Residential Costs with RSMeans data

2018

37th annual edition

Thomas Lane, Senior Editor

Chief Data Officer
Noam Reininger

Engineering Director
Bob Mewis, CCP

Contributing Editors
Christopher Babbitt
Sam Babbitt
Michelle Curran
Matthew Doheny (8)
Cheryl Elsmore
Linval Gentles
John Gomes (13, 41)
Derrick Hale, PE (2, 31, 32, 33, 34, 35, 44, 46)
Wafaa Hamitou (11, 12)

Joseph Kelble (14, 21, 22, 23, 25, 26, 27, 28, 48)
Charles Kibbee (1, 4)
Gerard Lafond, PE
Thomas Lane (6, 7)
Genevieve Medeiros
Elisa Mello
Ken Monty
Marilyn Phelan, AIA (9, 10)
Stephen C. Plotner (3, 5)
Callum Riley
Stephen Rosenberg
Jeff Sessions
Gabe Sirota

Matthew Sorrentino
Kevin Souza
Keegan Spraker
Tim Tonello
Jen Walsh
David Yazbek

Product Manager
Andrea Sillah

Production Manager
Debbie Panarelli

Production
Jonathan Forgit
Mary Lou Geary

Sharon Larsen
Sheryl Rose

Technical Support
Judy Abbruzzese
Gary L. Hoitt

Cover Design
Blaire Collins

Data Analytics
Tim Duggan
Todd Glowac
Matthew Kelliher-Gibson

Numbers in italics are the divisional responsibilities for each editor. Please contact the designated editor directly with any questions.

Gordian RSMeans data
Construction Publishers & Consultants
1099 Hingham Street, Suite 201
Rockland, MA 02370
United States of America
1-800-448-8182
www.RSMeans.com

Copyright 2017 by The Gordian Group Inc.
All rights reserved.
Cover photo © iStock.com/Maudib

Printed in the United States of America
ISSN 0896-8624
ISBN 978-1-946872-16-6

0178 | $241.99 per copy (in United States)
Price is subject to change without prior notice.

Related Data and Services

2018 Residential Costs with RSMeans data has been tirelessly researched and carefully compiled to provide construction cost data for single-family and small multi-family housing projects costing up to $1,000,000.

Our engineers recommend the following products and services to complement *Residential Costs with RSMeans data*:

Annual Cost Data Books
2018 Square Foot Costs with RSMeans data
2018 CPG Residential Repair and Remodeling Costs with RSMeans data

Reference Books
Estimating Building Costs
RSMeans Estimating Handbook
Green Building: Project Planning & Estimating
How to Estimate with RSMeans data
Plan Reading & Material Takeoff
Project Scheduling & Management for Construction
Universal Design Ideas for Style, Comfort & Safety

Seminars and In-House Training
Unit Price Estimating
Training for our online estimating solution
Practical Project Management for Construction Professionals
Scheduling with MSProject for Construction Professionals
Mechanical & Electrical Estimating

RSMeans data Online
For access to the latest cost data, an intuitive search, and an easy-to-use estimate builder, take advantage of the time savings available from our online application. To learn more visit: www.RSMeans.com/2018online.

Enterprise Solutions
Building owners, facility managers, building product manufacturers, and attorneys across the public and private sectors engage with RSMeans data Enterprise to solve unique challenges where trusted construction cost data is critical. To learn more visit: www.RSMeans.com/Enterprise.

Custom Built Data Sets
Building and Space Models: Quickly plan construction costs across multiple locations based on geography, project size, building system component, product options, and other variables for precise budgeting and cost control.

Predictive Analytics: Accurately plan future builds with custom graphical interactive dashboards, negotiate future costs of tenant build-outs, and identify and compare national account pricing.

Consulting
Building Product Manufacturing Analytics: Validate your claims and assist with new product launches.

Third-Party Legal Resources: Used in cases of construction cost or estimate disputes, construction product failure vs. installation failure, eminent domain, class action construction product liability, and more.

API
For resellers or internal application integration, RSMeans data is offered via API. Deliver Unit, Assembly, and Square Foot Model data within your interface. To learn more about how you can provide your customers with the latest in localized construction cost data visit: www.RSMeans.com/API.

Table of Contents

Foreword

The Value of RSMeans data from Gordian

Since 1942, RSMeans data has been the industry-standard materials, labor, and equipment cost information database for contractors, facility owners and managers, architects, engineers, and anyone else that requires the latest localized construction cost information. Over 75 years later, the objective remains the same: to provide facility and construction professionals with the most current and comprehensive construction cost database possible.

With the constant influx of new construction methods and materials, in addition to ever-changing labor and material costs, last year's cost data is not reliable for today's designs, estimates, or budgets. The RSMeans data engineers invest over 22,000 hours in cost research annually and apply real-world construction experience to identify and quantify new building products and methodologies, adjust productivity rates, and adjust costs to local market conditions across the nation. This unparalleled construction cost expertise is why so many facility and construction professionals rely on RSMeans data year over year.

About Gordian

Gordian originated in the spirit of innovation and a strong commitment to helping clients reach and exceed their construction goals. In 1982, Gordian's Chairman and Founder, Harry H. Mellon, created Job Order Contracting while serving as Chief Engineer at the Supreme Headquarters Allied Powers Europe. Job Order Contracting is a unique indefinite delivery/indefinite quantity (IDIQ) process, which enables facility owners to complete a substantial number of repair, maintenance, and construction projects with a single, competitively awarded contract. Realizing facility and infrastructure owners across various industries could greatly benefit from the time and cost saving advantages of this innovative construction procurement solution, he established Gordian in 1990.

Continuing the commitment to providing the most relevant and accurate facility and construction data, software, and expertise in the industry, Gordian enhanced the fortitude of its data with the acquisition of RSMeans in 2014. And in an effort to expand its facility management capabilities, Gordian acquired Sightlines, the leading provider of facilities benchmarking data and analysis, in 2015.

Our Offerings

Gordian is the leader in facility and construction cost data, software, and expertise for all phases of the building life cycle. From planning to design, procurement, construction, and operations, Gordian's solutions help clients maximize efficiency, optimize cost savings, and increase building quality with its highly specialized data engineers, software, and unique proprietary data sets.

Our Commitment

At Gordian, we do more than talk about the quality of our data and the usefulness of its application. We stand behind all of our RSMeans data—from historical cost indexes to construction materials and techniques—to craft current costs and predict future trends. If you have any questions about our products or services, please call us toll-free at 800-448-8182 or visit our website at www.gordian.com.

MasterFormat® 2014/ MasterFormat® 2016 Comparison Table

This table compares the 2014 edition of the Construction Specifications Institute's MasterFormat® to the expanded 2016 edition. For your convenience, all revised 2014 numbers and titles are listed along with the corresponding 2016 numbers and titles. In some cases, a designation of RSMeans is used to identify sections of data numbered exclusively in RSMeans products.

CSI 2014 MF ID	CSI 2014 MF Description	2014 Designation	CSI 2016 MF ID	CSI 2016 MF Description	2016 Designation
35 20 23.23	Hydraulic Dredging	CSI	35 24 13.13	Cutter Suction Dredging	RSMeans
35 20 23.13	Mechanical Dredging	CSI	35 24 23.13	Mechanical Dredging	CSI
35 20 23	Dredging	CSI	35 24 23	Clamshell Dredging	RSMeans
35 20 16.73	Slide Gates	RSMeans	35 22 73.16	Slide Gates	CSI
35 20 16.69	Knife Gates	RSMeans	35 22 69.16	Knife Gates	RSMeans
35 20 16.66	Flap Gates	RSMeans	35 22 66.16	Flap Gates	RSMeans
35 20 16.63	Canal Gates	RSMeans	35 22 63.16	Canal Gates	RSMeans
35 20 16.26	Hydraulic Sluice Gates	CSI	35 22 26.16	Hydraulic Sluice Gates	RSMeans
35 20 16	Hydraulic Gates	CSI	35 22 26	Sluice Gates	RSMeans
35 20	Waterway and Marine Construction and Equipment	CSI	35 22	Hydraulic Gates	RSMeans
33 72 33.46	Substation Converter Stations	CSI	33 78 33.46	Substation Converter Stations	RSMeans
33 72 33.36	Cable Trays For Utility Substations	CSI	26 05 36.36	Cable Trays For Utility Substations	CSI
33 72 33.33	Raceway/Boxes For Utility Substations	CSI	26 05 33.33	Raceway/Boxes For Utility Substations	CSI
33 52 43.13	Aviation Fuel Piping	CSI	33 52 13.43	Aviation Fuel Piping	RSMeans
33 52 16.13	Gasoline Piping	CSI	33 52 13.16	Gasoline Piping	RSMeans
33 52 13.14	Petroleum Products	RSMeans	33 52 13.19	Petroleum Products	RSMeans
33 51 33.10	Piping, Valves & Meters, Gas Distribution	RSMeans	33 59 33.10	Piping, Valves & Meters, Gas Distribution	CSI
33 51 33	Natural-Gas Metering	CSI	33 59 33	Natural-Gas Metering	CSI
33 51 13.30	Piping, Gas Service & Distribution, Steel	RSMeans	33 52 16.16	Piping, Gas Service & Distribution, Steel	RSMeans
33 51 13.20	Piping, Gas Service & Distribution, Steel	RSMeans	33 52 16.13	Steel Natural Gas Piping	RSMeans
33 51 13.10	Piping, Gas Service and Distribution, Polyethylene	RSMeans	33 52 16.20	Piping, Gas Service And Distribution, Polyethylene	CSI
33 49 23	Storm Drainage Water Retention Structures	CSI	33 46 23	Modular Buried Stormwater Storage Units	CSI
33 49 13	Storm Drainage Manholes, Frames, and Covers	CSI	33 05 61	Concrete Manholes	RSMeans
33 47 19	Water Ponds and Reservoirs	CSI	33 46 11	Stormwater Ponds	CSI
33 47 13.54	Garden Ponds	RSMeans	13 12 13.54	Garden Ponds	CSI
33 47 13.53	Reservoir Liners HDPE	CSI	31 05 19.53	Reservoir Liners HDPE	RSMeans
33 47 13	Pond and Reservoir Liners	CSI	31 05 19	Geosynthetics for Earthwork	RSMeans
33 46 26.10	Geotextiles For Subsurface Drainage	RSMeans	33 41 23.19	Geosynthetic Drainage Layers	RSMeans
33 46 26	Geotextile Subsurface Drainage Filtration	CSI	33 41 23	Drainage Layers	CSI
33 46 16	Subdrainage Piping	CSI	33 41 16	Subdrainage Piping	CSI
33 46	Subdrainage	CSI	33 41	Subdrainage	CSI
33 44 16	Utility Trench Drains	CSI	33 42 36	Stormwater Trench Drains	CSI
33 44 13	Utility Area Drains	CSI	33 42 33	Stormwater Curbside Drains and Inlets	RSMeans
33 42 16.15	Oval Arch Culverts	RSMeans	33 42 13.15	Oval Arch Culverts	RSMeans
33 42 16.13	Culverts & Box Trench Sections	CSI	33 42 13.14	Culverts & Box Trench Sections	RSMeans
33 41 13	Public Storm Utility Drainage Piping	CSI	33 42 11	Stormwater Gravity Piping	CSI
33 41	Storm Utility Drainage Piping	CSI	33 42	Stormwater Conveyance	CSI
33 36 50	Drainage Field Systems	RSMeans	33 34 51	Drainage Field System	CSI
33 36 33.13	Utility Septic Tank Tile Drainage Field	CSI	33 34 51.13	Utility Septic Tank Tile Drainage Field	RSMeans
33 36 19	Utility Septic Tank Effluent Filter	CSI	33 34 16	Septic Tank Effluent Filters	CSI
33 36 13.19	Polyethylene Utility Septic Tank	CSI	33 34 13.33	Polyethylene Septic Tanks	CSI
33 36 13.13	Concrete Utility Septic Tank	CSI	33 34 13.13	Concrete Septic Tanks	RSMeans
33 36 13	Utility Septic Tank and Effluent Wet Wells	CSI	33 34 13	Septic Tanks	CSI
33 36	Utility Septic Tanks	CSI	33 34	Onsite Wastewater Disposal	CSI
33 31 13	Public Sanitary Utility Sewerage Piping	CSI	33 31 11	Public Sanitary Sewerage Gravity Piping	CSI
33 21 13	Public Water Supply Wells	CSI	33 11 13	Potable Water Supply Wells	RSMeans
33 21	Water Supply Wells	CSI	33 11	Groundwater Sources	CSI
33 16 19.50	Elevated Water Storage Tanks	RSMeans	33 16 11.50	Elevated Water Storage Tanks	CSI
33 16 19	Elevated Water Utility Storage Tanks	CSI	33 16 11	Elevated Composite Water Storage Tanks	CSI
33 16 13.29	Wood Water Storage Tanks	RSMeans	33 16 59.29	Wood Water Storage Tanks	CSI
33 16 13.23	Plastic-Coated Fabric Pillow Water Tanks	RSMeans	33 16 56.23	Plastic-Coated Fabric Pillow Water Tanks	CSI
33 16 13.19	Horizontal Plastic Water Tanks	CSI	33 16 56.19	Horizontal Plastic Water Tanks	CSI
33 16 13.16	Prestressed Conc. Water Storage Tanks	CSI	33 16 36.16	Prestressed Conc. Water Storage Tanks	RSMeans
33 16 13.13	Steel Water Storage Tanks	CSI	33 16 23.13	Steel Water Storage Tanks	RSMeans
33 16 13	Aboveground Water Utility Storage Tanks	CSI	33 16 23	Ground-Level Steel Water Storage Tanks	CSI
33 12 19.10	Fire Hydrants	RSMeans	33 14 19.30	Fire Hydrants	CSI
33 12 16.20	Valves	RSMeans	33 14 19.20	Valves	CSI

v

Number	Title	Source	Number	Title	Source
33 12 16.10	Valves	RSMeans	33 14 19.10	Valves	CSI
33 12 16	Water Utility Distribution Valves	CSI	33 14 19	Valves and Hydrants for Water Utility Service	CSI
33 12 13.15	Tapping, Crosses and Sleeves	RSMeans	33 14 17.15	Tapping, Crosses and Sleeves	CSI
33 12 13	Water Service Connections	CSI	33 14 17	Site Water Utility Service Laterals	RSMeans
33 11 13	Public Water Utility Distribution Piping	CSI	33 14 13	Public Water Utility Distribution Piping	RSMeans
33 11	Water Utility Distribution Piping	CSI	33 14	Water Utility Transmission and Distribution	CSI
33 05 26	Utility Identification	CSI	33 05 97	Identification and Signage for Utilities	RSMeans
33 05 23.22	Directional Drilling	RSMeans	33 05 07.13	Utility Directional Drilling	RSMeans
33 05 23.20	Horizontal Boring	RSMeans	33 05 07.23	Utility Boring and Jacking	CSI
33 05 23.19	Microtunneling	CSI	33 05 07.36	Microtunneling	CSI
33 05 23	Trenchless Utility Installation	CSI	33 05 07	Trenchless Installation of Utility Piping	RSMeans
33 05 16	Utility Structures	CSI	33 05 63	Concrete Vaults and Chambers	RSMeans
33 01 30.71	Rehabilitation of Sewer Utilities	CSI	33 01 30.23	Pipe Bursting	RSMeans
33 01 30.16	TV Inspection of Sewer Pipelines	CSI	33 01 30.11	Television Inspection of Sewers	CSI
32 31 13.30	Fence, Chain Link, Gates & Posts	RSMeans	32 31 11.10	Gate Operators	CSI
31 71 21.10	Cut and Cover Tunnels	RSMeans	31 71 23.10	Cut and Cover Tunnels	CSI
31 71 21	Tunnel Excavation by Cut and Cover	RSMeans	31 71 23	Tunneling by Cut and Cover	CSI
28 46 13	Hard-Wired Detention Monitoring & Control Systems	CSI	28 52 11	Detention Monitoring and Control Systems	CSI
28 46	Electronic Detention Monitoring & Control Systems	CSI	28 52	Detention Security Systems	CSI
28 41 13	Building Systems	RSMeans	28 33 11	Electronic Structural Monitoring Systems	RSMeans
28 39 10	Notification Systems	RSMeans	28 47 12	Notification Systems	CSI
28 39	Mass Notification Systems	CSI	28 47	Mass Notification	RSMeans
28 33 33	Gas Detection Sensors	CSI	28 42 15	Gas Detection Sensors	CSI
28 33	Gas Detection and Alarm	CSI	28 42	Gas Detection and Alarm	RSMeans
28 32 33	Radiation Detection Sensors	CSI	28 41 15	Radiation Detection Sensors	CSI
28 31 49.50	Carbon-Monoxide Detectors	RSMeans	28 46 11.21	Carbon-Monoxide Detection Sensors	CSI
28 31 46.50	Smoke Detectors	RSMeans	28 46 11.27	Other Sensors	CSI
28 31 43	Fire Detection Sensors	CSI	28 46 11	Fire Sensors and Detectors	CSI
28 31 23	Fire Det. & Alarm Annunciation Panels & Fire Station	CSI	28 46 21	Fire Alarm	CSI
28 23 19.10	Digital Video Recorder (DVR)	RSMeans	28 05 19.11	Digital Video Recorders	RSMeans
28 23 19	Digital Video Recorders & Analog Recording Devices	CSI	28 05 19	Storage Appliances for Electronic Safety & Security	RSMeans
28 16 16	Intrusion Detection Systems Infrastructure	CSI	28 31 16	Intrusion Detection Systems Infrastructure	RSMeans
28 13 53.39	Security Access Full Body Imaging Machine	RSMeans	28 18 15.39	Security Access Full Body Imaging Machine	CSI
28 13 53.36	Security Access Debugging Kit	RSMeans	28 18 53.36	Security Access Debugging Kit	CSI
28 13 53.33	Security Access Counterfeit Money Detector	RSMeans	28 18 53.33	Security Access Counterfeit Money Detector	CSI
28 13 53.23	Security Access Explosive Detection Equipment	CSI	28 18 15.23	Security Access Explosive Detection Equipment	CSI
28 13 53.16	Security Access X-Ray Equipment	CSI	28 18 13.16	Security Access X-Ray Equipment	CSI
28 13 53.13	Security Access Metal Detectors	CSI	28 18 11.13	Security Access Metal Detectors	RSMeans
28 13 53	Security Access Detection	CSI	28 18 11	Security Access Metal Detectors	CSI
28 13 23.50	Vehicle Barriers	RSMeans	28 19 15.50	Vehicle Barriers	CSI
28 13 23	Access Control Remote Devices	RSMeans	28 19 15	Perimeter Vehicle Access Management Systems	RSMeans
28 13	Access Control	CSI	28 19	Access Control Vehicle Identification Systems	CSI
28 05 13.23	Fire Alarm Communications Conductors and Cables	CSI	27 15 01.19	Fire Alarm Communications Conductors and Cables	RSMeans
28 05 13.10	Alarm & Communications Cable	RSMeans	27 15 01.11	Conductors & Cables For Electronic Safety & Security	CSI
28 01 30.51	Maint. and Admin. of Elec. Detection and Alarm	CSI	28 01 80.51	Maint. & Administration of Fire Detection & Alarm	RSMeans
28 01 30	Operation and Maint. of Elec. Detection and Alarm	CSI	28 01 80	Operation and Maint. of Fire Detection and Alarm	RSMeans
26 56 19.20	Roadway Luminaire	RSMeans	26 56 21.20	Roadway Luminaire	RSMeans
26 56 19.10	Roadway Lighting Fixtures	RSMeans	26 56 21.10	LED Exterior Lighting	RSMeans
26 56 16.55	Parking LED Lighting	RSMeans	26 56 19.60	Parking LED Lighting	RSMeans
26 53 13.10	Exit Lighting Fixtures	RSMeans	26 52 13.16	Exit Signs	CSI
26 26 13.10	Power Distribution Unit	RSMeans	26 27 33.20	Power Distribution Unit	RSMeans
13 34 23.35	Geodesic Domes	RSMeans	13 33 13.35	Geodesic Domes	CSI
13 34 23.15	Domes	RSMeans	13 34 56.15	Domes	CSI
11 14 13.13	Portable Posts and Railings	CSI	11 14 19.13	Portable Posts and Railings	CSI
10 21 13.20	Plastic Toilet Compartment Components	RSMeans	10 21 14.19	Plastic Toilet Compartment Components	CSI
10 21 13.17	Plastic-Laminate Clad Toilet Compartment Components	RSMeans	10 21 14.16	Plastic-Laminate Clad Toilet Compartment Components	CSI
10 21 13.14	Metal Toilet Compartment Components	RSMeans	10 21 14.13	Metal Toilet Compartment Components	CSI
08 74 23.50	Security Access Control Accessories	RSMeans	28 15 11.19	Security Access Control Accessories	CSI
08 74 19.50	Biometric Identity Access	RSMeans	28 15 11.15	Biometric Identity Devices	RSMeans
08 74 16.50	Keypad Access	RSMeans	28 15 11.13	Keypads	RSMeans
08 74 13	Card Key Access Control Hardware	CSI	28 15 11	Integrated Credential Readers & Field Entry Mgmt	RSMeans
08 74	Access Control Hardware	CSI	28 15	Access Control Hardware Devices	CSI
08 56 63	Detention Windows	CSI	11 98 21	Detention Windows	RSMeans
08 34 63	Detention Doors and Frames	CSI	11 98 12	Detention Doors and Frames	RSMeans
07 72 73.10	Pitch Pockets, Variable Sizes	RSMeans	07 71 16.20	Pitch Pockets, Variable Sizes	CSI
05 53 13.70	Expanded Steel Grating, at Ground	RSMeans	05 53 19.20	Expanded Grating, Steel	CSI
02 85 33	Removal and Disposal of Materials with Mold	CSI	02 87 13.33	Removal and Disposal of Materials with Mold	RSMeans
02 85 16	Mold Remediation Preparation and Containment	CSI	02 87 13	Mold Remediation	RSMeans
02 85	Mold Remediation	CSI	02 87	Biohazard Remediation	CSI

For additional tools that help with the utilization of the Construction Specifications Institute's 2016 edition of MasterFormat® please visit the following website: http://www.masterformat.com/revisions/

How the Cost Data Is Built: An Overview

Unit Prices*
All cost data have been divided into 50 divisions according to the MasterFormat® system of classification and numbering.

Assemblies*
The cost data in this section have been organized in an "Assemblies" format. These assemblies are the functional elements of a building and are arranged according to the 7 elements of the UNIFORMAT II classification system. For a complete explanation of a typical "Assembly", see "RSMeans data: Assemblies—How They Work."

*Residential Models**
Model buildings for four classes of construction—economy, average, custom, and luxury—are developed and shown with complete costs per square foot.

*Commercial/Industrial/ Institutional Models**
This section contains complete costs for 77 typical model buildings expressed as costs per square foot.

*Green Commercial/Industrial/ Institutional Models**
This section contains complete costs for 25 green model buildings expressed as costs per square foot.

*References**
This section includes information on Equipment Rental Costs, Crew Listings, Historical Cost Indexes, City Cost Indexes, Location Factors, Reference Tables, and Change Orders, as well as a listing of abbreviations.

- **Equipment Rental Costs:** Included are the average costs to rent and operate hundreds of pieces of construction equipment.
- **Crew Listings:** This section lists all the crews referenced in the cost data. A crew is composed of more than one trade classification and/or the addition of power equipment to any trade classification. Power equipment is included in the cost of the crew. Costs are shown both with bare labor rates and with the installing contractor's overhead and profit added. For each, the total crew cost per eight-hour day and the composite cost per labor-hour are listed.

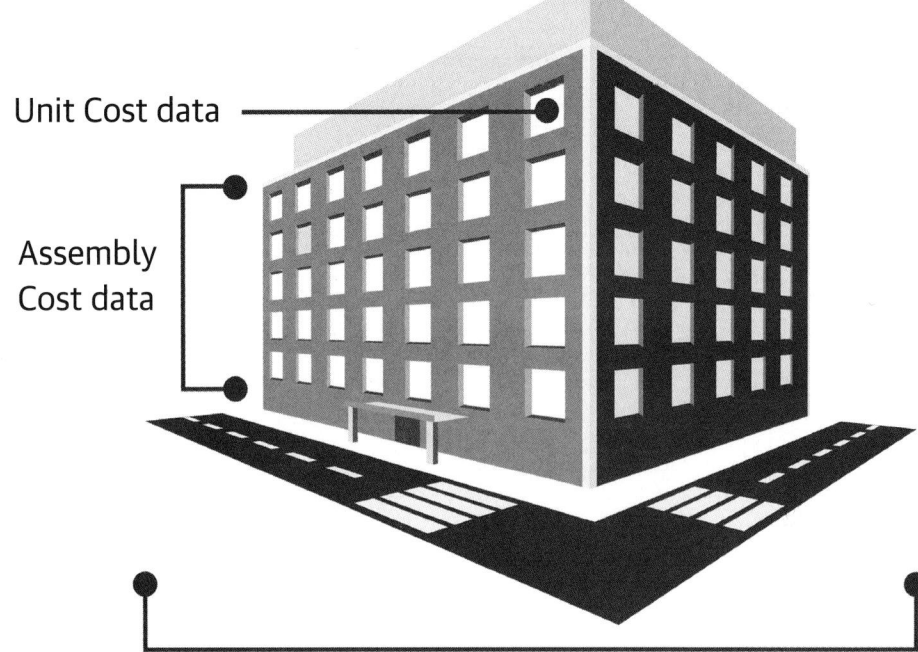

Unit Cost data

Assembly Cost data

Square Foot Models

- **Historical Cost Indexes**: These indexes provide you with data to adjust construction costs over time.
- **City Cost Indexes**: All costs in this data set are U.S. national averages. Costs vary by region. You can adjust for this by CSI Division to over 730 cities in 900+ 3-digit zip codes throughout the U.S. and Canada by using this data.
- **Location Factors**: You can adjust total project costs to over 730 cities in 900+ 3-digit zip codes throughout the U.S. and Canada by using the weighted number, which applies across all divisions.
- **Reference Tables**: At the beginning of selected major classifications in the Unit Prices are reference numbers indicators. These numbers refer you to related information in the Reference Section. In this section, you'll find reference tables, explanations, and estimating information that support how we develop the unit price data, technical data, and estimating procedures.
- **Change Orders**: This section includes information on the factors that influence the pricing of change orders.

- **Abbreviations**: A listing of abbreviations used throughout this information, along with the terms they represent, is included.

Index (printed versions only)
A comprehensive listing of all terms and subjects will help you quickly find what you need when you are not sure where it occurs in MasterFormat®.

Conclusion
This information is designed to be as comprehensive and easy to use as possible.

The Construction Specifications Institute (CSI) and Construction Specifications Canada (CSC) have produced the 2016 edition of MasterFormat®, a system of titles and numbers used extensively to organize construction information.

All unit prices in the RSMeans cost data are now arranged in the 50-division MasterFormat® 2016 system.

---

* Not all information is available in all data sets

Note: The material prices in RSMeans cost data
are "contractor's prices." They are the prices that
contractors can expect to pay at the lumberyards,
suppliers'/distributors' warehouses, etc. Small
orders of specialty items would be higher than
the costs shown, while very large orders, such as
truckload lots, would be less. The variation would
depend on the size, timing, and negotiating power
of the contractor. The labor costs are primarily
for new construction or major renovation rather
than repairs or minor alterations. With reasonable
exercise of judgment, the figures can be used for
any building work.

Estimating with RSMeans data: Unit Prices

Following these steps will allow you to complete an accurate estimate using RSMeans data Unit Prices.

1. Scope Out the Project

- Think through the project and identify the CSI divisions needed in your estimate.
- Identify the individual work tasks that will need to be covered in your estimate.
- The Unit Price data have been divided into 50 divisions according to CSI MasterFormat® 2016.
- In printed versions, the Unit Price Section Table of Contents on page 1 may also be helpful when scoping out your project.
- Experienced estimators find it helpful to begin with Division 2 and continue through completion. Division 1 can be estimated after the full project scope is known.

2. Quantify

- Determine the number of units required for each work task that you identified.
- Experienced estimators include an allowance for waste in their quantities. (Waste is not included in our Unit Price line items unless otherwise stated.)

3. Price the Quantities

- Use the search tools available to locate individual Unit Price line items for your estimate.
- Reference Numbers indicated within a Unit Price section refer to additional information that you may find useful.
- The crew indicates who is performing the work for that task. Crew codes are expanded in the Crew Listings in the Reference Section to include all trades and equipment that comprise the crew.
- The Daily Output is the amount of work the crew is expected to complete in one day.
- The Labor-Hours value is the amount of time it will take for the crew to install one unit of work.
- The abbreviated Unit designation indicates the unit of measure upon which the crew, productivity, and prices are based.
- Bare Costs are shown for materials, labor, and equipment needed to complete the Unit Price line item. Bare costs do not include waste, project overhead, payroll insurance, payroll taxes, main office overhead, or profit.
- The Total Incl O&P cost is the billing rate or invoice amount of the installing contractor or subcontractor who performs the work for the Unit Price line item.

4. Multiply

- Multiply the total number of units needed for your project by the Total Incl O&P cost for each Unit Price line item.
- Be careful that your take off unit of measure matches the unit of measure in the Unit column.
- The price you calculate is an estimate for a completed item of work.
- Keep scoping individual tasks, determining the number of units required for those tasks, matching each task with individual Unit Price line items, and multiplying quantities by Total Incl O&P costs.
- An estimate completed in this manner is priced as if a subcontractor, or set of subcontractors, is performing the work. The estimate does not yet include Project Overhead or Estimate Summary components such as general contractor markups on subcontracted work, general contractor office overhead and profit, contingency, and location factors.

5. Project Overhead

- Include project overhead items from Division 1-General Requirements.
- These items are needed to make the job run. They are typically, but not always, provided by the general contractor. Items include, but are not limited to, field personnel, insurance, performance bond, permits, testing, temporary utilities, field office and storage facilities, temporary scaffolding and platforms, equipment mobilization and demobilization, temporary roads and sidewalks, winter protection, temporary barricades and fencing, temporary security, temporary signs, field engineering and layout, final cleaning, and commissioning.
- Each item should be quantified and matched to individual Unit Price line items in Division 1, then priced and added to your estimate.
- An alternate method of estimating project overhead costs is to apply a percentage of the total project cost, usually 5% to 15% with an average of 10% (see General Conditions).
- Include other project related expenses in your estimate such as:
  - Rented equipment not itemized in the Crew Listings
  - Rubbish handling throughout the project (see section 02 41 19.19)

6. Estimate Summary

- Include sales tax as required by laws of your state or county.
- Include the general contractor's markup on self-performed work, usually 5% to 15% with an average of 10%.
- Include the general contractor's markup on subcontracted work, usually 5% to 15% with an average of 10%.
- Include the general contractor's main office overhead and profit:
  - RSMeans data provides general guidelines on the general contractor's main office overhead (see section 01 31 13.60 and Reference Number R013113-50).
  - Markups will depend on the size of the general contractor's operations, projected annual revenue, the level of risk, and the level of competition in the local area and for this project in particular.
- Include a contingency, usually 3% to 5%, if appropriate.
- Adjust your estimate to the project's location by using the City Cost Indexes or the Location Factors in the Reference Section:
  - Look at the rules in "How to Use the City Cost Indexes" to see how to apply the Indexes for your location.
  - When the proper Index or Factor has been identified for the project's location, convert it to a multiplier by dividing it by 100, then multiply that multiplier by your estimated total cost. The original estimated total cost will now be adjusted up or down from the national average to a total that is appropriate for your location.

Editors' Note:
We urge you to spend time reading and understanding the supporting material. An accurate estimate requires experience, knowledge, and careful calculation. The more you know about how we at RSMeans developed the data, the more accurate your estimate will be. In addition, it is important to take into consideration the reference material such as Equipment Listings, Crew Listings, City Cost Indexes, Location Factors, and Reference Tables.

How to Use the Cost Data: The Details

What's Behind the Numbers? The Development of Cost data

RSMeans data engineers continually monitor developments in the construction industry in order to ensure reliable, thorough, and up-to-date cost information. While overall construction costs may vary relative to general economic conditions, price fluctuations within the industry are dependent upon many factors. Individual price variations may, in fact, be opposite to overall economic trends. Therefore, costs are constantly tracked, and complete updates are performed yearly. Also, new items are frequently added in response to changes in materials and methods.

Costs in U.S. Dollars

All costs represent U.S. national averages and are given in U.S. dollars. The City Cost Index (CCI) with RSMeans data can be used to adjust costs to a particular location. The CCI for Canada can be used to adjust U.S. national averages to local costs in Canadian dollars. No exchange rate conversion is necessary because it has already been factored in.

G The processes or products identified by the green symbol in our publications have been determined to be environmentally responsible and/or resource-efficient solely by RSMeans data engineering staff. The inclusion of the green symbol does not represent compliance with any specific industry association or standard.

Material Costs

RSMeans data engineers contact manufacturers, dealers, distributors, and contractors all across the U.S. and Canada to determine national average material costs. If you have access to current material costs for your specific location, you may wish to make adjustments to reflect differences from the national average. Included within material costs are fasteners for a normal installation. RSMeans data engineers use manufacturers' recommendations, written specifications, and/or standard construction practices for the sizing and spacing of fasteners. Adjustments to material costs may be required for your specific application or location. The manufacturer's warranty is assumed. Extended warranties are not included in the material costs. **Material costs do not include sales tax.**

Labor Costs

Labor costs are based upon a mathematical average of trade-specific wages in 30 major U.S. cities. The type of wage (union, open shop, or residential) is identified on the inside back cover of printed publications or is selected by the estimator when using the electronic products. Markups for the wages can also be found on the inside back cover of printed publications and/or under the labor references found in the electronic products.

- If wage rates in your area vary from those used, or if rate increases are expected within a given year, labor costs should be adjusted accordingly.

Labor costs reflect productivity based on actual working conditions. In addition to actual installation, these figures include time spent during a normal weekday on tasks, such as material receiving and handling, mobilization at site, site movement, breaks, and cleanup.

Productivity data is developed over an extended period so as not to be influenced by abnormal variations and reflects a typical average.

Equipment Costs

Equipment costs include not only rental but also operating costs for equipment under normal use. The operating costs include parts and labor for routine servicing, such as the repair and replacement of pumps, filters, and worn lines. Normal operating expendables, such as fuel, lubricants, tires, and electricity (where applicable), are also included. Extraordinary operating expendables with highly variable wear patterns, such as diamond bits and blades, are excluded. These costs are included under materials. Equipment rental rates are obtained from industry sources throughout North America—contractors, suppliers, dealers, manufacturers, and distributors.

Rental rates can also be treated as reimbursement costs for contractor-owned equipment. Owned equipment costs include depreciation, loan payments, interest, taxes, insurance, storage, and major repairs.

Equipment costs do not include operators' wages.

Equipment Cost/Day—The cost of equipment required for each crew is included in the Crew Listings in the Reference Section (small tools that are considered essential everyday tools are not listed out separately). The Crew Listings itemize specialized tools and heavy equipment along with labor trades. The daily cost of itemized equipment included in a crew is based on dividing the weekly bare rental rate by 5 (number of working days per week), then adding the hourly operating cost times 8 (the number of hours per day). This Equipment Cost/Day is shown in the last column of the Equipment Rental Costs in the Reference Section.

Mobilization, Demobilization—The cost to move construction equipment from an equipment yard or rental company to the job site and back again is not included in equipment costs. Mobilization (to the site) and demobilization (from the site) costs can be found in the Unit Price Section. If a piece of equipment is already at the job site, it is not appropriate to utilize mobilization or demobilization costs again in an estimate.

Overhead and Profit

Total Cost including O&P for the installing contractor is shown in the last column of the Unit Price and/or Assemblies. This figure is the sum of the bare material cost plus 10% for profit, the bare labor cost plus total overhead and profit, and the bare equipment cost plus 10% for profit. Details for the calculation of overhead and profit on labor are shown on the inside back cover of the printed product and in the Reference Section of the electronic product.

General Conditions

Cost data in this data set are presented in two ways: Bare Costs and Total Cost including O&P (Overhead and Profit). General Conditions, or General Requirements, of the contract should also be added to the Total Cost including O&P when applicable. Costs for General Conditions are listed in Division 1 of the Unit Price Section and in the Reference Section.

General Conditions for the installing contractor may range from 0% to 10% of the Total Cost including O&P. For the general or prime contractor, costs for General Conditions may range from 5% to 15% of the Total Cost including O&P, with a figure of 10% as the most typical allowance. If applicable, the Assemblies and Models sections use costs that include the installing contractor's overhead and profit (O&P).

Factors Affecting Costs

Costs can vary depending upon a number of variables. Here's a listing of some factors that affect costs and points to consider.

Quality—The prices for materials and the workmanship upon which productivity is based represent sound construction work. They are also in line with industry standard and manufacturer specifications and are frequently used by federal, state, and local governments.

Overtime—We have made no allowance for overtime. If you anticipate premium time or work beyond normal working hours, be sure to make an appropriate adjustment to your labor costs.

Productivity—The productivity, daily output, and labor-hour figures for each line item are based on an eight-hour work day in daylight hours in moderate temperatures, and up to a 14' working height unless otherwise indicated. For work that extends beyond normal work hours or is performed under adverse conditions, productivity may decrease.

Size of Project—The size, scope of work, and type of construction project will have a significant impact on cost. Economies of scale can reduce costs for large projects. Unit costs can often run higher for small projects.

Location—Material prices are for metropolitan areas. However, in dense urban areas, traffic and site storage limitations may increase costs. Beyond a 20-mile radius of metropolitan areas, extra trucking or transportation charges may also increase the material costs slightly. On the other hand, lower wage rates may be in effect. Be sure to consider both of these factors when preparing an estimate, particularly if the job site is located in a central city or remote rural location. In addition, highly specialized subcontract items may require travel and per-diem expenses for mechanics.

Other Factors—

- season of year
- contractor management
- weather conditions
- local union restrictions
- building code requirements
- availability of:
  - adequate energy
  - skilled labor
  - building materials
- owner's special requirements/restrictions
- safety requirements
- environmental considerations
- access

Unpredictable Factors—General business conditions influence "in-place" costs of all items. Substitute materials and construction methods may have to be employed. These may affect the installed cost and/or life cycle costs. Such factors may be difficult to evaluate and cannot necessarily be predicted on the basis of the job's location in a particular section of the country. Thus, where these factors apply, you may find significant but unavoidable cost variations for which you will have to apply a measure of judgment to your estimate.

Rounding of Costs

In printed publications only, all unit prices in excess of $5.00 have been rounded to make them easier to use and still maintain adequate precision of the results.

How Subcontracted Items Affect Costs

A considerable portion of all large construction jobs is usually subcontracted. In fact, the percentage done by subcontractors is constantly increasing and may run over 90%. Since the workers employed by these companies do nothing else but install their particular product, they soon become experts in that line. The result is, installation by these firms is accomplished so efficiently that the total in-place cost, even with the general contractor's overhead and profit, is no more, and often less, than if the principal contractor had handled the installation. Companies that deal with construction specialties are anxious to have their product perform well and, consequently, the installation will be the best possible.

Contingencies

The allowance for contingencies generally provides for unforeseen construction difficulties. On alterations or repair jobs, 20% is not too much. If drawings are final and only field contingencies are being considered, 2% or 3% is probably sufficient, and often nothing needs to be added. Contractually, changes in plans will be covered by extras. The contractor should consider inflationary price trends and possible material shortages during the course of the job. These escalation factors are dependent upon both economic conditions and the anticipated time between the estimate and actual construction. If drawings are not complete or approved, or a budget cost is wanted, it is wise to add 5% to 10%. Contingencies, then, are a matter of judgment.

Important Estimating Considerations

The productivity, or daily output, of each craftsman or crew assumes a well-managed job where tradesmen with the proper tools and equipment, along with the appropriate construction materials, are present. Included are daily set-up and cleanup time, break time, and plan layout time. Unless otherwise indicated, time for material movement on site (for items

that can be transported by hand) of up to 200' into the building and to the first or second floor is also included. If material has to be transported by other means, over greater distances, or to higher floors, an additional allowance should be considered by the estimator.

While horizontal movement is typically a sole function of distances, vertical transport introduces other variables that can significantly impact productivity. In an occupied building, the use of elevators (assuming access, size, and required protective measures are acceptable) must be understood at the time of the estimate. For new construction, hoist wait and cycle times can easily be 15 minutes and may result in scheduled access extending beyond the normal work day. Finally, all vertical transport will impose strict weight limits likely to preclude the use of any motorized material handling.

The productivity, or daily output, also assumes installation that meets manufacturer/designer/ standard specifications. A time allowance for quality control checks, minor adjustments, and any task required to ensure the proper function or operation is also included. For items that require connections to services, time is included for positioning, leveling, securing the unit, and for making all the necessary connections (and start up where applicable), ensuring a complete installation. Estimating of the services themselves (electrical, plumbing, water, steam, hydraulics, dust collection, etc.) is separate.

In some cases, the estimator must consider the use of a crane and an appropriate crew for the installation of large or heavy items. For those situations where a crane is not included in the assigned crew and as part of the line item cost,

then equipment rental costs, mobilization and demobilization costs, and operator and support personnel costs must be considered.

Labor-Hours
The labor-hours expressed in this publication are derived by dividing the total daily labor-hours for the crew by the daily output. Based on average installation time and the assumptions listed above, the labor-hours include: direct labor, indirect labor, and nonproductive time. A typical day for a craftsman might include, but is not limited to:

- Direct Work
  - ☐ Measuring and layout
  - ☐ Preparing materials
  - ☐ Actual installation
  - ☐ Quality assurance/quality control
- Indirect Work
  - ☐ Reading plans or specifications
  - ☐ Preparing space
  - ☐ Receiving materials
  - ☐ Material movement
  - ☐ Giving or receiving instruction
  - ☐ Miscellaneous
- Non-Work
  - ☐ Chatting
  - ☐ Personal issues
  - ☐ Breaks
  - ☐ Interruptions (i.e., sickness, weather, material or equipment shortages, etc.)

If any of the items for a typical day do not apply to the particular work or project situation, the estimator should make any necessary adjustments.

Final Checklist
Estimating can be a straightforward process provided you remember the basics. Here's a checklist of some of the steps you should remember to complete before finalizing your estimate.

Did you remember to:
- factor in the City Cost Index for your locale?
- take into consideration which items have been marked up and by how much?
- mark up the entire estimate sufficiently for your purposes?
- read the background information on techniques and technical matters that could impact your project time span and cost?
- include all components of your project in the final estimate?
- double check your figures for accuracy?
- call RSMeans data engineers if you have any questions about your estimate or the data you've used? Remember, Gordian stands behind all of our products, including our extensive RSMeans data solutions. If you have any questions about your estimate, about the costs you've used from our data, or even about the technical aspects of the job that may affect your estimate, feel free to call the Gordian RSMeans editors at 1-800-448-8182.

Access Quarterly Data Updates

rsmeans.com/2018books

Square Foot Cost Section

Table of Contents

Introduction to the Square Foot Cost Section

The Square Foot Cost Section contains costs per square foot for four classes of construction in seven building types. Costs are listed for various exterior wall materials which are typical of the class and building type. There are cost tables for wings and ells with modification tables to adjust the base cost of each class of building. Non-standard items can easily be added to the standard structures.

Cost estimating for a residence is a three-step process:
1. Identification
2. Listing dimensions
3. Calculations

Guidelines and a sample cost estimating procedure are shown on the following pages.

Identification
To properly identify a residential building, the class of construction, type, and exterior wall material must be determined. The "Building Classes" information has drawings and guidelines for determining the class of construction. There are also detailed specifications and additional drawings at the beginning of each set of tables to further aid in proper building class and type identification.

Sketches for eight types of residential buildings and their configurations follow. Definitions of living area are next to each sketch. Sketches and definitions of garage types follow.

Living Area
Base cost tables are prepared as costs per square foot of living area. The living area of a residence is that area which is suitable and normally designed for full time living. It does not include basement recreation rooms or finished attics, although these areas are often considered full time living areas by the owners.

Living area is calculated from the exterior dimensions without the need to adjust for exterior wall thickness. When calculating the living area of a 1-1/2 story, two story, three story, or tri-level residence, overhangs and other differences in size and shape between floors must be considered.

Only the floor area with a ceiling height of seven feet or more in a 1-1/2 story residence is considered living area. In bi-levels and tri-levels, the areas that are below grade are considered living areas, even when these areas may not be completely finished.

Base Tables and Modifications
Base cost tables show the base cost per square foot without a basement, with one full bath and one full kitchen for economy and average homes, and an additional half bath for custom and luxury models. Adjustments for finished and unfinished basements are part of the base cost tables. Adjustments for multi-family residences, additional bathrooms, townhouses, alternative roofs, and air conditioning and heating systems are listed in Modifications, Adjustments, and Alternatives tables below the base cost tables.

Costs for other modifications, adjustments, and alternatives, including garages, breezeways, and site improvements, follow the base tables.

Listing of Dimensions
To use this section, only the dimensions used to calculate the horizontal area of the building and additions, modifications, adjustments, and alternatives are needed. The dimensions, normally the length and width, can come from drawings or field measurements. For ease in calculation, consider measuring in tenths of feet, i.e., 9'-6" = 9.5 ft. and 9'-4" = 9.3 ft.

In all cases, make a sketch of the building. Any protrusions or other variations in shape should be noted on the sketch with dimensions.

Calculations
The calculations portion of the estimate is a two-step activity:
1. The selection of appropriate costs from the tables
2. Computations

Selection of Appropriate Costs
To select the appropriate cost from the base tables, the following information is needed:
1. Class of construction
   - Economy
   - Average
   - Custom
   - Luxury
2. Type of residence
   - 1 story
   - 1-1/2 story
   - 2 story
   - 2-1/2 story
   - 3 story
   - Bi-level
   - Tri-level
3. Occupancy
   - One family
   - Two family
   - Three family
4. Building configuration
   - Detached
   - Town/Rowhouse
   - Semi-detached
5. Exterior wall construction
   - Wood frame
   - Brick veneer
   - Solid masonry
6. Living areas

Modifications are classified by class, type, and size.

Computations
The computation process should take the following sequence:
1. Multiply the base cost by the area.
2. Add or subtract the modifications, adjustments, and alternatives.
3. Apply the location modifier.

When selecting costs, interpolate or use the cost that most nearly matches the structure under study. This applies to size, exterior wall construction, and class.

How to Use the Residential Square Foot Cost Section

The following is a detailed explanation of a sample entry in the Residential Square Foot Cost Section. Each bold number below corresponds to the item being described in the following list with the appropriate component of the sample entry in parentheses. Prices listed are costs that include overhead and profit of the installing contractor. Total model costs include an additional markup for the general contractor's overhead and profit, and fees specific to the class of construction.

RESIDENTIAL	Average **1**	2 Story **2**

- **Simple design from standard plans**
- **Single family — 1 full bath, 1 kitchen**
- **No basement** **3**
- **Asphalt shingles on roof**
- **Hot air heat**
- **Gypsum wallboard interior finishes**
- **Materials and workmanship are average**

Note: The illustration shown may contain some optional components (for example: garages and/or fireplaces) whose costs are shown in the modifications, adjustments, & alternatives below or at the end of the square foot section.

Base cost per square foot of living area

Exterior Wall **4**	Living Area **5**										
	1000	1200	1400	1600	1800	2000	2200	2600	3000	3400	3800
Wood Siding - Wood Frame	149.75	135.30	128.30	123.35	118.45	113.25 **6**	109.75	103.15	96.85	93.90	91.20
Brick Veneer - Wood Frame	156.15	141.25	133.80	128.60	123.40	118.05	114.25	107.25	100.60	97.50	94.65
Stucco on Wood Frame	144.50	130.50	123.75	119.00	114.40	109.35	106.00	99.80	93.65	90.90	88.40
Solid Masonry	170.60	154.65	146.35	140.60	134.75	128.90	124.65	116.60	109.25	105.65	102.45
Finished Basement, Add **7**	23.25	22.90	22.10	21.65	21.05	20.70	20.30 **8**	19.55	18.95	18.60	18.30
Unfinished Basement, Add	9.20	8.55	8.05	7.75	7.40	7.15	6.95	6.45	6.15	5.90	5.70

Modifications

Add to the total cost

Upgrade Kitchen Cabinets **9**	$ + 5950
Solid Surface Countertops (Included)	
Full Bath - including plumbing, wall and floor finishes	+ 8050
Half Bath - including plumbing, wall and floor finishes	+ 4767
One Car Attached Garage	+ 15,086
One Car Detached Garage	+ 19,847
Fireplace & Chimney	+ 7671

Adjustments

For multi family - add to total cost

Additional Kitchen	$ + 9963
Additional Bath	+ 8050
Additional Entry & Exit	+ 1804
Separate Heating	+ 1650
Separate Electric	+ 1817

*For Townhouse/Rowhouse -
Multiply cost per square foot by*

Inner Unit **10**	.90
End Unit	.95

Alternatives

Add to or deduct from the cost per square foot of living area

Cedar Shake Roof	+ 1.75
Clay Tile Roof	+ 3.45
Slate Roof	+ 3.85
Upgrade Walls to Skim Coat Plaster **11**	+ .59
Upgrade Ceilings to Textured Finish	+ .60
Air Conditioning, in Heating Ductwork	+ 3
In Separate Ductwork	+ 5.81
Heating Systems, Hot Water	+ 1.52
Heat Pump	+ 1.59
Electric Heat	– .62
Not Heated	– 3.36

Additional upgrades or components

Kitchen Cabinets & Countertops	Page 93
Bathroom Vanities	94
Fireplaces & Chimneys	94
Windows, Skylights & Dormers	94
Appliances	95
Breezeways & Porches	95
Finished Attic	95
Garages	96
Site Improvements **12**	96
Wings & Ells	56

① Class of Construction (Average)

The class of construction depends upon the design and specifications of the plan. The four classes are economy, average, custom, and luxury.

② Type of Residence (2 Story)

The building type describes the number of stories or levels in the model. The seven building types are 1 story, 1-1/2 story, 2 story, 2-1/2 story, 3 story, bi-level, and tri-level.

③ Specification Highlights (Hot Air Heat)

These specifications include information concerning the components of the model, including the number of baths, roofing types, HVAC systems, materials, and workmanship. If the components listed are not appropriate, modifications can be made by consulting the information shown below or in the Assemblies section.

④ Exterior Wall System (Wood Siding-Wood Frame)

This section includes the types of exterior wall systems and the structural frames used. The exterior wall systems shown are typical of the class of construction and the building type shown.

⑤ Living Areas (2,000 S.F.)

The living area is that area of the residence which is suitable and normally designed for full time living. It does not include basement recreation rooms or finished attics. Living area is calculated from the exterior dimensions without the need to adjust for exterior wall thickness. When calculating the living area of a 1-1/2 story, 2 story, 3 story, or tri-level residence, overhangs and other differences in size and shape between floors must be considered. Only the floor area with a ceiling height of seven feet or more in a 1-1/2 story residence is considered living area. In bi-levels and tri-levels, the areas that are below grade are considered living areas, even when these areas may not be completely finished. A range of various living areas for the residential model is shown to aid in the selection of values from the matrix.

⑥ Base Costs per Square Foot of Living Area ($113.25)

Base cost tables show the cost per square foot of living area without a basement, with one full bath and one full kitchen for economy and average homes, and an additional half bath for custom and luxury models. When selecting costs, interpolate or use the cost that most nearly matches the residence under consideration for size, exterior

wall system, and class of construction. Prices listed are costs that include overhead and profit of the installing contractor, a general contractor markup, and an allowance for plans that vary by class of construction. For additional information on contractor overhead and architectural fees, see the Reference Section.

⑦ Basement Types (Finished)

The two types of basements are finished and unfinished. The specifications and components for both are shown under Building Classes in the Introduction to this section.

⑧ Additional Costs for Basements ($20.70 or $7.15)

These values indicate the additional cost per square foot of living area for either a finished or an unfinished basement.

⑨ Modifications and Adjustments (Upgrade Kitchen Cabinets $5,950)

Modifications and Adjustments are costs added to or subtracted from the total cost of the residence. The total cost of the residence is equal to the cost per square foot of living area times the living area. Typical modifications and adjustments include kitchens, baths, garages, and fireplaces.

⑩ Multiplier for Townhouse/Rowhouse (Inner Unit 0.90)

The multipliers shown adjust the base costs per square foot of living area for the common wall condition encountered in townhouses or rowhouses.

⑪ Alternatives (Skim Coat Plaster $0.59)

Alternatives are costs added to or subtracted from the base cost per square foot of living area. Typical alternatives include variations in kitchens, baths, roofing, and air conditioning and heating systems.

⑫ Additional Upgrades or Components (Wings & Ells)

Costs for additional upgrades or components, including wings or ells, breezeways, porches, finished attics, and site improvements, are shown at the end of each quality section and at the end of the Square Foot Section.

How to Use the Residential Square Foot Cost Section (Continued)

Average 2 Story

1 Living Area - 2000 S.F.
Perimeter - 135 L.F.

		Cost Per Square Foot Of Living Area			% of Total
		Mat.	**Inst.**	**Total**	**(rounded)**
1 Site Work	Site preparation for slab; 4' deep trench excavation for foundation wall.		1.32	1.32	1.2%
2 Foundation	Continuous reinforced concrete footing, 10" deep x 20" wide; damproofed and insulated 8" thick reinforced concrete block foundation wall, 4' deep; trowel finished 4" thick concrete slab on 4" crushed stone base and polyethylene vapor barrier.	3.58	4.72	8.30	7.3%
3 Framing	Exterior walls - 2" x 6" wood studs, 16" O.C.; 1/2" sheathing; gable end roof framing, 2" x 10" rafters, 16" O.C. with 1/2" plywood sheathing; 2" x 10" floor joists, 16" O.C. with bridging and 5/8" subflooring; 2" x 4" interior partitions.	7.41	10.02	17.43	15.4%
4 Exterior Walls	Beveled wood siding and housewrap on insulated wood frame walls; R38 attic insulation; double hung wood windows; flush solid core doors, frame and hardware, painted finish; aluminum storm and screen doors.	15.20	6.03	21.23	18.7%
5 Roofing	25 year asphalt roof shingles; #15 felt building paper; aluminum gutters, downspouts, drip edge and flashings.	1.13	1.40	2.53	2.2%
6 Interiors	Walls & ceilings, 1/2" taped & finished gypsum wallboard, primed & painted with 2 coats of finish paint; birch faced hollow core interior doors, frames & hardware, painted finish; medium weight carpeting with pad, 40%; sheet vinyl, 15%; oak hardwood, 40%; ceramic tile, 5%; hardwood tread stairway.	15.23	15.14	30.37	26.8%
7 Specialties	Average grade kitchen cabinets and countertop; stainless steel kitchen sink; 40 gallon electric water heater.	3.81	1.21	5.02	4.4%
8 Mechanical	Three fixture bathroom: bathtub, water closet, vanity and sink; gas fired hot air heating system.	3.44	3.35	6.79	6.0%
9 Electrical	200 amp electric service; wiring, duplex & GFI receptacles, wall switches, door bell, appliance circuits, fans and communications cabling; average grade lighting fixtures	1.46	2.37	3.83	3.4%
10 Overhead	Contractor's overhead and profit and plans.	8.69	7.74	16.43	14.5%
	Total	59.95	53.30	**113.25**	

Markers throughout table: 3, 4, 5, 6, 7, 8, 9

① Specifications

The parameters for an example dwelling from the previous pages are listed here. Included are the square foot dimensions of the proposed building. Living Area takes into account the number of floors and other factors needed to define a building's total square footage. Perimeter dimensions are defined in terms of linear feet.

② Building Type

This is a sketch of a cross section view through the dwelling. It is shown to help define the living area for the building type. For more information, see the Building Types in the Introduction.

③ Components (3 Framing)

This page contains the ten components needed to develop the complete square foot cost of the typical dwelling specified. All components are defined with a description of the materials and/or task involved. Use cost figures from each component to estimate the cost per square foot of that section of the project. The components listed here are typical of all sizes of residences from the facing page. Specific quantities of components required would vary with the size of the dwelling and the exterior wall system.

④ Materials (7.41)

This column gives the amount needed to develop the cost of materials. The figures given here are not bare costs. Ten percent has been added to bare material cost for profit.

⑤ Installation (10.02)

Installation includes labor and equipment costs. The labor rates included here incorporate the total overhead and profit costs for the installing contractor. The average mark-up used to create these figures is 72% over and above bare labor costs. The equipment rates include 10% for profit.

⑥ Total (17.43)

This column lists the sum of two figures. Use this total to determine the sum of material cost plus installation cost. The result is a convenient total cost for each of the ten components.

⑦ % of Total (rounded) (15.4%)

This column represents the percent of the total cost for this component group.

⑧ Overhead

The costs in components 1 through 9 include overhead and profit for the installing contractor. Item 10 is overhead and profit for the general contractor. This is typically a percentage mark-up of all other costs. The amount depends on the size and type of dwelling, building class, and economic conditions. An allowance for plans or design has been included where appropriate.

⑨ Bottom Line Total (113.25)

This figure is the complete square foot cost for the construction project and equals the sum of total material and total labor costs. To determine total project cost, multiply the bottom line total by the living area.

Building Classes

Economy Class

An economy class residence is usually built from stock plans. The materials and workmanship are sufficient to satisfy building codes. Low construction cost is more important than distinctive features. The overall shape of the foundation and structure is seldom other than square or rectangular.

An unfinished basement includes a 7' high, 8" thick foundation wall composed of either concrete block or cast-in-place concrete.

Included in the finished basement cost are inexpensive paneling or drywall as the interior finish on the foundation walls, a low cost sponge backed carpeting adhered to the concrete floor, a drywall ceiling, and overhead lighting.

Custom Class

A custom class residence is usually built from plans and specifications with enough features to give the building a distinct design. Materials and workmanship are generally above average with obvious attention given to construction details. Construction normally exceeds building code requirements.

An unfinished basement includes a 7'-6" high, 10" thick cast-in-place concrete foundation wall or a 7'-6" high, 12" thick concrete block foundation wall.

A finished basement includes painted drywall on insulated 2" × 4" wood furring as the interior finish to the concrete walls, a suspended ceiling, carpeting adhered to the concrete floor, overhead lighting, and heating.

Average Class

An average class residence is a simple design and built from standard plans. Materials and workmanship are average but often exceed minimum building codes. There are frequently special features that give the residence some distinctive characteristics.

An unfinished basement includes a 7'-6" high, 8" thick foundation wall composed of either cast-in-place concrete or concrete block.

Included in the finished basement are plywood paneling or drywall on furring that is fastened to the foundation walls, sponge backed carpeting adhered to the concrete floor, a suspended ceiling, overhead lighting, and heating.

Luxury Class

A luxury class residence is built from an architect's plan for a specific owner. It is unique both in design and workmanship. There are many special features, and construction usually exceeds all building codes. It is obvious that primary attention is placed on the owner's comfort and pleasure. Construction is supervised by an architect.

An unfinished basement includes an 8' high, 12" thick foundation wall that is composed of cast-in-place concrete or concrete block.

A finished basement includes painted drywall on 2" × 4" wood furring as the interior finish, a suspended ceiling, tackless carpet on wood subfloor with sleepers, overhead lighting, and heating.

Configurations

Detached House

This category of residence is a free-standing separate building with or without an attached garage. It has four complete walls.

Semi-Detached House

This category of residence has two side-by-side living units. The common wall is fireproof. Semi-detached residences can be treated as a rowhouse with two end units. Semi-detached residences can be any of the building types.

Town/Rowhouse

This category of residence has a number of attached units made up of inner units and end units. The units are joined by common walls. The inner units have only two exterior walls. The common walls are fireproof. The end units have three walls and a common wall. Town/rowhouses can be any of the building types.

Building Types

One Story

This is an example of a one-story dwelling. The living area of this type of residence is confined to the ground floor. The headroom in the attic is usually too low for use as a living area.

One-and-one-half Story

The living area on the upper level of this type of residence is 50% to 90% of the ground floor. This is made possible by a combination of this design's high-peaked roof and/or dormers. Only the upper level area with a ceiling height of seven feet or more is considered living area. The living area of this residence is the sum of the ground floor area plus the area on the second level with a ceiling height of seven feet or more.

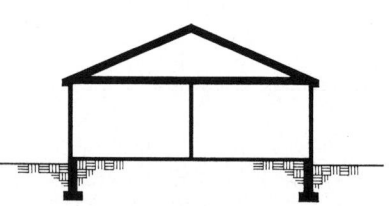

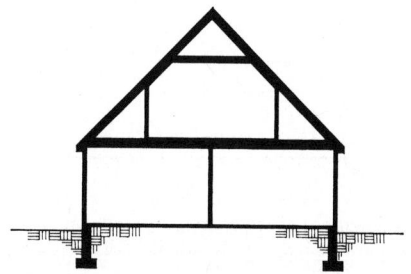

One Story with Finished Attic

The main living area in this type of residence is the ground floor. The upper level or attic area has sufficient headroom for use as a living area. This is made possible by a high-peaked roof. The living area in the attic is less than 50% of the ground floor. The living area of this type of residence is the ground floor area only. The finished attic is considered an adjustment.

Two Story

This type of residence has a second floor or upper level area which is equal or nearly equal to the ground floor area. The upper level of this type of residence can range from 90% to 110% of the ground floor area, depending on setbacks or overhangs. The living area is the sum of the ground floor area and the upper level floor area.

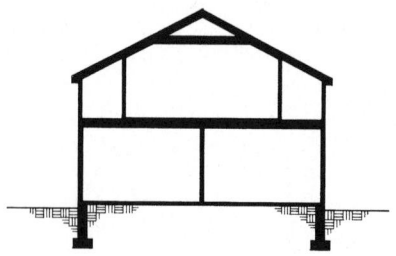

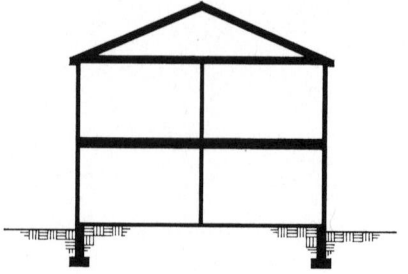

Two-and-one-half Story

This type of residence has two levels of equal or nearly equal area and a third level which has a living area that is 50% to 90% of the ground floor. This is made possible by a high-peaked roof, extended wall heights, and/or dormers. Only the upper level area with a ceiling height of seven feet or more is considered living area. The living area of this residence is the sum of the ground floor area, the second floor area, and the area on the third level with a ceiling height of seven feet or more.

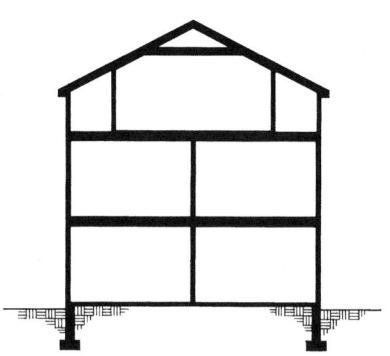

Bi-level

This type of residence has two living areas, one above the other. One area is about four feet below grade and the second is about four feet above grade. Both areas are equal in size. The lower level in this type of residence is designed and built to serve as a living area and not as a basement. Both levels have full ceiling heights. The living area is the sum of the lower level area and the upper level area.

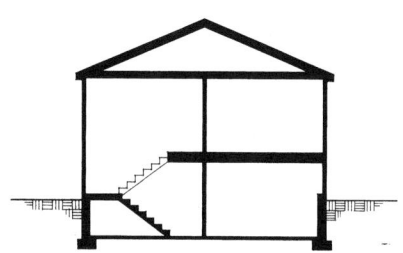

Three Story

This type of residence has three levels which are equal or nearly equal. As in the two story residence, the second and third floor areas may vary slightly depending on setbacks or overhangs. The living area is the sum of the ground floor area and the two upper level floor areas.

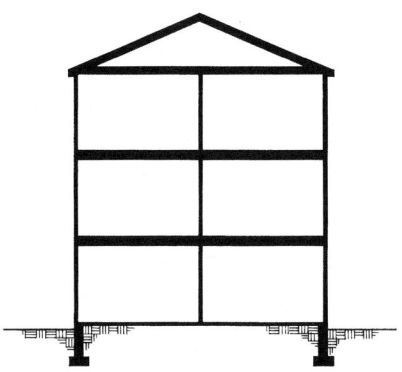

Tri-level

This type of residence has three levels of living area: one at grade level, one about four feet below grade, and one about four feet above grade. All levels are designed to serve as living areas. All levels have full ceiling heights. The living area is a sum of the areas of each of the three levels.

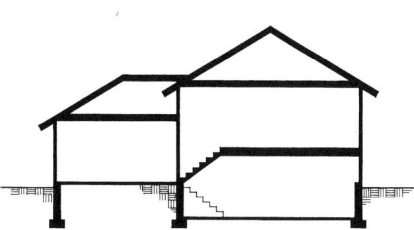

Garage Types

Attached Garage

Shares a common wall with the dwelling. Access is typically through a door between the dwelling and garage.

Basement Garage

Constructed under the roof of the dwelling but below the living area.

Built-In Garage

Constructed under the second floor living space and above the basement level of the dwelling. Reduces gross square feet of the living area.

Detached Garage

Constructed apart from the main dwelling. Shares no common area or wall with the dwelling.

Building Components

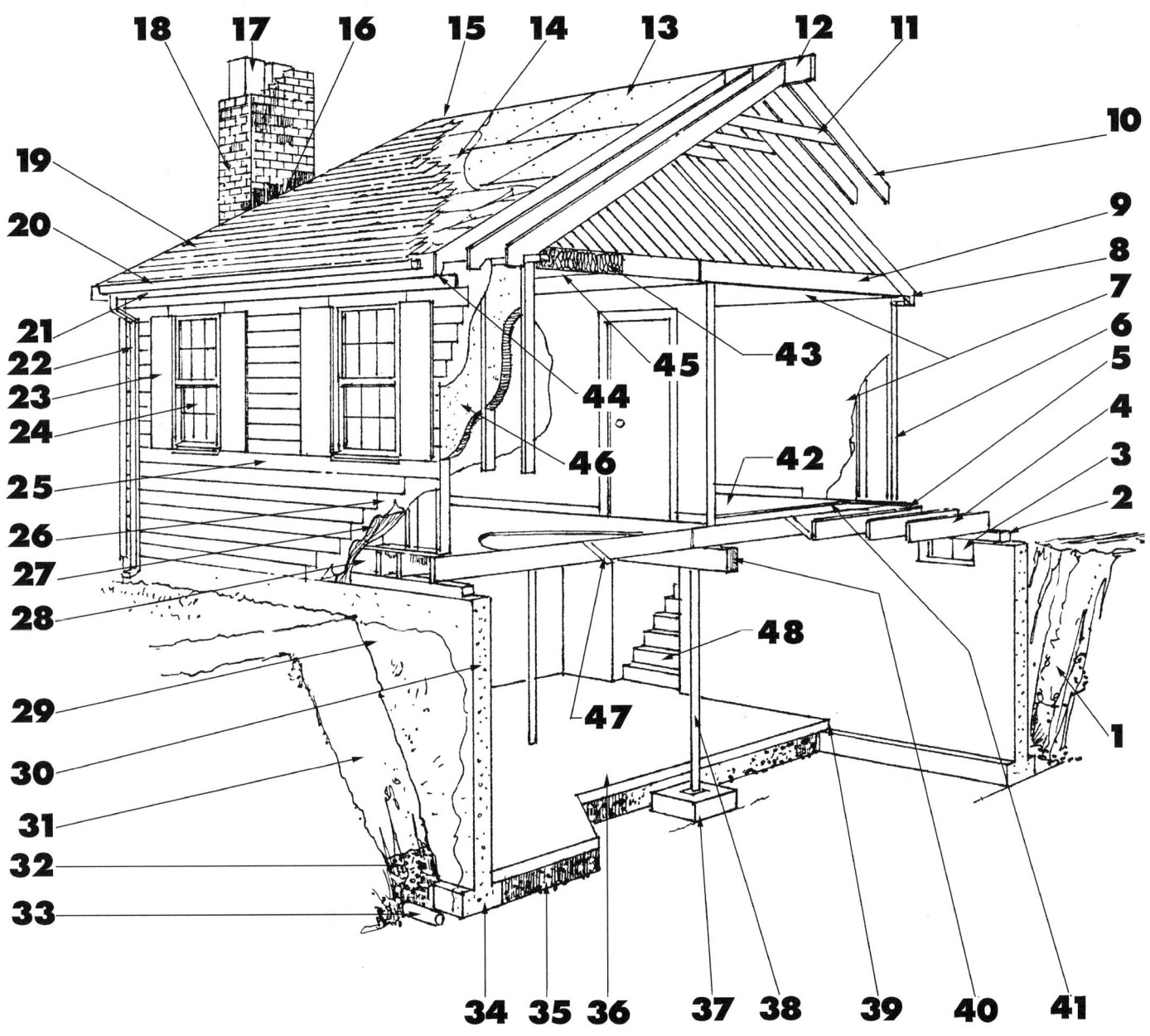

1. Excavation
2. Sill Plate
3. Basement Window
4. Floor Joist
5. Shoe Plate
6. Studs
7. Drywall
8. Plate
9. Ceiling Joists
10. Rafters
11. Collar Ties
12. Ridge Board
13. Roof Sheathing
14. Roof Felt
15. Roof Shingles
16. Flashing
17. Flue Lining
18. Chimney
19. Roof Shingles
20. Gutter
21. Fascia
22. Downspout
23. Shutter
24. Window
25. Wall Shingles
26. Weather Barrier
27. Wall Sheathing
28. Fire Stop
29. Dampproofing
30. Foundation Wall
31. Backfill
32. Drainage Stone
33. Drainage Tile
34. Wall Footing
35. Gravel
36. Concrete Slab
37. Column Footing
38. Pipe Column
39. Expansion Joint
40. Girder
41. Sub-floor
42. Finish Floor
43. Attic Insulation
44. Soffit
45. Ceiling Strapping
46. Wall Insulation
47. Cross Bridging
48. Bulkhead Stairs

Exterior Wall Construction

Typical Frame Construction

Typical wood frame construction consists of wood studs with insulation between them. A typical exterior surface is made up of sheathing, building paper, and exterior siding consisting of wood, vinyl, aluminum, or stucco over the wood sheathing.

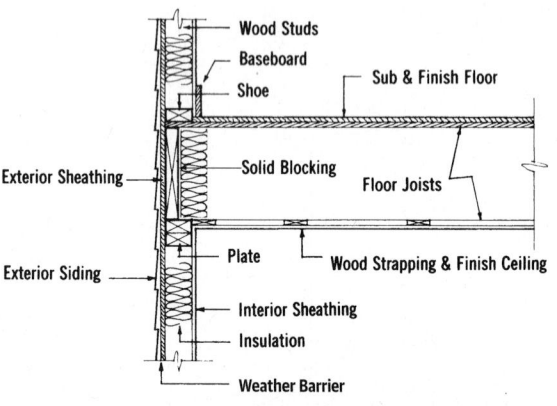

Brick Veneer

Typical brick veneer construction consists of wood studs with insulation between them. A typical exterior surface is sheathing, building paper, and an exterior of brick tied to the sheathing with metal strips.

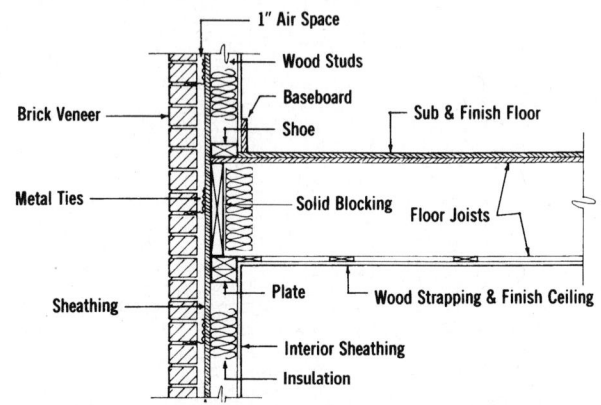

Stone

Typical solid masonry construction consists of a stone or block wall covered on the exterior with brick, stone, or other masonry.

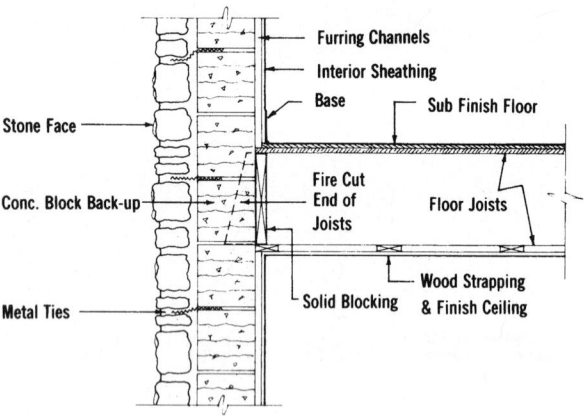

Residential Cost Estimate Worksheet

Worksheet Instructions

The residential cost estimate worksheet can be used as an outline for developing a residential construction or replacement cost. It is also useful for insurance appraisals. The design of the worksheet helps eliminate errors and omissions. To use the worksheet, follow the example below.

1. Fill out the owner's name, residence address, the estimator or appraiser's name, some type of project identifying number or code, and the date.

2. Determine from the plans, specifications, owner's description, photographs, or any other means possible the class of construction. The models in this data set use economy, average, custom, and luxury as classes. Fill in the appropriate box.

3. Fill in the appropriate box for the residence type, configuration, occupancy, and exterior wall. If you require clarification, the pages preceding this worksheet describe each of these.

4. Next, the living area of the residence must be established. The heated or air conditioned space of the residence, not including the basement, should be measured. It is easiest to break the structure up into separate components as shown in the example: the main house (A), a one-and-one-half story wing (B), and a one story wing (C). The breezeway (D), garage (E), and open covered porch (F) will be treated differently. Data entry blocks for the living area are included on the worksheet for your use. Keep each level of each component separate, and fill out the blocks as shown.

5. By using the information on the worksheet, find the model, wing, or ell in the following square foot cost pages that best matches the class, type, exterior finish, and size of the residence being estimated.

Use the Modifications, Adjustments, and Alternatives to determine the adjusted cost per square foot of living area for each component.

6. For each component, multiply the cost per square foot by the living area square footage. If the residence is a town/rowhouse, a multiplier should be applied based upon the configuration.

7. The second page of the residential cost estimate worksheet has space for the additional components of a house. The cost for additional bathrooms, finished attic space, breezeways, porches, fireplaces, appliance or cabinet upgrades, and garages should be added on this page. The information for each of these components is found with the model being used or in the Modifications, Adjustments, and Alternatives.

8. Add the total from page one of the estimate worksheet and the items listed on page two. The sum is the adjusted total building cost.

9. Depending on the use of the final estimated cost, one of the remaining two boxes should be filled out. Any additional items or exclusions should be added or subtracted at this time. The data contained in this data set are a national average. Construction costs are different throughout the country. To allow for this difference, a location factor based upon the first three digits of the residence's zip code must be applied. The location factor is a multiplier that increases or decreases the adjusted total building cost. Find the appropriate location factor and calculate the local cost. If depreciation is a concern, a dollar figure should be subtracted at this point.

10. No residence will match a model exactly. Many differences will be found. At this level of estimating, a variation of +/- 10% should be expected.

Adjustments Instructions

No residence matches a model exactly in shape, material, or specifications. The common differences are:

1. Two or more exterior wall systems:
   - Partial basement
   - Partly finished basement
2. Specifications or features that are between two classes
3. Crawl space instead of a basement

Examples

Below are quick examples. See pages 17–19 for complete examples of cost adjustments for these differences:

1. Residence "A" is an average one-story structure with 1,600 S.F. of living area and no basement. Three walls are wood siding on wood frame, and the fourth wall is brick veneer on wood frame. The brick veneer wall is 35% of the exterior wall area.

   Use page 38 to calculate the Base Cost per S.F. of Living Area. Wood Siding for 1,600 S.F. = $116.75 per S.F. and Brick Veneer for 1,600 S.F. = $120.60 per S.F.

   0.65 ($116.75) + 0.35 ($120.60) = $118.10 per S.F. of Living Area.

2a. Residence "B" is the same as Residence "A" but it has an unfinished basement under 50% of the building. To adjust the $118.10 per S.F. of living area for this partial basement, use page 38.

   $118.10 + 0.5 ($11.65) = $123.93 per S.F. of Living Area.

2b. Residence "C" is the same as Residence "A" but it has a full basement under the entire building. 640 S.F. or 40% of the basement area is finished.

   Using Page 38:

   $118.10 + 0.40 ($33.30) + 0.60 ($11.65) = $138.41 per S.F. of Living Area.

3. When specifications or features of a building are between classes, estimate the percent deviation, and use two tables to calculate the cost per S.F.

   A two-story residence with wood siding and 1,800 S.F. of living area has features 30% better than Average, but 70% less than Custom.

   From pages 42 and 64:

Custom 1,800 S.F. Base Cost	**=**	**$152.10 per S.F.**
Average 1,800 S.F. Base Cost	**=**	**$118.45 per S.F.**
DIFFERENCE	**=**	**$33.65 per S.F.**
Cost is $118.45 + 0.30 ($33.65)	**=**	**$128.55 per S.F. of Living Area.**

4. To add the cost of a crawl space, use the cost of an unfinished basement as a maximum. For specific costs of components to be added or deducted, such as vapor barrier, underdrain, and floor, see the "Assemblies" section (pages 97 to 281).

Model Residence Example

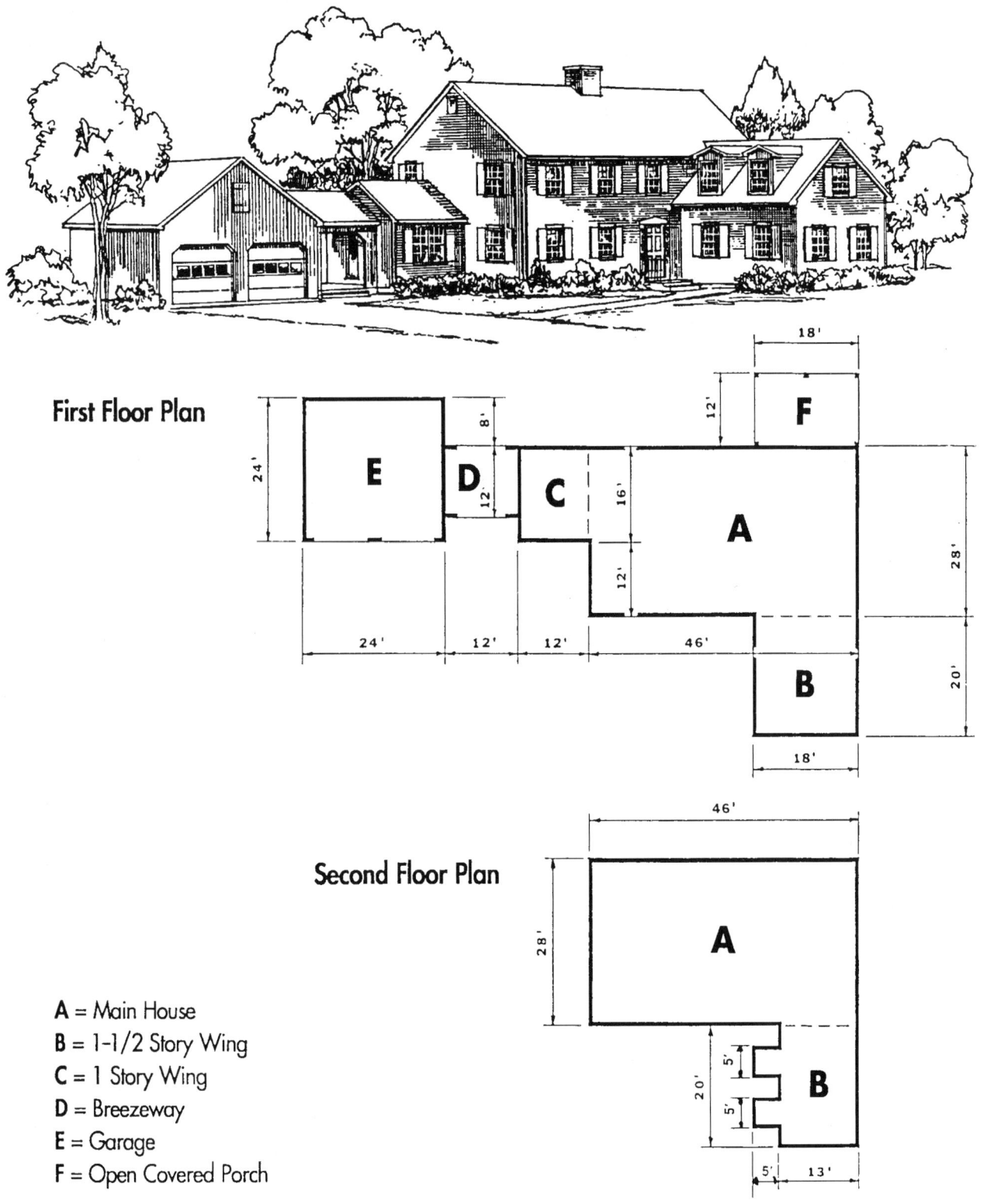

First Floor Plan

Second Floor Plan

A = Main House
B = 1-1/2 Story Wing
C = 1 Story Wing
D = Breezeway
E = Garage
F = Open Covered Porch

RESIDENTIAL COST ESTIMATE

OWNERS NAME:	**Albert Westenberg**	APPRAISER:	**Nicole Wojtowicz**
RESIDENCE ADDRESS:	**300 Sygiel Road**	PROJECT:	**# 55**
CITY, STATE, ZIP CODE:	**Three Rivers, MA 01080**	DATE:	**Jan. 1, 2018**

CLASS OF CONSTRUCTION	RESIDENCE TYPE	CONFIGURATION	EXTERIOR WALL SYSTEM
☐ ECONOMY	☐ 1 STORY	☑ DETACHED	☑ WOOD SIDING - WOOD FRAME
☑ AVERAGE	☐ 1 1/2 STORY	☐ TOWN/ROW HOUSE	☐ BRICK VENEER - WOOD FRAME
☐ CUSTOM	☑ 2 STORY	☐ SEMI-DETACHED	☐ STUCCO ON WOOD FRAME
☐ LUXURY	☐ 2 1/2 STORY		☐ PAINTED CONCRETE BLOCK
	☐ 3 STORY	OCCUPANCY	☐ SOLID MASONRY (AVERAGE & CUSTOM)
	☐ BI-LEVEL	☑ ONE FAMILY	☐ STONE VENEER - WOOD FRAME
	☐ TRI-LEVEL	☐ TWO FAMILY	☐ SOLID BRICK (LUXURY)
		☐ THREE FAMILY	☐ SOLID STONE (LUXURY)
		☐ OTHER	

* LIVING AREA (Main Building)			* LIVING AREA (Wing or Ell)	(**B**)		* LIVING AREA (Wing or Ell)	(**C**)
First Level	**1288**	S.F.	First Level	**360**	S.F.	First Level	**192**	S.F.
Second level	**1288**	S.F.	Second level	**310**	S.F.	Second level		S.F.
Third Level		S.F.	Third Level		S.F.	Third Level		S.F.
Total	**2576**	S.F.	Total	**670**	S.F.	Total	**192**	S.F.

* Basement Area is not part of living area.

MAIN BUILDING			COSTS PER S.F. LIVING AREA	
Cost per Square Foot of Living Area, from Page	**42**		$	103.15
Basement Addition: _____ % Finished,	**100**	% Unfinished	+	6.45
Roof Cover Adjustment: **Cedar Shake** Type, Page	**42**	(Add or Deduct)	( )	1.65
Central Air Conditioning: ☐ Separate Ducts ☑ Heating Ducts, Page	**42**		+	2.96
Heating System Adjustment: _____ Type, Page _____		(Add or Deduct)	( )	
Main Building: Adjusted Cost per S.F. of Living Area			$	# 114.21

MAIN BUILDING TOTAL COST	$ 114.21 /S.F.	X	2,576 S.F.	X	_____	=	$ 294,205
	Cost per S.F. Living Area		Living Area		Town/Row House Multiplier (Use 1 for Detached)		TOTAL COST

WING OR ELL (B)	1 - 1/2 STORY		COSTS PER S.F. LIVING AREA	
Cost per Square Foot of Living Area, from Page	**56**		$	98.20
Basement Addition: **100** % Finished,		% Unfinished	+	29.95
Roof Cover Adjustment: _____ Type, Page _____		(Add or Deduct)	( )	—
Central Air Conditioning: ☐ Separate Ducts ☑ Heating Ducts, Page	42		+	2.96
Heating System Adjustment: _____ Type, Page _____		(Add or Deduct)	( )	—
Wing or Ell (**B**): Adjusted Cost per S.F. of Living Area			$	# 131.11

WING OR ELL (B) TOTAL COST	$ 131.11 /S.F.	X	670 S.F.		=	$ 87,844
	Cost per S.F. Living Area		Living Area			TOTAL COST

WING OR ELL (C)	1 STORY	(WOOD SIDING)	COSTS PER S.F. LIVING AREA	
Cost per Square Foot of Living Area, from Page	# 56		$	# 150.10
Basement Addition: _____ % Finished,		% Unfinished	+	—
Roof Cover Adjustment: - Type, Page #		(Add or Deduct)	( )	—
Central Air Conditioning: ☐ Separate Ducts ☐ Heating Ducts, Page _____			+	—
Heating System Adjustment: _____ Type, Page _____		(Add or Deduct)	( )	—
Wing or Ell (C) Adjusted Cost per S.F. of Living Area			$	# 150.1

WING OR ELL (C) TOTAL COST	$ 150.10 /S.F.	X	192 S.F.		=	$ 28,819
	Cost per S.F. Living Area		Living Area			TOTAL COST

	TOTAL THIS PAGE	410,868

Page 1 of 2

RESIDENTIAL
COST ESTIMATE

					QUANTITY	UNIT COST		
Total Page 1							$	410,868
Additional Bathrooms: __2__ Full, __1__ Half 2 @ 7405 1 @ 4331							+	19,141
Finished Attic: __N/A__ Ft. x _____ Ft.						S.F.		
Breezeway: ☑ Open ☐ closed __12__ Ft. x __12__ Ft.					144	S.F. 40.07	+	5,770
Covered Porch: ☑ Open ☐ Enclosed __18__ Ft. x __12__ Ft.					216	S.F. 37.81	+	8,167
Fireplace: ☑ Interior Chimney ☐ Exterior Chimney ☑ No. of Flues (__2__) ☑ Additional Fireplaces **1 - 2nd Story**							+	12,746
Appliances:							+	—
Kitchen Cabinets Adjustments:				(±)				—
☑ Garage ☐ Carport: __2__ Car(s) Description **Wood, Attached** (±)								25,350
Miscellaneous:							+	

	ADJUSTED TOTAL BUILDING COST	$	482,042

REPLACEMENT COST		
ADJUSTED TOTAL BUILDING COST	$	482,042
Site Improvements		
(A) Paving & Sidewalks	$	
(B) Landscaping	$	
(C) Fences	$	
(D) Swimming Pools	$	
(E) Miscellaneous	$	
TOTAL	$	482,042
Location Factor	x	1.060
Location Replacement Cost	$	510,964
Depreciation	-$	52,592
LOCAL DEPRECIATED COST	$	458,372

INSURANCE COST		
ADJUSTED TOTAL BUILDING COST	$	
Insurance Exclusions		
(A) Footings, sitework, Underground Piping	-$	
(B) Architects Fees	-$	
Total Building Cost Less Exclusion	$	
Location Factor	x	
LOCAL INSURABLE REPLACEMENT COST	$	

SKETCH AND ADDITIONAL CALCULATIONS

1 Story

© Home Planners, Inc.

1-1/2 Story

2 Story

Bi-Level

Tri-Level

©Design Basics, Inc.

For customer support on your Residential Costs with RSMeans data, call 800.448.8182.

23

- **Mass produced from stock plans**
- **Single family — 1 full bath, 1 kitchen**
- **No basement**
- **Asphalt shingles on roof**
- **Hot air heat**
- **Gypsum wallboard interior finishes**
- **Materials and workmanship are sufficient to meet codes**

Note: The illustration shown may contain some optional components (for example: garages and/or fireplaces) whose costs are shown in the modifications, adjustments, & alternatives below or at the end of the square foot section.

©Home Planners, Inc.

Base cost per square foot of living area

Exterior Wall	Living Area										
	600	800	1000	1200	1400	1600	1800	2000	2400	2800	3200
Wood Siding - Wood Frame	147.60	133.20	122.20	113.55	105.85	100.95	98.60	95.25	88.80	84.10	80.75
Brick Veneer - Wood Frame	153.35	138.30	126.90	117.70	109.70	104.60	102.05	98.50	91.70	86.80	83.30
Stucco on Wood Frame	138.95	125.45	115.20	107.20	100.10	95.60	93.40	90.35	84.30	79.90	76.95
Painted Concrete Block	143.95	129.85	119.25	110.90	103.45	98.70	96.40	93.25	86.85	82.30	79.20
Finished Basement, Add	33.15	31.20	29.80	28.50	27.50	26.80	26.45	25.90	25.15	24.55	24.05
Unfinished Basement, Add	14.85	13.25	12.15	11.10	10.35	9.80	9.55	9.10	8.45	8.00	7.60

Modifications

Add to the total cost

Upgrade Kitchen Cabinets	$ + 1216
Solid Surface Countertops	+ 870
Full Bath - including plumbing, wall and floor finishes	+ 6440
Half Bath - including plumbing, wall and floor finishes	+ 3814
One Car Attached Garage	+ 13,994
One Car Detached Garage	+ 18,103
Fireplace & Chimney	+ 6624

Adjustments

For multi family - add to total cost

Additional Kitchen	$ + 5931
Additional Bath	+ 6440
Additional Entry & Exit	+ 1804
Separate Heating	+ 1650
Separate Electric	+ 1093

*For Townhouse/Rowhouse -
Multiply cost per square foot by*

Inner Unit	.95
End Unit	.97

Alternatives

Add to or deduct from the cost per square foot of living area

Composition Roll Roofing	– 1.05
Cedar Shake Roof	+ 4.20
Upgrade Walls and Ceilings to Skim Coat Plaster	+ .80
Upgrade Ceilings to Textured Finish	+ .60
Air Conditioning, in Heating Ductwork	+ 4.79
In Separate Ductwork	+ 7.05
Heating Systems, Hot Water	+ 1.60
Heat Pump	+ 1.32
Electric Heat	– 1.44
Not Heated	– 4.31

Additional upgrades or components

Kitchen Cabinets & Countertops	Page 93
Bathroom Vanities	94
Fireplaces & Chimneys	94
Windows, Skylights & Dormers	94
Appliances	95
Breezeways & Porches	95
Finished Attic	95
Garages	96
Site Improvements	96
Wings & Ells	34

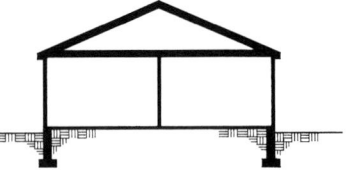

Economy 1 Story

Living Area - 1200 S.F.
Perimeter - 146 L.F.

		Cost Per Square Foot Of Living Area			% of Total
		Mat.	Inst.	Total	(rounded)
1 Site Work	Site preparation for slab; 4' deep trench excavation for foundation wall.		2.13	2.13	2.0%
2 Foundation	Continuous reinforced concrete footing, 10" deep x 20" wide; damproofed and insulated 8" thick reinforced concrete block foundation wall, 4' deep; trowel finished 4" thick concrete slab on 4" crushed stone base and polyethylene vapor barrier.	6.48	8.44	14.92	13.9%
3 Framing	Exterior walls - 2" x 4" wood studs, 16" O.C.; 1/2" sheathing; wood truss roof frame, 24" O.C. with 1/2" plywood sheating; 2" x 4" interior partitions.	5.66	7.69	13.35	12.5%
4 Exterior Walls	Metal lath reinforced stucco exterior on insulated wood frame walls; R38 attic insulation; sliding wood windows; flush solid core doors, frame and hardware, painted finish; aluminum storm and screen doors.	8.56	7.75	16.31	15.2%
5 Roofing	25 year asphalt roof shingles; #15 felt building paper; aluminum gutters, downspouts, drip edge and flashings.	2.20	2.71	4.91	4.6%
6 Interiors	Walls and ceilings, 1/2" taped and finished gypsum wallboard, primed and painted with 2 coats of finish paint; hollow core wood interior doors, frames and hardware, painted finish; lightweight carpeting with pad, 80%; sheet vinyl flooring, 20%.	12.08	13.19	25.27	23.6%
7 Specialties	Economy grade kitchen cabinets and countertops; stainless steel kitchen sink; 30 gallon electric water heater.	3.56	1.13	4.69	4.4%
8 Mechanical	Three fixture bathroom: bathtub, water closet and wall hung lavatory; gas fired hot air heating system.	4.57	3.91	8.48	7.9%
9 Electrical	100 amp electric service; wiring, duplex & GFI receptacles, wall switches, door bell, appliance circuits and communications cabling; economy grade lighting fixtures.	1.07	2.09	3.16	2.9%
10 Overhead	Contractor's overhead and profit.	6.62	7.36	13.98	13.0%
Total		50.80	56.40	**107.20**	

For customer support on your Residential Costs with RSMeans data, call 800.448.8182.

25

- **Mass produced from stock plans**
- **Single family — 1 full bath, 1 kitchen**
- **No basement**
- **Asphalt shingles on roof**
- **Hot air heat**
- **Gypsum wallboard interior finishes**
- **Materials and workmanship are sufficient to meet codes**

Note: The illustration shown may contain some optional components (for example: garages and/or fireplaces) whose costs are shown in the modifications, adjustments, & alternatives below or at the end of the square foot section.

Base cost per square foot of living area

Exterior Wall	Living Area										
	600	800	1000	1200	1400	1600	1800	2000	2400	2800	3200
Wood Siding - Wood Frame	173.65	144.00	129.25	122.20	117.10	109.25	105.45	101.55	93.35	90.25	86.80
Brick Veneer - Wood Frame	181.90	150.00	134.80	127.40	122.00	113.75	109.70	105.60	96.95	93.60	89.90
Stucco on Wood Frame	161.20	135.00	120.90	114.35	109.65	102.45	99.05	95.45	87.95	85.10	82.10
Painted Concrete Block	168.15	140.05	125.55	118.75	113.85	106.30	102.60	98.85	90.95	88.00	84.70
Finished Basement, Add	25.50	21.60	20.65	19.90	19.40	18.65	18.20	17.80	16.95	16.60	16.20
Unfinished Basement, Add	13.10	10.00	9.25	8.65	8.20	7.60	7.25	6.95	6.30	6.05	5.70

Modifications

Add to the total cost

Upgrade Kitchen Cabinets	$ + 1216
Solid Surface Countertops	+ 870
Full Bath - including plumbing, wall and floor finishes	+ 6440
Half Bath - including plumbing, wall and floor finishes	+ 3814
One Car Attached Garage	+ 13,994
One Car Detached Garage	+ 18,103
Fireplace & Chimney	+ 6624

Adjustments

For multi family - add to total cost

Additional Kitchen	$ + 5931
Additional Bath	+ 6440
Additional Entry & Exit	+ 1804
Separate Heating	+ 1650
Separate Electric	+ 1093

*For Townhouse/Rowhouse -
Multiply cost per square foot by*

Inner Unit	.95
End Unit	.97

Alternatives

Add to or deduct from the cost per square foot of living area

Composition Roll Roofing	– .75
Cedar Shake Roof	+ 3.05
Upgrade Walls and Ceilings to Skim Coat Plaster	+ .81
Upgrade Ceilings to Textured Finish	+ .60
Air Conditioning, in Heating Ductwork	+ 3.57
In Separate Ductwork	+ 6.19
Heating Systems, Hot Water	+ 1.52
Heat Pump	+ 1.45
Electric Heat	– 1.14
Not Heated	– 3.96

Additional upgrades or components

Kitchen Cabinets & Countertops	Page 93
Bathroom Vanities	94
Fireplaces & Chimneys	94
Windows, Skylights & Dormers	94
Appliances	95
Breezeways & Porches	95
Finished Attic	95
Garages	96
Site Improvements	96
Wings & Ells	34

		Cost Per Square Foot Of Living Area			% of Total
		Mat.	Inst.	Total	(rounded)
1 Site Work	Site preparation for slab; 4' deep trench excavation for foundation wall.		1.59	1.59	1.5%
2 Foundation	Continuous reinforced concrete footing, 10" deep x 20" wide; damproofed and insulated 8" thick reinforced concrete block foundation wall, 4' deep; trowel finished 4" thick concrete slab on 4" crushed stone base and polyethylene vapor barrier.	4.34	5.74	10.08	9.2%
3 Framing	Exterior walls - 2" x 4" wood studs, 16" O.C.; 1/2" sheathing; gable end roof framing, steep pitch 2" x 8" rafters, 16" O.C. with 1/2" plywood sheathing; 2" x 8" floor joists, 16" O.C. with bridging and 5/8" subflooring; 2" x 4" interior partitions.	6.56	10.15	16.71	15.3%
4 Exterior Walls	Beveled wood siding and housewrap on insulated wood frame walls; R38 attic insulation; sliding wood windows; flush solid core doors, frame and hardware, painted finish; aluminum storm and screen doors.	15.58	5.95	21.53	19.7%
5 Roofing	25 year asphalt roof shingles; #15 felt building paper; aluminum gutters, downspouts, drip edge and flashings.	1.60	1.96	3.56	3.3%
6 Interiors	Walls and ceilings, 1/2" taped and finished gypsum wallboard, primed and painted with 2 coats of finish paint; hollow core wood interior doors, frames and hardware, painted finish; lightweight carpeting with pad, 80%; sheet vinyl flooring, 20%; hardwood tread stairway.	13.86	14.06	27.92	25.6%
7 Specialties	Economy grade kitchen cabinets and countertops; stainless steel kitchen sink; 30 gallon electric water heater.	2.66	.84	3.50	3.2%
8 Mechanical	Three fixture bathroom: bathtub, water closet and wall hung lavatory; gas fired hot air heating system.	3.67	3.51	7.18	6.6%
9 Electrical	100 amp electric service; wiring, duplex & GFI receptacles, wall switches, door bell, appliance circuits and communications cabling; economy grade lighting fixtures.	1.01	1.92	2.93	2.7%
10 Overhead	Contractor's overhead and profit.	7.37	6.88	14.25	13.0%
Total		56.65	52.60	**109.25**	

For customer support on your Residential Costs with RSMeans data, call 800.448.8182.

27

- **Mass produced from stock plans**
- **Single family — 1 full bath, 1 kitchen**
- **No basement**
- **Asphalt shingles on roof**
- **Hot air heat**
- **Gypsum wallboard interior finishes**
- **Materials and workmanship are sufficient to meet codes**

Note: The illustration shown may contain some optional components (for example: garages and/or fireplaces) whose costs are shown in the modifications, adjustments, & alternatives below or at the end of the square foot section.

Base cost per square foot of living area

Exterior Wall	Living Area										
	1000	1200	1400	1600	1800	2000	2200	2600	3000	3400	3800
Wood Siding - Wood Frame	132.05	119.65	113.60	109.45	105.30	100.70	97.65	91.75	86.10	83.55	81.25
Brick Veneer - Wood Frame	138.00	125.15	118.70	114.35	109.90	105.15	101.85	95.60	89.70	86.90	84.45
Stucco on Wood Frame	123.05	111.35	105.80	102.00	98.35	93.95	91.20	86.00	80.80	78.45	76.40
Painted Concrete Block	128.55	116.40	110.55	106.55	102.60	98.05	95.10	89.50	84.05	81.55	79.35
Finished Basement, Add	17.40	16.60	16.00	15.65	15.20	14.95	14.60	14.05	13.65	13.40	13.10
Unfinished Basement, Add	8.05	7.45	6.95	6.65	6.30	6.10	5.85	5.35	5.10	4.85	4.65

Modifications

Add to the total cost

Upgrade Kitchen Cabinets	$ + 1216
Solid Surface Countertops	+ 870
Full Bath - including plumbing, wall and floor finishes	+ 6440
Half Bath - including plumbing, wall and floor finishes	+ 3814
One Car Attached Garage	+ 13,994
One Car Detached Garage	+ 18,103
Fireplace & Chimney	+ 7318

Adjustments

For multi family - add to total cost

Additional Kitchen	$ + 5931
Additional Bath	+ 6440
Additional Entry & Exit	+ 1804
Separate Heating	+ 1650
Separate Electric	+ 1093

For Townhouse/Rowhouse - Multiply cost per square foot by

Inner Unit	.93
End Unit	.96

Alternatives

Add to or deduct from the cost per square foot of living area

Composition Roll Roofing	– .55
Cedar Shake Roof	+ 2.10
Upgrade Walls and Ceilings to Skim Coat Plaster	+ .82
Upgrade Ceilings to Textured Finish	+ .60
Air Conditioning, in Heating Ductwork	+ 2.91
In Separate Ductwork	+ 5.68
Heating Systems, Hot Water	+ 1.48
Heat Pump	+ 1.53
Electric Heat	– 1
Not Heated	– 3.75

Additional upgrades or components

Kitchen Cabinets & Countertops	Page 93
Bathroom Vanities	94
Fireplaces & Chimneys	94
Windows, Skylights & Dormers	94
Appliances	95
Breezeways & Porches	95
Finished Attic	95
Garages	96
Site Improvements	96
Wings & Ells	34

		Cost Per Square Foot Of Living Area			% of Total
		Mat.	Inst.	Total	(rounded)
1 Site Work	Site preparation for slab; 4' deep trench excavation for foundation wall.		1.28	1.28	1.3%
2 Foundation	Continuous reinforced concrete footing, 10" deep x 20" wide; damproofed and insulated 8" thick reinforced concrete block foundation wall, 4' deep; trowel finished 4" thick concrete slab on 4" crushed stone base and polyethylene vapor barrier.	3.48	4.59	8.07	8.0%
3 Framing	Exterior walls - 2" x 4" wood studs, 16" O.C.; 1/2" sheathing; wood truss roof frame, 24" O.C. with 1/2" plywood sheating; 2" x 8" floor joists, 16" O.C. with bridging and 5/8" subflooring; 2" x 4" interior partitions.	6.04	8.94	14.98	14.9%
4 Exterior Walls	Beveled wood siding and housewrap on insulated wood frame walls; R38 attic insulation; sliding wood windows; flush solid core doors, frame and hardware, painted finish; aluminum storm and screen doors.	15.37	5.89	21.26	21.1%
5 Roofing	25 year asphalt roof shingles; #15 felt building paper; aluminum gutters, downspouts, drip edge and flashings.	1.10	1.36	2.46	2.4%
6 Interiors	Walls and ceilings, 1/2" taped and finished gypsum wallboard, primed and painted with 2 coats of finish paint; hollow core wood interior doors, frames and hardware, painted finish; lightweight carpeting with pad, 80%; sheet vinyl flooring, 20%; hardwood tread stairway.	13.53	13.98	27.51	27.3%
7 Specialties	Economy grade kitchen cabinets and countertops; stainless steel kitchen sink; 30 gallon electric water heater.	2.13	.68	2.81	2.8%
8 Mechanical	Three fixture bathroom: bathtub, water closet and wall hung lavatory; gas fired hot air heating system.	3.14	3.27	6.41	6.4%
9 Electrical	100 amp electric service; wiring, duplex & GFI receptacles, wall switches, door bell, appliance circuits and communications cabling; economy grade lighting fixtures.	.97	1.82	2.79	2.8%
10 Overhead	Contractor's overhead and profit.	6.84	6.29	13.13	13.0%
	Total	52.60	48.10	**100.70**	

For customer support on your Residential Costs with RSMeans data, call 800.448.8182.

29

- **Mass produced from stock plans**
- **Single family — 1 full bath, 1 kitchen**
- **No basement**
- **Asphalt shingles on roof**
- **Hot air heat**
- **Gypsum wallboard interior finishes**
- **Materials and workmanship are sufficient to meet codes**

Note: The illustration shown may contain some optional components (for example: garages and/or fireplaces) whose costs are shown in the modifications, adjustments, & alternatives below or at the end of the square foot section.

Base cost per square foot of living area

Exterior Wall	Living Area										
	1000	1200	1400	1600	1800	2000	2200	2600	3000	3400	3800
Wood Siding - Wood Frame	122.25	110.55	105.10	101.30	97.70	93.35	90.65	85.45	80.30	78.00	76.00
Brick Veneer - Wood Frame	126.75	114.75	109.05	105.05	101.20	96.70	93.85	88.45	83.00	80.55	78.40
Stucco on Wood Frame	115.40	104.20	99.20	95.65	92.35	88.20	85.75	81.05	76.25	74.10	72.30
Painted Concrete Block	119.55	108.00	102.75	99.05	95.50	91.30	88.70	83.70	78.65	76.45	74.50
Finished Basement, Add	17.40	16.60	16.00	15.65	15.20	14.95	14.60	14.05	13.65	13.40	13.10
Unfinished Basement, Add	8.05	7.45	6.95	6.65	6.30	6.10	5.85	5.35	5.10	4.85	4.65

Modifications

Add to the total cost

Upgrade Kitchen Cabinets	$ + 1216
Solid Surface Countertops	+ 870
Full Bath - including plumbing, wall and floor finishes	+ 6440
Half Bath - including plumbing, wall and floor finishes	+ 3814
One Car Attached Garage	+ 13,994
One Car Detached Garage	+ 18,103
Fireplace & Chimney	+ 6624

Adjustments

For multi family - add to total cost

Additional Kitchen	$ + 5931
Additional Bath	+ 6440
Additional Entry & Exit	+ 1804
Separate Heating	+ 1650
Separate Electric	+ 1093

For Townhouse/Rowhouse -
Multiply cost per square foot by

Inner Unit	.94
End Unit	.97

Alternatives

Add to or deduct from the cost per square foot of living area

Composition Roll Roofing	– .55
Cedar Shake Roof	+ 2.10
Upgrade Walls and Ceilings to Skim Coat Plaster	+ .77
Upgrade Ceilings to Textured Finish	+ .60
Air Conditioning, in Heating Ductwork	+ 2.91
In Separate Ductwork	+ 5.68
Heating Systems, Hot Water	+ 1.48
Heat Pump	+ 1.53
Electric Heat	– 1
Not Heated	– 3.75

Additional upgrades or components

Kitchen Cabinets & Countertops	Page 93
Bathroom Vanities	94
Fireplaces & Chimneys	94
Windows, Skylights & Dormers	94
Appliances	95
Breezeways & Porches	95
Finished Attic	95
Garages	96
Site Improvements	96
Wings & Ells	34

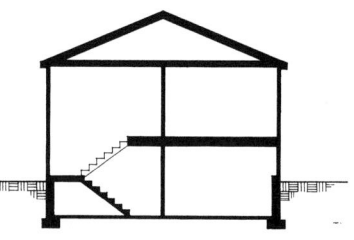

		Cost Per Square Foot Of Living Area			% of Total
		Mat.	Inst.	Total	(rounded)
1 Site Work	Site preparation for slab; 4' deep trench excavation for foundation wall.		1.28	1.28	1.4%
2 Foundation	Continuous reinforced concrete footing, 10" deep x 20" wide; damproofed and insulated 8" thick reinforced concrete block foundation wall, 4' deep; trowel finished 4" thick concrete slab on 4" crushed stone base and polyethylene vapor barrier.	3.48	4.59	8.07	8.6%
3 Framing	Exterior walls - 2" x 4" wood studs, 16" O.C.; 1/2" sheathing; wood truss roof frame, 24" O.C. with 1/2" plywood sheating; 2" x 8" floor joists, 16" O.C. with bridging and 5/8" subflooring; 2" x 4" interior partitions.	5.58	8.28	13.86	14.8%
4 Exterior Walls	Beveled wood siding and housewrap on insulated wood frame walls; R38 attic insulation; sliding wood windows; flush solid core doors, frame and hardware, painted finish; aluminum storm and screen doors.	12.06	4.66	16.72	17.9%
5 Roofing	25 year asphalt roof shingles; #15 felt building paper; aluminum gutters, downspouts, drip edge and flashings.	1.10	1.36	2.46	2.6%
6 Interiors	Walls and ceilings, 1/2" taped and finished gypsum wallboard, primed and painted with 2 coats of finish paint; hollow core wood interior doors, frames and hardware, painted finish; lightweight carpeting with pad, 80%; sheet vinyl flooring, 20%; hardwood tread stairway.	13.23	13.54	26.77	28.7%
7 Specialties	Economy grade kitchen cabinets and countertops; stainless steel kitchen sink; 30 gallon electric water heater.	2.13	.68	2.81	3.0%
8 Mechanical	Three fixture bathroom: bathtub, water closet and wall hung lavatory; gas fired hot air heating system.	3.14	3.27	6.41	6.9%
9 Electrical	100 amp electric service; wiring, duplex & GFI receptacles, wall switches, door bell, appliance circuits and communications cabling; economy grade lighting fixtures.	.97	1.82	2.79	3.0%
10 Overhead	Contractor's overhead and profit.	6.26	5.92	12.18	13.0%
Total		47.95	45.40	**93.35**	

- **Mass produced from stock plans**
- **Single family — 1 full bath, 1 kitchen**
- **No basement**
- **Asphalt shingles on roof**
- **Hot air heat**
- **Gypsum wallboard interior finishes**
- **Materials and workmanship are sufficient to meet codes**

Note: The illustration shown may contain some optional components (for example: garages and/or fireplaces) whose costs are shown in the modifications, adjustments, & alternatives below or at the end of the square foot section.

©Design Basics, Inc.

Base cost per square foot of living area

Exterior Wall	Living Area										
	1200	1500	1800	2000	2200	2400	2800	3200	3600	4000	4400
Wood Siding - Wood Frame	113.85	104.35	97.40	94.55	90.50	87.15	84.60	81.00	77.05	75.60	72.40
Brick Veneer - Wood Frame	118.00	108.15	100.80	97.85	93.60	90.10	87.40	83.60	79.45	77.95	74.55
Stucco on Wood Frame	107.45	98.55	92.20	89.60	85.80	82.75	80.35	77.10	73.40	72.00	69.05
Solid Masonry	111.20	102.00	95.25	92.50	88.55	85.35	82.80	79.40	75.50	74.15	71.05
Finished Basement, Add*	20.80	19.85	19.05	18.65	18.30	17.90	17.60	17.15	16.75	16.60	16.25
Unfinished Basement, Add*	8.85	8.10	7.45	7.10	6.85	6.55	6.30	5.95	5.65	5.50	5.25

*Basement under middle level only.

Modifications

Add to the total cost

Upgrade Kitchen Cabinets	$ + 1216
Solid Surface Countertops	+ 870
Full Bath - including plumbing, wall and floor finishes	+ 6440
Half Bath - including plumbing, wall and floor finishes	+ 3814
One Car Attached Garage	+ 13,994
One Car Detached Garage	+ 18,103
Fireplace & Chimney	+ 6624

Adjustments

For multi family - add to total cost

Additional Kitchen	$ + 5931
Additional Bath	+ 6440
Additional Entry & Exit	+ 1804
Separate Heating	+ 1650
Separate Electric	+ 1093

For Townhouse/Rowhouse -
Multiply cost per square foot by

Inner Unit	.93
End Unit	.96

Alternatives

Add to or deduct from the cost per square foot of living area

Composition Roll Roofing	– .75
Cedar Shake Roof	+ 3.05
Upgrade Walls and Ceilings to Skim Coat Plaster	+ .70
Upgrade Ceilings to Textured Finish	+ .60
Air Conditioning, in Heating Ductwork	+ 2.48
In Separate Ductwork	+ 5.28
Heating Systems, Hot Water	+ 1.42
Heat Pump	+ 1.59
Electric Heat	– .86
Not Heated	– 3.64

Additional upgrades or components

Kitchen Cabinets & Countertops	Page 93
Bathroom Vanities	94
Fireplaces & Chimneys	94
Windows, Skylights & Dormers	94
Appliances	95
Breezeways & Porches	95
Finished Attic	95
Garages	96
Site Improvements	96
Wings & Ells	34

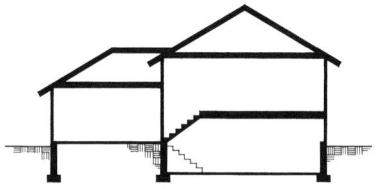

		Cost Per Square Foot Of Living Area			% of Total
		Mat.	Inst.	Total	(rounded)
1 Site Work	Site preparation for slab; 4' deep trench excavation for foundation wall.		1.06	1.06	1.2%
2 Foundation	Continuous reinforced concrete footing, 10" deep x 20" wide; damproofed and insulated 8" thick reinforced concrete block foundation wall, 4' deep; trowel finished 4" thick concrete slab on 4" crushed stone base and polyethylene vapor barrier.	3.90	4.98	8.88	10.2%
3 Framing	Exterior walls - 2" x 4" wood studs, 16" O.C.; 1/2" sheathing; wood truss roof frame, 24" O.C. with 1/2" plywood sheating; 2" x 8" floor joists, 16" O.C. with bridging and 5/8" subflooring; 2" x 4" interior partitions.	5.20	7.42	12.62	14.5%
4 Exterior Walls	Beveled wood siding and housewrap on insulated wood frame walls; R38 attic insulation; sliding wood windows; flush solid core doors, frame and hardware, painted finish; aluminum storm and screen doors.	10.62	4.05	14.67	16.8%
5 Roofing	25 year asphalt roof shingles; #15 felt building paper; aluminum gutters, downspouts, drip edge and flashings.	1.47	1.81	3.28	3.8%
6 Interiors	Walls and ceilings, 1/2" taped and finished gypsum wallboard, primed and painted with 2 coats of finish paint; hollow core wood interior doors, frames and hardware, painted finish; lightweight carpeting with pad, 80%; sheet vinyl flooring, 20%; hardwood tread stairway.	12.15	12.23	24.38	28.0%
7 Specialties	Economy grade kitchen cabinets and countertops; stainless steel kitchen sink; 30 gallon electric water heater.	1.78	.56	2.34	2.7%
8 Mechanical	Three fixture bathroom: bathtub, water closet and wall hung lavatory; gas fired hot air heating system.	2.77	3.11	5.88	6.7%
9 Electrical	100 amp electric service; wiring, duplex & GFI receptacles, wall switches, door bell, appliance circuits and communications cabling; economy grade lighting fixtures.	.95	1.75	2.70	3.1%
10 Overhead	Contractor's overhead and profit.	5.81	5.53	11.34	13.0%
Total		44.65	42.50	**87.15**	

1 Story — Base cost per square foot of living area

Exterior Wall	Living Area							
	50	100	200	300	400	500	600	700
Wood Siding - Wood Frame	196.40	149.15	128.70	105.90	99.25	95.20	92.60	93.10
Brick Veneer - Wood Frame	209.55	158.55	136.55	111.20	104.00	99.65	96.80	97.15
Stucco on Wood Frame	176.35	134.85	116.80	98.00	92.10	88.55	86.20	86.95
Painted Concrete Block	189.15	144.05	124.45	103.10	96.70	92.90	90.30	90.85
Finished Basement, Add	50.85	41.40	37.50	30.95	29.60	28.85	28.35	27.95
Unfinished Basement, Add	28.40	21.05	18.05	12.95	11.90	11.35	10.90	10.60

1-1/2 Story — Base cost per square foot of living area

Exterior Wall	Living Area							
	100	200	300	400	500	600	700	800
Wood Siding - Wood Frame	155.70	125.15	106.35	94.30	88.65	85.80	82.30	81.40
Brick Veneer - Wood Frame	167.50	134.60	114.20	100.45	94.30	91.15	87.35	86.40
Stucco on Wood Frame	137.85	110.85	94.45	85.05	80.05	77.70	74.70	73.80
Painted Concrete Block	149.30	120.00	102.05	90.95	85.55	82.90	79.60	78.65
Finished Basement, Add	33.95	30.00	27.35	24.50	23.70	23.20	22.70	22.65
Unfinished Basement, Add	17.45	14.45	12.35	10.15	9.50	9.10	8.75	8.70

2 Story — Base cost per square foot of living area

Exterior Wall	Living Area							
	100	200	400	600	800	1000	1200	1400
Wood Siding - Wood Frame	159.60	118.30	100.30	81.60	75.75	72.25	69.90	70.65
Brick Veneer - Wood Frame	172.85	127.70	108.20	86.90	80.45	76.65	74.10	74.70
Stucco on Wood Frame	139.65	104.00	88.40	73.65	68.65	65.55	63.55	64.50
Painted Concrete Block	152.45	113.20	96.05	78.80	73.20	69.85	67.60	68.45
Finished Basement, Add	25.45	20.75	18.80	15.55	14.85	14.45	14.20	14.05
Unfinished Basement, Add	14.25	10.55	9.05	6.50	5.95	5.70	5.45	5.30

Base costs do not include bathroom or kitchen facilities. Use Modifications/Adjustments/Alternatives on pages 93–96 where appropriate.

Did you know?

RSMeans data is available through our online application with 24/7 access:

- Search for unit prices by keyword
- Leverage the most up-to-date data
- Build and export estimates

Try it free for 30 days!
www.rsmeans.com/2018freetrial

1 Story

1-1/2 Story

2 Story

2-1/2 Story

Bi-Level

Tri-Level

For customer support on your Residential Costs with RSMeans data, call 800.448.8182.

37

- **Simple design from standard plans**
- **Single family — 1 full bath, 1 kitchen**
- **No basement**
- **Asphalt shingles on roof**
- **Hot air heat**
- **Gypsum wallboard interior finishes**
- **Materials and workmanship are average**

Note: The illustration shown may contain some optional components (for example: garages and/or fireplaces) whose costs are shown in the modifications, adjustments, & alternatives below or at the end of the square foot section.

•Home Planners, Inc.

Base cost per square foot of living area

Exterior Wall	Living Area										
	600	800	1000	1200	1400	1600	1800	2000	2400	2800	3200
Wood Siding - Wood Frame	172.50	154.65	141.65	131.25	122.50	116.75	113.60	109.85	102.45	97.05	93.30
Brick Veneer - Wood Frame	178.70	160.15	146.60	135.70	126.60	120.60	117.35	113.30	105.65	100.00	96.00
Stucco on Wood Frame	167.35	150.00	137.45	127.45	119.00	113.50	110.50	106.85	99.80	94.60	91.05
Solid Masonry	191.80	171.85	157.20	145.30	135.30	128.75	125.20	120.70	112.40	106.25	101.80
Finished Basement, Add	40.50	39.05	37.25	35.55	34.15	33.30	32.80	32.05	31.05	30.25	29.55
Unfinished Basement, Add	16.80	15.15	14.00	13.00	12.15	11.65	11.25	10.85	10.25	9.75	9.35

Modifications

Add to the total cost

Upgrade Kitchen Cabinets	$ + 5950
Solid Surface Countertops (Included)	
Full Bath - including plumbing, wall and floor finishes	+ 8050
Half Bath - including plumbing, wall and floor finishes	+ 4767
One Car Attached Garage	+ 15,086
One Car Detached Garage	+ 19,847
Fireplace & Chimney	+ 6942

Adjustments

For multi family - add to total cost

Additional Kitchen	$ + 9963
Additional Bath	+ 8050
Additional Entry & Exit	+ 1804
Separate Heating	+ 1650
Separate Electric	+ 1817

For Townhouse/Rowhouse - Multiply cost per square foot by

Inner Unit	.92
End Unit	.96

Alternatives

Add to or deduct from the cost per square foot of living area

Cedar Shake Roof	+ 3.50
Clay Tile Roof	+ 6.90
Slate Roof	+ 7.70
Upgrade Walls to Skim Coat Plaster	+ .49
Upgrade Ceilings to Textured Finish	+ .60
Air Conditioning, in Heating Ductwork	+ 4.95
In Separate Ductwork	+ 7.32
Heating Systems, Hot Water	+ 1.63
Heat Pump	+ 1.37
Electric Heat	– .79
Not Heated	– 3.61

Additional upgrades or components

Kitchen Cabinets & Countertops	Page 93
Bathroom Vanities	94
Fireplaces & Chimneys	94
Windows, Skylights & Dormers	94
Appliances	95
Breezeways & Porches	95
Finished Attic	95
Garages	96
Site Improvements	96
Wings & Ells	56

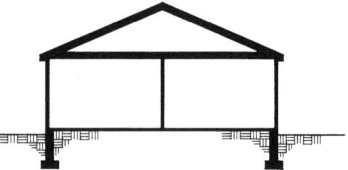

			Cost Per Square Foot Of Living Area			% of Total
			Mat.	Inst.	Total	(rounded)
1	**Site Work**	Site preparation for slab; 4' deep trench excavation for foundation wall.		1.63	1.63	1.4%
2	**Foundation**	Continuous reinforced concrete footing, 10" deep x 20" wide; damproofed and insulated 8" thick reinforced concrete block foundation wall, 4' deep; trowel finished 4" thick concrete slab on 4" crushed stone base and polyethylene vapor barrier.	6.02	7.67	13.69	11.7%
3	**Framing**	Exterior walls - 2" x 6" wood studs, 16" O.C.; 1/2" sheathing; gable end roof framing, 2" x 10" rafters, 16" O.C. with 1/2" plywood sheathing; 2" x 4" interior partitions.	7.05	9.74	16.79	14.4%
4	**Exterior Walls**	Beveled wood siding and housewrap on insulated wood frame walls; R38 attic insulation; double hung wood windows; flush solid core doors, frame and hardware, painted finish; aluminum storm and screen doors.	13.28	5.25	18.53	15.9%
5	**Roofing**	25 year asphalt roof shingles; #15 felt building paper; aluminum gutters, downspouts, drip edge and flashings.	2.26	2.79	5.05	4.3%
6	**Interiors**	Walls & ceilings, 1/2" taped & finished gypsum wallboard, primed & painted with 2 coats of finish paint; birch faced hollow core interior doors, frames & hardware, painted finish; medium weight carpeting with pad, 40%; sheet vinyl, 15%; oak hardwood, 40%; ceramic tile, 5%	12.89	13.23	26.12	22.4%
7	**Specialties**	Average grade kitchen cabinets and countertop; stainless steel kitchen sink; 40 gallon electric water heater.	4.76	1.50	6.26	5.4%
8	**Mechanical**	Three fixture bathroom: bathtub, water closet, vanity and sink; gas fired hot air heating system.	4.04	3.61	7.65	6.6%
9	**Electrical**	200 amp electric service; wiring, duplex & GFI receptacles, wall switches, door bell, appliance circuits, fans and communications cabling; average grade lighting fixtures.	1.53	2.54	4.07	3.5%
10	**Overhead**	Contractor's overhead and profit and plans.	8.82	8.14	16.96	14.5%
	Total		60.65	56.10	**116.75**	

For customer support on your Residential Costs with RSMeans data, call 800.448.8182.

39

- **Simple design from standard plans**
- **Single family — 1 full bath, 1 kitchen**
- **No basement**
- **Asphalt shingles on roof**
- **Hot air heat**
- **Gypsum wallboard interior finishes**
- **Materials and workmanship are average**

Note: The illustration shown may contain some optional components (for example: garages and/or fireplaces) whose costs are shown in the modifications, adjustments, & alternatives below or at the end of the square foot section.

©By Designer

Base cost per square foot of living area

Exterior Wall	Living Area										
	600	800	1000	1200	1400	1600	1800	2000	2400	2800	3200
Wood Siding - Wood Frame	197.00	163.20	145.95	137.45	131.35	122.40	118.05	113.50	104.30	100.65	96.70
Brick Veneer - Wood Frame	205.80	169.65	151.85	143.00	136.60	127.20	122.55	117.85	108.15	104.30	100.05
Stucco on Wood Frame	189.40	157.75	140.85	132.65	126.85	118.30	114.15	109.75	101.00	97.50	93.85
Solid Masonry	223.70	182.60	163.90	154.25	147.35	136.95	131.80	126.55	115.85	111.60	106.80
Finished Basement, Add	33.10	28.75	27.55	26.55	25.90	24.80	24.25	23.75	22.60	22.20	21.60
Unfinished Basement, Add	14.55	11.40	10.55	9.95	9.50	8.90	8.50	8.20	7.55	7.25	6.90

Modifications

Add to the total cost

Upgrade Kitchen Cabinets	$ + 5950
Solid Surface Countertops (Included)	
Full Bath - including plumbing, wall and floor finishes	+ 8050
Half Bath - including plumbing, wall and floor finishes	+ 4767
One Car Attached Garage	+ 15,086
One Car Detached Garage	+ 19,847
Fireplace & Chimney	+ 6942

Adjustments

For multi family - add to total cost

Additional Kitchen	$ + 9963
Additional Bath	+ 8050
Additional Entry & Exit	+ 1804
Separate Heating	+ 1650
Separate Electric	+ 1817

*For Townhouse/Rowhouse -
Multiply cost per square foot by*

Inner Unit	.92
End Unit	.96

Alternatives

Add to or deduct from the cost per square foot of living area

Cedar Shake Roof	+ 2.55
Clay Tile Roof	+ 5
Slate Roof	+ 5.60
Upgrade Walls to Skim Coat Plaster	+ .57
Upgrade Ceilings to Textured Finish	+ .60
Air Conditioning, in Heating Ductwork	+ 3.76
In Separate Ductwork	+ 6.40
Heating Systems, Hot Water	+ 1.55
Heat Pump	+ 1.51
Electric Heat	– .72
Not Heated	– 3.46

Additional upgrades or components

Kitchen Cabinets & Countertops	Page 93
Bathroom Vanities	94
Fireplaces & Chimneys	94
Windows, Skylights & Dormers	94
Appliances	95
Breezeways & Porches	95
Finished Attic	95
Garages	96
Site Improvements	96
Wings & Ells	56

		Cost Per Square Foot Of Living Area			% of Total
		Mat.	Inst.	Total	(rounded)
1 Site Work	Site preparation for slab; 4' deep trench excavation for foundation wall.		1.46	1.46	1.2%
2 Foundation	Continuous reinforced concrete footing, 10" deep x 20" wide; damproofed and insulated 8" thick reinforced concrete block foundation wall, 4' deep; trowel finished 4" thick concrete slab on 4" crushed stone base and polyethylene vapor barrier.	4.31	5.64	9.95	8.4%
3 Framing	Exterior walls - 2" x 6" wood studs, 16" O.C.; 1/2" sheathing; gable end roof framing, steep pitch 2" x 10" rafters, 16" O.C. with 1/2" plywood sheathing; 2" x 10" floor joists, 16" O.C. with bridging and 5/8" subflooring; 2" x 4" interior partitions.	7.80	10.60	18.40	15.6%
4 Exterior Walls	Beveled wood siding and housewrap on insulated wood frame walls; R38 attic insulation; double hung wood windows; flush solid core doors, frame and hardware, painted finish; aluminum storm and screen doors.	14.60	5.76	20.36	17.2%
5 Roofing	25 year asphalt roof shingles; #15 felt building paper; aluminum gutters, downspouts, drip edge and flashings.	1.64	2.01	3.65	3.1%
6 Interiors	Walls & ceilings, 1/2" taped & finished gypsum wallboard, primed & painted with 2 coats of finish paint; birch faced hollow core interior doors, frames & hardware, painted finish; medium weight carpeting with pad, 40%; sheet vinyl, 15%; oak hardwood, 40%; ceramic tile, 5%; hardwood tread stairway.	15.32	15.05	30.37	25.7%
7 Specialties	Average grade kitchen cabinets and countertop; stainless steel kitchen sink; 40 gallon electric water heater.	4.23	1.34	5.57	4.7%
8 Mechanical	Three fixture bathroom: bathtub, water closet, vanity and sink; gas fired hot air heating system.	3.71	3.46	7.17	6.1%
9 Electrical	200 amp electric service; wiring, duplex & GFI receptacles, wall switches, door bell, appliance circuits, fans and communications cabling; average grade lighting fixtures.	1.49	2.45	3.94	3.3%
10 Overhead	Contractor's overhead and profit and plans.	9.05	8.13	17.18	14.6%
	Total	62.15	55.90	**118.05**	

For customer support on your Residential Costs with RSMeans data, call 800.448.8182.

41

- Simple design from standard plans
- Single family — 1 full bath, 1 kitchen
- No basement
- Asphalt shingles on roof
- Hot air heat
- Gypsum wallboard interior finishes
- Materials and workmanship are average

Note: The illustration shown may contain some optional components (for example: garages and/or fireplaces) whose costs are shown in the modifications, adjustments, & alternatives below or at the end of the square foot section.

Base cost per square foot of living area

Exterior Wall	Living Area										
	1000	1200	1400	1600	1800	2000	2200	2600	3000	3400	3800
Wood Siding - Wood Frame	149.75	135.30	128.30	123.35	118.45	113.25	109.75	103.15	96.85	93.90	91.20
Brick Veneer - Wood Frame	156.15	141.25	133.80	128.60	123.40	118.05	114.25	107.25	100.60	97.50	94.65
Stucco on Wood Frame	144.50	130.50	123.75	119.00	114.40	109.35	106.00	99.80	93.65	90.90	88.40
Solid Masonry	170.60	154.65	146.35	140.60	134.75	128.90	124.65	116.60	109.25	105.65	102.45
Finished Basement, Add	23.25	22.90	22.10	21.65	21.05	20.70	20.30	19.55	18.95	18.60	18.30
Unfinished Basement, Add	9.20	8.55	8.05	7.75	7.40	7.15	6.95	6.45	6.15	5.90	5.70

Modifications

Add to the total cost

Upgrade Kitchen Cabinets	$ + 5950
Solid Surface Countertops (Included)	
Full Bath - including plumbing, wall and floor finishes	+ 8050
Half Bath - including plumbing, wall and floor finishes	+ 4767
One Car Attached Garage	+ 15,086
One Car Detached Garage	+ 19,847
Fireplace & Chimney	+ 7671

Adjustments

For multi family - add to total cost

Additional Kitchen	$ + 9963
Additional Bath	+ 8050
Additional Entry & Exit	+ 1804
Separate Heating	+ 1650
Separate Electric	+ 1817

For Townhouse/Rowhouse - Multiply cost per square foot by

Inner Unit	.90
End Unit	.95

Alternatives

Add to or deduct from the cost per square foot of living area

Cedar Shake Roof	+ 1.75
Clay Tile Roof	+ 3.45
Slate Roof	+ 3.85
Upgrade Walls to Skim Coat Plaster	+ .59
Upgrade Ceilings to Textured Finish	+ .60
Air Conditioning, in Heating Ductwork	+ 3
In Separate Ductwork	+ 5.81
Heating Systems, Hot Water	+ 1.52
Heat Pump	+ 1.59
Electric Heat	– .62
Not Heated	– 3.36

Additional upgrades or components

Kitchen Cabinets & Countertops	Page 93
Bathroom Vanities	94
Fireplaces & Chimneys	94
Windows, Skylights & Dormers	94
Appliances	95
Breezeways & Porches	95
Finished Attic	95
Garages	96
Site Improvements	96
Wings & Ells	56

			Cost Per Square Foot Of Living Area			% of Total
			Mat.	Inst.	Total	(rounded)
1	**Site Work**	Site preparation for slab; 4' deep trench excavation for foundation wall.		1.32	1.32	1.2%
2	**Foundation**	Continuous reinforced concrete footing, 10" deep x 20" wide; damproofed and insulated 8" thick reinforced concrete block foundation wall, 4' deep; trowel finished 4" thick concrete slab on 4" crushed stone base and polyethylene vapor barrier.	3.58	4.72	8.30	7.3%
3	**Framing**	Exterior walls - 2" x 6" wood studs, 16" O.C.; 1/2" sheathing; gable end roof framing, 2" x 10" rafters, 16" O.C. with 1/2" plywood sheathing; 2" x 10" floor joists, 16" O.C. with bridging and 5/8" subflooring; 2" x 4" interior partitions.	7.41	10.02	17.43	15.4%
4	**Exterior Walls**	Beveled wood siding and housewrap on insulated wood frame walls; R38 attic insulation; double hung wood windows; flush solid core doors, frame and hardware, painted finish; aluminum storm and screen doors.	15.20	6.03	21.23	18.7%
5	**Roofing**	25 year asphalt roof shingles; #15 felt building paper; aluminum gutters, downspouts, drip edge and flashings.	1.13	1.40	2.53	2.2%
6	**Interiors**	Walls & ceilings, 1/2" taped & finished gypsum wallboard, primed & painted with 2 coats of finish paint; birch faced hollow core interior doors, frames & hardware, painted finish; medium weight carpeting with pad, 40%; sheet vinyl, 15%; oak hardwood, 40%; ceramic tile, 5%; hardwood tread stairway.	15.23	15.14	30.37	26.8%
7	**Specialties**	Average grade kitchen cabinets and countertop; stainless steel kitchen sink; 40 gallon electric water heater.	3.81	1.21	5.02	4.4%
8	**Mechanical**	Three fixture bathroom: bathtub, water closet, vanity and sink; gas fired hot air heating system.	3.44	3.35	6.79	6.0%
9	**Electrical**	200 amp electric service; wiring, duplex & GFI receptacles, wall switches, door bell, appliance circuits, fans and communications cabling; average grade lighting fixtures	1.46	2.37	3.83	3.4%
10	**Overhead**	Contractor's overhead and profit and plans.	8.69	7.74	16.43	14.5%
		Total	59.95	53.30	**113.25**	

- **Simple design from standard plans**
- **Single family — 1 full bath, 1 kitchen**
- **No basement**
- **Asphalt shingles on roof**
- **Hot air heat**
- **Gypsum wallboard interior finishes**
- **Materials and workmanship are average**

Note: The illustration shown may contain some optional components (for example: garages and/or fireplaces) whose costs are shown in the modifications, adjustments, & alternatives below or at the end of the square foot section.

Base cost per square foot of living area

Exterior Wall	Living Area										
	1200	1400	1600	1800	2000	2400	2800	3200	3600	4000	4400
Wood Siding - Wood Frame	149.55	139.90	127.50	125.05	120.30	112.75	106.90	100.90	97.90	92.55	90.70
Brick Veneer - Wood Frame	156.50	146.20	133.30	130.85	125.70	117.65	111.60	105.20	102.00	96.30	94.40
Stucco on Wood Frame	143.70	134.60	122.65	120.20	115.80	108.60	103.00	97.35	94.55	89.40	87.70
Solid Masonry	171.65	159.90	145.95	143.50	137.65	128.40	121.95	114.50	110.75	104.50	102.35
Finished Basement, Add	19.70	19.30	18.55	18.45	17.95	17.20	16.85	16.25	15.90	15.55	15.40
Unfinished Basement, Add	7.65	7.05	6.65	6.50	6.25	5.75	5.55	5.20	5.00	4.75	4.65

Modifications

Add to the total cost

Upgrade Kitchen Cabinets	$ + 5950
Solid Surface Countertops (Included)	
Full Bath - including plumbing, wall and floor finishes	+ 8050
Half Bath - including plumbing, wall and floor finishes	+ 4767
One Car Attached Garage	+ 15,086
One Car Detached Garage	+ 19,847
Fireplace & Chimney	+ 8415

Adjustments

For multi family - add to total cost

Additional Kitchen	$ + 9963
Additional Bath	+ 8050
Additional Entry & Exit	+ 1804
Separate Heating	+ 1650
Separate Electric	+ 1817

For Townhouse/Rowhouse - Multiply cost per square foot by

Inner Unit	.90
End Unit	.95

Alternatives

Add to or deduct from the cost per square foot of living area

Cedar Shake Roof	+ 1.55
Clay Tile Roof	+ 3
Slate Roof	+ 3.35
Upgrade Walls to Skim Coat Plaster	+ .56
Upgrade Ceilings to Textured Finish	+ .60
Air Conditioning, in Heating Ductwork	+ 2.73
In Separate Ductwork	+ 5.62
Heating Systems, Hot Water	+ 1.38
Heat Pump	+ 1.63
Electric Heat	– 1.11
Not Heated	– 3.92

Additional upgrades or components

Kitchen Cabinets & Countertops	Page 93
Bathroom Vanities	94
Fireplaces & Chimneys	94
Windows, Skylights & Dormers	94
Appliances	95
Breezeways & Porches	95
Finished Attic	95
Garages	96
Site Improvements	96
Wings & Ells	56

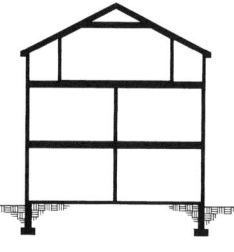

			Cost Per Square Foot Of Living Area			% of Total
			Mat.	Inst.	Total	(rounded)
1	**Site Work**	Site preparation for slab; 4' deep trench excavation for foundation wall.		.82	.82	0.8%
2	**Foundation**	Continuous reinforced concrete footing, 10" deep x 20" wide; damproofed and insulated 8" thick reinforced concrete block foundation wall, 4' deep; trowel finished 4" thick concrete slab on 4" crushed stone base and polyethylene vapor barrier.	2.55	3.34	5.89	5.8%
3	**Framing**	Exterior walls - 2" x 6" wood studs, 16" O.C.; 1/2" sheathing; gable end roof framing, steep pitch 2" x 10" rafters, 16" O.C. with 1/2" plywood sheathing; 2" x 10" floor joists, 16" O.C. with bridging and 5/8" subflooring; 2" x 4" interior partitions.	7.37	9.92	17.29	17.1%
4	**Exterior Walls**	Beveled wood siding and housewrap on insulated wood frame walls; R38 attic insulation; double hung wood windows; flush solid core doors, frame and hardware, painted finish; aluminum storm and screen doors.	13.05	5.12	18.17	18.0%
5	**Roofing**	25 year asphalt roof shingles; #15 felt building paper; aluminum gutters, downspouts, drip edge and flashings.	1.01	1.24	2.25	2.2%
6	**Interiors**	Walls & ceilings, 1/2" taped & finished gypsum wallboard, primed & painted with 2 coats of finish paint; birch faced hollow core interior doors, frames & hardware, painted finish; medium weight carpeting with pad, 40%; sheet vinyl, 15%; oak hardwood, 40%; ceramic tile, 5%; hardwood tread stairway.	15.05	14.64	29.69	29.4%
7	**Specialties**	Average grade kitchen cabinets and countertop; stainless steel kitchen sink; 40 gallon electric water heater.	2.39	.75	3.14	3.1%
8	**Mechanical**	Three fixture bathroom: bathtub, water closet, vanity and sink; gas fired hot air heating system.	2.52	2.99	5.51	5.5%
9	**Electrical**	200 amp electric service; wiring, duplex & GFI receptacles, wall switches, door bell, appliance circuits, fans and communications cabling; average grade lighting fixtures.	1.36	2.11	3.47	3.4%
10	**Overhead**	Contractor's overhead and profit and plans.	7.70	6.97	14.67	14.5%
	Total		53.00	47.90	**100.90**	

- **Simple design from standard plans**
- **Single family — 1 full bath, 1 kitchen**
- **No basement**
- **Asphalt shingles on roof**
- **Hot air heat**
- **Gypsum wallboard interior finishes**
- **Materials and workmanship are average**

Note: The illustration shown may contain some optional components (for example: garages and/or fireplaces) whose costs are shown in the modifications, adjustments, & alternatives below or at the end of the square foot section.

Base cost per square foot of living area

Exterior Wall	Living Area										
	1500	1800	2100	2500	3000	3500	4000	4500	5000	5500	6000
Wood Siding - Wood Frame	135.70	122.80	116.95	112.30	103.80	100.00	94.85	89.25	87.45	85.40	83.25
Brick Veneer - Wood Frame	141.95	128.65	122.35	117.45	108.55	104.40	98.90	93.00	91.15	88.90	86.50
Stucco on Wood Frame	130.55	118.10	112.55	108.00	99.95	96.35	91.55	86.20	84.50	82.55	80.55
Solid Masonry	156.60	142.20	135.05	129.55	119.50	114.85	108.35	101.70	99.55	97.05	94.15
Finished Basement, Add	16.85	16.65	16.15	15.75	15.15	14.85	14.40	14.00	13.90	13.70	13.45
Unfinished Basement, Add	6.20	5.80	5.45	5.25	4.90	4.70	4.40	4.15	4.05	3.95	3.80

Modifications

Add to the total cost

Upgrade Kitchen Cabinets	$ + 5950
Solid Surface Countertops (Included)	
Full Bath - including plumbing, wall and floor finishes	+ 8050
Half Bath - including plumbing, wall and floor finishes	+ 4767
One Car Attached Garage	+ 15,086
One Car Detached Garage	+ 19,847
Fireplace & Chimney	+ 8415

Adjustments

For multi family - add to total cost

Additional Kitchen	$ + 9963
Additional Bath	+ 8050
Additional Entry & Exit	+ 1804
Separate Heating	+ 1650
Separate Electric	+ 1817

For Townhouse/Rowhouse - Multiply cost per square foot by

Inner Unit	.88
End Unit	.94

Alternatives

Add to or deduct from the cost per square foot of living area

Cedar Shake Roof	+ 1.15
Clay Tile Roof	+ 2.30
Slate Roof	+ 2.55
Upgrade Walls to Skim Coat Plaster	+ .59
Upgrade Ceilings to Textured Finish	+ .60
Air Conditioning, in Heating Ductwork	+ 2.73
In Separate Ductwork	+ 5.62
Heating Systems, Hot Water	+ 1.38
Heat Pump	+ 1.63
Electric Heat	– .86
Not Heated	– 3.66

Additional upgrades or components

Kitchen Cabinets & Countertops	Page 93
Bathroom Vanities	94
Fireplaces & Chimneys	94
Windows, Skylights & Dormers	94
Appliances	95
Breezeways & Porches	95
Finished Attic	95
Garages	96
Site Improvements	96
Wings & Ells	56

Important: See the Reference Section for Location Factors (to adjust for your city) and Estimating Forms.

			Cost Per Square Foot Of Living Area			% of Total
			Mat.	Inst.	Total	(rounded)
1	**Site Work**	Site preparation for slab; 4' deep trench excavation for foundation wall.		.87	.87	0.8%
2	**Foundation**	Continuous reinforced concrete footing, 10" deep x 20" wide; damproofed and insulated 8" thick reinforced concrete block foundation wall, 4' deep; trowel finished 4" thick concrete slab on 4" crushed stone base and polyethylene vapor barrier.	2.38	3.15	5.53	5.3%
3	**Framing**	Exterior walls - 2" x 6" wood studs, 16" O.C.; 1/2" sheathing; gable end roof framing, 2" x 10" rafters, 16" O.C. with 1/2" plywood sheathing; 2" x 10" floor joists, 16" O.C. with bridging and 5/8" subflooring; 2" x 4" interior partitions.	7.30	9.80	17.10	16.5%
4	**Exterior Walls**	Beveled wood siding and housewrap on insulated wood frame walls; R38 attic insulation; double hung wood windows; flush solid core doors, frame and hardware, painted finish; aluminum storm and screen doors.	14.49	5.73	20.22	19.5%
5	**Roofing**	25 year asphalt roof shingles; #15 felt building paper; aluminum gutters, downspouts, drip edge and flashings.	.75	.93	1.68	1.6%
6	**Interiors**	Walls & ceilings, 1/2" taped & finished gypsum wallboard, primed & painted with 2 coats of finish paint; birch faced hollow core interior doors, frames & hardware, painted finish; medium weight carpeting with pad, 40%; sheet vinyl, 15%; oak hardwood, 40%; ceramic tile, 5%; hardwood tread stairway.	15.60	15.23	30.83	29.7%
7	**Specialties**	Average grade kitchen cabinets and countertop; stainless steel kitchen sink; 40 gallon electric water heater.	2.54	.80	3.34	3.2%
8	**Mechanical**	Three fixture bathroom: bathtub, water closet, vanity and sink; gas fired hot air heating system.	2.63	3.03	5.66	5.5%
9	**Electrical**	200 amp electric service; wiring, duplex & GFI receptacles, wall switches, door bell, appliance circuits, fans and communications cabling; average grade lighting fixtures.	1.37	2.14	3.51	3.4%
10	**Overhead**	Contractor's overhead and profit and plans.	7.99	7.07	15.06	14.5%
	Total		55.05	48.75	**103.80**	

- **Simple design from standard plans**
- **Single family — 1 full bath, 1 kitchen**
- **No basement**
- **Asphalt shingles on roof**
- **Hot air heat**
- **Gypsum wallboard interior finishes**
- **Materials and workmanship are average**

Note: The illustration shown may contain some optional components (for example: garages and/or fireplaces) whose costs are shown in the modifications, adjustments, & alternatives below or at the end of the square foot section.

Base cost per square foot of living area

Exterior Wall	Living Area										
	1000	1200	1400	1600	1800	2000	2200	2600	3000	3400	3800
Wood Siding - Wood Frame	139.50	125.85	119.40	114.85	110.45	105.55	102.45	96.55	90.70	88.05	85.70
Brick Veneer - Wood Frame	144.35	130.35	123.60	118.90	114.25	109.20	105.90	99.70	93.60	90.85	88.30
Stucco on Wood Frame	135.45	122.10	115.95	111.55	107.35	102.50	99.55	93.95	88.30	85.80	83.50
Solid Masonry	155.15	140.35	132.95	127.80	122.60	117.30	113.55	106.60	100.05	96.90	94.15
Finished Basement, Add	23.25	22.90	22.10	21.65	21.05	20.70	20.30	19.55	18.95	18.60	18.30
Unfinished Basement, Add	9.20	8.55	8.05	7.75	7.40	7.15	6.95	6.45	6.15	5.90	5.70

Modifications

Add to the total cost

Upgrade Kitchen Cabinets	$	+ 5950
Solid Surface Countertops (Included)		
Full Bath - including plumbing, wall and floor finishes		+ 8050
Half Bath - including plumbing, wall and floor finishes		+ 4767
One Car Attached Garage		+ 15,086
One Car Detached Garage		+ 19,847
Fireplace & Chimney		+ 6942

Adjustments

For multi family - add to total cost

Additional Kitchen	$	+ 9963
Additional Bath		+ 8050
Additional Entry & Exit		+ 1804
Separate Heating		+ 1650
Separate Electric		+ 1817

For Townhouse/Rowhouse - Multiply cost per square foot by

Inner Unit	.91
End Unit	.96

Alternatives

Add to or deduct from the cost per square foot of living area

Cedar Shake Roof	+ 1.75
Clay Tile Roof	+ 3.45
Slate Roof	+ 3.85
Upgrade Walls to Skim Coat Plaster	+ .54
Upgrade Ceilings to Textured Finish	+ .60
Air Conditioning, in Heating Ductwork	+ 3
In Separate Ductwork	+ 5.81
Heating Systems, Hot Water	+ 1.52
Heat Pump	+ 1.59
Electric Heat	– .62
Not Heated	– 3.36

Additional upgrades or components

Kitchen Cabinets & Countertops	Page 93
Bathroom Vanities	94
Fireplaces & Chimneys	94
Windows, Skylights & Dormers	94
Appliances	95
Breezeways & Porches	95
Finished Attic	95
Garages	96
Site Improvements	96
Wings & Ells	56

Important: See the Reference Section for Location Factors (to adjust for your city) and Estimating Forms.

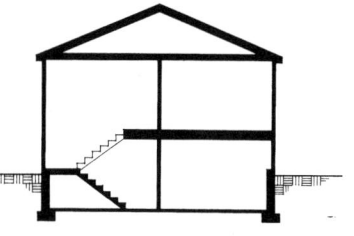

		Cost Per Square Foot Of Living Area			% of Total
		Mat.	Inst.	Total	(rounded)
1 Site Work	Site preparation for slab; 4' deep trench excavation for foundation wall.		1.32	1.32	1.3%
2 Foundation	Continuous reinforced concrete footing, 10" deep x 20" wide; damproofed and insulated 8" thick reinforced concrete block foundation wall, 4' deep; trowel finished 4" thick concrete slab on 4" crushed stone base and polyethylene vapor barrier.	3.58	4.72	8.30	7.9%
3 Framing	Exterior walls - 2" x 6" wood studs, 16" O.C.; 1/2" sheathing; gable end roof framing, 2" x 10" rafters, 16" O.C. with 1/2" plywood sheathing; 2" x 10" floor joists, 16" O.C. with bridging and 5/8" subflooring; 2" x 4" interior partitions.	6.86	9.27	16.13	15.3%
4 Exterior Walls	Beveled wood siding and housewrap on insulated wood frame walls; R38 attic insulation; double hung wood windows; flush solid core doors, frame and hardware, painted finish; aluminum storm and screen doors.	11.95	4.76	16.71	15.8%
5 Roofing	25 year roof asphalt shingles; #15 felt building paper; aluminum gutters, downspouts, drip edge and flashings.	1.13	1.40	2.53	2.4%
6 Interiors	Walls & ceilings, 1/2" taped & finished gypsum wallboard, primed & painted with 2 coats of finish paint; birch faced hollow core interior doors, frames & hardware, painted finish; medium weight carpeting with pad, 40%; sheet vinyl, 15%; oak hardwood, 40%; ceramic tile, 5%; hardwood tread stairway.	14.92	14.69	29.61	28.1%
7 Specialties	Average grade kitchen cabinets and countertop; stainless steel kitchen sink; 40 gallon electric water heater.	3.81	1.21	5.02	4.8%
8 Mechanical	Three fixture bathroom: bathtub, water closet, vanity and sink; gas fired hot air heating system.	3.44	3.35	6.79	6.4%
9 Electrical	200 amp electric service; wiring, duplex & GFI receptacles, wall switches, door bell, appliance circuits, fans and communications cabling; average grade lighting fixtures.	1.46	2.37	3.83	3.6%
10 Overhead	Contractor's overhead and profit and plans.	8.00	7.31	15.31	14.5%
Total		55.15	50.40	**105.55**	

- **Simple design from standard plans**
- **Single family — 1 full bath, 1 kitchen**
- **No basement**
- **Asphalt shingles on roof**
- **Hot air heat**
- **Gypsum wallboard interior finishes**
- **Materials and workmanship are average**

Note: The illustration shown may contain some optional components (for example: garages and/or fireplaces) whose costs are shown in the modifications, adjustments, & alternatives below or at the end of the square foot section.

●Design Basics, Inc.

Base cost per square foot of living area

Exterior Wall	Living Area										
	1200	1500	1800	2100	2400	2700	3000	3400	3800	4200	4600
Wood Siding - Wood Frame	129.80	118.85	110.75	103.65	99.15	96.60	93.65	90.95	86.65	83.10	81.35
Brick Veneer - Wood Frame	134.30	122.95	114.45	107.00	102.30	99.65	96.50	93.70	89.20	85.45	83.70
Stucco on Wood Frame	126.05	115.45	107.60	100.85	96.50	94.05	91.25	88.65	84.45	81.10	79.40
Solid Masonry	144.20	131.85	122.50	114.30	109.15	106.25	102.65	99.70	94.80	90.70	88.70
Finished Basement, Add*	26.70	26.20	25.05	24.20	23.55	23.25	22.80	22.45	21.95	21.60	21.35
Unfinished Basement, Add*	10.25	9.50	8.75	8.25	7.85	7.65	7.35	7.15	6.85	6.60	6.50

*Basement under middle level only.

Modifications

Add to the total cost

Upgrade Kitchen Cabinets	$ + 5950
Solid Surface Countertops (Included)	
Full Bath - including plumbing, wall and floor finishes	+ 8050
Half Bath - including plumbing, wall and floor finishes	+ 4767
One Car Attached Garage	+ 15,086
One Car Detached Garage	+ 19,847
Fireplace & Chimney	+ 6942

Adjustments

For multi family - add to total cost

Additional Kitchen	$ + 9963
Additional Bath	+ 8050
Additional Entry & Exit	+ 1804
Separate Heating	+ 1650
Separate Electric	+ 1817

For Townhouse/Rowhouse - Multiply cost per square foot by

Inner Unit	.90
End Unit	.95

Alternatives

Add to or deduct from the cost per square foot of living area

Cedar Shake Roof	+ 2.55
Clay Tile Roof	+ 5
Slate Roof	+ 5.60
Upgrade Walls to Skim Coat Plaster	+ .47
Upgrade Ceilings to Textured Finish	+ .60
Air Conditioning, in Heating Ductwork	+ 2.52
In Separate Ductwork	+ 5.47
Heating Systems, Hot Water	+ 1.47
Heat Pump	+ 1.66
Electric Heat	− .53
Not Heated	− 3.24

Additional upgrades or components

Kitchen Cabinets & Countertops	Page 93
Bathroom Vanities	94
Fireplaces & Chimneys	94
Windows, Skylights & Dormers	94
Appliances	95
Breezeways & Porches	95
Finished Attic	95
Garages	96
Site Improvements	96
Wings & Ells	56

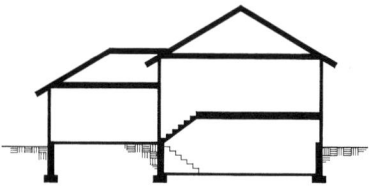

			Cost Per Square Foot Of Living Area			% of Total
			Mat.	Inst.	Total	(rounded)
1	**Site Work**	Site preparation for slab; 4' deep trench excavation for foundation wall.		1.09	1.09	1.1%
2	**Foundation**	Continuous reinforced concrete footing, 10" deep x 20" wide; damproofed and insulated 8" thick reinforced concrete block foundation wall, 4' deep; trowel finished 4" thick concrete slab on 4" crushed stone base and polyethylene vapor barrier.	4.01	5.11	9.12	9.2%
3	**Framing**	Exterior walls - 2" x 6" wood studs, 16" O.C.; 1/2" sheathing; gable end roof framing, 2" x 10" rafters, 16" O.C. with 1/2" plywood sheathing; 2" x 10" floor joists, 16" O.C. with bridging and 5/8" subflooring; 2" x 4" interior partitions.	6.59	8.90	15.49	15.6%
4	**Exterior Walls**	Beveled wood siding and housewrap on insulated wood frame walls; R38 attic insulation; double hung wood windows; flush solid core doors, frame and hardware, painted finish; aluminum storm and screen doors.	10.52	4.16	14.68	14.8%
5	**Roofing**	25 year asphalt roof shingles; #15 felt building paper; aluminum gutters, downspouts, drip edge and flashings.	1.51	1.86	3.37	3.4%
6	**Interiors**	Walls & ceilings, 1/2" taped & finished gypsum wallboard, primed & painted with 2 coats of finish paint; birch faced hollow core interior doors, frames & hardware, painted finish; medium weight carpeting with pad, 40%; sheet vinyl, 15%; oak hardwood, 40%; ceramic tile, 5%; hardwood tread stairway.	13.70	13.31	27.01	27.2%
7	**Specialties**	Average grade kitchen cabinets and countertop; stainless steel kitchen sink; 40 gallon electric water heater.	3.17	1.01	4.18	4.2%
8	**Mechanical**	Three fixture bathroom: bathtub, water closet, vanity and sink; gas fired hot air heating system.	2.92	3.19	6.11	6.2%
9	**Electrical**	200 amp electric service; wiring, duplex & GFI receptacles, wall switches, door bell, appliance circuits, fans and communications cabling; average grade lighting fixtures.	1.42	2.26	3.68	3.7%
10	**Overhead**	Contractor's overhead and profit and plans.	7.46	6.96	14.42	14.5%
		Total	51.30	47.85	**99.15**	

For customer support on your Residential Costs with RSMeans data, call 800.448.8182.

51

- **Post and beam frame**
- **Log exterior walls**
- **Simple design from standard plans**
- **Single family — 1 full bath, 1 kitchen**
- **No basement**
- **Asphalt shingles on roof**
- **Hot air heat**
- **Gypsum wallboard interior finishes**
- **Materials and workmanship are average**

Note: The illustration shown may contain some optional components (for example: garages and/or fireplaces) whose costs are shown in the modifications, adjustments, & alternatives below or at the end of the square foot section.

Base cost per square foot of living area

Exterior Wall	Living Area										
	600	800	1000	1200	1400	1600	1800	2000	2400	2800	3200
6" Log - Solid Wall	191.45	172.75	159.00	147.95	138.70	132.65	129.40	125.30	117.50	111.80	107.70
8" Log - Solid Wall	179.10	161.70	149.00	138.95	130.50	124.95	121.95	118.35	111.15	105.90	102.30
Finished Basement, Add	40.50	39.05	37.25	35.55	34.15	33.30	32.80	32.05	31.05	30.25	29.55
Unfinished Basement, Add	16.80	15.15	14.00	13.00	12.15	11.65	11.25	10.85	10.25	9.75	9.35

Modifications

Add to the total cost

Upgrade Kitchen Cabinets	$ + 5950
Solid Surface Countertops (Included)	
Full Bath - including plumbing, wall and floor finishes	+ 8050
Half Bath - including plumbing, wall and floor finishes	+ 4767
One Car Attached Garage	+ 15,086
One Car Detached Garage	+ 19,847
Fireplace & Chimney	+ 6942

Adjustments

For multi family - add to total cost

Additional Kitchen	$ + 9963
Additional Bath	+ 8050
Additional Entry & Exit	+ 1804
Separate Heating	+ 1650
Separate Electric	+ 1817

For Townhouse/Rowhouse - Multiply cost per square foot by

Inner Unit	.92
End Unit	.96

Alternatives

Add to or deduct from the cost per square foot of living area

Cedar Shake Roof	+ 3.50
Air Conditioning, in Heating Ductwork	+ 4.95
In Separate Ductwork	+ 7.31
Heating Systems, Hot Water	+ 1.63
Heat Pump	+ 1.36
Electric Heat	– .83
Not Heated	– 3.61

Additional upgrades or components

Kitchen Cabinets & Countertops	Page 93
Bathroom Vanities	94
Fireplaces & Chimneys	94
Windows, Skylights & Dormers	94
Appliances	95
Breezeways & Porches	95
Finished Attic	95
Garages	96
Site Improvements	96
Wings & Ells	56

			Cost Per Square Foot Of Living Area			% of Total
			Mat.	Inst.	Total	(rounded)
1	Site Work	Site preparation for slab; 4' deep trench excavation for foundation wall.		1.63	1.63	1.2%
2	Foundation	Continuous reinforced concrete footing, 10" deep x 20" wide; damproofed and insulated 8" thick reinforced concrete block foundation wall, 4' deep; trowel finished 4" thick concrete slab on 4" crushed stone base and polyethylene vapor barrier.	6.02	7.67	13.69	10.3%
3	Framing	Exterior walls - precut traditional log home, handcrafted white cedar or pine logs, delivery included; heavy timber roof framing with 2" thick tongue and groove decking, rigid insulation with 5/8" sheathing; 2" x 4" interior partitions.	26.51	16.29	42.80	32.3%
4	Exterior Walls	R38 attic insulation; double hung wood windows; flush solid core doors, frame and hardware, painted finish; aluminum storm and screen doors.	5.65	2.77	8.42	6.3%
5	Roofing	25 year asphalt roof shingles; #15 felt building paper; aluminum gutters, downspouts, drip edge and flashings.	2.26	2.79	5.05	3.8%
6	Interiors	Walls & ceilings, 1/2" taped & finished gypsum wallboard, primed & painted with 2 coats of finish paint; birch faced hollow core interior doors, frames & hardware, painted finish; medium weight carpeting with pad, 40%; sheet vinyl, 15%; oak hardwood, 40%; ceramic tile, 5%.	11.95	11.85	23.80	17.9%
7	Specialties	Average grade kitchen cabinets and countertop; stainless steel kitchen sink; 40 gallon electric water heater.	4.76	1.50	6.26	4.7%
8	Mechanical	Three fixture bathroom: bathtub, water closet, vanity and sink; gas fired hot air heating system.	4.04	3.61	7.65	5.8%
9	Electrical	200 amp electric service; wiring, duplex & GFI receptacles, wall switches, door bell, appliance circuits, fans and communications cabling; average grade lighting fixtures.	1.53	2.54	4.07	3.1%
10	Overhead	Contractor's overhead and profit and plans.	10.68	8.60	19.28	14.5%
	Total		73.40	59.25	**132.65**	

- **Post and beam frame**
- **Log exterior walls**
- **Simple design from standard plans**
- **Single family — 1 full bath, 1 kitchen**
- **No basement**
- **Asphalt shingles on roof**
- **Hot air heat**
- **Gypsum wallboard interior finishes**
- **Materials and workmanship are average**

Note: The illustration shown may contain some optional components (for example: garages and/or fireplaces) whose costs are shown in the modifications, adjustments, & alternatives below or at the end of the square foot section.

Base cost per square foot of living area

Exterior Wall	Living Area										
	1000	1200	1400	1600	1800	2000	2200	2600	3000	3400	3800
6" Log - Solid Wall	163.30	148.30	140.75	135.50	130.25	124.80	121.00	113.85	107.15	103.90	101.10
8" Log - Solid Wall	150.50	136.45	129.70	124.95	120.30	115.25	111.85	105.60	99.55	96.70	94.20
Finished Basement, Add	23.25	22.90	22.10	21.65	21.05	20.70	20.30	19.55	18.95	18.60	18.30
Unfinished Basement, Add	9.20	8.55	8.05	7.75	7.40	7.15	6.95	6.45	6.15	5.90	5.70

Modifications

Add to the total cost

Upgrade Kitchen Cabinets	$ + 5950
Solid Surface Countertops (Included)	
Full Bath - including plumbing, wall and floor finishes	+ 8050
Half Bath - including plumbing, wall and floor finishes	+ 4767
One Car Attached Garage	+ 15,086
One Car Detached Garage	+ 19,847
Fireplace & Chimney	+ 7671

Adjustments

For multi family - add to total cost

Additional Kitchen	$ + 9963
Additional Bath	+ 8050
Additional Entry & Exit	+ 1804
Separate Heating	+ 1650
Separate Electric	+ 1817

For Townhouse/Rowhouse - Multiply cost per square foot by

Inner Unit	.92
End Unit	.96

Alternatives

Add to or deduct from the cost per square foot of living area

Cedar Shake Roof	+ 1.75
Air Conditioning, in Heating Ductwork	+ 3
In Separate Ductwork	+ 5.81
Heating Systems, Hot Water	+ 1.52
Heat Pump	+ 1.59
Electric Heat	– .62
Not Heated	– 3.36

Additional upgrades or components

Kitchen Cabinets & Countertops	Page 93
Bathroom Vanities	94
Fireplaces & Chimneys	94
Windows, Skylights & Dormers	94
Appliances	95
Breezeways & Porches	95
Finished Attic	95
Garages	96
Site Improvements	96
Wings & Ells	56

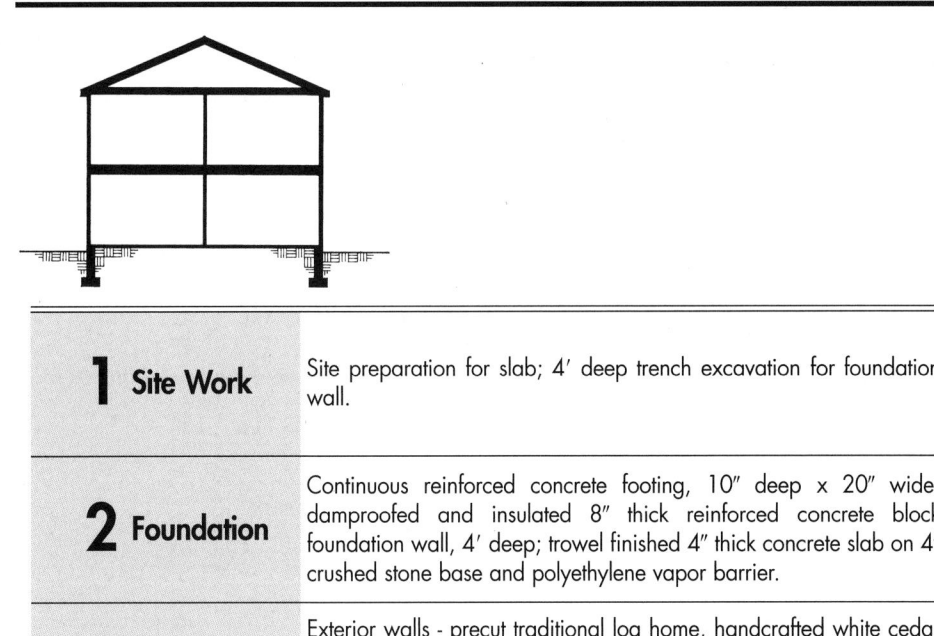

		Cost Per Square Foot Of Living Area			% of Total
		Mat.	Inst.	Total	(rounded)
1 Site Work	Site preparation for slab; 4' deep trench excavation for foundation wall.		1.32	1.32	1.1%
2 Foundation	Continuous reinforced concrete footing, 10" deep x 20" wide; damproofed and insulated 8" thick reinforced concrete block foundation wall, 4' deep; trowel finished 4" thick concrete slab on 4" crushed stone base and polyethylene vapor barrier.	3.58	4.72	8.30	6.7%
3 Framing	Exterior walls - precut traditional log home, handcrafted white cedar or pine logs, delivery included; heavy timber roof framing with 2" thick T. & G. decking, rigid insulation with 5/8" sheathing; heavy timber columns, beams & joists with 2" thick T. & G. decking; 2" x 4" interior partitions.	27.31	15.63	42.94	34.4%
4 Exterior Walls	R38 attic insulation; double hung wood windows; flush solid core doors, frame and hardware, painted finish; aluminum storm and screen doors.	5.67	2.92	8.59	6.9%
5 Roofing	25 year asphalt roof shingles; #15 felt building paper; aluminum gutters, downspouts, drip edge and flashings.	1.13	1.40	2.53	2.0%
6 Interiors	Walls & ceilings, 1/2" taped & finished gypsum wallboard, primed & painted with 2 coats of finish paint; birch faced hollow core interior doors, frames & hardware, painted finish; medium weight carpeting with pad, 40%; sheet vinyl, 15%; oak hardwood, 40%; ceramic tile, 5%; hardwood tread stairway.	14.01	13.34	27.35	21.9%
7 Specialties	Average grade kitchen cabinets and countertop; stainless steel kitchen sink; 40 gallon electric water heater.	3.81	1.21	5.02	4.0%
8 Mechanical	Three fixture bathroom: bathtub, water closet, vanity and sink; gas fired hot air heating system.	3.44	3.35	6.79	5.4%
9 Electrical	200 amp electric service; wiring, duplex & GFI receptacles, wall switches, door bell, appliance circuits, fans and communications cabling; average grade lighting fixtures.	1.46	2.37	3.83	3.1%
10 Overhead	Contractor's overhead and profit and plans.	10.29	7.84	18.13	14.5%
Total		70.70	54.10	**124.80**	

For customer support on your Residential Costs with RSMeans data, call 800.448.8182.

55

1 Story — Base cost per square foot of living area

Exterior Wall	Living Area							
	50	100	200	300	400	500	600	700
Wood Siding - Wood Frame	223.85	172.40	150.10	125.90	118.60	114.15	111.30	112.05
Brick Veneer - Wood Frame	224.60	169.10	145.10	118.10	110.20	105.55	102.40	103.00
Stucco on Wood Frame	212.20	164.00	143.05	121.10	114.25	110.10	107.45	108.35
Solid Masonry	275.45	209.35	180.85	146.40	137.05	131.40	129.55	129.45
Finished Basement, Add	65.20	53.90	48.45	39.35	37.50	36.40	35.65	35.20
Unfinished Basement, Add	30.75	23.20	20.05	14.80	13.75	13.15	12.70	12.40

1-1/2 Story — Base cost per square foot of living area

Exterior Wall	Living Area							
	100	200	300	400	500	600	700	800
Wood Siding - Wood Frame	181.10	145.85	124.70	111.65	105.30	102.10	98.20	97.10
Brick Veneer - Wood Frame	235.90	177.00	147.15	128.80	119.75	114.80	109.65	107.70
Stucco on Wood Frame	213.10	158.75	131.90	116.90	108.80	104.50	99.90	98.00
Solid Masonry	269.30	203.80	169.40	146.15	135.80	130.05	124.00	121.90
Finished Basement, Add	43.85	39.60	35.95	31.90	30.80	30.10	29.45	29.30
Unfinished Basement, Add	19.25	16.10	14.00	11.70	11.10	10.65	10.30	10.25

2 Story — Base cost per square foot of living area

Exterior Wall	Living Area							
	100	200	400	600	800	1000	1200	1400
Wood Siding - Wood Frame	180.70	135.45	115.75	95.80	89.40	85.45	82.90	83.95
Brick Veneer - Wood Frame	240.30	168.30	135.55	109.00	100.05	94.75	91.20	91.50
Stucco on Wood Frame	214.75	150.00	120.35	98.85	91.00	86.25	83.10	83.70
Solid Masonry	277.75	195.10	157.80	123.85	113.50	107.25	103.05	102.95
Finished Basement, Add	34.70	29.05	26.35	21.75	20.90	20.30	19.95	19.70
Unfinished Basement, Add	15.50	11.75	10.20	7.55	7.05	6.75	6.50	6.35

Base costs do not include bathroom or kitchen facilities. Use Modifications/Adjustments/Alternatives on pages 93–96 where appropriate.

Important: See the Reference Section for Location Factors (to adjust for your city) and Estimating Forms.

1 Story

1-1/2 Story

2 Story

2-1/2 Story

Bi-Level

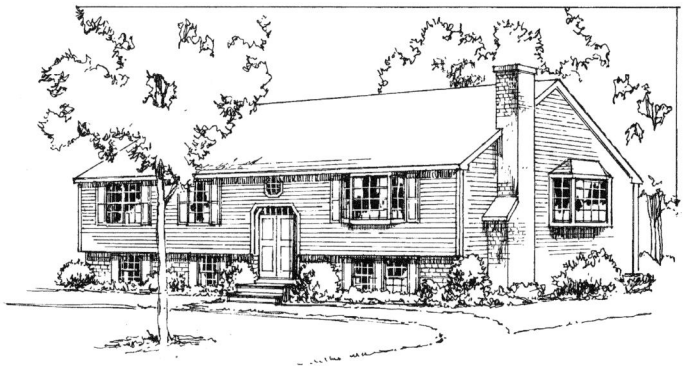

Tri-Level

For customer support on your Residential Costs with RSMeans data, call 800.448.8182.

59

- A distinct residence from designer's plans
- Single family — 1 full bath, 1 half bath, 1 kitchen
- No basement
- Asphalt shingles on roof
- Forced hot air heat/air conditioning
- Gypsum wallboard interior finishes
- Materials and workmanship are above average

Note: The illustration shown may contain some optional components (for example: garages and/or fireplaces) whose costs are shown in the modifications, adjustments, & alternatives below or at the end of the square foot section.

©Design Basics, Inc.

Base cost per square foot of living area

Exterior Wall	Living Area										
	800	1000	1200	1400	1600	1800	2000	2400	2800	3200	3600
Wood Siding - Wood Frame	216.45	195.05	178.40	165.10	155.95	150.70	144.80	134.00	126.05	120.45	114.80
Brick Veneer - Wood Frame	227.10	204.80	187.15	173.00	163.40	157.90	151.55	140.10	131.70	125.75	119.70
Stone Veneer - Wood Frame	236.95	213.70	195.20	180.35	170.30	164.55	157.80	145.80	137.00	130.60	124.25
Solid Masonry	237.50	214.20	195.70	180.80	170.75	164.90	158.15	146.15	137.25	130.90	124.50
Finished Basement, Add	60.80	60.55	57.95	55.85	54.50	53.65	52.55	51.00	49.75	48.70	47.75
Unfinished Basement, Add	26.70	25.20	23.90	22.80	22.10	21.65	21.10	20.25	19.60	19.05	18.55

Modifications

Add to the total cost

Upgrade Kitchen Cabinets	$	+ 1850
Solid Surface Countertops (Included)		
Full Bath - including plumbing, wall and floor finishes		+ 9660
Half Bath - including plumbing, wall and floor finishes		+ 5720
Two Car Attached Garage		+ 29,653
Two Car Detached Garage		+ 33,939
Fireplace & Chimney		+ 7266

Adjustments

For multi family - add to total cost

Additional Kitchen	$	+ 22,193
Additional Full Bath & Half Bath		+ 15,380
Additional Entry & Exit		+ 1804
Separate Heating & Air Conditioning		+ 7787
Separate Electric		+ 1817

For Townhouse/Rowhouse - Multiply cost per square foot by

Inner Unit	.90
End Unit	.95

Alternatives

Add to or deduct from the cost per square foot of living area

Cedar Shake Roof	+ 2.75
Clay Tile Roof	+ 6.10
Slate Roof	+ 6.90
Upgrade Ceilings to Textured Finish	+ .60
Air Conditioning, in Heating Ductwork	Base System
Heating Systems, Hot Water	+ 1.67
Heat Pump	+ 1.36
Electric Heat	− 2.33
Not Heated	− 4.52

Additional upgrades or components

Kitchen Cabinets & Countertops	Page 93
Bathroom Vanities	94
Fireplaces & Chimneys	94
Windows, Skylights & Dormers	94
Appliances	95
Breezeways & Porches	95
Finished Attic	95
Garages	96
Site Improvements	96
Wings & Ells	74

Important: See the Reference Section for Location Factors (to adjust for your city) and Estimating Forms.

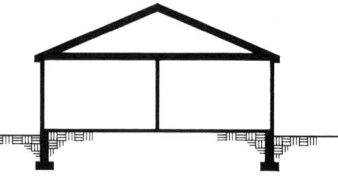

			Cost Per Square Foot Of Living Area			% of Total
			Mat.	Inst.	Total	(rounded)
1	**Site Work**	Site preparation for slab; 4' deep trench excavation for foundation wall.		1.23	1.23	0.9%
2	**Foundation**	Continuous reinforced concrete footing, 12" deep x 24" wide; damproofed and insulated 12" thick reinforced concrete block foundation wall, 4' deep; trowel finished 4" thick concrete slab on 4" crushed stone base and polyethylene vapor barrier.	6.68	8.12	14.80	11.0%
3	**Framing**	Exterior walls - 2" x 6" wood studs, 16" O.C.; 1/2" sheathing; gable end roof framing, 2" x 10" rafters, 16" O.C. with 1/2" plywood sheathing; 2" x 4" interior partitions.	6.77	9.29	16.06	12.0%
4	**Exterior Walls**	1" x 6" tongue and groove vertical wood siding and housewrap on insulated wood frame walls; R38 attic insulation; plastic clad double hung wood windows; raised panel exterior doors, frames and hardware, painted finish; wood storm and screen door.	11.93	4.13	16.06	12.0%
5	**Roofing**	Red cedar roof shingles, perfections; #15 felt building paper; aluminum gutters, downspouts and drip edge; copper flashings.	4.87	3.97	8.84	6.6%
6	**Interiors**	Skim coated 1/2" thick gypsum wallboard walls and ceilings, primed and painted with 2 coats; interior raised panel solid core doors, frames and hardware, painted finish; oak hardwood flooring, 70%; ceramic tile flooring, 30%.	14.80	13.51	28.31	21.1%
7	**Specialties**	Custom grade kitchen cabinets and countertops; double bowl kitchen sink; 75 gallon gas water heater.	8.61	1.63	10.24	7.6%
8	**Mechanical**	Three fixture bathroom: bathtub, water closet, vanity and sink; gas fired heating and air conditioning system.	7.39	3.69	11.08	8.3%
9	**Electrical**	200 amp electric service; wiring, duplex & GFI receptacles, wall switches, door bell, appliance circuits, air conditioning circuit, fans and communications cabling; custom grade lighting fixtures.	2.31	2.72	5.03	3.8%
10	**Overhead**	Contractor's overhead and profit and design.	12.69	9.66	22.35	16.7%
	Total		76.05	57.95	**134.00**	

For customer support on your Residential Costs with RSMeans data, call 800.448.8182.

61

- **A distinct residence from designer's plans**
- **Single family — 1 full bath, 1 half bath, 1 kitchen**
- **No basement**
- **Asphalt shingles on roof**
- **Forced hot air heat/air conditioning**
- **Gypsum wallboard interior finishes**
- **Materials and workmanship are above average**

Note: The illustration shown may contain some optional components (for example: garages and/or fireplaces) whose costs are shown in the modifications, adjustments, & alternatives below or at the end of the square foot section.

© Donald A. Gardner Architects, Inc.

Base cost per square foot of living area

Exterior Wall	Living Area										
	1000	1200	1400	1600	1800	2000	2400	2800	3200	3600	4000
Wood Siding - Wood Frame	195.55	181.35	171.25	159.45	152.45	145.80	133.30	127.50	122.10	118.05	112.45
Brick Veneer - Wood Frame	207.05	192.15	181.50	168.75	161.35	154.20	140.70	134.50	128.70	124.35	118.30
Stone Veneer - Wood Frame	217.70	202.15	191.00	177.35	169.55	161.95	147.55	141.05	134.65	130.10	123.70
Solid Masonry	218.20	202.55	191.50	177.75	169.85	162.35	147.85	141.35	134.90	130.40	123.90
Finished Basement, Add	40.35	40.65	39.50	37.95	37.00	36.15	34.50	33.75	32.90	32.45	31.75
Unfinished Basement, Add	17.95	17.15	16.65	15.75	15.30	14.90	14.05	13.65	13.15	12.95	12.60

Modifications

Add to the total cost

Upgrade Kitchen Cabinets	$ + 1850
Solid Surface Countertops (Included)	
Full Bath - including plumbing, wall and floor finishes	+ 9660
Half Bath - including plumbing, wall and floor finishes	+ 5720
Two Car Attached Garage	+ 29,653
Two Car Detached Garage	+ 33,939
Fireplace & Chimney	+ 7266

Adjustments

For multi family - add to total cost

Additional Kitchen	$ + 22,193
Additional Full Bath & Half Bath	+ 15,380
Additional Entry & Exit	+ 1804
Separate Heating & Air Conditioning	+ 7787
Separate Electric	+ 1817

For Townhouse/Rowhouse - Multiply cost per square foot by

Inner Unit	.90
End Unit	.95

Alternatives

Add to or deduct from the cost per square foot of living area

Cedar Shake Roof	+ 2
Clay Tile Roof	+ 4.40
Slate Roof	+ 5
Upgrade Ceilings to Textured Finish	+ .60
Air Conditioning, in Heating Ductwork	Base System
Heating Systems, Hot Water	+ 1.60
Heat Pump	+ 1.42
Electric Heat	- 2.05
Not Heated	- 4.17

Additional upgrades or components

Kitchen Cabinets & Countertops	Page 93
Bathroom Vanities	94
Fireplaces & Chimneys	94
Windows, Skylights & Dormers	94
Appliances	95
Breezeways & Porches	95
Finished Attic	95
Garages	96
Site Improvements	96
Wings & Ells	74

Important: See the Reference Section for Location Factors (to adjust for your city) and Estimating Forms.

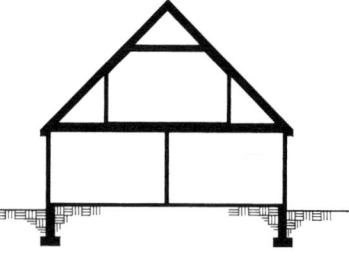

			Cost Per Square Foot Of Living Area			% of Total
			Mat.	Inst.	Total	(rounded)
1	**Site Work**	Site preparation for slab; 4' deep trench excavation for foundation wall.		1.06	1.06	0.8%
2	**Foundation**	Continuous reinforced concrete footing, 12" deep x 24" wide; damproofed and insulated 12" thick reinforced concrete block foundation wall, 4' deep; trowel finished 4" thick concrete slab on 4" crushed stone base and polyethylene vapor barrier.	4.68	5.80	10.48	8.2%
3	**Framing**	Exterior walls - 2" x 6" wood studs, 16" O.C.; 1/2" sheathing; gable end roof framing, steep pitch 2" x 10" rafters, 16" O.C. with 1/2" plywood sheathing; 2" x 10" floor joists, 16" O.C. with bridging and 5/8" subflooring; 2" x 4" interior partitions.	7.37	9.92	17.29	13.6%
4	**Exterior Walls**	1" x 6" tongue and groove vertical wood siding and housewrap on insulated wood frame walls; R38 attic insulation; plastic clad double hung wood windows; raised panel exterior doors, frames and hardware, painted finish; wood storm and screen door.	11.67	4.08	15.75	12.4%
5	**Roofing**	Red cedar roof shingles, perfections; #15 felt building paper; aluminum gutters, downspouts and drip edge; copper flashings.	3.53	2.89	6.42	5.0%
6	**Interiors**	Skim-coated 1/2" thick gypsum wallboard walls and ceilings, primed and painted with 2 coats; interior raised panel solid core doors, frames and hardware, painted finish; oak hardwood flooring, 70%; ceramic tile flooring, 30%; hardwood tread stairway.	16.80	14.92	31.72	24.9%
7	**Specialties**	Custom grade kitchen cabinets and countertops; double bowl kitchen sink; 75 gallon gas water heater.	7.38	1.40	8.78	6.9%
8	**Mechanical**	Three fixture bathroom: bathtub, water closet, vanity and sink; gas fired heating and air conditioning system.	6.38	3.45	9.83	7.7%
9	**Electrical**	200 amp electric service; wiring, duplex & GFI receptacles, wall switches, door bell, appliance circuits, air conditioning circuit, fans and communications cabling; custom grade lighting fixtures.	2.28	2.64	4.92	3.9%
10	**Overhead**	Contractor's overhead and profit and design.	12.01	9.24	21.25	16.7%
	Total		72.10	55.40	**127.50**	

- **A distinct residence from designer's plans**
- **Single family — 1 full bath, 1 half bath, 1 kitchen**
- **No basement**
- **Asphalt shingles on roof**
- **Forced hot air heat/air conditioning**
- **Gypsum wallboard interior finishes**
- **Materials and workmanship are above average**

Note: The illustration shown may contain some optional components (for example: garages and/or fireplaces) whose costs are shown in the modifications, adjustments, & alternatives below or at the end of the square foot section.

Base cost per square foot of living area

Exterior Wall	Living Area										
	1200	1400	1600	1800	2000	2400	2800	3200	3600	4000	4400
Wood Siding - Wood Frame	178.40	167.15	159.15	152.10	144.70	134.05	125.05	119.15	115.60	111.90	108.65
Brick Veneer - Wood Frame	189.80	177.85	169.40	161.70	154.00	142.40	132.65	126.30	122.50	118.30	114.90
Stone Veneer - Wood Frame	200.40	187.75	178.85	170.60	162.55	150.10	139.70	132.85	128.85	124.25	120.65
Solid Masonry	201.35	188.60	179.55	171.40	163.25	150.80	140.30	133.40	129.40	124.75	121.10
Finished Basement, Add	32.45	32.60	31.75	30.90	30.30	28.95	27.90	27.20	26.85	26.25	25.90
Unfinished Basement, Add	14.40	13.80	13.40	12.90	12.65	11.95	11.40	11.05	10.80	10.55	10.40

Modifications

Add to the total cost
Upgrade Kitchen Cabinets	$ + 1850
Solid Surface Countertops (Included)	
Full Bath - including plumbing, wall and floor finishes	+ 9660
Half Bath - including plumbing, wall and floor finishes	+ 5720
Two Car Attached Garage	+ 29,653
Two Car Detached Garage	+ 33,939
Fireplace & Chimney	+ 8202

Adjustments

For multi family - add to total cost
Additional Kitchen	$ + 22,193
Additional Full Bath & Half Bath	+ 15,380
Additional Entry & Exit	+ 1804
Separate Heating & Air Conditioning	+ 7787
Separate Electric	+ 1817

For Townhouse/Rowhouse - Multiply cost per square foot by
Inner Unit	.87
End Unit	.93

Alternatives

Add to or deduct from the cost per square foot of living area
Cedar Shake Roof	+ 1.35
Clay Tile Roof	+ 3.05
Slate Roof	+ 3.45
Upgrade Ceilings to Textured Finish	+ .60
Air Conditioning, in Heating Ductwork	Base System
Heating Systems, Hot Water	+ 1.55
Heat Pump	+ 1.58
Electric Heat	– 2.05
Not Heated	– 3.93

Additional upgrades or components

Kitchen Cabinets & Countertops	Page 93
Bathroom Vanities	94
Fireplaces & Chimneys	94
Windows, Skylights & Dormers	94
Appliances	95
Breezeways & Porches	95
Finished Attic	95
Garages	96
Site Improvements	96
Wings & Ells	74

			Cost Per Square Foot Of Living Area			% of Total
			Mat.	Inst.	Total	(rounded)
1	**Site Work**	Site preparation for slab; 4' deep trench excavation for foundation wall.		1.06	1.06	0.8%
2	**Foundation**	Continuous reinforced concrete footing, 12" deep x 24" wide; damproofed and insulated 12" thick reinforced concrete block foundation wall, 4' deep; trowel finished 4" thick concrete slab on 4" crushed stone base and polyethylene vapor barrier.	3.98	4.96	8.94	7.1%
3	**Framing**	Exterior walls - 2" x 6" wood studs, 16" O.C.; 1/2" sheathing; gable end roof framing, 2" x 10" rafters, 16" O.C. with 1/2" plywood sheathing; 2" x 10" floor joists, 16" O.C. with bridging and 5/8" subflooring; 2" x 4" interior partitions.	7.00	9.40	16.40	13.1%
4	**Exterior Walls**	1" x 6" tongue and groove vertical wood siding and housewrap on insulated wood frame walls; R38 attic insulation; plastic clad double hung wood windows; raised panel exterior doors, frames and hardware, painted finish; wood storm and screen door.	12.64	4.42	17.06	13.6%
5	**Roofing**	Red cedar roof shingles, perfections; #15 felt building paper; aluminum gutters, downspouts and drip edge; copper flashings.	2.44	1.99	4.43	3.5%
6	**Interiors**	Skim coated 1/2" thick gypsum wallboard walls and ceilings, primed and painted with 2 coats; interior raised panel solid core doors, frames and hardware, painted finish; oak hardwood flooring, 70%; ceramic tile flooring, 30%; hardwood tread stairway.	17.23	15.35	32.58	26.1%
7	**Specialties**	Custom grade kitchen cabinets and countertops; double bowl kitchen sink; 75 gallon gas water heater.	7.38	1.40	8.78	7.0%
8	**Mechanical**	Three fixture bathroom: bathtub, water closet, vanity and sink; gas fired heating and air conditioning system.	6.54	3.50	10.04	8.0%
9	**Electrical**	200 amp electric service; wiring, duplex & GFI receptacles, wall switches, door bell, appliance circuits, air conditioning circuit, fans and communications cabling; custom grade lighting fixtures.	2.28	2.64	4.92	3.9%
10	**Overhead**	Contractor's overhead and profit and design.	11.91	8.93	20.84	16.7%
	Total		71.40	53.65	**125.05**	

- **A distinct residence from designer's plans**
- **Single family — 1 full bath, 1 half bath, 1 kitchen**
- **No basement**
- **Asphalt shingles on roof**
- **Forced hot air heat/air conditioning**
- **Gypsum wallboard interior finishes**
- **Materials and workmanship are above average**

Note: The illustration shown may contain some optional components (for example: garages and/or fireplaces) whose costs are shown in the modifications, adjustments, & alternatives below or at the end of the square foot section.

Base cost per square foot of living area

Exterior Wall	Living Area										
	1500	1800	2100	2400	2800	3200	3600	4000	4500	5000	5500
Wood Siding - Wood Frame	175.50	157.75	147.20	140.75	132.60	124.75	120.50	113.70	110.25	106.95	103.70
Brick Veneer - Wood Frame	187.70	169.00	157.35	150.30	141.70	133.05	128.35	121.05	117.10	113.50	109.90
Stone Veneer - Wood Frame	199.00	179.40	166.60	159.10	150.15	140.70	135.55	127.75	123.45	119.55	115.75
Solid Masonry	199.80	180.15	167.30	159.75	150.75	141.25	136.10	128.15	123.90	120.00	116.15
Finished Basement, Add	25.75	25.75	24.40	23.75	23.20	22.30	21.75	21.20	20.80	20.40	20.10
Unfinished Basement, Add	11.55	10.95	10.25	9.95	9.70	9.15	8.90	8.60	8.35	8.15	8.00

Modifications

Add to the total cost

Upgrade Kitchen Cabinets	$ + 1850
Solid·Surface Countertops (Included)	
Full Bath - including plumbing, wall and floor finishes	+ 9660
Half Bath - including plumbing, wall and floor finishes	+ 5720
Two Car Attached Garage	+ 29,653
Two Car Detached Garage	+ 33,939
Fireplace & Chimney	+ 8202

Adjustments

For multi family - add to total cost

Additional Kitchen	$ + 22,193
Additional Full Bath & Half Bath	+ 15,380
Additional Entry & Exit	+ 1804
Separate Heating & Air Conditioning	+ 7787
Separate Electric	+ 1817

For Townhouse/Rowhouse - Multiply cost per square foot by

Inner Unit	.87
End Unit	.94

Alternatives

Add to or deduct from the cost per square foot of living area

Cedar Shake Roof	+ 1.20
Clay Tile Roof	+ 2.65
Slate Roof	+ 3
Upgrade Ceilings to Textured Finish	+ .60
Air Conditioning, in Heating Ductwork	Base System
Heating Systems, Hot Water	+ 1.40
Heat Pump	+ 1.63
Electric Heat	– 3.61
Not Heated	– 3.93

Additional upgrades or components

Kitchen Cabinets & Countertops	Page 93
Bathroom Vanities	94
Fireplaces & Chimneys	94
Windows, Skylights & Dormers	94
Appliances	95
Breezeways & Porches	95
Finished Attic	95
Garages	96
Site Improvements	96
Wings & Ells	74

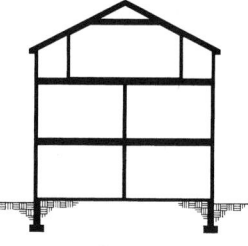

			Cost Per Square Foot Of Living Area			% of Total
			Mat.	Inst.	Total	(rounded)
1	**Site Work**	Site preparation for slab; 4' deep trench excavation for foundation wall.		.92	.92	0.7%
2	**Foundation**	Continuous reinforced concrete footing, 12" deep x 24" wide; damproofed and insulated 12" thick reinforced concrete block foundation wall, 4' deep; trowel finished 4" thick concrete slab on 4" crushed stone base and polyethylene vapor barrier.	3.28	4.12	7.40	5.9%
3	**Framing**	Exterior walls - 2" x 6" wood studs, 16" O.C.; 1/2" sheathing; gable end roof framing, steep pitch 2" x 10" rafters, 16" O.C. with 1/2" plywood sheathing; 2" x 10" floor joists, 16" O.C. with bridging and 5/8" subflooring; 2" x 4" interior partitions.	7.51	10.07	17.58	14.1%
4	**Exterior Walls**	1" x 6" tongue and groove vertical wood siding and housewrap on insulated wood frame walls; R38 attic insulation; plastic clad double hung wood windows; raised panel exterior doors, frames and hardware, painted finish; wood storm and screen door.	13.09	4.59	17.68	14.2%
5	**Roofing**	Red cedar roof shingles, perfections; #15 felt building paper; aluminum gutters, downspouts and drip edge; copper flashings.	2.17	1.77	3.94	3.2%
6	**Interiors**	Skim coated 1/2" thick gypsum wallboard walls and ceilings, primed and painted with 2 coats; interior raised panel solid core doors, frames and hardware, painted finish; oak hardwood flooring, 70%; ceramic tile flooring, 30%; hardwood tread stairway.	18.40	16.29	34.69	27.8%
7	**Specialties**	Custom grade kitchen cabinets and countertops; double bowl kitchen sink; 75 gallon gas water heater.	6.45	1.22	7.67	6.1%
8	**Mechanical**	Three fixture bathroom: bathtub, water closet, vanity and sink; gas fired heating and air conditioning system.	5.89	3.37	9.26	7.4%
9	**Electrical**	200 amp electric service; wiring, duplex & GFI receptacles, wall switches, door bell, appliance circuits, air conditioning circuit, fans and communications cabling; custom grade lighting fixtures.	2.26	2.58	4.84	3.9%
10	**Overhead**	Contractor's overhead and profit and design.	11.80	8.97	20.77	16.6%
	Total		70.85	53.90	**124.75**	

- **A distinct residence from designer's plans**
- **Single family — 1 full bath, 1 half bath, 1 kitchen**
- **No basement**
- **Asphalt shingles on roof**
- **Forced hot air heat/air conditioning**
- **Gypsum wallboard interior finishes**
- **Materials and workmanship are above average**

Note: The illustration shown may contain some optional components (for example: garages and/or fireplaces) whose costs are shown in the modifications, adjustments, & alternatives below or at the end of the square foot section.

Base cost per square foot of living area

Exterior Wall	Living Area										
	1500	1800	2100	2500	3000	3500	4000	4500	5000	5500	6000
Wood Siding - Wood Frame	172.25	154.80	146.20	138.90	127.90	122.40	115.70	108.90	106.25	103.60	100.85
Brick Veneer - Wood Frame	184.40	166.05	156.75	148.95	137.05	131.10	123.50	116.10	113.25	110.35	107.15
Stone Veneer - Wood Frame	195.65	176.45	166.50	158.25	145.50	139.05	130.80	122.85	119.75	116.60	113.00
Solid Masonry	196.70	177.45	167.45	159.05	146.25	139.75	131.45	123.50	120.30	117.15	113.60
Finished Basement, Add	22.55	22.55	21.65	21.10	20.20	19.65	18.90	18.30	18.10	17.85	17.50
Unfinished Basement, Add	10.10	9.60	9.15	8.85	8.40	8.15	7.75	7.45	7.35	7.20	7.00

Modifications

Add to the total cost

Upgrade Kitchen Cabinets	$ + 1850
Solid Surface Countertops (Included)	
Full Bath - including plumbing, wall and floor finishes	+ 9660
Half Bath - including plumbing, wall and floor finishes	+ 5720
Two Car Attached Garage	+ 29,653
Two Car Detached Garage	+ 33,939
Fireplace & Chimney	+ 9262

Adjustments

For multi family - add to total cost

Additional Kitchen	$ + 22,193
Additional Full Bath & Half Bath	+ 15,380
Additional Entry & Exit	+ 1804
Separate Heating & Air Conditioning	+ 7787
Separate Electric	+ 1817

For Townhouse/Rowhouse - Multiply cost per square foot by

Inner Unit	.85
End Unit	.93

Alternatives

Add to or deduct from the cost per square foot of living area

Cedar Shake Roof	+ .90
Clay Tile Roof	+ 2.05
Slate Roof	+ 2.30
Upgrade Ceilings to Textured Finish	+ .60
Air Conditioning, in Heating Ductwork	Base System
Heating Systems, Hot Water	+ 1.40
Heat Pump	+ 1.63
Electric Heat	− 3.61
Not Heated	− 3.81

Additional upgrades or components

Kitchen Cabinets & Countertops	Page 93
Bathroom Vanities	94
Fireplaces & Chimneys	94
Windows, Skylights & Dormers	94
Appliances	95
Breezeways & Porches	95
Finished Attic	95
Garages	96
Site Improvements	96
Wings & Ells	74

		Cost Per Square Foot Of Living Area			% of Total
		Mat.	Inst.	Total	(rounded)
1 Site Work	Site preparation for slab; 4' deep trench excavation for foundation wall.		.98	.98	0.8%
2 Foundation	Continuous reinforced concrete footing, 12" deep x 24" wide; damproofed and insulated 12" thick reinforced concrete block foundation wall, 4' deep; trowel finished 4" thick concrete slab on 4" crushed stone base and polyethylene vapor barrier.	3.07	3.90	6.97	5.4%
3 Framing	Exterior walls - 2" x 6" wood studs, 16" O.C.; 1/2" sheathing; gable end roof framing, 2" x 10" rafters, 16" O.C. with 1/2" plywood sheathing; 2" x 10" floor joists, 16" O.C. with bridging and 5/8" subflooring; 2" x 4" interior partitions.	7.44	9.95	17.39	13.6%
4 Exterior Walls	1" x 6" tongue and groove vertical wood siding and housewrap on insulated wood frame walls; R38 attic insulation; plastic clad double hung wood windows; raised panel exterior doors, frames and hardware, painted finish; wood storm and screen door.	14.55	5.12	19.67	15.4%
5 Roofing	Red cedar roof shingles, perfections; #15 felt building paper; aluminum gutters, downspouts and drip edge; copper flashings.	1.62	1.33	2.95	2.3%
6 Interiors	Skim coated 1/2" thick gypsum wallboard walls and ceilings, primed and painted with 2 coats; interior raised panel solid core doors, frames and hardware, painted finish; oak hardwood flooring, 70%; ceramic tile flooring, 30%; hardwood tread stairway.	19.08	16.89	35.97	28.1%
7 Specialties	Custom grade kitchen cabinets and countertops; double bowl kitchen sink; 75 gallon gas water heater.	6.87	1.31	8.18	6.4%
8 Mechanical	Three fixture bathroom: bathtub, water closet, vanity and sink; gas fired heating and air conditioning system.	6.19	3.43	9.62	7.5%
9 Electrical	200 amp electric service; wiring, duplex & GFI receptacles, wall switches, door bell, appliance circuits, air conditioning circuit, fans and communications cabling; custom grade lighting fixtures.	2.27	2.61	4.88	3.8%
10 Overhead	Contractor's overhead and profit and design.	12.21	9.08	21.29	16.6%
	Total	73.30	54.60	**127.90**	

- A distinct residence from designer's plans
- Single family — 1 full bath, 1 half bath, 1 kitchen
- No basement
- Asphalt shingles on roof
- Forced hot air heat/air conditioning
- Gypsum wallboard interior finishes
- Materials and workmanship are above average

Note: The illustration shown may contain some optional components (for example: garages and/or fireplaces) whose costs are shown in the modifications, adjustments, & alternatives below or at the end of the square foot section.

Base cost per square foot of living area

Exterior Wall	Living Area										
	1200	1400	1600	1800	2000	2400	2800	3200	3600	4000	4400
Wood Siding - Wood Frame	168.95	158.35	150.70	144.15	137.05	127.15	118.75	113.30	109.90	106.55	103.45
Brick Veneer - Wood Frame	177.60	166.45	158.50	151.45	144.10	133.50	124.55	118.70	115.15	111.45	108.20
Stone Veneer - Wood Frame	185.65	174.00	165.70	158.25	150.60	139.35	129.85	123.70	120.00	116.00	112.60
Solid Masonry	186.30	174.55	166.25	158.75	151.10	139.80	130.30	124.10	120.35	116.30	112.95
Finished Basement, Add	32.45	32.60	31.75	30.90	30.30	28.95	27.90	27.20	26.85	26.25	25.90
Unfinished Basement, Add	14.40	13.80	13.40	12.90	12.65	11.95	11.40	11.05	10.80	10.55	10.40

Modifications

Add to the total cost

Upgrade Kitchen Cabinets	$ + 1850
Solid Surface Countertops (Included)	
Full Bath - including plumbing, wall and floor finishes	+ 9660
Half Bath - including plumbing, wall and floor finishes	+ 5720
Two Car Attached Garage	+ 29,653
Two Car Detached Garage	+ 33,939
Fireplace & Chimney	+ 7266

Adjustments

For multi family - add to total cost

Additional Kitchen	$ + 22,193
Additional Full Bath & Half Bath	+ 15,380
Additional Entry & Exit	+ 1804
Separate Heating & Air Conditioning	+ 7787
Separate Electric	+ 1817

For Townhouse/Rowhouse - Multiply cost per square foot by

Inner Unit	.89
End Unit	.95

Alternatives

Add to or deduct from the cost per square foot of living area

Cedar Shake Roof	+ 1.35
Clay Tile Roof	+ 3.05
Slate Roof	+ 3.45
Upgrade Ceilings to Textured Finish	+ .60
Air Conditioning, in Heating Ductwork	Base System
Heating Systems, Hot Water	+ 1.55
Heat Pump	+ 1.58
Electric Heat	− 2.05
Not Heated	− 3.81

Additional upgrades or components

Kitchen Cabinets & Countertops	Page 93
Bathroom Vanities	94
Fireplaces & Chimneys	94
Windows, Skylights & Dormers	94
Appliances	95
Breezeways & Porches	95
Finished Attic	95
Garages	96
Site Improvements	96
Wings & Ells	74

Important: See the Reference Section for Location Factors (to adjust for your city) and Estimating Forms.

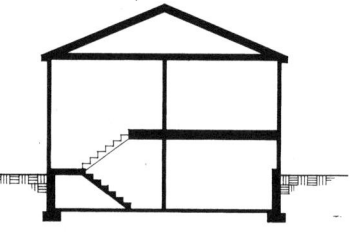

		Cost Per Square Foot Of Living Area			% of Total
		Mat.	Inst.	Total	(rounded)
1 Site Work	Site preparation for slab; 4' deep trench excavation for foundation wall.		1.06	1.06	0.9%
2 Foundation	Continuous reinforced concrete footing, 12" deep x 24" wide; damproofed and insulated 12" thick reinforced concrete block foundation wall, 4' deep; trowel finished 4" thick concrete slab on 4" crushed stone base and polyethylene vapor barrier.	3.98	4.96	8.94	7.5%
3 Framing	Exterior walls - 2" x 6" wood studs, 16" O.C.; 1/2" sheathing; gable end roof framing, 2" x 10" rafters, 16" O.C. with 1/2" plywood sheathing; 2" x 10" floor joists, 16" O.C. with bridging and 5/8" subflooring; 2" x 4" interior partitions.	6.54	8.77	15.31	12.9%
4 Exterior Walls	1" x 6" tongue and groove vertical wood siding and housewrap on insulated wood frame walls; R38 attic insulation; plastic clad double hung wood windows; raised panel exterior doors, frames and hardware, painted finish; wood storm and screen door.	10.08	3.50	13.58	11.4%
5 Roofing	Red cedar roof shingles, perfections; #15 felt building paper; aluminum gutters, downspouts and drip edge; copper flashings.	2.44	1.99	4.43	3.7%
6 Interiors	Skim coated 1/2" thick gypsum wallboard walls and ceilings, primed and painted with 2 coats; interior raised panel solid core doors, frames and hardware, painted finish; oak hardwood flooring, 70%; ceramic tile flooring, 30%; hardwood tread stairway.	16.95	14.94	31.89	26.9%
7 Specialties	Custom grade kitchen cabinets and countertops; double bowl kitchen sink; 75 gallon gas water heater.	7.38	1.40	8.78	7.4%
8 Mechanical	Three fixture bathroom: bathtub, water closet, vanity and sink; gas fired heating and air conditioning system.	6.54	3.50	10.04	8.5%
9 Electrical	200 amp electric service; wiring, duplex & GFI receptacles, wall switches, door bell, appliance circuits, air conditioning circuit, fans and communications cabling; custom grade lighting fixtures.	2.28	2.64	4.92	4.1%
10 Overhead	Contractor's overhead and profit and design.	11.26	8.54	19.80	16.7%
Total		67.45	51.30	**118.75**	

- **A distinct residence from designer's plans**
- **Single family — 1 full bath, 1 half bath, 1 kitchen**
- **No basement**
- **Asphalt shingles on roof**
- **Forced hot air heat/air conditioning**
- **Gypsum wallboard interior finishes**
- **Materials and workmanship are above average**

Note: The illustration shown may contain some optional components (for example: garages and/or fireplaces) whose costs are shown in the modifications, adjustments, & alternatives below or at the end of the square foot section.

©Design Basics, Inc.

Base cost per square foot of living area

Exterior Wall	Living Area										
	1200	1500	1800	2100	2400	2800	3200	3600	4000	4500	5000
Wood Siding - Wood Frame	174.65	157.95	145.55	135.35	128.45	123.35	117.50	111.45	108.75	103.10	99.90
Brick Veneer - Wood Frame	183.35	165.90	152.70	141.80	134.55	129.25	122.95	116.55	113.60	107.65	104.10
Stone Veneer - Wood Frame	191.45	173.20	159.25	147.85	140.20	134.60	127.95	121.20	118.15	111.75	108.00
Solid Masonry	192.05	173.75	159.75	148.25	140.60	135.00	128.30	121.50	118.45	112.10	108.30
Finished Basement, Add*	40.55	40.40	38.70	37.25	36.35	35.75	34.80	34.00	33.65	32.85	32.30
Unfinished Basement, Add*	17.85	16.85	15.95	15.20	14.75	14.40	13.95	13.50	13.30	12.90	12.60

*Basement under middle level only.

Modifications

Add to the total cost

Upgrade Kitchen Cabinets	$ + 1850
Solid Surface Countertops (Included)	
Full Bath - including plumbing, wall and floor finishes	+ 9660
Half Bath - including plumbing, wall and floor finishes	+ 5720
Two Car Attached Garage	+ 29,653
Two Car Detached Garage	+ 33,939
Fireplace & Chimney	+ 7266

Adjustments

For multi family - add to total cost

Additional Kitchen	$ + 22,193
Additional Full Bath & Half Bath	+ 15,380
Additional Entry & Exit	+ 1804
Separate Heating & Air Conditioning	+ 7787
Separate Electric	+ 1817

For Townhouse/Rowhouse - Multiply cost per square foot by

Inner Unit	.87
End Unit	.94

Alternatives

Add to or deduct from the cost per square foot of living area

Cedar Shake Roof	+ 2
Clay Tile Roof	+ 4.40
Slate Roof	+ 5
Upgrade Ceilings to Textured Finish	+ .60
Air Conditioning, in Heating Ductwork	Base System
Heating Systems, Hot Water	+ 1.50
Heat Pump	+ 1.65
Electric Heat	– 1.83
Not Heated	– 3.81

Additional upgrades or components

Kitchen Cabinets & Countertops	Page 93
Bathroom Vanities	94
Fireplaces & Chimneys	94
Windows, Skylights & Dormers	94
Appliances	95
Breezeways & Porches	95
Finished Attic	95
Garages	96
Site Improvements	96
Wings & Ells	74

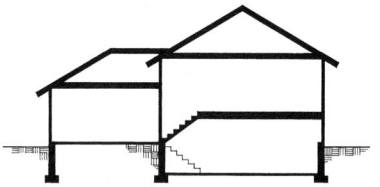

		Cost Per Square Foot Of Living Area			% of Total
		Mat.	Inst.	Total	(rounded)
1 Site Work	Site preparation for slab; 4' deep trench excavation for foundation wall.		.92	.92	0.8%
2 Foundation	Continuous reinforced concrete footing, 12" deep x 24" wide; damproofed and insulated 12" thick reinforced concrete block foundation wall, 4' deep; trowel finished 4" thick concrete slab on 4" crushed stone base and polyethylene vapor barrier.	4.67	5.71	10.38	8.8%
3 Framing	Exterior walls - 2" x 6" wood studs, 16" O.C.; 1/2" sheathing; gable end roof framing, 2" x 10" rafters, 16" O.C. with 1/2" plywood sheathing; 2" x 10" floor joists, 16" O.C. with bridging and 5/8" subflooring; 2" x 4" interior partitions.	6.52	8.85	15.37	13.1%
4 Exterior Walls	1" x 6" tongue and groove vertical wood siding and housewrap on insulated wood frame walls; R38 attic insulation; plastic clad double hung wood windows; raised panel exterior doors, frames and hardware, painted finish; wood storm and screen door.	9.97	3.47	13.44	11.4%
5 Roofing	Red cedar roof shingles, perfections; #15 felt building paper; aluminum gutters, downspouts and drip edge; copper flashings.	3.25	2.64	5.89	5.0%
6 Interiors	Skim coated 1/2" thick gypsum wallboard walls and ceilings, primed and painted with 2 coats; interior raised panel solid core doors, frames and hardware, painted finish; oak hardwood flooring, 70%; ceramic tile flooring, 30%; hardwood tread stairway.	15.94	14.21	30.15	25.7%
7 Specialties	Custom grade kitchen cabinets and countertops; double bowl kitchen sink; 75 gallon gas water heater.	6.45	1.22	7.67	6.5%
8 Mechanical	Three fixture bathroom: bathtub, water closet, vanity and sink; gas fired heating and air conditioning system.	5.89	3.37	9.26	7.9%
9 Electrical	200 amp electric service; wiring, duplex & GFI receptacles, wall switches, door bell, appliance circuits, air conditioning circuit, fans and communications cabling; custom grade lighting fixtures.	2.26	2.58	4.84	4.1%
10 Overhead	Contractor's overhead and profit and design.	11.00	8.58	19.58	16.7%
Total		65.95	51.55	**117.50**	

For customer support on your Residential Costs with RSMeans data, call 800.448.8182.

73

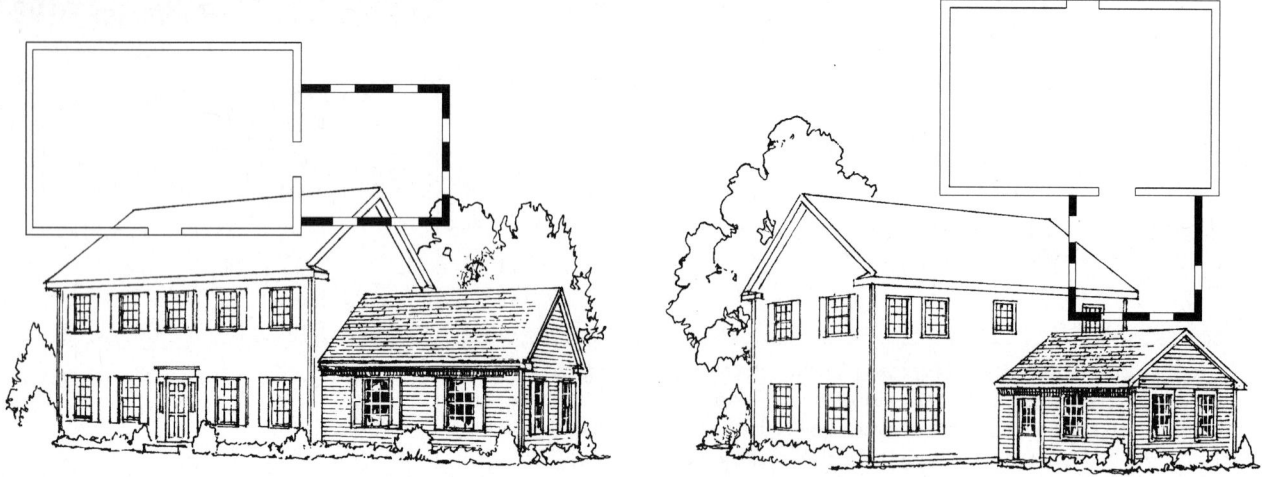

1 Story Base cost per square foot of living area

Exterior Wall	Living Area							
	50	100	200	300	400	500	600	700
Wood Siding - Wood Frame	257.45	199.95	175.05	148.30	140.05	135.15	131.90	132.85
Brick Veneer - Wood Frame	284.90	219.60	191.45	159.15	149.90	144.35	140.60	141.25
Stone Veneer - Wood Frame	310.35	237.75	206.55	169.25	158.95	152.80	148.70	149.05
Solid Masonry	316.10	241.85	209.95	171.50	161.00	154.75	150.45	150.80
Finished Basement, Add	98.20	83.60	75.70	62.50	59.85	58.30	57.25	56.50
Unfinished Basement, Add	72.90	50.40	40.10	30.80	28.35	26.90	25.90	25.20

1-1/2 Story Base cost per square foot of living area

Exterior Wall	Living Area							
	100	200	300	400	500	600	700	800
Wood Siding - Wood Frame	204.80	167.90	145.45	131.90	125.10	121.85	117.70	116.50
Brick Veneer - Wood Frame	229.35	187.55	161.80	144.70	136.85	133.00	128.20	126.95
Stone Veneer - Wood Frame	252.05	205.70	176.95	156.50	147.80	143.30	137.95	136.60
Solid Masonry	257.15	209.75	180.35	159.15	150.20	145.60	140.15	138.70
Finished Basement, Add	65.50	60.55	55.30	49.50	47.85	46.85	45.85	45.70
Unfinished Basement, Add	43.60	33.25	28.40	24.15	22.65	21.70	20.90	20.60

2 Story Base cost per square foot of living area

Exterior Wall	Living Area							
	100	200	400	600	800	1000	1200	1400
Wood Siding - Wood Frame	204.10	155.60	134.40	113.80	106.80	102.60	99.90	101.15
Brick Veneer - Wood Frame	231.65	175.25	150.75	124.65	116.65	111.80	108.60	109.60
Stone Veneer - Wood Frame	257.05	193.40	165.90	134.75	125.75	120.30	116.65	117.35
Solid Masonry	262.75	197.50	169.30	137.00	127.75	122.20	118.45	119.10
Finished Basement, Add	49.15	41.85	37.85	31.30	29.90	29.15	28.65	28.25
Unfinished Basement, Add	36.45	25.25	20.05	15.40	14.15	13.45	12.90	12.55

Base costs do not include bathroom or kitchen facilities. Use Modifications/Adjustments/Alternatives on pages 93–96 where appropriate.

1 Story

1-1/2 Story

2 Story

2-1/2 Story

Bi-Level

Tri-Level

For customer support on your Residential Costs with RSMeans data, call 800.448.8182.

77

- **Unique residence built from an architect's plan**
- **Single family — 1 full bath, 1 half bath, 1 kitchen**
- **No basement**
- **Cedar shakes on roof**
- **Forced hot air heat/air conditioning**
- **Gypsum wallboard interior finishes**
- **Many special features**
- **Extraordinary materials and workmanship**

Note: The illustration shown may contain some optional components (for example: garages and/or fireplaces) whose costs are shown in the modifications, adjustments, & alternatives below or at the end of the square foot section.

©Home Planners, Inc.

Base cost per square foot of living area

Exterior Wall	Living Area										
	1000	1200	1400	1600	1800	2000	2400	2800	3200	3600	4000
Wood Siding - Wood Frame	230.40	210.10	194.05	183.05	176.65	169.35	156.45	147.05	140.35	133.60	128.10
Brick Veneer - Wood Frame	241.20	219.85	202.95	191.35	184.65	176.90	163.30	153.40	146.25	139.10	133.20
Solid Brick	254.05	231.45	213.45	201.25	194.15	185.85	171.45	160.95	153.25	145.60	139.30
Solid Stone	261.80	238.50	219.90	207.25	199.90	191.25	176.40	165.45	157.45	149.55	143.00
Finished Basement, Add	59.80	64.35	61.70	60.05	59.00	57.60	55.65	54.15	52.80	51.70	50.70
Unfinished Basement, Add	27.00	25.30	24.00	23.10	22.65	21.95	20.90	20.10	19.45	18.90	18.40

Modifications

Add to the total cost

Upgrade Kitchen Cabinets	$ + 2604
Solid Surface Countertops (Included)	
Full Bath - including plumbing, wall and floor finishes	+ 11,593
Half Bath - including plumbing, wall and floor finishes	+ 6864
Two Car Attached Garage	+ 33,913
Two Car Detached Garage	+ 38,535
Fireplace & Chimney	+ 10,406

Adjustments

For multi family - add to total cost

Additional Kitchen	$ + 30,297
Additional Full Bath & Half Bath	+ 18,457
Additional Entry & Exit	+ 2418
Separate Heating & Air Conditioning	+ 7787
Separate Electric	+ 1817

For Townhouse/Rowhouse - Multiply cost per square foot by

Inner Unit	.90
End Unit	.95

Alternatives

Add to or deduct from the cost per square foot of living area

Heavyweight Asphalt Shingles	– 2.75
Clay Tile Roof	+ 3.40
Slate Roof	+ 4.20
Upgrade Ceilings to Textured Finish	+ .60
Air Conditioning, in Heating Ductwork	Base System
Heating Systems, Hot Water	+ 1.80
Heat Pump	+ 1.46
Electric Heat	– 2.05
Not Heated	– 4.92

Additional upgrades or components

Kitchen Cabinets & Countertops	Page 93
Bathroom Vanities	94
Fireplaces & Chimneys	94
Windows, Skylights & Dormers	94
Appliances	95
Breezeways & Porches	95
Finished Attic	95
Garages	96
Site Improvements	96
Wings & Ells	92

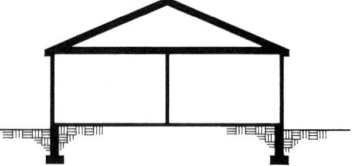

			Cost Per Square Foot Of Living Area			% of Total
			Mat.	Inst.	Total	(rounded)
1	**Site Work**	Site preparation for slab; 4' deep trench excavation for foundation wall.		1.14	1.14	0.8%
2	**Foundation**	Continuous reinforced concrete footing, 12" deep x 24" wide; damproofed and insulated 12" thick reinforced concrete block foundation wall, 4' deep; trowel finished 6" thick concrete slab on 4" crushed stone base and polyethylene vapor barrier.	7.90	8.41	16.31	11.1%
3	**Framing**	Exterior walls - 2" x 6" wood studs, 16" O.C.; 1/2" sheathing; gable end roof framing, 2" x 10" rafters, 16" O.C. with 1/2" plywood sheathing; 2" x 4" interior partitions.	7.11	9.73	16.84	11.5%
4	**Exterior Walls**	1" x 6" tongue and groove vertical wood siding and housewrap on insulated wood frame walls; R38 attic insulation; metal clad double hung wood windows; raised panel exterior doors, frame and hardware, painted finish; wood storm and screen door.	12.58	4.16	16.74	11.4%
5	**Roofing**	Red cedar roof shingles, perfections; #15 felt building paper; aluminum gutters, downspouts and drip edge; copper flashings.	5.27	4.30	9.57	6.5%
6	**Interiors**	Skim coated 1/2" thick gypsum wallboard walls and ceilings, primed and painted with 2 coats; interior raised panel solid core doors, frames and hardware, painted finish; oak hardwood flooring, 70%; ceramic tile flooring, 30%.	14.05	14.89	28.94	19.7%
7	**Specialties**	Luxury grade kitchen cabinets and countertops; double bowl kitchen sink; 75 gallon gas water heater.	9.75	1.83	11.58	7.9%
8	**Mechanical**	Three fixture bathroom: bathtub, water closet, vanity and sink; gas fired heating and air conditioning system.	7.35	3.85	11.20	7.6%
9	**Electrical**	200 amp electric service; wiring, duplex & GFI receptacles, wall switches, dimmer switches, door bell, appliance circuits, air conditioning circuit, fans and communications cabling; luxury grade lighting fixtures.	3.41	2.84	6.25	4.3%
10	**Overhead**	Contractor's overhead and profit and architect's fees.	16.18	12.30	28.48	19.4%
	Total		83.60	63.45	**147.05**	

- **Unique residence built from an architect's plan**
- **Single family — 1 full bath, 1 half bath, 1 kitchen**
- **No basement**
- **Cedar shakes on roof**
- **Forced hot air heat/air conditioning**
- **Gypsum wallboard interior finishes**
- **Many special features**
- **Extraordinary materials and workmanship**

Note: The illustration shown may contain some optional components (for example: garages and/or fireplaces) whose costs are shown in the modifications, adjustments, & alternatives below or at the end of the square foot section.

©Larry E. Belk Designs

Base cost per square foot of living area

Exterior Wall	Living Area										
	1000	1200	1400	1600	1800	2000	2400	2800	3200	3600	4000
Wood Siding - Wood Frame	231.70	214.10	201.60	187.40	179.00	170.90	156.00	148.85	142.40	137.50	130.85
Brick Veneer - Wood Frame	244.60	226.15	213.05	197.85	188.90	180.30	164.30	156.75	149.65	144.55	137.40
Solid Brick	259.20	239.85	226.10	209.70	200.10	190.95	173.75	165.65	157.90	152.45	144.80
Solid Stone	268.90	249.00	234.75	217.55	207.60	198.05	180.00	171.55	163.40	157.80	149.70
Finished Basement, Add	42.00	45.65	44.25	42.25	41.15	40.15	38.05	37.10	36.05	35.45	34.60
Unfinished Basement, Add	19.45	18.55	17.90	16.90	16.35	15.80	14.75	14.30	13.70	13.40	12.95

Modifications

Add to the total cost

Upgrade Kitchen Cabinets	$ + 2604
Solid Surface Countertops (Included)	
Full Bath - including plumbing, wall and floor finishes	+ 11,593
Half Bath - including plumbing, wall and floor finishes	+ 6864
Two Car Attached Garage	+ 33,913
Two Car Detached Garage	+ 38,535
Fireplace & Chimney	+ 10,406

Adjustments

For multi family - add to total cost

Additional Kitchen	$ + 30,297
Additional Full Bath & Half Bath	+ 18,457
Additional Entry & Exit	+ 2418
Separate Heating & Air Conditioning	+ 7787
Separate Electric	+ 1817

For Townhouse/Rowhouse - Multiply cost per square foot by

Inner Unit	.90
End Unit	.95

Alternatives

Add to or deduct from the cost per square foot of living area

Heavyweight Asphalt Shingles	– 2
Clay Tile Roof	+ 2.45
Slate Roof	+ 3.05
Upgrade Ceilings to Textured Finish	+ .60
Air Conditioning, in Heating Ductwork	Base System
Heating Systems, Hot Water	+ 1.72
Heat Pump	+ 1.62
Electric Heat	– 2.05
Not Heated	– 4.54

Additional upgrades or components

Kitchen Cabinets & Countertops	Page 93
Bathroom Vanities	94
Fireplaces & Chimneys	94
Windows, Skylights & Dormers	94
Appliances	95
Breezeways & Porches	95
Finished Attic	95
Garages	96
Site Improvements	96
Wings & Ells	92

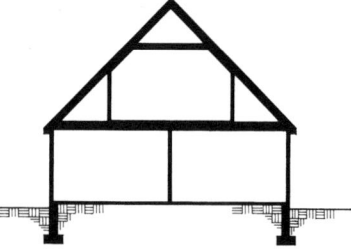

			Cost Per Square Foot Of Living Area			% of Total
			Mat.	Inst.	Total	(rounded)
1	**Site Work**	Site preparation for slab; 4' deep trench excavation for foundation wall.		1.14	1.14	0.8%
2	**Foundation**	Continuous reinforced concrete footing, 12" deep x 24" wide; damproofed and insulated 12" thick reinforced concrete block foundation wall, 4' deep; trowel finished 6" thick concrete slab on 4" crushed stone base and polyethylene vapor barrier.	5.72	6.36	12.08	8.1%
3	**Framing**	Exterior walls - 2" x 6" wood studs, 16" O.C.; 1/2" sheathing; gable end roof framing, steep pitch 2" x 10" rafters, 16" O.C. with 1/2" plywood sheathing; 2" x 12" floor joists, 16" O.C. with bridging and 5/8" subflooring; 2" x 4" interior partitions.	8.25	10.78	19.03	12.8%
4	**Exterior Walls**	1" x 6" tongue and groove vertical wood siding and housewrap on insulated wood frame walls; R38 attic insulation; metal clad double hung wood windows; raised panel exterior doors, frame and hardware, painted finish; wood storm and screen door.	13.48	4.51	17.99	12.1%
5	**Roofing**	Red cedar roof shingles, perfections; #15 felt building paper; aluminum gutters, downspouts and drip edge; copper flashings.	3.82	3.12	6.94	4.7%
6	**Interiors**	Skim coated 1/2" thick gypsum wallboard walls and ceilings, primed and painted with 2 coats; interior raised panel solid core doors, frames and hardware, painted finish; oak hardwood flooring, 70%; ceramic tile flooring, 30%; hardwood tread stairway.	16.77	17.07	33.84	22.7%
7	**Specialties**	Luxury grade kitchen cabinets and countertops; double bowl kitchen sink; 75 gallon gas water heater.	9.75	1.83	11.58	7.8%
8	**Mechanical**	Three fixture bathroom: bathtub, water closet, vanity and sink; gas fired heating and air conditioning system.	7.35	3.85	11.20	7.5%
9	**Electrical**	200 amp electric service; wiring, duplex & GFI receptacles, wall switches, dimmer switches, door bell, appliance circuits, air conditioning circuit, fans and communications cabling; luxury grade lighting fixtures.	3.41	2.84	6.25	4.2%
10	**Overhead**	Contractor's overhead and profit and architect's fees.	16.45	12.35	28.80	19.3%
		Total	85.00	63.85	**148.85**	

For customer support on your Residential Costs with RSMeans data, call 800.448.8182.

81

- **Unique residence built from an architect's plan**
- **Single family — 1 full bath, 1 half bath, 1 kitchen**
- **No basement**
- **Cedar shakes on roof**
- **Forced hot air heat/air conditioning**
- **Gypsum wallboard interior finishes**
- **Many special features**
- **Extraordinary materials and workmanship**

Note: The illustration shown may contain some optional components (for example: garages and/or fireplaces) whose costs are shown in the modifications, adjustments, & alternatives below or at the end of the square foot section.

Base cost per square foot of living area

Exterior Wall	Living Area										
	1200	1400	1600	1800	2000	2400	2800	3200	3600	4000	4400
Wood Siding - Wood Frame	210.20	196.60	186.80	178.25	169.45	156.60	145.95	138.90	134.60	130.10	126.35
Brick Veneer - Wood Frame	223.05	208.55	198.15	189.05	179.75	165.95	154.40	146.85	142.30	137.25	133.35
Solid Brick	239.30	223.70	212.65	202.70	192.95	177.80	165.20	156.95	152.10	146.45	142.20
Solid Stone	247.65	231.50	220.15	209.75	199.70	183.95	170.80	162.20	157.15	151.15	146.70
Finished Basement, Add	33.75	36.70	35.60	34.55	33.85	32.20	30.90	30.05	29.55	28.85	28.45
Unfinished Basement, Add	15.60	14.85	14.40	13.85	13.55	12.65	12.00	11.60	11.30	10.95	10.75

Modifications

Add to the total cost

Upgrade Kitchen Cabinets	$ + 2604
Solid Surface Countertops (Included)	
Full Bath - including plumbing, wall and floor finishes	+ 11,593
Half Bath - including plumbing, wall and floor finishes	+ 6864
Two Car Attached Garage	+ 33,913
Two Car Detached Garage	+ 38,535
Fireplace & Chimney	+ 11,407

Adjustments

For multi family - add to total cost

Additional Kitchen	$ + 30,297
Additional Full Bath & Half Bath	+ 18,457
Additional Entry & Exit	+ 2418
Separate Heating & Air Conditioning	+ 7787
Separate Electric	+ 1817

For Townhouse/Rowhouse -
Multiply cost per square foot by

Inner Unit	.86
End Unit	.93

Alternatives

Add to or deduct from the cost per square foot of living area

Heavyweight Asphalt Shingles	– 1.35
Clay Tile Roof	+ 1.70
Slate Roof	+ 2.10
Upgrade Ceilings to Textured Finish	+ .60
Air Conditioning, in Heating Ductwork	Base System
Heating Systems, Hot Water	+ 1.67
Heat Pump	+ 1.71
Electric Heat	– 1.86
Not Heated	– 4.29

Additional upgrades or components

Kitchen Cabinets & Countertops	Page 93
Bathroom Vanities	94
Fireplaces & Chimneys	94
Windows, Skylights & Dormers	94
Appliances	95
Breezeways & Porches	95
Finished Attic	95
Garages	96
Site Improvements	96
Wings & Ells	92

			Cost Per Square Foot Of Living Area			% of Total
			Mat.	**Inst.**	**Total**	**(rounded)**
1	**Site Work**	Site preparation for slab; 4' deep trench excavation for foundation wall.		.99	.99	0.7%
2	**Foundation**	Continuous reinforced concrete footing, 12" deep x 24" wide; damproofed and insulated 12" thick reinforced concrete block foundation wall, 4' deep; trowel finished 6" thick concrete slab on 4" crushed stone base and polyethylene vapor barrier.	4.63	5.17	9.80	7.1%
3	**Framing**	Exterior walls - 2" x 6" wood studs, 16" O.C.; 1/2" sheathing; gable end roof framing, 2" x 10" rafters, 16" O.C. with 1/2" plywood sheathing; 2" x 12" floor joists, 16" O.C. with bridging and 5/8" subflooring; 2" x 4" interior partitions.	7.82	10.05	17.87	12.9%
4	**Exterior Walls**	1" x 6" tongue and groove vertical wood siding and housewrap on insulated wood frame walls; R38 attic insulation; metal clad double hung wood windows; raised panel exterior doors, frame and hardware, painted finish; wood storm and screen door.	13.65	4.57	18.22	13.1%
5	**Roofing**	Red cedar roof shingles, perfections; #15 felt building paper; aluminum gutters, downspouts and drip edge; copper flashings.	2.63	2.15	4.78	3.4%
6	**Interiors**	Skim coated 1/2" thick gypsum wallboard walls and ceilings, primed and painted with 2 coats; interior raised panel solid core doors, frames and hardware, painted finish; oak hardwood flooring, 70%; ceramic tile flooring, 30%; hardwood tread stairway.	16.64	17.09	33.73	24.3%
7	**Specialties**	Luxury grade kitchen cabinets and countertops; double bowl kitchen sink; 75 gallon gas water heater.	8.54	1.59	10.13	7.3%
8	**Mechanical**	Three fixture bathroom: bathtub, water closet, vanity and sink; gas fired heating and air conditioning system.	6.60	3.73	10.33	7.4%
9	**Electrical**	200 amp electric service; wiring, duplex & GFI receptacles, wall switches, dimmer switches, door bell, appliance circuits, air conditioning circuit, fans and communications cabling; luxury grade lighting fixtures.	3.39	2.77	6.16	4.4%
10	**Overhead**	Contractor's overhead and profit and architect's fees.	15.35	11.54	26.89	19.4%
	Total		79.25	59.65	**138.90**	

- **Unique residence built from an architect's plan**
- **Single family — 1 full bath, 1 half bath, 1 kitchen**
- **No basement**
- **Cedar shakes on roof**
- **Forced hot air heat/air conditioning**
- **Gypsum wallboard interior finishes**
- **Many special features**
- **Extraordinary materials and workmanship**

Note: The illustration shown may contain some optional components (for example: garages and/or fireplaces) whose costs are shown in the modifications, adjustments, & alternatives below or at the end of the square foot section.

©Larry W. Garnett & Associates, Inc

Base cost per square foot of living area

Exterior Wall	Living Area										
	1500	1800	2100	2500	3000	3500	4000	4500	5000	5500	6000
Wood Siding - Wood Frame	205.60	184.70	172.00	162.55	149.90	140.90	132.25	128.10	124.20	120.30	116.40
Brick Veneer - Wood Frame	219.25	197.30	183.30	173.20	159.55	149.70	140.40	135.75	131.50	127.30	123.05
Solid Brick	235.90	212.70	197.05	186.15	171.30	160.40	150.30	145.15	140.50	135.80	131.25
Solid Stone	245.35	221.35	204.85	193.45	178.05	166.50	155.90	150.45	145.55	140.70	135.90
Finished Basement, Add	26.95	29.05	27.35	26.45	25.20	24.15	23.40	22.85	22.40	21.95	21.65
Unfinished Basement, Add	12.65	11.90	11.05	10.60	10.00	9.45	9.05	8.80	8.55	8.30	8.20

Modifications

Add to the total cost

Upgrade Kitchen Cabinets	$ + 2604
Solid Surface Countertops (Included)	
Full Bath - including plumbing, wall and floor finishes	+ 11,593
Half Bath - including plumbing, wall and floor finishes	+ 6864
Two Car Attached Garage	+ 33,913
Two Car Detached Garage	+ 38,535
Fireplace & Chimney	+ 12,482

Adjustments

For multi family - add to total cost

Additional Kitchen	$ + 30,297
Additional Full Bath & Half Bath	+ 18,457
Additional Entry & Exit	+ 2418
Separate Heating & Air Conditioning	+ 7787
Separate Electric	+ 1817

For Townhouse/Rowhouse - Multiply cost per square foot by

Inner Unit	.86
End Unit	.93

Alternatives

Add to or deduct from the cost per square foot of living area

Heavyweight Asphalt Shingles	– 1.20
Clay Tile Roof	+ 1.45
Slate Roof	+ 1.80
Upgrade Ceilings to Textured Finish	+ .60
Air Conditioning, in Heating Ductwork	Base System
Heating Systems, Hot Water	+ 1.51
Heat Pump	+ 1.76
Electric Heat	– 3.65
Not Heated	– 4.29

Additional upgrades or components

Kitchen Cabinets & Countertops	Page 93
Bathroom Vanities	94
Fireplaces & Chimneys	94
Windows, Skylights & Dormers	94
Appliances	95
Breezeways & Porches	95
Finished Attic	95
Garages	96
Site Improvements	96
Wings & Ells	92

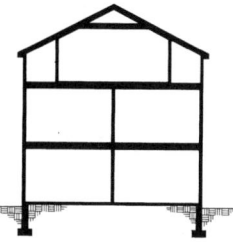

		Cost Per Square Foot Of Living Area			% of Total
		Mat.	Inst.	Total	(rounded)
1 Site Work	Site preparation for slab; 4' deep trench excavation for foundation wall.		1.06	1.06	0.7%
2 Foundation	Continuous reinforced concrete footing, 12" deep x 24" wide; damproofed and insulated 12" thick reinforced concrete block foundation wall, 4' deep; trowel finished 6" thick concrete slab on 4" crushed stone base and polyethylene vapor barrier.	4.07	4.69	8.76	5.8%
3 Framing	Exterior walls - 2" x 6" wood studs, 16" O.C.; 1/2" sheathing; gable end roof framing, steep pitch 2" x 10" rafters, 16" O.C. with 1/2" plywood sheathing; 2" x 12" floor joists, 16" O.C. with bridging and 5/8" subflooring; 2" x 4" interior partitions.	8.67	11.10	19.77	13.2%
4 Exterior Walls	1" x 6" tongue and groove vertical wood siding and housewrap on insulated wood frame walls; R38 attic insulation; metal clad double hung wood windows; raised panel exterior doors, frame and hardware, painted finish; wood storm and screen door.	15.84	5.32	21.16	14.1%
5 Roofing	Red cedar roof shingles, perfections; #15 felt building paper; aluminum gutters, downspouts and drip edge; copper flashings.	2.35	1.91	4.26	2.8%
6 Interiors	Skim coated 1/2" thick gypsum wallboard walls and ceilings, primed and painted with 2 coats; interior raised panel solid core doors, frames and hardware, painted finish; oak hardwood flooring, 70%; ceramic tile flooring, 30%; hardwood tread stairway.	19.14	19.02	38.16	25.5%
7 Specialties	Luxury grade kitchen cabinets and countertops; double bowl kitchen sink; 75 gallon gas water heater.	9.10	1.70	10.80	7.2%
8 Mechanical	Three fixture bathroom: bathtub, water closet, vanity and sink; gas fired heating and air conditioning system.	6.95	3.78	10.73	7.2%
9 Electrical	200 amp electric service; wiring, duplex & GFI receptacles, wall switches, dimmer switches, door bell, appliance circuits, air conditioning circuit, fans and communications cabling; luxury grade lighting fixtures.	3.40	2.81	6.21	4.1%
10 Overhead	Contractor's overhead and profit and architect's fees.	16.68	12.31	28.99	19.3%
Total		86.20	63.70	**149.90**	

- **Unique residence built from an architect's plan**
- **Single family — 1 full bath, 1 half bath, 1 kitchen**
- **No basement**
- **Cedar shakes on roof**
- **Forced hot air heat/air conditioning**
- **Gypsum wallboard interior finishes**
- **Many special features**
- **Extraordinary materials and workmanship**

Note: The illustration shown may contain some optional components (for example: garages and/or fireplaces) whose costs are shown in the modifications, adjustments, & alternatives below or at the end of the square foot section.

Base cost per square foot of living area

Exterior Wall	Living Area										
	1500	1800	2100	2500	3000	3500	4000	4500	5000	5500	6000
Wood Siding - Wood Frame	202.00	181.45	171.00	162.05	149.15	142.45	134.60	126.60	123.50	120.30	117.00
Brick Veneer - Wood Frame	215.60	194.05	182.80	173.30	159.35	152.10	143.30	134.70	131.30	127.80	124.05
Solid Brick	233.30	210.45	198.05	187.90	172.60	164.70	154.75	145.25	141.55	137.65	133.35
Solid Stone	241.85	218.40	205.50	195.00	179.05	170.75	160.30	150.40	146.50	142.40	137.80
Finished Basement, Add	23.65	25.50	24.45	23.70	22.55	21.85	21.00	20.25	20.00	19.65	19.20
Unfinished Basement, Add	11.05	10.40	9.95	9.55	9.00	8.70	8.20	7.85	7.70	7.55	7.30

Modifications

Add to the total cost

Upgrade Kitchen Cabinets	$ + 2604
Solid Surface Countertops (Included)	
Full Bath - including plumbing, wall and floor finishes	+ 11,593
Half Bath - including plumbing, wall and floor finishes	+ 6864
Two Car Attached Garage	+ 33,913
Two Car Detached Garage	+ 38,535
Fireplace & Chimney	+ 12,482

Adjustments

For multi family - add to total cost

Additional Kitchen	$ + 30,297
Additional Full Bath & Half Bath	+ 18,457
Additional Entry & Exit	+ 2418
Separate Heating & Air Conditioning	+ 7787
Separate Electric	+ 1817

For Townhouse/Rowhouse -
Multiply cost per square foot by

Inner Unit	.84
End Unit	.92

Alternatives

Add to or deduct from the cost per square foot of living area

Heavyweight Asphalt Shingles	– .90
Clay Tile Roof	+ 1.15
Slate Roof	+ 1.40
Upgrade Ceilings to Textured Finish	+ .60
Air Conditioning, in Heating Ductwork	Base System
Heating Systems, Hot Water	+ 1.51
Heat Pump	+ 1.76
Electric Heat	– 3.65
Not Heated	– 4.17

Additional upgrades or components

Kitchen Cabinets & Countertops	Page 93
Bathroom Vanities	94
Fireplaces & Chimneys	94
Windows, Skylights & Dormers	94
Appliances	95
Breezeways & Porches	95
Finished Attic	95
Garages	96
Site Improvements	96
Wings & Ells	92

			Cost Per Square Foot Of Living Area			% of Total
			Mat.	Inst.	Total	(rounded)
1	**Site Work**	Site preparation for slab; 4' deep trench excavation for foundation wall.		1.06	1.06	0.7%
2	**Foundation**	Continuous reinforced concrete footing, 12" deep x 24" wide; damproofed and insulated 12" thick reinforced concrete block foundation wall, 4' deep; trowel finished 6" thick concrete slab on 4" crushed stone base and polyethylene vapor barrier.	3.68	4.27	7.95	5.3%
3	**Framing**	Exterior walls - 2" x 6" wood studs, 16" O.C.; 1/2" sheathing; gable end roof framing, 2" x 10" rafters, 16" O.C. with 1/2" plywood sheathing; 2" x 12" floor joists, 16" O.C. with bridging and 5/8" subflooring; 2" x 4" interior partitions.	8.53	10.82	19.35	13.0%
4	**Exterior Walls**	1" x 6" tongue and groove vertical wood siding and housewrap on insulated wood frame walls; R38 attic insulation; metal clad double hung wood windows; raised panel exterior doors, frame and hardware, painted finish; wood storm and screen door.	16.88	5.69	22.57	15.1%
5	**Roofing**	Red cedar roof shingles, perfections; #15 felt building paper; aluminum gutters, downspouts and drip edge; copper flashings.	1.75	1.44	3.19	2.1%
6	**Interiors**	Skim coated 1/2" thick gypsum wallboard walls and ceilings, primed and painted with 2 coats; interior raised panel solid core doors, frames and hardware, painted finish; oak hardwood flooring, 70%; ceramic tile flooring, 30%; hardwood tread stairway.	19.22	19.20	38.42	25.8%
7	**Specialties**	Luxury grade kitchen cabinets and countertops; double bowl kitchen sink; 75 gallon gas water heater.	9.10	1.70	10.80	7.2%
8	**Mechanical**	Three fixture bathroom: bathtub, water closet, vanity and sink; gas fired heating and air conditioning system.	6.95	3.78	10.73	7.2%
9	**Electrical**	200 amp electric service; wiring, duplex & GFI receptacles, wall switches, dimmer switches, door bell, appliance circuits, air conditioning circuit, fans and communications cabling; luxury grade lighting fixtures.	3.40	2.81	6.21	4.2%
10	**Overhead**	Contractor's overhead and profit and architect's fees.	16.69	12.18	28.87	19.4%
		Total	86.20	62.95	**149.15**	

- **Unique residence built from an architect's plan**
- **Single family — 1 full bath, 1 half bath, 1 kitchen**
- **No basement**
- **Cedar shakes on roof**
- **Forced hot air heat/air conditioning**
- **Gypsum wallboard interior finishes**
- **Many special features**
- **Extraordinary materials and workmanship**

Note: The illustration shown may contain some optional components (for example: garages and/or fireplaces) whose costs are shown in the modifications, adjustments, & alternatives below or at the end of the square foot section.

Base cost per square foot of living area

Exterior Wall	Living Area										
	1200	1400	1600	1800	2000	2400	2800	3200	3600	4000	4400
Wood Siding - Wood Frame	199.15	186.25	176.90	169.00	160.50	148.60	138.60	132.00	128.05	123.95	120.30
Brick Veneer - Wood Frame	208.90	195.35	185.65	177.20	168.40	155.75	145.05	138.10	133.90	129.35	125.65
Solid Brick	221.00	206.60	196.45	187.35	178.20	164.50	153.15	145.60	141.15	136.20	132.25
Solid Stone	227.55	212.80	202.30	192.90	183.50	169.35	157.50	149.65	145.15	139.90	135.80
Finished Basement, Add	33.75	36.70	35.60	34.55	33.85	32.20	30.90	30.05	29.55	28.85	28.45
Unfinished Basement, Add	15.60	14.85	14.40	13.85	13.55	12.65	12.00	11.60	11.30	10.95	10.75

Modifications

Add to the total cost

Upgrade Kitchen Cabinets	$ + 2604
Solid Surface Countertops (Included)	
Full Bath - including plumbing, wall and floor finishes	+ 11,593
Half Bath - including plumbing, wall and floor finishes	+ 6864
Two Car Attached Garage	+ 33,913
Two Car Detached Garage	+ 38,535
Fireplace & Chimney	+ 10,406

Adjustments

For multi family - add to total cost

Additional Kitchen	$ + 30,297
Additional Full Bath & Half Bath	+ 18,457
Additional Entry & Exit	+ 2418
Separate Heating & Air Conditioning	+ 7787
Separate Electric	+ 1817

For Townhouse/Rowhouse - Multiply cost per square foot by

Inner Unit	.89
End Unit	.94

Alternatives

Add to or deduct from the cost per square foot of living area

Heavyweight Asphalt Shingles	– 1.35
Clay Tile Roof	+ 1.70
Slate Roof	+ 2.10
Upgrade Ceilings to Textured Finish	+ .60
Air Conditioning, in Heating Ductwork	Base System
Heating Systems, Hot Water	+ 1.67
Heat Pump	+ 1.71
Electric Heat	– 1.86
Not Heated	– 4.29

Additional upgrades or components

Kitchen Cabinets & Countertops	Page 93
Bathroom Vanities	94
Fireplaces & Chimneys	94
Windows, Skylights & Dormers	94
Appliances	95
Breezeways & Porches	95
Finished Attic	95
Garages	96
Site Improvements	96
Wings & Ells	92

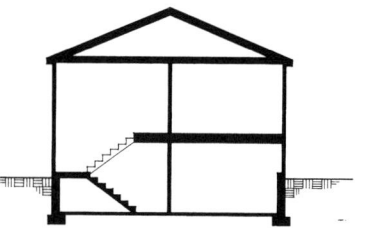

Luxury Bi-Level

Living Area - 3200 S.F.
Perimeter - 163 L.F.

			Cost Per Square Foot Of Living Area			% of Total
			Mat.	Inst.	Total	(rounded)
1	**Site Work**	Site preparation for slab; 4' deep trench excavation for foundation wall.		.99	.99	0.8%
2	**Foundation**	Continuous reinforced concrete footing, 12" deep x 24" wide; damproofed and insulated 12" thick reinforced concrete block foundation wall, 4' deep; trowel finished 6" thick concrete slab on 4" crushed stone base and polyethylene vapor barrier.	4.63	5.17	9.80	7.4%
3	**Framing**	Exterior walls - 2" x 6" wood studs, 16" O.C.; 1/2" sheathing; gable end roof framing, 2" x 10" rafters, 16" O.C. with 1/2" plywood sheathing; 2" x 12" floor joists, 16" O.C. with bridging and 5/8" subflooring; 2" x 4" interior partitions.	7.35	9.42	16.77	12.7%
4	**Exterior Walls**	1" x 6" tongue and groove vertical wood siding and housewrap on insulated wood frame walls; R38 attic insulation; metal clad double hung wood windows; raised panel exterior doors, frame and hardware, painted finish; wood storm and screen door.	10.84	3.62	14.46	11.0%
5	**Roofing**	Red cedar roof shingles, perfections: #15 felt building paper; aluminum gutters, downspouts and drip edge; copper flashings.	2.63	2.15	4.78	3.6%
6	**Interiors**	Skim coated 1/2" thick gypsum wallboard walls and ceilings, primed and painted with 2 coats; interior raised panel solid core doors, frames and hardware, painted finish; oak hardwood flooring, 70%; ceramic tile flooring, 30%; hardwood tread stairway.	16.36	16.68	33.04	25.0%
7	**Specialties**	Luxury grade kitchen cabinets and countertops; double bowl kitchen sink; 75 gallon gas water heater.	8.54	1.59	10.13	7.7%
8	**Mechanical**	Three fixture bathroom: bathtub, water closet, vanity and sink; gas fired heating and air conditioning system.	6.60	3.73	10.33	7.8%
9	**Electrical**	200 amp electric service; wiring, duplex & GFI receptacles, wall switches, dimmer switches, door bell, appliance circuits, air conditioning circuit, fans and communications cabling; luxury grade lighting fixtures	3.39	2.77	6.16	4.7%
10	**Overhead**	Contractor's overhead and profit and architect's fees.	14.46	11.08	25.54	19.3%
	Total		74.80	57.20	**132.00**	

For customer support on your Residential Costs with RSMeans data, call 800.448.8182.

89

- **Unique residence built from an architect's plan**
- **Single family — 1 full bath, 1 half bath, 1 kitchen**
- **No basement**
- **Cedar shakes on roof**
- **Forced hot air heat/air conditioning**
- **Gypsum wallboard interior finishes**
- **Many special features**
- **Extraordinary materials and workmanship**

Note: The illustration shown may contain some optional components (for example: garages and/or fireplaces) whose costs are shown in the modifications, adjustments, & alternatives below or at the end of the square foot section.

©Home Planners, Inc.

Base cost per square foot of living area

Exterior Wall	Living Area										
	1500	1800	2100	2400	2800	3200	3600	4000	4500	5000	5500
Wood Siding - Wood Frame	185.55	170.65	158.40	150.15	143.90	136.95	129.85	126.45	119.90	116.10	111.90
Brick Veneer - Wood Frame	194.40	178.65	165.70	156.90	150.45	143.05	135.50	131.90	124.90	120.80	116.35
Solid Brick	205.20	188.40	174.55	165.20	158.45	150.45	142.30	138.60	131.05	126.60	121.85
Solid Stone	211.35	193.90	179.60	170.00	162.95	154.60	146.25	142.40	134.60	129.85	124.95
Finished Basement, Add*	39.85	42.90	41.10	40.00	39.20	38.10	37.05	36.65	35.70	35.05	34.40
Unfinished Basement, Add*	18.00	16.90	16.05	15.40	15.05	14.50	13.95	13.70	13.20	12.90	12.60

*Basement under middle level only.

Modifications

Add to the total cost

Upgrade Kitchen Cabinets	$ + 2604
Solid Surface Countertops (Included)	
Full Bath - including plumbing, wall and floor finishes	+ 11,593
Half Bath - including plumbing, wall and floor finishes	+ 6864
Two Car Attached Garage	+ 33,913
Two Car Detached Garage	+ 38,535
Fireplace & Chimney	+ 10,406

Adjustments

For multi family - add to total cost

Additional Kitchen	$ + 30,297
Additional Full Bath & Half Bath	+ 18,457
Additional Entry & Exit	+ 2418
Separate Heating & Air Conditioning	+ 7787
Separate Electric	+ 1817

For Townhouse/Rowhouse - Multiply cost per square foot by

Inner Unit	.86
End Unit	.93

Alternatives

Add to or deduct from the cost per square foot of living area

Heavyweight Asphalt Shingles	– 2
Clay Tile Roof	+ 2.45
Slate Roof	+ 3.05
Upgrade Ceilings to Textured Finish	+ .60
Air Conditioning, in Heating Ductwork	Base System
Heating Systems, Hot Water	+ 1.62
Heat Pump	+ 1.78
Electric Heat	– 1.63
Not Heated	– 4.17

Additional upgrades or components

Kitchen Cabinets & Countertops	Page 93
Bathroom Vanities	94
Fireplaces & Chimneys	94
Windows, Skylights & Dormers	94
Appliances	95
Breezeways & Porches	95
Finished Attic	95
Garages	96
Site Improvements	96
Wings & Ells	92

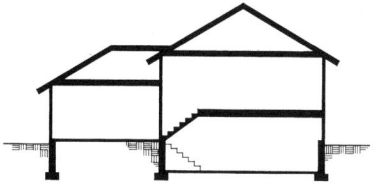

			Cost Per Square Foot Of Living Area			% of Total
			Mat.	Inst.	Total	(rounded)
1	**Site Work**	Site preparation for slab; 4' deep trench excavation for foundation wall.		.88	.88	0.7%
2	**Foundation**	Continuous reinforced concrete footing, 12" deep x 24" wide; damproofed and insulated 12" thick reinforced concrete block foundation wall, 4' deep; trowel finished 6" thick concrete slab on 4" crushed stone base and polyethylene vapor barrier.	5.52	5.94	11.46	8.8%
3	**Framing**	Exterior walls - 2" x 6" wood studs, 16" O.C.; 1/2" sheathing; gable end roof framing, 2" x 10" rafters, 16" O.C. with 1/2" plywood sheathing; 2" x 12" floor joists, 16" O.C. with bridging and 5/8" subflooring; 2" x 4" interior partitions.	7.11	9.27	16.38	12.6%
4	**Exterior Walls**	1" x 6" tongue and groove vertical wood siding and housewrap on insulated wood frame walls; R38 attic insulation; metal clad double hung wood windows; raised panel exterior doors, frame and hardware, painted finish; wood storm and screen door.	10.68	3.55	14.23	11.0%
5	**Roofing**	Red cedar roof shingles, perfections; #15 felt building paper; aluminum gutters, downspouts and drip edge; copper flashings.	3.52	2.86	6.38	4.9%
6	**Interiors**	Skim coated 1/2" thick gypsum wallboard walls and ceilings, primed and painted with 2 coats; interior raised panel solid core doors, frames and hardware, painted finish; oak hardwood flooring, 70%; ceramic tile flooring, 30%; hardwood tread stairway.	15.11	15.50	30.61	23.6%
7	**Specialties**	Luxury grade kitchen cabinets and countertops; double bowl kitchen sink; 75 gallon gas water heater.	7.60	1.42	9.02	6.9%
8	**Mechanical**	Three fixture bathroom: bathtub, water closet, vanity and sink; gas fired heating and air conditioning system.	6.03	3.60	9.63	7.4%
9	**Electrical**	200 amp electric service; wiring, duplex & GFI receptacles, wall switches, dimmer switches, door bell, appliance circuits, air conditioning circuit, fans and communications cabling; luxury grade lighting fixtures.	3.37	2.73	6.10	4.7%
10	**Overhead**	Contractor's overhead and profit and architect's fees.	14.16	11.00	25.16	19.4%
		Total	73.10	56.75	**129.85**	

1 Story — Base cost per square foot of living area

Exterior Wall	Living Area							
	50	100	200	300	400	500	600	700
Wood Siding - Wood Frame	281.20	217.20	189.35	159.55	150.40	145.00	141.35	142.40
Brick Veneer - Wood Frame	312.70	239.65	208.10	172.05	161.65	155.50	151.35	152.05
Solid Brick	358.90	272.70	235.65	190.40	178.20	170.90	166.00	166.20
Solid Stone	376.90	285.50	246.35	197.55	184.60	176.90	171.70	171.70
Finished Basement, Add	107.55	97.40	87.30	70.55	67.20	65.15	63.75	62.85
Unfinished Basement, Add	56.35	43.45	38.05	29.05	27.30	26.20	25.50	24.95

1-1/2 Story — Base cost per square foot of living area

Exterior Wall	Living Area							
	100	200	300	400	500	600	700	800
Wood Siding - Wood Frame	223.00	181.80	156.75	141.90	134.20	130.55	125.95	124.60
Brick Veneer - Wood Frame	251.15	204.30	175.50	156.50	147.70	143.30	137.95	136.55
Solid Brick	292.35	237.35	203.05	177.95	167.55	162.05	155.75	154.10
Solid Stone	308.40	250.20	213.75	186.25	175.20	169.35	162.60	160.90
Finished Basement, Add	70.70	69.85	63.15	55.75	53.75	52.40	51.15	51.00
Unfinished Basement, Add	35.80	30.40	26.80	22.90	21.80	21.10	20.45	20.35

2 Story — Base cost per square foot of living area

Exterior Wall	Living Area							
	100	200	400	600	800	1000	1200	1400
Wood Siding - Wood Frame	218.85	164.85	141.15	118.30	110.55	105.90	102.85	104.30
Brick Veneer - Wood Frame	250.40	187.35	159.95	130.75	121.75	116.40	112.85	113.90
Solid Brick	296.65	220.35	187.45	149.10	138.30	131.85	127.50	128.10
Solid Stone	314.60	233.20	198.15	156.30	144.70	137.80	133.15	133.60
Finished Basement, Add	53.80	48.75	43.70	35.25	33.60	32.60	31.95	31.45
Unfinished Basement, Add	28.15	21.70	19.05	14.55	13.65	13.10	12.75	12.50

Base costs do not include bathroom or kitchen facilities. Use Modifications/Adjustments/Alternatives on pages 93–96 where appropriate.

Kitchen cabinets - Base units, hardwood (Cost per Unit)

	Economy	Average	Custom	Luxury
24″ deep, 35″ high,				
One top drawer,				
One door below				
12″ wide	$ 278	$ 370	$ 492	$ 648
15″ wide	289	385	512	674
18″ wide	311	415	552	726
21″ wide	323	430	572	753
24″ wide	371	495	658	866
Four drawers				
12″ wide	293	390	519	683
15″ wide	296	395	525	691
18″ wide	323	430	572	753
24″ wide	353	470	625	823
Two top drawers,				
Two doors below				
27″ wide	401	535	712	936
30″ wide	435	580	771	1015
33″ wide	450	600	798	1050
36″ wide	465	620	825	1085
42″ wide	491	655	871	1146
48″ wide	525	700	931	1225
Range or sink base				
(Cost per unit)				
Two doors below				
30″ wide	368	490	652	858
33″ wide	390	520	692	910
36″ wide	409	545	725	954
42″ wide	428	570	758	998
48″ wide	450	600	798	1050
Corner Base Cabinet				
(Cost per unit)				
36″ wide	653	870	1157	1523
Lazy Susan (Cost per unit)				
With revolving door	825	1100	1463	1925

Kitchen cabinets - Wall cabinets, hardwood (Cost per Unit)

	Economy	Average	Custom	Luxury
12″ deep, 2 doors				
12″ high				
30″ wide	$ 255	$ 340	$ 452	$ 595
36″ wide	92	122	162	214
15″ high				
30″ wide	259	345	459	604
33″ wide	311	415	552	726
36″ wide	304	405	539	709
24″ high				
30″ wide	338	450	599	788
36″ wide	368	490	652	858
42″ wide	337	449	597	786
30″ high, 1 door				
12″ wide	248	330	439	578
15″ wide	259	345	459	604
18″ wide	281	375	499	656
24″ wide	323	430	572	753
30″ high, 2 doors				
27″ wide	353	470	625	823
30″ wide	375	500	665	875
36″ wide	424	565	751	989
42″ wide	454	605	805	1059
48″ wide	506	675	898	1181
Corner wall, 30″ high				
24″ wide	368	490	652	858
30″ wide	390	520	692	910
36″ wide	443	590	785	1033
Broom closet				
84″ high, 24″ deep				
18″ wide	686	915	1217	1601
Oven Cabinet				
84″ high, 24″ deep				
27″ wide	1031	1375	1829	2406

Kitchen countertops (Cost per L.F.)

	Economy	Average	Custom	Luxury
Solid Surface				
24″ wide, no backsplash	$ 110	$ 146	$ 194	$ 256
with backsplash	123	164	218	287
Stock plastic laminate, 24″ wide				
with backsplash	26	34	46	60
Custom plastic laminate, no splash				
7/8″ thick, alum. molding	40	53	71	94
1-1/4″ thick, no splash	44	59	78	103
Marble				
1/2″ - 3/4″ thick w/splash	61	81	108	143
Maple, laminated				
1-1/2″ thick w/splash	100	133	177	233
Stainless steel				
(per S.F.)	162	216	287	378
Cutting blocks, recessed				
16″ x 20″ x 1″ (each)	132	176	234	308

Vanity bases (Cost per Unit)

	Economy	Average	Custom	Luxury
2 door, 30" high, 21" deep				
24" wide	$ 315	$ 420	$ 559	$ 735
30" wide	379	505	672	884
36" wide	379	505	672	884
48" wide	488	650	865	1138

Solid surface vanity tops (Cost Each)

	Economy	Average	Custom	Luxury
Center bowl				
22" x 25"	$ 410	$ 443	$ 478	$ 516
22" x 31"	470	508	549	593
22" x 37"	540	583	630	680
22" x 49"	665	718	775	837

Fireplaces & Chimneys (Cost per Unit)

	1-1/2 Story	2 Story	3 Story
Economy (prefab metal)			
Exterior chimney & 1 fireplace	$ 6623	$ 7318	$ 8027
Interior chimney & 1 fireplace	6346	7056	7382
Average (masonry)			
Exterior chimney & 1 fireplace	6942	7671	8414
Interior chimney & 1 fireplace	6652	7396	7738
For more than 1 flue, add	472	804	1348
For more than 1 fireplace, add	4658	4658	4658
Custom (masonry)			
Exterior chimney & 1 fireplace	7266	8202	9262
Interior chimney & 1 fireplace	6813	7712	8311
For more than 1 flue, add	570	986	1651
For more than 1 fireplace, add	5219	5219	5219
Luxury (masonry)			
Exterior chimney & 1 fireplace	10406	11407	12482
Interior chimney & 1 fireplace	9927	10854	11489
For more than 1 flue, add	859	1434	2002
For more than 1 fireplace, add	8202	8202	8202

Windows and Skylights (Cost Each)

	Economy	Average	Custom	Luxury
Fixed Picture Windows				
3'-6" x 4'-0"	$ 375	$ 500	$ 665	$ 875
4'-0" x 6'-0"	648	864	1150	1513
5'-0" x 6'-0"	733	977	1300	1711
6'-0" x 6'-0"	761	1015	1350	1776
Bay/Bow Windows				
8'-0" x 5'-0"	874	1165	1550	2039
10'-0" x 5'-0"	930	1240	1650	2171
10'-0" x 6'-0"	1523	2030	2700	3553
12'-0" x 6'-0"	1988	2650	3525	4638
Palladian Windows				
3'-2" x 6'-4"		1522	2025	2664
4'-0" x 6'-0"		1672	2225	2928
5'-5" x 6'-10"		2293	3050	4013
8'-0" x 6'-0"		2763	3675	4836
Skylights				
46" x 21-1/2"	614	945	1145	1260
46" x 28"	631	970	1200	1320
57" x 44"	696	1070	1185	1304

Dormers (Cost/S.F. of plan area)

	Economy	Average	Custom	Luxury
Framing and Roofing Only				
Gable dormer, 2" x 6" roof frame	$ 33	$ 37	$ 41	$ 67
2" x 8" roof frame	34	38	42	70
Shed dormer, 2" x 6" roof frame	21	25	28	43
2" x 8" roof frame	23	26	29	44
2" x 10" roof frame	25	28	30	45

Appliances *(Cost per Unit)*

	Economy	Average	Custom	Luxury
Range				
30" free standing, 1 oven	$ 585	$ 1643	$ 2172	$ 2700
2 oven	1175	2438	3069	3700
30" built-in, 1 oven	970	1685	2043	2400
2 oven	1500	1725	1838	1950
21" free standing				
1 oven	590	748	827	905
Counter Top Ranges				
4 burner standard	450	1275	1688	2100
As above with griddle	1675	2913	3532	4150
Microwave Oven	238	504	637	770
Compactor				
4 to 1 compaction	855	1140	1283	1425
Deep Freeze				
15 to 23 C.F.	725	875	950	1025
30 C.F.	970	1073	1124	1175
Dehumidifier, portable, auto.				
15 pint	355	409	436	462
30 pint	415	478	509	540
Washing Machine, automatic	885	1443	1722	2000
Water Heater				
Electric, glass lined				
30 gal.	1125	1363	1482	1600
80 gal.	2150	2675	2938	3200
Water Heater, Gas, glass lined				
30 gal.	1975	2388	2594	2800
50 gal.	2325	2838	3094	3350
Dishwasher, built-in				
2 cycles	490	660	745	830
4 or more cycles	605	750	1188	1625
Dryer, automatic	720	1360	1680	2000
Garage Door Opener	545	650	703	755
Garbage Disposal	182	236	263	290
Heater, Electric, built-in				
1250 watt ceiling type	259	335	373	410
1250 watt wall type	350	370	380	390
Wall type w/blower				
1500 watt	340	391	417	442
3000 watt	705	811	864	917
Hood For Range, 2 speed				
30" wide	222	711	956	1200
42" wide	305	1353	1877	2400
Humidifier, portable				
7 gal. per day	162	187	199	211
15 gal. per day	210	242	258	273
Ice Maker, automatic				
13 lb. per day	1450	1668	1777	1885
51 lb. per day	1750	2013	2144	2275
Refrigerator, no frost				
10-12 C.F.	555	633	672	710
14-16 C.F.	690	835	908	980
18-20 C.F.	865	1433	1717	2000
21-29 C.F.	1275	2038	2419	2800
Sump Pump, 1/3 H.P.	330	428	477	525

Breezeway *(Cost per S.F.)*

Class	Type	Area (S.F.)			
		50	100	150	200
Economy	Open	$ 48.00	$ 39.84	$ 34.94	$ 32.49
	Enclosed	177.53	126.09	103.62	89.06
Average	Open	56.47	46.00	40.33	37.50
	Enclosed	181.46	131.09	108.53	97.01
Custom	Open	68.39	56.09	49.33	45.95
	Enclosed	244.30	175.68	144.85	129.18
Luxury	Open	76.43	62.94	55.78	52.20
	Enclosed	288.16	204.69	167.38	148.45

Porches *(Cost per S.F.)*

Class	Type	Area (S.F.)				
		25	50	100	200	300
Economy	Open	$ 85.19	$ 63.00	$ 47.28	$ 36.07	$ 32.28
	Enclosed	187.30	148.18	97.76	73.91	64.25
Average	Open	88.78	67.97	49.77	37.31	33.11
	Enclosed	205.26	162.13	104.14	76.50	65.59
Custom	Open	150.62	108.97	83.56	62.40	60.71
	Enclosed	249.90	199.63	131.69	99.36	105.77
Luxury	Open	163.59	118.89	90.71	66.13	60.04
	Enclosed	287.23	260.12	167.60	121.82	100.45

Finished attic *(Cost per S.F.)*

Class	Area (S.F.)				
	400	500	600	800	1000
Economy	$ 23.20	$ 22.42	$ 21.49	$ 21.13	$ 20.34
Average	35.76	34.98	34.12	33.69	32.75
Custom	45.75	44.73	43.71	43.04	42.16
Luxury	58.36	56.94	55.61	54.29	53.40

Alarm system *(Cost per System)*

	Burglar Alarm	Smoke Detector
Economy	$ 440	$ 73
Average	500	84
Custom	879	217
Luxury	1525	268

Sauna, prefabricated
(Cost per unit, including heater and controls—7' high)

Size	Cost
6' x 4'	$ 5400
6' x 5'	6350
6' x 6'	6675
6' x 9'	8425
8' x 10'	9875
8' x 12'	11,400
10' x 12'	13,200

Garages *

(Costs include exterior wall systems comparable with the quality of the residence. Included in the cost is an allowance for one personnel door, manual overhead door(s) and electrical fixture.)

Class	Type									
	Detached			Attached			Built-in		Basement	
	One Car	Two Car	Three Car	One Car	Two Car	Three Car	One Car	Two Car	One Car	Two Car
Economy										
Wood	$18,103	$27,607	$37,111	$13,994	$24,039	$33,543	$–1971	$–3941	$1832	$2524
Masonry	24,566	35,696	46,826	18,039	29,709	40,839	–2764	–5529		
Average										
Wood	19,847	29,790	39,733	15,086	25,569	35,512	–2185	–4370	2084	3028
Masonry	24,628	35,773	46,919	18,078	29,763	40,908	–2772	–4415		
Custom										
Wood	22,141	33,939	45,737	17,160	29,653	41,451	–3112	–2939	3046	4952
Masonry	26,881	39,871	52,861	20,126	33,811	46,801	–3694	–4103		
Luxury										
Wood	24,775	38,535	52,296	19,459	33,913	47,674	–3223	–3159	4121	6583
Masonry	32,126	47,735	63,344	24,059	40,362	55,971	–4125	–4965		

*See the Introduction to this section for definitions of garage types.

Swimming pools (Cost per S.F.)

Residential		
In-ground	$	40.50 - 97.50
Deck equipment		1.30
Paint pool, preparation & 3 coats (epoxy)		5.31
Rubber base paint		4.75
Pool Cover		1.91
Swimming Pool Heaters		
(not including wiring, external piping, base or pad)		
Gas		
155 MBH	$	3100.00
190 MBH		3700.00
500 MBH		13,100.
Electric		
15 KW 7200 gallon pool		2875.00
24 KW 9600 gallon pool		3350.00
54 KW 24,000 gallon pool		5250.00

Wood and coal stoves

Wood Only		
Free Standing (minimum)	$	2200
Fireplace Insert (minimum)		1871
Coal Only		
Free Standing	$	2114
Fireplace Insert		2314
Wood and Coal		
Free Standing	$	4348
Fireplace Insert		4456

Sidewalks (Cost per S.F.)

Concrete, 3000 psi with wire mesh	4" thick	$ 4.33
	5" thick	5.28
	6" thick	5.94
Precast concrete patio blocks (natural)	2" thick	6.95
Precast concrete patio blocks (colors)	2" thick	7.25
Flagstone, bluestone	1" thick	21.95
Flagstone, bluestone	1-1/2" thick	30.95
Slate (natural, irregular)	3/4" thick	19.60
Slate (random rectangular)	1/2" thick	30.55
Seeding		
Fine grading & seeding includes lime, fertilizer & seed	per S.Y.	3.11
Lawn Sprinkler System	per S.F.	1.14

Fencing (Cost per L.F.)

Chain Link, 4' high, galvanized	$ 17.75
Gate, 4' high (each)	238.00
Cedar Picket, 3' high, 2 rail	15.75
Gate (each)	235.00
3 Rail, 4' high	18.95
Gate (each)	253.00
Cedar Stockade, 3 Rail, 6' high	18.40
Gate (each)	256.00
Board & Battens, 2 sides 6' high, pine	35.50
6' high, cedar	39.00
No. 1 Cedar, basketweave, 6' high	37.00
Gate, 6' high (each)	315.00

Carport (Cost per S.F.)

Economy	$ 10.61
Average	15.79
Custom	23.17
Luxury	26.91

Assemblies Section

Table of Contents

RSMeans data: Assemblies— How They Work

The following is a detailed explanation of a sample Assemblies Cost Table. Included are an illustration and accompanying system descriptions. Additionally, related systems and price sheets may be included. Next to each bold number below is the item being described with the appropriate component of the sample entry in parentheses. General contractors should add an additional markup to the figures shown in the Assemblies section. Note: Throughout this section, the words assembly and system will be used interchangeably.

3 | FRAMING   12 | Gable End Roof Framing Systems

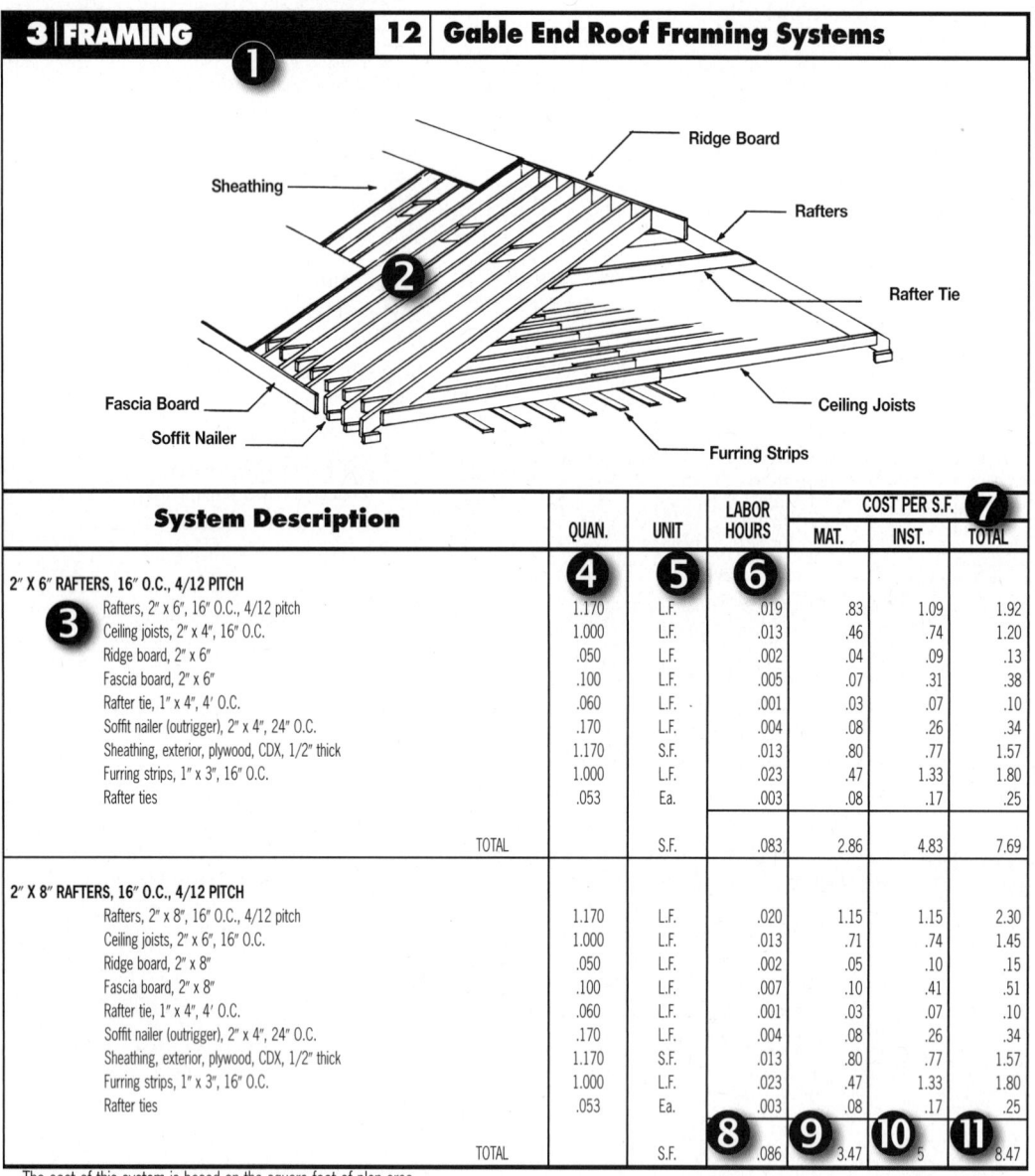

Labels: Ridge Board, Rafters, Rafter Tie, Sheathing, Ceiling Joists, Furring Strips, Fascia Board, Soffit Nailer

System Description	QUAN.	UNIT	LABOR HOURS	COST PER S.F. MAT.	COST PER S.F. INST.	COST PER S.F. TOTAL
2" X 6" RAFTERS, 16" O.C., 4/12 PITCH						
Rafters, 2" x 6", 16" O.C., 4/12 pitch	1.170	L.F.	.019	.83	1.09	1.92
Ceiling joists, 2" x 4", 16" O.C.	1.000	L.F.	.013	.46	.74	1.20
Ridge board, 2" x 6"	.050	L.F.	.002	.04	.09	.13
Fascia board, 2" x 6"	.100	L.F.	.005	.07	.31	.38
Rafter tie, 1" x 4", 4' O.C.	.060	L.F.	.001	.03	.07	.10
Soffit nailer (outrigger), 2" x 4", 24" O.C.	.170	L.F.	.004	.08	.26	.34
Sheathing, exterior, plywood, CDX, 1/2" thick	1.170	S.F.	.013	.80	.77	1.57
Furring strips, 1" x 3", 16" O.C.	1.000	L.F.	.023	.47	1.33	1.80
Rafter ties	.053	Ea.	.003	.08	.17	.25
TOTAL		S.F.	.083	2.86	4.83	7.69
2" X 8" RAFTERS, 16" O.C., 4/12 PITCH						
Rafters, 2" x 8", 16" O.C., 4/12 pitch	1.170	L.F.	.020	1.15	1.15	2.30
Ceiling joists, 2" x 6", 16" O.C.	1.000	L.F.	.013	.71	.74	1.45
Ridge board, 2" x 8"	.050	L.F.	.002	.05	.10	.15
Fascia board, 2" x 8"	.100	L.F.	.007	.10	.41	.51
Rafter tie, 1" x 4", 4' O.C.	.060	L.F.	.001	.03	.07	.10
Soffit nailer (outrigger), 2" x 4", 24" O.C.	.170	L.F.	.004	.08	.26	.34
Sheathing, exterior, plywood, CDX, 1/2" thick	1.170	S.F.	.013	.80	.77	1.57
Furring strips, 1" x 3", 16" O.C.	1.000	L.F.	.023	.47	1.33	1.80
Rafter ties	.053	Ea.	.003	.08	.17	.25
TOTAL		S.F.	.086	3.47	5	8.47

The cost of this system is based on the square foot of plan area.
All quantities have been adjusted accordingly.

Description	QUAN.	UNIT	LABOR HOURS	COST PER S.F. MAT.	COST PER S.F. INST.	COST PER S.F. TOTAL

① **System Identification**
(3 Framing 12)

Each Assemblies section has been assigned a unique identification number, component category, system number, and system description.

② **Illustration**

Included with most assemblies is an illustration with individual components labeled. Elements involved in the total system function are shown.

③ **System Description**
(2" x 6" Rafters, 16" O.C., 4/12 Pitch)

The components of a typical system are listed separately to show what has been included in the development of the total system price. Each assembly includes a brief outline of any special conditions to be used when pricing a system. Alternative components can also be found. Simply insert any chosen new element into the chart to develop a custom system.

④ **Quantities for Each Component**

Each material in a system is shown with the quantity required for the system unit. For example, there are 1.170 L.F. of rafter per S.F. of plan area.

⑤ **Unit of Measure for Each Component**

The abbreviated designation indicates the unit of measure, as defined by industry standards, upon which the individual component has been priced. In this example, items are priced by the linear foot (L.F.) or the square foot (S.F.).

⑥ **Labor-Hours**

This is the amount of time it takes to install the quantity of the individual component.

⑦ **Unit of Measure (Cost per S.F.)**

In the three right-hand columns, each cost figure is adjusted to agree with the unit of measure for the entire system. In this case, cost per S.F. is the common unit of measure.

⑧ **Labor-Hours (0.086)**

The labor-hours column shows the amount of time necessary to install the system per the unit of measure. For example, it takes 0.086 labor-hours to install one square foot (plan area) of this roof framing system.

⑨ **Materials (3.47)**

This column contains the material cost of each element. These cost figures include 10% for profit.

⑩ **Installation (5)**

This column contains labor and equipment costs. Labor rates include bare cost and the installing contractor's overhead and profit. On the average, the labor cost will be 69.5% over the bare labor cost. Equipment costs include 10% for profit.

⑪ **Totals (8.47)**

The figure in this column is the sum of the material and installation costs.

⑫ **Work Sheet**

Using the selective price sheet, it is possible to create estimates with alternative items for any number of systems.

Did you know?

RSMeans data is available through our online application with 24/7 access:

- Search for unit prices by keyword
- Leverage the most up-to-date data
- Build and export estimates

Try it free for 30 days!
www.rsmeans.com/2018freetrial

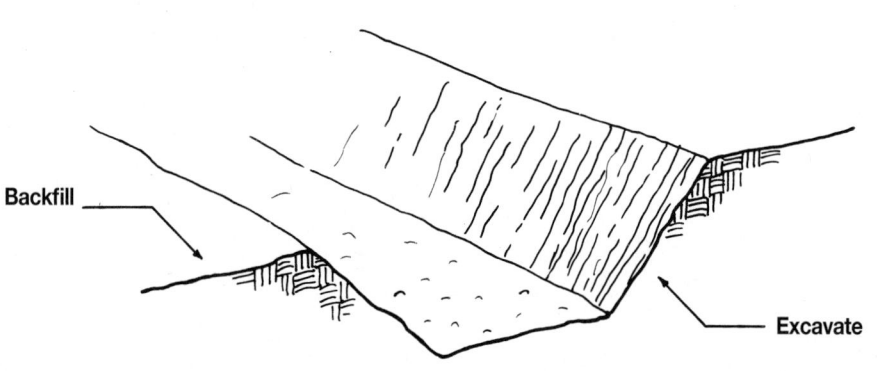

System Description	QUAN.	UNIT	LABOR HOURS	COST EACH		
				MAT.	INST.	TOTAL
BUILDING, 24′ X 38′, 4′ DEEP						
Cut & chip light trees to 6″ diam.	.190	Acre	9.120		783.75	783.75
Excavator, hydraulic, crawler mtd., 1 C.Y. cap. = 100 C.Y./hr.	174.000	C.Y.	3.480		360.18	360.18
Backfill, dozer, 4″ lifts, no compaction	87.000	C.Y.	.580		136.59	136.59
Rough grade, dozer, 30′ from building	87.000	C.Y.	.580		136.59	136.59
Mobilize and demobilize equipment	4.000	Ea.	12.000		950	950
TOTAL		Ea.	25.760		2,367.11	2,367.11
BUILDING, 26′ X 46′, 4′ DEEP						
Cut & chip light trees to 6″ diam.	.210	Acre	10.080		866.25	866.25
Excavator, hydraulic, crawler mtd., 1 C.Y. cap. = 100 C.Y./hr.	201.000	C.Y.	4.020		416.07	416.07
Backfill, dozer, 4″ lifts, no compaction	100.000	C.Y.	.667		157	157
Rough grade, dozer, 30′ from building	100.000	C.Y.	.667		157	157
Mobilize and demobilize equipment	4.000	Ea.	12.000		950	950
TOTAL		Ea.	27.434		2,546.32	2,546.32
BUILDING, 26′ X 60′, 4′ DEEP						
Cut & chip light trees to 6″ diam.	.240	Acre	11.520		990	990
Excavator, hydraulic, crawler mtd., 1 C.Y. cap. = 100 C.Y./hr.	240.000	C.Y.	4.800		496.80	496.80
Backfill, dozer, 4″ lifts, no compaction	120.000	C.Y.	.800		188.40	188.40
Rough grade, dozer, 30′ from building	120.000	C.Y.	.800		188.40	188.40
Mobilize and demobilize equipment	4.000	Ea.	12.000		950	950
TOTAL		Ea.	29.920		2,813.60	2,813.60
BUILDING, 30′ X 66′, 4′ DEEP						
Cut & chip light trees to 6″ diam.	.260	Acre	12.480		1,072.50	1,072.50
Excavator, hydraulic, crawler mtd., 1 C.Y. cap. = 100 C.Y./hr.	268.000	C.Y.	5.360		554.76	554.76
Backfill, dozer, 4″ lifts, no compaction	134.000	C.Y.	.894		210.38	210.38
Rough grade, dozer, 30′ from building	134.000	C.Y.	.894		210.38	210.38
Mobilize and demobilize equipment	4.000	Ea.	12.000		950	950
TOTAL		Ea.	31.628		2,998.02	2,998.02

The costs in this system are on a cost each basis.
Quantities are based on 1′-0″ clearance on each side of footing.

Description	QUAN.	UNIT	LABOR HOURS	COST EACH		
				MAT.	INST.	TOTAL

Footing Excavation Price Sheet	QUAN.	UNIT	LABOR HOURS	COST EACH		
				MAT.	INST.	TOTAL
Clear and grub, medium brush, 30' from building, 24' x 38'	.190	Acre	9.120		780	780
26' x 46'	.210	Acre	10.080		870	870
26' x 60'	.240	Acre	11.520		990	990
30' x 66'	.260	Acre	12.480		1,075	1,075
Light trees, to 6" dia. cut & chip, 24' x 38'	.190	Acre	9.120		780	780
26' x 46'	.210	Acre	10.080		870	870
26' x 60'	.240	Acre	11.520		990	990
30' x 66'	.260	Acre	12.480		1,075	1,075
Medium trees, to 10" dia. cut & chip, 24' x 38'	.190	Acre	13.029		1,125	1,125
26' x 46'	.210	Acre	14.400		1,250	1,250
26' x 60'	.240	Acre	16.457		1,425	1,425
30' x 66'	.260	Acre	17.829		1,550	1,550
Excavation, footing, 24' x 38', 2' deep	68.000	C.Y.	.906		141	141
4' deep	174.000	C.Y.	2.319		360	360
8' deep	384.000	C.Y.	5.119		795	795
26' x 46', 2' deep	79.000	C.Y.	1.053		164	164
4' deep	201.000	C.Y.	2.679		415	415
8' deep	404.000	C.Y.	5.385		835	835
26' x 60', 2' deep	94.000	C.Y.	1.253		195	195
4' deep	240.000	C.Y.	3.199		495	495
8' deep	483.000	C.Y.	6.438		1,000	1,000
30' x 66', 2' deep	105.000	C.Y.	1.400		217	217
4' deep	268.000	C.Y.	3.572		555	555
8' deep	539.000	C.Y.	7.185		1,125	1,125
Backfill, 24' x 38', 2" lifts, no compaction	34.000	C.Y.	.227		53.50	53.50
Compaction, air tamped, add	34.000	C.Y.	2.267		485	485
4" lifts, no compaction	87.000	C.Y.	.580		137	137
Compaction, air tamped, add	87.000	C.Y.	5.800		1,225	1,225
8" lifts, no compaction	192.000	C.Y.	1.281		300	300
Compaction, air tamped, add	192.000	C.Y.	12.801		2,725	2,725
26' x 46', 2" lifts, no compaction	40.000	C.Y.	.267		63	63
Compaction, air tamped, add	40.000	C.Y.	2.667		565	565
4" lifts, no compaction	100.000	C.Y.	.667		157	157
Compaction, air tamped, add	100.000	C.Y.	6.667		1,425	1,425
8" lifts, no compaction	202.000	C.Y.	1.347		315	315
Compaction, air tamped, add	202.000	C.Y.	13.467		2,875	2,875
26' x 60', 2" lifts, no compaction	47.000	C.Y.	.313		74	74
Compaction, air tamped, add	47.000	C.Y.	3.133		670	670
4" lifts, no compaction	120.000	C.Y.	.800		188	188
Compaction, air tamped, add	120.000	C.Y.	8.000		1,700	1,700
8" lifts, no compaction	242.000	C.Y.	1.614		380	380
Compaction, air tamped, add	242.000	C.Y.	16.134		3,450	3,450
30' x 66', 2" lifts, no compaction	53.000	C.Y.	.354		83	83
Compaction, air tamped, add	53.000	C.Y.	3.534		755	755
4" lifts, no compaction	134.000	C.Y.	.894		211	211
Compaction, air tamped, add	134.000	C.Y.	8.934		1,900	1,900
8" lifts, no compaction	269.000	C.Y.	1.794		425	425
Compaction, air tamped, add	269.000	C.Y.	17.934		3,825	3,825
Rough grade, 30' from building, 24' x 38'	87.000	C.Y.	.580		137	137
26' x 46'	100.000	C.Y.	.667		157	157
26' x 60'	120.000	C.Y.	.800		188	188
30' x 66'	134.000	C.Y.	.894		211	211
Mobilize and demobilize equipment	4.000	Ea.	12.000		950	950

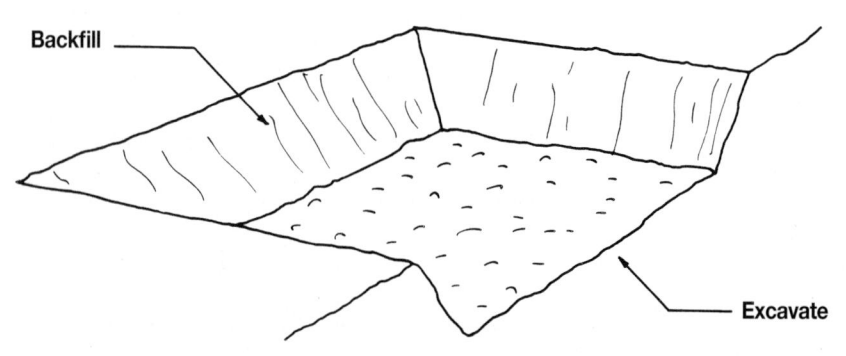

Backfill

Excavate

System Description	QUAN.	UNIT	LABOR HOURS	COST EACH		
				MAT.	INST.	TOTAL
BUILDING, 24' X 38', 8' DEEP						
Medium clearing	.190	Acre	2.027		284.05	284.05
Excavate, track loader, 1-1/2 C.Y. bucket	550.000	C.Y.	7.860		1,100	1,100
Backfill, dozer, 8" lifts, no compaction	180.000	C.Y.	1.201		282.60	282.60
Rough grade, dozer, 30' from building	280.000	C.Y.	1.868		439.60	439.60
Mobilize and demobilize equipment	4.000	Ea.	12.000		950	950
TOTAL		Ea.	24.956		3,056.25	3,056.25
BUILDING, 26' X 46', 8' DEEP						
Medium clearing	.210	Acre	2.240		313.95	313.95
Excavate, track loader, 1-1/2 C.Y. bucket	672.000	C.Y.	9.603		1,344	1,344
Backfill, dozer, 8" lifts, no compaction	220.000	C.Y.	1.467		345.40	345.40
Rough grade, dozer, 30' from building	340.000	C.Y.	2.268		533.80	533.80
Mobilize and demobilize equipment	4.000	Ea.	12.000		950	950
TOTAL		Ea.	27.578		3,487.15	3,487.15
BUILDING, 26' X 60', 8' DEEP						
Medium clearing	.240	Acre	2.560		358.80	358.80
Excavate, track loader, 1-1/2 C.Y. bucket	829.000	C.Y.	11.846		1,658	1,658
Backfill, dozer, 8" lifts, no compaction	270.000	C.Y.	1.801		423.90	423.90
Rough grade, dozer, 30' from building	420.000	C.Y.	2.801		659.40	659.40
Mobilize and demobilzie equipment	4.000	Ea.	12.000		950	950
TOTAL		Ea.	31.008		4,050.10	4,050.10
BUILDING, 30' X 66', 8' DEEP						
Medium clearing	.260	Acre	2.773		388.70	388.70
Excavate, track loader, 1-1/2 C.Y. bucket	990.000	C.Y.	14.147		1,980	1,980
Backfill dozer, 8" lifts, no compaction	320.000	C.Y.	2.134		502.40	502.40
Rough grade, dozer, 30' from building	500.000	C.Y.	3.335		785	785
Mobilize and demobilize equipment	4.000	Ea.	12.000		950	950
TOTAL		Ea.	34.389		4,606.10	4,606.10

The costs in this system are on a cost each basis.
Quantities are based on 1'-0" clearance beyond footing projection.

Description	QUAN.	UNIT	LABOR HOURS	COST EACH		
				MAT.	INST.	TOTAL

Foundation Excavation Price Sheet	QUAN.	UNIT	LABOR HOURS	COST EACH		
				MAT.	INST.	TOTAL
Clear & grub, medium brush, 30' from building, 24' x 38'	.190	Acre	2.027		284	284
26' x 46'	.210	Acre	2.240		315	315
26' x 60'	.240	Acre	2.560		360	360
30' x 66'	.260	Acre	2.773		390	390
Light trees, to 6" dia. cut & chip, 24' x 38'	.190	Acre	9.120		780	780
26' x 46'	.210	Acre	10.080		870	870
26' x 60'	.240	Acre	11.520		990	990
30' x 66'	.260	Acre	12.480		1,075	1,075
Medium trees, to 10" dia. cut & chip, 24' x 38'	.190	Acre	13.029		1,125	1,125
26' x 46'	.210	Acre	14.400		1,250	1,250
26' x 60'	.240	Acre	16.457		1,425	1,425
30' x 66'	.260	Acre	17.829		1,550	1,550
Excavation, basement, 24' x 38', 2' deep	98.000	C.Y.	1.400		197	197
4' deep	220.000	C.Y.	3.144		440	440
8' deep	550.000	C.Y.	7.860		1,100	1,100
26' x 46', 2' deep	123.000	C.Y.	1.758		246	246
4' deep	274.000	C.Y.	3.915		545	545
8' deep	672.000	C.Y.	9.603		1,350	1,350
26' x 60', 2' deep	157.000	C.Y.	2.244		315	315
4' deep	345.000	C.Y.	4.930		690	690
8' deep	829.000	C.Y.	11.846		1,650	1,650
30' x 66', 2' deep	192.000	C.Y.	2.744		385	385
4' deep	419.000	C.Y.	5.988		840	840
8' deep	990.000	C.Y.	14.147		1,975	1,975
Backfill, 24' x 38', 2" lifts, no compaction	32.000	C.Y.	.213		50.50	50.50
Compaction, air tamped, add	32.000	C.Y.	2.133		455	455
4" lifts, no compaction	72.000	C.Y.	.480		113	113
Compaction, air tamped, add	72.000	C.Y.	4.800		1,025	1,025
8" lifts, no compaction	180.000	C.Y.	1.201		283	283
Compaction, air tamped, add	180.000	C.Y.	12.001		2,550	2,550
26' x 46', 2" lifts, no compaction	40.000	C.Y.	.267		63	63
Compaction, air tamped, add	40.000	C.Y.	2.667		565	565
4" lifts, no compaction	90.000	C.Y.	.600		141	141
Compaction, air tamped, add	90.000	C.Y.	6.000		1,275	1,275
8" lifts, no compaction	220.000	C.Y.	1.467		345	345
Compacton, air tamped, add	220.000	C.Y.	14.667		3,125	3,125
26' x 60', 2" lifts, no compaction	50.000	C.Y.	.334		78.50	78.50
Compaction, air tamped, add	50.000	C.Y.	3.334		710	710
4" lifts, no compaction	110.000	C.Y.	.734		173	173
Compaction, air tamped, add	110.000	C.Y.	7.334		1,550	1,550
8" lifts, no compaction	270.000	C.Y.	1.801		425	425
Compaction, air tamped, add	270.000	C.Y.	18.001		3,825	3,825
30' x 66', 2" lifts, no compaction	60.000	C.Y.	.400		94	94
Compaction, air tamped, add	60.000	C.Y.	4.000		855	855
4" lifts, no compaction	130.000	C.Y.	.867		204	204
Compaction, air tamped, add	130.000	C.Y.	8.667		1,850	1,850
8" lifts, no compaction	320.000	C.Y.	2.134		505	505
Compaction, air tamped, add	320.000	C.Y.	21.334		4,550	4,550
Rough grade, 30' from building, 24' x 38'	280.000	C.Y.	1.868		440	440
26' x 46'	340.000	C.Y.	2.268		535	535
26' x 60'	420.000	C.Y.	2.801		660	660
30' x 66'	500.000	C.Y.	3.335		785	785
Mobilize and demobilize equipment	4.000	Ea.	12.000		950	950

For customer support on your Residential Costs with RSMeans data, call 800.448.8182.

105

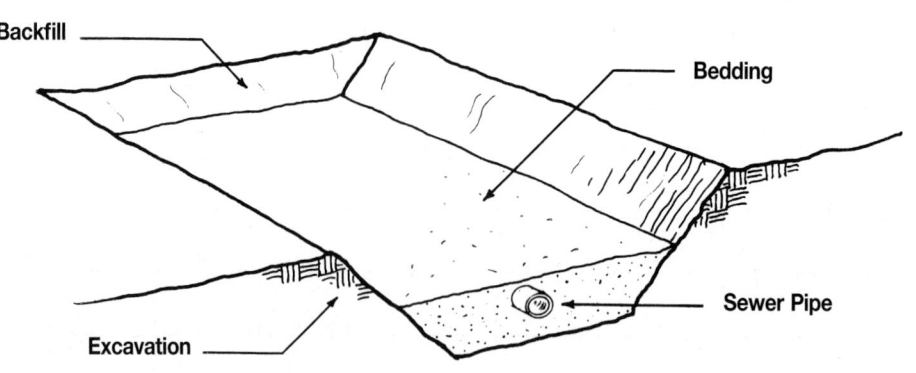

System Description	QUAN.	UNIT	LABOR HOURS	COST PER L.F.		
				MAT.	INST.	TOTAL
2′ DEEP						
Excavation, backhoe	.296	C.Y.	.032		2.34	2.34
Alternate pricing method, 4″ deep	.111	C.Y.	.044	3.66	2.25	5.91
Utility, sewer, 6″ cast iron	1.000	L.F.	.283	39.56	16.04	55.60
Compaction in 12″ layers, hand tamp, add to above	.185	C.Y.	.044		1.93	1.93
TOTAL		L.F.	.403	43.22	22.56	65.78
4′ DEEP						
Excavation, backhoe	.889	C.Y.	.095		7.02	7.02
Alternate pricing method, 4″ deep	.111	C.Y.	.044	3.66	2.25	5.91
Utility, sewer, 6″ cast iron	1.000	L.F.	.283	39.56	16.04	55.60
Compaction in 12″ layers, hand tamp, add to above	.778	C.Y.	.183		8.13	8.13
TOTAL		L.F.	.605	43.22	33.44	76.66
6′ DEEP						
Excavation, backhoe	1.770	C.Y.	.189		13.96	13.96
Alternate pricing method, 4″ deep	.111	C.Y.	.044	3.66	2.25	5.91
Utility, sewer, 6″ cast iron	1.000	L.F.	.283	39.56	16.04	55.60
Compaction in 12″ layers, hand tamp, add to above	1.660	C.Y.	.391		17.35	17.35
TOTAL		L.F.	.907	43.22	49.60	92.82
8′ DEEP						
Excavation, backhoe	2.960	C.Y.	.316		23.36	23.36
Alternate pricing method, 4″ deep	.111	C.Y.	.044	3.66	2.25	5.91
Utility, sewer, 6″ cast iron	1.000	L.F.	.283	39.56	16.04	55.60
Compaction in 12″ layers, hand tamp, add to above	2.850	C.Y.	.671		29.78	29.78
TOTAL		L.F.	1.314	43.22	71.43	114.65

The costs in this system are based on a cost per linear foot of trench, and based on 2′ wide at bottom of trench up to 6′ deep.

Description	QUAN.	UNIT	LABOR HOURS	COST PER L.F.		
				MAT.	INST.	TOTAL

Utility Trenching Price Sheet	QUAN.	UNIT	LABOR HOURS	COST PER UNIT		
				MAT.	INST.	TOTAL
Excavation, bottom of trench 2' wide, 2' deep	.296	C.Y.	.032		2.34	2.34
4' deep	.889	C.Y.	.095		7	7
6' deep	1.770	C.Y.	.142		11.15	11.15
8' deep	2.960	C.Y.	.105		21	21
Bedding, sand, bottom of trench 2' wide, no compaction, pipe, 2" diameter	.070	C.Y.	.028	2.31	1.42	3.73
4" diameter	.084	C.Y.	.034	2.77	1.71	4.48
6" diameter	.105	C.Y.	.042	3.47	2.13	5.60
8" diameter	.122	C.Y.	.049	4.03	2.48	6.51
Compacted, pipe, 2" diameter	.074	C.Y.	.030	2.44	1.50	3.94
4" diameter	.092	C.Y.	.037	3.04	1.87	4.91
6" diameter	.111	C.Y.	.044	3.66	2.25	5.91
8" diameter	.129	C.Y.	.052	4.26	2.62	6.88
3/4" stone, bottom of trench 2' wide, pipe, 4" diameter	.082	C.Y.	.033	2.71	1.66	4.37
6" diameter	.099	C.Y.	.040	3.27	2.01	5.28
3/8" stone, bottom of trench 2' wide, pipe, 4" diameter	.084	C.Y.	.034	2.77	1.71	4.48
6" diameter	.102	C.Y.	.041	3.37	2.07	5.44
Utilities, drainage & sewerage, corrugated plastic, 6" diameter	1.000	L.F.	.069	3.84	3.12	6.96
8" diameter	1.000	L.F.	.072	7.15	3.26	10.41
Concrete, non-reinforced, 6" diameter	1.000	L.F.	.181	8.30	9.85	18.15
8" diameter	1.000	L.F.	.214	9.15	11.70	20.85
PVC, SDR 35, 4" diameter	1.000	L.F.	.064	1.80	2.91	4.71
6" diameter	1.000	L.F.	.069	3.84	3.12	6.96
8" diameter	1.000	L.F.	.072	7.15	3.26	10.41
Gas & service, polyethylene, 1-1/4" diameter	1.000	L.F.	.059	1.80	3.09	4.89
Steel sched.40, 1" diameter	1.000	L.F.	.107	5.70	6.85	12.55
2" diameter	1.000	L.F.	.114	8.95	7.35	16.30
Sub-drainage, PVC, perforated, 3" diameter	1.000	L.F.	.064	1.80	2.91	4.71
4" diameter	1.000	L.F.	.064	1.80	2.91	4.71
5" diameter	1.000	L.F.	.069	3.84	3.12	6.96
6" diameter	1.000	L.F.	.069	3.84	3.12	6.96
Water service, copper, type K, 3/4"	1.000	L.F.	.083	7.10	5.45	12.55
1" diameter	1.000	L.F.	.093	10.40	6.10	16.50
PVC, 3/4"	1.000	L.F.	.121	4.74	7.90	12.64
1" diameter	1.000	L.F.	.134	7.45	8.80	16.25
Backfill, bottom of trench 2' wide no compact, 2' deep, pipe, 2" diameter	.226	L.F.	.053		2.36	2.36
4" diameter	.212	L.F.	.050		2.22	2.22
6" diameter	.185	L.F.	.044		1.93	1.93
4' deep, pipe, 2" diameter	.819	C.Y.	.193		8.55	8.55
4" diameter	.805	C.Y.	.189		8.40	8.40
6" diameter	.778	C.Y.	.183		8.15	8.15
6' deep, pipe, 2" diameter	1.700	C.Y.	.400		17.75	17.75
4" diameter	1.690	C.Y.	.398		17.65	17.65
6" diameter	1.660	C.Y.	.391		17.35	17.35
8' deep, pipe, 2" diameter	2.890	C.Y.	.680		30	30
4" diameter	2.870	C.Y.	.675		30	30
6" diameter	2.850	C.Y.	.671		30	30

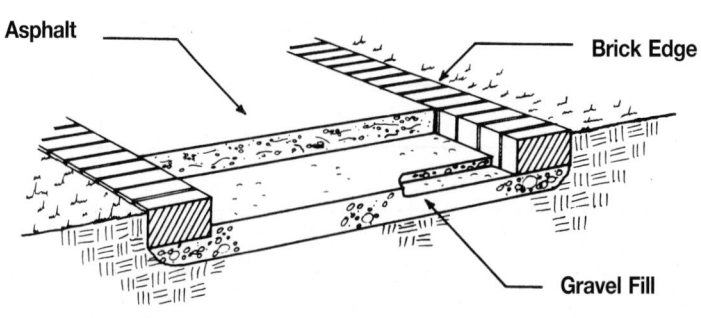

Asphalt

Brick Edge

Gravel Fill

System Description	QUAN.	UNIT	LABOR HOURS	COST PER S.F.		
				MAT.	INST.	TOTAL
ASPHALT SIDEWALK SYSTEM, 3' WIDE WALK						
Gravel fill, 4" deep	1.000	S.F.	.001	.29	.05	.34
Compact fill	.012	C.Y.			.02	.02
Handgrade	1.000	S.F.	.004		.18	.18
Walking surface, bituminous paving, 2" thick	1.000	S.F.	.007	.84	.38	1.22
Edging, brick, laid on edge	.670	L.F.	.079	2.05	4.19	6.24
TOTAL		S.F.	.091	3.18	4.82	8
CONCRETE SIDEWALK SYSTEM, 3' WIDE WALK						
Gravel fill, 4" deep	1.000	S.F.	.001	.29	.05	.34
Compact fill	.012	C.Y.			.02	.02
Handgrade	1.000	S.F.	.004		.18	.18
Walking surface, concrete, 4" thick	1.000	S.F.	.040	2.21	2.12	4.33
Edging, brick, laid on edge	.670	L.F.	.079	2.05	4.19	6.24
TOTAL		S.F.	.124	4.55	6.56	11.11
PAVERS, BRICK SIDEWALK SYSTEM, 3' WIDE WALK						
Sand base fill, 4" deep	1.000	S.F.	.001	.47	.08	.55
Compact fill	.012	C.Y.			.02	.02
Handgrade	1.000	S.F.	.004		.18	.18
Walking surface, brick pavers	1.000	S.F.	.160	3.14	8.45	11.59
Edging, redwood, untreated, 1" x 4"	.670	L.F.	.032	1.66	1.88	3.54
TOTAL		S.F.	.197	5.27	10.61	15.88

The costs in this system are based on a cost per square foot of sidewalk area. Concrete used is 3000 p.s.i.

Description	QUAN.	UNIT	LABOR HOURS	COST PER S.F.		
				MAT.	INST.	TOTAL

Sidewalk Price Sheet	QUAN.	UNIT	LABOR HOURS	COST PER S.F.		
				MAT.	INST.	TOTAL
Base, crushed stone, 3" deep	1.000	S.F.	.001	.33	.11	.44
6" deep	1.000	S.F.	.001	.66	.11	.77
9" deep	1.000	S.F.	.002	.97	.15	1.12
12" deep	1.000	S.F.	.002	1.66	.18	1.84
Bank run gravel, 6" deep	1.000	S.F.	.001	.44	.07	.51
9" deep	1.000	S.F.	.001	.64	.11	.75
12" deep	1.000	S.F.	.001	.88	.13	1.01
Compact base, 3" deep	.009	C.Y.	.001		.01	.01
6" deep	.019	C.Y.	.001		.03	.03
9" deep	.028	C.Y.	.001		.05	.05
Handgrade	1.000	S.F.	.004		.18	.18
Surface, brick, pavers dry joints, laid flat, running bond	1.000	S.F.	.160	3.14	8.45	11.59
Basket weave	1.000	S.F.	.168	4.29	8.90	13.19
Herringbone	1.000	S.F.	.174	4.29	9.20	13.49
Laid on edge, running bond	1.000	S.F.	.229	4.90	12.10	17
Mortar jts. laid flat, running bond	1.000	S.F.	.192	3.77	10.15	13.92
Basket weave	1.000	S.F.	.202	5.15	10.70	15.85
Herringbone	1.000	S.F.	.209	5.15	11.05	16.20
Laid on edge, running bond	1.000	S.F.	.274	5.90	14.50	20.40
Bituminous paving, 1-1/2" thick	1.000	S.F.	.006	.63	.28	.91
2" thick	1.000	S.F.	.007	.84	.38	1.22
2-1/2" thick	1.000	S.F.	.008	1.07	.41	1.48
Sand finish, 3/4" thick	1.000	S.F.	.001	.34	.12	.46
1" thick	1.000	S.F.	.001	.42	.15	.57
Concrete, reinforced, broom finish, 4" thick	1.000	S.F.	.040	2.21	2.12	4.33
5" thick	1.000	S.F.	.044	2.95	2.33	5.28
6" thick	1.000	S.F.	.047	3.45	2.49	5.94
Crushed stone, white marble, 3" thick	1.000	S.F.	.009	.53	.42	.95
Bluestone, 3" thick	1.000	S.F.	.009	.22	.42	.64
Flagging, bluestone, 1"	1.000	S.F.	.198	11.50	10.45	21.95
1-1/2"	1.000	S.F.	.188	21	9.95	30.95
Slate, natural cleft, 3/4"	1.000	S.F.	.174	10.40	9.20	19.60
Random rect., 1/2"	1.000	S.F.	.152	22.50	8.05	30.55
Granite blocks	1.000	S.F.	.174	21	9.20	30.20
Edging, corrugated aluminum, 4", 3' wide walk	.666	L.F.	.008	1.61	.48	2.09
4' wide walk	.500	L.F.	.006	1.21	.36	1.57
6", 3' wide walk	.666	L.F.	.010	2	.57	2.57
4' wide walk	.500	L.F.	.007	1.51	.43	1.94
Redwood-cedar-cypress, 1" x 4", 3' wide walk	.666	L.F.	.021	.85	1.24	2.09
4' wide walk	.500	L.F.	.016	.64	.93	1.57
2" x 4", 3' wide walk	.666	L.F.	.032	1.66	1.88	3.54
4' wide walk	.500	L.F.	.024	1.25	1.41	2.66
Brick, dry joints, 3' wide walk	.666	L.F.	.079	2.05	4.19	6.24
4' wide walk	.500	L.F.	.059	1.53	3.13	4.66
Mortar joints, 3' wide walk	.666	L.F.	.095	2.46	5.05	7.51
4' wide walk	.500	L.F.	.071	1.84	3.75	5.59

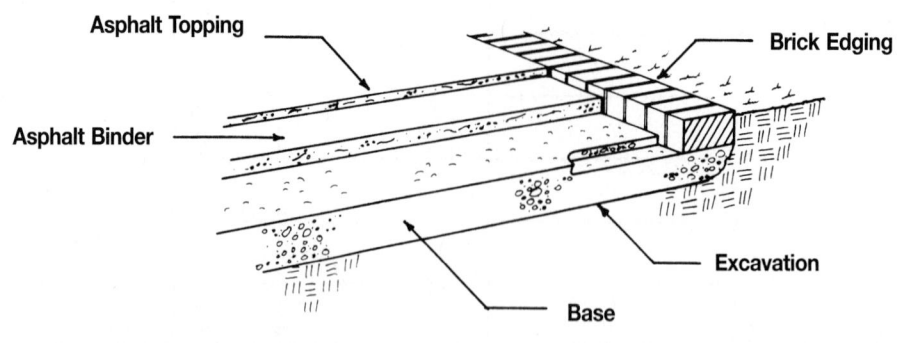

Asphalt Topping

Asphalt Binder

Brick Edging

Excavation

Base

System Description	QUAN.	UNIT	LABOR HOURS	COST PER S.F.		
				MAT.	INST.	TOTAL
ASPHALT DRIVEWAY TO 10′ WIDE						
Excavation, driveway to 10′ wide, 6″ deep	.019	C.Y.			.05	.05
Base, 6″ crushed stone	1.000	S.F.	.001	.66	.11	.77
Handgrade base	1.000	S.F.	.004		.18	.18
2″ thick base	1.000	S.F.	.002	.84	.20	1.04
1″ topping	1.000	S.F.	.001	.42	.15	.57
Edging, brick pavers	.200	L.F.	.024	.61	1.25	1.86
TOTAL		S.F.	.032	2.53	1.94	4.47
CONCRETE DRIVEWAY TO 10′ WIDE						
Excavation, driveway to 10′ wide, 6″ deep	.019	C.Y.			.05	.05
Base, 6″ crushed stone	1.000	S.F.	.001	.66	.11	.77
Handgrade base	1.000	S.F.	.004		.18	.18
Surface, concrete, 4″ thick	1.000	S.F.	.040	2.21	2.12	4.33
Edging, brick pavers	.200	L.F.	.024	.61	1.25	1.86
TOTAL		S.F.	.069	3.48	3.71	7.19
PAVERS, BRICK DRIVEWAY TO 10′ WIDE						
Excavation, driveway to 10′ wide, 6″ deep	.019	C.Y.			.05	.05
Base, 6″ sand	1.000	S.F.	.001	.75	.13	.88
Handgrade base	1.000	S.F.	.004		.18	.18
Surface, pavers, brick laid flat, running bond	1.000	S.F.	.160	3.14	8.45	11.59
Edging, redwood, untreated, 2″ x 4″	.200	L.F.	.010	.50	.56	1.06
TOTAL		S.F.	.175	4.39	9.37	13.76

Description	QUAN.	UNIT	LABOR HOURS	COST PER S.F.		
				MAT.	INST.	TOTAL

Driveway Price Sheet	QUAN.	UNIT	LABOR HOURS	COST PER S.F. MAT.	INST.	TOTAL
Excavation, by machine, 10' wide, 6" deep	.019	C.Y.	.001		.05	.05
12" deep	.037	C.Y.	.001		.08	.08
18" deep	.055	C.Y.	.001		.12	.12
20' wide, 6" deep	.019	C.Y.	.001		.05	.05
12" deep	.037	C.Y.	.001		.08	.08
18" deep	.055	C.Y.	.001		.12	.12
Base, crushed stone, 10' wide, 3" deep	1.000	S.F.	.001	.33	.06	.39
6" deep	1.000	S.F.	.001	.66	.11	.77
9" deep	1.000	S.F.	.002	.97	.15	1.12
20' wide, 3" deep	1.000	S.F.	.001	.33	.06	.39
6" deep	1.000	S.F.	.001	.66	.11	.77
9" deep	1.000	S.F.	.002	.97	.15	1.12
Bank run gravel, 10' wide, 3" deep	1.000	S.F.	.001	.22	.04	.26
6" deep	1.000	S.F.	.001	.44	.07	.51
9" deep	1.000	S.F.	.001	.64	.11	.75
20' wide, 3" deep	1.000	S.F.	.001	.22	.04	.26
6" deep	1.000	S.F.	.001	.44	.07	.51
9" deep	1.000	S.F.	.001	.64	.11	.75
Handgrade, 10' wide	1.000	S.F.	.004		.18	.18
20' wide	1.000	S.F.	.004		.18	.18
Surface, asphalt, 10' wide, 3/4" topping, 1" base	1.000	S.F.	.002	.99	.26	1.25
2" base	1.000	S.F.	.003	1.18	.32	1.50
1" topping, 1" base	1.000	S.F.	.002	1.07	.29	1.36
2" base	1.000	S.F.	.003	1.26	.35	1.61
20' wide, 3/4" topping, 1" base	1.000	S.F.	.002	.99	.26	1.25
2" base	1.000	S.F.	.003	1.18	.32	1.50
1" topping, 1" base	1.000	S.F.	.002	1.07	.29	1.36
2" base	1.000	S.F.	.003	1.26	.35	1.61
Concrete, 10' wide, 4" thick	1.000	S.F.	.040	2.21	2.12	4.33
6" thick	1.000	S.F.	.047	3.45	2.49	5.94
20' wide, 4" thick	1.000	S.F.	.040	2.21	2.12	4.33
6" thick	1.000	S.F.	.047	3.45	2.49	5.94
Paver, brick 10' wide dry joints, running bond, laid flat	1.000	S.F.	.160	3.14	8.45	11.59
Laid on edge	1.000	S.F.	.229	4.90	12.10	17
Mortar joints, laid flat	1.000	S.F.	.192	3.77	10.15	13.92
Laid on edge	1.000	S.F.	.274	5.90	14.50	20.40
20' wide, running bond, dry jts., laid flat	1.000	S.F.	.160	3.14	8.45	11.59
Laid on edge	1.000	S.F.	.229	4.90	12.10	17
Mortar joints, laid flat	1.000	S.F.	.192	3.77	10.15	13.92
Laid on edge	1.000	S.F.	.274	5.90	14.50	20.40
Crushed stone, 10' wide, white marble, 3"	1.000	S.F.	.009	.53	.42	.95
Bluestone, 3"	1.000	S.F.	.009	.22	.42	.64
20' wide, white marble, 3"	1.000	S.F.	.009	.53	.42	.95
Bluestone, 3"	1.000	S.F.	.009	.22	.42	.64
Soil cement, 10' wide	1.000	S.F.	.007	.39	1.01	1.40
20' wide	1.000	S.F.	.007	.39	1.01	1.40
Granite blocks, 10' wide	1.000	S.F.	.174	21	9.20	30.20
20' wide	1.000	S.F.	.174	21	9.20	30.20
Asphalt block, solid 1-1/4" thick	1.000	S.F.	.119	9.40	6.25	15.65
Solid 3" thick	1.000	S.F.	.123	13.20	6.50	19.70
Edging, brick, 10' wide	.200	L.F.	.024	.61	1.25	1.86
20' wide	.100	L.F.	.012	.31	.63	.94
Redwood, untreated 2" x 4", 10' wide	.200	L.F.	.010	.50	.56	1.06
20' wide	.100	L.F.	.005	.25	.28	.53
Granite, 4 1/2" x 12" straight, 10' wide	.200	L.F.	.032	1.67	2.05	3.72
20' wide	.100	L.F.	.016	.84	1.02	1.86
Finishes, asphalt sealer, 10' wide	1.000	S.F.	.023	.94	1.04	1.98
20' wide	1.000	S.F.	.023	.94	1.04	1.98
Concrete, exposed aggregate 10' wide	1.000	S.F.	.013	.22	.72	.94
20' wide	1.000	S.F.	.013	.22	.72	.94

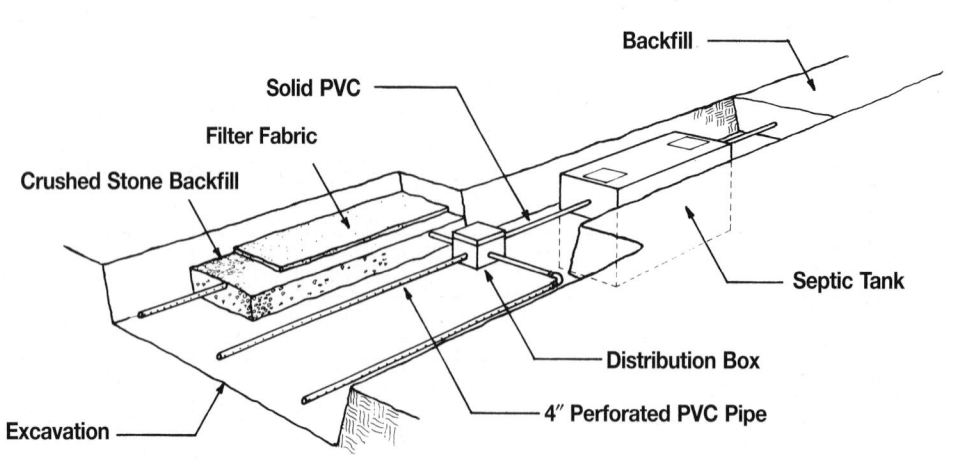

System Description	QUAN.	UNIT	LABOR HOURS	COST EACH		
				MAT.	INST.	TOTAL
SEPTIC SYSTEM WITH 600 S.F. LEACHING FIELD, 1000 GALLON TANK						
Mobilization	2.000	Ea.	16.000		1,356	1,356
Tank, precast, 1000 gallon	1.000	Ea.	3.500	1,150	184.80	1,334.80
Effluent filter	1.000	Ea.	1.000	43.50	60	103.50
Distribution box, precast	1.000	Ea.	1.000	102	44.50	146.50
Flow leveler	3.000	Ea.	.480	7.32	21.30	28.62
PVC pipe, 4" diameter	25.000	L.F.	1.600	45	72.75	117.75
Tee	1.000	Ea.	1.000	21.50	60	81.50
Elbow	2.000	Ea.	1.333	23.90	80	103.90
Viewport cap	1.000	Ea.	.333	9.05	19.95	29
Filter fabric	67.000	S.Y.	.447	112.56	20.10	132.66
Detectable marking tape	1.600	C.L.F.	.085	15.84	3.79	19.63
PVC perforated pipe, 4" diameter	135.000	L.F.	8.640	243	392.85	635.85
Excavation	160.000	C.Y.	17.654		1,744	1,744
Backfill	133.000	L.C.Y.	2.660		267.33	267.33
Spoil	55.000	L.C.Y.	3.056		304.15	304.15
Compaction	113.000	E.C.Y.	15.066		732.24	732.24
Stone fill	39.000	C.Y.	6.240	1,540.50	395.46	1,935.96
TOTAL		Ea.	80.094	3,314.17	5,759.22	9,073.39
SEPTIC SYSTEM WITH 750 S.F. LEACHING FIELD, 1500 GALLON TANK						
Mobilization	2.000	Ea.	16.000		1,356	1,356
Tank, precast, 1500 gallon	1.000	Ea.	4.000	1,775	211.50	1,986.50
Effluent filter	1.000	Ea.	1.000	43.50	60	103.50
Distribution box, precast	1.000	Ea.	1.000	102	44.50	146.50
Flow leveler	3.000	Ea.	.480	7.32	21.30	28.62
PVC pipe, 4" diameter	25.000	L.F.	1.600	45	72.75	117.75
Tee	125.000	Ea.	1.000	21.50	60	81.50
Elbow	2.000	Ea.	1.333	23.90	80	103.90
Viewport cap	1.000	Ea.	.333	9.05	19.95	29
Filter fabric	84.000	S.Y.	.560	141.12	25.20	166.32
Detectable marking tape	1.900	C.L.F.	.101	18.81	4.50	23.31
PVC perforated pipe, 4" diameter	165.000	L.F.	10.560	297	480.15	777.15
Excavation	199.000	C.Y.	21.958		2,169.10	2,169.10
Backfill	162.000	L.C.Y.	3.240		325.62	325.62
Spoil	73.500	L.C.Y.	4.084		406.46	406.46
Compaction	137.000	E.C.Y.	18.266		887.76	887.76
Stone fill	48.500	C.Y.	7.760	1,915.75	491.80	2,407.55
TOTAL		Ea.	93.275	4,399.95	6,716.59	11,116.54

The costs in this system include all necessary piping and excavation.

SEPTIC SYSTEM WITH 1000 S.F. LEACHING FIELD, 1500 GALLON TANK

System Description	QUAN.	UNIT	LABOR HOURS	COST EACH MAT.	COST EACH INST.	COST EACH TOTAL
Mobilization	2.000	Ea.	16.000		1,356	1,356
Tank, precast, 1500 gallon	1.000	Ea.	4.000	1,775	211.50	1,986.50
Effluent filter	1.000	Ea.	1.000	43.50	60	103.50
Distribution box, precast	1.000	Ea.	1.000	102	44.50	146.50
Flow leveler	4.000	Ea.	.640	9.76	28.40	38.16
PVC pipe, 4" diameter	25.000	L.F.	1.600	45	72.75	117.75
Tee	1.000	Ea.	1.000	21.50	60	81.50
Elbow	4.000	Ea.	2.667	47.80	160	207.80
Viewport cap	1.000	Ea.	.333	9.05	19.95	29
Filter fabric	111.000	S.Y.	.740	186.48	33.30	219.78
Detectable marking tape	2.400	C.L.F.	.128	23.76	5.69	29.45
PVC perforated pipe, 4" diameter	215.000	L.F.	13.760	387	625.65	1,012.65
Excavation	229.000	C.Y.	25.268		2,496.10	2,496.10
Backfill	178.000	L.C.Y.	3.560		357.78	357.78
Spoil	91.500	L.C.Y.	5.084		506	506
Compaction	151.000	E.C.Y.	20.133		978.48	978.48
Stone fill	64.000	C.Y.	10.240	2,528	648.96	3,176.96
TOTAL		Ea.	107.153	5,178.85	7,665.06	12,843.91

The costs in this system include all necessary piping and excavation.

Septic Systems Price Sheet	QUAN.	UNIT	LABOR HOURS	COST EACH MAT.	COST EACH INST.	COST EACH TOTAL
Tank, precast concrete, 1000 gallon	1.000	Ea.	3.500	1,150	185	1,335
1500 gallon	1.000	Ea.	4.000	1,775	212	1,987
Distribution box, concrete, 5 outlets	1.000	Ea.	1.000	102	44.50	146.50
12 outlets	1.000	Ea.	2.000	605	89	694
4" pipe, PVC, solid	25.000	L.F.	1.600	45	73	118
Tank and field excavation, 600 S.F. field	160.000	C.Y.	17.654		1,750	1,750
750 S.F. field	199.000	C.Y.	21.958		2,175	2,175
1000 S.F. field	229.000	C.Y.	25.268		2,500	2,500
Tank excavation only, 1000 gallon tank	20.000	C.Y.	2.206		218	218
1500 gallon tank	26.000	C.Y.	2.869		284	284
Backfill, crushed stone, 600 S.F. field	39.000	C.Y.	12.160	1,550	395	1,945
750 S.F. field	48.500	C.Y.	22.400	1,925	490	2,415
1000 S.F. field	64.000	C.Y.	.240	2,525	645	3,170
Backfill with excavated material, 600 S.F. field	133.000	L.C.Y.	.400		267	267
750 S.F. field	162.000	L.C.Y.	.367		325	325
1000 S.F. field	178.000	L.C.Y.	.280		355	355
Filter fabric, 600 S.F. field	67.000	S.Y.	2.376	113	20	133
750 S.F. field	84.000	S.Y.	4.860	141	25	166
1000 S.F. field	111.000	S.Y.	.740	186	33.50	219.50
4" pipe, PVC, perforated, 600 S.F. field	135.000	L.F.	9.280	243	395	638
750 S.F. field	165.000	L.F.	16.960	297	480	777
1000 S.F. field	215.000	L.F.	1.939	385	625	1,010
Pipe fittings, PVC, 600 S.F. field	2.000	Ea.	3.879	24	80	104
750 S.F. field	2.000	Ea.		24	80	104
1000 S.F. field	4.000	Ea.		48	160	208
Mobilization	2.000	Ea.	16.000		1,350	1,350
Effluent filter	1.000	Ea.	1.000	43.50	60	103.50
Flow leveler, 600 S.F. field	3.000	Ea.	.480	7.30	21.50	28.80
750 S.F. field	3.000	Ea.	.480	7.30	21.50	28.80
1000 S.F. field	4.000	Ea.	.640	9.75	28.50	38.25
Viewport cap	1.000	Ea.	.333	9.05	19.95	29
Detectable marking tape, 600 S.F. field	1.600	C.L.F.	.085	15.85	3.79	19.64
750 S.F. field	1.900	C.L.F.	.101	18.80	4.50	23.30
1000 S.F. field	2.400	C.L.F.	.128	24	5.70	29.70

System Description	QUAN.	UNIT	LABOR HOURS	COST PER UNIT		
				MAT.	INST.	TOTAL
Chain link fence						
Galv.9ga. wire, 1-5/8"post 10'O.C., 1-3/8"top rail, 2"corner post, 3'hi	1.000	L.F.	.130	11.85	5.90	17.75
4' high	1.000	L.F.	.141	11.30	6.45	17.75
6' high	1.000	L.F.	.209	13.50	9.50	23
Add for gate 3' wide 1-3/8" frame 3' high	1.000	Ea.	2.000	108	91	199
4' high	1.000	Ea.	2.400	129	109	238
6' high	1.000	Ea.	2.400	155	109	264
Add for gate 4' wide 1-3/8" frame 3' high	1.000	Ea.	2.667	123	121	244
4' high	1.000	Ea.	2.667	134	121	255
6' high	1.000	Ea.	3.000	160	137	297
Alum.9ga. wire, 1-5/8"post, 10'O.C., 1-3/8"top rail, 2"corner post,3'hi	1.000	L.F.	.130	9.65	5.90	15.55
4' high	1.000	L.F.	.141	10.20	6.45	16.65
6' high	1.000	L.F.	.209	13.50	9.50	23
Add for gate 3' wide 1-3/8" frame 3' high	1.000	Ea.	2.000	165	91	256
4' high	1.000	Ea.	2.400	176	109	285
6' high	1.000	Ea.	2.400	220	109	329
Add for gate 4' wide 1-3/8" frame 3' high	1.000	Ea.	2.400	170	109	279
4' high	1.000	Ea.	2.667	157	121	278
6' high	1.000	Ea.	3.000	212	137	349
Vinyl 9ga. wire, 1-5/8"post 10'O.C., 1-3/8"top rail, 2"corner post,3'hi	1.000	L.F.	.130	9.65	5.90	15.55
4' high	1.000	L.F.	.141	8	6.45	14.45
6' high	1.000	L.F.	.209	10.75	9.50	20.25
Add for gate 3' wide 1-3/8" frame 3' high	1.000	Ea.	2.000	118	91	209
4' high	1.000	Ea.	2.400	138	109	247
6' high	1.000	Ea.	2.400	171	109	280
Add for gate 4' wide 1-3/8" frame 3' high	1.000	Ea.	2.400	132	109	241
4' high	1.000	Ea.	2.667	143	121	264
6' high	1.000	Ea.	3.000	180	137	317
Tennis court, chain link fence, 10' high						
Galv.11ga.wire, 2"post 10'O.C., 1-3/8"top rail, 2-1/2"corner post	1.000	L.F.	.253	9.90	11.50	21.40
Add for gate 3' wide 1-3/8" frame	1.000	Ea.	2.400	255	109	364
Alum.11ga.wire, 2"post 10'O.C., 1-3/8"top rail, 2-1/2"corner post	1.000	L.F.	.253	12.10	11.50	23.60
Add for gate 3' wide 1-3/8" frame	1.000	Ea.	2.400	182	109	291
Vinyl 11ga.wire,2"post 10' O.C.,1-3/8"top rail,2-1/2"corner post	1.000	L.F.	.253	10.45	11.50	21.95
Add for gate 3' wide 1-3/8" frame	1.000	Ea.	2.400	360	109	469
Railings, commercial						
Aluminum balcony rail, 1-1/2" posts with pickets	1.000	L.F.	.164	86.50	12	98.50
With expanded metal panels	1.000	L.F.	.164	112	12	124
With porcelain enamel panel inserts	1.000	L.F.	.164	106	12	118
Mild steel, ornamental rounded top rail	1.000	L.F.	.164	100	12	112
As above, but pitch down stairs	1.000	L.F.	.183	130	13.35	143.35
Steel pipe, welded, 1-1/2" round, painted	1.000	L.F.	.160	34.50	11.70	46.20
Galvanized	1.000	L.F.	.160	48.50	11.70	60.20
Residential, stock units, mild steel, deluxe	1.000	L.F.	.102	18.50	7.45	25.95
Economy	1.000	L.F.	.102	13.80	7.45	21.25

System Description	QUAN.	UNIT	LABOR HOURS	COST PER UNIT		
				MAT.	INST.	TOTAL
Basketweave, 3/8"x4" boards, 2"x4" stringers on spreaders, 4"x4" posts						
No. 1 cedar, 6' high	1.000	L.F.	.150	28.50	8.40	36.90
Treated pine, 6' high	1.000	L.F.	.160	40.50	8.95	49.45
Board fence, 1"x4" boards, 2"x4" rails, 4"x4" posts						
Preservative treated, 2 rail, 3' high	1.000	L.F.	.166	11.30	9.25	20.55
4' high	1.000	L.F.	.178	13	9.95	22.95
3 rail, 5' high	1.000	L.F.	.185	13.65	10.30	23.95
6' high	1.000	L.F.	.192	16.85	10.70	27.55
Western cedar, No. 1, 2 rail, 3' high	1.000	L.F.	.166	13.55	9.25	22.80
3 rail, 4' high	1.000	L.F.	.178	12.95	9.95	22.90
5' high	1.000	L.F.	.185	15.20	10.30	25.50
6' high	1.000	L.F.	.192	16.25	10.70	26.95
No. 1 cedar, 2 rail, 3' high	1.000	L.F.	.166	14.65	9.25	23.90
4' high	1.000	L.F.	.178	16.20	9.95	26.15
3 rail, 5' high	1.000	L.F.	.185	19.15	10.30	29.45
6' high	1.000	L.F.	.192	23.50	10.70	34.20
Shadow box, 1"x6" boards, 2"x4" rails, 4"x4" posts						
Fir, pine or spruce, treated, 3 rail, 6' high	1.000	L.F.	.160	26.50	8.95	35.45
No. 1 cedar, 3 rail, 4' high	1.000	L.F.	.185	21	10.30	31.30
6' high	1.000	L.F.	.192	28.50	10.70	39.20
Open rail, split rails, No. 1 cedar, 2 rail, 3' high	1.000	L.F.	.150	10.75	8.40	19.15
3 rail, 4' high	1.000	L.F.	.160	13.35	8.95	22.30
No. 2 cedar, 2 rail, 3' high	1.000	L.F.	.150	8.95	8.40	17.35
3 rail, 4' high	1.000	L.F.	.160	8.75	8.95	17.70
Open rail, rustic rails, No. 1 cedar, 2 rail, 3' high	1.000	L.F.	.150	13.55	8.40	21.95
3 rail, 4' high	1.000	L.F.	.160	13	8.95	21.95
No. 2 cedar, 2 rail, 3' high	1.000	L.F.	.150	12.25	8.40	20.65
3 rail, 4' high	1.000	L.F.	.160	8.85	8.95	17.80
Rustic picket, molded pine pickets, 2 rail, 3' high	1.000	L.F.	.171	9.35	9.60	18.95
3 rail, 4' high	1.000	L.F.	.197	10.75	11	21.75
No. 1 cedar, 2 rail, 3' high	1.000	L.F.	.171	11.40	9.60	21
3 rail, 4' high	1.000	L.F.	.197	13.10	11	24.10
Picket fence, fir, pine or spruce, preserved, treated						
2 rail, 3' high	1.000	L.F.	.171	8.60	9.60	18.20
3 rail, 4' high	1.000	L.F.	.185	9.80	10.30	20.10
Western cedar, 2 rail, 3' high	1.000	L.F.	.171	9.90	9.60	19.50
3 rail, 4' high	1.000	L.F.	.185	10	10.30	20.30
No. 1 cedar, 2 rail, 3' high	1.000	L.F.	.171	15.05	9.60	24.65
3 rail, 4' high	1.000	L.F.	.185	19.60	10.30	29.90
Stockade, No. 1 cedar, 3-1/4" rails, 6' high	1.000	L.F.	.150	14.45	8.40	22.85
8' high	1.000	L.F.	.155	19.90	8.65	28.55
No. 2 cedar, treated rails, 6' high	1.000	L.F.	.150	14.80	8.40	23.20
Treated pine, treated rails, 6' high	1.000	L.F.	.150	15.35	8.40	23.75
Gates, No. 2 cedar, picket, 3'-6" wide 4' high	1.000	Ea.	2.667	87	149	236
No. 2 cedar, rustic round, 3' wide, 3' high	1.000	Ea.	2.667	113	149	262
No. 2 cedar, stockade screen, 3'-6" wide, 6' high	1.000	Ea.	3.000	98.50	168	266.50
General, wood, 3'-6" wide, 4' high	1.000	Ea.	2.400	119	135	254
6' high	1.000	Ea.	3.000	149	168	317

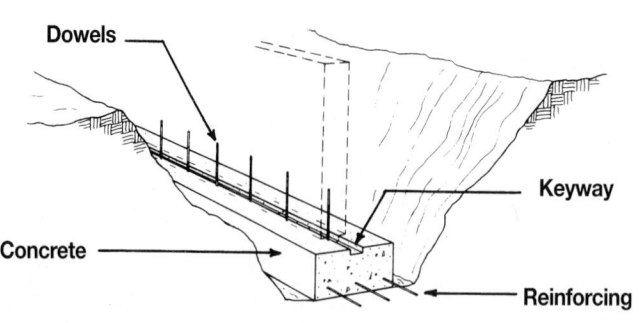

Dowels
Keyway
Concrete
Reinforcing

System Description	QUAN.	UNIT	LABOR HOURS	COST PER L.F.		
				MAT.	INST.	TOTAL
8″ THICK BY 18″ WIDE FOOTING						
Concrete, 3000 psi	.040	C.Y.		5.32		5.32
Place concrete, direct chute	.040	C.Y.	.016		.77	.77
Forms, footing, 4 uses	1.330	SFCA	.103	1	5.28	6.28
Reinforcing, 1/2″ diameter bars, 2 each	1.380	Lb.	.011	.73	.68	1.41
Keyway, 2″ x 4″, beveled, 4 uses	1.000	L.F.	.015	.24	.88	1.12
Dowels, 1/2″ diameter bars, 2′ long, 6′ O.C.	.166	Ea.	.006	.13	.36	.49
TOTAL		L.F.	.151	7.42	7.97	15.39
12″ THICK BY 24″ WIDE FOOTING						
Concrete, 3000 psi	.070	C.Y.		9.31		9.31
Place concrete, direct chute	.070	C.Y.	.028		1.35	1.35
Forms, footing, 4 uses	2.000	SFCA	.155	1.50	7.94	9.44
Reinforcing, 1/2″ diameter bars, 2 each	1.380	Lb.	.011	.73	.68	1.41
Keyway, 2″ x 4″, beveled, 4 uses	1.000	L.F.	.015	.24	.88	1.12
Dowels, 1/2″ diameter bars, 2′ long, 6′ O.C.	.166	Ea.	.006	.13	.36	.49
TOTAL		L.F.	.215	11.91	11.21	23.12
12″ THICK BY 36″ WIDE FOOTING						
Concrete, 3000 psi	.110	C.Y.		14.63		14.63
Place concrete, direct chute	.110	C.Y.	.044		2.12	2.12
Forms, footing, 4 uses	2.000	SFCA	.155	1.50	7.94	9.44
Reinforcing, 1/2″ diameter bars, 2 each	1.380	Lb.	.011	.73	.68	1.41
Keyway, 2″ x 4″, beveled, 4 uses	1.000	L.F.	.015	.24	.88	1.12
Dowels, 1/2″ diameter bars, 2′ long, 6′ O.C.	.166	Ea.	.006	.13	.36	.49
TOTAL		L.F.	.231	17.23	11.98	29.21

The footing costs in this system are on a cost per linear foot basis.

Description	QUAN.	UNIT	LABOR HOURS	COST PER L.F.		
				MAT.	INST.	TOTAL

For customer support on your Residential Costs with RSMeans data, call 800.448.8182.

Footing Price Sheet	QUAN.	UNIT	LABOR HOURS	COST PER L.F.		
				MAT.	INST.	TOTAL
Concrete, 8" thick by 18" wide footing						
2000 psi concrete	.040	C.Y.		5.10		5.10
2500 psi concrete	.040	C.Y.		5.20		5.20
3000 psi concrete	.040	C.Y.		5.30		5.30
3500 psi concrete	.040	C.Y.		5.50		5.50
4000 psi concrete	.040	C.Y.		5.60		5.60
12" thick by 24" wide footing						
2000 psi concrete	.070	C.Y.		8.90		8.90
2500 psi concrete	.070	C.Y.		9.10		9.10
3000 psi concrete	.070	C.Y.		9.30		9.30
3500 psi concrete	.070	C.Y.		9.60		9.60
4000 psi concrete	.070	C.Y.		9.80		9.80
12" thick by 36" wide footing						
2000 psi concrete	.110	C.Y.		13.95		13.95
2500 psi concrete	.110	C.Y.		14.30		14.30
3000 psi concrete	.110	C.Y.		14.65		14.65
3500 psi concrete	.110	C.Y.		15.05		15.05
4000 psi concrete	.110	C.Y.		15.40		15.40
Place concrete, 8" thick by 18" wide footing, direct chute	.040	C.Y.	.016		.77	.77
Pumped concrete	.040	C.Y.	.017		1.09	1.09
Crane & bucket	.040	C.Y.	.032		2.08	2.08
12" thick by 24" wide footing, direct chute	.070	C.Y.	.028		1.35	1.35
Pumped concrete	.070	C.Y.	.030		1.92	1.92
Crane & bucket	.070	C.Y.	.056		3.64	3.64
12" thick by 36" wide footing, direct chute	.110	C.Y.	.044		2.12	2.12
Pumped concrete	.110	C.Y.	.047		3.01	3.01
Crane & bucket	.110	C.Y.	.088		5.75	5.75
Forms, 8" thick footing, 1 use	1.330	SFCA	.140	3.07	7.20	10.27
4 uses	1.330	SFCA	.103	1	5.30	6.30
12" thick footing, 1 use	2.000	SFCA	.211	4.62	10.80	15.42
4 uses	2.000	SFCA	.155	1.50	7.95	9.45
Reinforcing, 3/8" diameter bar, 1 each	.400	Lb.	.003	.21	.20	.41
2 each	.800	Lb.	.006	.42	.39	.81
3 each	1.200	Lb.	.009	.64	.59	1.23
1/2" diameter bar, 1 each	.700	Lb.	.005	.37	.34	.71
2 each	1.380	Lb.	.011	.73	.68	1.41
3 each	2.100	Lb.	.016	1.11	1.03	2.14
5/8" diameter bar, 1 each	1.040	Lb.	.008	.55	.51	1.06
2 each	2.080	Lb.	.016	1.10	1.02	2.12
Keyway, beveled, 2" x 4", 1 use	1.000	L.F.	.030	.48	1.76	2.24
2 uses	1.000	L.F.	.023	.36	1.32	1.68
2" x 6", 1 use	1.000	L.F.	.032	.70	1.86	2.56
2 uses	1.000	L.F.	.024	.53	1.40	1.93
Dowels, 2 feet long, 6' O.C., 3/8" bar	.166	Ea.	.005	.07	.33	.40
1/2" bar	.166	Ea.	.006	.13	.36	.49
5/8" bar	.166	Ea.	.006	.20	.40	.60
3/4" bar	.166	Ea.	.006	.20	.40	.60

For customer support on your Residential Costs with RSMeans data, call 800.448.8182.

119

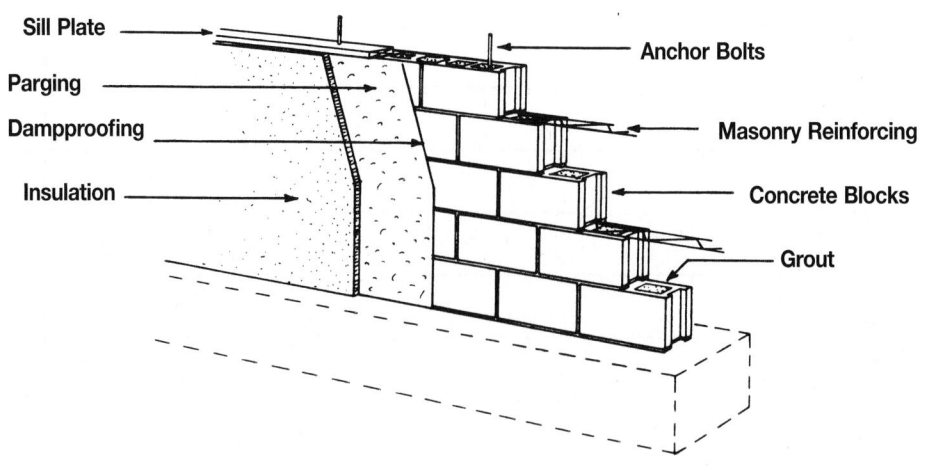

Sill Plate — Anchor Bolts

Parging — Masonry Reinforcing

Dampproofing —

Insulation — Concrete Blocks

Grout

System Description	QUAN.	UNIT	LABOR HOURS	COST PER S.F.		
				MAT.	INST.	TOTAL
8″ WALL, GROUTED, FULL HEIGHT						
Concrete block, 8″ x 16″ x 8″	1.000	S.F.	.094	3.42	5.10	8.52
Masonry reinforcing, every second course	.750	L.F.	.002	.18	.11	.29
Parging, plastering with portland cement plaster, 1 coat	1.000	S.F.	.014	.34	.80	1.14
Dampproofing, bituminous coating, 1 coat	1.000	S.F.	.012	.24	.66	.90
Insulation, 1″ rigid polystyrene	1.000	S.F.	.010	.63	.58	1.21
Grout, solid, pumped	1.000	S.F.	.059	1.43	3.10	4.53
Anchor bolts, 1/2″ diameter, 8″ long, 4′ O.C.	.060	Ea.	.004	.09	.21	.30
Sill plate, 2″ x 4″, treated	.250	L.F.	.007	.16	.42	.58
TOTAL		S.F.	.202	6.49	10.98	17.47
12″ WALL, GROUTED, FULL HEIGHT						
Concrete block, 8″ x 16″ x 12″	1.000	S.F.	.160	5.25	8.45	13.70
Masonry reinforcing, every second course	.750	L.F.	.003	.21	.17	.38
Parging, plastering with portland cement plaster, 1 coat	1.000	S.F.	.014	.34	.80	1.14
Dampproofing, bituminous coating, 1 coat	1.000	S.F.	.012	.24	.66	.90
Insulation, 1″ rigid polystyrene	1.000	S.F.	.010	.63	.58	1.21
Grout, solid, pumped	1.000	S.F.	.063	2.33	3.29	5.62
Anchor bolts, 1/2″ diameter, 8″ long, 4′ O.C.	.060	Ea.	.004	.09	.21	.30
Sill plate, 2″ x 4″, treated	.250	L.F.	.007	.16	.42	.58
TOTAL		S.F.	.273	9.25	14.58	23.83

The costs in this system are based on a square foot of wall. Do not subtract for window or door openings.

Description	QUAN.	UNIT	LABOR HOURS	COST PER S.F.		
				MAT.	INST.	TOTAL

Block Wall Price Sheet	QUAN.	UNIT	LABOR HOURS	COST PER S.F.		
				MAT.	INST.	TOTAL
Concrete, block, 8″ x 16″ x, 6″ thick	1.000	S.F.	.089	3.22	4.74	7.96
8″ thick	1.000	S.F.	.093	3.42	5.10	8.52
10″ thick	1.000	S.F.	.095	3.92	6.15	10.07
12″ thick	1.000	S.F.	.122	5.25	8.45	13.70
Solid block, 8″ x 16″ x, 6″ thick	1.000	S.F.	.091	3.29	4.90	8.19
8″ thick	1.000	S.F.	.096	4.65	5.20	9.85
10″ thick	1.000	S.F.	.096	4.65	5.20	9.85
12″ thick	1.000	S.F.	.126	6.70	7.25	13.95
Masonry reinforcing, wire strips, to 8″ wide, every course	1.500	L.F.	.004	.36	.23	.59
Every 2nd course	.750	L.F.	.002	.18	.11	.29
Every 3rd course	.500	L.F.	.001	.12	.08	.20
Every 4th course	.400	L.F.	.001	.10	.06	.16
Wire strips to 12″ wide, every course	1.500	L.F.	.006	.42	.35	.77
Every 2nd course	.750	L.F.	.003	.21	.17	.38
Every 3rd course	.500	L.F.	.002	.14	.12	.26
Every 4th course	.400	L.F.	.002	.11	.09	.20
Parging, plastering with portland cement plaster, 1 coat	1.000	S.F.	.014	.34	.80	1.14
2 coats	1.000	S.F.	.022	.53	1.24	1.77
Dampproofing, bituminous, brushed on, 1 coat	1.000	S.F.	.012	.24	.66	.90
2 coats	1.000	S.F.	.016	.49	.87	1.36
Sprayed on, 1 coat	1.000	S.F.	.010	.24	.53	.77
2 coats	1.000	S.F.	.016	.48	.87	1.35
Troweled on, 1/16″ thick	1.000	S.F.	.016	.40	.87	1.27
1/8″ thick	1.000	S.F.	.020	.71	1.09	1.80
1/2″ thick	1.000	S.F.	.023	2.32	1.25	3.57
Insulation, rigid, fiberglass, 1.5#/C.F., unfaced						
1-1/2″ thick R 6.2	1.000	S.F.	.008	.44	.46	.90
2″ thick R 8.5	1.000	S.F.	.008	.55	.46	1.01
3″ thick R 13	1.000	S.F.	.010	.66	.58	1.24
Perlite, 1″ thick R 2.77	1.000	S.F.	.010	.52	.58	1.10
2″ thick R 5.55	1.000	S.F.	.011	.86	.64	1.50
Polystyrene, extruded, 1″ thick R 5.4	1.000	S.F.	.010	.63	.58	1.21
2″ thick R 10.8	1.000	S.F.	.011	1.75	.64	2.39
Molded 1″ thick R 3.85	1.000	S.F.	.010	.30	.58	.88
2″ thick R 7.7	1.000	S.F.	.011	.89	.64	1.53
Grout, concrete block cores, 6″ thick	1.000	S.F.	.044	1.07	2.33	3.40
8″ thick	1.000	S.F.	.059	1.43	3.10	4.53
10″ thick	1.000	S.F.	.061	1.88	3.19	5.07
12″ thick	1.000	S.F.	.063	2.33	3.29	5.62
Anchor bolts, 2′ on center, 1/2″ diameter, 8″ long	.120	Ea.	.005	.19	.42	.61
12″ long	.120	Ea.	.005	.21	.42	.63
3/4″ diameter, 8″ long	.120	Ea.	.006	.62	.44	1.06
12″ long	.120	Ea.	.006	.77	.45	1.22
4′ on center, 1/2″ diameter, 8″ long	.060	Ea.	.002	.09	.21	.30
12″ long	.060	Ea.	.003	.10	.21	.31
3/4″ diameter, 8″ long	.060	Ea.	.003	.31	.22	.53
12″ long	.060	Ea.	.003	.38	.22	.60
Sill plates, treated, 2″ x 4″	.250	L.F.	.007	.16	.42	.58
4″ x 4″	.250	L.F.	.007	.33	.40	.73

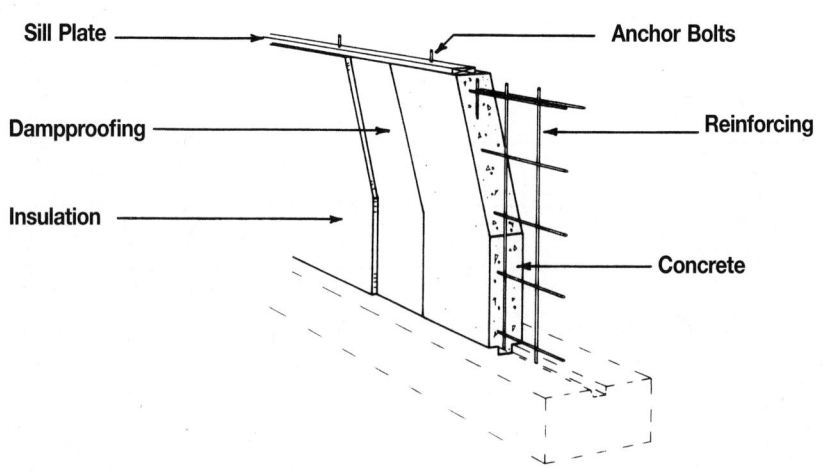

System Description	QUAN.	UNIT	LABOR HOURS	COST PER S.F.		
				MAT.	INST.	TOTAL
8″ THICK, POURED CONCRETE WALL						
Concrete, 8″ thick , 3000 psi	.025	C.Y.		3.33		3.33
Forms, prefabricated plywood, up to 8′ high	2.000	SFCA	.120	2.40	6.22	8.62
Reinforcing, light	.670	Lb.	.004	.36	.23	.59
Placing concrete, direct chute	.025	C.Y.	.013		.65	.65
Dampproofing, brushed on, 2 coats	1.000	S.F.	.016	.49	.87	1.36
Rigid insulation, 1″ polystyrene	1.000	S.F.	.010	.63	.58	1.21
Anchor bolts, 1/2″ diameter, 12″ long, 4′ O.C.	.060	Ea.	.004	.10	.21	.31
Sill plates, 2″ x 4″, treated	.250	L.F.	.007	.16	.42	.58
TOTAL		S.F.	.174	7.47	9.18	16.65
12″ THICK, POURED CONCRETE WALL						
Concrete, 12″ thick, 3000 psi	.040	C.Y.		5.32		5.32
Forms, prefabricated plywood, up to 8′ high	2.000	SFCA	.120	2.40	6.22	8.62
Reinforcing, light	1.000	Lb.	.005	.53	.34	.87
Placing concrete, direct chute	.040	C.Y.	.019		.92	.92
Dampproofing, brushed on, 2 coats	1.000	S.F.	.016	.49	.87	1.36
Rigid insulation, 1″ polystyrene	1.000	S.F.	.010	.63	.58	1.21
Anchor bolts, 1/2″ diameter, 12″ long, 4′ O.C.	.060	Ea.	.004	.10	.21	.31
Sill plates, 2″ x 4″ treated	.250	L.F.	.007	.16	.42	.58
TOTAL		S.F.	.181	9.63	9.56	19.19

The costs in this system are based on sq. ft. of wall. Do not subtract
for window and door openings. The costs assume a 4′ high wall.

Description	QUAN.	UNIT	LABOR HOURS	COST PER S.F.		
				MAT.	INST.	TOTAL

Concrete Wall Price Sheet

Concrete Wall Price Sheet	QUAN.	UNIT	LABOR HOURS	COST PER S.F.		
				MAT.	INST.	TOTAL
Formwork, prefabricated plywood, up to 8' high	2.000	SFCA	.081	2.40	6.20	8.60
Over 8' to 16' high	2.000	SFCA	.076	2.52	8.30	10.82
Job built forms, 1 use per month	2.000	SFCA	.320	6.05	13.50	19.55
4 uses per month	2.000	SFCA	.221	2.28	9.85	12.13
Reinforcing, 8" wall, light reinforcing	.670	Lb.	.004	.36	.23	.59
Heavy reinforcing	1.500	Lb.	.008	.80	.51	1.31
10" wall, light reinforcing	.850	Lb.	.005	.45	.29	.74
Heavy reinforcing	2.000	Lb.	.011	1.06	.68	1.74
12" wall light reinforcing	1.000	Lb.	.005	.53	.34	.87
Heavy reinforcing	2.250	Lb.	.012	1.19	.77	1.96
Placing concrete, 8" wall, direct chute	.025	C.Y.	.013		.65	.65
Pumped concrete	.025	C.Y.	.016		1.04	1.04
Crane & bucket	.025	C.Y.	.023		1.47	1.47
10" wall, direct chute	.030	C.Y.	.016		.77	.77
Pumped concrete	.030	C.Y.	.019		1.24	1.24
Crane & bucket	.030	C.Y.	.027		1.76	1.76
12" wall, direct chute	.040	C.Y.	.019		.92	.92
Pumped concrete	.040	C.Y.	.023		1.49	1.49
Crane & bucket	.040	C.Y.	.032		2.08	2.08
Dampproofing, bituminous, brushed on, 1 coat	1.000	S.F.	.012	.24	.66	.90
2 coats	1.000	S.F.	.016	.49	.87	1.36
Sprayed on, 1 coat	1.000	S.F.	.010	.24	.53	.77
2 coats	1.000	S.F.	.016	.48	.87	1.35
Troweled on, 1/16" thick	1.000	S.F.	.016	.40	.87	1.27
1/8" thick	1.000	S.F.	.020	.71	1.09	1.80
1/2" thick	1.000	S.F.	.023	2.32	1.25	3.57
Insulation rigid, fiberglass, 1.5#/C.F., unfaced						
1-1/2" thick, R 6.2	1.000	S.F.	.008	.44	.46	.90
2" thick, R 8.3	1.000	S.F.	.008	.55	.46	1.01
3" thick, R 12.4	1.000	S.F.	.010	.66	.58	1.24
Perlite, 1" thick R 2.77	1.000	S.F.	.010	.52	.58	1.10
2" thick R 5.55	1.000	S.F.	.011	.86	.64	1.50
Polystyrene, extruded, 1" thick R 5.40	1.000	S.F.	.010	.63	.58	1.21
2" thick R 10.8	1.000	S.F.	.011	1.75	.64	2.39
Molded, 1" thick R 3.85	1.000	S.F.	.010	.30	.58	.88
2" thick R 7.70	1.000	S.F.	.011	.89	.64	1.53
Anchor bolts, 2' on center, 1/2" diameter, 8" long	.120	Ea.	.005	.19	.42	.61
12" long	.120	Ea.	.005	.21	.42	.63
3/4" diameter, 8" long	.120	Ea.	.006	.62	.44	1.06
12" long	.120	Ea.	.006	.77	.45	1.22
Sill plates, treated lumber, 2" x 4"	.250	L.F.	.007	.16	.42	.58
4" x 4"	.250	L.F.	.007	.33	.40	.73

For customer support on your Residential Costs with RSMeans data, call 800.448.8182.

123

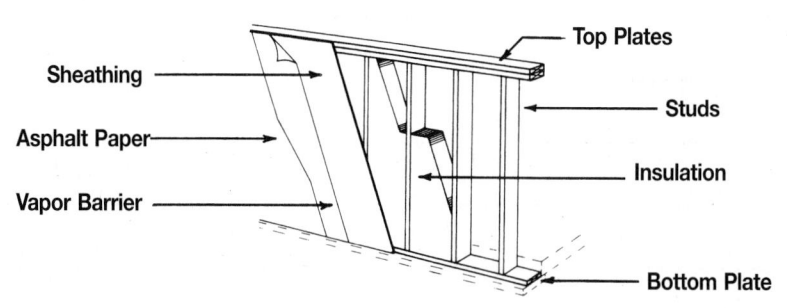

Sheathing — Top Plates
Asphalt Paper — Studs
Vapor Barrier — Insulation — Bottom Plate

System Description	QUAN.	UNIT	LABOR HOURS	COST PER S.F.		
				MAT.	INST.	TOTAL
2″ X 4″ STUDS, 16″ O.C., WALL						
Studs, 2″ x 4″, 16″ O.C., treated	1.000	L.F.	.015	.65	.85	1.50
Plates, double top plate, single bottom plate, treated, 2″ x 4″	.750	L.F.	.011	.49	.64	1.13
Sheathing, 1/2″, exterior grade, CDX, treated	1.000	S.F.	.014	.91	.83	1.74
Asphalt paper, 15# roll	1.100	S.F.	.002	.07	.14	.21
Vapor barrier, 4 mil polyethylene	1.000	S.F.	.002	.04	.13	.17
Fiberglass insulation, 3-1/2″ thick	1.000	S.F.	.007	.37	.40	.77
TOTAL		S.F.	.051	2.53	2.99	5.52
2″ X 6″ STUDS, 16″ O.C., WALL						
Studs, 2″ x 6″, 16″ O.C., treated	1.000	L.F.	.016	.81	.93	1.74
Plates, double top plate, single bottom plate, treated, 2″ x 6″	.750	L.F.	.012	.61	.70	1.31
Sheathing, 5/8″ exterior grade, CDX, treated	1.000	S.F.	.015	1.40	.89	2.29
Asphalt paper, 15# roll	1.100	S.F.	.002	.07	.14	.21
Vapor barrier, 4 mil polyethylene	1.000	S.F.	.002	.04	.13	.17
Fiberglass insulation, 6″ thick	1.000	S.F.	.007	.48	.40	.88
TOTAL		S.F.	.054	3.41	3.19	6.60
2″ X 8″ STUDS, 16″ O.C., WALL						
Studs, 2″ x 8″, 16″ O.C. treated	1.000	L.F.	.018	1.23	1.03	2.26
Plates, double top plate, single bottom plate, treated, 2″ x 8″	.750	L.F.	.013	.92	.77	1.69
Sheathing, 3/4″ exterior grade, CDX, treated	1.000	S.F.	.016	1.49	.95	2.44
Asphalt paper, 15# roll	1.100	S.F.	.002	.07	.14	.21
Vapor barrier, 4 mil polyethylene	1.000	S.F.	.002	.04	.13	.17
Fiberglass insulation, 9″ thick	1.000	S.F.	.006	.80	.34	1.14
TOTAL		S.F.	.057	4.55	3.36	7.91

The costs in this system are based on a sq. ft. of wall area. Do not
subtract for window or door openings. The costs assume a 4′ high wall.

Description	QUAN.	UNIT	LABOR HOURS	COST PER S.F.		
				MAT.	INST.	TOTAL

Wood Wall Foundation Price Sheet	QUAN.	UNIT	LABOR HOURS	COST PER S.F.		
				MAT.	INST.	TOTAL
Studs, treated, 2" x 4", 12" O.C.	1.250	L.F.	.018	.81	1.06	1.87
16" O.C.	1.000	L.F.	.015	.65	.85	1.50
2" x 6", 12" O.C.	1.250	L.F.	.020	1.01	1.16	2.17
16" O.C.	1.000	L.F.	.016	.81	.93	1.74
2" x 8", 12" O.C.	1.250	L.F.	.022	1.54	1.29	2.83
16" O.C.	1.000	L.F.	.018	1.23	1.03	2.26
Plates, treated double top single bottom, 2" x 4"	.750	L.F.	.011	.49	.64	1.13
2" x 6"	.750	L.F.	.012	.61	.70	1.31
2" x 8"	.750	L.F.	.013	.92	.77	1.69
Sheathing, treated exterior grade CDX, 1/2" thick	1.000	S.F.	.014	.91	.83	1.74
5/8" thick	1.000	S.F.	.015	1.40	.89	2.29
3/4" thick	1.000	S.F.	.016	1.49	.95	2.44
Asphalt paper, 15# roll	1.100	S.F.	.002	.07	.14	.21
Vapor barrier, polyethylene, 4 mil	1.000	S.F.	.002	.03	.13	.16
10 mil	1.000	S.F.	.002	.10	.13	.23
Insulation, rigid, fiberglass, 1.5#/C.F., unfaced	1.000	S.F.	.008	.37	.46	.83
1-1/2" thick, R 6.2	1.000	S.F.	.008	.44	.46	.90
2" thick, R 8.3	1.000	S.F.	.008	.55	.46	1.01
3" thick, R 12.4	1.000	S.F.	.010	.67	.59	1.26
Perlite 1" thick, R 2.77	1.000	S.F.	.010	.52	.58	1.10
2" thick, R 5.55	1.000	S.F.	.011	.86	.64	1.50
Polystyrene, extruded, 1" thick, R 5.40	1.000	S.F.	.010	.63	.58	1.21
2" thick, R 10.8	1.000	S.F.	.011	1.75	.64	2.39
Molded 1" thick, R 3.85	1.000	S.F.	.010	.30	.58	.88
2" thick, R 7.7	1.000	S.F.	.011	.89	.64	1.53
Non rigid, batts, fiberglass, paper backed, 3-1/2" thick roll, R 11	1.000	S.F.	.005	.37	.40	.77
6", R 19	1.000	S.F.	.006	.48	.40	.88
9", R 30	1.000	S.F.	.006	.80	.34	1.14
12", R 38	1.000	S.F.	.006	1.16	.34	1.50
Mineral fiber, paper backed, 3-1/2", R 13	1.000	S.F.	.005	.85	.29	1.14
6", R 19	1.000	S.F.	.005	1.33	.29	1.62
10", R 30	1.000	S.F.	.006	1.75	.34	2.09

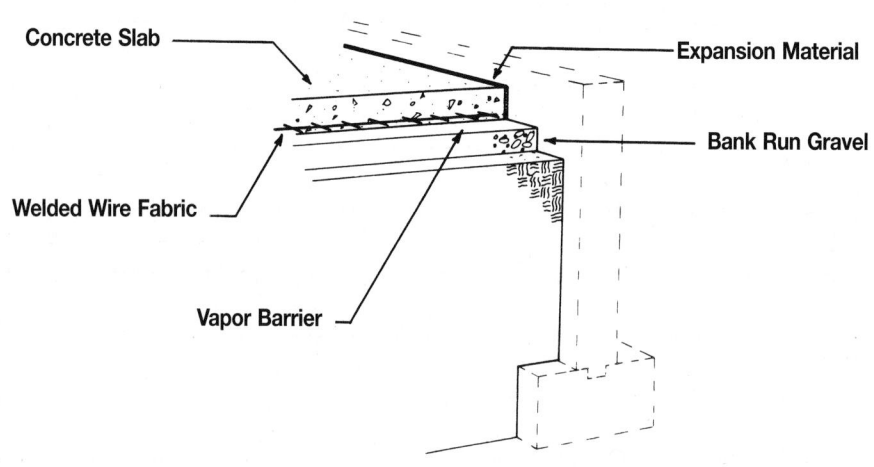

Concrete Slab — Expansion Material

Welded Wire Fabric — Bank Run Gravel

Vapor Barrier

System Description	QUAN.	UNIT	LABOR HOURS	COST PER S.F.		
				MAT.	INST.	TOTAL
4″ THICK SLAB						
Concrete, 4″ thick, 3000 psi concrete	.012	C.Y.		1.60		1.60
Place concrete, direct chute	.012	C.Y.	.005		.26	.26
Bank run gravel, 4″ deep	1.000	S.F.	.001	.33	.05	.38
Polyethylene vapor barrier, .006″ thick	1.000	S.F.	.002	.04	.13	.17
Edge forms, expansion material	.100	L.F.	.005	.03	.27	.30
Welded wire fabric, 6 x 6, 10/10 (W1.4/W1.4)	1.100	S.F.	.005	.19	.33	.52
Steel trowel finish	1.000	S.F.	.014		.76	.76
TOTAL		S.F.	.032	2.19	1.80	3.99
6″ THICK SLAB						
Concrete, 6″ thick, 3000 psi concrete	.019	C.Y.		2.53		2.53
Place concrete, direct chute	.019	C.Y.	.008		.40	.40
Bank run gravel, 4″ deep	1.000	S.F.	.001	.33	.05	.38
Polyethylene vapor barrier, .006″ thick	1.000	S.F.	.002	.04	.13	.17
Edge forms, expansion material	.100	L.F.	.005	.03	.27	.30
Welded wire fabric, 6 x 6, 10/10 (W1.4/W1.4)	1.100	S.F.	.005	.19	.33	.52
Steel trowel finish	1.000	S.F.	.014		.76	.76
TOTAL		S.F.	.035	3.12	1.94	5.06

The slab costs in this section are based on a cost per square foot of floor area.

Description	QUAN.	UNIT	LABOR HOURS	COST PER S.F.		
				MAT.	INST.	TOTAL

Floor Slab Price Sheet

Floor Slab Price Sheet	QUAN.	UNIT	LABOR HOURS	COST PER S.F. MAT.	COST PER S.F. INST.	COST PER S.F. TOTAL
Concrete, 4″ thick slab, 2000 psi concrete	.012	C.Y.		1.52		1.52
2500 psi concrete	.012	C.Y.		1.56		1.56
3000 psi concrete	.012	C.Y.		1.60		1.60
3500 psi concrete	.012	C.Y.		1.64		1.64
4000 psi concrete	.012	C.Y.		1.68		1.68
4500 psi concrete	.012	C.Y.		1.73		1.73
5″ thick slab, 2000 psi concrete	.015	C.Y.		1.91		1.91
2500 psi concrete	.015	C.Y.		1.95		1.95
3000 psi concrete	.015	C.Y.		2		2
3500 psi concrete	.015	C.Y.		2.06		2.06
4000 psi concrete	.015	C.Y.		2.10		2.10
4500 psi concrete	.015	C.Y.		2.16		2.16
6″ thick slab, 2000 psi concrete	.019	C.Y.		2.41		2.41
2500 psi concrete	.019	C.Y.		2.47		2.47
3000 psi concrete	.019	C.Y.		2.53		2.53
3500 psi concrete	.019	C.Y.		2.60		2.60
4000 psi concrete	.019	C.Y.		2.66		2.66
4500 psi concrete	.019	C.Y.		2.74		2.74
Place concrete, 4″ slab, direct chute	.012	C.Y.	.005		.26	.26
Pumped concrete	.012	C.Y.	.006		.38	.38
Crane & bucket	.012	C.Y.	.008		.51	.51
5″ slab, direct chute	.015	C.Y.	.007		.32	.32
Pumped concrete	.015	C.Y.	.007		.48	.48
Crane & bucket	.015	C.Y.	.010		.64	.64
6″ slab, direct chute	.019	C.Y.	.008		.40	.40
Pumped concrete	.019	C.Y.	.009		.61	.61
Crane & bucket	.019	C.Y.	.012		.81	.81
Gravel, bank run, 4″ deep	1.000	S.F.	.001	.33	.05	.38
6″ deep	1.000	S.F.	.001	.44	.07	.51
9″ deep	1.000	S.F.	.001	.64	.11	.75
12″ deep	1.000	S.F.	.001	.88	.13	1.01
3/4″ crushed stone, 3″ deep	1.000	S.F.	.001	.33	.06	.39
6″ deep	1.000	S.F.	.001	.66	.11	.77
9″ deep	1.000	S.F.	.002	.97	.15	1.12
12″ deep	1.000	S.F.	.002	1.66	.18	1.84
Vapor barrier polyethylene, .004″ thick	1.000	S.F.	.002	.03	.13	.16
.006″ thick	1.000	S.F.	.002	.04	.13	.17
Edge forms, expansion material, 4″ thick slab	.100	L.F.	.004	.02	.18	.20
6″ thick slab	.100	L.F.	.005	.03	.27	.30
Welded wire fabric 6 x 6, 10/10 (W1.4/W1.4)	1.100	S.F.	.005	.19	.33	.52
6 x 6, 6/6 (W2.9/W2.9)	1.100	S.F.	.006	.31	.40	.71
4 x 4, 10/10 (W1.4/W1.4)	1.100	S.F.	.006	.28	.36	.64
Finish concrete, screed finish	1.000	S.F.	.009		.31	.31
Float finish	1.000	S.F.	.011		.31	.31
Steel trowel, for resilient floor	1.000	S.F.	.013		.99	.99
For finished floor	1.000	S.F.	.015		.76	.76

System Description	QUAN.	UNIT	LABOR HOURS	COST PER S.F.		
				MAT.	INST.	TOTAL
2″ X 8″, 16″ O.C.						
Wood joists, 2″ x 8″, 16″ O.C.	1.000	L.F.	.015	.98	.85	1.83
Bridging, 1″ x 3″, 6′ O.C.	.080	Pr.	.005	.06	.29	.35
Box sills, 2″ x 8″	.150	L.F.	.002	.15	.13	.28
Concrete filled steel column, 4″ diameter	.125	L.F.	.002	.20	.14	.34
Girder, built up from three 2″ x 8″	.125	L.F.	.013	.37	.77	1.14
Sheathing, plywood, subfloor, 5/8″ CDX	1.000	S.F.	.012	.84	.69	1.53
Furring, 1″ x 3″, 16″ O.C.	1.000	L.F.	.023	.47	1.33	1.80
Joist hangers	.036	Ea.	.002	.05	.10	.15
TOTAL		S.F.	.074	3.12	4.30	7.42
2″ X 10″, 16″ O.C.						
Wood joists, 2″ x 10″, 16″ OC	1.000	L.F.	.018	1.55	1.03	2.58
Bridging, 1″ x 3″, 6′ OC	.080	Pr.	.005	.06	.29	.35
Box sills, 2″ x 10″	.150	L.F.	.003	.23	.15	.38
Concrete filled steel column, 4″ diameter	.125	L.F.	.002	.20	.14	.34
5/8″ thick	.125	L.F.	.014	.58	.83	1.41
Sheathing, plywood, subfloor, 5/8″ CDX	1.000	S.F.	.012	.84	.69	1.53
Furring, 1″ x 3″, 16″ OC	1.000	L.F.	.023	.47	1.33	1.80
Joist hangers	.036	Ea.	.002	.05	.10	.15
TOTAL		S.F.	.079	3.98	4.56	8.54
2″ X 12″, 16″ O.C.						
Wood joists, 2″ x 12″, 16″ O.C.	1.000	L.F.	.018	1.97	1.06	3.03
Bridging, 1″ x 3″, 6′ O.C.	.080	Pr.	.005	.06	.29	.35
Box sills, 2″ x 12″	.150	L.F.	.003	.30	.16	.46
Concrete filled steel column, 4″ diameter	.125	L.F.	.002	.20	.14	.34
Girder, built up from three 2″ x 12″	.125	L.F.	.015	.74	.87	1.61
Sheathing, plywood, subfloor, 5/8″ CDX	1.000	S.F.	.012	.84	.69	1.53
Furring, 1″ x 3″, 16″ O.C.	1.000	L.F.	.023	.47	1.33	1.80
Joist hangers	.036	Ea.	.002	.05	.10	.15
TOTAL		S.F.	.080	4.63	4.64	9.27

Floor costs on this page are given on a cost per square foot basis.

Description	QUAN.	UNIT	LABOR HOURS	COST PER S.F.		
				MAT.	INST.	TOTAL

Floor Framing Price Sheet (Wood)	QUAN.	UNIT	LABOR HOURS	COST PER S.F.		
				MAT.	INST.	TOTAL
Joists, #2 or better, pine, 2" x 4", 12" O.C.	1.250	L.F.	.016	.58	.93	1.51
16" O.C.	1.000	L.F.	.013	.46	.74	1.20
2" x 6", 12" O.C.	1.250	L.F.	.016	.89	.93	1.82
16" O.C.	1.000	L.F.	.013	.71	.74	1.45
2" x 8", 12" O.C.	1.250	L.F.	.018	1.23	1.06	2.29
16" O.C.	1.000	L.F.	.015	.98	.85	1.83
2" x 10", 12" O.C.	1.250	L.F.	.022	1.94	1.29	3.23
16" O.C.	1.000	L.F.	.018	1.55	1.03	2.58
2"x 12", 12" O.C.	1.250	L.F.	.023	2.46	1.33	3.79
16" O.C.	1.000	L.F.	.018	1.97	1.06	3.03
Bridging, wood 1" x 3", joists 12" O.C.	.100	Pr.	.006	.08	.36	.44
16" O.C.	.080	Pr.	.005	.06	.29	.35
Metal, galvanized, joists 12" O.C.	.100	Pr.	.006	.19	.36	.55
16" O.C.	.080	Pr.	.005	.15	.29	.44
Compression type, joists 12" O.C.	.100	Pr.	.004	.14	.23	.37
16" O.C.	.080	Pr.	.003	.11	.19	.30
Box sills, #2 or better, 2" x 4"	.150	L.F.	.002	.07	.11	.18
2" x 6"	.150	L.F.	.002	.11	.11	.22
2" x 8"	.150	L.F.	.002	.15	.13	.28
2" x 10"	.150	L.F.	.003	.23	.15	.38
2" x 12"	.150	L.F.	.003	.30	.16	.46
Girders, including lally columns, 3 pieces spiked together, 2" x 8"	.125	L.F.	.015	.57	.91	1.48
2" x 10"	.125	L.F.	.016	.78	.97	1.75
2" x 12"	.125	L.F.	.017	.94	1.01	1.95
Solid girders, 3" x 8"	.040	L.F.	.004	.33	.24	.57
3" x 10"	.040	L.F.	.004	.36	.25	.61
3" x 12"	.040	L.F.	.004	.40	.26	.66
4" x 8"	.040	L.F.	.004	.39	.25	.64
4" x 10"	.040	L.F.	.004	.42	.27	.69
4" x 12"	.040	L.F.	.004	.45	.28	.73
Steel girders, bolted & including fabrication, wide flange shapes						
12" deep, 14#/l.f.	.040	L.F.	.003	1.06	.28	1.34
10" deep, 15#/l.f.	.040	L.F.	.003	1.06	.28	1.34
8" deep, 10#/l.f.	.040	L.F.	.003	.71	.28	.99
6" deep, 9#/l.f.	.040	L.F.	.003	.64	.28	.92
5" deep, 16#/l.f.	.040	L.F.	.003	1.06	.28	1.34
Sheathing, plywood exterior grade CDX, 1/2" thick	1.000	S.F.	.011	.68	.66	1.34
5/8" thick	1.000	S.F.	.012	.84	.69	1.53
3/4" thick	1.000	S.F.	.013	1.02	.74	1.76
Boards, 1" x 8" laid regular	1.000	S.F.	.016	2.28	.93	3.21
Laid diagonal	1.000	S.F.	.019	2.28	1.09	3.37
1" x 10" laid regular	1.000	S.F.	.015	2.33	.85	3.18
Laid diagonal	1.000	S.F.	.018	2.33	1.03	3.36
Furring, 1" x 3", 12" O.C.	1.250	L.F.	.029	.59	1.66	2.25
16" O.C.	1.000	L.F.	.023	.47	1.33	1.80
24" O.C.	.750	L.F.	.017	.35	1	1.35

For customer support on your Residential Costs with RSMeans data, call 800.448.8182.

131

System Description	QUAN.	UNIT	LABOR HOURS	COST PER S.F.		
				MAT.	INST.	TOTAL
9-1/2″ COMPOSITE WOOD JOISTS, 16″ O.C.						
CWJ, 9-1/2″, 16″ O.C., 15′ span	1.000	L.F.	.018	1.95	1.03	2.98
Temp. strut line, 1″ x 4″, 8′ O.C.	.160	L.F.	.003	.09	.19	.28
CWJ rim joist, 9-1/2″	.150	L.F.	.003	.29	.15	.44
Concrete filled steel column, 4″ diameter	.125	L.F.	.002	.20	.14	.34
Girder, built up from three 2″ x 8″	.125	L.F.	.013	.37	.77	1.14
Sheathing, plywood, subfloor, 5/8″ CDX	1.000	S.F.	.012	.84	.69	1.53
TOTAL		S.F.	.051	3.74	2.97	6.71
11-1/2″ COMPOSITE WOOD JOISTS, 16″ O.C.						
CWJ, 11-1/2″, 16″ O.C., 18′ span	1.000	L.F.	.018	2.25	1.05	3.30
Temp. strut line, 1″ x 4″, 8′ O.C.	.160	L.F.	.003	.09	.19	.28
CWJ rim joist, 11-1/2″	.150	L.F.	.003	.34	.16	.50
Concrete filled steel column, 4″ diameter	.125	L.F.	.002	.20	.14	.34
Girder, built up from three 2″ x 10″	.125	L.F.	.014	.58	.83	1.41
Sheathing, plywood, subfloor, 5/8″ CDX	1.000	S.F.	.012	.84	.69	1.53
TOTAL		S.F.	.052	4.30	3.06	7.36
14″ COMPOSITE WOOD JOISTS, 16″ O.C.						
CWJ, 14″, 16″ O.C., 22′ span	1.000	L.F.	.020	2.83	1.13	3.96
Temp. strut line, 1″ x 4″, 8′ O.C.	.160	L.F.	.003	.09	.19	.28
CWJ rim joist, 14″	.150	L.F.	.003	.42	.17	.59
Concrete filled steel column, 4″ diameter	.600	L.F.	.002	.20	.14	.34
Girder, built up from three 2″ x 12″	.600	L.F.	.015	.74	.87	1.61
Sheathing, plywood, subfloor, 5/8″ CDX	1.000	S.F.	.012	.84	.69	1.53
TOTAL		S.F.	.055	5.12	3.19	8.31

Floor costs on this page are given on a cost per square foot basis.

Description	QUAN.	UNIT	LABOR HOURS	COST PER S.F.		
				MAT.	INST.	TOTAL

Floor Framing Price Sheet (Wood)

	QUAN.	UNIT	LABOR HOURS	COST PER S.F. MAT.	COST PER S.F. INST.	COST PER S.F. TOTAL
Composite wood joist 9-1/2" deep, 12" O.C.	1.250	L.F.	.022	2.44	1.28	3.72
16" O.C.	1.000	L.F.	.018	1.95	1.03	2.98
11-1/2" deep, 12" O.C.	1.250	L.F.	.023	2.81	1.31	4.12
16" O.C.	1.000	L.F.	.018	2.25	1.05	3.30
14" deep, 12" O.C.	1.250	L.F.	.024	3.53	1.41	4.94
16" O.C.	1.000	L.F.	.020	2.83	1.13	3.96
16 " deep, 12" O.C.	1.250	L.F.	.026	5.60	1.50	7.10
16" O.C.	1.000	L.F.	.021	4.48	1.20	5.68
CWJ rim joist, 9-1/2"	.150	L.F.	.003	.29	.15	.44
11-1/2"	.150	L.F.	.003	.34	.16	.50
14"	.150	L.F.	.003	.42	.17	.59
16"	.150	L.F.	.003	.67	.18	.85
Girders, including lally columns, 3 pieces spiked together, 2" x 8"	.125	L.F.	.015	.57	.91	1.48
2" x 10"	.125	L.F.	.016	.78	.97	1.75
2" x 12"	.125	L.F.	.017	.94	1.01	1.95
Solid girders, 3" x 8"	.040	L.F.	.004	.33	.24	.57
3" x 10"	.040	L.F.	.004	.36	.25	.61
3" x 12"	.040	L.F.	.004	.40	.26	.66
4" x 8"	.040	L.F.	.004	.39	.25	.64
4" x 10"	.040	L.F.	.004	.42	.27	.69
4" x 12"	.040	L.F.	.004	.45	.28	.73
Steel girders, bolted & including fabrication, wide flange shapes						
12" deep, 14#/l.f.	.040	L.F.	.061	24.50	6.50	31
10" deep, 15#/l.f.	.040	L.F.	.067	26.50	7.10	33.60
8" deep, 10#/l.f.	.040	L.F.	.067	17.75	7.10	24.85
6" deep, 9#/l.f.	.040	L.F.	.067	16	7.10	23.10
5" deep, 16#/l.f.	.040	L.F.	.064	25.50	6.90	32.40
Sheathing, plywood exterior grade CDX, 1/2" thick	1.000	S.F.	.011	.68	.66	1.34
5/8" thick	1.000	S.F.	.012	.84	.69	1.53
3/4" thick	1.000	S.F.	.013	1.02	.74	1.76
Boards, 1" x 8" laid regular	1.000	S.F.	.016	2.28	.93	3.21
Laid diagonal	1.000	S.F.	.019	2.28	1.09	3.37
1" x 10" laid regular	1.000	S.F.	.015	2.33	.85	3.18
Laid diagonal	1.000	S.F.	.018	2.33	1.03	3.36
Furring, 1" x 3", 12" O.C.	1.250	L.F.	.029	.59	1.66	2.25
16" O.C.	1.000	L.F.	.023	.47	1.33	1.80
24" O.C.	.750	L.F.	.017	.35	1	1.35

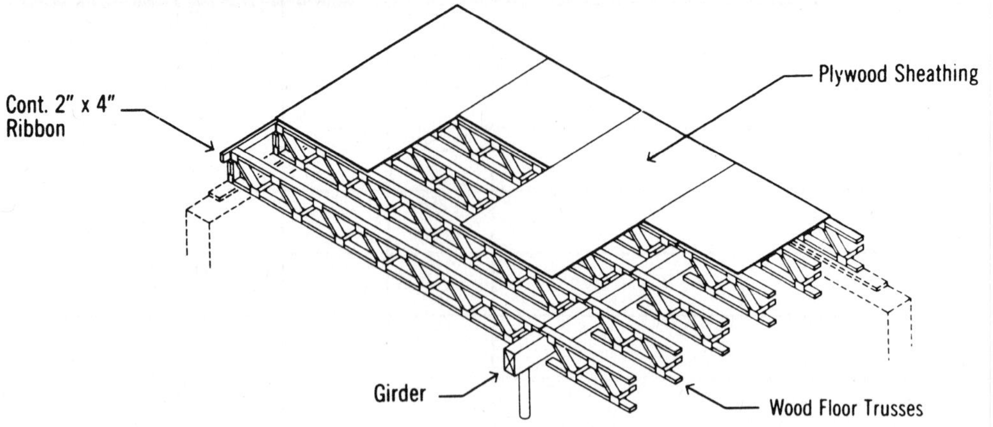

Cont. 2" x 4" Ribbon
Plywood Sheathing
Girder
Wood Floor Trusses

System Description	QUAN.	UNIT	LABOR HOURS	COST PER S.F.		
				MAT.	INST.	TOTAL
12" OPEN WEB JOISTS, 16" O.C.						
OWJ 12", 16" O.C., 21' span	1.000	L.F.	.018	3.98	1.05	5.03
Continuous ribbing, 2" x 4"	.150	L.F.	.002	.07	.11	.18
Concrete filled steel column, 4" diameter	.125	L.F.	.002	.20	.14	.34
Girder, built up from three 2" x 8"	.125	L.F.	.013	.37	.77	1.14
Sheathing, plywood, subfloor, 5/8" CDX	1.000	S.F.	.012	.84	.69	1.53
Furring, 1" x 3", 16" O.C.	1.000	L.F.	.023	.47	1.33	1.80
TOTAL		S.F.	.070	5.93	4.09	10.02
14" OPEN WEB WOOD JOISTS, 16" O.C.						
OWJ 14", 16" O.C., 22' span	1.000	L.F.	.020	4.30	1.13	5.43
Continuous ribbing, 2" x 4"	.150	L.F.	.002	.07	.11	.18
Concrete filled steel column, 4" diameter	.125	L.F.	.002	.20	.14	.34
Girder, built up from three 2" x 10"	.125	L.F.	.014	.58	.83	1.41
Sheathing, plywood, subfloor, 5/8" CDX	1.000	S.F.	.012	.84	.69	1.53
Furring, 1" x 3", 16" O.C.	1.000	L.F.	.023	.47	1.33	1.80
TOTAL		S.F.	.073	6.46	4.23	10.69
16" OPEN WEB WOOD JOISTS, 16" O.C.						
OWJ 16", 16" O.C., 24' span	1.000	L.F.	.021	4.25	1.20	5.45
Continuous ribbing, 2" x 4"	.150	L.F.	.002	.07	.11	.18
Concrete filled steel column, 4" diameter	.125	L.F.	.002	.20	.14	.34
Girder, built up from three 2" x 12"	.125	L.F.	.015	.74	.87	1.61
Sheathing, plywood, subfloor, 5/8" CDX	1.000	S.F.	.012	.84	.69	1.53
Furring, 1" x 3", 16" O.C.	1.000	L.F.	.023	.47	1.33	1.80
TOTAL		S.F.	.075	6.57	4.34	10.91

Floor costs on this page are given on a cost per square foot basis.

Description	QUAN.	UNIT	LABOR HOURS	COST PER S.F.		
				MAT.	INST.	TOTAL

Floor Framing Price Sheet (Wood)	QUAN.	UNIT	LABOR HOURS	COST PER S.F.		
				MAT.	INST.	TOTAL
Open web joists, 12" deep, 12" O.C.	1.250	L.F.	.023	4.97	1.31	6.28
16" O.C.	1.000	L.F.	.018	3.98	1.05	5.03
14" deep, 12" O.C.	1.250	L.F.	.024	5.40	1.41	6.81
16" O.C.	1.000	L.F.	.020	4.30	1.13	5.43
16 " deep, 12" O.C.	1.250	L.F.	.026	5.30	1.50	6.80
16" O.C.	1.000	L.F.	.021	4.25	1.20	5.45
18" deep, 12" O.C.	1.250	L.F.	.027	5.80	1.56	7.36
16" O.C.	1.000	L.F.	.022	4.65	1.25	5.90
Continuous ribbing, 2" x 4"	.150	L.F.	.002	.07	.11	.18
2" x 6"	.150	L.F.	.002	.11	.11	.22
2" x 8"	.150	L.F.	.002	.15	.13	.28
2" x 10"	.150	L.F.	.003	.23	.15	.38
2" x 12"	.150	L.F.	.003	.30	.16	.46
Girders, including lally columns, 3 pieces spiked together, 2" x 8"	.125	L.F.	.015	.57	.91	1.48
2" x 10"	.125	L.F.	.016	.78	.97	1.75
2" x 12"	.125	L.F.	.017	.94	1.01	1.95
Solid girders, 3" x 8"	.040	L.F.	.004	.33	.24	.57
3" x 10"	.040	L.F.	.004	.36	.25	.61
3" x 12"	.040	L.F.	.004	.40	.26	.66
4" x 8"	.040	L.F.	.004	.39	.25	.64
4" x 10"	.040	L.F.	.004	.42	.27	.69
4" x 12"	.040	L.F.	.004	.45	.28	.73
Steel girders, bolted & including fabrication, wide flange shapes						
12" deep, 14#/l.f.	.040	L.F.	.061	24.50	6.50	31
10" deep, 15#/l.f.	.040	L.F.	.067	26.50	7.10	33.60
8" deep, 10#/l.f.	.040	L.F.	.067	17.75	7.10	24.85
6" deep, 9#/l.f.	.040	L.F.	.067	16	7.10	23.10
5" deep, 16#/l.f.	.040	L.F.	.064	25.50	6.90	32.40
Sheathing, plywood exterior grade CDX, 1/2" thick	1.000	S.F.	.011	.68	.66	1.34
5/8" thick	1.000	S.F.	.012	.84	.69	1.53
3/4" thick	1.000	S.F.	.013	1.02	.74	1.76
Boards, 1" x 8" laid regular	1.000	S.F.	.016	2.28	.93	3.21
Laid diagonal	1.000	S.F.	.019	2.28	1.09	3.37
1" x 10" laid regular	1.000	S.F.	.015	2.33	.85	3.18
Laid diagonal	1.000	S.F.	.018	2.33	1.03	3.36
Furring, 1" x 3", 12" O.C.	1.250	L.F.	.029	.59	1.66	2.25
16" O.C.	1.000	L.F.	.023	.47	1.33	1.80
24" O.C.	.750	L.F.	.017	.35	1	1.35

Labels: Sheathing, Top Plates, Studs, Bottom Plate, Corner Bracing

System Description	QUAN.	UNIT	LABOR HOURS	COST PER S.F.		
				MAT.	INST.	TOTAL
2″ X 4″, 16″ O.C.						
2″ x 4″ studs, 16″ O.C.	1.000	L.F.	.015	.45	.85	1.30
Plates, 2″ x 4″, double top, single bottom	.375	L.F.	.005	.17	.32	.49
Corner bracing, let-in, 1″ x 6″	.063	L.F.	.003	.05	.20	.25
Sheathing, 1/2″ plywood, CDX	1.000	S.F.	.011	.68	.66	1.34
Framing connectors, holddowns	.013	Ea.	.013	.55	.75	1.30
TOTAL		S.F.	.047	1.90	2.78	4.68
2″ X 4″, 24″ O.C.						
2″ x 4″ studs, 24″ O.C.	.750	L.F.	.011	.34	.64	.98
Plates, 2″ x 4″, double top, single bottom	.375	L.F.	.005	.17	.32	.49
Corner bracing, let-in, 1″ x 6″	.063	L.F.	.002	.05	.13	.18
Sheathing, 1/2″ plywood, CDX	1.000	S.F.	.011	.68	.66	1.34
Framing connectors, holddowns	.013	Ea.	.013	.55	.75	1.30
TOTAL		S.F.	.042	1.79	2.50	4.29
2″ X 6″, 16″ O.C.						
2″ x 6″ studs, 16″ O.C.	1.000	L.F.	.016	.71	.93	1.64
Plates, 2″ x 6″, double top, single bottom	.375	L.F.	.006	.27	.35	.62
Corner bracing, let-in, 1″ x 6″	.063	L.F.	.003	.05	.20	.25
Sheathing, 1/2″ plywood, CDX	1.000	S.F.	.014	.68	.83	1.51
Framing connectors, holddowns	.013	Ea.	.013	.55	.75	1.30
TOTAL		S.F.	.052	2.26	3.06	5.32
2″ X 6″, 24″ O.C.						
2″ x 6″ studs, 24″ O.C.	.750	L.F.	.012	.53	.70	1.23
Plates, 2″ x 6″, double top, single bottom	.375	L.F.	.006	.27	.35	.62
Corner bracing, let-in, 1″ x 6″	.063	L.F.	.002	.05	.13	.18
Sheathing, 1/2″ plywood, CDX	1.000	S.F.	.011	.68	.66	1.34
Framing connectors, holddowns	.013	Ea.	.013	.55	.75	1.30
TOTAL		S.F.	.044	2.08	2.59	4.67

The wall costs on this page are given in cost per square foot of wall.
For window and door openings see below.

Description	QUAN.	UNIT	LABOR HOURS	COST PER S.F.		
				MAT.	INST.	TOTAL

Exterior Wall Framing Price Sheet	QUAN.	UNIT	LABOR HOURS	COST PER S.F.		
				MAT.	INST.	TOTAL
Studs, #2 or better, 2″ x 4″, 12″ O.C.	1.250	L.F.	.018	.56	1.06	1.62
16″ O.C.	1.000	L.F.	.015	.45	.85	1.30
24″ O.C.	.750	L.F.	.011	.34	.64	.98
32″ O.C.	.600	L.F.	.009	.27	.51	.78
2″ x 6″, 12″ O.C.	1.250	L.F.	.020	.89	1.16	2.05
16″ O.C.	1.000	L.F.	.016	.71	.93	1.64
24″ O.C.	.750	L.F.	.012	.53	.70	1.23
32″ O.C.	.600	L.F.	.010	.43	.56	.99
2″ x 8″, 12″ O.C.	1.250	L.F.	.025	1.65	1.45	3.10
16″ O.C.	1.000	L.F.	.020	1.32	1.16	2.48
24″ O.C.	.750	L.F.	.015	.99	.87	1.86
32″ O.C.	.600	L.F.	.012	.79	.70	1.49
Plates, #2 or better, double top, single bottom, 2″ x 4″	.375	L.F.	.005	.17	.32	.49
2″ x 6″	.375	L.F.	.006	.27	.35	.62
2″ x 8″	.375	L.F.	.008	.50	.44	.94
Corner bracing, let-in 1″ x 6″ boards, studs, 12″ O.C.	.070	L.F.	.004	.06	.22	.28
16″ O.C.	.063	L.F.	.003	.05	.20	.25
24″ O.C.	.063	L.F.	.002	.05	.13	.18
32″ O.C.	.057	L.F.	.002	.05	.12	.17
Let-in steel (″T″ shape), studs, 12″ O.C.	.070	L.F.	.001	.06	.06	.12
16″ O.C.	.063	L.F.	.001	.06	.05	.11
24″ O.C.	.063	L.F.	.001	.06	.05	.11
32″ O.C.	.057	L.F.	.001	.05	.04	.09
Sheathing, plywood CDX, 3/8″ thick	1.000	S.F.	.010	.66	.61	1.27
1/2″ thick	1.000	S.F.	.011	.68	.66	1.34
5/8″ thick	1.000	S.F.	.012	.84	.72	1.56
3/4″ thick	1.000	S.F.	.013	1.02	.77	1.79
Boards, 1″ x 6″, laid regular	1.000	S.F.	.025	1.86	1.43	3.29
Laid diagonal	1.000	S.F.	.027	1.86	1.59	3.45
1″ x 8″, laid regular	1.000	S.F.	.021	2.28	1.22	3.50
Laid diagonal	1.000	S.F.	.025	2.28	1.43	3.71
Wood fiber, regular, no vapor barrier, 1/2″ thick	1.000	S.F.	.013	.70	.77	1.47
5/8″ thick	1.000	S.F.	.013	.77	.77	1.54
Asphalt impregnated 25/32″ thick	1.000	S.F.	.013	.35	.77	1.12
1/2″ thick	1.000	S.F.	.013	.35	.77	1.12
Polystyrene, regular, 3/4″ thick	1.000	S.F.	.010	.63	.58	1.21
2″ thick	1.000	S.F.	.011	1.75	.64	2.39
Fiberglass, foil faced, 1″ thick	1.000	S.F.	.008	.92	.46	1.38
2″ thick	1.000	S.F.	.009	1.74	.52	2.26

Window & Door Openings	QUAN.	UNIT	LABOR HOURS	COST EACH		
				MAT.	INST.	TOTAL
The following costs are to be added to the total costs of the wall for each opening. Do not subtract the area of the openings.						
Headers, 2″ x 6″ double, 2′ long	4.000	L.F.	.178	2.84	10.30	13.14
3′ long	6.000	L.F.	.267	4.26	15.50	19.76
4′ long	8.000	L.F.	.356	5.70	20.50	26.20
5′ long	10.000	L.F.	.444	7.10	26	33.10
2″ x 8″ double, 4′ long	8.000	L.F.	.376	7.85	22	29.85
5′ long	10.000	L.F.	.471	9.80	27.50	37.30
6′ long	12.000	L.F.	.565	11.75	33	44.75
8′ long	16.000	L.F.	.753	15.70	43.50	59.20
2″ x 10″ double, 4′ long	8.000	L.F.	.400	12.40	23.50	35.90
6′ long	12.000	L.F.	.600	18.60	35	53.60
8′ long	16.000	L.F.	.800	25	46.50	71.50
10′ long	20.000	L.F.	1.000	31	58	89
2″ x 12″ double, 8′ long	16.000	L.F.	.853	31.50	49.50	81
12′ long	24.000	L.F.	1.280	47.50	74.50	122

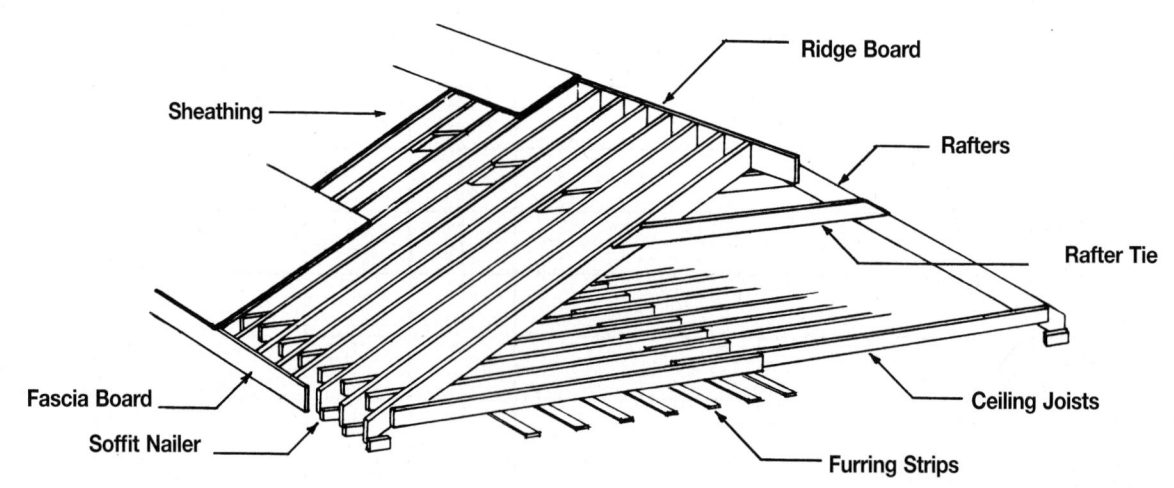

Sheathing
Ridge Board
Rafters
Rafter Tie
Fascia Board
Soffit Nailer
Ceiling Joists
Furring Strips

System Description	QUAN.	UNIT	LABOR HOURS	COST PER S.F.		
				MAT.	INST.	TOTAL
2″ X 6″ RAFTERS, 16″ O.C., 4/12 PITCH						
Rafters, 2″ x 6″, 16″ O.C., 4/12 pitch	1.170	L.F.	.019	.83	1.09	1.92
Ceiling joists, 2″ x 4″, 16″ O.C.	1.000	L.F.	.013	.46	.74	1.20
Ridge board, 2″ x 6″	.050	L.F.	.002	.04	.09	.13
Fascia board, 2″ x 6″	.100	L.F.	.005	.07	.31	.38
Rafter tie, 1″ x 4″, 4′ O.C.	.060	L.F.	.001	.03	.07	.10
Soffit nailer (outrigger), 2″ x 4″, 24″ O.C.	.170	L.F.	.004	.08	.26	.34
Sheathing, exterior, plywood, CDX, 1/2″ thick	1.170	S.F.	.013	.80	.77	1.57
Furring strips, 1″ x 3″, 16″ O.C.	1.000	L.F.	.023	.47	1.33	1.80
Rafter ties	.053	Ea.	.003	.08	.17	.25
TOTAL		S.F.	.083	2.86	4.83	7.69
2″ X 8″ RAFTERS, 16″ O.C., 4/12 PITCH						
Rafters, 2″ x 8″, 16″ O.C., 4/12 pitch	1.170	L.F.	.020	1.15	1.15	2.30
Ceiling joists, 2″ x 6″, 16″ O.C.	1.000	L.F.	.013	.71	.74	1.45
Ridge board, 2″ x 8″	.050	L.F.	.002	.05	.10	.15
Fascia board, 2″ x 8″	.100	L.F.	.007	.10	.41	.51
Rafter tie, 1″ x 4″, 4′ O.C.	.060	L.F.	.001	.03	.07	.10
Soffit nailer (outrigger), 2″ x 4″, 24″ O.C.	.170	L.F.	.004	.08	.26	.34
Sheathing, exterior, plywood, CDX, 1/2″ thick	1.170	S.F.	.013	.80	.77	1.57
Furring strips, 1″ x 3″, 16″ O.C.	1.000	L.F.	.023	.47	1.33	1.80
Rafter ties	.053	Ea.	.003	.08	.17	.25
TOTAL		S.F.	.086	3.47	5	8.47

The cost of this system is based on the square foot of plan area.
All quantities have been adjusted accordingly.

Description	QUAN.	UNIT	LABOR HOURS	COST PER S.F.		
				MAT.	INST.	TOTAL

Gable End Roof Framing Price Sheet	QUAN.	UNIT	LABOR HOURS	COST PER S.F. MAT.	INST.	TOTAL
Rafters, #2 or better, 16" O.C., 2" x 6", 4/12 pitch	1.170	L.F.	.019	.83	1.09	1.92
8/12 pitch	1.330	L.F.	.027	.94	1.54	2.48
2" x 8", 4/12 pitch	1.170	L.F.	.020	1.15	1.15	2.30
8/12 pitch	1.330	L.F.	.028	1.30	1.65	2.95
2" x 10", 4/12 pitch	1.170	L.F.	.030	1.81	1.73	3.54
8/12 pitch	1.330	L.F.	.043	2.06	2.50	4.56
24" O.C., 2" x 6", 4/12 pitch	.940	L.F.	.015	.67	.87	1.54
8/12 pitch	1.060	L.F.	.021	.75	1.23	1.98
2" x 8", 4/12 pitch	.940	L.F.	.016	.92	.92	1.84
8/12 pitch	1.060	L.F.	.023	1.04	1.31	2.35
2" x 10", 4/12 pitch	.940	L.F.	.024	1.46	1.39	2.85
8/12 pitch	1.060	L.F.	.034	1.64	1.99	3.63
Ceiling joist, #2 or better, 2" x 4", 16" O.C.	1.000	L.F.	.013	.46	.74	1.20
24" O.C.	.750	L.F.	.010	.35	.56	.91
2" x 6", 16" O.C.	1.000	L.F.	.013	.71	.74	1.45
24" O.C.	.750	L.F.	.010	.53	.56	1.09
2" x 8", 16" O.C.	1.000	L.F.	.015	.98	.85	1.83
24" O.C.	.750	L.F.	.011	.74	.64	1.38
2" x 10", 16" O.C.	1.000	L.F.	.018	1.55	1.03	2.58
24" O.C.	.750	L.F.	.013	1.16	.77	1.93
Ridge board, #2 or better, 1" x 6"	.050	L.F.	.001	.04	.08	.12
1" x 8"	.050	L.F.	.001	.07	.08	.15
1" x 10"	.050	L.F.	.002	.09	.09	.18
2" x 6"	.050	L.F.	.002	.04	.09	.13
2" x 8"	.050	L.F.	.002	.05	.10	.15
2" x 10"	.050	L.F.	.002	.08	.12	.20
Fascia board, #2 or better, 1" x 6"	.100	L.F.	.004	.05	.23	.28
1" x 8"	.100	L.F.	.005	.06	.26	.32
1" x 10"	.100	L.F.	.005	.07	.30	.37
2" x 6"	.100	L.F.	.006	.08	.33	.41
2" x 8"	.100	L.F.	.007	.10	.41	.51
2" x 10"	.100	L.F.	.004	.31	.21	.52
Rafter tie, #2 or better, 4' O.C., 1" x 4"	.060	L.F.	.001	.03	.07	.10
1" x 6"	.060	L.F.	.001	.04	.08	.12
2" x 4"	.060	L.F.	.002	.04	.09	.13
2" x 6"	.060	L.F.	.002	.06	.12	.18
Soffit nailer (outrigger), 2" x 4", 16" O.C.	.220	L.F.	.006	.10	.33	.43
24" O.C.	.170	L.F.	.004	.08	.26	.34
2" x 6", 16" O.C.	.220	L.F.	.006	.12	.38	.50
24" O.C.	.170	L.F.	.005	.09	.30	.39
Sheathing, plywood CDX, 4/12 pitch, 3/8" thick.	1.170	S.F.	.012	.77	.71	1.48
1/2" thick	1.170	S.F.	.013	.80	.77	1.57
5/8" thick	1.170	S.F.	.014	.98	.84	1.82
8/12 pitch, 3/8"	1.330	S.F.	.014	.88	.81	1.69
1/2" thick	1.330	S.F.	.015	.90	.88	1.78
5/8" thick	1.330	S.F.	.016	1.12	.96	2.08
Boards, 4/12 pitch roof, 1" x 6"	1.170	S.F.	.026	2.18	1.50	3.68
1" x 8"	1.170	S.F.	.021	2.67	1.24	3.91
8/12 pitch roof, 1" x 6"	1.330	S.F.	.029	2.47	1.70	4.17
1" x 8"	1.330	S.F.	.024	3.03	1.41	4.44
Furring, 1" x 3", 12" O.C.	1.200	L.F.	.027	.56	1.60	2.16
16" O.C.	1.000	L.F.	.023	.47	1.33	1.80
24" O.C.	.800	L.F.	.018	.38	1.06	1.44

For customer support on your Residential Costs with RSMeans data, call 800.448.8182.

139

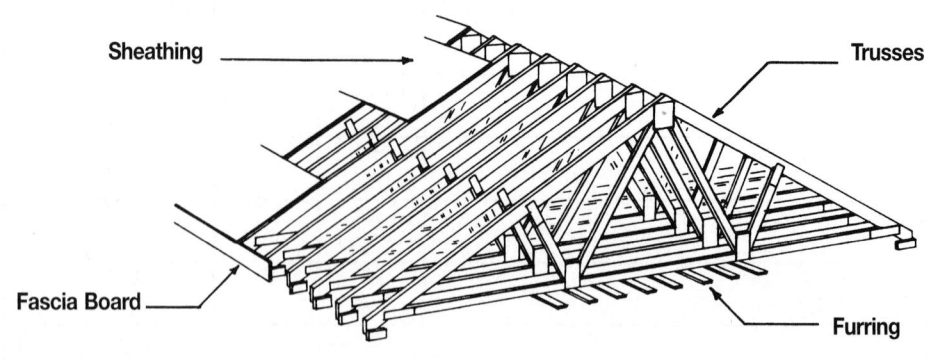

Sheathing

Trusses

Fascia Board

Furring

System Description	QUAN.	UNIT	LABOR HOURS	COST PER S.F.		
				MAT.	INST.	TOTAL
TRUSS, 16" O.C., 4/12 PITCH, 1' OVERHANG, 26' SPAN						
Truss, 40# loading, 16" O.C., 4/12 pitch, 26' span	.030	Ea.	.021	2.55	1.42	3.97
Fascia board, 2" x 6"	.100	L.F.	.005	.07	.31	.38
Sheathing, exterior, plywood, CDX, 1/2" thick	1.170	S.F.	.013	.80	.77	1.57
Furring, 1" x 3", 16" O.C.	1.000	L.F.	.023	.47	1.33	1.80
Rafter ties	.053	Ea.	.003	.08	.17	.25
TOTAL		S.F.	.065	3.97	4	7.97
TRUSS, 16" O.C., 8/12 PITCH, 1' OVERHANG, 26' SPAN						
Truss, 40# loading, 16" O.C., 8/12 pitch, 26' span	.030	Ea.	.023	3.57	1.55	5.12
Fascia board, 2" x 6"	.100	L.F.	.005	.07	.31	.38
Sheathing, exterior, plywood, CDX, 1/2" thick	1.330	S.F.	.015	.90	.88	1.78
Furring, 1" x 3", 16" O.C.	1.000	L.F.	.023	.47	1.33	1.80
Rafter ties	.053	Ea.	.003	.08	.17	.25
TOTAL		S.F.	.069	5.09	4.24	9.33
TRUSS, 24" O.C., 4/12 PITCH, 1' OVERHANG, 26' SPAN						
Truss, 40# loading, 24" O.C., 4/12 pitch, 26' span	.020	Ea.	.014	1.70	.94	2.64
Fascia board, 2" x 6"	.100	L.F.	.005	.07	.31	.38
Sheathing, exterior, plywood, CDX, 1/2" thick	1.170	S.F.	.013	.80	.77	1.57
Furring, 1" x 3", 16" O.C.	1.000	L.F.	.023	.47	1.33	1.80
Rafter ties	.035	Ea.	.002	.05	.11	.16
TOTAL		S.F.	.057	3.09	3.46	6.55
TRUSS, 24" O.C., 8/12 PITCH, 1' OVERHANG, 26' SPAN						
Truss, 40# loading, 24" O.C., 8/12 pitch, 26' span	.020	Ea.	.015	2.38	1.03	3.41
Fascia board, 2" x 6"	.100	L.F.	.005	.07	.31	.38
Sheathing, exterior, plywood, CDX, 1/2" thick	1.330	S.F.	.015	.90	.88	1.78
Furring, 1" x 3", 16" O.C.	1.000	L.F.	.023	.47	1.33	1.80
Rafter ties	.035	Ea.	.002	.05	.11	.16
TOTAL		S.F.	.060	3.87	3.66	7.53

The cost of this system is based on the square foot of plan area.
A one foot overhang is included.

Description	QUAN.	UNIT	LABOR HOURS	COST PER S.F.		
				MAT.	INST.	TOTAL

Truss Roof Framing Price Sheet	QUAN.	UNIT	LABOR HOURS	COST PER S.F.		
				MAT.	INST.	TOTAL
Truss, 40# loading, including 1' overhang, 4/12 pitch, 24' span, 16" O.C.	.033	Ea.	.022	2.85	1.47	4.32
24" O.C.	.022	Ea.	.015	1.90	.98	2.88
26' span, 16" O.C.	.030	Ea.	.021	2.55	1.42	3.97
24" O.C.	.020	Ea.	.014	1.70	.94	2.64
28' span, 16" O.C.	.027	Ea.	.020	2.57	1.36	3.93
24" O.C.	.019	Ea.	.014	1.81	.96	2.77
32' span, 16" O.C.	.024	Ea.	.019	3	1.28	4.28
24" O.C.	.016	Ea.	.013	2	.85	2.85
36' span, 16" O.C.	.022	Ea.	.019	3.43	1.28	4.71
24" O.C.	.015	Ea.	.013	2.34	.88	3.22
8/12 pitch, 24' span, 16" O.C.	.033	Ea.	.024	3.73	1.60	5.33
24" O.C.	.022	Ea.	.016	2.49	1.07	3.56
26' span, 16" O.C.	.030	Ea.	.023	3.57	1.55	5.12
24" O.C.	.020	Ea.	.015	2.38	1.03	3.41
28' span, 16" O.C.	.027	Ea.	.022	3.65	1.47	5.12
24" O.C.	.019	Ea.	.016	2.57	1.04	3.61
32' span, 16" O.C.	.024	Ea.	.021	3.62	1.43	5.05
24" O.C.	.016	Ea.	.014	2.42	.95	3.37
36' span, 16" O.C.	.022	Ea.	.021	4.20	1.43	5.63
24" O.C.	.015	Ea.	.015	2.87	.98	3.85
Fascia board, #2 or better, 1" x 6"	.100	L.F.	.004	.05	.23	.28
1" x 8"	.100	L.F.	.005	.06	.26	.32
1" x 10"	.100	L.F.	.005	.07	.30	.37
2" x 6"	.100	L.F.	.006	.08	.33	.41
2" x 8"	.100	L.F.	.007	.10	.41	.51
2" x 10"	.100	L.F.	.009	.16	.52	.68
Sheathing, plywood CDX, 4/12 pitch, 3/8" thick	1.170	S.F.	.012	.77	.71	1.48
1/2" thick	1.170	S.F.	.013	.80	.77	1.57
5/8" thick	1.170	S.F.	.014	.98	.84	1.82
8/12 pitch, 3/8" thick	1.330	S.F.	.014	.88	.81	1.69
1/2" thick	1.330	S.F.	.015	.90	.88	1.78
5/8" thick	1.330	S.F.	.016	1.12	.96	2.08
Boards, 4/12 pitch, 1" x 6"	1.170	S.F.	.026	2.18	1.50	3.68
1" x 8"	1.170	S.F.	.021	2.67	1.24	3.91
8/12 pitch, 1" x 6"	1.330	S.F.	.029	2.47	1.70	4.17
1" x 8"	1.330	S.F.	.024	3.03	1.41	4.44
Furring, 1" x 3", 12" O.C.	1.200	L.F.	.027	.56	1.60	2.16
16" O.C.	1.000	L.F.	.023	.47	1.33	1.80
24" O.C.	.800	L.F.	.018	.38	1.06	1.44
	.033	Ea.	.022	2.85	1.47	4.32

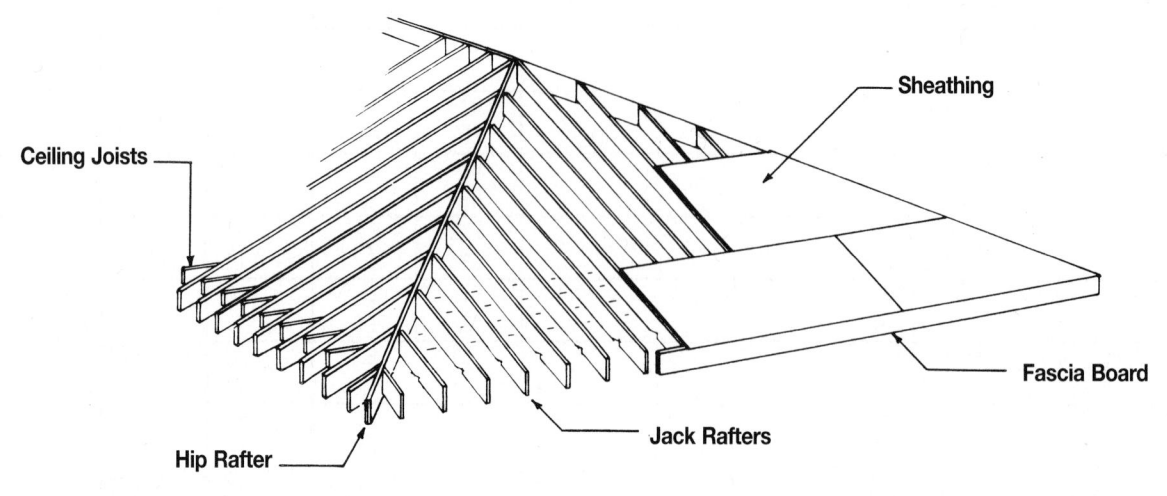

Ceiling Joists

Sheathing

Fascia Board

Jack Rafters

Hip Rafter

System Description	QUAN.	UNIT	LABOR HOURS	COST PER S.F.		
				MAT.	INST.	TOTAL
2″ X 6″, 16″ O.C., 4/12 PITCH						
Hip rafters, 2″ x 8″, 4/12 pitch	.160	L.F.	.004	.16	.21	.37
Jack rafters, 2″ x 6″, 16″ O.C., 4/12 pitch	1.430	L.F.	.038	1.02	2.22	3.24
Ceiling joists, 2″ x 6″, 16″ O.C.	1.000	L.F.	.013	.71	.74	1.45
Fascia board, 2″ x 8″	.220	L.F.	.016	.22	.91	1.13
Soffit nailer (outrigger), 2″ x 4″, 24″ O.C.	.220	L.F.	.006	.10	.33	.43
Sheathing, 1/2″ exterior plywood, CDX	1.570	S.F.	.018	1.07	1.04	2.11
Furring strips, 1″ x 3″, 16″ O.C.	1.000	L.F.	.023	.47	1.33	1.80
Rafter ties	.070	Ea.	.004	.11	.22	.33
TOTAL		S.F.	.122	3.86	7	10.86
2″ X 8″, 16″ O.C., 4/12 PITCH						
Hip rafters, 2″ x 10″, 4/12 pitch	.160	L.F.	.004	.25	.26	.51
Jack rafters, 2″ x 8″, 16″ O.C., 4/12 pitch	1.430	L.F.	.047	1.40	2.72	4.12
Ceiling joists, 2″ x 6″, 16″ O.C.	1.000	L.F.	.013	.71	.74	1.45
Fascia board, 2″ x 8″	.220	L.F.	.012	.17	.70	.87
Soffit nailer (outrigger), 2″ x 4″, 24″ O.C.	.220	L.F.	.006	.10	.33	.43
Sheathing, 1/2″ exterior plywood, CDX	1.570	S.F.	.018	1.07	1.04	2.11
Furring strips, 1″ x 3″, 16″ O.C.	1.000	L.F.	.023	.47	1.33	1.80
Rafter ties	.070	Ea.	.004	.11	.22	.33
TOTAL		S.F.	.127	4.28	7.34	11.62

The cost of this system is based on S.F. of plan area. Measurement is area under the hip roof only. See gable roof system for added costs.

Description	QUAN.	UNIT	LABOR HOURS	COST PER S.F.		
				MAT.	INST.	TOTAL

Hip Roof Framing Price Sheet	QUAN.	UNIT	LABOR HOURS	COST PER S.F.		
				MAT.	INST.	TOTAL
Hip rafters, #2 or better, 2" x 6", 4/12 pitch	.160	L.F.	.003	.11	.20	.31
8/12 pitch	.210	L.F.	.006	.15	.33	.48
2" x 8", 4/12 pitch	.160	L.F.	.004	.16	.21	.37
8/12 pitch	.210	L.F.	.006	.21	.36	.57
2" x 10", 4/12 pitch	.160	L.F.	.004	.25	.26	.51
8/12 pitch roof	.210	L.F.	.008	.33	.44	.77
Jack rafters, #2 or better, 16" O.C., 2" x 6", 4/12 pitch	1.430	L.F.	.038	1.02	2.22	3.24
8/12 pitch	1.800	L.F.	.061	1.28	3.53	4.81
2" x 8", 4/12 pitch	1.430	L.F.	.047	1.40	2.72	4.12
8/12 pitch	1.800	L.F.	.075	1.76	4.34	6.10
2" x 10", 4/12 pitch	1.430	L.F.	.051	2.22	2.96	5.18
8/12 pitch	1.800	L.F.	.082	2.79	4.79	7.58
24" O.C., 2" x 6", 4/12 pitch	1.150	L.F.	.031	.82	1.78	2.60
8/12 pitch	1.440	L.F.	.048	1.02	2.82	3.84
2" x 8", 4/12 pitch	1.150	L.F.	.038	1.13	2.19	3.32
8/12 pitch	1.440	L.F.	.060	1.41	3.47	4.88
2" x 10", 4/12 pitch	1.150	L.F.	.041	1.78	2.38	4.16
8/12 pitch	1.440	L.F.	.066	2.23	3.83	6.06
Ceiling joists, #2 or better, 2" x 4", 16" O.C.	1.000	L.F.	.013	.46	.74	1.20
24" O.C.	.750	L.F.	.010	.35	.56	.91
2" x 6", 16" O.C.	1.000	L.F.	.013	.71	.74	1.45
24" O.C.	.750	L.F.	.010	.53	.56	1.09
2" x 8", 16" O.C.	1.000	L.F.	.015	.98	.85	1.83
24" O.C.	.750	L.F.	.011	.74	.64	1.38
2" x 10", 16" O.C.	1.000	L.F.	.018	1.55	1.03	2.58
24" O.C.	.750	L.F.	.013	1.16	.77	1.93
Fascia board, #2 or better, 1" x 6"	.220	L.F.	.009	.12	.50	.62
1" x 8"	.220	L.F.	.010	.14	.58	.72
1" x 10"	.220	L.F.	.011	.15	.65	.80
2" x 6"	.220	L.F.	.013	.17	.73	.90
2" x 8"	.220	L.F.	.016	.22	.91	1.13
2" x 10"	.220	L.F.	.020	.34	1.13	1.47
Soffit nailer (outrigger), 2" x 4", 16" O.C.	.280	L.F.	.007	.13	.42	.55
24" O.C.	.220	L.F.	.006	.10	.33	.43
2" x 8", 16" O.C.	.280	L.F.	.007	.21	.39	.60
24" O.C.	.220	L.F.	.005	.17	.32	.49
Sheathing, plywood CDX, 4/12 pitch, 3/8" thick	1.570	S.F.	.016	1.04	.96	2
1/2" thick	1.570	S.F.	.018	1.07	1.04	2.11
5/8" thick	1.570	S.F.	.019	1.32	1.13	2.45
8/12 pitch, 3/8" thick	1.900	S.F.	.020	1.25	1.16	2.41
1/2" thick	1.900	S.F.	.022	1.29	1.25	2.54
5/8" thick	1.900	S.F.	.023	1.60	1.37	2.97
Boards, 4/12 pitch, 1" x 6" boards	1.450	S.F.	.032	2.70	1.86	4.56
1" x 8" boards	1.450	S.F.	.027	3.31	1.54	4.85
8/12 pitch, 1" x 6" boards	1.750	S.F.	.039	3.26	2.24	5.50
1" x 8" boards	1.750	S.F.	.032	3.99	1.86	5.85
Furring, 1" x 3", 12" O.C.	1.200	L.F.	.027	.56	1.60	2.16
16" O.C.	1.000	L.F.	.023	.47	1.33	1.80
24" O.C.	.800	L.F.	.018	.38	1.06	1.44

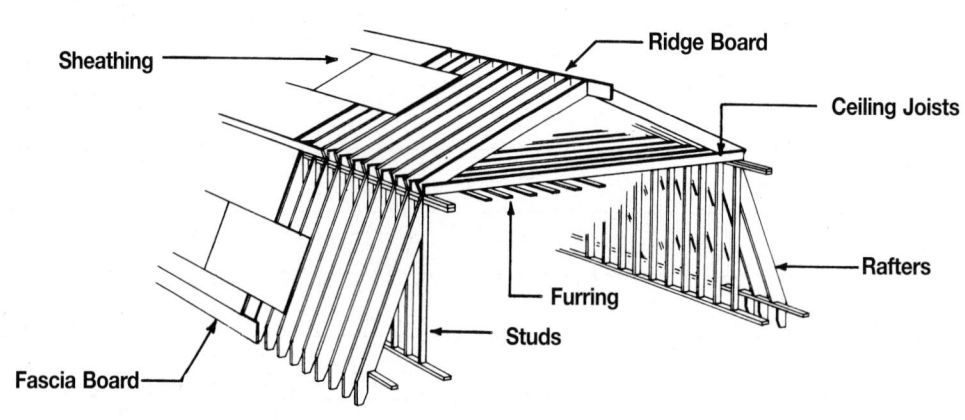

Sheathing · Ridge Board · Ceiling Joists · Rafters · Furring · Studs · Fascia Board

System Description	QUAN.	UNIT	LABOR HOURS	COST PER S.F.		
				MAT.	INST.	TOTAL
2" X 6" RAFTERS, 16" O.C.						
Roof rafters, 2" x 6", 16" O.C.	1.430	L.F.	.029	1.02	1.66	2.68
Ceiling joists, 2" x 6", 16" O.C.	.710	L.F.	.009	.50	.53	1.03
Stud wall, 2" x 4", 16" O.C., including plates	.790	L.F.	.012	.36	.73	1.09
Furring strips, 1" x 3", 16" O.C.	.710	L.F.	.016	.33	.94	1.27
Ridge board, 2" x 8"	.050	L.F.	.002	.05	.10	.15
Fascia board, 2" x 6"	.100	L.F.	.006	.08	.33	.41
Sheathing, exterior grade plywood, 1/2" thick	1.450	S.F.	.017	.99	.96	1.95
Rafter ties	.106	Ea.	.006	.16	.34	.50
TOTAL		S.F.	.097	3.49	5.59	9.08
2" X 8" RAFTERS, 16" O.C.						
Roof rafters, 2" x 8", 16" O.C.	1.430	L.F.	.031	1.40	1.77	3.17
Ceiling joists, 2" x 6", 16" O.C.	.710	L.F.	.009	.50	.53	1.03
Stud wall, 2" x 4", 16" O.C., including plates	.790	L.F.	.012	.36	.73	1.09
Furring strips, 1" x 3", 16" O.C.	.710	L.F.	.016	.33	.94	1.27
Ridge board, 2" x 8"	.050	L.F.	.002	.05	.10	.15
Fascia board, 2" x 8"	.100	L.F.	.007	.10	.41	.51
Sheathing, exterior grade plywood, 1/2" thick	1.450	S.F.	.017	.99	.96	1.95
Rafter ties	.106	Ea.	.006	.16	.34	.50
TOTAL		S.F.	.100	3.89	5.78	9.67

The cost of this system is based on the square foot of plan area on the first floor.

Description	QUAN.	UNIT	LABOR HOURS	COST PER S.F.		
				MAT.	INST.	TOTAL

Gambrel Roof Framing Price Sheet

	QUAN.	UNIT	LABOR HOURS	COST PER S.F. MAT.	COST PER S.F. INST.	COST PER S.F. TOTAL
Roof rafters, #2 or better, 2" x 6", 16" O.C.	1.430	L.F.	.029	1.02	1.66	2.68
24" O.C.	1.140	L.F.	.023	.81	1.32	2.13
2" x 8", 16" O.C.	1.430	L.F.	.031	1.40	1.77	3.17
24" O.C.	1.140	L.F.	.024	1.12	1.41	2.53
2" x 10", 16" O.C.	1.430	L.F.	.046	2.22	2.69	4.91
24" O.C.	1.140	L.F.	.037	1.77	2.14	3.91
Ceiling joist, #2 or better, 2" x 4", 16" O.C.	.710	L.F.	.009	.33	.53	.86
24" O.C.	.570	L.F.	.007	.26	.42	.68
2" x 6", 16" O.C.	.710	L.F.	.009	.50	.53	1.03
24" O.C.	.570	L.F.	.007	.40	.42	.82
2" x 8", 16" O.C.	.710	L.F.	.010	.70	.60	1.30
24" O.C.	.570	L.F.	.008	.56	.48	1.04
Stud wall, #2 or better, 2" x 4", 16" O.C.	.790	L.F.	.012	.36	.73	1.09
24" O.C.	.630	L.F.	.010	.29	.58	.87
2" x 6", 16" O.C.	.790	L.F.	.014	.56	.82	1.38
24" O.C.	.630	L.F.	.011	.45	.66	1.11
Furring, 1" x 3", 16" O.C.	.710	L.F.	.016	.33	.94	1.27
24" O.C.	.590	L.F.	.013	.28	.78	1.06
Ridge board, #2 or better, 1" x 6"	.050	L.F.	.001	.04	.08	.12
1" x 8"	.050	L.F.	.001	.07	.08	.15
1" x 10"	.050	L.F.	.002	.09	.09	.18
2" x 6"	.050	L.F.	.002	.04	.09	.13
2" x 8"	.050	L.F.	.002	.05	.10	.15
2" x 10"	.050	L.F.	.002	.08	.12	.20
Fascia board, #2 or better, 1" x 6"	.100	L.F.	.004	.05	.23	.28
1" x 8"	.100	L.F.	.005	.06	.26	.32
1" x 10"	.100	L.F.	.005	.07	.30	.37
2" x 6"	.100	L.F.	.006	.08	.33	.41
2" x 8"	.100	L.F.	.007	.10	.41	.51
2" x 10"	.100	L.F.	.009	.16	.52	.68
Sheathing, plywood, exterior grade CDX, 3/8" thick	1.450	S.F.	.015	.96	.88	1.84
1/2" thick	1.450	S.F.	.017	.99	.96	1.95
5/8" thick	1.450	S.F.	.018	1.22	1.04	2.26
3/4" thick	1.450	S.F.	.019	1.48	1.12	2.60
Boards, 1" x 6", laid regular	1.450	S.F.	.032	2.70	1.86	4.56
Laid diagonal	1.450	S.F.	.036	2.70	2.07	4.77
1" x 8", laid regular	1.450	S.F.	.027	3.31	1.54	4.85
Laid diagonal	1.450	S.F.	.032	3.31	1.86	5.17

For customer support on your Residential Costs with RSMeans data, call 800.448.8182.

145

System Description	QUAN.	UNIT	LABOR HOURS	COST PER S.F.		
				MAT.	INST.	TOTAL
2″ X 6″ RAFTERS, 16″ O.C.						
Roof rafters, 2″ x 6″, 16″ O.C.	1.210	L.F.	.033	.86	1.91	2.77
Rafter plates, 2″ x 6″, double top, single bottom	.364	L.F.	.010	.26	.58	.84
Ceiling joists, 2″ x 4″, 16″ O.C.	.920	L.F.	.012	.42	.68	1.10
Hip rafter, 2″ x 6″	.070	L.F.	.002	.05	.13	.18
Jack rafter, 2″ x 6″, 16″ O.C.	1.000	L.F.	.039	.71	2.27	2.98
Ridge board, 2″ x 6″	.018	L.F.	.001	.01	.03	.04
Sheathing, exterior grade plywood, 1/2″ thick	2.210	S.F.	.025	1.50	1.46	2.96
Furring strips, 1″ x 3″, 16″ O.C.	.920	L.F.	.021	.43	1.22	1.65
Rafter ties	.140	Ea.	.008	.21	.45	.66
TOTAL		S.F.	.151	4.45	8.73	13.18
2″ X 8″ RAFTERS, 16″ O.C.						
Roof rafters, 2″ x 8″, 16″ O.C.	1.210	L.F.	.036	1.19	2.08	3.27
Rafter plates, 2″ x 8″, double top, single bottom	.364	L.F.	.011	.36	.63	.99
Ceiling joists, 2″ x 6″, 16″ O.C.	.920	L.F.	.012	.65	.68	1.33
Hip rafter, 2″ x 8″	.070	L.F.	.002	.07	.14	.21
Jack rafter, 2″ x 8″, 16″ O.C.	1.000	L.F.	.048	.98	2.77	3.75
Ridge board, 2″ x 8″	.018	L.F.	.001	.02	.04	.06
Sheathing, exterior grade plywood, 1/2″ thick	2.210	S.F.	.025	1.50	1.46	2.96
Furring strips, 1″ x 3″, 16″ O.C.	.920	L.F.	.021	.43	1.22	1.65
Rafter ties	.140	Ea.	.008	.21	.45	.66
TOTAL		S.F.	.164	5.41	9.47	14.88

The cost of this system is based on the square foot of plan area.

Description	QUAN.	UNIT	LABOR HOURS	COST PER S.F.		
				MAT.	INST.	TOTAL

Mansard Roof Framing Price Sheet	QUAN.	UNIT	LABOR HOURS	COST PER S.F.		
				MAT.	INST.	TOTAL
Roof rafters, #2 or better, 2" x 6", 16" O.C.	1.210	L.F.	.033	.86	1.91	2.77
24" O.C.	.970	L.F.	.026	.69	1.53	2.22
2" x 8", 16" O.C.	1.210	L.F.	.036	1.19	2.08	3.27
24" O.C.	.970	L.F.	.029	.95	1.67	2.62
2" x 10", 16" O.C.	1.210	L.F.	.046	1.88	2.65	4.53
24" O.C.	.970	L.F.	.037	1.50	2.12	3.62
Rafter plates, #2 or better double top single bottom, 2" x 6"	.364	L.F.	.010	.26	.58	.84
2" x 8"	.364	L.F.	.011	.36	.63	.99
2" x 10"	.364	L.F.	.014	.56	.80	1.36
Ceiling joist, #2 or better, 2" x 4", 16" O.C.	.920	L.F.	.012	.42	.68	1.10
24" O.C.	.740	L.F.	.009	.34	.55	.89
2" x 6", 16" O.C.	.920	L.F.	.012	.65	.68	1.33
24" O.C.	.740	L.F.	.009	.53	.55	1.08
2" x 8", 16" O.C.	.920	L.F.	.013	.90	.78	1.68
24" O.C.	.740	L.F.	.011	.73	.63	1.36
Hip rafter, #2 or better, 2" x 6"	.070	L.F.	.002	.05	.13	.18
2" x 8"	.070	L.F.	.002	.07	.14	.21
2" x 10"	.070	L.F.	.003	.11	.17	.28
Jack rafter, #2 or better, 2" x 6", 16" O.C.	1.000	L.F.	.039	.71	2.27	2.98
24" O.C.	.800	L.F.	.031	.57	1.82	2.39
2" x 8", 16" O.C.	1.000	L.F.	.048	.98	2.77	3.75
24" O.C.	.800	L.F.	.038	.78	2.22	3
Ridge board, #2 or better, 1" x 6"	.018	L.F.	.001	.02	.03	.05
1" x 8"	.018	L.F.	.001	.03	.03	.06
1" x 10"	.018	L.F.	.001	.03	.03	.06
2" x 6"	.018	L.F.	.001	.01	.03	.04
2" x 8"	.018	L.F.	.001	.02	.04	.06
2" x 10"	.018	L.F.	.001	.03	.04	.07
Sheathing, plywood exterior grade CDX, 3/8" thick	2.210	S.F.	.023	1.46	1.35	2.81
1/2" thick	2.210	S.F.	.025	1.50	1.46	2.96
5/8" thick	2.210	S.F.	.027	1.86	1.59	3.45
3/4" thick	2.210	S.F.	.029	2.25	1.70	3.95
Boards, 1" x 6", laid regular	2.210	S.F.	.049	4.11	2.83	6.94
Laid diagonal	2.210	S.F.	.054	4.11	3.16	7.27
1" x 8", laid regular	2.210	S.F.	.040	5.05	2.34	7.39
Laid diagonal	2.210	S.F.	.049	5.05	2.83	7.88
Furring, 1" x 3", 12" O.C.	1.150	L.F.	.026	.54	1.53	2.07
24" O.C.	.740	L.F.	.017	.35	.98	1.33

For customer support on your Residential Costs with RSMeans data, call 800.448.8182.

147

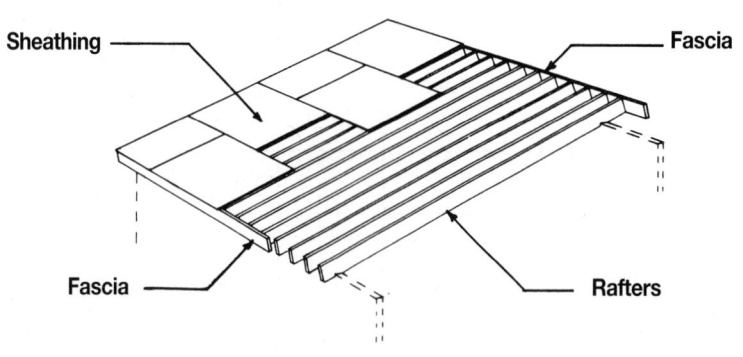

Sheathing · Fascia · Fascia · Rafters

System Description	QUAN.	UNIT	LABOR HOURS	COST PER S.F.		
				MAT.	INST.	TOTAL
2" X 6", 16" O.C., 4/12 PITCH						
Rafters, 2" x 6", 16" O.C., 4/12 pitch	1.170	L.F.	.019	.83	1.09	1.92
Fascia, 2" x 6"	.100	L.F.	.006	.08	.33	.41
Bridging, 1" x 3", 6' O.C.	.080	Pr.	.005	.06	.29	.35
Sheathing, exterior grade plywood, 1/2" thick	1.230	S.F.	.014	.84	.81	1.65
Rafter ties	.053	Ea.	.003	.08	.17	.25
TOTAL		S.F.	.047	1.89	2.69	4.58
2" X 6", 24" O.C., 4/12 PITCH						
Rafters, 2" x 6", 24" O.C., 4/12 pitch	.940	L.F.	.015	.67	.87	1.54
Fascia, 2" x 6"	.100	L.F.	.006	.08	.33	.41
Bridging, 1" x 3", 6' O.C.	.060	Pr.	.004	.05	.21	.26
Sheathing, exterior grade plywood, 1/2" thick	1.230	S.F.	.014	.84	.81	1.65
Rafter ties	.035	Ea.	.002	.05	.11	.16
TOTAL		S.F.	.041	1.69	2.33	4.02
2" X 8", 16" O.C., 4/12 PITCH						
Rafters, 2" x 8", 16" O.C., 4/12 pitch	1.170	L.F.	.020	1.15	1.15	2.30
Fascia, 2" x 8"	.100	L.F.	.007	.10	.41	.51
Bridging, 1" x 3", 6' O.C.	.080	Pr.	.005	.06	.29	.35
Sheathing, exterior grade plywood, 1/2" thick	1.230	S.F.	.014	.84	.81	1.65
Rafter ties	.053	Ea.	.003	.08	.17	.25
TOTAL		S.F.	.049	2.23	2.83	5.06
2" X 8", 24" O.C., 4/12 PITCH						
Rafters, 2" x 8", 24" O.C., 4/12 pitch	.940	L.F.	.016	.92	.92	1.84
Fascia, 2" x 8"	.100	L.F.	.007	.10	.41	.51
Bridging, 1" x 3", 6' O.C.	.060	Pr.	.004	.05	.21	.26
Sheathing, exterior grade plywood, 1/2" thick	1.230	S.F.	.014	.84	.81	1.65
Rafter ties	.035	Ea.	.002	.05	.11	.16
TOTAL		S.F.	.043	1.96	2.46	4.42

The cost of this system is based on the square foot of plan area.
A 1' overhang is assumed. No ceiling joists or furring are included.

Description	QUAN.	UNIT	LABOR HOURS	COST PER S.F.		
				MAT.	INST.	TOTAL

Shed/Flat Roof Framing Price Sheet	QUAN.	UNIT	LABOR HOURS	COST PER S.F.		
				MAT.	INST.	TOTAL
Rafters, #2 or better, 16" O.C., 2" x 4", 0 - 4/12 pitch	1.170	L.F.	.014	.62	.81	1.43
5/12 - 8/12 pitch	1.330	L.F.	.020	.71	1.16	1.87
2" x 6", 0 - 4/12 pitch	1.170	L.F.	.019	.83	1.09	1.92
5/12 - 8/12 pitch	1.330	L.F.	.027	.94	1.54	2.48
2" x 8", 0 - 4/12 pitch	1.170	L.F.	.020	1.15	1.15	2.30
5/12 - 8/12 pitch	1.330	L.F.	.028	1.30	1.65	2.95
2" x 10", 0 - 4/12 pitch	1.170	L.F.	.030	1.81	1.73	3.54
5/12 - 8/12 pitch	1.330	L.F.	.043	2.06	2.50	4.56
24" O.C., 2" x 4", 0 - 4/12 pitch	.940	L.F.	.011	.50	.66	1.16
5/12 - 8/12 pitch	1.060	L.F.	.021	.75	1.23	1.98
2" x 6", 0 - 4/12 pitch	.940	L.F.	.015	.67	.87	1.54
5/12 - 8/12 pitch	1.060	L.F.	.021	.75	1.23	1.98
2" x 8", 0 - 4/12 pitch	.940	L.F.	.016	.92	.92	1.84
5/12 - 8/12 pitch	1.060	L.F.	.023	1.04	1.31	2.35
2" x 10", 0 - 4/12 pitch	.940	L.F.	.024	1.46	1.39	2.85
5/12 - 8/12 pitch	1.060	L.F.	.034	1.64	1.99	3.63
Fascia, #2 or better,, 1" x 4"	.100	L.F.	.003	.04	.17	.21
1" x 6"	.100	L.F.	.004	.05	.23	.28
1" x 8"	.100	L.F.	.005	.06	.26	.32
1" x 10"	.100	L.F.	.005	.07	.30	.37
2" x 4"	.100	L.F.	.005	.07	.28	.35
2" x 6"	.100	L.F.	.006	.08	.33	.41
2" x 8"	.100	L.F.	.007	.10	.41	.51
2" x 10"	.100	L.F.	.009	.16	.52	.68
Bridging, wood 6' O.C., 1" x 3", rafters, 16" O.C.	.080	Pr.	.005	.06	.29	.35
24" O.C.	.060	Pr.	.004	.05	.21	.26
Metal, galvanized, rafters, 16" O.C.	.080	Pr.	.005	.15	.29	.44
24" O.C.	.060	Pr.	.003	.17	.20	.37
Compression type, rafters, 16" O.C.	.080	Pr.	.003	.11	.19	.30
24" O.C.	.060	Pr.	.002	.08	.14	.22
Sheathing, plywood, exterior grade, 3/8" thick, flat 0 - 4/12 pitch	1.230	S.F.	.013	.81	.75	1.56
5/12 - 8/12 pitch	1.330	S.F.	.014	.88	.81	1.69
1/2" thick, flat 0 - 4/12 pitch	1.230	S.F.	.014	.84	.81	1.65
5/12 - 8/12 pitch	1.330	S.F.	.015	.90	.88	1.78
5/8" thick, flat 0 - 4/12 pitch	1.230	S.F.	.015	1.03	.89	1.92
5/12 - 8/12 pitch	1.330	S.F.	.016	1.12	.96	2.08
3/4" thick, flat 0 - 4/12 pitch	1.230	S.F.	.016	1.25	.95	2.20
5/12 - 8/12 pitch	1.330	S.F.	.018	1.36	1.02	2.38
Boards, 1" x 6", laid regular, flat 0 - 4/12 pitch	1.230	S.F.	.027	2.29	1.57	3.86
5/12 - 8/12 pitch	1.330	S.F.	.041	2.47	2.38	4.85
Laid diagonal, flat 0 - 4/12 pitch	1.230	S.F.	.030	2.29	1.76	4.05
5/12 - 8/12 pitch	1.330	S.F.	.044	2.47	2.58	5.05
1" x 8", laid regular, flat 0 - 4/12 pitch	1.230	S.F.	.022	2.80	1.30	4.10
5/12 - 8/12 pitch	1.330	S.F.	.034	3.03	1.94	4.97
Laid diagonal, flat 0 - 4/12 pitch	1.230	S.F.	.027	2.80	1.57	4.37
5/12 - 8/12 pitch	1.330	S.F.	.044	2.47	2.58	5.05

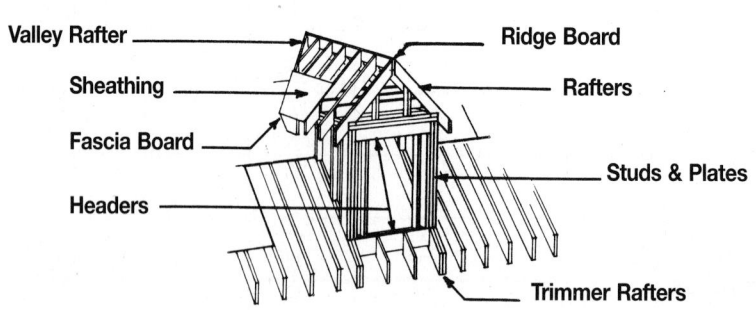

Valley Rafter — Ridge Board
Sheathing — Rafters
Fascia Board —
Headers — Studs & Plates
Trimmer Rafters

System Description	QUAN.	UNIT	LABOR HOURS	COST PER S.F.		
				MAT.	INST.	TOTAL
2" X 6", 16" O.C.						
Dormer rafter, 2" x 6", 16" O.C.	1.330	L.F.	.036	.94	2.10	3.04
Ridge board, 2" x 6"	.280	L.F.	.009	.20	.52	.72
Trimmer rafters, 2" x 6"	.880	L.F.	.014	.62	.82	1.44
Wall studs & plates, 2" x 4", 16" O.C.	3.160	L.F.	.056	1.45	3.25	4.70
Fascia, 2" x 6"	.220	L.F.	.012	.17	.70	.87
Valley rafter, 2" x 6", 16" O.C.	.280	L.F.	.009	.20	.51	.71
Cripple rafter, 2" x 6", 16" O.C.	.560	L.F.	.022	.40	1.27	1.67
Headers, 2" x 6", doubled	.670	L.F.	.030	.48	1.73	2.21
Ceiling joist, 2" x 4", 16" O.C.	1.000	L.F.	.013	.46	.74	1.20
Sheathing, exterior grade plywood, 1/2" thick	3.610	S.F.	.041	2.45	2.38	4.83
TOTAL		S.F.	.242	7.37	14.02	21.39
2" X 8", 16" O.C.						
Dormer rafter, 2" x 8", 16" O.C.	1.330	L.F.	.039	1.30	2.29	3.59
Ridge board, 2" x 8"	.280	L.F.	.010	.27	.58	.85
Trimmer rafter, 2" x 8"	.880	L.F.	.015	.86	.86	1.72
Wall studs & plates, 2" x 4", 16" O.C.	3.160	L.F.	.056	1.45	3.25	4.70
Fascia, 2" x 8"	.220	L.F.	.016	.22	.91	1.13
Valley rafter, 2" x 8", 16" O.C.	.280	L.F.	.010	.27	.55	.82
Cripple rafter, 2" x 8", 16" O.C.	.560	L.F.	.027	.55	1.55	2.10
Headers, 2" x 8", doubled	.670	L.F.	.032	.66	1.83	2.49
Ceiling joist, 2" x 4", 16" O.C.	1.000	L.F.	.013	.46	.74	1.20
Sheathing,, exterior grade plywood, 1/2" thick	3.610	S.F.	.041	2.45	2.38	4.83
TOTAL		S.F.	.259	8.49	14.94	23.43

The cost in this system is based on the square foot of plan area.
The measurement being the plan area of the dormer only.

Description	QUAN.	UNIT	LABOR HOURS	COST PER S.F.		
				MAT.	INST.	TOTAL

Gable Dormer Framing Price Sheet

Gable Dormer Framing Price Sheet	QUAN.	UNIT	LABOR HOURS	COST PER S.F. MAT.	COST PER S.F. INST.	COST PER S.F. TOTAL
Dormer rafters, #2 or better, 2" x 4", 16" O.C.	1.330	L.F.	.029	.76	1.68	2.44
24" O.C.	1.060	L.F.	.023	.60	1.34	1.94
2" x 6", 16" O.C.	1.330	L.F.	.036	.94	2.10	3.04
24" O.C.	1.060	L.F.	.029	.75	1.67	2.42
2" x 8", 16" O.C.	1.330	L.F.	.039	1.30	2.29	3.59
24" O.C.	1.060	L.F.	.031	1.04	1.82	2.86
Ridge board, #2 or better, 1" x 4"	.280	L.F.	.006	.19	.35	.54
1" x 6"	.280	L.F.	.007	.24	.43	.67
1" x 8"	.280	L.F.	.008	.40	.47	.87
2" x 4"	.280	L.F.	.007	.16	.42	.58
2" x 6"	.280	L.F.	.009	.20	.52	.72
2" x 8"	.280	L.F.	.010	.27	.58	.85
Trimmer rafters, #2 or better, 2" x 4"	.880	L.F.	.011	.50	.65	1.15
2" x 6"	.880	L.F.	.014	.62	.82	1.44
2" x 8"	.880	L.F.	.015	.86	.86	1.72
2" x 10"	.880	L.F.	.022	1.36	1.30	2.66
Wall studs & plates, #2 or better, 2" x 4" studs, 16" O.C.	3.160	L.F.	.056	1.45	3.25	4.70
24" O.C.	2.800	L.F.	.050	1.29	2.88	4.17
2" x 6" studs, 16" O.C.	3.160	L.F.	.063	2.24	3.67	5.91
24" O.C.	2.800	L.F.	.056	1.99	3.25	5.24
Fascia, #2 or better, 1" x 4"	.220	L.F.	.006	.09	.37	.46
1" x 6"	.220	L.F.	.008	.11	.45	.56
1" x 8"	.220	L.F.	.009	.13	.53	.66
2" x 4"	.220	L.F.	.011	.15	.62	.77
2" x 6"	.220	L.F.	.014	.19	.78	.97
2" x 8"	.220	L.F.	.016	.22	.91	1.13
Valley rafter, #2 or better, 2" x 4"	.280	L.F.	.007	.16	.41	.57
2" x 6"	.280	L.F.	.009	.20	.51	.71
2" x 8"	.280	L.F.	.010	.27	.55	.82
2" x 10"	.280	L.F.	.012	.43	.69	1.12
Cripple rafter, #2 or better, 2" x 4", 16" O.C.	.560	L.F.	.018	.32	1.02	1.34
24" O.C.	.450	L.F.	.014	.26	.82	1.08
2" x 6", 16" O.C.	.560	L.F.	.022	.40	1.27	1.67
24" O.C.	.450	L.F.	.018	.32	1.02	1.34
2" x 8", 16" O.C.	.560	L.F.	.027	.55	1.55	2.10
24" O.C.	.450	L.F.	.021	.44	1.25	1.69
Headers, #2 or better double header, 2" x 4"	.670	L.F.	.024	.38	1.39	1.77
2" x 6"	.670	L.F.	.030	.48	1.73	2.21
2" x 8"	.670	L.F.	.032	.66	1.83	2.49
2" x 10"	.670	L.F.	.034	1.04	1.95	2.99
Ceiling joist, #2 or better, 2" x 4", 16" O.C.	1.000	L.F.	.013	.46	.74	1.20
24" O.C.	.800	L.F.	.010	.37	.59	.96
2" x 6", 16" O.C.	1.000	L.F.	.013	.71	.74	1.45
24" O.C.	.800	L.F.	.010	.57	.59	1.16
Sheathing, plywood exterior grade, 3/8" thick	3.610	S.F.	.038	2.38	2.20	4.58
1/2" thick	3.610	S.F.	.041	2.45	2.38	4.83
5/8" thick	3.610	S.F.	.044	3.03	2.60	5.63
3/4" thick	3.610	S.F.	.048	3.68	2.78	6.46
Boards, 1" x 6", laid regular	3.610	S.F.	.089	6.70	5.15	11.85
Laid diagonal	3.610	S.F.	.099	6.70	5.75	12.45
1" x 8", laid regular	3.610	S.F.	.076	8.25	4.40	12.65
Laid diagonal	3.610	S.F.	.089	8.25	5.15	13.40

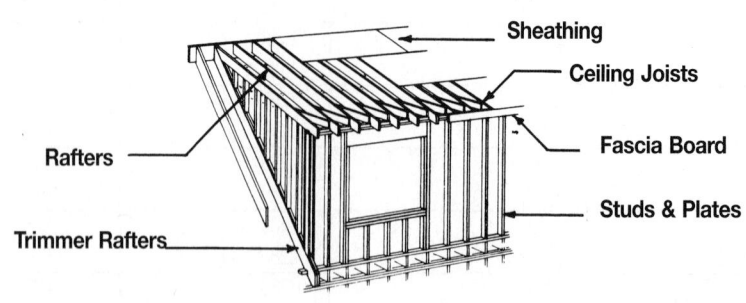

Labels: Sheathing, Ceiling Joists, Fascia Board, Studs & Plates, Rafters, Trimmer Rafters

System Description	QUAN.	UNIT	LABOR HOURS	COST PER S.F.		
				MAT.	INST.	TOTAL
2" X 6" RAFTERS, 16" O.C.						
Dormer rafter, 2" x 6", 16" O.C.	1.080	L.F.	.029	.77	1.71	2.48
Trimmer rafter, 2" x 6"	.400	L.F.	.006	.28	.37	.65
Studs & plates, 2" x 4", 16" O.C.	2.750	L.F.	.049	1.27	2.83	4.10
Fascia, 2" x 6"	.250	L.F.	.014	.19	.78	.97
Ceiling joist, 2" x 4", 16" O.C.	1.000	L.F.	.013	.46	.74	1.20
Sheathing, exterior grade plywood, CDX, 1/2" thick	2.940	S.F.	.034	2	1.94	3.94
TOTAL		S.F.	.145	4.97	8.37	13.34
2" X 8" RAFTERS, 16" O.C.						
Dormer rafter, 2" x 8", 16" O.C.	1.080	L.F.	.032	1.06	1.86	2.92
Trimmer rafter, 2" x 8"	.400	L.F.	.007	.39	.39	.78
Studs & plates, 2" x 4", 16" O.C.	2.750	L.F.	.049	1.27	2.83	4.10
Fascia, 2" x 8"	.250	L.F.	.018	.25	1.03	1.28
Ceiling joist, 2" x 6", 16" O.C.	1.000	L.F.	.013	.71	.74	1.45
Sheathing, exterior grade plywood, CDX, 1/2" thick	2.940	S.F.	.034	2	1.94	3.94
TOTAL		S.F.	.153	5.68	8.79	14.47
2" X 10" RAFTERS, 16" O.C.						
Dormer rafter, 2" x 10", 16" O.C.	1.080	L.F.	.041	1.67	2.37	4.04
Trimmer rafter, 2" x 10"	.400	L.F.	.010	.62	.59	1.21
Studs & plates, 2" x 4", 16" O.C.	2.750	L.F.	.049	1.27	2.83	4.10
Fascia, 2" x 10"	.250	L.F.	.022	.39	1.29	1.68
Ceiling joist, 2" x 6", 16" O.C.	1.000	L.F.	.013	.71	.74	1.45
Sheathing, exterior grade plywood, CDX, 1/2" thick	2.940	S.F.	.034	2	1.94	3.94
TOTAL		S.F.	.169	6.66	9.76	16.42

The cost in this system is based on the square foot of plan area.
The measurement is the plan area of the dormer only.

Description	QUAN.	UNIT	LABOR HOURS	COST PER S.F.		
				MAT.	INST.	TOTAL

Shed Dormer Framing Price Sheet

	QUAN.	UNIT	LABOR HOURS	COST PER S.F. MAT.	COST PER S.F. INST.	COST PER S.F. TOTAL
Dormer rafters, #2 or better, 2" x 4", 16" O.C.	1.080	L.F.	.023	.61	1.37	1.98
24" O.C.	.860	L.F.	.019	.49	1.09	1.58
2" x 6", 16" O.C.	1.080	L.F.	.029	.77	1.71	2.48
24" O.C.	.860	L.F.	.023	.61	1.36	1.97
2" x 8", 16" O.C.	1.080	L.F.	.032	1.06	1.86	2.92
24" O.C.	.860	L.F.	.025	.84	1.48	2.32
2" x 10", 16" O.C.	1.080	L.F.	.041	1.67	2.37	4.04
24" O.C.	.860	L.F.	.032	1.33	1.88	3.21
Trimmer rafter, #2 or better, 2" x 4"	.400	L.F.	.005	.23	.30	.53
2" x 6"	.400	L.F.	.006	.28	.37	.65
2" x 8"	.400	L.F.	.007	.39	.39	.78
2" x 10"	.400	L.F.	.010	.62	.59	1.21
Studs & plates, #2 or better, 2" x 4", 16" O.C.	2.750	L.F.	.049	1.27	2.83	4.10
24" O.C.	2.200	L.F.	.039	1.01	2.27	3.28
2" x 6", 16" O.C.	2.750	L.F.	.055	1.95	3.19	5.14
24" O.C.	2.200	L.F.	.044	1.56	2.55	4.11
Fascia, #2 or better, 1" x 4"	.250	L.F.	.006	.09	.37	.46
1" x 6"	.250	L.F.	.008	.11	.45	.56
1" x 8"	.250	L.F.	.009	.13	.53	.66
2" x 4"	.250	L.F.	.011	.15	.62	.77
2" x 6"	.250	L.F.	.014	.19	.78	.97
2" x 8"	.250	L.F.	.018	.25	1.03	1.28
Ceiling joist, #2 or better, 2" x 4", 16" O.C.	1.000	L.F.	.013	.46	.74	1.20
24" O.C.	.800	L.F.	.010	.37	.59	.96
2" x 6", 16" O.C.	1.000	L.F.	.013	.71	.74	1.45
24" O.C.	.800	L.F.	.010	.57	.59	1.16
2" x 8", 16" O.C.	1.000	L.F.	.015	.98	.85	1.83
24" O.C.	.800	L.F.	.012	.78	.68	1.46
Sheathing, plywood exterior grade, 3/8" thick	2.940	S.F.	.031	1.94	1.79	3.73
1/2" thick	2.940	S.F.	.034	2	1.94	3.94
5/8" thick	2.940	S.F.	.036	2.47	2.12	4.59
3/4" thick	2.940	S.F.	.039	3	2.26	5.26
Boards, 1" x 6", laid regular	2.940	S.F.	.072	5.45	4.20	9.65
Laid diagonal	2.940	S.F.	.080	5.45	4.67	10.12
1" x 8", laid regular	2.940	S.F.	.062	6.70	3.59	10.29
Laid diagonal	2.940	S.F.	.072	6.70	4.20	10.90

Window Openings

	QUAN.	UNIT	LABOR HOURS	COST EACH MAT.	COST EACH INST.	COST EACH TOTAL
The following are to be added to the total cost of the dormers for window openings. Do not subtract window area from the stud wall quantities.						
Headers, 2" x 6" doubled, 2' long	4.000	L.F.	.178	2.84	10.30	13.14
3' long	6.000	L.F.	.267	4.26	15.50	19.76
4' long	8.000	L.F.	.356	5.70	20.50	26.20
5' long	10.000	L.F.	.444	7.10	26	33.10
2" x 8" doubled, 4' long	8.000	L.F.	.376	7.85	22	29.85
5' long	10.000	L.F.	.471	9.80	27.50	37.30
6' long	12.000	L.F.	.565	11.75	33	44.75
8' long	16.000	L.F.	.753	15.70	43.50	59.20
2" x 10" doubled, 4' long	8.000	L.F.	.400	12.40	23.50	35.90
6' long	12.000	L.F.	.600	18.60	35	53.60
8' long	16.000	L.F.	.800	25	46.50	71.50
10' long	20.000	L.F.	1.000	31	58	89

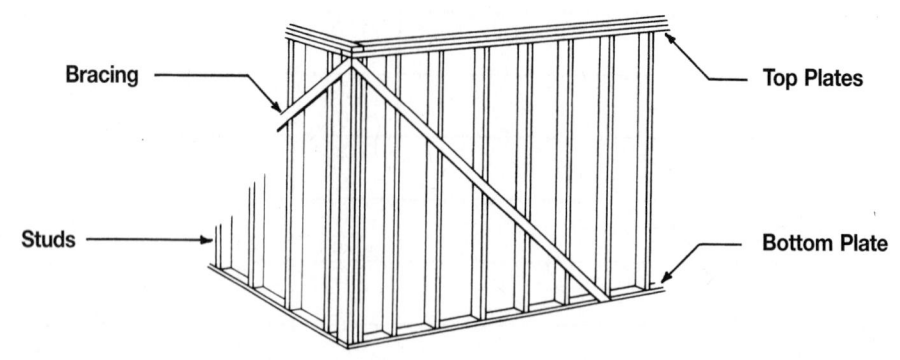

Bracing — Top Plates

Studs — Bottom Plate

System Description		QUAN.	UNIT	LABOR HOURS	COST PER S.F.		
					MAT.	INST.	TOTAL
2″ X 4″, 16″ O.C.							
2″ x 4″ studs, #2 or better, 16″ O.C.		1.000	L.F.	.015	.45	.85	1.30
Plates, double top, single bottom		.375	L.F.	.005	.17	.32	.49
Cross bracing, let-in, 1″ x 6″		.080	L.F.	.004	.07	.25	.32
	TOTAL		S.F.	.024	.69	1.42	2.11
2″ X 4″, 24″ O.C.							
2″ x 4″ studs, #2 or better, 24″ O.C.		.800	L.F.	.012	.36	.68	1.04
Plates, double top, single bottom		.375	L.F.	.005	.17	.32	.49
Cross bracing, let-in, 1″ x 6″		.080	L.F.	.003	.07	.16	.23
	TOTAL		S.F.	.020	.60	1.16	1.76
2″ X 6″, 16″ O.C.							
2″ x 6″ studs, #2 or better, 16″ O.C.		1.000	L.F.	.016	.71	.93	1.64
Plates, double top, single bottom		.375	L.F.	.006	.27	.35	.62
Cross bracing, let-in, 1″ x 6″		.080	L.F.	.004	.07	.25	.32
	TOTAL		S.F.	.026	1.05	1.53	2.58
2″ X 6″, 24″ O.C.							
2″ x 6″ studs, #2 or better, 24″ O.C.		.800	L.F.	.013	.57	.74	1.31
Plates, double top, single bottom		.375	L.F.	.006	.27	.35	.62
Cross bracing, let-in, 1″ x 6″		.080	L.F.	.003	.07	.16	.23
	TOTAL		S.F.	.022	.91	1.25	2.16

The costs in this system are based on a square foot of wall area. Do not subtract for door or window openings.

Description	QUAN.	UNIT	LABOR HOURS	COST PER S.F.		
				MAT.	INST.	TOTAL

Partition Framing Price Sheet

Partition Framing Price Sheet	QUAN.	UNIT	LABOR HOURS	COST PER S.F. MAT.	COST PER S.F. INST.	COST PER S.F. TOTAL
Wood studs, #2 or better, 2" x 4", 12" O.C.	1.250	L.F.	.018	.56	1.06	1.62
16" O.C.	1.000	L.F.	.015	.45	.85	1.30
24" O.C.	.800	L.F.	.012	.36	.68	1.04
32" O.C.	.650	L.F.	.009	.29	.55	.84
2" x 6", 12" O.C.	1.250	L.F.	.020	.89	1.16	2.05
16" O.C.	1.000	L.F.	.016	.71	.93	1.64
24" O.C.	.800	L.F.	.013	.57	.74	1.31
32" O.C.	.650	L.F.	.010	.46	.60	1.06
Plates, #2 or better double top single bottom, 2" x 4"	.375	L.F.	.005	.17	.32	.49
2" x 6"	.375	L.F.	.006	.27	.35	.62
2" x 8"	.375	L.F.	.005	.37	.32	.69
Cross bracing, let-in, 1" x 6" boards studs, 12" O.C.	.080	L.F.	.005	.09	.31	.40
16" O.C.	.080	L.F.	.004	.07	.25	.32
24" O.C.	.080	L.F.	.003	.07	.16	.23
32" O.C.	.080	L.F.	.002	.06	.13	.19
Let-in steel (T shaped) studs, 12" O.C.	.080	L.F.	.001	.09	.08	.17
16" O.C.	.080	L.F.	.001	.07	.06	.13
24" O.C.	.080	L.F.	.001	.07	.06	.13
32" O.C.	.080	L.F.	.001	.06	.05	.11
Steel straps studs, 12" O.C.	.080	L.F.	.001	.10	.07	.17
16" O.C.	.080	L.F.	.001	.09	.06	.15
24" O.C.	.080	L.F.	.001	.09	.06	.15
32" O.C.	.080	L.F.	.001	.09	.06	.15
Metal studs, load bearing 24" O.C., 20 ga. galv., 2-1/2" wide	1.000	S.F.	.015	.71	.87	1.58
3-5/8" wide	1.000	S.F.	.015	.84	.89	1.73
4" wide	1.000	S.F.	.016	.87	.91	1.78
6" wide	1.000	S.F.	.016	1.12	.92	2.04
16 ga., 2-1/2" wide	1.000	S.F.	.017	.82	.99	1.81
3-5/8" wide	1.000	S.F.	.017	.98	1.01	1.99
4" wide	1.000	S.F.	.018	1.03	1.04	2.07
6" wide	1.000	S.F.	.018	1.30	1.06	2.36
Non-load bearing 24" O.C., 25 ga. galv., 1-5/8" wide	1.000	S.F.	.011	.21	.61	.82
2-1/2" wide	1.000	S.F.	.011	.27	.62	.89
3-5/8" wide	1.000	S.F.	.011	.31	.63	.94
4" wide	1.000	S.F.	.011	.34	.63	.97
6" wide	1.000	S.F.	.011	.41	.64	1.05
20 ga., 2-1/2" wide	1.000	S.F.	.013	.33	.77	1.10
3-5/8" wide	1.000	S.F.	.014	.38	.79	1.17
4" wide	1.000	S.F.	.014	.46	.79	1.25
6" wide	1.000	S.F.	.014	.55	.80	1.35

Window & Door Openings

Window & Door Openings	QUAN.	UNIT	LABOR HOURS	COST EACH MAT.	COST EACH INST.	COST EACH TOTAL
The following costs are to be added to the total costs of the walls.						
Do not subtract openings from total wall area.						
Headers, 2" x 6" double, 2' long	4.000	L.F.	.178	2.84	10.30	13.14
3' long	6.000	L.F.	.267	4.26	15.50	19.76
4' long	8.000	L.F.	.356	5.70	20.50	26.20
5' long	10.000	L.F.	.444	7.10	26	33.10
2" x 8" double, 4' long	8.000	L.F.	.376	7.85	22	29.85
5' long	10.000	L.F.	.471	9.80	27.50	37.30
6' long	12.000	L.F.	.565	11.75	33	44.75
8' long	16.000	L.F.	.753	15.70	43.50	59.20
2" x 10" double, 4' long	8.000	L.F.	.400	12.40	23.50	35.90
6' long	12.000	L.F.	.600	18.60	35	53.60
8' long	16.000	L.F.	.800	25	46.50	71.50
10' long	20.000	L.F.	1.000	31	58	89
2" x 12" double, 8' long	16.000	L.F.	.853	31.50	49.50	81
12' long	24.000	L.F.	1.280	47.50	74.50	122

For customer support on your Residential Costs with RSMeans data, call 800.448.8182.

155

Did you know?

RSMeans data is available through our online application with 24/7 access:

- Search for unit prices by keyword
- Leverage the most up-to-date data
- Build and export estimates

Try it free for 30 days!
www.rsmeans.com/2018freetrial

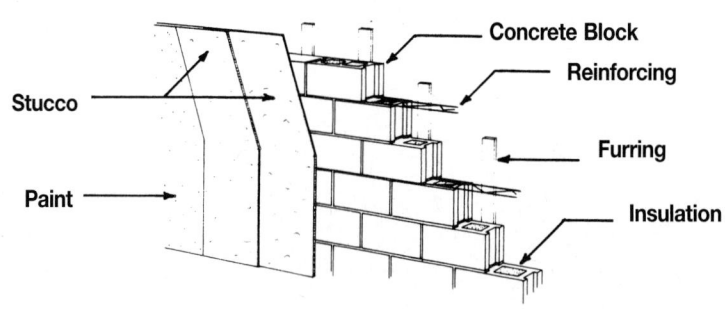

Stucco

Paint

Concrete Block

Reinforcing

Furring

Insulation

System Description	QUAN.	UNIT	LABOR HOURS	COST PER S.F.		
				MAT.	INST.	TOTAL
6″ THICK CONCRETE BLOCK WALL						
6″ thick concrete block, 6″ x 8″ x 16″	1.000	S.F.	.100	2.64	5.40	8.04
Masonry reinforcing, truss strips every other course	.625	L.F.	.002	.15	.09	.24
Furring, 1″ x 3″, 16″ O.C.	1.000	L.F.	.016	.51	.94	1.45
Masonry insulation, poured perlite	1.000	S.F.	.013	1.93	.77	2.70
Stucco, 2 coats	1.000	S.F.	.069	.73	3.86	4.59
Masonry paint, 2 coats	1.000	S.F.	.016	.25	.77	1.02
TOTAL		S.F.	.216	6.21	11.83	18.04
8″ THICK CONCRETE BLOCK WALL						
8″ thick concrete block, 8″ x 8″ x 16″	1.000	S.F.	.107	2.82	5.75	8.57
Masonry reinforcing, truss strips every other course	.625	L.F.	.002	.15	.09	.24
Furring, 1″ x 3″, 16″ O.C.	1.000	L.F.	.016	.51	.94	1.45
Masonry insulation, poured perlite	1.000	S.F.	.018	2.55	1.02	3.57
Stucco, 2 coats	1.000	S.F.	.069	.73	3.86	4.59
Masonry paint, 2 coats	1.000	S.F.	.016	.25	.77	1.02
TOTAL		S.F.	.228	7.01	12.43	19.44
12″ THICK CONCRETE BLOCK WALL						
12″ thick concrete block, 12″ x 8″ x 16″	1.000	S.F.	.141	4.69	7.45	12.14
Masonry reinforcing, truss strips every other course	.625	L.F.	.003	.18	.14	.32
Furring, 1″ x 3″, 16″ O.C.	1.000	L.F.	.016	.51	.94	1.45
Masonry insulation, poured perlite	1.000	S.F.	.026	3.77	1.51	5.28
Stucco, 2 coats	1.000	S.F.	.069	.73	3.86	4.59
Masonry paint, 2 coats	1.000	S.F.	.016	.25	.77	1.02
TOTAL		S.F.	.271	10.13	14.67	24.80

Costs for this system are based on a square foot of wall area. Do not subtract for window openings.

Description	QUAN.	UNIT	LABOR HOURS	COST PER S.F.		
				MAT.	INST.	TOTAL

Masonry Block Price Sheet	QUAN.	UNIT	LABOR HOURS	COST PER S.F.		
				MAT.	INST.	TOTAL
Block concrete, 8" x 16" regular, 4" thick	1.000	S.F.	.093	2.01	5	7.01
6" thick	1.000	S.F.	.100	2.64	5.40	8.04
8" thick	1.000	S.F.	.107	2.82	5.75	8.57
10" thick	1.000	S.F.	.111	3.36	6	9.36
12" thick	1.000	S.F.	.141	4.69	7.45	12.14
Solid block, 4" thick	1.000	S.F.	.096	2.29	5.20	7.49
6" thick	1.000	S.F.	.104	2.72	5.60	8.32
8" thick	1.000	S.F.	.111	4.05	6	10.05
10" thick	1.000	S.F.	.133	5.55	7	12.55
12" thick	1.000	S.F.	.148	6.15	7.80	13.95
Lightweight, 4" thick	1.000	S.F.	.093	2.01	5	7.01
6" thick	1.000	S.F.	.100	2.64	5.40	8.04
8" thick	1.000	S.F.	.107	2.82	5.75	8.57
10" thick	1.000	S.F.	.111	3.36	6	9.36
12" thick	1.000	S.F.	.141	4.69	7.45	12.14
Split rib profile, 4" thick	1.000	S.F.	.116	4.47	6.25	10.72
6" thick	1.000	S.F.	.123	5.05	6.65	11.70
8" thick	1.000	S.F.	.131	5.70	7.20	12.90
10" thick	1.000	S.F.	.157	6	8.35	14.35
12" thick	1.000	S.F.	.175	6.65	9.25	15.90
Masonry reinforcing, wire truss strips, every course, 8" block	1.375	L.F.	.004	.33	.21	.54
12" block	1.375	L.F.	.006	.39	.32	.71
Every other course, 8" block	.625	L.F.	.002	.15	.09	.24
12" block	.625	L.F.	.003	.18	.14	.32
Furring, wood, 1" x 3", 12" O.C.	1.250	L.F.	.020	.64	1.18	1.82
16" O.C.	1.000	L.F.	.016	.51	.94	1.45
24" O.C.	.800	L.F.	.013	.41	.75	1.16
32" O.C.	.640	L.F.	.010	.33	.60	.93
Steel, 3/4" channels, 12" O.C.	1.250	L.F.	.034	.44	1.89	2.33
16" O.C.	1.000	L.F.	.030	.39	1.68	2.07
24" O.C.	.800	L.F.	.023	.26	1.27	1.53
32" O.C.	.640	L.F.	.018	.21	1.02	1.23
Masonry insulation, vermiculite or perlite poured 4" thick	1.000	S.F.	.009	1.25	.50	1.75
6" thick	1.000	S.F.	.013	1.91	.77	2.68
8" thick	1.000	S.F.	.018	2.55	1.02	3.57
10" thick	1.000	S.F.	.021	3.09	1.24	4.33
12" thick	1.000	S.F.	.026	3.77	1.51	5.28
Block inserts polystyrene, 6" thick	1.000	S.F.		1.34		1.34
8" thick	1.000	S.F.		1.51		1.51
10" thick	1.000	S.F.		1.56		1.56
12" thick	1.000	S.F.		1.73		1.73
Stucco, 1 coat	1.000	S.F.	.057	.60	3.17	3.77
2 coats	1.000	S.F.	.069	.73	3.86	4.59
3 coats	1.000	S.F.	.081	.86	4.53	5.39
Painting, 1 coat	1.000	S.F.	.011	.16	.54	.70
2 coats	1.000	S.F.	.016	.25	.77	1.02
Primer & 1 coat	1.000	S.F.	.013	.23	.63	.86
2 coats	1.000	S.F.	.018	.32	.88	1.20
Lath, metal lath expanded 2.5 lb/S.Y., painted	1.000	S.F.	.010	.46	.58	1.04
Galvanized	1.000	S.F.	.012	.51	.64	1.15

For customer support on your Residential Costs with RSMeans data, call 800.448.8182.

159

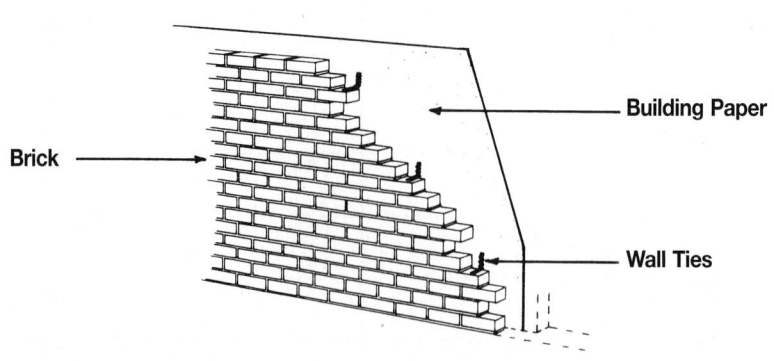

System Description	QUAN.	UNIT	LABOR HOURS	COST PER S.F.		
				MAT.	INST.	TOTAL
SELECT COMMON BRICK						
Brick, select common, running bond	1.000	S.F.	.174	4.75	9.40	14.15
Wall ties, 7/8" x 7", 22 gauge	1.000	Ea.	.008	.18	.44	.62
Building paper, spunbonded polypropylene	1.100	S.F.	.002	.17	.13	.30
Trim, pine, painted	.125	L.F.	.004	.08	.23	.31
TOTAL		S.F.	.188	5.18	10.20	15.38
RED FACED COMMON BRICK						
Brick, common, red faced, running bond	1.000	S.F.	.182	4.47	9.80	14.27
Wall ties, 7/8" x 7", 22 gauge	1.000	Ea.	.008	.18	.44	.62
Building paper, spundbonded polypropylene	1.100	S.F.	.002	.17	.13	.30
Trim, pine, painted	.125	L.F.	.004	.08	.23	.31
TOTAL		S.F.	.196	4.90	10.60	15.50
BUFF OR GREY FACE BRICK						
Brick, buff or grey	1.000	S.F.	.182	4.73	9.80	14.53
Wall ties, 7/8" x 7", 22 gauge	1.000	Ea.	.008	.18	.44	.62
Building paper, spundbonded polypropylene	1.100	S.F.	.002	.17	.13	.30
Trim, pine, painted	.125	L.F.	.004	.08	.23	.31
TOTAL		S.F.	.196	5.16	10.60	15.76
STONE WORK, ROUGH STONE, AVERAGE						
Field stone veneer	1.000	S.F.	.223	9.31	12.06	21.37
Wall ties, 7/8" x 7", 22 gauge	1.000	Ea.	.008	.18	.44	.62
Building paper, spundbonded polypropylene	1.000	S.F.	.002	.17	.13	.30
Trim, pine, painted	.125	L.F.	.004	.08	.23	.31
TOTAL		S.F.	.237	9.74	12.86	22.60

The costs in this system are based on a square foot of wall area. Do not subtract area for window & door openings.

Description	QUAN.	UNIT	LABOR HOURS	COST PER S.F.		
				MAT.	INST.	TOTAL

Brick/Stone Veneer Price Sheet	QUAN.	UNIT	LABOR HOURS	COST PER S.F.		
				MAT.	INST.	TOTAL
Brick						
Select common, running bond	1.000	S.F.	.174	4.75	9.40	14.15
Red faced, running bond	1.000	S.F.	.182	4.47	9.80	14.27
Buff or grey faced, running bond	1.000	S.F.	.182	4.73	9.80	14.53
Header every 6th course	1.000	S.F.	.216	5.20	11.65	16.85
English bond	1.000	S.F.	.286	6.65	15.40	22.05
Flemish bond	1.000	S.F.	.195	4.72	10.50	15.22
Common bond	1.000	S.F.	.267	5.95	14.40	20.35
Stack bond	1.000	S.F.	.182	4.73	9.80	14.53
Jumbo, running bond	1.000	S.F.	.092	6	4.96	10.96
Norman, running bond	1.000	S.F.	.125	7.05	6.75	13.80
Norwegian, running bond	1.000	S.F.	.107	6.25	5.75	12
Economy, running bond	1.000	S.F.	.129	4.90	6.95	11.85
Engineer, running bond	1.000	S.F.	.154	4.31	8.30	12.61
Roman, running bond	1.000	S.F.	.160	8	8.65	16.65
Utility, running bond	1.000	S.F.	.089	5.60	6	11.60
Glazed, running bond	1.000	S.F.	.190	14.30	10.25	24.55
Stone work, rough stone, average	1.000	S.F.	.179	9.30	12.05	21.35
Maximum	1.000	S.F.	.267	13.90	18	31.90
Wall ties, galvanized, corrugated 7/8" x 7", 22 gauge	1.000	Ea.	.008	.18	.44	.62
16 gauge	1.000	Ea.	.008	.33	.44	.77
Cavity wall, every 3rd course 6" long Z type, 1/4" diameter	1.330	L.F.	.010	.59	.57	1.16
3/16" diameter	1.330	L.F.	.010	.47	.57	1.04
8" long, Z type, 1/4" diameter	1.330	L.F.	.010	.68	.57	1.25
3/16" diameter	1.330	L.F.	.010	.37	.57	.94
Building paper, aluminum and kraft laminated foil, 1 side	1.000	S.F.	.002	.15	.13	.28
2 sides	1.000	S.F.	.002	.16	.13	.29
#15 asphalt paper	1.100	S.F.	.002	.07	.14	.21
Polyethylene, .002" thick	1.000	S.F.	.002	.02	.13	.15
.004" thick	1.000	S.F.	.002	.03	.13	.16
.006" thick	1.000	S.F.	.002	.04	.13	.17
.010" thick	1.000	S.F.	.002	.10	.13	.23
Trim, 1" x 4", cedar	.125	L.F.	.005	.17	.29	.46
Fir	.125	L.F.	.005	.11	.29	.40
Redwood	.125	L.F.	.005	.17	.29	.46
White pine	.125	L.F.	.005	.11	.29	.40

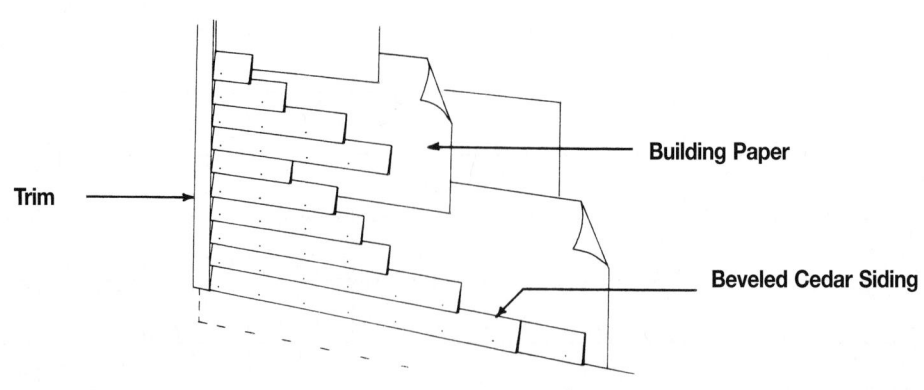

Trim

Building Paper

Beveled Cedar Siding

System Description	QUAN.	UNIT	LABOR HOURS	COST PER S.F.		
				MAT.	INST.	TOTAL
1/2″ X 6″ BEVELED CEDAR SIDING, ″A″ GRADE						
1/2″ x 6″ beveled cedar siding	1.000	S.F.	.027	4.86	1.58	6.44
Building wrap, spunbonded polypropylene	1.100	S.F.	.002	.17	.13	.30
Trim, cedar	.125	L.F.	.005	.17	.29	.46
Paint, primer & 2 coats	1.000	S.F.	.017	.24	.83	1.07
TOTAL		S.F.	.051	5.44	2.83	8.27
1/2″ X 8″ BEVELED CEDAR SIDING, ″A″ GRADE						
1/2″ x 8″ beveled cedar siding	1.000	S.F.	.024	8.25	1.41	9.66
Building wrap, spunbonded polypropylene	1.100	S.F.	.002	.17	.13	.30
Trim, cedar	.125	L.F.	.005	.17	.29	.46
Paint, primer & 2 coats	1.000	S.F.	.017	.24	.83	1.07
TOTAL		S.F.	.048	8.83	2.66	11.49
1″ X 4″ TONGUE & GROOVE, REDWOOD, VERTICAL GRAIN						
Redwood, clear, vertical grain, 1″ x 10″	1.000	S.F.	.020	5.46	1.14	6.60
Building wrap, spunbonded polypropylene	1.100	S.F.	.002	.17	.13	.30
Trim, redwood	.125	L.F.	.005	.17	.29	.46
Sealer, 1 coat, stain, 1 coat	1.000	S.F.	.013	.15	.64	.79
TOTAL		S.F.	.040	5.95	2.20	8.15
1″ X 6″ TONGUE & GROOVE, REDWOOD, VERTICAL GRAIN						
Redwood, clear, vertical grain, 1″ x 10″	1.000	S.F.	.020	5.61	1.17	6.78
Building wrap, spunbonded polypropylene	1.100	S.F.	.002	.17	.13	.30
Trim, redwood	.125	L.F.	.005	.17	.29	.46
Sealer, 1 coat, stain, 1 coat	1.000	S.F.	.013	.15	.64	.79
TOTAL		S.F.	.040	6.10	2.23	8.33

The costs in this system are based on a square foot of wall area.
Do not subtract area for door or window openings.

Description	QUAN.	UNIT	LABOR HOURS	COST PER S.F.		
				MAT.	INST.	TOTAL

Wood Siding Price Sheet

Wood Siding Price Sheet	QUAN.	UNIT	LABOR HOURS	COST PER S.F. MAT.	INST.	TOTAL
Siding, beveled cedar, "A" grade, 1/2" x 6"	1.000	S.F.	.028	4.86	1.58	6.44
1/2" x 8"	1.000	S.F.	.023	8.25	1.41	9.66
"B" grade, 1/2" x 6"	1.000	S.F.	.032	5.40	1.76	7.16
1/2" x 8"	1.000	S.F.	.029	9.15	1.57	10.72
Clear grade, 1/2" x 6"	1.000	S.F.	.028	6.10	1.98	8.08
1/2" x 8"	1.000	S.F.	.023	10.30	1.76	12.06
Redwood, clear vertical grain, 1/2" x 6"	1.000	S.F.	.036	5.30	1.58	6.88
1/2" x 8"	1.000	S.F.	.032	5.70	1.41	7.11
Clear all heart vertical grain, 1/2" x 6"	1.000	S.F.	.028	5.90	1.76	7.66
1/2" x 8"	1.000	S.F.	.023	6.35	1.57	7.92
Siding board & batten, cedar, "B" grade, 1" x 10"	1.000	S.F.	.031	5.25	1.11	6.36
1" x 12"	1.000	S.F.	.031	5.25	1.11	6.36
Redwood, clear vertical grain, 1" x 6"	1.000	S.F.	.043	5.35	1.58	6.93
1" x 8"	1.000	S.F.	.018	8.70	1.41	10.11
White pine, #2 & better, 1" x 10"	1.000	S.F.	.029	2.73	1.41	4.14
1" x 12"	1.000	S.F.	.029	2.73	1.41	4.14
Siding vertical, tongue & groove, cedar "B" grade, 1" x 4"	1.000	S.F.	.033	6.65	1.14	7.79
1" x 6"	1.000	S.F.	.024	6.85	1.17	8.02
1" x 8"	1.000	S.F.	.024	7.05	1.20	8.25
1" x 10"	1.000	S.F.	.021	7.25	1.24	8.49
"A" grade, 1" x 4"	1.000	S.F.	.033	6.10	1.04	7.14
1" x 6"	1.000	S.F.	.024	6.25	1.07	7.32
1" x 8"	1.000	S.F.	.024	6.40	1.10	7.50
1" x 10"	1.000	S.F.	.021	6.60	1.13	7.73
Clear vertical grain, 1" x 4"	1.000	S.F.	.033	5.60	.96	6.56
1" x 6"	1.000	S.F.	.024	5.75	.98	6.73
1" x 8"	1.000	S.F.	.024	5.90	1.01	6.91
1" x 10"	1.000	S.F.	.021	6.05	1.03	7.08
Redwood, clear vertical grain, 1" x 4"	1.000	S.F.	.033	5.45	1.14	6.59
1" x 6"	1.000	S.F.	.024	5.60	1.17	6.77
1" x 8"	1.000	S.F.	.024	5.80	1.20	7
1" x 10"	1.000	S.F.	.021	5.95	1.24	7.19
Clear all heart vertical grain, 1" x 4"	1.000	S.F.	.033	5	1.04	6.04
1" x 6"	1.000	S.F.	.024	5.15	1.07	6.22
1" x 8"	1.000	S.F.	.024	5.25	1.10	6.35
1" x 10"	1.000	S.F.	.021	5.40	1.13	6.53
White pine, 1" x 10"	1.000	S.F.	.024	4.05	1.24	5.29
Siding plywood, texture 1-11 cedar, 3/8" thick	1.000	S.F.	.024	1.40	1.38	2.78
5/8" thick	1.000	S.F.	.024	2.86	1.38	4.24
Redwood, 3/8" thick	1.000	S.F.	.024	1.40	1.38	2.78
5/8" thick	1.000	S.F.	.024	2.21	1.38	3.59
Fir, 3/8" thick	1.000	S.F.	.024	1.01	1.38	2.39
5/8" thick	1.000	S.F.	.024	1.51	1.38	2.89
Southern yellow pine, 3/8" thick	1.000	S.F.	.024	1.01	1.38	2.39
5/8" thick	1.000	S.F.	.024	1.58	1.38	2.96
Paper, #15 asphalt felt	1.100	S.F.	.002	.07	.14	.21
Trim, cedar	.125	L.F.	.005	.17	.29	.46
Fir	.125	L.F.	.005	.11	.29	.40
Redwood	.125	L.F.	.005	.17	.29	.46
White pine	.125	L.F.	.005	.11	.29	.40
Painting, primer, & 1 coat	1.000	S.F.	.013	.15	.64	.79
2 coats	1.000	S.F.	.017	.24	.83	1.07
Stain, sealer, & 1 coat	1.000	S.F.	.017	.17	.84	1.01
2 coats	1.000	S.F.	.019	.28	.91	1.19

For customer support on your Residential Costs with RSMeans data, call 800.448.8182.

163

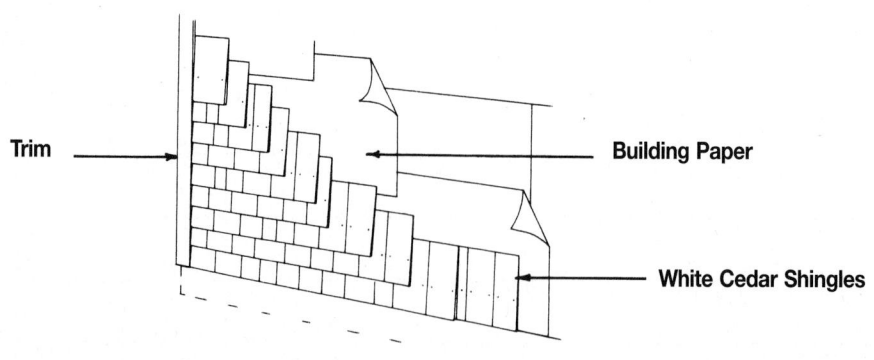

Trim — Building Paper — White Cedar Shingles

System Description	QUAN.	UNIT	LABOR HOURS	COST PER S.F.		
				MAT.	INST.	TOTAL
WHITE CEDAR SHINGLES, 5" EXPOSURE						
White cedar shingles, 16" long, grade "A", 5" exposure	1.000	S.F.	.033	2.17	1.94	4.11
Building wrap, spunbonded polypropylene	1.100	S.F.	.002	.17	.13	.30
Trim, cedar	.125	S.F.	.005	.17	.29	.46
Paint, primer & 1 coat	1.000	S.F.	.017	.17	.84	1.01
TOTAL		S.F.	.057	2.68	3.20	5.88
RESQUARED & REBUTTED PERFECTIONS, 5-1/2" EXPOSURE						
Resquared & rebutted perfections, 5-1/2" exposure	1.000	S.F.	.027	3.20	1.55	4.75
Building wrap, spunbonded polypropylene	1.100	S.F.	.002	.17	.13	.30
Trim, cedar	.125	S.F.	.005	.17	.29	.46
Stain, sealer & 1 coat	1.000	S.F.	.017	.17	.84	1.01
TOTAL		S.F.	.051	3.71	2.81	6.52
HAND-SPLIT SHAKES, 8-1/2" EXPOSURE						
Hand-split red cedar shakes, 18" long, 8-1/2" exposure	1.000	S.F.	.040	3	2.32	5.32
Building wrap, spunbonded polypropylene	1.100	S.F.	.002	.17	.13	.30
Trim, cedar	.125	S.F.	.005	.17	.29	.46
Stain, sealer & 1 coat	1.000	S.F.	.017	.17	.84	1.01
TOTAL		S.F.	.064	3.51	3.58	7.09

The costs in this system are based on a square foot of wall area.
Do not subtract area for door or window openings.

Description	QUAN.	UNIT	LABOR HOURS	COST PER S.F.		
				MAT.	INST.	TOTAL

Shingle Siding Price Sheet	QUAN.	UNIT	LABOR HOURS	COST PER S.F.		
				MAT.	INST.	TOTAL
Shingles wood, white cedar 16" long, "A" grade, 5" exposure	1.000	S.F.	.033	2.17	1.94	4.11
7" exposure	1.000	S.F.	.030	1.95	1.75	3.70
8-1/2" exposure	1.000	S.F.	.032	1.24	1.86	3.10
10" exposure	1.000	S.F.	.028	1.09	1.62	2.71
"B" grade, 5" exposure	1.000	S.F.	.040	1.89	2.32	4.21
7" exposure	1.000	S.F.	.028	1.32	1.62	2.94
8-1/2" exposure	1.000	S.F.	.024	1.13	1.39	2.52
10" exposure	1.000	S.F.	.020	.95	1.16	2.11
Fire retardant, "A" grade, 5" exposure	1.000	S.F.	.033	2.78	1.94	4.72
7" exposure	1.000	S.F.	.028	1.74	1.62	3.36
8-1/2" exposure	1.000	S.F.	.032	1.84	1.86	3.70
10" exposure	1.000	S.F.	.025	1.43	1.45	2.88
Fire retardant, 5" exposure	1.000	S.F.	.029	3.39	1.69	5.08
7" exposure	1.000	S.F.	.036	2.65	2.07	4.72
8-1/2" exposure	1.000	S.F.	.032	2.38	1.86	4.24
10" exposure	1.000	S.F.	.025	1.85	1.45	3.30
Resquared & rebutted, 5-1/2" exposure	1.000	S.F.	.027	3.20	1.55	4.75
7" exposure	1.000	S.F.	.024	2.88	1.40	4.28
8-1/2" exposure	1.000	S.F.	.021	2.56	1.24	3.80
10" exposure	1.000	S.F.	.019	2.24	1.09	3.33
Fire retardant, 5" exposure	1.000	S.F.	.027	3.81	1.55	5.36
7" exposure	1.000	S.F.	.024	3.42	1.40	4.82
8-1/2" exposure	1.000	S.F.	.021	3.04	1.24	4.28
10" exposure	1.000	S.F.	.023	2.07	1.33	3.40
Hand-split, red cedar, 24" long, 7" exposure	1.000	S.F.	.045	4.62	2.60	7.22
8-1/2" exposure	1.000	S.F.	.038	3.96	2.23	6.19
10" exposure	1.000	S.F.	.032	3.30	1.86	5.16
12" exposure	1.000	S.F.	.026	2.64	1.49	4.13
Fire retardant, 7" exposure	1.000	S.F.	.045	5.45	2.60	8.05
8-1/2" exposure	1.000	S.F.	.038	4.69	2.23	6.92
10" exposure	1.000	S.F.	.032	3.91	1.86	5.77
12" exposure	1.000	S.F.	.026	3.12	1.49	4.61
18" long, 5" exposure	1.000	S.F.	.068	5.10	3.94	9.04
7" exposure	1.000	S.F.	.048	3.60	2.78	6.38
8-1/2" exposure	1.000	S.F.	.040	3	2.32	5.32
10" exposure	1.000	S.F.	.036	2.70	2.09	4.79
Fire retardant, 5" exposure	1.000	S.F.	.068	6.15	3.94	10.09
7" exposure	1.000	S.F.	.048	4.33	2.78	7.11
8-1/2" exposure	1.000	S.F.	.040	3.61	2.32	5.93
10" exposure	1.000	S.F.	.036	3.24	2.09	5.33
Paper, #15 asphalt felt	1.100	S.F.	.002	.06	.13	.19
Trim, cedar	.125	S.F.	.005	.17	.29	.46
Fir	.125	S.F.	.005	.11	.29	.40
Redwood	.125	S.F.	.005	.17	.29	.46
White pine	.125	S.F.	.005	.11	.29	.40
Painting, primer, & 1 coat	1.000	S.F.	.013	.15	.64	.79
2 coats	1.000	S.F.	.017	.24	.83	1.07
Staining, sealer, & 1 coat	1.000	S.F.	.017	.17	.84	1.01
2 coats	1.000	S.F.	.019	.28	.91	1.19

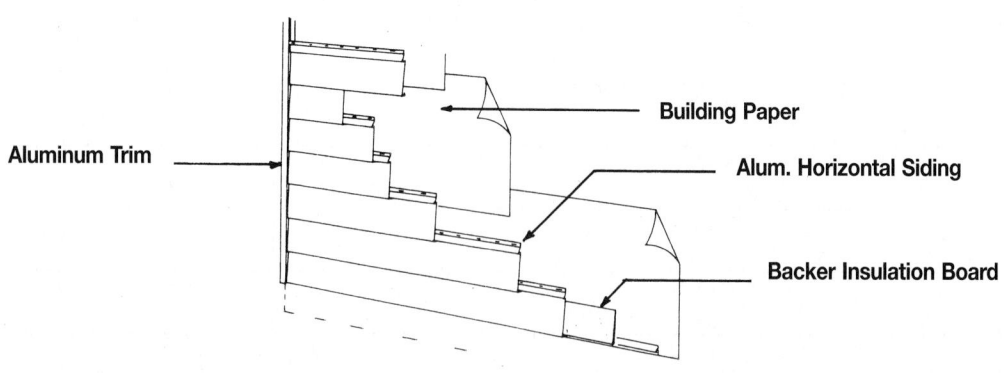

Aluminum Trim

Building Paper

Alum. Horizontal Siding

Backer Insulation Board

System Description	QUAN.	UNIT	LABOR HOURS	COST PER S.F.		
				MAT.	INST.	TOTAL
ALUMINUM CLAPBOARD SIDING, 8″ WIDE, WHITE						
Aluminum horizontal siding, 8″ clapboard	1.000	S.F.	.031	3.06	1.81	4.87
Backer, insulation board	1.000	S.F.	.008	.44	.46	.90
Trim, aluminum	.600	L.F.	.016	1.13	.91	2.04
Building wrap, spunbonded polypropylene	1.100	S.F.	.002	.17	.13	.30
TOTAL		S.F.	.057	4.80	3.31	8.11
ALUMINUM VERTICAL BOARD & BATTEN, WHITE						
Aluminum vertical board & batten	1.000	S.F.	.027	2.61	1.58	4.19
Backer insulation board	1.000	S.F.	.008	.44	.46	.90
Trim, aluminum	.600	L.F.	.016	1.13	.91	2.04
Building wrap, spunbonded polypropylene	1.100	S.F.	.002	.17	.13	.30
TOTAL		S.F.	.053	4.35	3.08	7.43
VINYL CLAPBOARD SIDING, 8″ WIDE, WHITE						
Vinyl siding, clabboard profile, smooth texture, .042 thick, single 8	1.000	S.F.	.032	.88	1.88	2.76
Backer, insulation board	1.000	S.F.	.008	.44	.46	.90
Vinyl siding, access., outside corner, woodgrain, 4″ face, 3/4″ pocket	.600	L.F.	.014	1.42	.80	2.22
Building wrap, spunbonded polypropylene	1.100	S.F.	.002	.17	.13	.30
TOTAL		S.F.	.056	2.91	3.27	6.18
VINYL VERTICAL BOARD & BATTEN, WHITE						
Vinyl siding, vertical pattern, .046 thick, double 5	1.000	S.F.	.029	1.72	1.69	3.41
Backer, insulation board	1.000	S.F.	.008	.44	.46	.90
Vinyl siding, access., outside corner, woodgrain, 4″ face, 3/4″ pocket	.600	L.F.	.014	1.42	.80	2.22
Building wrap, spunbonded polypropylene	1.100	S.F.	.002	.17	.13	.30
TOTAL		S.F.	.053	3.75	3.08	6.83

The costs in this system are on a square foot of wall basis.
Subtract openings from wall area.

Description	QUAN.	UNIT	LABOR HOURS	COST PER S.F.		
				MAT.	INST.	TOTAL

Metal & Plastic Siding Price Sheet	QUAN.	UNIT	LABOR HOURS	COST PER S.F.		
				MAT.	INST.	TOTAL
Siding, aluminum, .024" thick, smooth, 8" wide, white	1.000	S.F.	.031	3.06	1.81	4.87
Color	1.000	S.F.	.031	3.23	1.81	5.04
Double 4" pattern, 8" wide, white	1.000	S.F.	.031	3.14	1.81	4.95
Color	1.000	S.F.	.031	3.31	1.81	5.12
Double 5" pattern, 10" wide, white	1.000	S.F.	.029	3.03	1.69	4.72
Color	1.000	S.F.	.029	3.20	1.69	4.89
Embossed, single, 8" wide, white	1.000	S.F.	.031	3.07	1.81	4.88
Color	1.000	S.F.	.031	3.24	1.81	5.05
Double 4" pattern, 8" wide, white	1.000	S.F.	.031	3.08	1.81	4.89
Color	1.000	S.F.	.031	3.25	1.81	5.06
Double 5" pattern, 10" wide, white	1.000	S.F.	.029	3.10	1.69	4.79
Color	1.000	S.F.	.029	3.27	1.69	4.96
Alum siding with insulation board, smooth, 8" wide, white	1.000	S.F.	.031	2.72	1.81	4.53
Color	1.000	S.F.	.031	2.89	1.81	4.70
Double 4" pattern, 8" wide, white	1.000	S.F.	.031	2.70	1.81	4.51
Color	1.000	S.F.	.031	2.87	1.81	4.68
Double 5" pattern, 10" wide, white	1.000	S.F.	.029	2.71	1.69	4.40
Color	1.000	S.F.	.029	2.88	1.69	4.57
Embossed, single, 8" wide, white	1.000	S.F.	.031	3.14	1.81	4.95
Color	1.000	S.F.	.031	3.31	1.81	5.12
Double 4" pattern, 8" wide, white	1.000	S.F.	.031	3.16	1.81	4.97
Color	1.000	S.F.	.031	3.33	1.81	5.14
Double 5" pattern, 10" wide, white	1.000	S.F.	.029	3.16	1.69	4.85
Color	1.000	S.F.	.029	3.33	1.69	5.02
Aluminum, shake finish, 10" wide, white	1.000	S.F.	.029	3.41	1.69	5.10
Color	1.000	S.F.	.029	3.58	1.69	5.27
Aluminum, vertical, 12" wide, white	1.000	S.F.	.027	2.61	1.58	4.19
Color	1.000	S.F.	.027	2.78	1.58	4.36
Vinyl siding, 8" wide, smooth, white	1.000	S.F.	.032	.88	1.88	2.76
Color	1.000	S.F.	.032	1.05	1.88	2.93
10" wide, Dutch lap, smooth, white	1.000	S.F.	.029	1.20	1.69	2.89
Color	1.000	S.F.	.029	1.37	1.69	3.06
Double 4" pattern, 8" wide, white	1.000	S.F.	.032	.88	1.88	2.76
Color	1.000	S.F.	.032	1.05	1.88	2.93
Double 5" pattern, 10" wide, white	1.000	S.F.	.029	.88	1.69	2.57
Color	1.000	S.F.	.029	1.05	1.69	2.74
Embossed, single, 8" wide, white	1.000	S.F.	.032	1.51	1.88	3.39
Color	1.000	S.F.	.032	1.68	1.88	3.56
10" wide, white	1.000	S.F.	.029	1.81	1.69	3.50
Color	1.000	S.F.	.029	1.98	1.69	3.67
Double 4" pattern, 8" wide, white	1.000	S.F.	.032	1.18	1.88	3.06
Color	1.000	S.F.	.032	1.35	1.88	3.23
Double 5" pattern, 10" wide, white	1.000	S.F.	.029	1.18	1.69	2.87
Color	1.000	S.F.	.029	1.35	1.69	3.04
Vinyl, shake finish, 10" wide, white	1.000	S.F.	.029	4.09	2.32	6.41
Color	1.000	S.F.	.029	4.26	2.32	6.58
Vinyl, vertical, double 5" pattern, 10" wide, white	1.000	S.F.	.029	1.72	1.69	3.41
Color	1.000	S.F.	.029	1.89	1.69	3.58
Backer board, installed in siding panels 8" or 10" wide	1.000	S.F.	.008	.44	.46	.90
4' x 8' sheets, polystyrene, 3/4" thick	1.000	S.F.	.010	.63	.58	1.21
4' x 8' fiberboard, plain	1.000	S.F.	.008	.44	.46	.90
Trim, aluminum, white	.600	L.F.	.016	1.13	.91	2.04
Color	.600	L.F.	.016	1.22	.91	2.13
Vinyl, white	.600	L.F.	.014	1.42	.80	2.22
Color	.600	L.F.	.014	1.62	.80	2.42
Paper, #15 asphalt felt	1.100	S.F.	.002	.07	.14	.21
Kraft paper, plain	1.100	S.F.	.002	.17	.14	.31
Foil backed	1.100	S.F.	.002	.18	.14	.32

For customer support on your Residential Costs with RSMeans data, call 800.448.8182.

167

Description	QUAN.	UNIT	LABOR HOURS	COST PER S.F.		
				MAT.	INST.	TOTAL
Poured insulation, cellulose fiber, R3.8 per inch (1" thick)	1.000	S.F.	.003	.06	.19	.25
Fiberglass , R4.0 per inch (1" thick)	1.000	S.F.	.003	.06	.19	.25
Mineral wool, R3.0 per inch (1" thick)	1.000	S.F.	.003	.04	.19	.23
Polystyrene, R4.0 per inch (1" thick)	1.000	S.F.	.003	.13	.19	.32
Vermiculite, R2.7 per inch (1" thick)	1.000	S.F.	.003	.48	.19	.67
Perlite, R2.7 per inch (1" thick)	1.000	S.F.	.003	.48	.19	.67
Reflective insulation, aluminum foil reinforced with scrim	1.000	S.F.	.004	.17	.25	.42
Reinforced with woven polyolefin	1.000	S.F.	.004	.25	.25	.50
With single bubble air space, R8.8	1.000	S.F.	.005	.29	.31	.60
With double bubble air space, R9.8	1.000	S.F.	.005	.35	.31	.66
Rigid insulation, fiberglass, unfaced,						
1-1/2" thick, R6.2	1.000	S.F.	.008	.44	.46	.90
2" thick, R8.3	1.000	S.F.	.008	.55	.46	1.01
2-1/2" thick, R10.3	1.000	S.F.	.010	.66	.58	1.24
3" thick, R12.4	1.000	S.F.	.010	.66	.58	1.24
Foil faced, 1" thick, R4.3	1.000	S.F.	.008	.92	.46	1.38
1-1/2" thick, R6.2	1.000	S.F.	.008	1.38	.46	1.84
2" thick, R8.7	1.000	S.F.	.009	1.74	.52	2.26
2-1/2" thick, R10.9	1.000	S.F.	.010	2.05	.58	2.63
3" thick, R13.0	1.000	S.F.	.010	2.30	.58	2.88
Perlite, 1" thick R2.77	1.000	S.F.	.010	.52	.58	1.10
2" thick R5.55	1.000	S.F.	.011	.86	.64	1.50
Polystyrene, extruded, blue, 2.2#/C.F., 3/4" thick R4	1.000	S.F.	.010	.63	.58	1.21
1-1/2" thick R8.1	1.000	S.F.	.011	1.25	.64	1.89
2" thick R10.8	1.000	S.F.	.011	1.75	.64	2.39
Molded bead board, white, 1" thick R3.85	1.000	S.F.	.010	.30	.58	.88
1-1/2" thick, R5.6	1.000	S.F.	.011	.59	.64	1.23
2" thick, R7.7	1.000	S.F.	.011	.89	.64	1.53
Non-rigid insulation, batts						
Fiberglass, kraft faced, 3-1/2" thick, R13, 11" wide	1.000	S.F.	.005	.37	.40	.77
15" wide	1.000	S.F.	.005	.37	.40	.77
23" wide	1.000	S.F.	.005	.37	.40	.77
6" thick, R19, 11" wide	1.000	S.F.	.006	.48	.40	.88
15" wide	1.000	S.F.	.006	.48	.40	.88
23" wide	1.000	S.F.	.006	.48	.40	.88
9" thick, R30, 15" wide	1.000	S.F.	.006	.80	.34	1.14
23" wide	1.000	S.F.	.006	.80	.34	1.14
12" thick, R38, 15" wide	1.000	S.F.	.006	1.16	.34	1.50
23" wide	1.000	S.F.	.006	1.16	.34	1.50
Fiberglass, foil faced, 3-1/2" thick, R13, 15" wide	1.000	S.F.	.005	.52	.34	.86
23" wide	1.000	S.F.	.005	.52	.34	.86
6" thick, R19, 15" thick	1.000	S.F.	.005	.68	.29	.97
23" wide	1.000	S.F.	.005	.68	.29	.97
9" thick, R30, 15" wide	1.000	S.F.	.006	1.03	.34	1.37
23" wide	1.000	S.F.	.006	1.03	.34	1.37

Insulation Systems	QUAN.	UNIT	LABOR HOURS	COST PER S.F.		
				MAT.	INST.	TOTAL
Non-rigid insulation batts						
Fiberglass unfaced, 3-1/2" thick, R13, 15" wide	1.000	S.F.	.005	.36	.29	.65
23" wide	1.000	S.F.	.005	.36	.29	.65
6" thick, R19, 15" wide	1.000	S.F.	.006	.43	.34	.77
23" wide	1.000	S.F.	.006	.43	.34	.77
9" thick, R19, 15" wide	1.000	S.F.	.007	.66	.40	1.06
23" wide	1.000	S.F.	.007	.66	.40	1.06
12" thick, R38, 15" wide	1.000	S.F.	.007	.83	.40	1.23
23" wide	1.000	S.F.	.007	.83	.40	1.23
Mineral fiber batts, 3" thick, R11	1.000	S.F.	.005	.85	.29	1.14
3-1/2" thick, R13	1.000	S.F.	.005	.85	.29	1.14
6" thick, R19	1.000	S.F.	.005	1.33	.29	1.62
6-1/2" thick, R22	1.000	S.F.	.005	1.33	.29	1.62
10" thick, R30	1.000	S.F.	.006	1.75	.34	2.09
Non-rigid insulation batts						

For customer support on your Residential Costs with RSMeans data, call 800.448.8182.

169

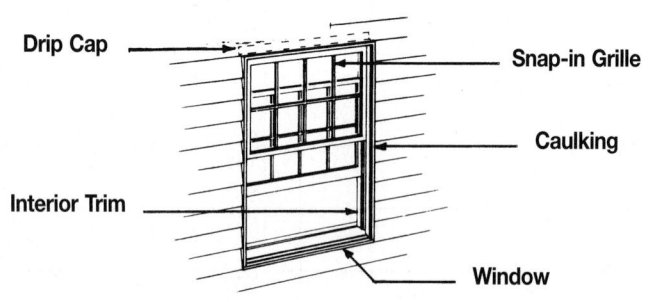

Drip Cap — Snap-in Grille

Caulking

Interior Trim — Window

System Description	QUAN.	UNIT	LABOR HOURS	COST EACH		
				MAT.	INST.	TOTAL
BUILDER'S QUALITY WOOD WINDOW 2' X 3', DOUBLE HUNG						
Window, primed, builder's quality, 2' x 3', insulating glass	1.000	Ea.	.800	264	46.50	310.50
Trim, interior casing	11.000	L.F.	.352	17.16	20.46	37.62
Paint, interior & exterior, primer & 2 coats	2.000	Face	1.778	2.80	86	88.80
Caulking	10.000	L.F.	.278	2.50	16.10	18.60
Snap-in grille	1.000	Set	.333	61.50	19.35	80.85
Drip cap, metal	2.000	L.F.	.040	1.28	2.32	3.60
TOTAL		Ea.	3.581	349.24	190.73	539.97
PLASTIC CLAD WOOD WINDOW 3' X 4', DOUBLE HUNG						
Window, plastic clad, premium, 3' x 4', insulating glass	1.000	Ea.	.889	440	51.50	491.50
Trim, interior casing	15.000	L.F.	.480	23.40	27.90	51.30
Paint, interior, primer & 2 coats	1.000	Face	.889	1.40	43	44.40
Caulking	14.000	L.F.	.389	3.50	22.54	26.04
Snap-in grille	1.000	Set	.333	61.50	19.35	80.85
TOTAL		Ea.	2.980	529.80	164.29	694.09
METAL CLAD WOOD WINDOW, 3' X 5', DOUBLE HUNG						
Window, metal clad, deluxe, 3' x 5', insulating glass	1.000	Ea.	1.000	425	58	483
Trim, interior casing	17.000	L.F.	.544	26.52	31.62	58.14
Paint, interior, primer & 2 coats	1.000	Face	.889	1.40	43	44.40
Caulking	16.000	L.F.	.444	4	25.76	29.76
Snap-in grille	1.000	Set	.235	154	13.65	167.65
Drip cap, metal	3.000	L.F.	.060	1.92	3.48	5.40
TOTAL		Ea.	3.172	612.84	175.51	788.35

The cost of this system is on a cost per each window basis.

Description	QUAN.	UNIT	LABOR HOURS	COST EACH		
				MAT.	INST.	TOTAL

Double Hung Window Price Sheet

Double Hung Window Price Sheet	QUAN.	UNIT	LABOR HOURS	MAT.	INST.	TOTAL
Windows, double-hung, builder's quality, 2' x 3', single glass	1.000	Ea.	.800	212	46.50	258.50
Insulating glass	1.000	Ea.	.800	264	46.50	310.50
3' x 4', single glass	1.000	Ea.	.889	315	51.50	366.50
Insulating glass	1.000	Ea.	.889	325	51.50	376.50
4' x 4'-6", single glass	1.000	Ea.	1.000	370	58	428
Insulating glass	1.000	Ea.	1.000	400	58	458
Plastic clad premium insulating glass, 2'-6" x 3'	1.000	Ea.	.800	370	46.50	416.50
3' x 3'-6"	1.000	Ea.	.800	375	46.50	421.50
3' x 4'	1.000	Ea.	.889	440	51.50	491.50
3' x 4'-6"	1.000	Ea.	.889	470	51.50	521.50
3' x 5'	1.000	Ea.	1.000	505	58	563
3'-6" x 6'	1.000	Ea.	1.000	565	58	623
Metal clad deluxe insulating glass, 2'-6" x 3'	1.000	Ea.	.800	315	46.50	361.50
3' x 3'-6"	1.000	Ea.	.800	360	46.50	406.50
3' x 4'	1.000	Ea.	.889	375	51.50	426.50
3' x 4'-6"	1.000	Ea.	.889	390	51.50	441.50
3' x 5'	1.000	Ea.	1.000	425	58	483
3'-6" x 6'	1.000	Ea.	1.000	510	58	568
Trim, interior casing, window 2' x 3'	11.000	L.F.	.367	17.15	20.50	37.65
2'-6" x 3'	12.000	L.F.	.400	18.70	22.50	41.20
3' x 3'-6"	14.000	L.F.	.467	22	26	48
3' x 4'	15.000	L.F.	.500	23.50	28	51.50
3' x 4'-6"	16.000	L.F.	.533	25	30	55
3' x 5'	17.000	L.F.	.567	26.50	31.50	58
3'-6" x 6'	20.000	L.F.	.667	31	37	68
4' x 4'-6"	18.000	L.F.	.600	28	33.50	61.50
Paint or stain, interior or exterior, 2' x 3' window, 1 coat	1.000	Face	.444	.47	21.50	21.97
2 coats	1.000	Face	.727	.94	35	35.94
Primer & 1 coat	1.000	Face	.727	.92	35	35.92
Primer & 2 coats	1.000	Face	.889	1.40	43	44.40
3' x 4' window, 1 coat	1.000	Face	.667	1.09	32	33.09
2 coats	1.000	Face	.667	1.81	32	33.81
Primer & 1 coat	1.000	Face	.727	1.43	35	36.43
Primer & 2 coats	1.000	Face	.889	1.40	43	44.40
4' x 4'-6" window, 1 coat	1.000	Face	.667	1.09	32	33.09
2 coats	1.000	Face	.667	1.81	32	33.81
Primer & 1 coat	1.000	Face	.727	1.43	35	36.43
Primer & 2 coats	1.000	Face	.889	1.40	43	44.40
Caulking, window, 2' x 3'	10.000	L.F.	.323	2.50	16.10	18.60
2'-6" x 3'	11.000	L.F.	.355	2.75	17.70	20.45
3' x 3'-6"	13.000	L.F.	.419	3.25	21	24.25
3' x 4'	14.000	L.F.	.452	3.50	22.50	26
3' x 4'-6"	15.000	L.F.	.484	3.75	24	27.75
3' x 5'	16.000	L.F.	.516	4	26	30
3'-6" x 6'	19.000	L.F.	.613	4.75	30.50	35.25
4' x 4'-6"	17.000	L.F.	.548	4.25	27.50	31.75
Grilles, glass size to, 16" x 24" per sash	1.000	Set	.333	61.50	19.35	80.85
32" x 32" per sash	1.000	Set	.235	154	13.65	167.65
Drip cap, aluminum, 2' long	2.000	L.F.	.040	1.28	2.32	3.60
3' long	3.000	L.F.	.060	1.92	3.48	5.40
4' long	4.000	L.F.	.080	2.56	4.64	7.20
Wood, 2' long	2.000	L.F.	.067	3.12	3.72	6.84
3' long	3.000	L.F.	.100	4.68	5.60	10.28
4' long	4.000	L.F.	.133	6.25	7.45	13.70

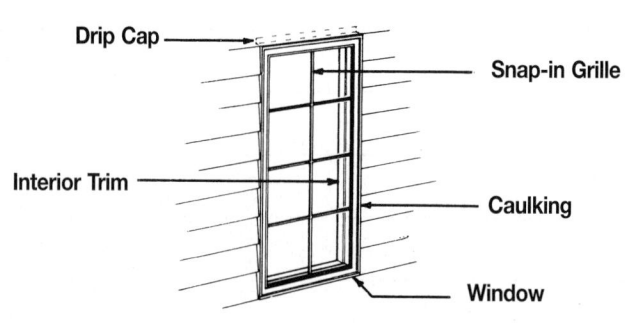

Drip Cap

Snap-in Grille

Interior Trim

Caulking

Window

System Description	QUAN.	UNIT	LABOR HOURS	COST EACH		
				MAT.	INST.	TOTAL
BUILDER'S QUALITY WINDOW, WOOD, 2' BY 3', CASEMENT						
Window, primed, builder's quality, 2' x 3', insulating glass	1.000	Ea.	.800	300	46.50	346.50
Trim, interior casing	11.000	L.F.	.352	17.16	20.46	37.62
Paint, interior & exterior, primer & 2 coats	2.000	Face	1.778	2.80	86	88.80
Caulking	10.000	L.F.	.278	2.50	16.10	18.60
Snap-in grille	1.000	Ea.	.267	37	15.50	52.50
Drip cap, metal	2.000	L.F.	.040	1.28	2.32	3.60
TOTAL		Ea.	3.515	360.74	186.88	547.62
PLASTIC CLAD WOOD WINDOW, 2' X 4', CASEMENT						
Window, plastic clad, premium, 2' x 4', insulating glass	1.000	Ea.	.889	365	51.50	416.50
Trim, interior casing	13.000	L.F.	.416	20.28	24.18	44.46
Paint, interior, primer & 2 coats	1.000	Ea.	.889	1.40	43	44.40
Caulking	12.000	L.F.	.333	3	19.32	22.32
Snap-in grille	1.000	Ea.	.267	37	15.50	52.50
TOTAL		Ea.	2.794	426.68	153.50	580.18
METAL CLAD WOOD WINDOW, 2' X 5', CASEMENT						
Window, metal clad, deluxe, 2' x 5', insulating glass	1.000	Ea.	1.000	375	58	433
Trim, interior casing	15.000	L.F.	.480	23.40	27.90	51.30
Paint, interior, primer & 2 coats	1.000	Ea.	.889	1.40	43	44.40
Caulking	14.000	L.F.	.389	3.50	22.54	26.04
Snap-in grille	1.000	Ea.	.250	51	14.55	65.55
Drip cap, metal	12.000	L.F.	.040	1.28	2.32	3.60
TOTAL		Ea.	3.048	455.58	168.31	623.89

The cost of this system is on a cost per each window basis.

Description	QUAN.	UNIT	LABOR HOURS	COST EACH		
				MAT.	INST.	TOTAL

Casement Window Price Sheet

Casement Window Price Sheet	QUAN.	UNIT	LABOR HOURS	COST EACH MAT.	COST EACH INST.	COST EACH TOTAL
Window, casement, builders quality, 2′ x 3′, double insulated glass	1.000	Ea.	.800	310	46.50	356.50
Insulating glass	1.000	Ea.	.800	300	46.50	346.50
2′ x 4′-6″, double insulated glass	1.000	Ea.	.727	1,000	42.50	1,042.50
Insulating glass	1.000	Ea.	.727	600	42.50	642.50
2′ x 6′, double insulated glass	1.000	Ea.	.889	490	58	548
Insulating glass	1.000	Ea.	.889	485	58	543
Plastic clad premium insulating glass, 2′ x 3′	1.000	Ea.	.800	297	46.50	343.50
2′ x 4′	1.000	Ea.	.889	455	51.50	506.50
2′ x 5′	1.000	Ea.	1.000	610	58	668
2′ x 6′	1.000	Ea.	1.000	445	58	503
Metal clad deluxe insulating glass, 2′ x 3′	1.000	Ea.	.800	325	46.50	371.50
2′ x 4′	1.000	Ea.	.889	310	51.50	361.50
2′ x 5′	1.000	Ea.	1.000	355	58	413
2′ x 6′	1.000	Ea.	1.000	405	58	463
Trim, interior casing, window 2′ x 3′	11.000	L.F.	.367	17.15	20.50	37.65
2′ x 4′	13.000	L.F.	.433	20.50	24	44.50
2′ x 4′-6″	14.000	L.F.	.467	22	26	48
2′ x 5′	15.000	L.F.	.500	23.50	28	51.50
2′ x 6′	17.000	L.F.	.567	26.50	31.50	58
Paint or stain, interior or exterior, 2′ x 3′ window, 1 coat	1.000	Face	.444	.47	21.50	21.97
2 coats	1.000	Face	.727	.94	35	35.94
Primer & 1 coat	1.000	Face	.727	.92	35	35.92
Primer & 2 coats	1.000	Face	.889	1.40	43	44.40
2′ x 4′ window, 1 coat	1.000	Face	.444	.47	21.50	21.97
2 coats	1.000	Face	.727	.94	35	35.94
Primer & 1 coat	1.000	Face	.727	.92	35	35.92
Primer & 2 coats	1.000	Face	.889	1.40	43	44.40
2′ x 6′ window, 1 coat	1.000	Face	.667	1.09	32	33.09
2 coats	1.000	Face	.667	1.81	32	33.81
Primer & 1 coat	1.000	Face	.727	1.43	35	36.43
Primer & 2 coats	1.000	Face	.889	1.40	43	44.40
Caulking, window, 2′ x 3′	10.000	L.F.	.323	2.50	16.10	18.60
2′ x 4′	12.000	L.F.	.387	3	19.30	22.30
2′ x 4′-6″	13.000	L.F.	.419	3.25	21	24.25
2′ x 5′	14.000	L.F.	.452	3.50	22.50	26
2′ x 6′	16.000	L.F.	.516	4	26	30
Grilles, glass size, to 20″ x 36″	1.000	Ea.	.267	37	15.50	52.50
To 20″ x 56″	1.000	Ea.	.250	51	14.55	65.55
Drip cap, metal, 2′ long	2.000	L.F.	.040	1.28	2.32	3.60
Wood, 2′ long	2.000	L.F.	.067	3.12	3.72	6.84

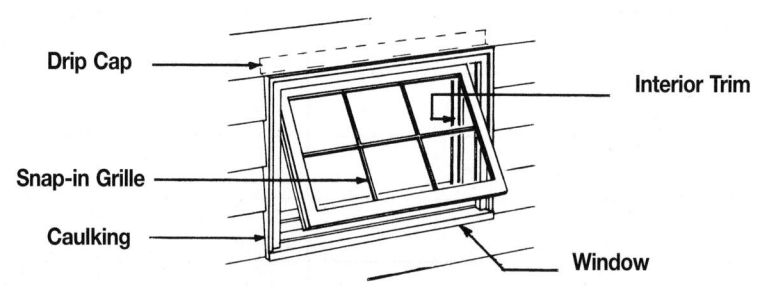

System Description	QUAN.	UNIT	LABOR HOURS	COST EACH		
				MAT.	INST.	TOTAL
BUILDER'S QUALITY WINDOW, WOOD, 34" X 22", AWNING						
Window, 34" x 22", insulating glass	1.000	Ea.	.800	330	46.50	376.50
Trim, interior casing	10.500	L.F.	.336	16.38	19.53	35.91
Paint, interior & exterior, primer & 2 coats	2.000	Face	1.778	2.80	86	88.80
Caulking	9.500	L.F.	.264	2.38	15.30	17.68
Snap-in grille	1.000	Ea.	.267	33.50	15.50	49
Drip cap, metal	3.000	L.F.	.060	1.92	3.48	5.40
TOTAL		Ea.	3.505	386.98	186.31	573.29
PLASTIC CLAD WOOD WINDOW, 40" X 28", AWNING						
Window, plastic clad, premium, 40" x 28", insulating glass	1.000	Ea.	.889	385	51.50	436.50
Trim interior casing	13.500	L.F.	.432	21.06	25.11	46.17
Paint, interior, primer & 2 coats	1.000	Face	.889	1.40	43	44.40
Caulking	12.500	L.F.	.347	3.13	20.13	23.26
Snap-in grille	1.000	Ea.	.267	33.50	15.50	49
TOTAL		Ea.	2.824	444.09	155.24	599.33
METAL CLAD WOOD WINDOW, 48" X 36", AWNING						
Window, metal clad, deluxe, 48" x 36", insulating glass	1.000	Ea.	1.000	410	58	468
Trim, interior casing	15.000	L.F.	.480	23.40	27.90	51.30
Paint, interior, primer & 2 coats	1.000	Face	.889	1.40	43	44.40
Caulking	14.000	L.F.	.389	3.50	22.54	26.04
Snap-in grille	1.000	Ea.	.250	48.50	14.55	63.05
Drip cap, metal	4.000	L.F.	.080	2.56	4.64	7.20
TOTAL		Ea.	3.088	489.36	170.63	659.99

The cost of this system is on a cost per each window basis.

Description	QUAN.	UNIT	LABOR HOURS	COST EACH		
				MAT.	INST.	TOTAL

Awning Window Price Sheet	QUAN.	UNIT	LABOR HOURS	COST EACH		
				MAT.	INST.	TOTAL
Windows, awning, builder's quality, 34" x 22", insulated glass	1.000	Ea.	.800	305	46.50	351.50
Low E glass	1.000	Ea.	.800	330	46.50	376.50
40" x 28", insulated glass	1.000	Ea.	.889	345	51.50	396.50
Low E glass	1.000	Ea.	.889	380	51.50	431.50
48" x 36", insulated glass	1.000	Ea.	1.000	515	58	573
Low E glass	1.000	Ea.	1.000	545	58	603
Plastic clad premium insulating glass, 34" x 22"	1.000	Ea.	.800	294	46.50	340.50
40" x 22"	1.000	Ea.	.800	325	46.50	371.50
36" x 28"	1.000	Ea.	.889	340	51.50	391.50
36" x 36"	1.000	Ea.	.889	385	51.50	436.50
48" x 28"	1.000	Ea.	1.000	410	58	468
60" x 36"	1.000	Ea.	1.000	570	58	628
Metal clad deluxe insulating glass, 34" x 22"	1.000	Ea.	.800	281	51.50	332.50
40" x 22"	1.000	Ea.	.800	325	51.50	376.50
36" x 25"	1.000	Ea.	.889	305	51.50	356.50
40" x 30"	1.000	Ea.	.889	375	51.50	426.50
48" x 28"	1.000	Ea.	1.000	385	58	443
60" x 36"	1.000	Ea.	1.000	410	58	468
Trim, interior casing window, 34" x 22"	10.500	L.F.	.350	16.40	19.55	35.95
40" x 22"	11.500	L.F.	.383	17.95	21.50	39.45
36" x 28"	12.500	L.F.	.417	19.50	23.50	43
40" x 28"	13.500	L.F.	.450	21	25	46
48" x 28"	14.500	L.F.	.483	22.50	27	49.50
48" x 36"	15.000	L.F.	.500	23.50	28	51.50
Paint or stain, interior or exterior, 34" x 22", 1 coat	1.000	Face	.444	.47	21.50	21.97
2 coats	1.000	Face	.727	.94	35	35.94
Primer & 1 coat	1.000	Face	.727	.92	35	35.92
Primer & 2 coats	1.000	Face	.889	1.40	43	44.40
36" x 28", 1 coat	1.000	Face	.444	.47	21.50	21.97
2 coats	1.000	Face	.727	.94	35	35.94
Primer & 1 coat	1.000	Face	.727	.92	35	35.92
Primer & 2 coats	1.000	Face	.889	1.40	43	44.40
48" x 36", 1 coat	1.000	Face	.667	1.09	32	33.09
2 coats	1.000	Face	.667	1.81	32	33.81
Primer & 1 coat	1.000	Face	.727	1.43	35	36.43
Primer & 2 coats	1.000	Face	.889	1.40	43	44.40
Caulking, window, 34" x 22"	9.500	L.F.	.306	2.38	15.30	17.68
40" x 22"	10.500	L.F.	.339	2.63	16.90	19.53
36" x 28"	11.500	L.F.	.371	2.88	18.50	21.38
40" x 28"	12.500	L.F.	.403	3.13	20	23.13
48" x 28"	13.500	L.F.	.436	3.38	21.50	24.88
48" x 36"	14.000	L.F.	.452	3.50	22.50	26
Grilles, glass size, to 28" by 16"	1.000	Ea.	.267	33.50	15.50	49
To 44" by 24"	1.000	Ea.	.250	48.50	14.55	63.05
Drip cap, aluminum, 3' long	3.000	L.F.	.060	1.92	3.48	5.40
3'-6" long	3.500	L.F.	.070	2.24	4.06	6.30
4' long	4.000	L.F.	.080	2.56	4.64	7.20
Wood, 3' long	3.000	L.F.	.100	4.68	5.60	10.28
3'-6" long	3.500	L.F.	.117	5.45	6.50	11.95
4' long	4.000	L.F.	.133	6.25	7.45	13.70

For customer support on your Residential Costs with RSMeans data, call 800.448.8182.

175

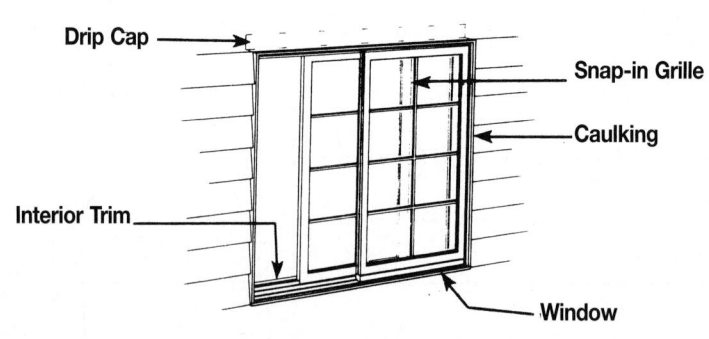

Drip Cap

Snap-in Grille

Caulking

Interior Trim

Window

System Description	QUAN.	UNIT	LABOR HOURS	COST EACH		
				MAT.	INST.	TOTAL
BUILDER'S QUALITY WOOD WINDOW, 3' X 2', SLIDING						
Window, primed, builder's quality, 3' x 3', insul. glass	1.000	Ea.	.800	350	46.50	396.50
Trim, interior casing	11.000	L.F.	.352	17.16	20.46	37.62
Paint, interior & exterior, primer & 2 coats	2.000	Face	1.778	2.80	86	88.80
Caulking	10.000	L.F.	.278	2.50	16.10	18.60
Snap-in grille	1.000	Set	.333	41.50	19.35	60.85
Drip cap, metal	3.000	L.F.	.060	1.92	3.48	5.40
TOTAL		Ea.	3.601	415.88	191.89	607.77
PLASTIC CLAD WOOD WINDOW, 4' X 3'-6", SLIDING						
Window, plastic clad, premium, 4' x 3'-6", insulating glass	1.000	Ea.	.889	760	51.50	811.50
Trim, interior casing	16.000	L.F.	.512	24.96	29.76	54.72
Paint, interior, primer & 2 coats	1.000	Face	.889	1.40	43	44.40
Caulking	17.000	L.F.	.472	4.25	27.37	31.62
Snap-in grille	1.000	Set	.333	41.50	19.35	60.85
TOTAL		Ea.	3.095	832.11	170.98	1,003.09
METAL CLAD WOOD WINDOW, 6' X 5', SLIDING						
Window, metal clad, deluxe, 6' x 5', insulating glass	1.000	Ea.	1.000	820	58	878
Trim, interior casing	23.000	L.F.	.736	35.88	42.78	78.66
Paint, interior, primer & 2 coats	1.000	Face	.889	1.40	43	44.40
Caulking	22.000	L.F.	.611	5.50	35.42	40.92
Snap-in grille	1.000	Set	.364	49.50	21	70.50
Drip cap, metal	6.000	L.F.	.120	3.84	6.96	10.80
TOTAL		Ea.	3.720	916.12	207.16	1,123.28

The cost of this system is on a cost per each window basis.

Description	QUAN.	UNIT	LABOR HOURS	COST EACH		
				MAT.	INST.	TOTAL

Sliding Window Price Sheet	QUAN.	UNIT	LABOR HOURS	COST EACH		
				MAT.	INST.	TOTAL
Windows, sliding, builder's quality, 3′ x 3′, single glass	1.000	Ea.	.800	320	46.50	366.50
Insulating glass	1.000	Ea.	.800	350	46.50	396.50
4′ x 3′-6″, single glass	1.000	Ea.	.889	420	51.50	471.50
Insulating glass	1.000	Ea.	.889	420	51.50	471.50
6′ x 5′, single glass	1.000	Ea.	1.000	555	58	613
Insulating glass	1.000	Ea.	1.000	605	58	663
Plastic clad premium insulating glass, 3′ x 3′	1.000	Ea.	.800	685	46.50	731.50
4′ x 3′-6″	1.000	Ea.	.889	760	51.50	811.50
5′ x 4′	1.000	Ea.	.889	1,000	51.50	1,051.50
6′ x 5′	1.000	Ea.	1.000	1,275	58	1,333
Metal clad deluxe insulating glass, 3′ x 3′	1.000	Ea.	.800	360	46.50	406.50
4′ x 3′-6″	1.000	Ea.	.889	440	51.50	491.50
5′ x 4′	1.000	Ea.	.889	525	51.50	576.50
6′ x 5′	1.000	Ea.	1.000	820	58	878
Trim, interior casing, window 3′ x 2′	11.000	L.F.	.367	17.15	20.50	37.65
3′ x 3′	13.000	L.F.	.433	20.50	24	44.50
4′ x 3′-6″	16.000	L.F.	.533	25	30	55
5′ x 4′	19.000	L.F.	.633	29.50	35.50	65
6′ x 5′	23.000	L.F.	.767	36	43	79
Paint or stain, interior or exterior, 3′ x 2′ window, 1 coat	1.000	Face	.444	.47	21.50	21.97
2 coats	1.000	Face	.727	.94	35	35.94
Primer & 1 coat	1.000	Face	.727	.92	35	35.92
Primer & 2 coats	1.000	Face	.889	1.40	43	44.40
4′ x 3′-6″ window, 1 coat	1.000	Face	.667	1.09	32	33.09
2 coats	1.000	Face	.667	1.81	32	33.81
Primer & 1 coat	1.000	Face	.727	1.43	35	36.43
Primer & 2 coats	1.000	Face	.889	1.40	43	44.40
6′ x 5′ window, 1 coat	1.000	Face	.889	2.67	43	45.67
2 coats	1.000	Face	1.333	4.87	64.50	69.37
Primer & 1 coat	1.000	Face	1.333	4.92	64.50	69.42
Primer & 2 coats	1.000	Face	1.600	7.45	77.50	84.95
Caulking, window, 3′ x 2′	10.000	L.F.	.323	2.50	16.10	18.60
3′ x 3′	12.000	L.F.	.387	3	19.30	22.30
4′ x 3′-6″	15.000	L.F.	.484	3.75	24	27.75
5′ x 4′	18.000	L.F.	.581	4.50	29	33.50
6′ x 5′	22.000	L.F.	.710	5.50	35.50	41
Grilles, glass size, to 14″ x 36″	1.000	Set	.333	41.50	19.35	60.85
To 36″ x 36″	1.000	Set	.364	49.50	21	70.50
Drip cap, aluminum, 3′ long	3.000	L.F.	.060	1.92	3.48	5.40
4′ long	4.000	L.F.	.080	2.56	4.64	7.20
5′ long	5.000	L.F.	.100	3.20	5.80	9
6′ long	6.000	L.F.	.120	3.84	6.95	10.79
Wood, 3′ long	3.000	L.F.	.100	4.68	5.60	10.28
4′ long	4.000	L.F.	.133	6.25	7.45	13.70
5′ long	5.000	L.F.	.167	7.80	9.30	17.10
6′ long	6.000	L.F.	.200	9.35	11.15	20.50

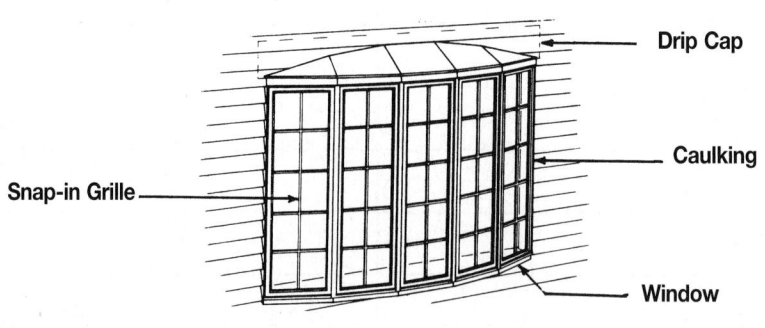

Drip Cap

Caulking

Snap-in Grille

Window

System Description	QUAN.	UNIT	LABOR HOURS	COST EACH		
				MAT.	INST.	TOTAL
AWNING TYPE BOW WINDOW, BUILDER'S QUALITY, 8′ X 5′						
Window, primed, builder's quality, 8′ x 5′, insulating glass	1.000	Ea.	1.600	1,450	93	1,543
Trim, interior casing	27.000	L.F.	.864	42.12	50.22	92.34
Paint, interior & exterior, primer & 1 coat	2.000	Face	3.200	14.90	155	169.90
Caulking	26.000	L.F.	.722	6.50	41.86	48.36
Snap-in grilles	1.000	Set	1.067	148	62	210
TOTAL		Ea.	7.453	1,661.52	402.08	2,063.60
CASEMENT TYPE BOW WINDOW, PLASTIC CLAD, 10′ X 6′						
Window, plastic clad, premium, 10′ x 6′, insulating glass	1.000	Ea.	2.286	3,075	133	3,208
Trim, interior casing	33.000	L.F.	1.056	51.48	61.38	112.86
Paint, interior, primer & 1 coat	1.000	Face	1.778	2.80	86	88.80
Caulking	32.000	L.F.	.889	8	51.52	59.52
Snap-in grilles	1.000	Set	1.333	185	77.50	262.50
TOTAL		Ea.	7.342	3,322.28	409.40	3,731.68
DOUBLE HUNG TYPE, METAL CLAD, 9′ X 5′						
Window, metal clad, deluxe, 9′ x 5′, insulating glass	1.000	Ea.	2.667	1,625	155	1,780
Trim, interior casing	29.000	L.F.	.928	45.24	53.94	99.18
Paint, interior, primer & 1 coat	1.000	Face	1.778	2.80	86	88.80
Caulking	28.000	L.F.	.778	7	45.08	52.08
Snap-in grilles	1.000	Set	1.067	148	62	210
TOTAL		Ea.	7.218	1,828.04	402.02	2,230.06

The cost of this system is on a cost per each window basis.

Description	QUAN.	UNIT	LABOR HOURS	COST EACH		
				MAT.	INST.	TOTAL

For customer support on your Residential Costs with RSMeans data, call 800.448.8182.

Bow/Bay Window Price Sheet	QUAN.	UNIT	LABOR HOURS	COST EACH MAT.	COST EACH INST.	COST EACH TOTAL
Windows, bow awning type, builder's quality, 8' x 5', insulating glass	1.000	Ea.	1.600	1,725	93	1,818
Low E glass	1.000	Ea.	1.600	1,450	93	1,543
12' x 6', insulating glass	1.000	Ea.	2.667	1,500	155	1,655
Low E glass	1.000	Ea.	2.667	1,600	155	1,755
Plastic clad premium insulating glass, 6' x 4'	1.000	Ea.	1.600	1,150	93	1,243
9' x 4'	1.000	Ea.	2.000	1,575	116	1,691
10' x 5'	1.000	Ea.	2.286	2,575	133	2,708
12' x 6'	1.000	Ea.	2.667	3,375	155	3,530
Metal clad deluxe insulating glass, 6' x 4'	1.000	Ea.	1.600	1,325	93	1,418
9' x 4'	1.000	Ea.	2.000	1,700	116	1,816
10' x 5'	1.000	Ea.	2.286	2,325	133	2,458
12' x 6'	1.000	Ea.	2.667	2,950	155	3,105
Bow casement type, builder's quality, 8' x 5', single glass	1.000	Ea.	1.600	2,100	93	2,193
Insulating glass	1.000	Ea.	1.600	2,550	93	2,643
12' x 6', single glass	1.000	Ea.	2.667	2,625	155	2,780
Insulating glass	1.000	Ea.	2.667	3,575	155	3,730
Plastic clad premium insulating glass, 8' x 5'	1.000	Ea.	1.600	1,975	93	2,068
10' x 5'	1.000	Ea.	2.000	2,650	116	2,766
10' x 6'	1.000	Ea.	2.286	3,075	133	3,208
12' x 6'	1.000	Ea.	2.667	3,700	155	3,855
Metal clad deluxe insulating glass, 8' x 5'	1.000	Ea.	1.600	1,875	93	1,968
10' x 5'	1.000	Ea.	2.000	2,000	116	2,116
10' x 6'	1.000	Ea.	2.286	2,375	133	2,508
12' x 6'	1.000	Ea.	2.667	3,225	155	3,380
Bow, double hung type, builder's quality, 8' x 4', single glass	1.000	Ea.	1.600	1,475	93	1,568
Insulating glass	1.000	Ea.	1.600	1,600	93	1,693
9' x 5', single glass	1.000	Ea.	2.667	1,600	155	1,755
Insulating glass	1.000	Ea.	2.667	1,675	155	1,830
Plastic clad premium insulating glass, 7' x 4'	1.000	Ea.	1.600	1,525	93	1,618
8' x 4'	1.000	Ea.	2.000	1,550	116	1,666
8' x 5'	1.000	Ea.	2.286	1,650	133	1,783
9' x 5'	1.000	Ea.	2.667	1,675	155	1,830
Metal clad deluxe insulating glass, 7' x 4'	1.000	Ea.	1.600	1,425	93	1,518
8' x 4'	1.000	Ea.	2.000	1,475	116	1,591
8' x 5'	1.000	Ea.	2.286	1,525	133	1,658
9' x 5'	1.000	Ea.	2.667	1,625	155	1,780
Trim, interior casing, window 7' x 4'	1.000	Ea.	.767	36	43	79
8' x 5'	1.000	Ea.	.900	42	50	92
10' x 6'	1.000	Ea.	1.100	51.50	61.50	113
12' x 6'	1.000	Ea.	1.233	57.50	69	126.50
Paint or stain, interior, or exterior, 7' x 4' window, 1 coat	1.000	Face	.889	2.67	43	45.67
Primer & 1 coat	1.000	Face	1.333	4.92	64.50	69.42
8' x 5' window, 1 coat	1.000	Face	.889	2.67	43	45.67
Primer & 1 coat	1.000	Face	1.333	4.92	64.50	69.42
10' x 6' window, 1 coat	1.000	Face	1.333	2.18	64	66.18
Primer & 1 coat	1.000	Face	1.778	2.80	86	88.80
12' x 6' window, 1 coat	1.000	Face	1.778	5.35	86	91.35
Primer & 1 coat	1.000	Face	2.667	9.85	129	138.85
Drip cap, vinyl moulded window, 7' long	1.000	Ea.	.533	125	24.50	149.50
8' long	1.000	Ea.	.533	143	28	171
Caulking, window, 7' x 4'	1.000	Ea.	.710	5.50	35.50	41
8' x 5'	1.000	Ea.	.839	6.50	42	48.50
10' x 6'	1.000	Ea.	1.032	8	51.50	59.50
12' x 6'	1.000	Ea.	1.161	9	58	67
Grilles, window, 7' x 4'	1.000	Set	.800	111	46.50	157.50
8' x 5'	1.000	Set	1.067	148	62	210
10' x 6'	1.000	Set	1.333	185	77.50	262.50
12' x 6'	1.000	Set	1.600	222	93	315

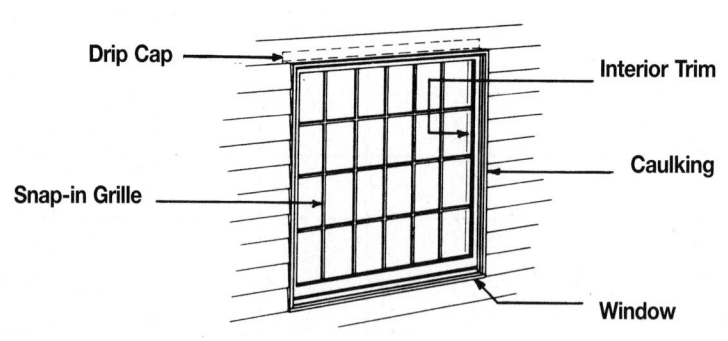

System Description	QUAN.	UNIT	LABOR HOURS	COST EACH		
				MAT.	INST.	TOTAL
BUILDER'S QUALITY PICTURE WINDOW, 4' X 4'						
Window, primed, builder's quality, 3'-0" x 4', insulating glass	1.000	Ea.	1.333	490	77.50	567.50
Trim, interior casing	17.000	L.F.	.544	26.52	31.62	58.14
Paint, interior & exterior, primer & 2 coats	2.000	Face	1.778	2.80	86	88.80
Caulking	16.000	L.F.	.444	4	25.76	29.76
Snap-in grille	1.000	Ea.	.267	141	15.50	156.50
Drip cap, metal	4.000	L.F.	.080	2.56	4.64	7.20
TOTAL		Ea.	4.446	666.88	241.02	907.90
PLASTIC CLAD WOOD WINDOW, 4'-6" X 6'-6"						
Window, plastic clad, prem., 4'-6" x 6'-6", insul. glass	1.000	Ea.	1.455	1,075	84.50	1,159.50
Trim, interior casing	23.000	L.F.	.736	35.88	42.78	78.66
Paint, interior, primer & 2 coats	1.000	Face	.889	1.40	43	44.40
Caulking	22.000	L.F.	.611	5.50	35.42	40.92
Snap-in grille	1.000	Ea.	.267	141	15.50	156.50
TOTAL		Ea.	3.958	1,258.78	221.20	1,479.98
METAL CLAD WOOD WINDOW, 6'-6" X 6'-6"						
Window, metal clad, deluxe, 6'-0" x 6'-0", insulating glass	1.000	Ea.	1.600	780	93	873
Trim interior casing	27.000	L.F.	.864	42.12	50.22	92.34
Paint, interior, primer & 2 coats	1.000	Face	1.600	7.45	77.50	84.95
Caulking	26.000	L.F.	.722	6.50	41.86	48.36
Snap-in grille	1.000	Ea.	.267	141	15.50	156.50
Drip cap, metal	6.500	L.F.	.130	4.16	7.54	11.70
TOTAL		Ea.	5.183	981.23	285.62	1,266.85

The cost of this system is on a cost per each window basis.

Description	QUAN.	UNIT	LABOR HOURS	COST EACH		
				MAT.	INST.	TOTAL

Fixed Window Price Sheet	QUAN.	UNIT	LABOR HOURS	COST EACH		
				MAT.	INST.	TOTAL
Window-picture, builder's quality, 4' x 4', single glass	1.000	Ea.	1.333	480	77.50	557.50
Insulating glass	1.000	Ea.	1.333	490	77.50	567.50
4' x 4'-6", single glass	1.000	Ea.	1.455	605	84.50	689.50
Insulating glass	1.000	Ea.	1.455	585	84.50	669.50
5' x 4', single glass	1.000	Ea.	1.455	640	84.50	724.50
Insulating glass	1.000	Ea.	1.455	665	84.50	749.50
6' x 4'-6", single glass	1.000	Ea.	1.600	700	93	793
Insulating glass	1.000	Ea.	1.600	705	93	798
Plastic clad premium insulating glass, 4' x 4'	1.000	Ea.	1.333	585	77.50	662.50
4'-6" x 6'-6"	1.000	Ea.	1.455	1,075	84.50	1,159.50
5'-6" x 6'-6"	1.000	Ea.	1.600	1,200	93	1,293
6'-6" x 6'-6"	1.000	Ea.	1.600	1,250	93	1,343
Metal clad deluxe insulating glass, 4' x 4'	1.000	Ea.	1.333	420	77.50	497.50
4'-6" x 6'-6"	1.000	Ea.	1.455	615	84.50	699.50
5'-6" x 6'-6"	1.000	Ea.	1.600	680	93	773
6'-6" x 6'-6"	1.000	Ea.	1.600	780	93	873
Trim, interior casing, window 4' x 4'	17.000	L.F.	.567	26.50	31.50	58
4'-6" x 4'-6"	19.000	L.F.	.633	29.50	35.50	65
5'-0" x 4'-0"	19.000	L.F.	.633	29.50	35.50	65
4'-6" x 6'-6"	23.000	L.F.	.767	36	43	79
5'-6" x 6'-6"	25.000	L.F.	.833	39	46.50	85.50
6'-6" x 6'-6"	27.000	L.F.	.900	42	50	92
Paint or stain, interior or exterior, 4' x 4' window, 1 coat	1.000	Face	.667	1.09	32	33.09
2 coats	1.000	Face	.667	1.81	32	33.81
Primer & 1 coat	1.000	Face	.727	1.43	35	36.43
Primer & 2 coats	1.000	Face	.889	1.40	43	44.40
4'-6" x 6'-6" window, 1 coat	1.000	Face	.667	1.09	32	33.09
2 coats	1.000	Face	.667	1.81	32	33.81
Primer & 1 coat	1.000	Face	.727	1.43	35	36.43
Primer & 2 coats	1.000	Face	.889	1.40	43	44.40
6'-6" x 6'-6" window, 1 coat	1.000	Face	.889	2.67	43	45.67
2 coats	1.000	Face	1.333	4.87	64.50	69.37
Primer & 1 coat	1.000	Face	1.333	4.92	64.50	69.42
Primer & 2 coats	1.000	Face	1.600	7.45	77.50	84.95
Caulking, window, 4' x 4'	1.000	Ea.	.516	4	26	30
4'-6" x 4'-6"	1.000	Ea.	.581	4.50	29	33.50
5'-0" x 4'-0"	1.000	Ea.	.581	4.50	29	33.50
4'-6" x 6'-6"	1.000	Ea.	.710	5.50	35.50	41
5'-6" x 6'-6"	1.000	Ea.	.774	6	38.50	44.50
6'-6" x 6'-6"	1.000	Ea.	.839	6.50	42	48.50
Grilles, glass size, to 48" x 48"	1.000	Ea.	.267	141	15.50	156.50
To 60" x 68"	1.000	Ea.	.286	216	16.60	232.60
Drip cap, aluminum, 4' long	4.000	L.F.	.080	2.56	4.64	7.20
4'-6" long	4.500	L.F.	.090	2.88	5.20	8.08
5' long	5.000	L.F.	.100	3.20	5.80	9
6' long	6.000	L.F.	.120	3.84	6.95	10.79
Wood, 4' long	4.000	L.F.	.133	6.25	7.45	13.70
4'-6" long	4.500	L.F.	.150	7	8.35	15.35
5' long	5.000	L.F.	.167	7.80	9.30	17.10
6' long	6.000	L.F.	.200	9.35	11.15	20.50

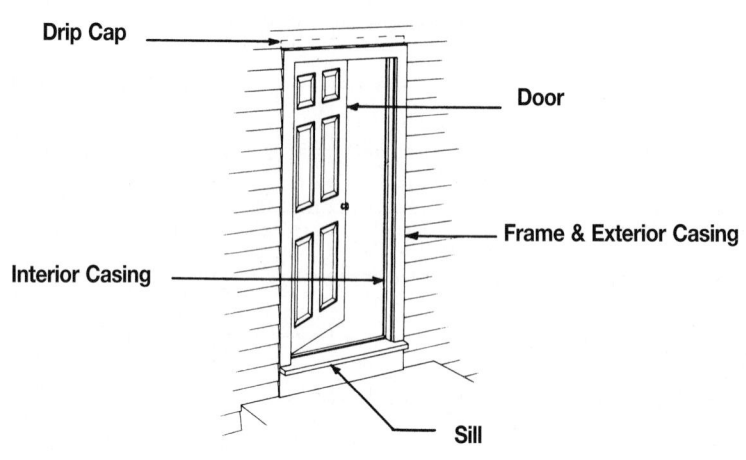

Drip Cap · Door · Frame & Exterior Casing · Interior Casing · Sill

System Description	QUAN.	UNIT	LABOR HOURS	COST EACH		
				MAT.	INST.	TOTAL
COLONIAL, 6 PANEL, 3′ X 6′-8″, WOOD						
Door, 3′ x 6′-8″ x 1-3/4″ thick, pine, 6 panel colonial	1.000	Ea.	1.067	615	62	677
Frame, 5-13/16″ deep, incl. exterior casing & drip cap	17.000	L.F.	.725	154.70	42.16	196.86
Interior casing, 2-1/2″ wide	18.000	L.F.	.576	28.08	33.48	61.56
Sill, 8/4 x 8″ deep	3.000	L.F.	.480	70.50	27.90	98.40
Butt hinges, brass, 4-1/2″ x 4-1/2″	1.500	Pr.		37.50		37.50
Average quality	1.000	Ea.	.571	57	33	90
Weatherstripping, metal, spring type, bronze	1.000	Set	1.053	27.50	61	88.50
Paint, interior & exterior, primer & 2 coats	2.000	Face	1.778	14.70	86	100.70
TOTAL		Ea.	6.250	1,004.98	345.54	1,350.52
SOLID CORE BIRCH, FLUSH, 3′ X 6′-8″						
Door, 3′ x 6′-8″, 1-3/4″ thick, birch, flush solid core	1.000	Ea.	1.067	167	62	229
Frame, 5-13/16″ deep, incl. exterior casing & drip cap	17.000	L.F.	.725	154.70	42.16	196.86
Interior casing, 2-1/2″ wide	18.000	L.F.	.576	28.08	33.48	61.56
Sill, 8/4 x 8″ deep	3.000	L.F.	.480	70.50	27.90	98.40
Butt hinges, brass, 4-1/2″ x 4-1/2″	1.500	Pr.		37.50		37.50
Average quality	1.000	Ea.	.571	57	33	90
Weatherstripping, metal, spring type, bronze	1.000	Set	1.053	27.50	61	88.50
Paint, Interior & exterior, primer & 2 coats	2.000	Face	1.778	13.80	86	99.80
TOTAL		Ea.	6.250	556.08	345.54	901.62

These systems are on a cost per each door basis.

Description	QUAN.	UNIT	LABOR HOURS	COST EACH		
				MAT.	INST.	TOTAL

Entrance Door Price Sheet

Entrance Door Price Sheet	QUAN.	UNIT	LABOR HOURS	COST EACH		
				MAT.	INST.	TOTAL
Door exterior wood 1-3/4" thick, pine, dutch door, 2'-8" x 6'-8" minimum	1.000	Ea.	1.333	765	77.50	842.50
Maximum	1.000	Ea.	1.600	1,075	93	1,168
3'-0" x 6'-8", minimum	1.000	Ea.	1.333	650	77.50	727.50
Maximum	1.000	Ea.	1.600	1,150	93	1,243
Colonial, 6 panel, 2'-8" x 6'-8"	1.000	Ea.	1.000	625	58	683
3'-0" x 6'-8"	1.000	Ea.	1.067	615	62	677
8 panel, 2'-6" x 6'-8"	1.000	Ea.	1.000	750	58	808
3'-0" x 6'-8"	1.000	Ea.	1.067	700	62	762
Flush, birch, solid core, 2'-8" x 6'-8"	1.000	Ea.	1.000	157	58	215
3'-0" x 6'-8"	1.000	Ea.	1.067	167	62	229
Porch door, 2'-8" x 6'-8"	1.000	Ea.	1.000	815	58	873
3'-0" x 6'-8"	1.000	Ea.	1.067	745	62	807
Hand carved mahogany, 2'-8" x 6'-8"	1.000	Ea.	1.067	2,175	62	2,237
3'-0" x 6'-8"	1.000	Ea.	1.067	1,550	62	1,612
Rosewood, 2'-8" x 6'-8"	1.000	Ea.	1.067	825	57.50	882.50
3'-0" x 6-8"	1.000	Ea.	1.067	760	62	822
Door, metal clad wood 1-3/8" thick raised panel, 2'-8" x 6'-8"	1.000	Ea.	1.067	450	54.50	504.50
3'-0" x 6'-8"	1.000	Ea.	1.067	320	62	382
Deluxe metal door, 3'-0" x 6'-8"	1.000	Ea.	1.231	320	62	382
3'-0" x 6'-8"	1.000	Ea.	1.231	320	62	382
Frame, pine, including exterior trim & drip cap, 5/4, x 4-9/16" deep	17.000	L.F.	.725	127	42	169
5-13/16" deep	17.000	L.F.	.725	155	42	197
6-9/16" deep	17.000	L.F.	.725	183	42	225
Safety glass lites, add	1.000	Ea.		94		94
Interior casing, 2'-8" x 6'-8" door	18.000	L.F.	.600	28	33.50	61.50
3'-0" x 6'-8" door	19.000	L.F.	.633	29.50	35.50	65
Sill, oak, 8/4 x 8" deep	3.000	L.F.	.480	70.50	28	98.50
8/4 x 10" deep	3.000	L.F.	.533	88.50	31	119.50
Butt hinges, steel plated, 4-1/2" x 4-1/2", plain	1.500	Pr.		37.50		37.50
Ball bearing	1.500	Pr.		58.50		58.50
Bronze, 4-1/2" x 4-1/2", plain	1.500	Pr.		43.50		43.50
Ball bearing	1.500	Pr.		61.50		61.50
Lockset, minimum	1.000	Ea.	.571	57	33	90
Maximum	1.000	Ea.	1.000	251	58	309
Weatherstripping, metal, interlocking, zinc	1.000	Set	2.667	52.50	155	207.50
Bronze	1.000	Set	2.667	66.50	155	221.50
Spring type, bronze	1.000	Set	1.053	27.50	61	88.50
Rubber, minimum	1.000	Set	1.053	10.95	61	71.95
Maximum	1.000	Set	1.143	14.05	66.50	80.55
Felt minimum	1.000	Set	.571	4.74	33	37.74
Maximum	1.000	Set	.615	5.50	36	41.50
Paint or stain, flush door, interior or exterior, 1 coat	2.000	Face	.941	5.20	45	50.20
2 coats	2.000	Face	1.455	10.40	70	80.40
Primer & 1 coat	2.000	Face	1.455	8.90	70	78.90
Primer & 2 coats	2.000	Face	1.778	13.80	86	99.80
Paneled door, interior & exterior, 1 coat	2.000	Face	1.143	5.55	55	60.55
2 coats	2.000	Face	2.000	11.10	97	108.10
Primer & 1 coat	2.000	Face	1.455	9.50	70	79.50
Primer & 2 coats	2.000	Face	1.778	14.70	86	100.70

For customer support on your Residential Costs with RSMeans data, call 800.448.8182.

183

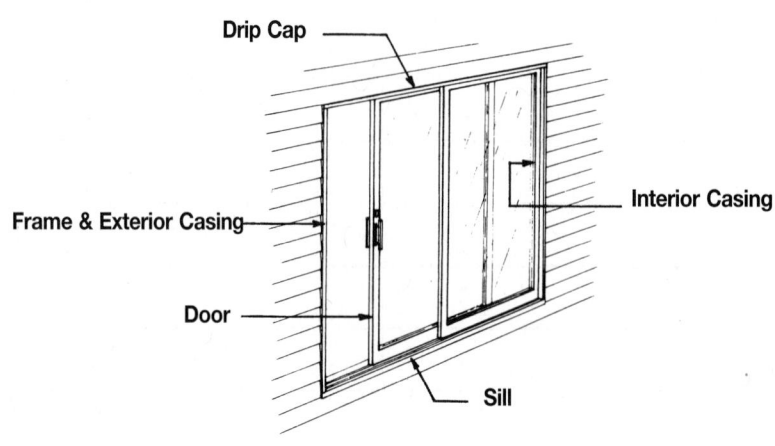

Drip Cap

Frame & Exterior Casing

Interior Casing

Door

Sill

System Description	QUAN.	UNIT	LABOR HOURS	COST EACH		
				MAT.	INST.	TOTAL
WOOD SLIDING DOOR, 8′ WIDE, PREMIUM						
Wood, 5/8″ thick tempered insul. glass, 8′ wide, premium	1.000	Ea.	5.333	2,025	310	2,335
Interior casing	22.000	L.F.	.704	34.32	40.92	75.24
Exterior casing	22.000	L.F.	.704	34.32	40.92	75.24
Sill, oak, 8/4 x 8″ deep	8.000	L.F.	1.280	188	74.40	262.40
Drip cap	8.000	L.F.	.160	5.12	9.28	14.40
Paint, interior & exterior, primer & 2 coats	2.000	Face	2.816	21.12	136.40	157.52
TOTAL		Ea.	10.997	2,307.88	611.92	2,919.80
ALUMINUM SLIDING DOOR, 8′ WIDE, PREMIUM						
Aluminum, 5/8″ tempered insul. glass, 8′ wide, premium	1.000	Ea.	5.333	1,875	310	2,185
Interior casing	22.000	L.F.	.704	34.32	40.92	75.24
Exterior casing	22.000	L.F.	.704	34.32	40.92	75.24
Sill, oak, 8/4 x 8″ deep	8.000	L.F.	1.280	188	74.40	262.40
Drip cap	8.000	L.F.	.160	5.12	9.28	14.40
Paint, interior & exterior, primer & 2 coats	2.000	Face	2.816	21.12	136.40	157.52
TOTAL		Ea.	10.997	2,157.88	611.92	2,769.80

The cost of this system is on a cost per each door basis.

Description	QUAN.	UNIT	LABOR HOURS	COST EACH		
				MAT.	INST.	TOTAL

Sliding Door Price Sheet	QUAN.	UNIT	LABOR HOURS	COST PER UNIT		
				MAT.	INST.	TOTAL
Sliding door, wood, 5/8" thick, tempered insul. glass, 6' wide, premium	1.000	Ea.	4.000	1,600	232	1,832
Economy	1.000	Ea.	4.000	1,325	232	1,557
8'wide, wood premium	1.000	Ea.	5.333	2,025	310	2,335
Economy	1.000	Ea.	5.333	1,675	310	1,985
12' wide, wood premium	1.000	Ea.	6.400	3,400	370	3,770
Economy	1.000	Ea.	6.400	2,725	370	3,095
Aluminum, 5/8" thick, tempered insul. glass, 6'wide, premium	1.000	Ea.	4.000	1,700	232	1,932
Economy	1.000	Ea.	4.000	905	232	1,137
8'wide, premium	1.000	Ea.	5.333	1,875	310	2,185
Economy	1.000	Ea.	5.333	1,625	310	1,935
12' wide, premium	1.000	Ea.	6.400	3,300	370	3,670
Economy	1.000	Ea.	6.400	1,725	370	2,095
Interior casing, 6' wide door	20.000	L.F.	.667	31	37	68
8' wide door	22.000	L.F.	.733	34.50	41	75.50
12' wide door	26.000	L.F.	.867	40.50	48.50	89
Exterior casing, 6' wide door	20.000	L.F.	.667	31	37	68
8' wide door	22.000	L.F.	.733	34.50	41	75.50
12' wide door	26.000	L.F.	.867	40.50	48.50	89
Sill, oak, 8/4 x 8" deep, 6' wide door	6.000	L.F.	.960	141	56	197
8' wide door	8.000	L.F.	1.280	188	74.50	262.50
12' wide door	12.000	L.F.	1.920	282	112	394
8/4 x 10" deep, 6' wide door	6.000	L.F.	1.067	177	62	239
8' wide door	8.000	L.F.	1.422	236	83	319
12' wide door	12.000	L.F.	2.133	355	124	479
Drip cap, 6' wide door	6.000	L.F.	.120	3.84	6.95	10.79
8' wide door	8.000	L.F.	.160	5.10	9.30	14.40
12' wide door	12.000	L.F.	.240	7.70	13.90	21.60
Paint or stain, interior & exterior, 6' wide door, 1 coat	2.000	Face	1.600	7.20	77.50	84.70
2 coats	2.000	Face	1.600	7.20	77.50	84.70
Primer & 1 coat	2.000	Face	1.778	12.80	85.50	98.30
Primer & 2 coats	2.000	Face	2.560	19.20	124	143.20
8' wide door, 1 coat	2.000	Face	1.760	7.90	85.50	93.40
2 coats	2.000	Face	1.760	7.90	85.50	93.40
Primer & 1 coat	2.000	Face	1.955	14.10	94	108.10
Primer & 2 coats	2.000	Face	2.816	21	136	157
12' wide door, 1 coat	2.000	Face	2.080	9.35	101	110.35
2 coats	2.000	Face	2.080	9.35	101	110.35
Primer & 1 coat	2.000	Face	2.311	16.65	111	127.65
Primer & 2 coats	2.000	Face	3.328	25	161	186
Aluminum door, trim only, interior & exterior, 6' door, 1 coat	2.000	Face	.800	3.60	39	42.60
2 coats	2.000	Face	.800	3.60	39	42.60
Primer & 1 coat	2.000	Face	.889	6.40	43	49.40
Primer & 2 coats	2.000	Face	1.280	9.60	62	71.60
8' wide door, 1 coat	2.000	Face	.880	3.96	42.50	46.46
2 coats	2.000	Face	.880	3.96	42.50	46.46
Primer & 1 coat	2.000	Face	.978	7.05	47	54.05
Primer & 2 coats	2.000	Face	1.408	10.55	68	78.55
12' wide door, 1 coat	2.000	Face	1.040	4.68	50.50	55.18
2 coats	2.000	Face	1.040	4.68	50.50	55.18
Primer & 1 coat	2.000	Face	1.155	8.30	55.50	63.80
Primer & 2 coats	2.000	Face	1.664	12.50	80.50	93

System Description	QUAN.	UNIT	LABOR HOURS	COST EACH		
				MAT.	INST.	TOTAL
OVERHEAD, SECTIONAL GARAGE DOOR, 9' X 7'						
Wood, overhead sectional door, std., incl. hardware, 9' x 7'	1.000	Ea.	2.000	1,125	116	1,241
Jamb & header blocking, 2" x 6"	25.000	L.F.	.901	17.75	52.25	70
Exterior trim	25.000	L.F.	.800	39	46.50	85.50
Paint, interior & exterior, primer & 2 coats	2.000	Face	3.556	29.40	172	201.40
Weatherstripping, molding type	1.000	Set	.736	35.88	42.78	78.66
Drip cap	9.000	L.F.	.180	5.76	10.44	16.20
TOTAL		Ea.	8.173	1,252.79	439.97	1,692.76
OVERHEAD, SECTIONAL GARAGE DOOR, 16' X 7'						
Wood, overhead sectional, std., incl. hardware, 16' x 7'	1.000	Ea.	2.667	1,850	155	2,005
Jamb & header blocking, 2" x 6"	30.000	L.F.	1.081	21.30	62.70	84
Exterior trim	30.000	L.F.	.960	46.80	55.80	102.60
Paint, interior & exterior, primer & 2 coats	2.000	Face	5.333	44.10	258	302.10
Weatherstripping, molding type	1.000	Set	.960	46.80	55.80	102.60
Drip cap	16.000	L.F.	.320	10.24	18.56	28.80
TOTAL		Ea.	11.321	2,019.24	605.86	2,625.10
OVERHEAD, SWING-UP TYPE, GARAGE DOOR, 16' X 7'						
Wood, overhead, swing-up, std., incl. hardware, 16' x 7'	1.000	Ea.	2.667	1,025	155	1,180
Jamb & header blocking, 2" x 6"	30.000	L.F.	1.081	21.30	62.70	84
Exterior trim	30.000	L.F.	.960	46.80	55.80	102.60
Paint, interior & exterior, primer & 2 coats	2.000	Face	5.333	44.10	258	302.10
Weatherstripping, molding type	1.000	Set	.960	46.80	55.80	102.60
Drip cap	16.000	L.F.	.320	10.24	18.56	28.80
TOTAL		Ea.	11.321	1,194.24	605.86	1,800.10

This system is on a cost per each door basis.

Description	QUAN.	UNIT	LABOR HOURS	COST EACH		
				MAT.	INST.	TOTAL

Resi Garage Door Price Sheet	QUAN.	UNIT	LABOR HOURS	COST EACH		
				MAT.	INST.	TOTAL
Overhead, sectional, including hardware, fiberglass, 9' x 7', standard	1.000	Ea.	3.030	1,125	186	1,311
Deluxe	1.000	Ea.	3.030	1,325	186	1,511
16' x 7', standard	1.000	Ea.	2.667	1,800	155	1,955
Deluxe	1.000	Ea.	2.667	2,475	155	2,630
Hardboard, 9' x 7', standard	1.000	Ea.	2.000	785	116	901
Deluxe	1.000	Ea.	2.000	945	116	1,061
16' x 7', standard	1.000	Ea.	2.667	1,425	155	1,580
Deluxe	1.000	Ea.	2.667	1,650	155	1,805
Metal, 9' x 7', standard	1.000	Ea.	3.030	975	116	1,091
Deluxe	1.000	Ea.	2.000	1,100	155	1,255
16' x 7', standard	1.000	Ea.	5.333	1,175	155	1,330
Deluxe	1.000	Ea.	2.667	1,600	186	1,786
Wood, 9' x 7', standard	1.000	Ea.	2.000	1,125	116	1,241
Deluxe	1.000	Ea.	2.000	2,475	116	2,591
16' x 7', standard	1.000	Ea.	2.667	1,850	155	2,005
Deluxe	1.000	Ea.	2.667	3,450	155	3,605
Overhead swing-up type including hardware, fiberglass, 9' x 7', standard	1.000	Ea.	2.000	1,150	116	1,266
Deluxe	1.000	Ea.	2.000	1,250	116	1,366
16' x 7', standard	1.000	Ea.	2.667	1,425	155	1,580
Deluxe	1.000	Ea.	2.667	1,825	155	1,980
Hardboard, 9' x 7', standard	1.000	Ea.	2.000	630	116	746
Deluxe	1.000	Ea.	2.000	745	116	861
16' x 7', standard	1.000	Ea.	2.667	765	155	920
Deluxe	1.000	Ea.	2.667	970	155	1,125
Metal, 9' x 7', standard	1.000	Ea.	2.000	695	116	811
Deluxe	1.000	Ea.	2.000	1,150	116	1,266
16' x 7', standard	1.000	Ea.	2.667	915	155	1,070
Deluxe	1.000	Ea.	2.667	1,250	155	1,405
Wood, 9' x 7', standard	1.000	Ea.	2.000	810	116	926
Deluxe	1.000	Ea.	2.000	1,325	116	1,441
16' x 7', standard	1.000	Ea.	2.667	1,025	155	1,180
Deluxe	1.000	Ea.	2.667	2,475	155	2,630
Jamb & header blocking, 2" x 6", 9' x 7' door	25.000	L.F.	.901	17.75	52.50	70.25
16' x 7' door	30.000	L.F.	1.081	21.50	62.50	84
2" x 8", 9' x 7' door	25.000	L.F.	1.000	24.50	58	82.50
16' x 7' door	30.000	L.F.	1.200	29.50	69.50	99
Exterior trim, 9' x 7' door	25.000	L.F.	.833	39	46.50	85.50
16' x 7' door	30.000	L.F.	1.000	47	56	103
Paint or stain, interior & exterior, 9' x 7' door, 1 coat	1.000	Face	2.286	11.10	110	121.10
2 coats	1.000	Face	4.000	22	194	216
Primer & 1 coat	1.000	Face	2.909	19.05	140	159.05
Primer & 2 coats	1.000	Face	3.556	29.50	172	201.50
16' x 7' door, 1 coat	1.000	Face	3.429	16.60	165	181.60
2 coats	1.000	Face	6.000	33.50	291	324.50
Primer & 1 coat	1.000	Face	4.364	28.50	210	238.50
Primer & 2 coats	1.000	Face	5.333	44	258	302
Weatherstripping, molding type, 9' x 7' door	1.000	Set	.767	36	43	79
16' x 7' door	1.000	Set	1.000	47	56	103
Drip cap, 9' door	9.000	L.F.	.180	5.75	10.45	16.20
16' door	16.000	L.F.	.320	10.25	18.55	28.80
Garage door opener, economy	1.000	Ea.	1.000	485	58	543
Deluxe, including remote control	1.000	Ea.	1.000	695	58	753

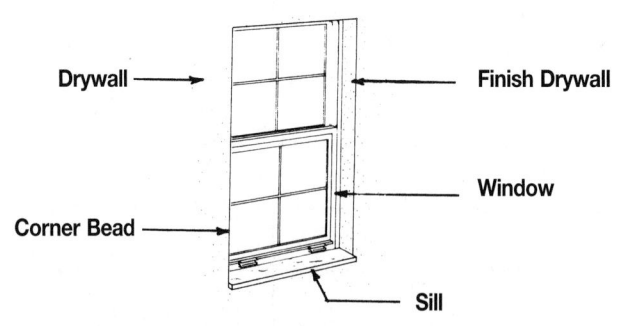

Drywall → ← Finish Drywall

← Window

Corner Bead → ← Sill

System Description	QUAN.	UNIT	LABOR HOURS	COST EACH		
				MAT.	INST.	TOTAL
SINGLE HUNG, 2′ X 3′ OPENING						
Window, 2′ x 3′ opening, enameled, insulating glass	1.000	Ea.	1.600	283	110	393
Blocking, 1″ x 3″ furring strip nailers	10.000	L.F.	.146	4.70	8.50	13.20
Drywall, 1/2″ thick, standard	5.000	S.F.	.040	1.85	2.30	4.15
Corner bead, 1″ x 1″, galvanized steel	8.000	L.F.	.160	1.36	9.28	10.64
Finish drywall, tape and finish corners inside and outside	16.000	L.F.	.269	1.76	15.68	17.44
Sill, slate	2.000	L.F.	.400	26.80	21.20	48
TOTAL		Ea.	2.615	319.47	166.96	486.43
SLIDING, 3′ X 2′ OPENING						
Window, 3′ x 2′ opening, enameled, insulating glass	1.000	Ea.	1.600	260	110	370
Blocking, 1″ x 3″ furring strip nailers	10.000	L.F.	.146	4.70	8.50	13.20
Drywall, 1/2″ thick, standard	5.000	S.F.	.040	1.85	2.30	4.15
Corner bead, 1″ x 1″, galvanized steel	7.000	L.F.	.140	1.19	8.12	9.31
Finish drywall, tape and finish corners inside and outside	14.000	L.F.	.236	1.54	13.72	15.26
Sill, slate	3.000	L.F.	.600	40.20	31.80	72
TOTAL		Ea.	2.762	309.48	174.44	483.92
AWNING, 3′-1″ X 3′-2″						
Window, 3′-1″ x 3′-2″ opening, enameled, insul. glass	1.000	Ea.	1.600	430	110	540
Blocking, 1″ x 3″ furring strip, nailers	12.500	L.F.	.182	5.88	10.63	16.51
Drywall, 1/2″ thick, standard	4.500	S.F.	.036	1.67	2.07	3.74
Corner bead, 1″ x 1″, galvanized steel	9.250	L.F.	.185	1.57	10.73	12.30
Finish drywall, tape and finish corners, inside and outside	18.500	L.F.	.312	2.04	18.13	20.17
Sill, slate	3.250	L.F.	.650	43.55	34.45	78
TOTAL		Ea.	2.965	484.71	186.01	670.72

Description	QUAN.	UNIT	LABOR HOURS	COST EACH		
				MAT.	INST.	TOTAL

Aluminum Window Price Sheet	QUAN.	UNIT	LABOR HOURS	COST EACH		
				MAT.	INST.	TOTAL
Window, aluminum, awning, 3'-1" x 3'-2", standard glass	1.000	Ea.	1.600	395	110	505
Insulating glass	1.000	Ea.	1.600	430	110	540
4'-5" x 5'-3", standard glass	1.000	Ea.	2.000	450	138	588
Insulating glass	1.000	Ea.	2.000	525	138	663
Casement, 3'-1" x 3'-2", standard glass	1.000	Ea.	1.600	420	110	530
Insulating glass	1.000	Ea.	1.600	570	110	680
Single hung, 2' x 3', standard glass	1.000	Ea.	1.600	233	110	343
Insulating glass	1.000	Ea.	1.600	283	110	393
2'-8" x 6'-8", standard glass	1.000	Ea.	2.000	405	138	543
Insulating glass	1.000	Ea.	2.000	530	138	668
3'-4" x 5'-0", standard glass	1.000	Ea.	1.778	335	122	457
Insulating glass	1.000	Ea.	1.778	375	122	497
Sliding, 3' x 2', standard glass	1.000	Ea.	1.600	242	110	352
Insulating glass	1.000	Ea.	1.600	260	110	370
5' x 3', standard glass	1.000	Ea.	1.778	370	122	492
Insulating glass	1.000	Ea.	1.778	430	122	552
8' x 4', standard glass	1.000	Ea.	2.667	395	184	579
Insulating glass	1.000	Ea.	2.667	635	184	819
Blocking, 1" x 3" furring, opening 3' x 2'	10.000	L.F.	.146	4.70	8.50	13.20
3' x 3'	12.500	L.F.	.182	5.90	10.65	16.55
3' x 5'	16.000	L.F.	.233	7.50	13.60	21.10
4' x 4'	16.000	L.F.	.233	7.50	13.60	21.10
4' x 5'	18.000	L.F.	.262	8.45	15.30	23.75
4' x 6'	20.000	L.F.	.291	9.40	17	26.40
4' x 8'	24.000	L.F.	.349	11.30	20.50	31.80
6'-8" x 2'-8"	19.000	L.F.	.276	8.95	16.15	25.10
Drywall, 1/2" thick, standard, opening 3' x 2'	5.000	S.F.	.040	1.85	2.30	4.15
3' x 3'	6.000	S.F.	.048	2.22	2.76	4.98
3' x 5'	8.000	S.F.	.064	2.96	3.68	6.64
4' x 4'	8.000	S.F.	.064	2.96	3.68	6.64
4' x 5'	9.000	S.F.	.072	3.33	4.14	7.47
4' x 6'	10.000	S.F.	.080	3.70	4.60	8.30
4' x 8'	12.000	S.F.	.096	4.44	5.50	9.94
6'-8" x 2'	9.500	S.F.	.076	3.52	4.37	7.89
Corner bead, 1" x 1", galvanized steel, opening 3' x 2'	7.000	L.F.	.140	1.19	8.10	9.29
3' x 3'	9.000	L.F.	.180	1.53	10.45	11.98
3' x 5'	11.000	L.F.	.220	1.87	12.75	14.62
4' x 4'	12.000	L.F.	.240	2.04	13.90	15.94
4' x 5'	13.000	L.F.	.260	2.21	15.10	17.31
4' x 6'	14.000	L.F.	.280	2.38	16.25	18.63
4' x 8'	16.000	L.F.	.320	2.72	18.55	21.27
6'-8" x 2'	15.000	L.F.	.300	2.55	17.40	19.95
Tape and finish corners, inside and outside, opening 3' x 2'	14.000	L.F.	.204	1.54	13.70	15.24
3' x 3'	18.000	L.F.	.262	1.98	17.65	19.63
3' x 5'	22.000	L.F.	.320	2.42	21.50	23.92
4' x 4'	24.000	L.F.	.349	2.64	23.50	26.14
4' x 5'	26.000	L.F.	.378	2.86	25.50	28.36
4' x 6'	28.000	L.F.	.407	3.08	27.50	30.58
4' x 8'	32.000	L.F.	.466	3.52	31.50	35.02
6'-8" x 2'	30.000	L.F.	.437	3.30	29.50	32.80
Sill, slate, 2' long	2.000	L.F.	.400	27	21	48
3' long	3.000	L.F.	.600	40	32	72
4' long	4.000	L.F.	.800	53.50	42.50	96
Wood, 1-5/8" x 6-1/4", 2' long	2.000	L.F.	.128	11.70	7.45	19.15
3' long	3.000	L.F.	.192	17.55	11.15	28.70
4' long	4.000	L.F.	.256	23.50	14.90	38.40

For customer support on your Residential Costs with RSMeans data, call 800.448.8182.

189

Aluminum Window

Aluminum Door

System Description	QUAN.	UNIT	LABOR HOURS	COST EACH		
				MAT.	INST.	TOTAL
Storm door, aluminum, combination, storm & screen, anodized, 2'-6" x 6'-8"	1.000	Ea.	1.067	232	62	294
2'-8" x 6'-8"	1.000	Ea.	1.143	242	66.50	308.50
3'-0" x 6'-8"	1.000	Ea.	1.143	210	66.50	276.50
Mill finish, 2'-6" x 6'-8"	1.000	Ea.	1.067	275	62	337
2'-8" x 6'-8"	1.000	Ea.	1.143	275	66.50	341.50
3'-0" x 6'-8"	1.000	Ea.	1.143	297	66.50	363.50
Painted, 2'-6" x 6'-8"	1.000	Ea.	1.067	246	62	308
2'-8" x 6'-8"	1.000	Ea.	1.143	263	66.50	329.50
3'-0" x 6'-8"	1.000	Ea.	1.143	345	66.50	411.50
Wood, combination, storm & screen, crossbuck, 2'-6" x 6'-9"	1.000	Ea.	1.455	370	84.50	454.50
2'-8" x 6'-9"	1.000	Ea.	1.600	345	93	438
3'-0" x 6'-9"	1.000	Ea.	1.778	360	103	463
Full lite, 2'-6" x 6'-9"	1.000	Ea.	1.455	360	84.50	444.50
2'-8" x 6'-9"	1.000	Ea.	1.600	365	93	458
3'-0" x 6'-9"	1.000	Ea.	1.778	365	103	468
Windows, aluminum, combination storm & screen, basement, 1'-10" x 1'-0"	1.000	Ea.	.533	39.50	31	70.50
2'-9" x 1'-6"	1.000	Ea.	.533	43	31	74
3'-4" x 2'-0"	1.000	Ea.	.533	50.50	31	81.50
Double hung, anodized, 2'-0" x 3'-5"	1.000	Ea.	.533	109	31	140
2'-6" x 5'-0"	1.000	Ea.	.571	132	33	165
4'-0" x 6'-0"	1.000	Ea.	.640	259	37	296
Painted, 2'-0" x 3'-5"	1.000	Ea.	.533	127	31	158
2'-6" x 5'-0"	1.000	Ea.	.571	193	33	226
4'-0" x 6'-0"	1.000	Ea.	.640	320	37	357
Fixed window, anodized, 4'-6" x 4'-6"	1.000	Ea.	.640	154	37	191
5'-8" x 4'-6"	1.000	Ea.	.800	171	46.50	217.50
Painted, 4'-6" x 4'-6"	1.000	Ea.	.640	154	37	191
5'-8" x 4'-6"	1.000	Ea.	.800	176	46.50	222.50

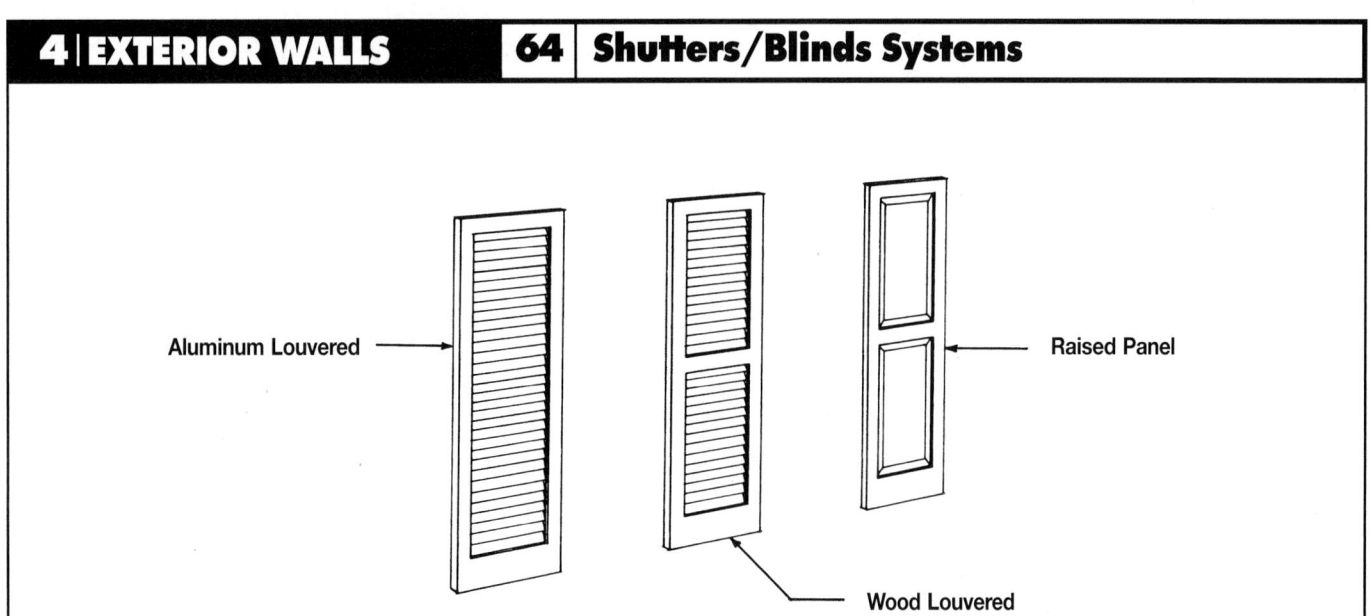

Aluminum Louvered →

← Raised Panel

Wood Louvered

System Description	QUAN.	UNIT	LABOR HOURS	COST PER PAIR		
				MAT.	INST.	TOTAL
Shutters, exterior blinds, aluminum, louvered, 1'-4" wide, 3"-0" long	1.000	Set	.800	211	46.50	257.50
4'-0" long	1.000	Set	.800	255	46.50	301.50
5'-4" long	1.000	Set	.800	300	46.50	346.50
6'-8" long	1.000	Set	.889	375	51.50	426.50
Wood, louvered, 1'-2" wide, 3'-3" long	1.000	Set	.800	259	46.50	305.50
4'-7" long	1.000	Set	.800	297	46.50	343.50
5'-3" long	1.000	Set	.800	355	46.50	401.50
1'-6" wide, 3'-3" long	1.000	Set	.800	280	46.50	326.50
4'-7" long	1.000	Set	.800	350	46.50	396.50
Polystyrene, louvered, 1'-2" wide, 3'-3" long	1.000	Set	.800	42	46.50	88.50
4'-7" long	1.000	Set	.800	49.50	46.50	96
5'-3" long	1.000	Set	.800	58.50	46.50	105
6'-8" long	1.000	Set	.889	76	51.50	127.50
Vinyl, louvered, 1'-2" wide, 4'-7" long	1.000	Set	.720	57	42	99
1'-4" x 6'-8" long	1.000	Set	.889	89.50	51.50	141

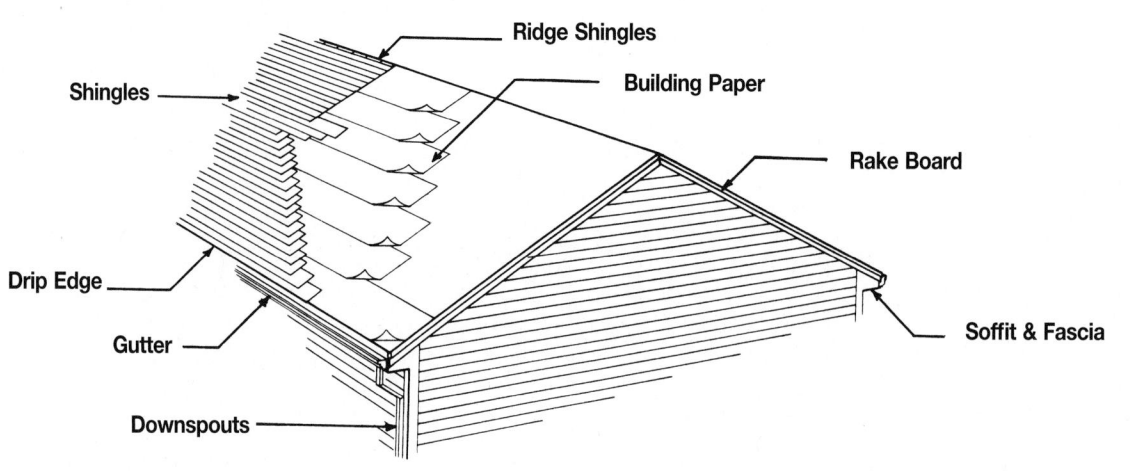

System Description	QUAN.	UNIT	LABOR HOURS	COST PER S.F.		
				MAT.	INST.	TOTAL
ASPHALT, ROOF SHINGLES, CLASS A						
Shingles, inorganic class A, 210-235 lb./sq., 4/12 pitch	1.160	S.F.	.017	.99	.95	1.94
Drip edge, metal, 5" wide	.150	L.F.	.003	.10	.17	.27
Building paper, #15 felt	1.300	S.F.	.002	.08	.10	.18
Ridge shingles, asphalt	.042	L.F.	.001	.11	.06	.17
Soffit & fascia, white painted aluminum, 1' overhang	.083	L.F.	.012	.41	.70	1.11
Rake trim, 1" x 6"	.040	L.F.	.002	.05	.09	.14
Rake trim, prime and paint	.040	L.F.	.002	.01	.09	.10
Gutter, seamless, aluminum painted	.083	L.F.	.005	.26	.34	.60
Downspouts, aluminum painted	.035	L.F.	.002	.08	.10	.18
Ridge vent	.042	L.F.	.002	.11	.11	.22
TOTAL		S.F.	.048	2.20	2.71	4.91
WOOD, CEDAR SHINGLES NO. 1 PERFECTIONS, 18" LONG						
Shingles, wood, cedar, No. 1 perfections, 4/12 pitch	1.160	S.F.	.035	3.34	2.03	5.37
Drip edge, metal, 5" wide	.150	L.F.	.003	.10	.17	.27
Building paper, #15 felt	1.300	S.F.	.002	.08	.10	.18
Ridge shingles, cedar	.042	L.F.	.001	.22	.07	.29
Soffit & fascia, white painted aluminum, 1' overhang	.083	L.F.	.012	.41	.70	1.11
Rake trim, 1" x 6"	.040	L.F.	.002	.05	.09	.14
Rake trim, prime and paint	.040	L.F.	.002	.01	.09	.10
Gutter, seamless, aluminum, painted	.083	L.F.	.005	.26	.34	.60
Downspouts, aluminum, painted	.035	L.F.	.002	.08	.10	.18
Ridge vent	.042	L.F.	.002	.11	.11	.22
TOTAL		S.F.	.066	4.66	3.80	8.46

The prices in these systems are based on a square foot of plan area.
All quantities have been adjusted accordingly.

Description	QUAN.	UNIT	LABOR HOURS	COST PER S.F.		
				MAT.	INST.	TOTAL

Gable End Roofing Price Sheet	QUAN.	UNIT	LABOR HOURS	COST PER S.F. MAT.	INST.	TOTAL
Shingles, asphalt, inorganic, class A, 210-235 lb./sq., 4/12 pitch	1.160	S.F.	.017	.99	.95	1.94
8/12 pitch	1.330	S.F.	.019	1.07	1.03	2.10
Laminated, multi-layered, 240-260 lb./sq., 4/12 pitch	1.160	S.F.	.021	1.45	1.16	2.61
8/12 pitch	1.330	S.F.	.023	1.57	1.26	2.83
Premium laminated, multi-layered, 260-300 lb./sq., 4/12 pitch	1.160	S.F.	.027	1.90	1.50	3.40
8/12 pitch	1.330	S.F.	.030	2.05	1.63	3.68
Clay tile, Spanish tile, red, 4/12 pitch	1.160	S.F.	.053	6.65	2.87	9.52
8/12 pitch	1.330	S.F.	.058	7.20	3.11	10.31
Mission tile, red, 4/12 pitch	1.160	S.F.	.083	5.45	2.87	8.32
8/12 pitch	1.330	S.F.	.090	5.90	3.11	9.01
French tile, red, 4/12 pitch	1.160	S.F.	.071	15.30	2.63	17.93
8/12 pitch	1.330	S.F.	.077	16.60	2.85	19.45
Slate, Buckingham, Virginia, black, 4/12 pitch	1.160	S.F.	.055	7.30	3.01	10.31
8/12 pitch	1.330	S.F.	.059	7.95	3.26	11.21
Vermont, black or grey, 4/12 pitch	1.160	S.F.	.055	6.40	3.01	9.41
8/12 pitch	1.330	S.F.	.059	6.95	3.26	10.21
Wood, No. 1 red cedar, 5X, 16" long, 5" exposure, 4/12 pitch	1.160	S.F.	.038	3.90	2.23	6.13
8/12 pitch	1.330	S.F.	.042	4.23	2.42	6.65
Fire retardant, 4/12 pitch	1.160	S.F.	.038	4.63	2.23	6.86
8/12 pitch	1.330	S.F.	.042	5	2.42	7.42
18" long, No.1 perfections, 5" exposure, 4/12 pitch	1.160	S.F.	.035	3.34	2.03	5.37
8/12 pitch	1.330	S.F.	.038	3.61	2.20	5.81
Fire retardant, 4/12 pitch	1.160	S.F.	.035	4.07	2.03	6.10
8/12 pitch	1.330	S.F.	.038	4.40	2.20	6.60
Resquared & rebutted, 18" long, 6" exposure, 4/12 pitch	1.160	S.F.	.032	3.84	1.86	5.70
8/12 pitch	1.330	S.F.	.035	4.16	2.02	6.18
Fire retardant, 4/12 pitch	1.160	S.F.	.032	4.57	1.86	6.43
8/12 pitch	1.330	S.F.	.035	4.95	2.02	6.97
Wood shakes hand split, 24" long, 10" exposure, 4/12 pitch	1.160	S.F.	.038	3.96	2.23	6.19
8/12 pitch	1.330	S.F.	.042	4.29	2.42	6.71
Fire retardant, 4/12 pitch	1.160	S.F.	.038	4.69	2.23	6.92
8/12 pitch	1.330	S.F.	.042	5.10	2.42	7.52
18" long, 8" exposure, 4/12 pitch	1.160	S.F.	.048	3.60	2.78	6.38
8/12 pitch	1.330	S.F.	.052	3.90	3.02	6.92
Fire retardant, 4/12 pitch	1.160	S.F.	.048	4.33	2.78	7.11
8/12 pitch	1.330	S.F.	.052	4.69	3.02	7.71
Drip edge, metal, 5" wide	.150	L.F.	.003	.10	.17	.27
8" wide	.150	L.F.	.003	.14	.17	.31
Building paper, #15 asphalt felt	1.300	S.F.	.002	.08	.10	.18
Ridge shingles, asphalt	.042	L.F.	.001	.11	.06	.17
Clay	.042	L.F.	.002	.23	.28	.51
Slate	.042	L.F.	.002	.47	.09	.56
Wood, shingles	.042	L.F.	.001	.22	.07	.29
Shakes	.042	L.F.	.001	.22	.07	.29
Soffit & fascia, aluminum, vented, 1' overhang	.083	L.F.	.012	.41	.70	1.11
2' overhang	.083	L.F.	.013	.60	.77	1.37
Vinyl, vented, 1' overhang	.083	L.F.	.011	.43	.64	1.07
2' overhang	.083	L.F.	.012	.57	.77	1.34
Wood, board fascia, plywood soffit, 1' overhang	.083	L.F.	.004	.03	.18	.21
2' overhang	.083	L.F.	.006	.04	.27	.31
Rake trim, painted, 1" x 6"	.040	L.F.	.004	.06	.18	.24
1" x 8"	.040	L.F.	.004	.23	.17	.40
Gutter, 5" box, aluminum, seamless, painted	.083	L.F.	.006	.26	.34	.60
Vinyl	.083	L.F.	.006	.15	.34	.49
Downspout, 2" x 3", aluminum, one story house	.035	L.F.	.001	.05	.09	.14
Two story house	.060	L.F.	.003	.09	.16	.25
Vinyl, one story house	.035	L.F.	.002	.08	.10	.18
Two story house	.060	L.F.	.003	.09	.16	.25

For customer support on your Residential Costs with RSMeans data, call 800.448.8182.

195

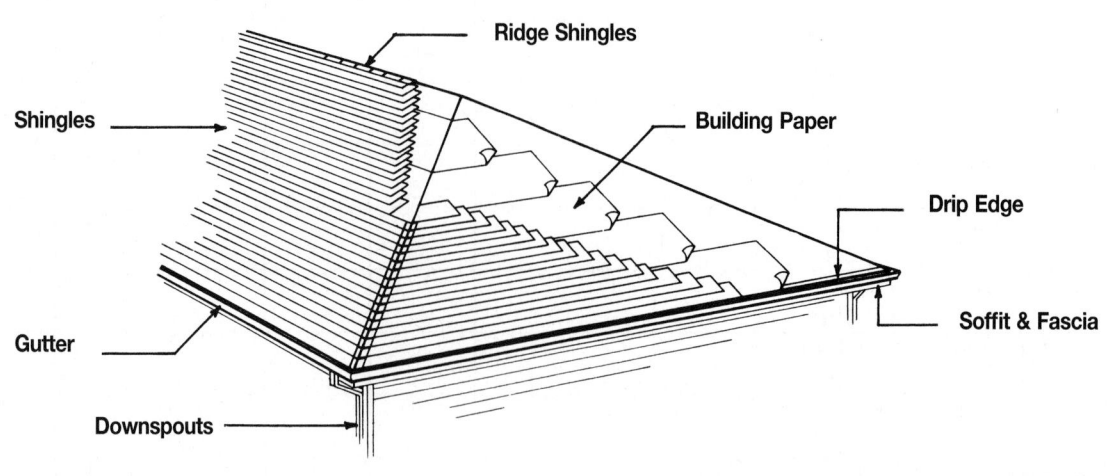

Ridge Shingles

Shingles

Building Paper

Drip Edge

Gutter

Soffit & Fascia

Downspouts

System Description	QUAN.	UNIT	LABOR HOURS	COST PER S.F.		
				MAT.	INST.	TOTAL
ASPHALT, ROOF SHINGLES, CLASS A						
Shingles, inorganic, class A, 210-235 lb./sq. 4/12 pitch	1.570	S.F.	.023	1.32	1.27	2.59
Drip edge, metal, 5″ wide	.122	L.F.	.002	.08	.14	.22
Building paper, #15 asphalt felt	1.800	S.F.	.002	.11	.14	.25
Ridge shingles, asphalt	.075	L.F.	.002	.19	.10	.29
Soffit & fascia, white painted aluminum, 1′ overhang	.120	L.F.	.017	.60	1.01	1.61
Gutter, seamless, aluminum, painted	.120	L.F.	.008	.38	.49	.87
Downspouts, aluminum, painted	.035	L.F.	.002	.08	.10	.18
Ridge vent	.028	L.F.	.001	.07	.08	.15
TOTAL		S.F.	.057	2.83	3.33	6.16
WOOD, CEDAR SHINGLES, NO. 1 PERFECTIONS, 18″ LONG						
Shingles, red cedar, No. 1 perfections, 5″ exp., 4/12 pitch	1.570	S.F.	.047	4.45	2.70	7.15
Drip edge, metal, 5″ wide	.122	L.F.	.002	.08	.14	.22
Building paper, #15 asphalt felt	1.800	S.F.	.002	.11	.14	.25
Ridge shingles, wood, cedar	.075	L.F.	.002	.39	.12	.51
Soffit & fascia, white painted aluminum, 1′ overhang	.120	L.F.	.017	.60	1.01	1.61
Gutter, seamless, aluminum, painted	.120	L.F.	.008	.38	.49	.87
Downspouts, aluminum, painted	.035	L.F.	.002	.08	.10	.18
Ridge vent	.028	L.F.	.001	.07	.08	.15
TOTAL		S.F.	.081	6.16	4.78	10.94

The prices in these systems are based on a square foot of plan area.
All quantities have been adjusted accordingly.

Description	QUAN.	UNIT	LABOR HOURS	COST PER S.F.		
				MAT.	INST.	TOTAL

Hip Roof - Roofing Price Sheet

	QUAN.	UNIT	LABOR HOURS	COST PER S.F. MAT.	COST PER S.F. INST.	COST PER S.F. TOTAL
Shingles, asphalt, inorganic, class A, 210-235 lb./sq., 4/12 pitch	1.570	S.F.	.023	1.32	1.27	2.59
8/12 pitch	1.850	S.F.	.028	1.57	1.51	3.08
Laminated, multi-layered, 240-260 lb./sq., 4/12 pitch	1.570	S.F.	.028	1.94	1.55	3.49
8/12 pitch	1.850	S.F.	.034	2.30	1.84	4.14
Prem. laminated, multi-layered, 260-300 lb./sq., 4/12 pitch	1.570	S.F.	.037	2.53	2	4.53
8/12 pitch	1.850	S.F.	.043	3	2.38	5.38
Clay tile, Spanish tile, red, 4/12 pitch	1.570	S.F.	.071	8.90	3.82	12.72
8/12 pitch	1.850	S.F.	.084	10.55	4.54	15.09
Mission tile, red, 4/12 pitch	1.570	S.F.	.111	7.30	3.82	11.12
8/12 pitch	1.850	S.F.	.132	8.65	4.54	13.19
French tile, red, 4/12 pitch	1.570	S.F.	.095	20.50	3.50	24
8/12 pitch	1.850	S.F.	.113	24	4.16	28.16
Slate, Buckingham, Virginia, black, 4/12 pitch	1.570	S.F.	.073	9.75	4.02	13.77
8/12 pitch	1.850	S.F.	.087	11.60	4.77	16.37
Vermont, black or grey, 4/12 pitch	1.570	S.F.	.073	8.55	4.02	12.57
8/12 pitch	1.850	S.F.	.087	10.15	4.77	14.92
Wood, red cedar, No.1 5X, 16" long, 5" exposure, 4/12 pitch	1.570	S.F.	.051	5.20	2.98	8.18
8/12 pitch	1.850	S.F.	.061	6.20	3.53	9.73
Fire retardant, 4/12 pitch	1.570	S.F.	.051	6.15	2.98	9.13
8/12 pitch	1.850	S.F.	.061	7.35	3.53	10.88
18" long, No.1 perfections, 5" exposure, 4/12 pitch	1.570	S.F.	.047	4.45	2.70	7.15
8/12 pitch	1.850	S.F.	.055	5.30	3.21	8.51
Fire retardant, 4/12 pitch	1.570	S.F.	.047	5.40	2.70	8.10
8/12 pitch	1.850	S.F.	.055	6.45	3.21	9.66
Resquared & rebutted, 18" long, 6" exposure, 4/12 pitch	1.570	S.F.	.043	5.10	2.48	7.58
8/12 pitch	1.850	S.F.	.051	6.10	2.95	9.05
Fire retardant, 4/12 pitch	1.570	S.F.	.043	6.10	2.48	8.58
8/12 pitch	1.850	S.F.	.051	7.25	2.95	10.20
Wood shakes hand split, 24" long, 10" exposure, 4/12 pitch	1.570	S.F.	.051	5.30	2.98	8.28
8/12 pitch	1.850	S.F.	.061	6.25	3.53	9.78
Fire retardant, 4/12 pitch	1.570	S.F.	.051	6.25	2.98	9.23
8/12 pitch	1.850	S.F.	.061	7.40	3.53	10.93
18" long, 8" exposure, 4/12 pitch	1.570	S.F.	.064	4.80	3.71	8.51
8/12 pitch	1.850	S.F.	.076	5.70	4.41	10.11
Fire retardant, 4/12 pitch	1.570	S.F.	.064	5.75	3.71	9.46
8/12 pitch	1.850	S.F.	.076	6.85	4.41	11.26
Drip edge, metal, 5" wide	.122	L.F.	.002	.08	.14	.22
8" wide	.122	L.F.	.002	.11	.14	.25
Building paper, #15 asphalt felt	1.800	S.F.	.002	.11	.14	.25
Ridge shingles, asphalt	.075	L.F.	.002	.19	.10	.29
Clay	.075	L.F.	.003	.41	.50	.91
Slate	.075	L.F.	.003	.83	.16	.99
Wood, shingles	.075	L.F.	.002	.39	.12	.51
Shakes	.075	L.F.	.002	.39	.12	.51
Soffit & fascia, aluminum, vented, 1' overhang	.120	L.F.	.017	.60	1.01	1.61
2' overhang	.120	L.F.	.019	.87	1.12	1.99
Vinyl, vented, 1' overhang	.120	L.F.	.016	.62	.93	1.55
2' overhang	.120	L.F.	.017	.83	1.12	1.95
Wood, board fascia, plywood soffit, 1' overhang	.120	L.F.	.004	.03	.18	.21
2' overhang	.120	L.F.	.006	.04	.27	.31
Gutter, 5" box, aluminum, seamless, painted	.120	L.F.	.008	.38	.49	.87
Vinyl	.120	L.F.	.009	.22	.48	.70
Downspout, 2" x 3", aluminum, one story house	.035	L.F.	.002	.08	.10	.18
Two story house	.060	L.F.	.003	.09	.16	.25
Vinyl, one story house	.035	L.F.	.001	.05	.09	.14
Two story house	.060	L.F.	.003	.09	.16	.25

For customer support on your Residential Costs with RSMeans data, call 800.448.8182.

197

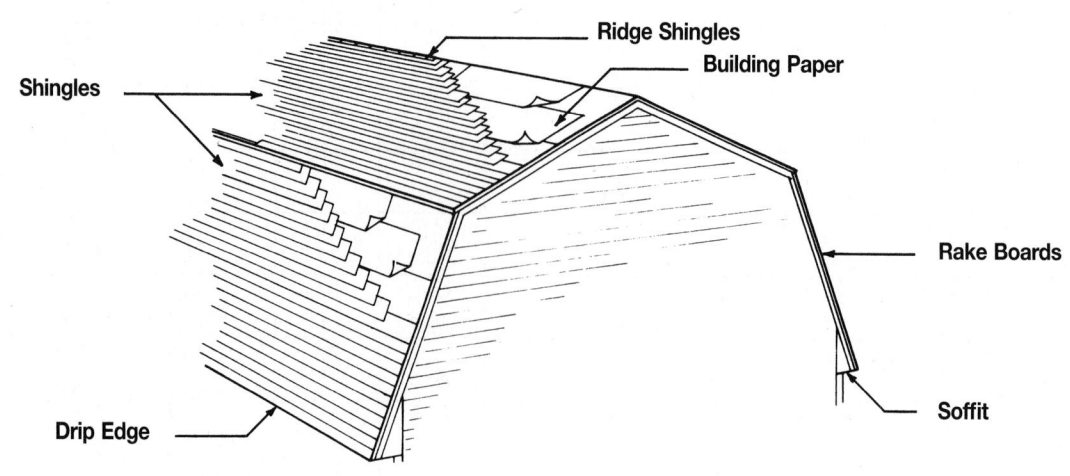

System Description	QUAN.	UNIT	LABOR HOURS	COST PER S.F.		
				MAT.	INST.	TOTAL
ASPHALT, ROOF SHINGLES, CLASS A						
Shingles, asphalt, inorganic, class A, 210-235 lb./sq.	1.450	S.F.	.022	1.24	1.19	2.43
Drip edge, metal, 5" wide	.146	L.F.	.003	.10	.17	.27
Building paper, #15 asphalt felt	1.500	S.F.	.002	.09	.11	.20
Ridge shingles, asphalt	.042	L.F.	.001	.11	.06	.17
Soffit & fascia, painted aluminum, 1' overhang	.083	L.F.	.012	.41	.70	1.11
Rake trim, 1" x 6"	.063	L.F.	.003	.07	.15	.22
Rake trim, prime and paint	.063	L.F.	.003	.02	.13	.15
Gutter, seamless, alumunum, painted	.083	L.F.	.005	.26	.34	.60
Downspouts, aluminum, painted	.042	L.F.	.002	.10	.12	.22
Ridge vent	.042	L.F.	.002	.11	.11	.22
TOTAL		S.F.	.055	2.51	3.08	5.59
WOOD, CEDAR SHINGLES, NO. 1 PERFECTIONS, 18" LONG						
Shingles, wood, red cedar, No. 1 perfections, 5" exposure	1.450	S.F.	.044	4.17	2.54	6.71
Drip edge, metal, 5" wide	.146	L.F.	.003	.10	.17	.27
Building paper, #15 asphalt felt	1.500	S.F.	.002	.09	.11	.20
Ridge shingles, wood	.042	L.F.	.001	.22	.07	.29
Soffit & fascia, white painted aluminum, 1' overhang	.083	L.F.	.012	.41	.70	1.11
Rake trim, 1" x 6"	.063	L.F.	.003	.07	.15	.22
Rake trim, prime and paint	.063	L.F.	.001	.02	.06	.08
Gutter, seamless, aluminum, painted	.083	L.F.	.005	.26	.34	.60
Downspouts, aluminum, painted	.042	L.F.	.002	.10	.12	.22
Ridge vent	.042	L.F.	.002	.11	.11	.22
TOTAL		S.F.	.075	5.55	4.37	9.92

The prices in this system are based on a square foot of plan area.
All quantities have been adjusted accordingly.

Description	QUAN.	UNIT	LABOR HOURS	COST PER S.F.		
				MAT.	INST.	TOTAL

Gambrel Roofing Price Sheet	QUAN.	UNIT	LABOR HOURS	COST PER S.F.		
				MAT.	INST.	TOTAL
Shingles, asphalt, standard, inorganic, class A, 210-235 lb./sq.	1.450	S.F.	.022	1.24	1.19	2.43
Laminated, multi-layered, 240-260 lb./sq.	1.450	S.F.	.027	1.82	1.46	3.28
Premium laminated, multi-layered, 260-300 lb./sq.	1.450	S.F.	.034	2.37	1.88	4.25
Slate, Buckingham, Virginia, black	1.450	S.F.	.069	9.15	3.77	12.92
Vermont, black or grey	1.450	S.F.	.069	8.05	3.77	11.82
Wood, red cedar, No.1 5X, 16" long, 5" exposure, plain	1.450	S.F.	.048	4.88	2.79	7.67
Fire retardant	1.450	S.F.	.048	5.80	2.79	8.59
18" long, No.1 perfections, 6" exposure, plain	1.450	S.F.	.044	4.17	2.54	6.71
Fire retardant	1.450	S.F.	.044	5.10	2.54	7.64
Resquared & rebutted, 18" long, 6" exposure, plain	1.450	S.F.	.040	4.80	2.33	7.13
Fire retardant	1.450	S.F.	.040	5.70	2.33	8.03
Shakes, hand split, 24" long, 10" exposure, plain	1.450	S.F.	.048	4.95	2.79	7.74
Fire retardant	1.450	S.F.	.048	5.85	2.79	8.64
18" long, 8" exposure, plain	1.450	S.F.	.060	4.50	3.48	7.98
Fire retardant	1.450	S.F.	.060	5.40	3.48	8.88
Drip edge, metal, 5" wide	.146	L.F.	.003	.10	.17	.27
8" wide	.146	L.F.	.003	.13	.17	.30
Building paper, #15 asphalt felt	1.500	S.F.	.002	.09	.11	.20
Ridge shingles, asphalt	.042	L.F.	.001	.11	.06	.17
Slate	.042	L.F.	.002	.47	.09	.56
Wood, shingles	.042	L.F.	.001	.22	.07	.29
Soffit & fascia, aluminum, vented, 1' overhang	.083	L.F.	.012	.41	.70	1.11
2' overhang	.083	L.F.	.013	.60	.77	1.37
Vinyl vented, 1' overhang	.083	L.F.	.011	.43	.64	1.07
2' overhang	.083	L.F.	.012	.57	.77	1.34
Wood board fascia, plywood soffit, 1' overhang	.083	L.F.	.004	.03	.18	.21
2' overhang	.083	L.F.	.006	.04	.27	.31
Rake trim, painted, 1" x 6"	.063	L.F.	.006	.09	.28	.37
1" x 8"	.063	L.F.	.007	.13	.37	.50
Gutter, 5" box, aluminum, seamless, painted	.083	L.F.	.006	.26	.34	.60
Vinyl	.083	L.F.	.006	.15	.34	.49
Downspout 2" x 3", aluminum, one story house	.042	L.F.	.002	.06	.11	.17
Two story house	.070	L.F.	.003	.11	.19	.30
Vinyl, one story house	.042	L.F.	.002	.06	.11	.17
Two story house	.070	L.F.	.003	.11	.19	.30

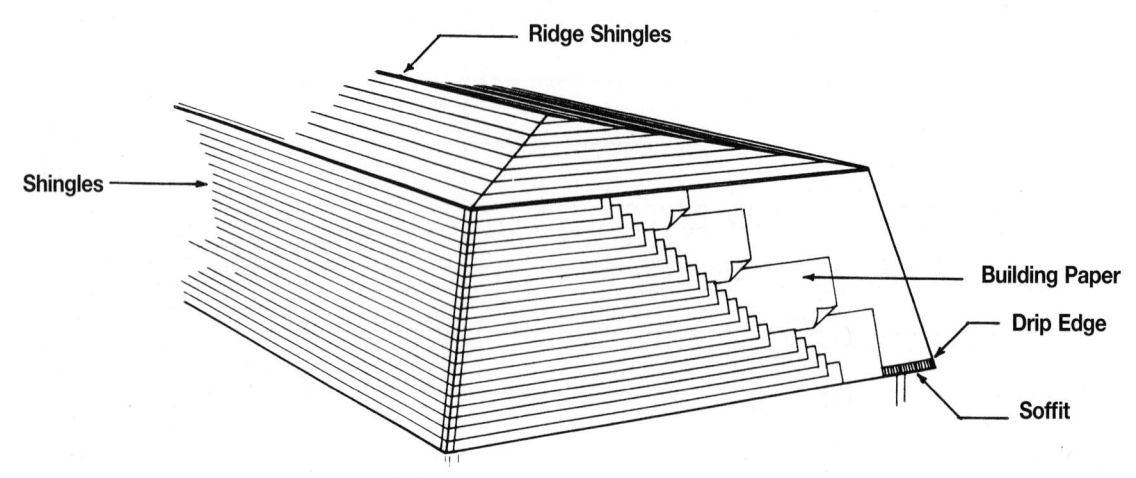

System Description	QUAN.	UNIT	LABOR HOURS	COST PER S.F.		
				MAT.	INST.	TOTAL
ASPHALT, ROOF SHINGLES, CLASS A						
Shingles, standard inorganic class A 210-235 lb./sq.	2.210	S.F.	.032	1.82	1.75	3.57
Drip edge, metal, 5″ wide	.122	L.F.	.002	.08	.14	.22
Building paper, #15 asphalt felt	2.300	S.F.	.003	.14	.17	.31
Ridge shingles, asphalt	.090	L.F.	.002	.23	.12	.35
Soffit & fascia, white painted aluminum, 1′ overhang	.122	L.F.	.018	.61	1.03	1.64
Gutter, seamless, aluminum, painted	.122	L.F.	.008	.39	.50	.89
Downspouts, aluminum, painted	.042	L.F.	.002	.10	.12	.22
Ridge vent	.028	L.F.	.001	.07	.08	.15
TOTAL		S.F.	.068	3.44	3.91	7.35
WOOD, CEDAR SHINGLES, NO. 1 PERFECTIONS, 18″ LONG						
Shingles, wood, red cedar, No. 1 perfections, 5″ exposure	2.210	S.F.	.064	6.12	3.72	9.84
Drip edge, metal, 5″ wide	.122	L.F.	.002	.08	.14	.22
Building paper, #15 asphalt felt	2.300	S.F.	.003	.14	.17	.31
Ridge shingles, wood	.090	L.F.	.003	.47	.15	.62
Soffit & fascia, white painted aluminum, 1′ overhang	.122	L.F.	.018	.61	1.03	1.64
Gutter, seamless, aluminum, painted	.122	L.F.	.008	.39	.50	.89
Downspouts, aluminum, painted	.042	L.F.	.002	.10	.12	.22
Ridge vent	.028	L.F.	.001	.07	.08	.15
TOTAL		S.F.	.101	7.98	5.91	13.89

The prices in these systems are based on a square foot of plan area.
All quantities have been adjusted accordingly.

Description	QUAN.	UNIT	LABOR HOURS	COST PER S.F.		
				MAT.	INST.	TOTAL

Mansard Roofing Price Sheet

	QUAN.	UNIT	LABOR HOURS	COST PER S.F.		
				MAT.	INST.	TOTAL
Shingles, asphalt, standard, inorganic, class A, 210-235 lb./sq.	2.210	S.F.	.032	1.82	1.75	3.57
Laminated, multi-layered, 240-260 lb./sq.	2.210	S.F.	.039	2.66	2.13	4.79
Premium laminated, multi-layered, 260-300 lb./sq.	2.210	S.F.	.050	3.48	2.75	6.23
Slate Buckingham, Virginia, black	2.210	S.F.	.101	13.40	5.50	18.90
Vermont, black or grey	2.210	S.F.	.101	11.75	5.50	17.25
Wood, red cedar, No.1 5X, 16" long, 5" exposure, plain	2.210	S.F.	.070	7.15	4.09	11.24
Fire retardant	2.210	S.F.	.070	8.50	4.09	12.59
18" long, No.1 perfections 6" exposure, plain	2.210	S.F.	.064	6.10	3.72	9.82
Fire retardant	2.210	S.F.	.064	7.45	3.72	11.17
Resquared & rebutted, 18" long, 6" exposure, plain	2.210	S.F.	.059	7.05	3.41	10.46
Fire retardant	2.210	S.F.	.059	8.35	3.41	11.76
Shakes, hand split, 24" long 10" exposure, plain	2.210	S.F.	.070	7.25	4.09	11.34
Fire retardant	2.210	S.F.	.070	8.60	4.09	12.69
18" long, 8" exposure, plain	2.210	S.F.	.088	6.60	5.10	11.70
Fire retardant	2.210	S.F.	.088	7.95	5.10	13.05
Drip edge, metal, 5" wide	.122	S.F.	.002	.08	.14	.22
8" wide	.122	S.F.	.002	.11	.14	.25
Building paper, #15 asphalt felt	2.300	S.F.	.003	.14	.17	.31
Ridge shingles, asphalt	.090	L.F.	.002	.23	.12	.35
Slate	.090	L.F.	.004	1	.20	1.20
Wood, shingles	.090	L.F.	.003	.47	.15	.62
Soffit & fascia, aluminum vented, 1' overhang	.122	L.F.	.018	.61	1.03	1.64
2' overhang	.122	L.F.	.020	.88	1.13	2.01
Vinyl vented, 1' overhang	.122	L.F.	.016	.63	.95	1.58
2' overhang	.122	L.F.	.018	.84	1.13	1.97
Wood board fascia, plywood soffit, 1' overhang	.122	L.F.	.013	.38	.71	1.09
2' overhang	.122	L.F.	.019	.53	1.09	1.62
Gutter, 5" box, aluminum, seamless, painted	.122	L.F.	.008	.39	.50	.89
Vinyl	.122	L.F.	.009	.22	.49	.71
Downspout 2" x 3", aluminum, one story house	.042	L.F.	.002	.06	.11	.17
Two story house	.070	L.F.	.003	.10	.18	.28
Vinyl, one story house	.042	L.F.	.002	.06	.11	.17
Two story house	.070	L.F.	.003	.10	.18	.28

For customer support on your Residential Costs with RSMeans data, call 800.448.8182.

201

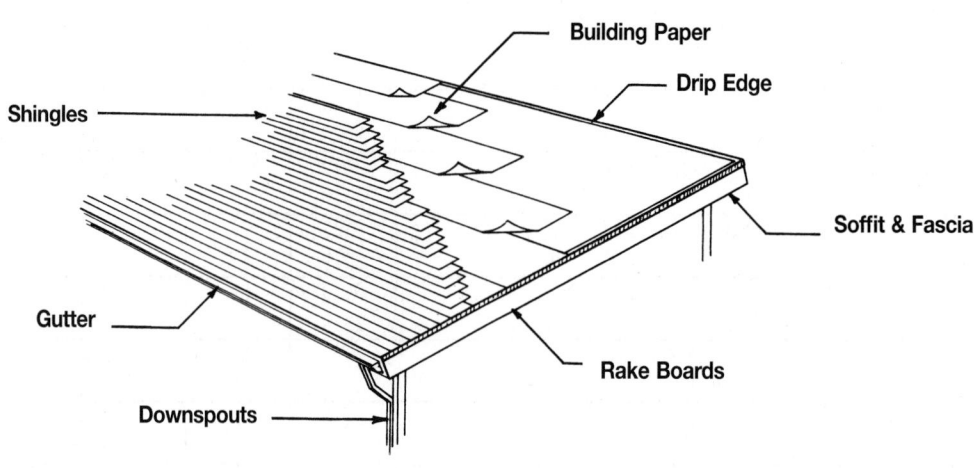

System Description	QUAN.	UNIT	LABOR HOURS	COST PER S.F.		
				MAT.	INST.	TOTAL
ASPHALT, ROOF SHINGLES, CLASS A						
Shingles, inorganic class A 210-235 lb./sq. 4/12 pitch	1.230	S.F.	.019	1.07	1.03	2.10
Drip edge, metal, 5" wide	.100	L.F.	.002	.07	.12	.19
Building paper, #15 asphalt felt	1.300	S.F.	.002	.08	.10	.18
Soffit & fascia, white painted aluminum, 1' overhang	.080	L.F.	.012	.40	.68	1.08
Rake trim, 1" x 6"	.043	L.F.	.002	.05	.10	.15
Rake trim, prime and paint	.043	L.F.	.002	.01	.09	.10
Gutter, seamless, aluminum, painted	.040	L.F.	.003	.13	.16	.29
Downspouts, painted aluminum	.020	L.F.	.001	.05	.06	.11
TOTAL		S.F.	.043	1.86	2.34	4.20
WOOD, CEDAR SHINGLES, NO. 1 PERFECTIONS, 18" LONG						
Shingles, red cedar, No. 1 perfections, 5" exp., 4/12 pitch	1.230	S.F.	.035	3.34	2.03	5.37
Drip edge, metal, 5" wide	.100	L.F.	.002	.07	.12	.19
Building paper, #15 asphalt felt	1.300	S.F.	.002	.08	.10	.18
Soffit & fascia, white painted aluminum, 1' overhang	.080	L.F.	.012	.40	.68	1.08
Rake trim, 1" x 6"	.043	L.F.	.002	.05	.10	.15
Rake trim, prime and paint	.043	L.F.	.001	.01	.04	.05
Gutter, seamless, aluminum, painted	.040	L.F.	.003	.13	.16	.29
Downspouts, painted aluminum	.020	L.F.	.001	.05	.06	.11
TOTAL		S.F.	.058	4.13	3.29	7.42

The prices in these systems are based on a square foot of plan area.
All quantities have been adjusted accordingly.

Description	QUAN.	UNIT	LABOR HOURS	COST PER S.F.		
				MAT.	INST.	TOTAL

Shed Roofing Price Sheet	QUAN.	UNIT	LABOR HOURS	COST PER S.F.		
				MAT.	INST.	TOTAL
Shingles, asphalt, inorganic, class A, 210-235 lb./sq., 4/12 pitch	1.230	S.F.	.017	.99	.95	1.94
8/12 pitch	1.330	S.F.	.019	1.07	1.03	2.10
Laminated, multi-layered, 240-260 lb./sq. 4/12 pitch	1.230	S.F.	.021	1.45	1.16	2.61
8/12 pitch	1.330	S.F.	.023	1.57	1.26	2.83
Premium laminated, multi-layered, 260-300 lb./sq. 4/12 pitch	1.230	S.F.	.027	1.90	1.50	3.40
8/12 pitch	1.330	S.F.	.030	2.05	1.63	3.68
Clay tile, Spanish tile, red, 4/12 pitch	1.230	S.F.	.053	6.65	2.87	9.52
8/12 pitch	1.330	S.F.	.058	7.20	3.11	10.31
Mission tile, red, 4/12 pitch	1.230	S.F.	.083	5.45	2.87	8.32
8/12 pitch	1.330	S.F.	.090	5.90	3.11	9.01
French tile, red, 4/12 pitch	1.230	S.F.	.071	15.30	2.63	17.93
8/12 pitch	1.330	S.F.	.077	16.60	2.85	19.45
Slate, Buckingham, Virginia, black, 4/12 pitch	1.230	S.F.	.055	7.30	3.01	10.31
8/12 pitch	1.330	S.F.	.059	7.95	3.26	11.21
Vermont, black or grey, 4/12 pitch	1.230	S.F.	.055	6.40	3.01	9.41
8/12 pitch	1.330	S.F.	.059	6.95	3.26	10.21
Wood, red cedar, No.1 5X, 16" long, 5" exposure, 4/12 pitch	1.230	S.F.	.038	3.90	2.23	6.13
8/12 pitch	1.330	S.F.	.042	4.23	2.42	6.65
Fire retardant, 4/12 pitch	1.230	S.F.	.038	4.63	2.23	6.86
8/12 pitch	1.330	S.F.	.042	5	2.42	7.42
18" long, 6" exposure, 4/12 pitch	1.230	S.F.	.035	3.34	2.03	5.37
8/12 pitch	1.330	S.F.	.038	3.61	2.20	5.81
Fire retardant, 4/12 pitch	1.230	S.F.	.035	4.07	2.03	6.10
8/12 pitch	1.330	S.F.	.038	4.40	2.20	6.60
Resquared & rebutted, 18" long, 6" exposure, 4/12 pitch	1.230	S.F.	.032	3.84	1.86	5.70
8/12 pitch	1.330	S.F.	.035	4.16	2.02	6.18
Fire retardant, 4/12 pitch	1.230	S.F.	.032	4.57	1.86	6.43
8/12 pitch	1.330	S.F.	.035	4.95	2.02	6.97
Wood shakes, hand split, 24" long, 10" exposure, 4/12 pitch	1.230	S.F.	.038	3.96	2.23	6.19
8/12 pitch	1.330	S.F.	.042	4.29	2.42	6.71
Fire retardant, 4/12 pitch	1.230	S.F.	.038	4.69	2.23	6.92
8/12 pitch	1.330	S.F.	.042	5.10	2.42	7.52
18" long, 8" exposure, 4/12 pitch	1.230	S.F.	.048	3.60	2.78	6.38
8/12 pitch	1.330	S.F.	.052	3.90	3.02	6.92
Fire retardant, 4/12 pitch	1.230	S.F.	.048	4.33	2.78	7.11
8/12 pitch	1.330	S.F.	.052	4.69	3.02	7.71
Drip edge, metal, 5" wide	.100	L.F.	.002	.07	.12	.19
8" wide	.100	L.F.	.002	.09	.12	.21
Building paper, #15 asphalt felt	1.300	S.F.	.002	.08	.10	.18
Soffit & fascia, aluminum vented, 1' overhang	.080	L.F.	.012	.40	.68	1.08
2' overhang	.080	L.F.	.013	.58	.74	1.32
Vinyl vented, 1' overhang	.080	L.F.	.011	.42	.62	1.04
2' overhang	.080	L.F.	.012	.55	.74	1.29
Wood board fascia, plywood soffit, 1' overhang	.080	L.F.	.010	.26	.52	.78
2' overhang	.080	L.F.	.014	.36	.79	1.15
Rake, trim, painted, 1" x 6"	.043	L.F.	.004	.06	.19	.25
1" x 8"	.043	L.F.	.004	.06	.19	.25
Gutter, 5" box, aluminum, seamless, painted	.040	L.F.	.003	.13	.16	.29
Vinyl	.040	L.F.	.003	.07	.16	.23
Downspout 2" x 3", aluminum, one story house	.020	L.F.	.001	.03	.05	.08
Two story house	.020	L.F.	.001	.05	.09	.14
Vinyl, one story house	.020	L.F.	.001	.03	.05	.08
Two story house	.020	L.F.	.001	.05	.09	.14

For customer support on your Residential Costs with RSMeans data, call 800.448.8182.

203

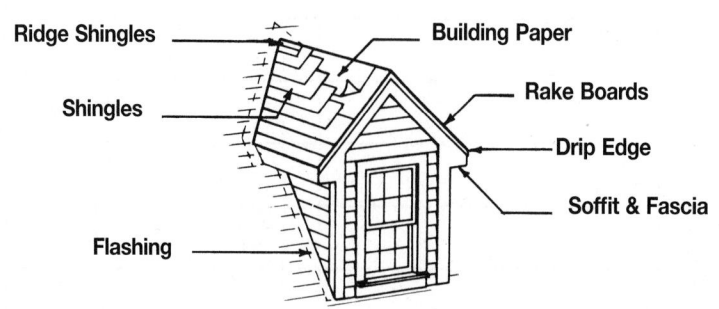

System Description	QUAN.	UNIT	LABOR HOURS	COST PER S.F.		
				MAT.	INST.	TOTAL
ASPHALT, ROOF SHINGLES, CLASS A						
Shingles, standard inorganic class A 210-235 lb./sq	1.400	S.F.	.020	1.16	1.11	2.27
Drip edge, metal, 5" wide	.220	L.F.	.004	.15	.26	.41
Building paper, #15 asphalt felt	1.500	S.F.	.002	.09	.11	.20
Ridge shingles, asphalt	.280	L.F.	.007	.70	.37	1.07
Soffit & fascia, aluminum, vented	.220	L.F.	.032	1.10	1.86	2.96
Flashing, aluminum, mill finish, .013" thick	1.500	S.F.	.083	1.37	4.52	5.89
TOTAL		S.F.	.148	4.57	8.23	12.80
WOOD, CEDAR, NO. 1 PERFECTIONS						
Shingles, red cedar, No.1 perfections, 18" long, 5" exp.	1.400	S.F.	.041	3.89	2.37	6.26
Drip edge, metal, 5" wide	.220	L.F.	.004	.15	.26	.41
Building paper, #15 asphalt felt	1.500	S.F.	.002	.09	.11	.20
Ridge shingles, wood	.280	L.F.	.008	1.47	.46	1.93
Soffit & fascia, aluminum, vented	.220	L.F.	.032	1.10	1.86	2.96
Flashing, aluminum, mill finish, .013" thick	1.500	S.F.	.083	1.37	4.52	5.89
TOTAL		S.F.	.170	8.07	9.58	17.65
SLATE, BUCKINGHAM, BLACK						
Shingles, Buckingham, Virginia, black	1.400	S.F.	.064	8.54	3.51	12.05
Drip edge, metal, 5" wide	.220	L.F.	.004	.15	.26	.41
Building paper, #15 asphalt felt	1.500	S.F.	.002	.09	.11	.20
Ridge shingles, slate	.280	L.F.	.011	3.11	.61	3.72
Soffit & fascia, aluminum, vented	.220	L.F.	.032	1.10	1.86	2.96
Flashing, copper, 16 oz.	1.500	S.F.	.104	13.35	5.69	19.04
TOTAL		S.F.	.217	26.34	12.04	38.38

The prices in these systems are based on a square foot of plan area under the dormer roof.

Description	QUAN.	UNIT	LABOR HOURS	COST PER S.F.		
				MAT.	INST.	TOTAL

Gable Dormer Roofing Price Sheet	QUAN.	UNIT	LABOR HOURS	COST PER S.F.		
				MAT.	INST.	TOTAL
Shingles, asphalt, standard, inorganic, class A, 210-235 lb./sq.	1.400	S.F.	.020	1.16	1.11	2.27
Laminated, multi-layered, 240-260 lb./sq.	1.400	S.F.	.025	1.69	1.36	3.05
Premium laminated, multi-layered, 260-300 lb./sq.	1.400	S.F.	.032	2.21	1.75	3.96
Clay tile, Spanish tile, red	1.400	S.F.	.062	7.75	3.35	11.10
Mission tile, red	1.400	S.F.	.097	6.35	3.35	9.70
French tile, red	1.400	S.F.	.083	17.85	3.07	20.92
Slate Buckingham, Virginia, black	1.400	S.F.	.064	8.55	3.51	12.06
Vermont, black or grey	1.400	S.F.	.064	7.50	3.51	11.01
Wood, red cedar, No.1 5X, 16" long, 5" exposure	1.400	S.F.	.045	4.55	2.60	7.15
Fire retardant	1.400	S.F.	.045	5.40	2.60	8
18" long, No.1 perfections, 5" exposure	1.400	S.F.	.041	3.89	2.37	6.26
Fire retardant	1.400	S.F.	.041	4.74	2.37	7.11
Resquared & rebutted, 18" long, 5" exposure	1.400	S.F.	.037	4.48	2.17	6.65
Fire retardant	1.400	S.F.	.037	5.35	2.17	7.52
Shakes hand split, 24" long, 10" exposure	1.400	S.F.	.045	4.62	2.60	7.22
Fire retardant	1.400	S.F.	.045	5.45	2.60	8.05
18" long, 8" exposure	1.400	S.F.	.056	4.20	3.25	7.45
Fire retardant	1.400	S.F.	.056	5.05	3.25	8.30
Drip edge, metal, 5" wide	.220	L.F.	.004	.15	.26	.41
8" wide	.220	L.F.	.004	.20	.26	.46
Building paper, #15 asphalt felt	1.500	S.F.	.002	.09	.11	.20
Ridge shingles, asphalt	.280	L.F.	.007	.70	.37	1.07
Clay	.280	L.F.	.011	1.53	1.85	3.38
Slate	.280	L.F.	.011	3.11	.61	3.72
Wood	.280	L.F.	.008	1.47	.46	1.93
Soffit & fascia, aluminum, vented	.220	L.F.	.032	1.10	1.86	2.96
Vinyl, vented	.220	L.F.	.029	1.14	1.71	2.85
Wood, board fascia, plywood soffit	.220	L.F.	.026	.70	1.42	2.12
Flashing, aluminum, .013" thick	1.500	S.F.	.083	1.37	4.52	5.89
.032" thick	1.500	S.F.	.083	2.24	4.52	6.76
.040" thick	1.500	S.F.	.083	3.78	4.52	8.30
.050" thick	1.500	S.F.	.083	4.55	4.52	9.07
Copper, 16 oz.	1.500	S.F.	.104	13.35	5.70	19.05
20 oz.	1.500	S.F.	.109	17.70	5.95	23.65
24 oz.	1.500	S.F.	.114	24.50	6.25	30.75
32 oz.	1.500	S.F.	.120	31.50	6.55	38.05

System Description	QUAN.	UNIT	LABOR HOURS	COST PER S.F.		
				MAT.	INST.	TOTAL
ASPHALT, ROOF SHINGLES, CLASS A						
Shingles, standard inorganic class A 210-235 lb./sq.	1.100	S.F.	.016	.91	.87	1.78
Drip edge, aluminum, 5″ wide	.250	L.F.	.005	.16	.29	.45
Building paper, #15 asphalt felt	1.200	S.F.	.002	.07	.09	.16
Soffit & fascia, aluminum, vented, 1′ overhang	.250	L.F.	.036	1.25	2.11	3.36
Flashing, aluminum, mill finish, 0.013″ thick	.800	L.F.	.044	.73	2.41	3.14
TOTAL		S.F.	.103	3.12	5.77	8.89
WOOD, CEDAR, NO. 1 PERFECTIONS, 18″ LONG						
Shingles, wood, red cedar, #1 perfections, 5″ exposure	1.100	S.F.	.032	3.06	1.86	4.92
Drip edge, aluminum, 5″ wide	.250	L.F.	.005	.16	.29	.45
Building paper, #15 asphalt felt	1.200	S.F.	.002	.07	.09	.16
Soffit & fascia, aluminum, vented, 1′ overhang	.250	L.F.	.036	1.25	2.11	3.36
Flashing, aluminum, mill finish, 0.013″ thick	.800	L.F.	.044	.73	2.41	3.14
TOTAL		S.F.	.119	5.27	6.76	12.03
SLATE, BUCKINGHAM, BLACK						
Shingles, slate, Buckingham, black	1.100	S.F.	.050	6.71	2.76	9.47
Drip edge, aluminum, 5″ wide	.250	L.F.	.005	.16	.29	.45
Building paper, #15 asphalt felt	1.200	S.F.	.002	.07	.09	.16
Soffit & fascia, aluminum, vented, 1′ overhang	.250	L.F.	.036	1.25	2.11	3.36
Flashing, copper, 16 oz.	.800	L.F.	.056	7.12	3.03	10.15
TOTAL		S.F.	.149	15.31	8.28	23.59

The prices in this system are based on a square foot of plan area under the dormer roof.

Description	QUAN.	UNIT	LABOR HOURS	COST PER S.F.		
				MAT.	INST.	TOTAL

Shed Dormer Roofing Price Sheet

	QUAN.	UNIT	LABOR HOURS	COST PER S.F.		
				MAT.	INST.	TOTAL
Shingles, asphalt, standard, inorganic, class A, 210-235 lb./sq.	1.100	S.F.	.016	.91	.87	1.78
Laminated, multi-layered, 240-260 lb./sq.	1.100	S.F.	.020	1.33	1.07	2.40
Premium laminated, multi-layered, 260-300 lb./sq.	1.100	S.F.	.025	1.74	1.38	3.12
Clay tile, Spanish tile, red	1.100	S.F.	.049	6.10	2.63	8.73
Mission tile, red	1.100	S.F.	.077	5	2.63	7.63
French tile, red	1.100	S.F.	.065	14.05	2.41	16.46
Slate Buckingham, Virginia, black	1.100	S.F.	.050	6.70	2.76	9.46
Vermont, black or grey	1.100	S.F.	.050	5.90	2.76	8.66
Wood, red cedar, No. 1 5X, 16" long, 5" exposure	1.100	S.F.	.035	3.58	2.05	5.63
Fire retardant	1.100	S.F.	.035	4.25	2.05	6.30
18" long, No.1 perfections, 5" exposure	1.100	S.F.	.032	3.06	1.86	4.92
Fire retardant	1.100	S.F.	.032	3.73	1.86	5.59
Resquared & rebutted, 18" long, 5" exposure	1.100	S.F.	.029	3.52	1.71	5.23
Fire retardant	1.100	S.F.	.029	4.19	1.71	5.90
Shakes hand split, 24" long, 10" exposure	1.100	S.F.	.035	3.63	2.05	5.68
Fire retardant	1.100	S.F.	.035	4.30	2.05	6.35
18" long, 8" exposure	1.100	S.F.	.044	3.30	2.55	5.85
Fire retardant	1.100	S.F.	.044	3.97	2.55	6.52
Drip edge, metal, 5" wide	.250	L.F.	.005	.16	.29	.45
8" wide	.250	L.F.	.005	.23	.29	.52
Building paper, #15 asphalt felt	1.200	S.F.	.002	.07	.09	.16
Soffit & fascia, aluminum, vented	.250	L.F.	.036	1.25	2.11	3.36
Vinyl, vented	.250	L.F.	.033	1.30	1.94	3.24
Wood, board fascia, plywood soffit	.250	L.F.	.030	.81	1.63	2.44
Flashing, aluminum, .013" thick	.800	L.F.	.044	.73	2.41	3.14
.032" thick	.800	L.F.	.044	1.19	2.41	3.60
.040" thick	.800	L.F.	.044	2.02	2.41	4.43
.050" thick	.800	L.F.	.044	2.42	2.41	4.83
Copper, 16 oz.	.800	L.F.	.056	7.10	3.03	10.13
20 oz.	.800	L.F.	.058	9.45	3.17	12.62
24 oz.	.800	L.F.	.061	13	3.32	16.32
32 oz.	.800	L.F.	.064	16.80	3.49	20.29

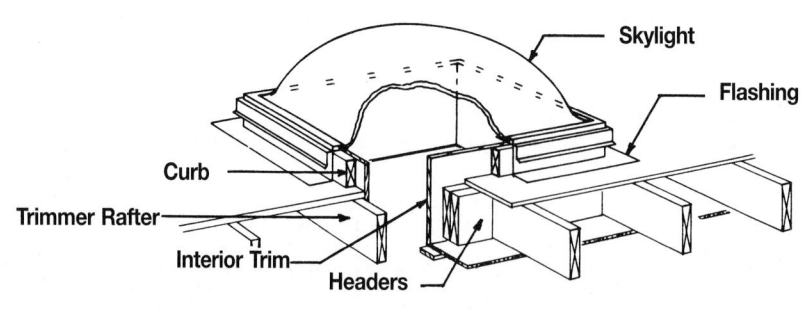

System Description	QUAN.	UNIT	LABOR HOURS	COST EACH		
				MAT.	INST.	TOTAL
SKYLIGHT, FIXED, 32″ X 32″						
Skylight, fixed bubble, insulating, 32″ x 32″	1.000	Ea.	1.422	234.66	76.80	311.46
Trimmer rafters, 2″ x 6″	28.000	L.F.	.448	19.88	26.04	45.92
Headers, 2″ x 6″	6.000	L.F.	.267	4.26	15.48	19.74
Curb, 2″ x 4″	12.000	L.F.	.154	5.52	8.88	14.40
Flashing, aluminum, .013″ thick	13.500	S.F.	.745	12.29	40.64	52.93
Moldings, casing, ogee, 11/16″ x 2-1/2″, pine	12.000	L.F.	.384	18.72	22.32	41.04
Trim primer coat, oil base, brushwork	12.000	L.F.	.148	.48	7.08	7.56
Trim paint, 1 coat, brushwork	12.000	L.F.	.148	.72	7.08	7.80
TOTAL		Ea.	3.716	296.53	204.32	500.85
SKYLIGHT, FIXED, 48″ X 48″						
Skylight, fixed bubble, insulating, 48″ x 48″	1.000	Ea.	1.296	560	70.08	630.08
Trimmer rafters, 2″ x 6″	28.000	L.F.	.448	19.88	26.04	45.92
Headers, 2″ x 6″	8.000	L.F.	.356	5.68	20.64	26.32
Curb, 2″ x 4″	16.000	L.F.	.205	7.36	11.84	19.20
Flashing, aluminum, .013″ thick	16.000	S.F.	.883	14.56	48.16	62.72
Moldings, casing, ogee, 11/16″ x 2-1/2″, pine	16.000	L.F.	.512	24.96	29.76	54.72
Trim primer coat, oil base, brushwork	16.000	L.F.	.197	.64	9.44	10.08
Trim paint, 1 coat, brushwork	16.000	L.F.	.197	.96	9.44	10.40
TOTAL		Ea.	4.094	634.04	225.40	859.44
SKYWINDOW, OPERATING, 24″ X 48″						
Skywindow, operating, thermopane glass, 24″ x 48″	1.000	Ea.	3.200	655	173	828
Trimmer rafters, 2″ x 6″	28.000	L.F.	.448	19.88	26.04	45.92
Headers, 2″ x 6″	8.000	L.F.	.267	4.26	15.48	19.74
Curb, 2″ x 4″	14.000	L.F.	.179	6.44	10.36	16.80
Flashing, aluminum, .013″ thick	14.000	S.F.	.772	12.74	42.14	54.88
Moldings, casing, ogee, 11/16″ x 2-1/2″, pine	14.000	L.F.	.448	21.84	26.04	47.88
Trim primer coat, oil base, brushwork	14.000	L.F.	.172	.56	8.26	8.82
Trim paint, 1 coat, brushwork	14.000	L.F.	.172	.84	8.26	9.10
TOTAL		Ea.	5.658	721.56	309.58	1,031.14

The prices in these systems are on a cost each basis.

Description	QUAN.	UNIT	LABOR HOURS	COST EACH		
				MAT.	INST.	TOTAL

Skylight/Skywindow Price Sheet	QUAN.	UNIT	LABOR HOURS	COST EACH		
				MAT.	INST.	TOTAL
Skylight, fixed bubble insulating, 24" x 24"	1.000	Ea.	.800	132	43	175
32" x 32"	1.000	Ea.	1.422	235	77	312
32" x 48"	1.000	Ea.	.864	375	46.50	421.50
48" x 48"	1.000	Ea.	1.296	560	70	630
Ventilating bubble insulating, 36" x 36"	1.000	Ea.	2.667	530	144	674
52" x 52"	1.000	Ea.	2.667	735	144	879
28" x 52"	1.000	Ea.	3.200	545	173	718
36" x 52"	1.000	Ea.	3.200	595	173	768
Skywindow, operating, thermopane glass, 24" x 48"	1.000	Ea.	3.200	655	173	828
32" x 48"	1.000	Ea.	3.556	685	192	877
Trimmer rafters, 2" x 6"	28.000	L.F.	.448	19.90	26	45.90
2" x 8"	28.000	L.F.	.472	27.50	27.50	55
2" x 10"	28.000	L.F.	.711	43.50	41.50	85
Headers, 24" window, 2" x 6"	4.000	L.F.	.178	2.84	10.30	13.14
2" x 8"	4.000	L.F.	.188	3.92	10.90	14.82
2" x 10"	4.000	L.F.	.200	6.20	11.65	17.85
32" window, 2" x 6"	6.000	L.F.	.267	4.26	15.50	19.76
2" x 8"	6.000	L.F.	.282	5.90	16.40	22.30
2" x 10"	6.000	L.F.	.300	9.30	17.45	26.75
48" window, 2" x 6"	8.000	L.F.	.356	5.70	20.50	26.20
2" x 8"	8.000	L.F.	.376	7.85	22	29.85
2" x 10"	8.000	L.F.	.400	12.40	23.50	35.90
Curb, 2" x 4", skylight, 24" x 24"	8.000	L.F.	.102	3.68	5.90	9.58
32" x 32"	12.000	L.F.	.154	5.50	8.90	14.40
32" x 48"	14.000	L.F.	.179	6.45	10.35	16.80
48" x 48"	16.000	L.F.	.205	7.35	11.85	19.20
Flashing, aluminum .013" thick, skylight, 24" x 24"	9.000	S.F.	.497	8.20	27	35.20
32" x 32"	13.500	S.F.	.745	12.30	40.50	52.80
32" x 48"	14.000	S.F.	.772	12.75	42	54.75
48" x 48"	16.000	S.F.	.883	14.55	48	62.55
Copper 16 oz., skylight, 24" x 24"	9.000	S.F.	.626	80	34	114
32" x 32"	13.500	S.F.	.939	120	51	171
32" x 48"	14.000	S.F.	.974	125	53	178
48" x 48"	16.000	S.F.	1.113	142	60.50	202.50
Trim, interior casing painted, 24" x 24"	8.000	L.F.	.347	13.70	18.70	32.40
32" x 32"	12.000	L.F.	.520	20.50	28	48.50
32" x 48"	14.000	L.F.	.607	24	33	57
48" x 48"	16.000	L.F.	.693	27.50	37.50	65

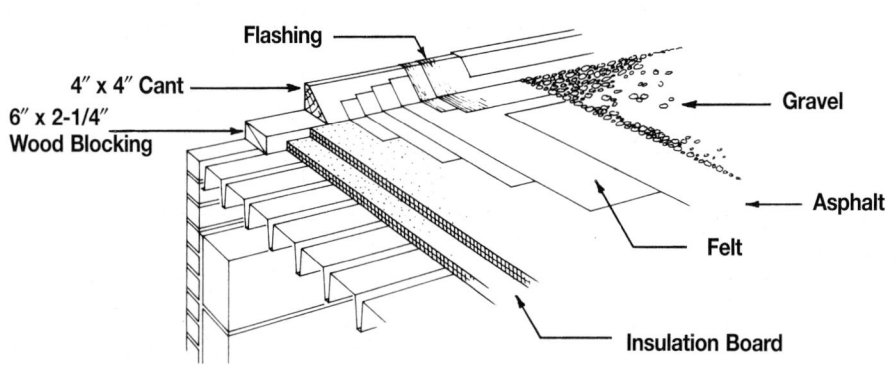

System Description	QUAN.	UNIT	LABOR HOURS	COST PER S.F.		
				MAT.	INST.	TOTAL
ASPHALT, ORGANIC, 4-PLY, INSULATED DECK						
Membrane, asphalt, 4-plies #15 felt, gravel surfacing	1.000	S.F.	.025	1.48	1.56	3.04
Insulation board, 2-layers of 1-1/16" glass fiber	2.000	S.F.	.012	2.36	.68	3.04
Roof deck insulation, fastening alternatives, coated screws, 4" long	1.000	S.F	.003	.12	.14	.26
Wood blocking, 2" x 6"	.040	L.F.	.004	.09	.25	.34
Treated 4" x 4" cant strip	.040	L.F.	.001	.08	.06	.14
Flashing, aluminum, 0.040" thick	.050	S.F.	.003	.13	.15	.28
TOTAL		S.F.	.048	4.26	2.84	7.10
ASPHALT, INORGANIC, 3-PLY, INSULATED DECK						
Membrane, asphalt, 3-plies type IV glass felt, gravel surfacing	1.000	S.F.	.028	1.47	1.72	3.19
Insulation board, 2-layers of 1-1/16" glass fiber	2.000	S.F.	.012	2.36	.68	3.04
Roof deck insulation, fastening alternatives, coated screws, 4" long	1.000	S.F	.003	.12	.14	.26
Wood blocking, 2" x 6"	.040	L.F.	.004	.09	.25	.34
Treated 4" x 4" cant strip	.040	L.F.	.001	.08	.06	.14
Flashing, aluminum, 0.040" thick	.050	S.F.	.003	.13	.15	.28
TOTAL		S.F.	.051	4.25	3	7.25
COAL TAR, ORGANIC, 4-PLY, INSULATED DECK						
Membrane, coal tar, 4-plies #15 felt, gravel surfacing	1.000	S.F.	.027	2.19	1.63	3.82
Insulation board, 2-layers of 1-1/16" glass fiber	2.000	S.F.	.012	2.36	.68	3.04
Roof deck insulation, fastening alternatives, coated screws, 4" long	1.000	S.F	.003	.12	.14	.26
Wood blocking, 2" x 6"	.040	L.F.	.004	.09	.25	.34
Treated 4" x 4" cant strip	.040	L.F.	.001	.08	.06	.14
Flashing, aluminum, 0.040" thick	.050	S.F.	.003	.13	.15	.28
TOTAL		S.F.	.050	4.97	2.91	7.88
COAL TAR, INORGANIC, 3-PLY, INSULATED DECK						
Membrane, coal tar, 3-plies type IV glass felt, gravel surfacing	1.000	S.F.	.029	1.81	1.81	3.62
Insulation board, 2-layers of 1-1/16" glass fiber	2.000	S.F.	.012	2.36	.68	3.04
Roof deck insulation, fastening alternatives, coated screws, 4" long	1.000	S.F	.003	.12	.14	.26
Wood blocking, 2" x 6"	.040	L.F.	.004	.09	.25	.34
Treated 4" x 4" cant strip	.040	L.F.	.001	.08	.06	.14
Flashing, aluminum, 0.040" thick	.050	S.F.	.003	.13	.15	.28
TOTAL		S.F.	.052	4.59	3.09	7.68

Built-Up Roofing Price Sheet	QUAN.	UNIT	LABOR HOURS	COST PER S.F.		
				MAT.	INST.	TOTAL
Membrane, asphalt, 4-plies #15 organic felt, gravel surfacing	1.000	S.F.	.025	1.48	1.56	3.04
Asphalt base sheet & 3-plies #15 asphalt felt	1.000	S.F.	.025	1.14	1.56	2.70
3-plies type IV glass fiber felt	1.000	S.F.	.028	1.47	1.72	3.19
4-plies type IV glass fiber felt	1.000	S.F.	.028	1.82	1.72	3.54
Coal tar, 4-plies #15 organic felt, gravel surfacing	1.000	S.F.	.027			
4-plies tarred felt	1.000	S.F.	.027	2.19	1.63	3.82
3-plies type IV glass fiber felt	1.000	S.F.	.029	1.81	1.81	3.62
4-plies type IV glass fiber felt	1.000	S.F.	.027	2.51	1.63	4.14
Roll, asphalt, 1-ply #15 organic felt, 2-plies mineral surfaced	1.000	S.F.	.021	.78	1.27	2.05
3-plies type IV glass fiber, 1-ply mineral surfaced	1.000	S.F.	.022	1.35	1.37	2.72
Insulation boards, glass fiber, 1-1/16" thick	1.000	S.F.	.008	1.30	.48	1.78
2-1/16" thick	1.000	S.F.	.010	1.79	.56	2.35
2-7/16" thick	1.000	S.F.	.010	2.01	.56	2.57
Expanded perlite, 1" thick	1.000	S.F.	.010	.71	.56	1.27
1-1/2" thick	1.000	S.F.	.010	1	.56	1.56
2" thick	1.000	S.F.	.011	1.31	.62	1.93
Fiberboard, 1" thick	1.000	S.F.	.010	.76	.56	1.32
1-1/2" thick	1.000	S.F.	.010	1.10	.56	1.66
2" thick	1.000	S.F.	.010	1.39	.56	1.95
Extruded polystyrene, 15 PSI compressive strength, 2" thick R10	1.000	S.F.	.006	.91	.41	1.32
3" thick R15	1.000	S.F.	.008	1.69	.48	2.17
4" thick R20	1.000	S.F.	.008	2.24	.48	2.72
Tapered for drainage	1.000	S.F.	.005	.68	.36	1.04
40 PSI compressive strength, 1" thick R5	1.000	S.F.	.005	1.08	.36	1.44
2" thick R10	1.000	S.F.	.006	1.94	.41	2.35
3" thick R15	1.000	S.F.	.008	2.75	.48	3.23
4" thick R20	1.000	S.F.	.008	3.57	.48	4.05
Fiberboard high density, 1/2" thick R1.3	1.000	S.F.	.008	.42	.48	.90
1" thick R2.5	1.000	S.F.	.010	.74	.56	1.30
1 1/2" thick R3.8	1.000	S.F.	.010	1.03	.56	1.59
Polyisocyanurate, 1 1/2" thick	1.000	S.F.	.006	.76	.41	1.17
2" thick	1.000	S.F.	.007	.93	.44	1.37
3 1/2" thick	1.000	S.F.	.008	2	.48	2.48
Tapered for drainage	1.000	S.F.	.006	.69	.38	1.07
Expanded polystyrene, 1" thick	1.000	S.F.	.005	.42	.36	.78
2" thick R10	1.000	S.F.	.006	.71	.41	1.12
3" thick R11	1.000	S.F.	.006	1.01	.41	1.42
Wood blocking, treated, 6" x 2" & 4" x 4" cant	.040	L.F.	.002	.14	.14	.28
6" x 4-1/2" & 4" x 4" cant	.040	L.F.	.005	.21	.31	.52
6" x 5" & 4" x 4" cant	.040	L.F.	.007	.24	.39	.63
Flashing, aluminum, 0.019" thick	.050	S.F.	.003	.08	.15	.23
0.032" thick	.050	S.F.	.003	.07	.15	.22
0.040" thick	.050	S.F.	.003	.13	.15	.28
Copper sheets, 16 oz., under 500 lbs.	.050	S.F.	.003	.45	.19	.64
Over 500 lbs.	.050	S.F.	.003	.45	.14	.59
20 oz., under 500 lbs.	.050	S.F.	.004	.59	.20	.79
Over 500 lbs.	.050	S.F.	.003	.56	.15	.71
Stainless steel, 32 gauge	.050	S.F.	.003	.18	.14	.32
28 gauge	.050	S.F.	.003	.26	.14	.40
26 gauge	.050	S.F.	.003	.25	.14	.39
24 gauge	.050	S.F.	.003	.28	.14	.42

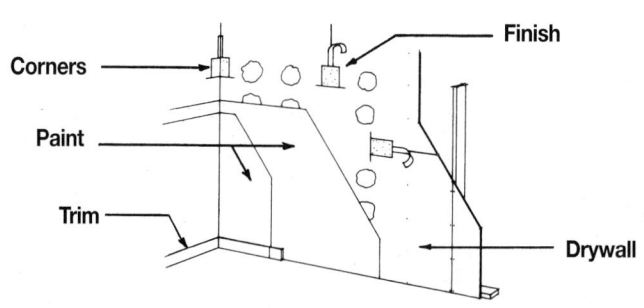

System Description	QUAN.	UNIT	LABOR HOURS	COST PER S.F.		
				MAT.	INST.	TOTAL
1/2″ DRYWALL, TAPED & FINISHED						
Gypsum wallboard, 1/2″ thick, standard	1.000	S.F.	.008	.37	.46	.83
Finish, taped & finished joints	1.000	S.F.	.008	.05	.46	.51
Corners, taped & finished, 32 L.F. per 12′ x 12′ room	.083	L.F.	.002	.01	.09	.10
Painting, primer & 2 coats	1.000	S.F.	.011	.22	.51	.73
Paint trim, to 6″ wide, primer + 1 coat enamel	.125	L.F.	.001	.02	.06	.08
Moldings, base, ogee profile, 9/16″ x 4-1/2, red oak	.125	L.F.	.005	.58	.26	.84
TOTAL		S.F.	.035	1.25	1.84	3.09
THINCOAT, SKIM-COAT, ON 1/2″ BACKER DRYWALL						
Gypsum wallboard, 1/2″ thick, thincoat backer	1.000	S.F.	.008	.37	.46	.83
Thincoat plaster	1.000	S.F.	.011	.12	.62	.74
Corners, taped & finished, 32 L.F. per 12′ x 12′ room	.083	L.F.	.002	.01	.09	.10
Painting, primer & 2 coats	1.000	S.F.	.011	.22	.51	.73
Paint trim, to 6″ wide, primer + 1 coat enamel	.125	L.F.	.001	.02	.06	.08
Moldings, base, ogee profile, 9/16″ x 4-1/2, red oak	.125	L.F.	.005	.58	.26	.84
TOTAL		S.F.	.038	1.32	2	3.32
5/8″ DRYWALL, TAPED & FINISHED						
Gypsum wallboard, 5/8″ thick, standard	1.000	S.F.	.008	.39	.46	.85
Finish, taped & finished joints	1.000	S.F.	.008	.05	.46	.51
Corners, taped & finished, 32 L.F. per 12′ x 12′ room	.083	L.F.	.002	.01	.09	.10
Painting, primer & 2 coats	1.000	S.F.	.011	.22	.51	.73
Moldings, base, ogee profile, 9/16″ x 4-1/2, red oak	.125	L.F.	.005	.58	.26	.84
Paint trim, to 6″ wide, primer + 1 coat enamel	.125	L.F.	.001	.02	.06	.08
TOTAL		S.F.	.035	1.27	1.84	3.11

The costs in this system are based on a square foot of wall.
Do not deduct for openings.

Description	QUAN.	UNIT	LABOR HOURS	COST PER S.F.		
				MAT.	INST.	TOTAL

Drywall & Thincoat Wall Price Sheet

	QUAN.	UNIT	LABOR HOURS	COST PER S.F. MAT.	COST PER S.F. INST.	COST PER S.F. TOTAL
Gypsum wallboard, 1/2" thick, standard	1.000	S.F.	.008	.37	.46	.83
Fire resistant	1.000	S.F.	.008	.41	.46	.87
Water resistant	1.000	S.F.	.008	.46	.46	.92
5/8" thick, standard	1.000	S.F.	.008	.39	.46	.85
Fire resistant	1.000	S.F.	.008	.40	.46	.86
Water resistant	1.000	S.F.	.008	.48	.46	.94
Gypsum wallboard backer for thincoat system, 1/2" thick	1.000	S.F.	.008	.37	.46	.83
5/8" thick	1.000	S.F.	.008	.39	.46	.85
Gypsum wallboard, taped & finished	1.000	S.F.	.008	.05	.46	.51
Texture spray	1.000	S.F.	.010	.04	.56	.60
Thincoat plaster, including tape	1.000	S.F.	.011	.12	.62	.74
Gypsum wallboard corners, taped & finished, 32 L.F. per 4' x 4' room	.250	L.F.	.004	.03	.25	.28
6' x 6' room	.110	L.F.	.002	.01	.11	.12
10' x 10' room	.100	L.F.	.001	.01	.10	.11
12' x 12' room	.083	L.F.	.001	.01	.08	.09
16' x 16' room	.063	L.F.	.001	.01	.06	.07
Thincoat system, 32 L.F. per 4' x 4' room	.250	L.F.	.003	.03	.16	.19
6' x 6' room	.110	L.F.	.001	.01	.06	.07
10' x 10' room	.100	L.F.	.001	.01	.06	.07
12' x 12' room	.083	L.F.	.001	.01	.05	.06
16' x 16' room	.063	L.F.	.001	.01	.04	.05
Painting, primer, & 1 coat	1.000	S.F.	.008	.14	.40	.54
& 2 coats	1.000	S.F.	.011	.22	.51	.73
Wallpaper, $7/double roll	1.000	S.F.	.013	.67	.60	1.27
$17/double roll	1.000	S.F.	.015	1.31	.72	2.03
$40/double roll	1.000	S.F.	.018	2.32	.89	3.21
Wallcovering, medium weight vinyl	1.000	S.F.	.017	1	.81	1.81
Tile, ceramic adhesive thin set, 4 1/4" x 4 1/4" tiles	1.000	S.F.	.084	2.87	4.08	6.95
6" x 6" tiles	1.000	S.F.	.080	3.93	4.43	8.36
Pregrouted sheets	1.000	S.F.	.067	6.20	3.23	9.43
Trim, painted or stained, baseboard	.125	L.F.	.006	.60	.32	.92
Base shoe	.125	L.F.	.005	.07	.30	.37
Chair rail	.125	L.F.	.005	.23	.28	.51
Cornice molding	.125	L.F.	.004	.16	.28	.44
Cove base, vinyl	.125	L.F.	.003	.18	.17	.35
Paneling, not including furring or trim						
Plywood, prefinished, 1/4" thick, 4' x 8' sheets, vert. grooves						
Birch faced, minimum	1.000	S.F.	.032	1.62	1.86	3.48
Average	1.000	S.F.	.038	1.32	2.21	3.53
Maximum	1.000	S.F.	.046	1.21	2.66	3.87
Mahogany, African	1.000	S.F.	.040	2.70	2.32	5.02
Philippine (lauan)	1.000	S.F.	.032	.72	1.86	2.58
Oak or cherry, minimum	1.000	S.F.	.032	1.47	1.86	3.33
Maximum	1.000	S.F.	.040	2.20	2.32	4.52
Rosewood	1.000	S.F.	.050	3.41	2.91	6.32
Teak	1.000	S.F.	.040	3.43	2.32	5.75
Chestnut	1.000	S.F.	.043	5.80	2.48	8.28
Pecan	1.000	S.F.	.040	2.75	2.32	5.07
Walnut, minimum	1.000	S.F.	.032	2.72	1.86	4.58
Maximum	1.000	S.F.	.040	2.71	2.32	5.03

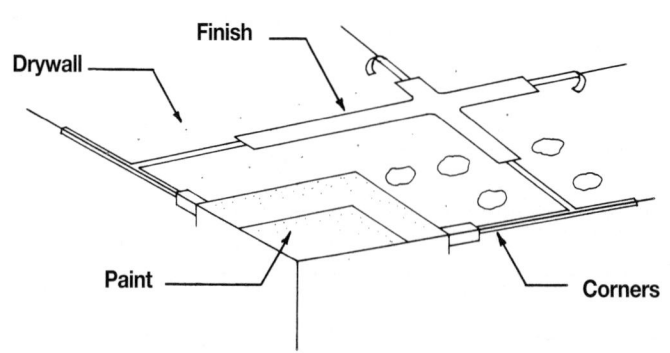

System Description	QUAN.	UNIT	LABOR HOURS	COST PER S.F.		
				MAT.	INST.	TOTAL
1/2″ GYPSUM WALLBOARD, TAPED & FINISHED						
Gypsum wallboard, 1/2″ thick, standard	1.000	S.F.	.008	.37	.46	.83
Finish, taped & finished	1.000	S.F.	.008	.05	.46	.51
Corners, taped & finished, 12′ x 12′ room	.333	L.F.	.006	.04	.32	.36
Paint, primer & 2 coats	1.000	S.F.	.011	.22	.51	.73
TOTAL		S.F.	.033	.68	1.75	2.43
THINCOAT, SKIM COAT ON 1/2″ GYPSUM WALLBOARD						
Gypsum wallboard, 1/2″ thick, thincoat backer	1.000	S.F.	.008	.37	.46	.83
Thincoat plaster	1.000	S.F.	.011	.12	.62	.74
Corners, taped & finished, 12′ x 12′ room	.333	L.F.	.006	.04	.32	.36
Paint, primer & 2 coats	1.000	S.F.	.011	.22	.51	.73
TOTAL		S.F.	.036	.75	1.91	2.66
WATER-RESISTANT GYPSUM WALLBOARD, 1/2″ THICK, TAPED & FINISHED						
Gypsum wallboard, 1/2″ thick, water-resistant	1.000	S.F.	.008	.46	.46	.92
Finish, taped & finished	1.000	S.F.	.008	.05	.46	.51
Corners, taped & finished, 12′ x 12′ room	.333	L.F.	.006	.04	.32	.36
Paint, primer & 2 coats	1.000	S.F.	.011	.22	.51	.73
TOTAL		S.F.	.033	.77	1.75	2.52
5/8″ GYPSUM WALLBOARD, TAPED & FINISHED						
Gypsum wallboard, 5/8″ thick, standard	1.000	S.F.	.008	.39	.46	.85
Finish, taped & finished	1.000	S.F.	.008	.05	.46	.51
Corners, taped & finished, 12′ x 12′ room	.333	L.F.	.006	.04	.32	.36
Paint, primer & 2 coats	1.000	S.F.	.011	.22	.51	.73
TOTAL		S.F.	.033	.70	1.75	2.45

The costs in this system are based on a square foot of ceiling.

Description	QUAN.	UNIT	LABOR HOURS	COST PER S.F.		
				MAT.	INST.	TOTAL

Drywall & Thincoat Ceilings Price Sheet

	QUAN.	UNIT	LABOR HOURS	COST PER S.F. MAT.	COST PER S.F. INST.	COST PER S.F. TOTAL
Gypsum wallboard ceilings, 1/2" thick, standard	1.000	S.F.	.008	.37	.46	.83
Fire resistant	1.000	S.F.	.008	.41	.46	.87
Water resistant	1.000	S.F.	.008	.46	.46	.92
5/8" thick, standard	1.000	S.F.	.008	.39	.46	.85
Fire resistant	1.000	S.F.	.008	.40	.46	.86
Water resistant	1.000	S.F.	.008	.48	.46	.94
Gypsum wallboard backer for thincoat ceiling system, 1/2" thick	1.000	S.F.	.016	.78	.92	1.70
5/8" thick	1.000	S.F.	.016	.80	.92	1.72
Gypsum wallboard ceilings, taped & finished	1.000	S.F.	.008	.05	.46	.51
Texture spray	1.000	S.F.	.010	.04	.56	.60
Thincoat plaster	1.000	S.F.	.011	.12	.62	.74
Corners taped & finished, 4' x 4' room	1.000	L.F.	.015	.11	.98	1.09
6' x 6' room	.667	L.F.	.010	.07	.66	.73
10' x 10' room	.400	L.F.	.006	.04	.39	.43
12' x 12' room	.333	L.F.	.005	.04	.32	.36
16' x 16' room	.250	L.F.	.003	.02	.18	.20
Thincoat system, 4' x 4' room	1.000	L.F.	.011	.12	.62	.74
6' x 6' room	.667	L.F.	.007	.08	.42	.50
10' x 10' room	.400	L.F.	.004	.05	.25	.30
12' x 12' room	.333	L.F.	.004	.04	.20	.24
16' x 16' room	.250	L.F.	.002	.02	.12	.14
Painting, primer & 1 coat	1.000	S.F.	.008	.14	.40	.54
& 2 coats	1.000	S.F.	.011	.22	.51	.73
Wallpaper, double roll, solid pattern, avg. workmanship	1.000	S.F.	.013	.67	.60	1.27
Basic pattern, avg. workmanship	1.000	S.F.	.015	1.31	.72	2.03
Basic pattern, quality workmanship	1.000	S.F.	.018	2.32	.89	3.21
Tile, ceramic adhesive thin set, 4 1/4" x 4 1/4" tiles	1.000	S.F.	.084	2.87	4.08	6.95
6" x 6" tiles	1.000	S.F.	.080	3.93	4.43	8.36
Pregrouted sheets	1.000	S.F.	.067	6.20	3.23	9.43

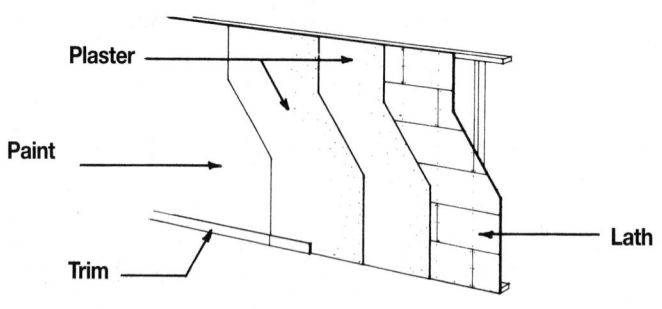

System Description	QUAN.	UNIT	LABOR HOURS	COST PER S.F.		
				MAT.	INST.	TOTAL
PLASTER ON GYPSUM LATH						
Plaster, gypsum or perlite, 2 coats	1.000	S.F.	.053	.46	2.99	3.45
Lath, 3/8" gypsum	1.000	S.F.	.010	.39	.58	.97
Corners, expanded metal, 32 L.F. per 12' x 12' room	.083	L.F.	.002	.01	.10	.11
Painting, primer & 2 coats	1.000	S.F.	.011	.22	.51	.73
Paint trim, to 6" wide, primer + 1 coat enamel	.125	L.F.	.001	.02	.06	.08
Moldings, base, ogee profile, 9/16" x 4-1/2, red oak	.125	L.F.	.005	.58	.26	.84
TOTAL		S.F.	.082	1.68	4.50	6.18
PLASTER ON METAL LATH						
Plaster, gypsum or perlite, 2 coats	1.000	S.F.	.053	.46	2.99	3.45
Lath, 2.5 Lb. diamond, metal	1.000	S.F.	.010	.46	.58	1.04
Corners, expanded metal, 32 L.F. per 12' x 12' room	.083	L.F.	.002	.01	.10	.11
Painting, primer & 2 coats	1.000	S.F.	.011	.22	.51	.73
Paint trim, to 6" wide, primer + 1 coat enamel	.125	L.F.	.001	.02	.06	.08
Moldings, base, ogee profile, 9/16" x 4-1/2, red oak	.125	L.F.	.005	.58	.26	.84
TOTAL		S.F.	.082	1.75	4.50	6.25
STUCCO ON METAL LATH						
Stucco, 2 coats	1.000	S.F.	.041	.41	2.29	2.70
Lath, 2.5 Lb. diamond, metal	1.000	S.F.	.010	.46	.58	1.04
Corners, expanded metal, 32 L.F. per 12' x 12' room	.083	L.F.	.002	.01	.10	.11
Painting, primer & 2 coats	1.000	S.F.	.011	.22	.51	.73
Paint trim, to 6" wide, primer + 1 coat enamel	.125	L.F.	.001	.02	.06	.08
Moldings, base, ogee profile, 9/16" x 4-1/2, red oak	.125	L.F.	.005	.58	.26	.84
TOTAL		S.F.	.070	1.70	3.80	5.50

The costs in these systems are based on a per square foot of wall area.
Do not deduct for openings.

Description	QUAN.	UNIT	LABOR HOURS	COST PER S.F.		
				MAT.	INST.	TOTAL

Plaster & Stucco Wall Price Sheet	QUAN.	UNIT	LABOR HOURS	COST PER S.F.		
				MAT.	INST.	TOTAL
Plaster, gypsum or perlite, 2 coats	1.000	S.F.	.053	.46	2.99	3.45
3 coats	1.000	S.F.	.065	.66	3.62	4.28
Lath, gypsum, standard, 3/8″ thick	1.000	S.F.	.010	.39	.58	.97
Fire resistant, 3/8″ thick	1.000	S.F.	.013	.31	.71	1.02
1/2″ thick	1.000	S.F.	.014	.35	.76	1.11
Metal, diamond, 2.5 Lb.	1.000	S.F.	.010	.46	.58	1.04
3.4 Lb.	1.000	S.F.	.012	.50	.66	1.16
Rib, 2.75 Lb.	1.000	S.F.	.012	.40	.66	1.06
3.4 Lb.	1.000	S.F.	.013	.52	.71	1.23
Corners, expanded metal, 32 L.F. per 4′ x 4′ room	.250	L.F.	.005	.04	.29	.33
6′ x 6′ room	.110	L.F.	.002	.02	.13	.15
10′ x 10′ room	.100	L.F.	.002	.02	.12	.14
12′ x 12′ room	.083	L.F.	.002	.01	.10	.11
16′ x 16′ room	.063	L.F.	.001	.01	.07	.08
Painting, primer & 1 coats	1.000	S.F.	.008	.14	.40	.54
Primer & 2 coats	1.000	S.F.	.011	.22	.51	.73
Wallpaper, low price double roll	1.000	S.F.	.013	.67	.60	1.27
Medium price double roll	1.000	S.F.	.015	1.31	.72	2.03
High price double roll	1.000	S.F.	.018	2.32	.89	3.21
Tile, ceramic thin set, 4-1/4″ x 4-1/4″ tiles	1.000	S.F.	.084	2.87	4.08	6.95
6″ x 6″ tiles	1.000	S.F.	.080	3.93	4.43	8.36
Pregrouted sheets	1.000	S.F.	.067	6.20	3.23	9.43
Trim, painted or stained, baseboard	.125	L.F.	.006	.60	.32	.92
Base shoe	.125	L.F.	.005	.07	.30	.37
Chair rail	.125	L.F.	.005	.23	.28	.51
Cornice molding	.125	L.F.	.004	.16	.28	.44
Cove base, vinyl	.125	L.F.	.003	.18	.17	.35
Paneling not including furring or trim						
Plywood, prefinished, 1/4″ thick, 4′ x 8′ sheets, vert. grooves						
Birch faced, minimum	1.000	S.F.	.032	1.62	1.86	3.48
Average	1.000	S.F.	.038	1.32	2.21	3.53
Maximum	1.000	S.F.	.046	1.21	2.66	3.87
Mahogany, African	1.000	S.F.	.040	2.70	2.32	5.02
Philippine (lauan)	1.000	S.F.	.032	.72	1.86	2.58
Oak or cherry, minimum	1.000	S.F.	.032	1.47	1.86	3.33
Maximum	1.000	S.F.	.040	2.20	2.32	4.52
Rosewood	1.000	S.F.	.050	3.41	2.91	6.32
Teak	1.000	S.F.	.040	3.43	2.32	5.75
Chestnut	1.000	S.F.	.043	5.80	2.48	8.28
Pecan	1.000	S.F.	.040	2.75	2.32	5.07
Walnut, minimum	1.000	S.F.	.032	2.72	1.86	4.58
Maximum	1.000	S.F.	.040	2.71	2.32	5.03

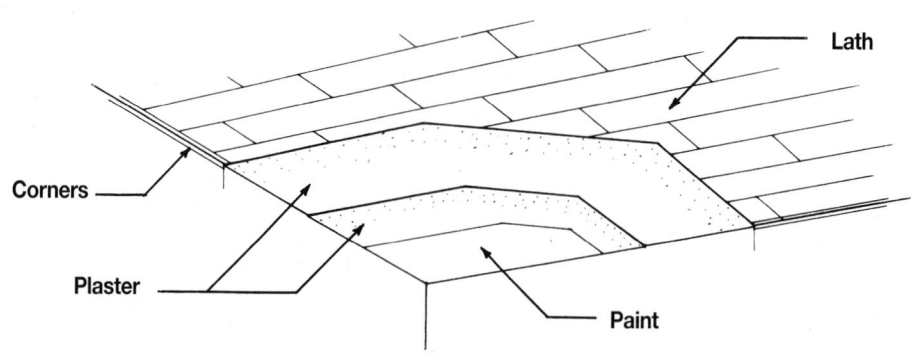

System Description	QUAN.	UNIT	LABOR HOURS	COST PER S.F.		
				MAT.	INST.	TOTAL
PLASTER ON GYPSUM LATH						
Plaster, gypsum or perlite, 2 coats	1.000	S.F.	.061	.46	3.39	3.85
Gypsum lath, plain or perforated, nailed, 3/8" thick	1.000	S.F.	.010	.39	.58	.97
Gypsum lath, ceiling installation adder	1.000	S.F.	.004		.23	.23
Corners, expanded metal, 12' x 12' room	.330	L.F.	.007	.06	.38	.44
Painting, primer & 2 coats	1.000	S.F.	.011	.22	.51	.73
TOTAL		S.F.	.093	1.13	5.09	6.22
PLASTER ON METAL LATH						
Plaster, gypsum or perlite, 2 coats	1.000	S.F.	.061	.46	3.39	3.85
Lath, 2.5 Lb. diamond, metal	1.000	S.F.	.012	.46	.66	1.12
Corners, expanded metal, 12' x 12' room	.330	L.F.	.007	.06	.38	.44
Painting, primer & 2 coats	1.000	S.F.	.011	.22	.51	.73
TOTAL		S.F.	.091	1.20	4.94	6.14
STUCCO ON GYPSUM LATH						
Stucco, 2 coats	1.000	S.F.	.041	.41	2.29	2.70
Gypsum lath, plain or perforated, nailed, 3/8" thick	1.000	S.F.	.010	.39	.58	.97
Gypsum lath, ceiling installation adder	1.000	S.F.	.004		.23	.23
Corners, expanded metal, 12' x 12' room	.330	L.F.	.007	.06	.38	.44
Painting, primer & 2 coats	1.000	S.F.	.011	.22	.51	.73
TOTAL		S.F.	.073	1.08	3.99	5.07
STUCCO ON METAL LATH						
Stucco, 2 coats	1.000	S.F.	.041	.41	2.29	2.70
Lath, 2.5 Lb. diamond, metal	1.000	S.F.	.012	.46	.66	1.12
Corners, expanded metal, 12' x 12' room	.330	L.F.	.007	.06	.38	.44
Painting, primer & 2 coats	1.000	S.F.	.011	.22	.51	.73
TOTAL		S.F.	.071	1.15	3.84	4.99

The costs in these systems are based on a square foot of ceiling area.

Description	QUAN.	UNIT	LABOR HOURS	COST PER S.F.		
				MAT.	INST.	TOTAL

Plaster & Stucco Ceiling Price Sheet	QUAN.	UNIT	LABOR HOURS	COST PER S.F.		
				MAT.	INST.	TOTAL
Plaster, gypsum or perlite, 2 coats	1.000	S.F.	.061	.46	3.39	3.85
3 coats	1.000	S.F.	.065	.66	3.62	4.28
Lath, gypsum, standard, 3/8" thick	1.000	S.F.	.014	.39	.81	1.20
Fire resistant, 3/8" thick	1.000	S.F.	.017	.31	.94	1.25
1/2" thick	1.000	S.F.	.018	.35	.99	1.34
Metal, diamond, 2.5 Lb.	1.000	S.F.	.012	.46	.66	1.12
3.4 Lb.	1.000	S.F.	.015	.50	.82	1.32
Rib, 2.75 Lb.	1.000	S.F.	.012	.40	.66	1.06
3.4 Lb.	1.000	S.F.	.013	.52	.71	1.23
Corners expanded metal, 4' x 4' room	1.000	L.F.	.020	.17	1.16	1.33
6' x 6' room	.667	L.F.	.013	.11	.78	.89
10' x 10' room	.400	L.F.	.008	.07	.46	.53
12' x 12' room	.333	L.F.	.007	.06	.38	.44
16' x 16' room	.250	L.F.	.004	.03	.22	.25
Painting, primer & 1 coat	1.000	S.F.	.008	.14	.40	.54
Primer & 2 coats	1.000	S.F.	.011	.22	.51	.73

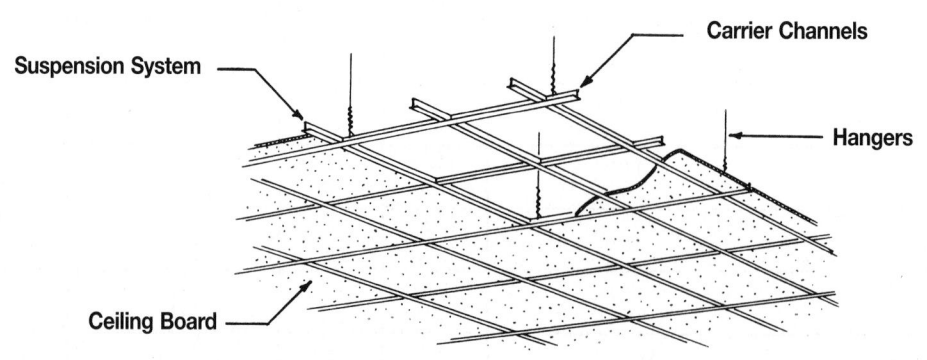

System Description	QUAN.	UNIT	LABOR HOURS	COST PER S.F.		
				MAT.	INST.	TOTAL
2′ X 2′ GRID, FILM FACED FIBERGLASS, 5/8″ THICK						
Suspension system, 2′ x 2′ grid, T bar	1.000	S.F.	.012	1.12	.72	1.84
Ceiling board, film faced fiberglass, 5/8″ thick	1.000	S.F.	.013	1.39	.74	2.13
Carrier channels, 1-1/2″ x 3/4″	1.000	S.F.	.017	.13	.99	1.12
Hangers, #12 wire	1.000	S.F.	.002	.01	.10	.11
TOTAL		S.F.	.044	2.65	2.55	5.20
2′ X 4′ GRID, FILM FACED FIBERGLASS, 5/8″ THICK						
Suspension system, 2′ x 4′ grid, T bar	1.000	S.F.	.010	.87	.58	1.45
Ceiling board, film faced fiberglass, 5/8″ thick	1.000	S.F.	.013	1.39	.74	2.13
Carrier channels, 1-1/2″ x 3/4″	1.000	S.F.	.017	.13	.99	1.12
Hangers, #12 wire	1.000	S.F.	.002	.01	.10	.11
TOTAL		S.F.	.042	2.40	2.41	4.81
2′ X 2′ GRID, MINERAL FIBER, REVEAL EDGE, 1″ THICK						
Suspension system, 2′ x 2′ grid, T bar	1.000	S.F.	.012	1.12	.72	1.84
Ceiling board, mineral fiber, reveal edge, 1″ thick	1.000	S.F.	.013	2.33	.77	3.10
Carrier channels, 1-1/2″ x 3/4″	1.000	S.F.	.017	.13	.99	1.12
Hangers, #12 wire	1.000	S.F.	.002	.01	.10	.11
TOTAL		S.F.	.044	3.59	2.58	6.17
2′ X 4′ GRID, MINERAL FIBER, REVEAL EDGE, 1″ THICK						
Suspension system, 2′ x 4′ grid, T bar	1.000	S.F.	.010	.87	.58	1.45
Ceiling board, mineral fiber, reveal edge, 1″ thick	1.000	S.F.	.013	2.33	.77	3.10
Carrier channels, 1-1/2″ x 3/4″	1.000	S.F.	.017	.13	.99	1.12
Hangers, #12 wire	1.000	S.F.	.002	.01	.10	.11
TOTAL		S.F.	.042	3.34	2.44	5.78

Description	QUAN.	UNIT	LABOR HOURS	COST PER S.F.		
				MAT.	INST.	TOTAL

Suspended Ceiling Price Sheet	QUAN.	UNIT	LABOR HOURS	COST PER S.F.		
				MAT.	INST.	TOTAL
Suspension systems, T bar, 2' x 2' grid	1.000	S.F.	.012	1.12	.72	1.84
2' x 4' grid	1.000	S.F.	.010	.87	.58	1.45
Concealed Z bar, 12" module	1.000	S.F.	.015	1.02	.89	1.91
Ceiling boards, fiberglass, film faced, 2' x 2' or 2' x 4', 5/8" thick	1.000	S.F.	.013	1.39	.74	2.13
3/4" thick	1.000	S.F.	.013	3.30	.77	4.07
3" thick thermal R11	1.000	S.F.	.018	3.85	1.03	4.88
Glass cloth faced, 3/4" thick	1.000	S.F.	.016	3.29	.93	4.22
1" thick	1.000	S.F.	.016	3.97	.96	4.93
1-1/2" thick, nubby face	1.000	S.F.	.017	3	.98	3.98
Mineral fiber boards, 5/8" thick, aluminum face 2' x 2'	1.000	S.F.	.013	4.30	.77	5.07
2' x 4'	1.000	S.F.	.012	4.30	.72	5.02
Standard faced, 2' x 2' or 2' x 4'	1.000	S.F.	.012	.96	.69	1.65
Plastic coated face, 2' x 2' or 2' x 4'	1.000	S.F.	.020	2.88	1.16	4.04
Fire rated, 2 hour rating, 5/8" thick	1.000	S.F.	.012	1.41	.69	2.10
Tegular edge, 2' x 2' or 2' x 4', 5/8" thick, fine textured	1.000	S.F.	.013	1.23	.99	2.22
Rough textured	1.000	S.F.	.015	1.47	.99	2.46
3/4" thick, fine textured	1.000	S.F.	.016	2.56	1.03	3.59
Rough textured	1.000	S.F.	.018	1.68	1.03	2.71
Luminous panels, prismatic, acrylic	1.000	S.F.	.020	3.22	1.16	4.38
Polystyrene	1.000	S.F.	.020	1.85	1.16	3.01
Flat or ribbed, acrylic	1.000	S.F.	.020	4.62	1.16	5.78
Polystyrene	1.000	S.F.	.020	2.60	1.16	3.76
Drop pan, white, acrylic	1.000	S.F.	.020	6.05	1.16	7.21
Polystyrene	1.000	S.F.	.020	4.94	1.16	6.10
Carrier channels, 4'-0" on center, 3/4" x 1-1/2"	1.000	S.F.	.017	.13	.99	1.12
1-1/2" x 3-1/2"	1.000	S.F.	.017	.35	.99	1.34
Hangers, #12 wire	1.000	S.F.	.002	.01	.10	.11

For customer support on your Residential Costs with RSMeans data, call 800.448.8182.

223

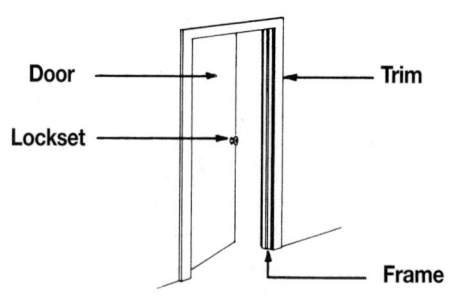

Door — Trim
Lockset —
Frame

System Description	QUAN.	UNIT	LABOR HOURS	COST EACH		
				MAT.	INST.	TOTAL
LAUAN, FLUSH DOOR, HOLLOW CORE						
Door, flush, lauan, hollow core, 2'-8" wide x 6'-8" high	1.000	Ea.	.889	77	51.50	128.50
Frame, pine, 4-5/8" jamb	17.000	L.F.	.725	90.10	42.16	132.26
Moldings, casing, ogee, 11/16" x 2-1/2", pine	34.000	L.F.	1.088	53.04	63.24	116.28
Paint trim, to 6" wide, primer + 1 coat enamel	34.000	L.F.	.340	5.10	16.32	21.42
Butt hinges, chrome, 3-1/2" x 3-1/2"	1.500	Pr.		45.75		45.75
Lockset, passage	1.000	Ea.	.500	41	29	70
Prime door & frame, oil, primer, brushwork	2.000	Face	1.600	8.76	77	85.76
Paint door and frame, oil, 2 coats	2.000	Face	2.667	10.40	129	139.40
TOTAL		Ea.	7.809	331.15	408.22	739.37
BIRCH, FLUSH DOOR, HOLLOW CORE						
Door, flush, birch, hollow core, 2'-8" wide x 6'-8" high	1.000	Ea.	.889	90	51.50	141.50
Frame, pine, 4-5/8" jamb	17.000	L.F.	.725	90.10	42.16	132.26
Moldings, casing, ogee, 11/16" x 2-1/2", pine	34.000	L.F.	1.088	53.04	63.24	116.28
Butt hinges, chrome, 3-1/2" x 3-1/2"	1.500	Pr.		45.75		45.75
Lockset, passage	1.000	Ea.	.500	41	29	70
Prime door & frame, oil, primer, brushwork	2.000	Face	1.600	8.76	77	85.76
Paint door and frame, oil, 2 coats	2.000	Face	2.667	10.40	129	139.40
TOTAL		Ea.	7.469	339.05	391.90	730.95
RAISED PANEL, SOLID, PINE DOOR						
Door, pine, raised panel, 2'-8" wide x 6'-8" high	1.000	Ea.	.889	284	51.50	335.50
Frame, pine, 4-5/8" jamb	17.000	L.F.	.725	90.10	42.16	132.26
Moldings, casing, ogee, 11/16" x 2-1/2", pine	34.000	L.F.	1.088	53.04	63.24	116.28
Butt hinges, bronze, 3-1/2" x 3-1/2"	1.500	Pr.		54		54
Lockset, passage	1.000	Ea.	.500	41	29	70
Prime door & frame, oil, primer, brushwork	2.000	Face	1.600	8.76	77	85.76
Paint door and frame, oil, 2 coats	2.000	Face	2.667	10.40	129	139.40
TOTAL		Ea.	7.469	541.30	391.90	933.20

The costs in these systems are based on a cost per each door.

Description	QUAN.	UNIT	LABOR HOURS	COST EACH		
				MAT.	INST.	TOTAL

For customer support on your Residential Costs with RSMeans data, call 800.448.8182.

Interior Door Price Sheet	QUAN.	UNIT	LABOR HOURS	COST EACH		
				MAT.	INST.	TOTAL
Door, hollow core, lauan 1-3/8" thick, 6'-8" high x 1'-6" wide	1.000	Ea.	.889	58.50	51.50	110
2'-0" wide	1.000	Ea.	.889	66	51.50	117.50
2'-6" wide	1.000	Ea.	.889	75	51.50	126.50
2'-8" wide	1.000	Ea.	.889	77	51.50	128.50
3'-0" wide	1.000	Ea.	.941	82.50	54.50	137
Birch 1-3/8" thick, 6'-8" high x 1'-6" wide	1.000	Ea.	.889	68	51.50	119.50
2'-0" wide	1.000	Ea.	.889	76	51.50	127.50
2'-6" wide	1.000	Ea.	.889	87	51.50	138.50
2'-8" wide	1.000	Ea.	.889	90	51.50	141.50
3'-0" wide	1.000	Ea.	.941	97	54.50	151.50
Louvered pine 1-3/8" thick, 6'-8" high x 1'-6" wide	1.000	Ea.	.842	163	49	212
2'-0" wide	1.000	Ea.	.889	180	51.50	231.50
2'-6" wide	1.000	Ea.	.889	219	51.50	270.50
2'-8" wide	1.000	Ea.	.889	230	51.50	281.50
3'-0" wide	1.000	Ea.	.941	241	54.50	295.50
Paneled pine 1-3/8" thick, 6'-8" high x 1'-6" wide	1.000	Ea.	.842	210	49	259
2'-0" wide	1.000	Ea.	.889	245	51.50	296.50
2'-6" wide	1.000	Ea.	.889	275	51.50	326.50
2'-8" wide	1.000	Ea.	.889	284	51.50	335.50
3'-0" wide	1.000	Ea.	.941	298	54.50	352.50
Frame, pine, 1'-6" thru 2'-0" wide door, 3-5/8" deep	16.000	L.F.	.683	73	39.50	112.50
4-5/8" deep	16.000	L.F.	.683	85	39.50	124.50
5-5/8" deep	16.000	L.F.	.683	97.50	39.50	137
2'-6" thru 3'0" wide door, 3-5/8" deep	17.000	L.F.	.725	77.50	42	119.50
4-5/8" deep	17.000	L.F.	.725	90	42	132
5-5/8" deep	17.000	L.F.	.725	104	42	146
Trim, casing, painted, both sides, 1'-6" thru 2'-6" wide door	32.000	L.F.	1.855	53	97.50	150.50
2'-6" thru 3'-0" wide door	34.000	L.F.	1.971	56.50	103	159.50
Butt hinges 3-1/2" x 3-1/2", steel plated, chrome	1.500	Pr.		46		46
Bronze	1.500	Pr.		54		54
Locksets, passage, minimum	1.000	Ea.	.500	41	29	70
Maximum	1.000	Ea.	.575	47	33.50	80.50
Privacy, miniumum	1.000	Ea.	.625	51.50	36.50	88
Maximum	1.000	Ea.	.675	55.50	39	94.50
Paint 2 sides, primer & 2 cts., flush door, 1'-6" to 2'-0" wide	2.000	Face	5.547	26	268	294
2'-6" thru 3'-0" wide	2.000	Face	6.933	32.50	335	367.50
Louvered door, 1'-6" thru 2'-0" wide	2.000	Face	6.400	23.50	310	333.50
2'-6" thru 3'-0" wide	2.000	Face	8.000	29.50	385	414.50
Paneled door, 1'-6" thru 2'-0" wide	2.000	Face	6.400	23.50	310	333.50
2'-6" thru 3'-0" wide	2.000	Face	8.000	29.50	385	414.50

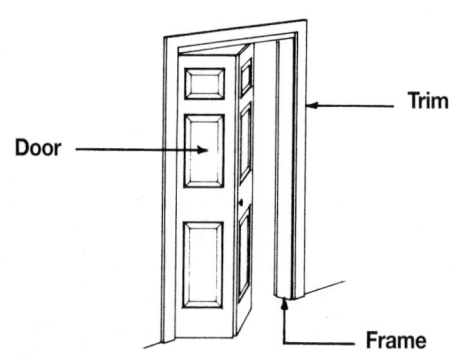

Trim

Door

Frame

System Description	QUAN.	UNIT	LABOR HOURS	COST EACH		
				MAT.	INST.	TOTAL
BI-PASSING, FLUSH, LAUAN, HOLLOW CORE, 4'-0" X 6'-8"						
Door, flush, lauan, hollow core, 4'-0" x 6'-8" opening	1.000	Ea.	1.333	184	77.50	261.50
Frame, pine, 4-5/8" jamb	18.000	L.F.	.768	95.40	44.64	140.04
Moldings, casing, ogee, 11/16" x 2-1/2", pine	36.000	L.F.	1.152	56.16	66.96	123.12
Prime door & frame, oil, primer, brushwork	2.000	Face	1.600	8.76	77	85.76
Paint door and frame, oil, 2 coats	2.000	Face	2.667	10.40	129	139.40
TOTAL		Ea.	7.520	354.72	395.10	749.82
BI-PASSING, FLUSH, BIRCH, HOLLOW CORE, 6'-0" X 6'-8"						
Door, flush, birch, hollow core, 6'-0" x 6'-8" opening	1.000	Ea.	1.600	370	93	463
Frame, pine, 4-5/8" jamb	19.000	L.F.	.811	100.70	47.12	147.82
Moldings, casing, ogee, 11/16" x 2-1/2", pine	38.000	L.F.	1.216	59.28	70.68	129.96
Prime door & frame, oil, primer, brushwork	2.000	Face	2.000	10.95	96.25	107.20
Paint door and frame, oil, 2 coats	2.000	Face	3.333	13	161.25	174.25
TOTAL		Ea.	8.960	553.93	468.30	1,022.23
BI-FOLD, PINE, PANELED, 3'-0" X 6'-8"						
Door, pine, paneled, 3'-0" x 6'-8" opening	1.000	Ea.	1.231	263	71.50	334.50
Frame, pine, 4-5/8" jamb	17.000	L.F.	.725	90.10	42.16	132.26
Moldings, casing, ogee, 11/16" x 2-1/2", pine	34.000	L.F.	1.088	53.04	63.24	116.28
Prime door & frame, oil, primer, brushwork	2.000	Face	1.600	8.76	77	85.76
Paint door and frame, oil, 2 coats	2.000	Face	2.667	10.40	129	139.40
TOTAL		Ea.	7.311	425.30	382.90	808.20
BI-FOLD, PINE, LOUVERED, 6'-0" X 6'-8"						
Door, pine, louvered, 6'-0" x 6'-8" opening	1.000	Ea.	1.600	340	93	433
Frame, pine, 4-5/8" jamb	19.000	L.F.	.811	100.70	47.12	147.82
Moldings, casing, ogee, 11/16" x 2-1/2", pine	38.000	L.F.	1.216	59.28	70.68	129.96
Prime door & frame, oil, primer, brushwork	2.500	Face	2.000	10.95	96.25	107.20
Paint door and frame, oil, 2 coats	2.500	Face	3.333	13	161.25	174.25
TOTAL		Ea.	8.960	523.93	468.30	992.23

The costs in this system are based on a cost per each door.

Description	QUAN.	UNIT	LABOR HOURS	COST EACH		
				MAT.	INST.	TOTAL

Closet Door Price Sheet	QUAN.	UNIT	LABOR HOURS	COST EACH		
				MAT.	INST.	TOTAL
Doors, bi-passing, pine, louvered, 4'-0" x 6'-8" opening	1.000	Ea.	1.333	530	77.50	607.50
6'-0" x 6'-8" opening	1.000	Ea.	1.600	790	93	883
Paneled, 4'-0" x 6'-8" opening	1.000	Ea.	1.333	540	77.50	617.50
6'-0" x 6'-8" opening	1.000	Ea.	1.600	925	93	1,018
Flush, birch, hollow core, 4'-0" x 6'-8" opening	1.000	Ea.	1.333	280	77.50	357.50
6'-0" x 6'-8" opening	1.000	Ea.	1.600	370	93	463
Flush, lauan, hollow core, 4'-0" x 6'-8" opening	1.000	Ea.	1.333	184	77.50	261.50
6'-0" x 6'-8" opening	1.000	Ea.	1.600	186	93	279
Bi-fold, pine, louvered, 3'-0" x 6'-8" opening	1.000	Ea.	1.231	263	71.50	334.50
6'-0" x 6'-8" opening	1.000	Ea.	1.600	340	93	433
Paneled, 3'-0" x 6'-8" opening	1.000	Ea.	1.231	263	71.50	334.50
6'-0" x 6'-8" opening	1.000	Ea.	1.600	340	93	433
Flush, birch, hollow core, 3'-0" x 6'-8" opening	1.000	Ea.	1.231	82	71.50	153.50
6'-0" x 6'-8" opening	1.000	Ea.	1.600	148	93	241
Flush, lauan, hollow core, 3'-0" x 6'8" opening	1.000	Ea.	1.231	320	71.50	391.50
6'-0" x 6'-8" opening	1.000	Ea.	1.600	490	93	583
Frame pine, 3'-0" door, 3-5/8" deep	17.000	L.F.	.725	77.50	42	119.50
4-5/8" deep	17.000	L.F.	.725	90	42	132
5-5/8" deep	17.000	L.F.	.725	104	42	146
4'-0" door, 3-5/8" deep	18.000	L.F.	.768	82.50	44.50	127
4-5/8" deep	18.000	L.F.	.768	95.50	44.50	140
5-5/8" deep	18.000	L.F.	.768	110	44.50	154.50
6'-0" door, 3-5/8" deep	19.000	L.F.	.811	87	47	134
4-5/8" deep	19.000	L.F.	.811	101	47	148
5-5/8" deep	19.000	L.F.	.811	116	47	163
Trim both sides, painted 3'-0" x 6'-8" door	34.000	L.F.	1.971	56.50	103	159.50
4'-0" x 6'-8" door	36.000	L.F.	2.086	60	109	169
6'-0" x 6'-8" door	38.000	L.F.	2.203	63	116	179
Paint 2 sides, primer & 2 cts., flush door & frame, 3' x 6'-8" opng	2.000	Face	2.914	14.35	155	169.35
4'-0" x 6'-8" opening	2.000	Face	3.886	19.15	206	225.15
6'-0" x 6'-8" opening	2.000	Face	4.857	24	258	282
Paneled door & frame, 3'-0" x 6'-8" opening	2.000	Face	6.000	22	290	312
4'-0" x 6'-8" opening	2.000	Face	8.000	29.50	385	414.50
6'-0" x 6'-8" opening	2.000	Face	10.000	36.50	485	521.50
Louvered door & frame, 3'-0" x 6'-8" opening	2.000	Face	6.000	22	290	312
4'-0" x 6'-8" opening	2.000	Face	8.000	29.50	385	414.50
6'-0" x 6'-8" opening	2.000	Face	10.000	36.50	485	521.50

System Description	QUAN.	UNIT	LABOR HOURS	COST PER S.F.		
				MAT.	INST.	TOTAL
Carpet, direct glue-down, nylon, level loop, 26 oz.	1.000	S.F.	.018	2.69	.65	3.34
32 oz.	1.000	S.F.	.018	4.82	.65	5.47
40 oz.	1.000	S.F.	.018	5.95	.65	6.60
Nylon, plush, 20 oz.	1.000	S.F.	.018	2.50	.65	3.15
24 oz.	1.000	S.F.	.018	2.38	.65	3.03
30 oz.	1.000	S.F.	.018	3.50	.65	4.15
42 oz.	1.000	S.F.	.022	5.55	.69	6.24
48 oz.	1.000	S.F.	.022	6.45	.69	7.14
54 oz.	1.000	S.F.	.022	7.25	.69	7.94
Olefin, 15 oz.	1.000	S.F.	.018	1.78	.65	2.43
22 oz.	1.000	S.F.	.018	1.81	.65	2.46
Tile, foam backed, needle punch	1.000	S.F.	.014	4.52	.77	5.29
Tufted loop or shag	1.000	S.F.	.014	3.62	.77	4.39
Wool, 36 oz., level loop	1.000	S.F.	.018	13.30	.69	13.99
32 oz., patterned	1.000	S.F.	.020	12.05	.69	12.74
48 oz., patterned	1.000	S.F.	.020	13.30	.69	13.99
Padding, sponge rubber cushion, minimum	1.000	S.F.	.006	.57	.32	.89
Maximum	1.000	S.F.	.006	1.07	.32	1.39
Felt, 32 oz. to 56 oz., minimum	1.000	S.F.	.006	.71	.32	1.03
Maximum	1.000	S.F.	.006	1.36	.32	1.68
Bonded urethane, 3/8" thick, minimum	1.000	S.F.	.006	.73	.32	1.05
Maximum	1.000	S.F.	.006	.98	.32	1.30
Prime urethane, 1/4" thick, minimum	1.000	S.F.	.006	.43	.32	.75
Maximum	1.000	S.F.	.006	.80	.32	1.12
Stairs, for stairs, add to above carpet prices	1.000	Riser	.267		14.60	14.60
Underlayment plywood, 3/8" thick	1.000	S.F.	.011	1.14	.62	1.76
1/2" thick	1.000	S.F.	.011	1.35	.64	1.99
5/8" thick	1.000	S.F.	.011	1.50	.66	2.16
3/4" thick	1.000	S.F.	.012	1.62	.72	2.34
Particle board, 3/8" thick	1.000	S.F.	.011	.45	.62	1.07
1/2" thick	1.000	S.F.	.011	.47	.64	1.11
5/8" thick	1.000	S.F.	.011	.61	.66	1.27
3/4" thick	1.000	S.F.	.012	.74	.72	1.46
Hardboard, 4' x 4', 0.215" thick	1.000	S.F.	.011	.75	.62	1.37

System Description	QUAN.	UNIT	LABOR HOURS	COST PER S.F.		
				MAT.	INST.	TOTAL
Resilient flooring, asphalt tile on concrete, 1/8" thick						
Color group B	1.000	S.F.	.020	1.57	1.09	2.66
Color group C & D	1.000	S.F.	.020	1.73	1.09	2.82
Asphalt tile on wood subfloor, 1/8" thick						
Color group B	1.000	S.F.	.020	1.86	1.09	2.95
Color group C & D	1.000	S.F.	.020	2.02	1.09	3.11
Vinyl composition tile, 12" x 12", 1/16" thick	1.000	S.F.	.016	1.34	.88	2.22
Embossed	1.000	S.F.	.016	2.93	.88	3.81
Marbleized	1.000	S.F.	.016	2.93	.88	3.81
Plain	1.000	S.F.	.016	3.78	.88	4.66
.080" thick, embossed	1.000	S.F.	.016	1.71	.88	2.59
Marbleized	1.000	S.F.	.016	3.37	.88	4.25
Plain	1.000	S.F.	.016	3.14	.88	4.02
1/8" thick, marbleized	1.000	S.F.	.016	2.68	.88	3.56
Plain	1.000	S.F.	.016	1.95	.88	2.83
Vinyl tile, 12" x 12", .050" thick, minimum	1.000	S.F.	.016	4.08	.88	4.96
Maximum	1.000	S.F.	.016	5.70	.88	6.58
1/8" thick, minimum	1.000	S.F.	.016	7.70	.88	8.58
Maximum	1.000	S.F.	.016	3.55	.88	4.43
1/8" thick, solid colors	1.000	S.F.	.016	6.80	.88	7.68
Florentine pattern	1.000	S.F.	.016	7.25	.88	8.13
Marbleized or travertine pattern	1.000	S.F.	.016	6.85	.88	7.73
Vinyl sheet goods, backed, .070" thick, minimum	1.000	S.F.	.032	4.73	1.75	6.48
Maximum	1.000	S.F.	.040	4.33	2.19	6.52
.093" thick, minimum	1.000	S.F.	.035	4.64	1.90	6.54
Maximum	1.000	S.F.	.040	7.10	2.19	9.29
.125" thick, minimum	1.000	S.F.	.035	4.26	1.90	6.16
Maximum	1.000	S.F.	.040	8.25	2.19	10.44
Wood, oak, finished in place, 25/32" x 2-1/2" clear	1.000	S.F.	.074	3.99	3.93	7.92
Select	1.000	S.F.	.074	4.95	3.93	8.88
No. 1 common	1.000	S.F.	.074	4.88	3.93	8.81
Prefinished, oak, 2-1/2" wide	1.000	S.F.	.047	5.75	2.73	8.48
3-1/4" wide	1.000	S.F.	.043	6.30	2.51	8.81
Ranch plank, oak, random width	1.000	S.F.	.055	8	3.21	11.21
Parquet, 5/16" thick, finished in place, oak, minimum	1.000	S.F.	.077	6.25	4.11	10.36
Maximum	1.000	S.F.	.107	11.10	5.85	16.95
Teak, minimum	1.000	S.F.	.077	7.20	4.11	11.31
Maximum	1.000	S.F.	.107	12.10	5.85	17.95
Sleepers, treated, 16" O.C., 1" x 2"	1.000	S.F.	.007	.33	.40	.73
1" x 3"	1.000	S.F.	.008	.54	.46	1
2" x 4"	1.000	S.F.	.011	.69	.62	1.31
2" x 6"	1.000	S.F.	.012	.90	.72	1.62
Subfloor, plywood, 1/2" thick	1.000	S.F.	.011	.68	.62	1.30
5/8" thick	1.000	S.F.	.012	.84	.69	1.53
3/4" thick	1.000	S.F.	.013	1.02	.74	1.76
Ceramic tile, color group 2, 1" x 1"	1.000	S.F.	.087	7.05	4.24	11.29
2" x 2" or 2" x 1"	1.000	S.F.	.084	6.90	4.08	10.98
Color group 1, 8" x 8"	1.000	S.F.	.064	5.80	2.58	8.38
12" x 12"	1.000	S.F.	.049	6.40	2.67	9.07
16" x 16"	1.000	S.F.	.029	8.25	2.77	11.02

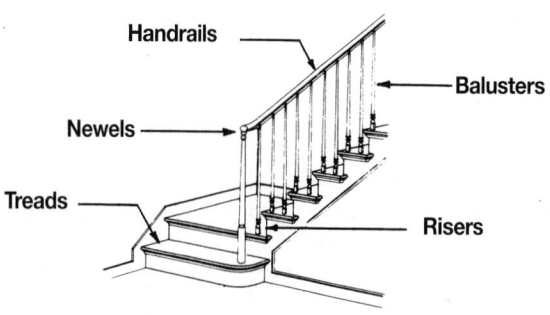

System Description	QUAN.	UNIT	LABOR HOURS	COST EACH		
				MAT.	INST.	TOTAL
7 RISERS, OAK TREADS, BOX STAIRS						
Treads, oak, 1-1/4" x 10" wide, 3' long	6.000	Ea.	2.667	636	156	792
Risers, 3/4" thick, beech	7.000	Ea.	.672	224.70	39.06	263.76
30" primed pine balusters	12.000	Ea.	1.000	48.84	58.08	106.92
Newels, 3" wide, plain, paint grade, square	2.000	Ea.	2.286	119	133	252
Handrails, oak laminated	7.000	L.F.	.933	294	54.25	348.25
Stringers, 2" x 10", 3 each	21.000	L.F.	.306	9.45	17.85	27.30
TOTAL		Ea.	7.864	1,331.99	458.24	1,790.23
14 RISERS, OAK TREADS, BOX STAIRS						
Treads, oak, 1-1/4" x 10" wide, 3' long	13.000	Ea.	5.778	1,378	338	1,716
Risers, 3/4" thick, beech	14.000	Ea.	1.344	449.40	78.12	527.52
30" primed pine balusters	26.000	Ea.	2.167	105.82	125.84	231.66
Newels, 3" wide, plain, paint grade, square	2.000	Ea.	2.286	119	133	252
Handrails, oak, laminated	14.000	L.F.	1.867	588	108.50	696.50
Stair stringers, 2" x 10"	42.000	L.F.	5.169	65.10	300.30	365.40
TOTAL		Ea.	18.611	2,705.32	1,083.76	3,789.08
14 RISERS, PINE TREADS, BOX STAIRS						
Treads, pine, 9-1/2" x 3/4" thick	13.000	Ea.	5.778	243.10	338	581.10
Risers, 3/4" thick, pine	14.000	Ea.	1.344	289.80	78.12	367.92
30" primed pine balusters	26.000	Ea.	2.167	105.82	125.84	231.66
Newels, 3" wide, plain, paint grade, square	2.000	Ea.	2.286	119	133	252
Handrails, oak, laminated	14.000	L.F.	1.867	588	108.50	696.50
Stair stringers, 2" x 10"	42.000	L.F.	5.169	65.10	300.30	365.40
TOTAL		Ea.	18.611	1,410.82	1,083.76	2,494.58

Description	QUAN.	UNIT	LABOR HOURS	COST EACH		
				MAT.	INST.	TOTAL

Stairway Price Sheet

Stairway Price Sheet	QUAN.	UNIT	LABOR HOURS	COST EACH		
				MAT.	INST.	TOTAL
Treads, oak, 1-1/16" x 9-1/2", 3' long, 7 riser stair	6.000	Ea.	2.667	635	156	791
14 riser stair	13.000	Ea.	5.778	1,375	340	1,715
1-1/16" x 11-1/2", 3' long, 7 riser stair	6.000	Ea.	2.667	695	156	851
14 riser stair	13.000	Ea.	5.778	1,500	340	1,840
Pine, 3/4" x 9-1/2", 3' long, 7 riser stair	6.000	Ea.	2.667	112	156	268
14 riser stair	13.000	Ea.	5.778	243	340	583
3/4" x 11-1/4", 3' long, 7 riser stair	6.000	Ea.	2.667	126	156	282
14 riser stair	13.000	Ea.	5.778	273	340	613
Risers, oak, 3/4" x 7-1/2" high, 7 riser stair	7.000	Ea.	2.625	150	39	189
14 riser stair	14.000	Ea.	5.250	300	78	378
Beech, 3/4" x 7-1/2" high, 7 riser stair	7.000	Ea.	2.625	225	39	264
14 riser stair	14.000	Ea.	5.250	450	78	528
Baluster, turned, 30" high, primed pine, 7 riser stair	12.000	Ea.	3.429	49	58	107
14 riser stair	26.000	Ea.	7.428	106	126	232
30" birch, 7 riser stair	12.000	Ea.	3.429	44.50	53	97.50
14 riser stair	26.000	Ea.	7.428	96	114	210
42" pine, 7 riser stair	12.000	Ea.	3.556	65	58	123
14 riser stair	26.000	Ea.	7.704	140	126	266
42" birch, 7 riser stair	12.000	Ea.	3.556	65	58	123
14 riser stair	26.000	Ea.	7.704	140	126	266
Newels, 3-1/4" wide, starting, 7 riser stair	2.000	Ea.	2.286	119	133	252
14 riser stair	2.000	Ea.	2.286	119	133	252
Landing, 7 riser stair	2.000	Ea.	3.200	224	186	410
14 riser stair	2.000	Ea.	3.200	224	186	410
Handrails, oak, laminated, 7 riser stair	7.000	L.F.	.933	294	54.50	348.50
14 riser stair	14.000	L.F.	1.867	590	109	699
Stringers, fir, 2" x 10" 7 riser stair	21.000	L.F.	2.585	32.50	150	182.50
14 riser stair	42.000	L.F.	5.169	65	300	365
2" x 12", 7 riser stair	21.000	L.F.	2.585	41.50	150	191.50
14 riser stair	42.000	L.F.	5.169	82.50	300	382.50

Special Stairways

Special Stairways	QUAN.	UNIT	LABOR HOURS	COST EACH		
				MAT.	INST.	TOTAL
Basement stairs, open risers	1.000	Flight	4.000	915	233	1,148
Spiral stairs, oak, 4'-6" diameter, prefabricated, 9' high	1.000	Flight	10.667	3,025	620	3,645
Aluminum, 5'-0" diameter stock unit	1.000	Flight	9.956	8,900	730	9,630
Custom unit	1.000	Flight	9.956	15,400	730	16,130
Cast iron, 4'-0" diameter, minimum	1.000	Flight	9.956	11,500	730	12,230
Maximum	1.000	Flight	17.920	20,000	1,300	21,300
Steel, industrial, pre-erected, 3'-6" wide, bar rail	1.000	Flight	7.724	8,550	825	9,375
Picket rail	1.000	Flight	7.724	9,525	825	10,350

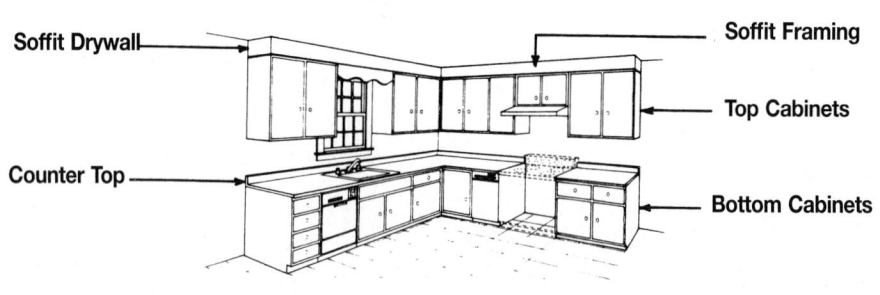

Soffit Drywall — Soffit Framing
Counter Top — Top Cabinets
Bottom Cabinets

System Description	QUAN.	UNIT	LABOR HOURS	COST PER L.F.		
				MAT.	INST.	TOTAL
KITCHEN, ECONOMY GRADE						
Top cabinets, economy grade	1.000	L.F.	.171	71.04	9.92	80.96
Bottom cabinets, economy grade	1.000	L.F.	.256	106.56	14.88	121.44
Square edge, plastic face countertop	1.000	L.F.	.267	37.50	15.50	53
Blocking, wood, 2″ x 4″	1.000	L.F.	.032	.46	1.86	2.32
Soffit, framing, wood, 2″ x 4″	4.000	L.F.	.071	1.84	4.12	5.96
Soffit drywall	2.000	S.F.	.047	.88	2.76	3.64
Drywall painting	2.000	S.F.	.013	.12	.78	.90
TOTAL		L.F.	.857	218.40	49.82	268.22
AVERAGE GRADE						
Top cabinets, average grade	1.000	L.F.	.213	88.80	12.40	101.20
Bottom cabinets, average grade	1.000	L.F.	.320	133.20	18.60	151.80
Solid surface countertop, solid color	1.000	L.F.	.800	77.50	46.50	124
Blocking, wood, 2″ x 4″	1.000	L.F.	.032	.46	1.86	2.32
Soffit framing, wood, 2″ x 4″	4.000	L.F.	.071	1.84	4.12	5.96
Soffit drywall	2.000	S.F.	.047	.88	2.76	3.64
Drywall painting	2.000	S.F.	.013	.12	.78	.90
TOTAL		L.F.	1.496	302.80	87.02	389.82
CUSTOM GRADE						
Top cabinets, custom grade	1.000	L.F.	.256	194	14.80	208.80
Bottom cabinets, custom grade	1.000	L.F.	.384	291	22.20	313.20
Solid surface countertop, premium patterned color	1.000	L.F.	1.067	143	62	205
Blocking, wood, 2″ x 4″	1.000	L.F.	.032	.46	1.86	2.32
Soffit framing, wood, 2″ x 4″	4.000	L.F.	.071	1.84	4.12	5.96
Soffit drywall	2.000	S.F.	.047	.88	2.76	3.64
Drywall painting	2.000	S.F.	.013	.12	.78	.90
TOTAL		L.F.	1.870	631.30	108.52	739.82

Description	QUAN.	UNIT	LABOR HOURS	COST PER L.F.		
				MAT.	INST.	TOTAL

Kitchen Price Sheet

Kitchen Price Sheet	QUAN.	UNIT	LABOR HOURS	COST PER L.F. MAT.	COST PER L.F. INST.	COST PER L.F. TOTAL
Top cabinets, economy grade	1.000	L.F.	.171	71	9.90	80.90
Average grade	1.000	L.F.	.213	89	12.40	101.40
Custom grade	1.000	L.F.	.256	194	14.80	208.80
Bottom cabinets, economy grade	1.000	L.F.	.256	107	14.90	121.90
Average grade	1.000	L.F.	.320	133	18.60	151.60
Custom grade	1.000	L.F.	.384	291	22	313
Counter top, laminated plastic, 7/8" thick, no splash	1.000	L.F.	.267	38	15.50	53.50
With backsplash	1.000	L.F.	.267	45	15.50	60.50
1-1/4" thick, no splash	1.000	L.F.	.286	42.50	16.60	59.10
With backsplash	1.000	L.F.	.286	47.50	16.60	64.10
Post formed, laminated plastic	1.000	L.F.	.267	13.85	15.50	29.35
Ceramic tile, with backsplash	1.000	L.F.	.427	21.50	9.30	30.80
Marble, with backsplash, minimum	1.000	L.F.	.471	54	27.50	81.50
Maximum	1.000	L.F.	.615	135	35.50	170.50
Maple, solid laminated, no backsplash	1.000	L.F.	.286	97	16.60	113.60
With backsplash	1.000	L.F.	.286	116	16.60	132.60
Solid Surface, with backsplash, minimum	1.000	L.F.	.842	91	49	140
Maximum	1.000	L.F.	1.067	143	62	205
Blocking, wood, 2" x 4"	1.000	L.F.	.032	.46	1.86	2.32
2" x 6"	1.000	L.F.	.036	.71	2.09	2.80
2" x 8"	1.000	L.F.	.040	.98	2.32	3.30
Soffit framing, wood, 2" x 3"	4.000	L.F.	.064	1.88	3.72	5.60
2" x 4"	4.000	L.F.	.071	1.84	4.12	5.96
Soffit, drywall, painted	2.000	S.F.	.060	1	3.54	4.54
Paneling, standard	2.000	S.F.	.064	3.24	3.72	6.96
Deluxe	2.000	S.F.	.091	2.42	5.30	7.72
Sinks, porcelain on cast iron, single bowl, 21" x 24"	1.000	Ea.	10.334	640	610	1,250
21" x 30"	1.000	Ea.	10.334	1,025	610	1,635
Double bowl, 20" x 32"	1.000	Ea.	10.810	720	635	1,355
Stainless steel, single bowl, 16" x 20"	1.000	Ea.	10.334	980	610	1,590
22" x 25"	1.000	Ea.	10.334	1,050	610	1,660
Double bowl, 20" x 32"	1.000	Ea.	10.810	900	635	1,535

Appliance Price Sheet	QUAN.	UNIT	LABOR HOURS	COST PER L.F.		
				MAT.	INST.	TOTAL
Range, free standing, minimum	1.000	Ea.	3.600	590	202	792
Maximum	1.000	Ea.	6.000	2,600	310	2,910
Built-in, minimum	1.000	Ea.	3.333	955	219	1,174
Maximum	1.000	Ea.	10.000	2,000	595	2,595
Counter top range, 4-burner, minimum	1.000	Ea.	3.333	435	219	654
Maximum	1.000	Ea.	4.667	2,000	305	2,305
Compactor, built-in, minimum	1.000	Ea.	2.215	780	134	914
Maximum	1.000	Ea.	3.282	1,300	196	1,496
Dishwasher, built-in, minimum	1.000	Ea.	6.735	570	440	1,010
Maximum	1.000	Ea.	9.235	745	610	1,355
Garbage disposer, minimum	1.000	Ea.	2.810	229	184	413
Maximum	1.000	Ea.	2.810	335	184	519
Microwave oven, minimum	1.000	Ea.	2.615	126	172	298
Maximum	1.000	Ea.	4.615	525	305	830
Range hood, ducted, minimum	1.000	Ea.	4.658	135	281	416
Maximum	1.000	Ea.	5.991	1,025	360	1,385
Ductless, minimum	1.000	Ea.	2.615	141	160	301
Maximum	1.000	Ea.	3.948	1,050	240	1,290
Refrigerator, 16 cu.ft., minimum	1.000	Ea.	2.000	605	89	694
Maximum	1.000	Ea.	3.200	970	142	1,112
16 cu.ft. with icemaker, minimum	1.000	Ea.	4.210	840	225	1,065
Maximum	1.000	Ea.	5.410	1,200	278	1,478
19 cu.ft., minimum	1.000	Ea.	2.667	590	118	708
Maximum	1.000	Ea.	4.667	1,025	207	1,232
19 cu.ft. with icemaker, minimum	1.000	Ea.	5.143	850	266	1,116
Maximum	1.000	Ea.	7.143	1,300	355	1,655
Sinks, porcelain on cast iron single bowl, 21" x 24"	1.000	Ea.	10.334	640	610	1,250
21" x 30"	1.000	Ea.	10.334	1,025	610	1,635
Double bowl, 20" x 32"	1.000	Ea.	10.810	720	635	1,355
Stainless steel, single bowl 16" x 20"	1.000	Ea.	10.334	980	610	1,590
22" x 25"	1.000	Ea.	10.334	1,050	610	1,660
Double bowl, 20" x 32"	1.000	Ea.	10.810	900	635	1,535
Water heater, electric, 30 gallon	1.000	Ea.	3.636	1,100	238	1,338
40 gallon	1.000	Ea.	4.000	1,175	262	1,437
Gas, 30 gallon	1.000	Ea.	4.000	1,950	262	2,212
75 gallon	1.000	Ea.	5.333	2,975	350	3,325
Wall, packaged terminal heater/air conditioner cabinet, wall sleeve, louver, electric heat, thermostat, manual changeover, 208V						
6000 BTUH cooling, 8800 BTU heating	1.000	Ea.	2.667	790	162	952
9000 BTUH cooling, 13,900 BTU heating	1.000	Ea.	3.200	1,025	194	1,219
12,000 BTUH cooling, 13,900 BTU heating	1.000	Ea.	4.000	1,600	243	1,843
15,000 BTUH cooling, 13,900 BTU heating	1.000	Ea.	5.333	1,550	325	1,875

System Description	QUAN.	UNIT	LABOR HOURS	COST EACH		
				MAT.	INST.	TOTAL
Curtain rods, stainless, 1″ diameter, 3′ long	1.000	Ea.	.615	30	36	66
5′ long	1.000	Ea.	.615	30	36	66
Grab bar, 1″ diameter, 12″ long	1.000	Ea.	.283	28	16.45	44.45
36″ long	1.000	Ea.	.340	32.50	19.55	52.05
1-1/4″ diameter, 12″ long	1.000	Ea.	.333	33	19.35	52.35
36″ long	1.000	Ea.	.400	38.50	23	61.50
1-1/2″ diameter, 12″ long	1.000	Ea.	.383	38	22.50	60.50
36″ long	1.000	Ea.	.460	44.50	26.50	71
Mirror, 18″ x 24″	1.000	Ea.	.400	52	23	75
72″ x 24″	1.000	Ea.	1.333	310	77.50	387.50
Medicine chest with mirror, 18″ x 24″	1.000	Ea.	.400	208	23	231
36″ x 24″	1.000	Ea.	.600	310	34.50	344.50
Toilet tissue dispenser, surface mounted, minimum	1.000	Ea.	.267	19.85	15.50	35.35
Maximum	1.000	Ea.	.400	30	23.50	53.50
Flush mounted, minimum	1.000	Ea.	.293	22	17.05	39.05
Maximum	1.000	Ea.	.427	32	25	57
Towel bar, 18″ long, minimum	1.000	Ea.	.278	37.50	16	53.50
Maximum	1.000	Ea.	.348	47	20	67
24″ long, minimum	1.000	Ea.	.313	42.50	18	60.50
Maximum	1.000	Ea.	.383	51.50	22	73.50
36″ long, minimum	1.000	Ea.	.381	57	22	79
Maximum	1.000	Ea.	.419	62.50	24	86.50

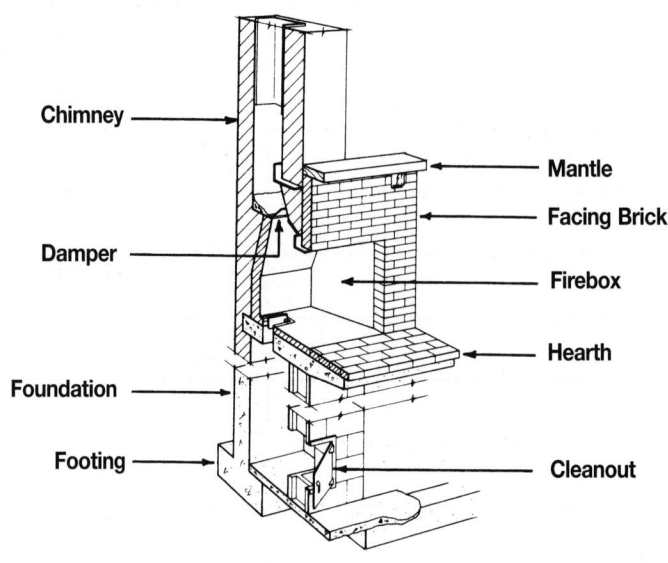

Chimney

Mantle

Facing Brick

Damper

Firebox

Hearth

Foundation

Footing

Cleanout

System Description	QUAN.	UNIT	LABOR HOURS	COST EACH		
				MAT.	INST.	TOTAL
MASONRY FIREPLACE						
Footing, 8" thick, concrete, 4' x 7'	.700	C.Y.	2.800	142.80	155.40	298.20
Foundation, concrete block, 32" x 60" x 4' deep	1.000	Ea.	5.275	193.20	284.40	477.60
Fireplace, brick firebox, 30" x 29" opening	1.000	Ea.	40.000	645	2,125	2,770
Damper, cast iron, 30" opening	1.000	Ea.	1.333	137	77.50	214.50
Facing brick, standard size brick, 6' x 5'	30.000	S.F.	5.217	142.50	282	424.50
Hearth, standard size brick, 3' x 6'	1.000	Ea.	8.000	234	425	659
Chimney, standard size brick, 8" x 12" flue, one story house	12.000	V.L.F.	12.000	528	636	1,164
Mantle, 4" x 8", wood	6.000	L.F.	1.333	53.10	77.40	130.50
Cleanout, cast iron, 8" x 8"	1.000	Ea.	.667	60	38.50	98.50
TOTAL		Ea.	76.625	2,135.60	4,101.20	6,236.80

The costs in this system are on a cost each basis.

Description	QUAN.	UNIT	LABOR HOURS	COST EACH		
				MAT.	INST.	TOTAL

Masonry Fireplace Price Sheet

	QUAN.	UNIT	LABOR HOURS	COST EACH		
				MAT.	INST.	TOTAL
Footing 8" thick, 3' x 6'	.440	C.Y.	1.326	90	97.50	187.50
4' x 7'	.700	C.Y.	2.110	143	156	299
5' x 8'	1.000	C.Y.	3.014	204	222	426
1' thick, 3' x 6'	.670	C.Y.	2.020	137	149	286
4' x 7'	1.030	C.Y.	3.105	210	229	439
5' x 8'	1.480	C.Y.	4.461	300	325	625
Foundation-concrete block, 24" x 48", 4' deep	1.000	Ea.	4.267	155	228	383
8' deep	1.000	Ea.	8.533	310	455	765
24" x 60", 4' deep	1.000	Ea.	4.978	180	265	445
8' deep	1.000	Ea.	9.956	360	530	890
32" x 48", 4' deep	1.000	Ea.	4.711	171	251	422
8' deep	1.000	Ea.	9.422	340	500	840
32" x 60", 4' deep	1.000	Ea.	5.333	193	284	477
8' deep	1.000	Ea.	10.845	395	580	975
32" x 72", 4' deep	1.000	Ea.	6.133	222	325	547
8' deep	1.000	Ea.	12.267	445	655	1,100
Fireplace, brick firebox 30" x 29" opening	1.000	Ea.	40.000	645	2,125	2,770
48" x 30" opening	1.000	Ea.	60.000	970	3,200	4,170
Steel fire box with registers, 25" opening	1.000	Ea.	26.667	1,200	1,425	2,625
48" opening	1.000	Ea.	44.000	2,050	2,350	4,400
Damper, cast iron, 30" opening	1.000	Ea.	1.333	137	77.50	214.50
36" opening	1.000	Ea.	1.556	160	90.50	250.50
Steel, 30" opening	1.000	Ea.	1.333	131	77.50	208.50
36" opening	1.000	Ea.	1.556	153	90.50	243.50
Facing for fireplace, standard size brick, 6' x 5'	30.000	S.F.	5.217	143	282	425
7' x 5'	35.000	S.F.	6.087	166	330	496
8' x 6'	48.000	S.F.	8.348	228	450	678
Fieldstone, 6' x 5'	30.000	S.F.	5.217	435	282	717
7' x 5'	35.000	S.F.	6.087	510	330	840
8' x 6'	48.000	S.F.	8.348	700	450	1,150
Sheetrock on metal, studs, 6' x 5'	30.000	S.F.	.980	26.50	56.50	83
7' x 5'	35.000	S.F.	1.143	31	66	97
8' x 6'	48.000	S.F.	1.568	42	90.50	132.50
Hearth, standard size brick, 3' x 6'	1.000	Ea.	8.000	234	425	659
3' x 7'	1.000	Ea.	9.280	271	495	766
3' x 8'	1.000	Ea.	10.640	310	565	875
Stone, 3' x 6'	1.000	Ea.	8.000	243	425	668
3' x 7'	1.000	Ea.	9.280	282	495	777
3' x 8'	1.000	Ea.	10.640	325	565	890
Chimney, standard size brick , 8" x 12" flue, one story house	12.000	V.L.F.	12.000	530	635	1,165
Two story house	20.000	V.L.F.	20.000	880	1,050	1,930
Mantle wood, beams, 4" x 8"	6.000	L.F.	1.333	53	77.50	130.50
4" x 10"	6.000	L.F.	1.371	71.50	80	151.50
Ornate, prefabricated, 6' x 3'-6" opening, minimum	1.000	Ea.	1.600	495	93	588
Maximum	1.000	Ea.	1.600	655	93	748
Cleanout, door and frame, cast iron, 8" x 8"	1.000	Ea.	.667	60	38.50	98.50
12" x 12"	1.000	Ea.	.800	98	46.50	144.50

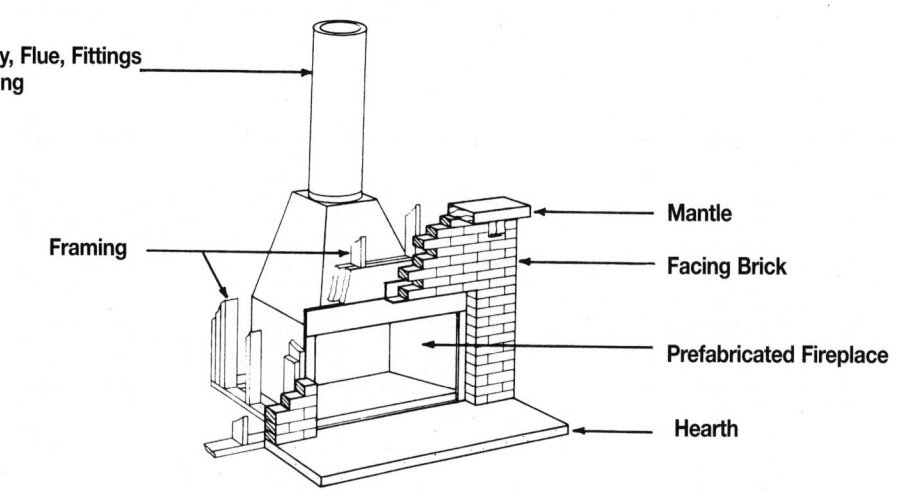

Chimney, Flue, Fittings & Framing

Framing

Mantle

Facing Brick

Prefabricated Fireplace

Hearth

System Description	QUAN.	UNIT	LABOR HOURS	COST EACH		
				MAT.	INST.	TOTAL
PREFABRICATED FIREPLACE						
Prefabricated fireplace, metal, painted	1.000	Ea.	6.154	1,775	360	2,135
Framing, 2" x 4" studs, 6' x 5'	35.000	L.F.	.509	15.75	29.75	45.50
Fire resistant gypsum drywall, unfinished	40.000	S.F.	.320	16.40	18.40	34.80
Drywall finishing adder	40.000	S.F.	.320	2	18.40	20.40
Facing, brick, standard size brick, 6' x 5'	30.000	S.F.	5.217	142.50	282	424.50
Hearth, standard size brick, 3' x 6'	1.000	Ea.	8.000	234	425	659
Chimney, one story house, framing, 2" x 4" studs	80.000	L.F.	1.164	36	68	104
Sheathing, plywood, 5/8" thick	32.000	S.F.	.758	91.52	44.16	135.68
Flue, 10" metal, insulated pipe	12.000	V.L.F.	4.000	570	229.20	799.20
Fittings, ceiling support	1.000	Ea.	.667	156	38	194
Fittings, joist shield	1.000	Ea.	.727	770	41.50	811.50
Fittings, roof flashing	1.000	Ea.	.667	440	38	478
Mantle beam, wood, 4" x 8"	6.000	L.F.	1.333	53.10	77.40	130.50
TOTAL		Ea.	29.836	4,302.27	1,669.81	5,972.08

The costs in this system are on a cost each basis.

Description	QUAN.	UNIT	LABOR HOURS	COST EACH		
				MAT.	INST.	TOTAL

Prefabricated Fireplace Price Sheet

	QUAN.	UNIT	LABOR HOURS	COST EACH		
				MAT.	INST.	TOTAL
Prefabricated fireplace, minimum	1.000	Ea.	6.154	1,775	360	2,135
Average	1.000	Ea.	8.000	2,000	465	2,465
Maximum	1.000	Ea.	8.889	3,575	515	4,090
Framing, 2" x 4" studs, fireplace, 6' x 5'	35.000	L.F.	.509	15.75	30	45.75
7' x 5'	40.000	L.F.	.582	18	34	52
8' x 6'	45.000	L.F.	.655	20.50	38.50	59
Sheetrock, 1/2" thick, fireplace, 6' x 5'	40.000	S.F.	.640	18.40	37	55.40
7' x 5'	45.000	S.F.	.720	20.50	41.50	62
8' x 6'	50.000	S.F.	.800	23	46	69
Facing for fireplace, brick, 6' x 5'	30.000	S.F.	5.217	143	282	425
7' x 5'	35.000	S.F.	6.087	166	330	496
8' x 6'	48.000	S.F.	8.348	228	450	678
Fieldstone, 6' x 5'	30.000	S.F.	5.217	420	282	702
7' x 5'	35.000	S.F.	6.087	490	330	820
8' x 6'	48.000	S.F.	8.348	670	450	1,120
Hearth, standard size brick, 3' x 6'	1.000	Ea.	8.000	234	425	659
3' x 7'	1.000	Ea.	9.280	271	495	766
3' x 8'	1.000	Ea.	10.640	310	565	875
Stone, 3' x 6'	1.000	Ea.	8.000	243	425	668
3' x 7'	1.000	Ea.	9.280	282	495	777
3' x 8'	1.000	Ea.	10.640	325	565	890
Chimney, framing, 2" x 4", one story house	80.000	L.F.	1.164	36	68	104
Two story house	120.000	L.F.	1.746	54	102	156
Sheathing, plywood, 5/8" thick	32.000	S.F.	.758	91.50	44	135.50
Stucco on plywood	32.000	S.F.	1.125	59.50	64.50	124
Flue, 10" metal pipe, insulated, one story house	12.000	V.L.F.	4.000	570	229	799
Two story house	20.000	V.L.F.	6.667	950	380	1,330
Fittings, ceiling support	1.000	Ea.	.667	156	38	194
Fittings joist sheild, one story house	1.000	Ea.	.667	770	41.50	811.50
Two story house	2.000	Ea.	1.333	1,550	83	1,633
Fittings roof flashing	1.000	Ea.	.667	440	38	478
Mantle, wood beam, 4" x 8"	6.000	L.F.	1.333	53	77.50	130.50
4" x 10"	6.000	L.F.	1.371	71.50	80	151.50
Ornate prefabricated, 6' x 3'-6" opening, minimum	1.000	Ea.	1.600	495	93	588
Maximum	1.000	Ea.	1.600	655	93	748

System Description	QUAN.	UNIT	LABOR HOURS	COST EACH		
				MAT.	INST.	TOTAL
Economy, lean to, shell only, not including 2' stub wall, fndtn, flrs, heat						
4' x 16'	1.000	Ea.	26.212	2,675	1,525	4,200
4' x 24'	1.000	Ea.	30.259	3,075	1,775	4,850
6' x 10'	1.000	Ea.	16.552	1,925	965	2,890
6' x 16'	1.000	Ea.	23.034	2,675	1,350	4,025
6' x 24'	1.000	Ea.	29.793	3,450	1,725	5,175
8' x 10'	1.000	Ea.	22.069	2,550	1,275	3,825
8' x 16'	1.000	Ea.	38.400	4,450	2,225	6,675
8' x 24'	1.000	Ea.	49.655	5,750	2,900	8,650
Free standing, 8' x 8'	1.000	Ea.	17.356	1,700	1,000	2,700
8' x 16'	1.000	Ea.	30.211	2,950	1,750	4,700
8' x 24'	1.000	Ea.	39.051	3,825	2,275	6,100
10' x 10'	1.000	Ea.	18.824	4,900	1,100	6,000
10' x 16'	1.000	Ea.	24.095	6,275	1,400	7,675
10' x 24'	1.000	Ea.	31.624	8,225	1,850	10,075
14' x 10'	1.000	Ea.	20.741	7,075	1,200	8,275
14' x 16'	1.000	Ea.	24.889	8,475	1,450	9,925
14' x 24'	1.000	Ea.	33.349	11,400	1,925	13,325
Standard,lean to,shell only,not incl.2'stub wall, fndtn,flrs, heat 4'x10'	1.000	Ea.	28.235	2,875	1,650	4,525
4' x 16'	1.000	Ea.	39.341	4,025	2,300	6,325
4' x 24'	1.000	Ea.	45.412	4,625	2,650	7,275
6' x 10'	1.000	Ea.	24.827	2,875	1,450	4,325
6' x 16'	1.000	Ea.	34.538	4,000	2,000	6,000
6' x 24'	1.000	Ea.	44.689	5,175	2,600	7,775
8' x 10'	1.000	Ea.	33.103	3,850	1,925	5,775
8' x 16'	1.000	Ea.	57.600	6,675	3,350	10,025
8' x 24'	1.000	Ea.	74.482	8,650	4,325	12,975
Free standing, 8' x 8'	1.000	Ea.	26.034	2,550	1,500	4,050
8' x 16'	1.000	Ea.	45.316	4,425	2,625	7,050
8' x 24'	1.000	Ea.	58.577	5,725	3,400	9,125
10' x 10'	1.000	Ea.	28.236	7,350	1,650	9,000
10' x 16'	1.000	Ea.	36.142	9,400	2,100	11,500
10' x 24'	1.000	Ea.	47.436	12,300	2,750	15,050
14' x 10'	1.000	Ea.	31.112	10,600	1,800	12,400
14' x 16'	1.000	Ea.	37.334	12,700	2,175	14,875
14' x 24'	1.000	Ea.	50.030	17,100	2,900	20,000
Deluxe,lean to,shell only,not incl.2'stub wall, fndtn, flrs or heat, 4'x10'	1.000	Ea.	20.645	4,650	1,200	5,850
4' x 16'	1.000	Ea.	33.032	7,425	1,925	9,350
4' x 24'	1.000	Ea.	49.548	11,100	2,875	13,975
6' x 10'	1.000	Ea.	30.968	6,950	1,800	8,750
6' x 16'	1.000	Ea.	49.548	11,100	2,875	13,975
6' x 24'	1.000	Ea.	74.323	16,700	4,325	21,025
8' x 10'	1.000	Ea.	41.290	9,275	2,400	11,675
8' x 16'	1.000	Ea.	66.065	14,800	3,850	18,650
8' x 24'	1.000	Ea.	99.097	22,300	5,750	28,050
Freestanding, 8' x 8'	1.000	Ea.	18.618	6,050	1,075	7,125
8' x 16'	1.000	Ea.	37.236	12,100	2,175	14,275
8' x 24'	1.000	Ea.	55.855	18,100	3,250	21,350
10' x 10'	1.000	Ea.	29.091	9,450	1,700	11,150
10' x 16'	1.000	Ea.	46.546	15,100	2,700	17,800
10' x 24'	1.000	Ea.	69.818	22,700	4,050	26,750
14' x 10'	1.000	Ea.	40.727	13,200	2,375	15,575
14' x 16'	1.000	Ea.	65.164	21,200	3,775	24,975
14' x 24'	1.000	Ea.	97.746	31,800	5,675	37,475

System Description	QUAN.	UNIT	LABOR HOURS	COST EACH		
				MAT.	INST.	TOTAL
Swimming pools, vinyl lined, metal sides, sand bottom, 12' x 28'	1.000	Ea.	50.177	7,800	3,075	10,875
12' x 32'	1.000	Ea.	55.366	8,600	3,400	12,000
12' x 36'	1.000	Ea.	60.061	9,325	3,675	13,000
16' x 32'	1.000	Ea.	66.798	10,400	4,100	14,500
16' x 36'	1.000	Ea.	71.190	11,100	4,375	15,475
16' x 40'	1.000	Ea.	74.703	11,600	4,575	16,175
20' x 36'	1.000	Ea.	77.860	12,100	4,775	16,875
20' x 40'	1.000	Ea.	82.135	12,800	5,050	17,850
20' x 44'	1.000	Ea.	90.348	14,000	5,550	19,550
24' x 40'	1.000	Ea.	98.562	15,300	6,050	21,350
24' x 44'	1.000	Ea.	108.418	16,800	6,650	23,450
24' x 48'	1.000	Ea.	118.274	18,400	7,250	25,650
Vinyl lined, concrete sides, 12' x 28'	1.000	Ea.	79.447	12,300	4,875	17,175
12' x 32'	1.000	Ea.	88.818	13,800	5,450	19,250
12' x 36'	1.000	Ea.	97.656	15,200	6,000	21,200
16' x 32'	1.000	Ea.	111.393	17,300	6,850	24,150
16' x 36'	1.000	Ea.	121.354	18,900	7,450	26,350
16' x 40'	1.000	Ea.	130.445	20,300	8,025	28,325
28' x 36'	1.000	Ea.	140.585	21,800	8,650	30,450
20' x 40'	1.000	Ea.	149.336	23,200	9,175	32,375
20' x 44'	1.000	Ea.	164.270	25,500	10,100	35,600
24' x 40'	1.000	Ea.	179.203	27,800	11,000	38,800
24' x 44'	1.000	Ea.	197.124	30,600	12,100	42,700
24' x 48'	1.000	Ea.	215.044	33,400	13,300	46,700
Gunite, bottom and sides, 12' x 28'	1.000	Ea.	129.767	18,000	7,950	25,950
12' x 32'	1.000	Ea.	142.164	19,700	8,700	28,400
12' x 36'	1.000	Ea.	153.028	21,200	9,375	30,575
16' x 32'	1.000	Ea.	167.743	23,200	10,300	33,500
16' x 36'	1.000	Ea.	176.421	24,400	10,800	35,200
16' x 40'	1.000	Ea.	182.368	25,300	11,200	36,500
20' x 36'	1.000	Ea.	187.949	26,000	11,500	37,500
20' x 40'	1.000	Ea.	179.200	34,400	11,000	45,400
20' x 44'	1.000	Ea.	197.120	37,800	12,100	49,900
24' x 40'	1.000	Ea.	215.040	41,300	13,200	54,500
24' x 44'	1.000	Ea.	273.244	37,900	16,700	54,600
24' x 48'	1.000	Ea.	298.077	41,300	18,200	59,500

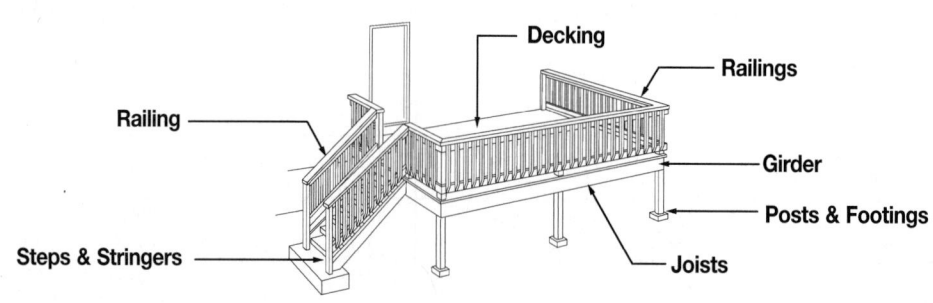

System Description	QUAN.	UNIT	LABOR HOURS	COST PER S.F.		
				MAT.	INST.	TOTAL
8' X 12' DECK, PRESSURE TREATED LUMBER, JOISTS 16" O.C.						
4" x 4" Posts	.500	L.F.	.021	.66	1.19	1.85
Ledger, bolted 4' O.C.	.125	L.F.	.005	.21	.30	.51
Joists, 2" x 10", 16" O.C.	1.080	L.F.	.019	1.72	1.11	2.83
5/4" x 6" decking	1.000	S.F.	.025	2.36	1.45	3.81
2" x 12" stair stringers	.375	L.F.	.046	.71	2.68	3.39
5/4" x 6" stair treads	.500	L.F.	.050	.55	2.90	3.45
4' x 4" handrail posts	.210	L.F.	.009	.28	.50	.78
2" x 6" handrail cap	.520	L.F.	.014	.43	.81	1.24
2" x 4" baluster support	1.040	L.F.	.028	.68	1.61	2.29
2" x 2" balusters	3.850	L.F.	.093	1.81	5.43	7.24
Post footings	.042	Ea.	.134	1.43	6.89	8.32
Concrete step	.010	Ea.	.016	.76	.93	1.69
Joist hangers	.210	Ea.	.010	.29	.59	.88
TOTAL		S.F.	.470	11.89	26.39	38.28
12' X 16' DECK, PRESSURE TREATED LUMBER, JOISTS 24" O.C.						
4" x 4" Posts	.312	L.F.	.013	.41	.74	1.15
Ledger, bolted 4' O.C.	.083	L.F.	.003	.12	.20	.32
Joists, 2" x 8", 16" O.C.	.810	L.F.	.014	1.29	.83	2.12
5/4" x 6" decking	1.000	S.F.	.025	2.36	1.45	3.81
2" x 12" stair stringers	.188	L.F.	.023	.36	1.34	1.70
5/4" x 6" stair treads	.250	L.F.	.025	1.73	1.45	3.18
4" x 4" handrail posts	.125	L.F.	.005	.17	.30	.47
2" x 6" handrail cap	.323	L.F.	.009	.26	.50	.76
2" x 4" baluster support	.650	L.F.	.017	.42	1.01	1.43
2" x 2" balusters	2.440	L.F.	.059	1.15	3.44	4.59
Post footings	.026	Ea.	.083	.88	4.26	5.14
Concrete step	.005	Ea.	.008	.38	.46	.84
Joist hangers	.135	Ea.	.007	.19	.38	.57
Deck support girder	.083	L.F.	.002	.26	.14	.40
TOTAL		S.F.	.293	9.98	16.50	26.48

The costs in this system are on a square foot basis.

12' X 24' DECK, REDWOOD OR CEDAR, JOISTS 16" O.C.

System Description	QUAN.	UNIT	LABOR HOURS	COST PER S.F. MAT.	COST PER S.F. INST.	COST PER S.F. TOTAL
4" x 4" redwood posts	.291	L.F.	.012	2.07	.69	2.76
Redwood joists, 2" x 8", 16" O.C.	.001	M.B.F.	.010	5.50	.58	6.08
5/4" x 6" redwood decking	1.000	S.F.	.025	5.25	1.45	6.70
Stair stringers	.125	L.F.	.015	.24	.89	1.13
Stringer trim	.083	L.F.	.003	.44	.19	.63
Stair treads	.135	S.F.	.003	.71	.20	.91
Handrail supports	.083	L.F.	.003	.59	.20	.79
Handrail cap	.152	L.F.	.004	1.25	.24	1.49
Baluster supports	.416	L.F.	.011	3.43	.64	4.07
Balusters	1.875	L.F.	.050	2.42	2.91	5.33
Post footings	.024	Ea.	.077	.82	3.94	4.76
Concrete step	.003	Ea.	.005	.23	.27	.50
Joist hangers	.132	Ea.	.006	.18	.37	.55
Deck support girder		M.B.F.	.003	1.50	.16	1.66
TOTAL		S.F.	.227	24.63	12.73	37.36

The costs in this system are on a square foot basis.

Wood Deck Price Sheet	QUAN.	UNIT	LABOR HOURS	COST PER S.F. MAT.	COST PER S.F. INST.	COST PER S.F. TOTAL
Decking, treated lumber, 1" x 4"	3.430	L.F.	.031	10.40	5.80	16.20
2" x 4"	3.430	L.F.	.033	7.55	5.30	12.85
2" x 6"	2.200	L.F.	.041	3.87	3.19	7.06
5/4" x 6"	2.200	L.F.	.027	5.20	3.19	8.39
Redwood or cedar, 1" x 4"	3.430	L.F.	.035	10.35	5.80	16.15
2" x 4"	3.430	L.F.	.036	21	5.30	26.30
2" x 6"	2.200	L.F.	.028	24.50	3.19	27.69
5/4" x 6"	2.200	L.F.	.027	15.20	3.19	18.39
Joists for deck, treated lumber, 2" x 8", 16" O.C.	1.000	L.F.	.015	1.27	.85	2.12
24" O.C.	.750	L.F.	.012	.95	.64	1.59
2" x 10", 16" O.C.	1.000	L.F.	.018	1.59	1.03	2.62
24" O.C.	.750	L.F.	.014	1.19	.77	1.96
Redwood or cedar, 2" x 8", 16" O.C.	1.000	L.F.	.015	5	.85	5.85
24" O.C.	.750	L.F.	.012	3.75	.64	4.39
2" x 10", 16" O.C.	1.000	L.F.	.018	8.35	1.03	9.38
24" O.C.	.750	L.F.	.014	6.25	.77	7.02
Girder for joists, treated lumber, 2" x 10", 8' x 12' deck	.250	L.F.	.002	.40	.21	.61
12' x 16' deck	.167	L.F.	.001	.26	.14	.40
12' x 24' deck	.167	L.F.	.001	.26	.14	.40
Redwood or cedar, 2" x 10", 8' x 12' deck	.250	L.F.	.002	2.09	.26	2.35
12' x 16' deck	.167	L.F.	.001	1.39	.17	1.56
12' x 24' deck	.167	L.F.	.001	1.39	.17	1.56
Posts, 4" x 4", including concrete footing, 8' x 12' deck	.500	L.F.	.022	2.05	7.90	9.95
12' x 16' deck	.312	L.F.	.017	1.29	5	6.29
12' x 24' deck	.291	L.F.	.017	1.20	4.63	5.83
Stairs 2" x 10" stringers, treated lumber, 8' x 12' deck	1.000	Set	.020	1.56	4.65	6.21
12' x 16' deck	1.000	Set	.012	.78	2.33	3.11
12' x 24' deck	1.000	Set	.008	.55	1.63	2.18
Redwood or cedar, 8' x 12' deck	1.000	Set	.040	3.90	4.65	8.55
12' x 16' deck	1.000	Set	.020	1.95	2.33	4.28
12' x 24' deck	1.000	Set	.012	1.37	1.63	3
Railings 2" x 4", treated lumber, 8' x 12' deck	.520	L.F.	.026	2.92	7.85	10.77
12' x 16' deck	.323	L.F.	.017	1.83	4.95	6.78
12' x 24' deck	.210	L.F.	.014	1.32	3.62	4.94
Redwood or cedar, 8' x 12' deck	1.000	L.F.	.009	10.60	8.35	18.95

For customer support on your Residential Costs with RSMeans data, call 800.448.8182.

245

Wood Deck Price Sheet	QUAN.	UNIT	LABOR HOURS	COST PER S.F.		
				MAT.	INST.	TOTAL
12' x 16' deck	.670	L.F.	.006	6.60	5.25	11.85
12' x 24' deck	.540	L.F.	.005	4.70	3.90	8.60
Alternative decking, wood/plastic composite, 5/4" x 6"	2.200	L.F.	.055	7.90	3.19	11.09
1" x 4" square edge fir	3.430	L.F.	.100	10.45	5.80	16.25
1" x 4" tongue and groove fir	3.430	L.F.	.122	5.70	7.10	12.80
1" x 4" mahogany	3.430	L.F.	.100	7.80	5.80	13.60
5/4" x 6" PVC	2.200	L.F.	.064	8.10	3.72	11.82

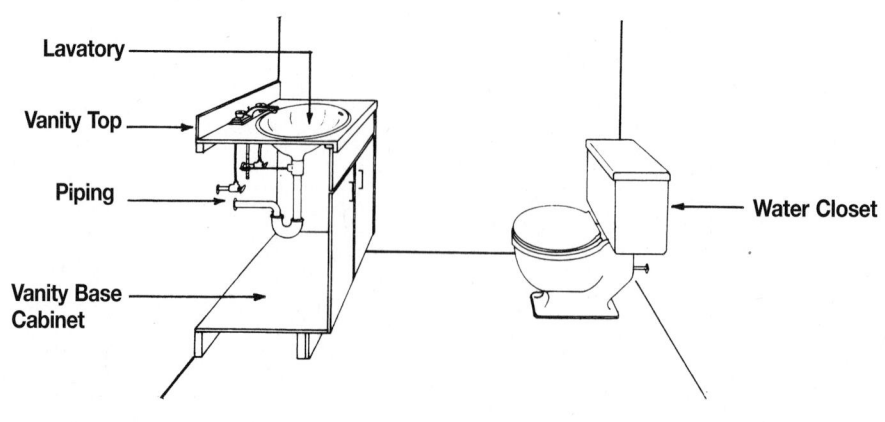

Lavatory

Vanity Top

Piping

Vanity Base
Cabinet

Water Closet

System Description	QUAN.	UNIT	LABOR HOURS	COST EACH		
				MAT.	INST.	TOTAL
LAVATORY INSTALLED WITH VANITY, PLUMBING IN 2 WALLS						
Water closet, floor mounted, 2 piece, close coupled, white	1.000	Ea.	3.019	253	178	431
Rough-in, vent, 2" diameter DWV piping	1.000	Ea.	.955	75.20	56.40	131.60
Waste, 4" diameter DWV piping	1.000	Ea.	.828	112.50	48.75	161.25
Supply, 1/2" diameter type "L" copper supply piping	1.000	Ea.	.593	22.08	38.70	60.78
Lavatory, 20" x 18", P.E. cast iron white	1.000	Ea.	2.500	330	147	477
Rough-in, vent, 1-1/2" diameter DWV piping	1.000	Ea.	.901	70.40	53.20	123.60
Waste, 2" diameter DWV piping	1.000	Ea.	.955	75.20	56.40	131.60
Supply, 1/2" diameter type "L" copper supply piping	1.000	Ea.	.988	36.80	64.50	101.30
Piping, supply, 1/2" diameter type "L" copper supply piping	10.000	L.F.	.988	36.80	64.50	101.30
Waste, 4" diameter DWV piping	7.000	L.F.	1.931	262.50	113.75	376.25
Vent, 2" diameter DWV piping	12.000	L.F.	2.866	225.60	169.20	394.80
Vanity base cabinet, 2 door, 30" wide	1.000	Ea.	1.000	460	58	518
Vanity top, plastic & laminated, square edge	2.670	L.F.	.712	121.49	41.39	162.88
TOTAL		Ea.	18.236	2,081.57	1,089.79	3,171.36
LAVATORY WITH WALL-HUNG LAVATORY, PLUMBING IN 2 WALLS						
Water closet, floor mounted, 2 piece close coupled, white	1.000	Ea.	3.019	253	178	431
Rough-in, vent, 2" diameter DWV piping	1.000	Ea.	.955	75.20	56.40	131.60
Waste, 4" diameter DWV piping	1.000	Ea.	.828	112.50	48.75	161.25
Supply, 1/2" diameter type "L" copper supply piping	1.000	Ea.	.593	22.08	38.70	60.78
Lavatory, 20" x 18", P.E. cast iron, wall hung, white	1.000	Ea.	2.000	270	118	388
Rough-in, vent, 1-1/2" diameter DWV piping	1.000	Ea.	.901	70.40	53.20	123.60
Waste, 2" diameter DWV piping	1.000	Ea.	.955	75.20	56.40	131.60
Supply, 1/2" diameter type "L" copper supply piping	1.000	Ea.	.988	36.80	64.50	101.30
Piping, supply, 1/2" diameter type "L" copper supply piping	10.000	L.F.	.988	36.80	64.50	101.30
Waste, 4" diameter DWV piping	7.000	L.F.	1.931	262.50	113.75	376.25
Vent, 2" diameter DWV piping	12.000	L.F.	2.866	225.60	169.20	394.80
Carrier, steel for studs, no arms	1.000	Ea.	1.143	63	75	138
TOTAL		Ea.	17.167	1,503.08	1,036.40	2,539.48

Description	QUAN.	UNIT	LABOR HOURS	COST EACH		
				MAT.	INST.	TOTAL

Two Fixture Lavatory Price Sheet

	QUAN.	UNIT	LABOR HOURS	COST EACH MAT.	COST EACH INST.	COST EACH TOTAL
Water closet, close coupled standard 2 piece, white	1.000	Ea.	3.019	253	178	431
Color	1.000	Ea.	3.019	460	178	638
One piece elongated bowl, white	1.000	Ea.	3.019	805	178	983
Color	1.000	Ea.	3.019	1,075	178	1,253
Low profile, one piece elongated bowl, white	1.000	Ea.	3.019	815	178	993
Color	1.000	Ea.	3.019	1,075	178	1,253
Rough-in for water closet						
1/2" copper supply, 4" cast iron waste, 2" cast iron vent	1.000	Ea.	2.376	210	144	354
4" PVC waste, 2" PVC vent	1.000	Ea.	2.678	93.50	162	255.50
4" copper waste , 2" copper vent	1.000	Ea.	2.520	283	157	440
3" cast iron waste, 1-1/2" cast iron vent	1.000	Ea.	2.244	157	136	293
3" PVC waste, 1-1/2" PVC vent	1.000	Ea.	2.388	84.50	150	234.50
3" copper waste, 1-1/2" copper vent	1.000	Ea.	2.524	211	152	363
1/2" PVC supply, 4" PVC waste, 2" PVC vent	1.000	Ea.	2.974	116	181	297
3" PVC waste, 1-1/2" PVC vent	1.000	Ea.	2.684	106	170	276
1/2" steel supply, 4" cast iron waste, 2" cast iron vent	1.000	Ea.	2.545	213	155	368
4" cast iron waste, 2" steel vent	1.000	Ea.	2.590	195	158	353
4" PVC waste, 2" PVC vent	1.000	Ea.	2.847	97	173	270
Lavatory, vanity top mounted, P.E. on cast iron 20" x 18" white	1.000	Ea.	2.500	330	147	477
Color	1.000	Ea.	2.500	535	147	682
Steel, enameled 10" x 17" white	1.000	Ea.	2.759	142	163	305
Color	1.000	Ea.	2.500	138	147	285
Vitreous china 20" x 16", white	1.000	Ea.	2.963	234	175	409
Color	1.000	Ea.	2.963	234	175	409
Wall hung, P.E. on cast iron, 20" x 18", white	1.000	Ea.	2.000	270	118	388
Color	1.000	Ea.	2.000	288	118	406
Vitreous china 19" x 17", white	1.000	Ea.	2.286	135	135	270
Color	1.000	Ea.	2.286	157	135	292
Rough-in supply waste and vent for lavatory						
1/2" copper supply, 2" cast iron waste, 1-1/2" cast iron vent	1.000	Ea.	2.844	182	174	356
2" PVC waste, 1-1/2" PVC vent	1.000	Ea.	2.962	109	187	296
2" copper waste, 1-1/2" copper vent	1.000	Ea.	2.308	166	151	317
1-1/2" PVC waste, 1-1/4" PVC vent	1.000	Ea.	2.639	108	173	281
1-1/2" copper waste, 1-1/4" copper vent	1.000	Ea.	2.114	137	138	275
1/2" PVC supply, 2" PVC waste, 1-1/2" PVC vent	1.000	Ea.	3.456	146	219	365
1-1/2" PVC waste, 1-1/4" PVC vent	1.000	Ea.	3.133	145	205	350
1/2" steel supply, 2" cast iron waste, 1-1/2" cast iron vent	1.000	Ea.	3.126	188	193	381
2" cast iron waste, 2" steel vent	1.000	Ea.	3.225	175	198	373
2" PVC waste, 1-1/2" PVC vent	1.000	Ea.	3.244	114	205	319
1-1/2" PVC waste, 1-1/4" PVC vent	1.000	Ea.	2.921	114	191	305
Piping, supply, 1/2" copper, type "L"	10.000	L.F.	.988	37	64.50	101.50
1/2" steel	10.000	L.F.	1.270	42.50	83	125.50
1/2" PVC	10.000	L.F.	1.482	73.50	97	170.50
Waste, 4" cast iron	7.000	L.F.	1.931	263	114	377
4" copper	7.000	L.F.	2.800	425	165	590
4" PVC	7.000	L.F.	2.333	78.50	138	216.50
Vent, 2" cast iron	12.000	L.F.	2.866	226	169	395
2" copper	12.000	L.F.	2.182	238	143	381
2" PVC	12.000	L.F.	3.254	114	192	306
2" steel	12.000	Ea.	3.000	171	177	348
Vanity base cabinet, 2 door, 24" x 30"	1.000	Ea.	1.000	460	58	518
24" x 36"	1.000	Ea.	1.200	445	69.50	514.50
Vanity top, laminated plastic, square edge 25" x 32"	2.670	L.F.	.712	121	41.50	162.50
25" x 38"	3.170	L.F.	.845	144	49	193
Post formed, laminated plastic, 25" x 32"	2.670	L.F.	.712	37	41.50	78.50
25" x 38"	3.170	L.F.	.845	44	49	93
Cultured marble, 25" x 32" with bowl	1.000	Ea.	2.500	183	147	330
25" x 38" with bowl	1.000	Ea.	2.500	216	147	363
Carrier for lavatory, steel for studs	1.000	Ea.	1.143	63	75	138
Wood 2" x 8" blocking	1.330	L.F.	.053	1.30	3.09	4.39

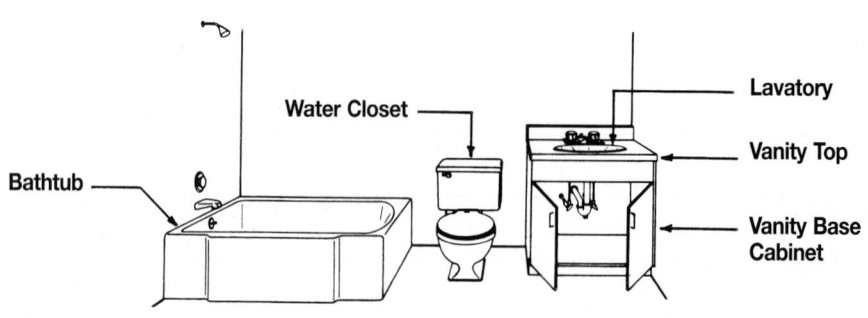

System Description	QUAN.	UNIT	LABOR HOURS	COST EACH		
				MAT.	INST.	TOTAL
BATHROOM INSTALLED WITH VANITY						
Water closet, floor mounted, 2 piece, close coupled, white	1.000	Ea.	3.019	253	178	431
Rough-in, waste, 4" diameter DWV piping	1.000	Ea.	.828	112.50	48.75	161.25
Vent, 2" diameter DWV piping	1.000	Ea.	.955	75.20	56.40	131.60
Supply, 1/2" diameter type "L" copper supply piping	1.000	Ea.	.593	22.08	38.70	60.78
Lavatory, 20" x 18", P.E. cast iron with accessories, white	1.000	Ea.	2.500	330	147	477
Rough-in, supply, 1/2" diameter type "L" copper supply piping	1.000	Ea.	.988	36.80	64.50	101.30
Waste, 1-1/2" diameter DWV piping	1.000	Ea.	1.803	140.80	106.40	247.20
Bathtub, P.E. cast iron, 5' long with accessories, white	1.000	Ea.	3.636	1,300	214	1,514
Rough-in, waste, 4" diameter DWV piping	1.000	Ea.	.828	112.50	48.75	161.25
Vent, 1-1/2" diameter DWV piping	1.000	Ea.	.593	49.40	38.80	88.20
Supply, 1/2" diameter type "L" copper supply piping	1.000	Ea.	.988	36.80	64.50	101.30
Piping, supply, 1/2" diameter type "L" copper supply piping	20.000	L.F.	1.975	73.60	129	202.60
Waste, 4" diameter DWV piping	9.000	L.F.	2.483	337.50	146.25	483.75
Vent, 2" diameter DWV piping	6.000	L.F.	1.500	85.50	88.50	174
Vanity base cabinet, 2 door, 30" wide	1.000	Ea.	1.000	460	58	518
Vanity top, plastic laminated square edge	2.670	L.F.	.712	100.13	41.39	141.52
TOTAL		Ea.	24.401	3,525.81	1,468.94	4,994.75
BATHROOM WITH WALL HUNG LAVATORY						
Water closet, floor mounted, 2 piece, close coupled, white	1.000	Ea.	3.019	253	178	431
Rough-in, vent, 2" diameter DWV piping	1.000	Ea.	.955	75.20	56.40	131.60
Waste, 4" diameter DWV piping	1.000	Ea.	.828	112.50	48.75	161.25
Supply, 1/2" diameter type "L" copper supply piping	1.000	Ea.	.593	22.08	38.70	60.78
Lavatory, 20" x 18" P.E. cast iron, wall hung, white	1.000	Ea.	2.000	270	118	388
Rough-in, waste, 1-1/2" diameter DWV piping	1.000	Ea.	1.803	140.80	106.40	247.20
Supply, 1/2" diameter type "L" copper supply piping	1.000	Ea.	.988	36.80	64.50	101.30
Bathtub, P.E. cast iron, 5' long with accessories, white	1.000	Ea.	3.636	1,300	214	1,514
Rough-in, waste, 4" diameter DWV piping	1.000	Ea.	.828	112.50	48.75	161.25
Supply, 1/2" diameter type "L" copper supply piping	1.000	Ea.	.988	36.80	64.50	101.30
Vent, 1-1/2" diameter DWV piping	1.000	Ea.	1.482	123.50	97	220.50
Piping, supply, 1/2" diameter type "L" copper supply piping	20.000	L.F.	1.975	73.60	129	202.60
Waste, 4" diameter DWV piping	9.000	L.F.	2.483	337.50	146.25	483.75
Vent, 2" diameter DWV piping	6.000	L.F.	1.500	85.50	88.50	174
Carrier, steel, for studs, no arms	1.000	Ea.	1.143	63	75	138
TOTAL		Ea.	24.221	3,042.78	1,473.75	4,516.53

The costs in this system are a cost each basis, all necessary piping
is included.

Three Fixture Bathroom Price Sheet	QUAN.	UNIT	LABOR HOURS	COST EACH		
				MAT.	INST.	TOTAL
Water closet, close coupled standard 2 piece, white	1.000	Ea.	3.019	253	178	431
Color	1.000	Ea.	3.019	460	178	638
One piece, elongated bowl, white	1.000	Ea.	3.019	805	178	983
Color	1.000	Ea.	3.019	1,075	178	1,253
Low profile, one piece elongated bowl, white	1.000	Ea.	3.019	815	178	993
Color	1.000	Ea.	3.019	1,075	178	1,253
Rough-in, for water closet						
1/2" copper supply, 4" cast iron waste, 2" cast iron vent	1.000	Ea.	2.376	210	144	354
4" PVC/DWV waste, 2" PVC vent	1.000	Ea.	2.678	93.50	162	255.50
4" copper waste, 2" copper vent	1.000	Ea.	2.520	283	157	440
3" cast iron waste, 1-1/2" cast iron vent	1.000	Ea.	2.244	157	136	293
3" PVC waste, 1-1/2" PVC vent	1.000	Ea.	2.388	84.50	150	234.50
3" copper waste, 1-1/2" copper vent	1.000	Ea.	2.014	152	126	278
1/2" PVC supply, 4" PVC waste, 2" PVC vent	1.000	Ea.	2.974	116	181	297
3" PVC waste, 1-1/2" PVC supply	1.000	Ea.	2.684	106	170	276
1/2" steel supply, 4" cast iron waste, 2" cast iron vent	1.000	Ea.	2.545	213	155	368
4" cast iron waste, 2" steel vent	1.000	Ea.	2.590	195	158	353
4" PVC waste, 2" PVC vent	1.000	Ea.	2.847	97	173	270
Lavatory, wall hung, P.E. cast iron 20" x 18", white	1.000	Ea.	2.000	270	118	388
Color	1.000	Ea.	2.000	288	118	406
Vitreous china 19" x 17", white	1.000	Ea.	2.286	135	135	270
Color	1.000	Ea.	2.286	157	135	292
Lavatory, for vanity top, P.E. cast iron 20" x 18"", white	1.000	Ea.	2.500	330	147	477
Color	1.000	Ea.	2.500	535	147	682
Steel, enameled 20" x 17", white	1.000	Ea.	2.759	142	163	305
Color	1.000	Ea.	2.500	138	147	285
Vitreous china 20" x 16", white	1.000	Ea.	2.963	234	175	409
Color	1.000	Ea.	2.963	234	175	409
Rough-in, for lavatory						
1/2" copper supply, 1-1/2" C.I. waste, 1-1/2" C.I. vent	1.000	Ea.	2.791	178	171	349
1-1/2" PVC waste, 1-1/4" PVC vent	1.000	Ea.	2.639	108	173	281
1/2" steel supply, 1-1/4" cast iron waste, 1-1/4" steel vent	1.000	Ea.	2.890	138	179	317
1-1/4" PVC waste, 1-1/4" PVC vent	1.000	Ea.	2.794	117	183	300
1/2" PVC supply, 1-1/2" PVC waste, 1-1/2" PVC vent	1.000	Ea.	3.260	142	213	355
Bathtub, P.E. cast iron, 5' long corner with fittings, white	1.000	Ea.	3.636	1,300	214	1,514
Color	1.000	Ea.	3.636	1,625	214	1,839
Rough-in, for bathtub						
1/2" copper supply, 4" cast iron waste, 1-1/2" copper vent	1.000	Ea.	2.409	199	152	351
4" PVC waste, 1-1/2" PVC vent	1.000	Ea.	2.877	104	182	286
1/2" steel supply, 4" cast iron waste, 1-1/2" steel vent	1.000	Ea.	2.898	183	179	362
4" PVC waste, 1-1/2" PVC vent	1.000	Ea.	3.159	110	200	310
1/2" PVC supply, 4" PVC waste, 1-1/2" PVC vent	1.000	Ea.	3.371	141	214	355
Piping, supply 1/2" copper	20.000	L.F.	1.975	73.50	129	202.50
1/2" steel	20.000	L.F.	2.540	85	166	251
1/2" PVC	20.000	L.F.	2.963	147	194	341
Piping, waste, 4" cast iron no hub	9.000	L.F.	2.483	340	146	486
4" PVC/DWV	9.000	L.F.	3.000	101	177	278
4" copper/DWV	9.000	L.F.	3.600	545	212	757
Piping, vent 2" cast iron no hub	6.000	L.F.	1.433	113	84.50	197.50
2" copper/DWV	6.000	L.F.	1.091	119	71.50	190.50
2" PVC/DWV	6.000	L.F.	1.627	57	96	153
2" steel, galvanized	6.000	L.F.	1.500	85.50	88.50	174
Vanity base cabinet, 2 door, 24" x 30"	1.000	Ea.	1.000	460	58	518
24" x 36"	1.000	Ea.	1.200	445	69.50	514.50
Vanity top, laminated plastic square edge 25" x 32"	2.670	L.F.	.712	100	41.50	141.50
25" x 38"	3.160	L.F.	.843	119	49	168
Cultured marble, 25" x 32", with bowl	1.000	Ea.	2.500	183	147	330
25" x 38", with bowl	1.000	Ea.	2.500	216	147	363
Carrier, for lavatory, steel for studs, no arms	1.000	Ea.	1.143	63	75	138
Wood, 2" x 8" blocking	1.300	L.F.	.052	1.27	3.02	4.29

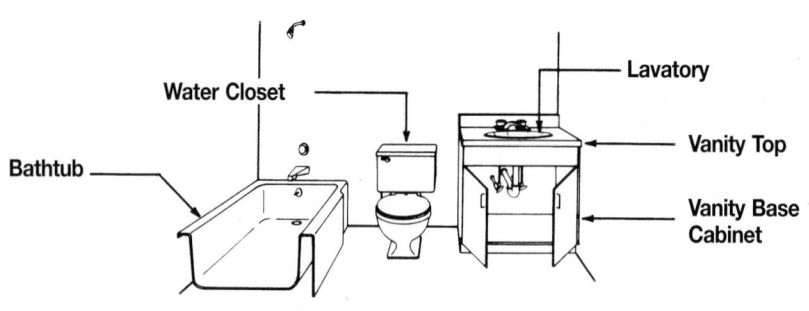

System Description	QUAN.	UNIT	LABOR HOURS	COST EACH		
				MAT.	INST.	TOTAL
BATHROOM WITH LAVATORY INSTALLED IN VANITY						
Water closet, floor mounted, 2 piece, close coupled, white	1.000	Ea.	3.019	253	178	431
Rough-in, waste, 4" diameter DWV piping	1.000	Ea.	.828	112.50	48.75	161.25
Vent, 2" diameter DWV piping	1.000	Ea.	.955	75.20	56.40	131.60
Supply, 1/2" diameter type "L" copper supply piping	1.000	Ea.	.593	22.08	38.70	60.78
Lavatory, 20" x 18", P.E. cast iron with accessories, white	1.000	Ea.	2.500	330	147	477
Rough-in, waste, 1-1/2" diameter DWV piping	1.000	Ea.	1.803	140.80	106.40	247.20
Supply, 1/2" diameter type "L" copper supply piping	1.000	Ea.	.988	36.80	64.50	101.30
Bathtub, P.E. cast iron 5' long with accessories, white	1.000	Ea.	3.636	1,300	214	1,514
Rough-in, waste, 4" diameter DWV piping	1.000	Ea.	.828	112.50	48.75	161.25
Vent, 1-1/2" diameter DWV piping	1.000	Ea.	.593	49.40	38.80	88.20
Supply, 1/2" diameter type "L" copper supply piping	1.000	Ea.	.988	36.80	64.50	101.30
Piping, supply, 1/2" diameter type "L" copper supply piping	10.000	L.F.	.988	36.80	64.50	101.30
Waste, 4" diameter DWV piping	6.000	L.F.	1.655	225	97.50	322.50
Vent, 2" diameter DWV piping	6.000	L.F.	1.500	85.50	88.50	174
Vanity base cabinet, 2 door, 30" wide	1.000	Ea.	1.000	460	58	518
Vanity top, plastic laminated square edge	2.670	L.F.	.712	100.13	41.39	141.52
TOTAL		Ea.	22.586	3,376.51	1,355.69	4,732.20
BATHROOM WITH WALL HUNG LAVATORY						
Water closet, floor mounted, 2 piece, close coupled, white	1.000	Ea.	3.019	253	178	431
Rough-in, vent, 2" diameter DWV piping	1.000	Ea.	.955	75.20	56.40	131.60
Waste, 4" diameter DWV piping	1.000	Ea.	.828	112.50	48.75	161.25
Supply, 1/2" diameter type "L" copper supply piping	1.000	Ea.	.593	22.08	38.70	60.78
Lavatory, 20" x 18" P.E. cast iron, wall hung, white	1.000	Ea.	2.000	270	118	388
Rough-in, waste, 1-1/2" diameter DWV piping	1.000	Ea.	1.803	140.80	106.40	247.20
Supply, 1/2" diameter type "L" copper supply piping	1.000	Ea.	.988	36.80	64.50	101.30
Bathtub, P.E. cast iron, 5' long with accessories, white	1.000	Ea.	3.636	1,300	214	1,514
Rough-in, waste, 4" diameter DWV piping	1.000	Ea.	.828	112.50	48.75	161.25
Supply, 1/2" diameter type "L" copper supply piping	1.000	Ea.	.988	36.80	64.50	101.30
Vent, 1-1/2" diameter DWV piping	1.000	Ea.	.593	49.40	38.80	88.20
Piping, supply, 1/2" diameter type "L" copper supply piping	10.000	L.F.	.988	36.80	64.50	101.30
Waste, 4" diameter DWV piping	6.000	L.F.	1.655	225	97.50	322.50
Vent, 2" diameter DWV piping	6.000	L.F.	1.500	85.50	88.50	174
Carrier, steel, for studs, no arms	1.000	Ea.	1.143	63	75	138
TOTAL		Ea.	21.517	2,819.38	1,302.30	4,121.68

The costs in this system are on a cost each basis. All necessary piping is included.

Three Fixture Bathroom Price Sheet	QUAN.	UNIT	LABOR HOURS	COST EACH		
				MAT.	INST.	TOTAL
Water closet, close coupled standard 2 piece, white	1.000	Ea.	3.019	253	178	431
Color	1.000	Ea.	3.019	460	178	638
One piece elongated bowl, white	1.000	Ea.	3.019	805	178	983
Color	1.000	Ea.	3.019	1,075	178	1,253
Low profile, one piece elongated bowl, white	1.000	Ea.	3.019	815	178	993
Color	1.000	Ea.	3.019	1,075	178	1,253
Rough-in for water closet						
1/2" copper supply, 4" cast iron waste, 2" cast iron vent	1.000	Ea.	2.376	210	144	354
4" PVC/DWV waste, 2" PVC vent	1.000	Ea.	2.678	93.50	162	255.50
4" carrier waste, 2" copper vent	1.000	Ea.	2.520	283	157	440
3" cast iron waste, 1-1/2" cast iron vent	1.000	Ea.	2.244	157	136	293
3" PVC waste, 1-1/2" PVC vent	1.000	Ea.	2.388	84.50	150	234.50
3" copper waste, 1-1/2" copper vent	1.000	Ea.	2.014	152	126	278
1/2" PVC supply, 4" PVC waste, 2" PVC vent	1.000	Ea.	2.974	116	181	297
3" PVC waste, 1-1/2" PVC supply	1.000	Ea.	2.684	106	170	276
1/2" steel supply, 4" cast iron waste, 2" cast iron vent	1.000	Ea.	2.545	213	155	368
4" cast iron waste, 2" steel vent	1.000	Ea.	2.590	195	158	353
4" PVC waste, 2" PVC vent	1.000	Ea.	2.847	97	173	270
Lavatory, wall hung, PE cast iron 20" x 18", white	1.000	Ea.	2.000	270	118	388
Color	1.000	Ea.	2.000	288	118	406
Vitreous china 19" x 17", white	1.000	Ea.	2.286	135	135	270
Color	1.000	Ea.	2.286	157	135	292
Lavatory, for vanity top, PE cast iron 20" x 18", white	1.000	Ea.	2.500	330	147	477
Color	1.000	Ea.	2.500	535	147	682
Steel enameled 20" x 17", white	1.000	Ea.	2.759	142	163	305
Color	1.000	Ea.	2.500	138	147	285
Vitreous china 20" x 16", white	1.000	Ea.	2.963	234	175	409
Color	1.000	Ea.	2.963	234	175	409
Rough-in for lavatory						
1/2" copper supply, 1-1/2" cast iron waste, 1-1/2" cast iron vent	1.000	Ea.	2.791	178	171	349
1-1/2" PVC waste, 1-1/4" PVC vent	1.000	Ea.	2.639	108	173	281
1/2" steel supply, 1-1/4" cast iron waste, 1-1/4" steel vent	1.000	Ea.	2.890	138	179	317
1-1/4" PVC waste, 1-1/4" PVC vent	1.000	Ea.	2.794	117	183	300
1/2" PVC supply, 1-1/2" PVC waste, 1-1/2" PVC vent	1.000	Ea.	3.260	142	213	355
Bathtub, PE cast iron, 5' long corner with fittings, white	1.000	Ea.	3.636	1,300	214	1,514
Color	1.000	Ea.	3.636	1,625	214	1,839
Rough-in for bathtub						
1/2" copper supply, 4" cast iron waste, 1-1/2" copper vent	1.000	Ea.	2.409	199	152	351
4" PVC waste, 1/2" PVC vent	1.000	Ea.	2.877	104	182	286
1/2" steel supply, 4" cast iron waste, 1-1/2" steel vent	1.000	Ea.	2.898	183	179	362
4" PVC waste, 1-1/2" PVC vent	1.000	Ea.	3.159	110	200	310
1/2" PVC supply, 4" PVC waste, 1-1/2" PVC vent	1.000	Ea.	3.371	141	214	355
Piping supply, 1/2" copper	10.000	L.F.	.988	37	64.50	101.50
1/2" steel	10.000	L.F.	1.270	42.50	83	125.50
1/2" PVC	10.000	L.F.	1.482	73.50	97	170.50
Piping waste, 4" cast iron no hub	6.000	L.F.	1.655	225	97.50	322.50
4" PVC/DWV	6.000	L.F.	2.000	67	118	185
4" copper/DWV	6.000	L.F.	2.400	365	141	506
Piping vent 2" cast iron no hub	6.000	L.F.	1.433	113	84.50	197.50
2" copper/DWV	6.000	L.F.	1.091	119	71.50	190.50
2" PVC/DWV	6.000	L.F.	1.627	57	96	153
2" steel, galvanized	6.000	L.F.	1.500	85.50	88.50	174
Vanity base cabinet, 2 door, 24" x 30"	1.000	Ea.	1.000	460	58	518
24" x 36"	1.000	Ea.	1.200	445	69.50	514.50
Vanity top, laminated plastic square edge 25" x 32"	2.670	L.F.	.712	100	41.50	141.50
25" x 38"	3.160	L.F.	.843	119	49	168
Cultured marble, 25" x 32", with bowl	1.000	Ea.	2.500	183	147	330
25" x 38", with bowl	1.000	Ea.	2.500	216	147	363
Carrier, for lavatory, steel for studs, no arms	1.000	Ea.	1.143	63	75	138
Wood, 2" x 8" blocking	1.300	L.F.	.052	1.27	3.02	4.29

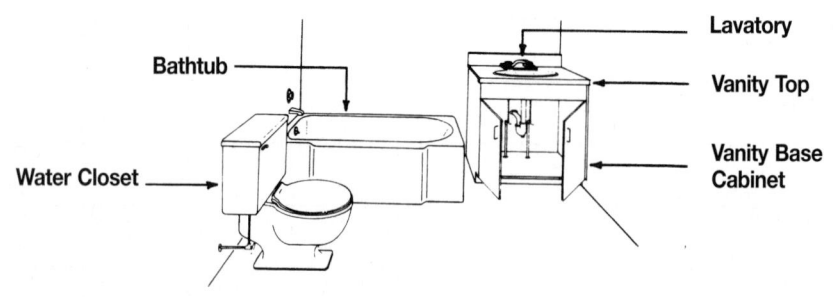

Bathtub

Water Closet

Lavatory

Vanity Top

Vanity Base Cabinet

System Description	QUAN.	UNIT	LABOR HOURS	COST EACH		
				MAT.	INST.	TOTAL
BATHROOM WITH LAVATORY INSTALLED IN VANITY						
Water closet, floor mounted, 2 piece, close coupled, white	1.000	Ea.	3.019	253	178	431
Rough-in, vent, 2" diameter DWV piping	1.000	Ea.	.955	75.20	56.40	131.60
Waste, 4" diameter DWV piping	1.000	Ea.	.828	112.50	48.75	161.25
Supply, 1/2" diameter type "L" copper supply piping	1.000	Ea.	.593	22.08	38.70	60.78
Lavatory, 20" x 18", PE cast iron with accessories, white	1.000	Ea.	2.500	330	147	477
Rough-in, vent, 1-1/2" diameter DWV piping	1.000	Ea.	1.803	140.80	106.40	247.20
Supply, 1/2" diameter type "L" copper supply piping	1.000	Ea.	.988	36.80	64.50	101.30
Bathtub, P.E. cast iron, 5' long with accessories, white	1.000	Ea.	3.636	1,300	214	1,514
Rough-in, waste, 4" diameter DWV piping	1.000	Ea.	.828	112.50	48.75	161.25
Supply, 1/2" diameter type "L" copper supply piping	1.000	Ea.	.988	36.80	64.50	101.30
Vent, 1-1/2" diameter DWV piping	1.000	Ea.	.593	49.40	38.80	88.20
Piping, supply, 1/2" diameter type "L" copper supply piping	32.000	L.F.	3.161	117.76	206.40	324.16
Waste, 4" diameter DWV piping	12.000	L.F.	3.310	450	195	645
Vent, 2" diameter DWV piping	6.000	L.F.	1.500	85.50	88.50	174
Vanity base cabinet, 2 door, 30" wide	1.000	Ea.	1.000	460	58	518
Vanity top, plastic laminated square edge	2.670	L.F.	.712	100.13	41.39	141.52
TOTAL		Ea.	26.414	3,682.47	1,595.09	5,277.56
BATHROOM WITH WALL HUNG LAVATORY						
Water closet, floor mounted, 2 piece, close coupled, white	1.000	Ea.	3.019	253	178	431
Rough-in, vent, 2" diameter DWV piping	1.000	Ea.	.955	75.20	56.40	131.60
Waste, 4" diameter DWV piping	1.000	Ea.	.828	112.50	48.75	161.25
Supply, 1/2" diameter type "L" copper supply piping	1.000	Ea.	.593	22.08	38.70	60.78
Lavatory, 20" x 18" P.E. cast iron, wall hung, white	1.000	Ea.	2.000	270	118	388
Rough-in, waste, 1-1/2" diameter DWV piping	1.000	Ea.	1.803	140.80	106.40	247.20
Supply, 1/2" diameter type "L" copper supply piping	1.000	Ea.	.988	36.80	64.50	101.30
Bathtub, P.E. cast iron, 5' long with accessories, white	1.000	Ea.	3.636	1,300	214	1,514
Rough-in, waste, 4" diameter DWV piping	1.000	Ea.	.828	112.50	48.75	161.25
Supply, 1/2" diameter type "L" copper supply piping	1.000	Ea.	.988	36.80	64.50	101.30
Vent, 1-1/2" diameter DWV piping	1.000	Ea.	.593	49.40	38.80	88.20
Piping, supply, 1/2" diameter type "L" copper supply piping	32.000	L.F.	3.161	117.76	206.40	324.16
Waste, 4" diameter DWV piping	12.000	L.F.	3.310	450	195	645
Vent, 2" diameter DWV piping	6.000	L.F.	1.500	85.50	88.50	174
Carrier steel, for studs, no arms	1.000	Ea.	1.143	63	75	138
TOTAL		Ea.	25.345	3,125.34	1,541.70	4,667.04

The costs in this system are on a cost each basis. All necessary piping is included.

Three Fixture Bathroom Price Sheet	QUAN.	UNIT	LABOR HOURS	COST EACH		
				MAT.	INST.	TOTAL
Water closet, close coupled, standard 2 piece, white	1.000	Ea.	3.019	253	178	431
Color	1.000	Ea.	3.019	460	178	638
One piece, elongated bowl, white	1.000	Ea.	3.019	805	178	983
Color	1.000	Ea.	3.019	1,075	178	1,253
Low profile, one piece, elongated bowl, white	1.000	Ea.	3.019	815	178	993
Color	1.000	Ea.	3.019	1,075	178	1,253
Rough-in, for water closet						
1/2" copper supply, 4" cast iron waste, 2" cast iron vent	1.000	Ea.	2.376	210	144	354
4" PVC/DWV waste, 2" PVC vent	1.000	Ea.	2.678	93.50	162	255.50
4" copper waste, 2" copper vent	1.000	Ea.	2.520	283	157	440
3" cast iron waste, 1-1/2" cast iron vent	1.000	Ea.	2.244	157	136	293
3" PVC waste, 1-1/2" PVC vent	1.000	Ea.	2.388	84.50	150	234.50
3" copper waste, 1-1/2" copper vent	1.000	Ea.	2.014	152	126	278
1/2" PVC supply, 4" PVC waste, 2" PVC vent	1.000	Ea.	2.974	116	181	297
3" PVC waste, 1-1/2" PVC supply	1.000	Ea.	2.684	106	170	276
1/2" steel supply, 4" cast iron waste, 2" cast iron vent	1.000	Ea.	2.545	213	155	368
4" cast iron waste, 2" steel vent	1.000	Ea.	2.590	195	158	353
4" PVC waste, 2" PVC vent	1.000	Ea.	2.847	97	173	270
Lavatory wall hung, P.E. cast iron, 20" x 18", white	1.000	Ea.	2.000	270	118	388
Color	1.000	Ea.	2.000	288	118	406
Vitreous china, 19" x 17", white	1.000	Ea.	2.286	135	135	270
Color	1.000	Ea.	2.286	157	135	292
Lavatory, for vanity top, P.E., cast iron, 20" x 18", white	1.000	Ea.	2.500	330	147	477
Color	1.000	Ea.	2.500	535	147	682
Steel, enameled, 20" x 17", white	1.000	Ea.	2.759	142	163	305
Color	1.000	Ea.	2.500	138	147	285
Vitreous china, 20" x 16", white	1.000	Ea.	2.963	234	175	409
Color	1.000	Ea.	2.963	234	175	409
Rough-in, for lavatory						
1/2" copper supply, 1-1/2" C.I. waste, 1-1/2" C.I. vent	1.000	Ea.	2.791	178	171	349
1-1/2" PVC waste, 1-1/4" PVC vent	1.000	Ea.	2.639	108	173	281
1/2" steel supply, 1-1/4" cast iron waste, 1-1/4" steel vent	1.000	Ea.	2.890	138	179	317
1-1/4" PVC waste, 1-1/4" PVC vent	1.000	Ea.	2.794	117	183	300
1/2" PVC supply, 1-1/2" PVC waste, 1-1/2" PVC vent	1.000	Ea.	3.260	142	213	355
Bathtub, P.E. cast iron, 5' long corner with fittings, white	1.000	Ea.	3.636	1,300	214	1,514
Color	1.000	Ea.	3.636	1,625	214	1,839
Rough-in, for bathtub						
1/2" copper supply, 4" cast iron waste, 1-1/2" copper vent	1.000	Ea.	2.409	199	152	351
4" PVC waste, 1/2" PVC vent	1.000	Ea.	2.877	104	182	286
1/2" steel supply, 4" cast iron waste, 1-1/2" steel vent	1.000	Ea.	2.898	183	179	362
4" PVC waste, 1-1/2" PVC vent	1.000	Ea.	3.159	110	200	310
1/2" PVC supply, 4" PVC waste, 1-1/2" PVC vent	1.000	Ea.	3.371	141	214	355
Piping, supply, 1/2" copper	32.000	L.F.	3.161	118	206	324
1/2" steel	32.000	L.F.	4.063	136	266	402
1/2" PVC	32.000	L.F.	4.741	235	310	545
Piping, waste, 4" cast iron no hub	12.000	L.F.	3.310	450	195	645
4" PVC/DWV	12.000	L.F.	4.000	134	236	370
4" copper/DWV	12.000	L.F.	4.800	725	282	1,007
Piping, vent, 2" cast iron no hub	6.000	L.F.	1.433	113	84.50	197.50
2" copper/DWV	6.000	L.F.	1.091	119	71.50	190.50
2" PVC/DWV	6.000	L.F.	1.627	57	96	153
2" steel, galvanized	6.000	L.F.	1.500	85.50	88.50	174
Vanity base cabinet, 2 door, 24" x 30"	1.000	Ea.	1.000	460	58	518
24" x 36"	1.000	Ea.	1.200	445	69.50	514.50
Vanity top, laminated plastic square edge, 25" x 32"	2.670	L.F.	.712	100	41.50	141.50
25" x 38"	3.160	L.F.	.843	119	49	168
Cultured marble, 25" x 32", with bowl	1.000	Ea.	2.500	183	147	330
25" x 38", with bowl	1.000	Ea.	2.500	216	147	363
Carrier, for lavatory, steel for studs, no arms	1.000	Ea.	1.143	63	75	138
Wood, 2" x 8" blocking	1.300	L.F.	.052	1.27	3.02	4.29

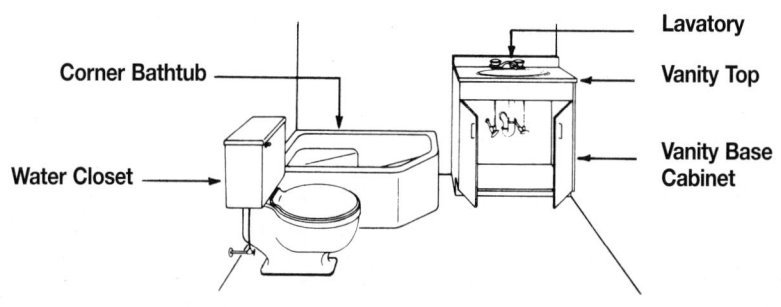

Corner Bathtub

Lavatory

Vanity Top

Water Closet

Vanity Base Cabinet

System Description	QUAN.	UNIT	LABOR HOURS	COST EACH		
				MAT.	INST.	TOTAL
BATHROOM WITH LAVATORY INSTALLED IN VANITY						
Water closet, floor mounted, 2 piece, close coupled, white	1.000	Ea.	3.019	253	178	431
Rough-in, vent, 2" diameter DWV piping	1.000	Ea.	.955	75.20	56.40	131.60
Waste, 4" diameter DWV piping	1.000	Ea.	.828	112.50	48.75	161.25
Supply, 1/2" diameter type "L" copper supply piping	1.000	Ea.	.593	22.08	38.70	60.78
Lavatory, 20" x 18", P.E. cast iron with fittings, white	1.000	Ea.	2.500	330	147	477
Rough-in, waste, 1-1/2" diameter DWV piping	1.000	Ea.	1.803	140.80	106.40	247.20
Supply, 1/2" diameter type "L" copper supply piping	1.000	Ea.	.988	36.80	64.50	101.30
Bathtub, P.E. cast iron, corner with fittings, white	1.000	Ea.	3.636	3,075	214	3,289
Rough-in, waste, 4" diameter DWV piping	1.000	Ea.	.828	112.50	48.75	161.25
Supply, 1/2" diameter type "L" copper supply piping	1.000	Ea.	.988	36.80	64.50	101.30
Vent, 1-1/2" diameter DWV piping	1.000	Ea.	.593	49.40	38.80	88.20
Piping, supply, 1/2" diameter type "L" copper supply piping	32.000	L.F.	3.161	117.76	206.40	324.16
Waste, 4" diameter DWV piping	12.000	L.F.	3.310	450	195	645
Vent, 2" diameter DWV piping	6.000	L.F.	1.500	85.50	88.50	174
Vanity base cabinet, 2 door, 30" wide	1.000	Ea.	1.000	460	58	518
Vanity top, plastic laminated, square edge	2.670	L.F.	.712	121.49	41.39	162.88
TOTAL		Ea.	26.414	5,478.83	1,595.09	7,073.92
BATHROOM WITH WALL HUNG LAVATORY						
Water closet, floor mounted, 2 piece, close coupled, white	1.000	Ea.	3.019	253	178	431
Rough-in, vent, 2" diameter DWV piping	1.000	Ea.	.955	75.20	56.40	131.60
Waste, 4" diameter DWV piping	1.000	Ea.	.828	112.50	48.75	161.25
Supply, 1/2" diameter type "L" copper supply piping	1.000	Ea.	.593	22.08	38.70	60.78
Lavatory, 20" x 18", P.E. cast iron, with fittings, white	1.000	Ea.	2.000	270	118	388
Rough-in, waste, 1-1/2" diameter DWV piping	1.000	Ea.	1.803	140.80	106.40	247.20
Supply, 1/2" diameter type "L" copper supply piping	1.000	Ea.	.988	36.80	64.50	101.30
Bathtub, P.E. cast iron, corner, with fittings, white	1.000	Ea.	3.636	3,075	214	3,289
Rough-in, waste, 4" diameter DWV piping	1.000	Ea.	.828	112.50	48.75	161.25
Supply, 1/2" diameter type "L" copper supply piping	1.000	Ea.	.988	36.80	64.50	101.30
Vent, 1-1/2" diameter DWV piping	1.000	Ea.	.593	49.40	38.80	88.20
Piping, supply, 1/2" diameter type "L" copper supply piping	32.000	L.F.	3.161	117.76	206.40	324.16
Waste, 4" diameter DWV piping	12.000	L.F.	3.310	450	195	645
Vent, 2" diameter DWV piping	6.000	L.F.	1.500	85.50	88.50	174
Carrier, steel, for studs, no arms	1.000	Ea.	1.143	63	75	138
TOTAL		Ea.	25.345	4,900.34	1,541.70	6,442.04

The costs in this system are on a cost each basis. All necessary piping is included.

Three Fixture Bathroom Price Sheet	QUAN.	UNIT	LABOR HOURS	COST EACH		
				MAT.	INST.	TOTAL
Water closet, close coupled, standard 2 piece, white	1.000	Ea.	3.019	253	178	431
Color	1.000	Ea.	3.019	460	178	638
One piece elongated bowl, white	1.000	Ea.	3.019	805	178	983
Color	1.000	Ea.	3.019	1,075	178	1,253
Low profile, one piece elongated bowl, white	1.000	Ea.	3.019	815	178	993
Color	1.000	Ea.	3.019	1,075	178	1,253
Rough-in, for water closet						
1/2" copper supply, 4" cast iron waste, 2" cast iron vent	1.000	Ea.	2.376	210	144	354
4" PVC/DWV waste, 2" PVC vent	1.000	Ea.	2.678	93.50	162	255.50
4" copper waste, 2" copper vent	1.000	Ea.	2.520	283	157	440
3" cast iron waste, 1-1/2" cast iron vent	1.000	Ea.	2.244	157	136	293
3" PVC waste, 1-1/2" PVC vent	1.000	Ea.	2.388	84.50	150	234.50
3" copper waste, 1-1/2" copper vent	1.000	Ea.	2.014	152	126	278
1/2" PVC supply, 4" PVC waste, 2" PVC vent	1.000	Ea.	2.974	116	181	297
3" PVC waste, 1-1/2" PVC supply	1.000	Ea.	2.684	106	170	276
1/2" steel supply, 4" cast iron waste, 2" cast iron vent	1.000	Ea.	2.545	213	155	368
4" cast iron waste, 2" steel vent	1.000	Ea.	2.590	195	158	353
4" PVC waste, 2" PVC vent	1.000	Ea.	2.847	97	173	270
Lavatory, wall hung P.E. cast iron 20" x 18", white	1.000	Ea.	2.000	270	118	388
Color	1.000	Ea.	2.000	288	118	406
Vitreous china 19" x 17", white	1.000	Ea.	2.286	135	135	270
Color	1.000	Ea.	2.286	157	135	292
Lavatory, for vanity top, P.E., cast iron, 20" x 18", white	1.000	Ea.	2.500	330	147	477
Color	1.000	Ea.	2.500	535	147	682
Steel enameled 20" x 17", white	1.000	Ea.	2.759	142	163	305
Color	1.000	Ea.	2.500	138	147	285
Vitreous china 20" x 16", white	1.000	Ea.	2.963	234	175	409
Color	1.000	Ea.	2.963	234	175	409
Rough-in, for lavatory						
1/2" copper supply, 1-1/2" cast iron waste, 1-1/2" cast iron vent	1.000	Ea.	2.791	178	171	349
1-1/2" PVC waste, 1-1/4" PVC vent	1.000	Ea.	2.639	108	173	281
1/2" steel supply, 1-1/4" cast iron waste, 1-1/4" steel vent	1.000	Ea.	2.890	138	179	317
1-1/4" PVC waste, 1-1/4" PVC vent	1.000	Ea.	2.794	117	183	300
1/2" PVC supply, 1-1/2" PVC waste, 1-1/2" PVC vent	1.000	Ea.	3.260	142	213	355
Bathtub, P.E. cast iron, corner with fittings, white	1.000	Ea.	3.636	3,075	214	3,289
Color	1.000	Ea.	4.000	2,975	236	3,211
Rough-in, for bathtub						
1/2" copper supply, 4" cast iron waste, 1-1/2" copper vent	1.000	Ea.	2.409	199	152	351
4" PVC waste, 1-1/2" PVC vent	1.000	Ea.	2.877	104	182	286
1/2" steel supply, 4" cast iron waste, 1-1/2" steel vent	1.000	Ea.	2.898	183	179	362
4" PVC waste, 1-1/2" PVC vent	1.000	Ea.	3.159	110	200	310
1/2" PVC supply, 4" PVC waste, 1-1/2" PVC vent	1.000	Ea.	3.371	141	214	355
Piping, supply, 1/2" copper	32.000	L.F.	3.161	118	206	324
1/2" steel	32.000	L.F.	4.063	136	266	402
1/2" PVC	32.000	L.F.	4.741	235	310	545
Piping, waste, 4" cast iron, no hub	12.000	L.F.	3.310	450	195	645
4" PVC/DWV	12.000	L.F.	4.000	134	236	370
4" copper/DWV	12.000	L.F.	4.800	725	282	1,007
Piping, vent 2" cast iron, no hub	6.000	L.F.	1.433	113	84.50	197.50
2" copper/DWV	6.000	L.F.	1.091	119	71.50	190.50
2" PVC/DWV	6.000	L.F.	1.627	57	96	153
2" steel, galvanized	6.000	L.F.	1.500	85.50	88.50	174
Vanity base cabinet, 2 door, 24" x 30"	1.000	Ea.	1.000	460	58	518
24" x 36"	1.000	Ea.	1.200	445	69.50	514.50
Vanity top, laminated plastic square edge 25" x 32"	2.670	L.F.	.712	121	41.50	162.50
25" x 38"	3.160	L.F.	.843	144	49	193
Cultured marble, 25" x 32", with bowl	1.000	Ea.	2.500	183	147	330
25" x 38", with bowl	1.000	Ea.	2.500	216	147	363
Carrier, for lavatory, steel for studs, no arms	1.000	Ea.	1.143	63	75	138
Wood, 2" x 8" blocking	1.300	L.F.	.053	1.30	3.09	4.39

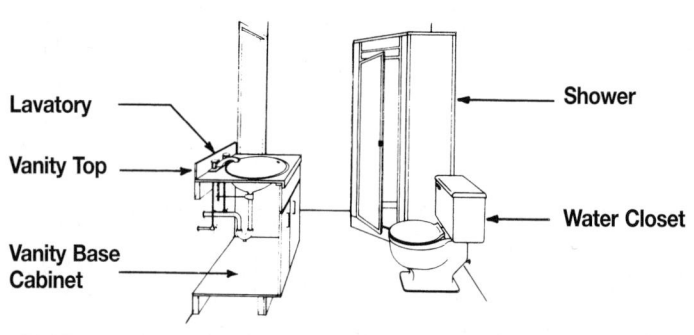

Lavatory

Vanity Top

Vanity Base Cabinet

Shower

Water Closet

System Description	QUAN.	UNIT	LABOR HOURS	COST EACH		
				MAT.	INST.	TOTAL
BATHROOM WITH SHOWER, LAVATORY INSTALLED IN VANITY						
Water closet, floor mounted, 2 piece, close coupled, white	1.000	Ea.	3.019	253	178	431
Rough-in, vent, 2" diameter DWV piping	1.000	Ea.	.955	75.20	56.40	131.60
Waste, 4" diameter DWV piping	1.000	Ea.	.828	112.50	48.75	161.25
Supply, 1/2" diameter type "L" copper supply piping	1.000	Ea.	.593	22.08	38.70	60.78
Lavatory, 20" x 18" P.E. cast iron with fittings, white	1.000	Ea.	2.500	330	147	477
Rough-in, waste, 1-1/2" diameter DWV piping	1.000	Ea.	1.803	140.80	106.40	247.20
Supply, 1/2" diameter type "L" copper supply piping	1.000	Ea.	.988	36.80	64.50	101.30
Shower, steel enameled, stone base, corner, white	1.000	Ea.	3.200	1,275	189	1,464
Shower mixing valve	1.000	Ea.	1.333	146	87.50	233.50
Shower door	1.000	Ea.	1.000	480	63.50	543.50
Rough-in, vent, 1-1/2" diameter DWV piping	1.000	Ea.	.225	17.60	13.30	30.90
Waste, 2" diameter DWV piping	1.000	Ea.	1.433	112.80	84.60	197.40
Supply, 1/2" diameter type "L" copper supply piping	1.000	Ea.	1.580	58.88	103.20	162.08
Piping, supply, 1/2" diameter type "L" copper supply piping	36.000	L.F.	4.148	154.56	270.90	425.46
Waste, 4" diameter DWV piping	7.000	L.F.	2.759	375	162.50	537.50
Vent, 2" diameter DWV piping	6.000	L.F.	2.250	128.25	132.75	261
Vanity base 2 door, 30" wide	1.000	Ea.	1.000	460	58	518
Vanity top, plastic laminated, square edge	2.170	L.F.	.712	120.15	41.39	161.54
TOTAL		Ea.	30.326	4,298.62	1,846.39	6,145.01
BATHROOM WITH SHOWER, WALL HUNG LAVATORY						
Water closet, floor mounted, close coupled	1.000	Ea.	3.019	253	178	431
Rough-in, vent, 2" diameter DWV piping	1.000	Ea.	.955	75.20	56.40	131.60
Waste, 4" diameter DWV piping	1.000	Ea.	.828	112.50	48.75	161.25
Supply, 1/2" diameter type "L" copper supply piping	1.000	Ea.	.593	22.08	38.70	60.78
Lavatory, 20" x 18" P.E. cast iron with fittings, white	1.000	Ea.	2.000	270	118	388
Rough-in, waste, 1-1/2" diameter DWV piping	1.000	Ea.	1.803	140.80	106.40	247.20
Supply, 1/2" diameter type "L" copper supply piping	1.000	Ea.	.988	36.80	64.50	101.30
Shower, steel enameled, stone base, white	1.000	Ea.	3.200	1,275	189	1,464
Mixing valve	1.000	Ea.	1.333	146	87.50	233.50
Shower door	1.000	Ea.	1.000	480	63.50	543.50
Rough-in, vent, 1-1/2" diameter DWV piping	1.000	Ea.	.225	17.60	13.30	30.90
Waste, 2" diameter DWV piping	1.000	Ea.	1.433	112.80	84.60	197.40
Supply, 1/2" diameter type "L" copper supply piping	1.000	Ea.	1.580	58.88	103.20	162.08
Piping, supply, 1/2" diameter type "L" copper supply piping	36.000	L.F.	4.148	154.56	270.90	425.46
Waste, 4" diameter DWV piping	7.000	L.F.	2.759	375	162.50	537.50
Vent, 2" diameter DWV piping	6.000	L.F.	2.250	128.25	132.75	261
Carrier, steel, for studs, no arms	1.000	Ea.	1.143	63	75	138
TOTAL		Ea.	29.257	3,721.47	1,793	5,514.47

The costs in this system are on a cost each basis. All necessary piping is included.

Three Fixture Bathroom Price Sheet	QUAN.	UNIT	LABOR HOURS	COST EACH MAT.	INST.	TOTAL
Water closet, close coupled, standard 2 piece, white	1.000	Ea.	3.019	253	178	431
Color	1.000	Ea.	3.019	460	178	638
One piece elongated bowl, white	1.000	Ea.	3.019	805	178	983
Color	1.000	Ea.	3.019	1,075	178	1,253
Low profile, one piece elongated bowl, white	1.000	Ea.	3.019	815	178	993
Color	1.000	Ea.	3.019	1,075	178	1,253
Rough-in, for water closet						
1/2" copper supply, 4" cast iron waste, 2" cast iron vent	1.000	Ea.	2.376	210	144	354
4" PVC/DWV waste, 2" PVC vent	1.000	Ea.	2.678	93.50	162	255.50
4" copper waste, 2" copper vent	1.000	Ea.	2.520	283	157	440
3" cast iron waste, 1-1/2" cast iron vent	1.000	Ea.	2.244	157	136	293
3" PVC waste, 1-1/2" PVC vent	1.000	Ea.	2.388	84.50	150	234.50
3" copper waste, 1-1/2" copper vent	1.000	Ea.	2.014	152	126	278
1/2" PVC supply, 4" PVC waste, 2" PVC vent	1.000	Ea.	2.974	116	181	297
3" PVC waste, 1-1/2" PVC supply	1.000	Ea.	2.684	106	170	276
1/2" steel supply, 4" cast iron waste, 2" cast iron vent	1.000	Ea.	2.545	213	155	368
4" cast iron waste, 2" steel vent	1.000	Ea.	2.590	195	158	353
4" PVC waste, 2" PVC vent	1.000	Ea.	2.847	97	173	270
Lavatory, wall hung, P.E. cast iron 20" x 18", white	1.000	Ea.	2.000	270	118	388
Color	1.000	Ea.	2.000	288	118	406
Vitreous china 19" x 17", white	1.000	Ea.	2.286	135	135	270
Color	1.000	Ea.	2.286	157	135	292
Lavatory, for vanity top, P.E. cast iron 20" x 18", white	1.000	Ea.	2.500	330	147	477
Color	1.000	Ea.	2.500	535	147	682
Steel enameled 20" x 17", white	1.000	Ea.	2.759	142	163	305
Color	1.000	Ea.	2.500	138	147	285
Vitreous china 20" x 16", white	1.000	Ea.	2.963	234	175	409
Color	1.000	Ea.	2.963	234	175	409
Rough-in, for lavatory						
1/2" copper supply, 1-1/2" cast iron waste, 1-1/2" cast iron vent	1.000	Ea.	2.791	178	171	349
1-1/2" PVC waste, 1-1/2" PVC vent	1.000	Ea.	2.639	108	173	281
1/2" steel supply, 1-1/4" cast iron waste, 1-1/4" steel vent	1.000	Ea.	2.890	138	179	317
1-1/4" PVC waste, 1-1/4" PVC vent	1.000	Ea.	2.921	114	191	305
1/2" PVC supply, 1-1/2" PVC waste, 1-1/2" PVC vent	1.000	Ea.	3.260	142	213	355
Shower, steel enameled stone base, 32" x 32", white	1.000	Ea.	8.000	1,275	189	1,464
Color	1.000	Ea.	7.822	1,950	173	2,123
36" x 36" white	1.000	Ea.	8.889	1,625	197	1,822
Color	1.000	Ea.	8.889	2,225	197	2,422
Rough-in, for shower						
1/2" copper supply, 4" cast iron waste, 1-1/2" copper vent	1.000	Ea.	3.238	189	201	390
4" PVC waste, 1-1/2" PVC vent	1.000	Ea.	3.429	124	214	338
1/2" steel supply, 4" cast iron waste, 1-1/2" steel vent	1.000	Ea.	3.665	188	229	417
4" PVC waste, 1-1/2" PVC vent	1.000	Ea.	3.881	133	243	376
1/2" PVC supply, 4" PVC waste, 1-1/2" PVC vent	1.000	Ea.	4.219	183	266	449
Piping, supply, 1/2" copper	36.000	L.F.	4.148	155	271	426
1/2" steel	36.000	L.F.	5.333	178	350	528
1/2" PVC	36.000	L.F.	6.222	310	405	715
Piping, waste, 4" cast iron no hub	7.000	L.F.	2.759	375	163	538
4" PVC/DWV	7.000	L.F.	3.333	112	197	309
4" copper/DWV	7.000	L.F.	4.000	605	235	840
Piping, vent, 2" cast iron no hub	6.000	L.F.	2.149	169	127	296
2" copper/DWV	6.000	L.F.	1.636	179	107	286
2" PVC/DWV	6.000	L.F.	2.441	85.50	144	229.50
2" steel, galvanized	6.000	L.F.	2.250	128	133	261
Vanity base cabinet, 2 door, 24" x 30"	1.000	Ea.	1.000	460	58	518
24" x 36"	1.000	Ea.	1.200	445	69.50	514.50
Vanity top, laminated plastic square edge, 25" x 32"	2.170	L.F.	.712	120	41.50	161.50
25" x 38"	2.670	L.F.	.845	143	49	192
Carrier, for lavatory, steel for studs, no arms	1.000	Ea.	1.143	63	75	138
Wood, 2" x 8" blocking	1.300	L.F.	.052	1.27	3.02	4.29

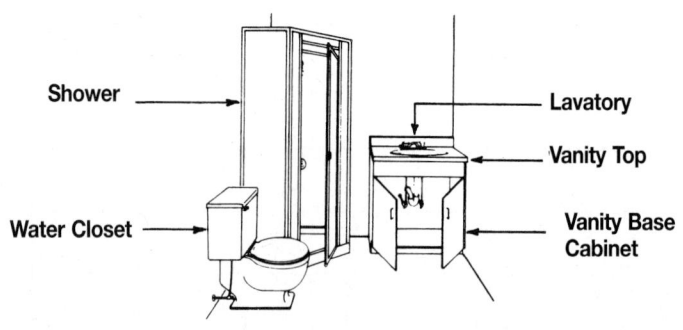

Shower — Lavatory — Vanity Top — Water Closet — Vanity Base Cabinet

System Description	QUAN.	UNIT	LABOR HOURS	COST EACH		
				MAT.	INST.	TOTAL
BATHROOM WITH LAVATORY INSTALLED IN VANITY						
Water closet, floor mounted, 2 piece, close coupled, white	1.000	Ea.	3.019	253	178	431
Rough-in, vent, 2" diameter DWV piping	1.000	Ea.	.955	75.20	56.40	131.60
Waste, 4" diameter DWV piping	1.000	Ea.	.828	112.50	48.75	161.25
Supply, 1/2" diameter type "L" copper supply piping	1.000	Ea.	.593	22.08	38.70	60.78
Lavatory, 20" x 18", P.E. cast iron with fittings, white	1.000	Ea.	2.500	330	147	477
Rough-in, waste, 1-1/2" diameter DWV piping	1.000	Ea.	1.803	140.80	106.40	247.20
Supply, 1/2" diameter type "L" copper supply piping	1.000	Ea.	.988	36.80	64.50	101.30
Shower, steel enameled, stone base, corner, white	1.000	Ea.	3.200	1,275	189	1,464
Mixing valve	1.000	Ea.	1.333	146	87.50	233.50
Shower door	1.000	Ea.	1.000	480	63.50	543.50
Rough-in, vent, 1-1/2" diameter DWV piping	1.000	Ea.	.225	17.60	13.30	30.90
Waste, 2" diameter DWV piping	1.000	Ea.	1.433	112.80	84.60	197.40
Supply, 1/2" diameter type "L" copper supply piping	1.000	Ea.	1.580	58.88	103.20	162.08
Piping, supply, 1/2" diameter type "L" copper supply piping	36.000	L.F.	3.556	132.48	232.20	364.68
Waste, 4" diameter DWV piping	7.000	L.F.	1.931	262.50	113.75	376.25
Vent, 2" diameter DWV piping	6.000	L.F.	1.500	85.50	88.50	174
Vanity base, 2 door, 30" wide	1.000	Ea.	1.000	460	58	518
Vanity top, plastic laminated, square edge	2.670	L.F.	.712	100.13	41.39	141.52
TOTAL		Ea.	28.156	4,101.27	1,714.69	5,815.96
BATHROOM, WITH WALL HUNG LAVATORY						
Water closet, floor mounted, 2 piece, close coupled, white	1.000	Ea.	3.019	253	178	431
Rough-in, vent, 2" diameter DWV piping	1.000	Ea.	.955	75.20	56.40	131.60
Waste, 4" diameter DWV piping	1.000	Ea.	.828	112.50	48.75	161.25
Supply, 1/2" diameter type "L" copper supply piping	1.000	Ea.	.593	22.08	38.70	60.78
Lavatory, wall hung, 20" x 18" P.E. cast iron with fittings, white	1.000	Ea.	2.000	270	118	388
Rough-in, waste, 1-1/2" diameter DWV piping	1.000	Ea.	1.803	140.80	106.40	247.20
Supply, 1/2" diameter type "L" copper supply piping	1.000	Ea.	.988	36.80	64.50	101.30
Shower, steel enameled, stone base, corner, white	1.000	Ea.	3.200	1,275	189	1,464
Mixing valve	1.000	Ea.	1.333	146	87.50	233.50
Shower door	1.000	Ea.	1.000	480	63.50	543.50
Rough-in, waste, 1-1/2" diameter DWV piping	1.000	Ea.	.225	17.60	13.30	30.90
Waste, 2" diameter DWV piping	1.000	Ea.	1.433	112.80	84.60	197.40
Supply, 1/2" diameter type "L" copper supply piping	1.000	Ea.	1.580	58.88	103.20	162.08
Piping, supply, 1/2" diameter type "L" copper supply piping	36.000	L.F.	3.556	132.48	232.20	364.68
Waste, 4" diameter DWV piping	7.000	L.F.	1.931	262.50	113.75	376.25
Vent, 2" diameter DWV piping	6.000	L.F.	1.500	85.50	88.50	174
Carrier, steel, for studs, no arms	1.000	Ea.	1.143	63	75	138
TOTAL		Ea.	27.087	3,544.14	1,661.30	5,205.44

The costs in this system are on a cost each basis. All necessary piping is included.

Three Fixture Bathroom Price Sheet	QUAN.	UNIT	LABOR HOURS	COST EACH MAT.	COST EACH INST.	COST EACH TOTAL
Water closet, close coupled, standard 2 piece, white	1.000	Ea.	3.019	253	178	431
Color	1.000	Ea.	3.019	460	178	638
One piece elongated bowl, white	1.000	Ea.	3.019	805	178	983
Color	1.000	Ea.	3.019	1,075	178	1,253
Low profile one piece elongated bowl, white	1.000	Ea.	3.019	815	178	993
Color	1.000	Ea.	3.623	1,075	178	1,253
Rough-in, for water closet						
1/2" copper supply, 4" cast iron waste, 2" cast iron vent	1.000	Ea.	2.376	210	144	354
4" P.V.C./DWV waste, 2" PVC vent	1.000	Ea.	2.678	93.50	162	255.50
4" copper waste, 2" copper vent	1.000	Ea.	2.520	283	157	440
3" cast iron waste, 1-1/2" cast iron vent	1.000	Ea.	2.244	157	136	293
3" PVC waste, 1-1/2" PVC vent	1.000	Ea.	2.388	84.50	150	234.50
3" copper waste, 1-1/2" copper vent	1.000	Ea.	2.014	152	126	278
1/2" P.V.C. supply, 4" P.V.C. waste, 2" P.V.C. vent	1.000	Ea.	2.974	116	181	297
3" P.V.C. waste, 1-1/2" P.V.C. vent	1.000	Ea.	2.684	106	170	276
1/2" steel supply, 4" cast iron waste, 2" cast iron vent	1.000	Ea.	2.545	213	155	368
4" cast iron waste, 2" steel vent	1.000	Ea.	2.590	195	158	353
4" P.V.C. waste, 2" P.V.C. vent	1.000	Ea.	2.847	97	173	270
Lavatory, wall hung P.E. cast iron 20" x 18", white	1.000	Ea.	2.000	270	118	388
Color	1.000	Ea.	2.000	288	118	406
Vitreous china 19" x 17", white	1.000	Ea.	2.286	135	135	270
Color	1.000	Ea.	2.286	157	135	292
Lavatory, for vanity top P.E. cast iron 20" x 18", white	1.000	Ea.	2.500	330	147	477
Color	1.000	Ea.	2.500	535	147	682
Steel enameled 20" x 17", white	1.000	Ea.	2.759	142	163	305
Color	1.000	Ea.	2.500	138	147	285
Vitreous china 20" x 16", white	1.000	Ea.	2.963	234	175	409
Color	1.000	Ea.	2.963	234	175	409
Rough-in, for lavatory						
1/2" copper supply, 1-1/2" cast iron waste, 1-1/2" cast iron vent	1.000	Ea.	2.791	178	171	349
1-1/2" P.V.C. waste, 1-1/2" P.V.C. vent	1.000	Ea.	2.639	108	173	281
1/2" steel supply, 1-1/2" cast iron waste, 1-1/4" steel vent	1.000	Ea.	2.890	138	179	317
1-1/2" P.V.C. waste, 1-1/4" P.V.C. vent	1.000	Ea.	2.921	114	191	305
1/2" P.V.C. supply, 1-1/2" P.V.C. waste, 1-1/2" P.V.C. vent	1.000	Ea.	3.260	142	213	355
Shower, steel enameled stone base, 32" x 32", white	1.000	Ea.	8.000	1,275	189	1,464
Color	1.000	Ea.	7.822	1,950	173	2,123
36" x 36", white	1.000	Ea.	8.889	1,625	197	1,822
Color	1.000	Ea.	8.889	2,225	197	2,422
Rough-in, for shower						
1/2" copper supply, 2" cast iron waste, 1-1/2" copper vent	1.000	Ea.	3.161	184	198	382
2" P.V.C. waste, 1-1/2" P.V.C. vent	1.000	Ea.	3.429	124	214	338
1/2" steel supply, 2" cast iron waste, 1-1/2" steel vent	1.000	Ea.	3.887	300	242	542
2" P.V.C. waste, 1-1/2" P.V.C. vent	1.000	Ea.	3.881	133	243	376
1/2" P.V.C. supply, 2" P.V.C. waste, 1-1/2" P.V.C. vent	1.000	Ea.	4.219	183	266	449
Piping, supply, 1/2" copper	36.000	L.F.	3.556	132	232	364
1/2" steel	36.000	L.F.	4.571	153	299	452
1/2" P.V.C.	36.000	L.F.	5.333	265	350	615
Waste, 4" cast iron, no hub	7.000	L.F.	1.931	263	114	377
4" P.V.C./DWV	7.000	L.F.	2.333	78.50	138	216.50
4" copper/DWV	7.000	L.F.	2.800	425	165	590
Vent, 2" cast iron, no hub	6.000	L.F.	1.091	119	71.50	190.50
2" copper/DWV	6.000	L.F.	1.091	119	71.50	190.50
2" P.V.C./DWV	6.000	L.F.	1.627	57	96	153
2" steel, galvanized	6.000	L.F.	1.500	85.50	88.50	174
Vanity base cabinet, 2 door, 24" x 30"	1.000	Ea.	1.000	460	58	518
24" x 36"	1.000	Ea.	1.200	445	69.50	514.50
Vanity top, laminated plastic square edge, 25" x 32"	2.670	L.F.	.712	100	41.50	141.50
25" x 38"	3.170	L.F.	.845	119	49	168
Carrier , for lavatory, steel, for studs, no arms	1.000	Ea.	1.143	63	75	138
Wood, 2" x 8" blocking	1.300	L.F.	.052	1.27	3.02	4.29

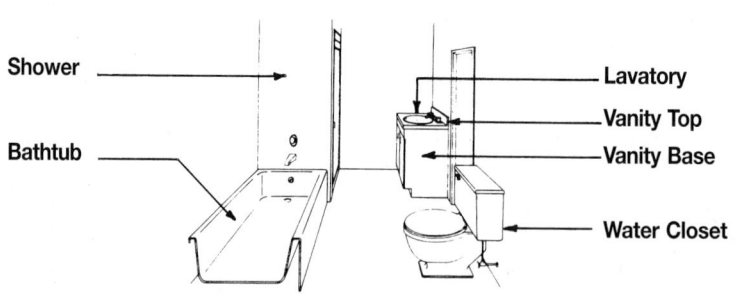

System Description	QUAN.	UNIT	LABOR HOURS	COST EACH		
				MAT.	INST.	TOTAL
BATHROOM WITH LAVATORY INSTALLED IN VANITY						
Water closet, floor mounted, 2 piece, close coupled, white	1.000	Ea.	3.019	253	178	431
Rough-in, vent, 2" diameter DWV piping	1.000	Ea.	.955	75.20	56.40	131.60
Waste, 4" diameter DWV piping	1.000	Ea.	.828	112.50	48.75	161.25
Supply, 1/2" diameter type "L" copper supply piping	1.000	Ea.	.593	22.08	38.70	60.78
Lavatory, 20" x 18" P.E. cast iron with fittings, white	1.000	Ea.	2.500	330	147	477
Shower, steel, enameled, stone base, corner, white	1.000	Ea.	3.333	1,925	197	2,122
Mixing valve	1.000	Ea.	1.333	146	87.50	233.50
Shower door	1.000	Ea.	1.000	480	63.50	543.50
Rough-in, waste, 1-1/2" diameter DWV piping	2.000	Ea.	4.507	352	266	618
Supply, 1/2" diameter type "L" copper supply piping	2.000	Ea.	3.161	117.76	206.40	324.16
Bathtub, P.E. cast iron, 5' long with fittings, white	1.000	Ea.	3.636	1,300	214	1,514
Rough-in, waste, 4" diameter DWV piping	1.000	Ea.	.828	112.50	48.75	161.25
Supply, 1/2" diameter type "L" copper supply piping	1.000	Ea.	.988	36.80	64.50	101.30
Vent, 1-1/2" diameter DWV piping	1.000	Ea.	.593	49.40	38.80	88.20
Piping, supply, 1/2" diameter type "L" copper supply piping	42.000	L.F.	4.148	154.56	270.90	425.46
Waste, 4" diameter DWV piping	10.000	L.F.	2.759	375	162.50	537.50
Vent, 2" diameter DWV piping	13.000	L.F.	3.250	185.25	191.75	377
Vanity base, 2 doors, 30" wide	1.000	Ea.	1.000	460	58	518
Vanity top, plastic laminated, square edge	2.670	L.F.	.712	100.13	41.39	141.52
TOTAL		Ea.	39.143	6,587.18	2,379.84	8,967.02
BATHROOM WITH WALL HUNG LAVATORY						
Water closet, floor mounted, 2 piece, close coupled, white	1.000	Ea.	3.019	253	178	431
Rough-in, vent, 2" diameter DWV piping	1.000	Ea.	.955	75.20	56.40	131.60
Waste, 4" diameter DWV piping	1.000	Ea.	.828	112.50	48.75	161.25
Supply, 1/2" diameter type "L" copper supply piping	1.000	Ea.	.593	22.08	38.70	60.78
Lavatory, 20" x 18" P.E. cast iron with fittings, white	1.000	Ea.	2.000	270	118	388
Shower, steel enameled, stone base, corner , white	1.000	Ea.	3.333	1,925	197	2,122
Mixing valve	1.000	Ea.	1.333	146	87.50	233.50
Shower door	1.000	Ea.	1.000	480	63.50	543.50
Rough-in, waste, 1-1/2" diameter DWV piping	2.000	Ea.	4.507	352	266	618
Supply, 1/2" diameter type "L" copper supply piping	2.000	Ea.	3.161	117.76	206.40	324.16
Bathtub, P.E. cast iron, 5' long with fittings, white	1.000	Ea.	3.636	1,300	214	1,514
Rough-in, waste, 4" diameter DWV piping	1.000	Ea.	.828	112.50	48.75	161.25
Supply, 1/2" diameter type "L" copper supply piping	1.000	Ea.	.988	36.80	64.50	101.30
Vent, 1-1/2" diameter copper DWV piping	1.000	Ea.	.593	49.40	38.80	88.20
Piping, supply, 1/2" diameter type "L" copper supply piping	42.000	L.F.	4.148	154.56	270.90	425.46
Waste, 4" diameter DWV piping	10.000	L.F.	2.759	375	162.50	537.50
Vent, 2" diameter DWV piping	13.000	L.F.	3.250	185.25	191.75	377
Carrier, steel, for studs, no arms	1.000	Ea.	1.143	63	75	138
TOTAL		Ea.	38.074	6,030.05	2,326.45	8,356.50

The costs in this system are on a cost each basis. All necessary piping is included.

Four Fixture Bathroom Price Sheet	QUAN.	UNIT	LABOR HOURS	COST EACH		
				MAT.	INST.	TOTAL
Water closet, close coupled, standard 2 piece, white	1.000	Ea.	3.019	253	178	431
Color	1.000	Ea.	3.019	460	178	638
One piece elongated bowl, white	1.000	Ea.	3.019	805	178	983
Color	1.000	Ea.	3.019	1,075	178	1,253
Low profile, one piece elongated bowl, white	1.000	Ea.	3.019	815	178	993
Color	1.000	Ea.	3.019	1,075	178	1,253
1/2" copper supply, 4" cast iron waste, 2" cast iron vent	1.000	Ea.	2.376	210	144	354
4" PVC/DWV waste, 2" PVC vent	1.000	Ea.	2.678	93.50	162	255.50
4" copper waste, 2" copper vent	1.000	Ea.	2.520	283	157	440
3" cast iron waste, 1-1/2" cast iron vent	1.000	Ea.	2.244	157	136	293
3" P.V.C. waste, 1-1/2" P.V.C. vent	1.000	Ea.	2.388	84.50	150	234.50
3" copper waste, 1-1/2" copper vent	1.000	Ea.	2.014	152	126	278
1/2" P.V.C. supply, 4" P.V.C. waste, 2" P.V.C. vent	1.000	Ea.	2.974	116	181	297
3" P.V.C. waste, 1-1/2" P.V.C. vent	1.000	Ea.	2.684	106	170	276
1/2" steel supply, 4" cast iron waste, 2" cast iron vent	1.000	Ea.	2.545	213	155	368
4" cast iron waste, 2" steel vent	1.000	Ea.	2.590	195	158	353
4" P.V.C. waste, 2" P.V.C. vent	1.000	Ea.	2.847	97	173	270
Lavatory, wall hung P.E. cast iron 20" x 18", white	1.000	Ea.	2.000	270	118	388
Color	1.000	Ea.	2.000	288	118	406
Vitreous china 19" x 17", white	1.000	Ea.	2.286	135	135	270
Color	1.000	Ea.	2.286	157	135	292
Lavatory for vanity top, P.E. cast iron 20" x 18", white	1.000	Ea.	2.500	330	147	477
Color	1.000	Ea.	2.500	535	147	682
Steel enameled, 20" x 17", white	1.000	Ea.	2.759	142	163	305
Color	1.000	Ea.	2.500	138	147	285
Vitreous china 20" x 16", white	1.000	Ea.	2.963	234	175	409
Color	1.000	Ea.	2.963	234	175	409
Shower, steel enameled stone base, 36" square, white	1.000	Ea.	8.889	1,925	197	2,122
Color	1.000	Ea.	8.889	1,975	197	2,172
Rough-in, for lavatory or shower						
1/2" copper supply, 1-1/2" cast iron waste, 1-1/2" cast iron vent	1.000	Ea.	3.834	235	236	471
1-1/2" P.V.C. waste, 1-1/4" P.V.C. vent	1.000	Ea.	3.675	147	241	388
1/2" steel supply, 1-1/4" cast iron waste, 1-1/4" steel vent	1.000	Ea.	4.103	198	255	453
1-1/4" P.V.C. waste, 1-1/4" P.V.C. vent	1.000	Ea.	3.937	161	258	419
1/2" P.V.C. supply, 1-1/2" P.V.C. waste, 1-1/2" P.V.C. vent	1.000	Ea.	4.592	203	300	503
Bathtub, P.E. cast iron, 5' long with fittings, white	1.000	Ea.	3.636	1,300	214	1,514
Color	1.000	Ea.	3.636	1,625	214	1,839
Steel, enameled 5' long with fittings, white	1.000	Ea.	2.909	555	172	727
Color	1.000	Ea.	2.909	555	172	727
Rough-in, for bathtub						
1/2" copper supply, 4" cast iron waste, 1-1/2" copper vent	1.000	Ea.	2.409	199	152	351
4" P.V.C. waste, 1-1/2" P.V.C. vent	1.000	Ea.	2.877	104	182	286
1/2" steel supply, 4" cast iron waste, 1-1/2" steel vent	1.000	Ea.	2.898	183	179	362
4" P.V.C. waste, 1-1/2" P.V.C. vent	1.000	Ea.	3.159	110	200	310
1/2" P.V.C. supply, 4" P.V.C. waste, 1-1/2" P.V.C. vent	1.000	Ea.	3.371	141	214	355
Piping, supply, 1/2" copper	42.000	L.F.	4.148	155	271	426
1/2" steel	42.000	L.F.	5.333	178	350	528
1/2" P.V.C.	42.000	L.F.	6.222	310	405	715
Waste, 4" cast iron, no hub	10.000	L.F.	2.759	375	163	538
4" P.V.C./DWV	10.000	L.F.	3.333	112	197	309
4" copper/DWV	10.000	Ea.	4.000	605	235	840
Vent 2" cast iron, no hub	13.000	L.F.	3.105	244	183	427
2" copper/DWV	13.000	L.F.	2.364	258	155	413
2" P.V.C./DWV	13.000	L.F.	3.525	124	208	332
2" steel, galvanized	13.000	L.F.	3.250	185	192	377
Vanity base cabinet, 2 doors, 30" wide	1.000	Ea.	1.000	460	58	518
Vanity top, plastic laminated, square edge	2.670	L.F.	.712	100	41.50	141.50
Carrier, steel for studs, no arms	1.000	Ea.	1.143	63	75	138
Wood, 2" x 8" blocking	1.300	L.F.	.052	1.27	3.02	4.29

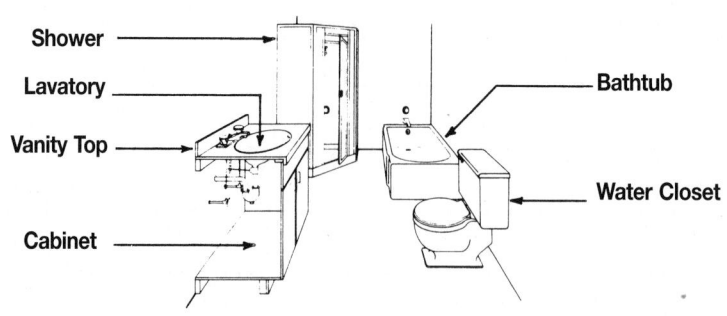

System Description	QUAN.	UNIT	LABOR HOURS	COST EACH		
				MAT.	INST.	TOTAL
BATHROOM WITH LAVATORY INSTALLED IN VANITY						
Water closet, floor mounted, 2 piece, close coupled, white	1.000	Ea.	3.019	253	178	431
Rough-in, vent, 2" diameter DWV piping	1.000	Ea.	.955	75.20	56.40	131.60
Waste, 4" diameter DWV piping	1.000	Ea.	.828	112.50	48.75	161.25
Supply, 1/2" diameter type "L" copper supply piping	1.000	Ea.	.593	22.08	38.70	60.78
Lavatory, 20" x 18" P.E. cast iron with fittings, white	1.000	Ea.	2.500	330	147	477
Shower, steel, enameled, stone base, corner, white	1.000	Ea.	3.333	1,925	197	2,122
Mixing valve	1.000	Ea.	1.333	146	87.50	233.50
Shower door	1.000	Ea.	1.000	480	63.50	543.50
Rough-in, waste, 1-1/2" diameter DWV piping	2.000	Ea.	4.507	352	266	618
Supply, 1/2" diameter type "L" copper supply piping	2.000	Ea.	3.161	117.76	206.40	324.16
Bathtub, P.E. cast iron, 5' long with fittings, white	1.000	Ea.	3.636	1,300	214	1,514
Rough-in, waste, 4" diameter DWV piping	1.000	Ea.	.828	112.50	48.75	161.25
Supply, 1/2" diameter type "L" copper supply piping	1.000	Ea.	.988	36.80	64.50	101.30
Vent, 1-1/2" diameter DWV piping	1.000	Ea.	.593	49.40	38.80	88.20
Piping, supply, 1/2" diameter type "L" copper supply piping	42.000	L.F.	4.939	184	322.50	506.50
Waste, 4" diameter DWV piping	10.000	L.F.	4.138	562.50	243.75	806.25
Vent, 2" diameter DWV piping	13.000	L.F.	4.500	256.50	265.50	522
Vanity base, 2 doors, 30" wide	1.000	Ea.	1.000	460	58	518
Vanity top, plastic laminated, square edge	2.670	L.F.	.712	120.15	41.39	161.54
TOTAL		Ea.	42.563	6,895.39	2,586.44	9,481.83
BATHROOM WITH WALL HUNG LAVATORY						
Water closet, floor mounted, 2 piece, close coupled, white	1.000	Ea.	3.019	253	178	431
Rough-in, vent, 2" diameter DWV piping	1.000	Ea.	.955	75.20	56.40	131.60
Waste, 4" diameter DWV piping	1.000	Ea.	.828	112.50	48.75	161.25
Supply, 1/2" diameter type "L" copper supply piping	1.000	Ea.	.593	22.08	38.70	60.78
Lavatory, 20" x 18" P.E. cast iron with fittings, white	1.000	Ea.	2.000	270	118	388
Shower, steel enameled, stone base, corner, white	1.000	Ea.	3.333	1,925	197	2,122
Mixing valve	1.000	Ea.	1.333	146	87.50	233.50
Shower door	1.000	Ea.	1.000	480	63.50	543.50
Rough-in, waste, 1-1/2" diameter DWV piping	2.000	Ea.	4.507	352	266	618
Supply, 1/2" diameter type "L" copper supply piping	2.000	Ea.	3.161	117.76	206.40	324.16
Bathtub, P.E. cast iron, 5" long with fittings, white	1.000	Ea.	3.636	1,300	214	1,514
Rough-in, waste, 4" diameter DWV piping	1.000	Ea.	.828	112.50	48.75	161.25
Supply, 1/2" diameter type "L" copper supply piping	1.000	Ea.	.988	36.80	64.50	101.30
Vent, 1-1/2" diameter DWV piping	1.000	Ea.	.593	49.40	38.80	88.20
Piping, supply, 1/2" diameter type "L" copper supply piping	42.000	L.F.	4.939	184	322.50	506.50
Waste, 4" diameter DWV piping	10.000	L.F.	4.138	562.50	243.75	806.25
Vent, 2" diameter DWV piping	13.000	L.F.	4.500	256.50	265.50	522
Carrier, steel for studs, no arms	1.000	Ea.	1.143	63	75	138
TOTAL		Ea.	41.494	6,318.24	2,533.05	8,851.29

The costs in this system are on a cost each basis. All necessary piping is included.

Four Fixture Bathroom Price Sheet	QUAN.	UNIT	LABOR HOURS	COST EACH		
				MAT.	INST.	TOTAL
Water closet, close coupled, standard 2 piece, white	1.000	Ea.	3.019	253	178	431
Color	1.000	Ea.	3.019	460	178	638
One piece, elongated bowl, white	1.000	Ea.	3.019	805	178	983
Color	1.000	Ea.	3.019	1,075	178	1,253
Low profile, one piece elongated bowl, white	1.000	Ea.	3.019	815	178	993
Color	1.000	Ea.	3.019	1,075	178	1,253
Rough-in, for water closet						
1/2" copper supply, 4" cast iron waste, 2" cast iron vent	1.000	Ea.	2.376	210	144	354
4" PVC/DWV waste, 2" PVC vent	1.000	Ea.	2.678	93.50	162	255.50
4" copper waste, 2" copper vent	1.000	Ea.	2.520	283	157	440
3" cast iron waste, 1-1/2" cast iron vent	1.000	Ea.	2.244	157	136	293
3" PVC waste, 1-1/2" PVC vent	1.000	Ea.	2.388	84.50	150	234.50
3" PVC waste, 1-1/2" PVC vent	1.000	Ea.	2.014	152	126	278
1/2" PVC supply, 4" PVC waste, 2" PVC vent	1.000	Ea.	2.974	116	181	297
3" PVC waste, 1-1/2" PVC vent	1.000	Ea.	2.684	106	170	276
1/2" steel supply, 4" cast iron waste, 2" cast iron vent	1.000	Ea.	2.545	213	155	368
4" cast iron waste, 2" steel vent	1.000	Ea.	2.590	195	158	353
4" PVC waste, 2" PVC vent	1.000	Ea.	2.847	97	173	270
Lavatory wall hung, P.E. cast iron 20" x 18", white	1.000	Ea.	2.000	270	118	388
Color	1.000	Ea.	2.000	288	118	406
Vitreous china 19" x 17", white	1.000	Ea.	2.286	135	135	270
Color	1.000	Ea.	2.286	157	135	292
Lavatory for vanity top, P.E. cast iron, 20" x 18", white	1.000	Ea.	2.500	330	147	477
Color	1.000	Ea.	2.500	535	147	682
Steel, enameled 20" x 17", white	1.000	Ea.	2.759	142	163	305
Color	1.000	Ea.	2.500	138	147	285
Vitreous china 20" x 16", white	1.000	Ea.	2.963	234	175	409
Color	1.000	Ea.	2.963	234	175	409
Shower, steel enameled, stone base 36" square, white	1.000	Ea.	8.889	1,925	197	2,122
Color	1.000	Ea.	8.889	1,975	197	2,172
Rough-in, for lavatory and shower						
1/2" copper supply, 1-1/2" cast iron waste, 1-1/2" cast iron vent	1.000	Ea.	7.668	470	470	940
1-1/2" PVC waste, 1-1/4" PVC vent	1.000	Ea.	7.352	295	480	775
1/2" steel supply, 1-1/4" cast iron waste, 1-1/4" steel vent	1.000	Ea.	8.205	395	510	905
1-1/4" PVC waste, 1-1/4" PVC vent	1.000	Ea.	7.873	325	515	840
1/2" PVC supply, 1-1/2" PVC waste, 1-1/2" PVC vent	1.000	Ea.	9.185	405	600	1,005
Bathtub, P.E. cast iron, 5' long with fittings, white	1.000	Ea.	3.636	1,300	214	1,514
Color	1.000	Ea.	3.636	1,625	214	1,839
Steel enameled, 5' long with fittings, white	1.000	Ea.	2.909	555	172	727
Color	1.000	Ea.	2.909	555	172	727
Rough-in, for bathtub						
1/2" copper supply, 4" cast iron waste, 1-1/2" copper vent	1.000	Ea.	2.409	199	152	351
4" PVC waste, 1-1/2" PVC vent	1.000	Ea.	2.877	104	182	286
1/2" steel supply, 4" cast iron waste, 1-1/2" steel vent	1.000	Ea.	2.898	183	179	362
4" PVC waste, 1-1/2" PVC vent	1.000	Ea.	3.159	110	200	310
1/2" PVC supply, 4" PVC waste, 1-1/2" PVC vent	1.000	Ea.	3.371	141	214	355
Piping supply, 1/2" copper	42.000	L.F.	4.148	155	271	426
1/2" steel	42.000	L.F.	5.333	178	350	528
1/2" PVC	42.000	L.F.	6.222	310	405	715
Piping, waste, 4" cast iron, no hub	10.000	L.F.	3.586	490	211	701
4" PVC/DWV	10.000	L.F.	4.333	146	255	401
4" copper/DWV	10.000	L.F.	5.200	785	305	1,090
Piping, vent, 2" cast iron, no hub	13.000	L.F.	3.105	244	183	427
2" copper/DWV	13.000	L.F.	2.364	258	155	413
2" PVC/DWV	13.000	L.F.	3.525	124	208	332
2" steel, galvanized	13.000	L.F.	3.250	185	192	377
Vanity base cabinet, 2 doors, 30" wide	1.000	Ea.	1.000	460	58	518
Vanity top, plastic laminated, square edge	3.160	L.F.	.843	119	49	168
Carrier, steel, for studs, no arms	1.000	Ea.	1.143	63	75	138
Wood, 2" x 8" blocking	1.300	L.F.	.052	1.27	3.02	4.29

For customer support on your Residential Costs with RSMeans data, call 800.448.8182.

265

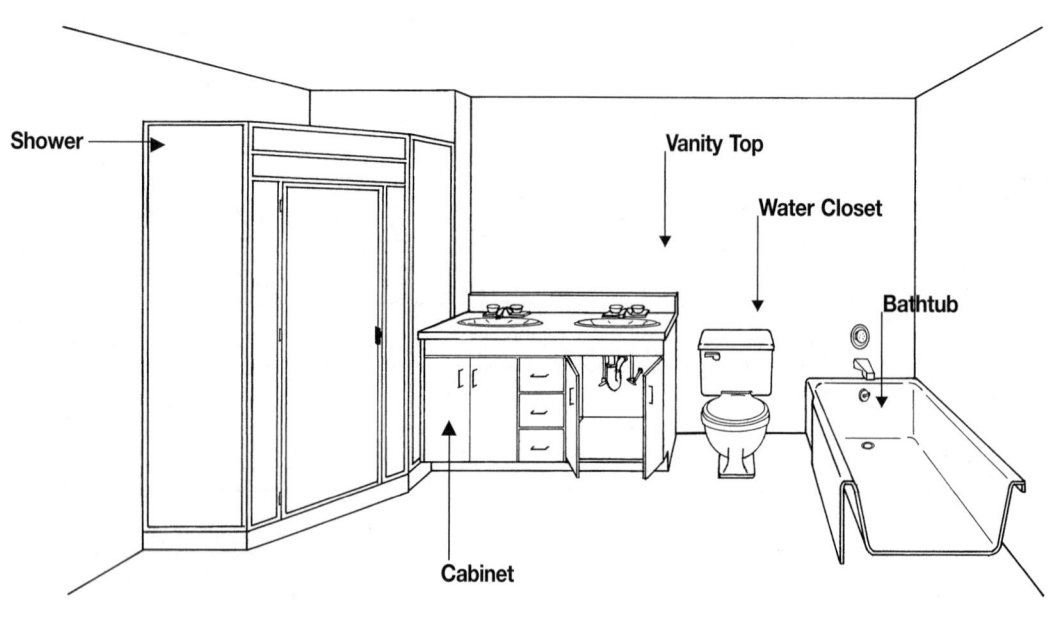

Shower · Vanity Top · Water Closet · Bathtub · Cabinet

System Description	QUAN.	UNIT	LABOR HOURS	COST EACH		
				MAT.	INST.	TOTAL
BATHROOM WITH SHOWER, BATHTUB, LAVATORIES IN VANITY						
Water closet, floor mounted, 1 piece, white	1.000	Ea.	3.019	815	178	993
Rough-in, vent, 2" diameter DWV piping	1.000	Ea.	.955	75.20	56.40	131.60
Waste, 4" diameter DWV piping	1.000	Ea.	.828	112.50	48.75	161.25
Supply, 1/2" diameter type "L" copper supply piping	1.000	Ea.	.593	22.08	38.70	60.78
Lavatory, 20" x 16", vitreous china oval, with fittings, white	2.000	Ea.	5.926	468	350	818
Shower, steel enameled, stone base, corner, white	1.000	Ea.	3.333	1,925	197	2,122
Mixing valve	1.000	Ea.	1.333	146	87.50	233.50
Shower door	1.000	Ea.	1.000	480	63.50	543.50
Rough-in, waste, 1-1/2" diameter DWV piping	3.000	Ea.	5.408	422.40	319.20	741.60
Supply, 1/2" diameter type "L" copper supply piping	3.000	Ea.	2.963	110.40	193.50	303.90
Bathtub, P.E. cast iron, 5' long with fittings, white	1.000	Ea.	3.636	1,300	214	1,514
Rough-in, waste, 4" diameter DWV piping	1.000	Ea.	1.103	150	65	215
Supply, 1/2" diameter type "L" copper supply piping	1.000	Ea.	.988	36.80	64.50	101.30
Vent, 1-1/2" diameter copper DWV piping	1.000	Ea.	.593	49.40	38.80	88.20
Piping, supply, 1/2" diameter type "L" copper supply piping	42.000	L.F.	4.148	154.56	270.90	425.46
Waste, 4" diameter DWV piping	10.000	L.F.	2.759	375	162.50	537.50
Vent, 2" diameter DWV piping	13.000	L.F.	3.250	185.25	191.75	377
Vanity base, 2 door, 24" x 48"	1.000	Ea.	1.400	585	81.50	666.50
Vanity top, plastic laminated, square edge	4.170	L.F.	1.112	156.38	64.64	221.02
TOTAL		Ea.	44.347	7,568.97	2,686.14	10,255.11

The costs in this system are on a cost each basis. All necessary piping is included.

Description	QUAN.	UNIT	LABOR HOURS	COST EACH		
				MAT.	INST.	TOTAL

Five Fixture Bathroom Price Sheet	QUAN.	UNIT	LABOR HOURS	COST EACH		
				MAT.	INST.	TOTAL
Water closet, close coupled, standard 2 piece, white	1.000	Ea.	3.019	253	178	431
Color	1.000	Ea.	3.019	460	178	638
One piece elongated bowl, white	1.000	Ea.	3.019	805	178	983
Color	1.000	Ea.	3.019	1,075	178	1,253
Low profile, one piece elongated bowl, white	1.000	Ea.	3.019	815	178	993
Color	1.000	Ea.	3.019	1,075	178	1,253
Rough-in, supply, waste and vent for water closet						
1/2" copper supply, 4" cast iron waste, 2" cast iron vent	1.000	Ea.	2.376	210	144	354
4" P.V.C./DWV waste, 2" P.V.C. vent	1.000	Ea.	2.678	93.50	162	255.50
4" copper waste, 2" copper vent	1.000	Ea.	2.520	283	157	440
3" cast iron waste, 1-1/2" cast iron vent	1.000	Ea.	2.244	157	136	293
3" P.V.C. waste, 1-1/2" P.V.C. vent	1.000	Ea.	2.388	84.50	150	234.50
3" copper waste, 1-1/2" copper vent	1.000	Ea.	2.014	152	126	278
1/2" P.V.C. supply, 4" P.V.C. waste, 2" P.V.C. vent	1.000	Ea.	2.974	116	181	297
3" P.V.C. waste, 1-1/2" P.V.C. supply	1.000	Ea.	2.684	106	170	276
1/2" steel supply, 4" cast iron waste, 2" cast iron vent	1.000	Ea.	2.545	213	155	368
4" cast iron waste, 2" steel vent	1.000	Ea.	2.590	195	158	353
4" P.V.C. waste, 2" P.V.C. vent	1.000	Ea.	2.847	97	173	270
Lavatory, wall hung, P.E. cast iron 20" x 18", white	2.000	Ea.	4.000	540	236	776
Color	2.000	Ea.	4.000	575	236	811
Vitreous china, 19" x 17", white	2.000	Ea.	4.571	270	270	540
Color	2.000	Ea.	4.571	315	270	585
Lavatory, for vanity top, P.E. cast iron, 20" x 18", white	2.000	Ea.	5.000	660	294	954
Color	2.000	Ea.	5.000	1,075	294	1,369
Steel enameled 20" x 17", white	2.000	Ea.	5.517	284	325	609
Color	2.000	Ea.	5.000	276	294	570
Vitreous china 20" x 16", white	2.000	Ea.	5.926	470	350	820
Color	2.000	Ea.	5.926	470	350	820
Shower, steel enameled, stone base 36" square, white	1.000	Ea.	8.889	1,925	197	2,122
Color	1.000	Ea.	8.889	1,975	197	2,172
Rough-in, for lavatory or shower						
1/2" copper supply, 1-1/2" cast iron waste, 1-1/2" cast iron vent	3.000	Ea.	8.371	535	515	1,050
1-1/2" P.V.C. waste, 1-1/4" P.V.C. vent	3.000	Ea.	7.916	325	520	845
1/2" steel supply, 1-1/4" cast iron waste, 1-1/4" steel vent	3.000	Ea.	8.670	415	535	950
1-1/4" P.V.C. waste, 1-1/4" P.V.C. vent	3.000	Ea.	8.381	350	550	900
1/2" P.V.C. supply, 1-1/2" P.V.C. waste, 1-1/2" P.V.C. vent	3.000	Ea.	9.778	425	640	1,065
Bathtub, P.E. cast iron 5' long with fittings, white	1.000	Ea.	3.636	1,300	214	1,514
Color	1.000	Ea.	3.636	1,625	214	1,839
Steel, enameled 5' long with fittings, white	1.000	Ea.	2.909	555	172	727
Color	1.000	Ea.	2.909	555	172	727
Rough-in, for bathtub						
1/2" copper supply, 4" cast iron waste, 1-1/2" copper vent	1.000	Ea.	2.684	236	168	404
4" P.V.C. waste, 1-1/2" P.V.C. vent	1.000	Ea.	3.210	116	201	317
1/2" steel supply, 4" cast iron waste, 1-1/2" steel vent	1.000	Ea.	3.173	221	195	416
4" P.V.C. waste, 1-1/2" P.V.C. vent	1.000	Ea.	3.492	121	220	341
1/2" P.V.C. supply, 4" P.V.C. waste, 1-1/2" P.V.C. vent	1.000	Ea.	3.704	152	234	386
Piping, supply, 1/2" copper	42.000	L.F.	4.148	155	271	426
1/2" steel	42.000	L.F.	5.333	178	350	528
1/2" P.V.C.	42.000	L.F.	6.222	310	405	715
Piping, waste, 4" cast iron, no hub	10.000	L.F.	2.759	375	163	538
4" P.V.C./DWV	10.000	L.F.	3.333	112	197	309
4" copper/DWV	10.000	L.F.	4.000	605	235	840
Piping, vent, 2" cast iron, no hub	13.000	L.F.	3.105	244	183	427
2" copper/DWV	13.000	L.F.	2.364	258	155	413
2" P.V.C./DWV	13.000	L.F.	3.525	124	208	332
2" steel, galvanized	13.000	L.F.	3.250	185	192	377
Vanity base cabinet, 2 doors, 24" x 48"	1.000	Ea.	1.400	585	81.50	666.50
Vanity top, plastic laminated, square edge	4.170	L.F.	1.112	156	64.50	220.50
Carrier, steel, for studs, no arms	1.000	Ea.	1.143	63	75	138
Wood, 2" x 8" blocking	1.300	L.F.	.052	1.27	3.02	4.29

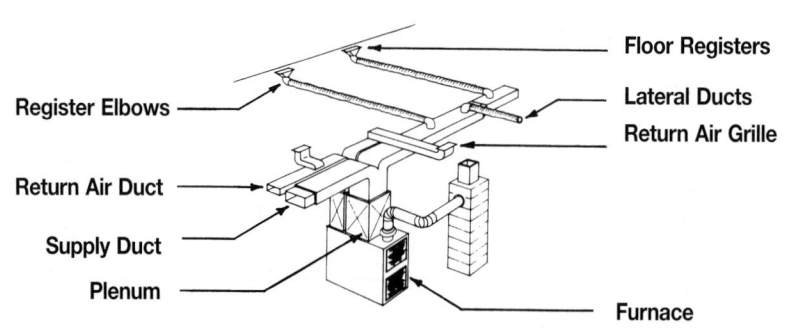

Floor Registers
Lateral Ducts
Return Air Grille
Register Elbows
Return Air Duct
Supply Duct
Plenum
Furnace

System Description	QUAN.	UNIT	LABOR HOURS	COST PER SYSTEM		
				MAT.	INST.	TOTAL
HEATING ONLY, GAS FIRED HOT AIR, ONE ZONE, 1200 S.F. BUILDING						
Furnace, gas, up flow	1.000	Ea.	5.000	815	287	1,102
Intermittent pilot	1.000	Ea.		273		273
Supply duct, rigid fiberglass	176.000	S.F.	12.068	174.24	718.08	892.32
Return duct, sheet metal, galvanized	158.000	Lb.	16.137	99.54	955.90	1,055.44
Lateral ducts, 6" flexible fiberglass	144.000	L.F.	8.862	492.48	508.32	1,000.80
Register, elbows	12.000	Ea.	6.400	195	366	561
Floor registers, enameled steel	12.000	Ea.	3.000	167.40	191.40	358.80
Floor grille, return air	2.000	Ea.	.727	62	46	108
Thermostat	1.000	Ea.	1.000	47	67.50	114.50
Plenum	1.000	Ea.	1.000	103	57.50	160.50
TOTAL		System	54.194	2,428.66	3,197.70	5,626.36
HEATING/COOLING, GAS FIRED FORCED AIR, ONE ZONE, 1200 S.F. BUILDING						
Furnace, including plenum, compressor, coil	1.000	Ea.	14.720	6,233	846.40	7,079.40
Intermittent pilot	1.000	Ea.		273		273
Supply duct, rigid fiberglass	176.000	S.F.	12.068	174.24	718.08	892.32
Return duct, sheet metal, galvanized	158.000	Lb.	16.137	99.54	955.90	1,055.44
Lateral duct, 6" flexible fiberglass	144.000	L.F.	8.862	492.48	508.32	1,000.80
Register elbows	12.000	Ea.	6.400	195	366	561
Floor registers, enameled steel	12.000	Ea.	3.000	167.40	191.40	358.80
Floor grille return air	2.000	Ea.	.727	62	46	108
Thermostat	1.000	Ea.	1.000	47	67.50	114.50
Refrigeration piping, 25 ft. (pre-charged)	1.000	Ea.		320		320
TOTAL		System	62.914	8,063.66	3,699.60	11,763.26

The costs in these systems are based on complete system basis. For larger buildings use the price sheet on the opposite page.

Description	QUAN.	UNIT	LABOR HOURS	COST PER SYSTEM		
				MAT.	INST.	TOTAL

Gas Heating/Cooling Price Sheet

Gas Heating/Cooling Price Sheet	QUAN.	UNIT	LABOR HOURS	COST EACH MAT.	COST EACH INST.	COST EACH TOTAL
Furnace, heating only, 100 MBH, area to 1200 S.F.	1.000	Ea.	5.000	815	287	1,102
120 MBH, area to 1500 S.F.	1.000	Ea.	5.000	815	287	1,102
160 MBH, area to 2000 S.F.	1.000	Ea.	5.714	1,700	330	2,030
200 MBH, area to 2400 S.F.	1.000	Ea.	6.154	3,500	355	3,855
Heating/cooling, 100 MBH heat, 36 MBH cool, to 1200 S.F.	1.000	Ea.	16.000	6,775	920	7,695
120 MBH heat, 42 MBH cool, to 1500 S.F.	1.000	Ea.	18.462	7,250	1,100	8,350
144 MBH heat, 47 MBH cool, to 2000 S.F.	1.000	Ea.	20.000	8,375	1,200	9,575
200 MBH heat, 60 MBH cool, to 2400 S.F.	1.000	Ea.	34.286	8,800	2,050	10,850
Intermittent pilot, 100 MBH furnace	1.000	Ea.		273		273
200 MBH furnace	1.000	Ea.		273		273
Supply duct, rectangular, area to 1200 S.F., rigid fiberglass	176.000	S.F.	12.068	174	720	894
Sheet metal insulated	228.000	Lb.	31.331	1,025	3,225	4,250
Area to 1500 S.F., rigid fiberglass	176.000	S.F.	12.068	174	720	894
Sheet metal insulated	228.000	Lb.	31.331	1,025	3,225	4,250
Area to 2400 S.F., rigid fiberglass	205.000	S.F.	14.057	203	835	1,038
Sheet metal insulated	271.000	Lb.	37.048	1,200	3,775	4,975
Round flexible, insulated 6" diameter, to 1200 S.F.	156.000	L.F.	9.600	535	550	1,085
To 1500 S.F.	184.000	L.F.	11.323	630	650	1,280
8" diameter, to 2000 S.F.	269.000	L.F.	23.911	1,125	1,375	2,500
To 2400 S.F.	248.000	L.F.	22.045	1,025	1,275	2,300
Return duct, sheet metal galvanized, to 1500 S.F.	158.000	Lb.	16.137	99.50	955	1,054.50
To 2400 S.F.	191.000	Lb.	19.507	120	1,150	1,270
Lateral ducts, flexible round 6" insulated, to 1200 S.F.	144.000	L.F.	8.862	490	510	1,000
To 1500 S.F.	172.000	L.F.	10.585	590	605	1,195
To 2000 S.F.	261.000	L.F.	16.062	895	920	1,815
To 2400 S.F.	300.000	L.F.	18.462	1,025	1,050	2,075
Spiral steel insulated, to 1200 S.F.	144.000	L.F.	20.067	1,600	3,175	4,775
To 1500 S.F.	172.000	L.F.	23.952	1,900	3,800	5,700
To 2000 S.F.	261.000	L.F.	36.352	2,875	5,750	8,625
To 2400 S.F.	300.000	L.F.	41.825	3,300	6,625	9,925
Rectangular sheet metal galvanized insulated, to 1200 S.F.	228.000	Lb.	39.056	1,875	4,975	6,850
To 1500 S.F.	344.000	Lb.	53.966	2,275	6,375	8,650
To 2000 S.F.	522.000	Lb.	81.926	3,450	9,700	13,150
To 2400 S.F.	600.000	Lb.	94.189	3,975	11,200	15,175
Register elbows, to 1500 S.F.	12.000	Ea.	6.400	195	365	560
To 2400 S.F.	14.000	Ea.	7.460	228	425	653
Floor registers, enameled steel w/damper, to 1500 S.F.	12.000	Ea.	3.000	167	191	358
To 2400 S.F.	14.000	Ea.	4.308	230	274	504
Return air grille, area to 1500 S.F. 12" x 12"	2.000	Ea.	.727	62	46	108
Area to 2400 S.F. 8" x 16"	2.000	Ea.	.444	40.50	28.50	69
Area to 2400 S.F. 8" x 16"	2.000	Ea.	.727	62	46	108
16" x 16"	1.000	Ea.	.364	39.50	23	62.50
Thermostat, manual, 1 set back	1.000	Ea.	1.000	47	67.50	114.50
Electric, timed, 1 set back	1.000	Ea.	1.000	84.50	67.50	152
2 set back	1.000	Ea.	1.000	251	67.50	318.50
Plenum, heating only, 100 M.B.H.	1.000	Ea.	1.000	103	57.50	160.50
120 MBH	1.000	Ea.	1.000	103	57.50	160.50
160 MBH	1.000	Ea.	1.000	103	57.50	160.50
200 MBH	1.000	Ea.	1.000	103	57.50	160.50
Refrigeration piping, 3/8"	25.000	L.F.		35.50		35.50
3/4"	25.000	L.F.		78		78
7/8"	25.000	L.F.		91		91
Refrigerant piping, 25 ft. (precharged)	1.000	Ea.		320		320
Diffusers, ceiling, 6" diameter, to 1500 S.F.	10.000	Ea.	4.444	121	285	406
To 2400 S.F.	12.000	Ea.	6.000	155	385	540
Floor, aluminum, adjustable, 2-1/4" x 12" to 1500 S.F.	12.000	Ea.	3.000	113	191	304
To 2400 S.F.	14.000	Ea.	3.500	132	223	355
Side wall, aluminum, adjustable, 8" x 4", to 1500 S.F.	12.000	Ea.	3.000	230	191	421
5" x 10" to 2400 S.F.	12.000	Ea.	3.692	305	235	540

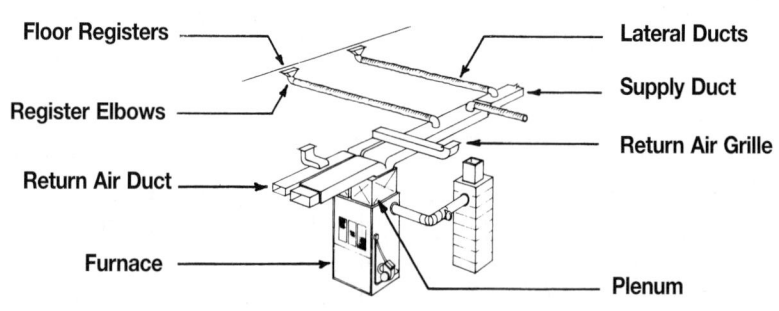

Floor Registers — Lateral Ducts
Register Elbows — Supply Duct
Return Air Duct — Return Air Grille
Furnace — Plenum

System Description	QUAN.	UNIT	LABOR HOURS	COST PER SYSTEM		
				MAT.	INST.	TOTAL
HEATING ONLY, OIL FIRED HOT AIR, ONE ZONE, 1200 S.F. BUILDING						
Furnace, oil fired, atomizing gun type burner	1.000	Ea.	4.571	3,150	262	3,412
3/8" diameter copper supply pipe	1.000	Ea.	2.759	123.30	180	303.30
Shut off valve	1.000	Ea.	.333	15.95	22	37.95
Oil tank, 275 gallon, on legs	1.000	Ea.	3.200	565	194	759
Supply duct, rigid fiberglass	176.000	S.F.	12.068	174.24	718.08	892.32
Return duct, sheet metal, galvanized	158.000	Lb.	16.137	99.54	955.90	1,055.44
Lateral ducts, 6" flexible fiberglass	144.000	L.F.	8.862	492.48	508.32	1,000.80
Register elbows	12.000	Ea.	6.400	195	366	561
Floor register, enameled steel	12.000	Ea.	3.000	167.40	191.40	358.80
Floor grille, return air	2.000	Ea.	.727	62	46	108
Thermostat	1.000	Ea.	1.000	47	67.50	114.50
TOTAL		System	59.057	5,091.91	3,511.20	8,603.11
HEATING/COOLING, OIL FIRED, FORCED AIR, ONE ZONE, 1200 S.F. BUILDING						
Furnace, including plenum, compressor, coil	1.000	Ea.	16.000	7,250	920	8,170
3/8" diameter copper supply pipe	1.000	Ea.	2.759	123.30	180	303.30
Shut off valve	1.000	Ea.	.333	15.95	22	37.95
Oil tank, 275 gallon on legs	1.000	Ea.	3.200	565	194	759
Supply duct, rigid fiberglass	176.000	S.F.	12.068	174.24	718.08	892.32
Return duct, sheet metal, galvanized	158.000	Lb.	16.137	99.54	955.90	1,055.44
Lateral ducts, 6" flexible fiberglass	144.000	L.F.	8.862	492.48	508.32	1,000.80
Register elbows	12.000	Ea.	6.400	195	366	561
Floor registers, enameled steel	12.000	Ea.	3.000	167.40	191.40	358.80
Floor grille, return air	2.000	Ea.	.727	62	46	108
Refrigeration piping (precharged)	25.000	L.F.		320		320
TOTAL		System	69.486	9,464.91	4,101.70	13,566.61

Description	QUAN.	UNIT	LABOR HOURS	COST PER SYSTEM		
				MAT.	INST.	TOTAL

Oil Fired Heating/Cooling Price Sheet

	QUAN.	UNIT	LABOR HOURS	COST EACH MAT.	COST EACH INST.	COST EACH TOTAL
Furnace, heating, 95.2 MBH, area to 1200 S.F.	1.000	Ea.	4.706	3,150	270	3,420
123.2 MBH, area to 1500 S.F.	1.000	Ea.	5.000	3,100	287	3,387
151.2 MBH, area to 2000 S.F.	1.000	Ea.	5.333	3,425	305	3,730
200 MBH, area to 2400 S.F.	1.000	Ea.	6.154	3,900	355	4,255
Heating/cooling, 95.2 MBH heat, 36 MBH cool, to 1200 S.F.	1.000	Ea.	16.000	7,250	920	8,170
112 MBH heat, 42 MBH cool, to 1500 S.F.	1.000	Ea.	24.000	10,900	1,375	12,275
151 MBH heat, 47 MBH cool, to 2000 S.F.	1.000	Ea.	20.800	9,425	1,200	10,625
184.8 MBH heat, 60 MBH cool, to 2400 S.F.	1.000	Ea.	24.000	5,200	1,425	6,625
Oil piping to furnace, 3/8" dia., copper	1.000	Ea.	3.412	284	220	504
Oil tank, on legs above ground, 275 gallons	1.000	Ea.	3.200	565	194	759
550 gallons	1.000	Ea.	5.926	4,775	360	5,135
Below ground, 275 gallons	1.000	Ea.	3.200	565	194	759
550 gallons	1.000	Ea.	5.926	4,775	360	5,135
1000 gallons	1.000	Ea.	6.400	7,825	390	8,215
Supply duct, rectangular, area to 1200 S.F., rigid fiberglass	176.000	S.F.	12.068	174	720	894
Sheet metal, insulated	228.000	Lb.	31.331	1,025	3,225	4,250
Area to 1500 S.F., rigid fiberglass	176.000	S.F.	12.068	174	720	894
Sheet metal, insulated	228.000	Lb.	31.331	1,025	3,225	4,250
Area to 2400 S.F., rigid fiberglass	205.000	S.F.	14.057	203	835	1,038
Sheet metal, insulated	271.000	Lb.	37.048	1,200	3,775	4,975
Round flexible, insulated, 6" diameter to 1200 S.F.	156.000	L.F.	9.600	535	550	1,085
To 1500 S.F.	184.000	L.F.	11.323	630	650	1,280
8" diameter to 2000 S.F.	269.000	L.F.	23.911	1,125	1,375	2,500
To 2400 S.F.	269.000	L.F.	22.045	1,025	1,275	2,300
Return duct, sheet metal galvanized, to 1500 S.F.	158.000	Lb.	16.137	99.50	955	1,054.50
To 2400 S.F.	191.000	Lb.	19.507	120	1,150	1,270
Lateral ducts, flexible round, 6", insulated to 1200 S.F.	144.000	L.F.	8.862	490	510	1,000
To 1500 S.F.	172.000	L.F.	10.585	590	605	1,195
To 2000 S.F.	261.000	L.F.	16.062	895	920	1,815
To 2400 S.F.	300.000	L.F.	18.462	1,025	1,050	2,075
Spiral steel, insulated to 1200 S.F.	144.000	L.F.	20.067	1,600	3,175	4,775
To 1500 S.F.	172.000	L.F.	23.952	1,900	3,800	5,700
To 2000 S.F.	261.000	L.F.	36.352	2,875	5,750	8,625
To 2400 S.F.	300.000	L.F.	41.825	3,300	6,625	9,925
Rectangular sheet metal galvanized insulated, to 1200 S.F.	288.000	Lb.	45.183	1,900	5,350	7,250
To 1500 S.F.	344.000	Lb.	53.966	2,275	6,375	8,650
To 2000 S.F.	522.000	Lb.	81.926	3,450	9,700	13,150
To 2400 S.F.	600.000	Lb.	94.189	3,975	11,200	15,175
Register elbows, to 1500 S.F.	12.000	Ea.	6.400	195	365	560
To 2400 S.F.	14.000	Ea.	7.470	228	425	653
Floor registers, enameled steel w/damper, to 1500 S.F.	12.000	Ea.	3.000	167	191	358
To 2400 S.F.	14.000	Ea.	4.308	230	274	504
Return air grille, area to 1500 S.F., 12" x 12"	2.000	Ea.	.727	62	46	108
12" x 24"	1.000	Ea.	.444	40.50	28.50	69
Area to 2400 S.F., 8" x 16"	2.000	Ea.	.727	62	46	108
16" x 16"	1.000	Ea.	.364	39.50	23	62.50
Thermostat, manual, 1 set back	1.000	Ea.	1.000	47	67.50	114.50
Electric, timed, 1 set back	1.000	Ea.	1.000	84.50	67.50	152
2 set back	1.000	Ea.	1.000	251	67.50	318.50
Refrigeration piping, 3/8"	25.000	L.F.		35.50		35.50
3/4"	25.000	L.F.		78		78
Diffusers, ceiling, 6" diameter, to 1500 S.F.	10.000	Ea.	4.444	121	285	406
To 2400 S.F.	12.000	Ea.	6.000	155	385	540
Floor, aluminum, adjustable, 2-1/4" x 12" to 1500 S.F.	12.000	Ea.	3.000	113	191	304
To 2400 S.F.	14.000	Ea.	3.500	132	223	355
Side wall, aluminum, adjustable, 8" x 4", to 1500 S.F.	12.000	Ea.	3.000	230	191	421
5" x 10" to 2400 S.F.	12.000	Ea.	3.692	305	235	540

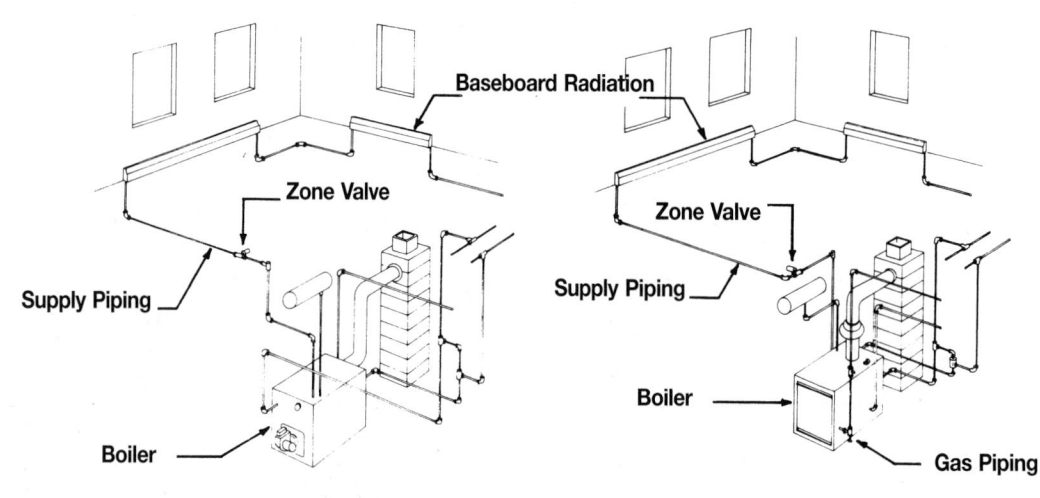

System Description	QUAN.	UNIT	LABOR HOURS	COST EACH		
				MAT.	INST.	TOTAL
OIL FIRED HOT WATER HEATING SYSTEM, AREA TO 1200 S.F.						
Boiler package, oil fired, 97 MBH, area to 1200 S.F. building	1.000	Ea.	15.000	2,075	875	2,950
3/8" diameter copper supply pipe	1.000	Ea.	2.759	123.30	180	303.30
Shut off valve	1.000	Ea.	.333	15.95	22	37.95
Oil tank, 275 gallon, with black iron filler pipe	1.000	Ea.	3.200	565	194	759
Supply piping, 3/4" copper tubing	176.000	L.F.	18.526	869.44	1,214.40	2,083.84
Supply fittings, copper 3/4"	36.000	Ea.	15.158	99.72	990	1,089.72
Supply valves, 3/4"	2.000	Ea.	.800	300	52	352
Baseboard radiation, 3/4"	106.000	L.F.	35.333	879.80	2,120	2,999.80
Zone valve	1.000	Ea.	.400	172	27	199
TOTAL		Ea.	91.509	5,100.21	5,674.40	10,774.61
OIL FIRED HOT WATER HEATING SYSTEM, AREA TO 2400 S.F.						
Boiler package, oil fired, 225 MBH, area to 2400 S.F. building	1.000	Ea.	25.105	3,650	1,475	5,125
3/8" diameter copper supply pipe	1.000	Ea.	2.759	123.30	180	303.30
Shut off valve	1.000	Ea.	.333	15.95	22	37.95
Oil tank, 550 gallon, with black iron pipe filler pipe	1.000	Ea.	5.926	4,775	360	5,135
Supply piping, 3/4" copper tubing	228.000	L.F.	23.999	1,126.32	1,573.20	2,699.52
Supply fittings, copper	46.000	Ea.	19.368	127.42	1,265	1,392.42
Supply valves	2.000	Ea.	.800	300	52	352
Baseboard radiation	212.000	L.F.	70.666	1,759.60	4,240	5,999.60
Zone valve	1.000	Ea.	.400	172	27	199
TOTAL		Ea.	149.356	12,049.59	9,194.20	21,243.79

The costs in this system are on a cost each basis. The costs represent total cost for the system based on a gross square foot of plan area.

Description	QUAN.	UNIT	LABOR HOURS	COST EACH		
				MAT.	INST.	TOTAL

Hot Water Heating Price Sheet	QUAN.	UNIT	LABOR HOURS	COST EACH		
				MAT.	INST.	TOTAL
Boiler, oil fired, 97 MBH, area to 1200 S.F.	1.000	Ea.	15.000	2,075	875	2,950
118 MBH, area to 1500 S.F.	1.000	Ea.	16.506	2,200	965	3,165
161 MBH, area to 2000 S.F.	1.000	Ea.	18.405	2,475	1,075	3,550
215 MBH, area to 2400 S.F.	1.000	Ea.	19.704	3,975	2,275	6,250
Oil piping, (valve & filter), 3/8" copper	1.000	Ea.	3.289	139	202	341
1/4" copper	1.000	Ea.	3.242	160	211	371
Oil tank, filler pipe and cap on legs, 275 gallon	1.000	Ea.	3.200	565	194	759
550 gallon	1.000	Ea.	5.926	4,775	360	5,135
Buried underground, 275 gallon	1.000	Ea.	3.200	565	194	759
550 gallon	1.000	Ea.	5.926	4,775	360	5,135
1000 gallon	1.000	Ea.	6.400	7,825	390	8,215
Supply piping copper, area to 1200 S.F., 1/2" tubing	176.000	L.F.	17.384	650	1,125	1,775
3/4" tubing	176.000	L.F.	18.526	870	1,225	2,095
Area to 1500 S.F., 1/2" tubing	186.000	L.F.	18.371	685	1,200	1,885
3/4" tubing	186.000	L.F.	19.578	920	1,275	2,195
Area to 2000 S.F., 1/2" tubing	204.000	L.F.	20.149	750	1,325	2,075
3/4" tubing	204.000	L.F.	21.473	1,000	1,400	2,400
Area to 2400 S.F., 1/2" tubing	228.000	L.F.	22.520	840	1,475	2,315
3/4" tubing	228.000	L.F.	23.999	1,125	1,575	2,700
Supply pipe fittings copper, area to 1200 S.F., 1/2"	36.000	Ea.	14.400	44.50	935	979.50
3/4"	36.000	Ea.	15.158	99.50	990	1,089.50
Area to 1500 S.F., 1/2"	40.000	Ea.	16.000	49.50	1,050	1,099.50
3/4"	40.000	Ea.	16.842	111	1,100	1,211
Area to 2000 S.F., 1/2"	44.000	Ea.	17.600	54.50	1,150	1,204.50
3/4"	44.000	Ea.	18.526	122	1,200	1,322
Area to 2400, S.F., 1/2"	46.000	Ea.	18.400	57	1,200	1,257
3/4"	46.000	Ea.	19.368	127	1,275	1,402
Supply valves, 1/2" pipe size	2.000	Ea.	.667	226	44	270
3/4"	2.000	Ea.	.800	300	52	352
Baseboard radiation, area to 1200 S.F., 1/2" tubing	106.000	L.F.	28.267	1,350	1,700	3,050
3/4" tubing	106.000	L.F.	35.333	880	2,125	3,005
Area to 1500 S.F., 1/2" tubing	134.000	L.F.	35.734	1,700	2,175	3,875
3/4" tubing	134.000	L.F.	44.666	1,100	2,675	3,775
Area to 2000 S.F., 1/2" tubing	178.000	L.F.	47.467	2,275	2,875	5,150
3/4" tubing	178.000	L.F.	59.333	1,475	3,550	5,025
Area to 2400 S.F., 1/2" tubing	212.000	L.F.	56.534	2,700	3,425	6,125
3/4" tubing	212.000	L.F.	70.666	1,750	4,250	6,000
Zone valves, 1/2" tubing	1.000	Ea.	.400	172	27	199
3/4" tubing	1.000	Ea.	.400	166	27	193

For customer support on your Residential Costs with RSMeans data, call 800.448.8182.

273

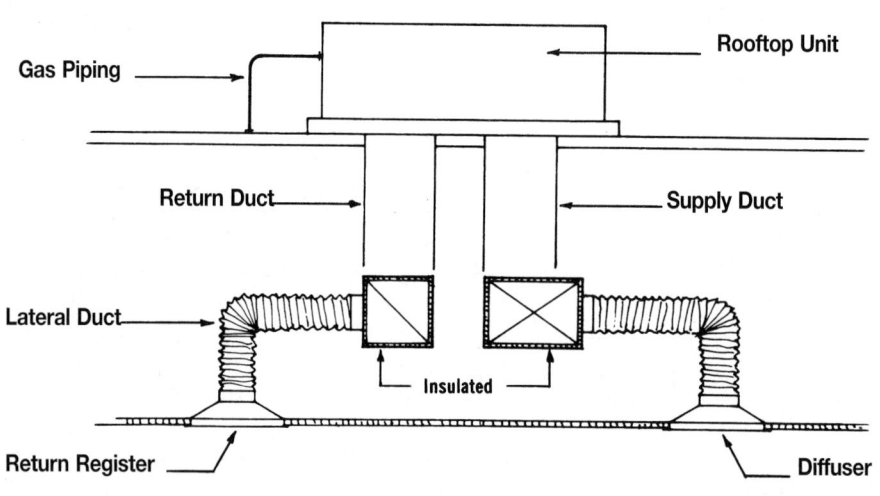

System Description	QUAN.	UNIT	LABOR HOURS	COST EACH		
				MAT.	INST.	TOTAL
ROOFTOP HEATING/COOLING UNIT, AREA TO 2000 S.F.						
Rooftop unit, single zone, electric cool, gas heat, to 2000 S.F.	1.000	Ea.	28.521	4,800	1,725	6,525
Gas piping	34.500	L.F.	5.207	166.29	341.55	507.84
Duct, supply and return, galvanized steel	38.000	Lb.	3.881	23.94	229.90	253.84
Insulation, ductwork	33.000	S.F.	6.286	164.67	344.85	509.52
Lateral duct, flexible duct 12" diameter, insulated	72.000	L.F.	11.520	417.60	662.40	1,080
Diffusers	4.000	Ea.	4.571	1,196	292	1,488
Return registers	1.000	Ea.	.727	98	46.50	144.50
TOTAL		Ea.	60.713	6,866.50	3,642.20	10,508.70
ROOFTOP HEATING/COOLING UNIT, AREA TO 5000 S.F.						
Rooftop unit, single zone, electric cool, gas heat, to 5000 S.F.	1.000	Ea.	42.032	14,400	2,450	16,850
Gas piping	86.250	L.F.	13.019	415.73	853.88	1,269.61
Duct supply and return, galvanized steel	95.000	Lb.	9.702	59.85	574.75	634.60
Insulation, ductwork	82.000	S.F.	15.619	409.18	856.90	1,266.08
Lateral duct, flexible duct, 12" diameter, insulated	180.000	L.F.	28.800	1,044	1,656	2,700
Diffusers	10.000	Ea.	11.429	2,990	730	3,720
Return registers	3.000	Ea.	2.182	294	139.50	433.50
TOTAL		Ea.	122.783	19,612.76	7,261.03	26,873.79

Description	QUAN.	UNIT	LABOR HOURS	COST EACH		
				MAT.	INST.	TOTAL

Rooftop Price Sheet	QUAN.	UNIT	LABOR HOURS	COST EACH		
				MAT.	INST.	TOTAL
Rooftop unit, single zone, electric cool, gas heat to 2000 S.F.	1.000	Ea.	28.521	4,800	1,725	6,525
Area to 3000 S.F.	1.000	Ea.	35.982	10,100	2,100	12,200
Area to 5000 S.F.	1.000	Ea.	42.032	14,400	2,450	16,850
Area to 10000 S.F.	1.000	Ea.	68.376	39,300	4,150	43,450
Gas piping, area 2000 through 4000 S.F.	34.500	L.F.	5.207	166	340	506
Area 5000 to 10000 S.F.	86.250	L.F.	13.019	415	855	1,270
Duct, supply and return, galvanized steel, to 2000 S.F.	38.000	Lb.	3.881	24	230	254
Area to 3000 S.F.	57.000	Lb.	5.821	36	345	381
Area to 5000 S.F.	95.000	Lb.	9.702	60	575	635
Area to 10000 S.F.	190.000	Lb.	19.405	120	1,150	1,270
Rigid fiberglass, area to 2000 S.F.	33.000	S.F.	2.263	32.50	135	167.50
Area to 3000 S.F.	49.000	S.F.	3.360	48.50	200	248.50
Area to 5000 S.F.	82.000	S.F.	5.623	81	335	416
Area to 10000 S.F.	164.000	S.F.	11.245	162	670	832
Insulation, supply and return, blanket type, area to 2000 S.F.	33.000	S.F.	1.508	165	345	510
Area to 3000 S.F.	49.000	S.F.	2.240	245	510	755
Area to 5000 S.F.	82.000	S.F.	3.748	410	855	1,265
Area to 10000 S.F.	164.000	S.F.	7.496	820	1,725	2,545
Lateral ducts, flexible round, 12″ insulated, to 2000 S.F.	72.000	L.F.	11.520	420	660	1,080
Area to 3000 S.F.	108.000	L.F.	17.280	625	995	1,620
Area to 5000 S.F.	180.000	L.F.	28.800	1,050	1,650	2,700
Area to 10000 S.F.	360.000	L.F.	57.600	2,100	3,300	5,400
Rectangular, galvanized steel, to 2000 S.F.	239.000	Lb.	24.409	151	1,450	1,601
Area to 3000 S.F.	360.000	Lb.	36.767	227	2,175	2,402
Area to 5000 S.F.	599.000	Lb.	61.176	375	3,625	4,000
Area to 10000 S.F.	998.000	Lb.	101.926	630	6,050	6,680
Diffusers, ceiling, 1 to 4 way blow, 24″ x 24″, to 2000 S.F.	4.000	Ea.	4.571	1,200	292	1,492
Area to 3000 S.F.	6.000	Ea.	6.857	1,800	440	2,240
Area to 5000 S.F.	10.000	Ea.	11.429	3,000	730	3,730
Area to 10000 S.F.	20.000	Ea.	22.857	5,975	1,450	7,425
Return grilles, 24″ x 24″, to 2000 S.F.	1.000	Ea.	.727	98	46.50	144.50
Area to 3000 S.F.	2.000	Ea.	1.455	196	93	289
Area to 5000 S.F.	3.000	Ea.	2.182	294	140	434
Area to 10000 S.F.	5.000	Ea.	3.636	490	233	723

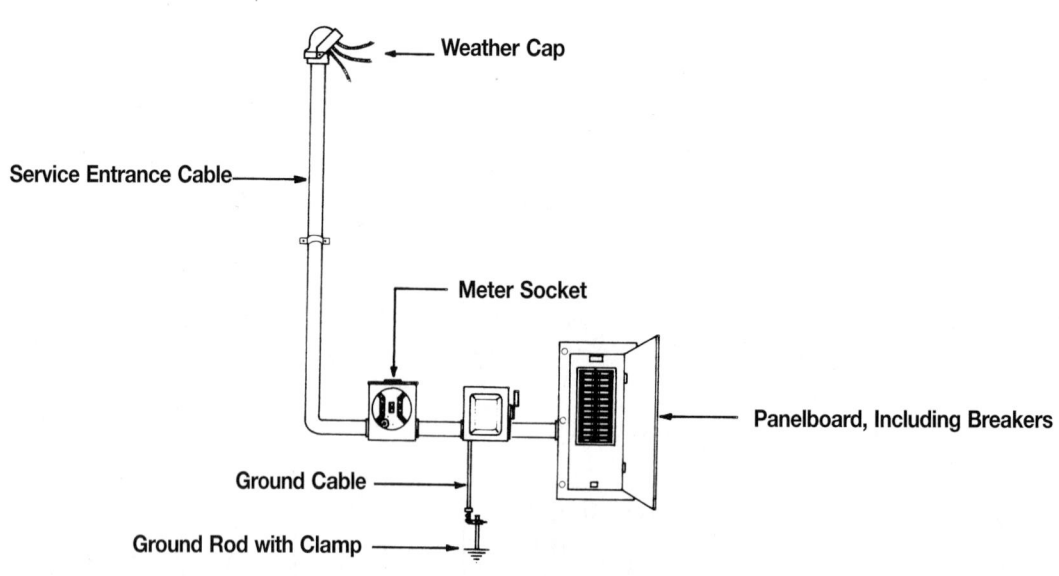

Weather Cap

Service Entrance Cable

Meter Socket

Panelboard, Including Breakers

Ground Cable

Ground Rod with Clamp

System Description	QUAN.	UNIT	LABOR HOURS	COST EACH		
				MAT.	INST.	TOTAL
100 AMP SERVICE						
Weather cap	1.000	Ea.	.667	8.90	44	52.90
Service entrance cable	10.000	L.F.	.762	36.60	50	86.60
Meter socket	1.000	Ea.	2.500	49.50	164	213.50
Ground rod with clamp	1.000	Ea.	1.455	22	95.50	117.50
Ground cable	5.000	L.F.	.250	7.35	16.40	23.75
Panel board, 12 circuit	1.000	Ea.	6.667	159	440	599
TOTAL		Ea.	12.301	283.35	809.90	1,093.25
200 AMP SERVICE						
Weather cap	1.000	Ea.	1.000	19.55	65.50	85.05
Service entrance cable	10.000	L.F.	1.143	32.50	75	107.50
Meter socket	1.000	Ea.	4.211	105	276	381
Ground rod with clamp	1.000	Ea.	1.818	46.50	119	165.50
Ground cable	10.000	L.F.	.500	14.70	32.80	47.50
3/4" EMT	5.000	L.F.	.308	6.20	20.20	26.40
Panel board, 24 circuit	1.000	Ea.	12.308	280	725	1,005
TOTAL		Ea.	21.288	504.45	1,313.50	1,817.95
400 AMP SERVICE						
Weather cap	1.000	Ea.	2.963	166	195	361
Service entrance cable	180.000	L.F.	5.760	464.40	378	842.40
Meter socket	1.000	Ea.	4.211	105	276	381
Ground rod with clamp	1.000	Ea.	2.000	54	131	185
Ground cable	20.000	L.F.	.485	47.60	31.80	79.40
3/4" Greenfield	20.000	L.F.	1.000	12.80	65.60	78.40
Current transformer cabinet	1.000	Ea.	6.154	197	405	602
Panel board, 42 circuit	1.000	Ea.	33.333	5,600	2,200	7,800
TOTAL		Ea.	55.906	6,646.80	3,682.40	10,329.20

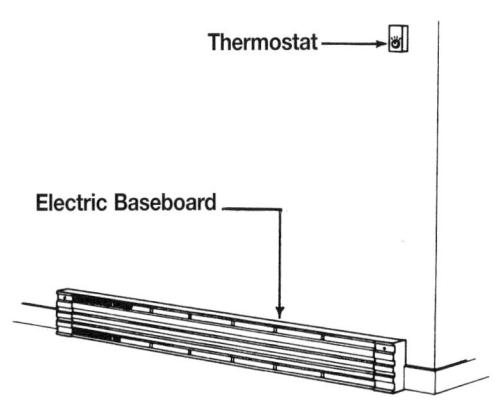

Thermostat

Electric Baseboard

System Description	QUAN.	UNIT	LABOR HOURS	COST EACH		
				MAT.	INST.	TOTAL
4' BASEBOARD HEATER						
Electric baseboard heater, 4' long	1.000	Ea.	1.194	41	78.50	119.50
Thermostat, integral	1.000	Ea.	.500	30	33	63
Romex, 12-3 with ground	40.000	L.F.	1.600	17.60	105.20	122.80
Panel board breaker, 15-50 amp	1.000	Ea.	.300	16.95	19.65	36.60
TOTAL		Ea.	3.594	105.55	236.35	341.90
6' BASEBOARD HEATER						
Electric baseboard heater, 6' long	1.000	Ea.	1.600	56.50	105	161.50
Thermostat, integral	1.000	Ea.	.500	30	33	63
Romex, 12-3 with ground	40.000	L.F.	1.600	17.60	105.20	122.80
Panel board breaker, 15-50 amp	1.000	Ea.	.400	22.60	26.20	48.80
TOTAL		Ea.	4.100	126.70	269.40	396.10
8' BASEBOARD HEATER						
Electric baseboard heater, 8' long	1.000	Ea.	2.000	69	131	200
Thermostat, integral	1.000	Ea.	.500	30	33	63
Romex, 12-3 with ground	40.000	L.F.	1.600	17.60	105.20	122.80
Panel board breaker, 15-50 amp	1.000	Ea.	.500	28.25	32.75	61
TOTAL		Ea.	4.600	144.85	301.95	446.80
10' BASEBOARD HEATER						
Electric baseboard heater, 10' long	1.000	Ea.	2.424	170	159	329
Thermostat, integral	1.000	Ea.	.500	30	33	63
Romex, 12-3 with ground	40.000	L.F.	1.600	17.60	105.20	122.80
Panel board breaker, 15-50 amp	1.000	Ea.	.750	42.38	49.13	91.51
TOTAL		Ea.	5.274	259.98	346.33	606.31

The costs in this system are on a cost each basis and include all necessary conduit fittings.

Description	QUAN.	UNIT	LABOR HOURS	COST EACH		
				MAT.	INST.	TOTAL

For customer support on your Residential Costs with RSMeans data, call 800.448.8182.

279

System Description	QUAN.	UNIT	LABOR HOURS	COST EACH		
				MAT.	INST.	TOTAL
Air conditioning receptacles						
Using non-metallic sheathed cable	1.000	Ea.	.800	19	52.50	71.50
Using BX cable	1.000	Ea.	.964	28.50	63.50	92
Using EMT conduit	1.000	Ea.	1.194	44	78.50	122.50
Disposal wiring						
Using non-metallic sheathed cable	1.000	Ea.	.889	21.50	58.50	80
Using BX cable	1.000	Ea.	1.067	30	70	100
Using EMT conduit	1.000	Ea.	1.333	47.50	87.50	135
Dryer circuit						
Using non-metallic sheathed cable	1.000	Ea.	1.455	30.50	95.50	126
Using BX cable	1.000	Ea.	1.739	40.50	114	154.50
Using EMT conduit	1.000	Ea.	2.162	54	142	196
Duplex receptacles						
Using non-metallic sheathed cable	1.000	Ea.	.615	19	40.50	59.50
Using BX cable	1.000	Ea.	.741	28.50	48.50	77
Using EMT conduit	1.000	Ea.	.920	44	60.50	104.50
Exhaust fan wiring						
Using non-metallic sheathed cable	1.000	Ea.	.800	12.20	52.50	64.70
Using BX cable	1.000	Ea.	.964	21.50	63.50	85
Using EMT conduit	1.000	Ea.	1.194	37	78.50	115.50
Furnace circuit & switch						
Using non-metallic sheathed cable	1.000	Ea.	1.333	27.50	87.50	115
Using BX cable	1.000	Ea.	1.600	38	105	143
Using EMT conduit	1.000	Ea.	2.000	50	131	181
Ground fault						
Using non-metallic sheathed cable	1.000	Ea.	1.000	56.50	65.50	122
Using BX cable	1.000	Ea.	1.212	66	79.50	145.50
Using EMT conduit	1.000	Ea.	1.481	83	97.50	180.50
Heater circuits						
Using non-metallic sheathed cable	1.000	Ea.	1.000	20.50	65.50	86
Using BX cable	1.000	Ea.	1.212	27	79.50	106.50
Using EMT conduit	1.000	Ea.	1.481	40.50	97.50	138
Lighting wiring						
Using non-metallic sheathed cable	1.000	Ea.	.500	23	33	56
Using BX cable	1.000	Ea.	.602	29	39.50	68.50
Using EMT conduit	1.000	Ea.	.748	38.50	49	87.50
Range circuits						
Using non-metallic sheathed cable	1.000	Ea.	2.000	77	131	208
Using BX cable	1.000	Ea.	2.424	114	159	273
Using EMT conduit	1.000	Ea.	2.963	95	195	290
Switches, single pole						
Using non-metallic sheathed cable	1.000	Ea.	.500	12.20	33	45.20
Using BX cable	1.000	Ea.	.602	21.50	39.50	61
Using EMT conduit	1.000	Ea.	.748	37	49	86
Switches, 3-way						
Using non-metallic sheathed cable	1.000	Ea.	.667	16.50	44	60.50
Using BX cable	1.000	Ea.	.800	23.50	52.50	76
Using EMT conduit	1.000	Ea.	1.333	40.50	87.50	128
Water heater						
Using non-metallic sheathed cable	1.000	Ea.	1.600	21	105	126
Using BX cable	1.000	Ea.	1.905	34.50	125	159.50
Using EMT conduit	1.000	Ea.	2.353	42	154	196
Weatherproof receptacle						
Using non-metallic sheathed cable	1.000	Ea.	1.333	140	87.50	227.50
Using BX cable	1.000	Ea.	1.600	147	105	252
Using EMT conduit	1.000	Ea.	2.000	162	131	293

System Description	QUAN.	UNIT	LABOR HOURS	COST EACH		
				MAT.	INST.	TOTAL
Fluorescent strip, 4' long, 1 light, average	1.000	Ea.	.941	34	62	96
Deluxe	1.000	Ea.	1.129	41	74.50	115.50
2 lights, average	1.000	Ea.	1.000	47.50	65.50	113
Deluxe	1.000	Ea.	1.200	57	78.50	135.50
8' long, 1 light, average	1.000	Ea.	1.194	60.50	78.50	139
Deluxe	1.000	Ea.	1.433	72.50	94	166.50
2 lights, average	1.000	Ea.	1.290	73	84.50	157.50
Deluxe	1.000	Ea.	1.548	87.50	101	188.50
Surface mounted, 4' x 1', economy	1.000	Ea.	.914	58	60	118
Average	1.000	Ea.	1.143	72.50	75	147.50
Deluxe	1.000	Ea.	1.371	87	90	177
4' x 2', economy	1.000	Ea.	1.208	75	79	154
Average	1.000	Ea.	1.509	93.50	99	192.50
Deluxe	1.000	Ea.	1.811	112	119	231
Recessed, 4'x 1', 2 lamps, economy	1.000	Ea.	1.123	45	73.50	118.50
Average	1.000	Ea.	1.404	56.50	92	148.50
Deluxe	1.000	Ea.	1.684	68	110	178
4' x 2', 4' lamps, economy	1.000	Ea.	1.362	55	89.50	144.50
Average	1.000	Ea.	1.702	68.50	112	180.50
Deluxe	1.000	Ea.	2.043	82	134	216
Incandescent, exterior, 150W, single spot	1.000	Ea.	.500	36.50	33	69.50
Double spot	1.000	Ea.	1.167	102	77	179
Recessed, 100W, economy	1.000	Ea.	.800	57	52.50	109.50
Average	1.000	Ea.	1.000	71	65.50	136.50
Deluxe	1.000	Ea.	1.200	85	78.50	163.50
150W, economy	1.000	Ea.	.800	92	52.50	144.50
Average	1.000	Ea.	1.000	115	65.50	180.50
Deluxe	1.000	Ea.	1.200	138	78.50	216.50
Surface mounted, 60W, economy	1.000	Ea.	.800	72	52.50	124.50
Average	1.000	Ea.	1.000	80	65.50	145.50
Deluxe	1.000	Ea.	1.194	113	78.50	191.50
Metal halide, recessed 2' x 2' 250W	1.000	Ea.	2.500	281	164	445
2' x 2', 400W	1.000	Ea.	2.759	405	181	586
Surface mounted, 2' x 2', 250W	1.000	Ea.	2.963	380	195	575
2' x 2', 400W	1.000	Ea.	3.333	450	219	669
High bay, single, unit, 400W	1.000	Ea.	3.478	455	228	683
Twin unit, 400W	1.000	Ea.	5.000	910	330	1,240
Low bay, 250W	1.000	Ea.	2.500	400	164	564

Unit Price Section

Table of Contents

Table of Contents (cont.)

RSMeans data: Unit Prices— How They Work

All RSMeans data: Unit Prices are organized in the same way.

03 30 Cast-In-Place Concrete

03 30 53 – Miscellaneous Cast-In-Place Concrete

① 03 30 53.40 Concrete In Place

03 30 53.40 Concrete In Place	Crew	Daily Output	Labor-Hours	Unit	Material	2018 Bare Costs Labor	Equipment	Total	Total Incl O&P
	④	**⑤**	**⑥**	**⑦**	**⑧**		**⑨**	**⑩**	**⑪**
0010 **CONCRETE IN PLACE**									
0020 Including forms (4 uses), Grade 60 rebar, concrete (Portland cement									
0050 Type I), placement and finishing unless otherwise indicated									
0500 Chimney foundations (5000 psi), over 5 C.Y.	C-14C	32.22	3.476	C.Y.	171	116	.80	287.80	380
0510 (3500 psi), under 5 C.Y.	"	23.71	4.724	"	198	157	1.09	356.09	480
3540 Equipment pad (3000 psi), 3' x 3' x 6" thick	C-14H	45	1.067	Ea.	44.50	37	.57	82.07	111
3550 4' x 4' x 6" thick		30	1.600		69	55	.85	124.85	168
3560 5' x 5' x 8" thick		18	2.667		126	92	1.41	219.41	293
3570 6' x 6' x 8" thick		14	3.429		173	118	1.82	292.82	390
3580 8' x 8' x 10" thick		8	6		370	207	3.18	580.18	755
3590 10' x 10' x 12" thick		5	9.600		645	330	5.10	980.10	1,275
3800 Footings (3000 psi), spread under 1 C.Y.	C-14C	28	4	C.Y.	185	133	.92	318.92	425
3825 1 C.Y. to 5 C.Y.		43	2.605		219	86.50	.60	306.10	385
3850 Over 5 C.Y.		75	1.493		203	49.50	.34	252.84	305

② (indicates line 3550) **③** (indicates lines 0510/3540 area)

It is important to understand the structure of RSMeans data: Unit Prices, so that you can find information easily and use it correctly.

① ### Line Numbers

Line Numbers consist of 12 characters, which identify a unique location in the database for each task. The first 6 or 8 digits conform to the Construction Specifications Institute MasterFormat® 2016. The remainder of the digits are a further breakdown in order to arrange items in understandable groups of similar tasks. Line numbers are consistent across all of our publications, so a line number in any of our products will always refer to the same item of work.

② ### Descriptions

Descriptions are shown in a hierarchical structure to make them readable. In order to read a complete description, read up through the indents to the top of the section. Include everything that is above and to the left that is not contradicted by information below. For instance, the complete description for line 03 30 53.40 3550 is "Concrete in place, including forms (4 uses), Grade 60 rebar, concrete (Portland cement Type 1), placement and finishing unless otherwise indicated; Equipment pad (3000 psi), 4' x 4' x 6" thick."

③ ### RSMeans data

When using **RSMeans data**, it is important to read through an entire section to ensure that you use the data that most closely matches your work. Note that sometimes there is additional information shown in the section that may improve your price. There are frequently lines that further describe, add to, or adjust data for specific situations.

④ ### Reference Information

Gordian's RSMeans engineers have created **reference** information to assist you in your estimate. **If** there is information that applies to a section, it will be indicated at the start of the section.

⑤ ### Crews

Crews include labor and/or equipment necessary to accomplish each task. In this case, Crew C-14H is used. Gordian's RSMeans staff selects a crew to represent the workers and equipment that are

typically used for that task. In this case, Crew C-14H consists of one carpenter foreman (outside), two carpenters, one rodman, one laborer, one cement finisher, and one gas engine vibrator. Details of all crews can be found in the Reference Section.

Crews - Residential

Crew C-14H	Bare Costs		Incl. Subs O&P		Cost Per Labor-Hour	
	Hr.	Daily	Hr.	Daily	Bare Costs	Incl. O&P
1 Carpenter Foreman (outside)	$36.95	$295.60	$61.45	$491.60	$34.51	$57.17
2 Carpenters	34.95	559.20	58.10	929.60		
1 Rodman (reinf.)	38.80	310.40	64.70	517.60		
1 Laborer	26.70	213.60	44.40	355.20		
1 Cement Finisher	34.70	277.60	56.30	450.40		
1 Gas Engine Vibrator		25.60		28.16	0.53	0.59
48 L.H., Daily Totals		$1682.00		$2772.56	$35.04	$57.76

6 ## Daily Output

The **Daily Output** is the amount of work that the crew can do in a normal 8-hour workday, including mobilization, layout, movement of materials, and cleanup. In this case, crew C-14H can install thirty 4' x 4' x 6" thick concrete pads in a day. Daily output is variable and based on many factors, including the size of the job, location, and environmental conditions. RSMeans data represents work done in daylight (or adequate lighting) and temperate conditions.

7 ## Labor-Hours

The figure in the **Labor-Hours** column is the amount of labor required to perform one unit of work—in this case the amount of labor required to construct one 4' x 4' equipment pad. This figure is calculated by dividing the number of hours of labor in the crew by the daily output (48 labor-hours divided by 30 pads = 1.6 hours of labor per pad). Multiply 1.6 times 60 to see the value in minutes: 60 x 1.6 = 96

minutes. Note: the labor-hour figure is not dependent on the crew size. A change in crew size will result in a corresponding change in daily output, but the labor-hours per unit of work will not change.

8 ## Unit of Measure

All RSMeans data: Unit Prices include the typical **Unit of Measure** used for estimating that item. For concrete-in-place the typical unit is cubic yards (C.Y.) or each (Ea.). For installing broadloom carpet it is square yard, and for gypsum board it is square foot. The estimator needs to take special care that the unit in the data matches the unit in the take-off. Unit conversions may be found in the Reference Section.

9 ## Bare Costs

Bare Costs are the costs of materials, labor, and equipment that the installing contractor pays. They represent the cost, in U.S. dollars, for one unit of work. They do not include any markups for profit or labor burden.

10 ## Bare Total

The **Total column** represents the total bare cost for the installing contractor in U.S. dollars. In this case, the sum of $69 for material + $55.00 for labor + $.85 for equipment is $124.85.

11 ## Total Incl O&P

The **Total Incl O&P column** is the total cost, including overhead and profit, that the installing contractor will charge the customer. This represents the cost of materials plus 10% profit, the cost of labor plus labor burden and 10% profit, and the cost of equipment plus 10% profit. It does not include the general contractor's overhead and profit. Note: See the inside back cover of the printed product or the Reference Section of the electronic product for details of how the labor burden is calculated.

National Average

*The RSMeans data in our print publications represents a "national average" cost. This data should be modified to the project location using the **City Cost Indexes** or **Location Factors** tables found in the Reference Section. Use the Location Factors to adjust estimate totals if the project covers multiple trades. Use the City Cost Indexes (CCI) for single trade projects or projects where a more detailed analysis is required. All figures in the two tables are derived from the same research. The last row of data in the CCI—the weighted average—is the same as the numbers reported for each location in the location factor table.*

RSMeans data: Unit Prices— How They Work (Continued)

Project Name: Pre-Engineered Steel Building			**Architect: As Shown**					
Location:	**Anywhere, USA**						**01/01/18**	**RESI**
Line Number	**Description**	**Qty**	**Unit**	**Material**	**Labor**	**Equipment**	**SubContract**	**Estimate Total**
03 30 53.40 3940	Strip footing, 12" x 24", reinforced	34	C.Y.	$5,406.00	$2,635.00	$18.36	$0.00	
03 30 53.40 3950	Strip footing, 12" x 36", reinforced	15	C.Y.	$2,295.00	$930.00	$6.45	$0.00	
03 11 13.65 3000	Concrete slab edge forms	500	L.F.	$145.00	$820.00	$0.00	$0.00	
03 22 11.10 0200	Welded wire fabric reinforcing	150	C.S.F.	$2,940.00	$3,000.00	$0.00	$0.00	
03 31 13.35 0300	Ready mix concrete, 4000 psi for slab on grade	278	C.Y.	$35,584.00	$0.00	$0.00	$0.00	
03 31 13.70 4300	Place, strike off & consolidate concrete slab	278	C.Y.	$0.00	$3,447.20	$130.66	$0.00	
03 35 13.30 0250	Machine float & trowel concrete slab	15,000	S.F.	$0.00	$6,750.00	$300.00	$0.00	
03 15 16.20 0140	Cut control joints in concrete slab	950	L.F.	$47.50	$294.50	$57.00	$0.00	
03 39 23.13 0300	Sprayed concrete curing membrane	150	C.S.F.	$1,815.00	$675.00	$0.00	$0.00	
Division 03	**Subtotal**			**$48,232.50**	**$18,551.70**	**$512.47**	**$0.00**	**$67,296.67**
08 36 13.10 2650	Manual 10' x 10' steel sectional overhead door	8	Ea.	$10,400.00	$2,480.00	$0.00	$0.00	
08 36 13.10 2860	Insulation and steel back panel for OH door	800	S.F.	$4,000.00	$0.00	$0.00	$0.00	
Division 08	**Subtotal**			**$14,400.00**	**$2,480.00**	**$0.00**	**$0.00**	**$16,880.00**
13 34 19.50 1100	Pre-Engineered Steel Building, 100' x 150' x 24'	15,000	SF Flr.	$0.00	$0.00	$0.00	$315,000.00	
13 34 19.50 6050	Framing for PESB door opening, 3' x 7'	4	Opng.	$0.00	$0.00	$0.00	$1,920.00	
13 34 19.50 6100	Framing for PESB door opening, 10' x 10'	8	Opng.	$0.00	$0.00	$0.00	$8,400.00	
13 34 19.50 6200	Framing for PESB window opening, 4' x 3'	6	Opng.	$0.00	$0.00	$0.00	$2,910.00	
13 34 19.50 5750	PESB door, 3' x 7', single leaf	4	Opng.	$2,620.00	$496.00	$0.00	$0.00	
13 34 19.50 7750	PESB sliding window, 4' x 3' with screen	6	Opng.	$2,550.00	$288.00	$45.30	$0.00	
13 34 19.50 6550	PESB gutter, eave type, 26 ga., painted	300	L.F.	$2,220.00	$582.00	$0.00	$0.00	
13 34 19.50 8650	PESB roof vent, 12" wide x 10' long	15	Ea.	$555.00	$2,325.00	$0.00	$0.00	
13 34 19.50 6900	PESB insulation, vinyl faced, 4" thick	27,400	S.F.	$13,152.00	$6,576.00	$0.00	$0.00	
Division 13	**Subtotal**			**$21,097.00**	**$10,267.00**	**$45.30**	**$328,230.00**	**$359,639.30**
			Subtotal	**$83,729.50**	**$31,298.70**	**$557.77**	**$328,230.00**	**$443,815.97**
Division 01	**General Requirements @ 7%**			5,861.07	2,190.91	39.04	22,976.10	
			Estimate Subtotal	$89,590.57	$33,489.61	$596.81	$351,206.10	$443,815.97
			Sales Tax @ 5%	4,479.53		29.84	8,780.15	
			Subtotal A	94,070.09	33,489.61	626.65	359,986.25	
			GC O & P	9,407.01	22,471.53	62.67	35,998.63	
			Subtotal B	103,477.10	55,961.14	689.32	395,984.88	$556,112.44
			Contingency @ 5%					27,805.62
			Subtotal C					$583,918.06
			Bond @ $12/1000 +10% O&P					7,707.72
			Subtotal D					$591,625.78
			Location Adjustment Factor		102.30			13,607.39
			Grand Total					$605,233.17

This estimate is based on an interactive spreadsheet. You are free to download it and adjust it to your methodology.
A copy of this spreadsheet is available at **www.RSMeans.com/2018books.**

Sample Estimate

This sample demonstrates the elements of an estimate, including a tally of the RSMeans data lines and a summary of the markups on a contractor's work to arrive at a total cost to the owner. The Location Factor with RSMeans data is added at the bottom of the estimate to adjust the cost of the work to a specific location.

---

1 Work Performed

The body of the estimate shows the RSMeans data selected, including the line number, a brief description of each item, its take-off unit and quantity, and the bare costs of materials, labor, and equipment. This estimate also includes a column titled "SubContract." This data is taken from the column "Total Incl O&P" and represents the total that a subcontractor would charge a general contractor for the work, including the sub's markup for overhead and profit.

2 Division 1, General Requirements

This is the first division numerically but the last division estimated. Division 1 includes project-wide needs provided by the general contractor. These requirements vary by project but may include temporary facilities and utilities, security, testing, project cleanup, etc. For small projects a percentage can be used—typically between 5% and 15% of project cost. For large projects the costs may be itemized and priced individually.

3 Sales Tax

If the work is subject to state or local sales taxes, the amount must be added to the estimate. Sales tax may be added to material costs, equipment costs, and subcontracted work. In this case, sales tax was added in all three categories. It was assumed that approximately half the subcontracted work would be material cost, so the tax was applied to 50% of the subcontract total.

4 GC O&P

This entry represents the general contractor's markup on material, labor, equipment, and subcontractor costs. Our standard markup on materials, equipment, and subcontracted work is 10%. In this estimate, the markup on the labor performed by the GC's workers uses "Skilled Workers Average" shown in Column F on the table "Installing Contractor's Overhead & Profit," which can be found on the inside back cover of the printed product or in the Reference Section of the electronic product.

5 Contingency

A factor for contingency may be added to any estimate to represent the cost of unknowns that may occur between the time that the estimate is performed and the time the project is constructed. The amount of the allowance will depend on the stage of design at which the estimate is done and the contractor's assessment of the risk involved. Refer to section 01 21 16.50 for contingency allowances.

6 Bonds

Bond costs should be added to the estimate. The figures here represent a typical performance bond, ensuring the owner that if the general contractor does not complete the obligations in the construction contract the bonding company will pay the cost for completion of the work.

7 Location Adjustment

Published prices are based on national average costs. If necessary, adjust the total cost of the project using a location factor from the "Location Factor" table or the "City Cost Index" table. Use location factors if the work is general, covering multiple trades. If the work is by a single trade (e.g., masonry) use the more specific data found in the "City Cost Indexes."

Estimating Tips

01 20 00 Price and Payment Procedures

- Allowances that should be added to estimates to cover contingencies and job conditions that are not included in the national average material and labor costs are shown in Section 01 21.

- When estimating historic preservation projects (depending on the condition of the existing structure and the owner's requirements), a 15%–20% contingency or allowance is recommended, regardless of the stage of the drawings.

01 30 00 Administrative Requirements

- Before determining a final cost estimate, it is a good practice to review all the items listed in Subdivisions 01 31 and 01 32 to make final adjustments for items that may need customizing to specific job conditions.

- Requirements for initial and periodic submittals can represent a significant cost to the General Requirements of a job. Thoroughly check the submittal specifications when estimating a project to determine any costs that should be included.

01 40 00 Quality Requirements

- All projects will require some degree of quality control. This cost is not included in the unit cost of construction listed in each division. Depending upon the terms of the contract, the various costs of inspection and testing can be the responsibility of either the owner or the contractor. Be sure to include the required costs in your estimate.

01 50 00 Temporary Facilities and Controls

- Barricades, access roads, safety nets, scaffolding, security, and many more requirements for the execution of a safe project are elements of direct cost. These costs can easily be overlooked when preparing an estimate. When looking through the major classifications of this subdivision, determine which items apply to each division in your estimate.

- Construction equipment rental costs can be found in the Reference Section in Section 01 54 33. Operators' wages are not included in equipment rental costs.

- Equipment mobilization and demobilization costs are not included in equipment rental costs and must be considered separately.

- The cost of small tools provided by the installing contractor for his workers is covered in the "Overhead" column on the "Installing Contractor's Overhead and Profit" table that lists labor trades, base rates, and markups and, therefore, is included in the "Total Incl. O&P" cost of any unit price line item.

01 70 00 Execution and Closeout Requirements

- When preparing an estimate, thoroughly read the specifications to determine the requirements for Contract Closeout. Final cleaning, record documentation, operation and maintenance data, warranties and bonds, and spare parts and maintenance materials can all be elements of cost for the completion of a contract. Do not overlook these in your estimate.

Reference Numbers

Reference numbers are shown at the beginning of some major classifications. These numbers refer to related items in the Reference Section. The reference information may be an estimating procedure, an alternate pricing method, or technical information.

Note: Not all subdivisions listed here necessarily appear. ■

Did you know?

RSMeans data is available through our online application with 24/7 access:

- Search for unit prices by keyword
- Leverage the most up-to-date data
- Build and export estimates

Try it free for 30 days!
www.rsmeans.com/2018freetrial

01 11 Summary of Work

01 11 31 – Professional Consultants

01 11 31.10 Architectural Fees		Crew	Daily Output	Labor-Hours	Unit	Material	2018 Bare Costs Labor	Equipment	Total	Total Incl O&P
0010	**ARCHITECTURAL FEES** R011110-10									
0020	For new construction									
0060	Minimum				Project				4.90%	4.90%
0090	Maximum								16%	16%
0100	For alteration work, to $500,000, add to new construction fee								50%	50%
0150	Over $500,000, add to new construction fee								25%	25%

01 11 31.20 Construction Management Fees										
0010	**CONSTRUCTION MANAGEMENT FEES**									
0060	For work to $100,000				Project				10%	10%
0070	To $250,000								9%	9%
0090	To $1,000,000								6%	6%

01 11 31.75 Renderings										
0010	**RENDERINGS** Color, matted, 20" x 30", eye level,									
0050	Average				Ea.	3,125			3,125	3,450

01 21 Allowances

01 21 16 – Contingency Allowances

01 21 16.50 Contingencies

		Crew	Daily Output	Labor-Hours	Unit	Material	2018 Bare Costs Labor	Equipment	Total	Total Incl O&P
0010	**CONTINGENCIES**, Add to estimate									
0020	Conceptual stage				Project				20%	20%
0150	Final working drawing stage				"				3%	3%

01 21 55 – Job Conditions Allowance

01 21 55.50 Job Conditions

		Crew	Daily Output	Labor-Hours	Unit	Material	Labor	Equipment	Total	Total Incl O&P
0010	**JOB CONDITIONS** Modifications to applicable									
8000	Remove and reset contents of small room	1 Clab	6.50	1.231	Room		33		33	54.50
8010	Average room		4.70	1.702			45.50		45.50	75.50
8020	Large room		3.50	2.286			61		61	101
8030	Extra large room		2.40	3.333			89		89	148

01 21 63 – Taxes

01 21 63.10 Taxes

		Crew	Daily Output	Labor-Hours	Unit	Material	Labor	Equipment	Total	Total Incl O&P
0010	**TAXES** R012909-80									
0020	Sales tax, State, average				%	5.08%				
0050	Maximum R012909-85					7.50%				
0200	Social Security, on first $118,500 of wages						7.65%			
0300	Unemployment, combined Federal and State, minimum R012909-86						.60%			
0350	Average						9.60%			
0400	Maximum						12%			

292

For customer support on your Residential Costs with RSMeans data, call 800.448.8182.

01 31 Project Management and Coordination

01 31 13 – Project Coordination

01 31 13.30 Insurance

		Crew	Daily Output	Labor-Hours	Unit	Material	2018 Bare Costs Labor	Equipment	Total	Total Incl O&P
0010	**INSURANCE** R013113-40									
0020	Builders risk, standard, minimum				Job				.24%	.24%
0050	Maximum R013113-50								.64%	.64%
0200	All-risk type, minimum								.25%	.25%
0250	Maximum R013113-60				▼				.62%	.62%
0400	Contractor's equipment floater, minimum				Value				.50%	.50%
0450	Maximum				"				1.50%	1.50%
0600	Public liability, average				Job				2.02%	2.02%
0800	Workers' compensation & employer's liability, average									
0850	by trade, carpentry, general				Payroll		13.05%			
0900	Clerical						.46%			
0950	Concrete						12.44%			
1000	Electrical						5.52%			
1050	Excavation						9.03%			
1100	Glazing						12.91%			
1150	Insulation						11.13%			
1200	Lathing						8.67%			
1250	Masonry						13.73%			
1300	Painting & decorating						11.68%			
1350	Pile driving						14.53%			
1400	Plastering						10.73%			
1450	Plumbing						6.94%			
1500	Roofing						30.73%			
1550	Sheet metal work (HVAC)						8.82%			
1600	Steel erection, structural						27.45%			
1650	Tile work, interior ceramic						8.85%			
1700	Waterproofing, brush or hand caulking						7.01%			
1800	Wrecking						20.85%			
2000	Range of 35 trades in 50 states, excl. wrecking & clerical, min.						1.41%			
2100	Average						12.25%			
2200	Maximum				▼		108.79%			

01 41 Regulatory Requirements

01 41 26 – Permit Requirements

01 41 26.50 Permits

		Crew	Daily Output	Labor-Hours	Unit	Material	2018 Bare Costs Labor	Equipment	Total	Total Incl O&P
0010	**PERMITS**									
0020	Rule of thumb, most cities, minimum				Job				.50%	.50%
0100	Maximum				"				2%	2%

01 54 Construction Aids

01 54 16 – Temporary Hoists

01 54 16.50 Weekly Forklift Crew

		Crew	Daily Output	Labor-Hours	Unit	Material	2018 Bare Costs Labor	Equipment	Total	Total Incl O&P
0010	**WEEKLY FORKLIFT CREW**									
0100	All-terrain forklift, 45' lift, 35' reach, 9000 lb. capacity	A-3P	.20	40	Week		1,425	2,425	3,850	5,000

For customer support on your Residential Costs with RSMeans data, call 800.448.8182.

293

01 54 Construction Aids

01 54 19 – Temporary Cranes

01 54 19.50 Daily Crane Crews

		Crew	Daily Output	Labor-Hours	Unit	Material	2018 Bare Costs Labor	Equipment	Total	Total Incl O&P
0010	**DAILY CRANE CREWS** for small jobs, portal to portal									
0100	12-ton truck-mounted hydraulic crane	A-3H	1	8	Day		296	630	926	1,175
0900	If crane is needed on a Saturday, Sunday or Holiday									
0910	At time-and-a-half, add				Day		50%			
0920	At double time, add				"		100%			

01 54 23 – Temporary Scaffolding and Platforms

01 54 23.60 Pump Staging

		Crew	Daily Output	Labor-Hours	Unit	Material	2018 Bare Costs Labor	Equipment	Total	Total Incl O&P
0010	**PUMP STAGING**, Aluminum									
1300	System in place, 50' working height, per use based on 50 uses	2 Carp	84.80	.189	C.S.F.	7.05	6.60		13.65	18.70
1400	100 uses R015423-20		84.80	.189		3.53	6.60		10.13	14.85
1500	150 uses		84.80	.189		2.36	6.60		8.96	13.55

01 54 23.70 Scaffolding

		Crew	Daily Output	Labor-Hours	Unit	Material	2018 Bare Costs Labor	Equipment	Total	Total Incl O&P
0010	**SCAFFOLDING** R015423-10									
0015	Steel tube, regular, no plank, labor only to erect & dismantle									
0091	Building exterior, wall face, 1 to 5 stories, 6'-4" x 5' frames	3 Clab	8	3	C.S.F.		80		80	133
0201	6 to 12 stories	4 Clab	8	4			107		107	178
0310	13 to 20 stories	5 Carp	8	5			175		175	291
0461	Building interior, wall face area, up to 16' high	3 Clab	12	2			53.50		53.50	89
0561	16' to 40' high		10	2.400			64		64	107
0801	Building interior floor area, up to 30' high		150	.160	C.C.F.		4.27		4.27	7.10
0901	Over 30' high	4 Clab	160	.200	"		5.35		5.35	8.90
0906	Complete system for face of walls, no plank, material only rent/mo				C.S.F.	33			33	36
0908	Interior spaces, no plank, material only rent/mo				C.C.F.	3.77			3.77	4.14
0910	Steel tubular, heavy duty shoring, buy									
0920	Frames 5' high 2' wide				Ea.	99.50			99.50	109
0925	5' high 4' wide					114			114	125
0930	6' high 2' wide					115			115	127
0935	6' high 4' wide					126			126	139
0940	Accessories									
0945	Cross braces				Ea.	20			20	22
0950	U-head, 8" x 8"					22			22	24
0955	J-head, 4" x 8"					16			16	17.60
0960	Base plate, 8" x 8"					17.70			17.70	19.45
0965	Leveling jack					39			39	42.50
1000	Steel tubular, regular, buy									
1100	Frames 3' high 5' wide				Ea.	91			91	100
1150	5' high 5' wide					107			107	117
1200	6'-4" high 5' wide					99.50			99.50	109
1350	7'-6" high 6' wide					170			170	187
1500	Accessories, cross braces					16			16	17.60
1550	Guardrail post					20			20	22
1600	Guardrail 7' section					8.05			8.05	8.85
1650	Screw jacks & plates					25.50			25.50	28
1700	Sidearm brackets					22.50			22.50	24.50
1750	8" casters					37			37	40.50
1800	Plank 2" x 10" x 16'-0"					66			66	72.50
1900	Stairway section					283			283	310
1910	Stairway starter bar					32			32	35
1920	Stairway inside handrail					53			53	58
1930	Stairway outside handrail					84			84	92.50
1940	Walk-thru frame guardrail					41.50			41.50	46

01 54 Construction Aids

01 54 23 – Temporary Scaffolding and Platforms

01 54 23.70 Scaffolding

		Crew	Daily Output	Labor-Hours	Unit	Material	2018 Bare Costs Labor	Equipment	Total	Total Incl O&P
2000	Steel tubular, regular, rent/mo.									
2100	Frames 3' high 5' wide				Ea.	4.38			4.38	4.82
2150	5' high 5' wide					4.38			4.38	4.82
2200	6'-4" high 5' wide					5.25			5.25	5.80
2250	7'-6" high 6' wide					9.75			9.75	10.75
2500	Accessories, cross braces					.88			.88	.97
2550	Guardrail post					.88			.88	.97
2600	Guardrail 7' section					.88			.88	.97
2650	Screw jacks & plates					1.75			1.75	1.93
2700	Sidearm brackets					1.75			1.75	1.93
2750	8" casters					7			7	7.70
2800	Outrigger for rolling tower					2.63			2.63	2.89
2850	Plank 2" x 10" x 16'-0"					9.75			9.75	10.75
2900	Stairway section					32.50			32.50	36
2940	Walk-thru frame guardrail				↓	2.19			2.19	2.41
3000	Steel tubular, heavy duty shoring, rent/mo.									
3250	5' high 2' & 4' wide				Ea.	8.30			8.30	9.15
3300	6' high 2' & 4' wide					8.30			8.30	9.15
3500	Accessories, cross braces					.88			.88	.97
3600	U-head, 8" x 8"					2.44			2.44	2.68
3650	J-head, 4" x 8"					2.44			2.44	2.68
3700	Base plate, 8" x 8"					.88			.88	.97
3750	Leveling jack					2.44			2.44	2.68
5700	Planks, 2" x 10" x 16'-0", labor only to erect & remove to 50' H	3 Carp	72	.333			11.65		11.65	19.35
5800	Over 50' high	4 Carp	80	.400	↓		14		14	23

01 54 23.80 Staging Aids

		Crew	Daily Output	Labor-Hours	Unit	Material	2018 Bare Costs Labor	Equipment	Total	Total Incl O&P
0010	**STAGING AIDS** and fall protection equipment									
0100	Sidewall staging bracket, tubular, buy				Ea.	57.50			57.50	63.50
0110	Cost each per day, based on 250 days use				Day	.23			.23	.25
0200	Guard post, buy				Ea.	54.50			54.50	60
0210	Cost each per day, based on 250 days use				Day	.22			.22	.24
0300	End guard chains, buy per pair				Pair	42			42	46
0310	Cost per set per day, based on 250 days use				Day	.23			.23	.25
1010	Cost each per day, based on 250 days use				"	.04			.04	.05
1100	Wood bracket, buy				Ea.	24.50			24.50	27
1110	Cost each per day, based on 250 days use				Day	.10			.10	.11
2010	Cost per pair per day, based on 250 days use					.51			.51	.56
3010	Cost each per day, based on 250 days use				↓	.23			.23	.25
3100	Aluminum scaffolding plank, 20" wide x 24' long, buy				Ea.	815			815	900
3110	Cost each per day, based on 250 days use				Day	3.27			3.27	3.59
4010	Cost each per day, based on 250 days use				"	.66			.66	.73
4100	Rope for safety line, 5/8" x 100' nylon, buy				Ea.	56			56	61.50
4110	Cost each per day, based on 250 days use				Day	.22			.22	.25
4200	Permanent U-Bolt roof anchor, buy				Ea.	30			30	33
4300	Temporary (one use) roof ridge anchor, buy				"	6.40			6.40	7
5000	Installation (setup and removal) of staging aids									
5010	Sidewall staging bracket	2 Carp	64	.250	Ea.		8.75		8.75	14.55
5020	Guard post with 2 wood rails	"	64	.250			8.75		8.75	14.55
5030	End guard chains, set	1 Carp	64	.125			4.37		4.37	7.25
5100	Roof shingling bracket		96	.083			2.91		2.91	4.84
5200	Ladder jack		64	.125			4.37		4.37	7.25
5300	Wood plank, 2" x 10" x 16'	2 Carp	80	.200	↓		7		7	11.60

For customer support on your Residential Costs with RSMeans data, call 800.448.8182.

295

01 54 Construction Aids

01 54 23 – Temporary Scaffolding and Platforms

01 54 23.80 Staging Aids	Crew	Daily Output	Labor-Hours	Unit	Material	2018 Bare Costs Labor	2018 Bare Costs Equipment	Total	Total Incl O&P
5310	Aluminum scaffold plank, 20" x 24'	2 Carp	40	.400	Ea.		14		14	23
5410	Safety rope	1 Carp	40	.200			7		7	11.60
5420	Permanent U-Bolt roof anchor (install only)	2 Carp	40	.400			14		14	23
5430	Temporary roof ridge anchor (install only)	1 Carp	64	.125	↓		4.37		4.37	7.25

01 54 36 – Equipment Mobilization

01 54 36.50 Mobilization

	01 54 36.50 Mobilization	Crew	Daily Output	Labor-Hours	Unit	Material	2018 Bare Costs Labor	2018 Bare Costs Equipment	Total	Total Incl O&P
0010	**MOBILIZATION** (Use line item again for demobilization) R015436-50									
0015	Up to 25 mi. haul dist. (50 mi. RT for mob/demob crew)									
1200	Small equipment, placed in rear of, or towed by pickup truck	A-3A	4	2	Ea.		71	31.50	102.50	152
1300	Equipment hauled on 3-ton capacity towed trailer	A-3Q	2.67	3			106	57	163	238
1400	20-ton capacity	B-34U	2	8			270	212	482	680
1500	40-ton capacity	B-34N	2	8			275	330	605	815
1700	Crane, truck-mounted, up to 75 ton (driver only)	1 Eqhv	4	2	↓		74		74	122
2500	For each additional 5 miles haul distance, add						10%	10%		
3000	For large pieces of equipment, allow for assembly/knockdown									
3100	For mob/demob of micro-tunneling equip, see Section 33 05 23.19									

01 56 Temporary Barriers and Enclosures

01 56 13 – Temporary Air Barriers

01 56 13.60 Tarpaulins

	01 56 13.60 Tarpaulins	Crew	Daily Output	Labor-Hours	Unit	Material	2018 Bare Costs Labor	2018 Bare Costs Equipment	Total	Total Incl O&P
0010	**TARPAULINS**									
0020	Cotton duck, 10-13.13 oz./S.Y., 6' x 8'				S.F.	.85			.85	.94
0050	30' x 30'					.59			.59	.65
0200	Reinforced polyethylene 3 mils thick, white					.04			.04	.04
0300	4 mils thick, white, clear or black					.13			.13	.14
0730	Polyester reinforced w/integral fastening system, 11 mils thick				↓	.19			.19	.21

01 56 16 – Temporary Dust Barriers

01 56 16.10 Dust Barriers, Temporary

	01 56 16.10 Dust Barriers, Temporary	Crew	Daily Output	Labor-Hours	Unit	Material	2018 Bare Costs Labor	2018 Bare Costs Equipment	Total	Total Incl O&P
0010	**DUST BARRIERS, TEMPORARY**									
0020	Spring loaded telescoping pole & head, to 12', erect and dismantle	1 Clab	240	.033	Ea.		.89		.89	1.48
0025	Cost per day (based upon 250 days)				Day	.29			.29	.31
0030	To 21', erect and dismantle	1 Clab	240	.033	Ea.		.89		.89	1.48
0035	Cost per day (based upon 250 days)				Day	.58			.58	.63
0040	Accessories, caution tape reel, erect and dismantle	1 Clab	480	.017	Ea.		.45		.45	.74
0045	Cost per day (based upon 250 days)				Day	.34			.34	.37
0060	Foam rail and connector, erect and dismantle	1 Clab	240	.033	Ea.		.89		.89	1.48
0065	Cost per day (based upon 250 days)				Day	.12			.12	.13
0070	Caution tape	1 Clab	384	.021	C.L.F.	3.13	.56		3.69	4.37
0080	Zipper, standard duty		60	.133	Ea.	7.30	3.56		10.86	13.90
0090	Heavy duty		48	.167	"	9.75	4.45		14.20	18.15
0100	Polyethylene sheet, 4 mil		37	.216	Sq.	2.58	5.75		8.33	12.45
0110	6 mil	↓	37	.216	"	3.73	5.75		9.48	13.70
1000	Dust partition, 6 mil polyethylene, 1" x 3" frame	2 Carp	2000	.008	S.F.	.32	.28		.60	.81
1080	2" x 4" frame	"	2000	.008	"	.35	.28		.63	.84

01 66 Product Storage and Handling Requirements

01 66 19 – Material Handling

01 66 19.10 Material Handling	Crew	Daily Output	Labor-Hours	Unit	Material	2018 Bare Costs Labor	Equipment	Total	Total Incl O&P
0010 **MATERIAL HANDLING**									
0020 Above 2nd story, via stairs, per C.Y. of material per floor	2 Clab	145	.110	C.Y.		2.95		2.95	4.90
0030 Via elevator, per C.Y. of material		240	.067			1.78		1.78	2.96
0050 Distances greater than 200', per C.Y. of material per each addl 200'		300	.053			1.42		1.42	2.37

01 71 Examination and Preparation

01 71 23 – Field Engineering

01 71 23.13 Construction Layout

	Crew	Daily Output	Labor-Hours	Unit	Material	2018 Bare Costs Labor	Equipment	Total	Total Incl O&P
0010 **CONSTRUCTION LAYOUT**									
1100 Crew for layout of building, trenching or pipe laying, 2 person crew	A-6	1	16	Day		560	50.50	610.50	985
1200 3 person crew	A-7	1	24	"		890	50.50	940.50	1,525

01 74 Cleaning and Waste Management

01 74 13 – Progress Cleaning

01 74 13.20 Cleaning Up

	Crew	Daily Output	Labor-Hours	Unit	Material	2018 Bare Costs Labor	Equipment	Total	Total Incl O&P
0010 **CLEANING UP**									
0020 After job completion, allow, minimum				Job				.30%	.30%
0040 Maximum				"				1%	1%

01 76 Protecting Installed Construction

01 76 13 – Temporary Protection of Installed Construction

01 76 13.20 Temporary Protection

	Crew	Daily Output	Labor-Hours	Unit	Material	2018 Bare Costs Labor	Equipment	Total	Total Incl O&P
0010 **TEMPORARY PROTECTION**									
0020 Flooring, 1/8" tempered hardboard, taped seams	2 Carp	1500	.011	S.F.	.45	.37		.82	1.12
0030 Peel away carpet protection	1 Clab	3200	.003	"	.13	.07		.20	.25

Division Notes

		CREW	DAILY OUTPUT	LABOR-HOURS	UNIT	BARE COSTS				TOTAL INCL O&P
						MAT.	LABOR	EQUIP.	TOTAL	

Estimating Tips
02 30 00 Subsurface Investigation
In preparing estimates on structures involving earthwork or foundations, all information concerning soil characteristics should be obtained. Look particularly for hazardous waste, evidence of prior dumping of debris, and previous stream beds.

02 40 00 Demolition and Structure Moving
The costs shown for selective demolition do not include rubbish handling or disposal. These items should be estimated separately using RSMeans data or other sources.

- Historic preservation often requires that the contractor remove materials from the existing structure, rehab them, and replace them. The estimator must be aware of any related measures and precautions that must be taken when doing selective demolition and cutting and patching. Requirements may include special handling and storage, as well as security.

- In addition to Subdivision 02 41 00, you can find selective demolition items in each division. Example: Roofing demolition is in Division 7.
- Absent of any other specific reference, an approximate demolish-in-place cost can be obtained by halving the new-install labor cost. To remove for reuse, allow the entire new-install labor figure.

02 40 00 Building Deconstruction
This section provides costs for the careful dismantling and recycling of most low-rise building materials.

02 50 00 Containment of Hazardous Waste
This section addresses on-site hazardous waste disposal costs.

02 80 00 Hazardous Material Disposal/Remediation
This subdivision includes information on hazardous waste handling, asbestos remediation, lead remediation, and mold remediation. See reference numbers

RO28213-20 and RO28319-60 for further guidance in using these unit price lines.

02 90 00 Monitoring Chemical Sampling, Testing Analysis
This section provides costs for on-site sampling and testing hazardous waste.

Reference Numbers
Reference numbers are shown at the beginning of some major classifications. These numbers refer to related items in the Reference Section. The reference information may be an estimating procedure, an alternate pricing method, or technical information.

Note: Not all subdivisions listed here necessarily appear. ■

Did you know?
RSMeans data is available through our online application with 24/7 access:
- Search for unit prices by keyword
- Leverage the most up-to-date data
- Build and export estimates

Try it free for 30 days!
www.rsmeans.com/2018freetrial

02 21 Surveys

02 21 13 – Site Surveys

02 21 13.09 Topographical Surveys

02 21 13.09 Topographical Surveys	Crew	Daily Output	Labor-Hours	Unit	Material	2018 Bare Costs Labor	2018 Bare Costs Equipment	Total	Total Incl O&P
0010 **TOPOGRAPHICAL SURVEYS**									
0020 Topographical surveying, conventional, minimum	A-7	3.30	7.273	Acre	21.50	269	15.25	305.75	485
0100 Maximum	A-8	.60	53.333	"	58.50	1,925	84.50	2,068	3,350

02 21 13.13 Boundary and Survey Markers

	Crew	Daily Output	Labor-Hours	Unit	Material	Labor	Equipment	Total	Total Incl O&P
0010 **BOUNDARY AND SURVEY MARKERS**									
0300 Lot location and lines, large quantities, minimum	A-7	2	12	Acre	34.50	445	25	504.50	805
0320 Average	"	1.25	19.200		58.50	710	40.50	809	1,275
0400 Small quantities, maximum	A-8	1	32	↓	73	1,150	50.50	1,273.50	2,050
0600 Monuments, 3' long	A-7	10	2.400	Ea.	31	89	5.05	125.05	188
0800 Property lines, perimeter, cleared land	"	1000	.024	L.F.	.07	.89	.05	1.01	1.62
0900 Wooded land	A-8	875	.037	"	.09	1.33	.06	1.48	2.36

02 32 Geotechnical Investigations

02 32 13 – Subsurface Drilling and Sampling

02 32 13.10 Boring and Exploratory Drilling

	Crew	Daily Output	Labor-Hours	Unit	Material	Labor	Equipment	Total	Total Incl O&P
0010 **BORING AND EXPLORATORY DRILLING**									
0020 Borings, initial field stake out & determination of elevations	A-6	1	16	Day		560	50.50	610.50	985
0100 Drawings showing boring details				Total		335		335	425
0200 Report and recommendations from P.E.						775		775	970
0300 Mobilization and demobilization	B-55	4	4	↓		117	249	366	465
0350 For over 100 miles, per added mile		450	.036	Mile		1.04	2.22	3.26	4.15
0600 Auger holes in earth, no samples, 2-1/2" diameter		78.60	.204	L.F.		5.95	12.70	18.65	24
0800 Cased borings in earth, with samples, 2-1/2" diameter		55.50	.288	"	16.65	8.40	17.95	43	52
1400 Borings, earth, drill rig and crew with truck mounted auger	↓	1	16	Day		465	995	1,460	1,875
1500 For inner city borings add, minimum								10%	10%
1510 Maximum								20%	20%

02 41 Demolition

02 41 13 – Selective Site Demolition

02 41 13.17 Demolish, Remove Pavement and Curb

	Crew	Daily Output	Labor-Hours	Unit	Material	Labor	Equipment	Total	Total Incl O&P
0010 **DEMOLISH, REMOVE PAVEMENT AND CURB** R024119-10									
5010 Pavement removal, bituminous roads, up to 3" thick	B-38	690	.035	S.Y.		1.03	1.63	2.66	3.50
5050 4"-6" thick		420	.057			1.69	2.68	4.37	5.75
5100 Bituminous driveways		640	.038			1.11	1.76	2.87	3.77
5200 Concrete to 6" thick, hydraulic hammer, mesh reinforced		255	.094			2.79	4.41	7.20	9.45
5300 Rod reinforced	↓	200	.120	↓		3.55	5.65	9.20	12.10
5600 With hand held air equipment, bituminous, to 6" thick	B-39	1900	.025	S.F.		.68	.12	.80	1.28
5700 Concrete to 6" thick, no reinforcing		1600	.030			.81	.15	.96	1.51
5800 Mesh reinforced		1400	.034			.93	.17	1.10	1.72
5900 Rod reinforced	↓	765	.063	↓		1.70	.31	2.01	3.16
6000 Curbs, concrete, plain	B-6	360	.067	L.F.		1.97	.87	2.84	4.22
6100 Reinforced		275	.087			2.58	1.14	3.72	5.55
6200 Granite		360	.067			1.97	.87	2.84	4.22
6300 Bituminous	↓	528	.045	↓		1.35	.59	1.94	2.88

02 41 13.23 Utility Line Removal

	Crew	Daily Output	Labor-Hours	Unit	Material	Labor	Equipment	Total	Total Incl O&P
0010 **UTILITY LINE REMOVAL**									
0015 No hauling, abandon catch basin or manhole	B-6	7	3.429	Ea.		101	44.50	145.50	217
0020 Remove existing catch basin or manhole, masonry		4	6			178	78	256	380
0030 Catch basin or manhole frames and covers, stored	↓	13	1.846	↓		54.50	24	78.50	117

For customer support on your Residential Costs with RSMeans data, call 800.448.8182.

02 41 Demolition

02 41 13 – Selective Site Demolition

02 41 13.23 Utility Line Removal

		Crew	Daily Output	Labor-Hours	Unit	Material	2018 Bare Costs Labor	Equipment	Total	Total Incl O&P
0040	Remove and reset	B-6	7	3.429	Ea.		101	44.50	145.50	217
2900	Pipe removal, sewer/water, no excavation, 12" diameter	↓	175	.137	L.F.		4.06	1.79	5.85	8.70
2930	15"-18" diameter	B-12Z	150	.160			4.82	9.40	14.22	18.35
2960	21"-24" diameter	"	120	.200	↓		6.05	11.75	17.80	23

02 41 13.30 Minor Site Demolition

		Crew	Daily Output	Labor-Hours	Unit	Material	2018 Bare Costs Labor	Equipment	Total	Total Incl O&P
0010	**MINOR SITE DEMOLITION** R024119-10									
1000	Masonry walls, block, solid	B-5	1800	.022	C.F.		.65	.80	1.45	1.95
1200	Brick, solid		900	.044			1.29	1.60	2.89	3.90
1400	Stone, with mortar		900	.044			1.29	1.60	2.89	3.90
1500	Dry set	↓	1500	.027	↓		.78	.96	1.74	2.34
4000	Sidewalk removal, bituminous, 2" thick	B-6	350	.069	S.Y.		2.03	.89	2.92	4.34
4010	2-1/2" thick		325	.074			2.19	.96	3.15	4.68
4050	Brick, set in mortar		185	.130			3.84	1.69	5.53	8.20
4100	Concrete, plain, 4"		160	.150			4.44	1.95	6.39	9.50
4110	Plain, 5"		140	.171			5.05	2.23	7.28	10.85
4120	Plain, 6"		120	.200			5.90	2.60	8.50	12.65
4200	Mesh reinforced, concrete, 4"		150	.160			4.74	2.08	6.82	10.15
4210	5" thick		131	.183			5.40	2.39	7.79	11.60
4220	6" thick	↓	112	.214	↓		6.35	2.79	9.14	13.55
4300	Slab on grade removal, plain	B-5	45	.889	C.Y.		26	32	58	78
4310	Mesh reinforced		33	1.212			35	43.50	78.50	107
4320	Rod reinforced	↓	25	1.600			46.50	57.50	104	141
4400	For congested sites or small quantities, add up to								200%	200%
4450	For disposal on site, add	B-11A	232	.069			2.18	5.50	7.68	9.65
4500	To 5 miles, add	B-34D	76	.105	↓		3.39	8.20	11.59	14.65

02 41 13.60 Selective Demolition Fencing

		Crew	Daily Output	Labor-Hours	Unit	Material	2018 Bare Costs Labor	Equipment	Total	Total Incl O&P
0010	**SELECTIVE DEMOLITION FENCING** R024119-10									
1600	Fencing, barbed wire, 3 strand	2 Clab	430	.037	L.F.		.99		.99	1.65
1650	5 strand	"	280	.057			1.53		1.53	2.54
1700	Chain link, posts & fabric, 8'-10' high, remove only	B-6	445	.054	↓		1.60	.70	2.30	3.42

02 41 16 – Structure Demolition

02 41 16.13 Building Demolition

		Crew	Daily Output	Labor-Hours	Unit	Material	2018 Bare Costs Labor	Equipment	Total	Total Incl O&P
0010	**BUILDING DEMOLITION** Large urban projects, incl. 20 mi. haul R024119-10									
0500	Small bldgs, or single bldgs, no salvage included, steel	B-3	14800	.003	C.F.		.10	.15	.25	.33
0600	Concrete	"	11300	.004	"		.13	.20	.33	.43
0605	Concrete, plain	B-5	33	1.212	C.Y.		35	43.50	78.50	107
0610	Reinforced		25	1.600			46.50	57.50	104	141
0615	Concrete walls		34	1.176			34	42.50	76.50	104
0620	Elevated slabs	↓	26	1.538	↓		44.50	55.50	100	135
0650	Masonry	B-3	14800	.003	C.F.		.10	.15	.25	.33
0700	Wood		14800	.003	"		.10	.15	.25	.33
1000	Demolition single family house, one story, wood 1600 S.F.		1	48	Ea.		1,475	2,300	3,775	4,950
1020	3200 S.F.		.50	96			2,925	4,575	7,500	9,875
1200	Demolition two family house, two story, wood 2400 S.F.		.67	71.964			2,200	3,425	5,625	7,400
1220	4200 S.F.		.38	128			3,900	6,100	10,000	13,200
1300	Demolition three family house, three story, wood 3200 S.F.		.50	96			2,925	4,575	7,500	9,875
1320	5400 S.F.	↓	.30	160	↓		4,875	7,625	12,500	16,500

02 41 16.17 Building Demolition Footings and Foundations

	02 41 16.17 Building Demolition Footings and Foundations	Crew	Daily Output	Labor-Hours	Unit	Material	2018 Bare Costs Labor	Equipment	Total	Total Incl O&P
0010	**BUILDING DEMOLITION FOOTINGS AND FOUNDATIONS** R024119-10									
0200	Floors, concrete slab on grade,									
0240	4" thick, plain concrete	B-13L	5000	.003	S.F.		.12	.39	.51	.63
0280	Reinforced, wire mesh		4000	.004			.15	.49	.64	.78
0300	Rods		4500	.004			.13	.44	.57	.70
0400	6" thick, plain concrete		4000	.004			.15	.49	.64	.78
0420	Reinforced, wire mesh		3200	.005			.19	.61	.80	.98
0440	Rods		3600	.004			.16	.54	.70	.87
1000	Footings, concrete, 1' thick, 2' wide		300	.053	L.F.		1.98	6.50	8.48	10.45
1080	1'-6" thick, 2' wide		250	.064			2.37	7.85	10.22	12.50
1120	3' wide		200	.080			2.96	9.80	12.76	15.65
1200	Average reinforcing, add								10%	10%
2000	Walls, block, 4" thick	B-13L	8000	.002	S.F.		.07	.24	.31	.39
2040	6" thick		6000	.003			.10	.33	.43	.52
2080	8" thick		4000	.004			.15	.49	.64	.78
2100	12" thick		3000	.005			.20	.65	.85	1.05
2400	Concrete, plain concrete, 6" thick		4000	.004			.15	.49	.64	.78
2420	8" thick		3500	.005			.17	.56	.73	.89
2440	10" thick		3000	.005			.20	.65	.85	1.05
2500	12" thick		2500	.006			.24	.78	1.02	1.25
2600	For average reinforcing, add								10%	10%
4000	For congested sites or small quantities, add up to								200%	200%
4200	Add for disposal, on site	B-11A	232	.069	C.Y.		2.18	5.50	7.68	9.65
4250	To five miles	B-30	220	.109	"		3.67	9	12.67	15.95

02 41 19 – Selective Demolition

02 41 19.13 Selective Building Demolition

0010	**SELECTIVE BUILDING DEMOLITION**									
0020	Costs related to selective demolition of specific building components									
0025	are included under Common Work Results (XX 05)									
0030	in the component's appropriate division.									

02 41 19.16 Selective Demolition, Cutout

	02 41 19.16 Selective Demolition, Cutout	Crew	Daily Output	Labor-Hours	Unit	Material	Labor	Equipment	Total	Total Incl O&P
0010	**SELECTIVE DEMOLITION, CUTOUT** R024119-10									
0020	Concrete, elev. slab, light reinforcement, under 6 C.F.	B-9	65	.615	C.F.		16.70	3.60	20.30	31.50
0050	Light reinforcing, over 6 C.F.		75	.533	"		14.45	3.12	17.57	27.50
0200	Slab on grade to 6" thick, not reinforced, under 8 S.F.		85	.471	S.F.		12.75	2.75	15.50	24
0250	8-16 S.F.		175	.229	"		6.20	1.34	7.54	11.75
0255	For over 16 S.F. see Line 02 41 16.17 0400									
0600	Walls, not reinforced, under 6 C.F.	B-9	60	.667	C.F.		18.05	3.90	21.95	34.50
0650	6-12 C.F.	"	80	.500	"		13.55	2.93	16.48	25.50
0655	For over 12 C.F. see Line 02 41 16.17 2500									
1000	Concrete, elevated slab, bar reinforced, under 6 C.F.	B-9	45	.889	C.F.		24	5.20	29.20	45.50
1050	Bar reinforced, over 6 C.F.		50	.800	"		21.50	4.68	26.18	41
1200	Slab on grade to 6" thick, bar reinforced, under 8 S.F.		75	.533	S.F.		14.45	3.12	17.57	27.50
1250	8-16 S.F.		150	.267	"		7.25	1.56	8.81	13.70
1255	For over 16 S.F. see Line 02 41 16.17 0440									
1400	Walls, bar reinforced, under 6 C.F.	B-9	50	.800	C.F.		21.50	4.68	26.18	41
1450	6-12 C.F.	"	70	.571	"		15.50	3.34	18.84	29.50
1455	For over 12 C.F. see Lines 02 41 16.17 2500 and 2600									
2000	Brick, to 4 S.F. opening, not including toothing									
2040	4" thick	B-9	30	1.333	Ea.		36	7.80	43.80	68.50
2060	8" thick		18	2.222			60	13	73	114

For customer support on your Residential Costs with RSMeans data, call 800.448.8182.

02 41 Demolition

02 41 19 – Selective Demolition

02 41 19.16 Selective Demolition, Cutout

		Crew	Daily Output	Labor-Hours	Unit	Material	2018 Bare Costs Labor	2018 Bare Costs Equipment	Total	Total Incl O&P
2080	12" thick	B-9	10	4	Ea.		108	23.50	131.50	206
2400	Concrete block, to 4 S.F. opening, 2" thick		35	1.143			31	6.70	37.70	59
2420	4" thick		30	1.333			36	7.80	43.80	68.50
2440	8" thick		27	1.481			40	8.65	48.65	76.50
2460	12" thick		24	1.667			45	9.75	54.75	86
2600	Gypsum block, to 4 S.F. opening, 2" thick		80	.500			13.55	2.93	16.48	25.50
2620	4" thick		70	.571			15.50	3.34	18.84	29.50
2640	8" thick		55	.727			19.70	4.25	23.95	37.50
2800	Terra cotta, to 4 S.F. opening, 4" thick		70	.571			15.50	3.34	18.84	29.50
2840	8" thick		65	.615			16.70	3.60	20.30	31.50
2880	12" thick	↓	50	.800	↓		21.50	4.68	26.18	41
3000	Toothing masonry cutouts, brick, soft old mortar	1 Brhe	40	.200	V.L.F.		5.75		5.75	9.55
3100	Hard mortar		30	.267			7.65		7.65	12.75
3200	Block, soft old mortar		70	.114			3.27		3.27	5.45
3400	Hard mortar	↓	50	.160	↓		4.58		4.58	7.65
6000	Walls, interior, not including re-framing,									
6010	openings to 5 S.F.									
6100	Drywall to 5/8" thick	1 Clab	24	.333	Ea.		8.90		8.90	14.80
6200	Paneling to 3/4" thick		20	.400			10.70		10.70	17.75
6300	Plaster, on gypsum lath		20	.400			10.70		10.70	17.75
6340	On wire lath	↓	14	.571	↓		15.25		15.25	25.50
7000	Wood frame, not including re-framing, openings to 5 S.F.									
7200	Floors, sheathing and flooring to 2" thick	1 Clab	5	1.600	Ea.		42.50		42.50	71
7310	Roofs, sheathing to 1" thick, not including roofing		6	1.333			35.50		35.50	59
7410	Walls, sheathing to 1" thick, not including siding	↓	7	1.143	↓		30.50		30.50	50.50

02 41 19.19 Selective Demolition

		Crew	Daily Output	Labor-Hours	Unit	Material	2018 Bare Costs Labor	2018 Bare Costs Equipment	Total	Total Incl O&P
0010	**SELECTIVE DEMOLITION**, Rubbish Handling R024119-10									
0020	The following are to be added to the demolition prices									
0600	Dumpster, weekly rental, 1 dump/week, 6 C.Y. capacity (2 tons)				Week	415			415	455
0700	10 C.Y. capacity (3 tons)					480			480	530
0725	20 C.Y. capacity (5 tons) R024119-20					565			565	625
0800	30 C.Y. capacity (7 tons)					730			730	800
0840	40 C.Y. capacity (10 tons)				↓	775			775	850
2000	Load, haul, dump and return, 0'-50' haul, hand carried	2 Clab	24	.667	C.Y.		17.80		17.80	29.50
2005	Wheeled		37	.432			11.55		11.55	19.20
2040	0'-100' haul, hand carried		16.50	.970			26		26	43
2045	Wheeled	↓	25	.640			17.10		17.10	28.50
2050	Forklift	A-3R	25	.320			11.35	5.95	17.30	25
2080	Haul and return, add per each extra 100' haul, hand carried	2 Clab	35.50	.451			12.05		12.05	20
2085	Wheeled		54	.296			7.90		7.90	13.15
2120	For travel in elevators, up to 10 floors, add		140	.114			3.05		3.05	5.05
2130	0'-50' haul, incl. up to 5 riser stairs, hand carried		23	.696			18.55		18.55	31
2135	Wheeled		35	.457			12.20		12.20	20.50
2140	6-10 riser stairs, hand carried		22	.727			19.40		19.40	32.50
2145	Wheeled		34	.471			12.55		12.55	21
2150	11-20 riser stairs, hand carried		20	.800			21.50		21.50	35.50
2155	Wheeled		31	.516			13.80		13.80	23
2160	21-40 riser stairs, hand carried		16	1			26.50		26.50	44.50
2165	Wheeled		24	.667			17.80		17.80	29.50
2170	0-100' haul, incl. 5 riser stairs, hand carried		15	1.067			28.50		28.50	47.50
2175	Wheeled		23	.696			18.55		18.55	31
2180	6-10 riser stairs, hand carried	↓	14	1.143			30.50		30.50	50.50

For customer support on your Residential Costs with RSMeans data, call 800.448.8182.

303

02 41 Demolition

02 41 19 – Selective Demolition

02 41 19.19 Selective Demolition

	02 41 19.19 Selective Demolition	Crew	Daily Output	Labor-Hours	Unit	Material	2018 Bare Costs Labor	Equipment	Total	Total Incl O&P
2185	Wheeled	2 Clab	21	.762	C.Y.		20.50		20.50	34
2190	11-20 riser stairs, hand carried		12	1.333			35.50		35.50	59
2195	Wheeled		18	.889			23.50		23.50	39.50
2200	21-40 riser stairs, hand carried		8	2			53.50		53.50	89
2205	Wheeled		12	1.333			35.50		35.50	59
2210	Haul and return, add per each extra 100' haul, hand carried		35.50	.451			12.05		12.05	20
2215	Wheeled		54	.296			7.90		7.90	13.15
2220	For each additional flight of stairs, up to 5 risers, add		550	.029	Flight		.78		.78	1.29
2225	6-10 risers, add		275	.058			1.55		1.55	2.58
2230	11-20 risers, add		138	.116			3.10		3.10	5.15
2235	21-40 risers, add		69	.232			6.20		6.20	10.30
3000	Loading & trucking, including 2 mile haul, chute loaded	B-16	45	.711	C.Y.		20.50	12.05	32.55	47
3040	Hand loading truck, 50' haul	"	48	.667			19.05	11.30	30.35	44
3080	Machine loading truck	B-17	120	.267			8.05	5.40	13.45	19.30
5000	Haul, per mile, up to 8 C.Y. truck	B-34B	1165	.007			.22	.47	.69	.87
5100	Over 8 C.Y. truck	"	1550	.005			.17	.35	.52	.66

02 41 19.20 Selective Demolition, Dump Charges

	02 41 19.20 Selective Demolition, Dump Charges	Crew	Daily Output	Labor-Hours	Unit	Material	2018 Bare Costs Labor	Equipment	Total	Total Incl O&P
0010	**SELECTIVE DEMOLITION, DUMP CHARGES** R024119-10									
0020	Dump charges, typical urban city, tipping fees only									
0100	Building construction materials				Ton	74			74	81
0200	Trees, brush, lumber					63			63	69.50
0300	Rubbish only					63			63	69.50
0500	Reclamation station, usual charge					74			74	81

02 41 19.21 Selective Demolition, Gutting

	02 41 19.21 Selective Demolition, Gutting	Crew	Daily Output	Labor-Hours	Unit	Material	2018 Bare Costs Labor	Equipment	Total	Total Incl O&P
0010	**SELECTIVE DEMOLITION, GUTTING** R024119-10									
0020	Building interior, including disposal, dumpster fees not included									
0500	Residential building									
0560	Minimum	B-16	400	.080	SF Flr.		2.29	1.36	3.65	5.30
0580	Maximum	"	360	.089	"		2.54	1.51	4.05	5.85
0900	Commercial building									
1000	Minimum	B-16	350	.091	SF Flr.		2.61	1.55	4.16	6.05
1020	Maximum	"	250	.128	"		3.66	2.17	5.83	8.45

02 83 Lead Remediation

02 83 19 – Lead-Based Paint Remediation

02 83 19.22 Preparation of Lead Containment Area

	02 83 19.22 Preparation of Lead Containment Area	Crew	Daily Output	Labor-Hours	Unit	Material	2018 Bare Costs Labor	Equipment	Total	Total Incl O&P
0010	**PREPARATION OF LEAD CONTAINMENT AREA**									
0020	Lead abatement work area, test kit, per swab	1 Skwk	16	.500	Ea.	3.12	17.90		21.02	33.50
0025	For dust barriers see Section 01 56 16.10									
0050	Caution sign	1 Skwk	48	.167	Ea.	8.30	5.95		14.25	19.10
0100	Pre-cleaning, HEPA vacuum and wet wipe, floor and wall surfaces	3 Skwk	5000	.005	S.F.	.01	.17		.18	.30
0105	Ceiling, 6'-11' high		4100	.006		.07	.21		.28	.43
0108	12'-15' high		3550	.007		.07	.24		.31	.48
0115	Over 15' high		3000	.008		.09	.29		.38	.58
0500	Cover surfaces with polyethylene sheeting									
0550	Floors, each layer, 6 mil	3 Skwk	8000	.003	S.F.	.04	.11		.15	.22
0560	Walls, each layer, 6 mil	"	6000	.004	"	.04	.14		.18	.28
0570	For heights above 14', add						20%			
2400	Post abatement cleaning of protective sheeting, HEPA vacuum & wet wipe	3 Skwk	5000	.005	S.F.	.01	.17		.18	.30
2450	Doff, bag and seal protective sheeting		12000	.002		.01	.07		.08	.13

02 83 Lead Remediation

02 83 19 – Lead-Based Paint Remediation

02 83 19.22 Preparation of Lead Containment Area	Crew	Daily Output	Labor-Hours	Unit	Material	2018 Bare Costs Labor	Equipment	Total	Total Incl O&P
2500 Post abatement cleaning, HEPA vacuum & wet wipe	3 Skwk	5000	.005	S.F.	.01	.17		.18	.30

02 83 19.23 Encapsulation of Lead-Based Paint

		Crew	Daily Output	Labor-Hours	Unit	Material	Labor	Equipment	Total	Total Incl O&P
0010	**ENCAPSULATION OF LEAD-BASED PAINT**									
0020	Interior, brushwork, trim, under 6"	1 Pord	240	.033	L.F.	2.32	.98		3.30	4.16
0030	6"-12" wide		180	.044		3.03	1.30		4.33	5.50
0040	Balustrades		300	.027		2.02	.78		2.80	3.51
0050	Pipe to 4" diameter		500	.016		1.21	.47		1.68	2.10
0060	To 8" diameter		375	.021		1.52	.63		2.15	2.70
0070	To 12" diameter		250	.032		2.22	.94		3.16	3.99
0080	To 16" diameter		170	.047		3.33	1.38		4.71	5.95
0090	Cabinets, ornate design		200	.040	S.F.	3.03	1.17		4.20	5.25
0100	Simple design		250	.032	"	2.32	.94		3.26	4.10
0110	Doors, 3' x 7', both sides, incl. frame & trim									
0120	Flush	1 Pord	6	1.333	Ea.	28.50	39		67.50	95.50
0130	French, 10-15 lite		3	2.667		6.05	78.50		84.55	136
0140	Panel		4	2		35.50	58.50		94	136
0150	Louvered		2.75	2.909		32.50	85.50		118	177
0160	Windows, per interior side, per 15 S.F.									
0170	1-6 lite	1 Pord	14	.571	Ea.	20	16.75		36.75	49.50
0180	7-10 lite		7.50	1.067		22	31.50		53.50	76
0190	12 lite		5.75	1.391		30.50	41		71.50	101
0200	Radiators		8	1		70.50	29.50		100	127
0210	Grilles, vents		275	.029	S.F.	2.02	.85		2.87	3.63
0220	Walls, roller, drywall or plaster		1000	.008		.61	.23		.84	1.06
0230	With spunbonded reinforcing fabric		720	.011		.71	.33		1.04	1.32
0240	Wood		800	.010		.71	.29		1	1.26
0250	Ceilings, roller, drywall or plaster		900	.009		.71	.26		.97	1.21
0260	Wood		700	.011		.81	.34		1.15	1.44
0270	Exterior, brushwork, gutters and downspouts		300	.027	L.F.	1.92	.78		2.70	3.40
0280	Columns		400	.020	S.F.	1.42	.59		2.01	2.53
0290	Spray, siding		600	.013	"	1.01	.39		1.40	1.75
0300	Miscellaneous									
0310	Electrical conduit, brushwork, to 2" diameter	1 Pord	500	.016	L.F.	1.21	.47		1.68	2.10
0320	Brick, block or concrete, spray		500	.016	S.F.	1.21	.47		1.68	2.10
0330	Steel, flat surfaces and tanks to 12"		500	.016		1.21	.47		1.68	2.10
0340	Beams, brushwork		400	.020		1.41	.59		2	2.52
0350	Trusses		400	.020		1.41	.59		2	2.52

02 83 19.26 Removal of Lead-Based Paint

		Crew	Daily Output	Labor-Hours	Unit	Material	Labor	Equipment	Total	Total Incl O&P
0010	**REMOVAL OF LEAD-BASED PAINT**									
0011	By chemicals, per application									
0050	Baseboard, to 6" wide	1 Pord	64	.125	L.F.	.80	3.67		4.47	6.95
0070	To 12" wide		32	.250	"	1.47	7.35		8.82	13.70
0200	Balustrades, one side		28	.286	S.F.	1.50	8.40		9.90	15.45
1400	Cabinets, simple design		32	.250		1.37	7.35		8.72	13.60
1420	Ornate design		25	.320		1.60	9.40		11	17.20
1600	Cornice, simple design		60	.133		1.60	3.91		5.51	8.20
1620	Ornate design		20	.400		5.55	11.75		17.30	25.50
2800	Doors, one side, flush		84	.095		1.73	2.80		4.53	6.50
2820	Two panel		80	.100		1.37	2.94		4.31	6.35
2840	Four panel		45	.178		1.43	5.20		6.63	10.15
2880	For trim, one side, add		64	.125	L.F.	.77	3.67		4.44	6.90
3000	Fence, picket, one side		30	.267	S.F.	1.37	7.85		9.22	14.40

For customer support on your Residential Costs with RSMeans data, call 800.448.8182.

305

02 83 Lead Remediation

02 83 19 – Lead-Based Paint Remediation

02 83 19.26 Removal of Lead-Based Paint

		Crew	Daily Output	Labor-Hours	Unit	Material	2018 Bare Costs Labor	Equipment	Total	Total Incl O&P
3200	Grilles, one side, simple design	1 Pord	30	.267	S.F.	1.37	7.85		9.22	14.40
3220	Ornate design		25	.320	↓	1.47	9.40		10.87	17.05
3240	Handrails		90	.089	L.F.	1.37	2.61		3.98	5.80
4400	Pipes, to 4" diameter		90	.089		1.73	2.61		4.34	6.20
4420	To 8" diameter		50	.160		3.47	4.70		8.17	11.55
4440	To 12" diameter		36	.222		5.20	6.50		11.70	16.45
4460	To 16" diameter		20	.400	↓	6.95	11.75		18.70	27
4500	For hangers, add		40	.200	Ea.	2.53	5.85		8.38	12.45
4800	Siding		90	.089	S.F.	1.32	2.61		3.93	5.75
5000	Trusses, open		55	.145	SF Face	2.02	4.27		6.29	9.25
6200	Windows, one side only, double-hung, 1/1 light, 24" x 48" high		4	2	Ea.	24	58.50		82.50	123
6220	30" x 60" high		3	2.667		32.50	78.50		111	165
6240	36" x 72" high		2.50	3.200		38.50	94		132.50	197
6280	40" x 80" high		2	4		47.50	117		164.50	245
6400	Colonial window, 6/6 light, 24" x 48" high		2	4		47.50	117		164.50	245
6420	30" x 60" high		1.50	5.333		62.50	157		219.50	325
6440	36" x 72" high		1	8		94	235		329	490
6480	40" x 80" high		1	8		94	235		329	490
6600	8/8 light, 24" x 48" high		2	4		47.50	117		164.50	245
6620	40" x 80" high		1	8		94	235		329	490
6800	12/12 light, 24" x 48" high		1	8		94	235		329	490
6820	40" x 80" high	↓	.75	10.667	↓	126	315		441	655
6840	Window frame & trim items, included in pricing above									

02 87 Biohazard Remediation

02 87 13 – Mold Remediation

02 87 13.33 Removal and Disposal of Materials With Mold

		Crew	Daily Output	Labor-Hours	Unit	Material	2018 Bare Costs Labor	Equipment	Total	Total Incl O&P
0010	**REMOVAL AND DISPOSAL OF MATERIALS WITH MOLD**									
0015	Demolition in mold contaminated area									
0200	Ceiling, including suspension system, plaster and lath	A-9	2100	.030	S.F.	.08	1.10		1.18	1.95
0210	Finished plaster, leaving wire lath		585	.109		.28	3.95		4.23	7
0220	Suspended acoustical tile		3500	.018		.05	.66		.71	1.17
0230	Concealed tile grid system		3000	.021		.05	.77		.82	1.36
0240	Metal pan grid system		1500	.043		.11	1.54		1.65	2.73
0250	Gypsum board		2500	.026		.07	.92		.99	1.63
0255	Plywood		2500	.026	↓	.07	.92		.99	1.63
0260	Lighting fixtures up to 2' x 4'	↓	72	.889	Ea.	2.26	32		34.26	57
0400	Partitions, non load bearing									
0410	Plaster, lath, and studs	A-9	690	.093	S.F.	.88	3.35		4.23	6.60
0450	Gypsum board and studs		1390	.046		.12	1.66		1.78	2.94
0465	Carpet & pad		1390	.046	↓	.12	1.66		1.78	2.94
0600	Pipe insulation, air cell type, up to 4" diameter pipe		900	.071	L.F.	.18	2.57		2.75	4.55
0610	4" to 8" diameter pipe		800	.080		.20	2.89		3.09	5.10
0620	10" to 12" diameter pipe		700	.091		.23	3.30		3.53	5.85
0630	14" to 16" diameter pipe		550	.116	↓	.30	4.20		4.50	7.45
0650	Over 16" diameter pipe	↓	650	.098	S.F.	.25	3.56		3.81	6.30
9000	For type B (supplied air) respirator equipment, add				%				10%	10%

Estimating Tips
General

- Carefully check all the plans and specifications. Concrete often appears on drawings other than structural drawings, including mechanical and electrical drawings for equipment pads. The cost of cutting and patching is often difficult to estimate. See Subdivision 03 81 for Concrete Cutting, Subdivision 02 41 19.16 for Cutout Demolition, Subdivision 03 05 05.10 for Concrete Demolition, and Subdivision 02 41 19.19 for Rubbish Handling (handling, loading, and hauling of debris).

- Always obtain concrete prices from suppliers near the job site. A volume discount can often be negotiated, depending upon competition in the area. Remember to add for waste, particularly for slabs and footings on grade.

03 10 00 Concrete Forming and Accessories

- A primary cost for concrete construction is forming. Most jobs today are constructed with prefabricated forms. The selection of the forms best suited for the job and the total square feet of forms required for efficient concrete forming and placing are key elements in estimating concrete construction. Enough forms must be available for erection to make efficient use of the concrete placing equipment and crew.

- Concrete accessories for forming and placing depend upon the systems used. Study the plans and specifications to ensure that all special accessory requirements have been included in the cost estimate, such as anchor bolts, inserts, and hangers.

- Included within costs for forms-in-place are all necessary bracing and shoring.

03 20 00 Concrete Reinforcing

- Ascertain that the reinforcing steel supplier has included all accessories, cutting, bending, and an allowance for lapping, splicing, and waste. A good rule of thumb is 10% for lapping, splicing, and waste. Also, 10% waste should be allowed for welded wire fabric.

- The unit price items in the subdivisions for Reinforcing In Place, Glass Fiber Reinforcing, and Welded Wire Fabric include the labor to install accessories such as beam and slab bolsters, high chairs, and bar ties and tie wire. The material cost for these accessories is not included; they may be obtained from the Accessories Subdivisions.

03 30 00 Cast-In-Place Concrete

- When estimating structural concrete, pay particular attention to requirements for concrete additives, curing methods, and surface treatments. Special consideration for climate, hot or cold, must be included in your estimate. Be sure to include requirements for concrete placing equipment and concrete finishing.

- For accurate concrete estimating, the estimator must consider each of the following major components individually: forms, reinforcing steel, ready-mix concrete, placement of the concrete, and finishing of the top surface. For faster estimating, Subdivision 03 30 53.40 for Concrete-In-Place can be used; here, various items of concrete work are presented that include the costs of all five major components (unless specifically stated otherwise).

03 40 00 Precast Concrete
03 50 00 Cast Decks and Underlayment

- The cost of hauling precast concrete structural members is often an important factor. For this reason, it is important to get a quote from the nearest supplier. It may become economically feasible to set up precasting beds on the site if the hauling costs are prohibitive.

Reference Numbers

Reference numbers are shown at the beginning of some major classifications. These numbers refer to related items in the Reference Section. The reference information may be an estimating procedure, an alternate pricing method, or technical information.

Note: Not all subdivisions listed here necessarily appear. ∎

Did you know?

RSMeans data is available through our online application with 24/7 access:

- Search for unit prices by keyword
- Leverage the most up-to-date data
- Build and export estimates

Try it free for 30 days!
www.rsmeans.com/2018freetrial

03 01 Maintenance of Concrete

03 01 30 – Maintenance of Cast-In-Place Concrete

03 01 30.64 Floor Patching		Crew	Daily Output	Labor-Hours	Unit	Material	2018 Bare Costs Labor	Equipment	Total	Total Incl O&P
0010	**FLOOR PATCHING**									
0012	Floor patching, 1/4" thick, small areas, regular	1 Cefi	170	.047	S.F.	3.42	1.63		5.05	6.40
0100	Epoxy	"	100	.080	"	8.55	2.78		11.33	13.90

03 05 Common Work Results for Concrete

03 05 13 – Basic Concrete Materials

03 05 13.85 Winter Protection

		Crew	Daily Output	Labor-Hours	Unit	Material	2018 Bare Costs Labor	Equipment	Total	Total Incl O&P
0010	**WINTER PROTECTION**									
0012	For heated ready mix, add				C.Y.	5.35			5.35	5.90
0100	Temporary heat to protect concrete, 24 hours	2 Clab	50	.320	M.S.F.	200	8.55		208.55	234
0200	Temporary shelter for slab on grade, wood frame/polyethylene sheeting									
0201	Build or remove, light framing for short spans	2 Carp	10	1.600	M.S.F.	305	56		361	430
0210	Large framing for long spans	"	3	5.333	"	410	186		596	760
0710	Electrically heated pads, 15 watts/S.F., 20 uses				S.F.	.56			.56	.61

03 11 Concrete Forming

03 11 13 – Structural Cast-In-Place Concrete Forming

03 11 13.25 Forms In Place, Columns

			Crew	Daily Output	Labor-Hours	Unit	Material	2018 Bare Costs Labor	Equipment	Total	Total Incl O&P
0010	**FORMS IN PLACE, COLUMNS**										
1500	Round fiber tube, recycled paper, 1 use, 8" diameter	G	C-1	155	.206	L.F.	2.88	6.35		9.23	13.75
1550	10" diameter	G		155	.206		3.61	6.35		9.96	14.55
1600	12" diameter	G		150	.213		4.05	6.55		10.60	15.40
1650	14" diameter	G		145	.221		4.29	6.80		11.09	16.05
1700	16" diameter	G		140	.229		5.30	7.05		12.35	17.60
1720	18" diameter	G		140	.229		6.20	7.05		13.25	18.60
5000	Job-built plywood, 8" x 8" columns, 1 use			165	.194	SFCA	2.66	5.95		8.61	12.90
5500	12" x 12" columns, 1 use			180	.178	"	2.58	5.45		8.03	12
7400	Steel framed plywood, based on 50 uses of purchased										
7420	forms, and 4 uses of bracing lumber										
7500	8" x 8" column		C-1	340	.094	SFCA	2.18	2.90		5.08	7.20
7550	10" x 10"			350	.091		1.89	2.82		4.71	6.75
7600	12" x 12"			370	.086		1.61	2.66		4.27	6.20

03 11 13.45 Forms In Place, Footings

		Crew	Daily Output	Labor-Hours	Unit	Material	2018 Bare Costs Labor	Equipment	Total	Total Incl O&P
0010	**FORMS IN PLACE, FOOTINGS**									
0020	Continuous wall, plywood, 1 use	C-1	375	.085	SFCA	6.60	2.63		9.23	11.65
0150	4 use	"	485	.066	"	2.15	2.03		4.18	5.75
1500	Keyway, 4 use, tapered wood, 2" x 4"	1 Carp	530	.015	L.F.	.22	.53		.75	1.12
1550	2" x 6"	"	500	.016	"	.32	.56		.88	1.28
5000	Spread footings, job-built lumber, 1 use	C-1	305	.105	SFCA	2.10	3.23		5.33	7.70
5150	4 use	"	414	.077	"	.68	2.38		3.06	4.72

03 11 13.50 Forms In Place, Grade Beam

		Crew	Daily Output	Labor-Hours	Unit	Material	2018 Bare Costs Labor	Equipment	Total	Total Incl O&P
0010	**FORMS IN PLACE, GRADE BEAM**									
0020	Job-built plywood, 1 use	C-2	530	.091	SFCA	2.97	2.82		5.79	7.95
0150	4 use	"	605	.079	"	.96	2.47		3.43	5.20

03 11 13.65 Forms In Place, Slab On Grade

			Crew	Daily Output	Labor-Hours	Unit	Material	2018 Bare Costs Labor	Equipment	Total	Total Incl O&P
0010	**FORMS IN PLACE, SLAB ON GRADE**										
1000	Bulkhead forms w/keyway, wood, 6" high, 1 use		C-1	510	.063	L.F.	1.05	1.93		2.98	4.38
1400	Bulkhead form for slab, 4-1/2" high, exp metal, incl keyway & stakes	G		1200	.027		.91	.82		1.73	2.37

308

03 11 Concrete Forming

03 11 13 – Structural Cast-In-Place Concrete Forming

03 11 13.65 Forms In Place, Slab On Grade

			Crew	Daily Output	Labor-Hours	Unit	Material	2018 Bare Costs Labor	2018 Bare Costs Equipment	Total	Total Incl O&P
1410	5-1/2" high	G	C-1	1100	.029	L.F.	1.14	.90		2.04	2.74
1420	7-1/2" high	G		960	.033		1.37	1.03		2.40	3.22
1430	9-1/2" high	G		840	.038		1.49	1.17		2.66	3.60
2000	Curb forms, wood, 6" to 12" high, on grade, 1 use			215	.149	SFCA	1.90	4.58		6.48	9.75
2150	4 use			275	.116	"	.62	3.58		4.20	6.65
3000	Edge forms, wood, 4 use, on grade, to 6" high			600	.053	L.F.	.29	1.64		1.93	3.06
3050	7" to 12" high			435	.074	SFCA	.69	2.26		2.95	4.54
4000	For slab blockouts, to 12" high, 1 use			200	.160	L.F.	.82	4.93		5.75	9.10
4100	Plastic (extruded), to 6" high, multiple use, on grade			800	.040	"	6.65	1.23		7.88	9.40
8760	Void form, corrugated fiberboard, 4" x 12", 4' long	G		3000	.011	S.F.	3.43	.33		3.76	4.32
8770	6" x 12", 4' long			3000	.011		4.10	.33		4.43	5.05
8780	1/4" thick hardboard protective cover for void form		2 Carp	1500	.011		.68	.37		1.05	1.37

03 11 13.85 Forms In Place, Walls

			Crew	Daily Output	Labor-Hours	Unit	Material	2018 Bare Costs Labor	2018 Bare Costs Equipment	Total	Total Incl O&P
0010	**FORMS IN PLACE, WALLS**										
0100	Box out for wall openings, to 16" thick, to 10 S.F.		C-2	24	2	Ea.	26.50	62		88.50	133
0150	Over 10 S.F. (use perimeter)		"	280	.171	L.F.	2.28	5.35		7.63	11.40
0250	Brick shelf, 4" w, add to wall forms, use wall area above shelf										
0260	1 use		C-2	240	.200	SFCA	2.47	6.20		8.67	13.10
0350	4 use			300	.160	"	.99	4.98		5.97	9.40
0500	Bulkhead, wood with keyway, 1 use, 2 piece			265	.181	L.F.	2.19	5.65		7.84	11.80
0600	Bulkhead forms with keyway, 1 piece expanded metal, 8" wall	G	C-1	1000	.032		1.37	.99		2.36	3.15
0610	10" wall	G	"	800	.040		1.49	1.23		2.72	3.69
2000	Wall, job-built plywood, to 8' high, 1 use		C-2	370	.130	SFCA	2.76	4.04		6.80	9.80
2050	2 use			435	.110		1.75	3.43		5.18	7.70
2100	3 use			495	.097		1.28	3.02		4.30	6.45
2150	4 use			505	.095		1.04	2.96		4	6.05
2400	Over 8' to 16' high, 1 use			280	.171		3.05	5.35		8.40	12.25
2450	2 use			345	.139		1.33	4.33		5.66	8.65
2500	3 use			375	.128		.95	3.98		4.93	7.70
2550	4 use			395	.122		.78	3.78		4.56	7.15
7800	Modular prefabricated plywood, based on 20 uses of purchased										
7820	forms, and 4 uses of bracing lumber										
7860	To 8' high		C-2	800	.060	SFCA	1.09	1.87		2.96	4.31
8060	Over 8' to 16' high		"	600	.080	"	1.15	2.49		3.64	5.40

03 11 19 – Insulating Concrete Forming

03 11 19.10 Insulating Forms, Left In Place

			Crew	Daily Output	Labor-Hours	Unit	Material	2018 Bare Costs Labor	2018 Bare Costs Equipment	Total	Total Incl O&P
0010	**INSULATING FORMS, LEFT IN PLACE**										
0020	S.F. is for exterior face, but includes forms for both faces (total R22)										
2000	4" wall, straight block, 16" x 48" (5.33 S.F.)	G	2 Carp	90	.178	Ea.	21	6.20		27.20	34
2010	90 corner block, exterior 16" x 38" x 22" (6.67 S.F.)	G		75	.213		25.50	7.45		32.95	40.50
2020	45 corner block, exterior 16" x 34" x 18" (5.78 S.F.)	G		75	.213		25	7.45		32.45	40
2100	6" wall, straight block, 16" x 48" (5.33 S.F.)	G		90	.178		22	6.20		28.20	34.50
2110	90 corner block, exterior 16" x 32" x 24" (6.22 S.F.)	G		75	.213		25	7.45		32.45	40
2120	45 corner block, exterior 16" x 26" x 18" (4.89 S.F.)	G		75	.213		24.50	7.45		31.95	39.50
2130	Brick ledge block, 16" x 48" (5.33 S.F.)	G		80	.200		27	7		34	41.50
2140	Taper top block, 16" x 48" (5.33 S.F.)	G		80	.200		25.50	7		32.50	39.50
2200	8" wall, straight block, 16" x 48" (5.33 S.F.)	G		90	.178		23	6.20		29.20	36
2210	90 corner block, exterior 16" x 34" x 26" (6.67 S.F.)	G		75	.213		31	7.45		38.45	46.50
2220	45 corner block, exterior 16" x 28" x 20" (5.33 S.F.)	G		75	.213		25.50	7.45		32.95	40.50
2230	Brick ledge block, 16" x 48" (5.33 S.F.)	G		80	.200		28	7		35	42.50
2240	Taper top block, 16" x 48" (5.33 S.F.)	G		80	.200		26.50	7		33.50	40.50

03 11 Concrete Forming

03 11 23 – Permanent Stair Forming

03 11 23.75 Forms In Place, Stairs

03 11 23.75 Forms In Place, Stairs	Crew	Daily Output	Labor-Hours	Unit	Material	2018 Bare Costs Labor	Equipment	Total	Total Incl O&P
0010 **FORMS IN PLACE, STAIRS**									
0015 (Slant length x width), 1 use	C-2	165	.291	S.F.	5.75	9.05		14.80	21.50
0150 4 use		190	.253		2.03	7.85		9.88	15.35
2000 Stairs, cast on sloping ground (length x width), 1 use		220	.218		2.34	6.80		9.14	13.95
2025 2 use		232	.207		1.29	6.45		7.74	12.15
2050 3 use		244	.197		.94	6.10		7.04	11.25
2100 4 use		256	.188		.76	5.85		6.61	10.60

03 15 Concrete Accessories

03 15 05 – Concrete Forming Accessories

03 15 05.12 Chamfer Strips

	Crew	Daily Output	Labor-Hours	Unit	Material	Labor	Equipment	Total	Total Incl O&P
0010 **CHAMFER STRIPS**									
5000 Wood, 1/2" wide	1 Carp	535	.015	L.F.	.14	.52		.66	1.02
5200 3/4" wide		525	.015		.16	.53		.69	1.07
5400 1" wide		515	.016		.29	.54		.83	1.22

03 15 05.15 Column Form Accessories

	Crew	Daily Output	Labor-Hours	Unit	Material	Labor	Equipment	Total	Total Incl O&P
0010 **COLUMN FORM ACCESSORIES**									
1000 Column clamps, adjustable to 24" x 24", buy ⑥				Set	172			172	189
1100 Rent per month ⑥				"	12.25			12.25	13.50

03 15 05.75 Sleeves and Chases

	Crew	Daily Output	Labor-Hours	Unit	Material	Labor	Equipment	Total	Total Incl O&P
0010 **SLEEVES AND CHASES**									
0100 Plastic, 1 use, 12" long, 2" diameter	1 Carp	100	.080	Ea.	1.89	2.80		4.69	6.75
0150 4" diameter		90	.089		5.30	3.11		8.41	11
0200 6" diameter		75	.107		9.30	3.73		13.03	16.45

03 15 05.80 Snap Ties

	Crew	Daily Output	Labor-Hours	Unit	Material	Labor	Equipment	Total	Total Incl O&P
0010 **SNAP TIES**, 8-1/4" L&W (Lumber and wedge)									
0100 2250 lb., w/flat washer, 8" wall ⑥				C	93			93	102
0150 10" wall ⑥					135			135	149
0200 12" wall ⑥					140			140	154
0500 With plastic cone, 8" wall ⑥					82			82	90
0550 10" wall ⑥					85			85	93.50
0600 12" wall ⑥					92			92	101

03 15 05.95 Wall and Foundation Form Accessories

	Crew	Daily Output	Labor-Hours	Unit	Material	Labor	Equipment	Total	Total Incl O&P
0010 **WALL AND FOUNDATION FORM ACCESSORIES**									
4000 Nail stakes, 3/4" diameter, 18" long ⑥				Ea.	1.98			1.98	2.18
4050 24" long ⑥					2.50			2.50	2.75
4200 30" long ⑥					3.16			3.16	3.48
4250 36" long ⑥					3.99			3.99	4.39

03 15 13 – Waterstops

03 15 13.50 Waterstops

	Crew	Daily Output	Labor-Hours	Unit	Material	Labor	Equipment	Total	Total Incl O&P
0010 **WATERSTOPS**, PVC and Rubber									
0020 PVC, ribbed 3/16" thick, 4" wide	1 Carp	155	.052	L.F.	1.43	1.80		3.23	4.57
0050 6" wide		145	.055		2.41	1.93		4.34	5.85
0500 With center bulb, 6" wide, 3/16" thick		135	.059		2.43	2.07		4.50	6.10
0550 3/8" thick		130	.062		4.43	2.15		6.58	8.45
0600 9" wide x 3/8" thick		125	.064		7.20	2.24		9.44	11.65

03 15 Concrete Accessories

03 15 16 – Concrete Construction Joints

03 15 16.20 Control Joints, Saw Cut

		Crew	Daily Output	Labor-Hours	Unit	Material	2018 Bare Costs Labor	Equipment	Total	Total Incl O&P
0010	**CONTROL JOINTS, SAW CUT**									
0100	Sawcut control joints in green concrete									
0120	1" depth	C-27	2000	.008	L.F.	.03	.28	.05	.36	.55
0140	1-1/2" depth		1800	.009		.05	.31	.06	.42	.62
0160	2" depth	↓	1600	.010	↓	.07	.35	.06	.48	.71
0180	Sawcut joint reservoir in cured concrete									
0182	3/8" wide x 3/4" deep, with single saw blade	C-27	1000	.016	L.F.	.05	.56	.10	.71	1.07
0184	1/2" wide x 1" deep, with double saw blades		900	.018		.10	.62	.11	.83	1.24
0186	3/4" wide x 1-1/2" deep, with double saw blades	↓	800	.020		.21	.69	.13	1.03	1.50
0190	Water blast joint to wash away laitance, 2 passes	C-29	2500	.003			.09	.03	.12	.17
0200	Air blast joint to blow out debris and air dry, 2 passes	C-28	2000	.004	↓		.14	.01	.15	.24
0300	For backer rod, see Section 07 91 23.10									
0340	For joint sealant, see Section 03 15 16.30 or 07 92 13.20									

03 15 16.30 Expansion Joints

				Crew	Daily Output	Labor-Hours	Unit	Material	2018 Bare Costs Labor	Equipment	Total	Total Incl O&P
0010	**EXPANSION JOINTS**											
0020	Keyed, cold, 24 ga., incl. stakes, 3-1/2" high		G	1 Carp	200	.040	L.F.	.85	1.40		2.25	3.26
0050	4-1/2" high		G		200	.040		.91	1.40		2.31	3.32
0100	5-1/2" high		G		195	.041		1.14	1.43		2.57	3.63
2000	Premolded, bituminous fiber, 1/2" x 6"				375	.021		.42	.75		1.17	1.70
2050	1" x 12"				300	.027		1.97	.93		2.90	3.72
2140	Concrete expansion joint, recycled paper and fiber, 1/2" x 6"		G		390	.021		.42	.72		1.14	1.65
2150	1/2" x 12"		G		360	.022		.84	.78		1.62	2.22
2500	Neoprene sponge, closed cell, 1/2" x 6"				375	.021		2.31	.75		3.06	3.78
2550	1" x 12"			↓	300	.027	↓	8.85	.93		9.78	11.25
5000	For installation in walls, add								75%			
5250	For installation in boxouts, add								25%			

03 15 19 – Cast-In Concrete Anchors

03 15 19.05 Anchor Bolt Accessories

		Crew	Daily Output	Labor-Hours	Unit	Material	2018 Bare Costs Labor	Equipment	Total	Total Incl O&P
0010	**ANCHOR BOLT ACCESSORIES**									
0015	For anchor bolts set in fresh concrete, see Section 03 15 19.10									
8150	Anchor bolt sleeve, plastic, 1" diameter bolts	1 Carp	60	.133	Ea.	13.40	4.66		18.06	22.50
8500	1-1/2" diameter		28	.286		19.75	10		29.75	38
8600	2" diameter		24	.333		19.45	11.65		31.10	41
8650	3" diameter	↓	20	.400	↓	35.50	14		49.50	62

03 15 19.10 Anchor Bolts

				Crew	Daily Output	Labor-Hours	Unit	Material	2018 Bare Costs Labor	Equipment	Total	Total Incl O&P
0010	**ANCHOR BOLTS**											
0015	Made from recycled materials											
0025	Single bolts installed in fresh concrete, no templates											
0030	Hooked w/nut and washer, 1/2" diameter, 8" long		G	1 Carp	132	.061	Ea.	1.42	2.12		3.54	5.10
0040	12" long		G		131	.061		1.58	2.13		3.71	5.30
0070	3/4" diameter, 8" long		G		127	.063		4.67	2.20		6.87	8.80
0080	12" long		G	↓	125	.064	↓	5.85	2.24		8.09	10.10

03 15 19.30 Inserts

				Crew	Daily Output	Labor-Hours	Unit	Material	2018 Bare Costs Labor	Equipment	Total	Total Incl O&P
0010	**INSERTS**											
1000	Inserts, slotted nut type for 3/4" bolts, 4" long		G	1 Carp	84	.095	Ea.	20	3.33		23.33	28
2100	6" long		G		84	.095		23	3.33		26.33	30.50
2150	8" long		G		84	.095		30.50	3.33		33.83	39
2200	Slotted, strap type, 4" long		G		84	.095		22	3.33		25.33	29.50
2300	8" long		G	↓	84	.095	↓	32.50	3.33		35.83	41.50
9950	For galvanized inserts, add								30%			

03 21 Reinforcement Bars

03 21 11 – Plain Steel Reinforcement Bars

03 21 11.60 Reinforcing In Place		Crew	Daily Output	Labor-Hours	Unit	Material	2018 Bare Costs Labor	Equipment	Total	Total Incl O&P
0010	**REINFORCING IN PLACE**, 50-60 ton lots, A615 Grade 60									
0020	Includes labor, but not material cost, to install accessories									
0030	Made from recycled materials									
0502	Footings, #4 to #7	G	4 Rodm	4200	.008	Lb.	.48	.30		.78	1.02
0550	#8 to #18	G		3.60	8.889	Ton	960	345		1,305	1,625
0602	Slab on grade, #3 to #7	G		4200	.008	Lb.	.48	.30		.78	1.02
0702	Walls, #3 to #7	G		6000	.005	"	.48	.21		.69	.87
0750	#8 to #18	G	↓	4	8	Ton	960	310		1,270	1,575
0900	For other than 50-60 ton lots									
1000	Under 10 ton job, #3 to #7, add						25%	10%		
1010	#8 to #18, add						20%	10%		
1050	10-50 ton job, #3 to #7, add						10%			
1060	#8 to #18, add						5%			
1100	60-100 ton job, #3 to #7, deduct						5%			
1110	#8 to #18, deduct						10%			
1150	Over 100 ton job, #3 to #7, deduct						10%			
1160	#8 to #18, deduct						15%			
2400	Dowels, 2 feet long, deformed, #3	G	2 Rodm	520	.031	Ea.	.40	1.19		1.59	2.43
2410	#4	G		480	.033		.70	1.29		1.99	2.93
2420	#5	G		435	.037		1.10	1.43		2.53	3.59
2430	#6	G	↓	360	.044	↓	1.58	1.72		3.30	4.62
2600	Dowel sleeves for CIP concrete, 2-part system									
2610	Sleeve base, plastic, for 5/8" smooth dowel sleeve, fasten to edge form		1 Rodm	200	.040	Ea.	.53	1.55		2.08	3.17
2615	Sleeve, plastic, 12" long, for 5/8" smooth dowel, snap onto base			400	.020		1.33	.78		2.11	2.75
2620	Sleeve base, for 3/4" smooth dowel sleeve			175	.046		.53	1.77		2.30	3.54
2625	Sleeve, 12" long, for 3/4" smooth dowel			350	.023		1.21	.89		2.10	2.81
2630	Sleeve base, for 1" smooth dowel sleeve			150	.053		.68	2.07		2.75	4.20
2635	Sleeve, 12" long, for 1" smooth dowel			300	.027		1.42	1.03		2.45	3.29
2700	Dowel caps, visual warning only, plastic, #3 to #8		2 Rodm	800	.020		.30	.78		1.08	1.62
2720	#8 to #18			750	.021		.74	.83		1.57	2.20
2750	Impalement protective, plastic, #4 to #9		↓	800	.020	↓	1.08	.78		1.86	2.48

03 22 Fabric and Grid Reinforcing

03 22 11 – Plain Welded Wire Fabric Reinforcing

03 22 11.10 Welded Wire Fabric

03 22 11.10 Welded Wire Fabric		Crew	Daily Output	Labor-Hours	Unit	Material	2018 Bare Costs Labor	Equipment	Total	Total Incl O&P
0011	**WELDED WIRE FABRIC**, 6 x 6 - W1.4 x W1.4 (10 x 10)	G	2 Rodm	3500	.005	S.F.	.15	.18		.33	.47
0301	6 x 6 - W2.9 x W2.9 (6 x 6) 42 lb./C.S.F.	G		2900	.006		.25	.21		.46	.64
0501	4 x 4 - W1.4 x W1.4 (10 x 10) 31 lb./C.S.F.	G	↓	3100	.005	↓	.23	.20		.43	.58
0750	Rolls									
0901	2 x 2 - #12 galv. for gunite reinforcing	G	2 Rodm	650	.025	S.F.	.70	.96		1.66	2.36

03 22 13 – Galvanized Welded Wire Fabric Reinforcing

03 22 13.10 Galvanized Welded Wire Fabric

0010	**GALVANIZED WELDED WIRE FABRIC**									
0100	Add to plain welded wire pricing for galvanized welded wire					Lb.	.23			.23	.25

03 22 16 – Epoxy-Coated Welded Wire Fabric Reinforcing

03 22 16.10 Epoxy-Coated Welded Wire Fabric

0010	**EPOXY-COATED WELDED WIRE FABRIC**									
0100	Add to plain welded wire pricing for epoxy-coated welded wire					Lb.	.39			.39	.43

03 23 Stressed Tendon Reinforcing

03 23 05 – Prestressing Tendons

03 23 05.50 Prestressing Steel

		Crew	Daily Output	Labor-Hours	Unit	Material	2018 Bare Costs Labor	Equipment	Total	Total Incl O&P
0010	**PRESTRESSING STEEL**									
3000	Slabs on grade, 0.5-inch diam. non-bonded strands, HDPE sheathed,									
3050	attached dead-end anchors, loose stressing-end anchors									
3100	25' x 30' slab, strands @ 36" OC, placing	2 Rodm	2940	.005	S.F.	.65	.21		.86	1.06
3105	Stressing	C-4A	3750	.004			.17	.01	.18	.29
3110	42" OC, placing	2 Rodm	3200	.005		.57	.19		.76	.95
3115	Stressing	C-4A	4040	.004			.15	.01	.16	.27
3120	48" OC, placing	2 Rodm	3510	.005		.50	.18		.68	.85
3125	Stressing	C-4A	4390	.004			.14	.01	.15	.25
3150	25' x 40' slab, strands @ 36" OC, placing	2 Rodm	3370	.005		.63	.18		.81	1
3155	Stressing	C-4A	4360	.004			.14	.01	.15	.25
3160	42" OC, placing	2 Rodm	3760	.004		.54	.17		.71	.87
3165	Stressing	C-4A	4820	.003			.13	.01	.14	.22
3170	48" OC, placing	2 Rodm	4090	.004		.48	.15		.63	.78
3175	Stressing	C-4A	5190	.003			.12	.01	.13	.21
3200	30' x 30' slab, strands @ 36" OC, placing	2 Rodm	3260	.005		.62	.19		.81	1
3205	Stressing	C-4A	4190	.004			.15	.01	.16	.26
3210	42" OC, placing	2 Rodm	3530	.005		.56	.18		.74	.91
3215	Stressing	C-4A	4500	.004			.14	.01	.15	.24
3220	48" OC, placing	2 Rodm	3840	.004		.50	.16		.66	.82
3225	Stressing	C-4A	4850	.003			.13	.01	.14	.22
3230	30' x 40' slab, strands @ 36" OC, placing	2 Rodm	3780	.004		.60	.16		.76	.93
3235	Stressing	C-4A	4920	.003			.13	.01	.14	.22
3240	42" OC, placing	2 Rodm	4190	.004		.52	.15		.67	.83
3245	Stressing	C-4A	5410	.003			.11	.01	.12	.20
3250	48" OC, placing	2 Rodm	4520	.004		.47	.14		.61	.75
3255	Stressing	C-4A	5790	.003			.11	.01	.12	.19
3260	30' x 50' slab, strands @ 36" OC, placing	2 Rodm	4300	.004		.57	.14		.71	.86
3265	Stressing	C-4A	5650	.003			.11	.01	.12	.19
3270	42" OC, placing	2 Rodm	4720	.003		.50	.13		.63	.77
3275	Stressing	C-4A	6150	.003			.10	.01	.11	.18
3280	48" OC, placing	2 Rodm	5240	.003		.44	.12		.56	.69
3285	Stressing	C-4A	6760	.002	▽		.09	.01	.10	.16

03 24 Fibrous Reinforcing

03 24 05 – Reinforcing Fibers

03 24 05.30 Synthetic Fibers

					Unit	Material			Total	Total Incl O&P
0010	**SYNTHETIC FIBERS**									
0100	Synthetic fibers, add to concrete				Lb.	4.71			4.71	5.20
0110	1-1/2 lb./C.Y.				C.Y.	7.30			7.30	8

03 24 05.70 Steel Fibers

						Unit	Material			Total	Total Incl O&P
0010	**STEEL FIBERS**										
0140	ASTM A850, Type V, continuously deformed, 1-1/2" long x 0.045" diam.										
0150	Add to price of ready mix concrete	G				Lb.	1.22			1.22	1.34
0205	Alternate pricing, dosing at 5 lb./C.Y., add to price of RMC	G				C.Y.	6.10			6.10	6.70
0210	10 lb./C.Y.	G					12.20			12.20	13.40
0215	15 lb./C.Y.	G					18.30			18.30	20
0220	20 lb./C.Y.	G					24.50			24.50	27
0225	25 lb./C.Y.	G					30.50			30.50	33.50
0230	30 lb./C.Y.	G					36.50			36.50	40.50
0235	35 lb./C.Y.	G					42.50			42.50	47

03 24 Fibrous Reinforcing

03 24 05 – Reinforcing Fibers

03 24 05.70 Steel Fibers		Crew	Daily Output	Labor-Hours	Unit	Material	2018 Bare Costs Labor	Equipment	Total	Total Incl O&P
0240	40 lb./C.Y.	G			C.Y.	49			49	53.50
0250	50 lb./C.Y.	G				61			61	67
0275	75 lb./C.Y.	G				91.50			91.50	101
0300	100 lb./C.Y.	G				122			122	134

03 30 Cast-In-Place Concrete

03 30 53 – Miscellaneous Cast-In-Place Concrete

03 30 53.40 Concrete In Place

		Crew	Daily Output	Labor-Hours	Unit	Material	2018 Bare Costs Labor	Equipment	Total	Total Incl O&P
0010	**CONCRETE IN PLACE**									
0020	Including forms (4 uses), Grade 60 rebar, concrete (Portland cement									
0050	Type I), placement and finishing unless otherwise indicated									
0500	Chimney foundations (5000 psi), over 5 C.Y.	C-14C	32.22	3.476	C.Y.	171	116	.80	287.80	380
0510	(3500 psi), under 5 C.Y.	"	23.71	4.724	"	198	157	1.09	356.09	480
3540	Equipment pad (3000 psi), 3' x 3' x 6" thick	C-14H	45	1.067	Ea.	44.50	37	.57	82.07	111
3550	4' x 4' x 6" thick		30	1.600		69	55	.85	124.85	168
3560	5' x 5' x 8" thick		18	2.667		126	92	1.41	219.41	293
3570	6' x 6' x 8" thick		14	3.429		173	118	1.82	292.82	390
3580	8' x 8' x 10" thick		8	6		370	207	3.18	580.18	755
3590	10' x 10' x 12" thick		5	9.600		645	330	5.10	980.10	1,275
3800	Footings (3000 psi), spread under 1 C.Y.	C-14C	28	4	C.Y.	185	133	.92	318.92	425
3825	1 C.Y. to 5 C.Y.		43	2.605		219	86.50	.60	306.10	385
3850	Over 5 C.Y.		75	1.493		203	49.50	.34	252.84	305
3900	Footings, strip (3000 psi), 18" x 9", unreinforced	C-14L	40	2.400		146	77.50	.65	224.15	291
3920	18" x 9", reinforced	C-14C	35	3.200		169	106	.74	275.74	365
3925	20" x 10", unreinforced	C-14L	45	2.133		142	69	.58	211.58	272
3930	20" x 10", reinforced	C-14C	40	2.800		161	93	.64	254.64	335
3935	24" x 12", unreinforced	C-14L	55	1.745		140	56.50	.47	196.97	248
3940	24" x 12", reinforced	C-14C	48	2.333		159	77.50	.54	237.04	305
3945	36" x 12", unreinforced	C-14L	70	1.371		136	44.50	.37	180.87	224
3950	36" x 12", reinforced	C-14C	60	1.867		153	62	.43	215.43	271
4000	Foundation mat (3000 psi), under 10 C.Y.		38.67	2.896		221	96.50	.67	318.17	405
4050	Over 20 C.Y.		56.40	1.986		196	66	.46	262.46	325
4520	Handicap access ramp (4000 psi), railing both sides, 3' wide	C-14H	14.58	3.292	L.F.	360	114	1.74	475.74	585
4525	5' wide		12.22	3.928		370	136	2.08	508.08	635
4530	With 6" curb and rails both sides, 3' wide		8.55	5.614		370	194	2.98	566.98	735
4535	5' wide		7.31	6.566		375	227	3.48	605.48	795
4650	Slab on grade (3500 psi), not including finish, 4" thick	C-14E	60.75	1.449	C.Y.	145	49.50	.42	194.92	243
4700	6" thick	"	92	.957	"	140	33	.28	173.28	209
4701	Thickened slab edge (3500 psi), for slab on grade poured									
4702	monolithically with slab; depth is in addition to slab thickness;									
4703	formed vertical outside edge, earthen bottom and inside slope									
4705	8" deep x 8" wide bottom, unreinforced	C-14L	2190	.044	L.F.	3.97	1.42	.01	5.40	6.70
4710	8" x 8", reinforced	C-14C	1670	.067		6.20	2.23	.02	8.45	10.55
4715	12" deep x 12" wide bottom, unreinforced	C-14L	1800	.053		8.15	1.73	.01	9.89	11.85
4720	12" x 12", reinforced	C-14C	1310	.086		12.25	2.84	.02	15.11	18.25
4725	16" deep x 16" wide bottom, unreinforced	C-14L	1440	.067		13.80	2.16	.02	15.98	18.80
4730	16" x 16", reinforced	C-14C	1120	.100		18.70	3.33	.02	22.05	26
4735	20" deep x 20" wide bottom, unreinforced	C-14L	1150	.083		21	2.70	.02	23.72	27.50
4740	20" x 20", reinforced	C-14C	920	.122		27	4.05	.03	31.08	36.50
4745	24" deep x 24" wide bottom, unreinforced	C-14L	930	.103		29.50	3.34	.03	32.87	38
4750	24" x 24", reinforced	C-14C	740	.151		38	5.05	.03	43.08	50

03 30 Cast-In-Place Concrete

03 30 53 – Miscellaneous Cast-In-Place Concrete

03 30 53.40 Concrete In Place	Crew	Daily Output	Labor-Hours	Unit	Material	2018 Bare Costs Labor	Equipment	Total	Total Incl O&P
4751	Slab on grade (3500 psi), incl. troweled finish, not incl. forms								
4760	or reinforcing, over 10,000 S.F., 4" thick	C-14F	3425	.021	S.F.	1.61	.68	.01	2.30	2.90
4820	6" thick	"	3350	.021	"	2.36	.69	.01	3.06	3.74
5000	Slab on grade (3000 psi), incl. broom finish, not incl. forms								
5001	or reinforcing, 4" thick	C-14G	2873	.019	S.F.	1.57	.62	.01	2.20	2.75
5010	6" thick		2590	.022		2.46	.68	.01	3.15	3.83
5020	8" thick		2320	.024		3.20	.76	.01	3.97	4.78
6800	Stairs (3500 psi), not including safety treads, free standing, 3'-6" wide	C-14H	83	.578	LF Nose	6.05	19.95	.31	26.31	40
6850	Cast on ground		125	.384	"	5.15	13.25	.20	18.60	28
7000	Stair landings, free standing		200	.240	S.F.	4.83	8.30	.13	13.26	19.15
7050	Cast on ground		475	.101	"	3.93	3.49	.05	7.47	10.20

03 31 Structural Concrete

03 31 13 – Heavyweight Structural Concrete

03 31 13.25 Concrete, Hand Mix

		Crew	Daily Output	Labor-Hours	Unit	Material	2018 Bare Costs Labor	Equipment	Total	Total Incl O&P
0010	**CONCRETE, HAND MIX** for small quantities or remote areas									
0050	Includes bulk local aggregate, bulk sand, bagged Portland									
0060	cement (Type I) and water, using gas powered cement mixer									
0125	2500 psi	C-30	135	.059	C.F.	3.80	1.58	1.19	6.57	8.10
0130	3000 psi		135	.059		4.11	1.58	1.19	6.88	8.45
0135	3500 psi		135	.059		4.29	1.58	1.19	7.06	8.65
0140	4000 psi		135	.059		4.50	1.58	1.19	7.27	8.90
0145	4500 psi		135	.059		4.75	1.58	1.19	7.52	9.15
0150	5000 psi		135	.059		5.05	1.58	1.19	7.82	9.55
0300	Using pre-bagged dry mix and wheelbarrow (80-lb. bag = 0.6 C.F.)									
0340	4000 psi	1 Clab	48	.167	C.F.	6.85	4.45		11.30	14.95

03 31 13.30 Concrete, Volumetric Site-Mixed

		Crew	Daily Output	Labor-Hours	Unit	Material	2018 Bare Costs Labor	Equipment	Total	Total Incl O&P
0010	**CONCRETE, VOLUMETRIC SITE-MIXED**									
0015	Mixed on-site in volumetric truck									
0020	Includes local aggregate, sand, Portland cement (Type I) and water									
0025	Excludes all additives and treatments									
0100	3000 psi, 1 C.Y. mixed and discharged				C.Y.	214			214	235
0110	2 C.Y.					158			158	174
0120	3 C.Y.					138			138	152
0130	4 C.Y.					125			125	137
0140	5 C.Y.					112			112	123
0200	For truck holding/waiting time past first 2 on-site hours, add				Hr.	93.50			93.50	103
0210	For trip charge beyond first 20 miles, each way, add				Mile	3.64			3.64	4
0220	For each additional increase of 500 psi, add				Ea.	4.17			4.17	4.59

03 31 13.35 Heavyweight Concrete, Ready Mix

		Crew	Daily Output	Labor-Hours	Unit	Material	2018 Bare Costs Labor	Equipment	Total	Total Incl O&P
0010	**HEAVYWEIGHT CONCRETE, READY MIX**, delivered									
0012	Includes local aggregate, sand, Portland cement (Type I) and water									
0015	Excludes all additives and treatments									
0020	2000 psi				C.Y.	115			115	127
0100	2500 psi					119			119	130
0150	3000 psi					121			121	133
0200	3500 psi					124			124	137
0300	4000 psi					128			128	140
0350	4500 psi					131			131	144
0400	5000 psi					134			134	148

03 31 13.35 Heavyweight Concrete, Ready Mix

		Crew	Daily Output	Labor-Hours	Unit	Material	2018 Bare Costs Labor	2018 Bare Costs Equipment	Total	Total Incl O&P
0411	6000 psi				C.Y.	138			138	152
0412	8000 psi					145			145	160
0413	10,000 psi					153			153	168
0414	12,000 psi					160			160	176
1000	For high early strength (Portland cement Type III), add					10%				
1300	For winter concrete (hot water), add					5.35			5.35	5.90
1410	For mid-range water reducer, add					3.59			3.59	3.95
1420	For high-range water reducer/superplasticizer, add					6.30			6.30	6.90
1430	For retarder, add					3.23			3.23	3.55
1440	For non-Chloride accelerator, add					6.40			6.40	7.05
1450	For Chloride accelerator, per 1%, add					3.86			3.86	4.25
1460	For fiber reinforcing, synthetic (1 lb./C.Y.), add					8			8	8.80
1500	For Saturday delivery, add					8.50			8.50	9.35
1510	For truck holding/waiting time past 1st hour per load, add				Hr.	105			105	116
1520	For short load (less than 4 C.Y.), add per load				Ea.	86			86	94.50
2000	For all lightweight aggregate, add				C.Y.	45%				

03 31 13.70 Placing Concrete

		Crew	Daily Output	Labor-Hours	Unit	Material	2018 Bare Costs Labor	2018 Bare Costs Equipment	Total	Total Incl O&P
0010	**PLACING CONCRETE**									
0020	Includes labor and equipment to place, level (strike off) and consolidate									
1900	Footings, continuous, shallow, direct chute	C-6	120	.400	C.Y.		11.35	.43	11.78	19.25
1950	Pumped	C-20	150	.427			12.45	6.20	18.65	27.50
2000	With crane and bucket	C-7	90	.800			23.50	11.90	35.40	52
2400	Footings, spread, under 1 C.Y., direct chute	C-6	55	.873			25	.93	25.93	42
2600	Over 5 C.Y., direct chute		120	.400			11.35	.43	11.78	19.25
2900	Foundation mats, over 20 C.Y., direct chute		350	.137			3.89	.15	4.04	6.60
4300	Slab on grade, up to 6" thick, direct chute		110	.436			12.40	.47	12.87	21
4350	Pumped	C-20	130	.492			14.35	7.20	21.55	32
4400	With crane and bucket	C-7	110	.655			19.35	9.75	29.10	42.50
4900	Walls, 8" thick, direct chute	C-6	90	.533			15.15	.57	15.72	25.50
4950	Pumped	C-20	100	.640			18.70	9.35	28.05	41.50
5000	With crane and bucket	C-7	80	.900			26.50	13.35	39.85	58.50
5050	12" thick, direct chute	C-6	100	.480			13.60	.51	14.11	23
5100	Pumped	C-20	110	.582			17	8.50	25.50	37.50
5200	With crane and bucket	C-7	90	.800			23.50	11.90	35.40	52
5600	Wheeled concrete dumping, add to placing costs above									
5610	Walking cart, 50' haul, add	C-18	32	.281	C.Y.		7.55	1.81	9.36	14.60
5620	150' haul, add		24	.375			10.10	2.41	12.51	19.45
5700	250' haul, add		18	.500			13.45	3.21	16.66	26
5800	Riding cart, 50' haul, add	C-19	80	.113			3.03	1.22	4.25	6.40
5810	150' haul, add		60	.150			4.04	1.62	5.66	8.50
5900	250' haul, add		45	.200			5.40	2.16	7.56	11.35

03 35 Concrete Finishing

03 35 13 – High-Tolerance Concrete Floor Finishing

03 35 13.30 Finishing Floors, High Tolerance

	03 35 13.30 Finishing Floors, High Tolerance	Crew	Daily Output	Labor-Hours	Unit	Material	2018 Bare Costs Labor	Equipment	Total	Total Incl O&P
0010	**FINISHING FLOORS, HIGH TOLERANCE**									
0012	Finishing of fresh concrete flatwork requires that concrete									
0013	first be placed, struck off & consolidated									
0015	Basic finishing for various unspecified flatwork									
0100	Bull float only	C-10	4000	.006	S.F.		.19		.19	.31
0125	Bull float & manual float		2000	.012			.38		.38	.63
0150	Bull float, manual float & broom finish, w/edging & joints		1850	.013			.42		.42	.68
0200	Bull float, manual float & manual steel trowel		1265	.019			.61		.61	.99
0210	For specified Random Access Floors in ACI Classes 1, 2, 3 and 4 to achieve									
0215	Composite Overall Floor Flatness and Levelness values up to FF35/FL25									
0250	Bull float, machine float & machine trowel (walk-behind)	C-10C	1715	.014	S.F.		.45	.02	.47	.76
0300	Power screed, bull float, machine float & trowel (walk-behind)	C-10D	2400	.010			.32	.05	.37	.57
0350	Power screed, bull float, machine float & trowel (ride-on)	C-10E	4000	.006			.19	.06	.25	.38

03 35 23 – Exposed Aggregate Concrete Finishing

03 35 23.30 Finishing Floors, Exposed Aggregate

	03 35 23.30 Finishing Floors, Exposed Aggregate	Crew	Daily Output	Labor-Hours	Unit	Material	Labor	Equipment	Total	Total Incl O&P
0010	**FINISHING FLOORS, EXPOSED AGGREGATE**									
1600	Exposed local aggregate finish, seeded on fresh concrete, 3 lb./S.F.	1 Cefi	625	.013	S.F.	.20	.44		.64	.94
1650	4 lb./S.F.	"	465	.017	"	.33	.60		.93	1.34

03 35 29 – Tooled Concrete Finishing

03 35 29.30 Finishing Floors, Tooled

	03 35 29.30 Finishing Floors, Tooled	Crew	Daily Output	Labor-Hours	Unit	Material	Labor	Equipment	Total	Total Incl O&P
0010	**FINISHING FLOORS, TOOLED**									
4400	Stair finish, fresh concrete, float finish	1 Cefi	275	.029	S.F.		1.01		1.01	1.64
4500	Steel trowel finish	"	200	.040	"		1.39		1.39	2.25

03 35 29.60 Finishing Walls

	03 35 29.60 Finishing Walls	Crew	Daily Output	Labor-Hours	Unit	Material	Labor	Equipment	Total	Total Incl O&P
0010	**FINISHING WALLS**									
0020	Break ties and patch voids	1 Cefi	540	.015	S.F.	.04	.51		.55	.88
0050	Burlap rub with grout	"	450	.018		.04	.62		.66	1.05
0300	Bush hammer, green concrete	B-39	1000	.048			1.30	.23	1.53	2.42
0350	Cured concrete	"	650	.074			2	.36	2.36	3.72
0500	Acid etch	1 Cefi	575	.014		.13	.48		.61	.93
0850	Grind form fins flush	1 Clab	700	.011	L.F.		.31		.31	.51

03 35 33 – Stamped Concrete Finishing

03 35 33.50 Slab Texture Stamping

	03 35 33.50 Slab Texture Stamping	Crew	Daily Output	Labor-Hours	Unit	Material	Labor	Equipment	Total	Total Incl O&P
0010	**SLAB TEXTURE STAMPING**									
0050	Stamping requires that concrete first be placed, struck off, consolidated,									
0060	bull floated and free of bleed water. Decorative stamping tasks include:									
0100	Step 1 - first application of dry shake colored hardener	1 Cefi	6400	.001	S.F.	.43	.04		.47	.54
0110	Step 2 - bull float		6400	.001			.04		.04	.07
0130	Step 3 - second application of dry shake colored hardener		6400	.001		.21	.04		.25	.30
0140	Step 4 - bull float, manual float & steel trowel	3 Cefi	1280	.019			.65		.65	1.06
0150	Step 5 - application of dry shake colored release agent	1 Cefi	6400	.001		.10	.04		.14	.18
0160	Step 6 - place, tamp & remove mats	3 Cefi	2400	.010		.81	.35		1.16	1.45
0170	Step 7 - touch up edges, mat joints & simulated grout lines	1 Cefi	1280	.006			.22		.22	.35
0300	Alternate stamping estimating method includes all tasks above	4 Cefi	800	.040		1.55	1.39		2.94	3.95
0400	Step 8 - pressure wash @ 3000 psi after 24 hours	1 Cefi	1600	.005			.17		.17	.28
0500	Step 9 - roll 2 coats cure/seal compound when dry	"	800	.010		.63	.35		.98	1.25

For customer support on your Residential Costs with RSMeans data, call 800.448.8182.

317

03 39 Concrete Curing

03 39 13 – Water Concrete Curing

03 39 13.50 Water Curing	Crew	Daily Output	Labor-Hours	Unit	Material	2018 Bare Costs Labor	Equipment	Total	Total Incl O&P
0011 **WATER CURING**									
0020 With burlap, 4 uses assumed, 7.5 oz.	2 Clab	5500	.003	S.F.	.15	.08		.23	.29
0101 10 oz.	"	5500	.003	"	.27	.08		.35	.42

03 39 23 – Membrane Concrete Curing

03 39 23.13 Chemical Compound Membrane Concrete Curing

	Crew	Daily Output	Labor-Hours	Unit	Material	Labor	Equipment	Total	Total Incl O&P
0010 **CHEMICAL COMPOUND MEMBRANE CONCRETE CURING**									
0301 Sprayed membrane curing compound	2 Clab	9500	.002	S.F.	.12	.04		.16	.20

03 39 23.23 Sheet Membrane Concrete Curing

	Crew	Daily Output	Labor-Hours	Unit	Material	Labor	Equipment	Total	Total Incl O&P
0010 **SHEET MEMBRANE CONCRETE CURING**									
0201 Curing blanket, burlap/poly, 2-ply	2 Clab	7000	.002	S.F.	.25	.06		.31	.38

03 41 Precast Structural Concrete

03 41 23 – Precast Concrete Stairs

03 41 23.50 Precast Stairs

	Crew	Daily Output	Labor-Hours	Unit	Material	Labor	Equipment	Total	Total Incl O&P
0010 **PRECAST STAIRS**									
0020 Precast concrete treads on steel stringers, 3' wide	C-12	75	.640	Riser	153	22	6.60	181.60	212
0300 Front entrance, 5' wide with 48" platform, 2 risers		16	3	Flight	625	103	31	759	895
0350 5 risers		12	4		995	137	41.50	1,173.50	1,375
0500 6' wide, 2 risers		15	3.200		695	110	33	838	985
0550 5 risers		11	4.364		1,100	150	45	1,295	1,500
0700 7' wide, 2 risers		14	3.429		885	117	35.50	1,037.50	1,200
1200 Basement entrance stairwell, 6 steps, incl. steel bulkhead door	B-51	22	2.182		1,875	61	8.55	1,944.55	2,150
1250 14 steps	"	11	4.364		3,125	122	17.15	3,264.15	3,650

03 48 Precast Concrete Specialties

03 48 43 – Precast Concrete Trim

03 48 43.40 Precast Lintels

	Crew	Daily Output	Labor-Hours	Unit	Material	Labor	Equipment	Total	Total Incl O&P
0010 **PRECAST LINTELS**, smooth gray, prestressed, stock units only									
0800 4" wide, 8" high, x 4' long	D-10	28	1.143	Ea.	32.50	39	13	84.50	115
0850 8' long		24	1.333		82	45.50	15.15	142.65	183
1000 6" wide, 8" high, x 4' long		26	1.231		48	42	14	104	138
1050 10' long		22	1.455		120	50	16.50	186.50	233

03 48 43.90 Precast Window Sills

	Crew	Daily Output	Labor-Hours	Unit	Material	Labor	Equipment	Total	Total Incl O&P
0010 **PRECAST WINDOW SILLS**									
0600 Precast concrete, 4" tapers to 3", 9" wide	D-1	70	.229	L.F.	19	7.25		26.25	33
0650 11" wide	"	60	.267	"	31	8.45		39.45	48

03 54 Cast Underlayment

03 54 13 – Gypsum Cement Underlayment

03 54 13.50 Poured Gypsum Underlayment	Crew	Daily Output	Labor-Hours	Unit	Material	2018 Bare Costs Labor	Equipment	Total	Total Incl O&P
0010 **POURED GYPSUM UNDERLAYMENT**									
0400 Underlayment, gypsum based, self-leveling 2500 psi, pumped, 1/2" thick	C-8	24000	.002	S.F.	.43	.07	.04	.54	.64
0500 3/4" thick		20000	.003		.65	.09	.04	.78	.91
0600 1" thick	↓	16000	.004		.87	.11	.06	1.04	1.20
1400 Hand placed, 1/2" thick	C-18	450	.020		.43	.54	.13	1.10	1.52
1500 3/4" thick	"	300	.030	↓	.65	.81	.19	1.65	2.27

03 63 Epoxy Grouting

03 63 05 – Grouting of Dowels and Fasteners

03 63 05.10 Epoxy Only

	Crew	Daily Output	Labor-Hours	Unit	Material	2018 Bare Costs Labor	Equipment	Total	Total Incl O&P
0010 **EPOXY ONLY**									
1500 Chemical anchoring, epoxy cartridge, excludes layout, drilling, fastener									
1530 For fastener 3/4" diam. x 6" embedment	2 Skwk	72	.222	Ea.	5.20	7.95		13.15	19.05
1535 1" diam. x 8" embedment		66	.242		7.80	8.70		16.50	23
1540 1-1/4" diam. x 10" embedment		60	.267		15.65	9.55		25.20	33
1545 1-3/4" diam. x 12" embedment		54	.296		26	10.60		36.60	46
1550 14" embedment		48	.333		31.50	11.95		43.45	54.50
1555 2" diam. x 12" embedment		42	.381		41.50	13.65		55.15	69
1560 18" embedment	↓	32	.500	↓	52	17.90		69.90	87.50

03 82 Concrete Boring

03 82 16 – Concrete Drilling

03 82 16.10 Concrete Impact Drilling

	Crew	Daily Output	Labor-Hours	Unit	Material	2018 Bare Costs Labor	Equipment	Total	Total Incl O&P
0010 **CONCRETE IMPACT DRILLING**									
0020 Includes bit cost, layout and set-up time, no anchors									
0050 Up to 4" deep in concrete/brick floors/walls									
0100 Holes, 1/4" diameter	1 Carp	75	.107	Ea.	.07	3.73		3.80	6.30
0150 For each additional inch of depth in same hole, add		430	.019		.02	.65		.67	1.10
0200 3/8" diameter		63	.127		.06	4.44		4.50	7.45
0250 For each additional inch of depth in same hole, add		340	.024		.01	.82		.83	1.39
0300 1/2" diameter		50	.160		.06	5.60		5.66	9.35
0350 For each additional inch of depth in same hole, add		250	.032		.01	1.12		1.13	1.88
0400 5/8" diameter		48	.167		.10	5.85		5.95	9.80
0450 For each additional inch of depth in same hole, add		240	.033		.03	1.16		1.19	1.97
0500 3/4" diameter		45	.178		.13	6.20		6.33	10.50
0550 For each additional inch of depth in same hole, add		220	.036		.03	1.27		1.30	2.15
0600 7/8" diameter		43	.186		.18	6.50		6.68	11
0650 For each additional inch of depth in same hole, add		210	.038		.05	1.33		1.38	2.26
0700 1" diameter		40	.200		.16	7		7.16	11.75
0750 For each additional inch of depth in same hole, add		190	.042		.04	1.47		1.51	2.49
0800 1-1/4" diameter		38	.211		.29	7.35		7.64	12.55
0850 For each additional inch of depth in same hole, add		180	.044		.07	1.55		1.62	2.66
0900 1-1/2" diameter		35	.229		.50	8		8.50	13.85
0950 For each additional inch of depth in same hole, add	↓	165	.048	↓	.12	1.69		1.81	2.96
1000 For ceiling installations, add						40%			

For customer support on your Residential Costs with RSMeans data, call 800.448.8182.

319

Division Notes

	CREW	DAILY OUTPUT	LABOR-HOURS	UNIT	BARE COSTS				TOTAL INCL O&P
					MAT.	LABOR	EQUIP.	TOTAL	

Estimating Tips
04 05 00 Common Work Results for Masonry

- The terms mortar and grout are often used interchangeably—and incorrectly. Mortar is used to bed masonry units, seal the entry of air and moisture, provide architectural appearance, and allow for size variations in the units. Grout is used primarily in reinforced masonry construction and to bond the masonry to the reinforcing steel. Common mortar types are M (2500 psi), S (1800 psi), N (750 psi), and O (350 psi), and they conform to ASTM C270. Grout is either fine or coarse and conforms to ASTM C476, and in-place strengths generally exceed 2500 psi. Mortar and grout are different components of masonry construction and are placed by entirely different methods. An estimator should be aware of their unique uses and costs.

- Mortar is included in all assembled masonry line items. The mortar cost, part of the assembled masonry material cost, includes all ingredients, all labor, and all equipment required. Please see reference number R040513-10.

- Waste, specifically the loss/droppings of mortar and the breakage of brick and block, is included in all unit cost lines that include mortar and masonry units in this division. A factor of 25% is added for mortar and 3% for brick and concrete masonry units.

- Scaffolding or staging is not included in any of the Division 4 costs. Refer to Subdivision 01 54 23 for scaffolding and staging costs.

04 20 00 Unit Masonry

- The most common types of unit masonry are brick and concrete masonry. The major classifications of brick are building brick (ASTM C62), facing brick (ASTM C216), glazed brick, fire brick, and pavers. Many varieties of texture and appearance can exist within these classifications, and the estimator would be wise to check local custom and availability within the project area. For repair and remodeling jobs, matching the existing brick may be the most important criteria.

- Brick and concrete block are priced by the piece and then converted into a price per square foot of wall. Openings less than two square feet are generally ignored by the estimator because any savings in units used are offset by the cutting and trimming required.

- It is often difficult and expensive to find and purchase small lots of historic brick. Costs can vary widely. Many design issues affect costs, selection of mortar mix, and repairs or replacement of masonry materials. Cleaning techniques must be reflected in the estimate.

- All masonry walls, whether interior or exterior, require bracing. The cost of bracing walls during construction should be included by the estimator, and this bracing must remain in place until permanent bracing is complete. Permanent bracing of masonry walls is accomplished by masonry itself, in the form of pilasters or abutting wall corners, or by anchoring the walls to the structural frame. Accessories in the form of anchors, anchor slots, and ties are used, but their supply and installation can be by different trades. For instance, anchor slots on spandrel beams and columns are supplied and welded in place by the steel fabricator, but the ties from the slots into the masonry are installed by the bricklayer. Regardless of the installation method, the estimator must be certain that these accessories are accounted for in pricing.

Reference Numbers

Reference numbers are shown at the beginning of some major classifications. These numbers refer to related items in the Reference Section. The reference information may be an estimating procedure, an alternate pricing method, or technical information.

Note: Not all subdivisions listed here necessarily appear. ■

Did you know?

RSMeans data is available through our online application with 24/7 access:

- Search for unit prices by keyword
- Leverage the most up-to-date data
- Build and export estimates

Try it free for 30 days!
www.rsmeans.com/2018freetrial

04 01 20 – Maintenance of Unit Masonry

04 01 20.20 Pointing Masonry

		Crew	Daily Output	Labor-Hours	Unit	Material	2018 Bare Costs Labor	2018 Bare Costs Equipment	Total	Total Incl O&P
0010	**POINTING MASONRY**									
0300	Cut and repoint brick, hard mortar, running bond	1 Bric	80	.100	S.F.	.56	3.47		4.03	6.40
0320	Common bond		77	.104		.56	3.61		4.17	6.65
0360	Flemish bond		70	.114		.59	3.97		4.56	7.30
0400	English bond		65	.123		.59	4.27		4.86	7.80
0600	Soft old mortar, running bond		100	.080		.56	2.78		3.34	5.25
0620	Common bond		96	.083		.56	2.89		3.45	5.45
0640	Flemish bond		90	.089		.59	3.08		3.67	5.80
0680	English bond		82	.098		.59	3.39		3.98	6.30
0700	Stonework, hard mortar		140	.057	L.F.	.74	1.98		2.72	4.13
0720	Soft old mortar		160	.050	"	.74	1.74		2.48	3.72
1000	Repoint, mask and grout method, running bond		95	.084	S.F.	.74	2.92		3.66	5.70
1020	Common bond		90	.089		.74	3.08		3.82	5.95
1040	Flemish bond		86	.093		.78	3.23		4.01	6.25
1060	English bond		77	.104		.78	3.61		4.39	6.90
2000	Scrub coat, sand grout on walls, thin mix, brushed		120	.067		3.36	2.31		5.67	7.55
2020	Troweled		98	.082		4.68	2.83		7.51	9.90

04 01 20.30 Pointing CMU

		Crew	Daily Output	Labor-Hours	Unit	Material	2018 Bare Costs Labor	2018 Bare Costs Equipment	Total	Total Incl O&P
0010	**POINTING CMU**									
0300	Cut and repoint block, hard mortar, running bond	1 Bric	190	.042	S.F.	.23	1.46		1.69	2.69
0310	Stacked bond		200	.040		.23	1.39		1.62	2.57
0600	Soft old mortar, running bond		230	.035		.23	1.21		1.44	2.27
0610	Stacked bond		245	.033		.23	1.13		1.36	2.14

04 01 30 – Unit Masonry Cleaning

04 01 30.60 Brick Washing

		Crew	Daily Output	Labor-Hours	Unit	Material	2018 Bare Costs Labor	2018 Bare Costs Equipment	Total	Total Incl O&P
0010	**BRICK WASHING**	R040130-10								
0012	Acid cleanser, smooth brick surface	1 Bric	560	.014	S.F.	.05	.50		.55	.89
0050	Rough brick		400	.020		.07	.69		.76	1.23
0060	Stone, acid wash		600	.013		.08	.46		.54	.86
1000	Muriatic acid, price per gallon in 5 gallon lots				Gal.	10.10			10.10	11.10

04 05 Common Work Results for Masonry

04 05 05 – Selective Demolition for Masonry

04 05 05.10 Selective Demolition

		Crew	Daily Output	Labor-Hours	Unit	Material	2018 Bare Costs Labor	2018 Bare Costs Equipment	Total	Total Incl O&P
0010	**SELECTIVE DEMOLITION**	R024119-10								
0200	Bond beams, 8" block with #4 bar	2 Clab	32	.500	L.F.		13.35		13.35	22
0300	Concrete block walls, unreinforced, 2" thick		1200	.013	S.F.		.36		.36	.59
0310	4" thick		1150	.014			.37		.37	.62
0320	6" thick		1100	.015			.39		.39	.65
0330	8" thick		1050	.015			.41		.41	.68
0340	10" thick		1000	.016			.43		.43	.71
0360	12" thick		950	.017			.45		.45	.75
0380	Reinforced alternate courses, 2" thick		1130	.014			.38		.38	.63
0390	4" thick		1080	.015			.40		.40	.66
0400	6" thick		1035	.015			.41		.41	.69
0410	8" thick		990	.016			.43		.43	.72
0420	10" thick		940	.017			.45		.45	.76
0430	12" thick		890	.018			.48		.48	.80
0440	Reinforced alternate courses & vertically 48" OC, 4" thick		900	.018			.47		.47	.79
0450	6" thick		850	.019			.50		.50	.84

04 05 05 – Selective Demolition for Masonry

04 05 05.10 Selective Demolition	Crew	Daily Output	Labor-Hours	Unit	Material	2018 Bare Costs Labor	2018 Bare Costs Equipment	Total	Total Incl O&P
0460	8" thick	2 Clab	800	.020	S.F.		.53		.53	.89
0480	10" thick		750	.021			.57		.57	.95
0490	12" thick	▼	700	.023	▼		.61		.61	1.02
1000	Chimney, 16" x 16", soft old mortar	1 Clab	55	.145	C.F.		3.88		3.88	6.45
1020	Hard mortar		40	.200			5.35		5.35	8.90
1030	16" x 20", soft old mortar		55	.145			3.88		3.88	6.45
1040	Hard mortar		40	.200			5.35		5.35	8.90
1050	16" x 24", soft old mortar		55	.145			3.88		3.88	6.45
1060	Hard mortar		40	.200			5.35		5.35	8.90
1080	20" x 20", soft old mortar		55	.145			3.88		3.88	6.45
1100	Hard mortar		40	.200			5.35		5.35	8.90
1110	20" x 24", soft old mortar		55	.145			3.88		3.88	6.45
1120	Hard mortar		40	.200			5.35		5.35	8.90
1140	20" x 32", soft old mortar		55	.145			3.88		3.88	6.45
1160	Hard mortar		40	.200			5.35		5.35	8.90
1200	48" x 48", soft old mortar		55	.145			3.88		3.88	6.45
1220	Hard mortar	▼	40	.200	▼		5.35		5.35	8.90
1250	Metal, high temp steel jacket, 24" diameter	E-2	130	.369	V.L.F.		14.35	13	27.35	39.50
1260	60" diameter	"	60	.800			31	28	59	85.50
1280	Flue lining, up to 12" x 12"	1 Clab	200	.040			1.07		1.07	1.78
1282	Up to 24" x 24"		150	.053			1.42		1.42	2.37
2000	Columns, 8" x 8", soft old mortar		48	.167			4.45		4.45	7.40
2020	Hard mortar		40	.200			5.35		5.35	8.90
2060	16" x 16", soft old mortar		16	.500			13.35		13.35	22
2100	Hard mortar		14	.571			15.25		15.25	25.50
2140	24" x 24", soft old mortar		8	1			26.50		26.50	44.50
2160	Hard mortar		6	1.333			35.50		35.50	59
2200	36" x 36", soft old mortar		4	2			53.50		53.50	89
2220	Hard mortar		3	2.667	▼		71		71	118
2230	Alternate pricing method, soft old mortar		30	.267	C.F.		7.10		7.10	11.85
2240	Hard mortar	▼	23	.348	"		9.30		9.30	15.45
3000	Copings, precast or masonry, to 8" wide									
3020	Soft old mortar	1 Clab	180	.044	L.F.		1.19		1.19	1.97
3040	Hard mortar	"	160	.050	"		1.34		1.34	2.22
3100	To 12" wide									
3120	Soft old mortar	1 Clab	160	.050	L.F.		1.34		1.34	2.22
3140	Hard mortar	"	140	.057	"		1.53		1.53	2.54
4000	Fireplace, brick, 30" x 24" opening									
4020	Soft old mortar	1 Clab	2	4	Ea.		107		107	178
4040	Hard mortar		1.25	6.400			171		171	284
4100	Stone, soft old mortar		1.50	5.333			142		142	237
4120	Hard mortar		1	8	▼		214		214	355
5000	Veneers, brick, soft old mortar		140	.057	S.F.		1.53		1.53	2.54
5020	Hard mortar		125	.064			1.71		1.71	2.84
5050	Glass block, up to 4" thick		500	.016			.43		.43	.71
5100	Granite and marble, 2" thick		180	.044			1.19		1.19	1.97
5120	4" thick		170	.047			1.26		1.26	2.09
5140	Stone, 4" thick		180	.044			1.19		1.19	1.97
5160	8" thick		175	.046	▼		1.22		1.22	2.03
5400	Alternate pricing method, stone, 4" thick		60	.133	C.F.		3.56		3.56	5.90
5420	8" thick	▼	85	.094	"		2.51		2.51	4.18

04 05 Common Work Results for Masonry

04 05 13 – Masonry Mortaring

04 05 13.10 Cement		Crew	Daily Output	Labor-Hours	Unit	Material	2018 Bare Costs Labor	Equipment	Total	Total Incl O&P
0010	**CEMENT** R040513-10									
0100	Masonry, 70 lb. bag, T.L. lots				Bag	13.80			13.80	15.20
0150	L.T.L. lots					14.65			14.65	16.10
0200	White, 70 lb. bag, T.L. lots					18.05			18.05	19.85
0250	L.T.L. lots					19.95			19.95	22

04 05 16 – Masonry Grouting

04 05 16.30 Grouting		Crew	Daily Output	Labor-Hours	Unit	Material	2018 Bare Costs Labor	Equipment	Total	Total Incl O&P
0010	**GROUTING**									
0011	Bond beams & lintels, 8" deep, 6" thick, 0.15 C.F./L.F.	D-4	1480	.027	L.F.	.75	.80	.09	1.64	2.26
0020	8" thick, 0.2 C.F./L.F.		1400	.029		1.22	.84	.09	2.15	2.85
0050	10" thick, 0.25 C.F./L.F.		1200	.033		1.26	.98	.11	2.35	3.14
0060	12" thick, 0.3 C.F./L.F.		1040	.038		1.51	1.13	.12	2.76	3.69
0200	Concrete block cores, solid, 4" thk., by hand, 0.067 C.F./S.F. of wall	D-8	1100	.036	S.F.	.34	1.17		1.51	2.33
0210	6" thick, pumped, 0.175 C.F./S.F.	D-4	720	.056		.88	1.64	.18	2.70	3.90
0250	8" thick, pumped, 0.258 C.F./S.F.		680	.059		1.30	1.73	.19	3.22	4.53
0300	10" thick, pumped, 0.340 C.F./S.F.		660	.061		1.71	1.79	.20	3.70	5.05
0350	12" thick, pumped, 0.422 C.F./S.F.		640	.063		2.12	1.84	.20	4.16	5.60

04 05 19 – Masonry Anchorage and Reinforcing

04 05 19.05 Anchor Bolts

04 05 19.05 Anchor Bolts		Crew	Daily Output	Labor-Hours	Unit	Material	2018 Bare Costs Labor	Equipment	Total	Total Incl O&P
0010	**ANCHOR BOLTS**									
0015	Installed in fresh grout in CMU bond beams or filled cores, no templates									
0020	Hooked, with nut and washer, 1/2" diameter, 8" long	1 Bric	132	.061	Ea.	1.42	2.10		3.52	5.10
0030	12" long		131	.061		1.58	2.12		3.70	5.30
0040	5/8" diameter, 8" long		129	.062		3.80	2.15		5.95	7.80
0050	12" long		127	.063		4.67	2.19		6.86	8.80
0060	3/4" diameter, 8" long		127	.063		4.67	2.19		6.86	8.80
0070	12" long		125	.064		5.85	2.22		8.07	10.10

04 05 19.16 Masonry Anchors

04 05 19.16 Masonry Anchors		Crew	Daily Output	Labor-Hours	Unit	Material	2018 Bare Costs Labor	Equipment	Total	Total Incl O&P
0010	**MASONRY ANCHORS**									
0020	For brick veneer, galv., corrugated, 7/8" x 7", 22 ga.	1 Bric	10.50	.762	C	16.05	26.50		42.55	61.50
0100	24 ga.		10.50	.762		10.15	26.50		36.65	55
0150	16 ga.		10.50	.762		30	26.50		56.50	77
0200	Buck anchors, galv., corrugated, 16 ga., 2" bend, 8" x 2"		10.50	.762		65	26.50		91.50	116
0250	8" x 3"		10.50	.762		67.50	26.50		94	119
0660	Cavity wall, Z-type, galvanized, 6" long, 1/8" diam.		10.50	.762		24	26.50		50.50	70
0670	3/16" diameter		10.50	.762		33	26.50		59.50	80
0680	1/4" diameter		10.50	.762		40.50	26.50		67	89
0850	8" long, 3/16" diameter		10.50	.762		26	26.50		52.50	72.50
0855	1/4" diameter		10.50	.762		47.50	26.50		74	96.50
1000	Rectangular type, galvanized, 1/4" diameter, 2" x 6"		10.50	.762		75.50	26.50		102	127
1050	4" x 6"		10.50	.762		91	26.50		117.50	144
1100	3/16" diameter, 2" x 6"		10.50	.762		48	26.50		74.50	97
1150	4" x 6"		10.50	.762		55	26.50		81.50	105
1500	Rigid partition anchors, plain, 8" long, 1" x 1/8"		10.50	.762		237	26.50		263.50	305
1550	1" x 1/4"		10.50	.762		279	26.50		305.50	350
1580	1-1/2" x 1/8"		10.50	.762		261	26.50		287.50	330
1600	1-1/2" x 1/4"		10.50	.762		325	26.50		351.50	405
1650	2" x 1/8"		10.50	.762		310	26.50		336.50	385
1700	2" x 1/4"		10.50	.762		405	26.50		431.50	490

04 05 19.26 Masonry Reinforcing Bars

		Crew	Daily Output	Labor-Hours	Unit	Material	2018 Bare Costs Labor	2018 Bare Costs Equipment	Total	Total Incl O&P
0010	**MASONRY REINFORCING BARS** R040519-50									
0015	Steel bars A615, placed horiz., #3 & #4 bars	1 Bric	450	.018	Lb.	.48	.62		1.10	1.56
0050	Placed vertical, #3 & #4 bars		350	.023		.48	.79		1.27	1.86
0060	#5 & #6 bars		650	.012		.48	.43		.91	1.24
0200	Joint reinforcing, regular truss, to 6" wide, mill std galvanized		30	.267	C.L.F.	24	9.25		33.25	42
0250	12" wide		20	.400		28	13.90		41.90	54
0400	Cavity truss with drip section, to 6" wide		30	.267		22.50	9.25		31.75	40.50
0450	12" wide		20	.400		26	13.90		39.90	51.50

04 21 13.13 Brick Veneer Masonry

		Crew	Daily Output	Labor-Hours	Unit	Material	2018 Bare Costs Labor	2018 Bare Costs Equipment	Total	Total Incl O&P
0010	**BRICK VENEER MASONRY**, T.L. lots, excl. scaff., grout & reinforcing									
0015	Material costs incl. 3% brick and 25% mortar waste R042110-10									
2000	Standard, sel. common, 4" x 2-2/3" x 8" (6.75/S.F.) R042110-20	D-8	230	.174	S.F.	4.32	5.60		9.92	14.15
2020	Red, 4" x 2-2/3" x 8", running bond		220	.182		4.06	5.85		9.91	14.25
2050	Full header every 6th course (7.88/S.F.) R042110-50		185	.216		4.73	7		11.73	16.85
2100	English, full header every 2nd course (10.13/S.F.)		140	.286		6.05	9.20		15.25	22
2150	Flemish, alternate header every course (9.00/S.F.)		150	.267		5.40	8.60		14	20.50
2200	Flemish, alt. header every 6th course (7.13/S.F.)		205	.195		4.29	6.30		10.59	15.20
2250	Full headers throughout (13.50/S.F.)		105	.381		8.05	12.30		20.35	29.50
2300	Rowlock course (13.50/S.F.)		100	.400		8.05	12.90		20.95	30.50
2350	Rowlock stretcher (4.50/S.F.)		310	.129		2.74	4.17		6.91	9.95
2400	Soldier course (6.75/S.F.)		200	.200		4.06	6.45		10.51	15.25
2450	Sailor course (4.50/S.F.)		290	.138		2.74	4.45		7.19	10.45
2600	Buff or gray face, running bond (6.75/S.F.)		220	.182		4.30	5.85		10.15	14.55
2700	Glazed face brick, running bond		210	.190		13	6.15		19.15	24.50
2750	Full header every 6th course (7.88/S.F.)		170	.235		15.15	7.60		22.75	29.50
3000	Jumbo, 6" x 4" x 12" running bond (3.00/S.F.)		435	.092		5.45	2.97		8.42	10.95
3050	Norman, 4" x 2-2/3" x 12" running bond (4.5/S.F.)		320	.125		6.40	4.04		10.44	13.80
3100	Norwegian, 4" x 3-1/5" x 12" (3.75/S.F.)		375	.107		5.65	3.44		9.09	12
3150	Economy, 4" x 4" x 8" (4.50/S.F.)		310	.129		4.46	4.17		8.63	11.85
3200	Engineer, 4" x 3-1/5" x 8" (5.63/S.F.)		260	.154		3.92	4.97		8.89	12.60
3250	Roman, 4" x 2" x 12" (6.00/S.F.)		250	.160		7.30	5.15		12.45	16.65
3300	S.C.R., 6" x 2-2/3" x 12" (4.50/S.F.)		310	.129		6.30	4.17		10.47	13.90
3350	Utility, 4" x 4" x 12" (3.00/S.F.)		360	.111		5.10	3.59		8.69	11.60
3360	For less than truck load lots, add					.10%				
3400	For cavity wall construction, add						15%			
3450	For stacked bond, add						10%			
3500	For interior veneer construction, add						15%			
3550	For curved walls, add						30%			

04 21 13.14 Thin Brick Veneer

		Crew	Daily Output	Labor-Hours	Unit	Material	2018 Bare Costs Labor	2018 Bare Costs Equipment	Total	Total Incl O&P
0010	**THIN BRICK VENEER**									
0015	Material costs incl. 3% brick and 25% mortar waste									
0020	On & incl. metal panel support sys, modular, 2-2/3" x 5/8" x 8", red	D-7	92	.174	S.F.	9.60	5.20		14.80	19.05
0100	Closure, 4" x 5/8" x 8"		110	.145		9.35	4.35		13.70	17.35
0110	Norman, 2-2/3" x 5/8" x 12"		110	.145		9.40	4.35		13.75	17.40
0120	Utility, 4" x 5/8" x 12"		125	.128		9.15	3.83		12.98	16.25
0130	Emperor, 4" x 3/4" x 16"		175	.091		10.30	2.73		13.03	15.80
0140	Super emperor, 8" x 3/4" x 16"		195	.082		10.60	2.45		13.05	15.70

For customer support on your Residential Costs with RSMeans data, call 800.448.8182.

325

04 21 Clay Unit Masonry

04 21 13 – Brick Masonry

04 21 13.14 Thin Brick Veneer

		Crew	Daily Output	Labor-Hours	Unit	Material	2018 Bare Costs Labor	Equipment	Total	Total Incl O&P
0150	For L shaped corners with 4" return, add				L.F.	9.25			9.25	10.20
0200	On masonry/plaster back-up, modular, 2-2/3" x 5/8" x 8", red	D-7	137	.117	S.F.	4.49	3.49		7.98	10.60
0210	Closure, 4" x 5/8" x 8"		165	.097		4.24	2.90		7.14	9.35
0220	Norman, 2-2/3" x 5/8" x 12"		165	.097		4.28	2.90		7.18	9.40
0230	Utility, 4" x 5/8" x 12"		185	.086		4	2.59		6.59	8.60
0240	Emperor, 4" x 3/4" x 16"		260	.062		5.15	1.84		6.99	8.70
0250	Super emperor, 8" x 3/4" x 16"		285	.056		5.50	1.68		7.18	8.75
0260	For L shaped corners with 4" return, add				L.F.	9.25			9.25	10.20
0270	For embedment into pre-cast concrete panels, add				S.F.	14.40			14.40	15.85

04 21 13.15 Chimney

		Crew	Daily Output	Labor-Hours	Unit	Material	2018 Bare Costs Labor	Equipment	Total	Total Incl O&P
0010	CHIMNEY, excludes foundation, scaffolding, grout and reinforcing									
0100	Brick, 16" x 16", 8" flue	D-1	18.20	.879	V.L.F.	25.50	28		53.50	74.50
0150	16" x 20" with one 8" x 12" flue		16	1		40	31.50		71.50	97
0200	16" x 24" with two 8" x 8" flues		14	1.143		58.50	36		94.50	125
0250	20" x 20" with one 12" x 12" flue		13.70	1.168		48	37		85	115
0300	20" x 24" with two 8" x 12" flues		12	1.333		66.50	42		108.50	144
0350	20" x 32" with two 12" x 12" flues		10	1.600		84.50	50.50		135	177

04 21 13.18 Columns

		Crew	Daily Output	Labor-Hours	Unit	Material	2018 Bare Costs Labor	Equipment	Total	Total Incl O&P
0010	COLUMNS, solid, excludes scaffolding, grout and reinforcing									
0050	Brick, 8" x 8", 9 brick/V.L.F.	D-1	56	.286	V.L.F.	5.30	9.05		14.35	21
0100	12" x 8", 13.5 brick/V.L.F.		37	.432		7.95	13.70		21.65	32
0200	12" x 12", 20 brick/V.L.F.		25	.640		11.80	20.50		32.30	47
0300	16" x 12", 27 brick/V.L.F.		19	.842		15.95	26.50		42.45	62
0400	16" x 16", 36 brick/V.L.F.		14	1.143		21.50	36		57.50	84
0500	20" x 16", 45 brick/V.L.F.		11	1.455		26.50	46		72.50	107
0600	20" x 20", 56 brick/V.L.F.		9	1.778		33	56.50		89.50	131
0700	24" x 20", 68 brick/V.L.F.		7	2.286		40	72.50		112.50	165
0800	24" x 24", 81 brick/V.L.F.		6	2.667		48	84.50		132.50	194
1000	36" x 36", 182 brick/V.L.F.		3	5.333		108	169		277	400

04 21 13.30 Oversized Brick

		Crew	Daily Output	Labor-Hours	Unit	Material	2018 Bare Costs Labor	Equipment	Total	Total Incl O&P
0010	OVERSIZED BRICK, excludes scaffolding, grout and reinforcing									
0100	Veneer, 4" x 2.25" x 16"	D-8	387	.103	S.F.	5.40	3.34		8.74	11.55
0102	8" x 2.25" x 16", multicell		265	.151		17.10	4.87		21.97	27
0105	4" x 2.75" x 16"		412	.097		5.60	3.13		8.73	11.40
0107	8" x 2.75" x 16", multicell		295	.136		17	4.38		21.38	26
0110	4" x 4" x 16"		460	.087		3.61	2.81		6.42	8.65
0120	4" x 8" x 16"		533	.075		4.26	2.42		6.68	8.75
0122	4" x 8" x 16" multicell		327	.122		16.05	3.95		20	24.50
0125	Loadbearing, 6" x 4" x 16", grouted and reinforced		387	.103		11.15	3.34		14.49	17.85
0130	8" x 4" x 16", grouted and reinforced		327	.122		12.25	3.95		16.20	20
0132	10" x 4" x 16", grouted and reinforced		327	.122		25	3.95		28.95	34
0135	6" x 8" x 16", grouted and reinforced		440	.091		14.40	2.93		17.33	21
0140	8" x 8" x 16", grouted and reinforced		400	.100		15.45	3.23		18.68	22.50
0145	Curtainwall/reinforced veneer, 6" x 4" x 16"		387	.103		16.15	3.34		19.49	23.50
0150	8" x 4" x 16"		327	.122		19.65	3.95		23.60	28
0152	10" x 4" x 16"		327	.122		27.50	3.95		31.45	37
0155	6" x 8" x 16"		440	.091		19.90	2.93		22.83	27
0160	8" x 8" x 16"		400	.100		28	3.23		31.23	36
0200	For 1 to 3 slots in face, add					15%				
0210	For 4 to 7 slots in face, add					25%				
0220	For bond beams, add					20%				
0230	For bullnose shapes, add					20%				

04 21 Clay Unit Masonry

04 21 13 – Brick Masonry

04 21 13.30 Oversized Brick

		Crew	Daily Output	Labor-Hours	Unit	Material	2018 Bare Costs Labor	Equipment	Total	Total Incl O&P
0240	For open end knockout, add				S.F.	10%				
0250	For white or gray color group, add					10%				
0260	For 135 degree corner, add				↓	250%				

04 21 13.32 Brick Veneer Masonry

		Crew	Daily Output	Labor-Hours	Unit	Material	Labor	Equipment	Total	Total Incl O&P
0010	**BRICK VENEER MASONRY**, for residential installations									
0020	Residential brick veneer, queen size	D-14	337	.095	S.F.	2.81	3.15		5.96	8.35
0030	Modular size	"	270	.119	"	3.62	3.93		7.55	10.55

04 21 13.35 Common Building Brick

		Crew	Daily Output	Labor-Hours	Unit	Material	Labor	Equipment	Total	Total Incl O&P
0010	**COMMON BUILDING BRICK**, C62, T.L. lots, material only R042110-20									
0020	Standard				M	570			570	625
0050	Select				"	535			535	585

04 21 13.45 Face Brick

		Crew	Daily Output	Labor-Hours	Unit	Material	Labor	Equipment	Total	Total Incl O&P
0010	**FACE BRICK** Material Only, C216, T.L. lots R042110-20									
0300	Standard modular, 4" x 2-2/3" x 8"				M	495			495	545
2170	For less than truck load lots, add					15			15	16.50
2180	For buff or gray brick, add				↓	16			16	17.60

04 22 Concrete Unit Masonry

04 22 10 – Concrete Masonry Units

04 22 10.11 Autoclave Aerated Concrete Block

			Crew	Daily Output	Labor-Hours	Unit	Material	Labor	Equipment	Total	Total Incl O&P
0010	**AUTOCLAVE AERATED CONCRETE BLOCK**, excl. scaffolding, grout & reinforcing										
0050	Solid, 4" x 8" x 24", incl. mortar	G	D-8	600	.067	S.F.	1.50	2.15		3.65	5.25
0060	6" x 8" x 24"	G		600	.067		2.25	2.15		4.40	6.05
0070	8" x 8" x 24"	G		575	.070		3	2.25		5.25	7.05
0080	10" x 8" x 24"	G		575	.070		3.66	2.25		5.91	7.75
0090	12" x 8" x 24"	G	↓	550	.073	↓	4.50	2.35		6.85	8.85

04 22 10.14 Concrete Block, Back-Up

		Crew	Daily Output	Labor-Hours	Unit	Material	Labor	Equipment	Total	Total Incl O&P
0010	**CONCRETE BLOCK, BACK-UP**, C90, 2000 psi R042210-20									
0020	Normal weight, 8" x 16" units, tooled joint 1 side									
0050	Not-reinforced, 2000 psi, 2" thick	D-8	475	.084	S.F.	1.60	2.72		4.32	6.30
0200	4" thick		460	.087		1.92	2.81		4.73	6.80
0300	6" thick		440	.091		2.50	2.93		5.43	7.65
0350	8" thick		400	.100		2.66	3.23		5.89	8.35
0400	10" thick	↓	330	.121		3.15	3.91		7.06	10
0450	12" thick	D-9	310	.155		4.36	4.91		9.27	13
1000	Reinforced, alternate courses, 4" thick	D-8	450	.089		2.10	2.87		4.97	7.10
1100	6" thick		430	.093		2.68	3		5.68	7.95
1150	8" thick		395	.101		2.88	3.27		6.15	8.60
1200	10" thick	↓	320	.125		3.32	4.04		7.36	10.40
1250	12" thick	D-9	300	.160	↓	4.54	5.05		9.59	13.45

04 22 10.16 Concrete Block, Bond Beam

		Crew	Daily Output	Labor-Hours	Unit	Material	Labor	Equipment	Total	Total Incl O&P
0010	**CONCRETE BLOCK, BOND BEAM**, C90, 2000 psi									
0020	Not including grout or reinforcing									
0125	Regular block, 6" thick	D-8	584	.068	L.F.	2.75	2.21		4.96	6.70
0130	8" high, 8" thick	"	565	.071		2.86	2.29		5.15	6.95
0150	12" thick	D-9	510	.094		4.05	2.98		7.03	9.45
0525	Lightweight, 6" thick	D-8	592	.068	↓	2.98	2.18		5.16	6.90

04 22 Concrete Unit Masonry

04 22 10 – Concrete Masonry Units

04 22 10.19 Concrete Block, Insulation Inserts

		Crew	Daily Output	Labor-Hours	Unit	Material	2018 Bare Costs Labor	Equipment	Total	Total Incl O&P
0010	**CONCRETE BLOCK, INSULATION INSERTS**									
0100	Styrofoam, plant installed, add to block prices									
0200	8" x 16" units, 6" thick				S.F.	1.22			1.22	1.34
0250	8" thick					1.37			1.37	1.51
0300	10" thick					1.42			1.42	1.56
0350	12" thick					1.57			1.57	1.73
0500	8" x 8" units, 8" thick					1.22			1.22	1.34
0550	12" thick					1.42			1.42	1.56

04 22 10.23 Concrete Block, Decorative

		Crew	Daily Output	Labor-Hours	Unit	Material	2018 Bare Costs Labor	Equipment	Total	Total Incl O&P
0010	**CONCRETE BLOCK, DECORATIVE**, C90, 2000 psi									
5000	Split rib profile units, 1" deep ribs, 8 ribs									
5100	8" x 16" x 4" thick	D-8	345	.116	S.F.	4.06	3.74		7.80	10.70
5150	6" thick		325	.123		4.61	3.97		8.58	11.70
5200	8" thick		300	.133		5.20	4.30		9.50	12.90
5250	12" thick	D-9	275	.175		6.05	5.55		11.60	15.90
5400	For special deeper colors, 4" thick, add					1.37			1.37	1.50
5450	12" thick, add					1.42			1.42	1.57
5600	For white, 4" thick, add					1.37			1.37	1.50
5650	6" thick, add					1.39			1.39	1.53
5700	8" thick, add					1.44			1.44	1.58
5750	12" thick, add					1.48			1.48	1.63

04 22 10.24 Concrete Block, Exterior

		Crew	Daily Output	Labor-Hours	Unit	Material	2018 Bare Costs Labor	Equipment	Total	Total Incl O&P
0010	**CONCRETE BLOCK, EXTERIOR**, C90, 2000 psi									
0020	Reinforced alt courses, tooled joints 2 sides									
0100	Normal weight, 8" x 16" x 6" thick	D-8	395	.101	S.F.	2.46	3.27		5.73	8.15
0200	8" thick		360	.111		3.80	3.59		7.39	10.20
0250	10" thick		290	.138		4.41	4.45		8.86	12.30
0300	12" thick	D-9	250	.192		5.15	6.10		11.25	15.80

04 22 10.26 Concrete Block Foundation Wall

		Crew	Daily Output	Labor-Hours	Unit	Material	2018 Bare Costs Labor	Equipment	Total	Total Incl O&P
0010	**CONCRETE BLOCK FOUNDATION WALL**, C90/C145									
0050	Normal-weight, cut joints, horiz joint reinf, no vert reinf.									
0200	Hollow, 8" x 16" x 6" thick	D-8	455	.088	S.F.	2.93	2.84		5.77	7.95
0250	8" thick		425	.094		3.11	3.04		6.15	8.50
0300	10" thick		350	.114		3.57	3.69		7.26	10.05
0350	12" thick	D-9	300	.160		4.78	5.05		9.83	13.70
0500	Solid, 8" x 16" block, 6" thick	D-8	440	.091		2.99	2.93		5.92	8.20
0550	8" thick	"	415	.096		4.23	3.11		7.34	9.85
0600	12" thick	D-9	350	.137		6.10	4.34		10.44	13.95

04 22 10.32 Concrete Block, Lintels

		Crew	Daily Output	Labor-Hours	Unit	Material	2018 Bare Costs Labor	Equipment	Total	Total Incl O&P
0010	**CONCRETE BLOCK, LINTELS**, C90, normal weight									
0100	Including grout and horizontal reinforcing									
0200	8" x 8" x 8", 1 #4 bar	D-4	300	.133	L.F.	4.03	3.93	.43	8.39	11.45
0250	2 #4 bars		295	.136		4.24	4	.44	8.68	11.80
0400	8" x 16" x 8", 1 #4 bar		275	.145		3.69	4.29	.47	8.45	11.70
0450	2 #4 bars		270	.148		3.91	4.37	.48	8.76	12.10
1000	12" x 8" x 8", 1 #4 bar		275	.145		5.50	4.29	.47	10.26	13.70
1100	2 #4 bars		270	.148		5.75	4.37	.48	10.60	14.10
1150	2 #5 bars		270	.148		6	4.37	.48	10.85	14.35
1200	2 #6 bars		265	.151		6.25	4.45	.49	11.19	14.85
1500	12" x 16" x 8", 1 #4 bar		250	.160		6.60	4.72	.52	11.84	15.65
1600	2 #3 bars		245	.163		6.60	4.81	.53	11.94	15.90

04 22 Concrete Unit Masonry

04 22 10 – Concrete Masonry Units

04 22 10.32 Concrete Block, Lintels

		Crew	Daily Output	Labor-Hours	Unit	Material	2018 Bare Costs Labor	Equipment	Total	Total Incl O&P
1650	2 #4 bars	D-4	245	.163	L.F.	6.80	4.81	.53	12.14	16.10
1700	2 #5 bars	↓	240	.167	↓	7.05	4.91	.54	12.50	16.55

04 22 10.33 Lintel Block

		Crew	Daily Output	Labor-Hours	Unit	Material	2018 Bare Costs Labor	Equipment	Total	Total Incl O&P
0010	**LINTEL BLOCK**									
3481	Lintel block 6" x 8" x 8"	D-1	300	.053	Ea.	1.36	1.69		3.05	4.32
3501	6" x 16" x 8"		275	.058		2.10	1.84		3.94	5.40
3521	8" x 8" x 8"		275	.058		1.21	1.84		3.05	4.41
3561	8" x 16" x 8"	↓	250	.064	↓	1.90	2.03		3.93	5.50

04 22 10.34 Concrete Block, Partitions

		Crew	Daily Output	Labor-Hours	Unit	Material	2018 Bare Costs Labor	Equipment	Total	Total Incl O&P
0010	**CONCRETE BLOCK, PARTITIONS**, excludes scaffolding									
1000	Lightweight block, tooled joints, 2 sides, hollow									
1100	Not reinforced, 8" x 16" x 4" thick	D-8	440	.091	S.F.	1.95	2.93		4.88	7.05
1150	6" thick		410	.098		2.75	3.15		5.90	8.30
1200	8" thick		385	.104		3.36	3.35		6.71	9.30
1250	10" thick	↓	370	.108		4.07	3.49		7.56	10.35
1300	12" thick	D-9	350	.137	↓	4.27	4.34		8.61	11.95
4000	Regular block, tooled joints, 2 sides, hollow									
4100	Not reinforced, 8" x 16" x 4" thick	D-8	430	.093	S.F.	1.82	3		4.82	7
4150	6" thick		400	.100		2.40	3.23		5.63	8.05
4200	8" thick		375	.107		2.57	3.44		6.01	8.55
4250	10" thick	↓	360	.111		3.05	3.59		6.64	9.35
4300	12" thick	D-9	340	.141	↓	4.26	4.47		8.73	12.15

04 23 Glass Unit Masonry

04 23 13 – Vertical Glass Unit Masonry

04 23 13.10 Glass Block

		Crew	Daily Output	Labor-Hours	Unit	Material	2018 Bare Costs Labor	Equipment	Total	Total Incl O&P
0010	**GLASS BLOCK**									
0100	Plain, 4" thick, under 1,000 S.F., 6" x 6"	D-8	115	.348	S.F.	26	11.25		37.25	48
0150	8" x 8"		160	.250		16.40	8.05		24.45	31.50
0160	end block		160	.250		60.50	8.05		68.55	80
0170	90 degree corner		160	.250		63.50	8.05		71.55	83
0180	45 degree corner		160	.250		54	8.05		62.05	72.50
0200	12" x 12"		175	.229		24	7.40		31.40	39
0210	4" x 8"		160	.250		30.50	8.05		38.55	47
0220	6" x 8"	↓	160	.250	↓	19.90	8.05		27.95	35.50
0700	For solar reflective blocks, add					100%				
1000	Thinline, plain, 3-1/8" thick, under 1,000 S.F., 6" x 6"	D-8	115	.348	S.F.	23.50	11.25		34.75	45
1050	8" x 8"		160	.250		13.15	8.05		21.20	28
1400	For cleaning block after installation (both sides), add	↓	1000	.040	↓	.16	1.29		1.45	2.34

329

For customer support on your Residential Costs with RSMeans data, call 800.448.8182.

04 24 Adobe Unit Masonry

04 24 16 – Manufactured Adobe Unit Masonry

04 24 16.06 Adobe Brick		Crew	Daily Output	Labor-Hours	Unit	Material	2018 Bare Costs Labor	Equipment	Total	Total Incl O&P
0010	**ADOBE BRICK**, Semi-stabilized, with cement mortar									
0060	Brick, 10" x 4" x 14", 2.6/S.F.	G	D-8	560	.071	S.F.	4.78	2.31		7.09	9.10
0080	12" x 4" x 16", 2.3/S.F.	G		580	.069		7	2.23		9.23	11.40
0100	10" x 4" x 16", 2.3/S.F.	G		590	.068		6.55	2.19		8.74	10.85
0120	8" x 4" x 16", 2.3/S.F.	G		560	.071		4.99	2.31		7.30	9.35
0140	4" x 4" x 16", 2.3/S.F.	G		540	.074		4.88	2.39		7.27	9.35
0160	6" x 4" x 16", 2.3/S.F.	G		540	.074		4.53	2.39		6.92	9
0180	4" x 4" x 12", 3.0/S.F.	G		520	.077		5.25	2.48		7.73	9.95
0200	8" x 4" x 12", 3.0/S.F.	G		520	.077		4.34	2.48		6.82	8.95

04 27 Multiple-Wythe Unit Masonry

04 27 10 – Multiple-Wythe Masonry

04 27 10.10 Cornices

		Crew	Daily Output	Labor-Hours	Unit	Material	Labor	Equipment	Total	Total Incl O&P
0010	**CORNICES**									
0110	Face bricks, 12 brick/S.F.	D-1	30	.533	SF Face	5.85	16.90		22.75	34.50
0150	15 brick/S.F.	"	23	.696	"	7	22		29	44.50

04 27 10.30 Brick Walls

		Crew	Daily Output	Labor-Hours	Unit	Material	Labor	Equipment	Total	Total Incl O&P
0010	**BRICK WALLS**, including mortar, excludes scaffolding									
0800	Face brick, 4" thick wall, 6.75 brick/S.F.	D-8	215	.186	S.F.	3.99	6		9.99	14.45
0850	Common brick, 4" thick wall, 6.75 brick/S.F.		240	.167		4.50	5.40		9.90	13.95
0900	8" thick, 13.50 brick/S.F.		135	.296		9.25	9.55		18.80	26
1000	12" thick, 20.25 brick/S.F.		95	.421		13.90	13.60		27.50	38
1050	16" thick, 27.00 brick/S.F.		75	.533		18.75	17.20		35.95	49.50
1200	Reinforced, face brick, 4" thick wall, 6.75 brick/S.F.		210	.190		4.15	6.15		10.30	14.80
1220	Common brick, 4" thick wall, 6.75 brick/S.F.		235	.170		4.66	5.50		10.16	14.35
1250	8" thick, 13.50 brick/S.F.		130	.308		9.55	9.95		19.50	27
1300	12" thick, 20.25 brick/S.F.		90	.444		14.40	14.35		28.75	40
1350	16" thick, 27.00 brick/S.F.		70	.571		19.40	18.45		37.85	52.50

04 27 10.40 Steps

		Crew	Daily Output	Labor-Hours	Unit	Material	Labor	Equipment	Total	Total Incl O&P
0010	**STEPS**									
0012	Entry steps, select common brick	D-1	.30	53.333	M	535	1,700		2,235	3,400

04 41 Dry-Placed Stone

04 41 10 – Dry Placed Stone

04 41 10.10 Rough Stone Wall

			Crew	Daily Output	Labor-Hours	Unit	Material	Labor	Equipment	Total	Total Incl O&P
0011	**ROUGH STONE WALL**, Dry										
0012	Dry laid (no mortar), under 18" thick	G	D-1	60	.267	C.F.	13.50	8.45		21.95	29
0100	Random fieldstone, under 18" thick	G	D-12	60	.533		13.50	16.90		30.40	43
0150	Over 18" thick	G	"	63	.508		16.20	16.10		32.30	45
0500	Field stone veneer	G	D-8	120	.333	S.F.	12.65	10.75		23.40	32
0510	Valley stone veneer	G		120	.333		12.65	10.75		23.40	32
0520	River stone veneer	G		120	.333		12.65	10.75		23.40	32
0600	Rubble stone walls, in mortar bed, up to 18" thick	G	D-11	75	.320	C.F.	16.30	10.45		26.75	35.50

04 43 Stone Masonry

04 43 10 – Masonry with Natural and Processed Stone

04 43 10.45 Granite		Crew	Daily Output	Labor-Hours	Unit	Material	2018 Bare Costs Labor	2018 Bare Costs Equipment	Total	Total Incl O&P
0010	**GRANITE**, cut to size									
2500	Steps, copings, etc., finished on more than one surface									
2550	Low price, gray, light gray, etc.	D-10	50	.640	C.F.	94	22	7.25	123.25	148
2575	Medium price, pink, brown, etc.		50	.640		122	22	7.25	151.25	179
2600	High price, red, black, etc.	↓	50	.640	↓	150	22	7.25	179.25	210
2800	Pavers, 4" x 4" x 4" blocks, split face and joints									
2850	Low price, gray, light gray, etc.	D-11	80	.300	S.F.	13.30	9.80		23.10	31
2875	Medium price, pink, brown, etc.		80	.300		21.50	9.80		31.30	40
2900	High price, red, black, etc.	↓	80	.300	↓	29.50	9.80		39.30	49
5000	Reclaimed or antique									
5010	Treads, up to 12" wide	D-10	100	.320	L.F.	42.50	10.95	3.64	57.09	69.50
5020	Up to 18" wide		100	.320		38.50	10.95	3.64	53.09	65
5030	Capstone, size varies		50	.640	↓	30.50	22	7.25	59.75	78
5040	Posts	↓	30	1.067	V.L.F.	30.50	36.50	12.10	79.10	108

04 43 10.55 Limestone

04 43 10.55 Limestone		Crew	Daily Output	Labor-Hours	Unit	Material	Labor	Equipment	Total	Total Incl O&P
0010	**LIMESTONE**, cut to size									
0020	Veneer facing panels									
0500	Texture finish, light stick, 4-1/2" thick, 5' x 12'	D-4	300	.133	S.F.	20.50	3.93	.43	24.86	29.50
0750	5" thick, 5' x 14' panels	D-10	275	.116		21.50	3.99	1.32	26.81	32
1000	Sugarcube finish, 2" thick, 3' x 5' panels		275	.116		29	3.99	1.32	34.31	40
1050	3" thick, 4' x 9' panels		275	.116		25.50	3.99	1.32	30.81	36.50
1200	4" thick, 5' x 11' panels		275	.116		31.50	3.99	1.32	36.81	42.50
1400	Sugarcube, textured finish, 4-1/2" thick, 5' x 12'		275	.116		32.50	3.99	1.32	37.81	44
1450	5" thick, 5' x 14' panels		275	.116	↓	34	3.99	1.32	39.31	45
2000	Coping, sugarcube finish, top & 2 sides		30	1.067	C.F.	68.50	36.50	12.10	117.10	150
2100	Sills, lintels, jambs, trim, stops, sugarcube finish, simple		20	1.600		68.50	55	18.20	141.70	187
2150	Detailed		20	1.600	↓	68.50	55	18.20	141.70	187
2300	Steps, extra hard, 14" wide, 6" rise	↓	50	.640	L.F.	25.50	22	7.25	54.75	72.50
3000	Quoins, plain finish, 6" x 12" x 12"	D-12	25	1.280	Ea.	41	40.50		81.50	114
3050	6" x 16" x 24"	"	25	1.280	"	55	40.50		95.50	129

04 43 10.60 Marble

04 43 10.60 Marble		Crew	Daily Output	Labor-Hours	Unit	Material	Labor	Equipment	Total	Total Incl O&P
0011	**MARBLE**, ashlar, split face, +/- 4" thick, random									
0040	Lengths 1' to 4' & heights 2" to 7-1/2", average	D-8	175	.229	S.F.	18.30	7.40		25.70	32.50
0100	Base, polished, 3/4" or 7/8" thick, polished, 6" high	D-10	65	.492	L.F.	12	16.90	5.60	34.50	47.50
1000	Facing, polished finish, cut to size, 3/4" to 7/8" thick									
1050	Carrara or equal	D-10	130	.246	S.F.	22.50	8.45	2.80	33.75	42
1100	Arabescato or equal	"	130	.246	"	39.50	8.45	2.80	50.75	60.50
2200	Window sills, 6" x 3/4" thick	D-1	85	.188	L.F.	11.10	5.95		17.05	22
2500	Flooring, polished tiles, 12" x 12" x 3/8" thick									
2510	Thin set, Giallo Solare or equal	D-11	90	.267	S.F.	17.50	8.70		26.20	34
2600	Sky Blue or equal		90	.267		16	8.70		24.70	32
2700	Mortar bed, Giallo Solare or equal		65	.369		17.50	12.05		29.55	39.50
2740	Sky Blue or equal	↓	65	.369		16	12.05		28.05	37.50
2780	Travertine, 3/8" thick, Sierra or equal	D-10	130	.246		9.40	8.45	2.80	20.65	27.50
2790	Silver or equal	"	130	.246	↓	26	8.45	2.80	37.25	45.50
3500	Thresholds, 3' long, 7/8" thick, 4" to 5" wide, plain	D-12	24	1.333	Ea.	35.50	42		77.50	110
3550	Beveled		24	1.333	"	71.50	42		113.50	150
3700	Window stools, polished, 7/8" thick, 5" wide		85	.376	L.F.	21.50	11.95		33.45	44

04 43 10.75 Sandstone or Brownstone

04 43 10.75 Sandstone or Brownstone		Crew	Daily Output	Labor-Hours	Unit	Material	Labor	Equipment	Total	Total Incl O&P
0011	**SANDSTONE OR BROWNSTONE**									
0100	Sawed face veneer, 2-1/2" thick, to 2' x 4' panels	D-10	130	.246	S.F.	21.50	8.45	2.80	32.75	40.50
0150	4" thick, to 3'-6" x 8' panels	↓	100	.320	↓	21.50	10.95	3.64	36.09	46

For customer support on your Residential Costs with RSMeans data, call 800.448.8182.

331

04 43 Stone Masonry

04 43 10 – Masonry with Natural and Processed Stone

04 43 10.75 Sandstone or Brownstone

		Crew	Daily Output	Labor-Hours	Unit	Material	2018 Bare Costs Labor	Equipment	Total	Total Incl O&P
0300	Split face, random sizes	D-10	100	.320	S.F.	13.40	10.95	3.64	27.99	37
0350	Cut stone trim (limestone)									
0360	Ribbon stone, 4" thick, 5' pieces	D-8	120	.333	Ea.	160	10.75		170.75	194
0370	Cove stone, 4" thick, 5' pieces		105	.381		161	12.30		173.30	198
0380	Cornice stone, 10" to 12" wide		90	.444		199	14.35		213.35	243
0390	Band stone, 4" thick, 5' pieces		145	.276		109	8.90		117.90	135
0410	Window and door trim, 3" to 4" wide		160	.250		93.50	8.05		101.55	117
0420	Key stone, 18" long	▼	60	.667	▼	93.50	21.50		115	139

04 43 10.80 Slate

		Crew	Daily Output	Labor-Hours	Unit	Material	2018 Bare Costs Labor	Equipment	Total	Total Incl O&P
0010	**SLATE**									
3100	Stair landings, 1" thick, black, clear	D-1	65	.246	S.F.	21	7.80		28.80	36.50
3200	Ribbon	"	65	.246	"	23.50	7.80		31.30	38.50
3500	Stair treads, sand finish, 1" thick x 12" wide									
3600	3 L.F. to 6 L.F.	D-10	120	.267	L.F.	25	9.15	3.03	37.18	46
3700	Ribbon, sand finish, 1" thick x 12" wide									
3750	To 6 L.F.	D-10	120	.267	L.F.	21	9.15	3.03	33.18	42

04 43 10.85 Window Sill

		Crew	Daily Output	Labor-Hours	Unit	Material	2018 Bare Costs Labor	Equipment	Total	Total Incl O&P
0010	**WINDOW SILL**									
0020	Bluestone, thermal top, 10" wide, 1-1/2" thick	D-1	85	.188	S.F.	9.60	5.95		15.55	20.50
0050	2" thick		75	.213	"	9.60	6.75		16.35	22
0100	Cut stone, 5" x 8" plain		48	.333	L.F.	12.55	10.55		23.10	31.50
0200	Face brick on edge, brick, 8" wide		80	.200		5.50	6.35		11.85	16.65
0400	Marble, 9" wide, 1" thick		85	.188		8.95	5.95		14.90	19.80
0900	Slate, colored, unfading, honed, 12" wide, 1" thick		85	.188		8.95	5.95		14.90	19.75
0950	2" thick	▼	70	.229	▼	8.95	7.25		16.20	22

04 51 Flue Liner Masonry

04 51 10 – Clay Flue Lining

04 51 10.10 Flue Lining

		Crew	Daily Output	Labor-Hours	Unit	Material	2018 Bare Costs Labor	Equipment	Total	Total Incl O&P
0010	**FLUE LINING,** including mortar									
0020	Clay, 8" x 8"	D-1	125	.128	V.L.F.	6	4.06		10.06	13.40
0100	8" x 12"		103	.155		8.75	4.92		13.67	17.80
0200	12" x 12"		93	.172		11.30	5.45		16.75	21.50
0300	12" x 18"		84	.190		22.50	6.05		28.55	35
0400	18" x 18"		75	.213		29	6.75		35.75	43.50
0500	20" x 20"		66	.242		43.50	7.70		51.20	60.50
0600	24" x 24"		56	.286		55.50	9.05		64.55	76
1000	Round, 18" diameter		66	.242		39.50	7.70		47.20	56.50
1100	24" diameter	▼	47	.340	▼	76	10.80		86.80	102

04 57 Masonry Fireplaces

04 57 10 – Brick or Stone Fireplaces

04 57 10.10 Fireplace

		Crew	Daily Output	Labor-Hours	Unit	Material	2018 Bare Costs Labor	Equipment	Total	Total Incl O&P
0010	**FIREPLACE**									
0100	Brick fireplace, not incl. foundations or chimneys									
0110	30" x 29" opening, incl. chamber, plain brickwork	D-1	.40	40	Ea.	590	1,275		1,865	2,775
0200	Fireplace box only (110 brick)	"	2	8	"	162	253		415	605
0300	For elaborate brickwork and details, add					35%	35%			
0400	For hearth, brick & stone, add	D-1	2	8	Ea.	213	253		466	660
0410	For steel, damper, cleanouts, add		4	4		17.55	127		144.55	231
0600	Plain brickwork, incl. metal circulator		.50	32		900	1,025		1,925	2,700
0800	Face brick only, standard size, 8" x 2-2/3" x 4"		.30	53.333	M	620	1,700		2,320	3,500
0900	Stone fireplace, fieldstone, add				SF Face	8.95			8.95	9.80
1000	Cut stone, add				"	8.40			8.40	9.25

04 72 Cast Stone Masonry

04 72 10 – Cast Stone Masonry Features

04 72 10.10 Coping

		Crew	Daily Output	Labor-Hours	Unit	Material	2018 Bare Costs Labor	Equipment	Total	Total Incl O&P
0010	**COPING**, stock units									
0050	Precast concrete, 10" wide, 4" tapers to 3-1/2", 8" wall	D-1	75	.213	L.F.	22.50	6.75		29.25	36.50
0100	12" wide, 3-1/2" tapers to 3", 10" wall		70	.229		24.50	7.25		31.75	39
0110	14" wide, 4" tapers to 3-1/2", 12" wall		65	.246		28	7.80		35.80	43.50
0150	16" wide, 4" tapers to 3-1/2", 14" wall		60	.267		30	8.45		38.45	47
0300	Limestone for 12" wall, 4" thick		90	.178		14.80	5.65		20.45	25.50
0350	6" thick		80	.200		22	6.35		28.35	35
0500	Marble, to 4" thick, no wash, 9" wide		90	.178		12.90	5.65		18.55	23.50
0550	12" wide		80	.200		18	6.35		24.35	30.50
0700	Terra cotta, 9" wide		90	.178		6.45	5.65		12.10	16.50
0750	12" wide		80	.200		8.80	6.35		15.15	20.50
0800	Aluminum, for 12" wall		80	.200		9.20	6.35		15.55	20.50

04 72 20 – Cultured Stone Veneer

04 72 20.10 Cultured Stone Veneer Components

		Crew	Daily Output	Labor-Hours	Unit	Material	2018 Bare Costs Labor	Equipment	Total	Total Incl O&P
0010	**CULTURED STONE VENEER COMPONENTS**									
0110	On wood frame and sheathing substrate, random sized cobbles, corner stones	D-8	70	.571	V.L.F.	10.40	18.45		28.85	42.50
0120	Field stones		140	.286	S.F.	7.55	9.20		16.75	23.50
0130	Random sized flats, corner stones		70	.571	V.L.F.	10.70	18.45		29.15	43
0140	Field stones		140	.286	S.F.	8.90	9.20		18.10	25
0150	Horizontal lined ledgestones, corner stones		75	.533	V.L.F.	10.40	17.20		27.60	40.50
0160	Field stones		150	.267	S.F.	7.55	8.60		16.15	22.50
0170	Random shaped flats, corner stones		65	.615	V.L.F.	10.40	19.85		30.25	44.50
0180	Field stones		150	.267	S.F.	7.55	8.60		16.15	22.50
0190	Random shaped/textured face, corner stones		65	.615	V.L.F.	10.40	19.85		30.25	44.50
0200	Field stones		130	.308	S.F.	7.55	9.95		17.50	25
0210	Random shaped river rock, corner stones		65	.615	V.L.F.	10.40	19.85		30.25	44.50
0220	Field stones		130	.308	S.F.	7.55	9.95		17.50	25
0240	On concrete or CMU substrate, random sized cobbles, corner stones		70	.571	V.L.F.	9.70	18.45		28.15	41.50
0250	Field stones		140	.286	S.F.	7.20	9.20		16.40	23.50
0260	Random sized flats, corner stones		70	.571	V.L.F.	10	18.45		28.45	42
0270	Field stones		140	.286	S.F.	8.55	9.20		17.75	25
0280	Horizontal lined ledgestones, corner stones		75	.533	V.L.F.	9.70	17.20		26.90	39.50
0290	Field stones		150	.267	S.F.	7.20	8.60		15.80	22.50
0300	Random shaped flats, corner stones		70	.571	V.L.F.	9.70	18.45		28.15	41.50
0310	Field stones		140	.286	S.F.	7.20	9.20		16.40	23.50

04 72 20.10 Cultured Stone Veneer Components	Crew	Daily Output	Labor-Hours	Unit	Material	2018 Bare Costs Labor	2018 Bare Costs Equipment	Total	Total Incl O&P
0320	Random shaped/textured face, corner stones	D-8	65	.615	V.L.F.	9.70	19.85		29.55	43.50
0330	Field stones		130	.308	S.F.	7.20	9.95		17.15	24.50
0340	Random shaped river rock, corner stones		65	.615	V.L.F.	9.70	19.85		29.55	43.50
0350	Field stones		130	.308	S.F.	7.20	9.95		17.15	24.50
0360	Cultured stone veneer, #15 felt weather resistant barrier	1 Clab	3700	.002	Sq.	5.35	.06		5.41	6
0390	Water table or window sill, 18" long	1 Bric	80	.100	Ea.	10	3.47		13.47	16.80

Estimating Tips
05 05 00 Common Work Results for Metals

- Nuts, bolts, washers, connection angles, and plates can add a significant amount to both the tonnage of a structural steel job and the estimated cost. As a rule of thumb, add 10% to the total weight to account for these accessories.

- Type 2 steel construction, commonly referred to as "simple construction," consists generally of field-bolted connections with lateral bracing supplied by other elements of the building, such as masonry walls or x-bracing. The estimator should be aware, however, that shop connections may be accomplished by welding or bolting. The method may be particular to the fabrication shop and may have an impact on the estimated cost.

05 10 00 Structural Steel

- Steel items can be obtained from two sources: a fabrication shop or a metals service center. Fabrication shops can fabricate items under more controlled conditions than crews in the field can. They are also more efficient and can produce items more economically. Metal service centers serve as a source of long mill shapes to both fabrication shops and contractors.

- Most line items in this structural steel subdivision, and most items in 05 50 00 Metal Fabrications, are indicated as being shop fabricated. The bare material cost for these shop fabricated items is the "Invoice Cost" from the shop and includes the mill base price of steel plus mill extras, transportation to the shop, shop drawings and detailing where warranted, shop fabrication and handling, sandblasting and a shop coat of primer paint, all necessary structural bolts, and delivery to the job site. The bare labor cost and bare equipment cost for these shop fabricated items are for field installation or erection.

- Line items in Subdivision 05 12 23.40 Lightweight Framing, and other items scattered in Division 5, are indicated as being field fabricated. The bare material cost for these field fabricated items is the "Invoice Cost" from the metals service center and includes the mill base price of steel plus mill extras, transportation to the metals service center, material handling, and delivery of long lengths of mill shapes to the job site. Material costs for structural bolts and welding rods should be added to the estimate. The bare labor cost and bare equipment cost for these items are for both field fabrication and field installation or erection, and include time for cutting, welding, and drilling in the fabricated metal items. Drilling into concrete and fasteners to fasten field fabricated items to other work are not included and should be added to the estimate.

05 20 00 Steel Joist Framing

- In any given project the total weight of open web steel joists is determined by the loads to be supported and the design. However, economies can be realized in minimizing the amount of labor used to place the joists. This is done by maximizing the joist spacing, and therefore minimizing the number of joists required to be installed on the job. Certain spacings and locations may be required by the design, but in other cases maximizing the spacing and keeping it as uniform as possible will keep the costs down.

05 30 00 Steel Decking

- The takeoff and estimating of a metal deck involves more than the area of the floor or roof and the type of deck specified or shown on the drawings. Many different sizes and types of openings may exist. Small openings for individual pipes or conduits may be drilled after the floor/roof is installed, but larger openings may require special deck lengths as well as reinforcing or structural support. The estimator should determine who will be supplying this reinforcing. Additionally, some deck terminations are part of the deck package, such as screed angles and pour stops, and others will be part of the steel contract, such as angles attached to structural members and cast-in-place angles and plates. The estimator must ensure that all pieces are accounted for in the complete estimate.

05 50 00 Metal Fabrications

- The most economical steel stairs are those that use common materials, standard details, and most importantly, a uniform and relatively simple method of field assembly. Commonly available A36/A992 channels and plates are very good choices for the main stringers of the stairs, as are angles and tees for the carrier members. Risers and treads are usually made by specialty shops, and it is most economical to use a typical detail in as many places as possible. The stairs should be pre-assembled and shipped directly to the site. The field connections should be simple and straightforward enough to be accomplished efficiently, and with minimum equipment and labor.

Reference Numbers

Reference numbers are shown at the beginning of some major classifications. These numbers refer to related items in the Reference Section. The reference information may be an estimating procedure, an alternate pricing method, or technical information.

Note: Not all subdivisions listed here necessarily appear. ■

Did you know?

RSMeans data is available through our online application with 24/7 access:

- Search for unit prices by keyword
- Leverage the most up-to-date data
- Build and export estimates

Try it free for 30 days!
www.rsmeans.com/2018freetrial

05 05 19.10 Chemical Anchors

		Crew	Daily Output	Labor-Hours	Unit	Material	2018 Bare Costs Labor	Equipment	Total	Total Incl O&P
0010	**CHEMICAL ANCHORS**									
0020	Includes layout & drilling									
1430	Chemical anchor, w/rod & epoxy cartridge, 3/4" diameter x 9-1/2" long	B-89A	27	.593	Ea.	10.85	18.50	4.15	33.50	47.50
1435	1" diameter x 11-3/4" long		24	.667		20	21	4.67	45.67	62
1440	1-1/4" diameter x 14" long		21	.762		38	24	5.35	67.35	87
1445	1-3/4" diameter x 15" long		20	.800		69.50	25	5.60	100.10	124
1450	18" long		17	.941		83	29.50	6.60	119.10	148
1455	2" diameter x 18" long		16	1		113	31.50	7	151.50	184
1460	24" long		15	1.067		147	33.50	7.50	188	225

05 05 19.20 Expansion Anchors

			Crew	Daily Output	Labor-Hours	Unit	Material	2018 Bare Costs Labor	Equipment	Total	Total Incl O&P
0010	**EXPANSION ANCHORS**										
0100	Anchors for concrete, brick or stone, no layout and drilling										
0200	Expansion shields, zinc, 1/4" diameter, 1-5/16" long, single	G	1 Carp	90	.089	Ea.	.45	3.11		3.56	5.65
0300	1-3/8" long, double	G		85	.094		.58	3.29		3.87	6.10
0400	3/8" diameter, 1-1/2" long, single	G		85	.094		.69	3.29		3.98	6.20
0500	2" long, double	G		80	.100		1.15	3.50		4.65	7.05
0600	1/2" diameter, 2-1/16" long, single	G		80	.100		1.26	3.50		4.76	7.20
0700	2-1/2" long, double	G		75	.107		2.13	3.73		5.86	8.55
0800	5/8" diameter, 2-5/8" long, single	G		75	.107		2.18	3.73		5.91	8.60
0900	2-3/4" long, double	G		70	.114		2.92	3.99		6.91	9.85
1000	3/4" diameter, 2-3/4" long, single	G		70	.114		3.29	3.99		7.28	10.25
1100	3-15/16" long, double	G		65	.123		5.45	4.30		9.75	13.15
2100	Hollow wall anchors for gypsum wall board, plaster or tile										
2500	3/16" diameter, short	G	1 Carp	150	.053	Ea.	.51	1.86		2.37	3.66
3000	Toggle bolts, bright steel, 1/8" diameter, 2" long	G		85	.094		.25	3.29		3.54	5.75
3100	4" long	G		80	.100		.29	3.50		3.79	6.10
3200	3/16" diameter, 3" long	G		80	.100		.33	3.50		3.83	6.15
3300	6" long	G		75	.107		.45	3.73		4.18	6.70
3400	1/4" diameter, 3" long	G		75	.107		.41	3.73		4.14	6.65
3500	6" long	G		70	.114		.59	3.99		4.58	7.30
3600	3/8" diameter, 3" long	G		70	.114		.86	3.99		4.85	7.60
3700	6" long	G		60	.133		1.50	4.66		6.16	9.40
3800	1/2" diameter, 4" long	G		60	.133		1.65	4.66		6.31	9.55
3900	6" long	G		50	.160		2.08	5.60		7.68	11.60
4000	Nailing anchors										
4100	Nylon nailing anchor, 1/4" diameter, 1" long		1 Carp	3.20	2.500	C	23	87.50		110.50	170
4200	1-1/2" long			2.80	2.857		25.50	100		125.50	195
4300	2" long			2.40	3.333		27.50	117		144.50	224
4400	Metal nailing anchor, 1/4" diameter, 1" long	G		3.20	2.500		22.50	87.50		110	170
4500	1-1/2" long	G		2.80	2.857		29	100		129	198
4600	2" long	G		2.40	3.333		35.50	117		152.50	233
5000	Screw anchors for concrete, masonry,										
5100	stone & tile, no layout or drilling included										
5700	Lag screw shields, 1/4" diameter, short	G	1 Carp	90	.089	Ea.	.47	3.11		3.58	5.65
5800	Long	G		85	.094		.54	3.29		3.83	6.05
5900	3/8" diameter, short	G		85	.094		.72	3.29		4.01	6.25
6000	Long	G		80	.100		.92	3.50		4.42	6.80
6100	1/2" diameter, short	G		80	.100		1.03	3.50		4.53	6.95
6200	Long	G		75	.107		1.37	3.73		5.10	7.70
6300	5/8" diameter, short	G		70	.114		1.42	3.99		5.41	8.20
6400	Long	G		65	.123		1.96	4.30		6.26	9.30
6600	Lead, #6 & #8, 3/4" long	G		260	.031		.19	1.08		1.27	2

05 05 Common Work Results for Metals

05 05 19 – Post-Installed Concrete Anchors

05 05 19.20 Expansion Anchors

		Crew	Daily Output	Labor-Hours	Unit	Material	2018 Bare Costs Labor	2018 Bare Costs Equipment	Total	Total Incl O&P
6700	#10 - #14, 1-1/2" long **G**	1 Carp	200	.040	Ea.	.38	1.40		1.78	2.74
6800	#16 & #18, 1-1/2" long **G**		160	.050		.41	1.75		2.16	3.36
6900	Plastic, #6 & #8, 3/4" long		260	.031		.05	1.08		1.13	1.85
7000	#8 & #10, 7/8" long		240	.033		.06	1.16		1.22	2.01
7100	#10 & #12, 1" long		220	.036		.07	1.27		1.34	2.19
7200	#14 & #16, 1-1/2" long		160	.050		.07	1.75		1.82	2.99
8950	Self-drilling concrete screw, hex washer head, 3/16" diam. x 1-3/4" long **G**		300	.027		.19	.93		1.12	1.76
8960	2-1/4" long **G**		250	.032		.21	1.12		1.33	2.09
8970	Phillips flat head, 3/16" diam. x 1-3/4" long **G**		300	.027		.19	.93		1.12	1.76
8980	2-1/4" long **G**		250	.032		.21	1.12		1.33	2.09

05 05 21 – Fastening Methods for Metal

05 05 21.15 Drilling Steel

		Crew	Daily Output	Labor-Hours	Unit	Material	2018 Bare Costs Labor	2018 Bare Costs Equipment	Total	Total Incl O&P
0010	**DRILLING STEEL**									
1910	Drilling & layout for steel, up to 1/4" deep, no anchor									
1920	Holes, 1/4" diameter	1 Sswk	112	.071	Ea.	.10	2.77		2.87	5.05
1925	For each additional 1/4" depth, add		336	.024		.10	.92		1.02	1.75
1930	3/8" diameter		104	.077		.09	2.98		3.07	5.40
1935	For each additional 1/4" depth, add		312	.026		.09	.99		1.08	1.87
1940	1/2" diameter		96	.083		.11	3.23		3.34	5.85
1945	For each additional 1/4" depth, add		288	.028		.11	1.08		1.19	2.03
1950	5/8" diameter		88	.091		.16	3.53		3.69	6.45
1955	For each additional 1/4" depth, add		264	.030		.16	1.18		1.34	2.27
1960	3/4" diameter		80	.100		.20	3.88		4.08	7.10
1965	For each additional 1/4" depth, add		240	.033		.20	1.29		1.49	2.51
1970	7/8" diameter		72	.111		.27	4.31		4.58	7.95
1975	For each additional 1/4" depth, add		216	.037		.27	1.44		1.71	2.84
1980	1" diameter		64	.125		.23	4.85		5.08	8.85
1985	For each additional 1/4" depth, add		192	.042		.23	1.62		1.85	3.12
1990	For drilling up, add						40%			

05 05 23 – Metal Fastenings

05 05 23.10 Bolts and Hex Nuts

		Crew	Daily Output	Labor-Hours	Unit	Material	2018 Bare Costs Labor	2018 Bare Costs Equipment	Total	Total Incl O&P
0010	**BOLTS & HEX NUTS**, Steel, A307									
0100	1/4" diameter, 1/2" long **G**	1 Sswk	140	.057	Ea.	.06	2.22		2.28	4
0200	1" long **G**		140	.057		.07	2.22		2.29	4.01
0300	2" long **G**		130	.062		.10	2.39		2.49	4.35
0400	3" long **G**		130	.062		.15	2.39		2.54	4.41
0500	4" long **G**		120	.067		.17	2.59		2.76	4.77
0600	3/8" diameter, 1" long **G**		130	.062		.14	2.39		2.53	4.40
0700	2" long **G**		130	.062		.18	2.39		2.57	4.44
0800	3" long **G**		120	.067		.24	2.59		2.83	4.85
0900	4" long **G**		120	.067		.30	2.59		2.89	4.92
1000	5" long **G**		115	.070		.38	2.70		3.08	5.20
1100	1/2" diameter, 1-1/2" long **G**		120	.067		.40	2.59		2.99	5.05
1200	2" long **G**		120	.067		.46	2.59		3.05	5.10
1300	4" long **G**		115	.070		.75	2.70		3.45	5.60
1400	6" long **G**		110	.073		1.05	2.82		3.87	6.15
1500	8" long **G**		105	.076		1.38	2.96		4.34	6.75
1600	5/8" diameter, 1-1/2" long **G**		120	.067		.85	2.59		3.44	5.55
1700	2" long **G**		120	.067		.94	2.59		3.53	5.60
1800	4" long **G**		115	.070		1.34	2.70		4.04	6.25
1900	6" long **G**		110	.073		1.72	2.82		4.54	6.90
2000	8" long **G**		105	.076		2.55	2.96		5.51	8.05

05 05 23.10 Bolts and Hex Nuts

			Crew	Daily Output	Labor-Hours	Unit	Material	2018 Bare Costs Labor	Equipment	Total	Total Incl O&P
2100	10" long	G	1 Sswk	100	.080	Ea.	3.20	3.10		6.30	9
2200	3/4" diameter, 2" long	G		120	.067		1.15	2.59		3.74	5.85
2300	4" long	G		110	.073		1.65	2.82		4.47	6.80
2400	6" long	G		105	.076		2.12	2.96		5.08	7.60
2500	8" long	G		95	.084		3.20	3.27		6.47	9.30
2600	10" long	G		85	.094		4.20	3.65		7.85	11.10
2700	12" long	G		80	.100		4.92	3.88		8.80	12.30
2800	1" diameter, 3" long	G		105	.076		2.89	2.96		5.85	8.40
2900	6" long	G		90	.089		4.20	3.45		7.65	10.70
3000	12" long	G	▼	75	.107	▼	7.50	4.14		11.64	15.60
3100	For galvanized, add						75%				
3200	For stainless, add						350%				

05 05 23.30 Lag Screws

			Crew	Daily Output	Labor-Hours	Unit	Material	2018 Bare Costs Labor	Equipment	Total	Total Incl O&P
0010	**LAG SCREWS**										
0020	Steel, 1/4" diameter, 2" long	G	1 Carp	200	.040	Ea.	.10	1.40		1.50	2.43
0100	3/8" diameter, 3" long	G		150	.053		.28	1.86		2.14	3.41
0200	1/2" diameter, 3" long	G		130	.062		.66	2.15		2.81	4.31
0300	5/8" diameter, 3" long	G	▼	120	.067	▼	1.29	2.33		3.62	5.30

05 05 23.50 Powder Actuated Tools and Fasteners

			Crew	Daily Output	Labor-Hours	Unit	Material	2018 Bare Costs Labor	Equipment	Total	Total Incl O&P
0010	**POWDER ACTUATED TOOLS & FASTENERS**										
0020	Stud driver, .22 caliber, single shot					Ea.	152			152	167
0100	.27 caliber, semi automatic, strip					"	460			460	505
0300	Powder load, single shot, .22 cal, power level 2, brown					C	5.65			5.65	6.25
0400	Strip, .27 cal, power level 4, red						8.15			8.15	9
0600	Drive pin, .300 x 3/4" long	G	1 Carp	4.80	1.667		4.39	58.50		62.89	102
0700	.300 x 3" long with washer	G	"	4	2	▼	13.40	70		83.40	131

05 05 23.55 Rivets

			Crew	Daily Output	Labor-Hours	Unit	Material	2018 Bare Costs Labor	Equipment	Total	Total Incl O&P
0010	**RIVETS**										
0100	Aluminum rivet & mandrel, 1/2" grip length x 1/8" diameter	G	1 Carp	4.80	1.667	C	7.70	58.50		66.20	105
0200	3/16" diameter	G		4	2		10.80	70		80.80	128
0300	Aluminum rivet, steel mandrel, 1/8" diameter	G		4.80	1.667		10.40	58.50		68.90	108
0400	3/16" diameter	G		4	2		16.65	70		86.65	134
0500	Copper rivet, steel mandrel, 1/8" diameter	G		4.80	1.667		9.95	58.50		68.45	108
0800	Stainless rivet & mandrel, 1/8" diameter	G		4.80	1.667		23	58.50		81.50	122
0900	3/16" diameter	G		4	2		37.50	70		107.50	158
1000	Stainless rivet, steel mandrel, 1/8" diameter	G		4.80	1.667		15.95	58.50		74.45	115
1100	3/16" diameter	G		4	2		26	70		96	145
1200	Steel rivet and mandrel, 1/8" diameter	G		4.80	1.667		7.55	58.50		66.05	105
1300	3/16" diameter	G	▼	4	2	▼	10.95	70		80.95	128
1400	Hand riveting tool, standard					Ea.	70			70	77
1500	Deluxe						395			395	435
1600	Power riveting tool, standard						540			540	595
1700	Deluxe					▼	1,650			1,650	1,825

338

For customer support on your Residential Costs with RSMeans data, call 800.448.8182.

05 12 Structural Steel Framing

05 12 23 – Structural Steel for Buildings

05 12 23.10 Ceiling Supports

		Crew	Daily Output	Labor-Hours	Unit	Material	2018 Bare Costs Labor	Equipment	Total	Total Incl O&P
0010	**CEILING SUPPORTS**									
1000	Entrance door/folding partition supports, shop fabricated	G E-4	60	.533	L.F.	27	21	1.64	49.64	68.50
1100	Linear accelerator door supports	G	14	2.286		123	90	7.05	220.05	300
1200	Lintels or shelf angles, hung, exterior hot dipped galv.	G	267	.120		18.40	4.71	.37	23.48	29
1250	Two coats primer paint instead of galv.	G	267	.120	↓	15.95	4.71	.37	21.03	26.50
1400	Monitor support, ceiling hung, expansion bolted	G	4	8	Ea.	425	315	24.50	764.50	1,050
1450	Hung from pre-set inserts	G	6	5.333		460	210	16.45	686.45	895
1600	Motor supports for overhead doors	G	4	8	↓	217	315	24.50	556.50	825
1700	Partition support for heavy folding partitions, without pocket	G	24	1.333	L.F.	61.50	52.50	4.11	118.11	165
1750	Supports at pocket only	G	12	2.667		123	105	8.20	236.20	330
2000	Rolling grilles & fire door supports	G	34	.941	↓	52.50	37	2.90	92.40	127
2100	Spider-leg light supports, expansion bolted to ceiling slab	G	8	4	Ea.	175	157	12.30	344.30	485
2150	Hung from pre-set inserts	G	12	2.667	"	188	105	8.20	301.20	400
2400	Toilet partition support	G	36	.889	L.F.	61.50	35	2.74	99.24	133
2500	X-ray travel gantry support	G	12	2.667	"	210	105	8.20	323.20	425

05 12 23.15 Columns, Lightweight

		Crew	Daily Output	Labor-Hours	Unit	Material	2018 Bare Costs Labor	Equipment	Total	Total Incl O&P
0010	**COLUMNS, LIGHTWEIGHT**									
8000	Lally columns, to 8', 3-1/2" diameter	2 Carp	24	.667	Ea.	46	23.50		69.50	89
8080	4" diameter	"	20	.800	"	61	28		89	114

05 12 23.17 Columns, Structural

		Crew	Daily Output	Labor-Hours	Unit	Material	2018 Bare Costs Labor	Equipment	Total	Total Incl O&P
0010	**COLUMNS, STRUCTURAL**									
0015	Made from recycled materials									
0020	Shop fab'd for 100-ton, 1-2 story project, bolted connections									
0800	Steel, concrete filled, extra strong pipe, 3-1/2" diameter	E-2	660	.073	L.F.	44.50	2.82	2.56	49.88	57
0830	4" diameter		780	.062		49.50	2.39	2.16	54.05	61
0890	5" diameter		1020	.047		59	1.83	1.65	62.48	70
0930	6" diameter		1200	.040		78	1.55	1.41	80.96	90.50
0940	8" diameter	↓	1100	.044	↓	78	1.70	1.53	81.23	90.50
1100	For galvanizing, add				Lb.	.25			.25	.28
1300	For web ties, angles, etc., add per added lb.	1 Sswk	945	.008		1.35	.33		1.68	2.06
1500	Steel pipe, extra strong, no concrete, 3" to 5" diameter	G E-2	16000	.003		1.35	.12	.11	1.58	1.80
1600	6" to 12" diameter	G	14000	.003		1.35	.13	.12	1.60	1.84
5100	Structural tubing, rect., 5" to 6" wide, light section	G	8000	.006		1.35	.23	.21	1.79	2.12
5200	Heavy section	G ↓	12000	.004	↓	1.35	.16	.14	1.65	1.90
8090	For projects 75 to 99 tons, add				%	10%				
8092	50 to 74 tons, add					20%				
8094	25 to 49 tons, add					30%	10%			
8096	10 to 24 tons, add					50%	25%			
8098	2 to 9 tons, add					75%	50%			
8099	Less than 2 tons, add				↓	100%	100%			

05 12 23.45 Lintels

		Crew	Daily Output	Labor-Hours	Unit	Material	2018 Bare Costs Labor	Equipment	Total	Total Incl O&P
0010	**LINTELS**									
0015	Made from recycled materials									
0020	Plain steel angles, shop fabricated, under 500 lb.	G	1 Bric	550	.015	Lb.	1.04	.50		1.54	1.98
0100	500 to 1,000 lb.	G	640	.013	"	1.01	.43		1.44	1.84
2000	Steel angles, 3-1/2" x 3", 1/4" thick, 2'-6" long	G	47	.170	Ea.	14.55	5.90		20.45	26
2100	4'-6" long	G	26	.308		26	10.70		36.70	47
2600	4" x 3-1/2", 1/4" thick, 5'-0" long	G	21	.381		33.50	13.20		46.70	58.50
2700	9'-0" long	G ↓	12	.667	↓	60	23		83	105

05 12 23 – Structural Steel for Buildings

05 12 23.65 Plates		Crew	Daily Output	Labor-Hours	Unit	Material	2018 Bare Costs Labor	Equipment	Total	Total Incl O&P
0010	**PLATES**									
0015	Made from recycled materials									
0020	For connections & stiffener plates, shop fabricated									
0050	1/8" thick (5.1 lb./S.F.)	G			S.F.	6.85			6.85	7.55
0100	1/4" thick (10.2 lb./S.F.)	G				13.75			13.75	15.10
0300	3/8" thick (15.3 lb./S.F.)	G				20.50			20.50	22.50
0400	1/2" thick (20.4 lb./S.F.)	G				27.50			27.50	30
0450	3/4" thick (30.6 lb./S.F.)	G				41			41	45.50
0500	1" thick (40.8 lb./S.F.)	G			▼	55			55	60.50
2000	Steel plate, warehouse prices, no shop fabrication									
2100	1/4" thick (10.2 lb./S.F.)	G			S.F.	7.30			7.30	8

05 12 23.79 Structural Steel

0010	**STRUCTURAL STEEL**									
0020	Shop fab'd for 100-ton, 1-2 story project, bolted conn's.									
0050	Beams, W 6 x 9	G	E-2	720	.067	L.F.	14.55	2.59	2.34	19.48	23
0100	W 8 x 10	G		720	.067		16.15	2.59	2.34	21.08	25
0200	Columns, W 6 x 15	G		540	.089		26.50	3.45	3.12	33.07	38.50
0250	W 8 x 31	G	▼	540	.089	▼	54.50	3.45	3.12	61.07	69
7990	For projects 75 to 99 tons, add					All	10%				
7992	50 to 75 tons, add						20%				
7994	25 to 49 tons, add						30%	10%			
7996	10 to 24 tons, add						50%	25%			
7998	2 to 9 tons, add						75%	50%			
7999	Less than 2 tons, add					▼	100%	100%			

05 31 23 – Steel Roof Decking

05 31 23.50 Roof Decking

0010	**ROOF DECKING**									
0015	Made from recycled materials									
2100	Open type, 1-1/2" deep, Type B, wide rib, galv., 22 ga., under 50 sq.	G	E-4	4500	.007	S.F.	2.38	.28	.02	2.68	3.14
2600	20 ga., under 50 squares	G		3865	.008		2.78	.33	.03	3.14	3.67
2900	18 ga., under 50 squares	G		3800	.008		3.58	.33	.03	3.94	4.56
3050	16 ga., under 50 squares	G	▼	3700	.009	▼	4.84	.34	.03	5.21	6

05 31 33 – Steel Form Decking

05 31 33.50 Form Decking

0010	**FORM DECKING**									
0015	Made from recycled materials									
6100	Slab form, steel, 28 ga., 9/16" deep, Type UFS, uncoated	G	E-4	4000	.008	S.F.	1.80	.31	.02	2.13	2.57
6200	Galvanized	G		4000	.008		1.59	.31	.02	1.92	2.34
6220	24 ga., 1" deep, Type UF1X, uncoated	G		3900	.008		1.72	.32	.03	2.07	2.49
6240	Galvanized	G		3900	.008		2.02	.32	.03	2.37	2.82
6300	24 ga., 1-5/16" deep, Type UFX, uncoated	G		3800	.008		1.83	.33	.03	2.19	2.63
6400	Galvanized	G		3800	.008		2.15	.33	.03	2.51	2.99
6500	22 ga., 1-5/16" deep, uncoated	G		3700	.009		2.32	.34	.03	2.69	3.18
6600	Galvanized	G		3700	.009		2.37	.34	.03	2.74	3.24
6700	22 ga., 2" deep, uncoated	G		3600	.009		3.02	.35	.03	3.40	3.97
6800	Galvanized	G	▼	3600	.009		2.96	.35	.03	3.34	3.91

340

For customer support on your Residential Costs with RSMeans data, call 800.448.8182.

05 41 Structural Metal Stud Framing

05 41 13 – Load-Bearing Metal Stud Framing

05 41 13.05 Bracing

		Crew	Daily Output	Labor-Hours	Unit	Material	2018 Bare Costs Labor	Equipment	Total	Total Incl O&P
0010	**BRACING**, shear wall X-bracing, per 10' x 10' bay, one face									
0015	Made of recycled materials									
0120	Metal strap, 20 ga. x 4" wide	[G] 2 Carp	18	.889	Ea.	18.90	31		49.90	72.50
0130	6" wide	[G]	18	.889		31.50	31		62.50	86
0160	18 ga. x 4" wide	[G]	16	1		33	35		68	94
0170	6" wide	[G]	16	1		48.50	35		83.50	112
0410	Continuous strap bracing, per horizontal row on both faces									
0420	Metal strap, 20 ga. x 2" wide, studs 12" OC	[G] 1 Carp	7	1.143	C.L.F.	57	40		97	130
0430	16" OC	[G]	8	1		57	35		92	121
0440	24" OC	[G]	10	.800		57	28		85	110
0450	18 ga. x 2" wide, studs 12" OC	[G]	6	1.333		79	46.50		125.50	165
0460	16" OC	[G]	7	1.143		79	40		119	154
0470	24" OC	[G]	8	1		79	35		114	145

05 41 13.10 Bridging

		Crew	Daily Output	Labor-Hours	Unit	Material	2018 Bare Costs Labor	Equipment	Total	Total Incl O&P
0010	**BRIDGING**, solid between studs w/1-1/4" leg track, per stud bay									
0015	Made from recycled materials									
0200	Studs 12" OC, 18 ga. x 2-1/2" wide	[G] 1 Carp	125	.064	Ea.	.95	2.24		3.19	4.77
0210	3-5/8" wide	[G]	120	.067		1.16	2.33		3.49	5.15
0220	4" wide	[G]	120	.067		1.22	2.33		3.55	5.20
0230	6" wide	[G]	115	.070		1.60	2.43		4.03	5.80
0240	8" wide	[G]	110	.073		1.97	2.54		4.51	6.40
0300	16 ga. x 2-1/2" wide	[G]	115	.070		1.21	2.43		3.64	5.35
0310	3-5/8" wide	[G]	110	.073		1.47	2.54		4.01	5.85
0320	4" wide	[G]	110	.073		1.57	2.54		4.11	5.95
0330	6" wide	[G]	105	.076		2.01	2.66		4.67	6.65
0340	8" wide	[G]	100	.080		2.51	2.80		5.31	7.40
1200	Studs 16" OC, 18 ga. x 2-1/2" wide	[G]	125	.064		1.22	2.24		3.46	5.05
1210	3-5/8" wide	[G]	120	.067		1.49	2.33		3.82	5.50
1220	4" wide	[G]	120	.067		1.57	2.33		3.90	5.60
1230	6" wide	[G]	115	.070		2.05	2.43		4.48	6.30
1240	8" wide	[G]	110	.073		2.52	2.54		5.06	7
1300	16 ga. x 2-1/2" wide	[G]	115	.070		1.55	2.43		3.98	5.75
1310	3-5/8" wide	[G]	110	.073		1.88	2.54		4.42	6.30
1320	4" wide	[G]	110	.073		2.01	2.54		4.55	6.45
1330	6" wide	[G]	105	.076		2.57	2.66		5.23	7.25
1340	8" wide	[G]	100	.080		3.22	2.80		6.02	8.20
2200	Studs 24" OC, 18 ga. x 2-1/2" wide	[G]	125	.064		1.77	2.24		4.01	5.65
2210	3-5/8" wide	[G]	120	.067		2.15	2.33		4.48	6.25
2220	4" wide	[G]	120	.067		2.27	2.33		4.60	6.35
2230	6" wide	[G]	115	.070		2.96	2.43		5.39	7.30
2240	8" wide	[G]	110	.073		3.65	2.54		6.19	8.25
2300	16 ga. x 2-1/2" wide	[G]	115	.070		2.24	2.43		4.67	6.50
2310	3-5/8" wide	[G]	110	.073		2.72	2.54		5.26	7.20
2320	4" wide	[G]	110	.073		2.91	2.54		5.45	7.45
2330	6" wide	[G]	105	.076		3.72	2.66		6.38	8.55
2340	8" wide	[G]	100	.080		4.65	2.80		7.45	9.75
3000	Continuous bridging, per row									
3100	16 ga. x 1-1/2" channel thru studs 12" OC	[G] 1 Carp	6	1.333	C.L.F.	51.50	46.50		98	135
3110	16" OC	[G]	7	1.143		51.50	40		91.50	124
3120	24" OC	[G]	8.80	.909		51.50	32		83.50	110
4100	2" x 2" angle x 18 ga., studs 12" OC	[G]	7	1.143		80.50	40		120.50	155
4110	16" OC	[G]	9	.889		80.50	31		111.50	140

05 41 Structural Metal Stud Framing

05 41 13 – Load-Bearing Metal Stud Framing

05 41 13.10 Bridging

		Crew	Daily Output	Labor-Hours	Unit	Material	2018 Bare Costs Labor	2018 Bare Costs Equipment	Total	Total Incl O&P
4120	24" OC	1 Carp	12	.667	C.L.F.	80.50	23.50		104	127
4200	16 ga., studs 12" OC	G	5	1.600		101	56		157	204
4210	16" OC	G	7	1.143		101	40		141	178
4220	24" OC	G	10	.800		101	28		129	158

05 41 13.25 Framing, Boxed Headers/Beams

		Crew	Daily Output	Labor-Hours	Unit	Material	2018 Bare Costs Labor	2018 Bare Costs Equipment	Total	Total Incl O&P
0010	**FRAMING, BOXED HEADERS/BEAMS**									
0015	Made from recycled materials									
0200	Double, 18 ga. x 6" deep	2 Carp	220	.073	L.F.	5.50	2.54		8.04	10.30
0210	8" deep	G	210	.076		6.05	2.66		8.71	11.10
0220	10" deep	G	200	.080		7.40	2.80		10.20	12.80
0230	12" deep	G	190	.084		8.05	2.94		10.99	13.75
0300	16 ga. x 8" deep	G	180	.089		7	3.11		10.11	12.85
0310	10" deep	G	170	.094		8.45	3.29		11.74	14.70
0320	12" deep	G	160	.100		9.15	3.50		12.65	15.90
0400	14 ga. x 10" deep	G	140	.114		9.75	3.99		13.74	17.35
0410	12" deep	G	130	.123		10.65	4.30		14.95	18.90
1210	Triple, 18 ga. x 8" deep	G	170	.094		8.75	3.29		12.04	15.10
1220	10" deep	G	165	.097		10.60	3.39		13.99	17.30
1230	12" deep	G	160	.100		11.60	3.50		15.10	18.55
1300	16 ga. x 8" deep	G	145	.110		10.15	3.86		14.01	17.55
1310	10" deep	G	140	.114		12.15	3.99		16.14	20
1320	12" deep	G	135	.119		13.25	4.14		17.39	21.50
1400	14 ga. x 10" deep	G	115	.139		13.25	4.86		18.11	22.50
1410	12" deep	G	110	.145		14.65	5.10		19.75	24.50

05 41 13.30 Framing, Stud Walls

		Crew	Daily Output	Labor-Hours	Unit	Material	2018 Bare Costs Labor	2018 Bare Costs Equipment	Total	Total Incl O&P
0010	**FRAMING, STUD WALLS** w/top & bottom track, no openings,									
0020	Headers, beams, bridging or bracing									
0025	Made from recycled materials									
4100	8' high walls, 18 ga. x 2-1/2" wide, studs 12" OC	2 Carp	54	.296	L.F.	9.15	10.35		19.50	27.50
4110	16" OC	G	77	.208		7.35	7.25		14.60	20
4120	24" OC	G	107	.150		5.50	5.25		10.75	14.75
4130	3-5/8" wide, studs 12" OC	G	53	.302		10.85	10.55		21.40	29.50
4140	16" OC	G	76	.211		8.70	7.35		16.05	22
4150	24" OC	G	105	.152		6.55	5.35		11.90	16.05
4160	4" wide, studs 12" OC	G	52	.308		11.30	10.75		22.05	30.50
4170	16" OC	G	74	.216		9.05	7.55		16.60	22.50
4180	24" OC	G	103	.155		6.80	5.45		12.25	16.55
4190	6" wide, studs 12" OC	G	51	.314		14.40	10.95		25.35	34
4200	16" OC	G	73	.219		11.55	7.65		19.20	25.50
4210	24" OC	G	101	.158		8.70	5.55		14.25	18.80
4220	8" wide, studs 12" OC	G	50	.320		17.45	11.20		28.65	38
4230	16" OC	G	72	.222		14	7.75		21.75	28.50
4240	24" OC	G	100	.160		10.60	5.60		16.20	21
4300	16 ga. x 2-1/2" wide, studs 12" OC	G	47	.340		10.85	11.90		22.75	32
4310	16" OC	G	68	.235		8.60	8.20		16.80	23
4320	24" OC	G	94	.170		6.35	5.95		12.30	16.85
4330	3-5/8" wide, studs 12" OC	G	46	.348		12.95	12.15		25.10	34
4340	16" OC	G	66	.242		10.25	8.45		18.70	25.50
4350	24" OC	G	92	.174		7.55	6.10		13.65	18.40
4360	4" wide, studs 12" OC	G	45	.356		13.55	12.45		26	35.50
4370	16" OC	G	65	.246		10.75	8.60		19.35	26
4380	24" OC	G	90	.178		7.95	6.20		14.15	19.05

05 41 13.30 Framing, Stud Walls		Crew	Daily Output	Labor-Hours	Unit	Material	2018 Bare Costs Labor	Equipment	Total	Total Incl O&P
4390	6" wide, studs 12" OC	G 2 Carp	44	.364	L.F.	17.05	12.70		29.75	40
4400	16" OC	G	64	.250		13.55	8.75		22.30	29.50
4410	24" OC	G	88	.182		10.05	6.35		16.40	21.50
4420	8" wide, studs 12" OC	G	43	.372		21	13		34	44.50
4430	16" OC	G	63	.254		16.60	8.90		25.50	33
4440	24" OC	G	86	.186		12.30	6.50		18.80	24.50
5100	10' high walls, 18 ga. x 2-1/2" wide, studs 12" OC	G	54	.296		11	10.35		21.35	29.50
5110	16" OC	G	77	.208		8.70	7.25		15.95	21.50
5120	24" OC	G	107	.150		6.40	5.25		11.65	15.75
5130	3-5/8" wide, studs 12" OC	G	53	.302		13	10.55		23.55	32
5140	16" OC	G	76	.211		10.30	7.35		17.65	23.50
5150	24" OC	G	105	.152		7.60	5.35		12.95	17.20
5160	4" wide, studs 12" OC	G	52	.308		13.55	10.75		24.30	33
5170	16" OC	G	74	.216		10.75	7.55		18.30	24.50
5180	24" OC	G	103	.155		7.95	5.45		13.40	17.75
5190	6" wide, studs 12" OC	G	51	.314		17.25	10.95		28.20	37
5200	16" OC	G	73	.219		13.70	7.65		21.35	28
5210	24" OC	G	101	.158		10.15	5.55		15.70	20.50
5220	8" wide, studs 12" OC	G	50	.320		21	11.20		32.20	41.50
5230	16" OC	G	72	.222		16.55	7.75		24.30	31
5240	24" OC	G	100	.160		12.30	5.60		17.90	23
5300	16 ga. x 2-1/2" wide, studs 12" OC	G	47	.340		13.10	11.90		25	34
5310	16" OC	G	68	.235		10.30	8.20		18.50	25
5320	24" OC	G	94	.170		7.45	5.95		13.40	18.10
5330	3-5/8" wide, studs 12" OC	G	46	.348		15.60	12.15		27.75	37
5340	16" OC	G	66	.242		12.25	8.45		20.70	27.50
5350	24" OC	G	92	.174		8.90	6.10		15	19.90
5360	4" wide, studs 12" OC	G	45	.356		16.35	12.45		28.80	38.50
5370	16" OC	G	65	.246		12.85	8.60		21.45	28.50
5380	24" OC	G	90	.178		9.35	6.20		15.55	20.50
5390	6" wide, studs 12" OC	G	44	.364		20.50	12.70		33.20	43.50
5400	16" OC	G	64	.250		16.15	8.75		24.90	32.50
5410	24" OC	G	88	.182		11.80	6.35		18.15	23.50
5420	8" wide, studs 12" OC	G	43	.372		25	13		38	49
5430	16" OC	G	63	.254		19.80	8.90		28.70	37
5440	24" OC	G	86	.186		14.45	6.50		20.95	26.50
6190	12' high walls, 18 ga. x 6" wide, studs 12" OC	G	41	.390		20	13.65		33.65	44.50
6200	16" OC	G	58	.276		15.80	9.65		25.45	33.50
6210	24" OC	G	81	.198		11.55	6.90		18.45	24
6220	8" wide, studs 12" OC	G	40	.400		24.50	14		38.50	49.50
6230	16" OC	G	57	.281		19.15	9.80		28.95	37.50
6240	24" OC	G	80	.200		14	7		21	27
6390	16 ga. x 6" wide, studs 12" OC	G	35	.457		24	16		40	53
6400	16" OC	G	51	.314		18.80	10.95		29.75	39
6410	24" OC	G	70	.229		13.55	8		21.55	28
6420	8" wide, studs 12" OC	G	34	.471		29.50	16.45		45.95	60
6430	16" OC	G	50	.320		23	11.20		34.20	44
6440	24" OC	G	69	.232		16.60	8.10		24.70	31.50
6530	14 ga. x 3-5/8" wide, studs 12" OC	G	34	.471		22.50	16.45		38.95	52.50
6540	16" OC	G	48	.333		17.75	11.65		29.40	39
6550	24" OC	G	65	.246		12.75	8.60		21.35	28.50
6560	4" wide, studs 12" OC	G	33	.485		24	16.95		40.95	54.50
6570	16" OC	G	47	.340		18.75	11.90		30.65	40.50

05 41 13 – Load-Bearing Metal Stud Framing

05 41 13.30 Framing, Stud Walls		Crew	Daily Output	Labor-Hours	Unit	Material	2018 Bare Costs Labor	Equipment	Total	Total Incl O&P
6580	24" OC	G 2 Carp	64	.250	L.F.	13.50	8.75		22.25	29.50
6730	12 ga. x 3-5/8" wide, studs 12" OC	G	31	.516		31.50	18.05		49.55	64.50
6740	16" OC	G	43	.372		24.50	13		37.50	48
6750	24" OC	G	59	.271		17.15	9.50		26.65	34.50
6760	4" wide, studs 12" OC	G	30	.533		33.50	18.65		52.15	68
6770	16" OC	G	42	.381		26	13.30		39.30	50.50
6780	24" OC	G	58	.276		18.30	9.65		27.95	36
7390	16' high walls, 16 ga. x 6" wide, studs 12" OC	G	33	.485		31	16.95		47.95	62
7400	16" OC	G	48	.333		24	11.65		35.65	46
7410	24" OC	G	67	.239		17.05	8.35		25.40	32.50
7420	8" wide, studs 12" OC	G	32	.500		38	17.50		55.50	71
7430	16" OC	G	47	.340		29.50	11.90		41.40	52.50
7440	24" OC	G	66	.242		21	8.45		29.45	37
7560	14 ga. x 4" wide, studs 12" OC	G	31	.516		31	18.05		49.05	64
7570	16" OC	G	45	.356		24	12.45		36.45	47
7580	24" OC	G	61	.262		17	9.15		26.15	34
7590	6" wide, studs 12" OC	G	30	.533		39	18.65		57.65	74
7600	16" OC	G	44	.364		30	12.70		42.70	54
7610	24" OC	G	60	.267		21.50	9.30		30.80	39
7760	12 ga. x 4" wide, studs 12" OC	G	29	.552		44	19.30		63.30	80
7770	16" OC	G	40	.400		33.50	14		47.50	60
7780	24" OC	G	55	.291		23.50	10.15		33.65	42.50
7790	6" wide, studs 12" OC	G	28	.571		55.50	19.95		75.45	94
7800	16" OC	G	39	.410		42.50	14.35		56.85	70.50
7810	24" OC	G	54	.296		29.50	10.35		39.85	49.50
8590	20' high walls, 14 ga. x 6" wide, studs 12" OC	G	29	.552		48	19.30		67.30	84.50
8600	16" OC	G	42	.381		37	13.30		50.30	62.50
8610	24" OC	G	57	.281		26	9.80		35.80	45
8620	8" wide, studs 12" OC	G	28	.571		52	19.95		71.95	90
8630	16" OC	G	41	.390		40	13.65		53.65	66.50
8640	24" OC	G	56	.286		28.50	10		38.50	47.50
8790	12 ga. x 6" wide, studs 12" OC	G	27	.593		68	20.50		88.50	110
8800	16" OC	G	37	.432		52	15.10		67.10	82.50
8810	24" OC	G	51	.314		36	10.95		46.95	58
8820	8" wide, studs 12" OC	G	26	.615		82.50	21.50		104	127
8830	16" OC	G	36	.444		63	15.55		78.55	95.50
8840	24" OC	G	50	.320		43.50	11.20		54.70	66.50

05 42 Cold-Formed Metal Joist Framing

05 42 13 – Cold-Formed Metal Floor Joist Framing

05 42 13.05 Bracing

		Crew	Daily Output	Labor-Hours	Unit	Material	2018 Bare Costs Labor	Equipment	Total	Total Incl O&P
0010	**BRACING**, continuous, per row, top & bottom									
0015	Made from recycled materials									
0120	Flat strap, 20 ga. x 2" wide, joists at 12" OC	G 1 Carp	4.67	1.713	C.L.F.	60	60		120	166
0130	16" OC	G	5.33	1.501		57.50	52.50		110	151
0140	24" OC	G	6.66	1.201		55.50	42		97.50	131
0150	18 ga. x 2" wide, joists at 12" OC	G	4	2		78.50	70		148.50	203
0160	16" OC	G	4.67	1.713		77	60		137	184
0170	24" OC	G	5.33	1.501		75.50	52.50		128	170

05 42 Cold-Formed Metal Joist Framing

05 42 13 – Cold-Formed Metal Floor Joist Framing

05 42 13.10 Bridging

05 42 13.10 Bridging		Crew	Daily Output	Labor-Hours	Unit	Material	2018 Bare Costs Labor	Equipment	Total	Total Incl O&P
0010	**BRIDGING**, solid between joists w/1-1/4" leg track, per joist bay									
0015	Made from recycled materials									
0230	Joists 12" OC, 18 ga. track x 6" wide	G	1 Carp	80	.100	Ea.	1.60	3.50		5.10	7.55
0240	8" wide	G		75	.107		1.97	3.73		5.70	8.35
0250	10" wide	G		70	.114		2.46	3.99		6.45	9.35
0260	12" wide	G		65	.123		2.78	4.30		7.08	10.20
0330	16 ga. track x 6" wide	G		70	.114		2.01	3.99		6	8.85
0340	8" wide	G		65	.123		2.51	4.30		6.81	9.90
0350	10" wide	G		60	.133		3.10	4.66		7.76	11.15
0360	12" wide	G		55	.145		3.58	5.10		8.68	12.40
0440	14 ga. track x 8" wide	G		60	.133		3.13	4.66		7.79	11.20
0450	10" wide	G		55	.145		3.89	5.10		8.99	12.75
0460	12" wide	G		50	.160		4.49	5.60		10.09	14.25
0550	12 ga. track x 10" wide	G		45	.178		5.70	6.20		11.90	16.60
0560	12" wide	G		40	.200		5.80	7		12.80	18
1230	16" OC, 18 ga. track x 6" wide	G		80	.100		2.05	3.50		5.55	8.05
1240	8" wide	G		75	.107		2.52	3.73		6.25	9
1250	10" wide	G		70	.114		3.15	3.99		7.14	10.10
1260	12" wide	G		65	.123		3.56	4.30		7.86	11.05
1330	16 ga. track x 6" wide	G		70	.114		2.57	3.99		6.56	9.50
1340	8" wide	G		65	.123		3.22	4.30		7.52	10.70
1350	10" wide	G		60	.133		3.98	4.66		8.64	12.10
1360	12" wide	G		55	.145		4.59	5.10		9.69	13.50
1440	14 ga. track x 8" wide	G		60	.133		4.01	4.66		8.67	12.15
1450	10" wide	G		55	.145		4.98	5.10		10.08	13.95
1460	12" wide	G		50	.160		5.75	5.60		11.35	15.65
1550	12 ga. track x 10" wide	G		45	.178		7.30	6.20		13.50	18.35
1560	12" wide	G		40	.200		7.45	7		14.45	19.80
2230	24" OC, 18 ga. track x 6" wide	G		80	.100		2.96	3.50		6.46	9.05
2240	8" wide	G		75	.107		3.65	3.73		7.38	10.20
2250	10" wide	G		70	.114		4.56	3.99		8.55	11.65
2260	12" wide	G		65	.123		5.15	4.30		9.45	12.80
2330	16 ga. track x 6" wide	G		70	.114		3.72	3.99		7.71	10.75
2340	8" wide	G		65	.123		4.65	4.30		8.95	12.25
2350	10" wide	G		60	.133		5.75	4.66		10.41	14.10
2360	12" wide	G		55	.145		6.65	5.10		11.75	15.75
2440	14 ga. track x 8" wide	G		60	.133		5.80	4.66		10.46	14.15
2450	10" wide	G		55	.145		7.20	5.10		12.30	16.40
2460	12" wide	G		50	.160		8.35	5.60		13.95	18.45
2550	12 ga. track x 10" wide	G		45	.178		10.55	6.20		16.75	22
2560	12" wide	G		40	.200		10.80	7		17.80	23.50

05 42 13.25 Framing, Band Joist

05 42 13.25 Framing, Band Joist		Crew	Daily Output	Labor-Hours	Unit	Material	2018 Bare Costs Labor	Equipment	Total	Total Incl O&P
0010	**FRAMING, BAND JOIST** (track) fastened to bearing wall									
0015	Made from recycled materials									
0220	18 ga. track x 6" deep	G	2 Carp	1000	.016	L.F.	1.30	.56		1.86	2.36
0230	8" deep	G		920	.017		1.61	.61		2.22	2.78
0240	10" deep	G		860	.019		2.01	.65		2.66	3.29
0320	16 ga. track x 6" deep	G		900	.018		1.64	.62		2.26	2.83
0330	8" deep	G		840	.019		2.05	.67		2.72	3.36
0340	10" deep	G		780	.021		2.53	.72		3.25	3.97
0350	12" deep	G		740	.022		2.92	.76		3.68	4.47
0430	14 ga. track x 8" deep	G		750	.021		2.55	.75		3.30	4.05

345

05 42 13 – Cold-Formed Metal Floor Joist Framing

05 42 13.25 Framing, Band Joist

		Crew	Daily Output	Labor-Hours	Unit	Material	2018 Bare Costs Labor	2018 Bare Costs Equipment	Total	Total Incl O&P
0440	10" deep	G 2 Carp	720	.022	L.F.	3.17	.78		3.95	4.78
0450	12" deep	G	700	.023		3.66	.80		4.46	5.35
0540	12 ga. track x 10" deep	G	670	.024		4.64	.83		5.47	6.50
0550	12" deep	G	650	.025		4.75	.86		5.61	6.65

05 42 13.30 Framing, Boxed Headers/Beams

		Crew	Daily Output	Labor-Hours	Unit	Material	2018 Bare Costs Labor	2018 Bare Costs Equipment	Total	Total Incl O&P
0010	**FRAMING, BOXED HEADERS/BEAMS**									
0015	Made from recycled materials									
0200	Double, 18 ga. x 6" deep	G 2 Carp	220	.073	L.F.	5.50	2.54		8.04	10.30
0210	8" deep	G	210	.076		6.05	2.66		8.71	11.10
0220	10" deep	G	200	.080		7.40	2.80		10.20	12.80
0230	12" deep	G	190	.084		8.05	2.94		10.99	13.75
0300	16 ga. x 8" deep	G	180	.089		7	3.11		10.11	12.85
0310	10" deep	G	170	.094		8.45	3.29		11.74	14.70
0320	12" deep	G	160	.100		9.15	3.50		12.65	15.90
0400	14 ga. x 10" deep	G	140	.114		9.75	3.99		13.74	17.35
0410	12" deep	G	130	.123		10.65	4.30		14.95	18.90
0500	12 ga. x 10" deep	G	110	.145		12.85	5.10		17.95	22.50
0510	12" deep	G	100	.160		14.15	5.60		19.75	25
1210	Triple, 18 ga. x 8" deep	G	170	.094		8.75	3.29		12.04	15.10
1220	10" deep	G	165	.097		10.60	3.39		13.99	17.30
1230	12" deep	G	160	.100		11.60	3.50		15.10	18.55
1300	16 ga. x 8" deep	G	145	.110		10.15	3.86		14.01	17.55
1310	10" deep	G	140	.114		12.15	3.99		16.14	20
1320	12" deep	G	135	.119		13.25	4.14		17.39	21.50
1400	14 ga. x 10" deep	G	115	.139		14.10	4.86		18.96	23.50
1410	12" deep	G	110	.145		15.50	5.10		20.60	25.50
1500	12 ga. x 10" deep	G	90	.178		18.75	6.20		24.95	31
1510	12" deep	G	85	.188		21	6.60		27.60	34

05 42 13.40 Framing, Joists

		Crew	Daily Output	Labor-Hours	Unit	Material	2018 Bare Costs Labor	2018 Bare Costs Equipment	Total	Total Incl O&P
0010	**FRAMING, JOISTS**, no band joists (track), web stiffeners, headers,									
0020	Beams, bridging or bracing									
0025	Made from recycled materials									
0030	Joists (2" flange) and fasteners, materials only									
0220	18 ga. x 6" deep	G			L.F.	1.69			1.69	1.86
0230	8" deep	G				1.98			1.98	2.18
0240	10" deep	G				2.34			2.34	2.58
0320	16 ga. x 6" deep	G				2.07			2.07	2.28
0330	8" deep	G				2.47			2.47	2.71
0340	10" deep	G				2.89			2.89	3.18
0350	12" deep	G				3.28			3.28	3.60
0430	14 ga. x 8" deep	G				3.10			3.10	3.41
0440	10" deep	G				3.57			3.57	3.93
0450	12" deep	G				4.06			4.06	4.47
0540	12 ga. x 10" deep	G				5.20			5.20	5.70
0550	12" deep	G				5.90			5.90	6.50
1010	Installation of joists to band joists, beams & headers, labor only									
1220	18 ga. x 6" deep	2 Carp	110	.145	Ea.		5.10		5.10	8.45
1230	8" deep		90	.178			6.20		6.20	10.35
1240	10" deep		80	.200			7		7	11.60
1320	16 ga. x 6" deep		95	.168			5.90		5.90	9.80
1330	8" deep		70	.229			8		8	13.30
1340	10" deep		60	.267			9.30		9.30	15.50

05 42 Cold-Formed Metal Joist Framing

05 42 13 – Cold-Formed Metal Floor Joist Framing

05 42 13.40 Framing, Joists		Crew	Daily Output	Labor-Hours	Unit	Material	2018 Bare Costs Labor	2018 Bare Costs Equipment	Total	Total Incl O&P
1350	12" deep	2 Carp	55	.291	Ea.		10.15		10.15	16.90
1430	14 ga. x 8" deep		65	.246			8.60		8.60	14.30
1440	10" deep		45	.356			12.45		12.45	20.50
1450	12" deep		35	.457			16		16	26.50
1540	12 ga. x 10" deep		40	.400			14		14	23
1550	12" deep	▼	30	.533	▼		18.65		18.65	31

05 42 13.45 Framing, Web Stiffeners

			Crew	Daily Output	Labor-Hours	Unit	Material	2018 Bare Costs Labor	2018 Bare Costs Equipment	Total	Total Incl O&P
0010	**FRAMING, WEB STIFFENERS** at joist bearing, fabricated from										
0020	Stud piece (1-5/8" flange) to stiffen joist (2" flange)										
0025	Made from recycled materials										
2120	For 6" deep joist, with 18 ga. x 2-1/2" stud	G	1 Carp	120	.067	Ea.	.92	2.33		3.25	4.88
2130	3-5/8" stud	G		110	.073		1.08	2.54		3.62	5.40
2140	4" stud	G		105	.076		1.12	2.66		3.78	5.65
2150	6" stud	G		100	.080		1.42	2.80		4.22	6.20
2160	8" stud	G		95	.084		1.71	2.94		4.65	6.75
2220	8" deep joist, with 2-1/2" stud	G		120	.067		1.23	2.33		3.56	5.25
2230	3-5/8" stud	G		110	.073		1.45	2.54		3.99	5.80
2240	4" stud	G		105	.076		1.50	2.66		4.16	6.10
2250	6" stud	G		100	.080		1.90	2.80		4.70	6.75
2260	8" stud	G		95	.084		2.29	2.94		5.23	7.40
2320	10" deep joist, with 2-1/2" stud	G		110	.073		1.53	2.54		4.07	5.90
2330	3-5/8" stud	G		100	.080		1.79	2.80		4.59	6.60
2340	4" stud	G		95	.084		1.86	2.94		4.80	6.95
2350	6" stud	G		90	.089		2.36	3.11		5.47	7.75
2360	8" stud	G		85	.094		2.84	3.29		6.13	8.55
2420	12" deep joist, with 2-1/2" stud	G		110	.073		1.84	2.54		4.38	6.25
2430	3-5/8" stud	G		100	.080		2.16	2.80		4.96	7.05
2440	4" stud	G		95	.084		2.24	2.94		5.18	7.35
2450	6" stud	G		90	.089		2.84	3.11		5.95	8.25
2460	8" stud	G		85	.094		3.42	3.29		6.71	9.20
3130	For 6" deep joist, with 16 ga. x 3-5/8" stud	G		100	.080		1.34	2.80		4.14	6.10
3140	4" stud	G		95	.084		1.40	2.94		4.34	6.45
3150	6" stud	G		90	.089		1.75	3.11		4.86	7.10
3160	8" stud	G		85	.094		2.14	3.29		5.43	7.80
3230	8" deep joist, with 3-5/8" stud	G		100	.080		1.80	2.80		4.60	6.65
3240	4" stud	G		95	.084		1.88	2.94		4.82	6.95
3250	6" stud	G		90	.089		2.35	3.11		5.46	7.75
3260	8" stud	G		85	.094		2.87	3.29		6.16	8.60
3330	10" deep joist, with 3-5/8" stud	G		85	.094		2.22	3.29		5.51	7.90
3340	4" stud	G		80	.100		2.32	3.50		5.82	8.35
3350	6" stud	G		75	.107		2.91	3.73		6.64	9.40
3360	8" stud	G		70	.114		3.55	3.99		7.54	10.55
3430	12" deep joist, with 3-5/8" stud	G		85	.094		2.68	3.29		5.97	8.40
3440	4" stud	G		80	.100		2.80	3.50		6.30	8.90
3450	6" stud	G		75	.107		3.50	3.73		7.23	10.05
3460	8" stud	G		70	.114		4.28	3.99		8.27	11.35
4230	For 8" deep joist, with 14 ga. x 3-5/8" stud	G		90	.089		2.22	3.11		5.33	7.60
4240	4" stud	G		85	.094		2.35	3.29		5.64	8.05
4250	6" stud	G		80	.100		2.95	3.50		6.45	9.05
4260	8" stud	G		75	.107		3.15	3.73		6.88	9.65
4330	10" deep joist, with 3-5/8" stud	G		75	.107		2.76	3.73		6.49	9.25
4340	4" stud	G	▼	70	.114	▼	2.91	3.99		6.90	9.85

05 42 Cold-Formed Metal Joist Framing

05 42 13 – Cold-Formed Metal Floor Joist Framing

05 42 13.45 Framing, Web Stiffeners

		Crew	Daily Output	Labor-Hours	Unit	Material	2018 Bare Costs Labor	Equipment	Total	Total Incl O&P
4350	6" stud	G 1 Carp	65	.123	Ea.	3.65	4.30		7.95	11.15
4360	8" stud	G	60	.133		3.90	4.66		8.56	12.05
4430	12" deep joist, with 3-5/8" stud	G	75	.107		3.32	3.73		7.05	9.85
4440	4" stud	G	70	.114		3.50	3.99		7.49	10.50
4450	6" stud	G	65	.123		4.40	4.30		8.70	12
4460	8" stud	G	60	.133		4.70	4.66		9.36	12.90
5330	For 10" deep joist, with 12 ga. x 3-5/8" stud	G	65	.123		3.97	4.30		8.27	11.50
5340	4" stud	G	60	.133		4.23	4.66		8.89	12.40
5350	6" stud	G	55	.145		5.35	5.10		10.45	14.35
5360	8" stud	G	50	.160		6.45	5.60		12.05	16.40
5430	12" deep joist, with 3-5/8" stud	G	65	.123		4.78	4.30		9.08	12.40
5440	4" stud	G	60	.133		5.10	4.66		9.76	13.35
5450	6" stud	G	55	.145		6.45	5.10		11.55	15.55
5460	8" stud	G	50	.160		7.80	5.60		13.40	17.85

05 42 23 – Cold-Formed Metal Roof Joist Framing

05 42 23.05 Framing, Bracing

		Crew	Daily Output	Labor-Hours	Unit	Material	2018 Bare Costs Labor	Equipment	Total	Total Incl O&P
0010	**FRAMING, BRACING**									
0015	Made from recycled materials									
0020	Continuous bracing, per row									
0100	16 ga. x 1-1/2" channel thru rafters/trusses @ 16" OC	G 1 Carp	4.50	1.778	C.L.F.	51.50	62		113.50	160
0120	24" OC	G	6	1.333		51.50	46.50		98	135
0300	2" x 2" angle x 18 ga., rafters/trusses @ 16" OC	G	6	1.333		80.50	46.50		127	166
0320	24" OC	G	8	1		80.50	35		115.50	147
0400	16 ga., rafters/trusses @ 16" OC	G	4.50	1.778		101	62		163	214
0420	24" OC	G	6.50	1.231		101	43		144	183

05 42 23.10 Framing, Bridging

		Crew	Daily Output	Labor-Hours	Unit	Material	2018 Bare Costs Labor	Equipment	Total	Total Incl O&P
0010	**FRAMING, BRIDGING**									
0015	Made from recycled materials									
0020	Solid, between rafters w/1-1/4" leg track, per rafter bay									
1200	Rafters 16" OC, 18 ga. x 4" deep	G 1 Carp	60	.133	Ea.	1.57	4.66		6.23	9.45
1210	6" deep	G	57	.140		2.05	4.91		6.96	10.40
1220	8" deep	G	55	.145		2.52	5.10		7.62	11.25
1230	10" deep	G	52	.154		3.15	5.40		8.55	12.40
1240	12" deep	G	50	.160		3.56	5.60		9.16	13.20
2200	24" OC, 18 ga. x 4" deep	G	60	.133		2.27	4.66		6.93	10.25
2210	6" deep	G	57	.140		2.96	4.91		7.87	11.40
2220	8" deep	G	.55	.145		3.65	5.10		8.75	12.45
2230	10" deep	G	52	.154		4.56	5.40		9.96	13.95
2240	12" deep	G	50	.160		5.15	5.60		10.75	14.95

05 42 23.50 Framing, Parapets

		Crew	Daily Output	Labor-Hours	Unit	Material	2018 Bare Costs Labor	Equipment	Total	Total Incl O&P
0010	**FRAMING, PARAPETS**									
0015	Made from recycled materials									
0100	3' high installed on 1st story, 18 ga. x 4" wide studs, 12" OC	G 2 Carp	100	.160	L.F.	5.70	5.60		11.30	15.55
0110	16" OC	G	150	.107		4.85	3.73		8.58	11.55
0120	24" OC	G	200	.080		4.01	2.80		6.81	9.05
0200	6" wide studs, 12" OC	G	100	.160		7.30	5.60		12.90	17.35
0210	16" OC	G	150	.107		6.25	3.73		9.98	13.05
0220	24" OC	G	200	.080		5.15	2.80		7.95	10.35
1100	Installed on 2nd story, 18 ga. x 4" wide studs, 12" OC	G	95	.168		5.70	5.90		11.60	16.05
1110	16" OC	G	145	.110		4.85	3.86		8.71	11.75
1120	24" OC	G	190	.084		4.01	2.94		6.95	9.30
1200	6" wide studs, 12" OC	G	95	.168		7.30	5.90		13.20	17.85

05 42 Cold-Formed Metal Joist Framing

05 42 23 – Cold-Formed Metal Roof Joist Framing

05 42 23.50 Framing, Parapets

			Crew	Daily Output	Labor-Hours	Unit	Material	2018 Bare Costs Labor	Equipment	Total	Total Incl O&P
1210	16" OC	G	2 Carp	145	.110	L.F.	6.25	3.86		10.11	13.25
1220	24" OC	G		190	.084		5.15	2.94		8.09	10.60
2100	Installed on gable, 18 ga. x 4" wide studs, 12" OC	G		85	.188		5.70	6.60		12.30	17.20
2110	16" OC	G		130	.123		4.85	4.30		9.15	12.50
2120	24" OC	G		170	.094		4.01	3.29		7.30	9.85
2200	6" wide studs, 12" OC	G		85	.188		7.30	6.60		13.90	19
2210	16" OC	G		130	.123		6.25	4.30		10.55	14
2220	24" OC	G		170	.094		5.15	3.29		8.44	11.15

05 42 23.60 Framing, Roof Rafters

			Crew	Daily Output	Labor-Hours	Unit	Material	2018 Bare Costs Labor	Equipment	Total	Total Incl O&P
0010	**FRAMING, ROOF RAFTERS**										
0015	Made from recycled materials										
0100	Boxed ridge beam, double, 18 ga. x 6" deep	G	2 Carp	160	.100	L.F.	5.50	3.50		9	11.85
0110	8" deep	G		150	.107		6.05	3.73		9.78	12.85
0120	10" deep	G		140	.114		7.40	3.99		11.39	14.80
0130	12" deep	G		130	.123		8.05	4.30		12.35	16
0200	16 ga. x 6" deep	G		150	.107		6.20	3.73		9.93	13.05
0210	8" deep	G		140	.114		7	3.99		10.99	14.35
0220	10" deep	G		130	.123		8.45	4.30		12.75	16.40
0230	12" deep	G		120	.133		9.15	4.66		13.81	17.85
1100	Rafters, 2" flange, material only, 18 ga. x 6" deep	G					1.69			1.69	1.86
1110	8" deep	G					1.98			1.98	2.18
1120	10" deep	G					2.34			2.34	2.58
1130	12" deep	G					2.70			2.70	2.97
1200	16 ga. x 6" deep	G					2.07			2.07	2.28
1210	8" deep	G					2.47			2.47	2.71
1220	10" deep	G					2.89			2.89	3.18
1230	12" deep	G					3.28			3.28	3.60
2100	Installation only, ordinary rafter to 4:12 pitch, 18 ga. x 6" deep		2 Carp	35	.457	Ea.		16		16	26.50
2110	8" deep			30	.533			18.65		18.65	31
2120	10" deep			25	.640			22.50		22.50	37
2130	12" deep			20	.800			28		28	46.50
2200	16 ga. x 6" deep			30	.533			18.65		18.65	31
2210	8" deep			25	.640			22.50		22.50	37
2220	10" deep			20	.800			28		28	46.50
2230	12" deep			15	1.067			37.50		37.50	62
8100	Add to labor, ordinary rafters on steep roofs							25%			
8110	Dormers & complex roofs							50%			
8200	Hip & valley rafters to 4:12 pitch							25%			
8210	Steep roofs							50%			
8220	Dormers & complex roofs							75%			
8300	Hip & valley jack rafters to 4:12 pitch							50%			
8310	Steep roofs							75%			
8320	Dormers & complex roofs							100%			

05 42 23.70 Framing, Soffits and Canopies

			Crew	Daily Output	Labor-Hours	Unit	Material	2018 Bare Costs Labor	Equipment	Total	Total Incl O&P
0010	**FRAMING, SOFFITS & CANOPIES**										
0015	Made from recycled materials										
0130	Continuous ledger track @ wall, studs @ 16" OC, 18 ga. x 4" wide	G	2 Carp	535	.030	L.F.	1.05	1.05		2.10	2.89
0140	6" wide	G		500	.032		1.36	1.12		2.48	3.36
0150	8" wide	G		465	.034		1.68	1.20		2.88	3.85
0160	10" wide	G		430	.037		2.10	1.30		3.40	4.47
0230	Studs @ 24" OC, 18 ga. x 4" wide	G		800	.020		1	.70		1.70	2.26
0240	6" wide	G		750	.021		1.30	.75		2.05	2.67

349

For customer support on your Residential Costs with RSMeans data, call 800.448.8182.

05 42 Cold-Formed Metal Joist Framing

05 42 23 – Cold-Formed Metal Roof Joist Framing

05 42 23.70 Framing, Soffits and Canopies

		Crew	Daily Output	Labor-Hours	Unit	Material	2018 Bare Costs Labor	Equipment	Total	Total Incl O&P
0250	8" wide	G 2 Carp	700	.023	L.F.	1.61	.80		2.41	3.10
0260	10" wide	G ↓	650	.025	↓	2.01	.86		2.87	3.64
1000	Horizontal soffit and canopy members, material only									
1030	1-5/8" flange studs, 18 ga. x 4" deep	G			L.F.	1.34			1.34	1.48
1040	6" deep	G				1.70			1.70	1.87
1050	8" deep	G				2.05			2.05	2.26
1140	2" flange joists, 18 ga. x 6" deep	G				1.93			1.93	2.13
1150	8" deep	G				2.27			2.27	2.49
1160	10" deep	G			↓	2.68			2.68	2.94
4030	Installation only, 18 ga., 1-5/8" flange x 4" deep	2 Carp	130	.123	Ea.		4.30		4.30	7.15
4040	6" deep		110	.145			5.10		5.10	8.45
4050	8" deep		90	.178			6.20		6.20	10.35
4140	2" flange, 18 ga. x 6" deep		110	.145			5.10		5.10	8.45
4150	8" deep		90	.178			6.20		6.20	10.35
4160	10" deep	↓	80	.200			7		7	11.60
6010	Clips to attach fascia to rafter tails, 2" x 2" x 18 ga. angle	G 1 Carp	120	.067		.95	2.33		3.28	4.91
6020	16 ga. angle	G "	100	.080	↓	1.20	2.80		4	5.95

05 44 Cold-Formed Metal Trusses

05 44 13 – Cold-Formed Metal Roof Trusses

05 44 13.60 Framing, Roof Trusses

		Crew	Daily Output	Labor-Hours	Unit	Material	2018 Bare Costs Labor	Equipment	Total	Total Incl O&P
0010	**FRAMING, ROOF TRUSSES**									
0015	Made from recycled materials									
0020	Fabrication of trusses on ground, Fink (W) or King Post, to 4:12 pitch									
0120	18 ga. x 4" chords, 16' span	G 2 Carp	12	1.333	Ea.	62.50	46.50		109	147
0130	20' span	G	11	1.455		78.50	51		129.50	171
0140	24' span	G	11	1.455		94	51		145	188
0150	28' span	G	10	1.600		110	56		166	214
0160	32' span	G	10	1.600		125	56		181	231
0250	6" chords, 28' span	G	9	1.778		139	62		201	256
0260	32' span	G	9	1.778		159	62		221	278
0270	36' span	G	8	2		179	70		249	315
0280	40' span	G	8	2		199	70		269	335
1120	5:12 to 8:12 pitch, 18 ga. x 4" chords, 16' span	G	10	1.600		71.50	56		127.50	172
1130	20' span	G	9	1.778		89.50	62		151.50	202
1140	24' span	G	9	1.778		108	62		170	221
1150	28' span	G	8	2		125	70		195	254
1160	32' span	G	8	2		143	70		213	274
1250	6" chords, 28' span	G	7	2.286		159	80		239	310
1260	32' span	G	7	2.286		182	80		262	335
1270	36' span	G	6	2.667		204	93		297	380
1280	40' span	G	6	2.667		227	93		320	405
2120	9:12 to 12:12 pitch, 18 ga. x 4" chords, 16' span	G	8	2		89.50	70		159.50	215
2130	20' span	G	7	2.286		112	80		192	256
2140	24' span	G	7	2.286		134	80		214	281
2150	28' span	G	6	2.667		157	93		250	325
2160	32' span	G	6	2.667		179	93		272	350
2250	6" chords, 28' span	G	5	3.200		199	112		311	405
2260	32' span	G	5	3.200		227	112		339	435
2270	36' span	G	4	4		256	140		396	515
2280	40' span	G ↓	4	4	↓	284	140		424	540

350

05 44 Cold-Formed Metal Trusses

05 44 13 – Cold-Formed Metal Roof Trusses

05 44 13.60 Framing, Roof Trusses		Crew	Daily Output	Labor-Hours	Unit	Material	2018 Bare Costs Labor	Equipment	Total	Total Incl O&P
5120	Erection only of roof trusses, to 4:12 pitch, 16' span	F-6	48	.833	Ea.		26.50	10.35	36.85	56
5130	20' span		46	.870			28	10.80	38.80	58.50
5140	24' span		44	.909			29	11.25	40.25	61
5150	28' span		42	.952			30.50	11.80	42.30	63.50
5160	32' span		40	1			32	12.40	44.40	66.50
5170	36' span		38	1.053			34	13.05	47.05	70.50
5180	40' span		36	1.111			35.50	13.80	49.30	74
5220	5:12 to 8:12 pitch, 16' span		42	.952			30.50	11.80	42.30	63.50
5230	20' span		40	1			32	12.40	44.40	66.50
5240	24' span		38	1.053			34	13.05	47.05	70.50
5250	28' span		36	1.111			35.50	13.80	49.30	74
5260	32' span		34	1.176			37.50	14.60	52.10	78.50
5270	36' span		32	1.250			40	15.50	55.50	83.50
5280	40' span		30	1.333			43	16.55	59.55	89
5320	9:12 to 12:12 pitch, 16' span		36	1.111			35.50	13.80	49.30	74
5330	20' span		34	1.176			37.50	14.60	52.10	78.50
5340	24' span		32	1.250			40	15.50	55.50	83.50
5350	28' span		30	1.333			43	16.55	59.55	89
5360	32' span		28	1.429			46	17.70	63.70	95.50
5370	36' span		26	1.538			49.50	19.10	68.60	103
5380	40' span		24	1.667			53.50	20.50	74	111

05 51 Metal Stairs

05 51 13 – Metal Pan Stairs

05 51 13.50 Pan Stairs

			Crew	Daily Output	Labor-Hours	Unit	Material	Labor	Equipment	Total	Total Incl O&P
0010	**PAN STAIRS**, shop fabricated, steel stringers										
0015	Made from recycled materials										
1700	Pre-erected, steel pan tread, 3'-6" wide, 2 line pipe rail	G	E-2	87	.552	Riser	555	21.50	19.40	595.90	670
1800	With flat bar picket rail	G	"	87	.552	"	615	21.50	19.40	655.90	740

05 51 23 – Metal Fire Escapes

05 51 23.50 Fire Escape Stairs

						Unit	Material			Total	Total Incl O&P
0010	**FIRE ESCAPE STAIRS**, portable										
0100	Portable ladder						Ea.	128			128	141

05 52 Metal Railings

05 52 13 – Pipe and Tube Railings

05 52 13.50 Railings, Pipe

			Crew	Daily Output	Labor-Hours	Unit	Material	Labor	Equipment	Total	Total Incl O&P
0010	**RAILINGS, PIPE**, shop fab'd, 3'-6" high, posts @ 5' OC										
0015	Made from recycled materials										
0020	Aluminum, 2 rail, satin finish, 1-1/4" diameter	G	E-4	160	.200	L.F.	52.50	7.85	.62	60.97	72.50
0030	Clear anodized	G		160	.200		64.50	7.85	.62	72.97	85.50
0040	Dark anodized	G		160	.200		71.50	7.85	.62	79.97	93
0080	1-1/2" diameter, satin finish	G		160	.200		61.50	7.85	.62	69.97	82.50
0090	Clear anodized	G		160	.200		69.50	7.85	.62	77.97	90.50
0100	Dark anodized	G		160	.200		76.50	7.85	.62	84.97	98.50
0140	Aluminum, 3 rail, 1-1/4" diam., satin finish	G		137	.234		68	9.20	.72	77.92	91.50
0150	Clear anodized	G		137	.234		84.50	9.20	.72	94.42	110
0160	Dark anodized	G		137	.234		93	9.20	.72	102.92	119
0200	1-1/2" diameter, satin finish	G		137	.234		80.50	9.20	.72	90.42	106

351

05 52 Metal Railings

05 52 13 – Pipe and Tube Railings

05 52 13.50 Railings, Pipe

		Crew	Daily Output	Labor-Hours	Unit	Material	2018 Bare Costs Labor	Equipment	Total	Total Incl O&P
0210	Clear anodized	G	E-4	137	.234	L.F.	91.50	9.20	.72	101.42	118
0220	Dark anodized	G		137	.234		101	9.20	.72	110.92	128
0500	Steel, 2 rail, on stairs, primed, 1-1/4" diameter	G		160	.200		29	7.85	.62	37.47	46.50
0520	1-1/2" diameter	G		160	.200		31.50	7.85	.62	39.97	49
0540	Galvanized, 1-1/4" diameter	G		160	.200		39	7.85	.62	47.47	57.50
0560	1-1/2" diameter	G		160	.200		44	7.85	.62	52.47	63
0580	Steel, 3 rail, primed, 1-1/4" diameter	G		137	.234		43	9.20	.72	52.92	64.50
0600	1-1/2" diameter	G		137	.234		45.50	9.20	.72	55.42	67
0620	Galvanized, 1-1/4" diameter	G		137	.234		60.50	9.20	.72	70.42	83.50
0640	1-1/2" diameter	G		137	.234		70	9.20	.72	79.92	94
0700	Stainless steel, 2 rail, 1-1/4" diam., #4 finish	G		137	.234		128	9.20	.72	137.92	157
0720	High polish	G		137	.234		206	9.20	.72	215.92	244
0740	Mirror polish	G		137	.234		258	9.20	.72	267.92	300
0760	Stainless steel, 3 rail, 1-1/2" diam., #4 finish	G		120	.267		192	10.50	.82	203.32	231
0770	High polish	G		120	.267		320	10.50	.82	331.32	370
0780	Mirror finish	G		120	.267		390	10.50	.82	401.32	445
0900	Wall rail, alum. pipe, 1-1/4" diam., satin finish	G		213	.150		25	5.90	.46	31.36	38.50
0905	Clear anodized	G		213	.150		31.50	5.90	.46	37.86	45.50
0910	Dark anodized	G		213	.150		37	5.90	.46	43.36	52
0915	1-1/2" diameter, satin finish	G		213	.150		28	5.90	.46	34.36	42
0920	Clear anodized	G		213	.150		35	5.90	.46	41.36	49.50
0925	Dark anodized	G		213	.150		44	5.90	.46	50.36	59.50
0930	Steel pipe, 1-1/4" diameter, primed	G		213	.150		17.35	5.90	.46	23.71	30
0935	Galvanized	G		213	.150		25	5.90	.46	31.36	38.50
0940	1-1/2" diameter	G		176	.182		17.90	7.15	.56	25.61	33
0945	Galvanized	G		213	.150		25	5.90	.46	31.36	38.50
0955	Stainless steel pipe, 1-1/2" diam., #4 finish	G		107	.299		102	11.75	.92	114.67	134
0960	High polish	G		107	.299		207	11.75	.92	219.67	250
0965	Mirror polish	G		107	.299		245	11.75	.92	257.67	292
2000	2-line pipe rail (1-1/2" T&B) with 1/2" pickets @ 4-1/2" OC,										
2005	attached handrail on brackets										
2010	42" high aluminum, satin finish, straight & level	G	E-4	120	.267	L.F.	268	10.50	.82	279.32	315
2050	42" high steel, primed, straight & level	G	"	120	.267		147	10.50	.82	158.32	181
4000	For curved and level rails, add						10%	10%			
4100	For sloped rails for stairs, add						30%	30%			

05 58 Formed Metal Fabrications

05 58 25 – Formed Lamp Posts

05 58 25.40 Lamp Posts

		Crew	Daily Output	Labor-Hours	Unit	Material	2018 Bare Costs Labor	Equipment	Total	Total Incl O&P
0010	**LAMP POSTS**										
0020	Aluminum, 7' high, stock units, post only	G	1 Carp	16	.500	Ea.	85.50	17.50		103	123
0100	Mild steel, plain	G	"	16	.500	"	63.50	17.50		81	98.50

05 71 Decorative Metal Stairs

05 71 13 – Fabricated Metal Spiral Stairs

05 71 13.50 Spiral Stairs

		Crew	Daily Output	Labor-Hours	Unit	Material	2018 Bare Costs Labor	2018 Bare Costs Equipment	Total	Total Incl O&P
0010	**SPIRAL STAIRS**									
1805	Shop fabricated, custom ordered									
1810	Aluminum, 5'-0" diameter, plain units	[G] E-4	45	.711	Riser	580	28	2.19	610.19	685
1820	Fancy units	[G]	45	.711		1,000	28	2.19	1,030.19	1,150
1900	Cast iron, 4'-0" diameter, plain units	[G]	45	.711		745	28	2.19	775.19	870
1920	Fancy units	[G]	25	1.280		1,300	50.50	3.94	1,354.44	1,525
3100	Spiral stair kits, 12 stacking risers to fit exact floor height									
3110	Steel, flat metal treads, primed, 3'-6" diameter	[G] 2 Carp	1.60	10	Flight	1,350	350		1,700	2,050
3120	4'-0" diameter	[G]	1.45	11.034		1,525	385		1,910	2,325
3130	4'-6" diameter	[G]	1.35	11.852		1,700	415		2,115	2,550
3140	5'-0" diameter	[G]	1.25	12.800		1,850	445		2,295	2,800
3310	Checkered plate tread, primed, 3'-6" diameter	[G]	1.45	11.034		1,600	385		1,985	2,400
3320	4'-0" diameter	[G]	1.35	11.852		1,800	415		2,215	2,675
3330	4'-6" diameter	[G]	1.25	12.800		1,975	445		2,420	2,900
3340	5'-0" diameter	[G]	1.15	13.913		2,150	485		2,635	3,175
3510	Red oak covers on flat metal treads, 3'-6" diameter		1.35	11.852		2,450	415		2,865	3,375
3520	4'-0" diameter		1.25	12.800		2,925	445		3,370	3,950
3530	4'-6" diameter		1.15	13.913		3,150	485		3,635	4,275
3540	5'-0" diameter		1.05	15.238		3,400	535		3,935	4,600

05 75 Decorative Formed Metal

05 75 13 – Columns

05 75 13.10 Aluminum Columns

		Crew	Daily Output	Labor-Hours	Unit	Material	2018 Bare Costs Labor	2018 Bare Costs Equipment	Total	Total Incl O&P
0010	**ALUMINUM COLUMNS**									
0015	Made from recycled materials									
0020	Aluminum, extruded, stock units, no cap or base, 6" diameter	[G] E-4	240	.133	L.F.	12.85	5.25	.41	18.51	24
0100	8" diameter	[G] "	170	.188	"	16.75	7.40	.58	24.73	32
0410	Caps and bases, plain, 6" diameter	[G]			Set	25			25	27.50
0420	8" diameter	[G]			"	31			31	34.50
0500	For square columns, add to column prices above				L.F.	50%				
0700	Residential, flat, 8' high, plain	[G] E-4	20	1.600	Ea.	101	63	4.93	168.93	228
0720	Fancy	[G]	20	1.600		196	63	4.93	263.93	335
0740	Corner type, plain	[G]	20	1.600		174	63	4.93	241.93	310
0760	Fancy	[G]	20	1.600		345	63	4.93	412.93	495

05 75 13.20 Columns, Ornamental

		Crew	Daily Output	Labor-Hours	Unit	Material	2018 Bare Costs Labor	2018 Bare Costs Equipment	Total	Total Incl O&P
0010	**COLUMNS, ORNAMENTAL**, shop fabricated									
6400	Mild steel, flat, 9" wide, stock units, painted, plain	[G] E-4	160	.200	V.L.F.	9.45	7.85	.62	17.92	25
6450	Fancy	[G]	160	.200		18.40	7.85	.62	26.87	35
6500	Corner columns, painted, plain	[G]	160	.200		16.35	7.85	.62	24.82	32.50
6550	Fancy	[G]	160	.200		32.50	7.85	.62	40.97	50

For customer support on your Residential Costs with RSMeans data, call 800.448.8182.

353

Division Notes

		CREW	DAILY OUTPUT	LABOR-HOURS	UNIT	BARE COSTS				TOTAL INCL O&P
						MAT.	LABOR	EQUIP.	TOTAL	

Estimating Tips
06 05 00 Common Work Results for Wood, Plastics, and Composites

- Common to any wood-framed structure are the accessory connector items such as screws, nails, adhesives, hangers, connector plates, straps, angles, and hold-downs. For typical wood-framed buildings, such as residential projects, the aggregate total for these items can be significant, especially in areas where seismic loading is a concern. For floor and wall framing, the material cost is based on 10 to 25 lbs. of accessory connectors per MBF. Hold-downs, hangers, and other connectors should be taken off by the piece.

  Included with material costs are fasteners for a normal installation. Gordian's RSMeans engineers use manufacturers' recommendations, written specifications, and/or standard construction practice for the sizing and spacing of fasteners. Prices for various fasteners are shown for informational purposes only. Adjustments should be made if unusual fastening conditions exist.

06 10 00 Carpentry

- Lumber is a traded commodity and therefore sensitive to supply and demand in the marketplace. Even with "budgetary" estimating of wood-framed projects, it is advisable to call local suppliers for the latest market pricing.

- The common quantity unit for wood-framed projects is "thousand board feet" (MBF). A board foot is a volume of wood—1" x 1' x 1' or 144 cubic inches. Board-foot quantities are generally calculated using nominal material dimensions— dressed sizes are ignored. Board foot per lineal foot of any stick of lumber can be calculated by dividing the nominal cross-sectional area by 12. As an example, 2,000 lineal feet of 2 x 12 equates to 4 MBF by dividing the nominal area, 2 x 12, by 12, which equals 2, and multiplying by 2,000 to give 4,000 board feet. This simple rule applies to all nominal dimensioned lumber.

- Waste is an issue of concern at the quantity takeoff for any area of construction. Framing lumber is sold in even foot lengths, i.e., 8', 10', 12', 14', 16' and depending on spans, wall heights, and the grade of lumber, waste is inevitable. A rule of thumb for lumber waste is 5%–10% depending on material quality and the complexity of the framing.

- Wood in various forms and shapes is used in many projects, even where the main structural framing is steel, concrete, or masonry. Plywood as a back-up partition material and 2x boards used as blocking and cant strips around roof edges are two common examples. The estimator should ensure that the costs of all wood materials are included in the final estimate.

06 20 00 Finish Carpentry

- It is necessary to consider the grade of workmanship when estimating labor costs for erecting millwork and an interior finish. In practice, there are three grades: premium, custom, and economy. The RSMeans daily output for base and case moldings is in the range of 200 to 250 L.F. per carpenter per day. This is appropriate for most average custom-grade projects. For premium projects, an adjustment to productivity of 25%–50% should be made, depending on the complexity of the job.

Reference Numbers

Reference numbers are shown at the beginning of some major classifications. These numbers refer to related items in the Reference Section. The reference information may be an estimating procedure, an alternate pricing method, or technical information.

Note: Not all subdivisions listed here necessarily appear. ∎

06 05 05.10 Selective Demolition Wood Framing	Crew	Daily Output	Labor-Hours	Unit	Material	2018 Bare Costs Labor	2018 Bare Costs Equipment	Total	Total Incl O&P
0010 SELECTIVE DEMOLITION WOOD FRAMING R024119-10									
0100 Timber connector, nailed, small	1 Clab	96	.083	Ea.		2.22		2.22	3.70
0110 Medium		60	.133			3.56		3.56	5.90
0120 Large		48	.167			4.45		4.45	7.40
0130 Bolted, small		48	.167			4.45		4.45	7.40
0140 Medium		32	.250			6.70		6.70	11.10
0150 Large	2 Clab	24	.333			8.90		8.90	14.80
2958 Beams, 2" x 6"	2 Clab	1100	.015	L.F.		.39		.39	.65
2960 2" x 8"		825	.019			.52		.52	.86
2965 2" x 10"		665	.024			.64		.64	1.07
2970 2" x 12"		550	.029			.78		.78	1.29
2972 2" x 14"		470	.034			.91		.91	1.51
2975 4" x 8"	B-1	413	.058			1.59		1.59	2.65
2980 4" x 10"		330	.073			1.99		1.99	3.31
2985 4" x 12"		275	.087			2.39		2.39	3.97
3000 6" x 8"		275	.087			2.39		2.39	3.97
3040 6" x 10"		220	.109			2.99		2.99	4.97
3080 6" x 12"		185	.130			3.55		3.55	5.90
3120 8" x 12"		140	.171			4.69		4.69	7.80
3160 10" x 12"		110	.218			5.95		5.95	9.95
3162 Alternate pricing method		1.10	21.818	M.B.F.		595		595	995
3170 Blocking, in 16" OC wall framing, 2" x 4"	1 Clab	600	.013	L.F.		.36		.36	.59
3172 2" x 6"		400	.020			.53		.53	.89
3174 In 24" OC wall framing, 2" x 4"		600	.013			.36		.36	.59
3176 2" x 6"		400	.020			.53		.53	.89
3178 Alt method, wood blocking removal from wood framing		.40	20	M.B.F.		535		535	890
3179 Wood blocking removal from steel framing		.36	22.222	"		595		595	985
3180 Bracing, let in, 1" x 3", studs 16" OC		1050	.008	L.F.		.20		.20	.34
3181 Studs 24" OC		1080	.007			.20		.20	.33
3182 1" x 4", studs 16" OC		1050	.008			.20		.20	.34
3183 Studs 24" OC		1080	.007			.20		.20	.33
3184 1" x 6", studs 16" OC		1050	.008			.20		.20	.34
3185 Studs 24" OC		1080	.007			.20		.20	.33
3186 2" x 3", studs 16" OC		800	.010			.27		.27	.44
3187 Studs 24" OC		830	.010			.26		.26	.43
3188 2" x 4", studs 16" OC		800	.010			.27		.27	.44
3189 Studs 24" OC		830	.010			.26		.26	.43
3190 2" x 6", studs 16" OC		800	.010			.27		.27	.44
3191 Studs 24" OC		830	.010			.26		.26	.43
3192 2" x 8", studs 16" OC		800	.010			.27		.27	.44
3193 Studs 24" OC		830	.010			.26		.26	.43
3194 "T" shaped metal bracing, studs at 16" OC		1060	.008			.20		.20	.34
3195 Studs at 24" OC		1200	.007			.18		.18	.30
3196 Metal straps, studs at 16" OC		1200	.007			.18		.18	.30
3197 Studs at 24" OC		1240	.006			.17		.17	.29
3200 Columns, round, 8' to 14' tall		40	.200	Ea.		5.35		5.35	8.90
3202 Dimensional lumber sizes	2 Clab	1.10	14.545	M.B.F.		390		390	645
3250 Blocking, between joists	1 Clab	320	.025	Ea.		.67		.67	1.11
3252 Bridging, metal strap, between joists		320	.025	Pr.		.67		.67	1.11
3254 Wood, between joists		320	.025	"		.67		.67	1.11
3260 Door buck, studs, header & access., 8' high 2" x 4" wall, 3' wide		32	.250	Ea.		6.70		6.70	11.10
3261 4' wide		32	.250			6.70		6.70	11.10
3262 5' wide		32	.250			6.70		6.70	11.10

06 05 05.10 Selective Demolition Wood Framing	Crew	Daily Output	Labor-Hours	Unit	Material	2018 Bare Costs Labor	Equipment	Total	Total Incl O&P
3263	6' wide	1 Clab	32	.250	Ea.		6.70		6.70	11.10
3264	8' wide		30	.267			7.10		7.10	11.85
3265	10' wide		30	.267			7.10		7.10	11.85
3266	12' wide		30	.267			7.10		7.10	11.85
3267	2" x 6" wall, 3' wide		32	.250			6.70		6.70	11.10
3268	4' wide		32	.250			6.70		6.70	11.10
3269	5' wide		32	.250			6.70		6.70	11.10
3270	6' wide		32	.250			6.70		6.70	11.10
3271	8' wide		30	.267			7.10		7.10	11.85
3272	10' wide		30	.267			7.10		7.10	11.85
3273	12' wide		30	.267			7.10		7.10	11.85
3274	Window buck, studs, header & access., 8' high 2" x 4" wall, 2' wide		24	.333			8.90		8.90	14.80
3275	3' wide		24	.333			8.90		8.90	14.80
3276	4' wide		24	.333			8.90		8.90	14.80
3277	5' wide		24	.333			8.90		8.90	14.80
3278	6' wide		24	.333			8.90		8.90	14.80
3279	7' wide		24	.333			8.90		8.90	14.80
3280	8' wide		22	.364			9.70		9.70	16.15
3281	10' wide		22	.364			9.70		9.70	16.15
3282	12' wide		22	.364			9.70		9.70	16.15
3283	2" x 6" wall, 2' wide		24	.333			8.90		8.90	14.80
3284	3' wide		24	.333			8.90		8.90	14.80
3285	4' wide		24	.333			8.90		8.90	14.80
3286	5' wide		24	.333			8.90		8.90	14.80
3287	6' wide		24	.333			8.90		8.90	14.80
3288	7' wide		24	.333			8.90		8.90	14.80
3289	8' wide		22	.364			9.70		9.70	16.15
3290	10' wide		22	.364			9.70		9.70	16.15
3291	12' wide		22	.364	▼		9.70		9.70	16.15
3360	Deck or porch decking		825	.010	L.F.		.26		.26	.43
3400	Fascia boards, 1" x 6"		500	.016			.43		.43	.71
3440	1" x 8"		450	.018			.47		.47	.79
3480	1" x 10"		400	.020			.53		.53	.89
3490	2" x 6"		450	.018			.47		.47	.79
3500	2" x 8"		400	.020			.53		.53	.89
3510	2" x 10"		350	.023	▼		.61		.61	1.02
3610	Furring, on wood walls or ceiling		4000	.002	S.F.		.05		.05	.09
3620	On masonry or concrete walls or ceiling		1200	.007	"		.18		.18	.30
3800	Headers over openings, 2 @ 2" x 6"		110	.073	L.F.		1.94		1.94	3.23
3840	2 @ 2" x 8"		100	.080			2.14		2.14	3.55
3880	2 @ 2" x 10"		90	.089	▼		2.37		2.37	3.95
3885	Alternate pricing method		.26	30.651	M.B.F.		820		820	1,350
3920	Joists, 1" x 4"		1250	.006	L.F.		.17		.17	.28
3930	1" x 6"		1135	.007			.19		.19	.31
3940	1" x 8"		1000	.008			.21		.21	.36
3950	1" x 10"		895	.009			.24		.24	.40
3960	1" x 12"		765	.010			.28		.28	.46
4200	2" x 4"	2 Clab	1000	.016			.43		.43	.71
4230	2" x 6"		970	.016			.44		.44	.73
4240	2" x 8"		940	.017			.45		.45	.76
4250	2" x 10"		910	.018			.47		.47	.78
4280	2" x 12"		880	.018			.49		.49	.81
4281	2" x 14"		850	.019	▼		.50		.50	.84

357

06 05 05.10 Selective Demolition Wood Framing	Crew	Daily Output	Labor-Hours	Unit	Material	2018 Bare Costs Labor	Equipment	Total	Total Incl O&P
4282	Composite joists, 9-1/2"	2 Clab	960	.017	L.F.		.45		.45	.74
4283	11-7/8"		930	.017			.46		.46	.76
4284	14"		897	.018			.48		.48	.79
4285	16"		865	.019			.49		.49	.82
4290	Wood joists, alternate pricing method		1.50	10.667	M.B.F.		285		285	475
4500	Open web joist, 12" deep		500	.032	L.F.		.85		.85	1.42
4505	14" deep		475	.034			.90		.90	1.50
4510	16" deep		450	.036			.95		.95	1.58
4520	18" deep		425	.038			1.01		1.01	1.67
4530	24" deep		400	.040			1.07		1.07	1.78
4550	Ledger strips, 1" x 2"	1 Clab	1200	.007			.18		.18	.30
4560	1" x 3"		1200	.007			.18		.18	.30
4570	1" x 4"		1200	.007			.18		.18	.30
4580	2" x 2"		1100	.007			.19		.19	.32
4590	2" x 4"		1000	.008			.21		.21	.36
4600	2" x 6"		1000	.008			.21		.21	.36
4601	2" x 8 or 2" x 10"		800	.010			.27		.27	.44
4602	4" x 6"		600	.013			.36		.36	.59
4604	4" x 8"		450	.018			.47		.47	.79
5400	Posts, 4" x 4"	2 Clab	800	.020			.53		.53	.89
5405	4" x 6"		550	.029			.78		.78	1.29
5410	4" x 8"		440	.036			.97		.97	1.61
5425	4" x 10"		390	.041			1.10		1.10	1.82
5430	4" x 12"		350	.046			1.22		1.22	2.03
5440	6" x 6"		400	.040			1.07		1.07	1.78
5445	6" x 8"		350	.046			1.22		1.22	2.03
5450	6" x 10"		320	.050			1.34		1.34	2.22
5455	6" x 12"		290	.055			1.47		1.47	2.45
5480	8" x 8"		300	.053			1.42		1.42	2.37
5500	10" x 10"		240	.067			1.78		1.78	2.96
5660	T&G floor planks		2	8	M.B.F.		214		214	355
5682	Rafters, ordinary, 16" OC, 2" x 4"		880	.018	S.F.		.49		.49	.81
5683	2" x 6"		840	.019			.51		.51	.85
5684	2" x 8"		820	.020			.52		.52	.87
5685	2" x 10"		820	.020			.52		.52	.87
5686	2" x 12"		810	.020			.53		.53	.88
5687	24" OC, 2" x 4"		1170	.014			.37		.37	.61
5688	2" x 6"		1117	.014			.38		.38	.64
5689	2" x 8"		1091	.015			.39		.39	.65
5690	2" x 10"		1091	.015			.39		.39	.65
5691	2" x 12"		1077	.015			.40		.40	.66
5795	Rafters, ordinary, 2" x 4" (alternate method)		862	.019	L.F.		.50		.50	.82
5800	2" x 6" (alternate method)		850	.019			.50		.50	.84
5840	2" x 8" (alternate method)		837	.019			.51		.51	.85
5855	2" x 10" (alternate method)		825	.019			.52		.52	.86
5865	2" x 12" (alternate method)		812	.020			.53		.53	.87
5870	Sill plate, 2" x 4"	1 Clab	1170	.007			.18		.18	.30
5871	2" x 6"		780	.010			.27		.27	.46
5872	2" x 8"		586	.014			.36		.36	.61
5873	Alternate pricing method		.78	10.256	M.B.F.		274		274	455
5885	Ridge board, 1" x 4"	2 Clab	900	.018	L.F.		.47		.47	.79
5886	1" x 6"		875	.018			.49		.49	.81
5887	1" x 8"		850	.019			.50		.50	.84

06 05 05 – Selective Demolition for Wood, Plastics, and Composites

06 05 05.10 Selective Demolition Wood Framing	Crew	Daily Output	Labor-Hours	Unit	Material	2018 Bare Costs Labor	Equipment	Total	Total Incl O&P
5888	1" x 10"	2 Clab	825	.019	L.F.		.52		.52	.86
5889	1" x 12"		800	.020			.53		.53	.89
5890	2" x 4"		900	.018			.47		.47	.79
5892	2" x 6"		875	.018			.49		.49	.81
5894	2" x 8"		850	.019			.50		.50	.84
5896	2" x 10"		825	.019			.52		.52	.86
5898	2" x 12"		800	.020			.53		.53	.89
6050	Rafter tie, 1" x 4"		1250	.013			.34		.34	.57
6052	1" x 6"		1135	.014			.38		.38	.63
6054	2" x 4"		1000	.016			.43		.43	.71
6056	2" x 6"		970	.016			.44		.44	.73
6070	Sleepers, on concrete, 1" x 2"	1 Clab	4700	.002			.05		.05	.08
6075	1" x 3"		4000	.002			.05		.05	.09
6080	2" x 4"		3000	.003			.07		.07	.12
6085	2" x 6"		2600	.003			.08		.08	.14
6086	Sheathing from roof, 5/16"	2 Clab	1600	.010	S.F.		.27		.27	.44
6088	3/8"		1525	.010			.28		.28	.47
6090	1/2"		1400	.011			.31		.31	.51
6092	5/8"		1300	.012			.33		.33	.55
6094	3/4"		1200	.013			.36		.36	.59
6096	Board sheathing from roof		1400	.011			.31		.31	.51
6100	Sheathing, from walls, 1/4"		1200	.013			.36		.36	.59
6110	5/16"		1175	.014			.36		.36	.60
6120	3/8"		1150	.014			.37		.37	.62
6130	1/2"		1125	.014			.38		.38	.63
6140	5/8"		1100	.015			.39		.39	.65
6150	3/4"		1075	.015			.40		.40	.66
6152	Board sheathing from walls		1500	.011			.28		.28	.47
6158	Subfloor/roof deck, with boards		2200	.007			.19		.19	.32
6159	Subfloor/roof deck, with tongue & groove boards		2000	.008			.21		.21	.36
6160	Plywood, 1/2" thick		768	.021			.56		.56	.92
6162	5/8" thick		760	.021			.56		.56	.93
6164	3/4" thick		750	.021			.57		.57	.95
6165	1-1/8" thick		720	.022			.59		.59	.99
6166	Underlayment, particle board, 3/8" thick	1 Clab	780	.010			.27		.27	.46
6168	1/2" thick		768	.010			.28		.28	.46
6170	5/8" thick		760	.011			.28		.28	.47
6172	3/4" thick		750	.011			.28		.28	.47
6200	Stairs and stringers, straight run	2 Clab	40	.400	Riser		10.70		10.70	17.75
6240	With platforms, winders or curves	"	26	.615	"		16.45		16.45	27.50
6300	Components, tread	1 Clab	110	.073	Ea.		1.94		1.94	3.23
6320	Riser		80	.100	"		2.67		2.67	4.44
6390	Stringer, 2" x 10"		260	.031	L.F.		.82		.82	1.37
6400	2" x 12"		260	.031			.82		.82	1.37
6410	3" x 10"		250	.032			.85		.85	1.42
6420	3" x 12"		250	.032			.85		.85	1.42
6590	Wood studs, 2" x 3"	2 Clab	3076	.005			.14		.14	.23
6600	2" x 4"		2000	.008			.21		.21	.36
6640	2" x 6"		1600	.010			.27		.27	.44
6720	Wall framing, including studs, plates and blocking, 2" x 4"	1 Clab	600	.013	S.F.		.36		.36	.59
6740	2" x 6"		480	.017	"		.45		.45	.74
6750	Headers, 2" x 4"		1125	.007	L.F.		.19		.19	.32
6755	2" x 6"		1125	.007			.19		.19	.32

For customer support on your Residential Costs with RSMeans data, call 800.448.8182.

359

06 05 05 – Selective Demolition for Wood, Plastics, and Composites

06 05 05.10 Selective Demolition Wood Framing

		Crew	Daily Output	Labor-Hours	Unit	Material	2018 Bare Costs Labor	Equipment	Total	Total Incl O&P
6760	2" x 8"	1 Clab	1050	.008	L.F.		.20		.20	.34
6765	2" x 10"		1050	.008			.20		.20	.34
6770	2" x 12"		1000	.008			.21		.21	.36
6780	4" x 10"		525	.015			.41		.41	.68
6785	4" x 12"		500	.016			.43		.43	.71
6790	6" x 8"		560	.014			.38		.38	.63
6795	6" x 10"		525	.015			.41		.41	.68
6797	6" x 12"	▼	500	.016	▼		.43		.43	.71
7000	Trusses									
7050	12' span	2 Clab	74	.216	Ea.		5.75		5.75	9.60
7150	24' span	F-3	66	.606			19.40	7.50	26.90	41
7200	26' span		64	.625			20	7.75	27.75	42
7250	28' span		62	.645			20.50	8	28.50	43.50
7300	30' span		58	.690			22	8.55	30.55	46.50
7350	32' span		56	.714			23	8.85	31.85	48
7400	34' span		54	.741			23.50	9.20	32.70	49.50
7450	36' span	▼	52	.769	▼		24.50	9.55	34.05	51.50
8000	Soffit, T&G wood	1 Clab	520	.015	S.F.		.41		.41	.68
8010	Hardboard, vinyl or aluminum	"	640	.013			.33		.33	.56
8030	Plywood	2 Carp	315	.051	▼		1.78		1.78	2.95
9500	See Section 02 41 19.19 for rubbish handling									

06 05 05.20 Selective Demolition Millwork and Trim

		Crew	Daily Output	Labor-Hours	Unit	Material	2018 Bare Costs Labor	Equipment	Total	Total Incl O&P
0010	**SELECTIVE DEMOLITION MILLWORK AND TRIM** R024119-10									
1000	Cabinets, wood, base cabinets, per L.F.	2 Clab	80	.200	L.F.		5.35		5.35	8.90
1020	Wall cabinets, per L.F.	"	80	.200	"		5.35		5.35	8.90
1060	Remove and reset, base cabinets	2 Carp	18	.889	Ea.		31		31	51.50
1070	Wall cabinets		20	.800			28		28	46.50
1072	Oven cabinet, 7' high	▼	11	1.455			51		51	84.50
1074	Cabinet door, up to 2' high	1 Clab	66	.121			3.24		3.24	5.40
1076	2' - 4' high	"	46	.174	▼		4.64		4.64	7.70
1100	Steel, painted, base cabinets	2 Clab	60	.267	L.F.		7.10		7.10	11.85
1120	Wall cabinets		60	.267	"		7.10		7.10	11.85
1200	Casework, large area		320	.050	S.F.		1.34		1.34	2.22
1220	Selective		200	.080	"		2.14		2.14	3.55
1500	Counter top, straight runs		200	.080	L.F.		2.14		2.14	3.55
1510	L, U or C shapes		120	.133	"		3.56		3.56	5.90
2000	Paneling, 4' x 8' sheets		2000	.008	S.F.		.21		.21	.36
2100	Boards, 1" x 4"		700	.023			.61		.61	1.02
2120	1" x 6"		750	.021			.57		.57	.95
2140	1" x 8"		800	.020	▼		.53		.53	.89
3000	Trim, baseboard, to 6" wide		1200	.013	L.F.		.36		.36	.59
3040	Greater than 6" and up to 12" wide	▼	1000	.016			.43		.43	.71
3080	Remove and reset, minimum	2 Carp	400	.040			1.40		1.40	2.32
3090	Maximum	"	300	.053			1.86		1.86	3.10
3100	Ceiling trim	2 Clab	1000	.016			.43		.43	.71
3120	Chair rail		1200	.013			.36		.36	.59
3140	Railings with balusters		240	.067	▼		1.78		1.78	2.96
3160	Wainscoting	▼	700	.023	S.F.		.61		.61	1.02
4000	Curtain rod	1 Clab	80	.100	L.F.		2.67		2.67	4.44

06 05 23 – Wood, Plastic, and Composite Fastenings

06 05 23.10 Nails	Crew	Daily Output	Labor-Hours	Unit	Material	2018 Bare Costs Labor	Equipment	Total	Total Incl O&P
0010 NAILS, material only, based upon 50# box purchase									
0020 Copper nails, plain				Lb.	11.15			11.15	12.30
0400 Stainless steel, plain					8.55			8.55	9.40
0500 Box, 3d to 20d, bright					1.43			1.43	1.57
0520 Galvanized					2.34			2.34	2.57
0600 Common, 3d to 60d, plain					1.15			1.15	1.27
0700 Galvanized					2.19			2.19	2.41
0800 Aluminum					10.80			10.80	11.90
1000 Annular or spiral thread, 4d to 60d, plain					3.19			3.19	3.51
1200 Galvanized					3.02			3.02	3.32
1400 Drywall nails, plain					1.78			1.78	1.96
1600 Galvanized					1.84			1.84	2.02
1800 Finish nails, 4d to 10d, plain					1.24			1.24	1.36
2000 Galvanized					1.81			1.81	1.99
2100 Aluminum					7.95			7.95	8.75
2300 Flooring nails, hardened steel, 2d to 10d, plain					3.54			3.54	3.89
2400 Galvanized					3.75			3.75	4.13
2500 Gypsum lath nails, 1-1/8", 13 ga. flathead, blued					3.32			3.32	3.65
2600 Masonry nails, hardened steel, 3/4" to 3" long, plain					2.34			2.34	2.57
2700 Galvanized					3.90			3.90	4.29
2900 Roofing nails, threaded, galvanized					2.76			2.76	3.04
3100 Aluminum					7.25			7.25	8
3300 Compressed lead head, threaded, galvanized					2.97			2.97	3.27
3600 Siding nails, plain shank, galvanized					2.53			2.53	2.78
3800 Aluminum					5.75			5.75	6.35
5000 Add to prices above for cement coating					.15			.15	.17
5200 Zinc or tin plating					.25			.25	.28
5500 Vinyl coated sinkers, 8d to 16d					2.56			2.56	2.82

06 05 23.50 Wood Screws

	Crew	Daily Output	Labor-Hours	Unit	Material	Labor	Equipment	Total	Total Incl O&P
0010 WOOD SCREWS									
0020 #8, 1" long, steel				C	4.60			4.60	5.05
0100 Brass					12.20			12.20	13.40
0200 #8, 2" long, steel					8			8	8.80
0300 Brass					24			24	26.50
0400 #10, 1" long, steel					3.30			3.30	3.63
0500 Brass					15.60			15.60	17.15
0600 #10, 2" long, steel					5.30			5.30	5.85
0700 Brass					27			27	29.50
0800 #10, 3" long, steel					8.85			8.85	9.75
1000 #12, 2" long, steel					7.30			7.30	8.05
1100 Brass					34.50			34.50	38
1500 #12, 3" long, steel					11.15			11.15	12.30
2000 #12, 4" long, steel					24.50			24.50	27

06 05 23.60 Timber Connectors

	Crew	Daily Output	Labor-Hours	Unit	Material	Labor	Equipment	Total	Total Incl O&P
0010 TIMBER CONNECTORS									
0020 Add up cost of each part for total cost of connection									
0100 Connector plates, steel, with bolts, straight	2 Carp	75	.213	Ea.	27.50	7.45		34.95	43
0110 Tee, 7 ga.		50	.320		34	11.20		45.20	56
0120 T- Strap, 14 ga., 12" x 8" x 2"		50	.320		34	11.20		45.20	56
0150 Anchor plates, 7 ga., 9" x 7"		75	.213		27.50	7.45		34.95	43
0200 Bolts, machine, sq. hd. with nut & washer, 1/2" diameter, 4" long	1 Carp	140	.057		.75	2		2.75	4.15
0300 7-1/2" long		130	.062		1.37	2.15		3.52	5.10

06 05 23.60 Timber Connectors	Crew	Daily Output	Labor-Hours	Unit	Material	2018 Bare Costs Labor	2018 Bare Costs Equipment	Total	Total Incl O&P
0500	3/4" diameter, 7-1/2" long	1 Carp	130	.062	Ea.	3.20	2.15		5.35	7.10
0610	Machine bolts, w/nut, washer, 3/4" diameter, 15" L, HD's & beam hangers		95	.084	▼	5.95	2.94		8.89	11.45
0720	Machine bolts, sq. hd. w/nut & wash		150	.053	Lb.	3.80	1.86		5.66	7.30
0800	Drilling bolt holes in timber, 1/2" diameter		450	.018	Inch		.62		.62	1.03
0900	1" diameter		350	.023	"		.80		.80	1.33
1100	Framing anchor, angle, 3" x 3" x 1-1/2", 12 ga.		175	.046	Ea.	2.43	1.60		4.03	5.35
1150	Framing anchors, 18 ga., 4-1/2" x 2-3/4"		175	.046		2.43	1.60		4.03	5.35
1160	Framing anchors, 18 ga., 4-1/2" x 3"		175	.046		2.43	1.60		4.03	5.35
1170	Clip anchors plates, 18 ga., 12" x 1-1/8"		175	.046		2.43	1.60		4.03	5.35
1250	Holdowns, 3 ga. base, 10 ga. body		8	1		38	35		73	100
1260	Holdowns, 7 ga. 11-1/16" x 3-1/4"		8	1		38	35		73	100
1270	Holdowns, 7 ga. 14-3/8" x 3-1/8"		8	1		38	35		73	100
1275	Holdowns, 12 ga. 8" x 2-1/2"		8	1		38	35		73	100
1300	Joist and beam hangers, 18 ga. galv., for 2" x 4" joist		175	.046		.75	1.60		2.35	3.49
1400	2" x 6" to 2" x 10" joist		165	.048		1.27	1.69		2.96	4.22
1600	16 ga. galv., 3" x 6" to 3" x 10" joist		160	.050		2.92	1.75		4.67	6.10
1700	3" x 10" to 3" x 14" joist		160	.050		4.32	1.75		6.07	7.65
1800	4" x 6" to 4" x 10" joist		155	.052		2.89	1.80		4.69	6.20
1900	4" x 10" to 4" x 14" joist		155	.052		4.89	1.80		6.69	8.40
2000	Two-2" x 6" to two-2" x 10" joists		150	.053		3.83	1.86		5.69	7.30
2100	Two-2" x 10" to two-2" x 14" joists		150	.053		4.29	1.86		6.15	7.80
2300	3/16" thick, 6" x 8" joist		145	.055		59.50	1.93		61.43	68.50
2400	6" x 10" joist		140	.057		62	2		64	71.50
2500	6" x 12" joist		135	.059		64.50	2.07		66.57	74.50
2700	1/4" thick, 6" x 14" joist	▼	130	.062		67	2.15		69.15	77
2900	Plywood clips, extruded aluminum H clip, for 3/4" panels					.22			.22	.24
3000	Galvanized 18 ga. back-up clip					.18			.18	.20
3200	Post framing, 16 ga. galv. for 4" x 4" base, 2 piece	1 Carp	130	.062		15.15	2.15		17.30	20.50
3300	Cap		130	.062		21	2.15		23.15	26.50
3500	Rafter anchors, 18 ga. galv., 1-1/2" wide, 5-1/4" long		145	.055		.45	1.93		2.38	3.71
3600	10-3/4" long		145	.055		1.37	1.93		3.30	4.72
3800	Shear plates, 2-5/8" diameter		120	.067		2.47	2.33		4.80	6.60
3900	4" diameter		115	.070		2.40	2.43		4.83	6.70
4000	Sill anchors, embedded in concrete or block, 25-1/2" long		115	.070		11.60	2.43		14.03	16.80
4100	Spike grids, 3" x 6"		120	.067		1.09	2.33		3.42	5.05
4400	Split rings, 2-1/2" diameter		120	.067		2.44	2.33		4.77	6.55
4500	4" diameter		110	.073		2.78	2.54		5.32	7.30
4550	Tie plate, 20 ga., 7" x 3-1/8"		110	.073		2.78	2.54		5.32	7.30
4560	5" x 4-1/8"		110	.073		2.78	2.54		5.32	7.30
4575	Twist straps, 18 ga., 12" x 1-1/4"		110	.073		2.78	2.54		5.32	7.30
4580	16" x 1-1/4"		110	.073		2.78	2.54		5.32	7.30
4600	Strap ties, 20 ga., 2-1/16" wide, 12-13/16" long		180	.044		.87	1.55		2.42	3.54
4700	16 ga., 1-3/8" wide, 12" long		180	.044		.87	1.55		2.42	3.54
4800	1-1/4" wide, 21-5/8" long		160	.050		2.70	1.75		4.45	5.90
5000	Toothed rings, 2-5/8" or 4" diameter		90	.089	▼	2.05	3.11		5.16	7.40
5200	Truss plates, nailed, 20 ga., up to 32' span	▼	17	.471	Truss	12.85	16.45		29.30	41.50
5400	Washers, 2" x 2" x 1/8"				Ea.	.46			.46	.51
5500	3" x 3" x 3/16"				"	1.32			1.32	1.45
6000	Angles and gussets, painted									
6012	7 ga., 3-1/4" x 3-1/4" x 2-1/2" long	1 Carp	1.90	4.211	C	1,075	147		1,222	1,425
6014	3-1/4" x 3-1/4" x 5" long		1.90	4.211		2,100	147		2,247	2,550
6016	3-1/4" x 3-1/4" x 7-1/2" long		1.85	4.324		3,975	151		4,126	4,600
6018	5-3/4" x 5-3/4" x 2-1/2" long	▼	1.85	4.324	▼	2,550	151		2,701	3,075

06 05 23.60 Timber Connectors		Crew	Daily Output	Labor-Hours	Unit	Material	2018 Bare Costs Labor	Equipment	Total	Total Incl O&P
6020	5-3/4" x 5-3/4" x 5" long	1 Carp	1.85	4.324	C	4,075	151		4,226	4,725
6022	5-3/4" x 5-3/4" x 7-1/2" long		1.80	4.444		6,025	155		6,180	6,875
6024	3 ga., 4-1/4" x 4-1/4" x 3" long		1.85	4.324		2,725	151		2,876	3,225
6026	4-1/4" x 4-1/4" x 6" long		1.85	4.324		5,850	151		6,001	6,675
6028	4-1/4" x 4-1/4" x 9" long		1.80	4.444		6,575	155		6,730	7,475
6030	7-1/4" x 7-1/4" x 3" long		1.80	4.444		4,700	155		4,855	5,425
6032	7-1/4" x 7-1/4" x 6" long		1.80	4.444		6,350	155		6,505	7,225
6034	7-1/4" x 7-1/4" x 9" long	↓	1.75	4.571	↓	14,200	160		14,360	15,900
6036	Gussets									
6038	7 ga., 8-1/8" x 8-1/8" x 2-3/4" long	1 Carp	1.80	4.444	C	4,850	155		5,005	5,575
6040	3 ga., 9-3/4" x 9-3/4" x 3-1/4" long	"	1.80	4.444	"	7,275	155		7,430	8,250
6101	Beam hangers, polymer painted									
6102	Bolted, 3 ga. (W x H x L)									
6104	3-1/4" x 9" x 12" top flange	1 Carp	1	8	C	19,100	280		19,380	21,500
6106	5-1/4" x 9" x 12" top flange		1	8		20,000	280		20,280	22,500
6108	5-1/4" x 11" x 11-3/4" top flange		1	8		20,800	280		21,080	23,300
6110	6-7/8" x 9" x 12" top flange		1	8		20,600	280		20,880	23,200
6112	6-7/8" x 11" x 13-1/2" top flange		1	8		24,300	280		24,580	27,300
6114	8-7/8" x 11" x 15-1/2" top flange	↓	1	8	↓	23,600	280		23,880	26,400
6116	Nailed, 3 ga. (W x H x L)									
6118	3-1/4" x 10-1/2" x 10" top flange	1 Carp	1.80	4.444	C	18,700	155		18,855	20,800
6120	3-1/4" x 10-1/2" x 12" top flange		1.80	4.444		21,400	155		21,555	23,900
6122	5-1/4" x 9-1/2" x 10" top flange		1.80	4.444		17,500	155		17,655	19,500
6124	5-1/4" x 9-1/2" x 12" top flange		1.80	4.444		29,800	155		29,955	33,100
6128	6-7/8" x 8-1/2" x 12" top flange	↓	1.80	4.444	↓	20,600	155		20,755	23,000
6134	Saddle hangers, glu-lam (W x H x L)									
6136	3-1/4" x 10-1/2" x 5-1/4" x 6" saddle	1 Carp	.50	16	C	14,900	560		15,460	17,300
6138	3-1/4" x 10-1/2" x 6-7/8" x 6" saddle		.50	16		15,700	560		16,260	18,100
6140	3-1/4" x 10-1/2" x 8-7/8" x 6" saddle		.50	16		17,400	560		17,960	20,100
6142	3-1/4" x 19-1/2" x 5-1/4" x 10-1/8" saddle		.40	20		14,900	700		15,600	17,600
6144	3-1/4" x 19-1/2" x 6-7/8" x 10-1/8" saddle		.40	20		15,700	700		16,400	18,400
6146	3-1/4" x 19-1/2" x 8-7/8" x 10-1/8" saddle		.40	20		16,500	700		17,200	19,300
6148	5-1/4" x 9-1/2" x 5-1/4" x 12" saddle		.50	16		17,800	560		18,360	20,400
6150	5-1/4" x 9-1/2" x 6-7/8" x 9" saddle		.50	16		19,500	560		20,060	22,300
6152	5-1/4" x 10-1/2" x spec x 12" saddle		.50	16		21,200	560		21,760	24,200
6154	5-1/4" x 18" x 5-1/4" x 12-1/8" saddle		.40	20		17,800	700		18,500	20,700
6156	5-1/4" x 18" x 6-7/8" x 12-1/8" saddle		.40	20		19,500	700		20,200	22,600
6158	5-1/4" x 18" x spec x 12-1/8" saddle		.40	20		21,200	700		21,900	24,500
6160	6-7/8" x 8-1/2" x 6-7/8" x 12" saddle		.50	16		20,000	560		20,560	22,900
6162	6-7/8" x 8-1/2" x 8-7/8" x 12" saddle		.50	16		20,700	560		21,260	23,600
6164	6-7/8" x 10-1/2" x spec x 12" saddle		.50	16		20,000	560		20,560	22,900
6166	6-7/8" x 18" x 6-7/8" x 13-3/4" saddle		.40	20		20,000	700		20,700	23,200
6168	6-7/8" x 18" x 8-7/8" x 13-3/4" saddle		.40	20		22,000	700		22,700	25,400
6170	6-7/8" x 18" x spec x 13-3/4" saddle		.40	20		22,300	700		23,000	25,700
6172	8-7/8" x 18" x spec x 15-3/4" saddle	↓	.40	20	↓	34,800	700		35,500	39,500
6201	Beam and purlin hangers, galvanized, 12 ga.									
6202	Purlin or joist size, 3" x 8"	1 Carp	1.70	4.706	C	1,850	164		2,014	2,325
6204	3" x 10"		1.70	4.706		2,125	164		2,289	2,625
6206	3" x 12"		1.65	4.848		2,300	169		2,469	2,800
6208	3" x 14"		1.65	4.848		2,450	169		2,619	2,950
6210	3" x 16"		1.65	4.848		2,575	169		2,744	3,125
6212	4" x 8"		1.65	4.848		1,875	169		2,044	2,325
6214	4" x 10"	↓	1.65	4.848	↓	2,125	169		2,294	2,625

For customer support on your Residential Costs with RSMeans data, call 800.448.8182.

363

06 05 23.60 Timber Connectors		Crew	Daily Output	Labor-Hours	Unit	Material	2018 Bare Costs Labor	Equipment	Total	Total Incl O&P
6216	4" x 12"	1 Carp	1.60	5	C	2,525	175		2,700	3,075
6218	4" x 14"		1.60	5		2,525	175		2,700	3,075
6220	4" x 16"		1.60	5		2,850	175		3,025	3,425
6222	6" x 8"		1.60	5		2,275	175		2,450	2,800
6224	6" x 10"		1.55	5.161		2,425	180		2,605	2,975
6226	6" x 12"		1.55	5.161		4,450	180		4,630	5,200
6228	6" x 14"		1.50	5.333		4,725	186		4,911	5,500
6230	6" x 16"		1.50	5.333		5,000	186		5,186	5,800
6250	Beam seats									
6252	Beam size, 5-1/4" wide									
6254	5" x 7" x 1/4"	1 Carp	1.80	4.444	C	7,500	155		7,655	8,500
6256	6" x 7" x 3/8"		1.80	4.444		8,425	155		8,580	9,525
6258	7" x 7" x 3/8"		1.80	4.444		9,000	155		9,155	10,200
6260	8" x 7" x 3/8"		1.80	4.444		10,600	155		10,755	12,000
6262	Beam size, 6-7/8" wide									
6264	5" x 9" x 1/4"	1 Carp	1.80	4.444	C	8,975	155		9,130	10,100
6266	6" x 9" x 3/8"		1.80	4.444		11,700	155		11,855	13,200
6268	7" x 9" x 3/8"		1.80	4.444		11,800	155		11,955	13,300
6270	8" x 9" x 3/8"		1.80	4.444		14,000	155		14,155	15,700
6272	Special beams, over 6-7/8" wide									
6274	5" x 10" x 3/8"	1 Carp	1.80	4.444	C	12,100	155		12,255	13,700
6276	6" x 10" x 3/8"		1.80	4.444		14,200	155		14,355	15,900
6278	7" x 10" x 3/8"		1.80	4.444		14,800	155		14,955	16,600
6280	8" x 10" x 3/8"		1.75	4.571		15,900	160		16,060	17,800
6282	5-1/4" x 12" x 5/16"		1.75	4.571		12,300	160		12,460	13,800
6284	6-1/2" x 12" x 3/8"		1.75	4.571		20,400	160		20,560	22,700
6286	5-1/4" x 16" x 5/16"		1.70	4.706		18,100	164		18,264	20,200
6288	6-1/2" x 16" x 3/8"		1.70	4.706		23,600	164		23,764	26,200
6290	5-1/4" x 20" x 5/16"		1.70	4.706		21,200	164		21,364	23,700
6292	6-1/2" x 20" x 3/8"		1.65	4.848		27,700	169		27,869	30,800
6300	Column bases									
6302	4" x 4", 16 ga.	1 Carp	1.80	4.444	C	715	155		870	1,050
6306	7 ga.		1.80	4.444		2,900	155		3,055	3,425
6308	4" x 6", 16 ga.		1.80	4.444		1,650	155		1,805	2,075
6312	7 ga.		1.80	4.444		2,850	155		3,005	3,375
6314	6" x 6", 16 ga.		1.75	4.571		1,875	160		2,035	2,350
6318	7 ga.		1.75	4.571		3,650	160		3,810	4,275
6320	6" x 8", 7 ga.		1.70	4.706		2,900	164		3,064	3,450
6322	6" x 10", 7 ga.		1.70	4.706		3,300	164		3,464	3,900
6324	6" x 12", 7 ga.		1.70	4.706		3,575	164		3,739	4,225
6326	8" x 8', 7 ga.		1.65	4.848		5,700	169		5,869	6,550
6330	8" x 10", 7 ga		1.65	4.848		7,225	169		7,394	8,225
6332	8" x 12", 7 ga.		1.60	5		7,850	175		8,025	8,925
6334	10" x 10", 3 ga.		1.60	5		8,000	175		8,175	9,125
6336	10" x 12", 3 ga.		1.60	5		9,200	175		9,375	10,400
6338	12" x 12", 3 ga.		1.55	5.161		10,000	180		10,180	11,300
6350	Column caps, painted, 3 ga.									
6352	3-1/4" x 3-5/8"	1 Carp	1.80	4.444	C	9,550	155		9,705	10,800
6354	3-1/4" x 5-1/2"		1.80	4.444		9,550	155		9,705	10,800
6356	3-5/8" x 3-5/8"		1.80	4.444		7,750	155		7,905	8,750
6358	3-5/8" x 5-1/2"		1.80	4.444		7,750	155		7,905	8,750
6360	5-1/4" x 5-1/2"		1.75	4.571		10,200	160		10,360	11,500
6362	5-1/4" x 7-1/2"		1.75	4.571		10,900	160		11,060	12,300

06 05 23 – Wood, Plastic, and Composite Fastenings

06 05 23.60 Timber Connectors	Crew	Daily Output	Labor-Hours	Unit	Material	2018 Bare Costs Labor	Equipment	Total	Total Incl O&P
6364	5-1/2" x 3-5/8"	1 Carp	1.75	4.571	C	10,200	160		10,360	11,500
6366	5-1/2" x 5-1/2"		1.75	4.571		10,200	160		10,360	11,500
6368	5-1/2" x 7-1/2"		1.70	4.706		11,700	164		11,864	13,200
6370	6-7/8" x 5-1/2"		1.70	4.706		12,200	164		12,364	13,800
6372	6-7/8" x 6-7/8"		1.70	4.706		12,200	164		12,364	13,800
6374	6-7/8" x 7-1/2"		1.70	4.706		12,200	164		12,364	13,800
6376	7-1/2" x 5-1/2"		1.65	4.848		12,800	169		12,969	14,400
6378	7-1/2" x 7-1/2"		1.65	4.848		12,800	169		12,969	14,400
6380	8-7/8" x 5-1/2"		1.60	5		13,500	175		13,675	15,200
6382	8-7/8" x 7-1/2"		1.60	5		13,500	175		13,675	15,200
6384	9-1/2" x 5-1/2"	↓	1.60	5	↓	18,300	175		18,475	20,400
6400	Floor tie anchors, polymer paint									
6402	10 ga., 3" x 37-1/2"	1 Carp	1.80	4.444	C	4,875	155		5,030	5,600
6404	3-1/2" x 45-1/2"		1.75	4.571		5,100	160		5,260	5,875
6406	3 ga., 3-1/2" x 56"	↓	1.70	4.706	↓	8,975	164		9,139	10,100
6410	Girder hangers									
6412	6" wall thickness, 4" x 6"	1 Carp	1.80	4.444	C	2,725	155		2,880	3,250
6414	4" x 8"		1.80	4.444		3,050	155		3,205	3,600
6416	8" wall thickness, 4" x 6"		1.80	4.444		3,075	155		3,230	3,625
6418	4" x 8"	↓	1.80	4.444	↓	3,475	155		3,630	4,075
6420	Hinge connections, polymer painted									
6422	3/4" thick top plate									
6424	5-1/4" x 12" w/5" x 5" top	1 Carp	1	8	C	32,900	280		33,180	36,600
6426	5-1/4" x 15" w/6" x 6" top		.80	10		34,900	350		35,250	39,000
6428	5-1/4" x 18" w/7" x 7" top		.70	11.429		36,700	400		37,100	41,100
6430	5-1/4" x 26" w/9" x 9" top	↓	.60	13.333	↓	39,100	465		39,565	43,800
6432	1" thick top plate									
6434	6-7/8" x 14" w/5" x 5" top	1 Carp	.80	10	C	40,000	350		40,350	44,600
6436	6-7/8" x 17" w/6" x 6" top		.80	10		44,500	350		44,850	49,500
6438	6-7/8" x 21" w/7" x 7" top		.70	11.429		48,500	400		48,900	54,000
6440	6-7/8" x 31" w/9" x 9" top	↓	.60	13.333	↓	53,000	465		53,465	59,500
6442	1-1/4" thick top plate									
6444	8-7/8" x 16" w/5" x 5" top	1 Carp	.60	13.333	C	49,900	465		50,365	56,000
6446	8-7/8" x 21" w/6" x 6" top		.50	16		55,000	560		55,560	61,500
6448	8-7/8" x 26" w/7" x 7" top		.40	20		62,000	700		62,700	69,500
6450	8-7/8" x 39" w/9" x 9" top	↓	.30	26.667	↓	77,500	930		78,430	86,500
6460	Holddowns									
6462	Embedded along edge									
6464	26" long, 12 ga.	1 Carp	.90	8.889	C	1,175	310		1,485	1,825
6466	35" long, 12 ga.		.85	9.412	C	1,625	330		1,955	2,350
6468	35" long, 10 ga.	↓	.85	9.412	↓	1,800	330		2,130	2,525
6470	Embedded away from edge									
6472	Medium duty, 12 ga.									
6474	18-1/2" long	1 Carp	.95	8.421	C	705	294		999	1,275
6476	23-3/4" long		.90	8.889		880	310		1,190	1,475
6478	28" long		.85	9.412		895	330		1,225	1,525
6480	35" long	↓	.85	9.412	↓	1,225	330		1,555	1,900
6482	Heavy duty, 10 ga.									
6484	28" long	1 Carp	.85	9.412	C	1,725	330		2,055	2,425
6486	35" long	"	.85	9.412	"	1,875	330		2,205	2,600
6490	Surface mounted (W x H)									
6492	2-1/2" x 5-3/4", 7 ga.	1 Carp	1	8	C	2,325	280		2,605	3,025
6494	2-1/2" x 8", 12 ga.	↓	1	8	↓	1,250	280		1,530	1,850

06 05 23 – Wood, Plastic, and Composite Fastenings

06 05 23.60 Timber Connectors		Crew	Daily Output	Labor-Hours	Unit	Material	2018 Bare Costs Labor	Equipment	Total	Total Incl O&P
6496	2-7/8" x 6-3/8", 7 ga.	1 Carp	1	8	C	4,775	280		5,055	5,725
6498	2-7/8" x 12-1/2", 3 ga.		1	8		4,050	280		4,330	4,925
6500	3-3/16" x 9-3/8", 10 ga.		1	8		4,325	280		4,605	5,225
6502	3-1/2" x 11-5/8", 3 ga.		1	8		4,875	280		5,155	5,850
6504	3-1/2" x 14-3/4", 3 ga.		1	8		12,100	280		12,380	13,800
6506	3-1/2" x 16-1/2", 3 ga.		1	8		14,100	280		14,380	16,000
6508	3-1/2" x 20-1/2", 3 ga.		.90	8.889		16,100	310		16,410	18,200
6510	3-1/2" x 24-1/2", 3 ga.		.90	8.889		22,400	310		22,710	25,200
6512	4-1/4" x 20-3/4", 3 ga.	↓	.90	8.889	↓	22,000	310		22,310	24,800
6520	Joist hangers									
6522	Sloped, field adjustable, 18 ga.									
6524	2" x 6"	1 Carp	1.65	4.848	C	455	169		624	780
6526	2" x 8"		1.65	4.848		970	169		1,139	1,350
6528	2" x 10" and up		1.65	4.848		1,125	169		1,294	1,500
6530	3" x 10" and up		1.60	5		1,175	175		1,350	1,600
6532	4" x 10" and up	↓	1.55	5.161	↓	1,350	180		1,530	1,775
6536	Skewed 45°, 16 ga.									
6538	2" x 4"	1 Carp	1.75	4.571	C	900	160		1,060	1,250
6540	2" x 6" or 2" x 8"		1.65	4.848		910	169		1,079	1,275
6542	2" x 10" or 2" x 12"		1.65	4.848		955	169		1,124	1,325
6544	2" x 14" or 2" x 16"		1.60	5		1,650	175		1,825	2,125
6546	(2) 2" x 6" or (2) 2" x 8"		1.60	5		1,500	175		1,675	1,950
6548	(2) 2" x 10" or (2) 2" x 12"		1.55	5.161		1,625	180		1,805	2,075
6550	(2) 2" x 14" or (2) 2" x 16"		1.50	5.333		2,550	186		2,736	3,100
6552	4" x 6" or 4" x 8"		1.60	5		1,400	175		1,575	1,850
6554	4" x 10" or 4" x 12"		1.55	5.161		1,500	180		1,680	1,950
6556	4" x 14" or 4" x 16"	↓	1.55	5.161	↓	2,300	180		2,480	2,825
6560	Skewed 45°, 14 ga.									
6562	(2) 2" x 6" or (2) 2" x 8"	1 Carp	1.60	5	C	1,725	175		1,900	2,200
6564	(2) 2" x 10" or (2) 2" x 12"		1.55	5.161		2,400	180		2,580	2,925
6566	(2) 2" x 14" or (2) 2" x 16"		1.50	5.333		3,525	186		3,711	4,175
6568	4" x 6" or 4" x 8"		1.60	5		2,050	175		2,225	2,550
6570	4" x 10" or 4" x 12"		1.55	5.161		2,175	180		2,355	2,675
6572	4" x 14" or 4" x 16"	↓	1.55	5.161	↓	2,900	180		3,080	3,500
6590	Joist hangers, heavy duty 12 ga., galvanized									
6592	2" x 4"	1 Carp	1.75	4.571	C	1,175	160		1,335	1,550
6594	2" x 6"		1.65	4.848		1,275	169		1,444	1,675
6595	2" x 6", 16 ga.		1.65	4.848		1,225	169		1,394	1,625
6596	2" x 8"		1.65	4.848		1,950	169		2,119	2,400
6597	2" x 8", 16 ga.		1.65	4.848		1,850	169		2,019	2,300
6598	2" x 10"		1.65	4.848		2,000	169		2,169	2,475
6600	2" x 12"		1.65	4.848		2,450	169		2,619	2,975
6602	2" x 14"		1.65	4.848		2,700	169		2,869	3,250
6604	2" x 16"		1.65	4.848		2,850	169		3,019	3,425
6606	3" x 4"		1.65	4.848		1,725	169		1,894	2,175
6608	3" x 6"		1.65	4.848		2,300	169		2,469	2,800
6610	3" x 8"		1.65	4.848		2,350	169		2,519	2,850
6612	3" x 10"		1.60	5		2,700	175		2,875	3,275
6614	3" x 12"		1.60	5		3,250	175		3,425	3,875
6616	3" x 14"		1.60	5		3,800	175		3,975	4,475
6618	3" x 16"		1.60	5		4,200	175		4,375	4,900
6620	(2) 2" x 4"		1.75	4.571		1,900	160		2,060	2,375
6622	(2) 2" x 6"	↓	1.60	5	↓	2,300	175		2,475	2,825

06 05 23.60 Timber Connectors		Crew	Daily Output	Labor-Hours	Unit	Material	2018 Bare Costs Labor	Equipment	Total	Total Incl O&P
6624	(2) 2" x 8"	1 Carp	1.60	5	C	2,350	175		2,525	2,875
6626	(2) 2" x 10"		1.55	5.161		2,550	180		2,730	3,100
6628	(2) 2" x 12"		1.55	5.161		3,250	180		3,430	3,875
6630	(2) 2" x 14"		1.50	5.333		3,475	186		3,661	4,100
6632	(2) 2" x 16"		1.50	5.333		3,500	186		3,686	4,150
6634	4" x 4"		1.65	4.848		1,450	169		1,619	1,875
6636	4" x 6"		1.60	5		1,600	175		1,775	2,075
6638	4" x 8"		1.60	5		1,850	175		2,025	2,350
6640	4" x 10"		1.55	5.161		2,275	180		2,455	2,800
6642	4" x 12"		1.55	5.161		2,425	180		2,605	2,975
6644	4" x 14"		1.55	5.161		2,900	180		3,080	3,500
6646	4" x 16"		1.55	5.161		3,175	180		3,355	3,800
6648	(3) 2" x 10"		1.50	5.333		3,325	186		3,511	3,950
6650	(3) 2" x 12"		1.50	5.333		3,700	186		3,886	4,375
6652	(3) 2" x 14"		1.45	5.517		4,200	193		4,393	4,950
6654	(3) 2" x 16"		1.45	5.517		4,275	193		4,468	5,025
6656	6" x 6"		1.60	5		1,725	175		1,900	2,200
6658	6" x 8"		1.60	5		1,950	175		2,125	2,450
6660	6" x 10"		1.55	5.161		2,325	180		2,505	2,875
6662	6" x 12"		1.55	5.161		2,650	180		2,830	3,225
6664	6" x 14"		1.50	5.333		3,325	186		3,511	3,950
6666	6" x 16"		1.50	5.333		3,925	186		4,111	4,600
6690	Knee braces, galvanized, 12 ga.									
6692	Beam depth, 10" x 15" x 5' long	1 Carp	1.80	4.444	C	5,100	155		5,255	5,850
6694	15" x 22-1/2" x 7' long		1.70	4.706		5,850	164		6,014	6,700
6696	22-1/2" x 28-1/2" x 8' long		1.60	5		6,275	175		6,450	7,200
6698	28-1/2" x 36" x 10' long		1.55	5.161		6,550	180		6,730	7,500
6700	36" x 42" x 12' long		1.50	5.333		7,225	186		7,411	8,250
6710	Mudsill anchors									
6714	2" x 4" or 3" x 4"	1 Carp	115	.070	C	177	2.43		179.43	199
6716	2" x 6" or 3" x 6"		115	.070		177	2.43		179.43	199
6718	Block wall, 13-1/4" long		115	.070		78.50	2.43		80.93	90
6720	21-1/4" long		115	.070		127	2.43		129.43	144
6730	Post bases, 12 ga. galvanized									
6732	Adjustable, 3-9/16" x 3-9/16"	1 Carp	1.30	6.154	C	975	215		1,190	1,425
6734	3-9/16" x 5-1/2"		1.30	6.154		1,950	215		2,165	2,500
6736	4" x 4"		1.30	6.154		915	215		1,130	1,350
6738	4" x 6"		1.30	6.154		1,325	215		1,540	1,800
6740	5-1/2" x 5-1/2"		1.30	6.154		1,725	215		1,940	2,250
6742	6" x 6"		1.30	6.154		3,425	215		3,640	4,100
6744	Elevated, 3-9/16" x 3-1/4"		1.30	6.154		1,100	215		1,315	1,550
6746	5-1/2" x 3-5/16"		1.30	6.154		1,475	215		1,690	1,975
6748	5-1/2" x 5"		1.30	6.154		2,250	215		2,465	2,825
6750	Regular, 3-9/16" x 3-3/8"		1.30	6.154		810	215		1,025	1,250
6752	4" x 3-3/8"		1.30	6.154		1,200	215		1,415	1,675
6754	18 ga., 5-1/4" x 3-1/8"		1.30	6.154		1,200	215		1,415	1,675
6755	5-1/2" x 3-3/8"		1.30	6.154		1,200	215		1,415	1,675
6756	5-1/2" x 5-3/8"		1.30	6.154		1,700	215		1,915	2,225
6758	6" x 3-3/8"		1.30	6.154		1,925	215		2,140	2,475
6760	6" x 5-3/8"		1.30	6.154		2,450	215		2,665	3,050
6762	Post combination cap/bases									
6764	3-9/16" x 3-9/16"	1 Carp	1.20	6.667	C	420	233		653	850
6766	3-9/16" x 5-1/2"		1.20	6.667		1,100	233		1,333	1,600

06 05 23.60 Timber Connectors		Crew	Daily Output	Labor-Hours	Unit	Material	2018 Bare Costs Labor	Equipment	Total	Total Incl O&P
6768	4" x 4"	1 Carp	1.20	6.667	C	2,225	233		2,458	2,800
6770	5-1/2" x 5-1/2"		1.20	6.667		1,100	233		1,333	1,600
6772	6" x 6"		1.20	6.667		4,050	233		4,283	4,850
6774	7-1/2" x 7-1/2"		1.20	6.667		4,225	233		4,458	5,025
6776	8" x 8"		1.20	6.667		4,450	233		4,683	5,275
6790	Post-beam connection caps									
6792	Beam size 3-9/16"									
6794	12 ga. post, 4" x 4"	1 Carp	1	8	C	2,500	280		2,780	3,225
6796	4" x 6"		1	8		3,750	280		4,030	4,600
6798	4" x 8"		1	8		5,550	280		5,830	6,575
6800	16 ga. post, 4" x 4"		1	8		1,100	280		1,380	1,700
6802	4" x 6"		1	8		1,925	280		2,205	2,600
6804	4" x 8"		1	8		3,575	280		3,855	4,400
6805	18 ga. post, 2- 7/8" x 3"		1	8		3,575	280		3,855	4,400
6806	Beam size 5-1/2"									
6808	12 ga. post, 6" x 4"	1 Carp	1	8	C	3,675	280		3,955	4,525
6810	6" x 6"		1	8		4,650	280		4,930	5,575
6812	6" x 8"		1	8		4,925	280		5,205	5,875
6816	16 ga. post, 6" x 4"		1	8		2,000	280		2,280	2,675
6818	6" x 6"		1	8		2,025	280		2,305	2,700
6820	Beam size 7-1/2"									
6822	12 ga. post, 8" x 4"	1 Carp	1	8	C	5,275	280		5,555	6,275
6824	8" x 6"		1	8		5,325	280		5,605	6,350
6826	8" x 8"		1	8		8,025	280		8,305	9,300
6840	Purlin anchors, embedded									
6842	Heavy duty, 10 ga.									
6844	Straight, 28" long	1 Carp	1.60	5	C	1,625	175		1,800	2,075
6846	35" long		1.50	5.333		1,700	186		1,886	2,175
6848	Twisted, 28" long		1.60	5		1,625	175		1,800	2,075
6850	35" long		1.50	5.333		1,700	186		1,886	2,175
6852	Regular duty, 12 ga.									
6854	Straight, 18-1/2" long	1 Carp	1.80	4.444	C	820	155		975	1,150
6856	23-3/4" long		1.70	4.706		1,025	164		1,189	1,400
6858	29" long		1.60	5		1,050	175		1,225	1,450
6860	35" long		1.50	5.333		1,450	186		1,636	1,900
6862	Twisted, 18" long		1.80	4.444		820	155		975	1,150
6866	28" long		1.60	5		980	175		1,155	1,375
6868	35" long		1.50	5.333		1,450	186		1,636	1,900
6870	Straight, plastic coated									
6872	23-1/2" long	1 Carp	1.60	5	C	2,025	175		2,200	2,525
6874	26-7/8" long		1.60	5		2,375	175		2,550	2,900
6876	32-1/2" long		1.50	5.333		2,525	186		2,711	3,075
6878	35-7/8" long		1.50	5.333		2,625	186		2,811	3,200
6890	Purlin hangers, painted									
6892	12 ga., 2" x 6"	1 Carp	1.80	4.444	C	1,700	155		1,855	2,125
6894	2" x 8"		1.80	4.444		1,850	155		2,005	2,275
6896	2" x 10"		1.80	4.444		2,000	155		2,155	2,450
6898	2" x 12"		1.75	4.571		2,150	160		2,310	2,625
6900	2" x 14"		1.75	4.571		2,275	160		2,435	2,800
6902	2" x 16"		1.75	4.571		2,425	160		2,585	2,950
6904	3" x 6"		1.70	4.706		1,700	164		1,864	2,150
6906	3" x 8"		1.70	4.706		1,850	164		2,014	2,325
6908	3" x 10"		1.70	4.706		2,000	164		2,164	2,475

06 05 23 – Wood, Plastic, and Composite Fastenings

06 05 23.60 Timber Connectors		Crew	Daily Output	Labor-Hours	Unit	Material	2018 Bare Costs Labor	Equipment	Total	Total Incl O&P
6910	3" x 12"	1 Carp	1.65	4.848	C	2,300	169		2,469	2,800
6912	3" x 14"		1.65	4.848		2,450	169		2,619	2,950
6914	3" x 16"		1.65	4.848		2,575	169		2,744	3,125
6916	4" x 6"		1.65	4.848		1,725	169		1,894	2,175
6918	4" x 8"		1.65	4.848		1,875	169		2,044	2,325
6920	4" x 10"		1.65	4.848		2,000	169		2,169	2,475
6922	4" x 12"		1.60	5		2,375	175		2,550	2,925
6924	4" x 14"		1.60	5		2,525	175		2,700	3,075
6926	4" x 16"		1.60	5		2,675	175		2,850	3,250
6928	6" x 6"		1.60	5		2,250	175		2,425	2,775
6930	6" x 8"		1.60	5		2,425	175		2,600	2,950
6932	6" x 10"		1.55	5.161		2,575	180		2,755	3,125
6934	double 2" x 6"		1.70	4.706		1,850	164		2,014	2,325
6936	double 2" x 8"		1.70	4.706		2,000	164		2,164	2,475
6938	double 2" x 10"		1.70	4.706		2,150	164		2,314	2,625
6940	double 2" x 12"		1.65	4.848		2,300	169		2,469	2,800
6942	double 2" x 14"		1.65	4.848		2,450	169		2,619	2,950
6944	double 2" x 16"		1.65	4.848		2,575	169		2,744	3,125
6960	11 ga., 4" x 6"		1.65	4.848		3,400	169		3,569	4,000
6962	4" x 8"		1.65	4.848		3,650	169		3,819	4,300
6964	4" x 10"		1.65	4.848		3,900	169		4,069	4,575
6966	6" x 6"		1.60	5		3,450	175		3,625	4,075
6968	6" x 8"		1.60	5		3,700	175		3,875	4,350
6970	6" x 10"		1.55	5.161		3,950	180		4,130	4,625
6972	6" x 12"		1.55	5.161		4,200	180		4,380	4,900
6974	6" x 14"		1.55	5.161		4,450	180		4,630	5,200
6976	6" x 16"		1.50	5.333		4,700	186		4,886	5,475
6978	7 ga., 8" x 6"		1.60	5		3,725	175		3,900	4,400
6980	8" x 8"		1.60	5		3,975	175		4,150	4,675
6982	8" x 10"		1.55	5.161		4,225	180		4,405	4,950
6984	8" x 12"		1.55	5.161		4,475	180		4,655	5,225
6986	8" x 14"		1.50	5.333		4,725	186		4,911	5,500
6988	8" x 16"		1.50	5.333		4,975	186		5,161	5,775
7000	Strap connectors, galvanized									
7002	12 ga., 2-1/16" x 36"	1 Carp	1.55	5.161	C	1,075	180		1,255	1,500
7004	2-1/16" x 47"		1.50	5.333		1,500	186		1,686	1,950
7005	10 ga., 2-1/16" x 72"		1.50	5.333		1,575	186		1,761	2,025
7006	7 ga., 2-1/16" x 34"		1.55	5.161		2,675	180		2,855	3,225
7008	2-1/16" x 45"		1.50	5.333		3,500	186		3,686	4,150
7010	3 ga., 3" x 32"		1.55	5.161		4,525	180		4,705	5,275
7012	3" x 41"		1.55	5.161		4,700	180		4,880	5,450
7014	3" x 50"		1.50	5.333		7,150	186		7,336	8,175
7016	3" x 59"		1.50	5.333		8,725	186		8,911	9,875
7018	3-1/2" x 68"		1.45	5.517		8,850	193		9,043	10,000
7030	Tension ties									
7032	19-1/8" long, 16 ga., 3/4" anchor bolt	1 Carp	1.80	4.444	C	1,250	155		1,405	1,625
7034	20" long, 12 ga., 1/2" anchor bolt		1.80	4.444		1,600	155		1,755	2,025
7036	20" long, 12 ga., 3/4" anchor bolt		1.80	4.444		1,600	155		1,755	2,025
7038	27-3/4" long, 12 ga., 3/4" anchor bolt		1.75	4.571		2,850	160		3,010	3,400
7050	Truss connectors, galvanized									
7052	Adjustable hanger									
7054	18 ga., 2" x 6"	1 Carp	1.65	4.848	C	485	169		654	815
7056	4" x 6"		1.65	4.848		640	169		809	980

06 05 Common Work Results for Wood, Plastics, and Composites

06 05 23 – Wood, Plastic, and Composite Fastenings

06 05 23.60 Timber Connectors

		Crew	Daily Output	Labor-Hours	Unit	Material	2018 Bare Costs Labor	Equipment	Total	Total Incl O&P
7058	16 ga., 4" x 10"	1 Carp	1.60	5	C	930	175		1,105	1,325
7060	(2) 2" x 10"	▼	1.60	5	▼	930	175		1,105	1,325
7062	Connectors to plate									
7064	16 ga., 2" x 4" plate	1 Carp	1.80	4.444	C	480	155		635	785
7066	2" x 6" plate	"	1.80	4.444	"	620	155		775	940
7068	Hip jack connector									
7070	14 ga.	1 Carp	1.50	5.333	C	2,950	186		3,136	3,550

06 05 23.80 Metal Bracing

		Crew	Daily Output	Labor-Hours	Unit	Material	2018 Bare Costs Labor	Equipment	Total	Total Incl O&P
0010	**METAL BRACING**									
0302	Let-in, "T" shaped, 22 ga. galv. steel, studs at 16" OC	1 Carp	580	.014	L.F.	.81	.48		1.29	1.69
0402	Studs at 24" OC		600	.013		.81	.47		1.28	1.66
0502	Steel straps, 16 ga. galv. steel, studs at 16" OC		600	.013		1.05	.47		1.52	1.93
0602	Studs at 24" OC	▼	620	.013	▼	1.05	.45		1.50	1.91

06 11 Wood Framing

06 11 10 – Framing with Dimensional, Engineered or Composite Lumber

06 11 10.01 Forest Stewardship Council Certification

		Crew	Daily Output	Labor-Hours	Unit	Material	2018 Bare Costs Labor	Equipment	Total	Total Incl O&P
0010	**FOREST STEWARDSHIP COUNCIL CERTIFICATION**									
0020	For Forest Stewardship Council (FSC) cert dimension lumber, add [G]					65%				

06 11 10.02 Blocking

		Crew	Daily Output	Labor-Hours	Unit	Material	2018 Bare Costs Labor	Equipment	Total	Total Incl O&P
0010	**BLOCKING**									
1790	Bolted to concrete									
1798	Ledger board, 2" x 4"	1 Carp	180	.044	L.F.	4.72	1.55		6.27	7.80
1800	2" x 6"		160	.050		4.94	1.75		6.69	8.35
1810	4" x 6"		140	.057		8.70	2		10.70	12.90
1820	4" x 8"	▼	120	.067	▼	9.80	2.33		12.13	14.60
1950	Miscellaneous, to wood construction									
2000	2" x 4"	1 Carp	250	.032	L.F.	.42	1.12		1.54	2.32
2005	Pneumatic nailed		305	.026		.42	.92		1.34	1.99
2050	2" x 6"		222	.036		.64	1.26		1.90	2.80
2055	Pneumatic nailed		271	.030		.65	1.03		1.68	2.44
2100	2" x 8"		200	.040		.89	1.40		2.29	3.30
2105	Pneumatic nailed		244	.033		.90	1.15		2.05	2.90
2150	2" x 10"		178	.045		1.41	1.57		2.98	4.16
2155	Pneumatic nailed		217	.037		1.42	1.29		2.71	3.70
2200	2" x 12"		151	.053		1.79	1.85		3.64	5.05
2205	Pneumatic nailed	▼	185	.043	▼	1.81	1.51		3.32	4.50
2300	To steel construction									
2320	2" x 4"	1 Carp	208	.038	L.F.	.42	1.34		1.76	2.69
2340	2" x 6"		180	.044		.64	1.55		2.19	3.29
2360	2" x 8"		158	.051		.90	1.77		2.67	3.93
2380	2" x 10"		136	.059		1.41	2.06		3.47	4.97
2400	2" x 12"	▼	109	.073	▼	1.80	2.57		4.37	6.25

06 11 10.04 Wood Bracing

		Crew	Daily Output	Labor-Hours	Unit	Material	2018 Bare Costs Labor	Equipment	Total	Total Incl O&P
0010	**WOOD BRACING**									
0012	Let-in, with 1" x 6" boards, studs @ 16" OC	1 Carp	150	.053	L.F.	.78	1.86		2.64	3.96
0202	Studs @ 24" OC	"	230	.035	"	.78	1.22		2	2.88

06 11 Wood Framing

06 11 10 – Framing with Dimensional, Engineered or Composite Lumber

06 11 10.06 Bridging

		Crew	Daily Output	Labor-Hours	Unit	Material	2018 Bare Costs Labor	Equipment	Total	Total Incl O&P
0010	**BRIDGING**									
0012	Wood, for joists 16" OC, 1" x 3"	1 Carp	130	.062	Pr.	.69	2.15		2.84	4.34
0017	Pneumatic nailed		170	.047		.76	1.64		2.40	3.57
0102	2" x 3" bridging		130	.062		.72	2.15		2.87	4.37
0107	Pneumatic nailed		170	.047		.75	1.64		2.39	3.56
0302	Steel, galvanized, 18 ga., for 2" x 10" joists at 12" OC		130	.062		1.71	2.15		3.86	5.45
0352	16" OC		135	.059		1.71	2.07		3.78	5.30
0402	24" OC		140	.057		2.57	2		4.57	6.15
0602	For 2" x 14" joists at 16" OC		130	.062		1.86	2.15		4.01	5.60
0902	Compression type, 16" OC, 2" x 8" joists		200	.040		1.26	1.40		2.66	3.70
1002	2" x 12" joists	▼	200	.040	▼	1.25	1.40		2.65	3.70

06 11 10.10 Beam and Girder Framing

		Crew	Daily Output	Labor-Hours	Unit	Material	2018 Bare Costs Labor	Equipment	Total	Total Incl O&P
0010	**BEAM AND GIRDER FRAMING** R061110-30									
1000	Single, 2" x 6"	2 Carp	700	.023	L.F.	.64	.80		1.44	2.04
1005	Pneumatic nailed		812	.020		.65	.69		1.34	1.86
1020	2" x 8"		650	.025		.89	.86		1.75	2.41
1025	Pneumatic nailed		754	.021		.90	.74		1.64	2.22
1040	2" x 10"		600	.027		1.41	.93		2.34	3.10
1045	Pneumatic nailed		696	.023		1.42	.80		2.22	2.90
1060	2" x 12"		550	.029		1.79	1.02		2.81	3.66
1065	Pneumatic nailed		638	.025		1.81	.88		2.69	3.45
1080	2" x 14"		500	.032		2.27	1.12		3.39	4.35
1085	Pneumatic nailed		580	.028		2.28	.96		3.24	4.11
1100	3" x 8"		550	.029		2.87	1.02		3.89	4.85
1120	3" x 10"		500	.032		3.62	1.12		4.74	5.85
1140	3" x 12"		450	.036		4.34	1.24		5.58	6.85
1160	3" x 14"	▼	400	.040		5.05	1.40		6.45	7.85
1170	4" x 6"	F-3	1100	.036		3.06	1.16	.45	4.67	5.80
1180	4" x 8"		1000	.040		4.15	1.28	.50	5.93	7.25
1200	4" x 10"		950	.042		4.96	1.35	.52	6.83	8.25
1220	4" x 12"		900	.044		5.50	1.42	.55	7.47	9.05
1240	4" x 14"		850	.047		6.40	1.51	.58	8.49	10.20
1250	6" x 8"		525	.076		7.90	2.44	.94	11.28	13.80
1260	6" x 10"		500	.080		6.90	2.56	.99	10.45	12.95
1290	8" x 12"	▼	300	.133		16.55	4.27	1.65	22.47	27
2000	Double, 2" x 6"	2 Carp	625	.026		1.29	.89		2.18	2.91
2005	Pneumatic nailed		725	.022		1.30	.77		2.07	2.71
2020	2" x 8"		575	.028		1.79	.97		2.76	3.59
2025	Pneumatic nailed		667	.024		1.81	.84		2.65	3.38
2040	2" x 10"		550	.029		2.82	1.02		3.84	4.79
2045	Pneumatic nailed		638	.025		2.85	.88		3.73	4.59
2060	2" x 12"		525	.030		3.59	1.07		4.66	5.70
2065	Pneumatic nailed		610	.026		3.62	.92		4.54	5.50
2080	2" x 14"		475	.034		4.53	1.18		5.71	6.95
2085	Pneumatic nailed		551	.029		4.56	1.01		5.57	6.70
3000	Triple, 2" x 6"		550	.029		1.93	1.02		2.95	3.81
3005	Pneumatic nailed		638	.025		1.95	.88		2.83	3.61
3020	2" x 8"		525	.030		2.68	1.07		3.75	4.72
3025	Pneumatic nailed		609	.026		2.71	.92		3.63	4.51
3040	2" x 10"		500	.032		4.23	1.12		5.35	6.50
3045	Pneumatic nailed		580	.028		4.27	.96		5.23	6.30
3060	2" x 12"	▼	475	.034	▼	5.40	1.18		6.58	7.85

For customer support on your Residential Costs with RSMeans data, call 800.448.8182.

371

06 11 Wood Framing

06 11 10 – Framing with Dimensional, Engineered or Composite Lumber

06 11 10.10 Beam and Girder Framing	Crew	Daily Output	Labor-Hours	Unit	Material	2018 Bare Costs Labor	Equipment	Total	Total Incl O&P
3065	Pneumatic nailed	2 Carp	551	.029	L.F.	5.45	1.01		6.46	7.65
3080	2" x 14"		450	.036		7.40	1.24		8.64	10.15
3085	Pneumatic nailed	↓	522	.031	↓	6.85	1.07		7.92	9.35

06 11 10.12 Ceiling Framing

		Crew	Daily Output	Labor-Hours	Unit	Material	Labor	Equipment	Total	Total Incl O&P
0010	**CEILING FRAMING**									
6000	Suspended, 2" x 3"	2 Carp	1000	.016	L.F.	.42	.56		.98	1.40
6050	2" x 4"		900	.018		.42	.62		1.04	1.49
6100	2" x 6"		800	.020		.64	.70		1.34	1.87
6150	2" x 8"	↓	650	.025	↓	.89	.86		1.75	2.41

06 11 10.14 Posts and Columns

		Crew	Daily Output	Labor-Hours	Unit	Material	Labor	Equipment	Total	Total Incl O&P
0010	**POSTS AND COLUMNS**									
0100	4" x 4"	2 Carp	390	.041	L.F.	1.83	1.43		3.26	4.39
0150	4" x 6"		275	.058		3.06	2.03		5.09	6.75
0200	4" x 8"		220	.073		4.15	2.54		6.69	8.80
0250	6" x 6"		215	.074		5.35	2.60		7.95	10.20
0300	6" x 8"		175	.091		7.90	3.20		11.10	14
0350	6" x 10"	↓	150	.107	↓	6.90	3.73		10.63	13.80

06 11 10.18 Joist Framing

		Crew	Daily Output	Labor-Hours	Unit	Material	Labor	Equipment	Total	Total Incl O&P
0010	**JOIST FRAMING**									
2000	Joists, 2" x 4"	2 Carp	1250	.013	L.F.	.42	.45		.87	1.20
2005	Pneumatic nailed		1438	.011		.42	.39		.81	1.12
2100	2" x 6"		1250	.013		.64	.45		1.09	1.45
2105	Pneumatic nailed		1438	.011		.65	.39		1.04	1.37
2150	2" x 8"		1100	.015		.89	.51		1.40	1.83
2155	Pneumatic nailed		1265	.013		.90	.44		1.34	1.73
2200	2" x 10"		900	.018		1.41	.62		2.03	2.58
2205	Pneumatic nailed		1035	.015		1.42	.54		1.96	2.46
2250	2" x 12"		875	.018		1.79	.64		2.43	3.03
2255	Pneumatic nailed		1006	.016		1.81	.56		2.37	2.91
2300	2" x 14"		770	.021		2.27	.73		3	3.70
2305	Pneumatic nailed		886	.018		2.28	.63		2.91	3.56
2350	3" x 6"		925	.017		2	.60		2.60	3.21
2400	3" x 10"		780	.021		3.62	.72		4.34	5.15
2450	3" x 12"		600	.027		4.34	.93		5.27	6.35
2500	4" x 6"		800	.020		3.06	.70		3.76	4.53
2550	4" x 10"		600	.027		4.96	.93		5.89	7
2600	4" x 12"		450	.036		5.50	1.24		6.74	8.10
2605	Sister joist, 2" x 6"		800	.020		.64	.70		1.34	1.87
2606	Pneumatic nailed		960	.017	↓	.65	.58		1.23	1.69
3000	Composite wood joist 9-1/2" deep		.90	17.778	M.L.F.	1,775	620		2,395	2,975
3010	11-1/2" deep		.88	18.182		2,050	635		2,685	3,300
3020	14" deep		.82	19.512		2,550	680		3,230	3,950
3030	16" deep		.78	20.513		4,075	715		4,790	5,675
4000	Open web joist 12" deep		.88	18.182	↓	3,600	635		4,235	5,025
4002	Per linear foot		880	.018	L.F.	3.60	.64		4.24	5
4004	Treated, per linear foot		880	.018	"	4.53	.64		5.17	6.05
4010	14" deep		.82	19.512	M.L.F.	3,900	680		4,580	5,425
4012	Per linear foot		820	.020	L.F.	3.90	.68		4.58	5.40
4014	Treated, per linear foot		820	.020	"	4.98	.68		5.66	6.65
4020	16" deep		.78	20.513	M.L.F.	3,850	715		4,565	5,450
4022	Per linear foot		780	.021	L.F.	3.86	.72		4.58	5.45
4024	Treated, per linear foot	↓	780	.021	"	5.10	.72		5.82	6.80

For customer support on your Residential Costs with RSMeans data, call 800.448.8182.

06 11 Wood Framing

06 11 10 – Framing with Dimensional, Engineered or Composite Lumber

06 11 10.18 Joist Framing	Crew	Daily Output	Labor-Hours	Unit	Material	2018 Bare Costs Labor	2018 Bare Costs Equipment	Total	Total Incl O&P
4030	18" deep	2 Carp	.74	21.622	M.L.F.	4,225	755		4,980	5,900
4032	Per linear foot		740	.022	L.F.	4.22	.76		4.98	5.90
4034	Treated, per linear foot		740	.022	"	5.60	.76		6.36	7.45
6000	Composite rim joist, 1-1/4" x 9-1/2"		.90	17.778	M.L.F.	2,075	620		2,695	3,300
6010	1-1/4" x 11-1/2"		.88	18.182		2,275	635		2,910	3,550
6020	1-1/4" x 14-1/2"		.82	19.512		3,000	680		3,680	4,425
6030	1-1/4" x 16-1/2"		.78	20.513		2,750	715		3,465	4,225

06 11 10.24 Miscellaneous Framing

	06 11 10.24 Miscellaneous Framing	Crew	Daily Output	Labor-Hours	Unit	Material	Labor	Equipment	Total	Total Incl O&P
0010	**MISCELLANEOUS FRAMING**									
2000	Firestops, 2" x 4"	2 Carp	780	.021	L.F.	.42	.72		1.14	1.65
2005	Pneumatic nailed		952	.017		.42	.59		1.01	1.45
2100	2" x 6"		600	.027		.64	.93		1.57	2.26
2105	Pneumatic nailed		732	.022		.65	.76		1.41	1.99
5000	Nailers, treated, wood construction, 2" x 4"		800	.020		.59	.70		1.29	1.81
5005	Pneumatic nailed		960	.017		.59	.58		1.17	1.62
5100	2" x 6"		750	.021		.74	.75		1.49	2.05
5105	Pneumatic nailed		900	.018		.74	.62		1.36	1.85
5120	2" x 8"		700	.023		1.12	.80		1.92	2.56
5125	Pneumatic nailed		840	.019		1.13	.67		1.80	2.35
5200	Steel construction, 2" x 4"		750	.021		.59	.75		1.34	1.89
5220	2" x 6"		700	.023		.74	.80		1.54	2.14
5240	2" x 8"		650	.025		1.12	.86		1.98	2.66
7000	Rough bucks, treated, for doors or windows, 2" x 6"		400	.040		.74	1.40		2.14	3.13
7005	Pneumatic nailed		480	.033		.74	1.16		1.90	2.76
7100	2" x 8"		380	.042		1.12	1.47		2.59	3.68
7105	Pneumatic nailed		456	.035		1.13	1.23		2.36	3.28
8000	Stair stringers, 2" x 10"		130	.123		1.41	4.30		5.71	8.70
8100	2" x 12"		130	.123		1.79	4.30		6.09	9.10
8150	3" x 10"		125	.128		3.62	4.47		8.09	11.45
8200	3" x 12"		125	.128		4.34	4.47		8.81	12.25
8870	Laminated structural lumber, 1-1/4" x 11-1/2"		130	.123		2.28	4.30		6.58	9.65
8880	1-1/4" x 14-1/2"		130	.123		2.97	4.30		7.27	10.40

06 11 10.26 Partitions

	06 11 10.26 Partitions	Crew	Daily Output	Labor-Hours	Unit	Material	Labor	Equipment	Total	Total Incl O&P
0010	**PARTITIONS**									
0020	Single bottom and double top plate, no waste, std. & better lumber									
0180	2" x 4" studs, 8' high, studs 12" OC	2 Carp	80	.200	L.F.	4.60	7		11.60	16.65
0185	12" OC, pneumatic nailed		96	.167		4.66	5.85		10.51	14.80
0200	16" OC		100	.160		3.77	5.60		9.37	13.45
0205	16" OC, pneumatic nailed		120	.133		3.81	4.66		8.47	11.95
0300	24" OC		125	.128		2.93	4.47		7.40	10.65
0305	24" OC, pneumatic nailed		150	.107		2.96	3.73		6.69	9.45
0380	10' high, studs 12" OC		80	.200		5.45	7		12.45	17.60
0385	12" OC, pneumatic nailed		96	.167		5.50	5.85		11.35	15.75
0400	16" OC		100	.160		4.39	5.60		9.99	14.15
0405	16" OC, pneumatic nailed		120	.133		4.44	4.66		9.10	12.65
0500	24" OC		125	.128		3.35	4.47		7.82	11.15
0505	24" OC, pneumatic nailed		150	.107		3.39	3.73		7.12	9.90
0580	12' high, studs 12" OC		65	.246		6.30	8.60		14.90	21
0585	12" OC, pneumatic nailed		78	.205		6.35	7.15		13.50	18.90
0600	16" OC		80	.200		5	7		12	17.10
0605	16" OC, pneumatic nailed		96	.167		5.10	5.85		10.95	15.30
0700	24" OC		100	.160		3.77	5.60		9.37	13.45

For customer support on your Residential Costs with RSMeans data, call 800.448.8182.

06 11 10 – Framing with Dimensional, Engineered or Composite Lumber

06 11 10.26 Partitions

		Crew	Daily Output	Labor-Hours	Unit	Material	2018 Bare Costs Labor	Equipment	Total	Total Incl O&P
0705	24" OC, pneumatic nailed	2 Carp	120	.133	L.F.	3.81	4.66		8.47	11.95
0780	2" x 6" studs, 8' high, studs 12" OC		70	.229		7.10	8		15.10	21
0785	12" OC, pneumatic nailed		84	.190		7.15	6.65		13.80	18.90
0800	16" OC		90	.178		5.80	6.20		12	16.70
0805	16" OC, pneumatic nailed		108	.148		5.85	5.20		11.05	15.05
0900	24" OC		115	.139		4.50	4.86		9.36	13.05
0905	24" OC, pneumatic nailed		138	.116		4.55	4.05		8.60	11.75
0980	10' high, studs 12" OC		70	.229		8.35	8		16.35	22.50
0985	12" OC, pneumatic nailed		84	.190		8.45	6.65		15.10	20.50
1000	16" OC		90	.178		6.75	6.20		12.95	17.80
1005	16" OC, pneumatic nailed		108	.148		6.85	5.20		12.05	16.10
1100	24" OC		115	.139		5.15	4.86		10.01	13.75
1105	24" OC, pneumatic nailed		138	.116		5.20	4.05		9.25	12.45
1180	12' high, studs 12" OC		55	.291		9.65	10.15		19.80	27.50
1185	12" OC, pneumatic nailed		66	.242		9.75	8.45		18.20	25
1200	16" OC		70	.229		7.70	8		15.70	22
1205	16" OC, pneumatic nailed		84	.190		7.80	6.65		14.45	19.65
1300	24" OC		90	.178		5.80	6.20		12	16.70
1305	24" OC, pneumatic nailed		108	.148		5.85	5.20		11.05	15.05
1400	For horizontal blocking, 2" x 4", add		600	.027		.42	.93		1.35	2.01
1500	2" x 6", add		600	.027		.64	.93		1.57	2.26
1600	For openings, add	▼	250	.064	▼		2.24		2.24	3.72
1702	Headers for above openings, material only, add				B.F.	.74			.74	.81

06 11 10.28 Porch or Deck Framing

		Crew	Daily Output	Labor-Hours	Unit	Material	2018 Bare Costs Labor	Equipment	Total	Total Incl O&P
0010	**PORCH OR DECK FRAMING**									
0100	Treated lumber, posts or columns, 4" x 4"	2 Carp	390	.041	L.F.	1.20	1.43		2.63	3.70
0110	4" x 6"		275	.058		1.94	2.03		3.97	5.50
0120	4" x 8"		220	.073		3.96	2.54		6.50	8.60
0130	Girder, single, 4" x 4"		675	.024		1.20	.83		2.03	2.70
0140	4" x 6"		600	.027		1.94	.93		2.87	3.69
0150	4" x 8"		525	.030		3.96	1.07		5.03	6.15
0160	Double, 2" x 4"		625	.026		1.21	.89		2.10	2.82
0170	2" x 6"		600	.027		1.52	.93		2.45	3.22
0180	2" x 8"		575	.028		2.30	.97		3.27	4.15
0190	2" x 10"		550	.029		2.88	1.02		3.90	4.86
0200	2" x 12"		525	.030		4.15	1.07		5.22	6.35
0210	Triple, 2" x 4"		575	.028		1.81	.97		2.78	3.61
0220	2" x 6"		550	.029		2.28	1.02		3.30	4.19
0230	2" x 8"		525	.030		3.45	1.07		4.52	5.55
0240	2" x 10"		500	.032		4.32	1.12		5.44	6.60
0250	2" x 12"		475	.034		6.20	1.18		7.38	8.80
0260	Ledger, bolted 4' OC, 2" x 4"		400	.040		.75	1.40		2.15	3.14
0270	2" x 6"		395	.041		.89	1.42		2.31	3.33
0280	2" x 8"		390	.041		1.27	1.43		2.70	3.78
0290	2" x 10"		385	.042		1.55	1.45		3	4.11
0300	2" x 12"		380	.042		2.17	1.47		3.64	4.84
0310	Joists, 2" x 4"		1250	.013		.60	.45		1.05	1.40
0320	2" x 6"		1250	.013		.76	.45		1.21	1.58
0330	2" x 8"		1100	.015		1.15	.51		1.66	2.12
0340	2" x 10"		900	.018		1.44	.62		2.06	2.62
0350	2" x 12"	▼	875	.018		1.73	.64		2.37	2.96
0360	Railings and trim, 1" x 4"	1 Carp	300	.027	▼	.51	.93		1.44	2.11

06 11 Wood Framing

06 11 10 – Framing with Dimensional, Engineered or Composite Lumber

06 11 10.28 Porch or Deck Framing	Crew	Daily Output	Labor-Hours	Unit	Material	2018 Bare Costs Labor	2018 Bare Costs Equipment	Total	Total Incl O&P
0370	2" x 2"	1 Carp	300	.027	L.F.	.43	.93		1.36	2.02
0380	2" x 4"		300	.027		.59	.93		1.52	2.20
0390	2" x 6"		300	.027		.74	.93		1.67	2.37
0400	Decking, 1" x 4"		275	.029	S.F.	2.76	1.02		3.78	4.72
0410	2" x 4"		300	.027		2	.93		2.93	3.75
0420	2" x 6"		320	.025		1.60	.87		2.47	3.21
0430	5/4" x 6"		320	.025		2.15	.87		3.02	3.81
0440	Balusters, square, 2" x 2"	2 Carp	660	.024	L.F.	.43	.85		1.28	1.88
0450	Turned, 2" x 2"		420	.038		.57	1.33		1.90	2.84
0460	Stair stringer, 2" x 10"		130	.123		1.44	4.30		5.74	8.75
0470	2" x 12"		130	.123		1.73	4.30		6.03	9.05
0480	Stair treads, 1" x 4"		140	.114		2.76	3.99		6.75	9.70
0490	2" x 4"		140	.114		.60	3.99		4.59	7.30
0500	2" x 6"		160	.100		.91	3.50		4.41	6.80
0510	5/4" x 6"		160	.100		1	3.50		4.50	6.90
0520	Turned handrail post, 4" x 4"		64	.250	Ea.	34	8.75		42.75	52
0530	Lattice panel, 4' x 8', 1/2"		1600	.010	S.F.	.72	.35		1.07	1.37
0535	3/4"		1600	.010	"	1.07	.35		1.42	1.75
0540	Cedar, posts or columns, 4" x 4"		390	.041	L.F.	3.69	1.43		5.12	6.45
0550	4" x 6"		275	.058		6.90	2.03		8.93	10.95
0560	4" x 8"		220	.073		11	2.54		13.54	16.35
0800	Decking, 1" x 4"		550	.029		2.75	1.02		3.77	4.71
0810	2" x 4"		600	.027		5.60	.93		6.53	7.70
0820	2" x 6"		640	.025		10.15	.87		11.02	12.60
0830	5/4" x 6"		640	.025		6.30	.87		7.17	8.35
0840	Railings and trim, 1" x 4"		600	.027		2.75	.93		3.68	4.57
0860	2" x 4"		600	.027		5.60	.93		6.53	7.70
0870	2" x 6"		600	.027		10.15	.93		11.08	12.70
0920	Stair treads, 1" x 4"		140	.114		2.75	3.99		6.74	9.65
0930	2" x 4"		140	.114		5.60	3.99		9.59	12.80
0940	2" x 6"		160	.100		10.15	3.50		13.65	16.95
0950	5/4" x 6"		160	.100		6.30	3.50		9.80	12.70
0980	Redwood, posts or columns, 4" x 4"		390	.041		6.45	1.43		7.88	9.50
0990	4" x 6"		275	.058		12.60	2.03		14.63	17.25
1000	4" x 8"		220	.073		23.50	2.54		26.04	30
1240	Decking, 1" x 4"	1 Carp	275	.029	S.F.	3.97	1.02		4.99	6.05
1260	2" x 6"		340	.024		7.50	.82		8.32	9.60
1270	5/4" x 6"		320	.025		4.77	.87		5.64	6.70
1280	Railings and trim, 1" x 4"	2 Carp	600	.027	L.F.	1.17	.93		2.10	2.84
1310	2" x 6"		600	.027		7.50	.93		8.43	9.80
1420	Alternative decking, wood/plastic composite, 5/4" x 6" [G]		640	.025		3.27	.87		4.14	5.05
1440	1" x 4" square edge fir		550	.029		2.77	1.02		3.79	4.73
1450	1" x 4" tongue and groove fir		450	.036		1.51	1.24		2.75	3.73
1460	1" x 4" mahogany		550	.029		2.07	1.02		3.09	3.96
1462	5/4" x 6" PVC		550	.029		3.36	1.02		4.38	5.40
1465	Framing, porch or deck, alt deck fastening, screws, add	1 Carp	240	.033	S.F.		1.16		1.16	1.94
1470	Accessories, joist hangers, 2" x 4"		160	.050	Ea.	.75	1.75		2.50	3.74
1480	2" x 6" through 2" x 12"		150	.053		1.27	1.86		3.13	4.50
1530	Post footing, incl excav, backfill, tube form & concrete, 4' deep, 8" diam.	F-7	12	2.667		17.55	82		99.55	156
1540	10" diameter		11	2.909		24	89.50		113.50	176
1550	12" diameter		10	3.200		30.50	98.50		129	198

For customer support on your Residential Costs with RSMeans data, call 800.448.8182.

375

06 11 10.30 Roof Framing

		Crew	Daily Output	Labor-Hours	Unit	Material	2018 Bare Costs Labor	Equipment	Total	Total Incl O&P
0010	**ROOF FRAMING**									
1900	Rough fascia, 2" x 6"	2 Carp	250	.064	L.F.	.64	2.24		2.88	4.43
2000	2" x 8"		225	.071		.89	2.49		3.38	5.10
2100	2" x 10"		180	.089		1.41	3.11		4.52	6.70
5002	Rafters, to 4 in 12 pitch, 2" x 6", ordinary		1000	.016		.64	.56		1.20	1.64
5021	On steep roofs		800	.020		.64	.70		1.34	1.87
5041	On dormers or complex roofs		590	.027		.64	.95		1.59	2.29
5062	2" x 8", ordinary		950	.017		.89	.59		1.48	1.96
5081	On steep roofs		750	.021		.89	.75		1.64	2.22
5101	On dormers or complex roofs		540	.030		.89	1.04		1.93	2.70
5122	2" x 10", ordinary		630	.025		1.41	.89		2.30	3.03
5141	On steep roofs		495	.032		1.41	1.13		2.54	3.43
5161	On dormers or complex roofs		425	.038		1.41	1.32		2.73	3.74
5182	2" x 12", ordinary		575	.028		1.79	.97		2.76	3.59
5201	On steep roofs		455	.035		1.79	1.23		3.02	4.01
5221	On dormers or complex roofs		395	.041		1.79	1.42		3.21	4.32
5250	Composite rafter, 9-1/2" deep		575	.028		1.76	.97		2.73	3.56
5260	11-1/2" deep		575	.028		2.05	.97		3.02	3.87
5301	Hip and valley rafters, 2" x 6", ordinary		760	.021		.64	.74		1.38	1.93
5321	On steep roofs		585	.027		.64	.96		1.60	2.30
5341	On dormers or complex roofs		510	.031		.64	1.10		1.74	2.53
5361	2" x 8", ordinary		720	.022		.89	.78		1.67	2.27
5381	On steep roofs		545	.029		.89	1.03		1.92	2.69
5401	On dormers or complex roofs		470	.034		.89	1.19		2.08	2.96
5421	2" x 10", ordinary		570	.028		1.41	.98		2.39	3.18
5441	On steep roofs		440	.036		1.41	1.27		2.68	3.66
5461	On dormers or complex roofs		380	.042		1.41	1.47		2.88	4
5481	2" x 12", ordinary		525	.030		1.79	1.07		2.86	3.74
5501	On steep roofs		410	.039		1.79	1.36		3.15	4.24
5521	On dormers or complex roofs		355	.045		1.79	1.58		3.37	4.59
5541	Hip and valley jacks, 2" x 6", ordinary		600	.027		.64	.93		1.57	2.26
5561	On steep roofs		475	.034		.64	1.18		1.82	2.67
5581	On dormers or complex roofs		410	.039		.64	1.36		2	2.98
5601	2" x 8", ordinary		490	.033		.89	1.14		2.03	2.88
5621	On steep roofs		385	.042		.89	1.45		2.34	3.39
5641	On dormers or complex roofs		335	.048		.89	1.67		2.56	3.75
5661	2" x 10", ordinary		450	.036		1.41	1.24		2.65	3.62
5681	On steep roofs		350	.046		1.41	1.60		3.01	4.21
5701	On dormers or complex roofs		305	.052		1.41	1.83		3.24	4.60
5721	2" x 12", ordinary		375	.043		1.79	1.49		3.28	4.45
5741	On steep roofs		295	.054		1.79	1.90		3.69	5.10
5762	On dormers or complex roofs		255	.063		1.79	2.19		3.98	5.60
5781	Rafter tie, 1" x 4", #3		800	.020		.51	.70		1.21	1.72
5791	2" x 4", #3		800	.020		.42	.70		1.12	1.62
5801	Ridge board, #2 or better, 1" x 6"		600	.027		.78	.93		1.71	2.41
5821	1" x 8"		550	.029		1.30	1.02		2.32	3.12
5841	1" x 10"		500	.032		1.69	1.12		2.81	3.72
5861	2" x 6"		500	.032		.64	1.12		1.76	2.57
5881	2" x 8"		450	.036		.89	1.24		2.13	3.05
5901	2" x 10"		400	.040		1.41	1.40		2.81	3.87
5921	Roof cants, split, 4" x 4"		650	.025		1.83	.86		2.69	3.44
5941	6" x 6"		600	.027		5.35	.93		6.28	7.45

06 11 Wood Framing

06 11 10 – Framing with Dimensional, Engineered or Composite Lumber

06 11 10.30 Roof Framing	Crew	Daily Output	Labor-Hours	Unit	Material	2018 Bare Costs Labor	Equipment	Total	Total Incl O&P
5961	Roof curbs, untreated, 2" x 6"	2 Carp	520	.031	L.F.	.64	1.08		1.72	2.50
5981	2" x 12"		400	.040		1.79	1.40		3.19	4.29
6001	Sister rafters, 2" x 6"		800	.020		.64	.70		1.34	1.87
6021	2" x 8"		640	.025		.88	.87		1.75	2.42
6041	2" x 10"		535	.030		1.41	1.05		2.46	3.29
6061	2" x 12"		455	.035		1.79	1.23		3.02	4.01

06 11 10.32 Sill and Ledger Framing

		Crew	Daily Output	Labor-Hours	Unit	Material	Labor	Equipment	Total	Total Incl O&P
0010	**SILL AND LEDGER FRAMING**									
0020	Extruded polystyrene sill sealer, 5-1/2" wide	1 Carp	1600	.005	L.F.	.16	.17		.33	.47
2002	Ledgers, nailed, 2" x 4"	2 Carp	755	.021		.42	.74		1.16	1.69
2052	2" x 6"		600	.027		.64	.93		1.57	2.26
2102	Bolted, not including bolts, 3" x 6"		325	.049		1.98	1.72		3.70	5.05
2152	3" x 12"		233	.069		4.31	2.40		6.71	8.75
2602	Mud sills, redwood, construction grade, 2" x 4"		895	.018		2.26	.62		2.88	3.53
2622	2" x 6"		780	.021		3.40	.72		4.12	4.93
4002	Sills, 2" x 4"		600	.027		.41	.93		1.34	2
4052	2" x 6"		550	.029		.63	1.02		1.65	2.39
4082	2" x 8"		500	.032		.88	1.12		2	2.83
4101	2" x 10"		450	.036		1.39	1.24		2.63	3.60
4121	2" x 12"		400	.040		1.78	1.40		3.18	4.27
4202	Treated, 2" x 4"		550	.029		.58	1.02		1.60	2.33
4222	2" x 6"		500	.032		.73	1.12		1.85	2.66
4242	2" x 8"		450	.036		1.11	1.24		2.35	3.29
4261	2" x 10"		400	.040		1.38	1.40		2.78	3.84
4281	2" x 12"		350	.046		2.01	1.60		3.61	4.87
4402	4" x 4"		450	.036		1.16	1.24		2.40	3.34
4422	4" x 6"		350	.046		1.88	1.60		3.48	4.73
4462	4" x 8"		300	.053		3.88	1.86		5.74	7.35
4480	4" x 10"		260	.062		5.60	2.15		7.75	9.75

06 11 10.34 Sleepers

		Crew	Daily Output	Labor-Hours	Unit	Material	Labor	Equipment	Total	Total Incl O&P
0010	**SLEEPERS**									
0100	On concrete, treated, 1" x 2"	2 Carp	2350	.007	L.F.	.30	.24		.54	.73
0150	1" x 3"		2000	.008		.49	.28		.77	1
0200	2" x 4"		1500	.011		.63	.37		1	1.31
0250	2" x 6"		1300	.012		.82	.43		1.25	1.62

06 11 10.36 Soffit and Canopy Framing

		Crew	Daily Output	Labor-Hours	Unit	Material	Labor	Equipment	Total	Total Incl O&P
0010	**SOFFIT AND CANOPY FRAMING**									
1002	Canopy or soffit framing, 1" x 4"	2 Carp	900	.018	L.F.	.51	.62		1.13	1.59
1021	1" x 6"		850	.019		.78	.66		1.44	1.95
1042	1" x 8"		750	.021		1.30	.75		2.05	2.67
1102	2" x 4"		620	.026		.42	.90		1.32	1.96
1121	2" x 6"		560	.029		.64	1		1.64	2.37
1142	2" x 8"		500	.032		.89	1.12		2.01	2.84
1202	3" x 4"		500	.032		1.20	1.12		2.32	3.18
1221	3" x 6"		400	.040		2	1.40		3.40	4.52
1242	3" x 10"		300	.053		3.62	1.86		5.48	7.10

06 11 10.38 Treated Lumber Framing Material

		Crew	Daily Output	Labor-Hours	Unit	Material	Labor	Equipment	Total	Total Incl O&P
0010	**TREATED LUMBER FRAMING MATERIAL**									
0100	2" x 4"				M.B.F.	870			870	960
0110	2" x 6"					725			725	800
0120	2" x 8"					830			830	915
0130	2" x 10"					830			830	915

For customer support on your Residential Costs with RSMeans data, call 800.448.8182.

377

06 11 Wood Framing

06 11 10 – Framing with Dimensional, Engineered or Composite Lumber

06 11 10.38 Treated Lumber Framing Material	Crew	Daily Output	Labor-Hours	Unit	Material	2018 Bare Costs Labor	2018 Bare Costs Equipment	Total	Total Incl O&P
0140	2" x 12"				M.B.F.	1,000			1,000	1,100
0200	4" x 4"					870			870	955
0210	4" x 6"					940			940	1,025
0220	4" x 8"					1,450			1,450	1,600

06 11 10.40 Wall Framing

	06 11 10.40 Wall Framing	Crew	Daily Output	Labor-Hours	Unit	Material	Labor	Equipment	Total	Total Incl O&P
0010	**WALL FRAMING** R061110-30									
0100	Door buck, studs, header, access, 8' high, 2" x 4" wall, 3' wide	1 Carp	32	.250	Ea.	17.55	8.75		26.30	34
0110	4' wide		32	.250		18.85	8.75		27.60	35
0120	5' wide		32	.250		22.50	8.75		31.25	39.50
0130	6' wide		32	.250		24.50	8.75		33.25	41.50
0140	8' wide		30	.267		36.50	9.30		45.80	55.50
0150	10' wide		30	.267		50	9.30		59.30	70.50
0160	12' wide		30	.267		68.50	9.30		77.80	91
0170	2" x 6" wall, 3' wide		32	.250		25	8.75		33.75	41.50
0180	4' wide		32	.250		26	8.75		34.75	43
0190	5' wide		32	.250		30	8.75		38.75	47.50
0200	6' wide		32	.250		31.50	8.75		40.25	49.50
0210	8' wide		30	.267		43.50	9.30		52.80	63.50
0220	10' wide		30	.267		57	9.30		66.30	78.50
0230	12' wide		30	.267		76	9.30		85.30	99
0240	Window buck, studs, header & access, 8' high 2" x 4" wall, 2' wide		24	.333		18.60	11.65		30.25	40
0250	3' wide		24	.333		21.50	11.65		33.15	43.50
0260	4' wide		24	.333		24	11.65		35.65	45.50
0270	5' wide		24	.333		27.50	11.65		39.15	50
0280	6' wide		24	.333		31	11.65		42.65	53.50
0290	7' wide		24	.333		41	11.65		52.65	64.50
0300	8' wide		22	.364		45.50	12.70		58.20	71
0310	10' wide		22	.364		60	12.70		72.70	87
0320	12' wide		22	.364		81	12.70		93.70	110
0330	2" x 6" wall, 2' wide		24	.333		27.50	11.65		39.15	50
0340	3' wide		24	.333		31	11.65		42.65	53.50
0350	4' wide		24	.333		33.50	11.65		45.15	56.50
0360	5' wide		24	.333		37.50	11.65		49.15	61
0370	6' wide		24	.333		41.50	11.65		53.15	65
0380	7' wide		24	.333		52.50	11.65		64.15	77
0390	8' wide		22	.364		57	12.70		69.70	84
0400	10' wide		22	.364		72.50	12.70		85.20	101
0410	12' wide		22	.364		95	12.70		107.70	126
2002	Headers over openings, 2" x 6"	2 Carp	360	.044	L.F.	.64	1.55		2.19	3.29
2007	2" x 6", pneumatic nailed		432	.037		.65	1.29		1.94	2.87
2052	2" x 8"		340	.047		.89	1.64		2.53	3.71
2057	2" x 8", pneumatic nailed		408	.039		.90	1.37		2.27	3.27
2101	2" x 10"		320	.050		1.41	1.75		3.16	4.46
2106	2" x 10", pneumatic nailed		384	.042		1.42	1.46		2.88	3.98
2152	2" x 12"		300	.053		1.79	1.86		3.65	5.05
2157	2" x 12", pneumatic nailed		360	.044		1.81	1.55		3.36	4.57
2180	4" x 8"		260	.062		4.15	2.15		6.30	8.15
2185	4" x 8", pneumatic nailed		312	.051		4.17	1.79		5.96	7.55
2191	4" x 10"		240	.067		4.96	2.33		7.29	9.30
2196	4" x 10", pneumatic nailed		288	.056		4.99	1.94		6.93	8.75
2202	4" x 12"		190	.084		5.50	2.94		8.44	10.95
2207	4" x 12", pneumatic nailed		228	.070		5.50	2.45		7.95	10.15

06 11 Wood Framing

06 11 10 – Framing with Dimensional, Engineered or Composite Lumber

06 11 10.40 Wall Framing		Crew	Daily Output	Labor-Hours	Unit	Material	2018 Bare Costs Labor	Equipment	Total	Total Incl O&P
2241	6" x 10"	2 Carp	165	.097	L.F.	6.90	3.39		10.29	13.25
2246	6" x 10", pneumatic nailed		198	.081		6.95	2.82		9.77	12.35
2251	6" x 12"		140	.114		8.70	3.99		12.69	16.20
2256	6" x 12", pneumatic nailed		168	.095		8.75	3.33		12.08	15.15
5002	Plates, untreated, 2" x 3"		850	.019		.42	.66		1.08	1.56
5007	2" x 3", pneumatic nailed		1020	.016		.43	.55		.98	1.38
5022	2" x 4"		800	.020		.42	.70		1.12	1.62
5027	2" x 4", pneumatic nailed		960	.017		.42	.58		1	1.44
5041	2" x 6"		750	.021		.64	.75		1.39	1.95
5045	2" x 6", pneumatic nailed		900	.018		.65	.62		1.27	1.75
5061	Treated, 2" x 3"		850	.019		.52	.66		1.18	1.66
5066	2" x 3", treated, pneumatic nailed		1020	.016		.52	.55		1.07	1.48
5081	2" x 4"		800	.020		.59	.70		1.29	1.81
5086	2" x 4", treated, pneumatic nailed		960	.017		.59	.58		1.17	1.62
5101	2" x 6"		750	.021		.74	.75		1.49	2.05
5106	2" x 6", treated, pneumatic nailed		900	.018		.74	.62		1.36	1.85
5122	Studs, 8' high wall, 2" x 3"		1200	.013		.42	.47		.89	1.24
5127	2" x 3", pneumatic nailed		1440	.011		.43	.39		.82	1.12
5142	2" x 4"		1100	.015		.41	.51		.92	1.30
5147	2" x 4", pneumatic nailed		1320	.012		.42	.42		.84	1.17
5162	2" x 6"		1000	.016		.64	.56		1.20	1.64
5167	2" x 6", pneumatic nailed		1200	.013		.65	.47		1.12	1.49
5182	3" x 4"		800	.020		1.20	.70		1.90	2.48
5187	3" x 4", pneumatic nailed		960	.017		1.21	.58		1.79	2.30
5201	Installed on second story, 2" x 3"		1170	.014		.42	.48		.90	1.26
5206	2" x 3", pneumatic nailed		1200	.013		.43	.47		.90	1.24
5221	2" x 4"		1015	.016		.42	.55		.97	1.38
5226	2" x 4", pneumatic nailed		1080	.015		.42	.52		.94	1.33
5241	2" x 6"		890	.018		.64	.63		1.27	1.75
5246	2" x 6", pneumatic nailed		1020	.016		.65	.55		1.20	1.63
5261	3" x 4"		800	.020		1.20	.70		1.90	2.48
5266	3" x 4", pneumatic nailed		960	.017		1.21	.58		1.79	2.30
5281	Installed on dormer or gable, 2" x 3"		1045	.015		.42	.54		.96	1.36
5286	2" x 3", pneumatic nailed		1254	.013		.43	.45		.88	1.21
5301	2" x 4"		905	.018		.42	.62		1.04	1.49
5306	2" x 4", pneumatic nailed		1086	.015		.42	.51		.93	1.33
5321	2" x 6"		800	.020		.64	.70		1.34	1.87
5326	2" x 6", pneumatic nailed		960	.017		.65	.58		1.23	1.69
5341	3" x 4"		700	.023		1.20	.80		2	2.65
5346	3" x 4", pneumatic nailed		840	.019		1.21	.67		1.88	2.44
5361	6' high wall, 2" x 3"		970	.016		.42	.58		1	1.43
5366	2" x 3", pneumatic nailed		1164	.014		.43	.48		.91	1.27
5381	2" x 4"		850	.019		.42	.66		1.08	1.55
5386	2" x 4", pneumatic nailed		1020	.016		.42	.55		.97	1.38
5401	2" x 6"		740	.022		.64	.76		1.40	1.97
5406	2" x 6", pneumatic nailed		888	.018		.65	.63		1.28	1.77
5421	3" x 4"		600	.027		1.20	.93		2.13	2.87
5426	3" x 4", pneumatic nailed		720	.022		1.21	.78		1.99	2.62
5441	Installed on second story, 2" x 3"		950	.017		.43	.59		1.02	1.45
5446	2" x 3", pneumatic nailed		1140	.014		.43	.49		.92	1.29
5461	2" x 4"		810	.020		.42	.69		1.11	1.61
5466	2" x 4", pneumatic nailed		972	.016		.42	.58		1	1.43
5481	2" x 6"		700	.023		.64	.80		1.44	2.04

For customer support on your Residential Costs with RSMeans data, call 800.448.8182.

379

06 11 Wood Framing

06 11 10 – Framing with Dimensional, Engineered or Composite Lumber

06 11 10.40 Wall Framing

		Crew	Daily Output	Labor-Hours	Unit	Material	2018 Bare Costs Labor	Equipment	Total	Total Incl O&P
5486	2" x 6", pneumatic nailed	2 Carp	840	.019	L.F.	.65	.67		1.32	1.83
5501	3" x 4"		550	.029		1.20	1.02		2.22	3.01
5506	3" x 4", pneumatic nailed		660	.024		1.21	.85		2.06	2.74
5521	Installed on dormer or gable, 2" x 3"		850	.019		.42	.66		1.08	1.56
5526	2" x 3", pneumatic nailed		1020	.016		.43	.55		.98	1.38
5541	2" x 4"		720	.022		.42	.78		1.20	1.75
5546	2" x 4", pneumatic nailed		864	.019		.42	.65		1.07	1.55
5561	2" x 6"		620	.026		.64	.90		1.54	2.21
5566	2" x 6", pneumatic nailed		744	.022		.65	.75		1.40	1.97
5581	3" x 4"		480	.033		1.20	1.16		2.36	3.26
5586	3" x 4", pneumatic nailed		576	.028		1.21	.97		2.18	2.94
5601	3' high wall, 2" x 3"		740	.022		.42	.76		1.18	1.73
5606	2" x 3", pneumatic nailed		888	.018		.43	.63		1.06	1.52
5621	2" x 4"		640	.025		.42	.87		1.29	1.91
5626	2" x 4", pneumatic nailed		768	.021		.42	.73		1.15	1.68
5641	2" x 6"		550	.029		.64	1.02		1.66	2.40
5646	2" x 6", pneumatic nailed		660	.024		.65	.85		1.50	2.13
5661	3" x 4"		440	.036		1.20	1.27		2.47	3.43
5666	3" x 4", pneumatic nailed		528	.030		1.21	1.06		2.27	3.09
5681	Installed on second story, 2" x 3"		700	.023		.42	.80		1.22	1.80
5686	2" x 3", pneumatic nailed		840	.019		.43	.67		1.10	1.58
5701	2" x 4"		610	.026		.42	.92		1.34	1.98
5706	2" x 4", pneumatic nailed		732	.022		.42	.76		1.18	1.74
5721	2" x 6"		520	.031		.64	1.08		1.72	2.50
5726	2" x 6", pneumatic nailed		624	.026		.65	.90		1.55	2.21
5741	3" x 4"		430	.037		1.20	1.30		2.50	3.48
5746	3" x 4", pneumatic nailed		516	.031		1.21	1.08		2.29	3.13
5761	Installed on dormer or gable, 2" x 3"		625	.026		.42	.89		1.31	1.96
5766	2" x 3", pneumatic nailed		750	.021		.43	.75		1.18	1.71
5781	2" x 4"		545	.029		.42	1.03		1.45	2.17
5786	2" x 4", pneumatic nailed		654	.024		.42	.85		1.27	1.89
5801	2" x 6"		465	.034		.64	1.20		1.84	2.71
5806	2" x 6", pneumatic nailed		558	.029		.65	1		1.65	2.39
5821	3" x 4"		380	.042		1.20	1.47		2.67	3.77
5826	3" x 4", pneumatic nailed	▼	456	.035	▼	1.21	1.23		2.44	3.37
8250	For second story & above, add						5%			
8300	For dormer & gable, add						15%			

06 11 10.42 Furring

		Crew	Daily Output	Labor-Hours	Unit	Material	2018 Bare Costs Labor	Equipment	Total	Total Incl O&P
0010	**FURRING**									
0012	Wood strips, 1" x 2", on walls, on wood	1 Carp	550	.015	L.F.	.27	.51		.78	1.14
0015	On wood, pneumatic nailed		710	.011		.27	.39		.66	.94
0300	On masonry		495	.016		.29	.56		.85	1.26
0400	On concrete		260	.031		.29	1.08		1.37	2.11
0600	1" x 3", on walls, on wood		550	.015		.43	.51		.94	1.32
0605	On wood, pneumatic nailed		710	.011		.43	.39		.82	1.12
0700	On masonry		495	.016		.46	.56		1.02	1.45
0800	On concrete		260	.031		.46	1.08		1.54	2.30
0850	On ceilings, on wood		350	.023		.43	.80		1.23	1.80
0855	On wood, pneumatic nailed		450	.018		.43	.62		1.05	1.50
0900	On masonry		320	.025		.46	.87		1.33	1.96
0950	On concrete	▼	210	.038	▼	.46	1.33		1.79	2.72

06 11 Wood Framing

06 11 10 – Framing with Dimensional, Engineered or Composite Lumber

06 11 10.44 Grounds

		Crew	Daily Output	Labor-Hours	Unit	Material	2018 Bare Costs Labor	Equipment	Total	Total Incl O&P
0010	**GROUNDS**									
0020	For casework, 1" x 2" wood strips, on wood	1 Carp	330	.024	L.F.	.27	.85		1.12	1.70
0100	On masonry		285	.028		.29	.98		1.27	1.95
0200	On concrete		250	.032		.29	1.12		1.41	2.18
0400	For plaster, 3/4" deep, on wood		450	.018		.27	.62		.89	1.32
0500	On masonry		225	.036		.29	1.24		1.53	2.39
0600	On concrete		175	.046		.29	1.60		1.89	2.98
0700	On metal lath		200	.040		.29	1.40		1.69	2.64

06 12 Structural Panels

06 12 10 – Structural Insulated Panels

06 12 10.10 OSB Faced Panels

			Crew	Daily Output	Labor-Hours	Unit	Material	2018 Bare Costs Labor	Equipment	Total	Total Incl O&P
0010	**OSB FACED PANELS**										
0100	Structural insul. panels, 7/16" OSB both faces, EPS insul., 3-5/8" T	G	F-3	2075	.019	S.F.	3.65	.62	.24	4.51	5.30
0110	5-5/8" thick	G		1725	.023		4.10	.74	.29	5.13	6.05
0120	7-3/8" thick	G		1425	.028		4.45	.90	.35	5.70	6.80
0130	9-3/8" thick	G		1125	.036		4.75	1.14	.44	6.33	7.65
0140	7/16" OSB one face, EPS insul., 3-5/8" thick	G		2175	.018		3.75	.59	.23	4.57	5.35
0150	5-5/8" thick	G		1825	.022		4.35	.70	.27	5.32	6.25
0160	7-3/8" thick	G		1525	.026		4.85	.84	.33	6.02	7.10
0170	9-3/8" thick	G		1225	.033		5.35	1.05	.40	6.80	8.10
0190	7/16" OSB - 1/2" GWB faces, EPS insul., 3-5/8" T	G		2075	.019		3.45	.62	.24	4.31	5.10
0200	5-5/8" thick	G		1725	.023		4.10	.74	.29	5.13	6.05
0210	7-3/8" thick	G		1425	.028		4.65	.90	.35	5.90	7
0220	9-3/8" thick	G		1125	.036		5.25	1.14	.44	6.83	8.20
0240	7/16" OSB - 1/2" MRGWB faces, EPS insul., 3-5/8" T	G		2075	.019		3.55	.62	.24	4.41	5.20
0250	5-5/8" thick	G		1725	.023		4.25	.74	.29	5.28	6.25
0260	7-3/8" thick	G		1425	.028		4.65	.90	.35	5.90	7
0270	9-3/8" thick	G		1125	.036		5.35	1.14	.44	6.93	8.30
0300	For 1/2" GWB added to OSB skin, add	G					1.40			1.40	1.54
0310	For 1/2" MRGWB added to OSB skin, add	G					1.40			1.40	1.54
0320	For one T1-11 skin, add to OSB-OSB	G					1.95			1.95	2.15
0330	For one 19/32" CDX skin, add to OSB-OSB	G					1.50			1.50	1.65
0500	Structural insulated panel, 7/16" OSB both sides, straw core										
0510	4-3/8" T, walls (w/sill, splines, plates)	G	F-6	2400	.017	S.F.	7.55	.53	.21	8.29	9.40
0520	Floors (w/splines)	G		2400	.017		7.55	.53	.21	8.29	9.40
0530	Roof (w/splines)	G		2400	.017		7.55	.53	.21	8.29	9.40
0550	7-7/8" T, walls (w/sill, splines, plates)	G		2400	.017		11.40	.53	.21	12.14	13.65
0560	Floors (w/splines)	G		2400	.017		11.40	.53	.21	12.14	13.65
0570	Roof (w/splines)	G		2400	.017		11.40	.53	.21	12.14	13.65

06 12 19 – Composite Shearwall Panels

06 12 19.10 Steel and Wood Composite Shearwall Panels

		Crew	Daily Output	Labor-Hours	Unit	Material	2018 Bare Costs Labor	Equipment	Total	Total Incl O&P
0010	**STEEL & WOOD COMPOSITE SHEARWALL PANELS**									
0020	Anchor bolts, 36" long (must be placed in wet concrete)	1 Carp	150	.053	Ea.	37	1.86		38.86	43.50
0030	On concrete, 2" x 4" & 2" x 6" walls, 7'-10' high, 360 lb. shear, 12" wide	2 Carp	8	2		460	70		530	620
0040	715 lb. shear, 15" wide		8	2		510	70		580	675
0050	1860 lb. shear, 18" wide		8	2		525	70		595	695
0060	2780 lb. shear, 21" wide		8	2		560	70		630	730
0070	3790 lb. shear, 24" wide		8	2		640	70		710	815
0080	2" x 6" walls, 11'-13' high, 1180 lb. shear, 18" wide		6	2.667		645	93		738	865

06 12 Structural Panels

06 12 19 – Composite Shearwall Panels

06 12 19.10 Steel and Wood Composite Shearwall Panels	Crew	Daily Output	Labor-Hours	Unit	Material	2018 Bare Costs Labor	Equipment	Total	Total Incl O&P
0090	1555 lb. shear, 21" wide	2 Carp	6	2.667	Ea.	720	93		813	945
0100	2280 lb. shear, 24" wide	↓	6	2.667	↓	805	93		898	1,050
0110	For installing above on wood floor frame, add									
0120	Coupler nuts, threaded rods, bolts, shear transfer plate kit	1 Carp	16	.500	Ea.	65	17.50		82.50	101
0130	Framing anchors, angle (2 required)	"	96	.083	"	2.43	2.91		5.34	7.50
0140	For blocking see Section 06 11 10.02									
0150	For installing above, first floor to second floor, wood floor frame, add									
0160	Add stack option to first floor wall panel				Ea.	71.50			71.50	78.50
0170	Threaded rods, bolts, shear transfer plate kit	1 Carp	16	.500		75	17.50		92.50	112
0180	Framing anchors, angle (2 required)	"	96	.083	↓	2.43	2.91		5.34	7.50
0190	For blocking see section 06 11 10.02									
0200	For installing stacked panels, balloon framing									
0210	Add stack option to first floor wall panel				Ea.	71.50			71.50	78.50
0220	Threaded rods, bolts kit	1 Carp	16	.500	"	45	17.50		62.50	78.50

06 13 Heavy Timber Construction

06 13 13 – Log Construction

06 13 13.10 Log Structures

		Crew	Daily Output	Labor-Hours	Unit	Material	2018 Bare Costs Labor	Equipment	Total	Total Incl O&P
0010	**LOG STRUCTURES**									
0020	Exterior walls, pine, D logs, with double T&G, 6" x 6"	2 Carp	500	.032	L.F.	4.97	1.12		6.09	7.30
0030	6" x 8"		375	.043		5.35	1.49		6.84	8.40
0040	8" x 6"		375	.043		4.97	1.49		6.46	7.95
0050	8" x 7"		322	.050		4.97	1.74		6.71	8.35
0060	Square/rectangular logs, with double T&G, 6" x 6"		500	.032		4.97	1.12		6.09	7.30
0070	6" x 8"		375	.043		4.97	1.49		6.46	7.95
0080	8" x 6"		375	.043		4.97	1.49		6.46	7.95
0090	8" x 7"		322	.050		4.97	1.74		6.71	8.35
0100	Round logs, with double T&G, 6" x 8"		375	.043		4.97	1.49		6.46	7.95
0110	8" x 7"		322	.050	↓	4.97	1.74		6.71	8.35
0120	Log siding, ship lapped, 2" x 6"		225	.071	S.F.	3.05	2.49		5.54	7.50
0130	2" x 8"		200	.080		2.56	2.80		5.36	7.45
0140	2" x 12"	↓	180	.089	↓	2.35	3.11		5.46	7.75
0150	Foam sealant strip, 3/8" x 3/8"	1 Carp	1920	.004	L.F.	.16	.15		.31	.42
0152	Chinking, 2" - 3" wide joint, 1/4" to 3/8" deep		600	.013		1.38	.47		1.85	2.29
0154	Caulking, 1/4" to 1/2" joint		900	.009		.50	.31		.81	1.07
0156	Backer rod, 1/4"	↓	900	.009	↓	.18	.31		.49	.72
0157	Penetrating wood preservative	1 Pord	2000	.004	S.F.	.15	.12		.27	.36
0158	Insect treatment	"	4000	.002	"	.36	.06		.42	.49
0160	Upper floor framing, pine, posts/columns, 4" x 6"	2 Carp	750	.021	L.F.	3.05	.75		3.80	4.59
0180	4" x 8"		562	.028		4.13	1		5.13	6.20
0190	6" x 6"		500	.032		5.35	1.12		6.47	7.70
0200	6" x 8"		375	.043		7.85	1.49		9.34	11.15
0210	8" x 8"		281	.057		9.45	1.99		11.44	13.70
0220	8" x 10"		225	.071		11.85	2.49		14.34	17.20
0230	Beams, 4" x 8"		562	.028		4.13	1		5.13	6.20
0240	4" x 10"		449	.036		4.93	1.25		6.18	7.50
0250	4" x 12"		375	.043		5.45	1.49		6.94	8.50
0260	6" x 8"		375	.043		7.85	1.49		9.34	11.15
0270	6" x 10"		300	.053		6.85	1.86		8.71	10.65
0280	6" x 12"		250	.064		8.65	2.24		10.89	13.20
0290	8" x 10"	↓	225	.071	↓	11.85	2.49		14.34	17.20

06 13 Heavy Timber Construction

06 13 13 – Log Construction

06 13 13.10 Log Structures

		Crew	Daily Output	Labor-Hours	Unit	Material	2018 Bare Costs Labor	Equipment	Total	Total Incl O&P
0300	8" x 12"	2 Carp	188	.085	L.F.	16.50	2.97		19.47	23
0310	Joists, 4" x 8"		562	.028		4.13	1		5.13	6.20
0320	4" x 10"		449	.036		4.93	1.25		6.18	7.50
0330	4" x 12"		375	.043		5.45	1.49		6.94	8.50
0340	6" x 8"		375	.043		7.85	1.49		9.34	11.15
0350	6" x 10"		300	.053		6.85	1.86		8.71	10.65
0360	6" x 12"		250	.064		8.65	2.24		10.89	13.20
0370	8" x 10"		225	.071		11.85	2.49		14.34	17.20
0380	8" x 12"		188	.085	▼	14.20	2.97		17.17	20.50
0390	Decking, 1" x 6" T&G		964	.017	S.F.	1.64	.58		2.22	2.76
0400	1" x 8" T&G		700	.023		1.49	.80		2.29	2.97
0410	2" x 6" T&G		482	.033	▼	3.51	1.16		4.67	5.80
0420	Gable end roof framing, rafters, 4" x 8"		562	.028	L.F.	4.13	1		5.13	6.20
0430	4" x 10"		449	.036		4.93	1.25		6.18	7.50
0450	4" x 12"		375	.043		5.45	1.49		6.94	8.50
0460	6" x 8"		375	.043		7.85	1.49		9.34	11.15
0470	6" x 10"		300	.053		6.85	1.86		8.71	10.65
0480	6" x 12"		250	.064		8.65	2.24		10.89	13.20
0490	8" x 10"		225	.071		11.85	2.49		14.34	17.20
0500	8" x 12"		188	.085		14.20	2.97		17.17	20.50
0510	Purlins, 4" x 8"		562	.028		4.13	1		5.13	6.20
0520	6" x 8"		375	.043	▼	7.85	1.49		9.34	11.15
0530	Roof decking, 1" x 6" T&G		640	.025	S.F.	1.64	.87		2.51	3.25
0540	1" x 8" T&G		430	.037		1.49	1.30		2.79	3.80
0550	2" x 6" T&G	▼	320	.050	▼	3.51	1.75		5.26	6.75

06 13 23 – Heavy Timber Framing

06 13 23.10 Heavy Framing

		Crew	Daily Output	Labor-Hours	Unit	Material	2018 Bare Costs Labor	Equipment	Total	Total Incl O&P
0010	**HEAVY FRAMING**									
0020	Beams, single 6" x 10"	2 Carp	1.10	14.545	M.B.F.	1,550	510		2,060	2,575
0100	Single 8" x 16"		1.20	13.333	"	1,925	465		2,390	2,900
0202	Built from 2" lumber, multiple 2" x 14"		900	.018	B.F.	.96	.62		1.58	2.09
0212	Built from 3" lumber, multiple 3" x 6"		700	.023		1.32	.80		2.12	2.78
0222	Multiple 3" x 8"		800	.020		1.43	.70		2.13	2.73
0232	Multiple 3" x 10"		900	.018		1.44	.62		2.06	2.61
0242	Multiple 3" x 12"		1000	.016		1.44	.56		2	2.51
0252	Built from 4" lumber, multiple 4" x 6"		800	.020		1.52	.70		2.22	2.84
0262	Multiple 4" x 8"		900	.018		1.55	.62		2.17	2.73
0272	Multiple 4" x 10"		1000	.016		1.48	.56		2.04	2.56
0282	Multiple 4" x 12"		1100	.015	▼	1.36	.51		1.87	2.35
0292	Columns, structural grade, 1500f, 4" x 4"		450	.036	L.F.	1.78	1.24		3.02	4.03
0302	6" x 6"		225	.071		4.17	2.49		6.66	8.70
0402	8" x 8"		240	.067		8	2.33		10.33	12.65
0502	10" x 10"		90	.178		13.50	6.20		19.70	25
0602	12" x 12"		70	.229	▼	18.55	8		26.55	34
0802	Floor planks, 2" thick, T&G, 2" x 6"		1050	.015	B.F.	1.61	.53		2.14	2.66
0902	2" x 10"		1100	.015		1.63	.51		2.14	2.64
1102	3" thick, 3" x 6"		1050	.015		1.62	.53		2.15	2.67
1202	3" x 10"		1100	.015		1.65	.51		2.16	2.67
1402	Girders, structural grade, 12" x 12"		800	.020		1.55	.70		2.25	2.86
1502	10" x 16"		1000	.016		2.55	.56		3.11	3.74
2302	Roof purlins, 4" thick, structural grade	▼	1050	.015	▼	1.55	.53		2.08	2.59

For customer support on your Residential Costs with RSMeans data, call 800.448.8182.

383

06 15 Wood Decking

06 15 16 – Wood Roof Decking

06 15 16.10 Solid Wood Roof Decking

		Crew	Daily Output	Labor-Hours	Unit	Material	2018 Bare Costs Labor	2018 Bare Costs Equipment	Total	Total Incl O&P
0010	**SOLID WOOD ROOF DECKING**									
0350	Cedar planks, 2" thick	2 Carp	350	.046	S.F.	6.65	1.60		8.25	10
0400	3" thick		320	.050		10	1.75		11.75	13.90
0500	4" thick		250	.064		13.35	2.24		15.59	18.35
0550	6" thick		200	.080		20	2.80		22.80	26.50
0650	Douglas fir, 2" thick		350	.046		2.78	1.60		4.38	5.70
0700	3" thick		320	.050		4.17	1.75		5.92	7.50
0800	4" thick		250	.064		5.55	2.24		7.79	9.80
0850	6" thick		200	.080		8.35	2.80		11.15	13.80
0950	Hemlock, 2" thick		350	.046		2.83	1.60		4.43	5.75
1000	3" thick		320	.050		4.25	1.75		6	7.60
1100	4" thick		250	.064		5.65	2.24		7.89	9.95
1150	6" thick		200	.080		8.50	2.80		11.30	14
1250	Western white spruce, 2" thick		350	.046		1.81	1.60		3.41	4.65
1300	3" thick		320	.050		2.71	1.75		4.46	5.90
1400	4" thick		250	.064		3.61	2.24		5.85	7.70
1450	6" thick		200	.080		5.40	2.80		8.20	10.60

06 15 23 – Laminated Wood Decking

06 15 23.10 Laminated Roof Deck

		Crew	Daily Output	Labor-Hours	Unit	Material	2018 Bare Costs Labor	2018 Bare Costs Equipment	Total	Total Incl O&P
0010	**LAMINATED ROOF DECK**									
0020	Pine or hemlock, 3" thick	2 Carp	425	.038	S.F.	5.75	1.32		7.07	8.55
0100	4" thick		325	.049		7.50	1.72		9.22	11.10
0300	Cedar, 3" thick		425	.038		7.05	1.32		8.37	9.95
0400	4" thick		325	.049		9.45	1.72		11.17	13.20
0600	Fir, 3" thick		425	.038		6	1.32		7.32	8.80
0700	4" thick		325	.049		7.55	1.72		9.27	11.15

06 16 Sheathing

06 16 13 – Insulating Sheathing

06 16 13.10 Insulating Sheathing

			Crew	Daily Output	Labor-Hours	Unit	Material	2018 Bare Costs Labor	2018 Bare Costs Equipment	Total	Total Incl O&P
0010	**INSULATING SHEATHING**										
0020	Expanded polystyrene, 1#/C.F. density, 3/4" thick, R2.89	G	2 Carp	1400	.011	S.F.	.36	.40		.76	1.06
0030	1" thick, R3.85	G		1300	.012		.43	.43		.86	1.19
0040	2" thick, R7.69	G		1200	.013		.70	.47		1.17	1.54
0050	Extruded polystyrene, 15 psi compressive strength, 1" thick, R5	G		1300	.012		.71	.43		1.14	1.50
0060	2" thick, R10	G		1200	.013		.87	.47		1.34	1.73
0070	Polyisocyanurate, 2#/C.F. density, 3/4" thick	G		1400	.011		.58	.40		.98	1.30
0080	1" thick	G		1300	.012		.59	.43		1.02	1.37
0090	1-1/2" thick	G		1250	.013		.74	.45		1.19	1.55
0100	2" thick	G		1200	.013		.90	.47		1.37	1.76

06 16 23 – Subflooring

06 16 23.10 Subfloor

		Crew	Daily Output	Labor-Hours	Unit	Material	2018 Bare Costs Labor	2018 Bare Costs Equipment	Total	Total Incl O&P
0010	**SUBFLOOR**									
0011	Plywood, CDX, 1/2" thick	2 Carp	1500	.011	SF Flr.	.62	.37		.99	1.30
0015	Pneumatic nailed		1860	.009		.62	.30		.92	1.18
0102	5/8" thick		1350	.012		.77	.41		1.18	1.53
0107	Pneumatic nailed		1674	.010		.77	.33		1.10	1.40
0202	3/4" thick		1250	.013		.92	.45		1.37	1.76
0207	Pneumatic nailed		1550	.010		.92	.36		1.28	1.62
0302	1-1/8" thick, 2-4-1 including underlayment		1050	.015		2.07	.53		2.60	3.17

06 16 Sheathing

06 16 23 – Subflooring

06 16 23.10 Subfloor

		Crew	Daily Output	Labor-Hours	Unit	Material	2018 Bare Costs Labor	Equipment	Total	Total Incl O&P
0440	With boards, 1" x 6", S4S, laid regular	2 Carp	900	.018	SF Flr.	1.70	.62		2.32	2.89
0452	1" x 8", laid regular		1000	.016		2.07	.56		2.63	3.21
0462	Laid diagonal		850	.019		2.07	.66		2.73	3.37
0502	1" x 10", laid regular		1100	.015		2.12	.51		2.63	3.18
0602	Laid diagonal		900	.018		2.12	.62		2.74	3.36
1500	OSB, 5/8" thick		1330	.012	S.F.	.55	.42		.97	1.31
1600	3/4" thick		1230	.013	"	.67	.45		1.12	1.50
8990	Subfloor adhesive, 3/8" bead	1 Carp	2300	.003	L.F.	.11	.12		.23	.32

06 16 26 – Underlayment

06 16 26.10 Wood Product Underlayment

		Crew	Daily Output	Labor-Hours	Unit	Material	2018 Bare Costs Labor	Equipment	Total	Total Incl O&P
0010	**WOOD PRODUCT UNDERLAYMENT**									
0015	Plywood, underlayment grade, 1/4" thick	2 Carp	1500	.011	S.F.	.94	.37		1.31	1.65
0018	Pneumatic nailed		1860	.009		.94	.30		1.24	1.53
0030	3/8" thick		1500	.011		1.04	.37		1.41	1.76
0070	Pneumatic nailed		1860	.009		1.04	.30		1.34	1.64
0102	1/2" thick		1450	.011		1.23	.39		1.62	1.99
0107	Pneumatic nailed		1798	.009		1.23	.31		1.54	1.87
0202	5/8" thick		1400	.011		1.36	.40		1.76	2.16
0207	Pneumatic nailed		1736	.009		1.36	.32		1.68	2.04
0302	3/4" thick		1300	.012		1.47	.43		1.90	2.34
0306	Pneumatic nailed		1612	.010		1.47	.35		1.82	2.20
0502	Particle board, 3/8" thick		1500	.011		.41	.37		.78	1.07
0507	Pneumatic nailed		1860	.009		.41	.30		.71	.95
0602	1/2" thick		1450	.011		.43	.39		.82	1.11
0607	Pneumatic nailed		1798	.009		.43	.31		.74	.99
0802	5/8" thick		1400	.011		.55	.40		.95	1.27
0807	Pneumatic nailed		1736	.009		.55	.32		.87	1.15
0902	3/4" thick		1300	.012		.67	.43		1.10	1.46
0907	Pneumatic nailed		1612	.010		.67	.35		1.02	1.32
1100	Hardboard, underlayment grade, 4' x 4', .215" thick [G]		1500	.011		.68	.37		1.05	1.37

06 16 33 – Wood Board Sheathing

06 16 33.10 Board Sheathing

		Crew	Daily Output	Labor-Hours	Unit	Material	2018 Bare Costs Labor	Equipment	Total	Total Incl O&P
0009	**BOARD SHEATHING**									
0010	Roof, 1" x 6" boards, laid horizontal	2 Carp	725	.022	S.F.	1.70	.77		2.47	3.14
0020	On steep roof		520	.031		1.70	1.08		2.78	3.65
0040	On dormers, hips, & valleys		480	.033		1.70	1.16		2.86	3.80
0050	Laid diagonal		650	.025		1.70	.86		2.56	3.29
0070	1" x 8" boards, laid horizontal		875	.018		2.07	.64		2.71	3.34
0080	On steep roof		635	.025		2.07	.88		2.95	3.74
0090	On dormers, hips, & valleys		580	.028		2.12	.96		3.08	3.93
0100	Laid diagonal		725	.022		2.07	.77		2.84	3.56
0110	Skip sheathing, 1" x 4", 7" OC	1 Carp	1200	.007		.63	.23		.86	1.09
0120	1" x 6", 9" OC		1450	.006		.78	.19		.97	1.18
0180	T&G sheathing/decking, 1" x 6"		1000	.008		1.80	.28		2.08	2.44
0190	2" x 6"		1000	.008		3.86	.28		4.14	4.71
0200	Walls, 1" x 6" boards, laid regular	2 Carp	650	.025		1.70	.86		2.56	3.29
0210	Laid diagonal		585	.027		1.70	.96		2.66	3.45
0220	1" x 8" boards, laid regular		765	.021		2.07	.73		2.80	3.50
0230	Laid diagonal		650	.025		2.07	.86		2.93	3.71

For customer support on your Residential Costs with RSMeans data, call 800.448.8182.

385

06 16 Sheathing

06 16 36 – Wood Panel Product Sheathing

06 16 36.10 Sheathing

		Crew	Daily Output	Labor-Hours	Unit	Material	2018 Bare Costs Labor	Equipment	Total	Total Incl O&P
0010	**SHEATHING** R061636-20									
0012	Plywood on roofs, CDX									
0032	5/16" thick	2 Carp	1600	.010	S.F.	.59	.35		.94	1.22
0037	Pneumatic nailed		1952	.008		.59	.29		.88	1.12
0052	3/8" thick		1525	.010		.60	.37		.97	1.27
0057	Pneumatic nailed		1860	.009		.60	.30		.90	1.16
0102	1/2" thick		1400	.011		.62	.40		1.02	1.34
0103	Pneumatic nailed		1708	.009		.62	.33		.95	1.22
0202	5/8" thick		1300	.012		.77	.43		1.20	1.56
0207	Pneumatic nailed		1586	.010		.77	.35		1.12	1.43
0302	3/4" thick		1200	.013		.92	.47		1.39	1.79
0307	Pneumatic nailed		1464	.011		.92	.38		1.30	1.66
0502	Plywood on walls, with exterior CDX, 3/8" thick		1200	.013		.60	.47		1.07	1.43
0507	Pneumatic nailed		1488	.011		.60	.38		.98	1.28
0603	1/2" thick		1125	.014		.62	.50		1.12	1.51
0608	Pneumatic nailed		1395	.011		.62	.40		1.02	1.35
0702	5/8" thick		1050	.015		.77	.53		1.30	1.73
0707	Pneumatic nailed		1302	.012		.77	.43		1.20	1.55
0803	3/4" thick		975	.016		.92	.57		1.49	1.97
0808	Pneumatic nailed		1209	.013		.92	.46		1.38	1.79
1000	For shear wall construction, add						20%			
1200	For structural 1 exterior plywood, add				S.F.	10%				
3000	Wood fiber, regular, no vapor barrier, 1/2" thick	2 Carp	1200	.013		.64	.47		1.11	1.47
3100	5/8" thick		1200	.013		.70	.47		1.17	1.54
3300	No vapor barrier, in colors, 1/2" thick		1200	.013		.81	.47		1.28	1.66
3400	5/8" thick		1200	.013		.85	.47		1.32	1.71
3600	With vapor barrier one side, white, 1/2" thick		1200	.013		.63	.47		1.10	1.46
3700	Vapor barrier 2 sides, 1/2" thick		1200	.013		.84	.47		1.31	1.69
3800	Asphalt impregnated, 25/32" thick		1200	.013		.32	.47		.79	1.12
3850	Intermediate, 1/2" thick		1200	.013		.32	.47		.79	1.12
4500	Oriented strand board, on roof, 7/16" thick [G]		1460	.011		.48	.38		.86	1.17
4505	Pneumatic nailed [G]		1780	.009		.48	.31		.79	1.05
4550	1/2" thick [G]		1400	.011		.48	.40		.88	1.19
4555	Pneumatic nailed [G]		1736	.009		.48	.32		.80	1.07
4600	5/8" thick [G]		1300	.012		.54	.43		.97	1.31
4605	Pneumatic nailed [G]		1586	.010		.54	.35		.89	1.18
4610	On walls, 7/16" thick [G]		1200	.013		.48	.47		.95	1.30
4615	Pneumatic nailed [G]		1488	.011		.48	.38		.86	1.15
4620	1/2" thick [G]		1195	.013		.48	.47		.95	1.31
4625	Pneumatic nailed [G]		1325	.012		.48	.42		.90	1.23
4630	5/8" thick [G]		1050	.015		.54	.53		1.07	1.48
4635	Pneumatic nailed [G]		1302	.012		.54	.43		.97	1.30
4700	Oriented strand board, factory laminated W.R. barrier, on roof, 1/2" thick [G]		1400	.011		.75	.40		1.15	1.49
4705	Pneumatic nailed [G]		1736	.009		.75	.32		1.07	1.37
4720	5/8" thick [G]		1300	.012		.91	.43		1.34	1.72
4725	Pneumatic nailed [G]		1586	.010		.91	.35		1.26	1.59
4730	5/8" thick, T&G [G]		1150	.014		1.09	.49		1.58	2.01
4735	Pneumatic nailed, T&G [G]		1400	.011		1.09	.40		1.49	1.86
4740	On walls, 7/16" thick [G]		1200	.013		.69	.47		1.16	1.53
4745	Pneumatic nailed [G]		1488	.011		.69	.38		1.07	1.38
4750	1/2" thick [G]		1195	.013		.75	.47		1.22	1.61
4755	Pneumatic nailed [G]		1325	.012		.75	.42		1.17	1.53

06 16 Sheathing

06 16 36 – Wood Panel Product Sheathing

06 16 36.10 Sheathing	Crew	Daily Output	Labor-Hours	Unit	Material	2018 Bare Costs Labor	Equipment	Total	Total Incl O&P
4800	Joint sealant tape, 3-1/2"	2 Carp	7600	.002	L.F.	.28	.07		.35	.43
4810	Joint sealant tape, 6"	↓	7600	.002	"	.41	.07		.48	.57

06 16 43 – Gypsum Sheathing

06 16 43.10 Gypsum Sheathing

		Crew	Daily Output	Labor-Hours	Unit	Material	2018 Bare Costs Labor	Equipment	Total	Total Incl O&P
0010	**GYPSUM SHEATHING**									
0020	Gypsum, weatherproof, 1/2" thick	2 Carp	1125	.014	S.F.	.46	.50		.96	1.34
0040	With embedded glass mats	"	1100	.015	"	.71	.51		1.22	1.63

06 17 Shop-Fabricated Structural Wood

06 17 33 – Wood I-Joists

06 17 33.10 Wood and Composite I-Joists

		Crew	Daily Output	Labor-Hours	Unit	Material	2018 Bare Costs Labor	Equipment	Total	Total Incl O&P
0010	**WOOD AND COMPOSITE I-JOISTS**									
0100	Plywood webs, incl. bridging & blocking, panels 24" OC									
1200	15' to 24' span, 50 psf live load	F-5	2400	.013	SF Flr.	1.99	.41		2.40	2.88
1300	55 psf live load		2250	.014		2.31	.44		2.75	3.27
1400	24' to 30' span, 45 psf live load		2600	.012		2.90	.38		3.28	3.82
1500	55 psf live load	↓	2400	.013	↓	4.62	.41		5.03	5.80

06 17 53 – Shop-Fabricated Wood Trusses

06 17 53.10 Roof Trusses

		Crew	Daily Output	Labor-Hours	Unit	Material	2018 Bare Costs Labor	Equipment	Total	Total Incl O&P
0010	**ROOF TRUSSES**									
5000	Common wood, 2" x 4" metal plate connected, 24" OC, 4/12 slope									
5010	1' overhang, 12' span	F-5	55	.582	Ea.	36.50	17.90		54.40	70
5050	20' span	F-6	62	.645		73.50	20.50	8	102	124
5100	24' span		60	.667		79	21.50	8.25	108.75	131
5150	26' span		57	.702		77	22.50	8.70	108.20	132
5200	28' span		53	.755		86.50	24	9.35	119.85	145
5240	30' span		51	.784		99	25	9.75	133.75	161
5250	32' span		50	.800		113	25.50	9.90	148.40	178
5280	34' span		48	.833		108	26.50	10.35	144.85	174
5350	8/12 pitch, 1' overhang, 20' span		57	.702		83.50	22.50	8.70	114.70	139
5400	24' span		55	.727		103	23.50	9	135.50	161
5450	26' span		52	.769		108	24.50	9.55	142.05	171
5500	28' span		49	.816		122	26	10.10	158.10	190
5550	32' span		45	.889		138	28.50	11	177.50	211
5600	36' span		41	.976		173	31.50	12.10	216.60	256
5650	38' span		40	1		191	32	12.40	235.40	277
5700	40' span	↓	40	1	↓	193	32	12.40	237.40	280

06 18 13.10 Laminated Beams

		Crew	Daily Output	Labor-Hours	Unit	Material	2018 Bare Costs Labor	Equipment	Total	Total Incl O&P
0010	**LAMINATED BEAMS**									
0050	3-1/2" x 18"	F-3	480	.083	L.F.	28	2.67	1.03	31.70	36.50
0100	5-1/4" x 11-7/8"		450	.089		27.50	2.85	1.10	31.45	36.50
0150	5-1/4" x 16"		360	.111		39	3.56	1.38	43.94	50.50
0200	5-1/4" x 18"		290	.138		42	4.42	1.71	48.13	55
0250	5-1/4" x 24"		220	.182		61	5.80	2.25	69.05	79
0300	7" x 11-7/8"		320	.125		42	4	1.55	47.55	54.50
0350	7" x 16"		260	.154		58.50	4.92	1.91	65.33	74.50
0400	7" x 18"		210	.190		68	6.10	2.36	76.46	88
0500	For premium appearance, add to L.F. prices					5%				
0550	For industrial type, deduct					15%				
0600	For stain and varnish, add					5%				
0650	For 3/4" laminations, add					25%				

06 18 13.20 Laminated Framing

		Crew	Daily Output	Labor-Hours	Unit	Material	2018 Bare Costs Labor	Equipment	Total	Total Incl O&P
0010	**LAMINATED FRAMING**									
0020	30 lb., short term live load, 15 lb. dead load									
0200	Straight roof beams, 20' clear span, beams 8' OC	F-3	2560	.016	SF Flr.	2.26	.50	.19	2.95	3.53
0300	Beams 16' OC		3200	.013		1.65	.40	.16	2.21	2.66
0500	40' clear span, beams 8' OC		3200	.013		4.29	.40	.16	4.85	5.55
0600	Beams 16' OC		3840	.010		3.54	.33	.13	4	4.59
0800	60' clear span, beams 8' OC	F-4	2880	.014		7.35	.44	.35	8.14	9.20
0900	Beams 16' OC	"	3840	.010		5.50	.33	.26	6.09	6.90
1100	Tudor arches, 30' to 40' clear span, frames 8' OC	F-3	1680	.024		9.60	.76	.30	10.66	12.15
1200	Frames 16' OC	"	2240	.018		7.50	.57	.22	8.29	9.45
1400	50' to 60' clear span, frames 8' OC	F-4	2200	.018		10.30	.58	.45	11.33	12.80
1500	Frames 16' OC		2640	.015		8.80	.49	.38	9.67	10.90
1700	Radial arches, 60' clear span, frames 8' OC		1920	.021		9.65	.67	.52	10.84	12.35
1800	Frames 16' OC		2880	.014		7.65	.44	.35	8.44	9.50
2000	100' clear span, frames 8' OC		1600	.025		9.95	.80	.62	11.37	12.95
2100	Frames 16' OC		2400	.017		8.80	.53	.41	9.74	11
2300	120' clear span, frames 8' OC		1440	.028		13.25	.89	.69	14.83	16.80
2400	Frames 16' OC		1920	.021		12.10	.67	.52	13.29	15
2600	Bowstring trusses, 20' OC, 40' clear span	F-3	2400	.017		6	.53	.21	6.74	7.70
2700	60' clear span	F-4	3600	.011		5.40	.36	.28	6.04	6.85
2800	100' clear span		4000	.010		7.65	.32	.25	8.22	9.20
2900	120' clear span		3600	.011		8.10	.36	.28	8.74	9.80
3100	For premium appearance, add to S.F. prices					5%				
3300	For industrial type, deduct					15%				
3500	For stain and varnish, add					5%				
3900	For 3/4" laminations, add to straight					25%				
4100	Add to curved					15%				
4300	Alternate pricing method: (use nominal footage of									
4310	components). Straight beams, camber less than 6"	F-3	3.50	11.429	M.B.F.	3,175	365	142	3,682	4,275
4400	Columns, including hardware		2	20		3,425	640	248	4,313	5,100
4600	Curved members, radius over 32'		2.50	16		3,500	510	198	4,208	4,925
4700	Radius 10' to 32'		3	13.333		3,475	425	165	4,065	4,700
4900	For complicated shapes, add maximum					100%				
5100	For pressure treating, add to straight					35%				
5200	Add to curved					45%				
6000	Laminated veneer members, southern pine or western species									
6050	1-3/4" wide x 5-1/2" deep	2 Carp	480	.033	L.F.	3.47	1.16		4.63	5.75
6100	9-1/2" deep		480	.033		4.25	1.16		5.41	6.60

06 18 Glued-Laminated Construction

06 18 13 – Glued-Laminated Beams

06 18 13.20 Laminated Framing	Crew	Daily Output	Labor-Hours	Unit	Material	2018 Bare Costs Labor	Equipment	Total	Total Incl O&P
6150	14" deep	2 Carp	450	.036	L.F.	7.35	1.24		8.59	10.15
6200	18" deep	↓	450	.036	↓	10.30	1.24		11.54	13.35
6300	Parallel strand members, southern pine or western species									
6350	1-3/4" wide x 9-1/4" deep	2 Carp	480	.033	L.F.	4.77	1.16		5.93	7.20
6400	11-1/4" deep		450	.036		4.94	1.24		6.18	7.50
6450	14" deep		400	.040		7.60	1.40		9	10.65
6500	3-1/2" wide x 9-1/4" deep		480	.033		16.10	1.16		17.26	19.65
6550	11-1/4" deep		450	.036		19.90	1.24		21.14	24
6600	14" deep		400	.040		22.50	1.40		23.90	27
6650	7" wide x 9-1/4" deep		450	.036		33.50	1.24		34.74	39
6700	11-1/4" deep		420	.038		42	1.33		43.33	48.50
6750	14" deep	↓	400	.040	↓	50	1.40		51.40	57.50
8000	Straight beams									
8102	20' span									
8104	3-1/8" x 9"	F-3	30	1.333	Ea.	149	42.50	16.55	208.05	253
8106	x 10-1/2"		30	1.333		174	42.50	16.55	233.05	280
8108	x 12"		30	1.333		199	42.50	16.55	258.05	310
8110	x 13-1/2"		30	1.333		224	42.50	16.55	283.05	335
8112	x 15"		29	1.379		249	44	17.10	310.10	365
8114	5-1/8" x 10-1/2"		30	1.333		285	42.50	16.55	344.05	405
8116	x 12"		30	1.333		325	42.50	16.55	384.05	450
8118	x 13-1/2"		30	1.333		365	42.50	16.55	424.05	495
8120	x 15"		29	1.379		410	44	17.10	471.10	540
8122	x 16-1/2"		29	1.379		450	44	17.10	511.10	585
8124	x 18"		29	1.379		490	44	17.10	551.10	630
8126	x 19-1/2"		29	1.379		530	44	17.10	591.10	675
8128	x 21"		28	1.429		570	45.50	17.70	633.20	725
8130	x 22-1/2"		28	1.429		610	45.50	17.70	673.20	765
8132	x 24"		28	1.429		650	45.50	17.70	713.20	810
8134	6-3/4" x 12"		29	1.379		430	44	17.10	491.10	560
8136	x 13-1/2"		29	1.379		485	44	17.10	546.10	620
8138	x 15"		29	1.379		535	44	17.10	596.10	680
8140	x 16-1/2"		28	1.429		590	45.50	17.70	653.20	745
8142	x 18"		28	1.429		645	45.50	17.70	708.20	805
8144	x 19-1/2"		28	1.429		700	45.50	17.70	763.20	865
8146	x 21"		27	1.481		750	47.50	18.35	815.85	925
8148	x 22-1/2"		27	1.481		805	47.50	18.35	870.85	985
8150	x 24"		27	1.481		860	47.50	18.35	925.85	1,050
8152	x 25-1/2"		27	1.481		915	47.50	18.35	980.85	1,100
8154	x 27"		26	1.538		965	49.50	19.10	1,033.60	1,175
8156	x 28-1/2"		26	1.538		1,025	49.50	19.10	1,093.60	1,225
8158	x 30"	↓	26	1.538	↓	1,075	49.50	19.10	1,143.60	1,275
8200	30' span									
8250	3-1/8" x 9"	F-3	30	1.333	Ea.	224	42.50	16.55	283.05	335
8252	x 10-1/2"		30	1.333		261	42.50	16.55	320.05	375
8254	x 12"		30	1.333		298	42.50	16.55	357.05	420
8256	x 13-1/2"		30	1.333		335	42.50	16.55	394.05	460
8258	x 15"		29	1.379		375	44	17.10	436.10	500
8260	5-1/8" x 10-1/2"		30	1.333		430	42.50	16.55	489.05	560
8262	x 12"		30	1.333		490	42.50	16.55	549.05	630
8264	x 13-1/2"		30	1.333		550	42.50	16.55	609.05	695
8266	x 15"		29	1.379		610	44	17.10	671.10	760
8268	x 16-1/2"	↓	29	1.379	↓	670	44	17.10	731.10	830

For customer support on your Residential Costs with RSMeans data, call 800.448.8182.

389

06 18 13.20 Laminated Framing	Crew	Daily Output	Labor-Hours	Unit	Material	2018 Bare Costs Labor	Equipment	Total	Total Incl O&P
8270	x 18"	F-3	29	1.379	Ea.	735	44	17.10	796.10	895
8272	x 19-1/2"		29	1.379		795	44	17.10	856.10	965
8274	x 21"		28	1.429		855	45.50	17.70	918.20	1,025
8276	x 22-1/2"		28	1.429		915	45.50	17.70	978.20	1,100
8278	x 24"		28	1.429		980	45.50	17.70	1,043.20	1,175
8280	6-3/4" x 12"		29	1.379		645	44	17.10	706.10	800
8282	x 13-1/2"		29	1.379		725	44	17.10	786.10	885
8284	x 15"		29	1.379		805	44	17.10	866.10	975
8286	x 16-1/2"		28	1.429		885	45.50	17.70	948.20	1,075
8288	x 18"		28	1.429		965	45.50	17.70	1,028.20	1,175
8290	x 19-1/2"		28	1.429		1,050	45.50	17.70	1,113.20	1,250
8292	x 21"		27	1.481		1,125	47.50	18.35	1,190.85	1,350
8294	x 22-1/2"		27	1.481		1,200	47.50	18.35	1,265.85	1,425
8296	x 24"		27	1.481		1,300	47.50	18.35	1,365.85	1,525
8298	x 25-1/2"		27	1.481		1,375	47.50	18.35	1,440.85	1,600
8300	x 27"		26	1.538		1,450	49.50	19.10	1,518.60	1,700
8302	x 28-1/2"		26	1.538		1,525	49.50	19.10	1,593.60	1,775
8304	x 30"	↓	26	1.538	↓	1,600	49.50	19.10	1,668.60	1,875
8400	40' span									
8402	3-1/8" x 9"	F-3	30	1.333	Ea.	298	42.50	16.55	357.05	420
8404	x 10-1/2"		30	1.333		350	42.50	16.55	409.05	475
8406	x 12"		30	1.333		400	42.50	16.55	459.05	525
8408	x 13-1/2"		30	1.333		445	42.50	16.55	504.05	580
8410	x 15"		29	1.379		495	44	17.10	556.10	635
8412	5-1/8" x 10-1/2"		30	1.333		570	42.50	16.55	629.05	720
8414	x 12"		30	1.333		650	42.50	16.55	709.05	805
8416	x 13-1/2"		30	1.333		735	42.50	16.55	794.05	895
8418	x 15"		29	1.379		815	44	17.10	876.10	985
8420	x 16-1/2"		29	1.379		895	44	17.10	956.10	1,075
8422	x 18"		29	1.379		980	44	17.10	1,041.10	1,175
8424	x 19-1/2"		29	1.379		1,050	44	17.10	1,111.10	1,275
8426	x 21"		28	1.429		1,150	45.50	17.70	1,213.20	1,350
8428	x 22-1/2"		28	1.429		1,225	45.50	17.70	1,288.20	1,450
8430	x 24"		28	1.429		1,300	45.50	17.70	1,363.20	1,525
8432	6-3/4" x 12"		29	1.379		860	44	17.10	921.10	1,025
8434	x 13-1/2"		29	1.379		965	44	17.10	1,026.10	1,175
8436	x 15"		29	1.379		1,075	44	17.10	1,136.10	1,275
8438	x 16-1/2"		28	1.429		1,175	45.50	17.70	1,238.20	1,400
8440	x 18"		28	1.429		1,300	45.50	17.70	1,363.20	1,525
8442	x 19-1/2"		28	1.429		1,400	45.50	17.70	1,463.20	1,625
8444	x 21"		27	1.481		1,500	47.50	18.35	1,565.85	1,750
8446	x 22-1/2"		27	1.481		1,600	47.50	18.35	1,665.85	1,875
8448	x 24"		27	1.481		1,725	47.50	18.35	1,790.85	2,000
8450	x 25-1/2"		27	1.481		1,825	47.50	18.35	1,890.85	2,100
8452	x 27"		26	1.538		1,925	49.50	19.10	1,993.60	2,225
8454	x 28-1/2"		26	1.538		2,050	49.50	19.10	2,118.60	2,350
8456	x 30"	↓	26	1.538	↓	2,150	49.50	19.10	2,218.60	2,450

For customer support on your Residential Costs with RSMeans data, call 800.448.8182.

06 22 Millwork

06 22 13 – Standard Pattern Wood Trim

06 22 13.10 Millwork

		Daily Output	Labor-Hours	Unit	Material	2018 Bare Costs Labor	Equipment	Total	Total Incl O&P
							Crew		
0010	**MILLWORK**								
0020	Rule of thumb, milled material equals rough lumber cost x 3								
1020	1" x 12", custom birch			L.F.	4.62			4.62	5.10
1040	Cedar				5.40			5.40	5.90
1060	Oak				5.25			5.25	5.80
1080	Redwood				4.71			4.71	5.20
1100	Southern yellow pine				4.06			4.06	4.47
1120	Sugar pine				5.85			5.85	6.45
1140	Teak				33.50			33.50	37
1160	Walnut				7.15			7.15	7.90
1180	White pine				5.30			5.30	5.80

06 22 13.15 Moldings, Base

		Crew	Daily Output	Labor-Hours	Unit	Material	Labor	Equipment	Total	Total Incl O&P
0010	**MOLDINGS, BASE**									
5100	Classic profile, 5/8" x 5-1/2", finger jointed and primed	1 Carp	250	.032	L.F.	1.59	1.12		2.71	3.61
5105	Poplar		240	.033		1.62	1.16		2.78	3.72
5110	Red oak		220	.036		2.49	1.27		3.76	4.85
5115	Maple		220	.036		3.75	1.27		5.02	6.25
5120	Cherry		220	.036		4.45	1.27		5.72	7
5125	3/4" x 7-1/2", finger jointed and primed		250	.032		2	1.12		3.12	4.06
5130	Poplar		240	.033		2.59	1.16		3.75	4.79
5135	Red oak		220	.036		3.61	1.27		4.88	6.10
5140	Maple		220	.036		4.99	1.27		6.26	7.60
5145	Cherry		220	.036		5.95	1.27		7.22	8.65
5150	Modern profile, 5/8" x 3-1/2", finger jointed and primed		250	.032		.96	1.12		2.08	2.91
5155	Poplar		240	.033		1.03	1.16		2.19	3.07
5160	Red oak		220	.036		1.70	1.27		2.97	3.98
5165	Maple		220	.036		2.63	1.27		3.90	5
5170	Cherry		220	.036		2.84	1.27		4.11	5.25
5175	Ogee profile, 7/16" x 3", finger jointed and primed		250	.032		.66	1.12		1.78	2.58
5180	Poplar		240	.033		.67	1.16		1.83	2.68
5185	Red oak		220	.036		.95	1.27		2.22	3.15
5200	9/16" x 3-1/2", finger jointed and primed		250	.032		.66	1.12		1.78	2.58
5205	Pine		240	.033		1.17	1.16		2.33	3.23
5210	Red oak		220	.036		2.80	1.27		4.07	5.20
5215	9/16" x 4-1/2", red oak		220	.036		4.19	1.27		5.46	6.70
5220	5/8" x 3-1/2", finger jointed and primed		250	.032		.98	1.12		2.10	2.94
5225	Poplar		240	.033		1.03	1.16		2.19	3.07
5230	Red oak		220	.036		1.70	1.27		2.97	3.98
5235	Maple		220	.036		2.63	1.27		3.90	5
5240	Cherry		220	.036		2.84	1.27		4.11	5.25
5245	5/8" x 4", finger jointed and primed		250	.032		1.25	1.12		2.37	3.23
5250	Poplar		240	.033		1.33	1.16		2.49	3.40
5255	Red oak		220	.036		1.91	1.27		3.18	4.21
5260	Maple		220	.036		2.94	1.27		4.21	5.35
5265	Cherry		220	.036		3.08	1.27		4.35	5.50
5270	Rectangular profile, oak, 3/8" x 1-1/4"		260	.031		1.24	1.08		2.32	3.16
5275	1/2" x 2-1/2"		255	.031		2.35	1.10		3.45	4.40
5280	1/2" x 3-1/2"		250	.032		2.78	1.12		3.90	4.92
5285	1" x 6"		240	.033		4.14	1.16		5.30	6.50
5290	1" x 8"		240	.033		5.15	1.16		6.31	7.60
5295	Pine, 3/8" x 1-3/4"		260	.031		.50	1.08		1.58	2.34
5300	7/16" x 2-1/2"		255	.031		.78	1.10		1.88	2.68

06 22 13 – Standard Pattern Wood Trim

06 22 13.15 Moldings, Base

		Crew	Daily Output	Labor-Hours	Unit	Material	2018 Bare Costs Labor	Equipment	Total	Total Incl O&P
5305	1" x 6"	1 Carp	240	.033	L.F.	.72	1.16		1.88	2.73
5310	1" x 8"		240	.033		.93	1.16		2.09	2.96
5315	Shoe, 1/2" x 3/4", primed		260	.031		.51	1.08		1.59	2.35
5320	Pine		240	.033		.34	1.16		1.50	2.31
5325	Poplar		240	.033		.41	1.16		1.57	2.39
5330	Red oak		220	.036		.55	1.27		1.82	2.71
5335	Maple		220	.036		.73	1.27		2	2.91
5340	Cherry		220	.036		.83	1.27		2.10	3.02
5345	11/16" x 1-1/2", pine		240	.033		.73	1.16		1.89	2.74
5350	Caps, 11/16" x 1-3/8", pine		240	.033		.58	1.16		1.74	2.58
5355	3/4" x 1-3/4", finger jointed and primed		260	.031		.77	1.08		1.85	2.64
5360	Poplar		240	.033		1.04	1.16		2.20	3.08
5365	Red oak		220	.036		1.25	1.27		2.52	3.48
5370	Maple		220	.036		1.63	1.27		2.90	3.90
5375	Cherry		220	.036		3.30	1.27		4.57	5.75
5380	Combination base & shoe, 9/16" x 3-1/2" & 1/2" x 3/4", pine		125	.064		1.51	2.24		3.75	5.40
5385	Three piece oak, 6" high		80	.100		5.95	3.50		9.45	12.35
5390	Including 3/4" x 1" base shoe		70	.114		6.45	3.99		10.44	13.75
5395	Flooring cant strip, 3/4" x 3/4", pre-finished pine		260	.031		.44	1.08		1.52	2.27
5400	For pre-finished, stain and clear coat, add					.55			.55	.61
5405	Clear coat only, add					.45			.45	.50

06 22 13.30 Moldings, Casings

		Crew	Daily Output	Labor-Hours	Unit	Material	2018 Bare Costs Labor	Equipment	Total	Total Incl O&P
0010	**MOLDINGS, CASINGS**									
0085	Apron, 9/16" x 2-1/2", pine	1 Carp	250	.032	L.F.	1.43	1.12		2.55	3.43
0090	5/8" x 2-1/2", pine		250	.032		1.71	1.12		2.83	3.74
0110	5/8" x 3-1/2", pine		220	.036		1.94	1.27		3.21	4.24
0300	Band, 11/16" x 1-1/8", pine		270	.030		.75	1.04		1.79	2.54
0310	11/16" x 1-1/2", finger jointed and primed		270	.030		.76	1.04		1.80	2.55
0320	Pine		270	.030		1	1.04		2.04	2.82
0330	11/16" x 1-3/4", finger jointed and primed		270	.030		.95	1.04		1.99	2.76
0350	Pine		270	.030		.95	1.04		1.99	2.76
0355	Beaded, 3/4" x 3-1/2", finger jointed and primed		220	.036		.98	1.27		2.25	3.19
0360	Poplar		220	.036		1.15	1.27		2.42	3.37
0365	Red oak		220	.036		1.70	1.27		2.97	3.98
0370	Maple		220	.036		2.63	1.27		3.90	5
0375	Cherry		220	.036		3.18	1.27		4.45	5.60
0380	3/4" x 4", finger jointed and primed		220	.036		1.22	1.27		2.49	3.45
0385	Poplar		220	.036		1.44	1.27		2.71	3.69
0390	Red oak		220	.036		2.02	1.27		3.29	4.33
0395	Maple		220	.036		2.65	1.27		3.92	5
0400	Cherry		220	.036		3.36	1.27		4.63	5.80
0405	3/4" x 5-1/2", finger jointed and primed		200	.040		1.35	1.40		2.75	3.80
0410	Poplar		200	.040		1.99	1.40		3.39	4.51
0415	Red oak		200	.040		2.93	1.40		4.33	5.55
0420	Maple		200	.040		3.85	1.40		5.25	6.55
0425	Cherry		200	.040		4.43	1.40		5.83	7.20
0430	Classic profile, 3/4" x 2-3/4", finger jointed and primed		250	.032		.86	1.12		1.98	2.80
0435	Poplar		250	.032		1.03	1.12		2.15	2.99
0440	Red oak		250	.032		1.51	1.12		2.63	3.52
0445	Maple		250	.032		2.16	1.12		3.28	4.23
0450	Cherry		250	.032		2.45	1.12		3.57	4.55
0455	Fluted, 3/4" x 3-1/2", poplar		220	.036		1.15	1.27		2.42	3.37

06 22 13 – Standard Pattern Wood Trim

06 22 13.30 Moldings, Casings	Crew	Daily Output	Labor-Hours	Unit	Material	2018 Bare Costs Labor	Equipment	Total	Total Incl O&P
0460	Red oak	1 Carp	220	.036	L.F.	1.70	1.27		2.97	3.98
0465	Maple		220	.036		2.63	1.27		3.90	5
0470	Cherry		220	.036		3.18	1.27		4.45	5.60
0475	3/4" x 4", poplar		220	.036		1.44	1.27		2.71	3.69
0480	Red oak		220	.036		2.02	1.27		3.29	4.33
0485	Maple		220	.036		2.65	1.27		3.92	5
0490	Cherry		220	.036		3.36	1.27		4.63	5.80
0495	3/4" x 5-1/2", poplar		200	.040		1.99	1.40		3.39	4.51
0500	Red oak		200	.040		2.93	1.40		4.33	5.55
0505	Maple		200	.040		3.85	1.40		5.25	6.55
0510	Cherry		200	.040		4.43	1.40		5.83	7.20
0515	3/4" x 7-1/2", poplar		190	.042		1.26	1.47		2.73	3.83
0520	Red oak		190	.042		3.93	1.47		5.40	6.75
0525	Maple		190	.042		5.65	1.47		7.12	8.65
0530	Cherry		190	.042		6.75	1.47		8.22	9.90
0535	3/4" x 9-1/2", poplar		180	.044		4.12	1.55		5.67	7.10
0540	Red oak		180	.044		6.55	1.55		8.10	9.80
0545	Maple		180	.044		9	1.55		10.55	12.50
0550	Cherry		180	.044		9.80	1.55		11.35	13.40
0555	Modern profile, 9/16" x 2-1/4", poplar		250	.032		.78	1.12		1.90	2.72
0560	Red oak		250	.032		.81	1.12		1.93	2.75
0565	11/16" x 2-1/2", finger jointed & primed		250	.032		.84	1.12		1.96	2.78
0570	Pine		250	.032		1.34	1.12		2.46	3.33
0575	3/4" x 2-1/2", poplar		250	.032		.91	1.12		2.03	2.86
0580	Red oak		250	.032		1.23	1.12		2.35	3.21
0585	Maple		250	.032		1.86	1.12		2.98	3.90
0590	Cherry		250	.032		2.56	1.12		3.68	4.67
0595	Mullion, 5/16" x 2", pine		270	.030		.90	1.04		1.94	2.71
0600	9/16" x 2-1/2", finger jointed and primed		250	.032		.98	1.12		2.10	2.94
0605	Pine		250	.032		1.34	1.12		2.46	3.33
0610	Red oak		250	.032		3.28	1.12		4.40	5.45
0615	1-1/16" x 3-3/4", red oak		220	.036		6.95	1.27		8.22	9.70
0620	Ogee, 7/16" x 2-1/2", poplar		250	.032		.74	1.12		1.86	2.67
0625	Red oak		250	.032		.94	1.12		2.06	2.89
0630	9/16" x 2-1/4", finger jointed and primed		250	.032		.55	1.12		1.67	2.46
0635	Poplar		250	.032		.59	1.12		1.71	2.51
0640	Red oak		250	.032		.81	1.12		1.93	2.75
0645	11/16" x 2-1/2", finger jointed and primed		250	.032		.72	1.12		1.84	2.65
0700	Pine		250	.032		1.42	1.12		2.54	3.42
0701	Red oak		250	.032		3.25	1.12		4.37	5.45
0730	11/16" x 3-1/2", finger jointed and primed		220	.036		1.37	1.27		2.64	3.62
0750	Pine		220	.036		1.79	1.27		3.06	4.08
0755	3/4" x 2-1/2", finger jointed and primed		250	.032		.77	1.12		1.89	2.71
0760	Poplar		250	.032		.91	1.12		2.03	2.86
0765	Red oak		250	.032		1.23	1.12		2.35	3.21
0770	Maple		250	.032		1.86	1.12		2.98	3.90
0775	Cherry		250	.032		2.34	1.12		3.46	4.43
0780	3/4" x 3-1/2", finger jointed and primed		220	.036		.98	1.27		2.25	3.19
0785	Poplar		220	.036		1.15	1.27		2.42	3.37
0790	Red oak		220	.036		1.70	1.27		2.97	3.98
0795	Maple		220	.036		2.63	1.27		3.90	5
0800	Cherry		220	.036		3.18	1.27		4.45	5.60
4700	Square profile, 1" x 1", teak		215	.037		2.24	1.30		3.54	4.62

393

06 22 Millwork

06 22 13 – Standard Pattern Wood Trim

06 22 13.30 Moldings, Casings	Crew	Daily Output	Labor-Hours	Unit	Material	2018 Bare Costs Labor	Equipment	Total	Total Incl O&P
4800	Rectangular profile, 1" x 3", teak	1 Carp	200	.040	L.F.	6.70	1.40		8.10	9.65

06 22 13.35 Moldings, Ceilings		Crew	Daily Output	Labor-Hours	Unit	Material	2018 Bare Costs Labor	Equipment	Total	Total Incl O&P
0010	**MOLDINGS, CEILINGS**									
0600	Bed, 9/16" x 1-3/4", pine	1 Carp	270	.030	L.F.	1.20	1.04		2.24	3.04
0650	9/16" x 2", pine		270	.030		1.19	1.04		2.23	3.03
0710	9/16" x 1-3/4", oak		270	.030		2.35	1.04		3.39	4.30
1200	Cornice, 9/16" x 1-3/4", pine		270	.030		1	1.04		2.04	2.82
1300	9/16" x 2-1/4", pine		265	.030		1.31	1.06		2.37	3.19
1350	Cove, 1/2" x 2-1/4", poplar		265	.030		1.07	1.06		2.13	2.92
1360	Red oak		265	.030		1.37	1.06		2.43	3.25
1370	Hard maple		265	.030		1.64	1.06		2.70	3.55
1380	Cherry		265	.030		2.27	1.06		3.33	4.24
2400	9/16" x 1-3/4", pine		270	.030		1	1.04		2.04	2.82
2401	Oak		270	.030		.91	1.04		1.95	2.72
2500	11/16" x 2-3/4", pine		265	.030		1.87	1.06		2.93	3.80
2510	Crown, 5/8" x 5/8", poplar		300	.027		.46	.93		1.39	2.05
2520	Red oak		300	.027		.54	.93		1.47	2.14
2530	Hard maple		300	.027		.73	.93		1.66	2.35
2540	Cherry		300	.027		.77	.93		1.70	2.40
2600	9/16" x 3-5/8", pine		250	.032		2.10	1.12		3.22	4.17
2700	11/16" x 4-1/4", pine		250	.032		2.97	1.12		4.09	5.10
2705	Oak		250	.032		6.30	1.12		7.42	8.80
2710	3/4" x 1-3/4", poplar		270	.030		.74	1.04		1.78	2.53
2720	Red oak		270	.030		1.09	1.04		2.13	2.92
2730	Hard maple		270	.030		1.44	1.04		2.48	3.30
2740	Cherry		270	.030		1.77	1.04		2.81	3.67
2750	3/4" x 2", poplar		270	.030		.97	1.04		2.01	2.79
2760	Red oak		270	.030		1.30	1.04		2.34	3.15
2770	Hard maple		270	.030		1.74	1.04		2.78	3.63
2780	Cherry		270	.030		1.94	1.04		2.98	3.85
2790	3/4" x 2-3/4", poplar		265	.030		1.05	1.06		2.11	2.90
2800	Red oak		265	.030		1.62	1.06		2.68	3.53
2810	Hard maple		265	.030		2.16	1.06		3.22	4.12
2820	Cherry		265	.030		2.48	1.06		3.54	4.47
2830	3/4" x 3-1/2", poplar		250	.032		1.34	1.12		2.46	3.33
2840	Red oak		250	.032		2.02	1.12		3.14	4.08
2850	Hard maple		250	.032		2.64	1.12		3.76	4.76
2860	Cherry		250	.032		3.09	1.12		4.21	5.25
2870	FJP poplar		250	.032		.99	1.12		2.11	2.95
2880	3/4" x 5", poplar		245	.033		1.99	1.14		3.13	4.09
2890	Red oak		245	.033		2.94	1.14		4.08	5.15
2900	Hard maple		245	.033		3.86	1.14		5	6.15
2910	Cherry		245	.033		4.45	1.14		5.59	6.80
2920	FJP poplar		245	.033		1.41	1.14		2.55	3.45
2930	3/4" x 6-1/4", poplar		240	.033		2.36	1.16		3.52	4.54
2940	Red oak		240	.033		3.57	1.16		4.73	5.85
2950	Hard maple		240	.033		4.66	1.16		5.82	7.10
2960	Cherry		240	.033		5.55	1.16		6.71	8.05
2970	7/8" x 8-3/4", poplar		220	.036		3.89	1.27		5.16	6.40
2980	Red oak		220	.036		6.05	1.27		7.32	8.75
2990	Hard maple		220	.036		8.40	1.27		9.67	11.30
3000	Cherry		220	.036		9.80	1.27		11.07	12.85

06 22 Millwork

06 22 13 – Standard Pattern Wood Trim

06 22 13.35 Moldings, Ceilings

		Crew	Daily Output	Labor-Hours	Unit	Material	2018 Bare Costs Labor	Equipment	Total	Total Incl O&P
3010	1" x 7-1/4", poplar	1 Carp	220	.036	L.F.	3.89	1.27		5.16	6.40
3020	Red oak		220	.036		6.05	1.27		7.32	8.75
3030	Hard maple		220	.036		8.40	1.27		9.67	11.30
3040	Cherry		220	.036		9.80	1.27		11.07	12.85
3050	1-1/16" x 4-1/4", poplar		250	.032		2.36	1.12		3.48	4.45
3060	Red oak		250	.032		2.78	1.12		3.90	4.91
3070	Hard maple		250	.032		3.67	1.12		4.79	5.90
3080	Cherry		250	.032		5.25	1.12		6.37	7.60
3090	Dentil crown, 3/4" x 5", poplar		250	.032		1.99	1.12		3.11	4.05
3100	Red oak		250	.032		2.94	1.12		4.06	5.10
3110	Hard maple		250	.032		3.86	1.12		4.98	6.10
3120	Cherry		250	.032		4.45	1.12		5.57	6.75
3130	Dentil piece for above, 1/2" x 1/2", poplar		300	.027		2.82	.93		3.75	4.65
3140	Red oak		300	.027		3.24	.93		4.17	5.10
3150	Hard maple		300	.027		3.68	.93		4.61	5.60
3160	Cherry		300	.027		4.01	.93		4.94	5.95

06 22 13.40 Moldings, Exterior

		Crew	Daily Output	Labor-Hours	Unit	Material	2018 Bare Costs Labor	Equipment	Total	Total Incl O&P
0010	**MOLDINGS, EXTERIOR**									
0100	Band board, cedar, rough sawn, 1" x 2"	1 Carp	300	.027	L.F.	.69	.93		1.62	2.31
0110	1" x 3"		300	.027		1.02	.93		1.95	2.68
0120	1" x 4"		250	.032		1.36	1.12		2.48	3.36
0130	1" x 6"		250	.032		2.05	1.12		3.17	4.11
0140	1" x 8"		225	.036		2.73	1.24		3.97	5.05
0150	1" x 10"		225	.036		3.39	1.24		4.63	5.80
0160	1" x 12"		200	.040		4.08	1.40		5.48	6.80
0240	STK, 1" x 2"		300	.027		.45	.93		1.38	2.04
0250	1" x 3"		300	.027		.49	.93		1.42	2.09
0260	1" x 4"		250	.032		.81	1.12		1.93	2.75
0270	1" x 6"		250	.032		1.32	1.12		2.44	3.31
0280	1" x 8"		225	.036		2.01	1.24		3.25	4.28
0290	1" x 10"		225	.036		2.77	1.24		4.01	5.10
0300	1" x 12"		200	.040		4.29	1.40		5.69	7.05
0310	Pine, #2, 1" x 2"		300	.027		.28	.93		1.21	1.86
0320	1" x 3"		300	.027		.45	.93		1.38	2.04
0330	1" x 4"		250	.032		.54	1.12		1.66	2.46
0340	1" x 6"		250	.032		.82	1.12		1.94	2.76
0350	1" x 8"		225	.036		1.33	1.24		2.57	3.53
0360	1" x 10"		225	.036		1.74	1.24		2.98	3.98
0370	1" x 12"		200	.040		2.13	1.40		3.53	4.66
0380	D & better, 1" x 2"		300	.027		.50	.93		1.43	2.10
0390	1" x 3"		300	.027		.64	.93		1.57	2.25
0400	1" x 4"		250	.032		.82	1.12		1.94	2.76
0410	1" x 6"		250	.032		1.07	1.12		2.19	3.03
0420	1" x 8"		225	.036		1.62	1.24		2.86	3.85
0430	1" x 10"		225	.036		2.24	1.24		3.48	4.53
0440	1" x 12"		200	.040		2.69	1.40		4.09	5.25
0450	Redwood, clear all heart, 1" x 2"		300	.027		.64	.93		1.57	2.25
0460	1" x 3"		300	.027		.95	.93		1.88	2.59
0470	1" x 4"		250	.032		1.20	1.12		2.32	3.18
0480	1" x 6"		252	.032		1.79	1.11		2.90	3.81
0490	1" x 8"		225	.036		2.37	1.24		3.61	4.67
0500	1" x 10"		225	.036		3.98	1.24		5.22	6.45

For customer support on your Residential Costs with RSMeans data, call 800.448.8182.

395

06 22 13.40 Moldings, Exterior		Crew	Daily Output	Labor-Hours	Unit	Material	2018 Bare Costs Labor	Equipment	Total	Total Incl O&P
0510	1" x 12"	1 Carp	200	.040	L.F.	4.77	1.40		6.17	7.55
0530	Corner board, cedar, rough sawn, 1" x 2"		225	.036		.69	1.24		1.93	2.83
0540	1" x 3"		225	.036		1.02	1.24		2.26	3.20
0550	1" x 4"		200	.040		1.36	1.40		2.76	3.82
0560	1" x 6"		200	.040		2.05	1.40		3.45	4.57
0570	1" x 8"		200	.040		2.73	1.40		4.13	5.30
0580	1" x 10"		175	.046		3.39	1.60		4.99	6.40
0590	1" x 12"		175	.046		4.08	1.60		5.68	7.15
0670	STK, 1" x 2"		225	.036		.45	1.24		1.69	2.56
0680	1" x 3"		225	.036		.49	1.24		1.73	2.61
0690	1" x 4"		200	.040		.79	1.40		2.19	3.19
0700	1" x 6"		200	.040		1.32	1.40		2.72	3.77
0710	1" x 8"		200	.040		2.01	1.40		3.41	4.53
0720	1" x 10"		175	.046		2.77	1.60		4.37	5.70
0730	1" x 12"		175	.046		4.29	1.60		5.89	7.35
0740	Pine, #2, 1" x 2"		225	.036		.28	1.24		1.52	2.38
0750	1" x 3"		225	.036		.45	1.24		1.69	2.56
0760	1" x 4"		200	.040		.54	1.40		1.94	2.92
0770	1" x 6"		200	.040		.82	1.40		2.22	3.22
0780	1" x 8"		200	.040		1.33	1.40		2.73	3.78
0790	1" x 10"		175	.046		1.74	1.60		3.34	4.57
0800	1" x 12"		175	.046		2.13	1.60		3.73	5
0810	D & better, 1" x 2"		225	.036		.50	1.24		1.74	2.62
0820	1" x 3"		225	.036		.64	1.24		1.88	2.77
0830	1" x 4"		200	.040		.82	1.40		2.22	3.22
0840	1" x 6"		200	.040		1.07	1.40		2.47	3.49
0850	1" x 8"		200	.040		1.62	1.40		3.02	4.10
0860	1" x 10"		175	.046		2.24	1.60		3.84	5.10
0870	1" x 12"		175	.046		2.69	1.60		4.29	5.60
0880	Redwood, clear all heart, 1" x 2"		225	.036		.64	1.24		1.88	2.77
0890	1" x 3"		225	.036		.95	1.24		2.19	3.11
0900	1" x 4"		200	.040		1.20	1.40		2.60	3.64
0910	1" x 6"		200	.040		1.79	1.40		3.19	4.29
0920	1" x 8"		200	.040		2.37	1.40		3.77	4.92
0930	1" x 10"		175	.046		3.98	1.60		5.58	7.05
0940	1" x 12"		175	.046		4.77	1.60		6.37	7.90
0950	Cornice board, cedar, rough sawn, 1" x 2"		330	.024		.69	.85		1.54	2.17
0960	1" x 3"		290	.028		1.02	.96		1.98	2.73
0970	1" x 4"		250	.032		1.36	1.12		2.48	3.36
0980	1" x 6"		250	.032		2.05	1.12		3.17	4.11
0990	1" x 8"		200	.040		2.73	1.40		4.13	5.30
1000	1" x 10"		180	.044		3.39	1.55		4.94	6.30
1010	1" x 12"		180	.044		4.08	1.55		5.63	7.05
1020	STK, 1" x 2"		330	.024		.45	.85		1.30	1.90
1030	1" x 3"		290	.028		.49	.96		1.45	2.14
1040	1" x 4"		250	.032		.81	1.12		1.93	2.75
1050	1" x 6"		250	.032		1.32	1.12		2.44	3.31
1060	1" x 8"		200	.040		2.01	1.40		3.41	4.53
1070	1" x 10"		180	.044		2.77	1.55		4.32	5.60
1080	1" x 12"		180	.044		4.29	1.55		5.84	7.30
1500	Pine, #2, 1" x 2"		330	.024		.28	.85		1.13	1.72
1510	1" x 3"		290	.028		.30	.96		1.26	1.93
1600	1" x 4"		250	.032		.54	1.12		1.66	2.46

06 22 13.40 Moldings, Exterior		Crew	Daily Output	Labor-Hours	Unit	Material	2018 Bare Costs Labor	Equipment	Total	Total Incl O&P
1700	1" x 6"	1 Carp	250	.032	L.F.	.82	1.12		1.94	2.76
1800	1" x 8"		200	.040		1.33	1.40		2.73	3.78
1900	1" x 10"		180	.044		1.74	1.55		3.29	4.49
2000	1" x 12"		180	.044		2.13	1.55		3.68	4.92
2020	D & better, 1" x 2"		330	.024		.50	.85		1.35	1.96
2030	1" x 3"		290	.028		.64	.96		1.60	2.30
2040	1" x 4"		250	.032		.82	1.12		1.94	2.76
2050	1" x 6"		250	.032		1.07	1.12		2.19	3.03
2060	1" x 8"		200	.040		1.62	1.40		3.02	4.10
2070	1" x 10"		180	.044		2.24	1.55		3.79	5.05
2080	1" x 12"		180	.044		2.69	1.55		4.24	5.55
2090	Redwood, clear all heart, 1" x 2"		330	.024		.64	.85		1.49	2.11
2100	1" x 3"		290	.028		.95	.96		1.91	2.64
2110	1" x 4"		250	.032		1.20	1.12		2.32	3.18
2120	1" x 6"		250	.032		1.79	1.12		2.91	3.83
2130	1" x 8"		200	.040		2.37	1.40		3.77	4.92
2140	1" x 10"		180	.044		3.98	1.55		5.53	6.95
2150	1" x 12"		180	.044		4.77	1.55		6.32	7.85
2160	3 piece, 1" x 2", 1" x 4", 1" x 6", rough sawn cedar		80	.100		4.11	3.50		7.61	10.30
2180	STK cedar		80	.100		2.57	3.50		6.07	8.65
2200	#2 pine		80	.100		1.64	3.50		5.14	7.60
2210	D & better pine		80	.100		2.38	3.50		5.88	8.40
2220	Clear all heart redwood		80	.100		3.62	3.50		7.12	9.80
2230	1" x 8", 1" x 10", 1" x 12", rough sawn cedar		65	.123		10.15	4.30		14.45	18.35
2240	STK cedar		65	.123		9	4.30		13.30	17.05
2300	#2 pine		65	.123		5.15	4.30		9.45	12.85
2320	D & better pine		65	.123		6.50	4.30		10.80	14.30
2330	Clear all heart redwood		65	.123		11.05	4.30		15.35	19.35
2340	Door/window casing, cedar, rough sawn, 1" x 2"		275	.029		.69	1.02		1.71	2.45
2350	1" x 3"		275	.029		1.02	1.02		2.04	2.82
2360	1" x 4"		250	.032		1.36	1.12		2.48	3.36
2370	1" x 6"		250	.032		2.05	1.12		3.17	4.11
2380	1" x 8"		230	.035		2.73	1.22		3.95	5
2390	1" x 10"		230	.035		3.39	1.22		4.61	5.75
2395	1" x 12"		210	.038		4.08	1.33		5.41	6.70
2410	STK, 1" x 2"		275	.029		.45	1.02		1.47	2.18
2420	1" x 3"		275	.029		.49	1.02		1.51	2.23
2430	1" x 4"		250	.032		.81	1.12		1.93	2.75
2440	1" x 6"		250	.032		1.32	1.12		2.44	3.31
2450	1" x 8"		230	.035		2.01	1.22		3.23	4.23
2460	1" x 10"		230	.035		2.77	1.22		3.99	5.05
2470	1" x 12"		210	.038		4.29	1.33		5.62	6.90
2550	Pine, #2, 1" x 2"		275	.029		.28	1.02		1.30	2
2560	1" x 3"		275	.029		.45	1.02		1.47	2.18
2570	1" x 4"		250	.032		.54	1.12		1.66	2.46
2580	1" x 6"		250	.032		.82	1.12		1.94	2.76
2590	1" x 8"		230	.035		1.33	1.22		2.55	3.48
2600	1" x 10"		230	.035		1.74	1.22		2.96	3.93
2610	1" x 12"		210	.038		2.13	1.33		3.46	4.55
2620	Pine, D & better, 1" x 2"		275	.029		.50	1.02		1.52	2.24
2630	1" x 3"		275	.029		.64	1.02		1.66	2.39
2640	1" x 4"		250	.032		.82	1.12		1.94	2.76
2650	1" x 6"		250	.032		1.07	1.12		2.19	3.03

06 22 13.40 Moldings, Exterior		Crew	Daily Output	Labor-Hours	Unit	Material	2018 Bare Costs Labor	Equipment	Total	Total Incl O&P
2660	1" x 8"	1 Carp	230	.035	L.F.	1.62	1.22		2.84	3.80
2670	1" x 10"		230	.035		2.24	1.22		3.46	4.48
2680	1" x 12"		210	.038		2.69	1.33		4.02	5.15
2690	Redwood, clear all heart, 1" x 2"		275	.029		.64	1.02		1.66	2.39
2695	1" x 3"		275	.029		.95	1.02		1.97	2.73
2710	1" x 4"		250	.032		1.20	1.12		2.32	3.18
2715	1" x 6"		250	.032		1.79	1.12		2.91	3.83
2730	1" x 8"		230	.035		2.37	1.22		3.59	4.62
2740	1" x 10"		230	.035		3.98	1.22		5.20	6.40
2750	1" x 12"		210	.038		4.77	1.33		6.10	7.45
3500	Bellyband, pine, 11/16" x 4-1/4"		250	.032		2.82	1.12		3.94	4.96
3610	Brickmold, pine, 1-1/4" x 2"		200	.040		2.10	1.40		3.50	4.63
3620	FJP, 1-1/4" x 2"		200	.040		.97	1.40		2.37	3.39
5100	Fascia, cedar, rough sawn, 1" x 2"		275	.029		.69	1.02		1.71	2.45
5110	1" x 3"		275	.029		1.02	1.02		2.04	2.82
5120	1" x 4"		250	.032		1.36	1.12		2.48	3.36
5200	1" x 6"		250	.032		2.05	1.12		3.17	4.11
5300	1" x 8"		230	.035		2.73	1.22		3.95	5
5310	1" x 10"		230	.035		3.39	1.22		4.61	5.75
5320	1" x 12"		210	.038		4.08	1.33		5.41	6.70
5400	2" x 4"		220	.036		1.07	1.27		2.34	3.28
5500	2" x 6"		220	.036		1.59	1.27		2.86	3.86
5600	2" x 8"		200	.040		2.12	1.40		3.52	4.66
5700	2" x 10"		180	.044		2.64	1.55		4.19	5.50
5800	2" x 12"		170	.047		6.30	1.64		7.94	9.65
6120	STK, 1" x 2"		275	.029		.45	1.02		1.47	2.18
6130	1" x 3"		275	.029		.49	1.02		1.51	2.23
6140	1" x 4"		250	.032		.81	1.12		1.93	2.75
6150	1" x 6"		250	.032		1.32	1.12		2.44	3.31
6160	1" x 8"		230	.035		2.01	1.22		3.23	4.23
6170	1" x 10"		230	.035		2.77	1.22		3.99	5.05
6180	1" x 12"		210	.038		4.29	1.33		5.62	6.90
6185	2" x 2"		260	.031		.74	1.08		1.82	2.60
6190	Pine, #2, 1" x 2"		275	.029		.28	1.02		1.30	2
6200	1" x 3"		275	.029		.45	1.02		1.47	2.18
6210	1" x 4"		250	.032		.54	1.12		1.66	2.46
6220	1" x 6"		250	.032		.82	1.12		1.94	2.76
6230	1" x 8"		230	.035		1.33	1.22		2.55	3.48
6240	1" x 10"		230	.035		1.74	1.22		2.96	3.93
6250	1" x 12"		210	.038		2.13	1.33		3.46	4.55
6260	D & better, 1" x 2"		275	.029		.50	1.02		1.52	2.24
6270	1" x 3"		275	.029		.64	1.02		1.66	2.39
6280	1" x 4"		250	.032		.82	1.12		1.94	2.76
6290	1" x 6"		250	.032		1.07	1.12		2.19	3.03
6300	1" x 8"		230	.035		1.62	1.22		2.84	3.80
6310	1" x 10"		230	.035		2.24	1.22		3.46	4.48
6312	1" x 12"		210	.038		2.69	1.33		4.02	5.15
6330	Southern yellow, 1-1/4" x 5"		240	.033		2.88	1.16		4.04	5.10
6340	1-1/4" x 6"		240	.033		2.57	1.16		3.73	4.76
6350	1-1/4" x 8"		215	.037		3.65	1.30		4.95	6.15
6360	1-1/4" x 12"		190	.042		5.35	1.47		6.82	8.35
6370	Redwood, clear all heart, 1" x 2"		275	.029		.64	1.02		1.66	2.39
6380	1" x 3"		275	.029		1.20	1.02		2.22	3.01

06 22 13 – Standard Pattern Wood Trim

06 22 13.40 Moldings, Exterior	Crew	Daily Output	Labor-Hours	Unit	Material	2018 Bare Costs Labor	Equipment	Total	Total Incl O&P
6390	1" x 4"	1 Carp	250	.032	L.F.	1.20	1.12		2.32	3.18
6400	1" x 6"		250	.032		1.79	1.12		2.91	3.83
6410	1" x 8"		230	.035		2.37	1.22		3.59	4.62
6420	1" x 10"		230	.035		3.98	1.22		5.20	6.40
6430	1" x 12"		210	.038		4.77	1.33		6.10	7.45
6440	1-1/4" x 5"		240	.033		1.87	1.16		3.03	4
6450	1-1/4" x 6"		240	.033		2.23	1.16		3.39	4.40
6460	1-1/4" x 8"		215	.037		3.54	1.30		4.84	6.05
6470	1-1/4" x 12"		190	.042		7.15	1.47		8.62	10.30
6580	Frieze, cedar, rough sawn, 1" x 2"		275	.029		.69	1.02		1.71	2.45
6590	1" x 3"		275	.029		1.02	1.02		2.04	2.82
6600	1" x 4"		250	.032		1.36	1.12		2.48	3.36
6610	1" x 6"		250	.032		2.05	1.12		3.17	4.11
6620	1" x 8"		250	.032		2.73	1.12		3.85	4.86
6630	1" x 10"		225	.036		3.39	1.24		4.63	5.80
6640	1" x 12"		200	.040		4.04	1.40		5.44	6.75
6650	STK, 1" x 2"		275	.029		.45	1.02		1.47	2.18
6660	1" x 3"		275	.029		.49	1.02		1.51	2.23
6670	1" x 4"		250	.032		.81	1.12		1.93	2.75
6680	1" x 6"		250	.032		1.32	1.12		2.44	3.31
6690	1" x 8"		250	.032		2.01	1.12		3.13	4.07
6700	1" x 10"		225	.036		2.77	1.24		4.01	5.10
6710	1" x 12"		200	.040		4.29	1.40		5.69	7.05
6790	Pine, #2, 1" x 2"		275	.029		.28	1.02		1.30	2
6800	1" x 3"		275	.029		.45	1.02		1.47	2.18
6810	1" x 4"		250	.032		.54	1.12		1.66	2.46
6820	1" x 6"		250	.032		.82	1.12		1.94	2.76
6830	1" x 8"		250	.032		1.33	1.12		2.45	3.32
6840	1" x 10"		225	.036		1.74	1.24		2.98	3.98
6850	1" x 12"		200	.040		2.13	1.40		3.53	4.66
6860	D & better, 1" x 2"		275	.029		.50	1.02		1.52	2.24
6870	1" x 3"		275	.029		.64	1.02		1.66	2.39
6880	1" x 4"		250	.032		.82	1.12		1.94	2.76
6890	1" x 6"		250	.032		1.07	1.12		2.19	3.03
6900	1" x 8"		250	.032		1.62	1.12		2.74	3.64
6910	1" x 10"		225	.036		2.24	1.24		3.48	4.53
6920	1" x 12"		200	.040		2.69	1.40		4.09	5.25
6930	Redwood, clear all heart, 1" x 2"		275	.029		.64	1.02		1.66	2.39
6940	1" x 3"		275	.029		.95	1.02		1.97	2.73
6950	1" x 4"		250	.032		1.20	1.12		2.32	3.18
6960	1" x 6"		250	.032		1.79	1.12		2.91	3.83
6970	1" x 8"		250	.032		2.37	1.12		3.49	4.46
6980	1" x 10"		225	.036		3.98	1.24		5.22	6.45
6990	1" x 12"		200	.040		4.77	1.40		6.17	7.55
7000	Grounds, 1" x 1", cedar, rough sawn		300	.027		.35	.93		1.28	1.94
7010	STK		300	.027		.28	.93		1.21	1.85
7020	Pine, #2		300	.027		.18	.93		1.11	1.75
7030	D & better		300	.027		.31	.93		1.24	1.89
7050	Redwood		300	.027		.39	.93		1.32	1.98
7060	Rake/verge board, cedar, rough sawn, 1" x 2"		225	.036		.69	1.24		1.93	2.83
7070	1" x 3"		225	.036		1.02	1.24		2.26	3.20
7080	1" x 4"		200	.040		1.36	1.40		2.76	3.82
7090	1" x 6"		200	.040		2.05	1.40		3.45	4.57

For customer support on your Residential Costs with RSMeans data, call 800.448.8182.

399

06 22 13.40 Moldings, Exterior		Crew	Daily Output	Labor-Hours	Unit	Material	2018 Bare Costs Labor	Equipment	Total	Total Incl O&P
7100	1" x 8"	1 Carp	190	.042	L.F.	2.73	1.47		4.20	5.45
7110	1" x 10"		190	.042		3.39	1.47		4.86	6.20
7120	1" x 12"		180	.044		4.08	1.55		5.63	7.05
7130	STK, 1" x 2"		225	.036		.45	1.24		1.69	2.56
7140	1" x 3"		225	.036		.49	1.24		1.73	2.61
7150	1" x 4"		200	.040		.81	1.40		2.21	3.21
7160	1" x 6"		200	.040		1.32	1.40		2.72	3.77
7170	1" x 8"		190	.042		2.01	1.47		3.48	4.66
7180	1" x 10"		190	.042		2.77	1.47		4.24	5.50
7190	1" x 12"		180	.044		4.29	1.55		5.84	7.30
7200	Pine, #2, 1" x 2"		225	.036		.28	1.24		1.52	2.38
7210	1" x 3"		225	.036		.45	1.24		1.69	2.56
7220	1" x 4"		200	.040		.54	1.40		1.94	2.92
7230	1" x 6"		200	.040		.82	1.40		2.22	3.22
7240	1" x 8"		190	.042		1.33	1.47		2.80	3.91
7250	1" x 10"		190	.042		1.74	1.47		3.21	4.36
7260	1" x 12"		180	.044		2.13	1.55		3.68	4.92
7340	D & better, 1" x 2"		225	.036		.50	1.24		1.74	2.62
7350	1" x 3"		225	.036		.64	1.24		1.88	2.77
7360	1" x 4"		200	.040		.82	1.40		2.22	3.22
7370	1" x 6"		200	.040		1.07	1.40		2.47	3.49
7380	1" x 8"		190	.042		1.62	1.47		3.09	4.23
7390	1" x 10"		190	.042		2.24	1.47		3.71	4.91
7400	1" x 12"		180	.044		2.69	1.55		4.24	5.55
7410	Redwood, clear all heart, 1" x 2"		225	.036		.64	1.24		1.88	2.77
7420	1" x 3"		225	.036		.95	1.24		2.19	3.11
7430	1" x 4"		200	.040		1.20	1.40		2.60	3.64
7440	1" x 6"		200	.040		1.79	1.40		3.19	4.29
7450	1" x 8"		190	.042		2.37	1.47		3.84	5.05
7460	1" x 10"		190	.042		3.98	1.47		5.45	6.80
7470	1" x 12"		180	.044		4.77	1.55		6.32	7.85
7480	2" x 4"		200	.040		2.30	1.40		3.70	4.85
7490	2" x 6"		182	.044		3.44	1.54		4.98	6.35
7500	2" x 8"		165	.048		4.57	1.69		6.26	7.80
7630	Soffit, cedar, rough sawn, 1" x 2"	2 Carp	440	.036		.69	1.27		1.96	2.87
7640	1" x 3"		440	.036		1.02	1.27		2.29	3.24
7650	1" x 4"		420	.038		1.36	1.33		2.69	3.71
7660	1" x 6"		420	.038		2.05	1.33		3.38	4.46
7670	1" x 8"		420	.038		2.73	1.33		4.06	5.20
7680	1" x 10"		400	.040		3.39	1.40		4.79	6.05
7690	1" x 12"		400	.040		4.08	1.40		5.48	6.80
7700	STK, 1" x 2"		440	.036		.45	1.27		1.72	2.60
7710	1" x 3"		440	.036		.49	1.27		1.76	2.65
7720	1" x 4"		420	.038		.81	1.33		2.14	3.10
7730	1" x 6"		420	.038		1.32	1.33		2.65	3.66
7740	1" x 8"		420	.038		2.01	1.33		3.34	4.42
7750	1" x 10"		400	.040		2.77	1.40		4.17	5.35
7760	1" x 12"		400	.040		4.29	1.40		5.69	7.05
7770	Pine, #2, 1" x 2"		440	.036		.28	1.27		1.55	2.42
7780	1" x 3"		440	.036		.45	1.27		1.72	2.60
7790	1" x 4"		420	.038		.54	1.33		1.87	2.81
7800	1" x 6"		420	.038		.82	1.33		2.15	3.11
7810	1" x 8"		420	.038		1.33	1.33		2.66	3.67

06 22 13 – Standard Pattern Wood Trim

06 22 13.40 Moldings, Exterior

		Crew	Daily Output	Labor-Hours	Unit	Material	2018 Bare Costs Labor	Equipment	Total	Total Incl O&P
7820	1" x 10"	2 Carp	400	.040	L.F.	1.74	1.40		3.14	4.23
7830	1" x 12"		400	.040		2.13	1.40		3.53	4.66
7840	D & better, 1" x 2"		440	.036		.50	1.27		1.77	2.66
7850	1" x 3"		440	.036		.64	1.27		1.91	2.81
7860	1" x 4"		420	.038		.82	1.33		2.15	3.11
7870	1" x 6"		420	.038		1.07	1.33		2.40	3.38
7880	1" x 8"		420	.038		1.62	1.33		2.95	3.99
7890	1" x 10"		400	.040		2.24	1.40		3.64	4.78
7900	1" x 12"		400	.040		2.69	1.40		4.09	5.25
7910	Redwood, clear all heart, 1" x 2"		440	.036		.64	1.27		1.91	2.81
7920	1" x 3"		440	.036		.95	1.27		2.22	3.15
7930	1" x 4"		420	.038		1.20	1.33		2.53	3.53
7940	1" x 6"		420	.038		1.79	1.33		3.12	4.18
7950	1" x 8"		420	.038		2.37	1.33		3.70	4.81
7960	1" x 10"		400	.040		3.98	1.40		5.38	6.70
7970	1" x 12"		400	.040		4.77	1.40		6.17	7.55
8050	Trim, crown molding, pine, 11/16" x 4-1/4"	1 Carp	250	.032		3.93	1.12		5.05	6.20
8060	Back band, 11/16" x 1-1/16"		250	.032		.90	1.12		2.02	2.85
8070	Insect screen frame stock, 1-1/16" x 1-3/4"		395	.020		2.23	.71		2.94	3.63
8080	Dentils, 2-1/2" x 2-1/2" x 4", 6" OC		30	.267		1.22	9.30		10.52	16.85
8100	Fluted, 5-1/2"		165	.048		5.05	1.69		6.74	8.35
8110	Stucco bead, 1-3/8" x 1-5/8"		250	.032		2.33	1.12		3.45	4.42

06 22 13.45 Moldings, Trim

		Crew	Daily Output	Labor-Hours	Unit	Material	2018 Bare Costs Labor	Equipment	Total	Total Incl O&P
0010	**MOLDINGS, TRIM**									
0200	Astragal, stock pine, 11/16" x 1-3/4"	1 Carp	255	.031	L.F.	1.24	1.10		2.34	3.18
0250	1-5/16" x 2-3/16"		240	.033		2.62	1.16		3.78	4.82
0800	Chair rail, stock pine, 5/8" x 2-1/2"		270	.030		1.54	1.04		2.58	3.41
0900	5/8" x 3-1/2"		240	.033		2.33	1.16		3.49	4.50
1000	Closet pole, stock pine, 1-1/8" diameter		200	.040		1.16	1.40		2.56	3.60
1100	Fir, 1-5/8" diameter		200	.040		2.05	1.40		3.45	4.58
1150	Corner, inside, 5/16" x 1"		225	.036		.34	1.24		1.58	2.44
1160	Outside, 1-1/16" x 1-1/16"		240	.033		1.36	1.16		2.52	3.44
1161	1-5/16" x 1-5/16"		240	.033		1.62	1.16		2.78	3.72
3300	Half round, stock pine, 1/4" x 1/2"		270	.030		.25	1.04		1.29	2
3350	1/2" x 1"		255	.031		.64	1.10		1.74	2.52
3400	Handrail, fir, single piece, stock, hardware not included									
3450	1-1/2" x 1-3/4"	1 Carp	80	.100	L.F.	2.10	3.50		5.60	8.10
3470	Pine, 1-1/2" x 1-3/4"		80	.100		1.97	3.50		5.47	7.95
3500	1-1/2" x 2-1/2"		76	.105		2.57	3.68		6.25	8.95
3600	Lattice, stock pine, 1/4" x 1-1/8"		270	.030		.36	1.04		1.40	2.12
3700	1/4" x 1-3/4"		250	.032		.61	1.12		1.73	2.53
3800	Miscellaneous, custom, pine, 1" x 1"		270	.030		.44	1.04		1.48	2.20
3850	1" x 2"		265	.030		.88	1.06		1.94	2.72
3900	1" x 3"		240	.033		1.32	1.16		2.48	3.39
4100	Birch or oak, nominal 1" x 1"		240	.033		.39	1.16		1.55	2.36
4200	Nominal 1" x 3"		215	.037		1.16	1.30		2.46	3.43
4400	Walnut, nominal 1" x 1"		215	.037		.60	1.30		1.90	2.82
4500	Nominal 1" x 3"		200	.040		1.79	1.40		3.19	4.29
4700	Teak, nominal 1" x 1"		215	.037		2.79	1.30		4.09	5.20
4800	Nominal 1" x 3"		200	.040		8.35	1.40		9.75	11.50
4900	Quarter round, stock pine, 1/4" x 1/4"		275	.029		.25	1.02		1.27	1.97
4950	3/4" x 3/4"		255	.031		.52	1.10		1.62	2.39

06 22 Millwork

06 22 13 – Standard Pattern Wood Trim

06 22 13.45 Moldings, Trim

		Crew	Daily Output	Labor-Hours	Unit	Material	2018 Bare Costs Labor	Equipment	Total	Total Incl O&P
5600	Wainscot moldings, 1-1/8" x 9/16", 2' high, minimum	1 Carp	76	.105	S.F.	12.35	3.68		16.03	19.65
5700	Maximum	▼	65	.123	"	16.25	4.30		20.55	25

06 22 13.50 Moldings, Window and Door

		Crew	Daily Output	Labor-Hours	Unit	Material	2018 Bare Costs Labor	Equipment	Total	Total Incl O&P
0010	**MOLDINGS, WINDOW AND DOOR**									
2800	Door moldings, stock, decorative, 1-1/8" wide, plain	1 Carp	17	.471	Set	48	16.45		64.45	80
2900	Detailed		17	.471	"	110	16.45		126.45	149
2960	Clear pine door jamb, no stops, 11/16" x 4-9/16"		240	.033	L.F.	4.93	1.16		6.09	7.35
3150	Door trim set, 1 head and 2 sides, pine, 2-1/2" wide		12	.667	Opng.	24	23.50		47.50	65
3170	3-1/2" wide		11	.727	"	30.50	25.50		56	76
3250	Glass beads, stock pine, 3/8" x 1/2"		275	.029	L.F.	.31	1.02		1.33	2.03
3270	3/8" x 7/8"		270	.030		.40	1.04		1.44	2.16
4850	Parting bead, stock pine, 3/8" x 3/4"		275	.029		.47	1.02		1.49	2.21
4870	1/2" x 3/4"		255	.031		.41	1.10		1.51	2.27
5000	Stool caps, stock pine, 11/16" x 3-1/2"		200	.040		2.17	1.40		3.57	4.71
5100	1-1/16" x 3-1/4"		150	.053	▼	3.40	1.86		5.26	6.85
5300	Threshold, oak, 3' long, inside, 5/8" x 3-5/8"		32	.250	Ea.	11.20	8.75		19.95	27
5400	Outside, 1-1/2" x 7-5/8"	▼	16	.500	"	47	17.50		64.50	80.50
5900	Window trim sets, including casings, header, stops,									
5910	stool and apron, 2-1/2" wide, FJP	1 Carp	13	.615	Opng.	32	21.50		53.50	71.50
5950	Pine		10	.800		38	28		66	88
6000	Oak	▼	6	1.333	▼	77	46.50		123.50	163

06 22 13.60 Moldings, Soffits

		Crew	Daily Output	Labor-Hours	Unit	Material	2018 Bare Costs Labor	Equipment	Total	Total Incl O&P
0010	**MOLDINGS, SOFFITS**									
0200	Soffits, pine, 1" x 4"	2 Carp	420	.038	L.F.	.51	1.33		1.84	2.77
0210	1" x 6"		420	.038		.78	1.33		2.11	3.07
0220	1" x 8"		420	.038		1.29	1.33		2.62	3.63
0230	1" x 10"		400	.040		1.68	1.40		3.08	4.17
0240	1" x 12"		400	.040		2.07	1.40		3.47	4.60
0250	STK cedar, 1" x 4"		420	.038		.77	1.33		2.10	3.06
0260	1" x 6"		420	.038		1.28	1.33		2.61	3.62
0270	1" x 8"		420	.038		1.97	1.33		3.30	4.38
0280	1" x 10"		400	.040		2.71	1.40		4.11	5.30
0290	1" x 12"		400	.040	▼	4.23	1.40		5.63	6.95
1000	Exterior AC plywood, 1/4" thick		400	.040	S.F.	1	1.40		2.40	3.42
1050	3/8" thick		400	.040		1.04	1.40		2.44	3.46
1100	1/2" thick	▼	400	.040		1.23	1.40		2.63	3.67
1150	Polyvinyl chloride, white, solid	1 Carp	230	.035		2.17	1.22		3.39	4.41
1160	Perforated	"	230	.035	▼	2.17	1.22		3.39	4.41
1170	Accessories, "J" channel 5/8"	2 Carp	700	.023	L.F.	.50	.80		1.30	1.88

06 25 Prefinished Paneling

06 25 13 – Prefinished Hardboard Paneling

06 25 13.10 Paneling, Hardboard

			Crew	Daily Output	Labor-Hours	Unit	Material	2018 Bare Costs Labor	Equipment	Total	Total Incl O&P
0010	**PANELING, HARDBOARD**										
0050	Not incl. furring or trim, hardboard, tempered, 1/8" thick	G	2 Carp	500	.032	S.F.	.45	1.12		1.57	2.36
0100	1/4" thick	G		500	.032		.64	1.12		1.76	2.56
0300	Tempered pegboard, 1/8" thick	G		500	.032		.44	1.12		1.56	2.34
0400	1/4" thick	G		500	.032		.70	1.12		1.82	2.63
0600	Untempered hardboard, natural finish, 1/8" thick	G		500	.032		.44	1.12		1.56	2.34
0700	1/4" thick	G	▼	500	.032	▼	.53	1.12		1.65	2.44

06 25 Prefinished Paneling

06 25 13 – Prefinished Hardboard Paneling

06 25 13.10 Paneling, Hardboard		Crew	Daily Output	Labor-Hours	Unit	Material	2018 Bare Costs Labor	Equipment	Total	Total Incl O&P
0900	Untempered pegboard, 1/8" thick ⒢	2 Carp	500	.032	S.F.	.48	1.12		1.60	2.39
1000	1/4" thick ⒢		500	.032		.48	1.12		1.60	2.39
1200	Plastic faced hardboard, 1/8" thick ⒢		500	.032		.62	1.12		1.74	2.54
1300	1/4" thick ⒢		500	.032		.84	1.12		1.96	2.78
1500	Plastic faced pegboard, 1/8" thick ⒢		500	.032		.62	1.12		1.74	2.54
1600	1/4" thick ⒢		500	.032		.80	1.12		1.92	2.74
1800	Wood grained, plain or grooved, 1/8" thick ⒢		500	.032		.65	1.12		1.77	2.58
1900	1/4" thick ⒢		425	.038		1.42	1.32		2.74	3.75
2100	Moldings, wood grained MDF		500	.032	L.F.	.41	1.12		1.53	2.31
2200	Pine		425	.038	"	1.41	1.32		2.73	3.74

06 25 16 – Prefinished Plywood Paneling

06 25 16.10 Paneling, Plywood

		Crew	Daily Output	Labor-Hours	Unit	Material	2018 Bare Costs Labor	Equipment	Total	Total Incl O&P
0010	**PANELING, PLYWOOD**									
2400	Plywood, prefinished, 1/4" thick, 4' x 8' sheets									
2410	with vertical grooves. Birch faced, economy	2 Carp	500	.032	S.F.	1.47	1.12		2.59	3.48
2420	Average		420	.038		1.20	1.33		2.53	3.53
2430	Custom		350	.046		1.10	1.60		2.70	3.87
2600	Mahogany, African		400	.040		2.45	1.40		3.85	5
2700	Philippine (Lauan)		500	.032		.65	1.12		1.77	2.58
2900	Oak		500	.032		1.34	1.12		2.46	3.33
3000	Cherry		400	.040		2	1.40		3.40	4.52
3200	Rosewood		320	.050		3.10	1.75		4.85	6.30
3400	Teak		400	.040		3.12	1.40		4.52	5.75
3600	Chestnut		375	.043		5.30	1.49		6.79	8.30
3800	Pecan		400	.040		2.50	1.40		3.90	5.05
3900	Walnut, average		500	.032		2.47	1.12		3.59	4.58
3950	Custom		400	.040		2.46	1.40		3.86	5.05
4000	Plywood, prefinished, 3/4" thick, stock grades, economy		320	.050		1.44	1.75		3.19	4.49
4100	Average		224	.071		4.93	2.50		7.43	9.55
4300	Architectural grade, custom		224	.071		5.40	2.50		7.90	10.05
4400	Luxury		160	.100		5.30	3.50		8.80	11.60
4600	Plywood, "A" face, birch, VC, 1/2" thick, natural		450	.036		1.38	1.24		2.62	3.59
4700	Select		450	.036		1.95	1.24		3.19	4.22
4900	Veneer core, 3/4" thick, natural		320	.050		2.25	1.75		4	5.40
5000	Select		320	.050		2.53	1.75		4.28	5.70
5200	Lumber core, 3/4" thick, natural		320	.050		3.14	1.75		4.89	6.35
5500	Plywood, knotty pine, 1/4" thick, A2 grade		450	.036		1.53	1.24		2.77	3.75
5600	A3 grade		450	.036		2.09	1.24		3.33	4.37
5800	3/4" thick, veneer core, A2 grade		320	.050		2.31	1.75		4.06	5.45
5900	A3 grade		320	.050		2.39	1.75		4.14	5.55
6100	Aromatic cedar, 1/4" thick, plywood		400	.040		2.25	1.40		3.65	4.80
6200	1/4" thick, particle board		400	.040		1.15	1.40		2.55	3.59

06 25 26 – Panel System

06 25 26.10 Panel Systems

		Crew	Daily Output	Labor-Hours	Unit	Material	2018 Bare Costs Labor	Equipment	Total	Total Incl O&P
0010	**PANEL SYSTEMS**									
0100	Raised panel, eng. wood core w/wood veneer, std., paint grade	2 Carp	300	.053	S.F.	12.15	1.86		14.01	16.45
0110	Oak veneer		300	.053		24	1.86		25.86	29.50
0120	Maple veneer		300	.053		30	1.86		31.86	36
0130	Cherry veneer		300	.053		37.50	1.86		39.36	44.50
0300	Class I fire rated, paint grade		300	.053		13.50	1.86		15.36	17.95
0310	Oak veneer		300	.053		30	1.86		31.86	36
0320	Maple veneer		300	.053		41	1.86		42.86	48

06 25 Prefinished Paneling

06 25 26 – Panel System

06 25 26.10 Panel Systems	Crew	Daily Output	Labor-Hours	Unit	Material	2018 Bare Costs Labor	Equipment	Total	Total Incl O&P
0330	Cherry veneer	2 Carp	300	.053	S.F.	49	1.86		50.86	57
0510	Beadboard, 5/8" MDF, standard, primed		300	.053		9.35	1.86		11.21	13.40
0520	Oak veneer, unfinished		300	.053		14.70	1.86		16.56	19.25
0530	Maple veneer, unfinished		300	.053		15.80	1.86		17.66	20.50
0610	Rustic paneling, 5/8" MDF, standard, maple veneer, unfinished	↓	300	.053	↓	18.50	1.86		20.36	23.50

06 26 Board Paneling

06 26 13 – Profile Board Paneling

06 26 13.10 Paneling, Boards

		Crew	Daily Output	Labor-Hours	Unit	Material	2018 Bare Costs Labor	Equipment	Total	Total Incl O&P
0010	**PANELING, BOARDS**									
6400	Wood board paneling, 3/4" thick, knotty pine	2 Carp	300	.053	S.F.	2.05	1.86		3.91	5.35
6500	Rough sawn cedar		300	.053		3.30	1.86		5.16	6.75
6700	Redwood, clear, 1" x 4" boards		300	.053		4.96	1.86		6.82	8.55
6900	Aromatic cedar, closet lining, boards	↓	275	.058		2.42	2.03		4.45	6.05
8950	On ceiling, wood board, install	1 Carp	225	.036	↓	3.30	1.24		4.54	5.70

06 43 Wood Stairs and Railings

06 43 13 – Wood Stairs

06 43 13.20 Prefabricated Wood Stairs

		Crew	Daily Output	Labor-Hours	Unit	Material	2018 Bare Costs Labor	Equipment	Total	Total Incl O&P
0010	**PREFABRICATED WOOD STAIRS**									
0100	Box stairs, prefabricated, 3'-0" wide									
0110	Oak treads, up to 14 risers	2 Carp	39	.410	Riser	99.50	14.35		113.85	133
0600	With pine treads for carpet, up to 14 risers	"	39	.410	"	64	14.35		78.35	94.50
1100	For 4' wide stairs, add				Flight	25%				
1550	Stairs, prefabricated stair handrail with balusters	1 Carp	30	.267	L.F.	82	9.30		91.30	106
1700	Basement stairs, prefabricated, pine treads									
1710	Pine risers, 3' wide, up to 14 risers	2 Carp	52	.308	Riser	64	10.75		74.75	88.50
4000	Residential, wood, oak treads, prefabricated		1.50	10.667	Flight	1,300	375		1,675	2,050
4200	Built in place	↓	.44	36.364	"	2,300	1,275		3,575	4,650
4400	Spiral, oak, 4'-6" diameter, unfinished, prefabricated,									
4500	incl. railing, 9' high	2 Carp	1.50	10.667	Flight	2,750	375		3,125	3,650

06 43 13.40 Wood Stair Parts

		Crew	Daily Output	Labor-Hours	Unit	Material	2018 Bare Costs Labor	Equipment	Total	Total Incl O&P
0010	**WOOD STAIR PARTS**									
0020	Pin top balusters, 1-1/4", oak, 34"	1 Carp	96	.083	Ea.	5.20	2.91		8.11	10.55
0030	38"		96	.083		5.85	2.91		8.76	11.30
0040	42"		96	.083		6.35	2.91		9.26	11.85
0050	Poplar, 34"		96	.083		3.37	2.91		6.28	8.55
0060	38"		96	.083		3.99	2.91		6.90	9.25
0070	42"		96	.083		8.80	2.91		11.71	14.50
0080	Maple, 34"		96	.083		4.90	2.91		7.81	10.25
0090	38"		96	.083		5.60	2.91		8.51	11.05
0100	42"		96	.083		6.50	2.91		9.41	12
0130	Primed, 34"		96	.083		3.70	2.91		6.61	8.90
0140	38"		96	.083		4.25	2.91		7.16	9.50
0150	42"		96	.083		4.89	2.91		7.80	10.25
0180	Box top balusters, 1-1/4", oak, 34"		60	.133		8.75	4.66		13.41	17.40
0190	38"		60	.133		11.85	4.66		16.51	21
0200	42"		60	.133		13	4.66		17.66	22
0210	Poplar, 34"	↓	60	.133	↓	7.50	4.66		12.16	16

06 43 13 – Wood Stairs

06 43 13.40 Wood Stair Parts	Crew	Daily Output	Labor-Hours	Unit	Material	2018 Bare Costs Labor	Equipment	Total	Total Incl O&P
0220	38"	1 Carp	60	.133	Ea.	8.10	4.66		12.76	16.65
0230	42"		60	.133		8.95	4.66		13.61	17.60
0240	Maple, 34"		60	.133		9.80	4.66		14.46	18.55
0250	38"		60	.133		10.75	4.66		15.41	19.60
0260	42"		60	.133		11.85	4.66		16.51	21
0290	Primed, 34"		60	.133		8.25	4.66		12.91	16.85
0300	38"		60	.133		9.50	4.66		14.16	18.20
0310	42"		60	.133		9.85	4.66		14.51	18.55
0340	Square balusters, cut from lineal stock, pine, 1-1/16" x 1-1/16"		180	.044	L.F.	1.31	1.55		2.86	4.02
0350	1-5/16" x 1-5/16"		180	.044		1.98	1.55		3.53	4.76
0360	1-5/8" x 1-5/8"		180	.044		2.68	1.55		4.23	5.55
0370	Turned newel, oak, 3-1/2" square, 48" high		8	1	Ea.	105	35		140	174
0380	62" high		8	1		101	35		136	169
0390	Poplar, 3-1/2" square, 48" high		8	1		46.50	35		81.50	110
0400	62" high		8	1		58	35		93	122
0410	Maple, 3-1/2" square, 48" high		8	1		65	35		100	130
0420	62" high		8	1		80	35		115	146
0430	Square newel, oak, 3-1/2" square, 48" high		8	1		55	35		90	119
0440	58" high		8	1		68.50	35		103.50	133
0450	Poplar, 3-1/2" square, 48" high		8	1		35.50	35		70.50	97
0460	58" high		8	1		43.50	35		78.50	106
0470	Maple, 3" square, 48" high		8	1		54	35		89	118
0480	58" high		8	1		68	35		103	133
0490	Railings, oak, economy		96	.083	L.F.	9.45	2.91		12.36	15.25
0500	Average		96	.083		13.50	2.91		16.41	19.70
0510	Custom		96	.083		17.95	2.91		20.86	24.50
0520	Maple, economy		96	.083		11.90	2.91		14.81	17.90
0530	Average		96	.083		13.50	2.91		16.41	19.70
0540	Custom		96	.083		20.50	2.91		23.41	27.50
0550	Oak, for bending rail, economy		48	.167		26	5.85		31.85	38
0560	Average		48	.167		29.50	5.85		35.35	42
0570	Custom		48	.167		33	5.85		38.85	46
0580	Maple, for bending rail, economy		48	.167		29	5.85		34.85	41
0590	Average		48	.167		31.50	5.85		37.35	44
0600	Custom		48	.167		35	5.85		40.85	47.50
0610	Risers, oak, 3/4" x 8", 36" long		80	.100	Ea.	13	3.50		16.50	20
0620	42" long		70	.114		15.15	3.99		19.14	23.50
0630	48" long		63	.127		17.30	4.44		21.74	26.50
0640	54" long		56	.143		19.50	4.99		24.49	30
0650	60" long		50	.160		21.50	5.60		27.10	33.50
0660	72" long		42	.190		26	6.65		32.65	39.50
0670	Poplar, 3/4" x 8", 36" long		80	.100		12.75	3.50		16.25	19.85
0680	42" long		71	.113		14.90	3.94		18.84	23
0690	48" long		63	.127		17	4.44		21.44	26
0700	54" long		56	.143		19.15	4.99		24.14	29.50
0710	60" long		50	.160		21.50	5.60		27.10	33
0720	72" long		42	.190		25.50	6.65		32.15	39
0730	Pine, 1" x 8", 36" long		80	.100		3.88	3.50		7.38	10.05
0740	42" long		70	.114		4.53	3.99		8.52	11.65
0750	48" long		63	.127		5.20	4.44		9.64	13.10
0760	54" long		56	.143		5.80	4.99		10.79	14.70
0770	60" long		50	.160		6.45	5.60		12.05	16.40
0780	72" long		42	.190		7.75	6.65		14.40	19.60

For customer support on your Residential Costs with RSMeans data, call 800.448.8182.

405

06 43 Wood Stairs and Railings

06 43 13 – Wood Stairs

06 43 13.40 Wood Stair Parts	Crew	Daily Output	Labor-Hours	Unit	Material	2018 Bare Costs Labor	Equipment	Total	Total Incl O&P
0790	Treads, oak, no returns, 1-1/32" x 11-1/2" x 36" long	1 Carp	32	.250	Ea.	29.50	8.75		38.25	47
0800	42" long		32	.250		34.50	8.75		43.25	52.50
0810	48" long		32	.250		39.50	8.75		48.25	57.50
0820	54" long		32	.250		44	8.75		52.75	63
0830	60" long		32	.250		49	8.75		57.75	68.50
0840	72" long		32	.250		59	8.75		67.75	79.50
0850	Mitred return one end, 1-1/32" x 11-1/2" x 36" long		24	.333		39	11.65		50.65	62.50
0860	42" long		24	.333		45.50	11.65		57.15	69.50
0870	48" long		24	.333		52	11.65		63.65	76.50
0880	54" long		24	.333		58.50	11.65		70.15	84
0890	60" long		24	.333		65	11.65		76.65	91
0900	72" long		24	.333		78	11.65		89.65	105
0910	Mitred return two ends, 1-1/32" x 11-1/2" x 36" long		12	.667		49	23.50		72.50	92.50
0920	42" long		12	.667		57	23.50		80.50	102
0930	48" long		12	.667		65.50	23.50		89	111
0940	54" long		12	.667		73.50	23.50		97	120
0950	60" long		12	.667		81.50	23.50		105	129
0960	72" long		12	.667		98	23.50		121.50	147
0970	Starting step, oak, 48", bullnose		8	1		172	35		207	247
0980	Double end bullnose		8	1		254	35		289	335
1030	Skirt board, pine, 1" x 10"		55	.145	L.F.	1.68	5.10		6.78	10.30
1040	1" x 12"		52	.154	"	2.07	5.40		7.47	11.25
1050	Oak landing tread, 1-1/16" thick		54	.148	S.F.	7.95	5.20		13.15	17.35
1060	Oak cove molding		96	.083	L.F.	1	2.91		3.91	5.95
1070	Oak stringer molding		96	.083	"	4	2.91		6.91	9.25
1090	Rail bolt, 5/16" x 3-1/2"		48	.167	Ea.	2.75	5.85		8.60	12.75
1100	5/16" x 4-1/2"		48	.167		2.75	5.85		8.60	12.75
1120	Newel post anchor		16	.500		13	17.50		30.50	43.50
1130	Tapered plug, 1/2"		240	.033		1	1.16		2.16	3.04
1140	1"		240	.033		.99	1.16		2.15	3.03

06 43 16 – Wood Railings

06 43 16.10 Wood Handrails and Railings

		Crew	Daily Output	Labor-Hours	Unit	Material	2018 Bare Costs Labor	Equipment	Total	Total Incl O&P
0010	**WOOD HANDRAILS AND RAILINGS**									
0020	Custom design, architectural grade, hardwood, plain	1 Carp	38	.211	L.F.	12.65	7.35		20	26
0100	Shaped		30	.267		53	9.30		62.30	74
0300	Stock interior railing with spindles 4" OC, 4' long		40	.200		46.50	7		53.50	63
0400	8' long		48	.167		46.50	5.85		52.35	61

06 44 Ornamental Woodwork

06 44 19 – Wood Grilles

06 44 19.10 Grilles

		Crew	Daily Output	Labor-Hours	Unit	Material	2018 Bare Costs Labor	Equipment	Total	Total Incl O&P
0010	**GRILLES** and panels, hardwood, sanded									
0020	2' x 4' to 4' x 8', custom designs, unfinished, economy	1 Carp	38	.211	S.F.	58	7.35		65.35	76.50
0050	Average		30	.267		68.50	9.30		77.80	91
0100	Custom		19	.421		72	14.70		86.70	104

06 44 33 – Wood Mantels

06 44 33.10 Fireplace Mantels

		Crew	Daily Output	Labor-Hours	Unit	Material	2018 Bare Costs Labor	Equipment	Total	Total Incl O&P
0010	**FIREPLACE MANTELS**									
0015	6" molding, 6' x 3'-6" opening, plain, paint grade	1 Carp	5	1.600	Opng.	450	56		506	590
0100	Ornate, oak		5	1.600		600	56		656	750

06 44 Ornamental Woodwork

06 44 33 – Wood Mantels

06 44 33.10 Fireplace Mantels	Crew	Daily Output	Labor-Hours	Unit	Material	2018 Bare Costs Labor	Equipment	Total	Total Incl O&P
0300 Prefabricated pine, colonial type, stock, deluxe	1 Carp	2	4	Opng.	1,825	140		1,965	2,225
0400 Economy	↓	3	2.667	↓	790	93		883	1,025

06 44 33.20 Fireplace Mantel Beam

0010 **FIREPLACE MANTEL BEAM**									
0020 Rough texture wood, 4" x 8"	1 Carp	36	.222	L.F.	8.05	7.75		15.80	22
0100 4" x 10"		35	.229	"	10.80	8		18.80	25
0300 Laminated hardwood, 2-1/4" x 10-1/2" wide, 6' long		5	1.600	Ea.	105	56		161	208
0400 8' long		5	1.600	"	146	56		202	253
0600 Brackets for above, rough sawn		12	.667	Pr.	10.10	23.50		33.60	49.50
0700 Laminated	↓	12	.667	"	19.50	23.50		43	60

06 44 39 – Wood Posts and Columns

06 44 39.10 Decorative Beams

0010 **DECORATIVE BEAMS**									
0020 Rough sawn cedar, non-load bearing, 4" x 4"	2 Carp	180	.089	L.F.	1.62	3.11		4.73	6.95
0100 4" x 6"		170	.094		1.77	3.29		5.06	7.40
0200 4" x 8"		160	.100		2.32	3.50		5.82	8.35
0300 4" x 10"		150	.107		4.04	3.73		7.77	10.65
0400 4" x 12"		140	.114		4.66	3.99		8.65	11.80
0500 8" x 8"		130	.123		4.77	4.30		9.07	12.40
0600 Plastic beam, "hewn finish", 6" x 2"		240	.067		3.44	2.33		5.77	7.65
0601 6" x 4"	↓	220	.073	↓	3.79	2.54		6.33	8.40

06 44 39.20 Columns

0010 **COLUMNS**									
0050 Aluminum, round colonial, 6" diameter	2 Carp	80	.200	V.L.F.	17.75	7		24.75	31
0100 8" diameter		62.25	.257		18.55	9		27.55	35.50
0200 10" diameter		55	.291		21	10.15		31.15	40.50
0250 Fir, stock units, hollow round, 6" diameter		80	.200		26	7		33	40
0300 8" diameter		80	.200		33.50	7		40.50	48
0350 10" diameter		70	.229		41	8		49	58.50
0360 12" diameter		65	.246		59	8.60		67.60	79.50
0400 Solid turned, to 8' high, 3-1/2" diameter		80	.200		10.15	7		17.15	23
0500 4-1/2" diameter		75	.213		12.10	7.45		19.55	26
0600 5-1/2" diameter		70	.229		16.35	8		24.35	31.50
0800 Square columns, built-up, 5" x 5"		65	.246		33.50	8.60		42.10	51.50
0900 Solid, 3-1/2" x 3-1/2"		130	.123		10.15	4.30		14.45	18.30
1600 Hemlock, tapered, T&G, 12" diam., 10' high		100	.160		47.50	5.60		53.10	61.50
1700 16' high		65	.246		79	8.60		87.60	101
1900 14" diameter, 10' high		100	.160		96	5.60		101.60	114
2000 18' high		65	.246		106	8.60		114.60	130
2200 18" diameter, 12' high		65	.246		173	8.60		181.60	205
2300 20' high		50	.320		124	11.20		135.20	155
2500 20" diameter, 14' high		40	.400		187	14		201	229
2600 20' high	↓	35	.457		177	16		193	222
2800 For flat pilasters, deduct				↓	33%				
3000 For splitting into halves, add				Ea.	110			110	121
4000 Rough sawn cedar posts, 4" x 4"	2 Carp	250	.064	V.L.F.	4.24	2.24		6.48	8.40
4100 4" x 6"		235	.068		8.30	2.38		10.68	13.05
4200 6" x 6"		220	.073		14	2.54		16.54	19.65
4300 8" x 8"	↓	200	.080	↓	18.70	2.80		21.50	25

06 48 Wood Frames

06 48 13 – Exterior Wood Door Frames

06 48 13.10 Exterior Wood Door Frames and Accessories	Crew	Daily Output	Labor-Hours	Unit	Material	2018 Bare Costs Labor	Equipment	Total	Total Incl O&P
0010 **EXTERIOR WOOD DOOR FRAMES AND ACCESSORIES**									
0400 Exterior frame, incl. ext. trim, pine, 5/4 x 4-9/16" deep	2 Carp	375	.043	L.F.	6.75	1.49		8.24	9.95
0420 5-3/16" deep		375	.043		8.30	1.49		9.79	11.60
0440 6-9/16" deep		375	.043		9.75	1.49		11.24	13.25
0600 Oak, 5/4 x 4-9/16" deep		350	.046		12.55	1.60		14.15	16.45
0620 5-3/16" deep		350	.046		13.80	1.60		15.40	17.80
0640 6-9/16" deep		350	.046		18.55	1.60		20.15	23
1000 Sills, 8/4 x 8" deep, oak, no horns		100	.160		7.15	5.60		12.75	17.15
1020 2" horns		100	.160		21.50	5.60		27.10	33
1040 3" horns		100	.160		21.50	5.60		27.10	33
1100 8/4 x 10" deep, oak, no horns		90	.178		6.55	6.20		12.75	17.55
1120 2" horns		90	.178		27	6.20		33.20	40
1140 3" horns		90	.178	↓	27	6.20		33.20	40
2000 Wood frame & trim, ext., colonial, 3' opng., fluted pilasters, flat head		22	.727	Ea.	505	25.50		530.50	600
2010 Dentil head		21	.762		600	26.50		626.50	705
2020 Ram's head		20	.800		650	28		678	760
2100 5'-4" opening, in-swing, fluted pilasters, flat head		17	.941		445	33		478	545
2120 Ram's head		15	1.067		1,325	37.50		1,362.50	1,500
2140 Out-swing, fluted pilasters, flat head		17	.941		480	33		513	580
2160 Ram's head		15	1.067		1,500	37.50		1,537.50	1,700
2400 6'-0" opening, in-swing, fluted pilasters, flat head		16	1		480	35		515	585
2420 Ram's head		10	1.600		1,500	56		1,556	1,750
2460 Out-swing, fluted pilasters, flat head		16	1		480	35		515	590
2480 Ram's head		10	1.600	↓	1,500	56		1,556	1,750
2600 For two sidelights, flat head, add		30	.533	Opng.	277	18.65		295.65	335
2620 Ram's head, add		20	.800	"	920	28		948	1,050
2700 Custom birch frame, 3'-0" opening		16	1	Ea.	232	35		267	315
2750 6'-0" opng.		16	1		365	35		400	460
2900 Exterior, modern, plain trim, 3' opng., in-swing, FJP		26	.615		48	21.50		69.50	89
2920 Fir		24	.667		56.50	23.50		80	101
2940 Oak	↓	22	.727	↓	66	25.50		91.50	115

06 48 16 – Interior Wood Door Frames

06 48 16.10 Interior Wood Door Jamb and Frames

	Crew	Daily Output	Labor-Hours	Unit	Material	Labor	Equipment	Total	Total Incl O&P
0010 **INTERIOR WOOD DOOR JAMB AND FRAMES**									
3000 Interior frame, pine, 11/16" x 3-5/8" deep	2 Carp	375	.043	L.F.	4.15	1.49		5.64	7.05
3020 4-9/16" deep		375	.043		4.80	1.49		6.29	7.80
3040 5-3/16" deep		375	.043		5.55	1.49		7.04	8.60
3200 Oak, 11/16" x 3-5/8" deep		350	.046		2.76	1.60		4.36	5.70
3220 4-9/16" deep		350	.046		11	1.60		12.60	14.75
3240 5-3/16" deep		350	.046		17.85	1.60		19.45	22.50
3400 Walnut, 11/16" x 3-5/8" deep		350	.046		9.10	1.60		10.70	12.65
3420 4-9/16" deep		350	.046		10.80	1.60		12.40	14.50
3440 5-3/16" deep		350	.046	↓	9.65	1.60		11.25	13.30
3600 Pocket door frame		16	1	Ea.	86	35		121	153
3800 Threshold, oak, 5/8" x 3-5/8" deep		200	.080	L.F.	3.57	2.80		6.37	8.60
3820 4-5/8" deep		190	.084		3.95	2.94		6.89	9.25
3840 5-5/8" deep	↓	180	.089	↓	6.65	3.11		9.76	12.45

06 49 19.10 Shutters, Exterior	Crew	Daily Output	Labor-Hours	Unit	Material	2018 Bare Costs Labor	Equipment	Total	Total Incl O&P
0010 **SHUTTERS, EXTERIOR**									
0012 Aluminum, louvered, 1'-4" wide, 3'-0" long	1 Carp	10	.800	Pr.	192	28		220	258
0200 4'-0" long		10	.800		232	28		260	300
0300 5'-4" long		10	.800		274	28		302	345
0400 6'-8" long		9	.889		340	31		371	425
1000 Pine, louvered, primed, each 1'-2" wide, 3'-3" long		10	.800		235	28		263	305
1100 4'-7" long		10	.800		270	28		298	345
1250 Each 1'-4" wide, 3'-0" long		10	.800		263	28		291	335
1350 5'-3" long		10	.800		320	28		348	400
1500 Each 1'-6" wide, 3'-3" long		10	.800		255	28		283	325
1600 4'-7" long		10	.800		315	28		343	395
1620 Cedar, louvered, 1'-2" wide, 5'-7" long		10	.800		320	28		348	395
1630 Each 1'-4" wide, 2'-2" long		10	.800		171	28		199	235
1640 3'-0" long		10	.800		219	28		247	288
1650 3'-3" long		10	.800		229	28		257	299
1660 3'-11" long		10	.800		265	28		293	340
1670 4'-3" long		10	.800		294	28		322	370
1680 5'-3" long		10	.800		335	28		363	415
1690 5'-11" long		10	.800		350	28		378	430
1700 Door blinds, 6'-9" long, each 1'-3" wide		9	.889		385	31		416	475
1710 1'-6" wide		9	.889		415	31		446	505
1720 Cedar, solid raised panel, each 1'-4" wide, 3'-3" long		10	.800		325	28		353	400
1730 3'-11" long		10	.800		325	28		353	405
1740 4'-3" long		10	.800		335	28		363	415
1750 4'-7" long		10	.800		370	28		398	455
1760 4'-11" long		10	.800		400	28		428	485
1770 5'-11" long		10	.800		515	28		543	615
1800 Door blinds, 6'-9" long, each 1'-3" wide		9	.889		490	31		521	590
1900 1'-6" wide		9	.889		580	31		611	690
2500 Polystyrene, solid raised panel, each 1'-4" wide, 3'-3" long		10	.800		82	28		110	137
2600 3'-11" long		10	.800		90	28		118	146
2700 4'-7" long		10	.800		102	28		130	159
2800 5'-3" long		10	.800		116	28		144	174
2900 6'-8" long		9	.889		144	31		175	210
4500 Polystyrene, louvered, each 1'-2" wide, 3'-3" long		10	.800		38.50	28		66.50	88.50
4600 4'-7" long		10	.800		45	28		73	96
4750 5'-3" long		10	.800		53.50	28		81.50	105
4850 6'-8" long		9	.889		69	31		100	128
6000 Vinyl, louvered, each 1'-2" x 4'-7" long		10	.800		58	28		86	110
6200 Each 1'-4" x 6'-8" long		9	.889		81.50	31		112.50	141
8000 PVC exterior rolling shutters									
8100 including crank control	1 Carp	8	1	Ea.	705	35		740	835
8500 Insulative - 6' x 6'-8" stock unit	"	8	1	"	1,000	35		1,035	1,150

For customer support on your Residential Costs with RSMeans data, call 800.448.8182.

409

06 51 Structural Plastic Shapes and Plates

06 51 13 – Plastic Lumber

06 51 13.10 Recycled Plastic Lumber

06 51 13.10 Recycled Plastic Lumber		Crew	Daily Output	Labor-Hours	Unit	Material	2018 Bare Costs Labor	2018 Bare Costs Equipment	Total	Total Incl O&P
0010	**RECYCLED PLASTIC LUMBER**									
4000	Sheeting, recycled plastic, black or white, 4' x 8' x 1/8"	G 2 Carp	1100	.015	S.F.	1.09	.51		1.60	2.05
4010	4' x 8' x 3/16"	G	1100	.015		1.51	.51		2.02	2.51
4020	4' x 8' x 1/4"	G	950	.017		1.81	.59		2.40	2.97
4030	4' x 8' x 3/8"	G	950	.017		3.02	.59		3.61	4.30
4040	4' x 8' x 1/2"	G	900	.018		4.03	.62		4.65	5.45
4050	4' x 8' x 5/8"	G	900	.018		6	.62		6.62	7.65
4060	4' x 8' x 3/4"	G	850	.019		7.50	.66		8.16	9.35
4070	Add for colors	G			Ea.	5%				
8500	100% recycled plastic, var colors, NLB, 2" x 2"	G			L.F.	1.90			1.90	2.09
8510	2" x 4"	G				3.90			3.90	4.29
8520	2" x 6"	G				6.10			6.10	6.75
8530	2" x 8"	G				8.60			8.60	9.50
8540	2" x 10"	G				12.25			12.25	13.50
8550	5/4" x 4"	G				4.58			4.58	5.05
8560	5/4" x 6"	G				5.70			5.70	6.25
8570	1" x 6"	G				2.83			2.83	3.11
8580	1/2" x 8"	G				3.32			3.32	3.65
8590	2" x 10" T&G	G				12.65			12.65	13.90
8600	3" x 10" T&G	G				18.95			18.95	21
8610	Add for premium colors	G				20%				

06 51 13.12 Structural Plastic Lumber

06 51 13.12 Structural Plastic Lumber		Crew	Daily Output	Labor-Hours	Unit	Material	2018 Bare Costs Labor	2018 Bare Costs Equipment	Total	Total Incl O&P
0010	**STRUCTURAL PLASTIC LUMBER**									
1320	Plastic lumber, posts or columns, 4" x 4"	2 Carp	390	.041	L.F.	11.10	1.43		12.53	14.65
1325	4" x 6"		275	.058		13.30	2.03		15.33	18.05
1330	4" x 8"		220	.073		19.55	2.54		22.09	25.50
1340	Girder, single, 4" x 4"		675	.024		11.10	.83		11.93	13.65
1345	4" x 6"		600	.027		13.30	.93		14.23	16.20
1350	4" x 8"		525	.030		19.55	1.07		20.62	23.50
1352	Double, 2" x 4"		625	.026		7.95	.89		8.84	10.25
1354	2" x 6"		600	.027		9.65	.93		10.58	12.15
1356	2" x 8"		575	.028		16.85	.97		17.82	20
1358	2" x 10"		550	.029		23.50	1.02		24.52	27.50
1360	2" x 12"		525	.030		25	1.07		26.07	29.50
1362	Triple, 2" x 4"		575	.028		11.90	.97		12.87	14.70
1364	2" x 6"		550	.029		14.45	1.02		15.47	17.60
1366	2" x 8"		525	.030		25	1.07		26.07	30
1368	2" x 10"		500	.032		35.50	1.12		36.62	41
1370	2" x 12"		475	.034		37.50	1.18		38.68	43
1372	Ledger, bolted 4' OC, 2" x 4"		400	.040		4.12	1.40		5.52	6.85
1374	2" x 6"		550	.029		4.89	1.02		5.91	7.10
1376	2" x 8"		390	.041		8.55	1.43		9.98	11.80
1378	2" x 10"		385	.042		11.90	1.45		13.35	15.50
1380	2" x 12"		380	.042		12.60	1.47		14.07	16.30
1382	Joists, 2" x 4"		1250	.013		3.97	.45		4.42	5.10
1384	2" x 6"		1250	.013		4.83	.45		5.28	6.05
1386	2" x 8"		1100	.015		8.40	.51		8.91	10.10
1388	2" x 10"		500	.032		11.90	1.12		13.02	14.95
1390	2" x 12"		875	.018		12.50	.64		13.14	14.80
1392	Railings and trim, 5/4" x 4"	1 Carp	300	.027		4.35	.93		5.28	6.35
1394	2" x 2"		300	.027		1.91	.93		2.84	3.65
1396	2" x 4"		300	.027		3.95	.93		4.88	5.90

06 51 Structural Plastic Shapes and Plates

06 51 13 – Plastic Lumber

06 51 13.12 Structural Plastic Lumber	Crew	Daily Output	Labor-Hours	Unit	Material	2018 Bare Costs Labor	Equipment	Total	Total Incl O&P
1398	2" x 6"	1 Carp	300	.027	L.F.	4.79	.93		5.72	6.80

06 63 Plastic Railings

06 63 10 – Plastic (PVC) Railings

06 63 10.10 Plastic Railings

		Crew	Daily Output	Labor-Hours	Unit	Material	Labor	Equipment	Total	Total Incl O&P
0010	**PLASTIC RAILINGS**									
0100	Horizontal PVC handrail with balusters, 3-1/2" wide, 36" high	1 Carp	96	.083	L.F.	29	2.91		31.91	36.50
0150	42" high		96	.083		28.50	2.91		31.41	36.50
0200	Angled PVC handrail with balusters, 3-1/2" wide, 36" high		72	.111		18.85	3.88		22.73	27.50
0250	42" high		72	.111		28.50	3.88		32.38	38
0300	Post sleeve for 4 x 4 post		96	.083		14.70	2.91		17.61	21
0400	Post cap for 4 x 4 post, flat profile		48	.167	Ea.	13	5.85		18.85	24
0450	Newel post style profile		48	.167		25.50	5.85		31.35	37.50
0500	Raised corbeled profile		48	.167		36.50	5.85		42.35	50
0550	Post base trim for 4 x 4 post		96	.083		20.50	2.91		23.41	27.50

06 65 Plastic Trim

06 65 10 – PVC Trim

06 65 10.10 PVC Trim, Exterior

		Crew	Daily Output	Labor-Hours	Unit	Material	Labor	Equipment	Total	Total Incl O&P
0010	**PVC TRIM, EXTERIOR**									
0100	Cornerboards, 5/4" x 6" x 6"	1 Carp	240	.033	L.F.	7.20	1.16		8.36	9.85
0110	Door/window casing, 1" x 4"		200	.040		1.39	1.40		2.79	3.85
0120	1" x 6"		200	.040		2.06	1.40		3.46	4.59
0130	1" x 8"		195	.041		2.72	1.43		4.15	5.35
0140	1" x 10"		195	.041		3.44	1.43		4.87	6.15
0150	1" x 12"		190	.042		4.22	1.47		5.69	7.10
0160	5/4" x 4"		195	.041		1.69	1.43		3.12	4.24
0170	5/4" x 6"		195	.041		2.67	1.43		4.10	5.30
0180	5/4" x 8"		190	.042		3.50	1.47		4.97	6.30
0190	5/4" x 10"		190	.042		4.50	1.47		5.97	7.40
0200	5/4" x 12"		185	.043		5.35	1.51		6.86	8.40
0210	Fascia, 1" x 4"		250	.032		1.39	1.12		2.51	3.39
0220	1" x 6"		250	.032		2.06	1.12		3.18	4.13
0230	1" x 8"		225	.036		2.72	1.24		3.96	5.05
0240	1" x 10"		225	.036		3.44	1.24		4.68	5.85
0250	1" x 12"		200	.040		4.22	1.40		5.62	6.95
0260	5/4" x 4"		240	.033		1.69	1.16		2.85	3.80
0270	5/4" x 6"		240	.033		2.67	1.16		3.83	4.88
0280	5/4" x 8"		215	.037		3.50	1.30		4.80	6
0290	5/4" x 10"		215	.037		4.50	1.30		5.80	7.10
0300	5/4" x 12"		190	.042		5.35	1.47		6.82	8.35
0310	Frieze, 1" x 4"		250	.032		1.39	1.12		2.51	3.39
0320	1" x 6"		250	.032		2.06	1.12		3.18	4.13
0330	1" x 8"		225	.036		2.72	1.24		3.96	5.05
0340	1" x 10"		225	.036		3.44	1.24		4.68	5.85
0350	1" x 12"		200	.040		4.22	1.40		5.62	6.95
0360	5/4" x 4"		240	.033		1.69	1.16		2.85	3.80
0370	5/4" x 6"		240	.033		2.67	1.16		3.83	4.88
0380	5/4" x 8"		215	.037		3.50	1.30		4.80	6

For customer support on your Residential Costs with RSMeans data, call 800.448.8182.

411

06 65 Plastic Trim

06 65 10 – PVC Trim

06 65 10.10 PVC Trim, Exterior		Crew	Daily Output	Labor-Hours	Unit	Material	2018 Bare Costs Labor	Equipment	Total	Total Incl O&P
0390	5/4" x 10"	1 Carp	215	.037	L.F.	4.50	1.30		5.80	7.10
0400	5/4" x 12"		190	.042		5.35	1.47		6.82	8.35
0410	Rake, 1" x 4"		200	.040		1.39	1.40		2.79	3.85
0420	1" x 6"		200	.040		2.06	1.40		3.46	4.59
0430	1" x 8"		190	.042		2.72	1.47		4.19	5.45
0440	1" x 10"		190	.042		3.44	1.47		4.91	6.25
0450	1" x 12"		180	.044		4.22	1.55		5.77	7.20
0460	5/4" x 4"		195	.041		1.69	1.43		3.12	4.24
0470	5/4" x 6"		195	.041		2.67	1.43		4.10	5.30
0480	5/4" x 8"		185	.043		3.50	1.51		5.01	6.35
0490	5/4" x 10"		185	.043		4.50	1.51		6.01	7.45
0500	5/4" x 12"		175	.046		5.35	1.60		6.95	8.55
0510	Rake trim, 1" x 4"		225	.036		1.39	1.24		2.63	3.60
0520	1" x 6"		225	.036		2.06	1.24		3.30	4.34
0560	5/4" x 4"		220	.036		1.69	1.27		2.96	3.97
0570	5/4" x 6"		220	.036		2.67	1.27		3.94	5.05
0610	Soffit, 1" x 4"	2 Carp	420	.038		1.39	1.33		2.72	3.74
0620	1" x 6"		420	.038		2.06	1.33		3.39	4.48
0630	1" x 8"		420	.038		2.72	1.33		4.05	5.20
0640	1" x 10"		400	.040		3.44	1.40		4.84	6.10
0650	1" x 12"		400	.040		4.22	1.40		5.62	6.95
0660	5/4" x 4"		410	.039		1.69	1.36		3.05	4.13
0670	5/4" x 6"		410	.039		2.67	1.36		4.03	5.20
0680	5/4" x 8"		410	.039		3.50	1.36		4.86	6.10
0690	5/4" x 10"		390	.041		4.50	1.43		5.93	7.35
0700	5/4" x 12"		390	.041		5.35	1.43		6.78	8.30

06 80 Composite Fabrications

06 80 10 – Composite Decking

06 80 10.10 Woodgrained Composite Decking

0010	WOODGRAINED COMPOSITE DECKING									
0100	Woodgrained composite decking, 1" x 6"	2 Carp	640	.025	L.F.	3.72	.87		4.59	5.55
0110	Grooved edge		660	.024		3.87	.85		4.72	5.65
0120	2" x 6"		640	.025		3.81	.87		4.68	5.65
0130	Encased, 1" x 6"		640	.025		3.81	.87		4.68	5.65
0140	Grooved edge		660	.024		3.96	.85		4.81	5.75
0150	2" x 6"		640	.025		5.35	.87		6.22	7.30

06 81 Composite Railings

06 81 10 – Encased Railings

06 81 10.10 Encased Composite Railings

0010	ENCASED COMPOSITE RAILINGS									
0100	Encased composite railing, 6' long, 36" high, incl. balusters	1 Carp	16	.500	Ea.	143	17.50		160.50	187
0110	42" high, incl. balusters		16	.500		207	17.50		224.50	257
0120	8' long, 36" high, incl. balusters		12	.667		165	23.50		188.50	221
0130	42" high, incl. balusters		12	.667		156	23.50		179.50	210
0140	Accessories, post sleeve, 4" x 4", 39" long		32	.250		25	8.75		33.75	42
0150	96" long		24	.333		77	11.65		88.65	104
0160	6" x 6", 39" long		32	.250		54	8.75		62.75	74

06 81 Composite Railings

06 81 10 – Encased Railings

06 81 10.10 Encased Composite Railings	Crew	Daily Output	Labor-Hours	Unit	Material	2018 Bare Costs Labor	Equipment	Total	Total Incl O&P
0170	96" long	1 Carp	24	.333	Ea.	155	11.65		166.65	189
0180	Accessories, post skirt, 4" x 4"		96	.083		3.99	2.91		6.90	9.25
0190	6" x 6"		96	.083		4.72	2.91		7.63	10.05
0200	Post cap, 4" x 4", flat		48	.167		6.65	5.85		12.50	17
0210	Pyramid		48	.167		5.70	5.85		11.55	15.95
0220	Post cap, 6" x 6", flat		48	.167		11.30	5.85		17.15	22
0230	Pyramid		48	.167		10.80	5.85		16.65	21.50

Division Notes

		CREW	DAILY OUTPUT	LABOR-HOURS	UNIT	BARE COSTS				TOTAL INCL O&P
						MAT.	LABOR	EQUIP.	TOTAL	

Estimating Tips

07 10 00 Dampproofing and Waterproofing

- Be sure of the job specifications before pricing this subdivision. The difference in cost between waterproofing and dampproofing can be great. Waterproofing will hold back standing water. Dampproofing prevents the transmission of water vapor. Also included in this section are vapor retarding membranes.

07 20 00 Thermal Protection

- Insulation and fireproofing products are measured by area, thickness, volume or R-value. Specifications may give only what the specific R-value should be in a certain situation. The estimator may need to choose the type of insulation to meet that R-value.

07 30 00 Steep Slope Roofing
07 40 00 Roofing and Siding Panels

- Many roofing and siding products are bought and sold by the square. One square is equal to an area that measures 100 square feet.

  This simple change in unit of measure could create a large error if the estimator is not observant. Accessories necessary for a complete installation must be figured into any calculations for both material and labor.

07 50 00 Membrane Roofing
07 60 00 Flashing and Sheet Metal
07 70 00 Roofing and Wall Specialties and Accessories

- The items in these subdivisions compose a roofing system. No one component completes the installation, and all must be estimated. Built-up or single-ply membrane roofing systems are made up of many products and installation trades. Wood blocking at roof perimeters or penetrations, parapet coverings, reglets, roof drains, gutters, downspouts, sheet metal flashing, skylights, smoke vents, and roof hatches all need to be considered along with the roofing material. Several different installation trades will need to work together on the roofing system. Inherent difficulties in the scheduling and coordination of various trades must be accounted for when estimating labor costs.

07 90 00 Joint Protection

- To complete the weather-tight shell, the sealants and caulkings must be estimated. Where different materials meet—at expansion joints, at flashing penetrations, and at hundreds of other locations throughout a construction project—caulking and sealants provide another line of defense against water penetration. Often, an entire system is based on the proper location and placement of caulking or sealants. The detailed drawings that are included as part of a set of architectural plans show typical locations for these materials. When caulking or sealants are shown at typical locations, this means the estimator must include them for all the locations where this detail is applicable. Be careful to keep different types of sealants separate, and remember to consider backer rods and primers if necessary.

Reference Numbers

Reference numbers are shown at the beginning of some major classifications. These numbers refer to related items in the Reference Section. The reference information may be an estimating procedure, an alternate pricing method, or technical information.

Note: Not all subdivisions listed here necessarily appear. ∎

Did you know?

RSMeans data is available through our online application with 24/7 access:

- Search for unit prices by keyword
- Leverage the most up-to-date data
- Build and export estimates

Try it free for 30 days!
www.rsmeans.com/2018freetrial

07 01 50.10 Roof Coatings		Crew	Daily Output	Labor-Hours	Unit	Material	2018 Bare Costs Labor	Equipment	Total	Total Incl O&P
0010	**ROOF COATINGS**									
0012	Asphalt, brush grade, material only				Gal.	8.85			8.85	9.75
0800	Glass fibered roof & patching cement, 5 gal.					8.40			8.40	9.25
1100	Roof patch & flashing cement, 5 gal.				↓	8.90			8.90	9.80

07 05 Common Work Results for Thermal and Moisture Protection

07 05 05 – Selective Demolition for Thermal and Moisture Protection

07 05 05.10 Selective Demo., Thermal and Moist. Protection

		Crew	Daily Output	Labor-Hours	Unit	Material	2018 Bare Costs Labor	Equipment	Total	Total Incl O&P
0010	**SELECTIVE DEMO., THERMAL AND MOISTURE PROTECTION**									
0020	Caulking/sealant, to 1" x 1" joint R024119-10	1 Clab	600	.013	L.F.		.36		.36	.59
0120	Downspouts, including hangers		350	.023	"		.61		.61	1.02
0220	Flashing, sheet metal		290	.028	S.F.		.74		.74	1.23
0420	Gutters, aluminum or wood, edge hung		240	.033	L.F.		.89		.89	1.48
0520	Built-in		100	.080	"		2.14		2.14	3.55
0620	Insulation, air/vapor barrier		3500	.002	S.F.		.06		.06	.10
0670	Batts or blankets	↓	1400	.006	C.F.		.15		.15	.25
0720	Foamed or sprayed in place	2 Clab	1000	.016	B.F.		.43		.43	.71
0770	Loose fitting	1 Clab	3000	.003	C.F.		.07		.07	.12
0870	Rigid board		3450	.002	B.F.		.06		.06	.10
1120	Roll roofing, cold adhesive		12	.667	Sq.		17.80		17.80	29.50
1170	Roof accessories, adjustable metal chimney flashing		9	.889	Ea.		23.50		23.50	39.50
1325	Plumbing vent flashing		32	.250	"		6.70		6.70	11.10
1375	Ridge vent strip, aluminum		310	.026	L.F.		.69		.69	1.15
1620	Skylight to 10 S.F.		8	1	Ea.		26.50		26.50	44.50
2120	Roof edge, aluminum soffit and fascia	↓	570	.014	L.F.		.37		.37	.62
2170	Concrete coping, up to 12" wide	2 Clab	160	.100			2.67		2.67	4.44
2220	Drip edge	1 Clab	1000	.008			.21		.21	.36
2270	Gravel stop		950	.008			.22		.22	.37
2370	Sheet metal coping, up to 12" wide	↓	240	.033	↓		.89		.89	1.48
2470	Roof insulation board, over 2" thick	B-2	7800	.005	B.F.		.14		.14	.23
2520	Up to 2" thick	"	3900	.010	S.F.		.28		.28	.46
2620	Roof ventilation, louvered gable vent	1 Clab	16	.500	Ea.		13.35		13.35	22
2670	Remove, roof hatch	G-3	15	2.133			69.50		69.50	115
2675	Rafter vents	1 Clab	960	.008	↓		.22		.22	.37
2720	Soffit vent and/or fascia vent		575	.014	L.F.		.37		.37	.62
2775	Soffit vent strip, aluminum, 3" to 4" wide		160	.050			1.34		1.34	2.22
2820	Roofing accessories, shingle moulding, to 1" x 4"	↓	1600	.005			.13		.13	.22
2870	Cant strip	B-2	2000	.020			.54		.54	.90
2920	Concrete block walkway	1 Clab	230	.035	↓		.93		.93	1.54
3070	Roofing, felt paper, #15		70	.114	Sq.		3.05		3.05	5.05
3125	#30 felt	↓	30	.267	"		7.10		7.10	11.85
3170	Asphalt shingles, 1 layer	B-2	3500	.011	S.F.		.31		.31	.52
3180	2 layers		1750	.023	"		.62		.62	1.03
3370	Modified bitumen		26	1.538	Sq.		41.50		41.50	69.50
3420	Built-up, no gravel, 3 ply		25	1.600			43.50		43.50	72
3470	4 ply		21	1.905	↓		51.50		51.50	86
3620	5 ply		1600	.025	S.F.		.68		.68	1.13
3720	5 ply, with gravel		890	.045			1.22		1.22	2.03
3725	Loose gravel removal		5000	.008			.22		.22	.36
3730	Embedded gravel removal		2000	.020			.54		.54	.90
3870	Fiberglass sheet		1200	.033	↓		.90		.90	1.50

07 05 Common Work Results for Thermal and Moisture Protection

07 05 05 – Selective Demolition for Thermal and Moisture Protection

07 05 05.10 Selective Demo., Thermal and Moist. Protection	Crew	Daily Output	Labor-Hours	Unit	Material	2018 Bare Costs Labor	Equipment	Total	Total Incl O&P
4120	Slate shingles	B-2	1900	.021	S.F.		.57		.57	.95
4170	Ridge shingles, clay or slate		2000	.020	L.F.		.54		.54	.90
4320	Single ply membrane, attached at seams		52	.769	Sq.		21		21	34.50
4370	Ballasted		75	.533			14.45		14.45	24
4420	Fully adhered	↓	39	1.026	↓		28		28	46
4550	Roof hatch, 2'-6" x 3'-0"	1 Clab	10	.800	Ea.		21.50		21.50	35.50
4670	Wood shingles	B-2	2200	.018	S.F.		.49		.49	.82
4820	Sheet metal roofing	"	2150	.019			.50		.50	.84
4970	Siding, horizontal wood clapboards	1 Clab	380	.021			.56		.56	.93
5025	Exterior insulation finish system	"	120	.067			1.78		1.78	2.96
5070	Tempered hardboard, remove and reset	1 Carp	380	.021			.74		.74	1.22
5120	Tempered hardboard sheet siding	"	375	.021	↓		.75		.75	1.24
5170	Metal, corner strips	1 Clab	850	.009	L.F.		.25		.25	.42
5225	Horizontal strips		444	.018	S.F.		.48		.48	.80
5320	Vertical strips		400	.020			.53		.53	.89
5520	Wood shingles		350	.023			.61		.61	1.02
5620	Stucco siding		360	.022			.59		.59	.99
5670	Textured plywood		725	.011			.29		.29	.49
5720	Vinyl siding		510	.016	↓		.42		.42	.70
5770	Corner strips		900	.009	L.F.		.24		.24	.39
5870	Wood, boards, vertical		400	.020	S.F.		.53		.53	.89
5880	Steel siding, corrugated/ribbed	↓	402.50	.020	"		.53		.53	.88
5920	Waterproofing, protection/drain board	2 Clab	3900	.004	B.F.		.11		.11	.18
5970	Over 1/2" thick		1750	.009	S.F.		.24		.24	.41
6020	To 1/2" thick	↓	2000	.008	"		.21		.21	.36

07 11 Dampproofing

07 11 13 – Bituminous Dampproofing

07 11 13.10 Bituminous Asphalt Coating

		Crew	Daily Output	Labor-Hours	Unit	Material	Labor	Equipment	Total	Total Incl O&P
0010	**BITUMINOUS ASPHALT COATING**									
0030	Brushed on, below grade, 1 coat	1 Rofc	665	.012	S.F.	.22	.36		.58	.90
0100	2 coat		500	.016		.44	.48		.92	1.36
0300	Sprayed on, below grade, 1 coat		830	.010		.22	.29		.51	.77
0400	2 coat	↓	500	.016	↓	.43	.48		.91	1.35
0500	Asphalt coating, with fibers				Gal.	8.40			8.40	9.25
0600	Troweled on, asphalt with fibers, 1/16" thick	1 Rofc	500	.016	S.F.	.37	.48		.85	1.27
0700	1/8" thick		400	.020	↓	.65	.60		1.25	1.80
1000	1/2" thick	↓	350	.023	↓	2.11	.68		2.79	3.57

07 11 16 – Cementitious Dampproofing

07 11 16.20 Cementitious Parging

		Crew	Daily Output	Labor-Hours	Unit	Material	Labor	Equipment	Total	Total Incl O&P
0010	**CEMENTITIOUS PARGING**									
0020	Portland cement, 2 coats, 1/2" thick	D-1	250	.064	S.F.	.37	2.03		2.40	3.80
0100	Waterproofed Portland cement, 1/2" thick, 2 coats	"	250	.064	"	5.80	2.03		7.83	9.75

For customer support on your Residential Costs with RSMeans data, call 800.448.8182.

417

07 19 Water Repellents

07 19 19 – Silicone Water Repellents

07 19 19.10 Silicone Based Water Repellents		Crew	Daily Output	Labor-Hours	Unit	Material	2018 Bare Costs Labor	Equipment	Total	Total Incl O&P
0010	**SILICONE BASED WATER REPELLENTS**									
0020	Water base liquid, roller applied	2 Rofc	7000	.002	S.F.	.42	.07		.49	.58
0200	Silicone or stearate, sprayed on CMU, 1 coat	1 Rofc	4000	.002		.39	.06		.45	.54
0300	2 coats	"	3000	.003	↓	.79	.08		.87	1.02

07 21 Thermal Insulation

07 21 13 – Board Insulation

07 21 13.10 Rigid Insulation

			Crew	Daily Output	Labor-Hours	Unit	Material	2018 Bare Costs Labor	Equipment	Total	Total Incl O&P
0010	**RIGID INSULATION**, for walls										
0040	Fiberglass, 1.5#/C.F., unfaced, 1" thick, R4.1	G	1 Carp	1000	.008	S.F.	.34	.28		.62	.83
0060	1-1/2" thick, R6.2	G		1000	.008		.40	.28		.68	.90
0080	2" thick, R8.3	G		1000	.008		.50	.28		.78	1.01
0120	3" thick, R12.4	G		800	.010		.60	.35		.95	1.24
0370	3#/C.F., unfaced, 1" thick, R4.3	G		1000	.008		.54	.28		.82	1.05
0390	1-1/2" thick, R6.5	G		1000	.008		.79	.28		1.07	1.33
0400	2" thick, R8.7	G		890	.009		1.07	.31		1.38	1.70
0420	2-1/2" thick, R10.9	G		800	.010		1.11	.35		1.46	1.80
0440	3" thick, R13	G		800	.010		1.62	.35		1.97	2.36
0520	Foil faced, 1" thick, R4.3	G		1000	.008		.84	.28		1.12	1.38
0540	1-1/2" thick, R6.5	G		1000	.008		1.25	.28		1.53	1.84
0560	2" thick, R8.7	G		890	.009		1.58	.31		1.89	2.26
0580	2-1/2" thick, R10.9	G		800	.010		1.86	.35		2.21	2.63
0600	3" thick, R13	G	↓	800	.010	↓	2.09	.35		2.44	2.88
1600	Isocyanurate, 4' x 8' sheet, foil faced, both sides										
1610	1/2" thick	G	1 Carp	800	.010	S.F.	.31	.35		.66	.92
1620	5/8" thick	G		800	.010		.48	.35		.83	1.11
1630	3/4" thick	G		800	.010		.45	.35		.80	1.08
1640	1" thick	G		800	.010		.61	.35		.96	1.25
1650	1-1/2" thick	G		730	.011		.67	.38		1.05	1.38
1660	2" thick	G		730	.011		.90	.38		1.28	1.63
1670	3" thick	G		730	.011		2.76	.38		3.14	3.68
1680	4" thick	G		730	.011		2.52	.38		2.90	3.41
1700	Perlite, 1" thick, R2.77	G		800	.010		.47	.35		.82	1.10
1750	2" thick, R5.55	G		730	.011		.78	.38		1.16	1.50
1900	Extruded polystyrene, 25 psi compressive strength, 1" thick, R5	G		800	.010		.57	.35		.92	1.21
1940	2" thick, R10	G		730	.011		1.14	.38		1.52	1.89
1960	3" thick, R15	G		730	.011		1.59	.38		1.97	2.39
2100	Expanded polystyrene, 1" thick, R3.85	G		800	.010		.27	.35		.62	.88
2120	2" thick, R7.69	G		730	.011		.54	.38		.92	1.23
2140	3" thick, R11.49	G	↓	730	.011	↓	.81	.38		1.19	1.53

07 21 13.13 Foam Board Insulation

			Crew	Daily Output	Labor-Hours	Unit	Material	2018 Bare Costs Labor	Equipment	Total	Total Incl O&P
0010	**FOAM BOARD INSULATION**										
0600	Polystyrene, expanded, 1" thick, R4	G	1 Carp	680	.012	S.F.	.27	.41		.68	.98
0700	2" thick, R8	G	"	675	.012	"	.54	.41		.95	1.28

07 21 16 – Blanket Insulation

07 21 16.10 Blanket Insulation for Floors/Ceilings

			Crew	Daily Output	Labor-Hours	Unit	Material	2018 Bare Costs Labor	Equipment	Total	Total Incl O&P
0010	**BLANKET INSULATION FOR FLOORS/CEILINGS**										
0020	Including spring type wire fasteners										
2000	Fiberglass, blankets or batts, paper or foil backing										
2100	3-1/2" thick, R13	G	1 Carp	700	.011	S.F.	.41	.40		.81	1.11

07 21 Thermal Insulation

07 21 16 – Blanket Insulation

07 21 16.10 Blanket Insulation for Floors/Ceilings

		Crew	Daily Output	Labor-Hours	Unit	Material	2018 Bare Costs Labor	Equipment	Total	Total Incl O&P
2150	6-1/4" thick, R19	G 1 Carp	600	.013	S.F.	.51	.47		.98	1.33
2210	9-1/2" thick, R30	G	500	.016		.73	.56		1.29	1.73
2220	12" thick, R38	G	475	.017		1.05	.59		1.64	2.14
3000	Unfaced, 3-1/2" thick, R13	G	600	.013		.33	.47		.80	1.13
3010	6-1/4" thick, R19	G	500	.016		.39	.56		.95	1.36
3020	9-1/2" thick, R30	G	450	.018		.60	.62		1.22	1.69
3030	12" thick, R38	G	425	.019		.75	.66		1.41	1.92

07 21 16.20 Blanket Insulation for Walls

		Crew	Daily Output	Labor-Hours	Unit	Material	2018 Bare Costs Labor	Equipment	Total	Total Incl O&P
0010	**BLANKET INSULATION FOR WALLS**									
0020	Kraft faced fiberglass, 3-1/2" thick, R11, 15" wide	G 1 Carp	1350	.006	S.F.	.32	.21		.53	.69
0030	23" wide	G	1600	.005		.32	.17		.49	.64
0060	R13, 11" wide	G	1150	.007		.34	.24		.58	.77
0080	15" wide	G	1350	.006		.34	.21		.55	.71
0100	23" wide	G	1600	.005		.34	.17		.51	.66
0110	R15, 11" wide	G	1150	.007		.49	.24		.73	.94
0120	15" wide	G	1350	.006		.49	.21		.70	.88
0140	6" thick, R19, 11" wide	G	1150	.007		.44	.24		.68	.88
0160	15" wide	G	1350	.006		.44	.21		.65	.82
0180	23" wide	G	1600	.005		.44	.17		.61	.77
0201	9" thick, R30, 15" wide	G	1350	.006		.73	.21		.94	1.14
0241	12" thick, R38, 15" wide	G	1350	.006		1.05	.21		1.26	1.50
0410	Foil faced fiberglass, 3-1/2" thick, R13, 11" wide	G	1150	.007		.47	.24		.71	.92
0420	15" wide	G	1350	.006		.47	.21		.68	.86
0442	R15, 11" wide	G	1150	.007		.48	.24		.72	.93
0444	15" wide	G	1350	.006		.48	.21		.69	.87
0461	6" thick, R19, 15" wide	G	1600	.005		.62	.17		.79	.97
0482	R21, 11" wide	G	1150	.007		.64	.24		.88	1.10
0501	9" thick, R30, 15" wide	G	1350	.006		.94	.21		1.15	1.37
0620	Unfaced fiberglass, 3-1/2" thick, R13, 11" wide	G	1150	.007		.33	.24		.57	.76
0821	15" wide	G	1600	.005		.33	.17		.50	.65
0832	R15, 11" wide	G	1150	.007		.46	.24		.70	.91
0834	15" wide	G	1350	.006		.46	.21		.67	.85
0861	6" thick, R19, 15" wide	G	1350	.006		.39	.21		.60	.77
0901	9" thick, R30, 15" wide	G	1150	.007		.60	.24		.84	1.06
0941	12" thick, R38, 15" wide	G	1150	.007		.75	.24		.99	1.23
1300	Wall or ceiling insulation, mineral wool batts									
1320	3-1/2" thick, R15	G 1 Carp	1600	.005	S.F.	.77	.17		.94	1.14
1340	5-1/2" thick, R23	G	1600	.005		1.21	.17		1.38	1.62
1380	7-1/4" thick, R30	G	1350	.006		1.60	.21		1.81	2.09
1700	Non-rigid insul., recycled blue cotton fiber, unfaced batts, R13, 16" wide	G	1600	.005		.99	.17		1.16	1.38
1710	R19, 16" wide	G	1600	.005		1.35	.17		1.52	1.78
1850	Friction fit wire insulation supports, 16" OC	G	960	.008	Ea.	.07	.29		.36	.56

07 21 19 – Foamed In Place Insulation

07 21 19.10 Masonry Foamed In Place Insulation

		Crew	Daily Output	Labor-Hours	Unit	Material	2018 Bare Costs Labor	Equipment	Total	Total Incl O&P
0010	**MASONRY FOAMED IN PLACE INSULATION**									
0100	Amino-plast foam, injected into block core, 6" block	G G-2A	6000	.004	Ea.	.17	.11	.10	.38	.49
0110	8" block	G	5000	.005		.20	.13	.12	.45	.58
0120	10" block	G	4000	.006		.25	.16	.15	.56	.73
0130	12" block	G	3000	.008		.34	.21	.20	.75	.98
0140	Injected into cavity wall	G	13000	.002	B.F.	.06	.05	.05	.16	.21
0150	Preparation, drill holes into mortar joint every 4 V.L.F., 5/8" diameter	1 Clab	960	.008	Ea.		.22		.22	.37
0160	7/8" diameter		680	.012			.31		.31	.52

For customer support on your Residential Costs with RSMeans data, call 800.448.8182.

419

07 21 Thermal Insulation

07 21 19 – Foamed In Place Insulation

07 21 19.10 Masonry Foamed In Place Insulation		Crew	Daily Output	Labor-Hours	Unit	Material	2018 Bare Costs Labor	Equipment	Total	Total Incl O&P
0170	Patch drilled holes, 5/8" diameter	1 Clab	1800	.004	Ea.	.04	.12		.16	.24
0180	7/8" diameter		1200	.007		.05	.18		.23	.36

07 21 23 – Loose-Fill Insulation

07 21 23.10 Poured Loose-Fill Insulation

			Crew	Daily Output	Labor-Hours	Unit	Material	Labor	Equipment	Total	Total Incl O&P
0010	**POURED LOOSE-FILL INSULATION**										
0020	Cellulose fiber, R3.8 per inch	G	1 Carp	200	.040	C.F.	.69	1.40		2.09	3.08
0021	4" thick	G		1000	.008	S.F.	.17	.28		.45	.64
0022	6" thick	G		800	.010	"	.28	.35		.63	.89
0080	Fiberglass wool, R4 per inch	G		200	.040	C.F.	.62	1.40		2.02	3
0081	4" thick	G		600	.013	S.F.	.21	.47		.68	1
0082	6" thick	G		400	.020	"	.30	.70		1	1.49
0100	Mineral wool, R3 per inch	G		200	.040	C.F.	.49	1.40		1.89	2.86
0101	4" thick	G		600	.013	S.F.	.16	.47		.63	.95
0102	6" thick	G		400	.020	"	.25	.70		.95	1.43
0300	Polystyrene, R4 per inch	G		200	.040	C.F.	1.42	1.40		2.82	3.88
0301	4" thick	G		600	.013	S.F.	.47	.47		.94	1.29
0302	6" thick	G		400	.020	"	.71	.70		1.41	1.94
0400	Perlite, R2.78 per inch	G		200	.040	C.F.	5.30	1.40		6.70	8.10
0401	4" thick	G		1000	.008	S.F.	1.76	.28		2.04	2.40
0402	6" thick	G		800	.010	"	2.65	.35		3	3.49

07 21 23.20 Masonry Loose-Fill Insulation

			Crew	Daily Output	Labor-Hours	Unit	Material	Labor	Equipment	Total	Total Incl O&P
0010	**MASONRY LOOSE-FILL INSULATION**, vermiculite or perlite										
0100	In cores of concrete block, 4" thick wall, .115 C.F./S.F.	G	D-1	4800	.003	S.F.	.61	.11		.72	.85
0700	Foamed in place, urethane in 2-5/8" cavity	G	G-2A	1035	.023		1.35	.61	.59	2.55	3.21
0800	For each 1" added thickness, add	G	"	2372	.010		.51	.27	.26	1.04	1.32

07 21 26 – Blown Insulation

07 21 26.10 Blown Insulation

			Crew	Daily Output	Labor-Hours	Unit	Material	Labor	Equipment	Total	Total Incl O&P
0010	**BLOWN INSULATION** Ceilings, with open access										
0020	Cellulose, 3-1/2" thick, R13	G	G-4	5000	.005	S.F.	.24	.13	.06	.43	.55
0030	5-3/16" thick, R19	G		3800	.006		.35	.17	.08	.60	.77
0050	6-1/2" thick, R22	G		3000	.008		.45	.22	.10	.77	.97
0100	8-11/16" thick, R30	G		2600	.009		.61	.25	.12	.98	1.22
0120	10-7/8" thick, R38	G		1800	.013		.78	.36	.17	1.31	1.65
1000	Fiberglass, 5.5" thick, R11	G		3800	.006		.21	.17	.08	.46	.61
1050	6" thick, R12	G		3000	.008		.30	.22	.10	.62	.80
1100	8.8" thick, R19	G		2200	.011		.37	.30	.14	.81	1.06
1200	10" thick, R22	G		1800	.013		.43	.36	.17	.96	1.28
1300	11.5" thick, R26	G		1500	.016		.52	.44	.20	1.16	1.52
1350	13" thick, R30	G		1400	.017		.60	.47	.22	1.29	1.68
1450	16" thick, R38	G		1145	.021		.77	.57	.27	1.61	2.08
1500	20" thick, R49	G		920	.026		1.01	.71	.33	2.05	2.67

07 21 27 – Reflective Insulation

07 21 27.10 Reflective Insulation Options

			Crew	Daily Output	Labor-Hours	Unit	Material	Labor	Equipment	Total	Total Incl O&P
0010	**REFLECTIVE INSULATION OPTIONS**										
0020	Aluminum foil on reinforced scrim	G	1 Carp	19	.421	C.S.F.	15	14.70		29.70	41
0100	Reinforced with woven polyolefin	G		19	.421		22	14.70		36.70	49
0500	With single bubble air space, R8.8	G		15	.533		26	18.65		44.65	60
0600	With double bubble air space, R9.8	G		15	.533		31.50	18.65		50.15	65.50

07 21 Thermal Insulation

07 21 29 – Sprayed Insulation

07 21 29.10 Sprayed-On Insulation

		Crew	Daily Output	Labor-Hours	Unit	Material	2018 Bare Costs Labor	2018 Bare Costs Equipment	Total	Total Incl O&P
0010	**SPRAYED-ON INSULATION**									
0300	Closed cell, spray polyurethane foam, 2 lb./C.F. density									
0310	1" thick	G	G-2A	6000	.004	S.F.	.51	.11	.10	.72	.86
0320	2" thick	G		3000	.008		1.03	.21	.20	1.44	1.73
0330	3" thick	G		2000	.012		1.54	.32	.31	2.17	2.59
0335	3-1/2" thick	G		1715	.014		1.80	.37	.36	2.53	3.02
0340	4" thick	G		1500	.016		2.05	.42	.41	2.88	3.45
0350	5" thick	G		1200	.020		2.57	.53	.51	3.61	4.31
0355	5-1/2" thick	G		1090	.022		2.82	.58	.56	3.96	4.75
0360	6" thick	G		1000	.024		3.08	.63	.61	4.32	5.20

07 22 Roof and Deck Insulation

07 22 16 – Roof Board Insulation

07 22 16.10 Roof Deck Insulation

		Crew	Daily Output	Labor-Hours	Unit	Material	2018 Bare Costs Labor	2018 Bare Costs Equipment	Total	Total Incl O&P
0010	**ROOF DECK INSULATION**, fastening excluded									
0016	Asphaltic cover board, fiberglass lined, 1/8" thick		1 Rofc	1400	.006	S.F.	.48	.17		.65	.84
0018	1/4" thick			1400	.006		.96	.17		1.13	1.37
0020	Fiberboard low density, 1/2" thick, R1.39	G		1300	.006		.35	.18		.53	.73
0030	1" thick, R2.78	G		1040	.008		.58	.23		.81	1.06
0080	1-1/2" thick, R4.17	G		1040	.008		.89	.23		1.12	1.40
0100	2" thick, R5.56	G		1040	.008		1.15	.23		1.38	1.69
0110	Fiberboard high density, 1/2" thick, R1.3	G		1300	.006		.27	.18		.45	.64
0120	1" thick, R2.5	G		1040	.008		.56	.23		.79	1.04
0130	1-1/2" thick, R3.8	G		1040	.008		.83	.23		1.06	1.33
0200	Fiberglass, 3/4" thick, R2.78	G		1300	.006		.62	.18		.80	1.02
0400	15/16" thick, R3.70	G		1300	.006		.82	.18		1	1.24
0460	1-1/16" thick, R4.17	G		1300	.006		1.07	.18		1.25	1.52
0600	1-5/16" thick, R5.26	G		1300	.006		1.42	.18		1.60	1.90
0650	2-1/16" thick, R8.33	G		1040	.008		1.52	.23		1.75	2.09
0700	2-7/16" thick, R10	G		1040	.008		1.72	.23		1.95	2.31
0800	Gypsum cover board, fiberglass mat facer, 1/4" thick			1400	.006		.48	.17		.65	.84
0810	1/2" thick			1300	.006		.59	.18		.77	.99
0820	5/8" thick			1200	.007		.62	.20		.82	1.04
0830	Primed fiberglass mat facer, 1/4" thick			1400	.006		.50	.17		.67	.86
0840	1/2" thick			1300	.006		.58	.18		.76	.98
0850	5/8" thick			1200	.007		.61	.20		.81	1.03
1650	Perlite, 1/2" thick, R1.32	G		1365	.006		.27	.18		.45	.62
1655	3/4" thick, R2.08	G		1040	.008		.38	.23		.61	.84
1660	1" thick, R2.78	G		1040	.008		.54	.23		.77	1.01
1670	1-1/2" thick, R4.17	G		1040	.008		.80	.23		1.03	1.30
1680	2" thick, R5.56	G		910	.009		1.08	.26		1.34	1.67
1685	2-1/2" thick, R6.67	G		910	.009		1.45	.26		1.71	2.08
1690	Tapered for drainage	G		1040	.008	B.F.	1.05	.23		1.28	1.58
1700	Polyisocyanurate, 2#/C.F. density, 3/4" thick	G		1950	.004	S.F.	.42	.12		.54	.68
1705	1" thick	G		1820	.004		.43	.13		.56	.71
1715	1-1/2" thick	G		1625	.005		.58	.15		.73	.91
1725	2" thick	G		1430	.006		.74	.17		.91	1.11
1735	2-1/2" thick	G		1365	.006		.99	.18		1.17	1.41
1745	3" thick	G		1300	.006		1.10	.18		1.28	1.55
1755	3-1/2" thick	G		1300	.006		1.71	.18		1.89	2.22

For customer support on your Residential Costs with RSMeans data, call 800.448.8182.

421

07 22 16 – Roof Board Insulation

07 22 16.10 Roof Deck Insulation		Crew	Daily Output	Labor-Hours	Unit	Material	2018 Bare Costs Labor	2018 Bare Costs Equipment	Total	Total Incl O&P
1765	Tapered for drainage	G	1 Rofc	1820	.004	B.F.	.52	.13		.65	.81
1900	Extruded polystyrene										
1910	15 psi compressive strength, 1" thick, R5	G	1 Rofc	1950	.004	S.F.	.55	.12		.67	.83
1920	2" thick, R10	G		1625	.005		.72	.15		.87	1.06
1930	3" thick, R15	G		1300	.006		1.43	.18		1.61	1.91
1932	4" thick, R20	G		1300	.006		1.93	.18		2.11	2.46
1934	Tapered for drainage	G		1950	.004	B.F.	.51	.12		.63	.78
1940	25 psi compressive strength, 1" thick, R5	G		1950	.004	S.F.	1.11	.12		1.23	1.44
1942	2" thick, R10	G		1625	.005		2.11	.15		2.26	2.59
1944	3" thick, R15	G		1300	.006		3.22	.18		3.40	3.88
1946	4" thick, R20	G		1300	.006		4.44	.18		4.62	5.20
1948	Tapered for drainage	G		1950	.004	B.F.	.56	.12		.68	.84
1950	40 psi compressive strength, 1" thick, R5	G		1950	.004	S.F.	.87	.12		.99	1.18
1952	2" thick, R10	G		1625	.005		1.65	.15		1.80	2.09
1954	3" thick, R15	G		1300	.006		2.39	.18		2.57	2.97
1956	4" thick, R20	G		1300	.006		3.13	.18		3.31	3.79
1958	Tapered for drainage	G		1820	.004	B.F.	.87	.13		1	1.20
1960	60 psi compressive strength, 1" thick, R5	G		1885	.004	S.F.	1.05	.13		1.18	1.39
1962	2" thick, R10	G		1560	.005		2	.15		2.15	2.47
1964	3" thick, R15	G		1270	.006		3.26	.19		3.45	3.92
1966	4" thick, R20	G		1235	.006		4.04	.19		4.23	4.80
1968	Tapered for drainage	G		1820	.004	B.F.	1	.13		1.13	1.34
2010	Expanded polystyrene, 1#/C.F. density, 3/4" thick, R2.89	G		1950	.004	S.F.	.20	.12		.32	.44
2020	1" thick, R3.85	G		1950	.004		.27	.12		.39	.52
2100	2" thick, R7.69	G		1625	.005		.54	.15		.69	.86
2110	3" thick, R11.49	G		1625	.005		.81	.15		.96	1.16
2120	4" thick, R15.38	G		1625	.005		1.08	.15		1.23	1.46
2130	5" thick, R19.23	G		1495	.005		1.35	.16		1.51	1.78
2140	6" thick, R23.26	G		1495	.005		1.62	.16		1.78	2.07
2150	Tapered for drainage			1950	.004	B.F.	.53	.12		.65	.80
2400	Composites with 2" EPS										
2410	1" fiberboard	G	1 Rofc	1325	.006	S.F.	1.45	.18		1.63	1.93
2420	7/16" oriented strand board	G		1040	.008		1.15	.23		1.38	1.69
2430	1/2" plywood	G		1040	.008		1.44	.23		1.67	2
2440	1" perlite	G		1040	.008		1.17	.23		1.40	1.71
2450	Composites with 1-1/2" polyisocyanurate										
2460	1" fiberboard	G	1 Rofc	1040	.008	S.F.	1.07	.23		1.30	1.60
2470	1" perlite	G		1105	.007		.96	.22		1.18	1.45
2480	7/16" oriented strand board	G		1040	.008		.84	.23		1.07	1.34
3000	Fastening alternatives, coated screws, 2" long			3744	.002	Ea.	.06	.06		.12	.19
3010	4" long			3120	.003		.11	.08		.19	.26
3020	6" long			2675	.003		.19	.09		.28	.37
3030	8" long			2340	.003		.28	.10		.38	.50
3040	10" long			1872	.004		.47	.13		.60	.75
3050	Pre-drill and drive wedge spike, 2-1/2"			1248	.006		.40	.19		.59	.79
3060	3-1/2"			1101	.007		.58	.22		.80	1.04
3070	4-1/2"			936	.009		.65	.26		.91	1.19
3075	3" galvanized deck plates			7488	.001		.08	.03		.11	.15
3080	Spot mop asphalt		G-1	295	.190	Sq.	5.80	5.30	1.74	12.84	18
3090	Full mop asphalt		"	192	.292		11.60	8.20	2.68	22.48	30.50
3110	Low-rise polyurethane adhesive, from 5 gallon kit, 12" OC beads		1 Rofc	45	.178		33	5.30		38.30	46
3120	6" OC beads			32	.250		66	7.50		73.50	86
3130	4" OC beads			30	.267		99	7.95		106.95	124

07 24 Exterior Insulation and Finish Systems

07 24 13 – Polymer-Based Exterior Insulation and Finish System

07 24 13.10 Exterior Insulation and Finish Systems		Crew	Daily Output	Labor-Hours	Unit	Material	2018 Bare Costs Labor	2018 Bare Costs Equipment	Total	Total Incl O&P
0010	**EXTERIOR INSULATION AND FINISH SYSTEMS**									
0095	Field applied, 1" EPS insulation	G	J-1	390	.103	S.F.	1.80	3.28	.34	5.42	7.70
0100	With 1/2" cement board sheathing	G		268	.149		2.59	4.77	.50	7.86	11.20
0105	2" EPS insulation	G		390	.103		2.07	3.28	.34	5.69	8
0110	With 1/2" cement board sheathing	G		268	.149		2.86	4.77	.50	8.13	11.50
0115	3" EPS insulation	G		390	.103		2.34	3.28	.34	5.96	8.30
0120	With 1/2" cement board sheathing	G		268	.149		3.13	4.77	.50	8.40	11.80
0125	4" EPS insulation	G		390	.103		2.61	3.28	.34	6.23	8.60
0130	With 1/2" cement board sheathing	G		268	.149		4.18	4.77	.50	9.45	12.95
0140	Premium finish add			1265	.032		.36	1.01	.11	1.48	2.18
0150	Heavy duty reinforcement add			914	.044		.60	1.40	.15	2.15	3.11
0160	2.5#/S.Y. metal lath substrate add		1 Lath	75	.107	S.Y.	3.10	3.67		6.77	9.35
0170	3.4#/S.Y. metal lath substrate add		"	75	.107	"	4.44	3.67		8.11	10.85
0180	Color or texture change		J-1	1265	.032	S.F.	.82	1.01	.11	1.94	2.68
0190	With substrate leveling base coat		1 Plas	530	.015		.83	.51		1.34	1.75
0210	With substrate sealing base coat		1 Pord	1224	.007		.12	.19		.31	.45
0370	V groove shape in panel face					L.F.	.68			.68	.75
0380	U groove shape in panel face					"	.85			.85	.94

07 25 Weather Barriers

07 25 10 – Weather Barriers or Wraps

07 25 10.10 Weather Barriers

		Crew	Daily Output	Labor-Hours	Unit	Material	2018 Bare Costs Labor	2018 Bare Costs Equipment	Total	Total Incl O&P
0010	**WEATHER BARRIERS**									
0400	Asphalt felt paper, #15	1 Carp	37	.216	Sq.	5.35	7.55		12.90	18.45
0401	Per square foot	"	3700	.002	S.F.	.05	.08		.13	.19
0450	Housewrap, exterior, spun bonded polypropylene									
0470	Small roll	1 Carp	3800	.002	S.F.	.15	.07		.22	.29
0480	Large roll	"	4000	.002	"	.14	.07		.21	.27
2100	Asphalt felt roof deck vapor barrier, class 1 metal decks	1 Rofc	37	.216	Sq.	22.50	6.45		28.95	36.50
2200	For all other decks	"	37	.216		16.95	6.45		23.40	30.50
2800	Asphalt felt, 50% recycled content, 15 lb., 4 sq./roll	1 Carp	36	.222		5.65	7.75		13.40	19.15
2810	30 lb., 2 sq./roll	"	36	.222		9	7.75		16.75	23
3000	Building wrap, spun bonded polyethylene	2 Carp	8000	.002	S.F.	.18	.07		.25	.32

07 26 Vapor Retarders

07 26 10 – Above-Grade Vapor Retarders

07 26 10.10 Building Paper

			Crew	Daily Output	Labor-Hours	Unit	Material	2018 Bare Costs Labor	2018 Bare Costs Equipment	Total	Total Incl O&P
0011	**BUILDING PAPER**, aluminum and kraft laminated, foil 1 side		1 Carp	3700	.002	S.F.	.14	.08		.22	.28
0101	Foil 2 sides	G		3700	.002		.14	.08		.22	.29
0601	Polyethylene vapor barrier, standard, 2 mil	G		3700	.002		.02	.08		.10	.15
0701	4 mil	G		3700	.002		.03	.08		.11	.16
0901	6 mil	G		3700	.002		.04	.08		.12	.17
1201	10 mil	G		3700	.002		.09	.08		.17	.23
1801	Reinf. waterproof, 2 mil polyethylene backing, 1 side			3700	.002		.10	.08		.18	.24
1901	2 sides			3700	.002		.13	.08		.21	.27

For customer support on your Residential Costs with RSMeans data, call 800.448.8182.

423

07 27 Air Barriers

07 27 13 – Modified Bituminous Sheet Air Barriers

07 27 13.10 Modified Bituminous Sheet Air Barrier

	07 27 13.10 Modified Bituminous Sheet Air Barrier	Crew	Daily Output	Labor-Hours	Unit	Material	2018 Bare Costs Labor	Equipment	Total	Total Incl O&P
0010	**MODIFIED BITUMINOUS SHEET AIR BARRIER**									
0100	SBS modified sheet laminated to polyethylene sheet, 40 mils, 4" wide	1 Carp	1200	.007	L.F.	.33	.23		.56	.75
0120	6" wide		1100	.007		.45	.25		.70	.91
0140	9" wide		1000	.008		.62	.28		.90	1.15
0160	12" wide		900	.009		.80	.31		1.11	1.40
0180	18" wide	2 Carp	1700	.009	S.F.	.75	.33		1.08	1.37
0200	36" wide	"	1800	.009		.73	.31		1.04	1.32
0220	Adhesive for above	1 Carp	1400	.006		.31	.20		.51	.68

07 27 26 – Fluid-Applied Membrane Air Barriers

07 27 26.10 Fluid Applied Membrane Air Barrier

	07 27 26.10 Fluid Applied Membrane Air Barrier	Crew	Daily Output	Labor-Hours	Unit	Material	Labor	Equipment	Total	Total Incl O&P
0010	**FLUID APPLIED MEMBRANE AIR BARRIER**									
0100	Spray applied vapor barrier, 25 S.F./gallon	1 Pord	1375	.006	S.F.	.01	.17		.18	.30

07 31 Shingles and Shakes

07 31 13 – Asphalt Shingles

07 31 13.10 Asphalt Roof Shingles

	07 31 13.10 Asphalt Roof Shingles	Crew	Daily Output	Labor-Hours	Unit	Material	Labor	Equipment	Total	Total Incl O&P
0010	**ASPHALT ROOF SHINGLES**									
0100	Standard strip shingles									
0150	Inorganic, class A, 25 year	1 Rofc	5.50	1.455	Sq.	75	43.50		118.50	162
0155	Pneumatic nailed		7	1.143		75	34		109	145
0200	30 year		5	1.600		95.50	48		143.50	192
0205	Pneumatic nailed		6.25	1.280		95.50	38.50		134	175
0250	Standard laminated multi-layered shingles									
0300	Class A, 240-260 lb./square	1 Rofc	4.50	1.778	Sq.	110	53		163	218
0305	Pneumatic nailed		5.63	1.422		110	42.50		152.50	199
0350	Class A, 250-270 lb./square		4	2		110	60		170	230
0355	Pneumatic nailed		5	1.600		110	48		158	208
0400	Premium, laminated multi-layered shingles									
0450	Class A, 260-300 lb./square	1 Rofc	3.50	2.286	Sq.	144	68.50		212.50	283
0455	Pneumatic nailed		4.37	1.831		144	54.50		198.50	258
0500	Class A, 300-385 lb./square		3	2.667		260	79.50		339.50	430
0505	Pneumatic nailed		3.75	2.133		260	64		324	400
0800	#15 felt underlayment		64	.125		5.35	3.74		9.09	12.70
0825	#30 felt underlayment		58	.138		10.30	4.12		14.42	18.85
0850	Self adhering polyethylene and rubberized asphalt underlayment		22	.364		78	10.85		88.85	105
0900	Ridge shingles		330	.024	L.F.	2.27	.72		2.99	3.82
0905	Pneumatic nailed		412.50	.019	"	2.27	.58		2.85	3.56
1000	For steep roofs (7 to 12 pitch or greater), add						50%			

07 31 26 – Slate Shingles

07 31 26.10 Slate Roof Shingles

	07 31 26.10 Slate Roof Shingles	Crew	Daily Output	Labor-Hours	Unit	Material	Labor	Equipment	Total	Total Incl O&P
0010	**SLATE ROOF SHINGLES** R073126-20									
0100	Buckingham Virginia black, 3/16" - 1/4" thick [G]	1 Rots	1.75	4.571	Sq.	555	137		692	860
0200	1/4" thick [G]		1.75	4.571		555	137		692	860
0900	Pennsylvania black, Bangor, #1 clear [G]		1.75	4.571		495	137		632	795
1200	Vermont, unfading, green, mottled green [G]		1.75	4.571		505	137		642	805
1300	Semi-weathering green & gray [G]		1.75	4.571		360	137		497	645
1400	Purple [G]		1.75	4.571		435	137		572	730
1500	Black or gray [G]		1.75	4.571		485	137		622	785
2700	Ridge shingles, slate		200	.040	L.F.	10.10	1.20		11.30	13.30

07 31 Shingles and Shakes

07 31 29 – Wood Shingles and Shakes

07 31 29.13 Wood Shingles

		Crew	Daily Output	Labor-Hours	Unit	Material	2018 Bare Costs Labor	Equipment	Total	Total Incl O&P
0010	**WOOD SHINGLES**									
0012	16" No. 1 red cedar shingles, 5" exposure, on roof	1 Carp	2.50	3.200	Sq.	295	112		407	510
0015	Pneumatic nailed		3.25	2.462		295	86		381	470
0200	7-1/2" exposure, on walls		2.05	3.902		196	136		332	445
0205	Pneumatic nailed		2.67	2.996		196	105		301	390
0300	18" No. 1 red cedar perfections, 5-1/2" exposure, on roof		2.75	2.909		253	102		355	445
0305	Pneumatic nailed		3.57	2.241		253	78.50		331.50	410
0500	7-1/2" exposure, on walls		2.25	3.556		186	124		310	410
0505	Pneumatic nailed		2.92	2.740		186	96		282	365
0600	Resquared and rebutted, 5-1/2" exposure, on roof		3	2.667		290	93		383	475
0605	Pneumatic nailed		3.90	2.051		290	71.50		361.50	440
0900	7-1/2" exposure, on walls		2.45	3.265		214	114		328	425
0905	Pneumatic nailed	▼	3.18	2.516		214	88		302	380
1000	Add to above for fire retardant shingles				▼	55			55	60.50
1060	Preformed ridge shingles	1 Carp	400	.020	L.F.	3.89	.70		4.59	5.45
2000	White cedar shingles, 16" long, extras, 5" exposure, on roof		2.40	3.333	Sq.	197	117		314	410
2005	Pneumatic nailed		3.12	2.564		197	89.50		286.50	365
2050	5" exposure on walls		2	4		197	140		337	450
2055	Pneumatic nailed		2.60	3.077		197	108		305	395
2100	7-1/2" exposure, on walls		2	4		141	140		281	385
2105	Pneumatic nailed		2.60	3.077		141	108		249	335
2150	"B" grade, 5" exposure on walls		2	4		172	140		312	420
2155	Pneumatic nailed		2.60	3.077		172	108		280	370
2300	For #15 organic felt underlayment on roof, 1 layer, add		64	.125		5.35	4.37		9.72	13.15
2400	2 layers, add	▼	32	.250		10.75	8.75		19.50	26.50
2600	For steep roofs (7/12 pitch or greater), add to above				▼		50%			
2700	Panelized systems, No.1 cedar shingles on 5/16" CDX plywood									
2800	On walls, 8' strips, 7" or 14" exposure	2 Carp	700	.023	S.F.	6.40	.80		7.20	8.40
3000	Ridge shakes or shingle, wood	1 Carp	280	.029	L.F.	4.75	1		5.75	6.90

07 31 29.16 Wood Shakes

		Crew	Daily Output	Labor-Hours	Unit	Material	2018 Bare Costs Labor	Equipment	Total	Total Incl O&P
0010	**WOOD SHAKES**									
1100	Hand-split red cedar shakes, 1/2" thick x 24" long, 10" exp. on roof	1 Carp	2.50	3.200	Sq.	300	112		412	515
1105	Pneumatic nailed		3.25	2.462		300	86		386	475
1110	3/4" thick x 24" long, 10" exp. on roof		2.25	3.556		300	124		424	535
1115	Pneumatic nailed		2.92	2.740		300	96		396	490
1200	1/2" thick, 18" long, 8-1/2" exp. on roof		2	4		275	140		415	530
1205	Pneumatic nailed		2.60	3.077		275	108		383	480
1210	3/4" thick x 18" long, 8-1/2" exp. on roof		1.80	4.444		275	155		430	560
1215	Pneumatic nailed		2.34	3.419		275	119		394	500
1255	10" exposure on walls		2	4		265	140		405	525
1260	10" exposure on walls, pneumatic nailed	▼	2.60	3.077		265	108		373	470
1700	Add to above for fire retardant shakes, 24" long					55			55	60.50
1800	18" long				▼	55			55	60.50
1810	Ridge shakes	1 Carp	350	.023	L.F.	5.75	.80		6.55	7.70

For customer support on your Residential Costs with RSMeans data, call 800.448.8182.

425

07 32 Roof Tiles

07 32 13 – Clay Roof Tiles

07 32 13.10 Clay Tiles

07 32 13.10 Clay Tiles	Crew	Daily Output	Labor-Hours	Unit	Material	2018 Bare Costs Labor	Equipment	Total	Total Incl O&P
0010 **CLAY TILES**, including accessories									
0300 Flat shingle, interlocking, 15", 166 pcs./sq., fireflashed blend	3 Rots	6	4	Sq.	470	120		590	735
0500 Terra cotta red	↓	6	4		520	120		640	790
0600 Roman pan and top, 18", 102 pcs./sq., fireflashed blend		5.50	4.364		505	131		636	795
0640 Terra cotta red	1 Rots	2.40	3.333		570	100		670	810
1100 Barrel mission tile, 18", 166 pcs./sq., fireflashed blend	3 Rots	5.50	4.364		415	131		546	695
1140 Terra cotta red		5.50	4.364		420	131		551	700
1700 Scalloped edge flat shingle, 14", 145 pcs./sq., fireflashed blend		6	4		1,150	120		1,270	1,500
1800 Terra cotta red	↓	6	4		1,050	120		1,170	1,375
3010 #15 felt underlayment	1 Rofc	64	.125		5.35	3.74		9.09	12.70
3020 #30 felt underlayment		58	.138		10.30	4.12		14.42	18.85
3040 Polyethylene and rubberized asph. underlayment	↓	22	.364	↓	78	10.85		88.85	105

07 32 16 – Concrete Roof Tiles

07 32 16.10 Concrete Tiles

	Crew	Daily Output	Labor-Hours	Unit	Material	2018 Bare Costs Labor	Equipment	Total	Total Incl O&P
0010 **CONCRETE TILES**									
0020 Corrugated, 13" x 16-1/2", 90 per sq., 950 lb./sq.									
0050 Earthtone colors, nailed to wood deck	1 Rots	1.35	5.926	Sq.	105	178		283	440
0150 Blues		1.35	5.926		105	178		283	440
0200 Greens	↓	1.35	5.926	↓	106	178		284	440
0250 Premium colors		1.35	5.926		106	178		284	440
0500 Shakes, 13" x 16-1/2", 90 per sq., 950 lb./sq.									
0600 All colors, nailed to wood deck	1 Rots	1.50	5.333	Sq.	126	160		286	430
1500 Accessory pieces, ridge & hip, 10" x 16-1/2", 8 lb. each	"	120	.067	Ea.	3.80	2		5.80	7.85
1700 Rake, 6-1/2" x 16-3/4", 9 lb. each					3.80			3.80	4.18
1800 Mansard hip, 10" x 16-1/2", 9.2 lb. each					3.80			3.80	4.18
1900 Hip starter, 10" x 16-1/2", 10.5 lb. each					10.50			10.50	11.55
2000 3 or 4 way apex, 10" each side, 11.5 lb. each				↓	12			12	13.20

07 33 Natural Roof Coverings

07 33 63 – Vegetated Roofing

07 33 63.10 Green Roof Systems

		Crew	Daily Output	Labor-Hours	Unit	Material	2018 Bare Costs Labor	Equipment	Total	Total Incl O&P
0010 **GREEN ROOF SYSTEMS**										
0020 Soil mixture for green roof 30% sand, 55% gravel, 15% soil										
0100 Hoist and spread soil mixture 4" depth up to 5 stories tall roof	G	B-13B	4000	.014	S.F.	.23	.41	.25	.89	1.20
0150 6" depth	G		2667	.021		.35	.62	.37	1.34	1.81
0200 8" depth	G		2000	.028		.46	.82	.50	1.78	2.42
0250 10" depth	G		1600	.035		.58	1.03	.62	2.23	3.01
0300 12" depth	G		1335	.042		.69	1.23	.74	2.66	3.62
0310 Alt. man-made soil mix, hoist & spread, 4" deep up to 5 stories tall roof	G	↓	4000	.014	↓	1.92	.41	.25	2.58	3.06
0350 Mobilization 55 ton crane to site	G	1 Eqhv	3.60	2.222	Ea.		82.50		82.50	136
0355 Hoisting cost to 5 stories per day (Avg. 28 picks per day)	G	B-13B	1	56	Day		1,650	995	2,645	3,825
0360 Mobilization or demobilization, 100 ton crane to site driver & escort	G	A-3E	2.50	6.400	Ea.		222	50.50	272.50	420
0365 Hoisting cost 6-10 stories per day (Avg. 21 picks per day)	G	B-13C	1	56	Day		1,650	1,875	3,525	4,775
0370 Hoist and spread soil mixture 4" depth 6-10 stories tall roof	G		4000	.014	S.F.	.23	.41	.47	1.11	1.44
0375 6" depth	G		2667	.021		.35	.62	.70	1.67	2.17
0380 8" depth	G		2000	.028		.46	.82	.94	2.22	2.90
0385 10" depth	G		1600	.035		.58	1.03	1.17	2.78	3.62
0390 12" depth	G		1335	.042	↓	.69	1.23	1.40	3.32	4.34
0400 Green roof edging treated lumber 4" x 4", no hoisting included	G	2 Carp	400	.040	L.F.	1.20	1.40		2.60	3.64
0410 4" x 6"	G		400	.040	↓	1.94	1.40		3.34	4.46

07 33 Natural Roof Coverings

07 33 63 – Vegetated Roofing

07 33 63.10 Green Roof Systems

07 33 63.10 Green Roof Systems		Crew	Daily Output	Labor-Hours	Unit	Material	2018 Bare Costs Labor	Equipment	Total	Total Incl O&P
0420	4" x 8"	G	2 Carp	360	.044	L.F.	3.96	1.55		5.51	6.95
0430	4" x 6" double stacked	G		300	.053		3.89	1.86		5.75	7.40
0500	Green roof edging redwood lumber 4" x 4", no hoisting included	G		400	.040		6.45	1.40		7.85	9.40
0510	4" x 6"	G		400	.040		12.60	1.40		14	16.15
0520	4" x 8"	G		360	.044		23.50	1.55		25.05	28.50
0530	4" x 6" double stacked	G	▼	300	.053	▼	25	1.86		26.86	30.50
0550	Components, not including membrane or insulation:										
0560	Fluid applied rubber membrane, reinforced, 215 mil thick	G	G-5	350	.114	S.F.	.30	3.12	.52	3.94	6.60
0570	Root barrier	G	2 Rofc	775	.021		.70	.62		1.32	1.90
0580	Moisture retention barrier and reservoir	G	"	900	.018		2.66	.53		3.19	3.90
0600	Planting sedum, light soil, potted, 2-1/4" diameter, 2 per S.F.	G	1 Clab	420	.019		6.10	.51		6.61	7.55
0610	1 per S.F.	G	"	840	.010		3.05	.25		3.30	3.78
0630	Planting sedum mat per S.F. including shipping (4000 S.F. min)	G	4 Clab	4000	.008		7.30	.21		7.51	8.40
0640	Installation sedum mat system (no soil required) per S.F. (4000 S.F. min)	G	"	4000	.008	▼	10.20	.21		10.41	11.55
0645	Note: pricing of sedum mats shipped in full truck loads (4000-5000 S.F.)										

07 41 Roof Panels

07 41 13 – Metal Roof Panels

07 41 13.10 Aluminum Roof Panels

07 41 13.10 Aluminum Roof Panels		Crew	Daily Output	Labor-Hours	Unit	Material	Labor	Equipment	Total	Total Incl O&P
0010	**ALUMINUM ROOF PANELS**										
0020	Corrugated or ribbed, .0155" thick, natural	G-3	1200	.027	S.F.	1	.87		1.87	2.54
0300	Painted	"	1200	.027	"	1.45	.87		2.32	3.04

07 41 33 – Plastic Roof Panels

07 41 33.10 Fiberglass Panels

07 41 33.10 Fiberglass Panels		Crew	Daily Output	Labor-Hours	Unit	Material	Labor	Equipment	Total	Total Incl O&P
0010	**FIBERGLASS PANELS**										
0012	Corrugated panels, roofing, 8 oz./S.F.	G-3	1000	.032	S.F.	2.47	1.04		3.51	4.45
0100	12 oz./S.F.		1000	.032		4.52	1.04		5.56	6.70
0300	Corrugated siding, 6 oz./S.F.		880	.036		1.99	1.18		3.17	4.16
0400	8 oz./S.F.		880	.036		2.47	1.18		3.65	4.69
0500	Fire retardant		880	.036		3.98	1.18		5.16	6.35
0600	12 oz. siding, textured		880	.036		3.87	1.18		5.05	6.25
0700	Fire retardant		880	.036		4.48	1.18		5.66	6.90
0900	Flat panels, 6 oz./S.F., clear or colors		880	.036		2.52	1.18		3.70	4.74
1100	Fire retardant, class A		880	.036		3.56	1.18		4.74	5.90
1300	8 oz./S.F., clear or colors	▼	880	.036	▼	2.47	1.18		3.65	4.69

07 42 Wall Panels

07 42 13 – Metal Wall Panels

07 42 13.20 Aluminum Siding

07 42 13.20 Aluminum Siding		Crew	Daily Output	Labor-Hours	Unit	Material	Labor	Equipment	Total	Total Incl O&P
0011	**ALUMINUM SIDING**										
6040	0.024" thick smooth white single 8" wide	2 Carp	515	.031	S.F.	2.78	1.09		3.87	4.87
6060	Double 4" pattern		515	.031		2.85	1.09		3.94	4.95
6080	Double 5" pattern		550	.029		2.75	1.02		3.77	4.72
6120	Embossed white, 8" wide		515	.031		2.79	1.09		3.88	4.88
6140	Double 4" pattern		515	.031		2.80	1.09		3.89	4.89
6160	Double 5" pattern		550	.029		2.82	1.02		3.84	4.79
6170	Vertical, embossed white, 12" wide		590	.027		2.83	.95		3.78	4.69
6320	0.019" thick, insulated, smooth white, 8" wide		515	.031		2.47	1.09		3.56	4.53
6340	Double 4" pattern		515	.031		2.45	1.09		3.54	4.51

For customer support on your Residential Costs with RSMeans data, call 800.448.8182.

427

07 42 Wall Panels

07 42 13 – Metal Wall Panels

07 42 13.20 Aluminum Siding

		Crew	Daily Output	Labor-Hours	Unit	Material	2018 Bare Costs Labor	Equipment	Total	Total Incl O&P
6360	Double 5" pattern	2 Carp	550	.029	S.F.	2.46	1.02		3.48	4.40
6400	Embossed white, 8" wide		515	.031		2.85	1.09		3.94	4.95
6420	Double 4" pattern		515	.031		2.87	1.09		3.96	4.97
6440	Double 5" pattern		550	.029		2.87	1.02		3.89	4.85
6500	Shake finish 10" wide white		550	.029		3.10	1.02		4.12	5.10
6600	Vertical pattern, 12" wide, white		590	.027		2.37	.95		3.32	4.19
6640	For colors add					.15			.15	.17
6700	Accessories, white									
6720	Starter strip 2-1/8"	2 Carp	610	.026	L.F.	.48	.92		1.40	2.05
6740	Sill trim		450	.036		.66	1.24		1.90	2.80
6760	Inside corner		610	.026		1.72	.92		2.64	3.41
6780	Outside corner post		610	.026		3.66	.92		4.58	5.55
6800	Door & window trim		440	.036		.63	1.27		1.90	2.80
6820	For colors add					.14			.14	.15
6900	Soffit & fascia 1' overhang solid	2 Carp	110	.145		4.51	5.10		9.61	13.40
6920	Vented		110	.145		4.54	5.10		9.64	13.45
6940	2' overhang solid		100	.160		6.60	5.60		12.20	16.55
6960	Vented		100	.160		6.60	5.60		12.20	16.55

07 42 13.30 Steel Siding

		Crew	Daily Output	Labor-Hours	Unit	Material	2018 Bare Costs Labor	Equipment	Total	Total Incl O&P
0010	**STEEL SIDING**									
0020	Beveled, vinyl coated, 8" wide	1 Carp	265	.030	S.F.	1.80	1.06		2.86	3.73
0050	10" wide	"	275	.029		1.95	1.02		2.97	3.84
0081	Galv., corrugated or ribbed, on steel frame, 30 ga.	G-3	775	.041		1.25	1.34		2.59	3.61
0101	28 ga.		775	.041		1.35	1.34		2.69	3.72
0301	26 ga.		775	.041		1.78	1.34		3.12	4.19
0401	24 ga.		775	.041		2.25	1.34		3.59	4.71
0601	22 ga.		775	.041		2.35	1.34		3.69	4.82
0701	Colored, corrugated/ribbed, on steel frame, 10 yr. finish, 28 ga.		775	.041		2.05	1.34		3.39	4.49
0901	26 ga.		775	.041		2.11	1.34		3.45	4.55
1001	24 ga.		775	.041		2.36	1.34		3.70	4.83

07 46 Siding

07 46 23 – Wood Siding

07 46 23.10 Wood Board Siding

		Crew	Daily Output	Labor-Hours	Unit	Material	2018 Bare Costs Labor	Equipment	Total	Total Incl O&P
0010	**WOOD BOARD SIDING**									
2000	Board & batten, cedar, "B" grade, 1" x 10"	1 Carp	375	.021	S.F.	6.60	.75		7.35	8.50
2200	Redwood, clear, vertical grain, 1" x 10"		375	.021		5.45	.75		6.20	7.20
2400	White pine, #2 & better, 1" x 10"		375	.021		3.68	.75		4.43	5.30
2410	White pine, #2 & better, 1" x 12"		420	.019		3.68	.67		4.35	5.15
3200	Wood, cedar bevel, A grade, 1/2" x 6"		295	.027		4.42	.95		5.37	6.45
3300	1/2" x 8"		330	.024		7.50	.85		8.35	9.65
3500	3/4" x 10", clear grade		375	.021		7.30	.75		8.05	9.25
3600	"B" grade		375	.021		3.99	.75		4.74	5.65
3800	Cedar, rough sawn, 1" x 4", A grade, natural		220	.036		7.30	1.27		8.57	10.15
3900	Stained		220	.036		7.45	1.27		8.72	10.30
4100	1" x 12", board & batten, #3 & Btr., natural		420	.019		4.76	.67		5.43	6.35
4200	Stained		420	.019		5.10	.67		5.77	6.70
4400	1" x 8" channel siding, #3 & Btr., natural		330	.024		4.74	.85		5.59	6.60
4500	Stained		330	.024		5	.85		5.85	6.90
4700	Redwood, clear, beveled, vertical grain, 1/2" x 4"		220	.036		4.81	1.27		6.08	7.40
4750	1/2" x 6"		295	.027		4.80	.95		5.75	6.90

07 46 Siding

07 46 23 – Wood Siding

07 46 23.10 Wood Board Siding	Crew	Daily Output	Labor-Hours	Unit	Material	2018 Bare Costs Labor	Equipment	Total	Total Incl O&P
4800	1/2" x 8"	1 Carp	330	.024	S.F.	5.20	.85		6.05	7.10
5000	3/4" x 10"		375	.021		4.95	.75		5.70	6.70
5200	Channel siding, 1" x 10", B grade		375	.021		4.60	.75		5.35	6.30
5250	Redwood, T&G boards, B grade, 1" x 4"		220	.036		7.50	1.27		8.77	10.35
5270	1" x 8"		330	.024		7.90	.85		8.75	10.10
5400	White pine, rough sawn, 1" x 8", natural		330	.024		2.48	.85		3.33	4.14
5500	Stained		330	.024		2.38	.85		3.23	4.03
5600	T&G, 1" x 8"	▼	330	.024	▼	2.51	.85		3.36	4.17

07 46 29 – Plywood Siding

07 46 29.10 Plywood Siding Options

		Crew	Daily Output	Labor-Hours	Unit	Material	Labor	Equipment	Total	Total Incl O&P
0010	**PLYWOOD SIDING OPTIONS**									
0900	Plywood, medium density overlaid, 3/8" thick	2 Carp	750	.021	S.F.	1.35	.75		2.10	2.73
1000	1/2" thick		700	.023		1.55	.80		2.35	3.04
1100	3/4" thick		650	.025		1.90	.86		2.76	3.52
1600	Texture 1-11, cedar, 5/8" thick, natural		675	.024		2.60	.83		3.43	4.24
1700	Factory stained		675	.024		2.87	.83		3.70	4.54
1900	Texture 1-11, fir, 5/8" thick, natural		675	.024		1.37	.83		2.20	2.89
2000	Factory stained		675	.024		1.91	.83		2.74	3.48
2050	Texture 1-11, S.Y.P., 5/8" thick, natural		675	.024		1.44	.83		2.27	2.96
2100	Factory stained		675	.024		1.51	.83		2.34	3.04
2200	Rough sawn cedar, 3/8" thick, natural		675	.024		1.27	.83		2.10	2.78
2300	Factory stained		675	.024		1.57	.83		2.40	3.11
2500	Rough sawn fir, 3/8" thick, natural		675	.024		.92	.83		1.75	2.39
2600	Factory stained		675	.024		1.09	.83		1.92	2.58
2800	Redwood, textured siding, 5/8" thick	▼	675	.024	▼	2.01	.83		2.84	3.59

07 46 33 – Plastic Siding

07 46 33.10 Vinyl Siding

		Crew	Daily Output	Labor-Hours	Unit	Material	Labor	Equipment	Total	Total Incl O&P
0010	**VINYL SIDING**									
3995	Clapboard profile, woodgrain texture, .048 thick, double 4	2 Carp	495	.032	S.F.	1.07	1.13		2.20	3.06
4000	Double 5		550	.029		1.07	1.02		2.09	2.87
4005	Single 8		495	.032		1.37	1.13		2.50	3.39
4010	Single 10		550	.029		1.65	1.02		2.67	3.50
4015	.044 thick, double 4		495	.032		1.05	1.13		2.18	3.04
4020	Double 5		550	.029		1.07	1.02		2.09	2.87
4025	.042 thick, double 4		495	.032		1.07	1.13		2.20	3.06
4030	Double 5		550	.029		1.07	1.02		2.09	2.87
4035	Cross sawn texture, .040 thick, double 4		495	.032		.73	1.13		1.86	2.69
4040	Double 5		550	.029		.67	1.02		1.69	2.43
4045	Smooth texture, .042 thick, double 4		495	.032		.80	1.13		1.93	2.76
4050	Double 5		550	.029		.80	1.02		1.82	2.57
4055	Single 8		495	.032		.80	1.13		1.93	2.76
4060	Cedar texture, .044 thick, double 4		495	.032		1.14	1.13		2.27	3.14
4065	Double 6		600	.027		1.15	.93		2.08	2.82
4070	Dutch lap profile, woodgrain texture, .048 thick, double 5		550	.029		1.09	1.02		2.11	2.89
4075	.044 thick, double 4.5		525	.030		1.09	1.07		2.16	2.97
4080	.042 thick, double 4.5		525	.030		.92	1.07		1.99	2.79
4085	.040 thick, double 4.5		525	.030		.73	1.07		1.80	2.58
4100	Shake profile, 10" wide		400	.040		3.71	1.40		5.11	6.40
4105	Vertical pattern, .046 thick, double 5		550	.029		1.56	1.02		2.58	3.41
4110	.044 thick, triple 3		550	.029		1.78	1.02		2.80	3.65
4115	.040 thick, triple 4		550	.029		1.68	1.02		2.70	3.54
4120	.040 thick, triple 2.66	▼	550	.029	▼	1.78	1.02		2.80	3.65

For customer support on your Residential Costs with RSMeans data, call 800.448.8182.

07 46 33 – Plastic Siding

07 46 33.10 Vinyl Siding

		Crew	Daily Output	Labor-Hours	Unit	Material	2018 Bare Costs Labor	Equipment	Total	Total Incl O&P
4125	Insulation, fan folded extruded polystyrene, 1/4"	2 Carp	2000	.008	S.F.	.29	.28		.57	.78
4130	3/8"		2000	.008	▼	.32	.28		.60	.81
4135	Accessories, J channel, 5/8" pocket		700	.023	L.F.	.51	.80		1.31	1.89
4140	3/4" pocket		695	.023		.56	.80		1.36	1.95
4145	1-1/4" pocket		680	.024		.86	.82		1.68	2.31
4150	Flexible, 3/4" pocket		600	.027		2.39	.93		3.32	4.18
4155	Under sill finish trim		500	.032		.56	1.12		1.68	2.47
4160	Vinyl starter strip		700	.023		.66	.80		1.46	2.05
4165	Aluminum starter strip		700	.023		.29	.80		1.09	1.65
4170	Window casing, 2-1/2" wide, 3/4" pocket		510	.031		1.71	1.10		2.81	3.70
4175	Outside corner, woodgrain finish, 4" face, 3/4" pocket		700	.023		2.15	.80		2.95	3.70
4180	5/8" pocket		700	.023		2.13	.80		2.93	3.68
4185	Smooth finish, 4" face, 3/4" pocket		700	.023		2.13	.80		2.93	3.68
4190	7/8" pocket		690	.023		2.03	.81		2.84	3.59
4195	1-1/4" pocket		700	.023		1.42	.80		2.22	2.90
4200	Soffit and fascia, 1' overhang, solid		120	.133		4.72	4.66		9.38	12.95
4205	Vented		120	.133		4.72	4.66		9.38	12.95
4207	18" overhang, solid		110	.145		5.50	5.10		10.60	14.50
4208	Vented		110	.145		5.50	5.10		10.60	14.50
4210	2' overhang, solid		100	.160		6.30	5.60		11.90	16.20
4215	Vented		100	.160		6.30	5.60		11.90	16.20
4217	3' overhang, solid		100	.160		7.85	5.60		13.45	17.90
4218	Vented	▼	100	.160	▼	7.85	5.60		13.45	17.90
4220	Colors for siding and soffits, add				S.F.	.15			.15	.17
4225	Colors for accessories and trim, add				L.F.	.31			.31	.34

07 46 33.20 Polypropylene Siding

		Crew	Daily Output	Labor-Hours	Unit	Material	2018 Bare Costs Labor	Equipment	Total	Total Incl O&P
0010	**POLYPROPYLENE SIDING**									
4090	Shingle profile, random grooves, double 7	2 Carp	400	.040	S.F.	3.24	1.40		4.64	5.90
4092	Cornerpost for above	1 Carp	365	.022	L.F.	13.10	.77		13.87	15.65
4095	Triple 5	2 Carp	400	.040	S.F.	3.24	1.40		4.64	5.90
4097	Cornerpost for above	1 Carp	365	.022	L.F.	12.30	.77		13.07	14.80
5000	Staggered butt, double 7"	2 Carp	400	.040	S.F.	3.70	1.40		5.10	6.40
5002	Cornerpost for above	1 Carp	365	.022	L.F.	13.10	.77		13.87	15.65
5010	Half round, double 6-1/4"	2 Carp	360	.044	S.F.	3.70	1.55		5.25	6.65
5020	Shake profile, staggered butt, double 9"	"	510	.031	"	3.70	1.10		4.80	5.90
5022	Cornerpost for above	1 Carp	365	.022	L.F.	9.80	.77		10.57	12
5030	Straight butt, double 7"	2 Carp	400	.040	S.F.	3.70	1.40		5.10	6.40
5032	Cornerpost for above	1 Carp	365	.022	L.F.	13	.77		13.77	15.55
6000	Accessories, J channel, 5/8" pocket	2 Carp	700	.023		.51	.80		1.31	1.89
6010	3/4" pocket		695	.023		.56	.80		1.36	1.95
6020	1-1/4" pocket		680	.024		.86	.82		1.68	2.31
6030	Aluminum starter strip	▼	700	.023	▼	.29	.80		1.09	1.65

07 46 46 – Fiber Cement Siding

07 46 46.10 Fiber Cement Siding

		Crew	Daily Output	Labor-Hours	Unit	Material	2018 Bare Costs Labor	Equipment	Total	Total Incl O&P
0010	**FIBER CEMENT SIDING**									
0020	Lap siding, 5/16" thick, 6" wide, 4-3/4" exposure, smooth texture	2 Carp	415	.039	S.F.	1.31	1.35		2.66	3.68
0025	Woodgrain texture		415	.039		1.31	1.35		2.66	3.68
0030	7-1/2" wide, 6-1/4" exposure, smooth texture		425	.038		1.63	1.32		2.95	3.99
0035	Woodgrain texture		425	.038		1.63	1.32		2.95	3.99
0040	8" wide, 6-3/4" exposure, smooth texture		425	.038		1.23	1.32		2.55	3.54
0045	Rough sawn texture		425	.038		1.23	1.32		2.55	3.54
0050	9-1/2" wide, 8-1/4" exposure, smooth texture	▼	440	.036	▼	1.29	1.27		2.56	3.53

For customer support on your Residential Costs with RSMeans data, call 800.448.8182.

07 46 Siding

07 46 46 – Fiber Cement Siding

07 46 46.10 Fiber Cement Siding

	07 46 46.10 Fiber Cement Siding	Crew	Daily Output	Labor-Hours	Unit	Material	2018 Bare Costs Labor	Equipment	Total	Total Incl O&P
0055	Woodgrain texture	2 Carp	440	.036	S.F.	1.29	1.27		2.56	3.53
0060	12" wide, 10-3/8" exposure, smooth texture		455	.035		2.09	1.23		3.32	4.34
0065	Woodgrain texture		455	.035		2.09	1.23		3.32	4.34
0070	Panel siding, 5/16" thick, smooth texture		750	.021		1.33	.75		2.08	2.70
0075	Stucco texture		750	.021		1.33	.75		2.08	2.70
0080	Grooved woodgrain texture		750	.021		1.33	.75		2.08	2.70
0085	V - grooved woodgrain texture		750	.021		1.33	.75		2.08	2.70
0088	Shingle siding, 48" x 15-1/4" panels, 7" exposure		700	.023	↓	4.19	.80		4.99	5.95
0090	Wood starter strip		400	.040	L.F.	.45	1.40		1.85	2.82
0200	Fiber cement siding, accessories, fascia, 5/4" x 3-1/2"		275	.058		.82	2.03		2.85	4.28
0210	5/4" x 5-1/2"		250	.064		1.98	2.24		4.22	5.90
0220	5/4" x 7-1/2"		225	.071		2.75	2.49		5.24	7.15
0230	5/4" x 9-1/2"	↓	210	.076	↓	3.30	2.66		5.96	8.05

07 46 73 – Soffit

07 46 73.10 Soffit Options

	07 46 73.10 Soffit Options	Crew	Daily Output	Labor-Hours	Unit	Material	2018 Bare Costs Labor	Equipment	Total	Total Incl O&P
0010	**SOFFIT OPTIONS**									
0012	Aluminum, residential, .020" thick	1 Carp	210	.038	S.F.	2.07	1.33		3.40	4.49
0100	Baked enamel on steel, 16 or 18 ga.		105	.076		6.05	2.66		8.71	11.10
0300	Polyvinyl chloride, white, solid		230	.035		2.17	1.22		3.39	4.41
0400	Perforated	↓	230	.035		2.17	1.22		3.39	4.41
0500	For colors, add				↓	.15			.15	.17

07 51 Built-Up Bituminous Roofing

07 51 13 – Built-Up Asphalt Roofing

07 51 13.10 Built-Up Roofing Components

	07 51 13.10 Built-Up Roofing Components	Crew	Daily Output	Labor-Hours	Unit	Material	2018 Bare Costs Labor	Equipment	Total	Total Incl O&P
0010	**BUILT-UP ROOFING COMPONENTS**									
0012	Asphalt saturated felt, #30, 2 sq./roll	1 Rofc	58	.138	Sq.	10.30	4.12		14.42	18.85
0200	#15, 4 sq./roll, plain or perforated, not mopped		58	.138		5.35	4.12		9.47	13.40
0250	Perforated		58	.138		5.35	4.12		9.47	13.40
0300	Roll roofing, smooth, #65		15	.533		10.40	15.95		26.35	40.50
0500	#90		12	.667		37	19.95		56.95	77
0520	Mineralized		12	.667		36	19.95		55.95	76
0540	D.C. (double coverage), 19" selvage edge	↓	10	.800	↓	48	24		72	96
0580	Adhesive (lap cement)				Gal.	8.45			8.45	9.30

07 51 13.20 Built-Up Roofing Systems

	07 51 13.20 Built-Up Roofing Systems	Crew	Daily Output	Labor-Hours	Unit	Material	2018 Bare Costs Labor	Equipment	Total	Total Incl O&P
0010	**BUILT-UP ROOFING SYSTEMS**									
0120	Asphalt flood coat with gravel/slag surfacing, not including									
0140	Insulation, flashing or wood nailers									
0200	Asphalt base sheet, 3 plies #15 asphalt felt, mopped	G-1	22	2.545	Sq.	104	71.50	23.50	199	270
0350	On nailable decks		21	2.667		107	75	24.50	206.50	281
0500	4 plies #15 asphalt felt, mopped		20	2.800		142	78.50	25.50	246	330
0550	On nailable decks	↓	19	2.947	↓	126	82.50	27	235.50	320
2000	Asphalt flood coat, smooth surface									
2200	Asphalt base sheet & 3 plies #15 asphalt felt, mopped	G-1	24	2.333	Sq.	110	65.50	21.50	197	264
2400	On nailable decks		23	2.435		102	68.50	22.50	193	261
2600	4 plies #15 asphalt felt, mopped		24	2.333		129	65.50	21.50	216	285
2700	On nailable decks	↓	23	2.435	↓	121	68.50	22.50	212	282
4500	Coal tar pitch with gravel/slag surfacing									
4600	4 plies #15 tarred felt, mopped	G-1	21	2.667	Sq.	199	75	24.50	298.50	380
4800	3 plies glass fiber felt (type IV), mopped	"	19	2.947	"	164	82.50	27	273.50	360

For customer support on your Residential Costs with RSMeans data, call 800.448.8182.

431

07 51 Built-Up Bituminous Roofing

07 51 13 – Built-Up Asphalt Roofing

07 51 13.30 Cants	Crew	Daily Output	Labor-Hours	Unit	Material	2018 Bare Costs Labor	Equipment	Total	Total Incl O&P
0010 **CANTS**									
0012 Lumber, treated, 4" x 4" cut diagonally	1 Rofc	325	.025	L.F.	1.91	.74		2.65	3.44
0300 Mineral or fiber, trapezoidal, 1" x 4" x 48"		325	.025		.30	.74		1.04	1.67
0400 1-1/2" x 5-5/8" x 48"	↓	325	.025	↓	.48	.74		1.22	1.87

07 52 Modified Bituminous Membrane Roofing

07 52 13 – Atactic-Polypropylene-Modified Bituminous Membrane Roofing

07 52 13.10 APP Modified Bituminous Membrane

07 52 13.10 APP Modified Bituminous Membrane	Crew	Daily Output	Labor-Hours	Unit	Material	2018 Bare Costs Labor	Equipment	Total	Total Incl O&P
0010 **APP MODIFIED BITUMINOUS MEMBRANE** R075213-30									
0020 Base sheet, #15 glass fiber felt, nailed to deck	1 Rofc	58	.138	Sq.	10.95	4.12		15.07	19.55
0030 Spot mopped to deck	G-1	295	.190		15.20	5.30	1.74	22.24	28.50
0040 Fully mopped to deck	"	192	.292		21	8.20	2.68	31.88	41
0050 #15 organic felt, nailed to deck	1 Rofc	58	.138		6.90	4.12		11.02	15.10
0060 Spot mopped to deck	G-1	295	.190		11.15	5.30	1.74	18.19	24
0070 Fully mopped to deck	"	192	.292	↓	16.95	8.20	2.68	27.83	36.50
2100 APP mod., smooth surf. cap sheet, poly. reinf., torched, 160 mils	G-5	2100	.019	S.F.	.76	.52	.09	1.37	1.88
2150 170 mils		2100	.019		.77	.52	.09	1.38	1.89
2200 Granule surface cap sheet, poly. reinf., torched, 180 mils		2000	.020		.96	.55	.09	1.60	2.15
2250 Smooth surface flashing, torched, 160 mils		1260	.032		.76	.87	.14	1.77	2.58
2300 170 mils		1260	.032		.77	.87	.14	1.78	2.59
2350 Granule surface flashing, torched, 180 mils	↓	1260	.032	↓	.96	.87	.14	1.97	2.80
2400 Fibrated aluminum coating	1 Rofc	3800	.002	↓	.08	.06		.14	.21
2450 Seam heat welding	"	205	.039	L.F.	.08	1.17		1.25	2.22

07 52 16 – Styrene-Butadiene-Styrene Modified Bituminous Membrane Roofing

07 52 16.10 SBS Modified Bituminous Membrane

07 52 16.10 SBS Modified Bituminous Membrane	Crew	Daily Output	Labor-Hours	Unit	Material	2018 Bare Costs Labor	Equipment	Total	Total Incl O&P
0010 **SBS MODIFIED BITUMINOUS MEMBRANE**									
0080 Mod. bit. rfng., SBS mod, gran surf. cap sheet, poly. reinf.									
0650 120 to 149 mils thick	G-1	2000	.028	S.F.	1.34	.78	.26	2.38	3.18
0750 150 to 160 mils	"	2000	.028		1.81	.78	.26	2.85	3.70
1150 For reflective granules, add									
1600 Smooth surface cap sheet, mopped, 145 mils	G-1	2100	.027		.82	.75	.24	1.81	2.53
1620 Lightweight base sheet, fiberglass reinforced, 35 to 47 mil		2100	.027		.29	.75	.24	1.28	1.95
1625 Heavyweight base/ply sheet, reinforced, 87 to 120 mil thick	↓	2100	.027		.93	.75	.24	1.92	2.65
1650 Granulated walkpad, 180 to 220 mils	1 Rofc	400	.020		1.88	.60		2.48	3.16
1700 Smooth surface flashing, 145 mils	G-1	1260	.044		.82	1.25	.41	2.48	3.62
1800 150 mils		1260	.044		.51	1.25	.41	2.17	3.28
1900 Granular surface flashing, 150 mils		1260	.044		.72	1.25	.41	2.38	3.51
2000 160 mils	↓	1260	.044		.75	1.25	.41	2.41	3.55
2010 Elastomeric asphalt primer	1 Rofc	2600	.003		.17	.09		.26	.36
2015 Roofing asphalt, 30 lb./square	G-1	19000	.003		.15	.08	.03	.26	.34
2020 Cold process adhesive, 20 to 30 mils thick	1 Rofc	750	.011		.25	.32		.57	.86
2025 Self adhering vapor retarder, 30 to 45 mils thick	G-5	2150	.019	↓	1.07	.51	.08	1.66	2.20
2050 Seam heat welding	1 Rofc	205	.039	L.F.	.08	1.17		1.25	2.22

07 57 Coated Foamed Roofing

07 57 13 – Sprayed Polyurethane Foam Roofing

07 57 13.10 Sprayed Polyurethane Foam Roofing (S.P.F.)	Crew	Daily Output	Labor-Hours	Unit	Material	2018 Bare Costs Labor	Equipment	Total	Total Incl O&P
0010 **SPRAYED POLYURETHANE FOAM ROOFING (S.P.F.)**									
0100 Primer for metal substrate (when required)	G-2A	3000	.008	S.F.	.49	.21	.20	.90	1.14
0200 Primer for non-metal substrate (when required)		3000	.008		.19	.21	.20	.60	.81
0300 Closed cell spray, polyurethane foam, 3 lb./C.F. density, 1", R6.7		15000	.002		.64	.04	.04	.72	.83
0400 2", R13.4		13125	.002		1.29	.05	.05	1.39	1.55
0500 3", R18.6		11485	.002		1.93	.06	.05	2.04	2.28
0550 4", R24.8		10080	.002		2.57	.06	.06	2.69	3.01
0700 Spray-on silicone coating		2500	.010		1.24	.25	.25	1.74	2.08
0800 Warranty 5-20 year manufacturer's								.15	.15
0900 Warranty 20 year, no dollar limit								.20	.20

07 58 Roll Roofing

07 58 10 – Asphalt Roll Roofing

07 58 10.10 Roll Roofing

		Crew	Daily Output	Labor-Hours	Unit	Material	Labor	Equipment	Total	Total Incl O&P
0010 **ROLL ROOFING**										
0100 Asphalt, mineral surface										
0200 1 ply #15 organic felt, 1 ply mineral surfaced										
0300 Selvage roofing, lap 19", nailed & mopped		G-1	27	2.074	Sq.	71	58	19.05	148.05	205
0400 3 plies glass fiber felt (type IV), 1 ply mineral surfaced										
0500 Selvage roofing, lapped 19", mopped		G-1	25	2.240	Sq.	123	63	20.50	206.50	272
0600 Coated glass fiber base sheet, 2 plies of glass fiber										
0700 Felt (type IV), 1 ply mineral surfaced selvage										
0800 Roofing, lapped 19", mopped		G-1	25	2.240	Sq.	131	63	20.50	214.50	281
0900 On nailable decks		"	24	2.333	"	120	65.50	21.50	207	275
1000 3 plies glass fiber felt (type III), 1 ply mineral surfaced										
1100 Selvage roofing, lapped 19", mopped		G-1	25	2.240	Sq.	123	63	20.50	206.50	272

07 61 Sheet Metal Roofing

07 61 13 – Standing Seam Sheet Metal Roofing

07 61 13.10 Standing Seam Sheet Metal Roofing, Field Fab.

		Crew	Daily Output	Labor-Hours	Unit	Material	Labor	Equipment	Total	Total Incl O&P
0010 **STANDING SEAM SHEET METAL ROOFING, FIELD FABRICATED**										
0400 Copper standing seam roofing, over 10 squares, 16 oz., 125 lb./sq.		1 Shee	1.30	6.154	Sq.	1,025	235		1,260	1,525
0600 18 oz., 140 lb./sq.		"	1.20	6.667		1,125	255		1,380	1,675
1200 For abnormal conditions or small areas, add						25%	100%			
1300 For lead-coated copper, add						25%				

07 61 16 – Batten Seam Sheet Metal Roofing

07 61 16.10 Batten Seam Sheet Metal Roofing, Field Fab.

		Crew	Daily Output	Labor-Hours	Unit	Material	Labor	Equipment	Total	Total Incl O&P
0010 **BATTEN SEAM SHEET METAL ROOFING, FIELD FABRICATED**										
0012 Copper batten seam roofing, over 10 sq., 16 oz., 130 lb./sq.		1 Shee	1.10	7.273	Sq.	1,275	278		1,553	1,875
0100 Zinc/copper alloy batten seam roofing, .020" thick			1.20	6.667		1,325	255		1,580	1,875
0200 Copper roofing, batten seam, over 10 sq., 18 oz., 145 lb./sq.			1	8		1,425	305		1,730	2,075
0800 Zinc, copper alloy roofing, batten seam, .027" thick			1.15	6.957		1,675	266		1,941	2,275
0900 .032" thick			1.10	7.273		1,900	278		2,178	2,575
1000 .040" thick			1.05	7.619		2,450	291		2,741	3,175

For customer support on your Residential Costs with RSMeans data, call 800.448.8182.

433

07 61 Sheet Metal Roofing

07 61 19 – Flat Seam Sheet Metal Roofing

07 61 19.10 Flat Seam Sheet Metal Roofing, Field Fabricated	Crew	Daily Output	Labor-Hours	Unit	Material	2018 Bare Costs Labor	Equipment	Total	Total Incl O&P
0010 **FLAT SEAM SHEET METAL ROOFING, FIELD FABRICATED**									
0900 Copper flat seam roofing, over 10 squares, 16 oz., 115 lb./sq.	1 Shee	1.20	6.667	Sq.	940	255		1,195	1,450
0950 18 oz., 130 lb./sq.		1.15	6.957		1,050	266		1,316	1,600
1000 20 oz., 145 lb./sq.		1.10	7.273		1,250	278		1,528	1,850
1008 Zinc flat seam roofing, .020" thick		1.20	6.667		1,125	255		1,380	1,675
1010 .027" thick		1.15	6.957		1,425	266		1,691	2,025
1020 .032" thick		1.12	7.143		1,625	273		1,898	2,225
1030 .040" thick		1.05	7.619		2,100	291		2,391	2,775
1100 Lead flat seam roofing, 5 lb./S.F.	↓	1.30	6.154	↓	1,400	235		1,635	1,950

07 62 Sheet Metal Flashing and Trim

07 62 10 – Sheet Metal Trim

07 62 10.10 Sheet Metal Cladding

	Crew	Daily Output	Labor-Hours	Unit	Material	Labor	Equipment	Total	Total Incl O&P
0010 **SHEET METAL CLADDING**									
0100 Aluminum, up to 6 bends, .032" thick, window casing	1 Carp	180	.044	S.F.	1.80	1.55		3.35	4.56
0200 Window sill		72	.111	L.F.	1.80	3.88		5.68	8.45
0300 Door casing		180	.044	S.F.	1.80	1.55		3.35	4.56
0400 Fascia		250	.032		1.80	1.12		2.92	3.84
0500 Rake trim		225	.036		1.80	1.24		3.04	4.05
0700 .024" thick, window casing		180	.044	↓	1.42	1.55		2.97	4.14
0800 Window sill		72	.111	L.F.	1.42	3.88		5.30	8
0900 Door casing		180	.044	S.F.	1.42	1.55		2.97	4.14
1000 Fascia		250	.032		1.42	1.12		2.54	3.42
1100 Rake trim		225	.036		1.42	1.24		2.66	3.63
1200 Vinyl coated aluminum, up to 6 bends, window casing		180	.044	↓	1.83	1.55		3.38	4.59
1300 Window sill		72	.111	L.F.	1.83	3.88		5.71	8.45
1400 Door casing		180	.044	S.F.	1.83	1.55		3.38	4.59
1500 Fascia		250	.032		1.83	1.12		2.95	3.87
1600 Rake trim	↓	225	.036	↓	1.83	1.24		3.07	4.08

07 65 Flexible Flashing

07 65 10 – Sheet Metal Flashing

07 65 10.10 Sheet Metal Flashing and Counter Flashing

	Crew	Daily Output	Labor-Hours	Unit	Material	Labor	Equipment	Total	Total Incl O&P
0010 **SHEET METAL FLASHING AND COUNTER FLASHING**									
0011 Including up to 4 bends									
0020 Aluminum, mill finish, .013" thick	1 Rofc	145	.055	S.F.	.83	1.65		2.48	3.92
0030 .016" thick		145	.055		.98	1.65		2.63	4.09
0060 .019" thick		145	.055		1.40	1.65		3.05	4.55
0100 .032" thick		145	.055		1.35	1.65		3	4.50
0200 .040" thick		145	.055		2.29	1.65		3.94	5.55
0300 .050" thick		145	.055	↓	2.75	1.65		4.40	6.05
0325 Mill finish 5" x 7" step flashing, .016" thick		1920	.004	Ea.	.15	.12		.27	.40
0350 Mill finish 12" x 12" step flashing, .016" thick	↓	1600	.005	"	.55	.15		.70	.88
0400 Painted finish, add				S.F.	.33			.33	.36
1600 Copper, 16 oz. sheets, under 1000 lb.	1 Rofc	115	.070		8.10	2.08		10.18	12.70
1900 20 oz. sheets, under 1000 lb.		110	.073		10.70	2.17		12.87	15.75
2200 24 oz. sheets, under 1000 lb.		105	.076		14.75	2.28		17.03	20.50
2500 32 oz. sheets, under 1000 lb.		100	.080	↓	19	2.39		21.39	25.50
2700 W shape for valleys, 16 oz., 24" wide	↓	100	.080	L.F.	16.40	2.39		18.79	22.50

07 65 Flexible Flashing

07 65 10 – Sheet Metal Flashing

07 65 10.10 Sheet Metal Flashing and Counter Flashing

07 65 10.10 Sheet Metal Flashing and Counter Flashing	Crew	Daily Output	Labor-Hours	Unit	Material	2018 Bare Costs Labor	Equipment	Total	Total Incl O&P
5800	Lead, 2.5 lb./S.F., up to 12" wide	1 Rofc	135	.059	S.F.	6.15	1.77		7.92	10
5900	Over 12" wide		135	.059		4.04	1.77		5.81	7.65
8900	Stainless steel sheets, 32 ga.		155	.052		3.35	1.54		4.89	6.50
9000	28 ga.		155	.052		4.66	1.54		6.20	7.95
9100	26 ga.		155	.052		4.50	1.54		6.04	7.75
9200	24 ga.		155	.052		5	1.54		6.54	8.30
9290	For mechanically keyed flashing, add					40%				
9320	Steel sheets, galvanized, 20 ga.	1 Rofc	130	.062	S.F.	1.27	1.84		3.11	4.75
9322	22 ga.		135	.059		1.25	1.77		3.02	4.61
9324	24 ga.		140	.057		.95	1.71		2.66	4.16
9326	26 ga.		148	.054		.83	1.62		2.45	3.86
9328	28 ga.		155	.052		.72	1.54		2.26	3.60
9340	30 ga.		160	.050		.60	1.50		2.10	3.39
9400	Terne coated stainless steel, .015" thick, 28 ga.		155	.052		8.10	1.54		9.64	11.70
9500	.018" thick, 26 ga.		155	.052		9.05	1.54		10.59	12.75
9600	Zinc and copper alloy (brass), .020" thick		155	.052		10.25	1.54		11.79	14.10
9700	.027" thick		155	.052		12.25	1.54		13.79	16.30
9800	.032" thick		155	.052		15.50	1.54		17.04	19.85
9900	.040" thick		155	.052		20.50	1.54		22.04	25.50

07 65 13 – Laminated Sheet Flashing

07 65 13.10 Laminated Sheet Flashing

		Crew	Daily Output	Labor-Hours	Unit	Material	Labor	Equipment	Total	Total Incl O&P
0010	**LAMINATED SHEET FLASHING**, Including up to 4 bends									
8550	Shower pan, 3 ply copper and fabric, 3 oz.	1 Rofc	155	.052	S.F.	4	1.54		5.54	7.20
8600	7 oz.	"	155	.052	"	4.73	1.54		6.27	8

07 65 19 – Plastic Sheet Flashing

07 65 19.10 Plastic Sheet Flashing and Counter Flashing

		Crew	Daily Output	Labor-Hours	Unit	Material	Labor	Equipment	Total	Total Incl O&P
0010	**PLASTIC SHEET FLASHING AND COUNTER FLASHING**									
7300	Polyvinyl chloride, black, 10 mil	1 Rofc	285	.028	S.F.	.26	.84		1.10	1.82
7400	20 mil		285	.028		.26	.84		1.10	1.82
7600	30 mil		285	.028		.33	.84		1.17	1.89
7700	60 mil		285	.028		.85	.84		1.69	2.47
8060	PVC tape, 5" x 45 mils, for joint covers, 100 L.F./roll				Ea.	177			177	194

07 65 23 – Rubber Sheet Flashing

07 65 23.10 Rubber Sheet Flashing and Counter Flashing

		Crew	Daily Output	Labor-Hours	Unit	Material	Labor	Equipment	Total	Total Incl O&P
0010	**RUBBER SHEET FLASHING AND COUNTER FLASHING**									
4810	EPDM 90 mils, 1" diameter pipe flashing	1 Rofc	32	.250	Ea.	19.95	7.50		27.45	35.50
4820	2" diameter		30	.267		19.90	7.95		27.85	36.50
4830	3" diameter		28	.286		21	8.55		29.55	39
4840	4" diameter		24	.333		28.50	9.95		38.45	49.50
4850	6" diameter		22	.364		28.50	10.85		39.35	51.50
8100	Rubber, butyl, 1/32" thick		285	.028	S.F.	2.22	.84		3.06	3.97
8200	1/16" thick		285	.028		3.27	.84		4.11	5.15
8300	Neoprene, cured, 1/16" thick		285	.028		2.60	.84		3.44	4.39
8400	1/8" thick		285	.028		6.10	.84		6.94	8.25

07 65 26 – Self-Adhering Sheet Flashing

07 65 26.10 Self-Adhering Sheet or Roll Flashing

		Crew	Daily Output	Labor-Hours	Unit	Material	Labor	Equipment	Total	Total Incl O&P
0010	**SELF-ADHERING SHEET OR ROLL FLASHING**									
0020	Self-adhered flashing, 25 mil cross laminated HDPE, 4" wide	1 Rofc	960	.008	L.F.	.20	.25		.45	.67
0040	6" wide		896	.009		.30	.27		.57	.82
0060	9" wide		832	.010		.45	.29		.74	1.02

For customer support on your Residential Costs with RSMeans data, call 800.448.8182.

435

07 65 Flexible Flashing

07 65 26 – Self-Adhering Sheet Flashing

07 65 26.10 Self-Adhering Sheet or Roll Flashing	Crew	Daily Output	Labor-Hours	Unit	Material	2018 Bare Costs Labor	Equipment	Total	Total Incl O&P
0080	12" wide	1 Rofc	768	.010	L.F.	.60	.31		.91	1.23

07 71 Roof Specialties

07 71 19 – Manufactured Gravel Stops and Fasciae

07 71 19.10 Gravel Stop

		Crew	Daily Output	Labor-Hours	Unit	Material	Labor	Equipment	Total	Total Incl O&P
0010	**GRAVEL STOP**									
0020	Aluminum, .050" thick, 4" face height, mill finish	1 Shee	145	.055	L.F.	6.55	2.11		8.66	10.70
0080	Duranodic finish		145	.055		7.40	2.11		9.51	11.65
0100	Painted		145	.055		7.50	2.11		9.61	11.75
1200	Copper, 16 oz., 3" face height		145	.055		24.50	2.11		26.61	30.50
1300	6" face height		135	.059		33.50	2.27		35.77	41
1350	Galv steel, 24 ga., 4" leg, plain, with continuous cleat, 4" face		145	.055		6.50	2.11		8.61	10.65
1360	6" face height		145	.055		6.55	2.11		8.66	10.70
1500	Polyvinyl chloride, 6" face height		135	.059		5.75	2.27		8.02	10.10
1800	Stainless steel, 24 ga., 6" face height	▼	135	.059	▼	15.80	2.27		18.07	21

07 71 19.30 Fascia

		Crew	Daily Output	Labor-Hours	Unit	Material	Labor	Equipment	Total	Total Incl O&P
0010	**FASCIA**									
0100	Aluminum, reverse board and batten, .032" thick, colored, no furring incl.	1 Shee	145	.055	S.F.	7	2.11		9.11	11.20
0200	Residential type, aluminum	1 Carp	200	.040	L.F.	2.02	1.40		3.42	4.54
0220	Vinyl	"	200	.040	"	2.17	1.40		3.57	4.71
0300	Steel, galv and enameled, stock, no furring, long panels	1 Shee	145	.055	S.F.	5.20	2.11		7.31	9.20
0600	Short panels	"	115	.070	"	5.20	2.66		7.86	10.15

07 71 23 – Manufactured Gutters and Downspouts

07 71 23.10 Downspouts

		Crew	Daily Output	Labor-Hours	Unit	Material	Labor	Equipment	Total	Total Incl O&P
0010	**DOWNSPOUTS**									
0020	Aluminum, embossed, .020" thick, 2" x 3"	1 Shee	190	.042	L.F.	.91	1.61		2.52	3.68
0100	Enameled		190	.042		1.36	1.61		2.97	4.18
0300	.024" thick, 2" x 3"		180	.044		2.12	1.70		3.82	5.15
0400	3" x 4"		140	.057		1.95	2.19		4.14	5.80
0600	Round, corrugated aluminum, 3" diameter, .020" thick		190	.042		2.08	1.61		3.69	4.97
0700	4" diameter, .025" thick		140	.057	▼	3.19	2.19		5.38	7.15
0900	Wire strainer, round, 2" diameter		155	.052	Ea.	1.74	1.97		3.71	5.20
1000	4" diameter		155	.052		2.38	1.97		4.35	5.90
1200	Rectangular, perforated, 2" x 3"		145	.055		2.32	2.11		4.43	6.05
1300	3" x 4"		145	.055	▼	3.33	2.11		5.44	7.15
1500	Copper, round, 16 oz., stock, 2" diameter		190	.042	L.F.	8.35	1.61		9.96	11.85
1600	3" diameter		190	.042		8.60	1.61		10.21	12.15
1800	4" diameter		145	.055		9.80	2.11		11.91	14.30
1900	5" diameter		130	.062		14.65	2.35		17	20
2100	Rectangular, corrugated copper, stock, 2" x 3"		190	.042		8.30	1.61		9.91	11.85
2200	3" x 4"		145	.055		9.35	2.11		11.46	13.80
2400	Rectangular, plain copper, stock, 2" x 3"		190	.042		11.25	1.61		12.86	15.10
2500	3" x 4"		145	.055	▼	14	2.11		16.11	18.90
2700	Wire strainers, rectangular, 2" x 3"		145	.055	Ea.	17.20	2.11		19.31	22.50
2800	3" x 4"		145	.055		17.95	2.11		20.06	23.50
3000	Round, 2" diameter		145	.055		6.50	2.11		8.61	10.65
3100	3" diameter		145	.055		7.25	2.11		9.36	11.50
3300	4" diameter		145	.055		12.25	2.11		14.36	17
3400	5" diameter		115	.070	▼	22.50	2.66		25.16	29.50
3600	Lead-coated copper, round, stock, 2" diameter		190	.042	L.F.	22.50	1.61		24.11	27.50

07 71 Roof Specialties

07 71 23 – Manufactured Gutters and Downspouts

07 71 23.10 Downspouts

		Crew	Daily Output	Labor-Hours	Unit	Material	2018 Bare Costs Labor	Equipment	Total	Total Incl O&P
3700	3" diameter	1 Shee	190	.042	L.F.	23	1.61		24.61	28
3900	4" diameter		145	.055		24	2.11		26.11	30
4000	5" diameter, corrugated		130	.062		24	2.35		26.35	30
4200	6" diameter, corrugated		105	.076		31.50	2.91		34.41	39.50
4300	Rectangular, corrugated, stock, 2" x 3"		190	.042		15.15	1.61		16.76	19.35
4500	Plain, stock, 2" x 3"		190	.042		24.50	1.61		26.11	29.50
4600	3" x 4"		145	.055		33	2.11		35.11	40
4800	Steel, galvanized, round, corrugated, 2" or 3" diameter, 28 ga.		190	.042		2.07	1.61		3.68	4.96
4900	4" diameter, 28 ga.		145	.055		2.08	2.11		4.19	5.80
5700	Rectangular, corrugated, 28 ga., 2" x 3"		190	.042		2.04	1.61		3.65	4.92
5800	3" x 4"		145	.055		2.20	2.11		4.31	5.95
6000	Rectangular, plain, 28 ga., galvanized, 2" x 3"		190	.042		3.83	1.61		5.44	6.90
6100	3" x 4"		145	.055		4.32	2.11		6.43	8.25
6300	Epoxy painted, 24 ga., corrugated, 2" x 3"		190	.042		2.40	1.61		4.01	5.30
6400	3" x 4"		145	.055	▼	2.91	2.11		5.02	6.70
6600	Wire strainers, rectangular, 2" x 3"		145	.055	Ea.	17.30	2.11		19.41	22.50
6700	3" x 4"		145	.055		19	2.11		21.11	24.50
6900	Round strainers, 2" or 3" diameter		145	.055		4.23	2.11		6.34	8.15
7000	4" diameter		145	.055	▼	6.25	2.11		8.36	10.40
8200	Vinyl, rectangular, 2" x 3"		210	.038	L.F.	1.90	1.46		3.36	4.52
8300	Round, 2-1/2"	▼	220	.036	"	1.38	1.39		2.77	3.84

07 71 23.20 Downspout Elbows

		Crew	Daily Output	Labor-Hours	Unit	Material	2018 Bare Costs Labor	Equipment	Total	Total Incl O&P
0010	**DOWNSPOUT ELBOWS**									
0020	Aluminum, embossed, 2" x 3", .020" thick	1 Shee	100	.080	Ea.	.93	3.06		3.99	6.10
0100	Enameled		100	.080		1.78	3.06		4.84	7.05
0200	Embossed, 3" x 4", .025" thick		100	.080		3.10	3.06		6.16	8.50
0300	Enameled		100	.080		3.93	3.06		6.99	9.40
0400	Embossed, corrugated, 3" diameter, .020" thick		100	.080		3.10	3.06		6.16	8.50
0500	4" diameter, .025" thick		100	.080		6.55	3.06		9.61	12.30
0600	Copper, 16 oz., 2" diameter		100	.080		9.45	3.06		12.51	15.50
0700	3" diameter		100	.080		9.10	3.06		12.16	15.10
0800	4" diameter		100	.080		13.95	3.06		17.01	20.50
1000	Rectangular, 2" x 3" corrugated		100	.080		9.25	3.06		12.31	15.25
1100	3" x 4" corrugated		100	.080		15.10	3.06		18.16	22
1300	Vinyl, 2-1/2" diameter, 45 or 75 degree bend		100	.080		3.98	3.06		7.04	9.50
1400	Tee Y junction	▼	75	.107	▼	12.95	4.08		17.03	21

07 71 23.30 Gutters

		Crew	Daily Output	Labor-Hours	Unit	Material	2018 Bare Costs Labor	Equipment	Total	Total Incl O&P
0010	**GUTTERS**									
0012	Aluminum, stock units, 5" K type, .027" thick, plain	1 Shee	125	.064	L.F.	2.80	2.45		5.25	7.15
0100	Enameled		125	.064		2.89	2.45		5.34	7.25
0300	5" K type, .032" thick, plain		125	.064		3.53	2.45		5.98	7.95
0400	Enameled		125	.064		3.52	2.45		5.97	7.95
0700	Copper, half round, 16 oz., stock units, 4" wide		125	.064		8.50	2.45		10.95	13.45
0900	5" wide		125	.064		7.10	2.45		9.55	11.90
1000	6" wide		118	.068		11.25	2.59		13.84	16.70
1200	K type, 16 oz., stock, 5" wide		125	.064		8.25	2.45		10.70	13.20
1300	6" wide		125	.064		9.05	2.45		11.50	14.05
1500	Lead coated copper, 16 oz., half round, stock, 4" wide		125	.064		15.55	2.45		18	21
1600	6" wide		118	.068		18.50	2.59		21.09	25
1800	K type, stock, 5" wide		125	.064		18.50	2.45		20.95	24.50
1900	6" wide		125	.064		18.50	2.45		20.95	24.50
2100	Copper clad stainless steel, K type, 5" wide		125	.064		7.75	2.45		10.20	12.65

07 71 23.30 Gutters		Crew	Daily Output	Labor-Hours	Unit	Material	2018 Bare Costs Labor	2018 Bare Costs Equipment	Total	Total Incl O&P
2200	6" wide	1 Shee	125	.064	L.F.	9.80	2.45		12.25	14.90
2400	Steel, galv, half round or box, 28 ga., 5" wide, plain		125	.064		2.18	2.45		4.63	6.50
2500	Enameled		125	.064		2.28	2.45		4.73	6.60
2700	26 ga., stock, 5" wide		125	.064		2.47	2.45		4.92	6.80
2800	6" wide		125	.064		2.50	2.45		4.95	6.85
3000	Vinyl, O.G., 4" wide	1 Carp	115	.070		1.34	2.43		3.77	5.50
3100	5" wide		115	.070		1.64	2.43		4.07	5.85
3200	4" half round, stock units		115	.070		1.39	2.43		3.82	5.55
3250	Joint connectors				Ea.	3.03			3.03	3.33
3300	Wood, clear treated cedar, fir or hemlock, 3" x 4"	1 Carp	100	.080	L.F.	11	2.80		13.80	16.75
3400	4" x 5"	"	100	.080	"	22	2.80		24.80	28.50
5000	Accessories, end cap, K type, aluminum 5"	1 Shee	625	.013	Ea.	.72	.49		1.21	1.61
5010	6"		625	.013		1.56	.49		2.05	2.54
5020	Copper, 5"		625	.013		3.49	.49		3.98	4.66
5030	6"		625	.013		3.54	.49		4.03	4.71
5040	Lead coated copper, 5"		625	.013		13	.49		13.49	15.10
5050	6"		625	.013		13.90	.49		14.39	16.05
5060	Copper clad stainless steel, 5"		625	.013		3.52	.49		4.01	4.69
5070	6"		625	.013		3.52	.49		4.01	4.69
5080	Galvanized steel, 5"		625	.013		1.36	.49		1.85	2.32
5090	6"		625	.013		2.34	.49		2.83	3.39
5100	Vinyl, 4"	1 Carp	625	.013		6.20	.45		6.65	7.55
5110	5"	"	625	.013		6.55	.45		7	7.95
5120	Half round, copper, 4"	1 Shee	625	.013		4.40	.49		4.89	5.65
5130	5"		625	.013		4.72	.49		5.21	6
5140	6"		625	.013		7.65	.49		8.14	9.25
5150	Lead coated copper, 5"		625	.013		14.65	.49		15.14	16.90
5160	6"		625	.013		22	.49		22.49	25
5170	Copper clad stainless steel, 5"		625	.013		4.53	.49		5.02	5.80
5180	6"		625	.013		4.53	.49		5.02	5.80
5190	Galvanized steel, 5"		625	.013		2.40	.49		2.89	3.46
5200	6"		625	.013		3	.49		3.49	4.12
5210	Outlet, aluminum, 2" x 3"		420	.019		.62	.73		1.35	1.89
5220	3" x 4"		420	.019		1.05	.73		1.78	2.37
5230	2-3/8" round		420	.019		.56	.73		1.29	1.83
5240	Copper, 2" x 3"		420	.019		7.20	.73		7.93	9.10
5250	3" x 4"		420	.019		8.15	.73		8.88	10.20
5260	2-3/8" round		420	.019		4.55	.73		5.28	6.20
5270	Lead coated copper, 2" x 3"		420	.019		26	.73		26.73	29.50
5280	3" x 4"		420	.019		29.50	.73		30.23	33
5290	2-3/8" round		420	.019		26	.73		26.73	29.50
5300	Copper clad stainless steel, 2" x 3"		420	.019		7.20	.73		7.93	9.10
5310	3" x 4"		420	.019		8.15	.73		8.88	10.20
5320	2-3/8" round		420	.019		4.55	.73		5.28	6.20
5330	Galvanized steel, 2" x 3"		420	.019		3.40	.73		4.13	4.95
5340	3" x 4"		420	.019		5.25	.73		5.98	6.95
5350	2-3/8" round		420	.019		4.25	.73		4.98	5.90
5360	K type mitres, aluminum		65	.123		3.38	4.71		8.09	11.55
5370	Copper		65	.123		13.15	4.71		17.86	22.50
5380	Lead coated copper		65	.123		55	4.71		59.71	68.50
5390	Copper clad stainless steel		65	.123		27.50	4.71		32.21	38
5400	Galvanized steel		65	.123		24	4.71		28.71	34.50
5420	Half round mitres, copper		65	.123		63.50	4.71		68.21	78

07 71 Roof Specialties

07 71 23 – Manufactured Gutters and Downspouts

07 71 23.30 Gutters

		Crew	Daily Output	Labor-Hours	Unit	Material	2018 Bare Costs Labor	Equipment	Total	Total Incl O&P
5430	Lead coated copper	1 Shee	65	.123	Ea.	90	4.71		94.71	107
5440	Copper clad stainless steel		65	.123		56.50	4.71		61.21	70
5450	Galvanized steel		65	.123		29	4.71		33.71	40
5460	Vinyl mitres and outlets		65	.123		10.45	4.71		15.16	19.35
5470	Sealant		940	.009	L.F.	.01	.33		.34	.55
5480	Soldering		96	.083	"	.26	3.19		3.45	5.60

07 71 23.35 Gutter Guard

		Crew	Daily Output	Labor-Hours	Unit	Material	2018 Bare Costs Labor	Equipment	Total	Total Incl O&P
0010	**GUTTER GUARD**									
0020	6" wide strip, aluminum mesh	1 Carp	500	.016	L.F.	2.58	.56		3.14	3.77
0100	Vinyl mesh	"	500	.016	"	2.78	.56		3.34	3.99

07 71 43 – Drip Edge

07 71 43.10 Drip Edge, Rake Edge, Ice Belts

		Crew	Daily Output	Labor-Hours	Unit	Material	2018 Bare Costs Labor	Equipment	Total	Total Incl O&P
0010	**DRIP EDGE, RAKE EDGE, ICE BELTS**									
0020	Aluminum, .016" thick, 5" wide, mill finish	1 Carp	400	.020	L.F.	.58	.70		1.28	1.80
0100	White finish		400	.020		.64	.70		1.34	1.86
0200	8" wide, mill finish		400	.020		1.51	.70		2.21	2.82
0300	Ice belt, 28" wide, mill finish		100	.080		7.85	2.80		10.65	13.25
0310	Vented, mill finish		400	.020		2.26	.70		2.96	3.65
0320	Painted finish		400	.020		2.54	.70		3.24	3.95
0400	Galvanized, 5" wide		400	.020		.61	.70		1.31	1.83
0500	8" wide, mill finish		400	.020		.83	.70		1.53	2.07
0510	Rake edge, aluminum, 1-1/2" x 1-1/2"		400	.020		.33	.70		1.03	1.52
0520	3-1/2" x 1-1/2"		400	.020		.45	.70		1.15	1.66

07 72 Roof Accessories

07 72 23 – Relief Vents

07 72 23.20 Vents

		Crew	Daily Output	Labor-Hours	Unit	Material	2018 Bare Costs Labor	Equipment	Total	Total Incl O&P
0010	**VENTS**									
0100	Soffit or eave, aluminum, mill finish, strips, 2-1/2" wide	1 Carp	200	.040	L.F.	.44	1.40		1.84	2.80
0200	3" wide		200	.040		.47	1.40		1.87	2.84
0300	Enamel finish, 3" wide		200	.040		.53	1.40		1.93	2.90
0400	Mill finish, rectangular, 4" x 16"		72	.111	Ea.	1.58	3.88		5.46	8.20
0500	8" x 16"		72	.111		2.50	3.88		6.38	9.20
2420	Roof ventilator	Q-9	16	1		47	34.50		81.50	110
2500	Vent, roof vent	1 Rofc	24	.333		23.50	9.95		33.45	44

07 72 26 – Ridge Vents

07 72 26.10 Ridge Vents and Accessories

		Crew	Daily Output	Labor-Hours	Unit	Material	2018 Bare Costs Labor	Equipment	Total	Total Incl O&P
0010	**RIDGE VENTS AND ACCESSORIES**									
0100	Aluminum strips, mill finish	1 Rofc	160	.050	L.F.	2.40	1.50		3.90	5.35
0150	Painted finish		160	.050	"	4.17	1.50		5.67	7.30
0200	Connectors		48	.167	Ea.	4.95	4.98		9.93	14.55
0300	End caps		48	.167	"	2.25	4.98		7.23	11.60
0400	Galvanized strips		160	.050	L.F.	3.71	1.50		5.21	6.80
0430	Molded polyethylene, shingles not included		160	.050	"	2.78	1.50		4.28	5.80
0440	End plugs		48	.167	Ea.	2.25	4.98		7.23	11.60
0450	Flexible roll, shingles not included		160	.050	L.F.	2.37	1.50		3.87	5.35
2300	Ridge vent strip, mill finish	1 Shee	155	.052	"	3.93	1.97		5.90	7.60

07 72 Roof Accessories

07 72 53 – Snow Guards

07 72 53.10 Snow Guard Options

		Crew	Daily Output	Labor-Hours	Unit	Material	2018 Bare Costs Labor	Equipment	Total	Total Incl O&P
0010	**SNOW GUARD OPTIONS**									
0100	Slate & asphalt shingle roofs, fastened with nails	1 Rofc	160	.050	Ea.	12.35	1.50		13.85	16.35
0200	Standing seam metal roofs, fastened with set screws		48	.167		17.55	4.98		22.53	28.50
0300	Surface mount for metal roofs, fastened with solder		48	.167	↓	7.45	4.98		12.43	17.30
0400	Double rail pipe type, including pipe	↓	130	.062	L.F.	34	1.84		35.84	41

07 72 80 – Vents

07 72 80.30 Vent Options

		Crew	Daily Output	Labor-Hours	Unit	Material	2018 Bare Costs Labor	Equipment	Total	Total Incl O&P
0010	**VENT OPTIONS**									
0800	Polystyrene baffles, 12" wide for 16" OC rafter spacing	1 Carp	90	.089	Ea.	.44	3.11		3.55	5.65
0900	For 24" OC rafter spacing	"	110	.073	"	.79	2.54		3.33	5.10

07 76 Roof Pavers

07 76 16 – Roof Decking Pavers

07 76 16.10 Roof Pavers and Supports

		Crew	Daily Output	Labor-Hours	Unit	Material	2018 Bare Costs Labor	Equipment	Total	Total Incl O&P
0010	**ROOF PAVERS AND SUPPORTS**									
1000	Roof decking pavers, concrete blocks, 2" thick, natural	1 Clab	115	.070	S.F.	3.51	1.86		5.37	6.95
1100	Colors		115	.070	"	3.77	1.86		5.63	7.25
1200	Support pedestal, bottom cap		960	.008	Ea.	3	.22		3.22	3.67
1300	Top cap		960	.008		4.80	.22		5.02	5.65
1400	Leveling shims, 1/16"		1920	.004		1.20	.11		1.31	1.51
1500	1/8"		1920	.004		1.20	.11		1.31	1.51
1600	Buffer pad		960	.008	↓	2.50	.22		2.72	3.12
1700	PVC legs (4" SDR 35)		2880	.003	Inch	.14	.07		.21	.27
2000	Alternate pricing method, system in place	↓	101	.079	S.F.	7.15	2.11		9.26	11.40

07 91 Preformed Joint Seals

07 91 13 – Compression Seals

07 91 13.10 Compression Seals

		Crew	Daily Output	Labor-Hours	Unit	Material	2018 Bare Costs Labor	Equipment	Total	Total Incl O&P
0010	**COMPRESSION SEALS**									
4900	O-ring type cord, 1/4"	1 Bric	472	.017	L.F.	.44	.59		1.03	1.46
4910	1/2"		440	.018		1.20	.63		1.83	2.37
4920	3/4"		424	.019		2.35	.65		3	3.68
4930	1"		408	.020		4.07	.68		4.75	5.60
4940	1-1/4"		384	.021		7.75	.72		8.47	9.70
4950	1-1/2"		368	.022		9.50	.75		10.25	11.70
4960	1-3/4"		352	.023		16.20	.79		16.99	19.15
4970	2"	↓	344	.023		22	.81		22.81	25.50

07 91 16 – Joint Gaskets

07 91 16.10 Joint Gaskets

		Crew	Daily Output	Labor-Hours	Unit	Material	2018 Bare Costs Labor	Equipment	Total	Total Incl O&P
0010	**JOINT GASKETS**									
4400	Joint gaskets, neoprene, closed cell w/adh, 1/8" x 3/8"	1 Bric	240	.033	L.F.	.33	1.16		1.49	2.29
4500	1/4" x 3/4"		215	.037		.63	1.29		1.92	2.85
4700	1/2" x 1"		200	.040		1.50	1.39		2.89	3.97
4800	3/4" x 1-1/2"	↓	165	.048	↓	1.63	1.68		3.31	4.60

440

For customer support on your Residential Costs with RSMeans data, call 800.448.8182.

07 91 Preformed Joint Seals

07 91 23 – Backer Rods

07 91 23.10 Backer Rods	Crew	Daily Output	Labor-Hours	Unit	Material	2018 Bare Costs Labor	Equipment	Total	Total Incl O&P
0010 **BACKER RODS**									
0030 Backer rod, polyethylene, 1/4" diameter	1 Bric	4.60	1.739	C.L.F.	2.37	60.50		62.87	104
0050 1/2" diameter		4.60	1.739		4.28	60.50		64.78	106
0070 3/4" diameter		4.60	1.739		6.75	60.50		67.25	108
0090 1" diameter	↓	4.60	1.739	↓	12.10	60.50		72.60	114

07 91 26 – Joint Fillers

07 91 26.10 Joint Fillers

07 91 26.10 Joint Fillers	Crew	Daily Output	Labor-Hours	Unit	Material	Labor	Equipment	Total	Total Incl O&P
0010 **JOINT FILLERS**									
4360 Butyl rubber filler, 1/4" x 1/4"	1 Bric	290	.028	L.F.	.22	.96		1.18	1.85
4365 1/2" x 1/2"		250	.032		.89	1.11		2	2.84
4370 1/2" x 3/4"		210	.038		1.34	1.32		2.66	3.68
4375 3/4" x 3/4"		230	.035		2.01	1.21		3.22	4.23
4380 1" x 1"	↓	180	.044	↓	2.68	1.54		4.22	5.50
4390 For coloring, add					12%				
4980 Polyethylene joint backing, 1/4" x 2"	1 Bric	2.08	3.846	C.L.F.	13.50	133		146.50	238
4990 1/4" x 6"		1.28	6.250	"	29	217		246	395
5600 Silicone, room temp vulcanizing foam seal, 1/4" x 1/2"		1312	.006	L.F.	.45	.21		.66	.85
5610 1/2" x 1/2"		656	.012		.90	.42		1.32	1.70
5620 1/2" x 3/4"		442	.018		1.35	.63		1.98	2.54
5630 3/4" x 3/4"		328	.024		2.03	.85		2.88	3.65
5640 1/8" x 1"		1312	.006		.45	.21		.66	.85
5650 1/8" x 3"		442	.018		1.35	.63		1.98	2.54
5670 1/4" x 3"		295	.027		2.71	.94		3.65	4.55
5680 1/4" x 6"		148	.054		5.40	1.88		7.28	9.10
5690 1/2" x 6"		82	.098		10.85	3.39		14.24	17.55
5700 1/2" x 9"		52.50	.152		16.25	5.30		21.55	27
5710 1/2" x 12"	↓	33	.242	↓	21.50	8.40		29.90	38

07 92 Joint Sealants

07 92 13 – Elastomeric Joint Sealants

07 92 13.20 Caulking and Sealant Options

07 92 13.20 Caulking and Sealant Options	Crew	Daily Output	Labor-Hours	Unit	Material	Labor	Equipment	Total	Total Incl O&P
0010 **CAULKING AND SEALANT OPTIONS**									
0050 Latex acrylic based, bulk				Gal.	29.50			29.50	32.50
0055 Bulk in place 1/4" x 1/4" bead	1 Bric	300	.027	L.F.	.09	.93		1.02	1.65
0060 1/4" x 3/8"		294	.027		.16	.94		1.10	1.75
0065 1/4" x 1/2"		288	.028		.21	.96		1.17	1.84
0075 3/8" x 3/8"		284	.028		.23	.98		1.21	1.89
0080 3/8" x 1/2"		280	.029		.31	.99		1.30	2
0085 3/8" x 5/8"		276	.029		.39	1.01		1.40	2.11
0095 3/8" x 3/4"		272	.029		.47	1.02		1.49	2.22
0100 1/2" x 1/2"		275	.029		.42	1.01		1.43	2.15
0105 1/2" x 5/8"		269	.030		.52	1.03		1.55	2.29
0110 1/2" x 3/4"		263	.030		.62	1.06		1.68	2.45
0115 1/2" x 7/8"		256	.031		.73	1.08		1.81	2.61
0120 1/2" x 1"		250	.032		.83	1.11		1.94	2.77
0125 3/4" x 3/4"		244	.033		.94	1.14		2.08	2.93
0130 3/4" x 1"		225	.036		1.25	1.23		2.48	3.43
0135 1" x 1"	↓	200	.040	↓	1.66	1.39		3.05	4.15
0190 Cartridges				Gal.	32.50			32.50	36
0200 11 fl. oz. cartridge				Ea.	2.81			2.81	3.09

For customer support on your Residential Costs with RSMeans data, call 800.448.8182.

441

07 92 Joint Sealants

07 92 13 – Elastomeric Joint Sealants

07 92 13.20 Caulking and Sealant Options	Crew	Daily Output	Labor-Hours	Unit	Material	2018 Bare Costs Labor	Equipment	Total	Total Incl O&P
0500	1/4" x 1/2"	1 Bric	288	.028	L.F.	.23	.96		1.19	1.86
0600	1/2" x 1/2"		275	.029		.46	1.01		1.47	2.19
0800	3/4" x 3/4"		244	.033		1.03	1.14		2.17	3.04
0900	3/4" x 1"		225	.036		1.38	1.23		2.61	3.58
1000	1" x 1"		200	.040		1.72	1.39		3.11	4.22
1400	Butyl based, bulk				Gal.	37			37	40.50
1500	Cartridges				"	40.50			40.50	44.50
1700	1/4" x 1/2", 154 L.F./gal.	1 Bric	288	.028	L.F.	.24	.96		1.20	1.87
1800	1/2" x 1/2", 77 L.F./gal.	"	275	.029	"	.48	1.01		1.49	2.22
2300	Polysulfide compounds, 1 component, bulk				Gal.	83			83	91
2400	Cartridges				"	128			128	141
2600	1 or 2 component, in place, 1/4" x 1/4", 308 L.F./gal.	1 Bric	300	.027	L.F.	.27	.93		1.20	1.85
2700	1/2" x 1/4", 154 L.F./gal.		288	.028		.54	.96		1.50	2.20
2900	3/4" x 3/8", 68 L.F./gal.		272	.029		1.22	1.02		2.24	3.05
3000	1" x 1/2", 38 L.F./gal.		250	.032		2.18	1.11		3.29	4.26
3200	Polyurethane, 1 or 2 component				Gal.	53.50			53.50	58.50
3300	Cartridges				"	67			67	73.50
3500	Bulk, in place, 1/4" x 1/4"	1 Bric	300	.027	L.F.	.17	.93		1.10	1.74
3655	1/2" x 1/4"		288	.028		.35	.96		1.31	1.99
3800	3/4" x 3/8"		272	.029		.79	1.02		1.81	2.57
3900	1" x 1/2"		250	.032		1.39	1.11		2.50	3.39
4100	Silicone rubber, bulk				Gal.	57			57	62.50
4200	Cartridges				"	52.50			52.50	57.50

07 92 19 – Acoustical Joint Sealants

07 92 19.10 Acoustical Sealant

		Crew	Daily Output	Labor-Hours	Unit	Material	2018 Bare Costs Labor	Equipment	Total	Total Incl O&P
0010	**ACOUSTICAL SEALANT**									
0020	Acoustical sealant, elastomeric, cartridges				Ea.	8.45			8.45	9.25
0025	In place, 1/4" x 1/4"	1 Bric	300	.027	L.F.	.34	.93		1.27	1.93
0030	1/4" x 1/2"		288	.028		.69	.96		1.65	2.37
0035	1/2" x 1/2"		275	.029		1.38	1.01		2.39	3.20
0040	1/2" x 3/4"		263	.030		2.07	1.06		3.13	4.03
0045	3/4" x 3/4"		244	.033		3.10	1.14		4.24	5.30
0050	1" x 1"		200	.040		5.50	1.39		6.89	8.35

Estimating Tips
08 10 00 Doors and Frames

All exterior doors should be addressed for their energy conservation (insulation and seals).

- Most metal doors and frames look alike, but there may be significant differences among them. When estimating these items, be sure to choose the line item that most closely compares to the specification or door schedule requirements regarding:
  - □ type of metal
  - □ metal gauge
  - □ door core material
  - □ fire rating
  - □ finish
- Wood and plastic doors vary considerably in price. The primary determinant is the veneer material. Lauan, birch, and oak are the most common veneers. Other variables include the following:
  - □ hollow or solid core
  - □ fire rating
  - □ flush or raised panel
  - □ finish
- Door pricing includes bore for cylindrical locksets and mortise for hinges.

08 30 00 Specialty Doors and Frames

- There are many varieties of special doors, and they are usually priced per each. Add frames, hardware, or operators required for a complete installation.

08 40 00 Entrances, Storefronts, and Curtain Walls

- Glazed curtain walls consist of the metal tube framing and the glazing material. The cost data in this subdivision is presented for the metal tube framing alone or the composite wall. If your estimate requires a detailed takeoff of the framing, be sure to add the glazing cost and any tints.

08 50 00 Windows

- Steel windows are unglazed and aluminum can be glazed or unglazed. Some metal windows are priced without glass. Refer to 08 80 00 Glazing for glass pricing. The grade C indicates commercial grade windows, usually ASTM C-35.
- All wood windows and vinyl are priced preglazed. The glazing is insulating glass. Add the cost of screens and grills if required and not already included.

08 70 00 Hardware

- Hardware costs add considerably to the cost of a door. The most efficient method to determine the hardware requirements for a project is to review the door and hardware schedule together. One type of door may have different hardware, depending on the door usage.
- Door hinges are priced by the pair, with most doors requiring 1-1/2 pairs per door. The hinge prices do not include installation labor, because it is included in door installation.

Hinges are classified according to the frequency of use, base material, and finish.

08 80 00 Glazing

- Different openings require different types of glass. The most common types are:
  - □ float
  - □ tempered
  - □ insulating
  - □ impact-resistant
  - □ ballistic-resistant
- Most exterior windows are glazed with insulating glass. Entrance doors and window walls, where the glass is less than 18" from the floor, are generally glazed with tempered glass. Interior windows and some residential windows are glazed with float glass.
- Coastal communities require the use of impact-resistant glass, dependent on wind speed.
- The insulation or 'u' value is a strong consideration, along with solar heat gain, to determine total energy efficiency.

Reference Numbers

Reference numbers are shown at the beginning of some major classifications. These numbers refer to related items in the Reference Section. The reference information may be an estimating procedure, an alternate pricing method, or technical information.

Note: Not all subdivisions listed here necessarily appear. ■

08 01 53 – Operation and Maintenance of Plastic Windows

08 01 53.81 Solid Vinyl Replacement Windows		Crew	Daily Output	Labor-Hours	Unit	Material	2018 Bare Costs Labor	Equipment	Total	Total Incl O&P
0010	**SOLID VINYL REPLACEMENT WINDOWS** R085313-20									
0020	Double-hung, insulated glass, up to 83 united inches	G	2 Carp	8	2	Ea.	173	70		243	305
0040	84 to 93	G		8	2		199	70		269	335
0060	94 to 101	G		6	2.667		199	93		292	375
0080	102 to 111	G		6	2.667		208	93		301	385
0100	112 to 120	G		6	2.667		222	93		315	400
0120	For each united inch over 120, add	G		800	.020	Inch	2.79	.70		3.49	4.23
0140	Casement windows, one operating sash, 42 to 60 united inches	G		8	2	Ea.	228	70		298	365
0160	61 to 70	G		8	2		259	70		329	400
0180	71 to 80	G		8	2		283	70		353	425
0200	81 to 96	G		8	2		298	70		368	445
0220	Two operating sash, 58 to 78 united inches	G		8	2		455	70		525	615
0240	79 to 88	G		8	2		490	70		560	650
0260	89 to 98	G		8	2		525	70		595	695
0280	99 to 108	G		6	2.667		555	93		648	765
0300	109 to 121	G		6	2.667		595	93		688	810
0320	Two operating, one fixed sash, 73 to 108 united inches	G		8	2		715	70		785	905
0340	109 to 118	G		8	2		755	70		825	945
0360	119 to 128	G		6	2.667		775	93		868	1,000
0380	129 to 138	G		6	2.667		820	93		913	1,050
0400	139 to 156	G		6	2.667		860	93		953	1,100
0420	Four operating sash, 98 to 118 united inches	G		8	2		1,025	70		1,095	1,250
0440	119 to 128	G		8	2		1,100	70		1,170	1,325
0460	129 to 138	G		6	2.667		1,175	93		1,268	1,425
0480	139 to 148	G		6	2.667		1,225	93		1,318	1,500
0500	149 to 168	G		6	2.667		1,300	93		1,393	1,600
0520	169 to 178	G		6	2.667		1,425	93		1,518	1,700
0560	Fixed picture window, up to 63 united inches	G		8	2		170	70		240	305
0580	64 to 83	G		8	2		200	70		270	335
0600	84 to 101	G		8	2		240	70		310	380
0620	For each united inch over 101, add	G		900	.018	Inch	2.85	.62		3.47	4.17
0800	Cellulose fiber insulation, poured into sash balance cavity	G	1 Carp	36	.222	C.F.	.69	7.75		8.44	13.65
0820	Silicone caulking at perimeter	G	"	800	.010	L.F.	.17	.35		.52	.77
2000	Impact resistant replacement windows										
2005	Laminated glass, 120 MPH rating, measure in united inches										
2010	Installation labor does not cover any rework of the window opening										
2020	Double-hung, insulated glass, up to 101 united inches		2 Carp	8	2	Ea.	520	70		590	685
2025	For each united inch over 101, add			80	.200	Inch	3.30	7		10.30	15.25
2100	Casement windows, impact resistant, up to 60 united inches			8	2	Ea.	480	70		550	640
2120	61 to 70			8	2		505	70		575	670
2130	71 to 80			8	2		535	70		605	700
2140	81 to 100			8	2		550	70		620	715
2150	For each united inch over 100, add			80	.200	Inch	4.65	7		11.65	16.70
2200	Awning windows, impact resistant, up to 60 united inches			8	2	Ea.	490	70		560	655
2220	61 to 70			8	2		515	70		585	680
2230	71 to 80			8	2		525	70		595	690
2240	For each united inch over 80, add			80	.200	Inch	4.65	7		11.65	16.70
2300	Picture windows, impact resistant, up to 63 united inches			8	2	Ea.	355	70		425	505
2320	63 to 83			8	2		385	70		455	535
2330	84 to 101			8	2		420	70		490	575
2340	For each united inch over 101, add			80	.200	Inch	3.30	7		10.30	15.25

08 05 05 – Selective Demolition for Openings

08 05 05.10 Selective Demolition Doors		Crew	Daily Output	Labor-Hours	Unit	Material	2018 Bare Costs Labor	Equipment	Total	Total Incl O&P
0010	**SELECTIVE DEMOLITION DOORS** R024119-10									
0200	Doors, exterior, 1-3/4" thick, single, 3' x 7' high	1 Clab	16	.500	Ea.		13.35		13.35	22
0210	3' x 8' high		10	.800			21.50		21.50	35.50
0215	Double, 3' x 8' high		6	1.333			35.50		35.50	59
0220	Double, 6' x 7' high		12	.667			17.80		17.80	29.50
0500	Interior, 1-3/8" thick, single, 3' x 7' high		20	.400			10.70		10.70	17.75
0520	Double, 6' x 7' high		16	.500			13.35		13.35	22
0700	Bi-folding, 3' x 6'-8" high		20	.400			10.70		10.70	17.75
0720	6' x 6'-8" high		18	.444			11.85		11.85	19.75
0900	Bi-passing, 3' x 6'-8" high		16	.500			13.35		13.35	22
0940	6' x 6'-8" high		14	.571			15.25		15.25	25.50
0960	Interior metal door 1-3/4" thick, 3'-0" x 6'-8"		18	.444			11.85		11.85	19.75
0980	Interior metal door 1-3/4" thick, 3'-0" x 7'-0"		18	.444			11.85		11.85	19.75
1000	Interior wood door 1-3/4" thick, 3'-0" x 6'-8"		20	.400			10.70		10.70	17.75
1020	Interior wood door 1-3/4" thick, 3'-0" x 7'-0"		20	.400			10.70		10.70	17.75
1100	Door demo, floor door	2 Sswk	5	3.200			124		124	220
1500	Remove and reset, hollow core	1 Carp	8	1			35		35	58
1520	Solid		6	1.333			46.50		46.50	77.50
2000	Frames, including trim, metal		8	1			35		35	58
2200	Wood	2 Carp	32	.500			17.50		17.50	29
2201	Alternate pricing method	1 Carp	200	.040	L.F.		1.40		1.40	2.32
3000	Special doors, counter doors	2 Carp	6	2.667	Ea.		93		93	155
3300	Glass, sliding, including frames		12	1.333			46.50		46.50	77.50
3400	Overhead, commercial, 12' x 12' high		4	4			140		140	232
3500	Residential, 9' x 7' high		8	2			70		70	116
3540	16' x 7' high		7	2.286			80		80	133
3600	Remove and reset, small		4	4			140		140	232
3620	Large		2.50	6.400			224		224	370
3660	Remove and reset elec. garage door opener	1 Carp	8	1			35		35	58
4000	Residential lockset, exterior		28	.286			10		10	16.60
4010	Residential lockset, exterior w/deadbolt		26	.308			10.75		10.75	17.90
4020	Residential lockset, interior		30	.267			9.30		9.30	15.50
4200	Deadbolt lock		32	.250			8.75		8.75	14.55
4224	Pocket door, no frame		8	1			35		35	58
5590	Remove mail slot	1 Clab	45	.178			4.75		4.75	7.90
5600	Remove door sidelight	1 Carp	6	1.333			46.50		46.50	77.50

08 05 05.20 Selective Demolition of Windows

		Crew	Daily Output	Labor-Hours	Unit	Material	Labor	Equipment	Total	Total Incl O&P
0010	**SELECTIVE DEMOLITION OF WINDOWS** R024119-10									
0200	Aluminum, including trim, to 12 S.F.	1 Clab	16	.500	Ea.		13.35		13.35	22
0240	To 25 S.F.		11	.727			19.40		19.40	32.50
0280	To 50 S.F.		5	1.600			42.50		42.50	71
0320	Storm windows/screens, to 12 S.F.		27	.296			7.90		7.90	13.15
0360	To 25 S.F.		21	.381			10.15		10.15	16.90
0400	To 50 S.F.		16	.500			13.35		13.35	22
0500	Screens, incl. aluminum frame, small		20	.400			10.70		10.70	17.75
0510	Large		16	.500			13.35		13.35	22
0600	Glass, up to 10 S.F./window		200	.040	S.F.		1.07		1.07	1.78
0620	Over 10 S.F./window		150	.053	"		1.42		1.42	2.37
2000	Wood, including trim, to 12 S.F.		22	.364	Ea.		9.70		9.70	16.15
2020	To 25 S.F.		18	.444			11.85		11.85	19.75
2060	To 50 S.F.		13	.615			16.45		16.45	27.50
2065	To 180 S.F.		8	1			26.50		26.50	44.50

For customer support on your Residential Costs with RSMeans data, call 800.448.8182.

445

08 05 Common Work Results for Openings

08 05 05 – Selective Demolition for Openings

08 05 05.20 Selective Demolition of Windows	Crew	Daily Output	Labor-Hours	Unit	Material	2018 Bare Costs Labor	2018 Bare Costs Equipment	Total	Total Incl O&P
4300	Remove bay/bow window	2 Carp	6	2.667	Ea.		93		93	155
4410	Remove skylight, plstc domes, flush/curb mtd	"	395	.041	S.F.		1.42		1.42	2.35
4420	Remove skylight, plstc/glass up to 2' x 3'	1 Carp	15	.533	Ea.		18.65		18.65	31
4440	Remove skylight, plstc/glass up to 4' x 6'	2 Carp	10	1.600			56		56	93
4480	Remove roof window up to 3' x 4'	1 Carp	8	1			35		35	58
4500	Remove roof window up to 4' x 6'	2 Carp	6	2.667			93		93	155
5020	Remove and reset window, up to a 2'x 2' window	1 Carp	6	1.333			46.50		46.50	77.50
5040	Up to a 3' x 3' window		4	2			70		70	116
5080	Up to a 4' x 5' window		2	4			140		140	232
6000	Screening only	1 Clab	4000	.002	S.F.		.05		.05	.09
9100	Window awning, residential	"	80	.100	L.F.		2.67		2.67	4.44

08 11 Metal Doors and Frames

08 11 63 – Metal Screen and Storm Doors and Frames

08 11 63.23 Aluminum Screen and Storm Doors and Frames

0010	ALUMINUM SCREEN AND STORM DOORS AND FRAMES									
0020	Combination storm and screen									
0420	Clear anodic coating, 2'-8" wide	2 Carp	14	1.143	Ea.	220	40		260	310
0440	3'-0" wide	"	14	1.143	"	191	40		231	277
0500	For 7'-0" door height, add					8%				
1020	Mill finish, 2'-8" wide	2 Carp	14	1.143	Ea.	250	40		290	340
1040	3'-0" wide	"	14	1.143		270	40		310	365
1100	For 7'-0" door, add					8%				
1520	White painted, 2'-8" wide	2 Carp	14	1.143		239	40		279	330
1540	3'-0" wide		14	1.143		315	40		355	410
1541	Storm door, painted, alum., insul., 6'-8" x 2'-6" wide		14	1.143		186	40		226	272
1545	2'-8" wide		14	1.143		291	40		331	385
1600	For 7'-0" door, add					8%				
1800	Aluminum screen door, 6'-8" x 2'-8" wide	2 Carp	14	1.143		199	40		239	286
1810	3'-0" wide	"	14	1.143		281	40		321	375
2000	Wood door & screen, see Section 08 14 33.20									

08 12 Metal Frames

08 12 13 – Hollow Metal Frames

08 12 13.13 Standard Hollow Metal Frames

0010	STANDARD HOLLOW METAL FRAMES										
0020	16 ga., up to 5-3/4" jamb depth										
0025	3'-0" x 6'-8" single	G	2 Carp	16	1	Ea.	121	35		156	191
0028	3'-6" wide, single	G		16	1		232	35		267	315
0030	4'-0" wide, single	G		16	1		268	35		303	355
0040	6'-0" wide, double	G		14	1.143		227	40		267	315
0045	8'-0" wide, double	G		14	1.143		227	40		267	315
0100	3'-0" x 7'-0" single	G		16	1		192	35		227	269
0110	3'-6" wide, single	G		16	1		194	35		229	271
0112	4'-0" wide, single	G		16	1		227	35		262	310
0140	6'-0" wide, double	G		14	1.143		227	40		267	315
0145	8'-0" wide, double	G		14	1.143		237	40		277	330
1000	16 ga., up to 4-7/8" deep, 3'-0" x 7'-0" single	G		16	1		181	35		216	257
1140	6'-0" wide, double	G		14	1.143		204	40		244	292

08 12 Metal Frames

08 12 13 – Hollow Metal Frames

08 12 13.13 Standard Hollow Metal Frames		Crew	Daily Output	Labor-Hours	Unit	Material	2018 Bare Costs Labor	Equipment	Total	Total Incl O&P
1200	16 ga., 8-3/4" deep, 3'-0" x 7'-0" single	**G** 2 Carp	16	1	Ea.	192	35		227	269
1240	6'-0" wide, double	**G**	14	1.143		256	40		296	350
2800	14 ga., up to 3-7/8" deep, 3'-0" x 7'-0" single	**G**	16	1		247	35		282	330
2840	6'-0" wide, double	**G**	14	1.143		278	40		318	370
3000	14 ga., up to 5-3/4" deep, 3'-0" x 6'-8" single	**G**	16	1		154	35		189	227
3002	3'-6" wide, single	**G**	16	1		239	35		274	320
3005	4'-0" wide, single	**G**	16	1		155	35		190	229
3600	up to 5-3/4" jamb depth, 4'-0" x 7'-0" single	**G**	15	1.067		237	37.50		274.50	325
3620	6'-0" wide, double	**G**	12	1.333		177	46.50		223.50	272
3640	8'-0" wide, double	**G**	12	1.333		282	46.50		328.50	390
3700	8'-0" high, 4'-0" wide, single	**G**	15	1.067		270	37.50		307.50	360
3740	8'-0" wide, double	**G**	12	1.333		320	46.50		366.50	430
4000	6-3/4" deep, 4'-0" x 7'-0" single	**G**	15	1.067		266	37.50		303.50	355
4020	6'-0" wide, double	**G**	12	1.333		300	46.50		346.50	410
4040	8'-0" wide, double	**G**	12	1.333		224	46.50		270.50	325
4100	8'-0" high, 4'-0" wide, single	**G**	15	1.067		176	37.50		213.50	255
4140	8'-0" wide, double	**G**	12	1.333		415	46.50		461.50	535
4400	8-3/4" deep, 4'-0" x 7'-0", single	**G**	15	1.067		385	37.50		422.50	480
4440	8'-0" wide, double	**G**	12	1.333		405	46.50		451.50	530
4500	4'-0" x 8'-0", single	**G**	15	1.067		440	37.50		477.50	545
4540	8'-0" wide, double	**G**	12	1.333		450	46.50		496.50	575
4900	For welded frames, add					51			51	56
5400	14 ga., "B" label, up to 5-3/4" deep, 4'-0" x 7'-0" single	**G** 2 Carp	15	1.067		167	37.50		204.50	246
5440	8'-0" wide, double	**G**	12	1.333		219	46.50		265.50	320
5800	6-3/4" deep, 7'-0" high, 4'-0" wide, single	**G**	15	1.067		173	37.50		210.50	253
5840	8'-0" wide, double	**G**	12	1.333		305	46.50		351.50	420
6200	8-3/4" deep, 4'-0" x 7'-0" single	**G**	15	1.067		247	37.50		284.50	335
6240	8'-0" wide, double	**G**	12	1.333		380	46.50		426.50	500
6300	For "A" label use same price as "B" label									
6400	For baked enamel finish, add					30%	15%			
6500	For galvanizing, add					20%				
6600	For hospital stop, add				Ea.	235			235	259
6620	For hospital stop, stainless steel, add				"	109			109	120
7900	Transom lite frames, fixed, add	2 Carp	155	.103	S.F.	43	3.61		46.61	53.50
8000	Movable, add	"	130	.123	"	55	4.30		59.30	67.50

08 13 Metal Doors

08 13 13 – Hollow Metal Doors

08 13 13.15 Metal Fire Doors

0010	**METAL FIRE DOORS** R081313-20									
0015	Steel, flush, "B" label, 90 minutes									
0020	Full panel, 20 ga., 2'-0" x 6'-8"	2 Carp	20	.800	Ea.	425	28		453	510
0040	2'-8" x 6'-8"		18	.889		440	31		471	535
0060	3'-0" x 6'-8"		17	.941		440	33		473	540
0080	3'-0" x 7'-0"		17	.941		455	33		488	555
0140	18 ga., 3'-0" x 6'-8"		16	1		505	35		540	615
0160	2'-8" x 7'-0"		17	.941		535	33		568	640
0180	3'-0" x 7'-0"		16	1		515	35		550	630
0200	4'-0" x 7'-0"		15	1.067		650	37.50		687.50	775
0220	For "A" label, 3 hour, 18 ga., use same price as "B" label									
0240	For vision lite, add				Ea.	163			163	179

For customer support on your Residential Costs with RSMeans data, call 800.448.8182.

447

08 13 Metal Doors

08 13 13 – Hollow Metal Doors

08 13 13.15 Metal Fire Doors

		Crew	Daily Output	Labor-Hours	Unit	Material	2018 Bare Costs Labor	Equipment	Total	Total Incl O&P
0520	Flush, "B" label, 90 minutes, egress core, 20 ga., 2'-0" x 6'-8"	2 Carp	18	.889	Ea.	665	31		696	785
0540	2'-8" x 6'-8"		17	.941		670	33		703	790
0560	3'-0" x 6'-8"		16	1		675	35		710	800
0580	3'-0" x 7'-0"		16	1		700	35		735	830
0640	Flush, "A" label, 3 hour, egress core, 18 ga., 3'-0" x 6'-8"		15	1.067		730	37.50		767.50	860
0660	2'-8" x 7'-0"		16	1		765	35		800	900
0680	3'-0" x 7'-0"		15	1.067		750	37.50		787.50	880
0700	4'-0" x 7'-0"		14	1.143		910	40		950	1,075

08 13 13.20 Residential Steel Doors

			Crew	Daily Output	Labor-Hours	Unit	Material	2018 Bare Costs Labor	Equipment	Total	Total Incl O&P
0010	**RESIDENTIAL STEEL DOORS**										
0020	Prehung, insulated, exterior										
0030	Embossed, full panel, 2'-8" x 6'-8"	G	2 Carp	17	.941	Ea.	405	33		438	505
0040	3'-0" x 6'-8"	G		15	1.067		290	37.50		327.50	380
0060	3'-0" x 7'-0"	G		15	1.067		350	37.50		387.50	445
0070	5'-4" x 6'-8", double	G		8	2		575	70		645	750
0220	Half glass, 2'-8" x 6'-8"	G		17	.941		330	33		363	420
0240	3'-0" x 6'-8"	G		16	1		330	35		365	425
0260	3'-0" x 7'-0"	G		16	1		400	35		435	500
0270	5'-4" x 6'-8", double	G		8	2		990	70		1,060	1,200
1320	Flush face, full panel, 2'-8" x 6'-8"	G		16	1		291	35		326	380
1340	3'-0" x 6'-8"	G		15	1.067		291	37.50		328.50	380
1360	3'-0" x 7'-0"	G		15	1.067		310	37.50		347.50	400
1380	5'-4" x 6'-8", double	G		8	2		720	70		790	905
1420	Half glass, 2'-8" x 6'-8"	G		17	.941		350	33		383	440
1440	3'-0" x 6'-8"	G		16	1		350	35		385	445
1460	3'-0" x 7'-0"	G		16	1		435	35		470	540
1480	5'-4" x 6'-8", double	G		8	2		665	70		735	845
1500	Sidelight, full lite, 1'-0" x 6'-8" with grille	G					345			345	375
1510	1'-0" x 6'-8", low E	G					365			365	405
1520	1'-0" x 6'-8", half lite	G					305			305	335
1530	1'-0" x 6'-8", half lite, low E	G					296			296	325
2300	Interior, residential, closet, bi-fold, 2'-0" x 6'-8"	G	2 Carp	16	1		222	35		257	300
2330	3'-0" wide	G		16	1		260	35		295	345
2360	4'-0" wide	G		15	1.067		280	37.50		317.50	370
2400	5'-0" wide	G		14	1.143		340	40		380	440
2420	6'-0" wide	G		13	1.231		300	43		343	400
2510	Bi-passing closet, incl. hardware, no frame or trim incl.										
2511	Mirrored, metal frame, 4'-0" x 6'-8"		2 Carp	10	1.600	Opng.	255	56		311	375
2512	5'-0" wide			10	1.600		263	56		319	380
2513	6'-0" wide			10	1.600		295	56		351	420
2514	7'-0" wide			9	1.778		285	62		347	420
2515	8'-0" wide			9	1.778		460	62		522	615
2611	Mirrored, metal, 4'-0" x 8'-0"			10	1.600		315	56		371	440
2612	5'-0" wide			10	1.600		325	56		381	450
2613	6'-0" wide			10	1.600		335	56		391	465
2614	7'-0" wide			9	1.778		355	62		417	495
2615	8'-0" wide			9	1.778		445	62		507	590

08 14 Wood Doors

08 14 13 – Carved Wood Doors

08 14 13.10 Types of Wood Doors, Carved

	08 14 13.10 Types of Wood Doors, Carved	Crew	Daily Output	Labor-Hours	Unit	Material	2018 Bare Costs Labor	Equipment	Total	Total Incl O&P
0010	**TYPES OF WOOD DOORS, CARVED**									
3000	Solid wood, 1-3/4" thick stile and rail									
3020	Mahogany, 3'-0" x 7'-0", six panel	2 Carp	14	1.143	Ea.	1,200	40		1,240	1,375
3030	With two lites		10	1.600		3,050	56		3,106	3,450
3040	3'-6" x 8'-0", six panel		10	1.600		1,625	56		1,681	1,900
3050	With two lites		8	2		2,850	70		2,920	3,275
3100	Pine, 3'-0" x 7'-0", six panel		14	1.143		510	40		550	625
3110	With two lites		10	1.600		805	56		861	980
3120	3'-6" x 8'-0", six panel		10	1.600		1,025	56		1,081	1,225
3130	With two lites		8	2		1,950	70		2,020	2,275
3200	Red oak, 3'-0" x 7'-0", six panel		14	1.143		1,325	40		1,365	1,525
3210	With two lites		10	1.600		2,525	56		2,581	2,875
3220	3'-6" x 8'-0", six panel		10	1.600		2,700	56		2,756	3,075
3230	With two lites	↓	8	2	↓	3,500	70		3,570	3,975
4000	Hand carved door, mahogany									
4020	3'-0" x 7'-0", simple design	2 Carp	14	1.143	Ea.	1,750	40		1,790	2,000
4030	Intricate design		11	1.455		3,675	51		3,726	4,100
4040	3'-6" x 8'-0", simple design		10	1.600		2,775	56		2,831	3,150
4050	Intricate design	↓	8	2		3,675	70		3,745	4,150
4400	For custom finish, add					575			575	635
4600	Side light, mahogany, 7'-0" x 1'-6" wide, 4 lites	2 Carp	18	.889		1,100	31		1,131	1,250
4610	6 lites		14	1.143		2,525	40		2,565	2,850
4620	8'-0" x 1'-6" wide, 4 lites		14	1.143		2,075	40		2,115	2,350
4630	6 lites		10	1.600		2,075	56		2,131	2,375
4640	Side light, oak, 7'-0" x 1'-6" wide, 4 lites		18	.889		1,200	31		1,231	1,375
4650	6 lites		14	1.143		2,100	40		2,140	2,375
4660	8'-0" x 1'-6" wide, 4 lites		14	1.143		1,250	40		1,290	1,450
4670	6 lites	↓	10	1.600	↓	2,100	56		2,156	2,400

08 14 16 – Flush Wood Doors

08 14 16.09 Smooth Wood Doors

	08 14 16.09 Smooth Wood Doors	Crew	Daily Output	Labor-Hours	Unit	Material	2018 Bare Costs Labor	Equipment	Total	Total Incl O&P
0010	**SMOOTH WOOD DOORS**									
0015	Flush, interior, hollow core									
0025	Lauan face, 1-3/8", 3'-0" x 6'-8"	2 Carp	17	.941	Ea.	55	33		88	115
0030	4'-0" x 6'-8"		16	1		126	35		161	197
0140	Birch face, 1-3/8", 2'-6" x 6'-8"		17	.941		113	33		146	180
0180	3'-0" x 6'-8"		17	.941		97	33		130	162
0200	4'-0" x 6'-8"		16	1		158	35		193	232
0202	1-3/4", 2'-0" x 6'-8"		17	.941		57.50	33		90.50	118
0204	2'-4" x 7'-0"		16	1		120	35		155	190
0206	2'-6" x 7'-0"		16	1		124	35		159	194
0208	2'-8" x 7'-0"		16	1		139	35		174	211
0210	3'-0" x 7'-0"		16	1		158	35		193	231
0212	3'-4" x 7'-0"		15	1.067	↓	220	37.50		257.50	305
0214	Pair of 3'-0" x 7'-0"	↓	9	1.778	Pr.	259	62		321	390
0480	For prefinishing, clear, add				Ea.	48.50			48.50	53
0500	For prefinishing, stain, add				"	60.50			60.50	66.50
0620	For dutch door with shelf, add					140%				
1320	M.D. overlay on hardboard, 1-3/8", 2'-0" x 6'-8"	2 Carp	17	.941	Ea.	123	33		156	190
1340	2'-6" x 6'-8"		17	.941		124	33		157	191
1380	3'-0" x 6'-8"		17	.941		130	33		163	198
1400	4'-0" x 6'-8"		16	1		185	35		220	262
1720	H.P. plastic laminate, 1-3/8", 2'-0" x 6'-8"	↓	16	1	↓	275	35		310	365

For customer support on your Residential Costs with RSMeans data, call 800.448.8182.

449

08 14 Wood Doors

08 14 16 – Flush Wood Doors

08 14 16.09 Smooth Wood Doors

		Crew	Daily Output	Labor-Hours	Unit	Material	2018 Bare Costs Labor	Equipment	Total	Total Incl O&P
1740	2'-6" x 6'-8"	2 Carp	16	1	Ea.	270	35		305	355
1780	3'-0" x 6'-8"	↓	15	1.067		299	37.50		336.50	390
1785	Door, plastic laminate, 3'-0" x 6'-8"					299			299	330
1800	4'-0" x 6'-8"	2 Carp	14	1.143		395	40		435	500
2020	Particle core, lauan face, 1-3/8", 2'-6" x 6'-8"		15	1.067		97	37.50		134.50	169
2040	3'-0" x 6'-8"		14	1.143		101	40		141	178
2120	Birch face, 1-3/8", 2'-6" x 6'-8"		15	1.067		110	37.50		147.50	183
2140	3'-0" x 6'-8"		14	1.143		121	40		161	200
3320	M.D. overlay on hardboard, 1-3/8", 2'-6" x 6'-8"		14	1.143		178	40		218	263
3340	3'-0" x 6'-8"		13	1.231		193	43		236	284
4000	Exterior, flush, solid core, birch, 1-3/4" x 2'-6" x 7'-0"		15	1.067		166	37.50		203.50	245
4020	2'-8" wide		15	1.067		247	37.50		284.50	335
4040	3'-0" wide	↓	14	1.143		217	40		257	305
4045	3'-0" x 8'-0"	1 Carp	8	1		475	35		510	585
4100	Oak faced 1-3/4" x 2'-6" x 7'-0"	2 Carp	15	1.067		222	37.50		259.50	305
4120	2'-8" wide		15	1.067		227	37.50		264.50	310
4140	3'-0" wide	↓	14	1.143		215	40		255	305
4160	Walnut faced, 1-3/4" x 3'-0" x 6'-8"	1 Carp	17	.471		510	16.45		526.45	590
4180	3'-6" wide	"	17	.471		615	16.45		631.45	705
4200	Walnut faced, 1-3/4" x 2'-6" x 7'-0"	2 Carp	15	1.067		325	37.50		362.50	420
4220	2'-8" wide		15	1.067		300	37.50		337.50	390
4240	3'-0" wide	↓	14	1.143		291	40		331	385
4250	3'-6" wide	1 Carp	14	.571		710	19.95		729.95	815
4260	3'-0" x 8'-0"		8	1		650	35		685	775
4270	3'-6" wide		8	1		775	35		810	915
4285	Cherry faced, flush, sc, 1-3/4" x 3'-0" x 8'-0" wide	↓	8	1	↓	475	35		510	585

08 14 16.10 Wood Doors Decorator

		Crew	Daily Output	Labor-Hours	Unit	Material	2018 Bare Costs Labor	Equipment	Total	Total Incl O&P
0010	**WOOD DOORS DECORATOR**									
1800	Exterior, flush, solid wood core, birch 1-3/4" x 2'-6" x 7'-0"	2 Carp	15	1.067	Ea.	315	37.50		352.50	405
1820	2'-8" wide		15	1.067		350	37.50		387.50	445
1840	3'-0" wide		14	1.143		355	40		395	455
1900	Oak faced, 1-3/4" x 2'-6" x 7'-0"		15	1.067		204	37.50		241.50	286
1920	2'-8" wide		15	1.067		440	37.50		477.50	545
1940	3'-0" wide		14	1.143		455	40		495	565
2100	Walnut faced, 1-3/4" x 2'-6" x 7'-0"		15	1.067		395	37.50		432.50	495
2120	2'-8" wide		15	1.067		410	37.50		447.50	515
2140	3'-0" wide	↓	14	1.143	↓	435	40		475	545

08 14 33 – Stile and Rail Wood Doors

08 14 33.10 Wood Doors Paneled

		Crew	Daily Output	Labor-Hours	Unit	Material	2018 Bare Costs Labor	Equipment	Total	Total Incl O&P
0010	**WOOD DOORS PANELED**									
0020	Interior, six panel, hollow core, 1-3/8" thick									
0040	Molded hardboard, 2'-0" x 6'-8"	2 Carp	17	.941	Ea.	62	33		95	123
0060	2'-6" x 6'-8"		17	.941		64	33		97	125
0070	2'-8" x 6'-8"		17	.941		67	33		100	128
0080	3'-0" x 6'-8"		17	.941		73.50	33		106.50	136
0140	Embossed print, molded hardboard, 2'-0" x 6'-8"		17	.941		64	33		97	125
0160	2'-6" x 6'-8"		17	.941		64	33		97	125
0180	3'-0" x 6'-8"		17	.941		73.50	33		106.50	136
0540	Six panel, solid, 1-3/8" thick, pine, 2'-0" x 6'-8"		15	1.067		155	37.50		192.50	233
0560	2'-6" x 6'-8"		14	1.143		154	40		194	237
0580	3'-0" x 6'-8"		13	1.231		148	43		191	235
1020	Two panel, bored rail, solid, 1-3/8" thick, pine, 1'-6" x 6'-8"	↓	16	1	↓	245	35		280	325

08 14 Wood Doors

08 14 33 – Stile and Rail Wood Doors

08 14 33.10 Wood Doors Paneled

		Crew	Daily Output	Labor-Hours	Unit	Material	2018 Bare Costs Labor	Equipment	Total	Total Incl O&P
1040	2'-0" x 6'-8"	2 Carp	15	1.067	Ea.	320	37.50		357.50	415
1060	2'-6" x 6'-8"		14	1.143		365	40		405	465
1340	Two panel, solid, 1-3/8" thick, fir, 2'-0" x 6'-8"		15	1.067		160	37.50		197.50	238
1360	2'-6" x 6'-8"		14	1.143		215	40		255	305
1380	3'-0" x 6'-8"		13	1.231		375	43		418	485
1740	Five panel, solid, 1-3/8" thick, fir, 2'-0" x 6'-8"		15	1.067		286	37.50		323.50	375
1760	2'-6" x 6'-8"		14	1.143		380	40		420	485
1780	3'-0" x 6'-8"		13	1.231		380	43		423	490
4190	Exterior, Knotty pine, paneled, 1-3/4", 3'-0" x 6'-8"		16	1		795	35		830	935
4195	Double 1-3/4", 3'-0" x 6'-8"		16	1		1,600	35		1,635	1,800
4200	Ash, paneled, 1-3/4", 3'-0" x 6'-8"		16	1		895	35		930	1,050
4205	Double 1-3/4", 3'-0" x 6'-8"		16	1		1,800	35		1,835	2,025
4210	Cherry, paneled, 1-3/4", 3'-0" x 6'-8"		16	1		1,025	35		1,060	1,200
4215	Double 1-3/4", 3'-0" x 6'-8"		16	1		2,075	35		2,110	2,325
4230	Ash, paneled, 1-3/4", 3'-0" x 8'-0"		16	1		1,075	35		1,110	1,250
4235	Double 1-3/4", 3'-0" x 8'-0"		16	1		2,175	35		2,210	2,425
4240	Hard maple, paneled, 1-3/4", 3'-0" x 8'-0"		16	1		1,175	35		1,210	1,350
4245	Double 1-3/4", 3'-0" x 8'-0"		16	1		2,375	35		2,410	2,650
4250	Cherry, paneled, 1-3/4", 3'-0" x 8'-0"		16	1		1,175	35		1,210	1,350
4255	Double 1-3/4", 3'-0" x 8'-0"		16	1		2,375	35		2,410	2,650

08 14 33.20 Wood Doors Residential

		Crew	Daily Output	Labor-Hours	Unit	Material	2018 Bare Costs Labor	Equipment	Total	Total Incl O&P
0010	**WOOD DOORS RESIDENTIAL**									
0200	Exterior, combination storm & screen, pine									
0260	2'-8" wide	2 Carp	10	1.600	Ea.	310	56		366	440
0280	3'-0" wide		9	1.778		325	62		387	465
0300	7'-1" x 3'-0" wide		9	1.778		355	62		417	495
0400	Full lite, 6'-9" x 2'-6" wide		11	1.455		325	51		376	445
0420	2'-8" wide		10	1.600		330	56		386	460
0440	3'-0" wide		9	1.778		335	62		397	470
0500	7'-1" x 3'-0" wide		9	1.778		365	62		427	505
0604	Door, screen, plain full		12	1.333		292	46.50		338.50	400
0614	Divided		12	1.333		420	46.50		466.50	540
0634	Decor full		12	1.333		505	46.50		551.50	635
0700	Dutch door, pine, 1-3/4" x 2'-8" x 6'-8", 6 panel		12	1.333		695	46.50		741.50	845
0720	Half glass		10	1.600		985	56		1,041	1,175
0800	3'-0" wide, 6 panel		12	1.333		590	46.50		636.50	730
0820	Half glass		10	1.600		1,050	56		1,106	1,250
1000	Entrance door, colonial, 1-3/4" x 6'-8" x 2'-8" wide		16	1		570	35		605	685
1020	6 panel pine, 3'-0" wide		15	1.067		560	37.50		597.50	675
1100	8 panel pine, 2'-8" wide		16	1		685	35		720	810
1120	3'-0" wide		15	1.067		635	37.50		672.50	760
1200	For tempered safety glass lites (min. of 2), add					85.50			85.50	94
1300	Flush, birch, solid core, 1-3/4" x 6'-8" x 2'-8" wide	2 Carp	16	1		142	35		177	215
1320	3'-0" wide		15	1.067		152	37.50		189.50	229
1350	7'-0" x 2'-8" wide		16	1		133	35		168	205
1360	3'-0" wide		15	1.067		155	37.50		192.50	233
1420	6'-8" x 3'-0" wide, fir		16	1		505	35		540	620
1720	Carved mahogany 3'-0" x 6'-8"		15	1.067		1,400	37.50		1,437.50	1,600
1760	Mahogany, 3'-0" x 6'-8"		15	1.067		690	37.50		727.50	820
1930	For dutch door with shelf, add					140%				
2700	Interior, closet, bi-fold, w/hardware, no frame or trim incl.									
2720	Flush, birch, 2'-6" x 6'-8"	2 Carp	13	1.231	Ea.	73	43		116	152

451

For customer support on your Residential Costs with RSMeans data, call 800.448.8182.

08 14 33.20 Wood Doors Residential		Crew	Daily Output	Labor-Hours	Unit	Material	2018 Bare Costs Labor	Equipment	Total	Total Incl O&P
2740	3'-0" wide	2 Carp	13	1.231	Ea.	74.50	43		117.50	154
2760	4'-0" wide		12	1.333		115	46.50		161.50	205
2780	5'-0" wide		11	1.455		114	51		165	211
2800	6'-0" wide		10	1.600		135	56		191	241
2804	Flush lauan 2'-0" x 6'-8"		14	1.143		50	40		90	122
2810	8'-0" wide		9	1.778		203	62		265	325
2817	6'-0" wide		9	1.778		149	62		211	266
2820	Flush, hardboard, primed, 6'-8" x 2'-6" wide		13	1.231		73.50	43		116.50	153
2840	3'-0" wide		13	1.231		85	43		128	165
2860	4'-0" wide		12	1.333		148	46.50		194.50	241
2880	5'-0" wide		11	1.455		177	51		228	280
2900	6'-0" wide		10	1.600		179	56		235	290
2920	Hardboard, primed 7'-0" x 4'-0" wide		12	1.333		193	46.50		239.50	290
2930	6'-0" wide		10	1.600		187	56		243	299
3000	Raised panel pine, 6'-6" or 6'-8" x 2'-6" wide		13	1.231		179	43		222	269
3020	3'-0" wide		13	1.231		289	43		332	390
3040	4'-0" wide		12	1.333		315	46.50		361.50	425
3060	5'-0" wide		11	1.455		405	51		456	530
3080	6'-0" wide		10	1.600		445	56		501	585
3180	Louvered, pine, 6'-6" or 6'-8" x 1'-6" wide		13	1.231		196	43		239	287
3190	2'-0" wide		14	1.143		175	40		215	260
3200	Louvered, pine, 6'-6" or 6'-8" x 2'-6" wide		13	1.231		154	43		197	241
3220	3'-0" wide	↓	13	1.231		239	43		282	335
3225	Door, interior louvered bi-fold, pine, 3'-0" x 6'-8"					239			239	263
3240	4'-0" wide	2 Carp	12	1.333		251	46.50		297.50	355
3260	5'-0" wide		11	1.455		278	51		329	390
3280	6'-0" wide		10	1.600		310	56		366	435
3290	8'-0" wide		10	1.600		470	56		526	615
3300	7'-0" x 3'-0" wide		12	1.333		281	46.50		327.50	390
3320	6'-0" wide	↓	10	1.600	↓	340	56		396	470
4400	Bi-passing closet, incl. hardware and frame, no trim incl.									
4420	Flush, lauan, 6'-8" x 4'-0" wide	2 Carp	12	1.333	Opng.	167	46.50		213.50	262
4440	5'-0" wide		11	1.455		183	51		234	286
4460	6'-0" wide		10	1.600		169	56		225	279
4600	Flush, birch, 6'-8" x 4'-0" wide		12	1.333		254	46.50		300.50	360
4620	5'-0" wide		11	1.455		223	51		274	330
4640	6'-0" wide		10	1.600		335	56		391	465
4800	Louvered, pine, 6'-8" x 4'-0" wide		12	1.333		485	46.50		531.50	610
4820	5'-0" wide		11	1.455		610	51		661	755
4840	6'-0" wide		10	1.600	↓	720	56		776	885
4900	Mirrored, 6'-8" x 4'-0" wide		12	1.333	Ea.	320	46.50		366.50	430
5000	Paneled, pine, 6'-8" x 4'-0" wide		12	1.333	Opng.	490	46.50		536.50	620
5020	5'-0" wide		11	1.455		630	51		681	775
5040	6'-0" wide		10	1.600		840	56		896	1,025
5042	8'-0" wide		12	1.333		980	46.50		1,026.50	1,150
5061	Hardboard, 6'-8" x 4'-0" wide		10	1.600		206	56		262	320
5062	5'-0" wide		10	1.600		207	56		263	320
5063	6'-0" wide	↓	10	1.600	↓	230	56		286	345
6100	Folding accordion, closet, including track and frame									
6200	Rigid PVC	2 Carp	10	1.600	Ea.	108	56		164	212
7310	Passage doors, flush, no frame included									
7320	Hardboard, hollow core, 1-3/8" x 6'-8" x 1'-6" wide	2 Carp	18	.889	Ea.	41.50	31		72.50	97.50
7330	2'-0" wide	↓	18	.889	↓	45	31		76	101

08 14 Wood Doors

08 14 33 – Stile and Rail Wood Doors

08 14 33.20 Wood Doors Residential	Crew	Daily Output	Labor-Hours	Unit	Material	2018 Bare Costs Labor	Equipment	Total	Total Incl O&P
7340	2'-6" wide	2 Carp	18	.889	Ea.	52.50	31		83.50	109
7350	2'-8" wide		18	.889		53	31		84	110
7360	3'-0" wide		17	.941		55	33		88	115
7420	Lauan, hollow core, 1-3/8" x 6'-8" x 1'-6" wide		18	.889		53	31		84	110
7440	2'-0" wide		18	.889		60	31		91	118
7450	2'-4" wide		18	.889		68	31		99	127
7460	2'-6" wide		18	.889		68	31		99	127
7480	2'-8" wide		18	.889		70	31		101	129
7500	3'-0" wide		17	.941		75	33		108	137
7540	2'-6" wide		16	1		83	35		118	150
7560	2'-8" wide		16	1		86	35		121	153
7580	3'-0" wide		16	1		90	35		125	157
7595	Pair of 3'-0" wide		9	1.778	Pr.	180	62		242	300
7700	Birch, hollow core, 1-3/8" x 6'-8" x 1'-6" wide		18	.889	Ea.	62	31		93	120
7720	2'-0" wide		18	.889		69	31		100	128
7740	2'-6" wide		18	.889		79	31		110	139
7760	2'-8" wide		18	.889		82	31		113	142
7780	3'-0" wide		17	.941		88	33		121	152
7790	2'-6" ash/oak door with hinges		18	.889		92	31		123	153
7910	2'-8" wide		16	1		78.50	35		113.50	145
7920	3'-0" wide		16	1		189	35		224	266
7940	Pair of 3'-0" wide		9	1.778	Pr.	380	62		442	520
8000	Pine louvered, 1-3/8" x 6'-8" x 1'-6" wide		19	.842	Ea.	148	29.50		177.50	212
8020	2'-0" wide		18	.889		164	31		195	232
8040	2'-6" wide		18	.889		199	31		230	271
8060	2'-8" wide		18	.889		209	31		240	282
8080	3'-0" wide		17	.941		219	33		252	296
8300	Pine paneled, 1-3/8" x 6'-8" x 1'-6" wide		19	.842		191	29.50		220.50	259
8320	2'-0" wide		18	.889		223	31		254	297
8330	2'-4" wide		18	.889		241	31		272	315
8340	2'-6" wide		18	.889		250	31		281	325
8360	2'-8" wide		18	.889		258	31		289	335
8380	3'-0" wide		17	.941		271	33		304	355
8450	French door, pine, 15 lites, 1-3/8" x 6'-8" x 2'-6" wide		18	.889		247	31		278	325
8470	2'-8" wide		18	.889		256	31		287	335
8490	3'-0" wide		17	.941		264	33		297	345
8804	Pocket door, 6 panel pine, 2'-6" x 6'-8" with frame		10.50	1.524		370	53.50		423.50	495
8814	2'-8" x 6'-8"		10.50	1.524		375	53.50		428.50	500
8824	3'-0" x 6'-8"		10.50	1.524		385	53.50		438.50	510
9000	Passage doors, flush, no frame, birch, solid core, 1-3/8" x 2'-4" x 7'-0"		16	1		125	35		160	196
9020	2'-8" wide		16	1		133	35		168	204
9040	3'-0" wide		16	1		146	35		181	219
9060	3'-4" wide		15	1.067		271	37.50		308.50	360
9080	Pair of 3'-0" wide		9	1.778	Pr.	291	62		353	425
9100	Lauan, solid core, 1-3/8" x 7'-0" x 2'-4" wide		16	1	Ea.	163	35		198	237
9120	2'-8" wide		16	1		147	35		182	220
9140	3'-0" wide		16	1		192	35		227	269
9160	3'-4" wide		15	1.067		206	37.50		243.50	289
9180	Pair of 3'-0" wide		9	1.778	Pr.	385	62		447	525
9200	Hardboard, solid core, 1-3/8" x 7'-0" x 2'-4" wide		16	1	Ea.	167	35		202	241
9220	2'-8" wide		16	1		174	35		209	249
9240	3'-0" wide		16	1		179	35		214	255
9260	3'-4" wide		15	1.067		340	37.50		377.50	435

For customer support on your Residential Costs with RSMeans data, call 800.448.8182.

453

08 14 Wood Doors

08 14 35 – Torrified Doors

08 14 35.10 Torrified Exterior Doors

		Crew	Daily Output	Labor-Hours	Unit	Material	2018 Bare Costs Labor	2018 Bare Costs Equipment	Total	Total Incl O&P
0010	**TORRIFIED EXTERIOR DOORS**									
0020	Wood doors made from torrified wood, exterior									
0030	All doors require a finish be applied, all glass is insulated									
0040	All doors require pilot holes for all fasteners									
0100	6 panel, paint grade poplar, 1-3/4" x 3'-0" x 6'-8"	2 Carp	12	1.333	Ea.	1,325	46.50		1,371.50	1,525
0120	Half glass 3'-0" x 6'-8"	"	12	1.333		1,425	46.50		1,471.50	1,625
0200	Side lite, full glass, 1-3/4" x 1'-2" x 6'-8"					930			930	1,025
0220	Side lite, half glass, 1-3/4" x 1'-2" x 6'-8"					910			910	1,000
0300	Raised face, 2 panel, paint grade poplar, 1-3/4" x 3'-0" x 7'-0"	2 Carp	12	1.333		1,325	46.50		1,371.50	1,525
0320	Side lite, raised face, half glass, 1-3/4" x 1'-2" x 7'-0"					1,050			1,050	1,175
0500	6 panel, Fir, 1-3/4" x 3'-0" x 6'-8"	2 Carp	12	1.333		1,875	46.50		1,921.50	2,125
0520	Half glass 3'-0" x 6'-8"	"	12	1.333		1,850	46.50		1,896.50	2,125
0600	Side lite, full glass, 1-3/4" x 1'-2" x 6'-8"					940			940	1,025
0620	Side lite, half glass, 1-3/4" x 1'-2" x 6'-8"					970			970	1,075
0700	6 panel, Mahogany, 1-3/4" x 3'-0" x 6'-8"	2 Carp	12	1.333		1,975	46.50		2,021.50	2,250
0800	Side lite, full glass, 1-3/4" x 1'-2" x 6'-8"					1,050			1,050	1,150
0820	Side lite, half glass, 1-3/4" x 1'-2" x 6'-8"					1,050			1,050	1,150

08 14 40 – Interior Cafe Doors

08 14 40.10 Cafe Style Doors

		Crew	Daily Output	Labor-Hours	Unit	Material	Labor	Equipment	Total	Total Incl O&P
0010	**CAFE STYLE DOORS**									
6520	Interior cafe doors, 2'-6" opening, stock, panel pine	2 Carp	16	1	Ea.	400	35		435	500
6540	3'-0" opening	"	16	1	"	445	35		480	545
6550	Louvered pine									
6560	2'-6" opening	2 Carp	16	1	Ea.	320	35		355	410
8000	3'-0" opening		16	1		335	35		370	430
8010	2'-6" opening, hardwood		16	1		360	35		395	455
8020	3'-0" opening		16	1		395	35		430	495

08 16 Composite Doors

08 16 13 – Fiberglass Doors

08 16 13.10 Entrance Doors, Fibrous Glass

			Crew	Daily Output	Labor-Hours	Unit	Material	Labor	Equipment	Total	Total Incl O&P
0010	**ENTRANCE DOORS, FIBROUS GLASS**										
0020	Exterior, fiberglass, door, 2'-8" wide x 6'-8" high	G	2 Carp	15	1.067	Ea.	281	37.50		318.50	370
0040	3'-0" wide x 6'-8" high	G		15	1.067		274	37.50		311.50	360
0060	3'-0" wide x 7'-0" high	G		15	1.067		490	37.50		527.50	595
0080	3'-0" wide x 6'-8" high, with two lites	G		15	1.067		340	37.50		377.50	435
0100	3'-0" wide x 8'-0" high, with two lites	G		15	1.067		525	37.50		562.50	640
0110	Half glass, 3'-0" wide x 6'-8" high	G		15	1.067		450	37.50		487.50	555
0120	3'-0" wide x 6'-8" high, low E	G		15	1.067		490	37.50		527.50	595
0130	3'-0" wide x 8'-0" high	G		15	1.067		585	37.50		622.50	705
0140	3'-0" wide x 8'-0" high, low E	G		15	1.067		675	37.50		712.50	805
0150	Side lights, 1'-0" wide x 6'-8" high	G					281			281	310
0160	1'-0" wide x 6'-8" high, low E	G					289			289	320
0180	1'-0" wide x 6'-8" high, full glass	G					340			340	370
0190	1'-0" wide x 6'-8" high, low E	G					360			360	395

08 16 Composite Doors

08 16 14 – French Doors

08 16 14.10 Exterior Doors With Glass Lites	Crew	Daily Output	Labor-Hours	Unit	Material	2018 Bare Costs Labor	Equipment	Total	Total Incl O&P
0010 **EXTERIOR DOORS WITH GLASS LITES**									
0020 French, Fir, 1-3/4", 3'-0" wide x 6'-8" high	2 Carp	12	1.333	Ea.	620	46.50		666.50	760
0025 Double		12	1.333		1,250	46.50		1,296.50	1,425
0030 Maple, 1-3/4", 3'-0" wide x 6'-8" high		12	1.333		695	46.50		741.50	845
0035 Double		12	1.333		1,400	46.50		1,446.50	1,600
0040 Cherry, 1-3/4", 3'-0" wide x 6'-8" high		12	1.333		810	46.50		856.50	970
0045 Double		12	1.333		1,625	46.50		1,671.50	1,850
0100 Mahogany, 1-3/4", 3'-0" wide x 8'-0" high		10	1.600		825	56		881	1,000
0105 Double		10	1.600		1,650	56		1,706	1,925
0110 Fir, 1-3/4", 3'-0" wide x 8'-0" high		10	1.600		1,225	56		1,281	1,450
0115 Double		10	1.600		2,475	56		2,531	2,825
0120 Oak, 1-3/4", 3'-0" wide x 8'-0" high		10	1.600		1,875	56		1,931	2,175
0125 Double	▼	10	1.600	▼	3,750	56		3,806	4,225

08 17 Integrated Door Opening Assemblies

08 17 23 – Integrated Wood Door Opening Assemblies

08 17 23.10 Pre-Hung Doors

	Crew	Daily Output	Labor-Hours	Unit	Material	2018 Bare Costs Labor	Equipment	Total	Total Incl O&P
0010 **PRE-HUNG DOORS**									
0300 Exterior, wood, comb. storm & screen, 6'-9" x 2'-6" wide	2 Carp	15	1.067	Ea.	248	37.50		285.50	335
0320 2'-8" wide		15	1.067		330	37.50		367.50	425
0340 3'-0" wide	▼	15	1.067	▼	310	37.50		347.50	400
0360 For 7'-0" high door, add					40			40	44
1600 Entrance door, flush, birch, solid core									
1620 4-5/8" solid jamb, 1-3/4" x 6'-8" x 2'-8" wide	2 Carp	16	1	Ea.	300	35		335	390
1640 3'-0" wide		16	1		405	35		440	505
1642 5-5/8" jamb	▼	16	1		345	35		380	440
1680 For 7'-0" high door, add				▼	25.50			25.50	28
2000 Entrance door, colonial, 6 panel pine									
2020 4-5/8" solid jamb, 1-3/4" x 6'-8" x 2'-8" wide	2 Carp	16	1	Ea.	660	35		695	785
2040 3'-0" wide	"	16	1		690	35		725	820
2060 For 7'-0" high door, add					56.50			56.50	62
2200 For 5-5/8" solid jamb, add					44.50			44.50	49
2230 French style, exterior, 1 lite, 1-3/4" x 3'-0" x 6'-8"	1 Carp	14	.571		655	19.95		674.95	755
2235 9 lites	"	14	.571		705	19.95		724.95	810
2245 15 lites	2 Carp	14	1.143	▼	735	40		775	875
2250 Double, 15 lites, 2'-0" x 6'-8", 4'-0" opening		7	2.286	Pr.	1,275	80		1,355	1,525
2260 2'-6" x 6'-8", 5'-0" opening		7	2.286		1,425	80		1,505	1,675
2280 3'-0" x 6'-8", 6'-0" opening		7	2.286	▼	1,550	80		1,630	1,825
2430 3'-0" x 7'-0", 15 lites		14	1.143	Ea.	1,000	40		1,040	1,175
2432 Two 3'-0" x 7'-0"		7	2.286	Pr.	2,075	80		2,155	2,400
2435 3'-0" x 8'-0"		14	1.143	Ea.	1,075	40		1,115	1,250
2437 Two, 3'-0" x 8'-0"	▼	7	2.286	Pr.	2,200	80		2,280	2,550
2500 Exterior, metal face, insulated, incl. jamb, brickmold and									
2520 Threshold, flush, 2'-8" x 6'-8"	2 Carp	16	1	Ea.	299	35		334	390
2550 3'-0" x 6'-8"		16	1		296	35		331	385
3500 Embossed, 6 panel, 2'-8" x 6'-8"		16	1		340	35		375	430
3550 3'-0" x 6'-8"		16	1		340	35		375	435
3600 2 narrow lites, 2'-8" x 6'-8"		16	1		310	35		345	400
3650 3'-0" x 6'-8"		16	1		315	35		350	410
3700 Half glass, 2'-8" x 6'-8"	▼	16	1	▼	335	35		370	430

08 17 Integrated Door Opening Assemblies

08 17 23 – Integrated Wood Door Opening Assemblies

08 17 23.10 Pre-Hung Doors		Crew	Daily Output	Labor-Hours	Unit	Material	2018 Bare Costs Labor	Equipment	Total	Total Incl O&P
3750	3'-0" x 6'-8"	2 Carp	16	1	Ea.	350	35		385	445
3800	2 top lites, 2'-8" x 6'-8"		16	1		325	35		360	420
3850	3'-0" x 6'-8"	↓	16	1	↓	365	35		400	460
4000	Interior, passage door, 4-5/8" solid jamb									
4370	Pine, louvered, 2'-8" x 6'-8"	2 Carp	17	.941	Ea.	205	33		238	281
4380	3'-0"		17	.941		213	33		246	289
4400	Lauan, flush, solid core, 1-3/8" x 6'-8" x 2'-6" wide		17	.941		194	33		227	268
4420	2'-8" wide		17	.941		194	33		227	268
4440	3'-0" wide		16	1		211	35		246	290
4600	Hollow core, 1-3/8" x 6'-8" x 2'-6" wide		17	.941		136	33		169	205
4620	2'-8" wide		17	.941		139	33		172	207
4640	3'-0" wide	↓	16	1		142	35		177	215
4700	For 7'-0" high door, add					41			41	45.50
5000	Birch, flush, solid core, 1-3/8" x 6'-8" x 2'-6" wide	2 Carp	17	.941		294	33		327	380
5020	2'-8" wide		17	.941		209	33		242	285
5040	3'-0" wide		16	1		320	35		355	415
5200	Hollow core, 1-3/8" x 6'-8" x 2'-6" wide		17	.941		242	33		275	320
5220	2'-8" wide		17	.941		273	33		306	355
5240	3'-0" wide	↓	16	1		264	35		299	350
5280	For 7'-0" high door, add					34.50			34.50	38
5500	Hardboard paneled, 1-3/8" x 6'-8" x 2'-6" wide	2 Carp	17	.941		152	33		185	222
5520	2'-8" wide		17	.941		163	33		196	234
5540	3'-0" wide		16	1		161	35		196	236
6000	Pine paneled, 1-3/8" x 6'-8" x 2'-6" wide		17	.941		276	33		309	360
6020	2'-8" wide		17	.941		293	33		326	380
6040	3'-0" wide	↓	16	1		300	35		335	390
7200	Prehung, bifold, mirrored, 6'-8" x 5'-0"	1 Carp	9	.889		405	31		436	495
7220	6'-8" x 6'-0"		9	.889		405	31		436	495
7240	6'-8" x 8'-0"		6	1.333		665	46.50		711.50	810
7600	Oak, 6 panel, 1-3/4" x 6'-8" x 3'-0"		17	.471		920	16.45		936.45	1,025
8500	Pocket door frame with lauan, flush, hollow core, 1-3/8" x 3'-0" x 6'-8"	↓	17	.471	↓	271	16.45		287.45	325

08 31 Access Doors and Panels

08 31 13 – Access Doors and Frames

08 31 13.20 Bulkhead/Cellar Doors

		Crew	Daily Output	Labor-Hours	Unit	Material	2018 Bare Costs Labor	Equipment	Total	Total Incl O&P
0010	**BULKHEAD/CELLAR DOORS**									
0020	Steel, not incl. sides, 44" x 62"	1 Carp	5.50	1.455	Ea.	655	51		706	805
0100	52" x 73"		5.10	1.569		825	55		880	995
0500	With sides and foundation plates, 57" x 45" x 24"		4.70	1.702		870	59.50		929.50	1,050
0600	42" x 49" x 51"	↓	4.30	1.860	↓	595	65		660	765

08 31 13.40 Kennel Doors

		Crew	Daily Output	Labor-Hours	Unit	Material	2018 Bare Costs Labor	Equipment	Total	Total Incl O&P
0010	**KENNEL DOORS**									
0020	2 way, swinging type, 13" x 19" opening	2 Carp	11	1.455	Opng.	90	51		141	184
0100	17" x 29" opening		11	1.455		130	51		181	228
0200	9" x 9" opening, electronic with accessories	↓	11	1.455	↓	149	51		200	249

08 32 Sliding Glass Doors

08 32 13 – Sliding Aluminum-Framed Glass Doors

08 32 13.10 Sliding Aluminum Doors

		Crew	Daily Output	Labor-Hours	Unit	Material	2018 Bare Costs Labor	2018 Bare Costs Equipment	Total	Total Incl O&P
0010	**SLIDING ALUMINUM DOORS**									
0350	Aluminum, 5/8" tempered insulated glass, 6' wide									
0400	Premium	2 Carp	4	4	Ea.	1,550	140		1,690	1,925
0450	Economy		4	4		820	140		960	1,125
0500	8' wide, premium		3	5.333		1,700	186		1,886	2,175
0550	Economy		3	5.333		1,475	186		1,661	1,925
0600	12' wide, premium		2.50	6.400		3,000	224		3,224	3,675
0650	Economy		2.50	6.400		1,575	224		1,799	2,100
4000	Aluminum, baked on enamel, temp glass, 6'-8" x 10'-0" wide		4	4		1,100	140		1,240	1,425
4020	Insulating glass, 6'-8" x 6'-0" wide		4	4		955	140		1,095	1,275
4040	8'-0" wide		3	5.333		1,125	186		1,311	1,525
4060	10'-0" wide		2	8		1,400	280		1,680	2,025
4080	Anodized, temp glass, 6'-8" x 6'-0" wide		4	4		455	140		595	730
4100	8'-0" wide		3	5.333		575	186		761	945
4120	10'-0" wide		2	8		660	280		940	1,200

08 32 19 – Sliding Wood-Framed Glass Doors

08 32 19.10 Sliding Wood Doors

		Crew	Daily Output	Labor-Hours	Unit	Material	Labor	Equipment	Total	Total Incl O&P
0010	**SLIDING WOOD DOORS**									
0020	Wood, tempered insul. glass, 6' wide, premium	2 Carp	4	4	Ea.	1,450	140		1,590	1,825
0100	Economy		4	4		1,200	140		1,340	1,550
0150	8' wide, wood, premium		3	5.333		1,850	186		2,036	2,325
0200	Economy		3	5.333		1,525	186		1,711	1,975
0235	10' wide, wood, premium		2.50	6.400		2,675	224		2,899	3,300
0240	Economy		2.50	6.400		2,275	224		2,499	2,875
0250	12' wide, wood, premium		2.50	6.400		3,100	224		3,324	3,775
0300	Economy		2.50	6.400		2,475	224		2,699	3,100

08 32 19.15 Sliding Glass Vinyl-Clad Wood Doors

			Crew	Daily Output	Labor-Hours	Unit	Material	Labor	Equipment	Total	Total Incl O&P
0010	**SLIDING GLASS VINYL-CLAD WOOD DOORS**										
0020	Glass, sliding vinyl-clad, insul. glass, 6'-0" x 6'-8"	G	2 Carp	4	4	Opng.	1,525	140		1,665	1,900
0025	6'-0" x 6'-10" high	G		4	4		1,700	140		1,840	2,075
0030	6'-0" x 8'-0" high	G		4	4		2,050	140		2,190	2,475
0050	5'-0" x 6'-8" high	G		4	4		1,575	140		1,715	1,975
0100	8'-0" x 6'-10" high	G		4	4		2,025	140		2,165	2,450
0104	8'-0" x 6'-8" high	G		4	4		2,025	140		2,165	2,450
0150	8'-0" x 8'-0" high	G		4	4		2,350	140		2,490	2,825
0500	4 leaf, 9'-0" x 6'-10" high	G		3	5.333		3,350	186		3,536	4,000
0550	9'-0" x 8'-0" high	G		3	5.333		3,925	186		4,111	4,600
0600	12'-0" x 6'-10" high	G		3	5.333		4,025	186		4,211	4,725

08 36 Panel Doors

08 36 13 – Sectional Doors

08 36 13.20 Residential Garage Doors

		Crew	Daily Output	Labor-Hours	Unit	Material	Labor	Equipment	Total	Total Incl O&P
0010	**RESIDENTIAL GARAGE DOORS**									
0050	Hinged, wood, custom, double door, 9' x 7'	2 Carp	4	4	Ea.	875	140		1,015	1,200
0070	16' x 7'		3	5.333		1,300	186		1,486	1,725
0200	Overhead, sectional, incl. hardware, fiberglass, 9' x 7', standard		5	3.200		1,025	112		1,137	1,300
0220	Deluxe		5	3.200		1,200	112		1,312	1,500
0300	16' x 7', standard		6	2.667		1,625	93		1,718	1,950
0320	Deluxe		6	2.667		2,250	93		2,343	2,625
0500	Hardboard, 9' x 7', standard		8	2		715	70		785	900

For customer support on your Residential Costs with RSMeans data, call 800.448.8182.

457

08 36 Panel Doors

08 36 13 – Sectional Doors

08 36 13.20 Residential Garage Doors		Crew	Daily Output	Labor-Hours	Unit	Material	2018 Bare Costs Labor	Equipment	Total	Total Incl O&P
0520	Deluxe	2 Carp	8	2	Ea.	860	70		930	1,050
0600	16' x 7', standard		6	2.667		1,300	93		1,393	1,575
0620	Deluxe		6	2.667		1,500	93		1,593	1,800
0700	Metal, 9' x 7', standard		8	2		885	70		955	1,100
0720	Deluxe		6	2.667		995	93		1,088	1,250
0800	16' x 7', standard		6	2.667		1,075	93		1,168	1,325
0820	Deluxe		5	3.200		1,450	112		1,562	1,775
0900	Wood, 9' x 7', standard		8	2		1,025	70		1,095	1,250
0920	Deluxe		8	2		2,250	70		2,320	2,600
1000	16' x 7', standard		6	2.667		1,675	93		1,768	2,000
1020	Deluxe		6	2.667		3,125	93		3,218	3,600
1800	Door hardware, sectional	1 Carp	4	2		360	70		430	515
1810	Door tracks only		4	2		170	70		240	305
1820	One side only		7	1.143		128	40		168	207
4000	For electric operator, economy, add		8	1		440	35		475	545
4100	Deluxe, including remote control		8	1		635	35		670	755
4500	For transmitter/receiver control, add to operator				Total	113			113	124
4600	Transmitters, additional				"	63.50			63.50	70
6000	Replace section, on sectional door, fiberglass, 9' x 7'	1 Carp	4	2	Ea.	635	70		705	810
6020	16' x 7'		3.50	2.286		760	80		840	970
6200	Hardboard, 9' x 7'		4	2		176	70		246	310
6220	16' x 7'		3.50	2.286		238	80		318	395
6300	Metal, 9' x 7'		4	2		225	70		295	365
6320	16' x 7'		3.50	2.286		330	80		410	500
6500	Wood, 9' x 7'		4	2		125	70		195	253
6520	16' x 7'		3.50	2.286		250	80		330	410
7010	Garage doors, row of lites					126			126	138

08 51 Metal Windows

08 51 13 – Aluminum Windows

08 51 13.20 Aluminum Windows

		Crew	Daily Output	Labor-Hours	Unit	Material	2018 Bare Costs Labor	Equipment	Total	Total Incl O&P
0010	**ALUMINUM WINDOWS**, incl. frame and glazing, commercial grade									
1000	Stock units, casement, 3'-1" x 3'-2" opening	2 Sswk	10	1.600	Ea.	380	62		442	530
1040	Insulating glass	"	10	1.600		520	62		582	680
1050	Add for storms					122			122	134
1600	Projected, with screen, 3'-1" x 3'-2" opening	2 Sswk	10	1.600		360	62		422	505
1650	Insulating glass	"	10	1.600		390	62		452	540
1700	Add for storms					119			119	131
2000	4'-5" x 5'-3" opening	2 Sswk	8	2		410	77.50		487.50	590
2050	Insulating glass	"	8	2		480	77.50		557.50	665
2100	Add for storms					128			128	141
2500	Enamel finish windows, 3'-1" x 3'-2"	2 Sswk	10	1.600		365	62		427	515
2550	Insulating glass		10	1.600		315	62		377	460
2600	4'-5" x 5'-3"		8	2		415	77.50		492.50	595
2700	Insulating glass		8	2		415	77.50		492.50	595
3000	Single-hung, 2' x 3' opening, enameled, standard glazed		10	1.600		212	62		274	345
3100	Insulating glass		10	1.600		257	62		319	395
3300	2'-8" x 6'-8" opening, standard glazed		8	2		370	77.50		447.50	545
3400	Insulating glass		8	2		480	77.50		557.50	670
3700	3'-4" x 5'-0" opening, standard glazed		9	1.778		305	69		374	455
3800	Insulating glass		9	1.778		340	69		409	495

For customer support on your Residential Costs with RSMeans data, call 800.448.8182.

08 51 Metal Windows

08 51 13 – Aluminum Windows

08 51 13.20 Aluminum Windows

		Crew	Daily Output	Labor-Hours	Unit	Material	2018 Bare Costs Labor	Equipment	Total	Total Incl O&P
4000	Sliding aluminum, 3' x 2' opening, standard glazed	2 Sswk	10	1.600	Ea.	220	62		282	350
4100	Insulating glass		10	1.600		236	62		298	370
4300	5' x 3' opening, standard glazed		9	1.778		335	69		404	490
4400	Insulating glass		9	1.778		390	69		459	550
4600	8' x 4' opening, standard glazed		6	2.667		355	103		458	580
4700	Insulating glass		6	2.667		575	103		678	820
5000	9' x 5' opening, standard glazed		4	4		540	155		695	870
5100	Insulating glass		4	4		850	155		1,005	1,200
5500	Sliding, with thermal barrier and screen, 6' x 4', 2 track		8	2		725	77.50		802.50	935
5700	4 track	▼	8	2		910	77.50		987.50	1,150
6000	For above units with bronze finish, add					15%				
6200	For installation in concrete openings, add				▼	8%				

08 51 13.30 Impact Resistant Aluminum Windows

		Crew	Daily Output	Labor-Hours	Unit	Material	2018 Bare Costs Labor	Equipment	Total	Total Incl O&P
0010	**IMPACT RESISTANT ALUMINUM WINDOWS**, incl. frame and glazing									
0100	Single-hung, impact resistant, 2'-8" x 5'-0"	2 Carp	9	1.778	Ea.	1,250	62		1,312	1,475
0120	3'-0" x 5'-0"		9	1.778		1,350	62		1,412	1,600
0130	4'-0" x 5'-0"		9	1.778		1,450	62		1,512	1,700
0250	Horizontal slider, impact resistant, 5'-5" x 5'-2"	▼	9	1.778	▼	1,625	62		1,687	1,875

08 51 23 – Steel Windows

08 51 23.40 Basement Utility Windows

		Crew	Daily Output	Labor-Hours	Unit	Material	2018 Bare Costs Labor	Equipment	Total	Total Incl O&P
0010	**BASEMENT UTILITY WINDOWS**									
0015	1'-3" x 2'-8"	1 Carp	16	.500	Ea.	143	17.50		160.50	187
1100	1'-7" x 2'-8"		16	.500		147	17.50		164.50	191
1200	1'-11" x 2'-8"	▼	14	.571	▼	151	19.95		170.95	199

08 51 66 – Metal Window Screens

08 51 66.10 Screens

		Crew	Daily Output	Labor-Hours	Unit	Material	2018 Bare Costs Labor	Equipment	Total	Total Incl O&P
0010	**SCREENS**									
0020	For metal sash, aluminum or bronze mesh, flat screen	2 Sswk	1200	.013	S.F.	4.50	.52		5.02	5.85
0500	Wicket screen, inside window	"	1000	.016	"	6.85	.62		7.47	8.65
0600	Residential, aluminum mesh and frame, 2' x 3'	2 Carp	32	.500	Ea.	17.45	17.50		34.95	48
0610	Rescreen		50	.320		13.90	11.20		25.10	34
0620	3' x 5'		32	.500		58.50	17.50		76	93
0630	Rescreen		45	.356		36	12.45		48.45	60
0640	4' x 8'		25	.640		89.50	22.50		112	136
0650	Rescreen		40	.400		52.50	14		66.50	80.50
0660	Patio door		25	.640	▼	206	22.50		228.50	264
0680	Rescreening	▼	1600	.010	S.F.	2.63	.35		2.98	3.47
1000	Screens for solar louvers	2 Sswk	160	.100	"	25.50	3.88		29.38	35

08 52 Wood Windows

08 52 10 – Plain Wood Windows

08 52 10.10 Wood Windows

		Crew	Daily Output	Labor-Hours	Unit	Material	2018 Bare Costs Labor	Equipment	Total	Total Incl O&P
0010	**WOOD WINDOWS**, including frame, screens and grilles									
0020	Residential, stock units									
0050	Awning type, double insulated glass, 2'-10" x 1'-9" opening	2 Carp	12	1.333	Opng.	230	46.50		276.50	330
0100	2'-10" x 6'-0" opening	1 Carp	8	1		555	35		590	670
0200	4'-0" x 3'-6" single pane		10	.800	▼	375	28		403	460
0300	6' x 5' single pane	▼	8	1	Ea.	510	35		545	620
1000	Casement, 2'-0" x 3'-4" high	2 Carp	20	.800		258	28		286	330
1020	2'-0" x 4'-0"	▼	18	.889	▼	273	31		304	350

For customer support on your Residential Costs with RSMeans data, call 800.448.8182.

459

08 52 10.10 Wood Windows

		Crew	Daily Output	Labor-Hours	Unit	Material	2018 Bare Costs Labor	2018 Bare Costs Equipment	Total	Total Incl O&P
1040	2'-0" x 5'-0"	2 Carp	17	.941	Ea.	315	33		348	405
1060	2'-0" x 6'-0"		16	1		315	35		350	405
1080	4'-0" x 3'-4"		15	1.067		590	37.50		627.50	710
1100	4'-0" x 4'-0"		15	1.067		665	37.50		702.50	790
1120	4'-0" x 5'-0"		14	1.143		750	40		790	890
1140	4'-0" x 6'-0"		12	1.333		845	46.50		891.50	1,000
1600	Casement units, 8' x 5', with screens, double insulated glass		2.50	6.400	Opng.	1,500	224		1,724	2,025
1700	Low E glass		2.50	6.400		1,625	224		1,849	2,175
2300	Casements, including screens, 2'-0" x 3'-4", double insulated glass		11	1.455		275	51		326	390
2400	Low E glass		11	1.455		275	51		326	390
2600	2 lite, 4'-0" x 4'-0", double insulated glass		9	1.778		500	62		562	655
2700	Low E glass		9	1.778		515	62		577	670
2900	3 lite, 5'-2" x 5'-0", double insulated glass		7	2.286		760	80		840	970
3000	Low E glass		7	2.286		805	80		885	1,025
3200	4 lite, 7'-0" x 5'-0", double insulated glass		6	2.667		1,075	93		1,168	1,350
3300	Low E glass		6	2.667		1,150	93		1,243	1,425
3500	5 lite, 8'-6" x 5'-0", double insulated glass		5	3.200		1,450	112		1,562	1,750
3600	Low E glass		5	3.200		1,450	112		1,562	1,750
3800	For removable wood grilles, diamond pattern, add				Leaf	39			39	43
3900	Rectangular pattern, add				"	39			39	43
4000	Bow, fixed lites, 8' x 5', double insulated glass	2 Carp	3	5.333	Opng.	1,450	186		1,636	1,875
4100	Low E glass	"	3	5.333	"	1,975	186		2,161	2,475
4150	6'-0" x 5'-0"	1 Carp	8	1	Ea.	1,275	35		1,310	1,450
4300	Fixed lites, 9'-9" x 5'-0", double insulated glass	2 Carp	2	8	Opng.	995	280		1,275	1,575
4400	Low E glass	"	2	8	"	1,075	280		1,355	1,675
5000	Bow, casement, 8'-1" x 4'-8" high	3 Carp	8	3	Ea.	1,600	105		1,705	1,925
5020	9'-6" x 4'-8"		8	3		1,725	105		1,830	2,075
5040	8'-1" x 5'-1"		8	3		1,925	105		2,030	2,300
5060	9'-6" x 5'-1"		6	4		1,950	140		2,090	2,375
5080	8'-1" x 6'-0"		6	4		1,925	140		2,065	2,350
5100	9'-6" x 6'-0"		6	4		2,050	140		2,190	2,475
5800	Skylights, hatches, vents, and sky roofs, see Section 08 62 13.00									

08 52 10.20 Awning Window

		Crew	Daily Output	Labor-Hours	Unit	Material	2018 Bare Costs Labor	2018 Bare Costs Equipment	Total	Total Incl O&P
0010	**AWNING WINDOW,** Including frame, screens and grilles									
0100	34" x 22", insulated glass	1 Carp	10	.800	Ea.	276	28		304	350
0200	Low E glass		10	.800		298	28		326	375
0300	40" x 28", insulated glass		9	.889		315	31		346	395
0400	Low E glass		9	.889		345	31		376	430
0500	48" x 36", insulated glass		8	1		470	35		505	575
0600	Low E glass		8	1		495	35		530	605

08 52 10.30 Wood Windows

		Crew	Daily Output	Labor-Hours	Unit	Material	2018 Bare Costs Labor	2018 Bare Costs Equipment	Total	Total Incl O&P
0010	**WOOD WINDOWS,** double-hung									
0020	Including frame, double insulated glass, screens and grilles									
0040	Double-hung, 2'-2" x 3'-4" high	2 Carp	15	1.067	Ea.	215	37.50		252.50	299
0060	2'-2" x 4'-4"		14	1.143		234	40		274	325
0080	2'-6" x 3'-4"		13	1.231		225	43		268	320
0100	2'-6" x 4'-0"		12	1.333		233	46.50		279.50	335
0120	2'-6" x 4'-8"		12	1.333		253	46.50		299.50	355
0140	2'-10" x 3'-4"		10	1.600		227	56		283	345
0160	2'-10" x 4'-0"		10	1.600		252	56		308	370
0180	3'-7" x 3'-4"		9	1.778		259	62		321	390
0200	3'-7" x 5'-4"		9	1.778		294	62		356	430

08 52 Wood Windows

08 52 10 – Plain Wood Windows

08 52 10.30 Wood Windows

		Crew	Daily Output	Labor-Hours	Unit	Material	2018 Bare Costs Labor	Equipment	Total	Total Incl O&P
0220	3'-10" x 5'-4"	2 Carp	8	2	Ea.	520	70		590	685
3800	Triple glazing for above, add				▼	25%				

08 52 10.40 Casement Window

			Crew	Daily Output	Labor-Hours	Unit	Material	2018 Bare Costs Labor	Equipment	Total	Total Incl O&P
0010	**CASEMENT WINDOW**, including frame, screen and grilles R085216-10										
0100	2'-0" x 3'-0" H, double insulated glass	G	1 Carp	10	.800	Ea.	282	28		310	355
0150	Low E glass	G		10	.800		275	28		303	345
0200	2'-0" x 4'-6" high, double insulated glass	G		9	.889		385	31		416	475
0250	Low E glass	G		9	.889		385	31		416	475
0260	Casement 4'-2" x 4'-2" double insulated glass	G		11	.727		920	25.50		945.50	1,050
0270	4'-0" x 4'-0" Low E glass	G		11	.727		550	25.50		575.50	645
0290	6'-4" x 5'-7" Low E glass	G		9	.889		1,175	31		1,206	1,350
0300	2'-4" x 6'-0" high, double insulated glass	G		8	1		445	35		480	550
0350	Low E glass	G		8	1		440	35		475	545
0522	Vinyl-clad, premium, double insulated glass, 2'-0" x 3'-0"	G		10	.800		285	28		313	360
0524	2'-0" x 4'-0"	G		9	.889		330	31		361	415
0525	2'-0" x 5'-0"	G		8	1		380	35		415	480
0528	2'-0" x 6'-0"	G		8	1		400	35		435	505
0600	3'-0" x 5'-0"	G		8	1		700	35		735	825
0700	4'-0" x 3'-0"	G		8	1		765	35		800	900
0710	4'-0" x 4'-0"	G		8	1		655	35		690	780
0720	4'-8" x 4'-0"	G		8	1		720	35		755	855
0730	4'-8" x 5'-0"	G		6	1.333		825	46.50		871.50	985
0740	4'-8" x 6'-0"	G		6	1.333		930	46.50		976.50	1,100
0750	6'-0" x 4'-0"	G		6	1.333		845	46.50		891.50	1,000
0800	6'-0" x 5'-0"	G	▼	6	1.333		930	46.50		976.50	1,100
0900	5'-6" x 5'-6"	G	2 Carp	15	1.067	▼	1,475	37.50		1,512.50	1,675
2000	Bay, casement units, 8' x 5', w/screens, double insulated glass			2.50	6.400	Opng.	1,625	224		1,849	2,175
2100	Low E glass		▼	2.50	6.400	"	1,725	224		1,949	2,250
3020	Vinyl-clad, premium, double insulated glass, multiple leaf units										
3080	Single unit, 1'-6" x 5'-0"	G	2 Carp	20	.800	Ea.	330	28		358	405
3100	2'-0" x 2'-0"	G		20	.800		225	28		253	295
3140	2'-0" x 2'-6"	G		20	.800		285	28		313	360
3220	2'-0" x 3'-6"	G		20	.800		290	28		318	365
3260	2'-0" x 4'-0"	G		19	.842		330	29.50		359.50	415
3300	2'-0" x 4'-6"	G		19	.842		335	29.50		364.50	420
3340	2'-0" x 5'-0"	G		18	.889		380	31		411	470
3460	2'-4" x 3'-0"	G		20	.800		290	28		318	365
3500	2'-4" x 4'-0"	G		19	.842		360	29.50		389.50	445
3540	2'-4" x 5'-0"	G		18	.889		425	31		456	520
3700	Double unit, 2'-8" x 5'-0"	G		18	.889		600	31		631	710
3740	2'-8" x 6'-0"	G		17	.941		695	33		728	820
3840	3'-0" x 4'-6"	G		18	.889		540	31		571	645
3860	3'-0" x 5'-0"	G		17	.941		700	33		733	820
3880	3'-0" x 6'-0"	G		17	.941		755	33		788	885
3980	3'-4" x 2'-6"	G		19	.842		460	29.50		489.50	555
4000	3'-4" x 3'-0"	G		12	1.333		465	46.50		511.50	590
4030	3'-4" x 4'-0"	G		18	.889		560	31		591	665
4050	3'-4" x 5'-0"	G		12	1.333		700	46.50		746.50	850
4100	3'-4" x 6'-0"	G		11	1.455		755	51		806	915
4200	3'-6" x 3'-0"	G		18	.889		540	31		571	645
4340	4'-0" x 3'-0"	G		18	.889		520	31		551	625
4380	4'-0" x 3'-6"	G	▼	17	.941		560	33		593	670

For customer support on your Residential Costs with RSMeans data, call 800.448.8182.

461

08 52 10 – Plain Wood Windows

08 52 10.40 Casement Window		Crew	Daily Output	Labor-Hours	Unit	Material	2018 Bare Costs Labor	Equipment	Total	Total Incl O&P
4420	4'-0" x 4'-0" G	2 Carp	16	1	Ea.	655	35		690	780
4460	4'-0" x 4'-4" G		16	1		660	35		695	785
4540	4'-0" x 5'-0" G		16	1		715	35		750	845
4580	4'-0" x 6'-0" G		15	1.067		815	37.50		852.50	955
4740	4'-8" x 3'-0" G		18	.889		580	31		611	690
4780	4'-8" x 3'-6" G		17	.941		620	33		653	735
4820	4'-8" x 4'-0" G		16	1		720	35		755	855
4860	4'-8" x 5'-0" G		15	1.067		825	37.50		862.50	965
4900	4'-8" x 6'-0" G		15	1.067		930	37.50		967.50	1,075
5060	5'-0" x 5'-0" G		15	1.067		1,075	37.50		1,112.50	1,250
5100	Triple unit, 5'-6" x 3'-0" G		17	.941		735	33		768	865
5140	5'-6" x 3'-6" G		16	1		825	35		860	970
5180	5'-6" x 4'-6" G		15	1.067		890	37.50		927.50	1,050
5220	5'-6" x 5'-6" G		15	1.067		1,150	37.50		1,187.50	1,325
5300	6'-0" x 4'-6" G		15	1.067		900	37.50		937.50	1,050
5850	5'-0" x 3'-0" G		12	1.333		620	46.50		666.50	760
5900	5'-0" x 4'-0" G		11	1.455		960	51		1,011	1,125
6100	5'-0" x 5'-6" G		10	1.600		1,100	56		1,156	1,325
6150	5'-0" x 6'-0" G		10	1.600		1,250	56		1,306	1,450
6200	6'-0" x 3'-0" G		12	1.333		1,200	46.50		1,246.50	1,400
6250	6'-0" x 3'-4" G		12	1.333		675	46.50		721.50	825
6300	6'-0" x 4'-0" G		11	1.455		860	51		911	1,025
6350	6'-0" x 5'-0" G		10	1.600		950	56		1,006	1,150
6400	6'-0" x 6'-0" G		10	1.600		1,200	56		1,256	1,425
6500	Quadruple unit, 7'-0" x 4'-0" G		9	1.778		1,150	62		1,212	1,350
6700	8'-0" x 4'-6" G		9	1.778		1,425	62		1,487	1,675
6950	6'-8" x 4'-0" G		10	1.600		1,125	56		1,181	1,325
7000	6'-8" x 6'-0" G	▼	10	1.600		1,500	56		1,556	1,750
8190	For installation, add per leaf				▼		15%			
8200	For multiple leaf units, deduct for stationary sash									
8220	2' high				Ea.	24			24	26.50
8240	4'-6" high					27			27	30
8260	6' high				▼	36.50			36.50	40

08 52 10.50 Double-Hung

	08 52 10.50 Double-Hung	Crew	Daily Output	Labor-Hours	Unit	Material	Labor	Equipment	Total	Total Incl O&P
0010	**DOUBLE-HUNG**, Including frame, screens and grilles									
0100	2'-0" x 3'-0" high, low E insul. glass G	1 Carp	10	.800	Ea.	193	28		221	259
0200	3'-0" x 4'-0" high, double insulated glass G		9	.889		288	31		319	365
0300	4'-0" x 4'-6" high, low E insulated glass G	▼	8	1	▼	335	35		370	430

08 52 10.55 Picture Window

	08 52 10.55 Picture Window	Crew	Daily Output	Labor-Hours	Unit	Material	Labor	Equipment	Total	Total Incl O&P
0010	**PICTURE WINDOW**, Including frame and grilles									
0100	3'-6" x 4'-0" high, double insulated glass	2 Carp	12	1.333	Ea.	435	46.50		481.50	560
0150	Low E glass		12	1.333		445	46.50		491.50	570
0200	4'-0" x 4'-6" high, double insulated glass		11	1.455		550	51		601	690
0250	Low E glass		11	1.455		530	51		581	670
0300	5'-0" x 4'-0" high, double insulated glass		11	1.455		580	51		631	725
0350	Low E glass		11	1.455		605	51		656	750
0400	6'-0" x 4'-6" high, double insulated glass		10	1.600		640	56		696	795
0450	Low E glass	▼	10	1.600	▼	640	56		696	800

08 52 10.65 Wood Sash

	08 52 10.65 Wood Sash	Crew	Daily Output	Labor-Hours	Unit	Material	Labor	Equipment	Total	Total Incl O&P
0010	**WOOD SASH**, Including glazing but not trim									
0050	Custom, 5'-0" x 4'-0", 1" double glazed, 3/16" thick lites	2 Carp	3.20	5	Ea.	233	175		408	545
0100	1/4" thick lites	▼	5	3.200	▼	263	112		375	475

For customer support on your Residential Costs with RSMeans data, call 800.448.8182.

08 52 Wood Windows

08 52 10 – Plain Wood Windows

08 52 10.65 Wood Sash

		Crew	Daily Output	Labor-Hours	Unit	Material	2018 Bare Costs Labor	Equipment	Total	Total Incl O&P
0200	1" thick, triple glazed	2 Carp	5	3.200	Ea.	430	112		542	660
0300	7'-0" x 4'-6" high, 1" double glazed, 3/16" thick lites		4.30	3.721		430	130		560	685
0400	1/4" thick lites		4.30	3.721		490	130		620	750
0500	1" thick, triple glazed		4.30	3.721		560	130		690	830
0600	8'-6" x 5'-0" high, 1" double glazed, 3/16" thick lites		3.50	4.571		580	160		740	905
0700	1/4" thick lites		3.50	4.571		640	160		800	965
0800	1" thick, triple glazed	↓	3.50	4.571	↓	675	160		835	1,000
0900	Window frames only, based on perimeter length				L.F.	4.13			4.13	4.54

08 52 10.70 Sliding Windows

			Crew	Daily Output	Labor-Hours	Unit	Material	2018 Bare Costs Labor	Equipment	Total	Total Incl O&P
0010	**SLIDING WINDOWS**										
0100	3'-0" x 3'-0" high, double insulated	G	1 Carp	10	.800	Ea.	293	28		321	365
0120	Low E glass	G		10	.800		320	28		348	395
0200	4'-0" x 3'-6" high, double insulated	G		9	.889		380	31		411	470
0220	Low E glass	G		9	.889		385	31		416	470
0300	6'-0" x 5'-0" high, double insulated	G		8	1		505	35		540	615
0320	Low E glass	G	↓	8	1	↓	550	35		585	665
6000	Sliding, insulating glass, including screens,										
6100	3'-0" x 3'-0"		2 Carp	6.50	2.462	Ea.	330	86		416	510
6200	4'-0" x 3'-6"			6.30	2.540		335	89		424	520
6300	5'-0" x 4'-0"		↓	6	2.667	↓	415	93		508	615

08 52 13 – Metal-Clad Wood Windows

08 52 13.10 Awning Windows, Metal-Clad

		Crew	Daily Output	Labor-Hours	Unit	Material	2018 Bare Costs Labor	Equipment	Total	Total Incl O&P
0010	**AWNING WINDOWS, METAL-CLAD**									
2000	Metal-clad, awning deluxe, double insulated glass, 34" x 22"	1 Carp	9	.889	Ea.	255	31		286	335
2050	36" x 25"		9	.889		276	31		307	355
2100	40" x 22"		9	.889		295	31		326	375
2150	40" x 30"		9	.889		340	31		371	425
2200	48" x 28"		8	1		350	35		385	445
2250	60" x 36"	↓	8	1	↓	375	35		410	470

08 52 13.20 Casement Windows, Metal-Clad

			Crew	Daily Output	Labor-Hours	Unit	Material	2018 Bare Costs Labor	Equipment	Total	Total Incl O&P
0010	**CASEMENT WINDOWS, METAL-CLAD**										
0100	Metal-clad, deluxe, dbl. insul. glass, 2'-0" x 3'-0" high	G	1 Carp	10	.800	Ea.	294	28		322	370
0120	2'-0" x 4'-0" high	G		9	.889		320	31		351	400
0130	2'-0" x 5'-0" high	G		8	1		340	35		375	435
0140	2'-0" x 6'-0" high	G		8	1		380	35		415	480
0150	Casement window, metal-clad, double insul. glass, 3'-6" x 3'-6"	G	↓	8.90	.899		505	31.50		536.50	605
0300	Metal-clad, casement, bldrs mdl, 6'-0" x 4'-0", dbl. insul. glass, 3 panels		2 Carp	10	1.600		1,200	56		1,256	1,425
0310	9'-0" x 4'-0", 4 panels			8	2		1,550	70		1,620	1,825
0320	10'-0" x 5'-0", 5 panels			7	2.286		2,100	80		2,180	2,450
0330	12'-0" x 6'-0", 6 panels		↓	6	2.667	↓	2,700	93		2,793	3,100

08 52 13.30 Double-Hung Windows, Metal-Clad

			Crew	Daily Output	Labor-Hours	Unit	Material	2018 Bare Costs Labor	Equipment	Total	Total Incl O&P
0010	**DOUBLE-HUNG WINDOWS, METAL-CLAD**										
0100	Metal-clad, deluxe, dbl. insul. glass, 2'-6" x 3'-0" high	G	1 Carp	10	.800	Ea.	285	28		313	360
0120	3'-0" x 3'-6" high	G		10	.800		325	28		353	405
0140	3'-0" x 4'-0" high	G		9	.889		340	31		371	425
0160	3'-0" x 4'-6" high	G		9	.889		355	31		386	440
0180	3'-0" x 5'-0" high	G		8	1		385	35		420	485
0200	3'-6" x 6'-0" high	G	↓	8	1	↓	465	35		500	570

08 52 13 – Metal-Clad Wood Windows

08 52 13.35 Picture and Sliding Windows Metal-Clad

		Crew	Daily Output	Labor-Hours	Unit	Material	2018 Bare Costs Labor	Equipment	Total	Total Incl O&P
0010	**PICTURE AND SLIDING WINDOWS METAL-CLAD**									
2000	Metal-clad, dlx picture, dbl. insul. glass, 4'-0" x 4'-0" high	2 Carp	12	1.333	Ea.	380	46.50		426.50	500
2100	4'-0" x 6'-0" high		11	1.455		560	51		611	700
2200	5'-0" x 6'-0" high		10	1.600		620	56		676	775
2300	6'-0" x 6'-0" high		10	1.600		710	56		766	875
2400	Metal-clad, dlx sliding, dbl. insul. glass, 3'-0" x 3'-0" high G	1 Carp	10	.800		330	28		358	405
2420	4'-0" x 3'-6" high G		9	.889		400	31		431	490
2440	5'-0" x 4'-0" high G		9	.889		480	31		511	575
2460	6'-0" x 5'-0" high G		8	1		745	35		780	880

08 52 13.40 Bow and Bay Windows, Metal-Clad

		Crew	Daily Output	Labor-Hours	Unit	Material	2018 Bare Costs Labor	Equipment	Total	Total Incl O&P
0010	**BOW AND BAY WINDOWS, METAL-CLAD**									
0100	Metal-clad, deluxe, dbl. insul. glass, 8'-0" x 5'-0" high, 4 panels	2 Carp	10	1.600	Ea.	1,700	56		1,756	1,975
0120	10'-0" x 5'-0" high, 5 panels		8	2		1,825	70		1,895	2,125
0140	10'-0" x 6'-0" high, 5 panels		7	2.286		2,150	80		2,230	2,500
0160	12'-0" x 6'-0" high, 6 panels		6	2.667		2,925	93		3,018	3,375
0400	Double-hung, bldrs. model, bay, 8' x 4' high, dbl. insul. glass		10	1.600		1,350	56		1,406	1,575
0440	Low E glass		10	1.600		1,450	56		1,506	1,700
0460	9'-0" x 5'-0" high, dbl. insul. glass		6	2.667		1,450	93		1,543	1,750
0480	Low E glass		6	2.667		1,525	93		1,618	1,825
0500	Metal-clad, deluxe, dbl. insul. glass, 7'-0" x 4'-0" high		10	1.600		1,300	56		1,356	1,525
0520	8'-0" x 4'-0" high		8	2		1,350	70		1,420	1,600
0540	8'-0" x 5'-0" high		7	2.286		1,375	80		1,455	1,650
0560	9'-0" x 5'-0" high		6	2.667		1,475	93		1,568	1,775

08 52 16 – Plastic-Clad Wood Windows

08 52 16.10 Bow Window

		Crew	Daily Output	Labor-Hours	Unit	Material	2018 Bare Costs Labor	Equipment	Total	Total Incl O&P
0010	**BOW WINDOW** including frames, screens, and grilles									
0020	End panels operable									
1000	Bow type, casement, wood, bldrs. mdl., 8' x 5' dbl. insul. glass, 4 panel	2 Carp	10	1.600	Ea.	1,575	56		1,631	1,825
1050	Low E glass		10	1.600		1,325	56		1,381	1,550
1100	10'-0" x 5'-0", dbl. insul. glass, 6 panels		6	2.667		1,350	93		1,443	1,650
1200	Low E glass, 6 panels		6	2.667		1,450	93		1,543	1,750
1300	Vinyl-clad, bldrs. model, dbl. insul. glass, 6'-0" x 4'-0", 3 panel		10	1.600		1,050	56		1,106	1,250
1340	9'-0" x 4'-0", 4 panel		8	2		1,425	70		1,495	1,700
1380	10'-0" x 6'-0", 5 panels		7	2.286		2,350	80		2,430	2,700
1420	12'-0" x 6'-0", 6 panels		6	2.667		3,075	93		3,168	3,525
2000	Bay window, 8' x 5', dbl. insul. glass		10	1.600		1,925	56		1,981	2,200
2050	Low E glass		10	1.600		2,325	56		2,381	2,650
2100	12'-0" x 6'-0", dbl. insul. glass, 6 panels		6	2.667		2,400	93		2,493	2,775
2200	Low E glass		6	2.667		3,250	93		3,343	3,725
2280	6'-0" x 4'-0"		11	1.455		1,250	51		1,301	1,450
2300	Vinyl-clad, premium, dbl. insul. glass, 8'-0" x 5'-0"		10	1.600		1,800	56		1,856	2,075
2340	10'-0" x 5'-0"		8	2		2,400	70		2,470	2,775
2380	10'-0" x 6'-0"		7	2.286		2,800	80		2,880	3,200
2420	12'-0" x 6'-0"		6	2.667		3,350	93		3,443	3,850
2430	14'-0" x 3'-0"		7	2.286		3,050	80		3,130	3,475
2440	14'-0" x 6'-0"		5	3.200		5,100	112		5,212	5,775
3300	Vinyl-clad, premium, dbl. insul. glass, 7'-0" x 4'-6"		10	1.600		1,400	56		1,456	1,625
3340	8'-0" x 4'-6"		8	2		1,425	70		1,495	1,675
3380	8'-0" x 5'-0"		7	2.286		1,500	80		1,580	1,775
3420	9'-0" x 5'-0"		6	2.667		1,525	93		1,618	1,825

464

For customer support on your Residential Costs with RSMeans data, call 800.448.8182.

08 52 Wood Windows

08 52 16 – Plastic-Clad Wood Windows

08 52 16.15 Awning Window Vinyl-Clad

		Crew	Daily Output	Labor-Hours	Unit	Material	2018 Bare Costs Labor	Equipment	Total	Total Incl O&P
0010	**AWNING WINDOW VINYL-CLAD** including frames, screens, and grilles									
0200	Vinyl-clad, premium, double insulated glass, 24" x 17"	1 Carp	12	.667	Ea.	213	23.50		236.50	274
0210	24" x 28"		11	.727		261	25.50		286.50	330
0250	36" x 17"		9	.889		268	31		299	345
0280	36" x 28"		9	.889		305	31		336	390
0300	36" x 36"		9	.889		350	31		381	435
0320	36" x 40"		9	.889		400	31		431	490
0340	40" x 22"		10	.800		295	28		323	370
0360	48" x 28"		8	1		370	35		405	470
0380	60" x 36"		8	1		520	35		555	630

08 52 16.20 Half Round, Vinyl-Clad

		Crew	Daily Output	Labor-Hours	Unit	Material	2018 Bare Costs Labor	Equipment	Total	Total Incl O&P
0010	**HALF ROUND, VINYL-CLAD**, double insulated glass, incl. grille									
0800	14" height x 24" base	2 Carp	9	1.778	Ea.	420	62		482	565
1040	15" height x 25" base		8	2		405	70		475	560
1060	16" height x 28" base		7	2.286		430	80		510	605
1080	17" height x 29" base		7	2.286		445	80		525	620
2000	19" height x 33" base	1 Carp	6	1.333		490	46.50		536.50	615
2100	20" height x 35" base		6	1.333		485	46.50		531.50	610
2200	21" height x 37" base		6	1.333		495	46.50		541.50	625
2250	23" height x 41" base	2 Carp	6	2.667		535	93		628	740
2300	26" height x 48" base		6	2.667		555	93		648	765
2350	30" height x 56" base		6	2.667		645	93		738	865
3000	36" height x 67" base	1 Carp	4	2		1,150	70		1,220	1,375
3040	38" height x 71" base	2 Carp	5	3.200		1,000	112		1,112	1,275
3050	40" height x 75" base	"	5	3.200		1,325	112		1,437	1,625
5000	Elliptical, 71" x 16"	1 Carp	11	.727		995	25.50		1,020.50	1,150
5100	95" x 21"	"	10	.800		1,425	28		1,453	1,625

08 52 16.30 Palladian Windows

		Crew	Daily Output	Labor-Hours	Unit	Material	2018 Bare Costs Labor	Equipment	Total	Total Incl O&P
0010	**PALLADIAN WINDOWS**									
0020	Vinyl-clad, double insulated glass, including frame and grilles									
0040	3'-2" x 2'-6" high	2 Carp	11	1.455	Ea.	1,275	51		1,326	1,475
0060	3'-2" x 4'-10"		11	1.455		1,750	51		1,801	2,000
0080	3'-2" x 6'-4"		10	1.600		1,750	56		1,806	2,025
0100	4'-0" x 4'-0"		10	1.600		1,575	56		1,631	1,825
0120	4'-0" x 5'-4"	3 Carp	10	2.400		1,875	84		1,959	2,225
0140	4'-0" x 6'-0"		9	2.667		1,875	93		1,968	2,225
0160	4'-0" x 7'-4"		9	2.667		2,125	93		2,218	2,500
0180	5'-5" x 4'-10"		9	2.667		2,300	93		2,393	2,675
0200	5'-5" x 6'-10"		9	2.667		2,650	93		2,743	3,050
0220	5'-5" x 7'-9"		9	2.667		2,800	93		2,893	3,250
0240	6'-0" x 7'-11"		8	3		3,575	105		3,680	4,125
0260	8'-0" x 6'-0"		8	3		3,200	105		3,305	3,675

08 52 16.35 Double-Hung Window

			Crew	Daily Output	Labor-Hours	Unit	Material	2018 Bare Costs Labor	Equipment	Total	Total Incl O&P
0010	**DOUBLE-HUNG WINDOW** including frames, screens, and grilles										
0300	Vinyl-clad, premium, double insulated glass, 2'-6" x 3'-0"	G	1 Carp	10	.800	Ea.	340	28		368	415
0305	2'-6" x 4'-0"	G		10	.800		375	28		403	460
0400	3'-0" x 3'-6"	G		10	.800		345	28		373	420
0500	3'-0" x 4'-0"	G		9	.889		400	31		431	490
0600	3'-0" x 4'-6"	G		9	.889		425	31		456	520
0700	3'-0" x 5'-0"	G		8	1		460	35		495	565
0790	3'-4" x 5'-0"	G		8	1		450	35		485	555

For customer support on your Residential Costs with RSMeans data, call 800.448.8182.

465

08 52 Wood Windows

08 52 16 – Plastic-Clad Wood Windows

08 52 16.35 Double-Hung Window

		Crew	Daily Output	Labor-Hours	Unit	Material	2018 Bare Costs Labor	2018 Bare Costs Equipment	Total	Total Incl O&P
0800	3'-6" x 6'-0" **G**	1 Carp	8	1	Ea.	515	35		550	625
0820	4'-0" x 5'-0" **G**		7	1.143		570	40		610	695
0830	4'-0" x 6'-0" **G**		7	1.143		725	40		765	860

08 52 16.40 Transom Windows

		Crew	Daily Output	Labor-Hours	Unit	Material	2018 Bare Costs Labor	2018 Bare Costs Equipment	Total	Total Incl O&P
0010	**TRANSOM WINDOWS**									
0050	Vinyl-clad, premium, dbl. insul. glass, 32" x 8"	1 Carp	16	.500	Ea.	191	17.50		208.50	239
0100	36" x 8"		16	.500		206	17.50		223.50	256
0110	36" x 12"		16	.500		221	17.50		238.50	272
1000	Vinyl-clad, premium, dbl. insul. glass, 4'-0" x 4'-0"	2 Carp	12	1.333		530	46.50		576.50	665
1100	4'-0" x 6'-0"		11	1.455		975	51		1,026	1,150
1200	5'-0" x 6'-0"		10	1.600		1,075	56		1,131	1,300
1300	6'-0" x 6'-0"		10	1.600		1,150	56		1,206	1,350

08 52 16.45 Trapezoid Windows

		Crew	Daily Output	Labor-Hours	Unit	Material	2018 Bare Costs Labor	2018 Bare Costs Equipment	Total	Total Incl O&P
0010	**TRAPEZOID WINDOWS**									
0100	Vinyl-clad, including frame and exterior trim									
0900	20" base x 44" leg x 53" leg	2 Carp	13	1.231	Ea.	400	43		443	510
1000	24" base x 90" leg x 102" leg		8	2		700	70		770	885
3000	36" base x 40" leg x 22" leg		12	1.333		440	46.50		486.50	565
3010	36" base x 44" leg x 25" leg		13	1.231		465	43		508	580
3050	36" base x 26" leg x 48" leg		.9	1.778		470	62		532	620
3100	36" base x 42" legs, 50" peak		9	1.778		555	62		617	715
3200	36" base x 60" leg x 81" leg		11	1.455		730	51		781	885
4320	44" base x 23" leg x 56" leg		11	1.455		570	51		621	710
4350	44" base x 59" leg x 92" leg		10	1.600		880	56		936	1,075
4500	46" base x 15" leg x 46" leg		8	2		440	70		510	600
4550	46" base x 16" leg x 48" leg		8	2		465	70		535	625
4600	46" base x 50" leg x 80" leg		7	2.286		700	80		780	905
6600	66" base x 12" leg x 42" leg		8	2		600	70		670	775
6650	66" base x 12" legs, 28" peak		9	1.778		480	62		542	635
6700	68" base x 23" legs, 31" peak		8	2		600	70		670	775

08 52 16.70 Vinyl-Clad, Premium, DBL. Insul. Glass

		Crew	Daily Output	Labor-Hours	Unit	Material	2018 Bare Costs Labor	2018 Bare Costs Equipment	Total	Total Incl O&P
0010	**VINYL-CLAD, PREMIUM, DBL. INSUL. GLASS**									
1000	Sliding, 3'-0" x 3'-0" **G**	1 Carp	10	.800	Ea.	625	28		653	730
1020	4'-0" x 1'-11" **G**		11	.727		600	25.50		625.50	705
1040	4'-0" x 3'-0" **G**		10	.800		750	28		778	870
1050	4'-0" x 3'-6" **G**		9	.889		690	31		721	810
1090	4'-0" x 5'-0" **G**		9	.889		940	31		971	1,075
1100	5'-0" x 4'-0" **G**		9	.889		920	31		951	1,050
1120	5'-0" x 5'-0" **G**		8	1		1,075	35		1,110	1,250
1140	6'-0" x 4'-0" **G**		8	1		1,125	35		1,160	1,275
1150	6'-0" x 5'-0" **G**		8	1		1,175	35		1,210	1,325

08 52 50 – Window Accessories

08 52 50.10 Window Grille or Muntin

		Crew	Daily Output	Labor-Hours	Unit	Material	2018 Bare Costs Labor	2018 Bare Costs Equipment	Total	Total Incl O&P
0010	**WINDOW GRILLE OR MUNTIN**, snap in type									
0020	Standard pattern interior grilles									
2000	Wood, awning window, glass size, 28" x 16" high	1 Carp	30	.267	Ea.	30.50	9.30		39.80	49
2060	44" x 24" high		32	.250		44	8.75		52.75	63
2100	Casement, glass size, 20" x 36" high		30	.267		33.50	9.30		42.80	52.50
2180	20" x 56" high		32	.250		46.50	8.75		55.25	65.50
2200	Double-hung, glass size, 16" x 24" high		24	.333	Set	56	11.65		67.65	81
2280	32" x 32" high		34	.235	"	140	8.20		148.20	168

08 52 Wood Windows

08 52 50 – Window Accessories

08 52 50.10 Window Grille or Muntin		Crew	Daily Output	Labor-Hours	Unit	Material	2018 Bare Costs Labor	Equipment	Total	Total Incl O&P
2500	Picture, glass size, 48" x 48" high	1 Carp	30	.267	Ea.	128	9.30		137.30	157
2580	60" x 68" high		28	.286	"	196	10		206	233
2600	Sliding, glass size, 14" x 36" high		24	.333	Set	37.50	11.65		49.15	61
2680	36" x 36" high	▼	22	.364	"	45	12.70		57.70	70.50

08 52 66 – Wood Window Screens

08 52 66.10 Wood Screens

		Crew	Daily Output	Labor-Hours	Unit	Material	2018 Bare Costs Labor	Equipment	Total	Total Incl O&P
0010	**WOOD SCREENS**									
0020	Over 3 S.F., 3/4" frames	2 Carp	375	.043	S.F.	4.75	1.49		6.24	7.75
0100	1-1/8" frames	"	375	.043	"	8.45	1.49		9.94	11.80

08 52 69 – Wood Storm Windows

08 52 69.10 Storm Windows

			Crew	Daily Output	Labor-Hours	Unit	Material	2018 Bare Costs Labor	Equipment	Total	Total Incl O&P
0010	**STORM WINDOWS**, aluminum residential										
0300	Basement, mill finish, incl. fiberglass screen										
0320	1'-10" x 1'-0" high	G	2 Carp	30	.533	Ea.	36	18.65		54.65	70.50
0340	2'-9" x 1'-6" high	G		30	.533		39	18.65		57.65	74
0360	3'-4" x 2'-0" high	G	▼	30	.533	▼	46	18.65		64.65	81.50
1600	Double-hung, combination, storm & screen										
1700	Custom, clear anodic coating, 2'-0" x 3'-5" high		2 Carp	30	.533	Ea.	99	18.65		117.65	140
1720	2'-6" x 5'-0" high			28	.571		120	19.95		139.95	165
1740	4'-0" x 6'-0" high			25	.640		235	22.50		257.50	296
1800	White painted, 2'-0" x 3'-5" high			30	.533		115	18.65		133.65	158
1820	2'-6" x 5'-0" high			28	.571		175	19.95		194.95	226
1840	4'-0" x 6'-0" high			25	.640		290	22.50		312.50	355
2000	Clear anodic coating, 2'-0" x 3'-5" high	G		30	.533		97	18.65		115.65	138
2020	2'-6" x 5'-0" high	G		28	.571		124	19.95		143.95	169
2040	4'-0" x 6'-0" high	G		25	.640		135	22.50		157.50	186
2400	White painted, 2'-0" x 3'-5" high	G		30	.533		95	18.65		113.65	136
2420	2'-6" x 5'-0" high	G		28	.571		100	19.95		119.95	143
2440	4'-0" x 6'-0" high	G		25	.640		115	22.50		137.50	164
2600	Mill finish, 2'-0" x 3'-5" high	G		30	.533		90	18.65		108.65	130
2620	2'-6" x 5'-0" high	G		28	.571		95	19.95		114.95	138
2640	4'-0" x 6'-8" high	G	▼	25	.640	▼	115	22.50		137.50	164
4000	Picture window, storm, 1 lite, white or bronze finish										
4020	4'-6" x 4'-6" high		2 Carp	25	.640	Ea.	140	22.50		162.50	191
4040	5'-8" x 4'-6" high			20	.800		155	28		183	218
4400	Mill finish, 4'-6" x 4'-6" high			25	.640		140	22.50		162.50	191
4420	5'-8" x 4'-6" high		▼	20	.800	▼	160	28		188	223
4600	3 lite, white or bronze finish										
4620	4'-6" x 4'-6" high		2 Carp	25	.640	Ea.	165	22.50		187.50	219
4640	5'-8" x 4'-6" high			20	.800		175	28		203	240
4800	Mill finish, 4'-6" x 4'-6" high			25	.640		160	22.50		182.50	213
4820	5'-8" x 4'-6" high		▼	20	.800	▼	165	28		193	229
6000	Sliding window, storm, 2 lite, white or bronze finish										
6020	3'-4" x 2'-7" high		2 Carp	28	.571	Ea.	143	19.95		162.95	190
6040	4'-4" x 3'-3" high			25	.640		160	22.50		182.50	213
6060	5'-4" x 6'-0" high		▼	20	.800	▼	216	28		244	285
9000	Interior storm window										
9100	Storm window interior glass		1 Glaz	107	.075	S.F.	6.20	2.54		8.74	11.05

08 53 13.10 Solid Vinyl Windows		Crew	Daily Output	Labor-Hours	Unit	Material	2018 Bare Costs Labor	Equipment	Total	Total Incl O&P
0010	**SOLID VINYL WINDOWS**									
0020	Double-hung, including frame and screen, 2'-0" x 2'-6"	2 Carp	15	1.067	Ea.	205	37.50		242.50	288
0040	2'-0" x 3'-6"		14	1.143		215	40		255	305
0060	2'-6" x 4'-6"		13	1.231		256	43		299	355
0080	3'-0" x 4'-0"		10	1.600		227	56		283	345
0100	3'-0" x 4'-6"		9	1.778		292	62		354	425
0120	3'-6" x 4'-6"		8	2		310	70		380	455
0140	3'-6" x 6'-0"		7	2.286		385	80		465	560

08 53 13.20 Vinyl Single-Hung Windows			Crew	Daily Output	Labor-Hours	Unit	Material	Labor	Equipment	Total	Total Incl O&P
0010	**VINYL SINGLE-HUNG WINDOWS**, insulated glass										
0020	Grids, low E, J fin, extension jambs										
0130	25" x 41"	G	2 Carp	20	.800	Ea.	210	28		238	278
0140	25" x 49"	G		18	.889		220	31		251	294
0150	25" x 57"	G		17	.941		230	33		263	310
0160	25" x 65"	G		16	1		250	35		285	335
0170	29" x 41"	G		18	.889		210	31		241	283
0180	29" x 53"	G		18	.889		230	31		261	305
0190	29" x 57"	G		17	.941		240	33		273	320
0200	29" x 65"	G		16	1		250	35		285	335
0210	33" x 41"	G		20	.800		225	28		253	295
0220	33" x 53"	G		18	.889		245	31		276	320
0230	33" x 57"	G		17	.941		250	33		283	330
0240	33" x 65"	G		16	1		260	35		295	345
0250	37" x 41"	G		20	.800		250	28		278	320
0260	37" x 53"	G		18	.889		275	31		306	355
0270	37" x 57"	G		17	.941		285	33		318	370
0280	37" x 65"	G		16	1		300	35		335	390
0500	Vinyl-clad, premium, double insulated glass, circle, 24" diameter		1 Carp	6	1.333		490	46.50		536.50	615
1000	2'-4" diameter			6	1.333		565	46.50		611.50	700
1500	2'-11" diameter			6	1.333		655	46.50		701.50	800

08 53 13.30 Vinyl Double-Hung Windows			Crew	Daily Output	Labor-Hours	Unit	Material	Labor	Equipment	Total	Total Incl O&P
0010	**VINYL DOUBLE-HUNG WINDOWS**, insulated glass										
0100	Grids, low E, J fin, ext. jambs, 21" x 53"	G	2 Carp	18	.889	Ea.	220	31		251	294
0102	21" x 37"	G		18	.889		230	31		261	305
0104	21" x 41"	G		18	.889		240	31		271	315
0106	21" x 49"	G		18	.889		250	31		281	325
0110	21" x 57"	G		17	.941		270	33		303	350
0120	21" x 65"	G		16	1		285	35		320	375
0128	25" x 37"	G		20	.800		240	28		268	310
0130	25" x 41"	G		20	.800		250	28		278	320
0140	25" x 49"	G		18	.889		260	31		291	340
0145	25" x 53"	G		18	.889		275	31		306	355
0150	25" x 57"	G		17	.941		280	33		313	365
0160	25" x 65"	G		16	1		295	35		330	385
0162	25" x 69"	G		16	1		300	35		335	390
0164	25" x 77"	G		16	1		325	35		360	420
0168	29" x 37"	G		18	.889		250	31		281	325
0170	29" x 41"	G		18	.889		260	31		291	340
0172	29" x 49"	G		18	.889		270	31		301	350
0180	29" x 53"	G		18	.889		280	31		311	360
0190	29" x 57"	G		17	.941		290	33		323	375
0200	29" x 65"	G		16	1		310	35		345	400

08 53 Plastic Windows

08 53 13 – Vinyl Windows

08 53 13.30 Vinyl Double-Hung Windows

			Crew	Daily Output	Labor-Hours	Unit	Material	2018 Bare Costs Labor	Equipment	Total	Total Incl O&P
0202	29" x 69"	G	2 Carp	16	1	Ea.	315	35		350	405
0205	29" x 77"	G		16	1		340	35		375	435
0208	33" x 37"	G		20	.800		265	28		293	340
0210	33" x 41"	G		20	.800		270	28		298	345
0215	33" x 49"	G		20	.800		285	28		313	360
0220	33" x 53"	G		18	.889		290	31		321	370
0230	33" x 57"	G		17	.941		300	33		333	385
0240	33" x 65"	G		16	1		320	35		355	410
0242	33" x 69"	G		16	1		330	35		365	425
0246	33" x 77"	G		16	1		350	35		385	445
0250	37" x 41"	G		20	.800		295	28		323	370
0255	37" x 49"	G		20	.800		297	28		325	370
0260	37" x 53"	G		18	.889		320	31		351	400
0270	37" x 57"	G		17	.941		340	33		373	430
0280	37" x 65"	G		16	1		360	35		395	455
0282	37" x 69"	G		16	1		365	35		400	460
0286	37" x 77"	G		16	1		380	35		415	480
0300	Solid vinyl, average quality, double insulated glass, 2'-0" x 3'-0"	G	1 Carp	10	.800		291	28		319	365
0310	3'-0" x 4'-0"	G		9	.889		214	31		245	287
0330	Premium, double insulated glass, 2'-6" x 3'-0"	G		10	.800		276	28		304	350
0340	3'-0" x 3'-6"	G		9	.889		310	31		341	390
0350	3'-0" x 4'-0"	G		9	.889		330	31		361	415
0360	3'-0" x 4'-6"	G		9	.889		335	31		366	420
0370	3'-0" x 5'-0"	G		8	1		360	35		395	455
0380	3'-6" x 6'-0"	G		8	1		395	35		430	495

08 53 13.40 Vinyl Casement Windows

			Crew	Daily Output	Labor-Hours	Unit	Material	2018 Bare Costs Labor	Equipment	Total	Total Incl O&P
0010	**VINYL CASEMENT WINDOWS**, insulated glass										
0015	Grids, low E, J fin, extension jambs, screens										
0100	One lite, 21" x 41"	G	2 Carp	20	.800	Ea.	315	28		343	390
0110	21" x 47"	G		20	.800		330	28		358	410
0120	21" x 53"	G		20	.800		365	28		393	445
0128	24" x 35"	G		19	.842		297	29.50		326.50	375
0130	24" x 41"	G		19	.842		320	29.50		349.50	400
0140	24" x 47"	G		19	.842		355	29.50		384.50	440
0150	24" x 53"	G		19	.842		380	29.50		409.50	470
0158	28" x 35"	G		19	.842		310	29.50		339.50	390
0160	28" x 41"	G		19	.842		340	29.50		369.50	425
0170	28" x 47"	G		19	.842		375	29.50		404.50	465
0180	28" x 53"	G		19	.842		405	29.50		434.50	495
0184	28" x 59"	G		19	.842		415	29.50		444.50	510
0188	Two lites, 33" x 35"	G		18	.889		495	31		526	595
0190	33" x 41"	G		18	.889		530	31		561	635
0200	33" x 47"	G		18	.889		565	31		596	670
0210	33" x 53"	G		18	.889		580	31		611	690
0212	33" x 59"	G		18	.889		635	31		666	750
0215	33" x 72"	G		18	.889		665	31		696	785
0220	41" x 41"	G		18	.889		550	31		581	655
0230	41" x 47"	G		18	.889		610	31		641	720
0240	41" x 53"	G		17	.941		650	33		683	770
0242	41" x 59"	G		17	.941		690	33		723	810
0246	41" x 72"	G		17	.941		720	33		753	845
0250	47" x 41"	G		17	.941		585	33		618	695

For customer support on your Residential Costs with RSMeans data, call 800.448.8182.

469

08 53 Plastic Windows

08 53 13 – Vinyl Windows

08 53 13.40 Vinyl Casement Windows

			Crew	Daily Output	Labor-Hours	Unit	Material	2018 Bare Costs Labor	Equipment	Total	Total Incl O&P
0260	47" x 47"	G	2 Carp	17	.941	Ea.	625	33		658	745
0270	47" x 53"	G		17	.941		665	33		698	790
0272	47" x 59"	G		17	.941		725	33		758	850
0280	56" x 41"	G		15	1.067		625	37.50		662.50	750
0290	56" x 47"	G		15	1.067		665	37.50		702.50	795
0300	56" x 53"	G		15	1.067		725	37.50		762.50	855
0302	56" x 59"	G		15	1.067		750	37.50		787.50	885
0310	56" x 72"	G		15	1.067		810	37.50		847.50	950
0340	Solid vinyl, premium, double insulated glass, 2'-0" x 3'-0" high	G	1 Carp	10	.800		270	28		298	345
0360	2'-0" x 4'-0" high	G		9	.889		299	31		330	380
0380	2'-0" x 5'-0" high	G		8	1		335	35		370	430

08 53 13.50 Vinyl Picture Windows

			Crew	Daily Output	Labor-Hours	Unit	Material	Labor	Equipment	Total	Total Incl O&P
0010	**VINYL PICTURE WINDOWS**, insulated glass										
0120	Grids, low E, J fin, ext. jambs, 47" x 35"		2 Carp	12	1.333	Ea.	305	46.50		351.50	415
0130	47" x 41"			12	1.333		400	46.50		446.50	520
0140	47" x 47"			12	1.333		350	46.50		396.50	465
0150	47" x 53"			11	1.455		375	51		426	500
0160	71" x 35"			11	1.455		400	51		451	525
0170	71" x 41"			11	1.455		420	51		471	545
0180	71" x 47"			11	1.455		450	51		501	580

08 53 13.60 Vinyl Half Round Windows

			Crew	Daily Output	Labor-Hours	Unit	Material	Labor	Equipment	Total	Total Incl O&P
0010	**VINYL HALF ROUND WINDOWS**, Including grille, J fin, low E, ext. jambs										
0100	10" height x 20" base		2 Carp	8	2	Ea.	475	70		545	640
0110	15" height x 30" base			8	2		480	70		550	645
0120	17" height x 34" base			7	2.286		395	80		475	570
0130	19" height x 38" base			7	2.286		440	80		520	620
0140	19" height x 33" base			7	2.286		575	80		655	770
0150	24" height x 48" base		1 Carp	6	1.333		500	46.50		546.50	630
0160	25" height x 50" base		"	6	1.333		775	46.50		821.50	935
0170	30" height x 60" base		2 Carp	6	2.667		795	93		888	1,025

08 54 Composite Windows

08 54 13 – Fiberglass Windows

08 54 13.10 Fiberglass Single-Hung Windows

			Crew	Daily Output	Labor-Hours	Unit	Material	Labor	Equipment	Total	Total Incl O&P
0010	**FIBERGLASS SINGLE-HUNG WINDOWS**										
0100	Grids, low E, 18" x 24"	G	2 Carp	18	.889	Ea.	335	31		366	420
0110	18" x 40"	G		17	.941		340	33		373	430
0130	24" x 40"	G		20	.800		360	28		388	440
0230	36" x 36"	G		17	.941		370	33		403	460
0250	36" x 48"	G		20	.800		405	28		433	490
0260	36" x 60"	G		18	.889		445	31		476	540
0280	36" x 72"	G		16	1		470	35		505	575
0290	48" x 40"	G		16	1		470	35		505	575

08 54 13.30 Fiberglass Slider Windows

			Crew	Daily Output	Labor-Hours	Unit	Material	Labor	Equipment	Total	Total Incl O&P
0010	**FIBERGLASS SLIDER WINDOWS**										
0100	Grids, low E, 36" x 24"	G	2 Carp	20	.800	Ea.	340	28		368	420
0110	36" x 36"	G	"	20	.800	"	390	28		418	475

08 54 Composite Windows

08 54 13 – Fiberglass Windows

08 54 13.50 Fiberglass Bay Windows		Crew	Daily Output	Labor-Hours	Unit	Material	2018 Bare Costs Labor	Equipment	Total	Total Incl O&P
0010	**FIBERGLASS BAY WINDOWS**									
0150	48" x 36"	[G] 2 Carp	11	1.455	Ea.	1,100	51		1,151	1,275

08 61 Roof Windows

08 61 13 – Metal Roof Windows

08 61 13.10 Roof Windows

		Crew	Daily Output	Labor-Hours	Unit	Material	Labor	Equipment	Total	Total Incl O&P
0010	**ROOF WINDOWS**, fixed high perf tmpd glazing, metallic framed									
0020	46" x 21-1/2", Flashed for shingled roof	1 Carp	8	1	Ea.	281	35		316	370
0100	46" x 28"		8	1		305	35		340	395
0125	57" x 44"		6	1.333		375	46.50		421.50	495
0130	72" x 28"		7	1.143		375	40		415	480
0150	Fixed, laminated tempered glazing, 46" x 21-1/2"		8	1		465	35		500	570
0175	46" x 28"		8	1		515	35		550	625
0200	57" x 44"		6	1.333		485	46.50		531.50	610
0500	Vented flashing set for shingled roof, 46" x 21-1/2"		7	1.143		465	40		505	575
0525	46" x 28"		6	1.333		515	46.50		561.50	645
0550	57" x 44"		5	1.600		645	56		701	805
0560	72" x 28"		5	1.600		645	56		701	805
0575	Flashing set for low pitched roof, 46" x 21-1/2"		7	1.143		535	40		575	655
0600	46" x 28"		7	1.143		585	40		625	710
0625	57" x 44"		5	1.600		730	56		786	895
0650	Flashing set for curb, 46" x 21-1/2"		7	1.143		620	40		660	745
0675	46" x 28"		7	1.143		670	40		710	805
0700	57" x 44"		5	1.600		825	56		881	1,000

08 61 16 – Wood Roof Windows

08 61 16.16 Roof Windows, Wood Framed

		Crew	Daily Output	Labor-Hours	Unit	Material	Labor	Equipment	Total	Total Incl O&P
0010	**ROOF WINDOWS, WOOD FRAMED**									
5600	Roof window incl. frame, flashing, double insulated glass & screens,									
5610	complete unit, 22" x 38"	2 Carp	3	5.333	Ea.	755	186		941	1,150
5650	2'-5" x 3'-8"		3.20	5		945	175		1,120	1,350
5700	3'-5" x 4'-9"		3.40	4.706		1,050	164		1,214	1,425

08 62 Unit Skylights

08 62 13 – Domed Unit Skylights

08 62 13.10 Domed Skylights

		Crew	Daily Output	Labor-Hours	Unit	Material	Labor	Equipment	Total	Total Incl O&P
0010	**DOMED SKYLIGHTS**									
0020	Skylight, fixed dome type, 22" x 22"	[G] G-3	12	2.667	Ea.	216	86.50		302.50	380
0030	22" x 46"	[G]	10	3.200		269	104		373	470
0040	30" x 30"	[G]	12	2.667		286	86.50		372.50	460
0050	30" x 46"	[G]	10	3.200		380	104		484	595
0110	Fixed, double glazed, 22" x 27"	[G]	12	2.667		275	86.50		361.50	450
0120	22" x 46"	[G]	10	3.200		293	104		397	495
0130	44" x 46"	[G]	10	3.200		430	104		534	650
0210	Operable, double glazed, 22" x 27"	[G]	12	2.667		405	86.50		491.50	590
0220	22" x 46"	[G]	10	3.200		465	104		569	690
0230	44" x 46"	[G]	10	3.200		925	104		1,029	1,200

For customer support on your Residential Costs with RSMeans data, call 800.448.8182.

471

08 62 13 – Domed Unit Skylights

08 62 13.20 Skylights

		Crew	Daily Output	Labor-Hours	Unit	Material	2018 Bare Costs Labor	Equipment	Total	Total Incl O&P
0010	**SKYLIGHTS**, flush or curb mounted									
2120	Ventilating insulated plexiglass dome with									
2130	curb mounting, 36" x 36"	G-3	12	2.667	Ea.	480	86.50		566.50	675
2150	52" x 52"		12	2.667		670	86.50		756.50	880
2160	28" x 52"		10	3.200		495	104		599	720
2170	36" x 52"		10	3.200		545	104		649	770
2180	For electric opening system, add					315			315	350
2210	Operating skylight, with thermopane glass, 24" x 48"	G-3	10	3.200		595	104		699	830
2220	32" x 48"	"	9	3.556		620	115		735	875
2310	Non venting insulated plexiglass dome skylight with									
2320	Flush mount 22" x 46"	G-3	15.23	2.101	Ea.	335	68		403	480
2330	30" x 30"		16	2		310	65		375	450
2340	46" x 46"		13.91	2.301		570	74.50		644.50	750
2350	Curb mount 22" x 46"		15.23	2.101		365	68		433	515
2360	30" x 30"		16	2		405	65		470	560
2370	46" x 46"		13.91	2.301		625	74.50		699.50	815
2381	Non-insulated flush mount 22" x 46"		15.23	2.101		227	68		295	365
2382	30" x 30"		16	2		204	65		269	335
2383	46" x 46"		13.91	2.301		385	74.50		459.50	550
2384	Curb mount 22" x 46"		15.23	2.101		191	68		259	325
2385	30" x 30"		16	2		188	65		253	315
4000	Skylight, solar tube kit, incl. dome, flashing, diffuser, 1 pipe, 10" diam.	1 Carp	2	4		300	140		440	560
4010	14" diam.		2	4		400	140		540	670
4020	21" diam.		2	4		490	140		630	770
4030	Accessories for, 1' long x 9" diam. pipe		24	.333		50	11.65		61.65	74.50
4040	2' long x 9" diam. pipe		24	.333		43	11.65		54.65	67
4050	4' long x 9" diam. pipe		20	.400		75	14		89	106
4060	1' long x 13" diam. pipe		24	.333		70	11.65		81.65	96.50
4070	2' long x 13" diam. pipe		24	.333		55	11.65		66.65	80
4080	4' long x 13" diam. pipe		20	.400		102	14		116	135
4090	6.5" turret ext. for 21" diam. pipe		16	.500		110	17.50		127.50	150
4100	12' long x 21" diam. flexible pipe		12	.667		90	23.50		113.50	138
4110	45 degree elbow, 10"		16	.500		155	17.50		172.50	200
4120	14"		16	.500		85	17.50		102.50	123
4130	Interior decorative ring, 9"		20	.400		45	14		59	72.50
4140	13"		20	.400		60	14		74	89

08 71 Door Hardware

08 71 20 – Hardware

08 71 20.15 Hardware

		Crew	Daily Output	Labor-Hours	Unit	Material	2018 Bare Costs Labor	Equipment	Total	Total Incl O&P
0010	**HARDWARE**									
0020	Average hardware cost									
1000	Door hardware, apartment, interior	1 Carp	4	2	Door	505	70		575	670
1300	Average, door hardware, motel/hotel interior, with access card		4	2	"	660	70		730	840
2100	Pocket door		6	1.333	Ea.	111	46.50		157.50	200
4000	Door knocker, bright brass		32	.250		78.50	8.75		87.25	101
4100	Mail slot, bright brass, 2" x 11"		25	.320		60.50	11.20		71.70	85
4200	Peep hole, add to price of door					12.55			12.55	13.80

08 71 Door Hardware

08 71 20 – Hardware

08 71 20.40 Lockset	Crew	Daily Output	Labor-Hours	Unit	Material	2018 Bare Costs Labor	Equipment	Total	Total Incl O&P
0010 **LOCKSET**, Standard duty									
0020 Non-keyed, passage, w/sect. trim	1 Carp	12	.667	Ea.	78	23.50		101.50	125
0100 Privacy		12	.667		80	23.50		103.50	127
0400 Keyed, single cylinder function		10	.800		151	28		179	214
0500 Lever handled, keyed, single cylinder function	↓	10	.800	↓	127	28		155	187

08 71 20.41 Dead Locks

	Crew	Daily Output	Labor-Hours	Unit	Material	Labor	Equipment	Total	Total Incl O&P
0010 **DEAD LOCKS**									
1203 Deadlock night latch	1 Carp	7.70	1.039	Ea.	50	36.50		86.50	116
1420 Deadbolt lock, single cylinder		10	.800		46.50	28		74.50	97.50
1440 Double cylinder	↓	10	.800	↓	64	28		92	117

08 71 20.42 Mortise Locksets

	Crew	Daily Output	Labor-Hours	Unit	Material	Labor	Equipment	Total	Total Incl O&P
0010 **MORTISE LOCKSETS**, Comm., wrought knobs & full escutcheon trim									
0015 Assumes mortise is cut									
0020 Non-keyed, passage, Grade 3	1 Carp	9	.889	Ea.	157	31		188	225
0030 Grade 1		8	1		425	35		460	525
0040 Privacy set, Grade 3		9	.889		172	31		203	242
0050 Grade 1		8	1		465	35		500	570
0100 Keyed, office/entrance/apartment, Grade 2		8	1		201	35		236	279
0110 Grade 1		7	1.143		530	40		570	645
0120 Single cylinder, typical, Grade 3		8	1		193	35		228	270
0130 Grade 1		7	1.143		525	40		565	645
0200 Hotel, room, Grade 3		7	1.143		192	40		232	278
0210 Grade 1 (see also Section 08 71 20.15)		6	1.333		530	46.50		576.50	660
0300 Double cylinder, Grade 3		8	1		229	35		264	310
0310 Grade 1		7	1.143		545	40		585	665
1000 Wrought knobs and sectional trim, non-keyed, passage, Grade 3		10	.800		133	28		161	193
1010 Grade 1		9	.889		425	31		456	520
1040 Privacy, Grade 3		10	.800		153	28		181	216
1050 Grade 1		9	.889		480	31		511	580
1100 Keyed, entrance, office/apartment, Grade 3		9	.889		225	31		256	299
1103 Install lockset		6.92	1.156		225	40.50		265.50	315
1110 Grade 1		8	1		550	35		585	660
1120 Single cylinder, Grade 3		9	.889		230	31		261	305
1130 Grade 1	↓	8	1	↓	510	35		545	620
2000 Cast knobs and full escutcheon trim									
2010 Non-keyed, passage, Grade 3	1 Carp	9	.889	Ea.	280	31		311	360
2020 Grade 1		8	1		385	35		420	485
2040 Privacy, Grade 3		9	.889		325	31		356	410
2050 Grade 1		8	1		450	35		485	555
2120 Keyed, single cylinder, Grade 3		8	1		335	35		370	430
2123 Mortise lock		6.15	1.301		335	45.50		380.50	445
2130 Grade 1	↓	7	1.143	↓	535	40		575	655

08 71 20.50 Door Stops

	Crew	Daily Output	Labor-Hours	Unit	Material	Labor	Equipment	Total	Total Incl O&P
0010 **DOOR STOPS**									
0020 Holder & bumper, floor or wall	1 Carp	32	.250	Ea.	36	8.75		44.75	54
1300 Wall bumper, 4" diameter, with rubber pad, aluminum		32	.250		12.70	8.75		21.45	28.50
1600 Door bumper, floor type, aluminum		32	.250		3.32	8.75		12.07	18.20
1620 Brass		32	.250		11.15	8.75		19.90	27
1630 Bronze		32	.250		19.20	8.75		27.95	35.50
1900 Plunger type, door mounted		32	.250		30.50	8.75		39.25	48.50
2520 Wall type, aluminum	↓	32	.250		32	8.75		40.75	49.50

08 71 20 – Hardware

08 71 20.50 Door Stops	Crew	Daily Output	Labor-Hours	Unit	Material	2018 Bare Costs Labor	2018 Bare Costs Equipment	Total	Total Incl O&P
2540	Plunger type, aluminum	1 Carp	32	.250	Ea.	30.50	8.75		39.25	48.50
2560	Brass		32	.250		46	8.75		54.75	65
3020	Floor mounted, US3		3	2.667		335	93		428	525

08 71 20.60 Entrance Locks

		Crew	Daily Output	Labor-Hours	Unit	Material	Labor	Equipment	Total	Total Incl O&P
0010	**ENTRANCE LOCKS**									
0015	Cylinder, grip handle deadlocking latch	1 Carp	9	.889	Ea.	181	31		212	251
0020	Deadbolt		8	1		195	35		230	272
0100	Push and pull plate, dead bolt		8	1		234	35		269	315
0900	For handicapped lever, add					152			152	167

08 71 20.65 Thresholds

		Crew	Daily Output	Labor-Hours	Unit	Material	Labor	Equipment	Total	Total Incl O&P
0010	**THRESHOLDS**									
0011	Threshold 3' long saddles aluminum	1 Carp	48	.167	L.F.	10.40	5.85		16.25	21
0100	Aluminum, 8" wide, 1/2" thick		12	.667	Ea.	50	23.50		73.50	93.50
0500	Bronze		60	.133	L.F.	45.50	4.66		50.16	58
0600	Bronze, panic threshold, 5" wide, 1/2" thick		12	.667	Ea.	162	23.50		185.50	217
0700	Rubber, 1/2" thick, 5-1/2" wide		20	.400		41	14		55	68
0800	2-3/4" wide		20	.400		49.50	14		63.50	77
1950	ADA compliant thresholds									
2000	Threshold, wood oak 3-1/2" wide x 24" long	1 Carp	12	.667	Ea.	11.90	23.50		35.40	51.50
2010	3-1/2" wide x 36" long		12	.667		16	23.50		39.50	56
2020	3-1/2" wide x 48" long		12	.667		21.50	23.50		45	62
2030	4-1/2" wide x 24" long		12	.667		13.75	23.50		37.25	53.50
2040	4-1/2" wide x 36" long		12	.667		20	23.50		43.50	60.50
2050	4-1/2" wide x 48" long		12	.667		26.50	23.50		50	67.50
2060	6-1/2" wide x 24" long		12	.667		19.10	23.50		42.60	59.50
2070	6-1/2" wide x 36" long		12	.667		28.50	23.50		52	70
2080	6-1/2" wide x 48" long		12	.667		38	23.50		61.50	80.50
2090	Threshold, wood cherry 3-1/2" wide x 24" long		12	.667		14.50	23.50		38	54.50
2100	3-1/2" wide x 36" long		12	.667		28.50	23.50		52	70
2110	3-1/2" wide x 48" long		12	.667		35.50	23.50		59	77.50
2120	4-1/2" wide x 24" long		12	.667		18.50	23.50		42	59
2130	4-1/2" wide x 36" long		12	.667		34.50	23.50		58	76.50
2140	4-1/2" wide x 48" long		12	.667		43.50	23.50		67	86.50
2150	6-1/2" wide x 24" long		12	.667		26.50	23.50		50	67.50
2160	6-1/2" wide x 36" long		12	.667		46.50	23.50		70	89.50
2170	6-1/2" wide x 48" long		12	.667		59.50	23.50		83	104
2180	Threshold, wood walnut 3-1/2" wide x 24" long		12	.667		17.50	23.50		41	58
2190	3-1/2" wide x 36" long		12	.667		33.50	23.50		57	75.50
2200	3-1/2" wide x 48" long		12	.667		42	23.50		65.50	84.50
2210	4-1/2" wide x 24" long		12	.667		25	23.50		48.50	66
2220	4-1/2" wide x 36" long		12	.667		44	23.50		67.50	87
2230	4-1/2" wide x 48" long		12	.667		56	23.50		79.50	100
2240	6-1/2" wide x 24" long		12	.667		42	23.50		65.50	84.50
2250	6-1/2" wide x 36" long		12	.667		69	23.50		92.50	115
2260	6-1/2" wide x 48" long		12	.667		90	23.50		113.50	138
2300	Threshold, aluminum 4" wide x 36" long		12	.667		25	23.50		48.50	66
2310	4" wide x 48" long		12	.667		30	23.50		53.50	71.50
2320	4" wide x 72" long		12	.667		39	23.50		62.50	81.50
2330	5" wide x 36" long		12	.667		33	23.50		56.50	75
2340	5" wide x 48" long		12	.667		40	23.50		63.50	82.50
2350	5" wide x 72" long		12	.667		67	23.50		90.50	112
2360	6" wide x 36" long		12	.667		39	23.50		62.50	81.50

08 71 Door Hardware

08 71 20 – Hardware

08 71 20.65 Thresholds

		Crew	Daily Output	Labor-Hours	Unit	Material	2018 Bare Costs Labor	Equipment	Total	Total Incl O&P
2370	6" wide x 48" long	1 Carp	12	.667	Ea.	48	23.50		71.50	91.50
2380	6" wide x 72" long		12	.667		67	23.50		90.50	112
2390	7" wide x 36" long		12	.667		49	23.50		72.50	92.50
2400	7" wide x 48" long		12	.667		60	23.50		83.50	105
2410	7" wide x 72" long		12	.667		91	23.50		114.50	139
2500	Threshold, ramp, aluminum or rubber 24" x 24"	↓	12	.667	↓	181	23.50		204.50	238

08 71 20.75 Door Hardware Accessories

		Crew	Daily Output	Labor-Hours	Unit	Material	2018 Bare Costs Labor	Equipment	Total	Total Incl O&P
0010	**DOOR HARDWARE ACCESSORIES**									
1000	Knockers, brass, standard	1 Carp	16	.500	Ea.	64	17.50		81.50	99.50
1100	Deluxe	"	10	.800	"	90.50	28		118.50	146
2000	Torsion springs for overhead doors									
2050	1-3/4" diam., 32" long, 0.243" spring	1 Carp	8	1	Ea.	49	35		84	112
2060	1-3/4" diam., 32" long, 0.250" spring		8	1		49	35		84	112
2070	2" diam., 32" long, 0.243" spring		8	1		42	35		77	104
2080	2" diam., 32" long, 0.253" spring		8	1		50	35		85	113
4500	Rubber door silencers	↓	540	.015	↓	.39	.52		.91	1.29

08 71 20.90 Hinges

		Crew	Daily Output	Labor-Hours	Unit	Material	2018 Bare Costs Labor	Equipment	Total	Total Incl O&P
0010	**HINGES**									
0012	Full mortise, avg. freq., steel base, USP, 4-1/2" x 4-1/2"				Pr.	39.50			39.50	43.50
0100	5" x 5", USP					59			59	65
0200	6" x 6", USP					124			124	137
0400	Brass base, 4-1/2" x 4-1/2", US10					64			64	70
0500	5" x 5", US10					71.50			71.50	78.50
0600	6" x 6", US10					164			164	180
0800	Stainless steel base, 4-1/2" x 4-1/2", US32				↓	77.50			77.50	85
0900	For non removable pin, add (security item)				Ea.	5.55			5.55	6.10
0910	For floating pin, driven tips, add					3.40			3.40	3.74
0930	For hospital type tip on pin, add					14.20			14.20	15.60
0940	For steeple type tip on pin, add				↓	19.30			19.30	21
0950	Full mortise, high frequency, steel base, 3-1/2" x 3-1/2", US26D				Pr.	31			31	34
1000	4-1/2" x 4-1/2", USP					66.50			66.50	73.50
1100	5" x 5", USP					52.50			52.50	58
1200	6" x 6", USP					139			139	153
1400	Brass base, 3-1/2" x 3-1/2", US4					52.50			52.50	57.50
1430	4-1/2" x 4-1/2", US10					76			76	83.50
1500	5" x 5", US10					127			127	139
1600	6" x 6", US10					171			171	188
1800	Stainless steel base, 4-1/2" x 4-1/2", US32					104			104	115
1810	5" x 4-1/2", US32				↓	140			140	154
1930	For hospital type tip on pin, add				Ea.	14.60			14.60	16.05
1950	Full mortise, low frequency, steel base, 3-1/2" x 3-1/2", US26D				Pr.	28			28	31
2000	4-1/2" x 4-1/2", USP					23.50			23.50	25.50
2100	5" x 5", USP					49			49	54
2200	6" x 6", USP					94			94	104
2300	4-1/2" x 4-1/2", US3					17.70			17.70	19.45
2310	5" x 5", US3					42.50			42.50	46.50
2400	Brass bass, 4-1/2" x 4-1/2", US10					55			55	60.50
2500	5" x 5", US10					79			79	86.50
2800	Stainless steel base, 4-1/2" x 4-1/2", US32					76.50			76.50	84.50
8000	Install hinge	1 Carp	34	.235	↓		8.20		8.20	13.65

08 71 Door Hardware

08 71 20 – Hardware

08 71 20.91 Special Hinges

	Crew	Daily Output	Labor-Hours	Unit	Material	2018 Bare Costs Labor	Equipment	Total	Total Incl O&P
0010 **SPECIAL HINGES**									
8000 Continuous hinges									
8010 Steel, piano, 2" x 72"	1 Carp	20	.400	Ea.	23	14		37	48.50
8020 Brass, piano, 1-1/16" x 30"		30	.267		9	9.30		18.30	25.50
8030 Acrylic, piano, 1-3/4" x 12"	↓	40	.200	↓	16	7		23	29

08 71 20.92 Mortised Hinges

	Crew	Daily Output	Labor-Hours	Unit	Material	2018 Bare Costs Labor	Equipment	Total	Total Incl O&P
0010 **MORTISED HINGES**									
0200 Average frequency, steel plated, ball bearing, 3-1/2" x 3-1/2"				Pr.	27.50			27.50	30.50
0300 Bronze, ball bearing					33			33	36
0900 High frequency, steel plated, ball bearing					102			102	112
1100 Bronze, ball bearing					102			102	112
1300 Average frequency, steel plated, ball bearing, 4-1/2" x 4-1/2"					35.50			35.50	39
1500 Bronze, ball bearing, to 36" wide					37			37	41
1700 Low frequency, steel, plated, plain bearing					22.50			22.50	25
1900 Bronze, plain bearing				↓	26.50			26.50	29

08 71 20.95 Kick Plates

	Crew	Daily Output	Labor-Hours	Unit	Material	2018 Bare Costs Labor	Equipment	Total	Total Incl O&P
0010 **KICK PLATES**									
0020 Stainless steel, .050", 16 ga., 8" x 28", US32	1 Carp	15	.533	Ea.	41	18.65		59.65	76
0080 Mop/Kick, 4" x 28"		15	.533		37	18.65		55.65	71.50
0090 4" x 30"		15	.533		39	18.65		57.65	74
0100 4" x 34"		15	.533		44	18.65		62.65	79.50
0110 6" x 28"		15	.533		42	18.65		60.65	77
0120 6" x 30"		15	.533		51	18.65		69.65	87
0130 6" x 34"	↓	15	.533	↓	57	18.65		75.65	94

08 71 21 – Astragals

08 71 21.10 Exterior Mouldings, Astragals

	Crew	Daily Output	Labor-Hours	Unit	Material	2018 Bare Costs Labor	Equipment	Total	Total Incl O&P
0010 **EXTERIOR MOULDINGS, ASTRAGALS**									
4170 Astragal for double doors, aluminum	1 Carp	4	2	Opng.	37.50	70		107.50	157
4174 Bronze	"	4	2	"	51	70		121	173

08 71 25 – Weatherstripping

08 71 25.10 Mechanical Seals, Weatherstripping

	Crew	Daily Output	Labor-Hours	Unit	Material	2018 Bare Costs Labor	Equipment	Total	Total Incl O&P
0010 **MECHANICAL SEALS, WEATHERSTRIPPING**									
1000 Doors, wood frame, interlocking, for 3' x 7' door, zinc	1 Carp	3	2.667	Opng.	47.50	93		140.50	208
1100 Bronze		3	2.667		60.50	93		153.50	222
1300 6' x 7' opening, zinc		2	4		58.50	140		198.50	296
1400 Bronze		2	4	↓	69.50	140		209.50	310
1500 Vinyl V strip	↓	6.40	1.250	Ea.	12.90	43.50		56.40	86.50
1700 Wood frame, spring type, bronze									
1800 3' x 7' door	1 Carp	7.60	1.053	Opng.	25	37		62	88.50
1900 6' x 7' door		7	1.143		31	40		71	101
1920 Felt, 3' x 7' door		14	.571		4.31	19.95		24.26	37.50
1930 6' x 7' door		13	.615		4.99	21.50		26.49	41.50
1950 Rubber, 3' x 7' door		7.60	1.053		9.95	37		46.95	72
1960 6' x 7' door	↓	7	1.143	↓	12.75	40		52.75	80.50
2200 Metal frame, spring type, bronze									
2300 3' x 7' door	1 Carp	3	2.667	Opng.	49.50	93		142.50	209
2400 6' x 7' door	"	2.50	3.200	"	53	112		165	245
2500 For stainless steel, spring type, add					133%				
2700 Metal frame, extruded sections, 3' x 7' door, aluminum	1 Carp	3	2.667	Opng.	28	93		121	186
2800 Bronze	↓	3	2.667	↓	82.50	93		175.50	246

476

08 71 Door Hardware

08 71 25 – Weatherstripping

08 71 25.10 Mechanical Seals, Weatherstripping	Crew	Daily Output	Labor-Hours	Unit	Material	2018 Bare Costs Labor	Equipment	Total	Total Incl O&P
3100	6' x 7' door, aluminum	1 Carp	1.50	5.333	Opng.	35	186		221	350
3200	Bronze	↓	1.50	5.333	↓	140	186		326	465
3500	Threshold weatherstripping									
3650	Door sweep, flush mounted, aluminum	1 Carp	25	.320	Ea.	20.50	11.20		31.70	41
3700	Vinyl		25	.320		18.35	11.20		29.55	38.50
5000	Garage door bottom weatherstrip, 12' aluminum, clear		14	.571		25.50	19.95		45.45	61
5010	Bronze		14	.571		91.50	19.95		111.45	134
5050	Bottom protection, rubber		14	.571		41.50	19.95		61.45	78.50
5100	Threshold	↓	14	.571	↓	74.50	19.95		94.45	115

08 75 Window Hardware

08 75 10 – Window Handles and Latches

08 75 10.10 Handles and Latches

		Crew	Daily Output	Labor-Hours	Unit	Material	2018 Bare Costs Labor	Equipment	Total	Total Incl O&P
0010	**HANDLES AND LATCHES**									
1000	Handles, surface mounted, aluminum	1 Carp	24	.333	Ea.	5.45	11.65		17.10	25.50
1020	Brass		24	.333		5.75	11.65		17.40	25.50
1040	Chrome		24	.333		8.20	11.65		19.85	28.50
1200	Window handles window crank ADA		24	.333		16	11.65		27.65	37
1500	Recessed, aluminum		12	.667		3.39	23.50		26.89	42
1520	Brass		12	.667		5.10	23.50		28.60	44
1540	Chrome		12	.667		4.24	23.50		27.74	43
2000	Latches, aluminum		20	.400		4.05	14		18.05	27.50
2020	Brass		20	.400		5.35	14		19.35	29
2040	Chrome		20	.400		3.95	14		17.95	27.50

08 75 10.15 Window Opening Control

		Crew	Daily Output	Labor-Hours	Unit	Material	2018 Bare Costs Labor	Equipment	Total	Total Incl O&P
0010	**WINDOW OPENING CONTROL**									
0015	Window stops									
0020	For double-hung window	1 Carp	35	.229	Ea.	38.50	8		46.50	56
0030	Cam action for sliding window		35	.229		4.92	8		12.92	18.70
0040	Thumb screw for sliding window	↓	35	.229	↓	4.65	8		12.65	18.40
0100	Window guards, child safety bars for single or double-hung									
0110	14" to 17" wide max vert opening 26"	1 Carp	16	.500	Ea.	70	17.50		87.50	106
0120	17" to 23" wide max vert opening 26"		16	.500		74	17.50		91.50	111
0130	23" to 36" wide max vert opening 26"		16	.500		79	17.50		96.50	116
0140	35" to 58" wide max vert opening 26"		16	.500		94	17.50		111.50	132
0150	58" to 90" wide max vert opening 26"		16	.500		125	17.50		142.50	167
0160	73" to 120" wide max vert opening 26"	↓	16	.500	↓	227	17.50		244.50	279

08 75 30 – Weatherstripping

08 75 30.10 Mechanical Weather Seals

		Crew	Daily Output	Labor-Hours	Unit	Material	2018 Bare Costs Labor	Equipment	Total	Total Incl O&P
0010	**MECHANICAL WEATHER SEALS**, Window, double-hung, 3' x 5'									
0020	Zinc	1 Carp	7.20	1.111	Opng.	22	39		61	88.50
0100	Bronze		7.20	1.111		42	39		81	111
0200	Vinyl V strip		7	1.143		10.65	40		50.65	78.50
0500	As above but heavy duty, zinc		4.60	1.739		20.50	61		81.50	124
0600	Bronze	↓	4.60	1.739	↓	75.50	61		136.50	184

For customer support on your Residential Costs with RSMeans data, call 800.448.8182.

477

08 79 Hardware Accessories

08 79 20 – Door Accessories

08 79 20.10 Door Hardware Accessories	Crew	Daily Output	Labor-Hours	Unit	Material	2018 Bare Costs Labor	Equipment	Total	Total Incl O&P
0010 **DOOR HARDWARE ACCESSORIES**									
0140 Door bolt, surface, 4"	1 Carp	32	.250	Ea.	15.55	8.75		24.30	31.50
0160 Door latch	"	12	.667	"	8.75	23.50		32.25	48
0200 Sliding closet door									
0220 Track and hanger, single	1 Carp	10	.800	Ea.	67	28		95	120
0240 Double		8	1		81.50	35		116.50	148
0260 Door guide, single		48	.167		31	5.85		36.85	43.50
0280 Double		48	.167		41	5.85		46.85	54.50
0600 Deadbolt and lock cover plate, brass or stainless steel		30	.267		30	9.30		39.30	48.50
0620 Hole cover plate, brass or chrome		35	.229		8.20	8		16.20	22.50
2240 Mortise lockset, passage, lever handle		9	.889		161	31		192	229
4000 Security chain, standard		18	.444		12.95	15.55		28.50	40.50
4100 Deluxe	▼	18	.444	▼	45.50	15.55		61.05	76

08 81 Glass Glazing

08 81 10 – Float Glass

08 81 10.10 Various Types and Thickness of Float Glass

	Crew	Daily Output	Labor-Hours	Unit	Material	2018 Bare Costs Labor	Equipment	Total	Total Incl O&P
0010 **VARIOUS TYPES AND THICKNESS OF FLOAT GLASS**									
0020 3/16" plain	2 Glaz	130	.123	S.F.	6.80	4.18		10.98	14.40
0200 Tempered, clear		130	.123		7.25	4.18		11.43	14.90
0300 Tinted		130	.123		6.80	4.18		10.98	14.40
0600 1/4" thick, clear, plain		120	.133		9.90	4.53		14.43	18.40
0700 Tinted		120	.133		9.45	4.53		13.98	17.90
0800 Tempered, clear		120	.133		6.80	4.53		11.33	15
0900 Tinted		120	.133		11.70	4.53		16.23	20.50
1600 3/8" thick, clear, plain		75	.213		11.20	7.25		18.45	24.50
1700 Tinted		75	.213		16.55	7.25		23.80	30
1800 Tempered, clear		75	.213		17.70	7.25		24.95	31.50
1900 Tinted		75	.213		20	7.25		27.25	34
2200 1/2" thick, clear, plain		55	.291		19.05	9.90		28.95	37.50
2300 Tinted		55	.291		29	9.90		38.90	48.50
2400 Tempered, clear		55	.291		26.50	9.90		36.40	45.50
2500 Tinted		55	.291		28	9.90		37.90	47
2800 5/8" thick, clear, plain		45	.356		29	12.05		41.05	52
2900 Tempered, clear	▼	45	.356	▼	33	12.05		45.05	56.50

08 81 25 – Glazing Variables

08 81 25.10 Applications of Glazing

	Crew	Daily Output	Labor-Hours	Unit	Material	2018 Bare Costs Labor	Equipment	Total	Total Incl O&P
0010 **APPLICATIONS OF GLAZING**									
0600 For glass replacement, add				S.F.		100%			
0700 For gasket settings, add				L.F.	6.10			6.10	6.75
0900 For sloped glazing, add				S.F.		26%			
2000 Fabrication, polished edges, 1/4" thick				Inch	.59			.59	.65
2100 1/2" thick					1.38			1.38	1.52
2500 Mitered edges, 1/4" thick					1.38			1.38	1.52
2600 1/2" thick				▼	2.29			2.29	2.52

08 81 Glass Glazing

08 81 30 – Insulating Glass

08 81 30.10 Reduce Heat Transfer Glass

		Crew	Daily Output	Labor-Hours	Unit	Material	2018 Bare Costs Labor	Equipment	Total	Total Incl O&P
0010	**REDUCE HEAT TRANSFER GLASS**									
0015	2 lites 1/8" float, 1/2" thk under 15 S.F.									
0100	Tinted	[G] 2 Glaz	95	.168	S.F.	14.50	5.70		20.20	25.50
0280	Double glazed, 5/8" thk unit, 3/16" float, 15 to 30 S.F., clear		90	.178		13.80	6.05		19.85	25
0400	1" thk, dbl. glazed, 1/4" float, 30 to 70 S.F., clear	[G]	75	.213		17.35	7.25		24.60	31
0500	Tinted	[G]	75	.213		23.50	7.25		30.75	38
2000	Both lites, light & heat reflective	[G]	85	.188		31.50	6.40		37.90	45.50
2500	Heat reflective, film inside, 1" thick unit, clear	[G]	85	.188		27.50	6.40		33.90	41
2600	Tinted	[G]	85	.188		30	6.40		36.40	44
3000	Film on weatherside, clear, 1/2" thick unit	[G]	95	.168		19.75	5.70		25.45	31
3100	5/8" thick unit	[G]	90	.178		20.50	6.05		26.55	32.50
3200	1" thick unit	[G]	85	.188		27.50	6.40		33.90	40.50
5000	Spectrally selective film, on ext., blocks solar gain/allows 70% of light	[G] ↓	95	.168	↓	15.20	5.70		20.90	26

08 81 40 – Plate Glass

08 81 40.10 Plate Glass

		Crew	Daily Output	Labor-Hours	Unit	Material	2018 Bare Costs Labor	Equipment	Total	Total Incl O&P
0010	**PLATE GLASS** Twin ground, polished,									
0015	3/16" thick, material				S.F.	5.70			5.70	6.25
0020	3/16" thick	2 Glaz	100	.160		5.70	5.45		11.15	15.25
0100	1/4" thick		94	.170		7.80	5.80		13.60	18.15
0200	3/8" thick		60	.267		13.35	9.05		22.40	29.50
0300	1/2" thick	↓	40	.400	↓	26	13.60		39.60	51

08 81 55 – Window Glass

08 81 55.10 Sheet Glass

		Crew	Daily Output	Labor-Hours	Unit	Material	2018 Bare Costs Labor	Equipment	Total	Total Incl O&P
0010	**SHEET GLASS** (window), clear float, stops, putty bed									
0015	1/8" thick, clear float	2 Glaz	480	.033	S.F.	3.92	1.13		5.05	6.20
0500	3/16" thick, clear		480	.033		6.30	1.13		7.43	8.80
0600	Tinted		480	.033		7.85	1.13		8.98	10.50
0700	Tempered	↓	480	.033	↓	9.75	1.13		10.88	12.55

08 83 Mirrors

08 83 13 – Mirrored Glass Glazing

08 83 13.10 Mirrors

		Crew	Daily Output	Labor-Hours	Unit	Material	2018 Bare Costs Labor	Equipment	Total	Total Incl O&P
0010	**MIRRORS**, No frames, wall type, 1/4" plate glass, polished edge									
0100	Up to 5 S.F.	2 Glaz	125	.128	S.F.	9.80	4.35		14.15	18
0200	Over 5 S.F.		160	.100		9.55	3.40		12.95	16.10
0500	Door type, 1/4" plate glass, up to 12 S.F.		160	.100		9.15	3.40		12.55	15.65
1000	Float glass, up to 10 S.F., 1/8" thick		160	.100		6.10	3.40		9.50	12.30
1100	3/16" thick		150	.107		7.50	3.62		11.12	14.25
1500	12" x 12" wall tiles, square edge, clear		195	.082		2.61	2.79		5.40	7.50
1600	Veined		195	.082		6.55	2.79		9.34	11.80
2010	Bathroom, unframed, laminated	↓	160	.100	↓	14.75	3.40		18.15	22

For customer support on your Residential Costs with RSMeans data, call 800.448.8182.

479

08 87 Glazing Surface Films

08 87 26 – Bird Control Film

08 87 26.10 Bird Control Film

	08 87 26.10 Bird Control Film	Crew	Daily Output	Labor-Hours	Unit	Material	2018 Bare Costs Labor	Equipment	Total	Total Incl O&P
0010	**BIRD CONTROL FILM**									
0050	Patterned, adhered to glass	2 Glaz	180	.089	S.F.	6.75	3.02		9.77	12.45
0200	Decals small, adhered to glass	1 Glaz	50	.160	Ea.	2.48	5.45		7.93	11.75
0250	Decals large, adhered to glass		50	.160	"	6.95	5.45		12.40	16.65
0300	Decals set of 4, adhered to glass		25	.320	Set	6.95	10.85		17.80	25.50
0400	Bird control film tape		10	.800	Roll	19.95	27		46.95	67

08 87 33 – Electrically Tinted Window Film

08 87 33.20 Window Film

	08 87 33.20 Window Film	Crew	Daily Output	Labor-Hours	Unit	Material	2018 Bare Costs Labor	Equipment	Total	Total Incl O&P
0010	**WINDOW FILM** adhered on glass (glass not included)									
0015	Film is pre-wired and can be trimmed									
0100	Window film 4 S.F. adhered on glass	1 Glaz	200	.040	S.F.	224	1.36		225.36	249
0120	6 S.F.		180	.044		335	1.51		336.51	375
0160	9 S.F.		170	.047		505	1.60		506.60	560
0180	10 S.F.		150	.053		560	1.81		561.81	620
0200	12 S.F.		144	.056		670	1.89		671.89	740
0220	14 S.F.		140	.057		780	1.94		781.94	860
0240	16 S.F.		128	.063		890	2.12		892.12	985
0260	18 S.F.		126	.063		1,000	2.16		1,002.16	1,100
0280	20 S.F.		120	.067		1,125	2.26		1,127.26	1,225

08 87 53 – Security Films On Glass

08 87 53.10 Security Film Adhered On Glass

	08 87 53.10 Security Film Adhered On Glass	Crew	Daily Output	Labor-Hours	Unit	Material	2018 Bare Costs Labor	Equipment	Total	Total Incl O&P
0010	**SECURITY FILM ADHERED ON GLASS** (glass not included)									
0020	Security film, clear, 7 mil	1 Glaz	200	.040	S.F.	1.09	1.36		2.45	3.45
0030	8 mil		200	.040		1.97	1.36		3.33	4.42
0040	9 mil		200	.040		1.08	1.36		2.44	3.44
0050	10 mil		200	.040		1.87	1.36		3.23	4.31
0060	12 mil		200	.040		1.84	1.36		3.20	4.27
0075	15 mil		200	.040		2.32	1.36		3.68	4.80
0100	Security film, sealed with structural adhesive, 7 mil		180	.044		5.55	1.51		7.06	8.60
0110	8 mil		180	.044		6.45	1.51		7.96	9.55
0140	14 mil		180	.044		7.05	1.51		8.56	10.25

08 91 Louvers

08 91 19 – Fixed Louvers

08 91 19.10 Aluminum Louvers

	08 91 19.10 Aluminum Louvers	Crew	Daily Output	Labor-Hours	Unit	Material	2018 Bare Costs Labor	Equipment	Total	Total Incl O&P
0010	**ALUMINUM LOUVERS**									
0020	Aluminum with screen, residential, 8" x 8"	1 Carp	38	.211	Ea.	20	7.35		27.35	34.50
0100	12" x 12"		38	.211		16	7.35		23.35	30
0200	12" x 18"		35	.229		21	8		29	37
0250	14" x 24"		30	.267		29	9.30		38.30	47.50
0300	18" x 24"		27	.296		32	10.35		42.35	52
0500	24" x 30"		24	.333		61.50	11.65		73.15	87
0700	Triangle, adjustable, small		20	.400		58.50	14		72.50	87.50
0800	Large		15	.533		80.50	18.65		99.15	120
2100	Midget, aluminum, 3/4" deep, 1" diameter		85	.094		.86	3.29		4.15	6.40
2150	3" diameter		60	.133		2.94	4.66		7.60	11
2200	4" diameter		50	.160		5.60	5.60		11.20	15.45
2250	6" diameter		30	.267		3.78	9.30		13.08	19.65

08 95 Vents

08 95 13 – Soffit Vents

08 95 13.10 Wall Louvers

		Crew	Daily Output	Labor-Hours	Unit	Material	2018 Bare Costs Labor	Equipment	Total	Total Incl O&P
0010	**WALL LOUVERS**									
2400	Under eaves vent, aluminum, mill finish, 16" x 4"	1 Carp	48	.167	Ea.	1.89	5.85		7.74	11.80
2500	16" x 8"	"	48	.167	"	2.61	5.85		8.46	12.55

08 95 16 – Wall Vents

08 95 16.10 Louvers

		Crew	Daily Output	Labor-Hours	Unit	Material	2018 Bare Costs Labor	Equipment	Total	Total Incl O&P
0010	**LOUVERS**									
0020	Redwood, 2'-0" diameter, full circle	1 Carp	16	.500	Ea.	275	17.50		292.50	335
0100	Half circle		16	.500		214	17.50		231.50	264
0200	Octagonal		16	.500		194	17.50		211.50	242
0300	Triangular, 5/12 pitch, 5'-0" at base		16	.500		555	17.50		572.50	640
1000	Rectangular, 1'-4" x 1'-3"		16	.500		25.50	17.50		43	57
1100	Rectangular, 1'-4" x 1'-8"		16	.500		38.50	17.50		56	71
1200	1'-4" x 2'-2"		15	.533		44	18.65		62.65	79.50
1300	1'-9" x 2'-2"		15	.533		52.50	18.65		71.15	89
1400	2'-3" x 2'-2"		14	.571		70	19.95		89.95	110
1700	2'-4" x 2'-11"		13	.615		70	21.50		91.50	113
2000	Aluminum, 12" x 16"		25	.320		22.50	11.20		33.70	43
2010	16" x 20"		25	.320		30	11.20		41.20	51.50
2020	24" x 30"		25	.320		61.50	11.20		72.70	86
2100	6' triangle		12	.667		181	23.50		204.50	238
3100	Round, 2'-2" diameter		16	.500		179	17.50		196.50	226
7000	Vinyl gable vent, 8" x 8"		38	.211		17	7.35		24.35	31
7020	12" x 12"		38	.211		30	7.35		37.35	45.50
7080	12" x 18"		35	.229		38.50	8		46.50	56
7200	18" x 24"		30	.267		55.50	9.30		64.80	76.50

Division Notes

	CREW	DAILY OUTPUT	LABOR-HOURS	UNIT	BARE COSTS				TOTAL INCL O&P
					MAT.	LABOR	EQUIP.	TOTAL	

Estimating Tips
General

- Room Finish Schedule: A complete set of plans should contain a room finish schedule. If one is not available, it would be well worth the time and effort to obtain one.

09 20 00 Plaster and Gypsum Board

- Lath is estimated by the square yard plus a 5% allowance for waste. Furring, channels, and accessories are measured by the linear foot. An extra foot should be allowed for each accessory miter or stop.

- Plaster is also estimated by the square yard. Deductions for openings vary by preference, from zero deduction to 50% of all openings over 2 feet in width. The estimator should allow one extra square foot for each linear foot of horizontal interior or exterior angle located below the ceiling level. Also, double the areas of small radius work.

- Drywall accessories, studs, track, and acoustical caulking are all measured by the linear foot. Drywall taping is figured by the square foot. Gypsum wallboard is estimated by the square foot. No material deductions should be made for door or window openings under 32 S.F.

09 60 00 Flooring

- Tile and terrazzo areas are taken off on a square foot basis. Trim and base materials are measured by the linear foot. Accent tiles are listed per each. Two basic methods of installation are used. Mud set is approximately 30% more expensive than thin set. The cost of grout is included with tile unit price lines unless otherwise noted. In terrazzo work, be sure to include the linear footage of embedded decorative strips, grounds, machine rubbing, and power cleanup.

- Wood flooring is available in strip, parquet, or block configuration. The latter two types are set in adhesives with quantities estimated by the square foot. The laying pattern will influence labor costs and material waste. In addition to the material and labor for laying wood floors, the estimator must make allowances for sanding and finishing these areas, unless the flooring is prefinished.

- Sheet flooring is measured by the square yard. Roll widths vary, so consideration should be given to use the most economical width, as waste must be figured into the total quantity. Consider also the installation methods available—direct glue down or stretched. Direct glue-down installation is assumed with sheet carpet unit price lines unless otherwise noted.

09 70 00 Wall Finishes

- Wall coverings are estimated by the square foot. The area to be covered is measured—length by height of the wall above the baseboards—to calculate the square footage of each wall. This figure is divided by the number of square feet in the single roll which is being used. Deduct, in full, the areas of openings such as doors and windows. Where a pattern match is required allow 25%–30% waste.

09 80 00 Acoustic Treatment

- Acoustical systems fall into several categories. The takeoff of these materials should be by the square foot of area with a 5% allowance for waste. Do not forget about scaffolding, if applicable, when estimating these systems.

09 90 00 Painting and Coating

- A major portion of the work in painting involves surface preparation. Be sure to include cleaning, sanding, filling, and masking costs in the estimate.

- Protection of adjacent surfaces is not included in painting costs. When considering the method of paint application, an important factor is the amount of protection and masking required. These must be estimated separately and may be the determining factor in choosing the method of application.

Reference Numbers

Reference numbers are shown at the beginning of some major classifications. These numbers refer to related items in the Reference Section. The reference information may be an estimating procedure, an alternate pricing method, or technical information.

Note: Not all subdivisions listed here necessarily appear. ■

Did you know?

RSMeans data is available through our online application with 24/7 access:

- Search for unit prices by keyword
- Leverage the most up-to-date data
- Build and export estimates

Try it free for 30 days!
www.rsmeans.com/2018freetrial

09 01 Maintenance of Finishes

09 01 70 – Maintenance of Wall Finishes

09 01 70.10 Gypsum Wallboard Repairs

		Crew	Daily Output	Labor-Hours	Unit	Material	2018 Bare Costs Labor	Equipment	Total	Total Incl O&P
0010	**GYPSUM WALLBOARD REPAIRS**									
0100	Fill and sand, pin/nail holes	1 Carp	960	.008	Ea.		.29		.29	.48
0110	Screw head pops		480	.017			.58		.58	.97
0120	Dents, up to 2" square		48	.167		.01	5.85		5.86	9.70
0130	2" to 4" square		24	.333		.03	11.65		11.68	19.40
0140	Cut square, patch, sand and finish, holes, up to 2" square		12	.667		.03	23.50		23.53	38.50
0150	2" to 4" square		11	.727		.09	25.50		25.59	42.50
0160	4" to 8" square		10	.800		.24	28		28.24	47
0170	8" to 12" square		8	1		.48	35		35.48	58.50
0180	12" to 32" square		6	1.333		1.60	46.50		48.10	79.50
0210	16" by 48"		5	1.600		2.76	56		58.76	96
0220	32" by 48"		4	2		4.52	70		74.52	121
0230	48" square		3.50	2.286		6.40	80		86.40	140
0240	60" square		3.20	2.500		9.95	87.50		97.45	156
0500	Skim coat surface with joint compound		1600	.005	S.F.	.03	.17		.20	.32
0510	Prepare, retape and refinish joints		60	.133	L.F.	.63	4.66		5.29	8.45

09 05 Common Work Results for Finishes

09 05 05 – Selective Demolition for Finishes

09 05 05.10 Selective Demolition, Ceilings

		Crew	Daily Output	Labor-Hours	Unit	Material	2018 Bare Costs Labor	Equipment	Total	Total Incl O&P
0010	**SELECTIVE DEMOLITION, CEILINGS** R024119-10									
0200	Ceiling, gypsum wall board, furred and nailed or screwed	2 Clab	800	.020	S.F.		.53		.53	.89
1000	Plaster, lime and horse hair, on wood lath, incl. lath		700	.023			.61		.61	1.02
1200	Suspended ceiling, mineral fiber, 2' x 2' or 2' x 4'		1500	.011			.28		.28	.47
1250	On suspension system, incl. system		1200	.013			.36		.36	.59
1500	Tile, wood fiber, 12" x 12", glued		900	.018			.47		.47	.79
1540	Stapled		1500	.011			.28		.28	.47
2000	Wood, tongue and groove, 1" x 4"		1000	.016			.43		.43	.71
2040	1" x 8"		1100	.015			.39		.39	.65
2400	Plywood or wood fiberboard, 4' x 8' sheets		1200	.013			.36		.36	.59
2500	Remove & refinish textured ceiling	1 Plas	222	.036		.03	1.22		1.25	2.03

09 05 05.20 Selective Demolition, Flooring

		Crew	Daily Output	Labor-Hours	Unit	Material	2018 Bare Costs Labor	Equipment	Total	Total Incl O&P
0010	**SELECTIVE DEMOLITION, FLOORING** R024119-10									
0200	Brick with mortar	2 Clab	475	.034	S.F.		.90		.90	1.50
0400	Carpet, bonded, including surface scraping		2000	.008			.21		.21	.36
0480	Tackless		9000	.002			.05		.05	.08
0550	Carpet tile, releasable adhesive		5000	.003			.09		.09	.14
0560	Permanent adhesive		1850	.009			.23		.23	.38
0800	Resilient, sheet goods		1400	.011			.31		.31	.51
0850	Vinyl or rubber cove base	1 Clab	1000	.008	L.F.		.21		.21	.36
0860	Vinyl or rubber cove base, molded corner	"	1000	.008	Ea.		.21		.21	.36
0870	For glued and caulked installation, add to labor						50%			
0900	Vinyl composition tile, 12" x 12"	2 Clab	1000	.016	S.F.		.43		.43	.71
2000	Tile, ceramic, thin set		675	.024			.63		.63	1.05
2020	Mud set		625	.026			.68		.68	1.14
3000	Wood, block, on end	1 Carp	400	.020			.70		.70	1.16
3200	Parquet		450	.018			.62		.62	1.03
3400	Strip flooring, interior, 2-1/4" x 25/32" thick		325	.025			.86		.86	1.43
3500	Exterior, porch flooring, 1" x 4"		220	.036			1.27		1.27	2.11
3800	Subfloor, tongue and groove, 1" x 6"		325	.025			.86		.86	1.43
3820	1" x 8"		430	.019			.65		.65	1.08

09 05 Common Work Results for Finishes

09 05 05 - Selective Demolition for Finishes

09 05 05.20 Selective Demolition, Flooring

		Crew	Daily Output	Labor-Hours	Unit	Material	2018 Bare Costs Labor	Equipment	Total	Total Incl O&P
3840	1" x 10"	1 Carp	520	.015	S.F.		.54		.54	.89
4000	Plywood, nailed		600	.013			.47		.47	.77
4100	Glued and nailed		400	.020			.70		.70	1.16
4200	Hardboard, 1/4" thick	↓	760	.011	↓		.37		.37	.61
9050	For grinding concrete floors, see Section 03 35 43.10									

09 05 05.30 Selective Demolition, Walls and Partitions

		Crew	Daily Output	Labor-Hours	Unit	Material	2018 Bare Costs Labor	Equipment	Total	Total Incl O&P
0010	**SELECTIVE DEMOLITION, WALLS AND PARTITIONS** R024119-10									
0020	Walls, concrete, reinforced	B-39	120	.400	C.F.		10.80	1.95	12.75	20
0025	Plain	"	160	.300	"		8.10	1.46	9.56	15.10
1000	Gypsum wallboard, nailed or screwed	1 Clab	1000	.008	S.F.		.21		.21	.36
1010	2 layers		400	.020			.53		.53	.89
1500	Fiberboard, nailed		900	.009			.24		.24	.39
1568	Plenum barrier, sheet lead	↓	300	.027			.71		.71	1.18
2200	Metal or wood studs, finish 2 sides, fiberboard	B-1	520	.046			1.26		1.26	2.10
2250	Lath and plaster		260	.092			2.53		2.53	4.20
2300	Gypsum wallboard		520	.046			1.26		1.26	2.10
2350	Plywood	↓	450	.053			1.46		1.46	2.43
2800	Paneling, 4' x 8' sheets	1 Clab	475	.017			.45		.45	.75
3000	Plaster, lime and horsehair, on wood lath		400	.020			.53		.53	.89
3020	On metal lath		335	.024	↓		.64		.64	1.06
3450	Plaster, interior gypsum, acoustic, or cement		60	.133	S.Y.		3.56		3.56	5.90
3500	Stucco, on masonry		145	.055			1.47		1.47	2.45
3510	Commercial 3-coat		80	.100			2.67		2.67	4.44
3520	Interior stucco		25	.320	↓		8.55		8.55	14.20
3760	Tile, ceramic, on walls, thin set		300	.027	S.F.		.71		.71	1.18
3765	Mud set	↓	250	.032	"		.85		.85	1.42

09 22 Supports for Plaster and Gypsum Board

09 22 03 - Fastening Methods for Finishes

09 22 03.20 Drilling Plaster/Drywall

		Crew	Daily Output	Labor-Hours	Unit	Material	2018 Bare Costs Labor	Equipment	Total	Total Incl O&P
0010	**DRILLING PLASTER/DRYWALL**									
1100	Drilling & layout for drywall/plaster walls, up to 1" deep, no anchor									
1200	Holes, 1/4" diameter	1 Carp	150	.053	Ea.	.01	1.86		1.87	3.11
1300	3/8" diameter		140	.057		.01	2		2.01	3.33
1400	1/2" diameter		130	.062		.01	2.15		2.16	3.59
1500	3/4" diameter		120	.067		.02	2.33		2.35	3.89
1600	1" diameter		110	.073		.02	2.54		2.56	4.25
1700	1-1/4" diameter		100	.080		.04	2.80		2.84	4.69
1800	1-1/2" diameter	↓	90	.089		.06	3.11		3.17	5.20
1900	For ceiling installations, add				↓		40%			

09 22 13 - Metal Furring

09 22 13.13 Metal Channel Furring

		Crew	Daily Output	Labor-Hours	Unit	Material	2018 Bare Costs Labor	Equipment	Total	Total Incl O&P
0010	**METAL CHANNEL FURRING**									
0030	Beams and columns, 7/8" hat channels, galvanized, 12" OC	1 Lath	155	.052	S.F.	.44	1.78		2.22	3.35
0050	16" OC		170	.047		.36	1.62		1.98	3.01
0070	24" OC		185	.043		.24	1.49		1.73	2.67
0100	Ceilings, on steel, 7/8" hat channels, galvanized, 12" OC		210	.038		.40	1.31		1.71	2.56
0300	16" OC		290	.028		.36	.95		1.31	1.93
0400	24" OC		420	.019		.24	.66		.90	1.32
0600	1-5/8" hat channels, galvanized, 12" OC	↓	190	.042		.53	1.45		1.98	2.92

For customer support on your Residential Costs with RSMeans data, call 800.448.8182.

485

09 22 13 – Metal Furring

09 22 13.13 Metal Channel Furring		Crew	Daily Output	Labor-Hours	Unit	Material	2018 Bare Costs Labor	Equipment	Total	Total Incl O&P
0700	16" OC	1 Lath	260	.031	S.F.	.47	1.06		1.53	2.23
0900	24" OC		390	.021		.32	.71		1.03	1.49
0930	7/8" hat channels with sound isolation clips, 12" OC		120	.067		1.72	2.29		4.01	5.60
0940	16" OC		100	.080		1.29	2.75		4.04	5.85
0950	24" OC		165	.048		.86	1.67		2.53	3.64
0960	1-5/8" hat channels, galvanized, 12" OC		110	.073		1.85	2.50		4.35	6.10
0970	16" OC		100	.080		1.38	2.75		4.13	5.95
0980	24" OC		155	.052		.92	1.78		2.70	3.88
1000	Walls, 7/8" hat channels, galvanized, 12" OC		235	.034		.40	1.17		1.57	2.33
1200	16" OC		265	.030		.36	1.04		1.40	2.07
1300	24" OC		350	.023		.24	.79		1.03	1.53
1500	1-5/8" hat channels, galvanized, 12" OC		210	.038		.53	1.31		1.84	2.70
1600	16" OC		240	.033		.47	1.15		1.62	2.37
1800	24" OC		305	.026		.32	.90		1.22	1.81
1920	7/8" hat channels with sound isolation clips, 12" OC		125	.064		1.72	2.20		3.92	5.45
1940	16" OC		100	.080		1.29	2.75		4.04	5.85
1950	24" OC		150	.053		.86	1.83		2.69	3.91
1960	1-5/8" hat channels, galvanized, 12" OC		115	.070		1.85	2.39		4.24	5.90
1970	16" OC		95	.084		1.38	2.90		4.28	6.20
1980	24" OC		140	.057		.92	1.97		2.89	4.19
3000	Z Furring, walls, 1" deep, 25 ga., 24" OC		350	.023		1.38	.79		2.17	2.78
3010	48" OC		700	.011		.69	.39		1.08	1.40
3020	1-1/2" deep, 24" OC		345	.023		1.61	.80		2.41	3.06
3030	48" OC		695	.012		.80	.40		1.20	1.52
3040	2" deep, 24" OC		340	.024		1.93	.81		2.74	3.43
3050	48" OC		690	.012		.97	.40		1.37	1.71
3060	1" deep, 20 ga., 24" OC		350	.023		2.28	.79		3.07	3.77
3070	48" OC		700	.011		1.14	.39		1.53	1.89
3080	1-1/2" deep, 24" OC		345	.023		2.62	.80		3.42	4.17
3090	48" OC		695	.012		1.31	.40		1.71	2.08
4000	2" deep, 24" OC		340	.024		3.22	.81		4.03	4.85
4010	48" OC		690	.012		1.61	.40		2.01	2.42

09 22 16 – Non-Structural Metal Framing

09 22 16.13 Non-Structural Metal Stud Framing

0010	NON-STRUCTURAL METAL STUD FRAMING									
1600	Non-load bearing, galv., 8' high, 25 ga. 1-5/8" wide, 16" OC	1 Carp	619	.013	S.F.	.27	.45		.72	1.05
1610	24" OC		950	.008		.20	.29		.49	.71
1620	2-1/2" wide, 16" OC		613	.013		.36	.46		.82	1.15
1630	24" OC		938	.009		.27	.30		.57	.79
1640	3-5/8" wide, 16" OC		600	.013		.40	.47		.87	1.21
1650	24" OC		925	.009		.30	.30		.60	.83
1660	4" wide, 16" OC		594	.013		.45	.47		.92	1.27
1670	24" OC		925	.009		.33	.30		.63	.87
1680	6" wide, 16" OC		588	.014		.53	.48		1.01	1.37
1690	24" OC		906	.009		.40	.31		.71	.95
1700	20 ga. studs, 1-5/8" wide, 16" OC		494	.016		.34	.57		.91	1.32
1710	24" OC		763	.010		.26	.37		.63	.89
1720	2-1/2" wide, 16" OC		488	.016		.44	.57		1.01	1.43
1730	24" OC		750	.011		.33	.37		.70	.98
1740	3-5/8" wide, 16" OC		481	.017		.50	.58		1.08	1.52
1750	24" OC		738	.011		.37	.38		.75	1.04
1760	4" wide, 16" OC		475	.017		.60	.59		1.19	1.64

09 22 16 – Non-Structural Metal Framing

09 22 16.13 Non-Structural Metal Stud Framing	Crew	Daily Output	Labor-Hours	Unit	Material	2018 Bare Costs Labor	Equipment	Total	Total Incl O&P
1770	24" OC	1 Carp	738	.011	S.F.	.45	.38		.83	1.13
1780	6" wide, 16" OC		469	.017		.71	.60		1.31	1.78
1790	24" OC		725	.011		.54	.39		.93	1.23
2000	Non-load bearing, galv., 10' high, 25 ga. 1-5/8" wide, 16" OC		495	.016		.25	.56		.81	1.22
2100	24" OC		760	.011		.19	.37		.56	.82
2200	2-1/2" wide, 16" OC		490	.016		.34	.57		.91	1.32
2250	24" OC		750	.011		.25	.37		.62	.89
2300	3-5/8" wide, 16" OC		480	.017		.38	.58		.96	1.39
2350	24" OC		740	.011		.28	.38		.66	.94
2400	4" wide, 16" OC		475	.017		.42	.59		1.01	1.45
2450	24" OC		740	.011		.31	.38		.69	.97
2500	6" wide, 16" OC		470	.017		.50	.59		1.09	1.54
2550	24" OC		725	.011		.37	.39		.76	1.05
2600	20 ga. studs, 1-5/8" wide, 16" OC		395	.020		.32	.71		1.03	1.54
2650	24" OC		610	.013		.24	.46		.70	1.02
2700	2-1/2" wide, 16" OC		390	.021		.41	.72		1.13	1.64
2750	24" OC		600	.013		.30	.47		.77	1.10
2800	3-5/8" wide, 16" OC		385	.021		.47	.73		1.20	1.73
2850	24" OC		590	.014		.35	.47		.82	1.17
2900	4" wide, 16" OC		380	.021		.57	.74		1.31	1.85
2950	24" OC		590	.014		.42	.47		.89	1.25
3000	6" wide, 16" OC		375	.021		.68	.75		1.43	1.98
3050	24" OC		580	.014		.50	.48		.98	1.35
3060	Non-load bearing, galv., 12' high, 25 ga. 1-5/8" wide, 16" OC		413	.019		.24	.68		.92	1.40
3070	24" OC		633	.013		.18	.44		.62	.93
3080	2-1/2" wide, 16" OC		408	.020		.32	.69		1.01	1.50
3090	24" OC		625	.013		.24	.45		.69	1
3100	3-5/8" wide, 16" OC		400	.020		.36	.70		1.06	1.56
3110	24" OC		617	.013		.26	.45		.71	1.04
3120	4" wide, 16" OC		396	.020		.40	.71		1.11	1.61
3130	24" OC		617	.013		.30	.45		.75	1.07
3140	6" wide, 16" OC		392	.020		.48	.71		1.19	1.72
3150	24" OC		604	.013		.35	.46		.81	1.16
3160	20 ga. studs, 1-5/8" wide, 16" OC		329	.024		.31	.85		1.16	1.75
3170	24" OC		508	.016		.23	.55		.78	1.17
3180	2-1/2" wide, 16" OC		325	.025		.39	.86		1.25	1.86
3190	24" OC		500	.016		.29	.56		.85	1.25
3200	3-5/8" wide, 16" OC		321	.025		.45	.87		1.32	1.95
3210	24" OC		492	.016		.33	.57		.90	1.30
3220	4" wide, 16" OC		317	.025		.55	.88		1.43	2.07
3230	24" OC		492	.016		.40	.57		.97	1.38
3240	6" wide, 16" OC		313	.026		.65	.89		1.54	2.20
3250	24" OC		483	.017		.47	.58		1.05	1.48
3260	Non-load bearing, galv., 16' high, 25 ga. 4" wide, 12" OC		195	.041		.52	1.43		1.95	2.95
3270	16" OC		275	.029		.41	1.02		1.43	2.14
3280	24" OC		400	.020		.30	.70		1	1.49
3290	6" wide, 12" OC		190	.042		.62	1.47		2.09	3.13
3300	16" OC		280	.029		.49	1		1.49	2.19
3310	24" OC		400	.020		.35	.70		1.05	1.55
3320	20 ga. studs, 2-1/2" wide, 12" OC		180	.044		.50	1.55		2.05	3.13
3330	16" OC		254	.032		.40	1.10		1.50	2.27
3340	24" OC		390	.021		.29	.72		1.01	1.51
3350	3-5/8" wide, 12" OC		170	.047		.58	1.64		2.22	3.36

09 22 16 – Non-Structural Metal Framing

09 22 16.13 Non-Structural Metal Stud Framing

		Crew	Daily Output	Labor-Hours	Unit	Material	2018 Bare Costs Labor	Equipment	Total	Total Incl O&P
3360	16" OC	1 Carp	251	.032	S.F.	.45	1.11		1.56	2.35
3370	24" OC		384	.021		.33	.73		1.06	1.57
3380	4" wide, 12" OC		170	.047		.70	1.64		2.34	3.50
3390	16" OC		247	.032		.55	1.13		1.68	2.49
3400	24" OC		384	.021		.40	.73		1.13	1.65
3410	6" wide, 12" OC		175	.046		.83	1.60		2.43	3.57
3420	16" OC		245	.033		.65	1.14		1.79	2.62
3430	24" OC		400	.020		.48	.70		1.18	1.69
3440	Non-load bearing, galv., 20' high, 25 ga. 6" wide, 12" OC		125	.064		.60	2.24		2.84	4.38
3450	16" OC		220	.036		.47	1.27		1.74	2.63
3460	24" OC		360	.022		.34	.78		1.12	1.67
3470	20 ga. studs, 4" wide, 12" OC		120	.067		.69	2.33		3.02	4.62
3480	16" OC		215	.037		.54	1.30		1.84	2.75
3490	6" wide, 12" OC		115	.070		.81	2.43		3.24	4.93
3500	16" OC		215	.037		.64	1.30		1.94	2.86
3510	24" OC		331	.024		.46	.84		1.30	1.91
5000	For load bearing studs, see Section 05 41 13.30									

09 22 26 – Suspension Systems

09 22 26.13 Ceiling Suspension Systems

		Crew	Daily Output	Labor-Hours	Unit	Material	2018 Bare Costs Labor	Equipment	Total	Total Incl O&P
0010	**CEILING SUSPENSION SYSTEMS** for gypsum board or plaster									
8000	Suspended ceilings, including carriers									
8200	1-1/2" carriers, 24" OC with:									
8300	7/8" channels, 16" OC	1 Lath	275	.029	S.F.	.58	1		1.58	2.25
8320	24" OC		310	.026		.46	.89		1.35	1.94
8400	1-5/8" channels, 16" OC		205	.039		.69	1.34		2.03	2.93
8420	24" OC		250	.032		.53	1.10		1.63	2.37
8600	2" carriers, 24" OC with:									
8700	7/8" channels, 16" OC	1 Lath	250	.032	S.F.	.66	1.10		1.76	2.50
8720	24" OC		285	.028		.54	.97		1.51	2.15
8800	1-5/8" channels, 16" OC		190	.042		.77	1.45		2.22	3.19
8820	24" OC		225	.036		.62	1.22		1.84	2.66

09 22 36 – Lath

09 22 36.13 Gypsum Lath

		Crew	Daily Output	Labor-Hours	Unit	Material	2018 Bare Costs Labor	Equipment	Total	Total Incl O&P
0011	**GYPSUM LATH** Plain or perforated, nailed, 3/8" thick	1 Lath	765	.010	S.F.	.35	.36		.71	.97
0101	1/2" thick, nailed		720	.011		.28	.38		.66	.93
0301	Clipped to steel studs, 3/8" thick		675	.012		.35	.41		.76	1.05
0401	1/2" thick		630	.013		.28	.44		.72	1.02
0601	Firestop gypsum base, to steel studs, 3/8" thick		630	.013		.28	.44		.72	1.02
0701	1/2" thick		585	.014		.32	.47		.79	1.11
0901	Foil back, to steel studs, 3/8" thick		675	.012		.36	.41		.77	1.06
1001	1/2" thick		630	.013		.38	.44		.82	1.13
1501	For ceiling installations, add		1950	.004			.14		.14	.23
1601	For columns and beams, add		1550	.005			.18		.18	.29

09 22 36.23 Metal Lath

			Crew	Daily Output	Labor-Hours	Unit	Material	2018 Bare Costs Labor	Equipment	Total	Total Incl O&P
0010	**METAL LATH**	R092000-50									
3601	2.5 lb. diamond painted, on wood framing, on walls		1 Lath	765	.010	S.F.	.42	.36		.78	1.04
3701	On ceilings			675	.012		.42	.41		.83	1.12
4201	3.4 lb. diamond painted, wired to steel framing, on walls			675	.012		.45	.41		.86	1.16
4301	On ceilings			540	.015		.45	.51		.96	1.32
5101	Rib lath, painted, wired to steel, on walls, 2.75 lb.			675	.012		.36	.41		.77	1.06
5201	3.4 lb.			630	.013		.47	.44		.91	1.23

09 22 Supports for Plaster and Gypsum Board

09 22 36 – Lath

09 22 36.23 Metal Lath	Crew	Daily Output	Labor-Hours	Unit	Material	2018 Bare Costs Labor	Equipment	Total	Total Incl O&P
5701	Suspended ceiling system, incl. 3.4 lb. diamond lath, painted	1 Lath	135	.059	S.F.	.46	2.04		2.50	3.81
5801	Galvanized	↓	135	.059	↓	.50	2.04		2.54	3.85

09 22 36.83 Accessories, Plaster

		Crew	Daily Output	Labor-Hours	Unit	Material	Labor	Equipment	Total	Total Incl O&P
0010	**ACCESSORIES, PLASTER**									
0020	Casing bead, expanded flange, galvanized	1 Lath	2.70	2.963	C.L.F.	52.50	102		154.50	223
0200	Foundation weep screed, galvanized	"	2.70	2.963		53	102		155	224
0900	Channels, cold rolled, 16 ga., 3/4" deep, galvanized					39.50			39.50	43.50
1620	Corner bead, expanded bullnose, 3/4" radius, #10, galvanized	1 Lath	2.60	3.077		26.50	106		132.50	200
1650	#1, galvanized		2.55	3.137		49	108		157	229
1670	Expanded wing, 2-3/4" wide, #1, galvanized		2.65	3.019		40.50	104		144.50	213
1700	Inside corner (corner rite), 3" x 3", painted		2.60	3.077		21	106		127	194
1750	Strip-ex, 4" wide, painted		2.55	3.137		23.50	108		131.50	201
1800	Expansion joint, 3/4" grounds, limited expansion, galv., 1 piece		2.70	2.963		73	102		175	246
2100	Extreme expansion, galvanized, 2 piece	↓	2.60	3.077	↓	141	106		247	325

09 23 Gypsum Plastering

09 23 13 – Acoustical Gypsum Plastering

09 23 13.10 Perlite or Vermiculite Plaster

		Crew	Daily Output	Labor-Hours	Unit	Material	Labor	Equipment	Total	Total Incl O&P
0010	**PERLITE OR VERMICULITE PLASTER** R092000-50									
0020	In 100 lb. bags, under 200 bags				Bag	18.20			18.20	20
0301	2 coats, no lath included, on walls	J-1	830	.048	S.F.	.66	1.54	.16	2.36	3.43
0401	On ceilings		710	.056		.66	1.80	.19	2.65	3.89
0901	3 coats, no lath included, on walls		665	.060		.71	1.92	.20	2.83	4.15
1001	On ceilings	↓	565	.071		.71	2.26	.24	3.21	4.75
1700	For irregular or curved surfaces, add to above				S.Y.		30%			
1800	For columns and beams, add to above						50%			
1900	For soffits, add to ceiling prices				↓		40%			

09 23 20 – Gypsum Plaster

09 23 20.10 Gypsum Plaster On Walls and Ceilings

		Crew	Daily Output	Labor-Hours	Unit	Material	Labor	Equipment	Total	Total Incl O&P
0010	**GYPSUM PLASTER ON WALLS AND CEILINGS** R092000-50									
0020	80# bag, less than 1 ton				Bag	15.80			15.80	17.40
0302	2 coats, no lath included, on walls	J-1	750	.053	S.F.	.42	1.70	.18	2.30	3.45
0402	On ceilings		660	.061		.42	1.94	.20	2.56	3.85
0903	3 coats, no lath included, on walls		620	.065		.60	2.06	.21	2.87	4.28
1002	On ceilings	↓	560	.071	↓	1.06	2.28	.24	3.58	5.15
1600	For irregular or curved surfaces, add						30%			
1800	For columns & beams, add						50%			

09 24 Cement Plastering

09 24 23 – Cement Stucco

09 24 23.40 Stucco

		Crew	Daily Output	Labor-Hours	Unit	Material	Labor	Equipment	Total	Total Incl O&P
0010	**STUCCO** R092000-50									
0011	3 coats 1" thick, float finish, with mesh, on wood frame	J-2	470	.102	S.F.	1.09	3.30	.28	4.67	6.90
0101	On masonry construction	J-1	495	.081		.79	2.58	.27	3.64	5.40
0151	2 coats, 5/8" thick, float finish, no lath incl.	"	980	.041		.37	1.30	.14	1.81	2.70
0301	For trowel finish, add	1 Plas	1530	.005	↓		.18		.18	.29
0600	For coloring, add	J-1	685	.058	S.Y.	.42	1.87	.19	2.48	3.73
0700	For special texture, add	↓	200	.200	"	1.46	6.40	.67	8.53	12.85

For customer support on your Residential Costs with RSMeans data, call 800.448.8182.

489

09 24 Cement Plastering

09 24 23 – Cement Stucco

09 24 23.40 Stucco	Crew	Daily Output	Labor-Hours	Unit	Material	2018 Bare Costs Labor	Equipment	Total	Total Incl O&P
1001	Stucco, with bonding agent, 3 coats, on walls	J-1	1800	.022	S.F.	.48	.71	.07	1.26	1.77
1201	Ceilings		1620	.025		.38	.79	.08	1.25	1.79
1301	Beams		720	.056		.38	1.77	.19	2.34	3.52
1501	Columns		900	.044		.38	1.42	.15	1.95	2.90
1601	Mesh, galvanized, nailed to wood, 1.8 lb.	1 Lath	540	.015		.82	.51		1.33	1.72
1801	3.6 lb.		495	.016		.48	.56		1.04	1.42
1901	Wired to steel, galvanized, 1.8 lb.		477	.017		.82	.58		1.40	1.83
2101	3.6 lb.		450	.018		.48	.61		1.09	1.51

09 25 Other Plastering

09 25 23 – Lime Based Plastering

09 25 23.10 Venetian Plaster

		Crew	Daily Output	Labor-Hours	Unit	Material	2018 Bare Costs Labor	Equipment	Total	Total Incl O&P
0010	**VENETIAN PLASTER**									
0100	Walls, 1 coat primer, roller applied	1 Plas	950	.008	S.F.	.17	.29		.46	.66
0210	For pigment, light colors add per S.F. plaster					.02			.02	.02
0220	For pigment, dark colors add per S.F. plaster					.04			.04	.04
0300	For sealer/wax coat incl. burnishing, add	1 Plas	300	.027		.38	.90		1.28	1.89

09 26 Veneer Plastering

09 26 13 – Gypsum Veneer Plastering

09 26 13.20 Blueboard

		Crew	Daily Output	Labor-Hours	Unit	Material	2018 Bare Costs Labor	Equipment	Total	Total Incl O&P
0010	**BLUEBOARD** For use with thin coat									
0100	plaster application see Section 09 26 13.80									
1000	3/8" thick, on walls or ceilings, standard, no finish included	2 Carp	1900	.008	S.F.	.34	.29		.63	.86
1100	With thin coat plaster finish		875	.018		.45	.64		1.09	1.56
1400	On beams, columns, or soffits, standard, no finish included		675	.024		.39	.83		1.22	1.81
1450	With thin coat plaster finish		475	.034		.50	1.18		1.68	2.51
3000	1/2" thick, on walls or ceilings, standard, no finish included		1900	.008		.35	.29		.64	.88
3100	With thin coat plaster finish		875	.018		.46	.64		1.10	1.57
3300	Fire resistant, no finish included		1900	.008		.35	.29		.64	.88
3400	With thin coat plaster finish		875	.018		.46	.64		1.10	1.57
3450	On beams, columns, or soffits, standard, no finish included		675	.024		.40	.83		1.23	1.82
3500	With thin coat plaster finish		475	.034		.52	1.18		1.70	2.53
3700	Fire resistant, no finish included		675	.024		.40	.83		1.23	1.82
3800	With thin coat plaster finish		475	.034		.52	1.18		1.70	2.53
5000	5/8" thick, on walls or ceilings, fire resistant, no finish included		1900	.008		.36	.29		.65	.89
5100	With thin coat plaster finish		875	.018		.47	.64		1.11	1.58
5500	On beams, columns, or soffits, no finish included		675	.024		.41	.83		1.24	1.84
5600	With thin coat plaster finish		475	.034		.53	1.18		1.71	2.54
6000	For high ceilings, over 8' high, add		3060	.005			.18		.18	.30
6500	For distribution costs 3 stories and above, add per story		6100	.003			.09		.09	.15

09 26 13.80 Thin Coat Plaster

		Crew	Daily Output	Labor-Hours	Unit	Material	2018 Bare Costs Labor	Equipment	Total	Total Incl O&P
0010	**THIN COAT PLASTER** R092000-50									
0012	1 coat veneer, not incl. lath	J-1	3600	.011	S.F.	.11	.35	.04	.50	.74
1000	In 50 lb. bags				Bag	15.20			15.20	16.75

09 28 Backing Boards and Underlayments

09 28 13 – Cementitious Backing Boards

09 28 13.10 Cementitious Backerboard	Crew	Daily Output	Labor-Hours	Unit	Material	2018 Bare Costs Labor	Equipment	Total	Total Incl O&P
0010 **CEMENTITIOUS BACKERBOARD**									
0070 Cementitious backerboard, on floor, 3' x 4' x 1/2" sheets	2 Carp	525	.030	S.F.	.86	1.07		1.93	2.72
0080 3' x 5' x 1/2" sheets		525	.030		.79	1.07		1.86	2.64
0090 3' x 6' x 1/2" sheets		525	.030		.78	1.07		1.85	2.63
0100 3' x 4' x 5/8" sheets		525	.030		.95	1.07		2.02	2.82
0110 3' x 5' x 5/8" sheets		525	.030		.95	1.07		2.02	2.81
0120 3' x 6' x 5/8" sheets		525	.030		.95	1.07		2.02	2.82
0150 On wall, 3' x 4' x 1/2" sheets		350	.046		.86	1.60		2.46	3.61
0160 3' x 5' x 1/2" sheets		350	.046		.79	1.60		2.39	3.53
0170 3' x 6' x 1/2" sheets		350	.046		.78	1.60		2.38	3.52
0180 3' x 4' x 5/8" sheets		350	.046		.95	1.60		2.55	3.71
0190 3' x 5' x 5/8" sheets		350	.046		.95	1.60		2.55	3.70
0200 3' x 6' x 5/8" sheets		350	.046		.95	1.60		2.55	3.71
0250 On counter, 3' x 4' x 1/2" sheets		180	.089		.86	3.11		3.97	6.10
0260 3' x 5' x 1/2" sheets		180	.089		.79	3.11		3.90	6
0270 3' x 6' x 1/2" sheets		180	.089		.78	3.11		3.89	6
0300 3' x 4' x 5/8" sheets		180	.089		.95	3.11		4.06	6.20
0310 3' x 5' x 5/8" sheets		180	.089		.95	3.11		4.06	6.20
0320 3' x 6' x 5/8" sheets		180	.089		.95	3.11		4.06	6.20

09 29 Gypsum Board

09 29 10 – Gypsum Board Panels

09 29 10.20 Taping and Finishing

	Crew	Daily Output	Labor-Hours	Unit	Material	Labor	Equipment	Total	Total Incl O&P
0010 **TAPING AND FINISHING**									
3600 For taping and finishing joints, add	2 Carp	2000	.008	S.F.	.05	.28		.33	.51
4500 For thin coat plaster instead of taping, add	J-1	3600	.011	"	.11	.35	.04	.50	.74

09 29 10.30 Gypsum Board

	Crew	Daily Output	Labor-Hours	Unit	Material	Labor	Equipment	Total	Total Incl O&P
0010 **GYPSUM BOARD** on walls & ceilings R092910-10									
0100 Nailed or screwed to studs unless otherwise noted									
0110 1/4" thick, on walls or ceilings, standard, no finish included	2 Carp	1330	.012	S.F.	.37	.42		.79	1.11
0115 1/4" thick, on walls or ceilings, flexible, no finish included		1050	.015		.53	.53		1.06	1.47
0117 1/4" thick, on columns or soffits, flexible, no finish included		1050	.015		.53	.53		1.06	1.47
0130 1/4" thick, standard, no finish included, less than 800 S.F.		510	.031		.37	1.10		1.47	2.23
0150 3/8" thick, on walls, standard, no finish included		2000	.008		.36	.28		.64	.86
0200 On ceilings, standard, no finish included		1800	.009		.36	.31		.67	.92
0250 On beams, columns, or soffits, no finish included		675	.024		.36	.83		1.19	1.78
0300 1/2" thick, on walls, standard, no finish included		2000	.008		.34	.28		.62	.83
0350 Taped and finished (level 4 finish)		965	.017		.39	.58		.97	1.39
0390 With compound skim coat (level 5 finish)		775	.021		.44	.72		1.16	1.68
0400 Fire resistant, no finish included		2000	.008		.37	.28		.65	.87
0450 Taped and finished (level 4 finish)		965	.017		.42	.58		1	1.42
0490 With compound skim coat (level 5 finish)		775	.021		.47	.72		1.19	1.72
0500 Water resistant, no finish included		2000	.008		.42	.28		.70	.92
0550 Taped and finished (level 4 finish)		965	.017		.47	.58		1.05	1.47
0590 With compound skim coat (level 5 finish)		775	.021		.52	.72		1.24	1.77
0600 Prefinished, vinyl, clipped to studs		900	.018		.51	.62		1.13	1.59
0700 Mold resistant, no finish included		2000	.008		.44	.28		.72	.94
0710 Taped and finished (level 4 finish)		965	.017		.49	.58		1.07	1.50
0720 With compound skim coat (level 5 finish)		775	.021		.54	.72		1.26	1.79
1000 On ceilings, standard, no finish included		1800	.009		.34	.31		.65	.89
1050 Taped and finished (level 4 finish)		765	.021		.39	.73		1.12	1.65

For customer support on your Residential Costs with RSMeans data, call 800.448.8182.

491

09 29 Gypsum Board

09 29 10 – Gypsum Board Panels

09 29 10.30 Gypsum Board	Crew	Daily Output	Labor-Hours	Unit	Material	2018 Bare Costs Labor	Equipment	Total	Total Incl O&P
1090	With compound skim coat (level 5 finish)	2 Carp	610	.026	S.F.	.44	.92		1.36	2
1100	Fire resistant, no finish included		1800	.009		.37	.31		.68	.93
1150	Taped and finished (level 4 finish)		765	.021		.42	.73		1.15	1.68
1195	With compound skim coat (level 5 finish)		610	.026		.47	.92		1.39	2.04
1200	Water resistant, no finish included		1800	.009		.42	.31		.73	.98
1250	Taped and finished (level 4 finish)		765	.021		.47	.73		1.20	1.73
1290	With compound skim coat (level 5 finish)		610	.026		.52	.92		1.44	2.09
1310	Mold resistant, no finish included		1800	.009		.44	.31		.75	1
1320	Taped and finished (level 4 finish)		765	.021		.49	.73		1.22	1.76
1330	With compound skim coat (level 5 finish)		610	.026		.54	.92		1.46	2.11
1350	Sag resistant, no finish included		1600	.010		.36	.35		.71	.98
1360	Taped and finished (level 4 finish)		765	.021		.41	.73		1.14	1.67
1370	With compound skim coat (level 5 finish)		610	.026		.46	.92		1.38	2.02
1500	On beams, columns, or soffits, standard, no finish included		675	.024		.39	.83		1.22	1.81
1550	Taped and finished (level 4 finish)		540	.030		.39	1.04		1.43	2.15
1590	With compound skim coat (level 5 finish)		475	.034		.44	1.18		1.62	2.44
1600	Fire resistant, no finish included		675	.024		.37	.83		1.20	1.79
1650	Taped and finished (level 4 finish)		540	.030		.42	1.04		1.46	2.18
1690	With compound skim coat (level 5 finish)		475	.034		.47	1.18		1.65	2.48
1700	Water resistant, no finish included		675	.024		.48	.83		1.31	1.91
1750	Taped and finished (level 4 finish)		540	.030		.47	1.04		1.51	2.23
1790	With compound skim coat (level 5 finish)		475	.034		.52	1.18		1.70	2.53
1800	Mold resistant, no finish included		675	.024		.51	.83		1.34	1.94
1810	Taped and finished (level 4 finish)		540	.030		.49	1.04		1.53	2.26
1820	With compound skim coat (level 5 finish)		475	.034		.54	1.18		1.72	2.55
1850	Sag resistant, no finish included		675	.024		.41	.83		1.24	1.84
1860	Taped and finished (level 4 finish)		540	.030		.41	1.04		1.45	2.17
1870	With compound skim coat (level 5 finish)		475	.034		.46	1.18		1.64	2.46
2000	5/8" thick, on walls, standard, no finish included		2000	.008		.35	.28		.63	.85
2050	Taped and finished (level 4 finish)		965	.017		.40	.58		.98	1.40
2090	With compound skim coat (level 5 finish)		775	.021		.45	.72		1.17	1.69
2100	Fire resistant, no finish included		2000	.008		.36	.28		.64	.86
2150	Taped and finished (level 4 finish)		965	.017		.41	.58		.99	1.41
2195	With compound skim coat (level 5 finish)		775	.021		.46	.72		1.18	1.70
2200	Water resistant, no finish included		2000	.008		.44	.28		.72	.94
2250	Taped and finished (level 4 finish)		965	.017		.49	.58		1.07	1.50
2290	With compound skim coat (level 5 finish)		775	.021		.54	.72		1.26	1.79
2300	Prefinished, vinyl, clipped to studs		900	.018		.79	.62		1.41	1.90
2510	Mold resistant, no finish included		2000	.008		.49	.28		.77	1
2520	Taped and finished (level 4 finish)		965	.017		.54	.58		1.12	1.55
2530	With compound skim coat (level 5 finish)		775	.021		.59	.72		1.31	1.85
3000	On ceilings, standard, no finish included		1800	.009		.35	.31		.66	.91
3050	Taped and finished (level 4 finish)		765	.021		.40	.73		1.13	1.66
3090	With compound skim coat (level 5 finish)		615	.026		.45	.91		1.36	2
3100	Fire resistant, no finish included		1800	.009		.36	.31		.67	.92
3150	Taped and finished (level 4 finish)		765	.021		.41	.73		1.14	1.67
3190	With compound skim coat (level 5 finish)		615	.026		.46	.91		1.37	2.01
3200	Water resistant, no finish included		1800	.009		.44	.31		.75	1
3250	Taped and finished (level 4 finish)		765	.021		.49	.73		1.22	1.76
3290	With compound skim coat (level 5 finish)		615	.026		.54	.91		1.45	2.10
3300	Mold resistant, no finish included		1800	.009		.49	.31		.80	1.06
3310	Taped and finished (level 4 finish)		765	.021		.54	.73		1.27	1.81
3320	With compound skim coat (level 5 finish)		615	.026		.59	.91		1.50	2.16

09 29 Gypsum Board

09 29 10 – Gypsum Board Panels

09 29 10.30 Gypsum Board	Crew	Daily Output	Labor-Hours	Unit	Material	2018 Bare Costs Labor	Equipment	Total	Total Incl O&P
3500	On beams, columns, or soffits, no finish included	2 Carp	675	.024	S.F.	.40	.83		1.23	1.82
3550	Taped and finished (level 4 finish)		475	.034		.46	1.18		1.64	2.46
3590	With compound skim coat (level 5 finish)		380	.042		.52	1.47		1.99	3.02
3600	Fire resistant, no finish included		675	.024		.41	.83		1.24	1.84
3650	Taped and finished (level 4 finish)		475	.034		.47	1.18		1.65	2.47
3690	With compound skim coat (level 5 finish)		380	.042		.46	1.47		1.93	2.95
3700	Water resistant, no finish included		675	.024		.51	.83		1.34	1.94
3750	Taped and finished (level 4 finish)		475	.034		.54	1.18		1.72	2.55
3790	With compound skim coat (level 5 finish)		380	.042		.56	1.47		2.03	3.07
3800	Mold resistant, no finish included		675	.024		.56	.83		1.39	2
3810	Taped and finished (level 4 finish)		475	.034		.59	1.18		1.77	2.61
3820	With compound skim coat (level 5 finish)		380	.042		.62	1.47		2.09	3.13
4000	Fireproofing, beams or columns, 2 layers, 1/2" thick, incl finish		330	.048		.83	1.69		2.52	3.74
4010	Mold resistant		330	.048		.97	1.69		2.66	3.89
4050	5/8" thick		300	.053		.81	1.86		2.67	3.99
4060	Mold resistant		300	.053		1.07	1.86		2.93	4.28
4100	3 layers, 1/2" thick		225	.071		1.25	2.49		3.74	5.50
4110	Mold resistant		225	.071		1.46	2.49		3.95	5.75
4150	5/8" thick		210	.076		1.22	2.66		3.88	5.75
4160	Mold resistant		210	.076		1.61	2.66		4.27	6.20
5200	For work over 8' high, add		3060	.005			.18		.18	.30
5270	For textured spray, add	2 Lath	1600	.010		.04	.34		.38	.60
5350	For finishing inner corners, add	2 Carp	950	.017	L.F.	.10	.59		.69	1.09
5355	For finishing outer corners, add	"	1250	.013		.23	.45		.68	.99
5500	For acoustical sealant, add per bead	1 Carp	500	.016		.04	.56		.60	.98
5550	Sealant, 1 quart tube				Ea.	7.05			7.05	7.80
6000	Gypsum sound dampening panels									
6010	1/2" thick on walls, multi-layer, lightweight, no finish included	2 Carp	1500	.011	S.F.	2.16	.37		2.53	3
6015	Taped and finished (level 4 finish)		725	.022		2.21	.77		2.98	3.71
6020	With compound skim coat (level 5 finish)		580	.028		2.26	.96		3.22	4.08
6025	5/8" thick on walls, for wood studs, no finish included		1500	.011		2.21	.37		2.58	3.05
6030	Taped and finished (level 4 finish)		725	.022		2.26	.77		3.03	3.76
6035	With compound skim coat (level 5 finish)		580	.028		2.31	.96		3.27	4.14
6040	For metal stud, no finish included		1500	.011		2.28	.37		2.65	3.13
6045	Taped and finished (level 4 finish)		725	.022		2.33	.77		3.10	3.84
6050	With compound skim coat (level 5 finish)		580	.028		2.38	.96		3.34	4.22
6055	Abuse resist, no finish included		1500	.011		4.10	.37		4.47	5.15
6060	Taped and finished (level 4 finish)		725	.022		4.15	.77		4.92	5.85
6065	With compound skim coat (level 5 finish)		580	.028		4.20	.96		5.16	6.20
6070	Shear rated, no finish included		1500	.011		5.15	.37		5.52	6.25
6075	Taped and finished (level 4 finish)		725	.022		5.20	.77		5.97	7
6080	With compound skim coat (level 5 finish)		580	.028		5.25	.96		6.21	7.35
6085	For SCIF applications, no finish included		1500	.011		5.15	.37		5.52	6.25
6090	Taped and finished (level 4 finish)		725	.022		5.20	.77		5.97	7
6095	With compound skim coat (level 5 finish)		580	.028		5.25	.96		6.21	7.35
6100	1-3/8" thick on walls, THX certified, no finish included		1500	.011		9.15	.37		9.52	10.65
6105	Taped and finished (level 4 finish)		725	.022		9.20	.77		9.97	11.40
6110	With compound skim coat (level 5 finish)		580	.028		9.25	.96		10.21	11.75
6115	5/8" thick on walls, score & snap installation, no finish included		2000	.008		1.96	.28		2.24	2.62
6120	Taped and finished (level 4 finish)		965	.017		2.01	.58		2.59	3.17
6125	With compound skim coat (level 5 finish)		775	.021		2.06	.72		2.78	3.46
7020	5/8" thick on ceilings, for wood joists, no finish included		1200	.013		2.21	.47		2.68	3.20
7025	Taped and finished (level 4 finish)		510	.031		2.26	1.10		3.36	4.30

For customer support on your Residential Costs with RSMeans data, call 800.448.8182.

493

09 29 Gypsum Board

09 29 10 – Gypsum Board Panels

09 29 10.30 Gypsum Board		Crew	Daily Output	Labor-Hours	Unit	Material	2018 Bare Costs Labor	Equipment	Total	Total Incl O&P
7030	With compound skim coat (level 5 finish)	2 Carp	410	.039	S.F.	2.31	1.36		3.67	4.81
7035	For metal joists, no finish included		1200	.013		2.28	.47		2.75	3.28
7040	Taped and finished (level 4 finish)		510	.031		2.33	1.10		3.43	4.38
7045	With compound skim coat (level 5 finish)		410	.039		2.38	1.36		3.74	4.89
7050	Abuse resist, no finish included		1200	.013		4.10	.47		4.57	5.30
7055	Taped and finished (level 4 finish)		510	.031		4.15	1.10		5.25	6.40
7060	With compound skim coat (level 5 finish)		410	.039		4.20	1.36		5.56	6.90
7065	Shear rated, no finish included		1200	.013		5.15	.47		5.62	6.40
7070	Taped and finished (level 4 finish)		510	.031		5.20	1.10		6.30	7.50
7075	With compound skim coat (level 5 finish)		410	.039		5.25	1.36		6.61	8
7080	For SCIF applications, no finish included		1200	.013		5.15	.47		5.62	6.40
7085	Taped and finished (level 4 finish)		510	.031		5.20	1.10		6.30	7.50
7090	With compound skim coat (level 5 finish)		410	.039		5.25	1.36		6.61	8
8010	5/8" thick on ceilings, score & snap installation, no finish included		1600	.010		1.96	.35		2.31	2.74
8015	Taped and finished (level 4 finish)		680	.024		2.01	.82		2.83	3.58
8020	With compound skim coat (level 5 finish)		545	.029		2.06	1.03		3.09	3.97

09 29 15 – Gypsum Board Accessories

09 29 15.10 Accessories, Drywall

		Crew	Daily Output	Labor-Hours	Unit	Material	2018 Bare Costs Labor	Equipment	Total	Total Incl O&P
0011	**ACCESSORIES, DRYWALL** Casing bead, galvanized steel	1 Carp	290	.028	L.F.	.24	.96		1.20	1.86
0101	Vinyl		290	.028		.23	.96		1.19	1.85
0401	Corner bead, galvanized steel, 1-1/4" x 1-1/4"		350	.023		.17	.80		.97	1.52
0411	1-1/4" x 1-1/4", 10' long		35	.229	Ea.	1.70	8		9.70	15.15
0601	Vinyl corner bead		400	.020	L.F.	.20	.70		.90	1.38
0901	Furring channel, galv. steel, 7/8" deep, standard		260	.031		.34	1.08		1.42	2.17
1001	Resilient		260	.031		.26	1.08		1.34	2.08
1101	J trim, galvanized steel, 1/2" wide		300	.027			.93		.93	1.55
1121	5/8" wide		300	.027		.32	.93		1.25	1.91
1160	Screws #6 x 1" A				M	10.85			10.85	11.95
1170	#6 x 1-5/8" A				"	14.40			14.40	15.80
1501	Z stud, galvanized steel, 1-1/2" wide	1 Carp	260	.031	L.F.	.42	1.08		1.50	2.25

09 30 Tiling

09 30 13 – Ceramic Tiling

09 30 13.45 Ceramic Tile Accessories

		Crew	Daily Output	Labor-Hours	Unit	Material	2018 Bare Costs Labor	Equipment	Total	Total Incl O&P
0010	**CERAMIC TILE ACCESSORIES**									
0100	Spacers, 1/8"				C	1.98			1.98	2.18
1310	Sealer for natural stone tile, installed	1 Tilf	650	.012	S.F.	.05	.42		.47	.73

09 30 29 – Metal Tiling

09 30 29.10 Metal Tile

		Crew	Daily Output	Labor-Hours	Unit	Material	2018 Bare Costs Labor	Equipment	Total	Total Incl O&P
0010	**METAL TILE** 4' x 4' sheet, 24 ga., tile pattern, nailed									
0200	Stainless steel	2 Carp	512	.031	S.F.	28	1.09		29.09	33
0400	Aluminized steel	"	512	.031	"	19	1.09		20.09	23

09 30 95 – Tile & Stone Setting Materials and Specialties

09 30 95.10 Moisture Resistant, Anti-Fracture Membrane

		Crew	Daily Output	Labor-Hours	Unit	Material	2018 Bare Costs Labor	Equipment	Total	Total Incl O&P
0010	**MOISTURE RESISTANT, ANTI-FRACTURE MEMBRANE**									
0200	Elastomeric membrane, 1/16" thick	D-7	275	.058	S.F.	1.12	1.74		2.86	4.05

09 31 13.10 Thin-Set Ceramic Tile	Crew	Daily Output	Labor-Hours	Unit	Material	2018 Bare Costs Labor	2018 Bare Costs Equipment	Total	Total Incl O&P
0010 THIN-SET CERAMIC TILE									
0020 Backsplash, average grade tiles	1 Tilf	50	.160	S.F.	2.79	5.40		8.19	11.80
0022 Custom grade tiles		50	.160		5.60	5.40		11	14.90
0024 Luxury grade tiles		50	.160		11.15	5.40		16.55	21
0026 Economy grade tiles	▼	50	.160	▼	2.56	5.40		7.96	11.55
0100 Base, using 1' x 4" high piece with 1" x 1" tiles	D-7	128	.125	L.F.	4.99	3.74		8.73	11.55
0300 For 6" high base, 1" x 1" tile face, add					1.22			1.22	1.35
0400 For 2" x 2" tile face, add to above					.72			.72	.79
0700 Cove base, 4-1/4" x 4-1/4"	D-7	128	.125		4.19	3.74		7.93	10.65
1000 6" x 4-1/4" high		137	.117		4.32	3.49		7.81	10.40
1300 Sanitary cove base, 6" x 4-1/4" high		124	.129		4.64	3.86		8.50	11.35
1600 6" x 6" high		117	.137	▼	5.30	4.09		9.39	12.45
1800 Bathroom accessories, average (soap dish, toothbrush holder)		82	.195	Ea.	9.80	5.85		15.65	20.50
1900 Bathtub, 5', rec. 4-1/4" x 4-1/4" tile wainscot, adhesive set 6' high		2.90	5.517		175	165		340	460
2100 7' high wainscot		2.50	6.400		204	191		395	535
2200 8' high wainscot		2.20	7.273	▼	233	217		450	605
2500 Bullnose trim, 4-1/4" x 4-1/4"		128	.125	L.F.	4.17	3.74		7.91	10.65
2800 2" x 6"		124	.129	"	4.22	3.86		8.08	10.90
3300 Ceramic tile, porcelain type, 1 color, color group 2, 1" x 1"		183	.087	S.F.	6.40	2.61		9.01	11.30
3310 2" x 2" or 2" x 1"	▼	190	.084		6.25	2.52		8.77	11
3350 For random blend, 2 colors, add					1			1	1.10
3360 4 colors, add					1.50			1.50	1.65
4300 Specialty tile, 4-1/4" x 4-1/4" x 1/2", decorator finish	D-7	183	.087		12.70	2.61		15.31	18.25
4500 Add for epoxy grout, 1/16" joint, 1" x 1" tile		800	.020		.68	.60		1.28	1.72
4600 2" x 2" tile		820	.020		.62	.58		1.20	1.63
4610 Add for epoxy grout, 1/8" joint, 8" x 8" x 3/8" tile, add	▼	900	.018	▼	1.42	.53		1.95	2.42
4800 Pregrouted sheets, walls, 4-1/4" x 4-1/4", 6" x 4-1/4"									
4810 and 8-1/2" x 4-1/4", 4 S.F. sheets, silicone grout	D-7	240	.067	S.F.	5.60	1.99		7.59	9.45
5100 Floors, unglazed, 2 S.F. sheets,									
5110 urethane adhesive	D-7	180	.089	S.F.	2.03	2.66		4.69	6.55
5400 Walls, interior, 4-1/4" x 4-1/4" tile		190	.084		2.61	2.52		5.13	6.95
5500 6" x 4-1/4" tile		190	.084		3.15	2.52		5.67	7.55
5700 8-1/2" x 4-1/4" tile		190	.084		5.65	2.52		8.17	10.35
5800 6" x 6" tile		175	.091		3.57	2.73		6.30	8.35
5810 8" x 8" tile		170	.094		5.30	2.81		8.11	10.40
5820 12" x 12" tile		160	.100		4.66	2.99		7.65	10
5830 16" x 16" tile		150	.107		5.15	3.19		8.34	10.80
6000 Decorated wall tile, 4-1/4" x 4-1/4", color group 1		270	.059		3.68	1.77		5.45	6.90
6100 Color group 4		180	.089		52.50	2.66		55.16	62
9300 Ceramic tiles, recycled glass, standard colors, 2" x 2" thru 6" x 6" [G]		190	.084		22	2.52		24.52	28.50
9310 6" x 6" [G]		175	.091		22	2.73		24.73	29
9320 8" x 8" [G]		170	.094		23.50	2.81		26.31	30.50
9330 12" x 12" [G]		160	.100		23.50	2.99		26.49	31
9340 Earthtones, 2" x 2" to 4" x 8" [G]		190	.084		26	2.52		28.52	32.50
9350 6" x 6" [G]		175	.091		26	2.73		28.73	33
9360 8" x 8" [G]		170	.094		27	2.81		29.81	34
9370 12" x 12" [G]		160	.100		27	2.99		29.99	34.50
9380 Deep colors, 2" x 2" to 4" x 8" [G]		190	.084		30.50	2.52		33.02	38
9390 6" x 6" [G]		175	.091		30.50	2.73		33.23	38.50
9400 8" x 8" [G]		170	.094		32	2.81		34.81	40
9410 12" x 12" [G]	▼	160	.100	▼	32	2.99		34.99	40.50

09 31 Thin-Set Tiling

09 31 33 – Thin-Set Stone Tiling

09 31 33.10 Tiling, Thin-Set Stone		Crew	Daily Output	Labor-Hours	Unit	Material	2018 Bare Costs Labor	Equipment	Total	Total Incl O&P
0010	**TILING, THIN-SET STONE**									
3000	Floors, natural clay, random or uniform, color group 1	D-7	183	.087	S.F.	4.31	2.61		6.92	9
3100	Color group 2		183	.087		5.95	2.61		8.56	10.80
3255	Floors, glazed, 6" x 6", color group 1		300	.053		5.25	1.59		6.84	8.40
3260	8" x 8" tile		300	.053		5.25	1.59		6.84	8.40
3270	12" x 12" tile		290	.055		5.85	1.65		7.50	9.05
3280	16" x 16" tile		280	.057		7.50	1.71		9.21	11
3281	18" x 18" tile		270	.059		7.10	1.77		8.87	10.70
3282	20" x 20" tile		260	.062		7.05	1.84		8.89	10.75
3283	24" x 24" tile		250	.064		9	1.91		10.91	13
3285	Border, 6" x 12" tile		200	.080		11.05	2.39		13.44	16.10
3290	3" x 12" tile		200	.080		10.60	2.39		12.99	15.55

09 32 Mortar-Bed Tiling

09 32 13 – Mortar-Bed Ceramic Tiling

09 32 13.10 Ceramic Tile

		Crew	Daily Output	Labor-Hours	Unit	Material	2018 Bare Costs Labor	Equipment	Total	Total Incl O&P
0010	**CERAMIC TILE**									
0050	Base, using 1' x 4" high pc. with 1" x 1" tiles	D-7	82	.195	L.F.	5.25	5.85		11.10	15.25
0600	Cove base, 4-1/4" x 4-1/4" high		91	.176		4.33	5.25		9.58	13.25
0900	6" x 4-1/4" high		100	.160		4.46	4.78		9.24	12.65
1200	Sanitary cove base, 6" x 4-1/4" high		93	.172		4.78	5.15		9.93	13.60
1500	6" x 6" high		84	.190		5.45	5.70		11.15	15.20
2400	Bullnose trim, 4-1/4" x 4-1/4"		82	.195		4.27	5.85		10.12	14.15
2700	2" x 6" bullnose trim		84	.190		4.29	5.70		9.99	13.95
6210	Wall tile, 4-1/4" x 4-1/4", better grade	1 Tilf	50	.160	S.F.	9.50	5.40		14.90	19.20
6240	2" x 2"		50	.160		6.50	5.40		11.90	15.90
6250	6" x 6"		55	.145		10.15	4.91		15.06	19.10
6260	8" x 8"		60	.133		9.40	4.50		13.90	17.65
6600	Crystalline glazed, 4-1/4" x 4-1/4", plain	D-7	100	.160		4.52	4.78		9.30	12.70
6700	4-1/4" x 4-1/4", scored tile		100	.160		6.50	4.78		11.28	14.90
6900	6" x 6" plain		93	.172		5.75	5.15		10.90	14.70
7000	For epoxy grout, 1/16" joints, 4-1/4" tile, add		800	.020		.41	.60		1.01	1.42
7200	For tile set in dry mortar, add		1735	.009			.28		.28	.45
7300	For tile set in Portland cement mortar, add		290	.055		.18	1.65		1.83	2.87

09 32 16 – Mortar-Bed Quarry Tiling

09 32 16.10 Quarry Tile

		Crew	Daily Output	Labor-Hours	Unit	Material	2018 Bare Costs Labor	Equipment	Total	Total Incl O&P
0010	**QUARRY TILE**									
0100	Base, cove or sanitary, to 5" high, 1/2" thick	D-7	110	.145	L.F.	6.40	4.35		10.75	14.10
0300	Bullnose trim, red, 6" x 6" x 1/2" thick		120	.133		5.40	3.99		9.39	12.35
0400	4" x 4" x 1/2" thick		110	.145		5	4.35		9.35	12.55
0600	4" x 8" x 1/2" thick, using 8" as edge		130	.123		5.35	3.68		9.03	11.85
0700	Floors, 1,000 S.F. lots, red, 4" x 4" x 1/2" thick		120	.133	S.F.	8.70	3.99		12.69	16.05
0900	6" x 6" x 1/2" thick		140	.114		8.25	3.42		11.67	14.65
1000	4" x 8" x 1/2" thick		130	.123		6.60	3.68		10.28	13.25
1300	For waxed coating, add					.76			.76	.84
1500	For non-standard colors, add					.46			.46	.51
1600	For abrasive surface, add					.52			.52	.57
1800	Brown tile, imported, 6" x 6" x 3/4"	D-7	120	.133		7.20	3.99		11.19	14.35
1900	8" x 8" x 1"		110	.145		9.55	4.35		13.90	17.55
2100	For thin set mortar application, deduct		700	.023			.68		.68	1.11

09 32 Mortar-Bed Tiling

09 32 16 – Mortar-Bed Quarry Tiling

09 32 16.10 Quarry Tile

		Crew	Daily Output	Labor-Hours	Unit	Material	2018 Bare Costs Labor	Equipment	Total	Total Incl O&P
2700	Stair tread, 6" x 6" x 3/4", plain	D-7	50	.320	S.F.	7.20	9.55		16.75	23.50
2800	Abrasive		47	.340		8.55	10.20		18.75	26
3000	Wainscot, 6" x 6" x 1/2", thin set, red		105	.152		6.30	4.56		10.86	14.30
3100	Non-standard colors		105	.152		6.40	4.56		10.96	14.40
3300	Window sill, 6" wide, 3/4" thick		90	.178	L.F.	8.65	5.30		13.95	18.10
3400	Corners		80	.200	Ea.	6	6		12	16.30

09 32 23 – Mortar-Bed Glass Mosaic Tiling

09 32 23.10 Glass Mosaics

		Crew	Daily Output	Labor-Hours	Unit	Material	2018 Bare Costs Labor	Equipment	Total	Total Incl O&P
0010	**GLASS MOSAICS** 3/4" tile on 12" sheets, standard grout									
1020	1" tile on 12" sheets, opalescent finish	D-7	73	.219	S.F.	18.40	6.55		24.95	31
1040	1" x 2" tile on 12" sheet, blend		73	.219		21	6.55		27.55	33.50
1060	2" tile on 12" sheet, blend		73	.219		17.60	6.55		24.15	30
1080	5/8" x random tile, linear, on 12" sheet, blend		73	.219		25	6.55		31.55	37.50
1600	Dots on 12" sheet		73	.219		25	6.55		31.55	38
1700	For glass mosaic tiles set in dry mortar, add		290	.055		.45	1.65		2.10	3.17
1720	For glass mosaic tiles set in Portland cement mortar, add		290	.055		.01	1.65		1.66	2.68
1730	For polyblend sanded tile grout		96.15	.166	Lb.	2.19	4.98		7.17	10.45

09 34 Waterproofing-Membrane Tiling

09 34 13 – Waterproofing-Membrane Ceramic Tiling

09 34 13.10 Ceramic Tile Waterproofing Membrane

		Crew	Daily Output	Labor-Hours	Unit	Material	2018 Bare Costs Labor	Equipment	Total	Total Incl O&P
0010	**CERAMIC TILE WATERPROOFING MEMBRANE**									
0020	On floors, including thinset									
0030	Fleece laminated polyethylene grid, 1/8" thick	D-7	250	.064	S.F.	2.28	1.91		4.19	5.60
0040	5/16" thick	"	250	.064	"	2.60	1.91		4.51	5.95
0050	On walls, including thinset									
0060	Fleece laminated polyethylene sheet, 8 mil thick	D-7	480	.033	S.F.	2.28	1		3.28	4.11
0070	Accessories, including thinset									
0080	Joint and corner sheet, 4 mils thick, 5" wide	1 Tilf	240	.033	L.F.	1.35	1.12		2.47	3.30
0090	7-1/4" wide		180	.044		1.71	1.50		3.21	4.31
0100	10" wide		120	.067		2.08	2.25		4.33	5.95
0110	Pre-formed corners, inside		32	.250	Ea.	7.85	8.45		16.30	22.50
0120	Outside		32	.250		7.65	8.45		16.10	22
0130	2" flanged floor drain with 6" stainless steel grate		16	.500		370	16.90		386.90	440
0140	EPS, sloped shower floor		480	.017	S.F.	5.55	.56		6.11	7
0150	Curb		32	.250	L.F.	14.05	8.45		22.50	29

09 35 Chemical-Resistant Tiling

09 35 13 – Chemical-Resistant Ceramic Tiling

09 35 13.10 Chemical-Resistant Ceramic Tiling

		Crew	Daily Output	Labor-Hours	Unit	Material	2018 Bare Costs Labor	Equipment	Total	Total Incl O&P
0010	**CHEMICAL-RESISTANT CERAMIC TILING**									
0100	4-1/4" x 4-1/4" x 1/4", 1/8" joint	D-7	130	.123	S.F.	12.05	3.68		15.73	19.20
0200	6" x 6" x 1/2" thick		120	.133		9.75	3.99		13.74	17.15
0300	8" x 8" x 1/2" thick		110	.145		10.90	4.35		15.25	19.05
0400	4-1/4" x 4-1/4" x 1/4", 1/4" joint		130	.123		12.80	3.68		16.48	20
0500	6" x 6" x 1/2" thick		120	.133		10.85	3.99		14.84	18.40
0600	8" x 8" x 1/2" thick		110	.145		11.55	4.35		15.90	19.75
0700	4-1/4" x 4-1/4" x 1/4", 3/8" joint		130	.123		13.50	3.68		17.18	21
0800	6" x 6" x 1/2" thick		120	.133		11.80	3.99		15.79	19.40

497

For customer support on your Residential Costs with RSMeans data, call 800.448.8182.

09 35 Chemical-Resistant Tiling

09 35 13 – Chemical-Resistant Ceramic Tiling

09 35 13.10 Chemical-Resistant Ceramic Tiling	Crew	Daily Output	Labor-Hours	Unit	Material	2018 Bare Costs Labor	Equipment	Total	Total Incl O&P
0900	8"x 8" x 1/2" thick	D-7	110	.145	S.F.	12.70	4.35		17.05	21

09 35 16 – Chemical-Resistant Quarry Tiling

09 35 16.10 Chemical-Resistant Quarry Tiling

		Crew	Daily Output	Labor-Hours	Unit	Material	Labor	Equipment	Total	Total Incl O&P
0010	**CHEMICAL-RESISTANT QUARRY TILING**									
0100	4"x 8" x 1/2" thick, 1/8" joint	D-7	130	.123	S.F.	11.20	3.68		14.88	18.25
0200	6"x 6" x 1/2" thick		120	.133		11.25	3.99		15.24	18.80
0300	8"x 8" x 1/2" thick		110	.145		10.35	4.35		14.70	18.45
0400	4"x 8" x 1/2" thick, 1/4" joint		130	.123		12.40	3.68		16.08	19.60
0500	6"x 6" x 1/2" thick		120	.133		12.35	3.99		16.34	20
0600	8"x 8" x 1/2" thick		110	.145		11	4.35		15.35	19.15
0700	4"x 8" x 1/2" thick, 3/8" joint		130	.123		13.50	3.68		17.18	21
0800	6"x 6" x 1/2" thick		120	.133		13.30	3.99		17.29	21
0900	8"x 8" x 1/2" thick	▼	110	.145	▼	12.15	4.35		16.50	20.50

09 51 Acoustical Ceilings

09 51 13 – Acoustical Panel Ceilings

09 51 13.10 Ceiling, Acoustical Panel

		Crew	Daily Output	Labor-Hours	Unit	Material	Labor	Equipment	Total	Total Incl O&P
0010	**CEILING, ACOUSTICAL PANEL**									
0100	Fiberglass boards, film faced, 2' x 2' or 2' x 4', 5/8" thick	1 Carp	625	.013	S.F.	1.26	.45		1.71	2.13
0120	3/4" thick		600	.013		3	.47		3.47	4.07
0130	3" thick, thermal, R11	▼	450	.018	▼	3.50	.62		4.12	4.88

09 51 14 – Acoustical Fabric-Faced Panel Ceilings

09 51 14.10 Ceiling, Acoustical Fabric-Faced Panel

		Crew	Daily Output	Labor-Hours	Unit	Material	Labor	Equipment	Total	Total Incl O&P
0010	**CEILING, ACOUSTICAL FABRIC-FACED PANEL**									
0100	Glass cloth faced fiberglass, 3/4" thick	1 Carp	500	.016	S.F.	2.99	.56		3.55	4.22
0120	1" thick		485	.016		3.61	.58		4.19	4.93
0130	1-1/2" thick, nubby face	▼	475	.017	▼	2.73	.59		3.32	3.98

09 51 23 – Acoustical Tile Ceilings

09 51 23.10 Suspended Acoustic Ceiling Tiles

		Crew	Daily Output	Labor-Hours	Unit	Material	Labor	Equipment	Total	Total Incl O&P
0010	**SUSPENDED ACOUSTIC CEILING TILES**, not including									
0100	suspension system									
1110	Mineral fiber tile, lay-in, 2' x 2' or 2' x 4', 5/8" thick, fine texture	1 Carp	625	.013	S.F.	.79	.45		1.24	1.61
1115	Rough textured		625	.013		.75	.45		1.20	1.57
1125	3/4" thick, fine textured		600	.013		2.11	.47		2.58	3.09
1130	Rough textured		600	.013		1.73	.47		2.20	2.67
1135	Fissured		600	.013		2.17	.47		2.64	3.16
1150	Tegular, 5/8" thick, fine textured		470	.017		1.12	.59		1.71	2.22
1155	Rough textured		470	.017		1.34	.59		1.93	2.46
1165	3/4" thick, fine textured		450	.018		2.33	.62		2.95	3.59
1170	Rough textured		450	.018		1.53	.62		2.15	2.71
1175	Fissured	▼	450	.018		2.21	.62		2.83	3.46
1185	For plastic film face, add					.81			.81	.89
1190	For fire rating, add					.50			.50	.55
3750	Wood fiber in cementitious binder, 2' x 2' or 4', painted, 1" thick	1 Carp	600	.013		2.12	.47		2.59	3.10
3760	2" thick		550	.015		3.71	.51		4.22	4.93
3770	2-1/2" thick		500	.016		4.63	.56		5.19	6.05
3780	3" thick	▼	450	.018	▼	5.55	.62		6.17	7.20

09 51 Acoustical Ceilings

09 51 23 – Acoustical Tile Ceilings

09 51 23.30 Suspended Ceilings, Complete

	Crew	Daily Output	Labor-Hours	Unit	Material	2018 Bare Costs Labor	Equipment	Total	Total Incl O&P
0010 **SUSPENDED CEILINGS, COMPLETE**, incl. standard									
0100 suspension system but not incl. 1-1/2" carrier channels									
0600 Fiberglass ceiling board, 2' x 4' x 5/8", plain faced	1 Carp	500	.016	S.F.	2.05	.56		2.61	3.19
0700 Offices, 2' x 4' x 3/4"		380	.021		3.79	.74		4.53	5.40
1800 Tile, Z bar suspension, 5/8" mineral fiber tile		150	.053		2.14	1.86		4	5.45
1900 3/4" mineral fiber tile	↓	150	.053	↓	2.42	1.86		4.28	5.75

09 51 53 – Direct-Applied Acoustical Ceilings

09 51 53.10 Ceiling Tile

	Crew	Daily Output	Labor-Hours	Unit	Material	2018 Bare Costs Labor	Equipment	Total	Total Incl O&P
0010 **CEILING TILE**, stapled or cemented									
0100 12" x 12" or 12" x 24", not including furring									
0600 Mineral fiber, vinyl coated, 5/8" thick	1 Carp	300	.027	S.F.	2.34	.93		3.27	4.12
0700 3/4" thick		300	.027		3.05	.93		3.98	4.91
0900 Fire rated, 3/4" thick, plain faced		300	.027		1.42	.93		2.35	3.11
1000 Plastic coated face		300	.027		2.10	.93		3.03	3.86
1200 Aluminum faced, 5/8" thick, plain	↓	300	.027		1.83	.93		2.76	3.56
3300 For flameproofing, add					.07			.07	.08
3400 For sculptured 3 dimensional, add					.32			.32	.35
3900 For ceiling primer, add					.12			.12	.13
4000 For ceiling cement, add					.40			.40	.44

09 53 Acoustical Ceiling Suspension Assemblies

09 53 23 – Metal Acoustical Ceiling Suspension Assemblies

09 53 23.30 Ceiling Suspension Systems

	Crew	Daily Output	Labor-Hours	Unit	Material	2018 Bare Costs Labor	Equipment	Total	Total Incl O&P
0010 **CEILING SUSPENSION SYSTEMS** for boards and tile									
0050 Class A suspension system, 15/16" T bar, 2' x 4' grid	1 Carp	800	.010	S.F.	.79	.35		1.14	1.45
0300 2' x 2' grid	"	650	.012		1.02	.43		1.45	1.84
0350 For 9/16" grid, add					.16			.16	.18
0360 For fire rated grid, add					.09			.09	.10
0370 For colored grid, add					.21			.21	.23
0400 Concealed Z bar suspension system, 12" module	1 Carp	520	.015		.93	.54		1.47	1.91
0600 1-1/2" carrier channels, 4' OC, add		470	.017		.12	.59		.71	1.12
0650 1-1/2" x 3-1/2" channels	↓	470	.017	↓	.32	.59		.91	1.34
0700 Carrier channels for ceilings with									
0900 recessed lighting fixtures, add	1 Carp	460	.017	S.F.	.22	.61		.83	1.25
5000 Wire hangers, #12 wire	"	300	.027	Ea.	.07	.93		1	1.63

09 54 Specialty Ceilings

09 54 16 – Luminous Ceilings

09 54 16.10 Ceiling, Luminous

	Crew	Daily Output	Labor-Hours	Unit	Material	2018 Bare Costs Labor	Equipment	Total	Total Incl O&P
0010 **CEILING, LUMINOUS**									
0020 Translucent lay-in panels, 2' x 2'	1 Carp	500	.016	S.F.	23	.56		23.56	26
0030 2' x 6'	"	500	.016	"	17.80	.56		18.36	20.50

09 54 23 – Linear Metal Ceilings

09 54 23.10 Metal Ceilings

	Crew	Daily Output	Labor-Hours	Unit	Material	2018 Bare Costs Labor	Equipment	Total	Total Incl O&P
0010 **METAL CEILINGS**									
0015 Solid alum. planks, 3-1/4" x 12', open reveal	1 Carp	500	.016	S.F.	2.35	.56		2.91	3.52
0020 Closed reveal		500	.016		3	.56		3.56	4.23
0030 7-1/4" x 12', open reveal	↓	500	.016	↓	4	.56		4.56	5.35

09 54 Specialty Ceilings

09 54 23 – Linear Metal Ceilings

09 54 23.10 Metal Ceilings		Crew	Daily Output	Labor-Hours	Unit	Material	2018 Bare Costs Labor	Equipment	Total	Total Incl O&P
0040	Closed reveal	1 Carp	500	.016	S.F.	5.10	.56		5.66	6.55
0050	Metal, open cell, 2' x 2', 6" cell		500	.016		8.70	.56		9.26	10.50
0060	8" cell		500	.016		9.60	.56		10.16	11.50
0070	2' x 4', 6" cell		500	.016		5.70	.56		6.26	7.20
0080	8" cell		500	.016		5.70	.56		6.26	7.20

09 61 Flooring Treatment

09 61 19 – Concrete Floor Staining

09 61 19.40 Floors, Interior

09 61 19.40 Floors, Interior		Crew	Daily Output	Labor-Hours	Unit	Material	Labor	Equipment	Total	Total Incl O&P
0010	**FLOORS, INTERIOR**									
0300	Acid stain and sealer									
0310	Stain, one coat	1 Pord	650	.012	S.F.	.14	.36		.50	.74
0320	Two coats		570	.014		.28	.41		.69	.99
0330	Acrylic sealer, one coat		2600	.003		.22	.09		.31	.40
0340	Two coats		1400	.006		.45	.17		.62	.77

09 62 Specialty Flooring

09 62 19 – Laminate Flooring

09 62 19.10 Floating Floor

09 62 19.10 Floating Floor			Crew	Daily Output	Labor-Hours	Unit	Material	Labor	Equipment	Total	Total Incl O&P
0010	**FLOATING FLOOR**										
8300	Floating floor, laminate, wood pattern strip, complete		1 Clab	133	.060	S.F.	4.32	1.61		5.93	7.40
8310	Components, T&G wood composite strips						3.83			3.83	4.22
8320	Film						.17			.17	.18
8330	Foam						.26			.26	.29
8340	Adhesive						.43			.43	.47
8350	Installation kit						.19			.19	.21
8360	Trim, 2" wide x 3' long					L.F.	4.30			4.30	4.73
8370	Reducer moulding					"	5.70			5.70	6.25

09 62 23 – Bamboo Flooring

09 62 23.10 Flooring, Bamboo

09 62 23.10 Flooring, Bamboo			Crew	Daily Output	Labor-Hours	Unit	Material	Labor	Equipment	Total	Total Incl O&P
0010	**FLOORING, BAMBOO**										
8600	Flooring, wood, bamboo strips, unfinished, 5/8" x 4" x 3'	G	1 Carp	255	.031	S.F.	5.70	1.10		6.80	8.05
8610	5/8" x 4" x 4'	G		275	.029		5.90	1.02		6.92	8.20
8620	5/8" x 4" x 6'	G		295	.027		6.45	.95		7.40	8.70
8630	Finished, 5/8" x 4" x 3'	G		255	.031		6.25	1.10		7.35	8.70
8640	5/8" x 4" x 4'	G		275	.029		6.55	1.02		7.57	8.90
8650	5/8" x 4" x 6'	G		295	.027		4.99	.95		5.94	7.10
8660	Stair treads, unfinished, 1-1/16" x 11-1/2" x 4'	G		18	.444	Ea.	54.50	15.55		70.05	86
8670	Finished, 1-1/16" x 11-1/2" x 4'	G		18	.444		83	15.55		98.55	118
8680	Stair risers, unfinished, 5/8" x 7-1/2" x 4'	G		18	.444		20	15.55		35.55	48
8690	Finished, 5/8" x 7-1/2" x 4'	G		18	.444		38	15.55		53.55	68
8700	Stair nosing, unfinished, 6' long	G		16	.500		44.50	17.50		62	78
8710	Finished, 6' long	G		16	.500		42	17.50		59.50	75.50

09 62 29 – Cork Flooring

09 62 29.10 Cork Tile Flooring

09 62 29.10 Cork Tile Flooring			Crew	Daily Output	Labor-Hours	Unit	Material	Labor	Equipment	Total	Total Incl O&P
0010	**CORK TILE FLOORING**										
2200	Cork tile, standard finish, 1/8" thick	G	1 Tilf	315	.025	S.F.	5.25	.86		6.11	7.15
2250	3/16" thick	G		315	.025		5.80	.86		6.66	7.75

09 62 Specialty Flooring

09 62 29 – Cork Flooring

09 62 29.10 Cork Tile Flooring

	09 62 29.10 Cork Tile Flooring		Crew	Daily Output	Labor-Hours	Unit	Material	2018 Bare Costs Labor	Equipment	Total	Total Incl O&P
2300	5/16" thick	G	1 Tilf	315	.025	S.F.	6.45	.86		7.31	8.50
2350	1/2" thick	G		315	.025		7	.86		7.86	9.10
2500	Urethane finish, 1/8" thick	G		315	.025		5.25	.86		6.11	7.20
2550	3/16" thick	G		315	.025		7.30	.86		8.16	9.45
2600	5/16" thick	G		315	.025		7.25	.86		8.11	9.35
2650	1/2" thick	G		315	.025		7.20	.86		8.06	9.30

09 63 Masonry Flooring

09 63 13 – Brick Flooring

09 63 13.10 Miscellaneous Brick Flooring

	09 63 13.10 Miscellaneous Brick Flooring		Crew	Daily Output	Labor-Hours	Unit	Material	2018 Bare Costs Labor	Equipment	Total	Total Incl O&P
0010	**MISCELLANEOUS BRICK FLOORING**										
0020	Acid-proof shales, red, 8" x 3-3/4" x 1-1/4" thick		D-7	.43	37.209	M	715	1,125		1,840	2,575
0050	2-1/4" thick		D-1	.40	40	"	1,050	1,275		2,325	3,275
0260	Cast ceramic, pressed, 4" x 8" x 1/2", unglazed		D-7	100	.160	S.F.	6.85	4.78		11.63	15.30
0270	Glazed			100	.160		9.15	4.78		13.93	17.80
0280	Hand molded flooring, 4" x 8" x 3/4", unglazed			95	.168		9.05	5.05		14.10	18.10
0290	Glazed			95	.168		11.40	5.05		16.45	20.50
0300	8" hexagonal, 3/4" thick, unglazed			85	.188		9.95	5.65		15.60	20
0310	Glazed			85	.188		17.95	5.65		23.60	29
0450	Acid-proof joints, 1/4" wide		D-1	65	.246		1.58	7.80		9.38	14.80
0500	Pavers, 8" x 4", 1" to 1-1/4" thick, red		D-7	95	.168		3.99	5.05		9.04	12.55
0510	Ironspot		"	95	.168		5.65	5.05		10.70	14.35
0540	1-3/8" to 1-3/4" thick, red		D-1	95	.168		3.85	5.35		9.20	13.15
0560	Ironspot			95	.168		5.60	5.35		10.95	15.05
0580	2-1/4" thick, red			90	.178		3.92	5.65		9.57	13.70
0590	Ironspot			90	.178		6.10	5.65		11.75	16.10
0700	Paver, adobe brick, 6" x 12", 1/2" joint	G		42	.381		1.44	12.05		13.49	21.50
0710	Mexican red, 12" x 12"	G	1 Tilf	48	.167		1.85	5.65		7.50	11.15
0720	Saltillo, 12" x 12"	G	"	48	.167		1.49	5.65		7.14	10.75
0800	For sidewalks and patios with pavers, see Section 32 14 16.10										
0870	For epoxy joints, add		D-1	600	.027	S.F.	3.05	.84		3.89	4.77
0880	For Furan underlayment, add		"	600	.027		2.53	.84		3.37	4.19
0890	For waxed surface, steam cleaned, add		A-1H	1000	.008		.21	.21	.07	.49	.67

09 63 40 – Stone Flooring

09 63 40.10 Marble

	09 63 40.10 Marble	Crew	Daily Output	Labor-Hours	Unit	Material	2018 Bare Costs Labor	Equipment	Total	Total Incl O&P
0010	**MARBLE**									
0020	Thin gauge tile, 12" x 6", 3/8", white Carara	D-7	60	.267	S.F.	17	7.95		24.95	31.50
0100	Travertine		60	.267		8.80	7.95		16.75	22.50
0200	12" x 12" x 3/8", thin set, floors		60	.267		11.10	7.95		19.05	25
0300	On walls		52	.308		9.85	9.20		19.05	26
1000	Marble threshold, 4" wide x 36" long x 5/8" thick, white		60	.267	Ea.	11.10	7.95		19.05	25

09 63 40.20 Slate Tile

	09 63 40.20 Slate Tile	Crew	Daily Output	Labor-Hours	Unit	Material	2018 Bare Costs Labor	Equipment	Total	Total Incl O&P
0010	**SLATE TILE**									
0020	Vermont, 6" x 6" x 1/4" thick, thin set	D-7	180	.089	S.F.	7.70	2.66		10.36	12.75

For customer support on your Residential Costs with RSMeans data, call 800.448.8182.

501

09 64 Wood Flooring

09 64 23 – Wood Parquet Flooring

09 64 23.10 Wood Parquet

		Crew	Daily Output	Labor-Hours	Unit	Material	2018 Bare Costs Labor	Equipment	Total	Total Incl O&P
0010	**WOOD PARQUET** flooring									
5200	Parquetry, 5/16" thk, no finish, oak, plain pattern	1 Carp	160	.050	S.F.	5.45	1.75		7.20	8.90
5300	Intricate pattern		100	.080		9.90	2.80		12.70	15.50
5500	Teak, plain pattern		160	.050		6.35	1.75		8.10	9.85
5600	Intricate pattern		100	.080		10.75	2.80		13.55	16.50
5650	13/16" thick, select grade oak, plain pattern		160	.050		11.85	1.75		13.60	15.90
5700	Intricate pattern		100	.080		17.35	2.80		20.15	24
5800	Custom parquetry, including finish, plain pattern		100	.080		17.10	2.80		19.90	23.50
5900	Intricate pattern		50	.160		25	5.60		30.60	37
6700	Parquetry, prefinished white oak, 5/16" thick, plain pattern		160	.050		8	1.75		9.75	11.70
6800	Intricate pattern		100	.080		8.65	2.80		11.45	14.15
7000	Walnut or teak, parquetry, plain pattern		160	.050		8.50	1.75		10.25	12.25
7100	Intricate pattern	▼	100	.080	▼	13.65	2.80		16.45	19.65
7200	Acrylic wood parquet blocks, 12" x 12" x 5/16",									
7210	Irradiated, set in epoxy	1 Carp	160	.050	S.F.	10.45	1.75		12.20	14.40

09 64 29 – Wood Strip and Plank Flooring

09 64 29.10 Wood

		Crew	Daily Output	Labor-Hours	Unit	Material	2018 Bare Costs Labor	Equipment	Total	Total Incl O&P
0010	**WOOD**									
0020	Fir, vertical grain, 1" x 4", not incl. finish, grade B & better	1 Carp	255	.031	S.F.	3.45	1.10		4.55	5.60
0100	Grade C & better		255	.031		3.25	1.10		4.35	5.40
0300	Flat grain, 1" x 4", not incl. finish, grade B & better		255	.031		3.94	1.10		5.04	6.15
0400	Grade C & better		255	.031		3.79	1.10		4.89	6
4000	Maple, strip, 25/32" x 2-1/4", not incl. finish, select		170	.047		4.93	1.64		6.57	8.15
4100	#2 & better		170	.047		4.84	1.64		6.48	8.05
4300	33/32" x 3-1/4", not incl. finish, #1 grade		170	.047		5.65	1.64		7.29	8.95
4400	#2 & better	▼	170	.047	▼	5	1.64		6.64	8.25
4600	Oak, white or red, 25/32" x 2-1/4", not incl. finish									
4700	#1 common	1 Carp	170	.047	S.F.	3.42	1.64		5.06	6.50
4900	Select quartered, 2-1/4" wide		170	.047		4.29	1.64		5.93	7.45
5000	Clear		170	.047		4.23	1.64		5.87	7.40
6100	Prefinished, white oak, prime grade, 2-1/4" wide		170	.047		5.25	1.64		6.89	8.50
6200	3-1/4" wide		185	.043		5.70	1.51		7.21	8.80
6400	Ranch plank		145	.055		7.30	1.93		9.23	11.20
6500	Hardwood blocks, 9" x 9", 25/32" thick		160	.050		7.45	1.75		9.20	11.05
7400	Yellow pine, 3/4" x 3-1/8", T&G, C & better, not incl. finish	▼	200	.040		1.67	1.40		3.07	4.16
7500	Refinish wood floor, sand, 2 coats poly, wax, soft wood	1 Clab	400	.020		.21	.53		.74	1.12
7600	Hardwood		130	.062		.21	1.64		1.85	2.96
7800	Sanding and finishing, 2 coats polyurethane	▼	295	.027	▼	.21	.72		.93	1.43
7900	Subfloor and underlayment, see Section 06 16									
8015	Transition molding, 2-1/4" wide, 5' long	1 Carp	19.20	.417	Ea.	18.65	14.55		33.20	44.50

09 65 Resilient Flooring

09 65 10 – Resilient Tile Underlayment

09 65 10.10 Latex Underlayment	Crew	Daily Output	Labor-Hours	Unit	Material	2018 Bare Costs Labor	Equipment	Total	Total Incl O&P
0010 **LATEX UNDERLAYMENT**									
3600 Latex underlayment, 1/8" thk., cementitious for resilient flooring	1 Tilf	160	.050	S.F.	1.25	1.69		2.94	4.12
4000 Liquid, fortified				Gal.	32			32	35

09 65 13 – Resilient Base and Accessories

09 65 13.13 Resilient Base

	Crew	Daily Output	Labor-Hours	Unit	Material	Labor	Equipment	Total	Total Incl O&P
0010 **RESILIENT BASE**									
0690 1/8" vinyl base, 2-1/2" H, straight or cove, standard colors	1 Tilf	315	.025	L.F.	.70	.86		1.56	2.16
0700 4" high		315	.025		1.19	.86		2.05	2.70
0710 6" high		315	.025		1.47	.86		2.33	3.01
0720 Corners, 2-1/2" high		315	.025	Ea.	2.21	.86		3.07	3.82
0730 4" high		315	.025		2.52	.86		3.38	4.16
0740 6" high		315	.025		2.84	.86		3.70	4.51
0800 1/8" rubber base, 2-1/2" H, straight or cove, standard colors		315	.025	L.F.	1.10	.86		1.96	2.60
1100 4" high		315	.025		1.28	.86		2.14	2.80
1110 6" high		315	.025		1.90	.86		2.76	3.48
1150 Corners, 2-1/2" high		315	.025	Ea.	2.46	.86		3.32	4.10
1153 4" high		315	.025		2.54	.86		3.40	4.18
1155 6" high		315	.025		3.11	.86		3.97	4.81
1450 For premium color/finish add					50%				
1500 Millwork profile	1 Tilf	315	.025	L.F.	6.25	.86		7.11	8.30

09 65 13.37 Vinyl Transition Strips

	Crew	Daily Output	Labor-Hours	Unit	Material	Labor	Equipment	Total	Total Incl O&P
0010 **VINYL TRANSITION STRIPS**									
0100 Various mats. to various mats., adhesive applied, 1/4" to 1/8"	1 Tilf	315	.025	L.F.	1.47	.86		2.33	3.01
0105 0.08" to 1/8"		315	.025		1.33	.86		2.19	2.85
0110 0.08" to 1/4"		315	.025		1.45	.86		2.31	2.99
0115 1/4" to 3/8"		315	.025		1.35	.86		2.21	2.88
0120 1/4" to 1/2"		315	.025		1.35	.86		2.21	2.88
0125 1/4" to 0.08"		315	.025		1.45	.86		2.31	2.99
0200 Vinyl wheeled trans. strips, carpet to var. mats., 1/4" to 1/8" x 2-1/2"		315	.025		4.71	.86		5.57	6.60
0205 1/4" to 1/8" x 4"		315	.025		6.10	.86		6.96	8.10
0210 Various mats. to various mats. 1/4" to 0.08" x 2-1/2"		315	.025		5.05	.86		5.91	6.95
0215 Carpet to various materials, 1/4" to flush x 2-1/2"		315	.025		3.93	.86		4.79	5.70
0220 1/4" to flush x 4"		315	.025		6.05	.86		6.91	8.05
0225 Various materials to resilient, 3/8" to 1/8" x 2-1/2"		315	.025		3.93	.86		4.79	5.70
0230 Carpet to various materials, 3/8" to 1/4" x 2-1/2"		315	.025		5.40	.86		6.26	7.35
0235 1/4" to 1/4" x 2-1/2"		315	.025		6.05	.86		6.91	8.05
0240 Various materials to resilient, 1/8" to 1/8" x 2-1/2"		315	.025		4.57	.86		5.43	6.45
0245 Various materials to var. mats., 1/8" to flush x 2-1/2"		315	.025		3.06	.86		3.92	4.76
0250 3/8" to flush x 4"		315	.025		6.15	.86		7.01	8.20
0255 1/2" to flush x 4"		315	.025		8.25	.86		9.11	10.50
0260 Various materials to resilient, 1/8" to 0.08" x 2-1/2"		315	.025		3.55	.86		4.41	5.30
0265 0.08" to 0.08" x 2-1/2"		315	.025		3.37	.86		4.23	5.10
0270 3/8" to 0.08" x 2-1/2"		315	.025		3.37	.86		4.23	5.10

09 65 16 – Resilient Sheet Flooring

09 65 16.10 Rubber and Vinyl Sheet Flooring

		Crew	Daily Output	Labor-Hours	Unit	Material	Labor	Equipment	Total	Total Incl O&P
0010 **RUBBER AND VINYL SHEET FLOORING**										
5500 Linoleum, sheet goods	G	1 Tilf	360	.022	S.F.	3.59	.75		4.34	5.15
5900 Rubber, sheet goods, 36" wide, 1/8" thick			120	.067		8.25	2.25		10.50	12.75
5950 3/16" thick			100	.080		10.20	2.70		12.90	15.65
6000 1/4" thick			90	.089		12.05	3		15.05	18.10
8000 Vinyl sheet goods, backed, .065" thick, plain pattern/colors			250	.032		4.30	1.08		5.38	6.50

For customer support on your Residential Costs with RSMeans data, call 800.448.8182.

503

09 65 Resilient Flooring

09 65 16 – Resilient Sheet Flooring

09 65 16.10 Rubber and Vinyl Sheet Flooring

09 65 16.10 Rubber and Vinyl Sheet Flooring	Crew	Daily Output	Labor-Hours	Unit	Material	2018 Bare Costs Labor	Equipment	Total	Total Incl O&P
8050	Intricate pattern/colors	1 Tilf	200	.040	S.F.	3.94	1.35		5.29	6.50
8100	.080" thick, plain pattern/colors		230	.035		4.22	1.17		5.39	6.55
8150	Intricate pattern/colors		200	.040		6.45	1.35		7.80	9.30
8200	.125" thick, plain pattern/colors		230	.035		3.87	1.17		5.04	6.15
8250	Intricate pattern/colors		200	.040		7.50	1.35		8.85	10.45
8700	Adhesive cement, 1 gallon per 200 to 300 S.F.				Gal.	31.50			31.50	34.50
8800	Asphalt primer, 1 gallon per 300 S.F.					14.95			14.95	16.45
8900	Emulsion, 1 gallon per 140 S.F.					18.95			18.95	21

09 65 19 – Resilient Tile Flooring

09 65 19.19 Vinyl Composition Tile Flooring

		Crew	Daily Output	Labor-Hours	Unit	Material	2018 Bare Costs Labor	Equipment	Total	Total Incl O&P
0010	**VINYL COMPOSITION TILE FLOORING**									
7000	Vinyl composition tile, 12" x 12", 1/16" thick	1 Tilf	500	.016	S.F.	1.22	.54		1.76	2.22
7050	Embossed		500	.016		2.66	.54		3.20	3.81
7100	Marbleized		500	.016		2.66	.54		3.20	3.81
7150	Solid		500	.016		3.44	.54		3.98	4.66
7200	3/32" thick, embossed		500	.016		1.55	.54		2.09	2.59
7250	Marbleized		500	.016		3.06	.54		3.60	4.25
7300	Solid		500	.016		2.85	.54		3.39	4.02
7350	1/8" thick, marbleized		500	.016		2.44	.54		2.98	3.56
7400	Solid		500	.016		1.77	.54		2.31	2.83
7450	Conductive		500	.016		5.90	.54		6.44	7.40

09 65 19.23 Vinyl Tile Flooring

		Crew	Daily Output	Labor-Hours	Unit	Material	2018 Bare Costs Labor	Equipment	Total	Total Incl O&P
0010	**VINYL TILE FLOORING**									
7500	Vinyl tile, 12" x 12", 3/32" thick, standard colors/patterns	1 Tilf	500	.016	S.F.	3.71	.54		4.25	4.96
7550	1/8" thick, standard colors/patterns		500	.016		5.20	.54		5.74	6.60
7600	1/8" thick, premium colors/patterns		500	.016		7	.54		7.54	8.60
7650	Solid colors		500	.016		3.23	.54		3.77	4.43
7700	Marbleized or Travertine pattern		500	.016		6.20	.54		6.74	7.70
7750	Florentine pattern		500	.016		6.60	.54		7.14	8.15
7800	Premium colors/patterns		500	.016		6.25	.54		6.79	7.75

09 65 19.33 Rubber Tile Flooring

		Crew	Daily Output	Labor-Hours	Unit	Material	2018 Bare Costs Labor	Equipment	Total	Total Incl O&P
0010	**RUBBER TILE FLOORING**									
6050	Rubber tile, marbleized colors, 12" x 12", 1/8" thick	1 Tilf	400	.020	S.F.	5.80	.68		6.48	7.45
6100	3/16" thick		400	.020		8	.68		8.68	9.90
6300	Special tile, plain colors, 1/8" thick		400	.020		8.15	.68		8.83	10.05
6350	3/16" thick		400	.020		9.75	.68		10.43	11.80

09 65 33 – Conductive Resilient Flooring

09 65 33.10 Conductive Rubber and Vinyl Flooring

		Crew	Daily Output	Labor-Hours	Unit	Material	2018 Bare Costs Labor	Equipment	Total	Total Incl O&P
0010	**CONDUCTIVE RUBBER AND VINYL FLOORING**									
1700	Conductive flooring, rubber tile, 1/8" thick	1 Tilf	315	.025	S.F.	7.05	.86		7.91	9.15
1800	Homogeneous vinyl tile, 1/8" thick	"	315	.025	"	6.80	.86		7.66	8.90

09 66 Terrazzo Flooring

09 66 13 – Portland Cement Terrazzo Flooring

09 66 13.10 Portland Cement Terrazzo	Crew	Daily Output	Labor-Hours	Unit	Material	2018 Bare Costs Labor	Equipment	Total	Total Incl O&P
0010 **PORTLAND CEMENT TERRAZZO**, cast-in-place									
4300 Stone chips, onyx gemstone, per 50 lb. bag				Bag	18.60			18.60	20.50

09 66 16 – Terrazzo Floor Tile

09 66 16.13 Portland Cement Terrazzo Floor Tile

	Crew	Daily Output	Labor-Hours	Unit	Material	Labor	Equipment	Total	Total Incl O&P
0010 **PORTLAND CEMENT TERRAZZO FLOOR TILE**									
1200 Floor tiles, non-slip, 1" thick, 12" x 12"	D-1	60	.267	S.F.	24.50	8.45		32.95	41
1300 1-1/4" thick, 12" x 12"		60	.267		25.50	8.45		33.95	42
1500 16" x 16"		50	.320		27.50	10.15		37.65	47.50
1600 1-1/2" thick, 16" x 16"	↓	45	.356	↓	25	11.25		36.25	47

09 66 16.30 Terrazzo, Precast

	Crew	Daily Output	Labor-Hours	Unit	Material	Labor	Equipment	Total	Total Incl O&P
0010 **TERRAZZO, PRECAST**									
0020 Base, 6" high, straight	1 Mstz	70	.114	L.F.	12.35	3.82		16.17	19.75
0100 Cove		60	.133		16.50	4.45		20.95	25.50
0300 8" high, straight		60	.133		14.75	4.45		19.20	23.50
0400 Cove	↓	50	.160		21.50	5.35		26.85	32.50
0600 For white cement, add					.55			.55	.61
0700 For 16 ga. zinc toe strip, add					2.18			2.18	2.40
0900 Curbs, 4" x 4" high	1 Mstz	40	.200		40	6.70		46.70	55
1000 8" x 8" high		30	.267	↓	46.50	8.90		55.40	65.50
4800 Wainscot, 12" x 12" x 1" tiles		12	.667	S.F.	8.75	22.50		31.25	45.50
4900 16" x 16" x 1-1/2" tiles	↓	8	1	"	17.45	33.50		50.95	73

09 68 Carpeting

09 68 05 – Carpet Accessories

09 68 05.11 Flooring Transition Strip

	Crew	Daily Output	Labor-Hours	Unit	Material	Labor	Equipment	Total	Total Incl O&P
0010 **FLOORING TRANSITION STRIP**									
0107 Clamp down brass divider, 12' strip, vinyl to carpet	1 Tilf	31.25	.256	Ea.	14.65	8.65		23.30	30
0117 Vinyl to hard surface	"	31.25	.256	"	14.65	8.65		23.30	30

09 68 10 – Carpet Pad

09 68 10.10 Commercial Grade Carpet Pad

	Crew	Daily Output	Labor-Hours	Unit	Material	Labor	Equipment	Total	Total Incl O&P
0010 **COMMERCIAL GRADE CARPET PAD**									
9001 Sponge rubber pad, 20 oz./sq. yd.	1 Tilf	1350	.006	S.F.	.52	.20		.72	.89
9101 40 to 62 oz./sq. yd.		1350	.006		.97	.20		1.17	1.39
9201 Felt pad, 20 to 32 oz./sq. yd.		1350	.006		.64	.20		.84	1.03
9301 Maximum		1350	.006		1.24	.20		1.44	1.68
9401 Bonded urethane pad, 2.7 density		1350	.006		.66	.20		.86	1.05
9501 13.0 density		1350	.006		.89	.20		1.09	1.30
9601 Prime urethane pad, 2.7 density		1350	.006		.39	.20		.59	.75
9701 13.0 density	↓	1350	.006	↓	.73	.20		.93	1.12

09 68 13 – Tile Carpeting

09 68 13.10 Carpet Tile

	Crew	Daily Output	Labor-Hours	Unit	Material	Labor	Equipment	Total	Total Incl O&P
0010 **CARPET TILE**									
0100 Tufted nylon, 18" x 18", hard back, 20 oz.	1 Tilf	80	.100	S.Y.	27	3.38		30.38	35
0110 26 oz.		80	.100		25.50	3.38		28.88	33.50
0200 Cushion back, 20 oz.		80	.100		25	3.38		28.38	33
0210 26 oz.		80	.100		30	3.38		33.38	39
6000 Electrostatic dissipative carpet tile, 24" x 24", 24 oz.		80	.100		37.50	3.38		40.88	46.50
6100 Electrostatic dissapative carpet tile for access floors, 24" x 24", 24 oz.	↓	80	.100	↓	47	3.38		50.38	57

For customer support on your Residential Costs with RSMeans data, call 800.448.8182.

505

09 68 Carpeting

09 68 16 – Sheet Carpeting

09 68 16.10 Sheet Carpet

		Crew	Daily Output	Labor-Hours	Unit	Material	2018 Bare Costs Labor	2018 Bare Costs Equipment	Total	Total Incl O&P
0010	**SHEET CARPET**									
0701	Nylon, level loop, 26 oz., light to medium traffic	1 Tilf	675	.012	S.F.	2.44	.40		2.84	3.34
0901	32 oz., medium traffic		675	.012		4.38	.40		4.78	5.45
1101	40 oz., medium to heavy traffic		675	.012		5.40	.40		5.80	6.60
2101	Nylon, plush, 20 oz., light traffic		675	.012		2.27	.40		2.67	3.15
2801	24 oz., light to medium traffic		675	.012		2.16	.40		2.56	3.03
2901	30 oz., medium traffic		675	.012		3.18	.40		3.58	4.15
3001	36 oz., medium traffic		675	.012		4.02	.40		4.42	5.05
3101	42 oz., medium to heavy traffic		630	.013		5.05	.43		5.48	6.25
3201	46 oz., medium to heavy traffic		630	.013		5.85	.43		6.28	7.15
3301	54 oz., heavy traffic		630	.013		6.60	.43		7.03	7.95
3501	Olefin, 15 oz., light traffic		675	.012		1.62	.40		2.02	2.43
3651	22 oz., light traffic		675	.012		1.65	.40		2.05	2.46
4501	50 oz., medium to heavy traffic, level loop		630	.013		12.10	.43		12.53	14
4701	32 oz., medium to heavy traffic, patterned		630	.013		10.95	.43		11.38	12.75
4901	48 oz., heavy traffic, patterned	↓	630	.013	↓	12.10	.43		12.53	14
5000	For less than full roll (approx. 1500 S.F.), add					25%				
5100	For small rooms, less than 12' wide, add						25%			
5200	For large open areas (no cuts), deduct						25%			
5600	For bound carpet baseboard, add	1 Tilf	300	.027	L.F.	1.81	.90		2.71	3.45
5610	For stairs, not incl. price of carpet, add	"	30	.267	Riser		9		9	14.60
8950	For tackless, stretched installation, add padding from 09 68 10.10 to above									
9850	For brand-named specific fiber, add				S.Y.	25%				

09 68 20 – Athletic Carpet

09 68 20.10 Indoor Athletic Carpet

		Crew	Daily Output	Labor-Hours	Unit	Material	2018 Bare Costs Labor	2018 Bare Costs Equipment	Total	Total Incl O&P
0010	**INDOOR ATHLETIC CARPET**									
3700	Polyethylene, in rolls, no base incl., landscape surfaces	1 Tilf	275	.029	S.F.	4.08	.98		5.06	6.10
3800	Nylon action surface, 1/8" thick		275	.029		3.94	.98		4.92	5.90
3900	1/4" thick		275	.029		5.70	.98		6.68	7.85
4000	3/8" thick	↓	275	.029	↓	7.15	.98		8.13	9.45

09 72 Wall Coverings

09 72 13 – Cork Wall Coverings

09 72 13.10 Covering, Cork Wall

		Crew	Daily Output	Labor-Hours	Unit	Material	2018 Bare Costs Labor	2018 Bare Costs Equipment	Total	Total Incl O&P
0010	**COVERING, CORK WALL**									
0600	Cork tiles, light or dark, 12" x 12" x 3/16"	1 Pape	240	.033	S.F.	4.26	.98		5.24	6.30
0700	5/16" thick		235	.034		3.33	1		4.33	5.30
0900	1/4" basket weave		240	.033		3.39	.98		4.37	5.35
1000	1/2" natural, non-directional pattern		240	.033		6.80	.98		7.78	9.05
1100	3/4" natural, non-directional pattern		240	.033		11.60	.98		12.58	14.35
1200	Granular surface, 12" x 36", 1/2" thick		385	.021		1.30	.61		1.91	2.43
1300	1" thick		370	.022		1.66	.64		2.30	2.88
1500	Polyurethane coated, 12" x 12" x 3/16" thick		240	.033		4.03	.98		5.01	6.05
1600	5/16" thick		235	.034		6	1		7	8.25
1800	Cork wallpaper, paperbacked, natural		480	.017		1.60	.49		2.09	2.57
1900	Colors	↓	480	.017	↓	2.84	.49		3.33	3.93

09 72 Wall Coverings

09 72 16 – Vinyl-Coated Fabric Wall Coverings

09 72 16.13 Flexible Vinyl Wall Coverings

09 72 16.13 Flexible Vinyl Wall Coverings		Crew	Daily Output	Labor-Hours	Unit	Material	2018 Bare Costs Labor	Equipment	Total	Total Incl O&P
0010	**FLEXIBLE VINYL WALL COVERINGS**									
3000	Vinyl wall covering, fabric-backed, lightweight, type 1 (12-15 oz./S.Y.)	1 Pape	640	.013	S.F.	1.07	.37		1.44	1.78
3300	Medium weight, type 2 (20-24 oz./S.Y.)		480	.017		.91	.49		1.40	1.81
3400	Heavy weight, type 3 (28 oz./S.Y.)	↓	435	.018	↓	1.40	.54		1.94	2.43
3600	Adhesive, 5 gal. lots (18 S.Y./gal.)				Gal.	12.05			12.05	13.25

09 72 16.16 Rigid-Sheet Vinyl Wall Coverings

		Crew	Daily Output	Labor-Hours	Unit	Material	Labor	Equipment	Total	Total Incl O&P
0010	**RIGID-SHEET VINYL WALL COVERINGS**									
0100	Acrylic, modified, semi-rigid PVC, .028" thick	2 Carp	330	.048	S.F.	1.30	1.69		2.99	4.25
0110	.040" thick	"	320	.050	"	1.85	1.75		3.60	4.95

09 72 19 – Textile Wall Coverings

09 72 19.10 Textile Wall Covering

		Crew	Daily Output	Labor-Hours	Unit	Material	Labor	Equipment	Total	Total Incl O&P
0010	**TEXTILE WALL COVERING**, including sizing; add 10-30% waste @ takeoff									
0020	Silk	1 Pape	640	.013	S.F.	4.72	.37		5.09	5.80
0030	Cotton		640	.013		7.05	.37		7.42	8.40
0040	Linen		640	.013		1.92	.37		2.29	2.71
0050	Blend	↓	640	.013	↓	3.26	.37		3.63	4.19
0060	Linen wall covering, paper backed									
0070	Flame treatment				S.F.	1.07			1.07	1.18
0080	Stain resistance treatment					1.92			1.92	2.11
0100	Grass cloths with lining paper	G	1 Pape	400	.020		1.38	.59		1.97	2.49
0110	Premium texture/color	G	"	350	.023	↓	3.15	.67		3.82	4.58

09 72 20 – Natural Fiber Wall Covering

09 72 20.10 Natural Fiber Wall Covering

		Crew	Daily Output	Labor-Hours	Unit	Material	Labor	Equipment	Total	Total Incl O&P
0010	**NATURAL FIBER WALL COVERING**, including sizing; add 10-30% waste @ takeoff									
0015	Bamboo	1 Pape	640	.013	S.F.	2.22	.37		2.59	3.04
0030	Burlap		640	.013		1.88	.37		2.25	2.67
0045	Jute		640	.013		1.33	.37		1.70	2.06
0060	Sisal	↓	640	.013	↓	1.53	.37		1.90	2.28

09 72 23 – Wallpapering

09 72 23.10 Wallpaper

		Crew	Daily Output	Labor-Hours	Unit	Material	Labor	Equipment	Total	Total Incl O&P
0010	**WALLPAPER** including sizing; add 10-30% waste @ takeoff R097223-10									
0050	Aluminum foil	1 Pape	275	.029	S.F.	1.03	.86		1.89	2.54
0100	Copper sheets, .025" thick, vinyl backing		240	.033		5.50	.98		6.48	7.65
0300	Phenolic backing	↓	240	.033	↓	7.10	.98		8.08	9.45
2400	Gypsum-based, fabric-backed, fire resistant									
2500	for masonry walls, 21 oz./S.Y.	1 Pape	800	.010	S.F.	.77	.29		1.06	1.33
2600	Average		720	.011		1.26	.33		1.59	1.93
2700	Small quantities		640	.013		.68	.37		1.05	1.35
3700	Wallpaper, average workmanship, solid pattern, low cost paper		640	.013		.61	.37		.98	1.27
3900	Basic patterns (matching required), avg. cost paper		535	.015		1.19	.44		1.63	2.03
4000	Paper at $85 per double roll, quality workmanship	↓	435	.018	↓	2.11	.54		2.65	3.21

For customer support on your Residential Costs with RSMeans data, call 800.448.8182.

09 74 Flexible Wood Sheets

09 74 16 – Flexible Wood Veneers

09 74 16.10 Veneer, Flexible Wood	Crew	Daily Output	Labor-Hours	Unit	Material	2018 Bare Costs Labor	Equipment	Total	Total Incl O&P
0010 **VENEER, FLEXIBLE WOOD**									
0100 Flexible wood veneer, 1/32" thick, plain woods	1 Pape	100	.080	S.F.	2.41	2.35		4.76	6.50
0110 Exotic woods	"	95	.084	"	3.64	2.48		6.12	8.05

09 91 Painting

09 91 03 – Paint Restoration

09 91 03.20 Sanding

	Crew	Daily Output	Labor-Hours	Unit	Material	2018 Bare Costs Labor	Equipment	Total	Total Incl O&P
0010 **SANDING** and puttying interior trim, compared to									
0100 Painting 1 coat, on quality work				L.F.		100%			
0300 Medium work						50%			
0400 Industrial grade						25%			
0500 Surface protection, placement and removal									
0510 Surface protection, placement and removal, basic drop cloths	1 Pord	6400	.001	S.F.		.04		.04	.06
0520 Masking with paper		800	.010		.07	.29		.36	.56
0530 Volume cover up (using plastic sheathing or building paper)		16000	.001			.01		.01	.02

09 91 03.30 Exterior Surface Preparation

	Crew	Daily Output	Labor-Hours	Unit	Material	2018 Bare Costs Labor	Equipment	Total	Total Incl O&P
0010 **EXTERIOR SURFACE PREPARATION**									
0015 Doors, per side, not incl. frames or trim									
0020 Scrape & sand									
0030 Wood, flush	1 Pord	616	.013	S.F.		.38		.38	.63
0040 Wood, detail		496	.016			.47		.47	.78
0050 Wood, louvered		280	.029			.84		.84	1.38
0060 Wood, overhead		616	.013			.38		.38	.63
0070 Wire brush									
0080 Metal, flush	1 Pord	640	.013	S.F.		.37		.37	.60
0090 Metal, detail		520	.015			.45		.45	.74
0100 Metal, louvered		360	.022			.65		.65	1.07
0110 Metal or fibr., overhead		640	.013			.37		.37	.60
0120 Metal, roll up		560	.014			.42		.42	.69
0130 Metal, bulkhead		640	.013			.37		.37	.60
0140 Power wash, based on 2500 lb. operating pressure									
0150 Metal, flush	A-1H	2240	.004	S.F.		.10	.03	.13	.20
0160 Metal, detail		2120	.004			.10	.04	.14	.21
0170 Metal, louvered		2000	.004			.11	.04	.15	.22
0180 Metal or fibr., overhead		2400	.003			.09	.03	.12	.18
0190 Metal, roll up		2400	.003			.09	.03	.12	.18
0200 Metal, bulkhead		2200	.004			.10	.03	.13	.20
0400 Windows, per side, not incl. trim									
0410 Scrape & sand									
0420 Wood, 1-2 lite	1 Pord	320	.025	S.F.		.73		.73	1.21
0430 Wood, 3-6 lite		280	.029			.84		.84	1.38
0440 Wood, 7-10 lite		240	.033			.98		.98	1.61
0450 Wood, 12 lite		200	.040			1.17		1.17	1.93
0460 Wood, Bay/Bow		320	.025			.73		.73	1.21
0470 Wire brush									
0480 Metal, 1-2 lite	1 Pord	480	.017	S.F.		.49		.49	.81
0490 Metal, 3-6 lite		400	.020			.59		.59	.97
0500 Metal, Bay/Bow		480	.017			.49		.49	.81
0510 Power wash, based on 2500 lb. operating pressure									
0520 1-2 lite	A-1H	4400	.002	S.F.		.05	.02	.07	.10
0530 3-6 lite		4320	.002			.05	.02	.07	.10

508

09 91 Painting

09 91 03 – Paint Restoration

09 91 03.30 Exterior Surface Preparation	Crew	Daily Output	Labor-Hours	Unit	Material	2018 Bare Costs Labor	Equipment	Total	Total Incl O&P
0540	7-10 lite	A-1H	4240	.002	S.F.		.05	.02	.07	.10
0550	12 lite		4160	.002			.05	.02	.07	.11
0560	Bay/Bow	▼	4400	.002	▼		.05	.02	.07	.10
0600	Siding, scrape and sand, light=10-30%, med.=30-70%									
0610	Heavy=70-100% of surface to sand									
0650	Texture 1-11, light	1 Pord	480	.017	S.F.		.49		.49	.81
0660	Med.		440	.018			.53		.53	.88
0670	Heavy		360	.022			.65		.65	1.07
0680	Wood shingles, shakes, light		440	.018			.53		.53	.88
0690	Med.		360	.022			.65		.65	1.07
0700	Heavy		280	.029			.84		.84	1.38
0710	Clapboard, light		520	.015			.45		.45	.74
0720	Med.		480	.017			.49		.49	.81
0730	Heavy	▼	400	.020	▼		.59		.59	.97
0740	Wire brush									
0750	Aluminum, light	1 Pord	600	.013	S.F.		.39		.39	.64
0760	Med.		520	.015			.45		.45	.74
0770	Heavy	▼	440	.018	▼		.53		.53	.88
0780	Pressure wash, based on 2500 lb. operating pressure									
0790	Stucco	A-1H	3080	.003	S.F.		.07	.02	.09	.15
0800	Aluminum or vinyl		3200	.003			.07	.02	.09	.14
0810	Siding, masonry, brick & block	▼	2400	.003	▼		.09	.03	.12	.18
1300	Miscellaneous, wire brush									
1310	Metal, pedestrian gate	1 Pord	100	.080	S.F.		2.35		2.35	3.86

09 91 03.40 Interior Surface Preparation	Crew	Daily Output	Labor-Hours	Unit	Material	2018 Bare Costs Labor	Equipment	Total	Total Incl O&P
0010	**INTERIOR SURFACE PREPARATION**									
0020	Doors, per side, not incl. frames or trim									
0030	Scrape & sand									
0040	Wood, flush	1 Pord	616	.013	S.F.		.38		.38	.63
0050	Wood, detail		496	.016			.47		.47	.78
0060	Wood, louvered	▼	280	.029	▼		.84		.84	1.38
0070	Wire brush									
0080	Metal, flush	1 Pord	640	.013	S.F.		.37		.37	.60
0090	Metal, detail		520	.015			.45		.45	.74
0100	Metal, louvered	▼	360	.022	▼		.65		.65	1.07
0110	Hand wash									
0120	Wood, flush	1 Pord	2160	.004	S.F.		.11		.11	.18
0130	Wood, detail		2000	.004			.12		.12	.19
0140	Wood, louvered		1360	.006			.17		.17	.28
0150	Metal, flush		2160	.004			.11		.11	.18
0160	Metal, detail		2000	.004			.12		.12	.19
0170	Metal, louvered	▼	1360	.006	▼		.17		.17	.28
0400	Windows, per side, not incl. trim									
0410	Scrape & sand									
0420	Wood, 1-2 lite	1 Pord	360	.022	S.F.		.65		.65	1.07
0430	Wood, 3-6 lite		320	.025			.73		.73	1.21
0440	Wood, 7-10 lite		280	.029			.84		.84	1.38
0450	Wood, 12 lite		240	.033			.98		.98	1.61
0460	Wood, Bay/Bow	▼	360	.022	▼		.65		.65	1.07
0470	Wire brush									
0480	Metal, 1-2 lite	1 Pord	520	.015	S.F.		.45		.45	.74
0490	Metal, 3-6 lite	▼	440	.018	▼		.53		.53	.88

For customer support on your Residential Costs with RSMeans data, call 800.448.8182.

509

09 91 Painting

09 91 03 – Paint Restoration

09 91 03.40 Interior Surface Preparation

		Crew	Daily Output	Labor-Hours	Unit	Material	2018 Bare Costs Labor	Equipment	Total	Total Incl O&P
0500	Metal, Bay/Bow	1 Pord	520	.015	S.F.		.45		.45	.74
0600	Walls, sanding, light=10-30%, medium=30-70%,									
0610	heavy=70-100% of surface to sand									
0650	Walls, sand									
0660	Gypsum board or plaster, light	1 Pord	3077	.003	S.F.		.08		.08	.13
0670	Gypsum board or plaster, medium		2160	.004			.11		.11	.18
0680	Gypsum board or plaster, heavy		923	.009			.25		.25	.42
0690	Wood, T&G, light		2400	.003			.10		.10	.16
0700	Wood, T&G, medium		1600	.005			.15		.15	.24
0710	Wood, T&G, heavy	↓	800	.010	↓		.29		.29	.48
0720	Walls, wash									
0730	Gypsum board or plaster	1 Pord	3200	.003	S.F.		.07		.07	.12
0740	Wood, T&G		3200	.003			.07		.07	.12
0750	Masonry, brick & block, smooth		2800	.003			.08		.08	.14
0760	Masonry, brick & block, coarse	↓	2000	.004	↓		.12		.12	.19
8000	For chemical washing, see Section 04 01 30									

09 91 03.41 Scrape After Fire Damage

		Crew	Daily Output	Labor-Hours	Unit	Material	2018 Bare Costs Labor	Equipment	Total	Total Incl O&P
0010	**SCRAPE AFTER FIRE DAMAGE**									
0050	Boards, 1" x 4"	1 Pord	336	.024	L.F.		.70		.70	1.15
0060	1" x 6"		260	.031			.90		.90	1.49
0070	1" x 8"		207	.039			1.13		1.13	1.87
0080	1" x 10"		174	.046			1.35		1.35	2.22
0500	Framing, 2" x 4"		265	.030			.89		.89	1.46
0510	2" x 6"		221	.036			1.06		1.06	1.75
0520	2" x 8"		190	.042			1.24		1.24	2.03
0530	2" x 10"		165	.048			1.42		1.42	2.34
0540	2" x 12"		144	.056			1.63		1.63	2.68
1000	Heavy framing, 3" x 4"		226	.035			1.04		1.04	1.71
1010	4" x 4"		210	.038			1.12		1.12	1.84
1020	4" x 6"		191	.042			1.23		1.23	2.02
1030	4" x 8"		165	.048			1.42		1.42	2.34
1040	4" x 10"		144	.056			1.63		1.63	2.68
1060	4" x 12"		131	.061	↓		1.79		1.79	2.95
2900	For sealing, light damage		825	.010	S.F.	.15	.28		.43	.64
2920	Heavy damage	↓	460	.017	"	.33	.51		.84	1.20

09 91 13 – Exterior Painting

09 91 13.30 Fences

		Crew	Daily Output	Labor-Hours	Unit	Material	2018 Bare Costs Labor	Equipment	Total	Total Incl O&P
0010	**FENCES**	R099100-20								
0100	Chain link or wire metal, one side, water base									
0110	Roll & brush, first coat	1 Pord	960	.008	S.F.	.07	.24		.31	.48
0120	Second coat		1280	.006		.07	.18		.25	.37
0130	Spray, first coat		2275	.004		.07	.10		.17	.25
0140	Second coat	↓	2600	.003	↓	.07	.09		.16	.23
0150	Picket, water base									
0160	Roll & brush, first coat	1 Pord	865	.009	S.F.	.08	.27		.35	.53
0170	Second coat		1050	.008		.08	.22		.30	.45
0180	Spray, first coat		2275	.004		.08	.10		.18	.25
0190	Second coat	↓	2600	.003	↓	.08	.09		.17	.23
0200	Stockade, water base									
0210	Roll & brush, first coat	1 Pord	1040	.008	S.F.	.08	.23		.31	.45
0220	Second coat		1200	.007		.08	.20		.28	.40
0230	Spray, first coat	↓	2275	.004	↓	.08	.10		.18	.25

09 91 Painting

09 91 13 – Exterior Painting

09 91 13.30 Fences

	Crew	Daily Output	Labor-Hours	Unit	Material	2018 Bare Costs Labor	Equipment	Total	Total Incl O&P
0240 Second coat	1 Pord	2600	.003	S.F.	.08	.09		.17	.23

09 91 13.42 Miscellaneous, Exterior

	Crew	Daily Output	Labor-Hours	Unit	Material	2018 Bare Costs Labor	Equipment	Total	Total Incl O&P
0010 **MISCELLANEOUS, EXTERIOR** R099100-20									
0100 Railing, ext., decorative wood, incl. cap & baluster									
0110 Newels & spindles @ 12" OC									
0120 Brushwork, stain, sand, seal & varnish									
0130 First coat	1 Pord	90	.089	L.F.	.88	2.61		3.49	5.25
0140 Second coat	"	120	.067	"	.88	1.96		2.84	4.18
0150 Rough sawn wood, 42" high, 2" x 2" verticals, 6" OC									
0160 Brushwork, stain, each coat	1 Pord	90	.089	L.F.	.31	2.61		2.92	4.63
0170 Wrought iron, 1" rail, 1/2" sq. verticals									
0180 Brushwork, zinc chromate, 60" high, bars 6" OC									
0190 Primer	1 Pord	130	.062	L.F.	.88	1.81		2.69	3.94
0200 Finish coat		130	.062		1.13	1.81		2.94	4.21
0210 Additional coat	↓	190	.042	↓	1.32	1.24		2.56	3.48
0220 Shutters or blinds, single panel, 2' x 4', paint all sides									
0230 Brushwork, primer	1 Pord	20	.400	Ea.	.69	11.75		12.44	20
0240 Finish coat, exterior latex		20	.400		.57	11.75		12.32	19.95
0250 Primer & 1 coat, exterior latex		13	.615		1.11	18.05		19.16	30.50
0260 Spray, primer		35	.229		1.01	6.70		7.71	12.15
0270 Finish coat, exterior latex		35	.229		1.21	6.70		7.91	12.40
0280 Primer & 1 coat, exterior latex	↓	20	.400	↓	1.09	11.75		12.84	20.50
0290 For louvered shutters, add				S.F.	10%				
0300 Stair stringers, exterior, metal									
0310 Roll & brush, zinc chromate, to 14", each coat	1 Pord	320	.025	L.F.	.38	.73		1.11	1.62
0320 Rough sawn wood, 4" x 12"									
0330 Roll & brush, exterior latex, each coat	1 Pord	215	.037	L.F.	.08	1.09		1.17	1.89
0340 Trellis/lattice, 2" x 2" @ 3" OC with 2" x 8" supports									
0350 Spray, latex, per side, each coat	1 Pord	475	.017	S.F.	.08	.49		.57	.90
0450 Decking, ext., sealer, alkyd, brushwork, sealer coat		1140	.007		.10	.21		.31	.45
0460 1st coat		1140	.007		.12	.21		.33	.47
0470 2nd coat		1300	.006		.09	.18		.27	.39
0500 Paint, alkyd, brushwork, primer coat		1140	.007		.11	.21		.32	.46
0510 1st coat		1140	.007		.14	.21		.35	.49
0520 2nd coat		1300	.006		.10	.18		.28	.41
0600 Sand paint, alkyd, brushwork, 1 coat	↓	150	.053	↓	.14	1.57		1.71	2.73

09 91 13.60 Siding Exterior

	Crew	Daily Output	Labor-Hours	Unit	Material	2018 Bare Costs Labor	Equipment	Total	Total Incl O&P
0010 **SIDING EXTERIOR**, Alkyd (oil base)									
0450 Steel siding, oil base, paint 1 coat, brushwork	2 Pord	2015	.008	S.F.	.11	.23		.34	.50
0500 Spray		4550	.004		.17	.10		.27	.36
0800 Paint 2 coats, brushwork		1300	.012		.23	.36		.59	.84
1000 Spray		2750	.006		.15	.17		.32	.45
1200 Stucco, rough, oil base, paint 2 coats, brushwork		1300	.012		.23	.36		.59	.84
1400 Roller		1625	.010		.24	.29		.53	.74
1600 Spray		2925	.005		.25	.16		.41	.54
1800 Texture 1-11 or clapboard, oil base, primer coat, brushwork		1300	.012		.14	.36		.50	.75
2000 Spray		4550	.004		.14	.10		.24	.33
2100 Paint 1 coat, brushwork		1300	.012		.16	.36		.52	.77
2200 Spray		4550	.004		.16	.10		.26	.35
2400 Paint 2 coats, brushwork		810	.020		.33	.58		.91	1.31
2600 Spray		2600	.006		.36	.18		.54	.70
3000 Stain 1 coat, brushwork	↓	1520	.011	↓	.10	.31		.41	.62

For customer support on your Residential Costs with RSMeans data, call 800.448.8182.

511

09 91 13.60 Siding Exterior

		Crew	Daily Output	Labor-Hours	Unit	Material	2018 Bare Costs Labor	2018 Bare Costs Equipment	Total	Total Incl O&P
3200	Spray	2 Pord	5320	.003	S.F.	.11	.09		.20	.28
3400	Stain 2 coats, brushwork		950	.017		.20	.49		.69	1.04
4000	Spray		3050	.005		.23	.15		.38	.50
4200	Wood shingles, oil base primer coat, brushwork		1300	.012		.13	.36		.49	.74
4400	Spray		3900	.004		.12	.12		.24	.34
4600	Paint 1 coat, brushwork		1300	.012		.14	.36		.50	.74
4800	Spray		3900	.004		.17	.12		.29	.39
5000	Paint 2 coats, brushwork		810	.020		.27	.58		.85	1.25
5200	Spray		2275	.007		.26	.21		.47	.63
5800	Stain 1 coat, brushwork		1500	.011		.10	.31		.41	.63
6000	Spray		3900	.004		.10	.12		.22	.31
6500	Stain 2 coats, brushwork		950	.017		.20	.49		.69	1.04
7000	Spray		2660	.006		.28	.18		.46	.60
8000	For latex paint, deduct					10%				
8100	For work over 12' H, from pipe scaffolding, add						15%			
8200	For work over 12' H, from extension ladder, add						25%			
8300	For work over 12' H, from swing staging, add						35%			

09 91 13.62 Siding, Misc.

		Crew	Daily Output	Labor-Hours	Unit	Material	2018 Bare Costs Labor	2018 Bare Costs Equipment	Total	Total Incl O&P
0010	**SIDING, MISC.**, latex paint R099100-10									
0100	Aluminum siding									
0110	Brushwork, primer	2 Pord	2275	.007	S.F.	.06	.21		.27	.41
0120	Finish coat, exterior latex		2275	.007		.06	.21		.27	.40
0130	Primer & 1 coat exterior latex		1300	.012		.13	.36		.49	.74
0140	Primer & 2 coats exterior latex		975	.016		.19	.48		.67	1
0150	Mineral fiber shingles									
0160	Brushwork, primer	2 Pord	1495	.011	S.F.	.14	.31		.45	.68
0170	Finish coat, industrial enamel		1495	.011		.18	.31		.49	.72
0180	Primer & 1 coat enamel		810	.020		.32	.58		.90	1.30
0190	Primer & 2 coats enamel		540	.030		.50	.87		1.37	1.98
0200	Roll, primer		1625	.010		.16	.29		.45	.65
0210	Finish coat, industrial enamel		1625	.010		.19	.29		.48	.69
0220	Primer & 1 coat enamel		975	.016		.35	.48		.83	1.18
0230	Primer & 2 coats enamel		650	.025		.55	.72		1.27	1.79
0240	Spray, primer		3900	.004		.12	.12		.24	.34
0250	Finish coat, industrial enamel		3900	.004		.16	.12		.28	.38
0260	Primer & 1 coat enamel		2275	.007		.28	.21		.49	.65
0270	Primer & 2 coats enamel		1625	.010		.45	.29		.74	.97
0280	Waterproof sealer, first coat		4485	.004		.12	.10		.22	.30
0290	Second coat		5235	.003		.11	.09		.20	.27
0300	Rough wood incl. shingles, shakes or rough sawn siding									
0310	Brushwork, primer	2 Pord	1280	.013	S.F.	.14	.37		.51	.75
0320	Finish coat, exterior latex		1280	.013		.10	.37		.47	.71
0330	Primer & 1 coat exterior latex		960	.017		.24	.49		.73	1.08
0340	Primer & 2 coats exterior latex		700	.023		.34	.67		1.01	1.48
0350	Roll, primer		2925	.005		.19	.16		.35	.47
0360	Finish coat, exterior latex		2925	.005		.12	.16		.28	.39
0370	Primer & 1 coat exterior latex		1790	.009		.31	.26		.57	.77
0380	Primer & 2 coats exterior latex		1300	.012		.43	.36		.79	1.07
0390	Spray, primer		3900	.004		.16	.12		.28	.37
0400	Finish coat, exterior latex		3900	.004		.09	.12		.21	.30
0410	Primer & 1 coat exterior latex		2600	.006		.25	.18		.43	.58
0420	Primer & 2 coats exterior latex		2080	.008		.35	.23		.58	.75

09 91 Painting

09 91 13 – Exterior Painting

09 91 13.62 Siding, Misc.

		Crew	Daily Output	Labor-Hours	Unit	Material	2018 Bare Costs Labor	2018 Bare Costs Equipment	Total	Total Incl O&P
0430	Waterproof sealer, first coat	2 Pord	4485	.004	S.F.	.21	.10		.31	.40
0440	Second coat	▼	4485	.004	▼	.12	.10		.22	.30
0450	Smooth wood incl. butt, T&G, beveled, drop or B&B siding									
0460	Brushwork, primer	2 Pord	2325	.007	S.F.	.10	.20		.30	.44
0470	Finish coat, exterior latex		1280	.013		.10	.37		.47	.71
0480	Primer & 1 coat exterior latex		800	.020		.20	.59		.79	1.19
0490	Primer & 2 coats exterior latex		630	.025		.31	.75		1.06	1.57
0500	Roll, primer		2275	.007		.11	.21		.32	.46
0510	Finish coat, exterior latex		2275	.007		.11	.21		.32	.46
0520	Primer & 1 coat exterior latex		1300	.012		.22	.36		.58	.83
0530	Primer & 2 coats exterior latex		975	.016		.33	.48		.81	1.16
0540	Spray, primer		4550	.004		.09	.10		.19	.27
0550	Finish coat, exterior latex		4550	.004		.09	.10		.19	.27
0560	Primer & 1 coat exterior latex		2600	.006		.18	.18		.36	.50
0570	Primer & 2 coats exterior latex		1950	.008		.28	.24		.52	.70
0580	Waterproof sealer, first coat		5230	.003		.12	.09		.21	.28
0590	Second coat	▼	5980	.003	▼	.12	.08		.20	.26
0600	For oil base paint, add					10%				

09 91 13.70 Doors and Windows, Exterior

		Crew	Daily Output	Labor-Hours	Unit	Material	2018 Bare Costs Labor	2018 Bare Costs Equipment	Total	Total Incl O&P
0010	**DOORS AND WINDOWS, EXTERIOR** R099100-10									
0100	Door frames & trim, only									
0110	Brushwork, primer R099100-20	1 Pord	512	.016	L.F.	.06	.46		.52	.82
0120	Finish coat, exterior latex		512	.016		.07	.46		.53	.83
0130	Primer & 1 coat, exterior latex		300	.027		.13	.78		.91	1.44
0140	Primer & 2 coats, exterior latex	▼	265	.030	▼	.21	.89		1.10	1.69
0150	Doors, flush, both sides, incl. frame & trim									
0160	Roll & brush, primer	1 Pord	10	.800	Ea.	4.80	23.50		28.30	44
0170	Finish coat, exterior latex		10	.800		5.40	23.50		28.90	44.50
0180	Primer & 1 coat, exterior latex		7	1.143		10.20	33.50		43.70	66.50
0190	Primer & 2 coats, exterior latex		5	1.600		15.65	47		62.65	94.50
0200	Brushwork, stain, sealer & 2 coats polyurethane	▼	4	2	▼	30	58.50		88.50	130
0210	Doors, French, both sides, 10-15 lite, incl. frame & trim									
0220	Brushwork, primer	1 Pord	6	1.333	Ea.	2.40	39		41.40	67
0230	Finish coat, exterior latex		6	1.333		2.71	39		41.71	67.50
0240	Primer & 1 coat, exterior latex		3	2.667		5.10	78.50		83.60	135
0250	Primer & 2 coats, exterior latex		2	4		7.65	117		124.65	201
0260	Brushwork, stain, sealer & 2 coats polyurethane	▼	2.50	3.200	▼	11.10	94		105.10	167
0270	Doors, louvered, both sides, incl. frame & trim									
0280	Brushwork, primer	1 Pord	7	1.143	Ea.	4.80	33.50		38.30	60.50
0290	Finish coat, exterior latex		7	1.143		5.40	33.50		38.90	61
0300	Primer & 1 coat, exterior latex		4	2		10.20	58.50		68.70	108
0310	Primer & 2 coats, exterior latex		3	2.667		15.35	78.50		93.85	146
0320	Brushwork, stain, sealer & 2 coats polyurethane	▼	4.50	1.778	▼	30	52		82	119
0330	Doors, panel, both sides, incl. frame & trim									
0340	Roll & brush, primer	1 Pord	6	1.333	Ea.	4.80	39		43.80	70
0350	Finish coat, exterior latex		6	1.333		5.40	39		44.40	70.50
0360	Primer & 1 coat, exterior latex		3	2.667		10.20	78.50		88.70	140
0370	Primer & 2 coats, exterior latex		2.50	3.200		15.35	94		109.35	172
0380	Brushwork, stain, sealer & 2 coats polyurethane	▼	3	2.667	▼	30	78.50		108.50	162
0400	Windows, per ext. side, based on 15 S.F.									
0410	1 to 6 lite									
0420	Brushwork, primer	1 Pord	13	.615	Ea.	.95	18.05		19	30.50

For customer support on your Residential Costs with RSMeans data, call 800.448.8182.

513

09 91 Painting

09 91 13 – Exterior Painting

09 91 13.70 Doors and Windows, Exterior

		Crew	Daily Output	Labor-Hours	Unit	Material	2018 Bare Costs Labor	Equipment	Total	Total Incl O&P
0430	Finish coat, exterior latex	1 Pord	13	.615	Ea.	1.07	18.05		19.12	30.50
0440	Primer & 1 coat, exterior latex		8	1		2.02	29.50		31.52	50.50
0450	Primer & 2 coats, exterior latex		6	1.333		3.03	39		42.03	68
0460	Stain, sealer & 1 coat varnish		7	1.143		4.38	33.50		37.88	60
0470	7 to 10 lite									
0480	Brushwork, primer	1 Pord	11	.727	Ea.	.95	21.50		22.45	36
0490	Finish coat, exterior latex		11	.727		1.07	21.50		22.57	36
0500	Primer & 1 coat, exterior latex		7	1.143		2.02	33.50		35.52	57
0510	Primer & 2 coats, exterior latex		5	1.600		3.03	47		50.03	81
0520	Stain, sealer & 1 coat varnish		6	1.333		4.38	39		43.38	69.50
0530	12 lite									
0540	Brushwork, primer	1 Pord	10	.800	Ea.	.95	23.50		24.45	39.50
0550	Finish coat, exterior latex		10	.800		1.07	23.50		24.57	39.50
0560	Primer & 1 coat, exterior latex		6	1.333		2.02	39		41.02	66.50
0570	Primer & 2 coats, exterior latex		5	1.600		3.03	47		50.03	81
0580	Stain, sealer & 1 coat varnish		6	1.333		4.28	39		43.28	69
0590	For oil base paint, add					10%				

09 91 13.80 Trim, Exterior

		Crew	Daily Output	Labor-Hours	Unit	Material	2018 Bare Costs Labor	Equipment	Total	Total Incl O&P
0010	**TRIM, EXTERIOR** R099100-10									
0100	Door frames & trim (see Doors, interior or exterior)									
0110	Fascia, latex paint, one coat coverage									
0120	1" x 4", brushwork	1 Pord	640	.013	L.F.	.02	.37		.39	.62
0130	Roll		1280	.006		.02	.18		.20	.33
0140	Spray		2080	.004		.02	.11		.13	.21
0150	1" x 6" to 1" x 10", brushwork		640	.013		.08	.37		.45	.68
0160	Roll		1230	.007		.08	.19		.27	.40
0170	Spray		2100	.004		.06	.11		.17	.25
0180	1" x 12", brushwork		640	.013		.08	.37		.45	.68
0190	Roll		1050	.008		.08	.22		.30	.46
0200	Spray		2200	.004		.06	.11		.17	.25
0210	Gutters & downspouts, metal, zinc chromate paint									
0220	Brushwork, gutters, 5", first coat	1 Pord	640	.013	L.F.	.40	.37		.77	1.04
0230	Second coat		960	.008		.38	.24		.62	.81
0240	Third coat		1280	.006		.30	.18		.48	.64
0250	Downspouts, 4", first coat		640	.013		.40	.37		.77	1.04
0260	Second coat		960	.008		.38	.24		.62	.81
0270	Third coat		1280	.006		.30	.18		.48	.64
0280	Gutters & downspouts, wood									
0290	Brushwork, gutters, 5", primer	1 Pord	640	.013	L.F.	.06	.37		.43	.67
0300	Finish coat, exterior latex		640	.013		.06	.37		.43	.67
0310	Primer & 1 coat exterior latex		400	.020		.13	.59		.72	1.12
0320	Primer & 2 coats exterior latex		325	.025		.21	.72		.93	1.42
0330	Downspouts, 4", primer		640	.013		.06	.37		.43	.67
0340	Finish coat, exterior latex		640	.013		.06	.37		.43	.67
0350	Primer & 1 coat exterior latex		400	.020		.13	.59		.72	1.12
0360	Primer & 2 coats exterior latex		325	.025		.10	.72		.82	1.30
0370	Molding, exterior, up to 14" wide									
0380	Brushwork, primer	1 Pord	640	.013	L.F.	.08	.37		.45	.68
0390	Finish coat, exterior latex		640	.013		.08	.37		.45	.68
0400	Primer & 1 coat exterior latex		400	.020		.16	.59		.75	1.15
0410	Primer & 2 coats exterior latex		315	.025		.16	.75		.91	1.41
0420	Stain & fill		1050	.008		.12	.22		.34	.51

09 91 Painting

09 91 13 – Exterior Painting

09 91 13.80 Trim, Exterior

	Crew	Daily Output	Labor-Hours	Unit	Material	2018 Bare Costs Labor	Equipment	Total	Total Incl O&P
0430	Shellac	1 Pord	1850	.004	L.F.	.15	.13		.28	.38
0440	Varnish	↓	1275	.006	↓	.11	.18		.29	.42

09 91 13.90 Walls, Masonry (CMU), Exterior

		Crew	Daily Output	Labor-Hours	Unit	Material	Labor	Equipment	Total	Total Incl O&P
0010	**WALLS, MASONRY (CMU), EXTERIOR**									
0360	Concrete masonry units (CMU), smooth surface									
0370	Brushwork, latex, first coat	1 Pord	640	.013	S.F.	.06	.37		.43	.67
0380	Second coat		960	.008		.05	.24		.29	.45
0390	Waterproof sealer, first coat		736	.011		.27	.32		.59	.83
0400	Second coat		1104	.007		.27	.21		.48	.65
0410	Roll, latex, paint, first coat		1465	.005		.07	.16		.23	.34
0420	Second coat		1790	.004		.06	.13		.19	.28
0430	Waterproof sealer, first coat		1680	.005		.27	.14		.41	.53
0440	Second coat		2060	.004		.27	.11		.38	.49
0450	Spray, latex, paint, first coat		1950	.004		.06	.12		.18	.26
0460	Second coat		2600	.003		.05	.09		.14	.20
0470	Waterproof sealer, first coat		2245	.004		.27	.10		.37	.47
0480	Second coat	↓	2990	.003	↓	.27	.08		.35	.43
0490	Concrete masonry unit (CMU), porous									
0500	Brushwork, latex, first coat	1 Pord	640	.013	S.F.	.12	.37		.49	.73
0510	Second coat		960	.008		.06	.24		.30	.47
0520	Waterproof sealer, first coat		736	.011		.27	.32		.59	.83
0530	Second coat		1104	.007		.27	.21		.48	.65
0540	Roll latex, first coat		1465	.005		.09	.16		.25	.36
0550	Second coat		1790	.004		.06	.13		.19	.28
0560	Waterproof sealer, first coat		1680	.005		.27	.14		.41	.53
0570	Second coat		2060	.004		.27	.11		.38	.49
0580	Spray latex, first coat		1950	.004		.07	.12		.19	.27
0590	Second coat		2600	.003		.05	.09		.14	.20
0600	Waterproof sealer, first coat		2245	.004		.27	.10		.37	.47
0610	Second coat	↓	2990	.003	↓	.27	.08		.35	.43

09 91 23 – Interior Painting

09 91 23.20 Cabinets and Casework

		Crew	Daily Output	Labor-Hours	Unit	Material	Labor	Equipment	Total	Total Incl O&P
0010	**CABINETS AND CASEWORK**									
1000	Primer coat, oil base, brushwork	1 Pord	650	.012	S.F.	.07	.36		.43	.66
2000	Paint, oil base, brushwork, 1 coat		650	.012		.12	.36		.48	.72
2500	2 coats		400	.020		.23	.59		.82	1.22
3000	Stain, brushwork, wipe off		650	.012		.10	.36		.46	.70
4000	Shellac, 1 coat, brushwork		650	.012		.13	.36		.49	.73
4500	Varnish, 3 coats, brushwork, sand after 1st coat	↓	325	.025		.27	.72		.99	1.49
5000	For latex paint, deduct				↓	10%				

09 91 23.33 Doors and Windows, Interior Alkyd (Oil Base)

		Crew	Daily Output	Labor-Hours	Unit	Material	Labor	Equipment	Total	Total Incl O&P
0010	**DOORS AND WINDOWS, INTERIOR ALKYD (OIL BASE)**									
0500	Flush door & frame, 3' x 7', oil, primer, brushwork	1 Pord	10	.800	Ea.	3.98	23.50		27.48	43
1000	Paint, 1 coat		10	.800		4.16	23.50		27.66	43
1200	2 coats		6	1.333		4.71	39		43.71	69.50
1400	Stain, brushwork, wipe off		18	.444		2.15	13.05		15.20	24
1600	Shellac, 1 coat, brushwork		25	.320		2.64	9.40		12.04	18.35
1800	Varnish, 3 coats, brushwork, sand after 1st coat		9	.889		5.75	26		31.75	49.50
2000	Panel door & frame, 3' x 7', oil, primer, brushwork		6	1.333		2.50	39		41.50	67.50
2200	Paint, 1 coat		6	1.333		4.16	39		43.16	69
2400	2 coats		3	2.667		10.85	78.50		89.35	141
2600	Stain, brushwork, panel door, 3' x 7', not incl. frame		16	.500		2.15	14.70		16.85	26.50

09 91 23.33 Doors and Windows, Interior Alkyd (Oil Base)

		Daily Output	Labor-Hours	Unit	Material	2018 Bare Costs Labor	Equipment	Total	Total Incl O&P
		Crew							
2800	Shellac, 1 coat, brushwork	1 Pord	22	.364	Ea.	2.64	10.65		13.29	20.50
3000	Varnish, 3 coats, brushwork, sand after 1st coat	▼	7.50	1.067	▼	5.75	31.50		37.25	58
3020	French door, incl. 3' x 7', 6 lites, frame & trim									
3022	Paint, 1 coat, over existing paint	1 Pord	5	1.600	Ea.	8.30	47		55.30	86.50
3024	2 coats, over existing paint		5	1.600		16.15	47		63.15	95.50
3026	Primer & 1 coat		3.50	2.286		13.35	67		80.35	125
3028	Primer & 2 coats		3	2.667		21.50	78.50		100	153
3032	Varnish or polyurethane, 1 coat		5	1.600		7.95	47		54.95	86.50
3034	2 coats, sanding between	▼	3	2.667	▼	15.95	78.50		94.45	147
4400	Windows, including frame and trim, per side									
4600	Colonial type, 6/6 lites, 2' x 3', oil, primer, brushwork	1 Pord	14	.571	Ea.	.39	16.75		17.14	28
5800	Paint, 1 coat		14	.571		.66	16.75		17.41	28
6000	2 coats		9	.889		1.28	26		27.28	44.50
6200	3' x 5' opening, 6/6 lites, primer coat, brushwork		12	.667		.99	19.55		20.54	33
6400	Paint, 1 coat		12	.667		1.64	19.55		21.19	34
6600	2 coats		7	1.143		3.19	33.50		36.69	58.50
6800	4' x 8' opening, 6/6 lites, primer coat, brushwork		8	1		2.11	29.50		31.61	51
7000	Paint, 1 coat		8	1		3.50	29.50		33	52.50
7200	2 coats		5	1.600		6.80	47		53.80	85
8000	Single lite type, 2' x 3', oil base, primer coat, brushwork		33	.242		.39	7.10		7.49	12.15
8200	Paint, 1 coat		33	.242		.66	7.10		7.76	12.40
8400	2 coats		20	.400		1.28	11.75		13.03	20.50
8600	3' x 5' opening, primer coat, brushwork		20	.400		.99	11.75		12.74	20.50
8800	Paint, 1 coat		20	.400		1.64	11.75		13.39	21
8900	2 coats		13	.615		3.19	18.05		21.24	33
9200	4' x 8' opening, primer coat, brushwork		14	.571		2.11	16.75		18.86	30
9400	Paint, 1 coat		14	.571		3.50	16.75		20.25	31.50
9600	2 coats	▼	8	1	▼	6.80	29.50		36.30	56

09 91 23.35 Doors and Windows, Interior Latex

		Crew	Daily Output	Labor-Hours	Unit	Material	2018 Bare Costs Labor	Equipment	Total	Total Incl O&P
0010	**DOORS & WINDOWS, INTERIOR LATEX** R099100-10									
0100	Doors, flush, both sides, incl. frame & trim									
0110	Roll & brush, primer	1 Pord	10	.800	Ea.	4.08	23.50		27.58	43
0120	Finish coat, latex		10	.800		5.55	23.50		29.05	44.50
0130	Primer & 1 coat latex		7	1.143		9.65	33.50		43.15	65.50
0140	Primer & 2 coats latex		5	1.600		14.90	47		61.90	94
0160	Spray, both sides, primer		20	.400		4.30	11.75		16.05	24
0170	Finish coat, latex		20	.400		5.85	11.75		17.60	25.50
0180	Primer & 1 coat latex		11	.727		10.20	21.50		31.70	46
0190	Primer & 2 coats latex	▼	8	1	▼	15.75	29.50		45.25	66
0200	Doors, French, both sides, 10-15 lite, incl. frame & trim									
0210	Roll & brush, primer	1 Pord	6	1.333	Ea.	2.04	39		41.04	67
0220	Finish coat, latex		6	1.333		2.78	39		41.78	67.50
0230	Primer & 1 coat latex		3	2.667		4.82	78.50		83.32	134
0240	Primer & 2 coats latex	▼	2	4	▼	7.45	117		124.45	201
0260	Doors, louvered, both sides, incl. frame & trim									
0270	Roll & brush, primer	1 Pord	7	1.143	Ea.	4.08	33.50		37.58	59.50
0280	Finish coat, latex		7	1.143		5.55	33.50		39.05	61
0290	Primer & 1 coat, latex		4	2		9.40	58.50		67.90	107
0300	Primer & 2 coats, latex		3	2.667		15.20	78.50		93.70	146
0320	Spray, both sides, primer		20	.400		4.30	11.75		16.05	24
0330	Finish coat, latex		20	.400		5.85	11.75		17.60	25.50
0340	Primer & 1 coat, latex	▼	11	.727		10.20	21.50		31.70	46

09 91 23.35 Doors and Windows, Interior Latex

		Crew	Daily Output	Labor- Hours	Unit	Material	2018 Bare Costs Labor	Equipment	Total	Total Incl O&P
0350	Primer & 2 coats, latex	1 Pord	8	1	Ea.	16.10	29.50		45.60	66
0360	Doors, panel, both sides, incl. frame & trim									
0370	Roll & brush, primer	1 Pord	6	1.333	Ea.	4.30	39		43.30	69
0380	Finish coat, latex		6	1.333		5.55	39		44.55	70.50
0390	Primer & 1 coat, latex		3	2.667		9.65	78.50		88.15	140
0400	Primer & 2 coats, latex		2.50	3.200		15.20	94		109.20	172
0420	Spray, both sides, primer		10	.800		4.30	23.50		27.80	43
0430	Finish coat, latex		10	.800		5.85	23.50		29.35	45
0440	Primer & 1 coat, latex		5	1.600		10.20	47		57.20	88.50
0450	Primer & 2 coats, latex		4	2		16.10	58.50		74.60	114
0460	Windows, per interior side, based on 15 S.F.									
0470	1 to 6 lite									
0480	Brushwork, primer	1 Pord	13	.615	Ea.	.81	18.05		18.86	30.50
0490	Finish coat, enamel		13	.615		1.10	18.05		19.15	30.50
0500	Primer & 1 coat enamel		8	1		1.90	29.50		31.40	50.50
0510	Primer & 2 coats enamel		6	1.333		3	39		42	68
0530	7 to 10 lite									
0540	Brushwork, primer	1 Pord	11	.727	Ea.	.81	21.50		22.31	36
0550	Finish coat, enamel		11	.727		1.10	21.50		22.60	36
0560	Primer & 1 coat enamel		7	1.143		1.90	33.50		35.40	57
0570	Primer & 2 coats enamel		5	1.600		3	47		50	81
0590	12 lite									
0600	Brushwork, primer	1 Pord	10	.800	Ea.	.81	23.50		24.31	39.50
0610	Finish coat, enamel		10	.800		1.10	23.50		24.60	39.50
0620	Primer & 1 coat enamel		6	1.333		1.90	39		40.90	66.50
0630	Primer & 2 coats enamel		5	1.600		3	47		50	81
0650	For oil base paint, add					10%				

09 91 23.39 Doors and Windows, Interior Latex, Zero Voc

			Crew	Daily Output	Labor- Hours	Unit	Material	2018 Bare Costs Labor	Equipment	Total	Total Incl O&P
0010	**DOORS & WINDOWS, INTERIOR LATEX, ZERO VOC**										
0100	Doors flush, both sides, incl. frame & trim										
0110	Roll & brush, primer	G	1 Pord	10	.800	Ea.	6.30	23.50		29.80	45.50
0120	Finish coat, latex	G		10	.800		6.95	23.50		30.45	46
0130	Primer & 1 coat latex	G		7	1.143		13.25	33.50		46.75	69.50
0140	Primer & 2 coats latex	G		5	1.600		19.80	47		66.80	99.50
0160	Spray, both sides, primer	G		20	.400		6.65	11.75		18.40	26.50
0170	Finish coat, latex	G		20	.400		7.30	11.75		19.05	27.50
0180	Primer & 1 coat latex	G		11	.727		14.05	21.50		35.55	50.50
0190	Primer & 2 coats latex	G		8	1		21	29.50		50.50	71.50
0200	Doors, French, both sides, 10-15 lite, incl. frame & trim										
0210	Roll & brush, primer	G	1 Pord	6	1.333	Ea.	3.15	39		42.15	68
0220	Finish coat, latex	G		6	1.333		3.48	39		42.48	68.50
0230	Primer & 1 coat latex	G		3	2.667		6.65	78.50		85.15	136
0240	Primer & 2 coats latex	G		2	4		9.90	117		126.90	204
0360	Doors, panel, both sides, incl. frame & trim										
0370	Roll & brush, primer	G	1 Pord	6	1.333	Ea.	6.65	39		45.65	72
0380	Finish coat, latex	G		6	1.333		6.95	39		45.95	72
0390	Primer & 1 coat, latex	G		3	2.667		13.25	78.50		91.75	144
0400	Primer & 2 coats, latex	G		2.50	3.200		20	94		114	177
0420	Spray, both sides, primer	G		10	.800		6.65	23.50		30.15	46
0430	Finish coat, latex	G		10	.800		7.30	23.50		30.80	46.50
0440	Primer & 1 coat, latex	G		5	1.600		14.05	47		61.05	93
0450	Primer & 2 coats, latex	G		4	2		21.50	58.50		80	120

For customer support on your Residential Costs with RSMeans data, call 800.448.8182.

517

09 91 Painting

09 91 23 – Interior Painting

09 91 23.39 Doors and Windows, Interior Latex, Zero Voc

		Crew	Daily Output	Labor-Hours	Unit	Material	2018 Bare Costs Labor	Equipment	Total	Total Incl O&P
0460	Windows, per interior side, based on 15 S.F.									
0470	1 to 6 lite									
0480	Brushwork, primer ☐G	1 Pord	13	.615	Ea.	1.24	18.05		19.29	31
0490	Finish coat, enamel ☐G		13	.615		1.37	18.05		19.42	31
0500	Primer & 1 coat enamel ☐G		8	1		2.62	29.50		32.12	51.50
0510	Primer & 2 coats enamel ☐G		6	1.333		3.99	39		42.99	69

09 91 23.40 Floors, Interior

		Crew	Daily Output	Labor-Hours	Unit	Material	2018 Bare Costs Labor	Equipment	Total	Total Incl O&P
0010	**FLOORS, INTERIOR**									
0100	Concrete paint, latex									
0110	Brushwork									
0120	1st coat	1 Pord	975	.008	S.F.	.16	.24		.40	.57
0130	2nd coat		1150	.007		.10	.20		.30	.45
0140	3rd coat		1300	.006		.08	.18		.26	.39
0150	Roll									
0160	1st coat	1 Pord	2600	.003	S.F.	.21	.09		.30	.38
0170	2nd coat		3250	.002		.12	.07		.19	.26
0180	3rd coat		3900	.002		.09	.06		.15	.20
0190	Spray									
0200	1st coat	1 Pord	2600	.003	S.F.	.18	.09		.27	.35
0210	2nd coat		3250	.002		.10	.07		.17	.23
0220	3rd coat		3900	.002		.08	.06		.14	.19

09 91 23.52 Miscellaneous, Interior

		Crew	Daily Output	Labor-Hours	Unit	Material	2018 Bare Costs Labor	Equipment	Total	Total Incl O&P
0010	**MISCELLANEOUS, INTERIOR**									
2400	Floors, conc./wood, oil base, primer/sealer coat, brushwork	2 Pord	1950	.008	S.F.	.09	.24		.33	.50
2450	Roller		5200	.003		.09	.09		.18	.25
2600	Spray		6000	.003		.09	.08		.17	.23
2650	Paint 1 coat, brushwork		1950	.008		.11	.24		.35	.52
2800	Roller		5200	.003		.11	.09		.20	.27
2850	Spray		6000	.003		.12	.08		.20	.26
3000	Stain, wood floor, brushwork, 1 coat		4550	.004		.10	.10		.20	.28
3200	Roller		5200	.003		.11	.09		.20	.27
3250	Spray		6000	.003		.11	.08		.19	.25
3400	Varnish, wood floor, brushwork		4550	.004		.09	.10		.19	.27
3450	Roller		5200	.003		.10	.09		.19	.26
3600	Spray		6000	.003		.10	.08		.18	.24
3800	Grilles, per side, oil base, primer coat, brushwork	1 Pord	520	.015		.13	.45		.58	.88
3850	Spray		1140	.007		.14	.21		.35	.49
3880	Paint 1 coat, brushwork		520	.015		.22	.45		.67	.98
3900	Spray		1140	.007		.24	.21		.45	.61
3920	Paint 2 coats, brushwork		325	.025		.43	.72		1.15	1.66
3940	Spray		650	.012		.49	.36		.85	1.13
4500	Louvers, 1 side, primer, brushwork		524	.015		.09	.45		.54	.84
4520	Paint 1 coat, brushwork		520	.015		.10	.45		.55	.85
4530	Spray		1140	.007		.11	.21		.32	.46
4540	Paint 2 coats, brushwork		325	.025		.19	.72		.91	1.40
4550	Spray		650	.012		.21	.36		.57	.83
4560	Paint 3 coats, brushwork		270	.030		.29	.87		1.16	1.75
4570	Spray		500	.016		.32	.47		.79	1.12
5000	Pipe, 1"-4" diameter, primer or sealer coat, oil base, brushwork	2 Pord	1250	.013	L.F.	.09	.38		.47	.72
5100	Spray		2165	.007		.09	.22		.31	.46
5200	Paint 1 coat, brushwork		1250	.013		.12	.38		.50	.75
5300	Spray		2165	.007		.10	.22		.32	.47

09 91 Painting

09 91 23 – Interior Painting

09 91 23.52 Miscellaneous, Interior

		Crew	Daily Output	Labor-Hours	Unit	Material	2018 Bare Costs Labor	Equipment	Total	Total Incl O&P
5350	Paint 2 coats, brushwork	2 Pord	775	.021	L.F.	.20	.61		.81	1.22
5400	Spray		1240	.013		.23	.38		.61	.87
5450	5"-8" diameter, primer or sealer coat, brushwork		620	.026		.19	.76		.95	1.46
5500	Spray		1085	.015		.31	.43		.74	1.05
5550	Paint 1 coat, brushwork		620	.026		.31	.76		1.07	1.59
5600	Spray		1085	.015		.34	.43		.77	1.09
5650	Paint 2 coats, brushwork		385	.042		.41	1.22		1.63	2.46
5700	Spray	↓	620	.026	↓	.45	.76		1.21	1.75
6600	Radiators, per side, primer, brushwork	1 Pord	520	.015	S.F.	.09	.45		.54	.84
6620	Paint, 1 coat		520	.015		.08	.45		.53	.83
6640	2 coats		340	.024		.19	.69		.88	1.35
6660	3 coats	↓	283	.028	↓	.29	.83		1.12	1.69
7000	Trim, wood, incl. puttying, under 6" wide									
7200	Primer coat, oil base, brushwork	1 Pord	650	.012	L.F.	.03	.36		.39	.63
7250	Paint, 1 coat, brushwork		650	.012		.05	.36		.41	.65
7400	2 coats		400	.020		.11	.59		.70	1.09
7450	3 coats		325	.025		.16	.72		.88	1.36
7500	Over 6" wide, primer coat, brushwork		650	.012		.07	.36		.43	.66
7550	Paint, 1 coat, brushwork		650	.012		.11	.36		.47	.71
7600	2 coats		400	.020		.21	.59		.80	1.20
7650	3 coats		325	.025	↓	.32	.72		1.04	1.54
8000	Cornice, simple design, primer coat, oil base, brushwork		650	.012	S.F.	.07	.36		.43	.66
8250	Paint, 1 coat		650	.012		.11	.36		.47	.71
8300	2 coats		400	.020		.21	.59		.80	1.20
8350	Ornate design, primer coat		350	.023		.07	.67		.74	1.17
8400	Paint, 1 coat		350	.023		.11	.67		.78	1.22
8450	2 coats		400	.020		.21	.59		.80	1.20
8600	Balustrades, primer coat, oil base, brushwork		520	.015		.07	.45		.52	.81
8650	Paint, 1 coat		520	.015		.11	.45		.56	.86
8700	2 coats		325	.025		.21	.72		.93	1.42
8900	Trusses and wood frames, primer coat, oil base, brushwork		800	.010		.07	.29		.36	.55
8950	Spray		1200	.007		.07	.20		.27	.40
9000	Paint 1 coat, brushwork		750	.011		.11	.31		.42	.64
9200	Spray		1200	.007		.12	.20		.32	.45
9220	Paint 2 coats, brushwork		500	.016		.21	.47		.68	1
9240	Spray		600	.013		.24	.39		.63	.90
9260	Stain, brushwork, wipe off		600	.013		.10	.39		.49	.75
9280	Varnish, 3 coats, brushwork	↓	275	.029		.27	.85		1.12	1.71
9350	For latex paint, deduct				↓	10%				

09 91 23.72 Walls and Ceilings, Interior

		Crew	Daily Output	Labor-Hours	Unit	Material	2018 Bare Costs Labor	Equipment	Total	Total Incl O&P
0010	**WALLS AND CEILINGS, INTERIOR**									
0100	Concrete, drywall or plaster, latex, primer or sealer coat									
0200	Smooth finish, brushwork	1 Pord	1150	.007	S.F.	.06	.20		.26	.41
0240	Roller		1350	.006		.06	.17		.23	.36
0280	Spray		2750	.003		.05	.09		.14	.20
0300	Sand finish, brushwork		975	.008		.06	.24		.30	.47
0340	Roller		1150	.007		.06	.20		.26	.41
0380	Spray		2275	.004		.05	.10		.15	.23
0400	Paint 1 coat, smooth finish, brushwork		1200	.007		.07	.20		.27	.40
0440	Roller		1300	.006		.07	.18		.25	.38
0480	Spray		2275	.004		.06	.10		.16	.24
0500	Sand finish, brushwork		1050	.008		.07	.22		.29	.44

For customer support on your Residential Costs with RSMeans data, call 800.448.8182.

519

09 91 23.72 Walls and Ceilings, Interior	Crew	Daily Output	Labor-Hours	Unit	Material	2018 Bare Costs Labor	Equipment	Total	Total Incl O&P
0540	Roller	1 Pord	1600	.005	S.F.	.07	.15		.22	.32
0580	Spray		2100	.004		.02	.11		.13	.21
0800	Paint 2 coats, smooth finish, brushwork		680	.012		.14	.35		.49	.73
0840	Roller		800	.010		.14	.29		.43	.64
0880	Spray		1625	.005		.13	.14		.27	.39
0900	Sand finish, brushwork		605	.013		.14	.39		.53	.80
0940	Roller		1020	.008		.14	.23		.37	.54
0980	Spray		1700	.005		.13	.14		.27	.38
1200	Paint 3 coats, smooth finish, brushwork		510	.016		.22	.46		.68	1
1240	Roller		650	.012		.22	.36		.58	.83
1280	Spray		850	.009		.20	.28		.48	.67
1300	Sand finish, brushwork		454	.018		.32	.52		.84	1.20
1340	Roller		680	.012		.34	.35		.69	.94
1380	Spray		1133	.007		.29	.21		.50	.66
1600	Glaze coating, 2 coats, spray, clear		1200	.007		.56	.20		.76	.94
1640	Multicolor		1200	.007		.85	.20		1.05	1.26
1660	Painting walls, complete, including surface prep, primer &									
1670	2 coats finish, on drywall or plaster, with roller	1 Pord	325	.025	S.F.	.21	.72		.93	1.42
1700	For oil base paint, add					10%				
1800	For ceiling installations, add						25%			
2000	Masonry or concrete block, primer/sealer, latex paint									
2100	Primer, smooth finish, brushwork	1 Pord	1000	.008	S.F.	.15	.23		.38	.56
2110	Roller		1150	.007		.11	.20		.31	.46
2180	Spray		2400	.003		.10	.10		.20	.27
2200	Sand finish, brushwork		850	.009		.11	.28		.39	.57
2210	Roller		975	.008		.11	.24		.35	.52
2280	Spray		2050	.004		.10	.11		.21	.30
2400	Finish coat, smooth finish, brush		1100	.007		.09	.21		.30	.44
2410	Roller		1300	.006		.09	.18		.27	.39
2480	Spray		2400	.003		.08	.10		.18	.24
2500	Sand finish, brushwork		950	.008		.09	.25		.34	.50
2510	Roller		1090	.007		.09	.22		.31	.44
2580	Spray		2040	.004		.08	.12		.20	.27
2800	Primer plus one finish coat, smooth brush		525	.015		.30	.45		.75	1.07
2810	Roller		615	.013		.19	.38		.57	.84
2880	Spray		1200	.007		.17	.20		.37	.51
2900	Sand finish, brushwork		450	.018		.19	.52		.71	1.07
2910	Roller		515	.016		.19	.46		.65	.96
2980	Spray		1025	.008		.17	.23		.40	.57
3200	Primer plus 2 finish coats, smooth, brush		355	.023		.28	.66		.94	1.40
3210	Roller		415	.019		.28	.57		.85	1.24
3280	Spray		800	.010		.25	.29		.54	.75
3300	Sand finish, brushwork		305	.026		.28	.77		1.05	1.58
3310	Roller		350	.023		.28	.67		.95	1.41
3380	Spray		675	.012		.25	.35		.60	.84
3600	Glaze coating, 3 coats, spray, clear		900	.009		.80	.26		1.06	1.31
3620	Multicolor		900	.009		1.05	.26		1.31	1.58
4000	Block filler, 1 coat, brushwork		425	.019		.13	.55		.68	1.05
4100	Silicone, water repellent, 2 coats, spray		2000	.004		.45	.12		.57	.68
4120	For oil base paint, add					10%				
8200	For work 8'-15' H, add						10%			
8300	For work over 15' H, add						20%			
8400	For light textured surfaces, add						10%			

09 91 23 – Interior Painting

09 91 23.72 Walls and Ceilings, Interior		Crew	Daily Output	Labor-Hours	Unit	Material	2018 Bare Costs Labor	2018 Bare Costs Equipment	Total	Total Incl O&P
8410	Heavy textured, add				S.F.		25%			

09 91 23.74 Walls and Ceilings, Interior, Zero VOC Latex

		Crew	Daily Output	Labor-Hours	Unit	Material	Labor	Equipment	Total	Total Incl O&P
0010	**WALLS AND CEILINGS, INTERIOR, ZERO VOC LATEX**									
0100	Concrete, dry wall or plaster, latex, primer or sealer coat									
0200	Smooth finish, brushwork	G 1 Pord	1150	.007	S.F.	.08	.20		.28	.43
0240	Roller	G	1350	.006		.08	.17		.25	.38
0280	Spray	G	2750	.003		.06	.09		.15	.21
0300	Sand finish, brushwork	G	975	.008		.08	.24		.32	.49
0340	Roller	G	1150	.007		.09	.20		.29	.44
0380	Spray	G	2275	.004		.07	.10		.17	.24
0400	Paint 1 coat, smooth finish, brushwork	G	1200	.007		.09	.20		.29	.42
0440	Roller	G	1300	.006		.09	.18		.27	.40
0480	Spray	G	2275	.004		.08	.10		.18	.26
0500	Sand finish, brushwork	G	1050	.008		.09	.22		.31	.47
0540	Roller	G	1600	.005		.09	.15		.24	.34
0580	Spray	G	2100	.004		.08	.11		.19	.27
0800	Paint 2 coats, smooth finish, brushwork	G	680	.012		.18	.35		.53	.77
0840	Roller	G	800	.010		.19	.29		.48	.68
0880	Spray	G	1625	.005		.16	.14		.30	.42
0900	Sand finish, brushwork	G	605	.013		.18	.39		.57	.83
0940	Roller	G	1020	.008		.19	.23		.42	.58
0980	Spray	G	1700	.005		.16	.14		.30	.41
1200	Paint 3 coats, smooth finish, brushwork	G	510	.016		.26	.46		.72	1.05
1240	Roller	G	650	.012		.28	.36		.64	.90
1280	Spray	G	850	.009		.24	.28		.52	.71
1800	For ceiling installations, add	G					25%			
8200	For work 8'- 15' H, add						10%			
8300	For work over 15' H, add						20%			

09 91 23.75 Dry Fall Painting

		Crew	Daily Output	Labor-Hours	Unit	Material	Labor	Equipment	Total	Total Incl O&P
0010	**DRY FALL PAINTING**									
0100	Sprayed on walls, gypsum board or plaster									
0220	One coat	1 Pord	2600	.003	S.F.	.09	.09		.18	.24
0250	Two coats		1560	.005		.17	.15		.32	.44
0280	Concrete or textured plaster, one coat		1560	.005		.09	.15		.24	.34
0310	Two coats		1300	.006		.17	.18		.35	.49
0340	Concrete block, one coat		1560	.005		.09	.15		.24	.34
0370	Two coats		1300	.006		.17	.18		.35	.49
0400	Wood, one coat		877	.009		.09	.27		.36	.53
0430	Two coats		650	.012		.17	.36		.53	.78
0440	On ceilings, gypsum board or plaster									
0470	One coat	1 Pord	1560	.005	S.F.	.09	.15		.24	.34
0500	Two coats		1300	.006		.17	.18		.35	.49
0530	Concrete or textured plaster, one coat		1560	.005		.09	.15		.24	.34
0560	Two coats		1300	.006		.17	.18		.35	.49
0570	Structural steel, bar joists or metal deck, one coat		1560	.005		.09	.15		.24	.34
0580	Two coats		1040	.008		.17	.23		.40	.56

09 93 Staining and Transparent Finishing

09 93 23 – Interior Staining and Finishing

09 93 23.10 Varnish		Crew	Daily Output	Labor-Hours	Unit	Material	2018 Bare Costs Labor	Equipment	Total	Total Incl O&P
0010	**VARNISH**									
0012	1 coat + sealer, on wood trim, brush, no sanding included	1 Pord	400	.020	S.F.	.07	.59		.66	1.05
0020	1 coat + sealer, on wood trim, brush, no sanding included, no VOC		400	.020		.22	.59		.81	1.21
0100	Hardwood floors, 2 coats, no sanding included, roller	↓	1890	.004	↓	.15	.12		.27	.37

09 96 High-Performance Coatings

09 96 56 – Epoxy Coatings

09 96 56.20 Wall Coatings

09 96 56.20 Wall Coatings		Crew	Daily Output	Labor-Hours	Unit	Material	2018 Bare Costs Labor	Equipment	Total	Total Incl O&P
0010	**WALL COATINGS**									
0100	Acrylic glazed coatings, matte	1 Pord	525	.015	S.F.	.35	.45		.80	1.13
0200	Gloss		305	.026		.73	.77		1.50	2.07
0300	Epoxy coatings, solvent based		525	.015		.45	.45		.90	1.24
0400	Water based		170	.047		.32	1.38		1.70	2.62
0600	Exposed aggregate, troweled on, 1/16" to 1/4", solvent based		235	.034		.70	1		1.70	2.41
0700	Water based (epoxy or polyacrylate)		130	.062		1.50	1.81		3.31	4.62
0900	1/2" to 5/8" aggregate, solvent based		130	.062		1.35	1.81		3.16	4.46
1000	Water based		80	.100		2.35	2.94		5.29	7.40
1200	1" aggregate size, solvent based		90	.089		2.39	2.61		5	6.90
1300	Water based		55	.145		3.64	4.27		7.91	11.05
1500	Exposed aggregate, sprayed on, 1/8" aggregate, solvent based		295	.027		.57	.80		1.37	1.94
1600	Water based	↓	145	.055	↓	1.18	1.62		2.80	3.96

09 97 Special Coatings

09 97 35 – Dry Erase Coatings

09 97 35.10 Dry Erase Coatings

09 97 35.10 Dry Erase Coatings		Crew	Daily Output	Labor-Hours	Unit	Material	2018 Bare Costs Labor	Equipment	Total	Total Incl O&P
0010	**DRY ERASE COATINGS**									
0020	Dry erase coatings, clear, roller applied	1 Pord	1325	.006	S.F.	2.08	.18		2.26	2.58

For customer support on your Residential Costs with RSMeans data, call 800.448.8182.

Estimating Tips
General

- The items in this division are usually priced per square foot or each.

- Many items in Division 10 require some type of support system or special anchors that are not usually furnished with the item. The required anchors must be added to the estimate in the appropriate division.

- Some items in Division 10, such as lockers, may require assembly before installation. Verify the amount of assembly required. Assembly can often exceed installation time.

10 20 00 Interior Specialties

- Support angles and blocking are not included in the installation of toilet compartments, shower/dressing compartments, or cubicles. Appropriate line items from Division 5 or 6 may need to be added to support the installations.

- Toilet partitions are priced by the stall. A stall consists of a side wall, pilaster, and door with hardware. Toilet tissue holders and grab bars are extra.

- The required acoustical rating of a folding partition can have a significant impact on costs. Verify the sound transmission coefficient rating of the panel priced against the specification requirements.

- Grab bar installation does not include supplemental blocking or backing to support the required load. When grab bars are installed at an existing facility, provisions must be made to attach the grab bars to a solid structure.

Reference Numbers

Reference numbers are shown at the beginning of some major classifications. These numbers refer to related items in the Reference Section. The reference information may be an estimating procedure, an alternate pricing method, or technical information.

Note: Not all subdivisions listed here necessarily appear. ■

Did you know?

RSMeans data is available through our online application with 24/7 access:

■ Search for unit prices by keyword

■ Leverage the most up-to-date data

■ Build and export estimates

Try it free for 30 days!
www.rsmeans.com/2018freetrial

10 28 Toilet, Bath, and Laundry Accessories

10 28 13 – Toilet Accessories

10 28 13.13 Commercial Toilet Accessories

		Crew	Daily Output	Labor-Hours	Unit	Material	2018 Bare Costs Labor	Equipment	Total	Total Incl O&P
0010	**COMMERCIAL TOILET ACCESSORIES**									
0200	Curtain rod, stainless steel, 5' long, 1" diameter	1 Carp	13	.615	Ea.	27.50	21.50		49	66
0300	1-1/4" diameter		13	.615		30	21.50		51.50	69
0800	Grab bar, straight, 1-1/4" diameter, stainless steel, 18" long		24	.333		30	11.65		41.65	52.50
1100	36" long		20	.400		35	14		49	61.50
1105	42" long		20	.400		36.50	14		50.50	63
1120	Corner, 36" long		20	.400		92.50	14		106.50	125
3000	Mirror, with stainless steel 3/4" square frame, 18" x 24"		20	.400		47	14		61	75
3100	36" x 24"		15	.533		123	18.65		141.65	166
3300	72" x 24"		6	1.333		282	46.50		328.50	390
4300	Robe hook, single, regular		96	.083		19.45	2.91		22.36	26.50
4400	Heavy duty, concealed mounting		56	.143		22.50	4.99		27.49	33.50
6400	Towel bar, stainless steel, 18" long		23	.348		42.50	12.15		54.65	67
6500	30" long		21	.381		52	13.30		65.30	79
7400	Tumbler holder, for tumbler only		30	.267		25	9.30		34.30	43
7410	Tumbler holder, recessed		20	.400		8	14		22	32
7500	Soap, tumbler & toothbrush		30	.267		19.60	9.30		28.90	37
7510	Tumbler & toothbrush holder		20	.400		13.45	14		27.45	38

10 28 16 – Bath Accessories

10 28 16.20 Medicine Cabinets

		Crew	Daily Output	Labor-Hours	Unit	Material	2018 Bare Costs Labor	Equipment	Total	Total Incl O&P
0010	**MEDICINE CABINETS**									
0020	With mirror, sst frame, 16" x 22", unlighted	1 Carp	14	.571	Ea.	99	19.95		118.95	142
0100	Wood frame		14	.571		136	19.95		155.95	183
0300	Sliding mirror doors, 20" x 16" x 4-3/4", unlighted		7	1.143		124	40		164	204
0400	24" x 19" x 8-1/2", lighted		5	1.600		206	56		262	320
0600	Triple door, 30" x 32", unlighted, plywood body		7	1.143		350	40		390	450
0700	Steel body		7	1.143		375	40		415	475
0900	Oak door, wood body, beveled mirror, single door		7	1.143		175	40		215	259
1000	Double door		6	1.333		370	46.50		416.50	485

10 28 19 – Tub and Shower Enclosures

10 28 19.10 Partitions, Shower

		Crew	Daily Output	Labor-Hours	Unit	Material	2018 Bare Costs Labor	Equipment	Total	Total Incl O&P
0010	**PARTITIONS, SHOWER** floor mounted, no plumbing									
0400	Cabinet, one piece, fiberglass, 32" x 32"	2 Carp	5	3.200	Ea.	625	112		737	875
0420	36" x 36"		5	3.200		695	112		807	950
0440	36" x 48"		5	3.200		1,400	112		1,512	1,700
0460	Acrylic, 32" x 32"		5	3.200		360	112		472	580
0480	36" x 36"		5	3.200		1,075	112		1,187	1,350
0500	36" x 48"		5	3.200		1,300	112		1,412	1,600
0520	Shower door for above, clear plastic, 24" wide	1 Carp	8	1		182	35		217	258
0540	28" wide		8	1		249	35		284	330
0560	Tempered glass, 24" wide		8	1		239	35		274	320
0580	28" wide		8	1		281	35		316	370
2400	Glass stalls, with doors, no receptors, chrome on brass	2 Shee	3	5.333		1,650	204		1,854	2,150
2700	Anodized aluminum	"	4	4		1,325	153		1,478	1,700
3200	Receptors, precast terrazzo, 32" x 32"	2 Marb	14	1.143		355	40		395	455
3300	48" x 34"		9.50	1.684		470	58.50		528.50	620
3500	Plastic, simulated terrazzo receptor, 32" x 32"		14	1.143		167	40		207	251
3600	32" x 48"		12	1.333		278	46.50		324.50	385
3800	Precast concrete, colors, 32" x 32"		14	1.143		251	40		291	345
3900	48" x 48"		8	2		273	69.50		342.50	415
4100	Shower doors, economy plastic, 24" wide	1 Shee	9	.889		137	34		171	208
4200	Tempered glass door, economy		8	1		293	38.50		331.50	385

10 28 Toilet, Bath, and Laundry Accessories

10 28 19 – Tub and Shower Enclosures

10 28 19.10 Partitions, Shower	Crew	Daily Output	Labor-Hours	Unit	Material	2018 Bare Costs Labor	Equipment	Total	Total Incl O&P
4400	Folding, tempered glass, aluminum frame	1 Shee	6	1.333	Ea.	450	51		501	580
4700	Deluxe, tempered glass, chrome on brass frame, 42" to 44"		8	1		440	38.50		478.50	545
4800	39" to 48" wide		1	8		670	305		975	1,250
4850	On anodized aluminum frame, obscure glass		2	4		580	153		733	895
4900	Clear glass		1	8		670	305		975	1,250
5100	Shower enclosure, tempered glass, anodized alum. frame									
5120	2 panel & door, corner unit, 32" x 32"	1 Shee	2	4	Ea.	965	153		1,118	1,325
5140	Neo-angle corner unit, 16" x 24" x 16"	"	2	4		1,125	153		1,278	1,500
5200	Shower surround, 3 wall, polypropylene, 32" x 32"	1 Carp	4	2		620	70		690	795
5220	PVC, 32" x 32"		4	2		395	70		465	550
5240	Fiberglass		4	2		400	70		470	555
5250	2 wall, polypropylene, 32" x 32"		4	2		310	70		380	460
5270	PVC		4	2		380	70		450	530
5290	Fiberglass		4	2		380	70		450	535
5300	Tub doors, tempered glass & frame, obscure glass	1 Shee	8	1		224	38.50		262.50	310
5400	Clear glass		6	1.333		520	51		571	660
5600	Chrome plated, brass frame, obscure glass		8	1		295	38.50		333.50	390
5700	Clear glass		6	1.333		725	51		776	880
5900	Tub/shower enclosure, temp. glass, alum. frame, obscure glass		2	4		400	153		553	695
6200	Clear glass		1.50	5.333		835	204		1,039	1,250
6500	On chrome-plated brass frame, obscure glass		2	4		550	153		703	860
6600	Clear glass		1.50	5.333		1,175	204		1,379	1,650
6800	Tub surround, 3 wall, polypropylene	1 Carp	4	2		252	70		322	395
6900	PVC		4	2		375	70		445	525
7000	Fiberglass, obscure glass		4	2		390	70		460	545
7100	Clear glass		3	2.667		665	93		758	885

10 28 23 – Laundry Accessories

10 28 23.13 Built-In Ironing Boards

		Crew	Daily Output	Labor-Hours	Unit	Material	Labor	Equipment	Total	Total Incl O&P
0010	**BUILT-IN IRONING BOARDS**									
0020	Including cabinet, board & light, 42"	1 Carp	2	4	Ea.	445	140		585	720

10 31 Manufactured Fireplaces

10 31 13 – Manufactured Fireplace Chimneys

10 31 13.10 Fireplace Chimneys

		Crew	Daily Output	Labor-Hours	Unit	Material	Labor	Equipment	Total	Total Incl O&P
0010	**FIREPLACE CHIMNEYS**									
0500	Chimney dbl. wall, all stainless, over 8'-6", 7" diam., add to fireplace	1 Carp	33	.242	V.L.F.	87.50	8.45		95.95	110
0600	10" diameter, add to fireplace		32	.250		112	8.75		120.75	139
0700	12" diameter, add to fireplace		31	.258		172	9		181	205
0800	14" diameter, add to fireplace		30	.267		227	9.30		236.30	266
1000	Simulated brick chimney top, 4' high, 16" x 16"		10	.800	Ea.	455	28		483	550
1100	24" x 24"		7	1.143	"	560	40		600	680

10 31 13.20 Chimney Accessories

		Crew	Daily Output	Labor-Hours	Unit	Material	Labor	Equipment	Total	Total Incl O&P
0010	**CHIMNEY ACCESSORIES**									
0020	Chimney screens, galv., 13" x 13" flue	1 Bric	8	1	Ea.	58.50	34.50		93	122
0050	24" x 24" flue		5	1.600		124	55.50		179.50	230
0200	Stainless steel, 13" x 13" flue		8	1		97.50	34.50		132	165
0250	20" x 20" flue		5	1.600		152	55.50		207.50	260
2400	Squirrel and bird screens, galvanized, 8" x 8" flue		16	.500		56	17.35		73.35	90.50
2450	13" x 13" flue		12	.667		66.50	23		89.50	112

For customer support on your Residential Costs with RSMeans data, call 800.448.8182.

525

10 31 Manufactured Fireplaces

10 31 16 – Manufactured Fireplace Forms

10 31 16.10 Fireplace Forms	Crew	Daily Output	Labor-Hours	Unit	Material	2018 Bare Costs Labor	Equipment	Total	Total Incl O&P
0010 **FIREPLACE FORMS**									
1800 Fireplace forms, no accessories, 32" opening	1 Bric	3	2.667	Ea.	745	92.50		837.50	975
1900 36" opening		2.50	3.200		950	111		1,061	1,225
2000 40" opening		2	4		1,275	139		1,414	1,625
2100 78" opening	↓	1.50	5.333	↓	1,850	185		2,035	2,325

10 31 23 – Prefabricated Fireplaces

10 31 23.10 Fireplace, Prefabricated

	Crew	Daily Output	Labor-Hours	Unit	Material	2018 Bare Costs Labor	Equipment	Total	Total Incl O&P
0010 **FIREPLACE, PREFABRICATED**, free standing or wall hung									
0100 With hood & screen, painted	1 Carp	1.30	6.154	Ea.	1,625	215		1,840	2,125
0150 Average		1	8		1,825	280		2,105	2,475
0200 Stainless steel		.90	8.889	↓	3,250	310		3,560	4,100
1500 Simulated logs, gas fired, 40,000 BTU, 2' long, manual safety pilot		7	1.143	Set	545	40		585	665
1600 Adjustable flame remote pilot		6	1.333		1,275	46.50		1,321.50	1,475
1700 Electric, 1,500 BTU, 1'-6" long, incandescent flame		7	1.143		260	40		300	355
1800 1,500 BTU, LED flame		6	1.333	↓	345	46.50		391.50	460
2000 Fireplace, built-in, 36" hearth, radiant		1.30	6.154	Ea.	720	215		935	1,150
2100 Recirculating, small fan		1	8		925	280		1,205	1,500
2150 Large fan		.90	8.889		2,075	310		2,385	2,825
2200 42" hearth, radiant		1.20	6.667		970	233		1,203	1,450
2300 Recirculating, small fan		.90	8.889		1,250	310		1,560	1,925
2350 Large fan		.80	10		1,450	350		1,800	2,175
2400 48" hearth, radiant		1.10	7.273		2,350	254		2,604	3,025
2500 Recirculating, small fan		.80	10		2,475	350		2,825	3,300
2550 Large fan		.70	11.429		2,575	400		2,975	3,500
3000 See through, including doors		.80	10		2,375	350		2,725	3,175
3200 Corner (2 wall)	↓	1	8	↓	3,150	280		3,430	3,950

10 32 Fireplace Specialties

10 32 13 – Fireplace Dampers

10 32 13.10 Dampers

	Crew	Daily Output	Labor-Hours	Unit	Material	2018 Bare Costs Labor	Equipment	Total	Total Incl O&P
0010 **DAMPERS**									
0800 Damper, rotary control, steel, 30" opening	1 Bric	6	1.333	Ea.	119	46.50		165.50	209
0850 Cast iron, 30" opening		6	1.333		125	46.50		171.50	215
1200 Steel plate, poker control, 60" opening		8	1		320	34.50		354.50	415
1250 84" opening, special order		5	1.600		585	55.50		640.50	740
1400 "Universal" type, chain operated, 32" x 20" opening		8	1		250	34.50		284.50	335
1450 48" x 24" opening	↓	5	1.600	↓	375	55.50		430.50	505

10 32 23 – Fireplace Doors

10 32 23.10 Doors

	Crew	Daily Output	Labor-Hours	Unit	Material	2018 Bare Costs Labor	Equipment	Total	Total Incl O&P
0010 **DOORS**									
0400 Cleanout doors and frames, cast iron, 8" x 8"	1 Bric	12	.667	Ea.	54.50	23		77.50	98.50
0450 12" x 12"		10	.800		89	28		117	145
0500 18" x 24"		8	1		150	34.50		184.50	223
0550 Cast iron frame, steel door, 24" x 30"		5	1.600		315	55.50		370.50	440
1600 Dutch oven door and frame, cast iron, 12" x 15" opening		13	.615		131	21.50		152.50	181
1650 Copper plated, 12" x 15" opening	↓	13	.615	↓	257	21.50		278.50	320

10 35 Stoves

10 35 13 – Heating Stoves

10 35 13.10 Wood Burning Stoves	Crew	Daily Output	Labor-Hours	Unit	Material	2018 Bare Costs Labor	Equipment	Total	Total Incl O&P
0010 **WOOD BURNING STOVES**									
0015 Cast iron, less than 1,500 S.F.	2 Carp	1.30	12.308	Ea.	1,350	430		1,780	2,200
0020 1,500 to 2,000 S.F.		1	16		2,075	560		2,635	3,200
0030 greater than 2,000 S.F.		.80	20		2,800	700		3,500	4,225
0050 For gas log lighter, add					47			47	52

10 44 Fire Protection Specialties

10 44 16 – Fire Extinguishers

10 44 16.13 Portable Fire Extinguishers	Crew	Daily Output	Labor-Hours	Unit	Material	Labor	Equipment	Total	Total Incl O&P
0010 **PORTABLE FIRE EXTINGUISHERS**									
0140 CO_2, with hose and "H" horn, 10 lb.				Ea.	300			300	330
1000 Dry chemical, pressurized									
1040 Standard type, portable, painted, 2-1/2 lb.				Ea.	41			41	45
1080 10 lb.					85.50			85.50	94
1100 20 lb.					138			138	152
1120 30 lb.					425			425	465
2000 ABC all purpose type, portable, 2-1/2 lb.					23.50			23.50	26
2080 9-1/2 lb.					51.50			51.50	56.50

10 55 Postal Specialties

10 55 23 – Mail Boxes

10 55 23.10 Mail Boxes	Crew	Daily Output	Labor-Hours	Unit	Material	Labor	Equipment	Total	Total Incl O&P
0011 **MAIL BOXES**									
1900 Letter slot, residential	1 Carp	20	.400	Ea.	80.50	14		94.50	112
2400 Residential, galv. steel, small 20" x 7" x 9"	1 Clab	16	.500		203	13.35		216.35	245
2410 With galv. steel post, 54" long		6	1.333		250	35.50		285.50	335
2420 Large, 24" x 12" x 15"		16	.500		200	13.35		213.35	242
2430 With galv. steel post, 54" long		6	1.333		243	35.50		278.50	325
2440 Decorative, polyethylene, 22" x 10" x 10"		16	.500		60	13.35		73.35	88
2450 With alum. post, decorative, 54" long		6	1.333		200	35.50		235.50	279

10 56 Storage Assemblies

10 56 13 – Metal Storage Shelving

10 56 13.10 Shelving	Crew	Daily Output	Labor-Hours	Unit	Material	Labor	Equipment	Total	Total Incl O&P
0010 **SHELVING**									
0020 Metal, industrial, cross-braced, 3' W, 12" D	1 Sswk	175	.046	SF Shlf	8.10	1.77		9.87	12.05
0100 24" D		330	.024		6.10	.94		7.04	8.35
2200 Wide span, 1600 lb. capacity per shelf, 6' W, 24" D		380	.021		7.15	.82		7.97	9.35
2400 36" D		440	.018		6.10	.71		6.81	7.95
3000 Residential, vinyl covered wire, wardrobe, 12" D	1 Carp	195	.041	L.F.	15.25	1.43		16.68	19.15
3100 16" D		195	.041		11.60	1.43		13.03	15.15
3200 Standard, 6" D		195	.041		4.61	1.43		6.04	7.45
3300 9" D		195	.041		6.60	1.43		8.03	9.65
3400 12" D		195	.041		7.90	1.43		9.33	11.10
3500 16" D		195	.041		18	1.43		19.43	22
3600 20" D		195	.041		21.50	1.43		22.93	26
3700 Support bracket		80	.100	Ea.	7	3.50		10.50	13.50

For customer support on your Residential Costs with RSMeans data, call 800.448.8182.

527

10 57 Wardrobe and Closet Specialties

10 57 23 – Closet and Utility Shelving

10 57 23.19 Wood Closet and Utility Shelving	Crew	Daily Output	Labor-Hours	Unit	Material	2018 Bare Costs Labor	Equipment	Total	Total Incl O&P
0010 **WOOD CLOSET AND UTILITY SHELVING**									
0020 Pine, clear grade, no edge band, 1" x 8"	1 Carp	115	.070	L.F.	3.52	2.43		5.95	7.90
0100 1" x 10"		110	.073		4.38	2.54		6.92	9.05
0200 1" x 12"		105	.076		5.30	2.66		7.96	10.25
0450 1" x 18"		95	.084		7.90	2.94		10.84	13.60
0460 1" x 24"		85	.094		10.55	3.29		13.84	17.05
0600 Plywood, 3/4" thick with lumber edge, 12" wide		75	.107		1.91	3.73		5.64	8.30
0700 24" wide		70	.114	↓	3.38	3.99		7.37	10.35
0900 Bookcase, clear grade pine, shelves 12" OC, 8" deep, per S.F. shelf		70	.114	S.F.	11.45	3.99		15.44	19.25
1000 12" deep shelves		65	.123	"	17.15	4.30		21.45	26
1200 Adjustable closet rod and shelf, 12" wide, 3' long		20	.400	Ea.	12.60	14		26.60	37
1300 8' long		15	.533	"	24.50	18.65		43.15	58
1500 Prefinished shelves with supports, stock, 8" wide		75	.107	L.F.	5.60	3.73		9.33	12.35
1600 10" wide	↓	70	.114	"	5.40	3.99		9.39	12.60

10 73 Protective Covers

10 73 16 – Canopies

10 73 16.10 Canopies, Residential

	Crew	Daily Output	Labor-Hours	Unit	Material	2018 Bare Costs Labor	Equipment	Total	Total Incl O&P
0010 **CANOPIES, RESIDENTIAL** Prefabricated									
0500 Carport, free standing, baked enamel, alum., .032", 40 psf									
0520 16' x 8', 4 posts	2 Carp	3	5.333	Ea.	4,850	186		5,036	5,625
0600 20' x 10', 6 posts		2	8		4,975	280		5,255	5,950
0605 30' x 10', 8 posts	↓	2	8		7,450	280		7,730	8,675
1000 Door canopies, extruded alum., .032", 42" projection, 4' wide	1 Carp	8	1		207	35		242	285
1020 6' wide	"	6	1.333		291	46.50		337.50	400
1040 8' wide	2 Carp	9	1.778		435	62		497	585
1060 10' wide		7	2.286		510	80		590	700
1080 12' wide	↓	5	3.200		585	112		697	830
1200 54" projection, 4' wide	1 Carp	8	1		232	35		267	315
1220 6' wide	"	6	1.333		296	46.50		342.50	405
1240 8' wide	2 Carp	9	1.778		355	62		417	500
1260 10' wide		7	2.286		815	80		895	1,025
1280 12' wide	↓	5	3.200		880	112		992	1,150
1300 Painted, add					20%				
1310 Bronze anodized, add					50%				
3000 Window awnings, aluminum, window 3' high, 4' wide	1 Carp	10	.800		160	28		188	223
3020 6' wide	"	8	1		192	35		227	269
3040 9' wide	2 Carp	9	1.778		495	62		557	650
3060 12' wide	"	5	3.200		495	112		607	730
3100 Window, 4' high, 4' wide	1 Carp	10	.800		289	28		317	365
3120 6' wide	"	8	1		355	35		390	450
3140 9' wide	2 Carp	9	1.778		625	62		687	795
3160 12' wide	"	5	3.200		710	112		822	965
3200 Window, 6' high, 4' wide	1 Carp	10	.800		340	28		368	420
3220 6' wide	"	8	1		390	35		425	490
3240 9' wide	2 Carp	9	1.778		955	62		1,017	1,150
3260 12' wide	"	5	3.200		1,450	112		1,562	1,775
3400 Roll-up aluminum, 2'-6" wide	1 Carp	14	.571		208	19.95		227.95	261
3420 3' wide		12	.667		213	23.50		236.50	273
3440 4' wide		10	.800		267	28		295	340
3460 6' wide	↓	8	1	↓	286	35		321	375

10 73 Protective Covers

10 73 16 – Canopies

10 73 16.10 Canopies, Residential	Crew	Daily Output	Labor-Hours	Unit	Material	2018 Bare Costs Labor	Equipment	Total	Total Incl O&P
3480	9' wide	2 Carp	9	1.778	Ea.	410	62		472	555
3500	12' wide	"	5	3.200		520	112		632	755
3600	Window awnings, canvas, 24" drop, 3' wide	1 Carp	30	.267	L.F.	58	9.30		67.30	79.50
3620	4' wide		40	.200		46	7		53	62
3700	30" drop, 3' wide		30	.267		60	9.30		69.30	81.50
3720	4' wide		40	.200		50	7		57	66.50
3740	5' wide		45	.178		46	6.20		52.20	61.50
3760	6' wide		48	.167		42.50	5.85		48.35	56.50
3780	8' wide		48	.167		39.50	5.85		45.35	53
3800	10' wide		50	.160		41.50	5.60		47.10	55

10 74 Manufactured Exterior Specialties

10 74 23 – Cupolas

10 74 23.10 Wood Cupolas

		Crew	Daily Output	Labor-Hours	Unit	Material	2018 Bare Costs Labor	Equipment	Total	Total Incl O&P
0010	**WOOD CUPOLAS**									
0020	Stock units, pine, painted, 18" sq., 28" high, alum. roof	1 Carp	4.10	1.951	Ea.	285	68		353	430
0100	Copper roof		3.80	2.105		291	73.50		364.50	440
0300	23" square, 33" high, aluminum roof		3.70	2.162		440	75.50		515.50	610
0400	Copper roof		3.30	2.424		590	84.50		674.50	790
0600	30" square, 37" high, aluminum roof		3.70	2.162		600	75.50		675.50	785
0700	Copper roof		3.30	2.424		715	84.50		799.50	930
0900	Hexagonal, 31" wide, 46" high, copper roof		4	2		945	70		1,015	1,175
1000	36" wide, 50" high, copper roof		3.50	2.286		1,675	80		1,755	1,950
1200	For deluxe stock units, add to above					25%				
1400	For custom built units, add to above					50%	50%			

10 74 33 – Weathervanes

10 74 33.10 Residential Weathervanes

		Crew	Daily Output	Labor-Hours	Unit	Material	2018 Bare Costs Labor	Equipment	Total	Total Incl O&P
0010	**RESIDENTIAL WEATHERVANES**									
0020	Residential types, 18" to 24"	1 Carp	8	1	Ea.	134	35		169	206
0100	24" to 48"	"	2	4	"	1,825	140		1,965	2,225

10 74 46 – Window Wells

10 74 46.10 Area Window Wells

		Crew	Daily Output	Labor-Hours	Unit	Material	2018 Bare Costs Labor	Equipment	Total	Total Incl O&P
0010	**AREA WINDOW WELLS**, Galvanized steel									
0020	20 ga., 3'-2" wide, 1' deep	1 Sswk	29	.276	Ea.	16.95	10.70		27.65	37.50
0100	2' deep		23	.348		31	13.50		44.50	58.50
0300	16 ga., 3'-2" wide, 1' deep		29	.276		23	10.70		33.70	44.50
0400	3' deep		23	.348		47	13.50		60.50	75.50
0600	Welded grating for above, 15 lb., painted		45	.178		91.50	6.90		98.40	113
0700	Galvanized		45	.178		124	6.90		130.90	149
0900	Translucent plastic cap for above		60	.133		20.50	5.15		25.65	31.50

529

10 75 Flagpoles

10 75 16 – Ground-Set Flagpoles

10 75 16.10 Flagpoles	Crew	Daily Output	Labor-Hours	Unit	Material	2018 Bare Costs Labor	Equipment	Total	Total Incl O&P
0010 **FLAGPOLES**, ground set									
0050 Not including base or foundation									
0100 Aluminum, tapered, ground set 20' high	K-1	2	8	Ea.	1,100	266	119	1,485	1,775
0200 25' high		1.70	9.412		1,175	315	140	1,630	1,950
0300 30' high		1.50	10.667		1,400	355	159	1,914	2,275
0500 40' high		1.20	13.333		3,025	445	198	3,668	4,275

Estimating Tips
General

- The items in this division are usually priced per square foot or each. Many of these items are purchased by the owner for installation by the contractor. Check the specifications for responsibilities and include time for receiving, storage, installation, and mechanical and electrical hookups in the appropriate divisions.

- Many items in Division 11 require some type of support system that is not usually furnished with the item. Examples of these systems include blocking for the attachment of casework and support angles for ceiling-hung projection screens. The required blocking or supports must be added to the estimate in the appropriate division.

- Some items in Division 11 may require assembly or electrical hookups. Verify the amount of assembly required or the need for a hard electrical connection and add the appropriate costs.

Reference Numbers

Reference numbers are shown at the beginning of some major classifications. These numbers refer to related items in the Reference Section. The reference information may be an estimating procedure, an alternate pricing method, or technical information.

Note: Not all subdivisions listed here necessarily appear. ■

Did you know?

RSMeans data is available through our online application with 24/7 access:

- Search for unit prices by keyword
- Leverage the most up-to-date data
- Build and export estimates

Try it free for 30 days!
www.rsmeans.com/2018freetrial

11 30 13.15 Cooking Equipment

		Crew	Daily Output	Labor-Hours	Unit	Material	2018 Bare Costs Labor	Equipment	Total	Total Incl O&P
0010	**COOKING EQUIPMENT**									
0020	Cooking range, 30" free standing, 1 oven, minimum	2 Clab	10	1.600	Ea.	470	42.50		512.50	585
0050	Maximum		4	4		2,300	107		2,407	2,700
0150	2 oven, minimum		10	1.600		1,000	42.50		1,042.50	1,175
0200	Maximum		10	1.600		3,300	42.50		3,342.50	3,700
0350	Built-in, 30" wide, 1 oven, minimum	1 Elec	6	1.333		800	53.50		853.50	970
0400	Maximum	2 Carp	2	8		1,750	280		2,030	2,400
0500	2 oven, conventional, minimum		4	4		1,150	140		1,290	1,500
0550	1 conventional, 1 microwave, maximum		2	8		2,275	280		2,555	2,975
0700	Free standing, 1 oven, 21" wide range, minimum	2 Clab	10	1.600		475	42.50		517.50	590
0750	21" wide, maximum	"	4	4		660	107		767	905
0900	Countertop cooktops, 4 burner, standard, minimum	1 Elec	6	1.333		325	53.50		378.50	450
0950	Maximum		3	2.667		1,750	107		1,857	2,100
1050	As above, but with grill and griddle attachment, minimum		6	1.333		1,450	53.50		1,503.50	1,675
1100	Maximum		3	2.667		3,625	107		3,732	4,150
1250	Microwave oven, minimum		4	2		97	80.50		177.50	238
1300	Maximum		2	4		460	161		621	770
5380	Oven, built-in, standard		4	2		900	80.50		980.50	1,125
5390	Deluxe		2	4		3,050	161		3,211	3,625

11 30 13.16 Refrigeration Equipment

		Crew	Daily Output	Labor-Hours	Unit	Material	2018 Bare Costs Labor	Equipment	Total	Total Incl O&P
0010	**REFRIGERATION EQUIPMENT**									
2000	Deep freeze, 15 to 23 C.F., minimum	2 Clab	10	1.600	Ea.	595	42.50		637.50	725
2050	Maximum		5	3.200		805	85.50		890.50	1,025
2200	30 C.F., minimum		8	2		800	53.50		853.50	970
2250	Maximum		3	5.333		860	142		1,002	1,175
5200	Icemaker, automatic, 20 lbs./day	1 Plum	7	1.143		1,250	45.50		1,295.50	1,450
5350	51 lbs./day	"	2	4		1,375	159		1,534	1,750
5450	Refrigerator, no frost, 6 C.F.	2 Clab	15	1.067		370	28.50		398.50	460
5500	Refrigerator, no frost, 10 C.F. to 12 C.F., minimum		10	1.600		440	42.50		482.50	555
5600	Maximum		6	2.667		535	71		606	710
5750	14 C.F. to 16 C.F., minimum		9	1.778		555	47.50		602.50	690
5800	Maximum		5	3.200		765	85.50		850.50	980
5950	18 C.F. to 20 C.F., minimum		8	2		705	53.50		758.50	865
6000	Maximum		4	4		1,650	107		1,757	2,000
6150	21 C.F. to 29 C.F., minimum		7	2.286		1,075	61		1,136	1,275
6200	Maximum		3	5.333		2,350	142		2,492	2,800
6790	Energy-star qualified, 18 C.F., minimum G	2 Carp	4	4		535	140		675	820
6795	Maximum G		2	8		1,600	280		1,880	2,225
6797	21.7 C.F., minimum G		4	4		1,400	140		1,540	1,775
6799	Maximum G		4	4		1,875	140		2,015	2,300

11 30 13.17 Kitchen Cleaning Equipment

		Crew	Daily Output	Labor-Hours	Unit	Material	2018 Bare Costs Labor	Equipment	Total	Total Incl O&P
0010	**KITCHEN CLEANING EQUIPMENT**									
2750	Dishwasher, built-in, 2 cycles, minimum	L-1	4	2.500	Ea.	296	99.50		395.50	490
2800	Maximum		2	5		455	199		654	830
2950	4 or more cycles, minimum		4	2.500		400	99.50		499.50	605
2960	Average		4	2.500		535	99.50		634.50	750
3000	Maximum		2	5		1,175	199		1,374	1,625
3100	Energy-star qualified, minimum G		4	2.500		380	99.50		479.50	585
3110	Maximum G		2	5		1,700	199		1,899	2,200

11 30 Residential Equipment

11 30 13 – Residential Appliances

11 30 13.18 Waste Disposal Equipment

		Crew	Daily Output	Labor-Hours	Unit	Material	2018 Bare Costs Labor	Equipment	Total	Total Incl O&P
0010	**WASTE DISPOSAL EQUIPMENT**									
1750	Compactor, residential size, 4 to 1 compaction, minimum	1 Carp	5	1.600	Ea.	690	56		746	855
1800	Maximum	"	3	2.667		1,150	93		1,243	1,425
3300	Garbage disposal, sink type, minimum	L-1	10	1		106	40		146	182
3350	Maximum	"	10	1		204	40		244	290

11 30 13.19 Kitchen Ventilation Equipment

		Crew	Daily Output	Labor-Hours	Unit	Material	2018 Bare Costs Labor	Equipment	Total	Total Incl O&P
0010	**KITCHEN VENTILATION EQUIPMENT**									
4150	Hood for range, 2 speed, vented, 30" wide, minimum	L-3	5	2	Ea.	93	72		165	222
4200	Maximum		3	3.333		910	120		1,030	1,200
4300	42" wide, minimum		5	2		170	72		242	305
4330	Custom		5	2		1,650	72		1,722	1,925
4350	Maximum		3	3.333		2,000	120		2,120	2,400
4500	For ventless hood, 2 speed, add					17.45			17.45	19.20
4650	For vented 1 speed, deduct from maximum					63			63	69

11 30 13.24 Washers

		Crew	Daily Output	Labor-Hours	Unit	Material	2018 Bare Costs Labor	Equipment	Total	Total Incl O&P
0010	**WASHERS**									
6650	Washing machine, automatic, minimum	1 Plum	3	2.667	Ea.	645	106		751	885
6700	Maximum		1	8		1,350	320		1,670	2,000
6750	Energy star qualified, front loading, minimum [G]		3	2.667		700	106		806	945
6760	Maximum [G]		1	8		1,800	320		2,120	2,500
6764	Top loading, minimum [G]		3	2.667		640	106		746	880
6766	Maximum [G]		3	2.667		1,000	106		1,106	1,275

11 30 13.25 Dryers

		Crew	Daily Output	Labor-Hours	Unit	Material	2018 Bare Costs Labor	Equipment	Total	Total Incl O&P
0010	**DRYERS**									
6770	Electric, front loading, energy-star qualified, minimum [G]	L-2	3	5.333	Ea.	405	164		569	720
6780	Maximum [G]	"	2	8		1,475	246		1,721	2,000
7450	Vent kits for dryers	1 Carp	10	.800		43.50	28		71.50	94.50

11 30 15 – Miscellaneous Residential Appliances

11 30 15.13 Sump Pumps

		Crew	Daily Output	Labor-Hours	Unit	Material	2018 Bare Costs Labor	Equipment	Total	Total Incl O&P
0010	**SUMP PUMPS**									
6400	Cellar drainer, pedestal, 1/3 HP, molded PVC base	1 Plum	3	2.667	Ea.	140	106		246	330
6450	Solid brass	"	2	4	"	240	159		399	525
6460	Sump pump, see also Section 22 14 29.16									

11 30 15.23 Water Heaters

		Crew	Daily Output	Labor-Hours	Unit	Material	2018 Bare Costs Labor	Equipment	Total	Total Incl O&P
0010	**WATER HEATERS**									
6900	Electric, glass lined, 30 gallon, minimum	L-1	5	2	Ea.	900	79.50		979.50	1,125
6950	Maximum		3	3.333		1,250	133		1,383	1,600
7100	80 gallon, minimum		2	5		1,675	199		1,874	2,150
7150	Maximum		1	10		2,325	400		2,725	3,200
7180	Gas, glass lined, 30 gallon, minimum	2 Plum	5	3.200		1,600	127		1,727	1,975
7220	Maximum		3	5.333		2,225	212		2,437	2,800
7260	50 gallon, minimum		2.50	6.400		1,725	255		1,980	2,325
7300	Maximum		1.50	10.667		2,400	425		2,825	3,350
7310	Water heater, see also Section 22 33 30.13									

11 30 15.43 Air Quality

		Crew	Daily Output	Labor-Hours	Unit	Material	2018 Bare Costs Labor	Equipment	Total	Total Incl O&P
0010	**AIR QUALITY**									
2450	Dehumidifier, portable, automatic, 15 pint	1 Elec	4	2	Ea.	205	80.50		285.50	355
2550	40 pint		3.75	2.133		250	85.50		335.50	415
3550	Heater, electric, built-in, 1250 watt, ceiling type, minimum		4	2		117	80.50		197.50	259
3600	Maximum		3	2.667		213	107		320	410

For customer support on your Residential Costs with RSMeans data, call 800.448.8182.

533

11 30 Residential Equipment

11 30 15 – Miscellaneous Residential Appliances

11 30 15.43 Air Quality

		Crew	Daily Output	Labor-Hours	Unit	Material	2018 Bare Costs Labor	Equipment	Total	Total Incl O&P
3700	Wall type, minimum	1 Elec	4	2	Ea.	197	80.50		277.50	350
3750	Maximum		3	2.667		194	107		301	390
3900	1500 watt wall type, with blower		4	2		188	80.50		268.50	340
3950	3000 watt		3	2.667		480	107		587	705
4850	Humidifier, portable, 8 gallons/day					148			148	162
5000	15 gallons/day					191			191	210

11 30 33 – Retractable Stairs

11 30 33.10 Disappearing Stairway

		Crew	Daily Output	Labor-Hours	Unit	Material	2018 Bare Costs Labor	Equipment	Total	Total Incl O&P
0010	**DISAPPEARING STAIRWAY** No trim included									
0020	One piece, yellow pine, 8'-0" ceiling	2 Carp	4	4	Ea.	227	140		367	480
0030	9'-0" ceiling		4	4		269	140		409	530
0040	10'-0" ceiling		3	5.333		251	186		437	585
0050	11'-0" ceiling		3	5.333		325	186		511	665
0060	12'-0" ceiling		3	5.333		335	186		521	680
0100	Custom grade, pine, 8'-6" ceiling, minimum	1 Carp	4	2		219	70		289	355
0150	Average		3.50	2.286		243	80		323	400
0200	Maximum		3	2.667		291	93		384	475
0500	Heavy duty, pivoted, from 7'-7" to 12'-10" floor to floor		3	2.667		1,375	93		1,468	1,675
0600	16'-0" ceiling		2	4		1,575	140		1,715	1,950
0800	Economy folding, pine, 8'-6" ceiling		4	2		177	70		247	310
0900	9'-6" ceiling		4	2		205	70		275	340
1100	Automatic electric, aluminum, floor to floor height, 8' to 9'	2 Carp	1	16		9,350	560		9,910	11,200

11 32 Unit Kitchens

11 32 13 – Metal Unit Kitchens

11 32 13.10 Commercial Unit Kitchens

		Crew	Daily Output	Labor-Hours	Unit	Material	2018 Bare Costs Labor	Equipment	Total	Total Incl O&P
0010	**COMMERCIAL UNIT KITCHENS**									
1500	Combination range, refrigerator and sink, 30" wide, minimum	L-1	2	5	Ea.	1,225	199		1,424	1,675
1550	Maximum		1	10		1,075	400		1,475	1,825
1570	60" wide, average		1.40	7.143		1,275	285		1,560	1,875
1590	72" wide, average		1.20	8.333		1,700	330		2,030	2,425

11 41 Foodservice Storage Equipment

11 41 13 – Refrigerated Food Storage Cases

11 41 13.30 Wine Cellar

		Crew	Daily Output	Labor-Hours	Unit	Material	2018 Bare Costs Labor	Equipment	Total	Total Incl O&P
0010	**WINE CELLAR**, refrigerated, Redwood interior, carpeted, walk-in type									
0020	6'-8" high, including racks									
0200	80" W x 48" D for 900 bottles	2 Carp	1.50	10.667	Ea.	4,200	375		4,575	5,250
0250	80" W x 72" D for 1300 bottles		1.33	12.030		5,125	420		5,545	6,325
0300	80" W x 94" D for 1900 bottles		1.17	13.675		6,200	480		6,680	7,625

11 81 Facility Maintenance Equipment

11 81 19 – Vacuum Cleaning Systems

11 81 19.10 Vacuum Cleaning	Crew	Daily Output	Labor-Hours	Unit	Material	2018 Bare Costs Labor	Equipment	Total	Total Incl O&P
0010 **VACUUM CLEANING**									
0020 Central, 3 inlet, residential	1 Skwk	.90	8.889	Total	1,150	320		1,470	1,800
0400 5 inlet system, residential		.50	16		1,675	575		2,250	2,775
0600 7 inlet system, commercial		.40	20		1,875	715		2,590	3,275
0800 9 inlet system, residential	↓	.30	26.667		4,025	955		4,980	6,025
4010 Rule of thumb: First 1200 S.F., installed								1,425	1,575
4020 For each additional S.F., add				S.F.				.26	.26

Division Notes

	CREW	DAILY OUTPUT	LABOR-HOURS	UNIT	BARE COSTS				TOTAL INCL O&P
					MAT.	LABOR	EQUIP.	TOTAL	

Estimating Tips
General

- The items in this division are usually priced per square foot or each. Most of these items are purchased by the owner and installed by the contractor. Do not assume the items in Division 12 will be purchased and installed by the contractor. Check the specifications for responsibilities and include receiving, storage, installation, and mechanical and electrical hookups in the appropriate divisions.

- Some items in this division require some type of support system that is not usually furnished with the item. Examples of these systems include blocking for the attachment of casework and heavy drapery rods. The required blocking must be added to the estimate in the appropriate division.

Reference Numbers

Reference numbers are shown at the beginning of some major classifications. These numbers refer to related items in the Reference Section. The reference information may be an estimating procedure, an alternate pricing method, or technical information.

Note: Not all subdivisions listed here necessarily appear. ■

Did you know?

RSMeans data is available through our online application with 24/7 access:

- Search for unit prices by keyword
- Leverage the most up-to-date data
- Build and export estimates

Try it free for 30 days!
www.rsmeans.com/2018freetrial

12 21 Window Blinds

12 21 13 – Horizontal Louver Blinds

12 21 13.13 Metal Horizontal Louver Blinds	Crew	Daily Output	Labor-Hours	Unit	Material	2018 Bare Costs Labor	2018 Bare Costs Equipment	Total	Total Incl O&P
0010 **METAL HORIZONTAL LOUVER BLINDS**									
0020 Horizontal, 1" aluminum slats, solid color, stock	1 Carp	590	.014	S.F.	5.90	.47		6.37	7.30

12 21 13.33 Vinyl Horizontal Louver Blinds

	Crew	Daily Output	Labor-Hours	Unit	Material	Labor	Equipment	Total	Total Incl O&P
0010 **VINYL HORIZONTAL LOUVER BLINDS**									
0015 1" composite, 48" wide, 48" high	1 Carp	30	.267	Ea.	31	9.30		40.30	49.50
0020 72" high		30	.267		38	9.30		47.30	57.50
0030 60" wide, 60" high		30	.267		45.50	9.30		54.80	65.50
0040 72" high		30	.267		45.50	9.30		54.80	65.50
0050 72" wide x 72" high		30	.267		54.50	9.30		63.80	75.50

12 22 Curtains and Drapes

12 22 16 – Drapery Track and Accessories

12 22 16.10 Drapery Hardware

	Crew	Daily Output	Labor-Hours	Unit	Material	Labor	Equipment	Total	Total Incl O&P
0010 **DRAPERY HARDWARE**									
0030 Standard traverse, per foot, minimum	1 Carp	59	.136	L.F.	6.15	4.74		10.89	14.70
0100 Maximum		51	.157	"	15	5.50		20.50	25.50
0200 Decorative traverse, 28" to 48", minimum		22	.364	Ea.	25	12.70		37.70	48.50
0220 Maximum		21	.381		56	13.30		69.30	83.50
0300 48" to 84", minimum		20	.400		27.50	14		41.50	53
0320 Maximum		19	.421		71.50	14.70		86.20	103
0400 66" to 120", minimum		18	.444		37.50	15.55		53.05	67
0420 Maximum		17	.471		109	16.45		125.45	148
0500 84" to 156", minimum		16	.500		48	17.50		65.50	82
0520 Maximum		15	.533		149	18.65		167.65	195
0600 130" to 240", minimum		14	.571		70	19.95		89.95	110
0620 Maximum		13	.615		206	21.50		227.50	262
0700 Slide rings, each, minimum					1.27			1.27	1.40
0720 Maximum					2.13			2.13	2.34
4000 Traverse rods, adjustable, 28" to 48"	1 Carp	22	.364		32	12.70		44.70	56
4020 48" to 84"		20	.400		40.50	14		54.50	67.50
4040 66" to 120"		18	.444		49.50	15.55		65.05	80
4060 84" to 156"		16	.500		56	17.50		73.50	90.50
4080 100" to 180"		14	.571		65.50	19.95		85.45	105
4100 228" to 312"		13	.615		90.50	21.50		112	136
4500 Curtain rod, 28" to 48", single		22	.364		10	12.70		22.70	32
4510 Double		22	.364		17	12.70		29.70	39.50
4520 48" to 86", single		20	.400		17.10	14		31.10	42
4530 Double		20	.400		28.50	14		42.50	54.50
4540 66" to 120", single		18	.444		28.50	15.55		44.05	57.50
4550 Double		18	.444		44.50	15.55		60.05	75
4600 Valance, pinch pleated fabric, 12" deep, up to 54" long, minimum					41			41	45
4610 Maximum					102			102	112
4620 Up to 77" long, minimum					72.50			72.50	79.50
4630 Maximum					165			165	181
5000 Stationary rods, first 2'					8.35			8.35	9.20

12 23 Interior Shutters

12 23 10 – Wood Interior Shutters

12 23 10.10 Wood Interior Shutters		Crew	Daily Output	Labor-Hours	Unit	Material	2018 Bare Costs Labor	Equipment	Total	Total Incl O&P
0010	**WOOD INTERIOR SHUTTERS**, louvered									
0200	Two panel, 27" wide, 36" high	1 Carp	5	1.600	Set	160	56		216	269
0300	33" wide, 36" high		5	1.600		206	56		262	320
0500	47" wide, 36" high		5	1.600		276	56		332	400
1000	Four panel, 27" wide, 36" high		5	1.600		158	56		214	267
1100	33" wide, 36" high		5	1.600		203	56		259	315
1300	47" wide, 36" high		5	1.600		271	56		327	390
1400	Plantation shutters, 16" x 48"		5	1.600	Ea.	181	56		237	292
1450	16" x 96"		4	2		297	70		367	440
1460	36" x 96"		3	2.667		655	93		748	875

12 23 10.13 Wood Panels

		Crew	Daily Output	Labor-Hours	Unit	Material	2018 Bare Costs Labor	Equipment	Total	Total Incl O&P
0010	**WOOD PANELS**									
3000	Wood folding panels with movable louvers, 7" x 20" each	1 Carp	17	.471	Pr.	92	16.45		108.45	129
3300	8" x 28" each		17	.471		92	16.45		108.45	129
3450	9" x 36" each		17	.471		106	16.45		122.45	145
3600	10" x 40" each		17	.471		115	16.45		131.45	155
4000	Fixed louver type, stock units, 8" x 20" each		17	.471		96.50	16.45		112.95	134
4150	10" x 28" each		17	.471		81.50	16.45		97.95	118
4300	12" x 36" each		17	.471		96.50	16.45		112.95	134
4450	18" x 40" each		17	.471		138	16.45		154.45	180
5000	Insert panel type, stock, 7" x 20" each		17	.471		27	16.45		43.45	57.50
5150	8" x 28" each		17	.471		50	16.45		66.45	82
5300	9" x 36" each		17	.471		63	16.45		79.45	97
5450	10" x 40" each		17	.471		67.50	16.45		83.95	102
5600	Raised panel type, stock, 10" x 24" each		17	.471		115	16.45		131.45	155
5650	12" x 26" each		17	.471		115	16.45		131.45	155
5700	14" x 30" each		17	.471		127	16.45		143.45	168
5750	16" x 36" each		17	.471		141	16.45		157.45	183
6000	For custom built pine, add					22%				
6500	For custom built hardwood blinds, add					42%				

12 24 Window Shades

12 24 13 – Roller Window Shades

12 24 13.10 Shades

			Crew	Daily Output	Labor-Hours	Unit	Material	2018 Bare Costs Labor	Equipment	Total	Total Incl O&P
0010	**SHADES**										
0020	Basswood, roll-up, stain finish, 3/8" slats		1 Carp	300	.027	S.F.	19.75	.93		20.68	23
5011	Insulative shades	G		125	.064		15.60	2.24		17.84	21
6011	Solar screening, fiberglass	G		85	.094		7.65	3.29		10.94	13.85
8011	Interior insulative shutter										
8111	Stock unit, 15" x 60"	G	1 Carp	17	.471	Pr.	15.75	16.45		32.20	45

For customer support on your Residential Costs with RSMeans data, call 800.448.8182.

539

12 32 Manufactured Wood Casework

12 32 23 – Hardwood Casework

12 32 23.10 Manufactured Wood Casework, Stock Units	Crew	Daily Output	Labor-Hours	Unit	Material	2018 Bare Costs Labor	Equipment	Total	Total Incl O&P
0010 **MANUFACTURED WOOD CASEWORK, STOCK UNITS**									
0700 Kitchen base cabinets, hardwood, not incl. counter tops,									
0710 24" deep, 35" high, prefinished									
0800 One top drawer, one door below, 12" wide	2 Carp	24.80	.645	Ea.	300	22.50		322.50	370
0820 15" wide		24	.667		315	23.50		338.50	385
0840 18" wide		23.30	.687		340	24		364	415
0860 21" wide		22.70	.705		350	24.50		374.50	430
0880 24" wide		22.30	.717		415	25		440	495
1000 Four drawers, 12" wide		24.80	.645		315	22.50		337.50	390
1020 15" wide		24	.667		320	23.50		343.50	395
1040 18" wide		23.30	.687		355	24		379	430
1060 24" wide		22.30	.717		390	25		415	470
1200 Two top drawers, two doors below, 27" wide		22	.727		445	25.50		470.50	535
1220 30" wide		21.40	.748		485	26		511	580
1240 33" wide		20.90	.766		505	27		532	600
1260 36" wide		20.30	.788		525	27.50		552.50	620
1280 42" wide		19.80	.808		550	28		578	655
1300 48" wide		18.90	.847		590	29.50		619.50	700
1500 Range or sink base, two doors below, 30" wide		21.40	.748		405	26		431	490
1520 33" wide		20.90	.766		430	27		457	520
1540 36" wide		20.30	.788		455	27.50		482.50	545
1560 42" wide		19.80	.808		475	28		503	570
1580 48" wide	▼	18.90	.847		500	29.50		529.50	600
1800 For sink front units, deduct					179			179	197
2000 Corner base cabinets, 36" wide, standard	2 Carp	18	.889		745	31		776	870
2100 Lazy Susan with revolving door	"	16.50	.970	▼	955	34		989	1,100
4000 Kitchen wall cabinets, hardwood, 12" deep with two doors									
4050 12" high, 30" wide	2 Carp	24.80	.645	Ea.	272	22.50		294.50	340
4100 36" wide		24	.667		75.50	23.50		99	122
4400 15" high, 30" wide		24	.667		277	23.50		300.50	345
4420 33" wide		23.30	.687		340	24		364	415
4440 36" wide		22.70	.705		330	24.50		354.50	405
4450 42" wide		22.70	.705		375	24.50		399.50	450
4700 24" high, 30" wide		23.30	.687		370	24		394	450
4720 36" wide		22.70	.705		405	24.50		429.50	490
4740 42" wide		22.30	.717		278	25		303	345
5000 30" high, one door, 12" wide		22	.727		261	25.50		286.50	330
5020 15" wide		21.40	.748		273	26		299	345
5040 18" wide		20.90	.766		300	27		327	375
5060 24" wide		20.30	.788		350	27.50		377.50	430
5300 Two doors, 27" wide		19.80	.808		390	28		418	470
5320 30" wide		19.30	.829		410	29		439	500
5340 36" wide		18.80	.851		465	29.50		494.50	565
5360 42" wide		18.50	.865		505	30		535	605
5380 48" wide		18.40	.870		570	30.50		600.50	675
6000 Corner wall, 30" high, 24" wide		18	.889		400	31		431	490
6050 30" wide		17.20	.930		425	32.50		457.50	520
6100 36" wide		16.50	.970		485	34		519	590
6500 Revolving Lazy Susan		15.20	1.053		127	37		164	201
7000 Broom cabinet, 84" high, 24" deep, 18" wide		10	1.600		745	56		801	915
7500 Oven cabinets, 84" high, 24" deep, 27" wide		8	2	▼	1,125	70		1,195	1,375
7750 Valance board trim	▼	396	.040	L.F.	18.05	1.41		19.46	22.50
7780 Toe kick trim	1 Carp	256	.031	"	3.45	1.09		4.54	5.60

12 32 Manufactured Wood Casework

12 32 23 – Hardwood Casework

12 32 23.10 Manufactured Wood Casework, Stock Units	Crew	Daily Output	Labor-Hours	Unit	Material	2018 Bare Costs Labor	Equipment	Total	Total Incl O&P
7790	Base cabinet corner filler	1 Carp	16	.500	Ea.	48.50	17.50		66	82
7800	Cabinet filler, 3" x 24"		20	.400		19.05	14		33.05	44
7810	3" x 30"		20	.400		24	14		38	49
7820	3" x 42"		18	.444		33.50	15.55		49.05	62.50
7830	3" x 80"		16	.500		63.50	17.50		81	99
7850	Cabinet panel		50	.160	S.F.	10.35	5.60		15.95	20.50
9000	For deluxe models of all cabinets, add					40%				
9500	For custom built in place, add					25%	10%			
9558	Rule of thumb, kitchen cabinets not including									
9560	appliances & counter top, minimum	2 Carp	30	.533	L.F.	202	18.65		220.65	253
9600	Maximum	"	25	.640	"	440	22.50		462.50	520

12 32 23.30 Manufactured Wood Casework Vanities

		Crew	Daily Output	Labor-Hours	Unit	Material	Labor	Equipment	Total	Total Incl O&P
0010	**MANUFACTURED WOOD CASEWORK VANITIES**									
8000	Vanity bases, 2 doors, 30" high, 21" deep, 24" wide	2 Carp	20	.800	Ea.	350	28		378	430
8050	30" wide		16	1		420	35		455	520
8100	36" wide		13.33	1.200		405	42		447	515
8150	48" wide		11.43	1.400		530	49		579	665
9000	For deluxe models of all vanities, add to above					40%				
9500	For custom built in place, add to above					25%	10%			

12 32 23.35 Manufactured Wood Casework Hardware

		Crew	Daily Output	Labor-Hours	Unit	Material	Labor	Equipment	Total	Total Incl O&P
0010	**MANUFACTURED WOOD CASEWORK HARDWARE**									
1000	Catches, minimum	1 Carp	235	.034	Ea.	1.40	1.19		2.59	3.52
1020	Average		119.40	.067		4.20	2.34		6.54	8.50
1040	Maximum		80	.100		8.05	3.50		11.55	14.65
2000	Door/drawer pulls, handles									
2200	Handles and pulls, projecting, metal, minimum	1 Carp	48	.167	Ea.	5.20	5.85		11.05	15.40
2220	Average		42	.190		7.90	6.65		14.55	19.75
2240	Maximum		36	.222		10.65	7.75		18.40	24.50
2300	Wood, minimum		48	.167		5.30	5.85		11.15	15.55
2320	Average		42	.190		7.10	6.65		13.75	18.85
2340	Maximum		36	.222		9.80	7.75		17.55	23.50
2600	Flush, metal, minimum		48	.167		5.30	5.85		11.15	15.55
2620	Average		42	.190		7.25	6.65		13.90	19.05
2640	Maximum		36	.222		9.90	7.75		17.65	24
3000	Drawer tracks/glides, minimum		48	.167	Pr.	8.95	5.85		14.80	19.55
3020	Average		32	.250		16	8.75		24.75	32
3040	Maximum		24	.333		26	11.65		37.65	48
4000	Cabinet hinges, minimum		160	.050		3.10	1.75		4.85	6.30
4020	Average		95.24	.084		5.30	2.94		8.24	10.70
4040	Maximum		68	.118		14.10	4.11		18.21	22.50

For customer support on your Residential Costs with RSMeans data, call 800.448.8182.

541

12 34 Manufactured Plastic Casework

12 34 16 – Manufactured Solid-Plastic Casework

12 34 16.10 Outdoor Casework		Crew	Daily Output	Labor-Hours	Unit	Material	2018 Bare Costs Labor	Equipment	Total	Total Incl O&P
0010	**OUTDOOR CASEWORK**									
0020	Cabinet, base, sink/range, 36"	2 Carp	20.30	.788	Ea.	480	27.50		507.50	570
0100	Base, 36"		20.30	.788		4,600	27.50		4,627.50	5,100
0200	Filler strip, 1" x 30"	↓	158	.101	↓	43.50	3.54		47.04	54

12 36 Countertops

12 36 16 – Metal Countertops

12 36 16.10 Stainless Steel Countertops

		Crew	Daily Output	Labor-Hours	Unit	Material	Labor	Equipment	Total	Total Incl O&P
0010	**STAINLESS STEEL COUNTERTOPS**									
3200	Stainless steel, custom	1 Carp	24	.333	S.F.	179	11.65		190.65	216

12 36 19 – Wood Countertops

12 36 19.10 Maple Countertops

		Crew	Daily Output	Labor-Hours	Unit	Material	Labor	Equipment	Total	Total Incl O&P
0010	**MAPLE COUNTERTOPS**									
2900	Solid, laminated, 1-1/2" thick, no splash	1 Carp	28	.286	L.F.	88.50	10		98.50	114
3000	With square splash		28	.286	"	105	10		115	133
3400	Recessed cutting block with trim, 16" x 20" x 1"		8	1	Ea.	108	35		143	176
3411	Replace cutting block only	↓	16	.500	"	89	17.50		106.50	127

12 36 23 – Plastic Countertops

12 36 23.13 Plastic-Laminate-Clad Countertops

		Crew	Daily Output	Labor-Hours	Unit	Material	Labor	Equipment	Total	Total Incl O&P
0010	**PLASTIC-LAMINATE-CLAD COUNTERTOPS**									
0020	Stock, 24" wide w/backsplash, minimum	1 Carp	30	.267	L.F.	17.25	9.30		26.55	34.50
0100	Maximum		25	.320		39	11.20		50.20	61.50
0300	Custom plastic, 7/8" thick, aluminum molding, no splash		30	.267		34.50	9.30		43.80	53.50
0400	Cove splash		30	.267		41	9.30		50.30	60.50
0600	1-1/4" thick, no splash		28	.286		39	10		49	59
0700	Square splash		28	.286		43.50	10		53.50	64
0900	Square edge, plastic face, 7/8" thick, no splash		30	.267		34	9.30		43.30	53
1000	With splash	↓	30	.267		41	9.30		50.30	61
1200	For stainless channel edge, 7/8" thick, add					3.66			3.66	4.03
1300	1-1/4" thick, add					4.36			4.36	4.80
1500	For solid color suede finish, add				↓	5.80			5.80	6.40
1700	For end splash, add				Ea.	21			21	23
1901	For cut outs, standard, add, minimum	1 Carp	32	.250			8.75		8.75	14.55
2000	Maximum		8	1		7	35		42	65.50
2010	Cut out in backsplash for elec. wall outlet		38	.211			7.35		7.35	12.25
2020	Cut out for sink		20	.400			14		14	23
2030	Cut out for stove top		18	.444	↓		15.55		15.55	26
2100	Postformed, including backsplash and front edge		30	.267	L.F.	12.60	9.30		21.90	29.50
2110	Mitred, add		12	.667	Ea.		23.50		23.50	38.50
2200	Built-in place, 25" wide, plastic laminate	↓	25	.320	L.F.	60	11.20		71.20	84.50

12 36 33 – Tile Countertops

12 36 33.10 Ceramic Tile Countertops

		Crew	Daily Output	Labor-Hours	Unit	Material	Labor	Equipment	Total	Total Incl O&P
0010	**CERAMIC TILE COUNTERTOPS**									
2300	Ceramic tile mosaic	1 Carp	25	.320	L.F.	39.50	11.20		50.70	62

12 36 40 – Stone Countertops

12 36 40.10 Natural Stone Countertops

		Crew	Daily Output	Labor-Hours	Unit	Material	Labor	Equipment	Total	Total Incl O&P
0010	**NATURAL STONE COUNTERTOPS**									
2500	Marble, stock, with splash, 1/2" thick, minimum	1 Bric	17	.471	L.F.	49	16.35		65.35	81.50
2700	3/4" thick, maximum	↓	13	.615	↓	123	21.50		144.50	171

12 36 Countertops

12 36 40 – Stone Countertops

12 36 40.10 Natural Stone Countertops	Crew	Daily Output	Labor-Hours	Unit	Material	2018 Bare Costs Labor	Equipment	Total	Total Incl O&P
2800	Granite, average, 1-1/4" thick, 24" wide, no splash	1 Bric	13.01	.615	L.F.	165	21.50		186.50	218

12 36 61 – Simulated Stone Countertops

12 36 61.16 Solid Surface Countertops

		Crew	Daily Output	Labor-Hours	Unit	Material	Labor	Equipment	Total	Total Incl O&P
0010	**SOLID SURFACE COUNTERTOPS**, Acrylic polymer									
2000	Pricing for order of 1-50 L.F.									
2100	25" wide, solid colors	2 Carp	20	.800	L.F.	70.50	28		98.50	124
2200	Patterned colors		20	.800		90	28		118	146
2300	Premium patterned colors		20	.800		119	28		147	178
2400	With silicone attached 4" backsplash, solid colors		19	.842		82.50	29.50		112	140
2500	Patterned colors		19	.842		105	29.50		134.50	164
2600	Premium patterned colors		19	.842		130	29.50		159.50	192
2700	With hard seam attached 4" backsplash, solid colors		15	1.067		82.50	37.50		120	153
2800	Patterned colors		15	1.067		105	37.50		142.50	177
2900	Premium patterned colors		15	1.067		130	37.50		167.50	205
3800	Sinks, pricing for order of 1-50 units									
3900	Single bowl, hard seamed, solid colors, 13" x 17"	1 Carp	2	4	Ea.	510	140		650	790
4000	10" x 15"		4.55	1.758		235	61.50		296.50	360
4100	Cutouts for sinks		5.25	1.524			53.50		53.50	88.50

12 36 61.17 Solid Surface Vanity Tops

		Crew	Daily Output	Labor-Hours	Unit	Material	Labor	Equipment	Total	Total Incl O&P
0010	**SOLID SURFACE VANITY TOPS**									
0015	Solid surface, center bowl, 17" x 19"	1 Carp	12	.667	Ea.	181	23.50		204.50	238
0020	19" x 25"		12	.667		185	23.50		208.50	242
0030	19" x 31"		12	.667		216	23.50		239.50	277
0040	19" x 37"		12	.667		252	23.50		275.50	315
0050	22" x 25"		10	.800		330	28		358	410
0060	22" x 31"		10	.800		385	28		413	470
0070	22" x 37"		10	.800		450	28		478	540
0080	22" x 43"		10	.800		520	28		548	615
0090	22" x 49"		10	.800		560	28		588	665
0110	22" x 55"		8	1		405	35		440	505
0120	22" x 61"		8	1		460	35		495	565
0220	Double bowl, 22" x 61"		8	1		515	35		550	630
0230	Double bowl, 22" x 73"		8	1		980	35		1,015	1,125
0240	For aggregate colors, add					35%				
0250	For faucets and fittings, see Section 22 41 39.10									

12 36 61.19 Quartz Agglomerate Countertops

		Crew	Daily Output	Labor-Hours	Unit	Material	Labor	Equipment	Total	Total Incl O&P
0010	**QUARTZ AGGLOMERATE COUNTERTOPS**									
0100	25" wide, 4" backsplash, color group A, minimum	2 Carp	15	1.067	L.F.	67.50	37.50		105	136
0110	Maximum		15	1.067		94	37.50		131.50	166
0120	Color group B, minimum		15	1.067		72	37.50		109.50	141
0130	Maximum		15	1.067		98.50	37.50		136	170
0140	Color group C, minimum		15	1.067		81	37.50		118.50	151
0150	Maximum		15	1.067		115	37.50		152.50	189
0160	Color group D, minimum		15	1.067		88	37.50		125.50	159
0170	Maximum		15	1.067		119	37.50		156.50	193

For customer support on your Residential Costs with RSMeans data, call 800.448.8182.

543

Division Notes

		CREW	DAILY OUTPUT	LABOR-HOURS	UNIT	BARE COSTS				TOTAL INCL O&P
						MAT.	LABOR	EQUIP.	TOTAL	

Estimating Tips
General

- The items and systems in this division are usually estimated, purchased, supplied, and installed as a unit by one or more subcontractors. The estimator must ensure that all parties are operating from the same set of specifications and assumptions, and that all necessary items are estimated and will be provided. Many times the complex items and systems are covered, but the more common ones, such as excavation or a crane, are overlooked for the very reason that everyone assumes nobody could miss them. The estimator should be the central focus and be able to ensure that all systems are complete.

- Another area where problems can develop in this division is at the interface between systems. The estimator must ensure, for instance, that anchor bolts, nuts, and washers are estimated and included for the air-supported structures and pre-engineered buildings to be bolted to their foundations. Utility supply is a common area where essential items or pieces of equipment can be missed or overlooked, because each subcontractor may feel it is another's responsibility. The estimator should also be aware of certain items which may be supplied as part of a package but installed by others, and ensure that the installing contractor's estimate includes the cost of installation. Conversely, the estimator must also ensure that items are not costed by two different subcontractors, resulting in an inflated overall estimate.

13 30 00 Special Structures

- The foundations and floor slab, as well as rough mechanical and electrical, should be estimated, as this work is required for the assembly and erection of the structure. Generally, as noted in the data set, the pre-engineered building comes as a shell. Pricing is based on the size and structural design parameters stated in the reference section. Additional features, such as windows and doors with their related structural framing, must also be included by the estimator. Here again, the estimator must have a clear understanding of the scope of each portion of the work and all the necessary interfaces.

Reference Numbers

Reference numbers are shown at the beginning of some major classifications. These numbers refer to related items in the Reference Section. The reference information may be an estimating procedure, an alternate pricing method, or technical information.

Note: Not all subdivisions listed here necessarily appear. ■

Did you know?

RSMeans data is available through our online application with 24/7 access:

- Search for unit prices by keyword
- Leverage the most up-to-date data
- Build and export estimates

Try it free for 30 days!
www.rsmeans.com/2018freetrial

13 11 Swimming Pools

13 11 13 – Below-Grade Swimming Pools

13 11 13.50 Swimming Pools

		Crew	Daily Output	Labor-Hours	Unit	Material	2018 Bare Costs Labor	2018 Bare Costs Equipment	Total	Total Incl O&P
0010	**SWIMMING POOLS** Residential in-ground, vinyl lined									
0020	Concrete sides, w/equip, sand bottom	B-52	300	.187	SF Surf	26	5.55	2.01	33.56	40.50
0100	Metal or polystyrene sides R131113-20	B-14	410	.117		22	3.33	.76	26.09	30.50
0200	Add for vermiculite bottom				↓	1.67			1.67	1.84
0500	Gunite bottom and sides, white plaster finish									
0600	12' x 30' pool	B-52	145	.386	SF Surf	49	11.50	4.15	64.65	77
0720	16' x 32' pool		155	.361		44	10.75	3.88	58.63	70.50
0750	20' x 40' pool	↓	250	.224	↓	39.50	6.65	2.41	48.56	56.50
0810	Concrete bottom and sides, tile finish									
0820	12' x 30' pool	B-52	80	.700	SF Surf	49.50	21	7.55	78.05	97
0830	16' x 32' pool		95	.589		40.50	17.50	6.35	64.35	81
0840	20' x 40' pool	↓	130	.431	↓	32.50	12.80	4.63	49.93	62
1600	For water heating system, see Section 23 52 28.10									
1700	Filtration and deck equipment only, as % of total				Total				20%	20%
1800	Deck equipment, rule of thumb, 20' x 40' pool				SF Pool				1.18	1.30
3000	Painting pools, preparation + 3 coats, 20' x 40' pool, epoxy	2 Pord	.33	48.485	Total	1,725	1,425		3,150	4,250
3100	Rubber base paint, 18 gallons	"	.33	48.485	"	1,325	1,425		2,750	3,800

13 11 23 – On-Grade Swimming Pools

13 11 23.50 Swimming Pools

		Crew	Daily Output	Labor-Hours	Unit	Material	2018 Bare Costs Labor	2018 Bare Costs Equipment	Total	Total Incl O&P
0010	**SWIMMING POOLS** Residential above ground, steel construction									
0100	Round, 15' diam.	B-80A	3	8	Ea.	785	214	79.50	1,078.50	1,300
0120	18' diam.		2.50	9.600		930	256	95	1,281	1,550
0140	21' diam.		2	12		1,075	320	119	1,514	1,850
0160	24' diam.		1.80	13.333		1,175	355	132	1,662	2,000
0180	27' diam.		1.50	16		1,500	425	159	2,084	2,550
0200	30' diam.		1	24		1,650	640	238	2,528	3,125
0220	Oval, 12' x 24'		2.30	10.435		1,600	279	104	1,983	2,350
0240	15' x 30'		1.80	13.333		2,525	355	132	3,012	3,500
0260	18' x 33'	↓	1	24		2,775	640	238	3,653	4,375

13 11 46 – Swimming Pool Accessories

13 11 46.50 Swimming Pool Equipment

		Crew	Daily Output	Labor-Hours	Unit	Material	2018 Bare Costs Labor	2018 Bare Costs Equipment	Total	Total Incl O&P
0010	**SWIMMING POOL EQUIPMENT**									
0020	Diving stand, stainless steel, 3 meter	2 Carp	.40	40	Ea.	16,700	1,400		18,100	20,700
0300	1 meter		2.70	5.926		10,100	207		10,307	11,400
0600	Diving boards, 16' long, aluminum		2.70	5.926		4,325	207		4,532	5,100
0700	Fiberglass		2.70	5.926		3,375	207		3,582	4,050
0800	14' long, aluminum		2.70	5.926		3,975	207		4,182	4,725
0850	Fiberglass	↓	2.70	5.926		3,350	207		3,557	4,025
1100	Bulkhead, movable, PVC, 8'-2" wide	2 Clab	8	2		2,425	53.50		2,478.50	2,775
1120	7'-9" wide		8	2		2,075	53.50		2,128.50	2,400
1140	7'-3" wide		8	2		2,125	53.50		2,178.50	2,425
1160	6'-9" wide	↓	8	2		2,125	53.50		2,178.50	2,425
1200	Ladders, heavy duty, stainless steel, 2 tread	2 Carp	7	2.286		880	80		960	1,100
1500	4 tread	"	6	2.667		775	93		868	1,000
2100	Lights, underwater, 12 volt, with transformer, 300 watt	1 Elec	1	8		360	320		680	920
2200	110 volt, 500 watt, standard	"	1	8	↓	295	320		615	850
3000	Pool covers, reinforced vinyl	3 Clab	1800	.013	S.F.	1.20	.36		1.56	1.91
3100	Vinyl, for winter, 400 S.F. max pool surface		3200	.008		.22	.20		.42	.57
3200	With water tubes, 400 S.F. max pool surface		3000	.008		.37	.21		.58	.77
3250	Sealed air bubble polyethylene solar blanket, 16 mils					.37			.37	.41
3300	Slides, tubular, fiberglass, aluminum handrails & ladder, 5'-0", straight	2 Carp	1.60	10	Ea.	3,700	350		4,050	4,625
3320	8'-0", curved	"	3	5.333	"	7,225	186		7,411	8,250

13 12 Fountains

13 12 13 – Exterior Fountains

13 12 13.10 Outdoor Fountains	Crew	Daily Output	Labor-Hours	Unit	Material	2018 Bare Costs Labor	Equipment	Total	Total Incl O&P
0010 **OUTDOOR FOUNTAINS**									
0100 Outdoor fountain, 48" high with bowl and figures	2 Clab	2	8	Ea.	570	214		784	980
0200 Commercial, concrete or cast stone, 40-60" H, simple		2	8		775	214		989	1,200
0220 Average		2	8		1,350	214		1,564	1,825
0240 Ornate		2	8		2,625	214		2,839	3,225
0260 Metal, 72" high		2	8		1,200	214		1,414	1,675
0280 90" high		2	8		1,875	214		2,089	2,425
0300 120" high		2	8		5,000	214		5,214	5,850
0320 Resin or fiberglass, 40-60" H, wall type		2	8		480	214		694	885
0340 Waterfall type		2	8		1,000	214		1,214	1,450

13 12 23 – Interior Fountains

13 12 23.10 Indoor Fountains

	Crew	Daily Output	Labor-Hours	Unit	Material	Labor	Equipment	Total	Total Incl O&P
0010 **INDOOR FOUNTAINS**									
0100 Commercial, floor type, resin or fiberglass, lighted, cascade type	2 Clab	2	8	Ea.	286	214		500	670
0120 Tiered type		2	8		285	214		499	670
0140 Waterfall type		2	8		275	214		489	655

13 17 Tubs and Pools

13 17 13 – Hot Tubs

13 17 13.10 Redwood Hot Tub System

	Crew	Daily Output	Labor-Hours	Unit	Material	Labor	Equipment	Total	Total Incl O&P
0010 **REDWOOD HOT TUB SYSTEM**									
7050 4' diameter x 4' deep	Q-1	1	16	Ea.	3,225	575		3,800	4,500
7150 6' diameter x 4' deep		.80	20		4,950	715		5,665	6,600
7200 8' diameter x 4' deep		.80	20		7,250	715		7,965	9,150

13 17 33 – Whirlpool Tubs

13 17 33.10 Whirlpool Bath

	Crew	Daily Output	Labor-Hours	Unit	Material	Labor	Equipment	Total	Total Incl O&P
0010 **WHIRLPOOL BATH**									
6000 Whirlpool, bath with vented overflow, molded fiberglass									
6100 66" x 36" x 24"	Q-1	1	16	Ea.	1,025	575		1,600	2,075
6400 72" x 36" x 21"		1	16		1,250	575		1,825	2,325
6500 60" x 34" x 21"		1	16		1,175	575		1,750	2,250
6600 72" x 42" x 23"		1	16		1,275	575		1,850	2,350

13 24 Special Activity Rooms

13 24 16 – Saunas

13 24 16.50 Saunas and Heaters

	Crew	Daily Output	Labor-Hours	Unit	Material	Labor	Equipment	Total	Total Incl O&P
0010 **SAUNAS AND HEATERS**									
0020 Prefabricated, incl. heater & controls, 7' high, 6' x 4', C/C	L-7	2.20	11.818	Ea.	5,125	355		5,480	6,250
0050 6' x 4', C/P		2	13		4,325	390		4,715	5,400
0400 6' x 5', C/C		2	13		5,950	390		6,340	7,200
0450 6' x 5', C/P		2	13		5,175	390		5,565	6,350
0600 6' x 6', C/C		1.80	14.444		7,000	435		7,435	8,425
0650 6' x 6', C/P		1.80	14.444		5,425	435		5,860	6,675
0800 6' x 9', C/C		1.60	16.250		7,900	490		8,390	9,525
0850 6' x 9', C/P		1.60	16.250		6,900	490		7,390	8,425
1000 8' x 12', C/C		1.10	23.636		10,500	715		11,215	12,800
1050 8' x 12', C/P		1.10	23.636		9,300	715		10,015	11,400
1200 8' x 8', C/C		1.40	18.571		8,275	560		8,835	10,000

For customer support on your Residential Costs with RSMeans data, call 800.448.8182.

547

13 24 Special Activity Rooms

13 24 16 – Saunas

13 24 16.50 Saunas and Heaters

		Crew	Daily Output	Labor-Hours	Unit	Material	2018 Bare Costs Labor	Equipment	Total	Total Incl O&P
1250	8' x 8', C/P	L-7	1.40	18.571	Ea.	7,375	560		7,935	9,025
1400	8' x 10', C/C		1.20	21.667		9,050	655		9,705	11,100
1450	8' x 10', C/P		1.20	21.667		7,975	655		8,630	9,875
1600	10' x 12', C/C		1	26		12,200	785		12,985	14,700
1650	10' x 12', C/P		1	26		10,800	785		11,585	13,200
1700	Door only, cedar, 2'x6', with 1' x 4' tempered insulated glass window	2 Carp	3.40	4.706		620	164		784	955
1800	Prehung, incl. jambs, pulls & hardware	"	12	1.333		680	46.50		726.50	830
2500	Heaters only (incl. above), wall mounted, to 200 C.F.					775			775	855
2750	To 300 C.F.					1,050			1,050	1,150
3000	Floor standing, to 720 C.F., 10,000 watts, w/controls	1 Elec	3	2.667		2,500	107		2,607	2,925
3250	To 1,000 C.F., 16,000 watts	"	3	2.667		3,625	107		3,732	4,150

13 24 26 – Steam Baths

13 24 26.50 Steam Baths and Components

		Crew	Daily Output	Labor-Hours	Unit	Material	2018 Bare Costs Labor	Equipment	Total	Total Incl O&P
0010	**STEAM BATHS AND COMPONENTS**									
0020	Heater, timer & head, single, to 140 C.F.	1 Plum	1.20	6.667	Ea.	2,325	265		2,590	2,975
0500	To 300 C.F.	"	1.10	7.273		2,600	289		2,889	3,325
2700	Conversion unit for residential tub, including door					3,750			3,750	4,100

13 34 Fabricated Engineered Structures

13 34 13 – Glazed Structures

13 34 13.13 Greenhouses

		Crew	Daily Output	Labor-Hours	Unit	Material	2018 Bare Costs Labor	Equipment	Total	Total Incl O&P
0010	**GREENHOUSES**, Shell only, stock units, not incl. 2' stub walls,									
0020	foundation, floors, heat or compartments									
0300	Residential type, free standing, 8'-6" long x 7'-6" wide	2 Carp	59	.271	SF Flr.	24	9.50		33.50	42.50
0400	10'-6" wide		85	.188		44.50	6.60		51.10	60
0600	13'-6" wide		108	.148		46	5.20		51.20	59
0700	17'-0" wide		160	.100		46	3.50		49.50	56.50
0900	Lean-to type, 3'-10" wide		34	.471		44	16.45		60.45	75.50
1000	6'-10" wide		58	.276		29	9.65		38.65	48
1100	Wall mounted to existing window, 3' x 3'	1 Carp	4	2	Ea.	2,125	70		2,195	2,450
1120	4' x 5'	"	3	2.667	"	2,325	93		2,418	2,700
1200	Deluxe quality, free standing, 7'-6" wide	2 Carp	55	.291	SF Flr.	86	10.15		96.15	111
1220	10'-6" wide		81	.198		64.50	6.90		71.40	82.50
1240	13'-6" wide		104	.154		53.50	5.40		58.90	68
1260	17'-0" wide		150	.107		49	3.73		52.73	59.50
1400	Lean-to type, 3'-10" wide		31	.516		105	18.05		123.05	146
1420	6'-10" wide		55	.291		70.50	10.15		80.65	94.50
1440	8'-0" wide		97	.165		60	5.75		65.75	75.50

13 34 13.19 Swimming Pool Enclosures

		Crew	Daily Output	Labor-Hours	Unit	Material	2018 Bare Costs Labor	Equipment	Total	Total Incl O&P
0010	**SWIMMING POOL ENCLOSURES** Translucent, free standing									
0020	not including foundations, heat or light									
0200	Economy	2 Carp	200	.080	SF Hor.	51	2.80		53.80	60.50
0600	Deluxe	"	70	.229	"	93	8		101	115

13 34 63 – Natural Fiber Construction

13 34 63.50 Straw Bale Construction

		Crew	Daily Output	Labor-Hours	Unit	Material	2018 Bare Costs Labor	Equipment	Total	Total Incl O&P
0010	**STRAW BALE CONSTRUCTION**									
2020	Straw bales in walls w/modified post and beam frame [G]	2 Carp	320	.050	S.F.	6.35	1.75		8.10	9.85

Estimating Tips
General
- Many products in Division 14 will require some type of support or blocking for installation not included with the item itself. Examples are supports for conveyors or tube systems, attachment points for lifts, and footings for hoists or cranes. Add these supports in the appropriate division.

14 10 00 Dumbwaiters
14 20 00 Elevators
- Dumbwaiters and elevators are estimated and purchased in a method similar to buying a car. The manufacturer has a base unit with standard features. Added to this base unit price will be whatever options the owner or specifications require. Increased load capacity, additional vertical travel, additional stops, higher speed, and cab finish options are items to be considered. When developing an estimate for dumbwaiters and elevators, remember that some items needed by the installers may have to be included as part of the general contract.

Examples are:
- ☐ shaftway
- ☐ rail support brackets
- ☐ machine room
- ☐ electrical supply
- ☐ sill angles
- ☐ electrical connections
- ☐ pits
- ☐ roof penthouses
- ☐ pit ladders

Check the job specifications and drawings before pricing.
- Installation of elevators and handicapped lifts in historic structures can require significant additional costs. The associated structural requirements may involve cutting into and repairing finishes, moldings, flooring, etc. The estimator must account for these special conditions.

14 30 00 Escalators and Moving Walks
- Escalators and moving walks are specialty items installed by specialty contractors. There are numerous options associated with these items. For specific options, contact a manufacturer or contractor. In a method similar to estimating dumbwaiters and elevators, you should verify the extent of general contract work and add items as necessary.

14 40 00 Lifts
14 90 00 Other Conveying Equipment
- Products such as correspondence lifts, chutes, and pneumatic tube systems, as well as other items specified in this subdivision, may require trained installers. The general contractor might not have any choice as to who will perform the installation or when it will be performed. Long lead times are often required for these products, making early decisions in scheduling necessary.

Reference Numbers
Reference numbers are shown at the beginning of some major classifications. These numbers refer to related items in the Reference Section. The reference information may be an estimating procedure, an alternate pricing method, or technical information.

Note: Not all subdivisions listed here necessarily appear. ∎

---

14 21 Electric Traction Elevators

14 21 33 – Electric Traction Residential Elevators

14 21 33.20 Residential Elevators	Crew	Daily Output	Labor-Hours	Unit	Material	2018 Bare Costs Labor	Equipment	Total	Total Incl O&P
0010 **RESIDENTIAL ELEVATORS**									
7000 Residential, cab type, 1 floor, 2 stop, economy model	2 Elev	.20	80	Ea.	10,800	4,350		15,150	18,900
7100 Custom model		.10	160		18,200	8,675		26,875	34,200
7200 2 floor, 3 stop, economy model		.12	133		16,000	7,225		23,225	29,400
7300 Custom model		.06	267		26,100	14,500		40,600	52,500

14 42 Wheelchair Lifts

14 42 13 – Inclined Wheelchair Lifts

14 42 13.10 Inclined Wheelchair Lifts and Stairclimbers

	Crew	Daily Output	Labor-Hours	Unit	Material	2018 Bare Costs Labor	Equipment	Total	Total Incl O&P
0010 **INCLINED WHEELCHAIR LIFTS AND STAIRCLIMBERS**									
7700 Stair climber (chair lift), single seat, minimum	2 Elev	1	16	Ea.	5,025	870		5,895	6,950
7800 Maximum	"	.20	80	"	6,875	4,350		11,225	14,700

Estimating Tips

Pipe for fire protection and all uses is located in Subdivisions 21 11 13 and 22 11 13.

The labor adjustment factors listed in Subdivision 22 01 02.20 also apply to Division 21.

Many, but not all, areas in the U.S. require backflow protection in the fire system. Insurance underwriters may have specific requirements for the type of materials to be installed or design requirements based on the hazard to be protected. Local jurisdictions may have requirements not covered by code. It is advisable to be aware of any special conditions.

For your reference, the following is a list of the most applicable Fire Codes and Standards which may be purchased from the NFPA, 1 Batterymarch Park, Quincy, MA 02169-7471.

- NFPA 1: Uniform Fire Code
- NFPA 10: Portable Fire Extinguishers
- NFPA 11: Low-, Medium-, and High-Expansion Foam
- NFPA 12: Carbon Dioxide Extinguishing Systems (Also companion 12A)
- NFPA 13: Installation of Sprinkler Systems (Also companion 13D, 13E, and 13R)
- NFPA 14: Installation of Standpipe and Hose Systems
- NFPA 15: Water Spray Fixed Systems for Fire Protection
- NFPA 16: Installation of Foam-Water Sprinkler and Foam-Water Spray Systems
- NFPA 17: Dry Chemical Extinguishing Systems (Also companion 17A)
- NFPA 18: Wetting Agents
- NFPA 20: Installation of Stationary Pumps for Fire Protection
- NFPA 22: Water Tanks for Private Fire Protection
- NFPA 24: Installation of Private Fire Service Mains and their Appurtenances
- NFPA 25: Inspection, Testing and Maintenance of Water-Based Fire Protection

Reference Numbers

Reference numbers are shown at the beginning of some major classifications. These numbers refer to related items in the Reference Section. The reference information may be an estimating procedure, an alternate pricing method, or technical information.

Note: Not all subdivisions listed here necessarily appear. ■

---

21 05 23 – General-Duty Valves for Water-Based Fire-Suppression Piping

21 05 23.50 General-Duty Valves	Crew	Daily Output	Labor-Hours	Unit	Material	2018 Bare Costs Labor	Equipment	Total	Total Incl O&P
0010 **GENERAL-DUTY VALVES**, for water-based fire suppression									
6200 Valves and components									
6210 Wet alarm, includes									
6220 retard chamber, trim, gauges, alarm line strainer									
6260 3" size	Q-12	3	5.333	Ea.	1,775	187		1,962	2,250
6280 4" size	"	2	8		1,925	280		2,205	2,550
6300 6" size	Q-13	4	8	↓	1,825	280		2,105	2,475
6400 Dry alarm, includes									
6405 retard chamber, trim, gauges, alarm line strainer									
6410 1-1/2" size	Q-12	3	5.333	Ea.	4,500	187		4,687	5,250
6420 2" size		3	5.333		4,500	187		4,687	5,250
6430 3" size		3	5.333		4,575	187		4,762	5,350
6440 4" size	↓	2	8		4,900	280		5,180	5,825
6450 6" size	Q-13	3	10.667		5,650	375		6,025	6,825
6460 8" size	"	3	10.667	↓	8,300	375		8,675	9,750
6500 Check, swing, C.I. body, brass fittings, auto. ball drip									
6520 4" size	Q-12	3	5.333	Ea.	380	187		567	725
6800 Check, wafer, butterfly type, C.I. body, bronze fittings									
6820 4" size	Q-12	4	4	Ea.	1,275	140		1,415	1,625
8700 Floor control valve, includes trim and gauges, 2" size		6	2.667		890	93.50		983.50	1,125
8710 2-1/2" size		6	2.667		960	93.50		1,053.50	1,200
8720 3" size		6	2.667		1,000	93.50		1,093.50	1,250
8730 4" size		6	2.667		1,000	93.50		1,093.50	1,250
8740 6" size		5	3.200		1,000	112		1,112	1,275
8800 Flow control valve, includes trim and gauges, 2" size		2	8		4,725	280		5,005	5,650
8820 3" size	↓	1.50	10.667		5,675	375		6,050	6,875
8840 4" size	Q-13	2.80	11.429		5,675	400		6,075	6,900
8860 6" size	"	2	16		7,725	560		8,285	9,425
9200 Pressure operated relief valve, brass body	1 Spri	18	.444	↓	625	17.30		642.30	720
9600 Waterflow indicator, vane type, with recycling retard and									
9610 two single pole retard switches, 2" thru 6" pipe size	1 Spri	8	1	Ea.	163	39		202	244

21 05 53 – Identification For Fire-Suppression Piping and Equipment

21 05 53.50 Identification

	Crew	Daily Output	Labor-Hours	Unit	Material	2018 Bare Costs Labor	Equipment	Total	Total Incl O&P
0010 **IDENTIFICATION**, for fire suppression piping and equipment									
3010 Plates and escutcheons for identification of fire dept. service/connections									
3100 Wall mount, round, aluminum									
3110 4"	1 Plum	96	.083	Ea.	22.50	3.32		25.82	30.50
3120 6"	"	96	.083	"	60	3.32		63.32	71.50
3200 Wall mount, round, cast brass									
3210 2-1/2"	1 Plum	70	.114	Ea.	54	4.55		58.55	67
3220 3"		70	.114		63	4.55		67.55	77
3230 4"		70	.114		108	4.55		112.55	127
3240 6"	↓	70	.114	↓	126	4.55		130.55	147
3250 For polished brass, add									
3260 For rough chrome, add									
3270 For polished chrome, add									
3300 Wall mount, square, cast brass									
3310 2-1/2"	1 Plum	70	.114	Ea.	144	4.55		148.55	166
3320 3"	"	70	.114	"	153	4.55		157.55	176
3330 For polished brass, add									
3340 For rough chrome, add									
3350 For polished chrome, add									

21 05 Common Work Results for Fire Suppression

21 05 53 – Identification For Fire-Suppression Piping and Equipment

21 05 53.50 Identification	Crew	Daily Output	Labor-Hours	Unit	Material	2018 Bare Costs Labor	2018 Bare Costs Equipment	Total	Total Incl O&P
3400 Wall mount, cast brass, multiple outlets									
3410 rect. 2 way	Q-1	5	3.200	Ea.	200	115		315	410
3420 rect. 3 way		4	4		470	143		613	750
3430 rect. 4 way		4	4		585	143		728	880
3440 square 4 way		4	4		575	143		718	870
3450 rect. 6 way	▼	3	5.333	▼	800	191		991	1,200
3460 For polished brass, add									
3470 For rough chrome, add									
3480 For polished chrome, add									
3500 Base mount, free standing fdc, cast brass									
3510 4"	1 Plum	60	.133	Ea.	108	5.30		113.30	128
3520 6"	"	60	.133	"	144	5.30		149.30	167
3530 For polished brass, add									
3540 For rough chrome, add									
3550 For polished chrome, add									

21 11 Facility Fire-Suppression Water-Service Piping

21 11 13 – Facility Water Distribution Piping

21 11 13.16 Pipe, Plastic

	Crew	Daily Output	Labor-Hours	Unit	Material	2018 Bare Costs Labor	2018 Bare Costs Equipment	Total	Total Incl O&P
0010 **PIPE, PLASTIC**									
0020 CPVC, fire suppression (C-UL-S, FM, NFPA 13, 13D & 13R)									
0030 Socket joint, no couplings or hangers									
0100 SDR 13.5 (ASTM F442)									
0120 3/4" diameter	Q-12	420	.038	L.F.	1.67	1.33		3	4.04
0130 1" diameter		340	.047		2.58	1.65		4.23	5.55
0140 1-1/4" diameter		260	.062		4.10	2.15		6.25	8.05
0150 1-1/2" diameter		190	.084		5.65	2.95		8.60	11.10
0160 2" diameter		140	.114		9	4		13	16.50
0170 2-1/2" diameter		130	.123		16.65	4.31		20.96	25.50
0180 3" diameter	▼	120	.133	▼	25.50	4.67		30.17	35.50

21 11 13.18 Pipe Fittings, Plastic

	Crew	Daily Output	Labor-Hours	Unit	Material	2018 Bare Costs Labor	2018 Bare Costs Equipment	Total	Total Incl O&P
0010 **PIPE FITTINGS, PLASTIC**									
0020 CPVC, fire suppression (C-UL-S, FM, NFPA 13, 13D & 13R)									
0030 Socket joint									
0100 90° elbow									
0120 3/4"	1 Plum	26	.308	Ea.	1.78	12.25		14.03	22
0130 1"		22.70	.352		3.91	14.05		17.96	27.50
0140 1-1/4"		20.20	.396		4.94	15.75		20.69	31.50
0150 1-1/2"	▼	18.20	.440		7	17.50		24.50	36.50
0160 2"	Q-1	33.10	.483		8.70	17.30		26	38
0170 2-1/2"		24.20	.661		16.75	23.50		40.25	57.50
0180 3"	▼	20.80	.769	▼	23	27.50		50.50	70.50
0200 45° elbow									
0210 3/4"	1 Plum	26	.308	Ea.	2.45	12.25		14.70	22.50
0220 1"		22.70	.352		2.87	14.05		16.92	26
0230 1-1/4"		20.20	.396		4.15	15.75		19.90	30.50
0240 1-1/2"	▼	18.20	.440		5.80	17.50		23.30	35.50
0250 2"	Q-1	33.10	.483		7.20	17.30		24.50	36.50
0260 2-1/2"		24.20	.661		12.95	23.50		36.45	53
0270 3"	▼	20.80	.769	▼	18.55	27.50		46.05	66
0300 Tee									

For customer support on your Residential Costs with RSMeans data, call 800.448.8182.

553

21 11 13.18 Pipe Fittings, Plastic		Crew	Daily Output	Labor-Hours	Unit	Material	2018 Bare Costs Labor	Equipment	Total	Total Incl O&P
0310	3/4"	1 Plum	17.30	.462	Ea.	2.45	18.40		20.85	33
0320	1"		15.20	.526		4.82	21		25.82	40
0330	1-1/4"		13.50	.593		7.25	23.50		30.75	47
0340	1-1/2"		12.10	.661		10.70	26.50		37.20	55.50
0350	2"	Q-1	20	.800		15.80	28.50		44.30	64.50
0360	2-1/2"		16.20	.988		25.50	35.50		61	86.50
0370	3"		13.90	1.151		40	41		81	112
0400	Tee, reducing x any size									
0420	1"	1 Plum	15.20	.526	Ea.	4.09	21		25.09	39
0430	1-1/4"		13.50	.593		7.50	23.50		31	47.50
0440	1-1/2"		12.10	.661		9.10	26.50		35.60	53.50
0450	2"	Q-1	20	.800		15.50	28.50		44	64
0460	2-1/2"		16.20	.988		19.95	35.50		55.45	80.50
0470	3"		13.90	1.151		23	41		64	93.50
0500	Coupling									
0510	3/4"	1 Plum	26	.308	Ea.	1.71	12.25		13.96	22
0520	1"		22.70	.352		2.26	14.05		16.31	25.50
0530	1-1/4"		20.20	.396		3.29	15.75		19.04	29.50
0540	1-1/2"		18.20	.440		4.70	17.50		22.20	34
0550	2"	Q-1	33.10	.483		6.35	17.30		23.65	35.50
0560	2-1/2"		24.20	.661		9.70	23.50		33.20	49.50
0570	3"		20.80	.769		12.60	27.50		40.10	59.50
0600	Coupling, reducing									
0610	1" x 3/4"	1 Plum	22.70	.352	Ea.	2.26	14.05		16.31	25.50
0620	1-1/4" x 1"		20.20	.396		3.42	15.75		19.17	30
0630	1-1/2" x 3/4"		18.20	.440		5.15	17.50		22.65	34.50
0640	1-1/2" x 1"		18.20	.440		4.94	17.50		22.44	34.50
0650	1-1/2" x 1-1/4"		18.20	.440		4.70	17.50		22.20	34
0660	2" x 1"	Q-1	33.10	.483		6.60	17.30		23.90	36
0670	2" x 1-1/2"	"	33.10	.483		6.35	17.30		23.65	35.50
0700	Cross									
0720	3/4"	1 Plum	13	.615	Ea.	3.84	24.50		28.34	44.50
0730	1"		11.30	.708		4.82	28		32.82	52
0740	1-1/4"		10.10	.792		6.65	31.50		38.15	59.50
0750	1-1/2"		9.10	.879		9.20	35		44.20	67.50
0760	2"	Q-1	16.60	.964		15	34.50		49.50	73.50
0770	2-1/2"	"	12.10	1.322		33	47.50		80.50	114
0800	Cap									
0820	3/4"	1 Plum	52	.154	Ea.	1.04	6.10		7.14	11.25
0830	1"		45	.178		1.47	7.10		8.57	13.25
0840	1-1/4"		40	.200		2.38	7.95		10.33	15.70
0850	1-1/2"		36.40	.220		3.29	8.75		12.04	18
0860	2"	Q-1	66	.242		4.94	8.70		13.64	19.75
0870	2-1/2"		48.40	.331		7.15	11.85		19	27.50
0880	3"		41.60	.385		11.50	13.80		25.30	35
0900	Adapter, sprinkler head, female w/metal thd. insert (s x FNPT)									
0920	3/4" x 1/2"	1 Plum	52	.154	Ea.	4.61	6.10		10.71	15.15
0930	1" x 1/2"		45	.178		4.86	7.10		11.96	17
0940	1" x 3/4"		45	.178		7.65	7.10		14.75	20

21 11 16.50 Fire Hydrants for Buildings

		Crew	Daily Output	Labor-Hours	Unit	Material	2018 Bare Costs Labor	Equipment	Total	Total Incl O&P
0010	**FIRE HYDRANTS FOR BUILDINGS**									
3750	Hydrants, wall, w/caps, single, flush, polished brass									
3800	2-1/2" x 2-1/2"	Q-12	5	3.200	Ea.	248	112		360	460
3840	2-1/2" x 3"		5	3.200		500	112		612	735
3860	3" x 3"	↓	4.80	3.333	↓	405	117		522	640
3900	For polished chrome, add					20%				
3950	Double, flush, polished brass									
4000	2-1/2" x 2-1/2" x 4"	Q-12	5	3.200	Ea.	765	112		877	1,025
4040	2-1/2" x 2-1/2" x 6"		4.60	3.478		875	122		997	1,150
4080	3" x 3" x 4"		4.90	3.265		1,125	114		1,239	1,450
4120	3" x 3" x 6"	↓	4.50	3.556	↓	1,600	124		1,724	1,950
4200	For polished chrome, add					10%				
4350	Double, projecting, polished brass									
4400	2-1/2" x 2-1/2" x 4"	Q-12	5	3.200	Ea.	296	112		408	510
4450	2-1/2" x 2-1/2" x 6"	"	4.60	3.478	"	605	122		727	865
4460	Valve control, dbl. flush/projecting hydrant, cap &									
4470	chain, extension rod & cplg., escutcheon, polished brass	Q-12	8	2	Ea.	272	70		342	415
4480	Four-way square, flush, polished brass									
4540	2-1/2" (4) x 6"	Q-12	3.60	4.444	Ea.	3,575	156		3,731	4,200

21 11 19.50 Connections for the Fire-Department

		Crew	Daily Output	Labor-Hours	Unit	Material	2018 Bare Costs Labor	Equipment	Total	Total Incl O&P
0010	**CONNECTIONS FOR THE FIRE-DEPARTMENT**									
4000	Storz type, with cap and chain									
6000	Roof manifold, horiz., brass, without valves & caps									
6040	2-1/2" x 2-1/2" x 4"	Q-12	4.80	3.333	Ea.	194	117		311	405
6060	2-1/2" x 2-1/2" x 6"		4.60	3.478		221	122		343	445
6080	2-1/2" x 2-1/2" x 2-1/2" x 4"		4.60	3.478		325	122		447	560
6090	2-1/2" x 2-1/2" x 2-1/2" x 6"	↓	4.60	3.478	↓	335	122		457	570
7000	Sprinkler line tester, cast brass					38			38	42
7140	Standpipe connections, wall, w/plugs & chains									
7160	Single, flush, brass, 2-1/2" x 2-1/2"	Q-12	5	3.200	Ea.	188	112		300	390
7180	2-1/2" x 3"	"	5	3.200	"	196	112		308	400
7240	For polished chrome, add					15%				
7280	Double, flush, polished brass									
7300	2-1/2" x 2-1/2" x 4"	Q-12	5	3.200	Ea.	760	112		872	1,025
7330	2-1/2" x 2-1/2" x 6"		4.60	3.478		890	122		1,012	1,175
7340	3" x 3" x 4"		4.90	3.265		1,025	114		1,139	1,325
7370	3" x 3" x 6"	↓	4.50	3.556	↓	1,250	124		1,374	1,575
7400	For polished chrome, add					15%				
7440	For sill cock combination, add				Ea.	96			96	105
7580	Double projecting, polished brass									
7600	2-1/2" x 2-1/2" x 4"	Q-12	5	3.200	Ea.	535	112		647	770
7630	2-1/2" x 2-1/2" x 6"	"	4.60	3.478	"	890	122		1,012	1,175
7680	For polished chrome, add					15%				
7900	Three way, flush, polished brass									
7920	2-1/2" (3) x 4"	Q-12	4.80	3.333	Ea.	1,950	117		2,067	2,350
7930	2-1/2" (3) x 6"	"	4.60	3.478		2,000	122		2,122	2,400
8000	For polished chrome, add				↓	9%				
8020	Three way, projecting, polished brass									
8040	2-1/2" (3) x 4"	Q-12	4.80	3.333	Ea.	850	117		967	1,125
8070	2-1/2" (3) x 6"	"	4.60	3.478	↓	1,750	122		1,872	2,125
8100	For polished chrome, add					12%				

For customer support on your Residential Costs with RSMeans data, call 800.448.8182.

555

21 11 Facility Fire-Suppression Water-Service Piping

21 11 19 – Fire-Department Connections

		Crew	Daily Output	Labor-Hours	Unit	Material	2018 Bare Costs Labor	Equipment	Total	Total Incl O&P
21 11 19.50	**Connections for the Fire-Department**									
8200	Four way, square, flush, polished brass,									
8240	2-1/2" (4) x 6"	Q-12	3.60	4.444	Ea.	1,500	156		1,656	1,900
8300	For polished chrome, add				"	10%				
8550	Wall, vertical, flush, cast brass									
8600	Two way, 2-1/2" x 2-1/2" x 4"	Q-12	5	3.200	Ea.	395	112		507	620
8660	Four way, 2-1/2" (4) x 6"		3.80	4.211		1,325	147		1,472	1,700
8680	Six way, 2-1/2" (6) x 6"		3.40	4.706		1,550	165		1,715	2,000
8700	For polished chrome, add					10%				
8800	Free standing siamese unit, polished brass, two way									
8820	2-1/2" x 2-1/2" x 4"	Q-12	2.50	6.400	Ea.	735	224		959	1,175
8850	2-1/2" x 2-1/2" x 6"		2	8		745	280		1,025	1,275
8860	3" x 3" x 4"		2.50	6.400		500	224		724	920
8890	3" x 3" x 6"		2	8		1,325	280		1,605	1,900
8940	For polished chrome, add					12%				
9100	Free standing siamese unit, polished brass, three way									
9120	2-1/2" x 2-1/2" x 2-1/2" x 6"	Q-12	2	8	Ea.	950	280		1,230	1,500
9160	For polished chrome, add				"	15%				

21 12 Fire-Suppression Standpipes

21 12 19 – Fire-Suppression Hose Racks

		Crew	Daily Output	Labor-Hours	Unit	Material	2018 Bare Costs Labor	Equipment	Total	Total Incl O&P
21 12 19.50	**Fire Hose Racks**									
0010	**FIRE HOSE RACKS**									
2600	Hose rack, swinging, for 1-1/2" diameter hose,									
2620	Enameled steel, 50' and 75' lengths of hose	Q-12	20	.800	Ea.	69	28		97	122
2640	100' and 125' lengths of hose		20	.800		90	28		118	145
2680	Chrome plated, 50' and 75' lengths of hose		20	.800		79.50	28		107.50	134
2700	100' and 125' lengths of hose		20	.800		139	28		167	199
2780	For hose rack nipple, 1-1/2" polished brass, add					35			35	38.50
2820	2-1/2" polished brass, add					54.50			54.50	60
2840	1-1/2" polished chrome, add					34.50			34.50	38
2860	2-1/2" polished chrome, add					71			71	78

21 12 23 – Fire-Suppression Hose Valves

		Crew	Daily Output	Labor-Hours	Unit	Material	2018 Bare Costs Labor	Equipment	Total	Total Incl O&P
21 12 23.70	**Fire Hose Valves**									
0010	**FIRE HOSE VALVES**									
0020	Angle, combination pressure adjust/restricting, rough brass									
0030	1-1/2"	1 Spri	12	.667	Ea.	96	26		122	149
0040	2-1/2"	"	7	1.143	"	208	44.50		252.50	305
0042	Nonpressure adjustable/restricting, rough brass									
0044	1-1/2"	1 Spri	12	.667	Ea.	48	26		74	95.50
0046	2-1/2"	"	7	1.143	"	95.50	44.50		140	179
0050	For polished brass, add					30%				
0060	For polished chrome, add					40%				
1000	Ball drip, automatic, rough brass, 1/2"	1 Spri	20	.400	Ea.	23.50	15.55		39.05	51
1010	3/4"	"	20	.400	"	18.70	15.55		34.25	46
1100	Ball, 175 lb., sprinkler system, FM/UL, threaded, bronze									
1120	Slow close									
1150	1" size	1 Spri	19	.421	Ea.	256	16.40		272.40	310
1160	1-1/4" size		15	.533		276	21		297	340
1170	1-1/2" size		13	.615		395	24		419	475
1180	2" size		11	.727		445	28.50		473.50	530

21 12 Fire-Suppression Standpipes

21 12 23 – Fire-Suppression Hose Valves

21 12 23.70 Fire Hose Valves		Crew	Daily Output	Labor-Hours	Unit	Material	2018 Bare Costs Labor	Equipment	Total	Total Incl O&P
1190	2-1/2" size	Q-12	15	1.067	Ea.	595	37.50		632.50	715
1230	For supervisory switch kit, all sizes									
1240	One circuit, add	1 Spri	48	.167	Ea.	176	6.50		182.50	205
1280	Quarter turn for trim									
1300	1/2" size	1 Spri	22	.364	Ea.	34.50	14.15		48.65	61.50
1310	3/4" size		20	.400		37	15.55		52.55	66.50
1320	1" size		19	.421		41	16.40		57.40	72.50
1330	1-1/4" size		15	.533		67.50	21		88.50	109
1340	1-1/2" size		13	.615		84.50	24		108.50	133
1350	2" size		11	.727		101	28.50		129.50	158
1400	Caps, polished brass with chain, 3/4"					44.50			44.50	49
1420	1"					57			57	62.50
1440	1-1/2"					20.50			20.50	22.50
1460	2-1/2"					31			31	34
1480	3"					38.50			38.50	42
1900	Escutcheon plate, for angle valves, polished brass, 1-1/2"					15.20			15.20	16.70
1920	2-1/2"					25			25	27.50
1940	3"					31.50			31.50	34.50
1980	For polished chrome, add					15%				
3000	Gate, hose, wheel handle, N.R.S., rough brass, 1-1/2"	1 Spri	12	.667		154	26		180	212
3040	2-1/2", 300 lb.	"	7	1.143		222	44.50		266.50	320
3080	For polished brass, add					40%				
3090	For polished chrome, add					50%				
5000	Pressure reducing rough brass, 1-1/2"	1 Spri	12	.667		330	26		356	405
5020	2-1/2"	"	7	1.143		385	44.50		429.50	500
5080	For polished brass, add					105%				
5090	For polished chrome, add					140%				

21 13 Fire-Suppression Sprinkler Systems

21 13 13 – Wet-Pipe Sprinkler Systems

21 13 13.50 Wet-Pipe Sprinkler System Components

		Crew	Daily Output	Labor-Hours	Unit	Material	2018 Bare Costs Labor	Equipment	Total	Total Incl O&P
0010	**WET-PIPE SPRINKLER SYSTEM COMPONENTS**									
1100	Alarm, electric pressure switch (circuit closer)	1 Spri	26	.308	Ea.	105	11.95		116.95	136
1140	For explosion proof, max 20 psi, contacts close or open		26	.308		660	11.95		671.95	745
1220	Water motor, complete with gong		4	2		445	78		523	620
1900	Flexible sprinkler head connectors									
1910	Braided stainless steel hose with mounting bracket									
1920	1/2" and 3/4" outlet size									
1940	40" length	1 Spri	30	.267	Ea.	76	10.35		86.35	101
1960	60" length	"	22	.364	"	87.50	14.15		101.65	120
1982	May replace hard-pipe armovers									
1984	for wet and pre-action systems.									
2000	Release, emergency, manual, for hydraulic or pneumatic system	1 Spri	12	.667	Ea.	235	26		261	300
2060	Release, thermostatic, for hydraulic or pneumatic release line		20	.400		730	15.55		745.55	825
2200	Sprinkler cabinets, 6 head capacity		16	.500		85.50	19.45		104.95	126
2260	12 head capacity		16	.500		90.50	19.45		109.95	132
2340	Sprinkler head escutcheons, standard, brass tone, 1" size		40	.200		3.09	7.80		10.89	16.25
2360	Chrome, 1" size		40	.200		3.29	7.80		11.09	16.45
2400	Recessed type, bright brass		40	.200		10.90	7.80		18.70	25
2440	Chrome or white enamel		40	.200		3.63	7.80		11.43	16.85
2600	Sprinkler heads, not including supply piping									

21 13 13 – Wet-Pipe Sprinkler Systems

21 13 13.50 Wet-Pipe Sprinkler System Components

21 13 13.50 Wet-Pipe Sprinkler System Components	Crew	Daily Output	Labor-Hours	Unit	Material	2018 Bare Costs Labor	Equipment	Total	Total Incl O&P
3700 Standard spray, pendent or upright, brass, 135°F to 286°F									
3720 1/2" NPT, 3/8" orifice	1 Spri	16	.500	Ea.	16.20	19.45		35.65	50
3730 1/2" NPT, 7/16" orifice		16	.500		16	19.45		35.45	49.50
3732 1/2" NPT, 7/16" orifice, chrome		16	.500		16.80	19.45		36.25	50.50
3740 1/2" NPT, 1/2" orifice		16	.500		10.40	19.45		29.85	43.50
3760 1/2" NPT, 17/32" orifice		16	.500		13.50	19.45		32.95	47
3780 3/4" NPT, 17/32" orifice		16	.500		12.45	19.45		31.90	45.50
3840 For chrome, add					3.70			3.70	4.07
4200 Sidewall, vertical brass, 135°F to 286°F									
4240 1/2" NPT, 1/2" orifice	1 Spri	16	.500	Ea.	26.50	19.45		45.95	61.50
4280 3/4" NPT, 17/32" orifice	"	16	.500		74	19.45		93.45	113
4360 For satin chrome, add					3.70			3.70	4.07
4500 Sidewall, horizontal, brass, 135°F to 286°F									
4520 1/2" NPT, 1/2" orifice	1 Spri	16	.500	Ea.	26.50	19.45		45.95	61.50
5600 Concealed, complete with cover plate									
5620 1/2" NPT, 1/2" orifice, 135°F to 212°F	1 Spri	9	.889	Ea.	25.50	34.50		60	85
6025 Residential sprinkler components (one and two family)									
6026 Water motor alarm with strainer	1 Spri	4	2	Ea.	445	78		523	620
6027 Fast response, glass bulb, 135°F to 155°F									
6028 1/2" NPT, pendent, brass	1 Spri	16	.500	Ea.	28.50	19.45		47.95	63.50
6029 1/2" NPT, sidewall, brass		16	.500		28.50	19.45		47.95	63.50
6030 1/2" NPT, pendent, brass, extended coverage		16	.500		25	19.45		44.45	59.50
6031 1/2" NPT, sidewall, brass, extended coverage		16	.500		25	19.45		44.45	59.50
6032 3/4" NPT sidewall, brass, extended coverage		16	.500		25.50	19.45		44.95	60
6033 For chrome, add					15%				
6034 For polyester/teflon coating add					20%				
6100 Sprinkler head wrenches, standard head				Ea.	26			26	28.50
6120 Recessed head					39.50			39.50	43.50
6160 Tamper switch (valve supervisory switch)	1 Spri	16	.500		238	19.45		257.45	294
6165 Flow switch (valve supervisory switch)	"	16	.500		238	19.45		257.45	294

21 13 16 – Dry-Pipe Sprinkler Systems

21 13 16.50 Dry-Pipe Sprinkler System Components

	Crew	Daily Output	Labor-Hours	Unit	Material	2018 Bare Costs Labor	Equipment	Total	Total Incl O&P
0010 **DRY-PIPE SPRINKLER SYSTEM COMPONENTS**									
0600 Accelerator	1 Spri	8	1	Ea.	820	39		859	965
0800 Air compressor for dry pipe system, automatic, complete									
0820 30 gal. system capacity, 3/4 HP	1 Spri	1.30	6.154	Ea.	1,150	239		1,389	1,650
0860 30 gal. system capacity, 1 HP		1.30	6.154		1,425	239		1,664	1,950
0960 Air pressure maintenance control		24	.333		375	12.95		387.95	430
1600 Dehydrator package, incl. valves and nipples		12	.667		770	26		796	890
2600 Sprinkler heads, not including supply piping									
2640 Dry, pendent, 1/2" orifice, 3/4" or 1" NPT									
2660 1/2" to 6" length	1 Spri	14	.571	Ea.	132	22		154	182
2670 6-1/4" to 8" length		14	.571		137	22		159	187
2680 8-1/4" to 12" length		14	.571		143	22		165	194
2800 For each inch or fraction, add					3.70			3.70	4.07
6330 Valves and components									
6340 Alarm test/shut off valve, 1/2"	1 Spri	20	.400	Ea.	23.50	15.55		39.05	51
8000 Dry pipe air check valve, 3" size	Q-12	2	8		1,925	280		2,205	2,550
8200 Dry pipe valve, incl. trim and gauges, 3" size		2	8		2,800	280		3,080	3,525
8220 4" size		1	16		3,075	560		3,635	4,325
8240 6" size	Q-13	2	16		3,625	560		4,185	4,900
8280 For accelerator trim with gauges, add	1 Spri	8	1		258	39		297	350

Estimating Tips
22 10 00 Plumbing
Piping and Pumps

This subdivision is primarily basic pipe and related materials. The pipe may be used by any of the mechanical disciplines, i.e., plumbing, fire protection, heating, and air conditioning.

Note: CPVC plastic piping approved for fire protection is located in 21 11 13.

- The labor adjustment factors listed in Subdivision 22 01 02.20 apply throughout Divisions 21, 22, and 23. CAUTION: the correct percentage may vary for the same items. For example, the percentage add for the basic pipe installation should be based on the maximum height that the installer must install for that particular section. If the pipe is to be located 14' above the floor but it is suspended on threaded rod from beams, the bottom flange of which is 18' high (4' rods), then the height is actually 18' and the add is 20%. The pipe cover, however, does not have to go above the 14', and so the add should be 10%.

- Most pipe is priced first as straight pipe with a joint (coupling, weld, etc.) every 10' and a hanger usually every 10'. There are exceptions with hanger spacing such as for cast iron pipe (5')

and plastic pipe (3 per 10'). Following each type of pipe there are several lines listing sizes and the amount to be subtracted to delete couplings and hangers. This is for pipe that is to be buried or supported together on trapeze hangers. The reason that the couplings are deleted is that these runs are usually long, and frequently longer lengths of pipe are used. By deleting the couplings, the estimator is expected to look up and add back the correct reduced number of couplings.

- When preparing an estimate, it may be necessary to approximate the fittings. Fittings usually run between 25% and 50% of the cost of the pipe. The lower percentage is for simpler runs, and the higher number is for complex areas, such as mechanical rooms.

- For historic restoration projects, the systems must be as invisible as possible, and pathways must be sought for pipes, conduit, and ductwork. While installations in accessible spaces (such as basements and attics) are relatively straightforward to estimate, labor costs may be more difficult to determine when delivery systems must be concealed.

22 40 00 Plumbing Fixtures

- Plumbing fixture costs usually require two lines: the fixture itself and its "rough-in, supply, and waste."

- In the Assemblies Section (Plumbing D2010) for the desired fixture, the System Components Group at the center of the page shows the fixture on the first line. The rest of the list (fittings, pipe, tubing, etc.) will total up to what we refer to in the Unit Price section as "Rough-in, supply, waste, and vent." Note that for most fixtures we allow a nominal 5' of tubing to reach from the fixture to a main or riser.

- Remember that gas- and oil-fired units need venting.

Reference Numbers

Reference numbers are shown at the beginning of some major classifications. These numbers refer to related items in the Reference Section. The reference information may be an estimating procedure, an alternate pricing method, or technical information.

Note: Not all subdivisions listed here necessarily appear. ■

Did you know?

RSMeans data is available through our online application with 24/7 access:

- Search for unit prices by keyword
- Leverage the most up-to-date data
- Build and export estimates

Try it free for 30 days!
www.rsmeans.com/2018freetrial

22 05 Common Work Results for Plumbing

22 05 05 – Selective Demolition for Plumbing

22 05 05.10 Plumbing Demolition

		Crew	Daily Output	Labor-Hours	Unit	Material	2018 Bare Costs Labor	Equipment	Total	Total Incl O&P
0010	**PLUMBING DEMOLITION**									
1020	Fixtures, including 10' piping									
1101	Bathtubs, cast iron	1 Clab	4	2	Ea.		53.50		53.50	89
1121	Fiberglass		6	1.333			35.50		35.50	59
1141	Steel	↓	5	1.600			42.50		42.50	71
1200	Lavatory, wall hung	1 Plum	10	.800			32		32	52.50
1221	Counter top	1 Clab	16	.500			13.35		13.35	22
1301	Sink, single compartment		16	.500			13.35		13.35	22
1321	Double	↓	10	.800			21.50		21.50	35.50
1400	Water closet, floor mounted	1 Plum	8	1			40		40	65.50
1421	Wall mounted	1 Clab	7	1.143	↓		30.50		30.50	50.50
2001	Piping, metal, to 1-1/2" diameter		200	.040	L.F.		1.07		1.07	1.78
2051	2" thru 3-1/2" diameter		150	.053			1.42		1.42	2.37
2101	4" thru 6" diameter	↓	100	.080			2.14		2.14	3.55
2160	Plastic pipe with fittings, up thru 1-1/2" diameter	1 Plum	250	.032			1.27		1.27	2.10
2162	2" thru 3" diameter	"	200	.040			1.59		1.59	2.62
2164	4" thru 6" diameter	Q-1	200	.080			2.87		2.87	4.72
2166	8" thru 14" diameter		150	.107			3.82		3.82	6.30
2168	16" diameter	↓	100	.160	↓		5.75		5.75	9.45
3000	Submersible sump pump	1 Plum	24	.333	Ea.		13.25		13.25	22
6000	Remove and reset fixtures, easy access		6	1.333			53		53	87.50
6100	Difficult access	↓	4	2	↓		79.50		79.50	131

22 05 23 – General-Duty Valves for Plumbing Piping

22 05 23.20 Valves, Bronze

		Crew	Daily Output	Labor-Hours	Unit	Material	2018 Bare Costs Labor	Equipment	Total	Total Incl O&P
0010	**VALVES, BRONZE**									
1750	Check, swing, class 150, regrinding disc, threaded									
1860	3/4"	1 Plum	20	.400	Ea.	86.50	15.90		102.40	121
1870	1"	"	19	.421	"	141	16.75		157.75	183
2850	Gate, N.R.S., soldered, 125 psi									
2940	3/4"	1 Plum	20	.400	Ea.	64	15.90		79.90	96
2950	1"	"	19	.421	"	74.50	16.75		91.25	110
5600	Relief, pressure & temperature, self-closing, ASME, threaded									
5640	3/4"	1 Plum	28	.286	Ea.	209	11.35		220.35	249
5650	1"		24	.333		335	13.25		348.25	390
5660	1-1/4"	↓	20	.400	↓	685	15.90		700.90	780
6400	Pressure, water, ASME, threaded									
6440	3/4"	1 Plum	28	.286	Ea.	144	11.35		155.35	177
6450	1"	"	24	.333	"	296	13.25		309.25	345
6900	Reducing, water pressure									
6920	300 psi to 25-75 psi, threaded or sweat									
6940	1/2"	1 Plum	24	.333	Ea.	440	13.25		453.25	505
6950	3/4"		20	.400		450	15.90		465.90	520
6960	1"	↓	19	.421	↓	700	16.75		716.75	800
8350	Tempering, water, sweat connections									
8400	1/2"	1 Plum	24	.333	Ea.	106	13.25		119.25	138
8440	3/4"	"	20	.400	"	139	15.90		154.90	179
8650	Threaded connections									
8700	1/2"	1 Plum	24	.333	Ea.	148	13.25		161.25	185
8740	3/4"	"	20	.400	"	875	15.90		890.90	985
8800	Water heater water & gas safety shut off									
8810	Protection against a leaking water heater									
8814	Shut off valve	1 Plum	16	.500	Ea.	185	19.90		204.90	237

22 05 Common Work Results for Plumbing

22 05 23 – General-Duty Valves for Plumbing Piping

22 05 23.20 Valves, Bronze		Crew	Daily Output	Labor-Hours	Unit	Material	2018 Bare Costs Labor	Equipment	Total	Total Incl O&P
8818	Water heater dam	1 Plum	32	.250	Ea.	31.50	9.95		41.45	51
8822	Gas control wiring harness	↓	32	.250	↓	19.75	9.95		29.70	38
8830	Whole house flood safety shut off									
8834	Connections									
8838	3/4" NPT	1 Plum	12	.667	Ea.	935	26.50		961.50	1,075
8842	1" NPT		11	.727		960	29		989	1,100
8846	1-1/4" NPT	↓	10	.800	↓	995	32		1,027	1,150

22 05 29 – Hangers and Supports for Plumbing Piping and Equipment

22 05 29.10 Hangers & Supp. for Plumb'g/HVAC Pipe/Equip.

		Crew	Daily Output	Labor-Hours	Unit	Material	2018 Bare Costs Labor	Equipment	Total	Total Incl O&P
0010	**HANGERS AND SUPPORTS FOR PLUMB'G/HVAC PIPE/EQUIP.**									
8000	Pipe clamp, plastic, 1/2" CTS	1 Plum	80	.100	Ea.	.23	3.98		4.21	6.80
8010	3/4" CTS		73	.110		.24	4.36		4.60	7.45
8020	1" CTS		68	.118		.53	4.68		5.21	8.30
8080	Economy clamp, 1/4" CTS		175	.046		.05	1.82		1.87	3.05
8090	3/8" CTS		168	.048		.05	1.90		1.95	3.18
8100	1/2" CTS		160	.050		.05	1.99		2.04	3.34
8110	3/4" CTS		145	.055		.05	2.20		2.25	3.67
8200	Half clamp, 1/2" CTS		80	.100		.07	3.98		4.05	6.65
8210	3/4" CTS		73	.110		.10	4.36		4.46	7.30
8300	Suspension clamp, 1/2" CTS		80	.100		.23	3.98		4.21	6.80
8310	3/4" CTS		73	.110		.24	4.36		4.60	7.45
8320	1" CTS		68	.118		.53	4.68		5.21	8.30
8400	Insulator, 1/2" CTS		80	.100		.32	3.98		4.30	6.90
8410	3/4" CTS		73	.110		.33	4.36		4.69	7.55
8420	1" CTS		68	.118		.37	4.68		5.05	8.10
8500	J hook clamp with nail, 1/2" CTS		240	.033		.10	1.33		1.43	2.29
8501	3/4" CTS	↓	240	.033	↓	.10	1.33		1.43	2.29

22 05 48 – Vibration and Seismic Controls for Plumbing Piping and Equipment

22 05 48.10 Seismic Bracing Supports

		Crew	Daily Output	Labor-Hours	Unit	Material	2018 Bare Costs Labor	Equipment	Total	Total Incl O&P
0010	**SEISMIC BRACING SUPPORTS**									
0020	Clamps									
0030	C-clamp, for mounting on steel beam									
0040	3/8" threaded rod	1 Skwk	160	.050	Ea.	3.93	1.79		5.72	7.30
0050	1/2" threaded rod		160	.050		4.88	1.79		6.67	8.35
0060	5/8" threaded rod		160	.050		5.30	1.79		7.09	8.85
0070	3/4" threaded rod	↓	160	.050	↓	6.80	1.79		8.59	10.50
0100	Brackets									
0110	Beam side or wall malleable iron									
0120	3/8" threaded rod	1 Skwk	48	.167	Ea.	4.07	5.95		10.02	14.45
0130	1/2" threaded rod		48	.167		3.78	5.95		9.73	14.10
0140	5/8" threaded rod		48	.167		11	5.95		16.95	22
0150	3/4" threaded rod		48	.167		18.90	5.95		24.85	31
0160	7/8" threaded rod	↓	48	.167	↓	12.50	5.95		18.45	23.50
0170	For concrete installation, add						30%			
0180	Wall, welded steel									
0190	0 size 12" wide 18" deep	1 Skwk	34	.235	Ea.	178	8.40		186.40	209
0200	1 size 18" wide 24" deep		34	.235		211	8.40		219.40	246
0210	2 size 24" wide 30" deep	↓	34	.235	↓	279	8.40		287.40	320
0300	Rod, carbon steel									
0310	Continuous thread									
0320	1/4" thread	1 Skwk	144	.056	L.F.	2.18	1.99		4.17	5.70
0330	3/8" thread	↓	144	.056	↓	2.33	1.99		4.32	5.90

22 05 48 – Vibration and Seismic Controls for Plumbing Piping and Equipment

22 05 48.10 Seismic Bracing Supports	Crew	Daily Output	Labor-Hours	Unit	Material	2018 Bare Costs Labor	2018 Bare Costs Equipment	Total	Total Incl O&P
0340	1/2" thread	1 Skwk	144	.056	L.F.	3.67	1.99		5.66	7.35
0350	5/8" thread		144	.056		5.20	1.99		7.19	9
0360	3/4" thread		144	.056		9.15	1.99		11.14	13.40
0370	7/8" thread		144	.056		11.50	1.99		13.49	15.95
0380	For galvanized, add					30%				
0400	Channel, steel									
0410	3/4" x 1-1/2"	1 Skwk	80	.100	L.F.	2.61	3.58		6.19	8.85
0420	1-1/2" x 1-1/2"		70	.114		3.41	4.09		7.50	10.60
0430	1-7/8" x 1-1/2"		60	.133		16.30	4.77		21.07	26
0440	3" x 1-1/2"		50	.160		18.25	5.75		24	29.50
0450	Spring nuts									
0460	3/8"	1 Skwk	100	.080	Ea.	1.59	2.86		4.45	6.55
0470	1/2"	"	80	.100	"	1.67	3.58		5.25	7.85
0500	Welding, field									
0510	Cleaning and welding plates, bars, or rods									
0520	To existing beams, columns, or trusses									
0530	1" weld	1 Skwk	144	.056	Ea.	.23	1.99		2.22	3.57
0540	2" weld		72	.111		.46	3.98		4.44	7.15
0550	3" weld		54	.148		.69	5.30		5.99	9.60
0560	4" weld		36	.222		.92	7.95		8.87	14.30
0570	5" weld		30	.267		1.15	9.55		10.70	17.20
0580	6" weld		24	.333		1.38	11.95		13.33	21.50
0600	Vibration absorbers									
0610	Hangers, neoprene flex									
0620	10-120 lb. capacity	1 Skwk	8	1	Ea.	25.50	36		61.50	88
0630	75-550 lb. capacity		8	1		40.50	36		76.50	105
0640	250-1,100 lb. capacity		6	1.333		82.50	47.50		130	170
0650	1,000-4,000 lb. capacity		6	1.333		147	47.50		194.50	242

22 05 76 – Facility Drainage Piping Cleanouts

22 05 76.10 Cleanouts

		Crew	Daily Output	Labor-Hours	Unit	Material	2018 Bare Costs Labor	2018 Bare Costs Equipment	Total	Total Incl O&P
0010	**CLEANOUTS**									
0060	Floor type									
0080	Round or square, scoriated nickel bronze top									
0100	2" pipe size	1 Plum	10	.800	Ea.	184	32		216	256
0120	3" pipe size		8	1		249	40		289	340
0140	4" pipe size		6	1.333		277	53		330	395

22 05 76.20 Cleanout Tees

		Crew	Daily Output	Labor-Hours	Unit	Material	2018 Bare Costs Labor	2018 Bare Costs Equipment	Total	Total Incl O&P
0010	**CLEANOUT TEES**									
0100	Cast iron, B&S, with countersunk plug									
0220	3" pipe size	1 Plum	3.60	2.222	Ea.	150	88.50		238.50	310
0240	4" pipe size	"	3.30	2.424	"	232	96.50		328.50	415
0500	For round smooth access cover, same price									
4000	Plastic, tees and adapters. Add plugs									
4010	ABS, DWV									
4020	Cleanout tee, 1-1/2" pipe size	1 Plum	15	.533	Ea.	12.20	21		33.20	48.50

22 07 Plumbing Insulation

22 07 16 – Plumbing Equipment Insulation

22 07 16.10 Insulation for Plumbing Equipment		Crew	Daily Output	Labor-Hours	Unit	Material	2018 Bare Costs Labor	Equipment	Total	Total Incl O&P
0010	**INSULATION FOR PLUMBING EQUIPMENT**									
2900	Domestic water heater wrap kit									
2920	1-1/2" with vinyl jacket, 20 to 60 gal.	G 1 Plum	8	1	Ea.	16.50	40		56.50	83.50

22 07 19 – Plumbing Piping Insulation

22 07 19.10 Piping Insulation

		Crew	Daily Output	Labor-Hours	Unit	Material	Labor	Equipment	Total	Total Incl O&P
0010	**PIPING INSULATION**									
0230	Insulated protectors (ADA)									
0235	For exposed piping under sinks or lavatories									
0240	Vinyl coated foam, velcro tabs									
0245	P Trap, 1-1/4" or 1-1/2"	1 Plum	32	.250	Ea.	15.90	9.95		25.85	34
0260	Valve and supply cover									
0265	1/2", 3/8", and 7/16" pipe size	1 Plum	32	.250	Ea.	15.60	9.95		25.55	33.50
0285	1-1/4" pipe size	"	32	.250	"	12.90	9.95		22.85	30.50
0600	Pipe covering (price copper tube one size less than IPS)									
6600	Fiberglass, with all service jacket									
6840	1" wall, 1/2" iron pipe size	G Q-14	240	.067	L.F.	.85	2.16		3.01	4.60
6860	3/4" iron pipe size	G	230	.070		.92	2.26		3.18	4.83
6870	1" iron pipe size	G	220	.073		.99	2.36		3.35	5.10
6900	2" iron pipe size	G	200	.080		1.53	2.60		4.13	6.05
7879	Rubber tubing, flexible closed cell foam									
8100	1/2" wall, 1/4" iron pipe size	G 1 Asbe	90	.089	L.F.	.72	3.20		3.92	6.20
8130	1/2" iron pipe size	G	89	.090		.84	3.24		4.08	6.40
8140	3/4" iron pipe size	G	89	.090		.95	3.24		4.19	6.55
8150	1" iron pipe size	G	88	.091		.82	3.28		4.10	6.45
8170	1-1/2" iron pipe size	G	87	.092		1.49	3.31		4.80	7.25
8180	2" iron pipe size	G	86	.093		1.91	3.35		5.26	7.75
8300	3/4" wall, 1/4" iron pipe size	G	90	.089		.89	3.20		4.09	6.40
8330	1/2" iron pipe size	G	89	.090		1.09	3.24		4.33	6.70
8340	3/4" iron pipe size	G	89	.090		1.54	3.24		4.78	7.20
8350	1" iron pipe size	G	88	.091		1.77	3.28		5.05	7.50
8380	2" iron pipe size	G	86	.093		3.04	3.35		6.39	9
8444	1" wall, 1/2" iron pipe size	G	86	.093		2.32	3.35		5.67	8.20
8445	3/4" iron pipe size	G	84	.095		2.81	3.43		6.24	8.90
8446	1" iron pipe size	G	84	.095		3.39	3.43		6.82	9.55
8447	1-1/4" iron pipe size	G	82	.098		3.75	3.52		7.27	10.10
8448	1-1/2" iron pipe size	G	82	.098		4.88	3.52		8.40	11.30
8449	2" iron pipe size	G	80	.100		6.10	3.61		9.71	12.80
8450	2-1/2" iron pipe size	G	80	.100		7.45	3.61		11.06	14.30
8456	Rubber insulation tape, 1/8" x 2" x 30'	G			Ea.	21			21	23

22 11 Facility Water Distribution

22 11 13 – Facility Water Distribution Piping

22 11 13.23 Pipe/Tube, Copper

		Crew	Daily Output	Labor-Hours	Unit	Material	Labor	Equipment	Total	Total Incl O&P
0010	**PIPE/TUBE, COPPER**, Solder joints									
1000	Type K tubing, couplings & clevis hanger assemblies 10' OC									
1180	3/4" diameter	1 Plum	74	.108	L.F.	8.40	4.30		12.70	16.30
1200	1" diameter	"	66	.121	"	12.30	4.82		17.12	21.50
2000	Type L tubing, couplings & clevis hanger assemblies 10' OC									
2140	1/2" diameter	1 Plum	81	.099	L.F.	3.34	3.93		7.27	10.15
2160	5/8" diameter		79	.101		5.75	4.03		9.78	13

For customer support on your Residential Costs with RSMeans data, call 800.448.8182.

563

22 11 13.23 Pipe/Tube, Copper

		Crew	Daily Output	Labor-Hours	Unit	Material	2018 Bare Costs Labor	2018 Bare Costs Equipment	Total	Total Incl O&P
2180	3/4" diameter	1 Plum	76	.105	L.F.	4.49	4.19		8.68	11.85
2200	1" diameter		68	.118		6.90	4.68		11.58	15.30
2220	1-1/4" diameter	▼	58	.138	▼	11.45	5.50		16.95	21.50
3000	Type M tubing, couplings & clevis hanger assemblies 10' OC									
3140	1/2" diameter	1 Plum	84	.095	L.F.	3.70	3.79		7.49	10.30
3180	3/4" diameter		78	.103		5.20	4.08		9.28	12.45
3200	1" diameter		70	.114		8.20	4.55		12.75	16.50
3220	1-1/4" diameter		60	.133		10.75	5.30		16.05	20.50
3240	1-1/2" diameter		54	.148		13.85	5.90		19.75	25
3260	2" diameter	▼	44	.182	▼	20.50	7.25		27.75	34.50
4000	Type DWV tubing, couplings & clevis hanger assemblies 10' OC									
4100	1-1/4" diameter	1 Plum	60	.133	L.F.	11.60	5.30		16.90	21.50
4120	1-1/2" diameter		54	.148		11.25	5.90		17.15	22
4140	2" diameter	▼	44	.182		18.05	7.25		25.30	32
4160	3" diameter	Q-1	58	.276		24.50	9.90		34.40	43.50
4180	4" diameter	"	40	.400	▼	55	14.35		69.35	84

22 11 13.25 Pipe/Tube Fittings, Copper

		Crew	Daily Output	Labor-Hours	Unit	Material	2018 Bare Costs Labor	2018 Bare Costs Equipment	Total	Total Incl O&P
0010	**PIPE/TUBE FITTINGS, COPPER**, Wrought unless otherwise noted									
0040	Solder joints, copper x copper									
0070	90° elbow, 1/4"	1 Plum	22	.364	Ea.	3.15	14.45		17.60	27.50
0100	1/2"		20	.400		1.13	15.90		17.03	27
0120	3/4"		19	.421		2.52	16.75		19.27	30.50
0250	45° elbow, 1/4"		22	.364		8	14.45		22.45	33
0280	1/2"		20	.400		2.83	15.90		18.73	29
0290	5/8"		19	.421		12	16.75		28.75	40.50
0300	3/4"		19	.421		3.56	16.75		20.31	31.50
0310	1"		16	.500		8.95	19.90		28.85	43
0320	1-1/4"		15	.533		16.20	21		37.20	53
0450	Tee, 1/4"		14	.571		7.70	22.50		30.20	46
0480	1/2"		13	.615		1.92	24.50		26.42	42.50
0490	5/8"		12	.667		16.45	26.50		42.95	61.50
0500	3/4"		12	.667		6.35	26.50		32.85	50.50
0510	1"		10	.800		14.10	32		46.10	68
0520	1-1/4"		9	.889		22	35.50		57.50	82
0612	Tee, reducing on the outlet, 1/4"		15	.533		14.70	21		35.70	51
0613	3/8"		15	.533		14.25	21		35.25	50.50
0614	1/2"		14	.571		13.15	22.50		35.65	52
0615	5/8"		13	.615		26.50	24.50		51	69.50
0616	3/4"		12	.667		8.50	26.50		35	53
0617	1"		11	.727		23.50	29		52.50	73.50
0618	1-1/4"		10	.800		28.50	32		60.50	83.50
0619	1-1/2"		9	.889		30	35.50		65.50	91
0620	2"	▼	8	1		49	40		89	120
0621	2-1/2"	Q-1	9	1.778		136	63.50		199.50	254
0622	3"		8	2		140	71.50		211.50	272
0623	4"		6	2.667		283	95.50		378.50	465
0624	5"	▼	5	3.200		1,550	115		1,665	1,900
0625	6"	Q-2	7	3.429		2,000	118		2,118	2,400
0626	8"	"	6	4		8,125	138		8,263	9,175
0630	Tee, reducing on the run, 1/4"	1 Plum	15	.533		19	21		40	56
0631	3/8"		15	.533		26.50	21		47.50	64
0632	1/2"	▼	14	.571	▼	18.10	22.50		40.60	57.50

22 11 13.25 Pipe/Tube Fittings, Copper

		Crew	Daily Output	Labor-Hours	Unit	Material	2018 Bare Costs Labor	Equipment	Total	Total Incl O&P
0633	5/8"	1 Plum	13	.615	Ea.	28.50	24.50		53	72
0634	3/4"		12	.667		19.80	26.50		46.30	65.50
0635	1"		11	.727		23	29		52	73
0636	1-1/4"		10	.800		36.50	32		68.50	92.50
0637	1-1/2"		9	.889		62	35.50		97.50	126
0638	2"	▼	8	1		82	40		122	156
0639	2-1/2"	Q-1	9	1.778		175	63.50		238.50	298
0640	3"		8	2		256	71.50		327.50	400
0641	4"		6	2.667		520	95.50		615.50	730
0642	5"	▼	5	3.200		1,475	115		1,590	1,800
0643	6"	Q-2	7	3.429		2,225	118		2,343	2,650
0644	8"	"	6	4		8,875	138		9,013	9,975
0650	Coupling, 1/4"	1 Plum	24	.333		.95	13.25		14.20	23
0680	1/2"		22	.364		.86	14.45		15.31	25
0690	5/8"		21	.381		3.43	15.15		18.58	29
0700	3/4"		21	.381		2.35	15.15		17.50	27.50
0710	1"		18	.444		4.66	17.70		22.36	34
0715	1-1/4"	▼	17	.471	▼	6.10	18.75		24.85	37.50
2000	DWV, solder joints, copper x copper									
2030	90° elbow, 1-1/4"	1 Plum	13	.615	Ea.	16.50	24.50		41	58.50
2050	1-1/2"		12	.667		22	26.50		48.50	67.50
2070	2"	▼	10	.800		35	32		67	91
2090	3"	Q-1	10	1.600		84.50	57.50		142	188
2100	4"	"	9	1.778		410	63.50		473.50	555
2250	Tee, sanitary, 1-1/4"	1 Plum	9	.889		25.50	35.50		61	86
2270	1-1/2"		8	1		31.50	40		71.50	100
2290	2"	▼	7	1.143		49.50	45.50		95	129
2310	3"	Q-1	7	2.286		188	82		270	340
2330	4"	"	6	2.667		455	95.50		550.50	655
2400	Coupling, 1-1/4"	1 Plum	14	.571		6.85	22.50		29.35	45
2420	1-1/2"		13	.615		8.50	24.50		33	50
2440	2"	▼	11	.727		11.80	29		40.80	60.50
2460	3"	Q-1	11	1.455		27.50	52		79.50	116
2480	4"	"	10	1.600	▼	60.50	57.50		118	161

22 11 13.44 Pipe, Steel

		Crew	Daily Output	Labor-Hours	Unit	Material	2018 Bare Costs Labor	Equipment	Total	Total Incl O&P
0010	**PIPE, STEEL**									
0050	Schedule 40, threaded, with couplings, and clevis hanger									
0060	assemblies sized for covering, 10' OC									
0540	Black, 1/4" diameter	1 Plum	66	.121	L.F.	5.55	4.82		10.37	14.05
0560	1/2" diameter		63	.127		3.82	5.05		8.87	12.50
0570	3/4" diameter		61	.131		4.18	5.20		9.38	13.20
0580	1" diameter	▼	53	.151		4.38	6		10.38	14.70
0590	1-1/4" diameter	Q-1	89	.180		5.15	6.45		11.60	16.25
0600	1-1/2" diameter		80	.200		5.70	7.15		12.85	18.05
0610	2" diameter	▼	64	.250	▼	11.95	8.95		20.90	28

22 11 13.45 Pipe Fittings, Steel, Threaded

		Crew	Daily Output	Labor-Hours	Unit	Material	2018 Bare Costs Labor	Equipment	Total	Total Incl O&P
0010	**PIPE FITTINGS, STEEL, THREADED**									
5000	Malleable iron, 150 lb.									
5020	Black									
5040	90° elbow, straight									
5090	3/4"	1 Plum	14	.571	Ea.	4.41	22.50		26.91	42.50
5100	1"	"	13	.615	▼	7.70	24.50		32.20	49

22 11 13.45 Pipe Fittings, Steel, Threaded		Crew	Daily Output	Labor-Hours	Unit	Material	2018 Bare Costs Labor	Equipment	Total	Total Incl O&P
5120	1-1/2"	Q-1	20	.800	Ea.	16.65	28.50		45.15	65.50
5130	2"	"	18	.889		28.50	32		60.50	84
5450	Tee, straight									
5500	3/4"	1 Plum	9	.889	Ea.	7.05	35.50		42.55	66
5510	1"	"	8	1		12.05	40		52.05	79
5520	1-1/4"	Q-1	14	1.143		19.50	41		60.50	89
5530	1-1/2"		13	1.231		24	44		68	99
5540	2"		11	1.455		41.50	52		93.50	132
5650	Coupling									
5700	3/4"	1 Plum	18	.444	Ea.	5.95	17.70		23.65	35.50
5710	1"	"	15	.533		8.90	21		29.90	45
5720	1-1/4"	Q-1	26	.615		11.50	22		33.50	49
5730	1-1/2"		24	.667		15.55	24		39.55	56.50
5740	2"		21	.762		23	27.50		50.50	70.50

22 11 13.74 Pipe, Plastic

		Crew	Daily Output	Labor-Hours	Unit	Material	Labor	Equipment	Total	Total Incl O&P
0010	**PIPE, PLASTIC**									
1800	PVC, couplings 10' OC, clevis hanger assemblies, 3 per 10'									
1820	Schedule 40									
1860	1/2" diameter	1 Plum	54	.148	L.F.	5.10	5.90		11	15.30
1870	3/4" diameter		51	.157		5.60	6.25		11.85	16.40
1880	1" diameter		46	.174		8.75	6.90		15.65	21
1890	1-1/4" diameter		42	.190		9.50	7.60		17.10	23
1900	1-1/2" diameter		36	.222		9.60	8.85		18.45	25
1910	2" diameter	Q-1	59	.271		11.10	9.70		20.80	28
1920	2-1/2" diameter		56	.286		10.60	10.25		20.85	28.50
1930	3" diameter		53	.302		12.90	10.80		23.70	32
1940	4" diameter		48	.333		30.50	11.95		42.45	53
4100	DWV type, schedule 40, couplings 10' OC, clevis hanger assy's, 3 per 10'									
4210	ABS, schedule 40, foam core type									
4212	Plain end black									
4214	1-1/2" diameter	1 Plum	39	.205	L.F.	8	8.15		16.15	22
4216	2" diameter	Q-1	62	.258		8.55	9.25		17.80	24.50
4218	3" diameter		56	.286		8.35	10.25		18.60	26
4220	4" diameter		51	.314		25	11.25		36.25	46
4222	6" diameter		42	.381		18.95	13.65		32.60	43.50
4240	To delete coupling & hangers, subtract									
4244	1-1/2" diam. to 6" diam.					43%	48%			
4400	PVC									
4410	1-1/4" diameter	1 Plum	42	.190	L.F.	8.50	7.60		16.10	22
4420	1-1/2" diameter	"	36	.222		7.75	8.85		16.60	23
4460	2" diameter	Q-1	59	.271		8.65	9.70		18.35	25.50
4470	3" diameter		53	.302		8.55	10.80		19.35	27
4480	4" diameter		48	.333		10.20	11.95		22.15	31
5300	CPVC, socket joint, couplings 10' OC, clevis hanger assemblies, 3 per 10'									
5302	Schedule 40									
5304	1/2" diameter	1 Plum	54	.148	L.F.	5.85	5.90		11.75	16.15
5305	3/4" diameter		51	.157		6.90	6.25		13.15	17.85
5306	1" diameter		46	.174		10.50	6.90		17.40	23
5307	1-1/4" diameter		42	.190		11.85	7.60		19.45	25.50
5308	1-1/2" diameter		36	.222		11.75	8.85		20.60	27.50
5309	2" diameter	Q-1	59	.271		14.80	9.70		24.50	32.50
5360	CPVC, threaded, couplings 10' OC, clevis hanger assemblies, 3 per 10'									

566

For customer support on your Residential Costs with RSMeans data, call 800.448.8182.

22 11 13.74 Pipe, Plastic	Crew	Daily Output	Labor-Hours	Unit	Material	2018 Bare Costs Labor	Equipment	Total	Total Incl O&P
5380 Schedule 40									
5460 1/2" diameter	1 Plum	54	.148	L.F.	6.70	5.90		12.60	17.05
5470 3/4" diameter		51	.157		8.35	6.25		14.60	19.45
5480 1" diameter		46	.174		12	6.90		18.90	24.50
5490 1-1/4" diameter		42	.190		13	7.60		20.60	27
5500 1-1/2" diameter	↓	36	.222		12.75	8.85		21.60	28.50
5510 2" diameter	Q-1	59	.271	↓	16.05	9.70		25.75	33.50
6500 Residential installation, plastic pipe									
6510 Couplings 10' OC, strap hangers 3 per 10'									
6520 PVC, Schedule 40									
6530 1/2" diameter	1 Plum	138	.058	L.F.	1.03	2.31		3.34	4.93
6540 3/4" diameter		128	.063		1.19	2.49		3.68	5.40
6550 1" diameter		119	.067		1.76	2.68		4.44	6.35
6560 1-1/4" diameter		111	.072		2.14	2.87		5.01	7.05
6570 1-1/2" diameter	↓	104	.077		2.33	3.06		5.39	7.60
6580 2" diameter	Q-1	197	.081		2.94	2.91		5.85	8
6590 2-1/2" diameter		162	.099		4.83	3.54		8.37	11.15
6600 4" diameter	↓	123	.130	↓	8.10	4.66		12.76	16.55
6700 PVC, DWV, Schedule 40									
6720 1-1/4" diameter	1 Plum	100	.080	L.F.	2.04	3.18		5.22	7.50
6730 1-1/2" diameter	"	94	.085		1.46	3.39		4.85	7.15
6740 2" diameter	Q-1	178	.090		2.06	3.22		5.28	7.55
6760 4" diameter	"	110	.145	↓	5.80	5.20		11	14.95
7280 PEX, flexible, no couplings or hangers									
7282 Note: For labor costs add 25% to the couplings and fittings labor total.									
7285 For fittings see section 23 83 16.10 7000									
7300 Non-barrier type, hot/cold tubing rolls									
7310 1/4" diameter x 100'				L.F.	.49			.49	.54
7350 3/8" diameter x 100'					.55			.55	.61
7360 1/2" diameter x 100'					.61			.61	.67
7370 1/2" diameter x 500'					.61			.61	.67
7380 1/2" diameter x 1000'					.61			.61	.67
7400 3/4" diameter x 100'					.92			.92	1.01
7410 3/4" diameter x 500'					1.11			1.11	1.22
7420 3/4" diameter x 1000'					1.11			1.11	1.22
7460 1" diameter x 100'					1.90			1.90	2.09
7470 1" diameter x 300'					1.90			1.90	2.09
7480 1" diameter x 500'					1.90			1.90	2.09
7500 1-1/4" diameter x 100'					3.23			3.23	3.55
7510 1-1/4" diameter x 300'					3.23			3.23	3.55
7540 1-1/2" diameter x 100'					4.40			4.40	4.84
7550 1-1/2" diameter x 300'				↓	4.40			4.40	4.84
7596 Most sizes available in red or blue									
7700 Non-barrier type, hot/cold tubing straight lengths									
7710 1/2" diameter x 20'				L.F.	.61			.61	.67
7750 3/4" diameter x 20'					1.11			1.11	1.22
7760 1" diameter x 20'					1.90			1.90	2.09
7770 1-1/4" diameter x 20'					3.23			3.23	3.55
7780 1-1/2" diameter x 20'					4.40			4.40	4.84
7790 2" diameter				↓	8.60			8.60	9.45
7796 Most sizes available in red or blue									

For customer support on your Residential Costs with RSMeans data, call 800.448.8182.

567

22 11 13.76 Pipe Fittings, Plastic	Crew	Daily Output	Labor-Hours	Unit	Material	2018 Bare Costs Labor	Equipment	Total	Total Incl O&P
0010 PIPE FITTINGS, PLASTIC									
2700 PVC (white), schedule 40, socket joints									
2760 90° elbow, 1/2"	1 Plum	33.30	.240	Ea.	.50	9.55		10.05	16.30
2770 3/4"		28.60	.280		.57	11.15		11.72	18.95
2780 1"		25	.320		1.02	12.75		13.77	22
2790 1-1/4"		22.20	.360		1.80	14.35		16.15	25.50
2800 1-1/2"	▼	20	.400		1.94	15.90		17.84	28
2810 2"	Q-1	36.40	.440		3.04	15.75		18.79	29.50
2820 2-1/2"		26.70	.599		9.25	21.50		30.75	45.50
2830 3"		22.90	.699		11.10	25		36.10	53
2840 4"	▼	18.20	.879		19.80	31.50		51.30	74
3180 Tee, 1/2"	1 Plum	22.20	.360		.63	14.35		14.98	24
3190 3/4"		19	.421		.73	16.75		17.48	28.50
3200 1"		16.70	.479		1.36	19.05		20.41	33
3210 1-1/4"		14.80	.541		2.12	21.50		23.62	38
3220 1-1/2"	▼	13.30	.602		2.57	24		26.57	42.50
3230 2"	Q-1	24.20	.661		3.75	23.50		27.25	43
3240 2-1/2"		17.80	.899		12.35	32		44.35	66.50
3250 3"		15.20	1.053		16.25	37.50		53.75	80
3260 4"	▼	12.10	1.322		29.50	47.50		77	111
3380 Coupling, 1/2"	1 Plum	33.30	.240		.33	9.55		9.88	16.10
3390 3/4"		28.60	.280		.46	11.15		11.61	18.80
3400 1"		25	.320		.80	12.75		13.55	22
3410 1-1/4"		22.20	.360		1.11	14.35		15.46	24.50
3420 1-1/2"	▼	20	.400		1.19	15.90		17.09	27.50
3430 2"	Q-1	36.40	.440		1.81	15.75		17.56	28
3440 2-1/2"		26.70	.599		4.01	21.50		25.51	40
3450 3"		22.90	.699		6.30	25		31.30	48
3460 4"	▼	18.20	.879	▼	9.10	31.50		40.60	62
4500 DWV, ABS, non pressure, socket joints									
4540 1/4 bend, 1-1/4"	1 Plum	20.20	.396	Ea.	4.46	15.75		20.21	31
4560 1-1/2"	"	18.20	.440		3.44	17.50		20.94	33
4570 2"	Q-1	33.10	.483	▼	5.30	17.30		22.60	34.50
4650 1/8 bend, same as 1/4 bend									
4800 Tee, sanitary									
4820 1-1/4"	1 Plum	13.50	.593	Ea.	5.90	23.50		29.40	45.50
4830 1-1/2"	"	12.10	.661		5.10	26.50		31.60	49
4840 2"	Q-1	20	.800	▼	7.85	28.50		36.35	55.50
5000 DWV, PVC, schedule 40, socket joints									
5040 1/4 bend, 1-1/4"	1 Plum	20.20	.396	Ea.	8.80	15.75		24.55	35.50
5060 1-1/2"	"	18.20	.440		2.51	17.50		20.01	32
5070 2"	Q-1	33.10	.483		3.95	17.30		21.25	33
5080 3"		20.80	.769		11.60	27.50		39.10	58.50
5090 4"	▼	16.50	.970		23	34.50		57.50	82
5110 1/4 bend, long sweep, 1-1/2"	1 Plum	18.20	.440		5.80	17.50		23.30	35.50
5112 2"	Q-1	33.10	.483		6.50	17.30		23.80	35.50
5114 3"		20.80	.769		14.90	27.50		42.40	62
5116 4"	▼	16.50	.970		28.50	34.50		63	88
5250 Tee, sanitary 1-1/4"	1 Plum	13.50	.593		9.40	23.50		32.90	49.50
5254 1-1/2"	"	12.10	.661		4.37	26.50		30.87	48.50
5255 2"	Q-1	20	.800		6.45	28.50		34.95	54
5256 3"	▼	13.90	1.151	▼	16.95	41		57.95	86.50

22 11 13 – Facility Water Distribution Piping

22 11 13.76 Pipe Fittings, Plastic		Crew	Daily Output	Labor-Hours	Unit	Material	2018 Bare Costs Labor	2018 Bare Costs Equipment	Total	Total Incl O&P
5257	4"	Q-1	11	1.455	Ea.	31	52		83	121
5259	6"	↓	6.70	2.388		125	85.50		210.50	279
5261	8"	Q-2	6.20	3.871		277	134		411	525
5264	2" x 1-1/2"	Q-1	22	.727		5.70	26		31.70	49.50
5266	3" x 1-1/2"		15.50	1.032		12.40	37		49.40	74.50
5268	4" x 3"		12.10	1.322		36.50	47.50		84	118
5271	6" x 4"	↓	6.90	2.319		121	83		204	270
5314	Combination Y & 1/8 bend, 1-1/2"	1 Plum	12.10	.661		10.65	26.50		37.15	55.50
5315	2"	Q-1	20	.800		13.30	28.50		41.80	61.50
5317	3"		13.90	1.151		29	41		70	100
5318	4"	↓	11	1.455	↓	58	52		110	150
5324	Combination Y & 1/8 bend, reducing									
5325	2" x 2" x 1-1/2"	Q-1	22	.727	Ea.	15	26		41	59.50
5327	3" x 3" x 1-1/2"		15.50	1.032		27	37		64	90.50
5328	3" x 3" x 2"		15.30	1.046		19.95	37.50		57.45	83.50
5329	4" x 4" x 2"	↓	12.20	1.311		30	47		77	111
5331	Wye, 1-1/4"	1 Plum	13.50	.593		12.05	23.50		35.55	52.50
5332	1-1/2"	"	12.10	.661		8	26.50		34.50	52.50
5333	2"	Q-1	20	.800		7.85	28.50		36.35	55.50
5334	3"		13.90	1.151		21	41		62	91.50
5335	4"		11	1.455		38.50	52		90.50	129
5336	6"	↓	6.70	2.388		112	85.50		197.50	264
5337	8"	Q-2	6.20	3.871		197	134		331	435
5341	2" x 1-1/2"	Q-1	22	.727		9.65	26		35.65	53.50
5342	3" x 1-1/2"		15.50	1.032		14.30	37		51.30	76.50
5343	4" x 3"		12.10	1.322		31.50	47.50		79	113
5344	6" x 4"	↓	6.90	2.319		85	83		168	231
5345	8" x 6"	Q-2	6.40	3.750		186	129		315	415
5347	Double wye, 1-1/2"	1 Plum	9.10	.879		18.05	35		53.05	77.50
5348	2"	Q-1	16.60	.964		20	34.50		54.50	79
5349	3"		10.40	1.538		42	55		97	137
5350	4"	↓	8.25	1.939	↓	85	69.50		154.50	208
5353	Double wye, reducing									
5354	2" x 2" x 1-1/2" x 1-1/2"	Q-1	16.80	.952	Ea.	18.40	34		52.40	76
5355	3" x 3" x 2" x 2"		10.60	1.509		31	54		85	124
5356	4" x 4" x 3" x 3"		8.45	1.893		67.50	68		135.50	186
5357	6" x 6" x 4" x 4"	↓	7.25	2.207		236	79		315	390
5374	Coupling, 1-1/4"	1 Plum	20.20	.396		5.70	15.75		21.45	32.50
5376	1-1/2"	"	18.20	.440		1.17	17.50		18.67	30.50
5378	2"	Q-1	33.10	.483		1.61	17.30		18.91	30.50
5380	3"		20.80	.769		5.65	27.50		33.15	51.50
5390	4"		16.50	.970		9.55	34.50		44.05	67.50
5410	Reducer bushing, 2" x 1-1/4"		36.50	.438		3.35	15.70		19.05	29.50
5412	3" x 1-1/2"		27.30	.586		9.90	21		30.90	45.50
5414	4" x 2"		18.20	.879		17.20	31.50		48.70	71
5416	6" x 4"	↓	11.10	1.441		44	51.50		95.50	134
5418	8" x 6"	Q-2	10.20	2.353	↓	86.50	81		167.50	229
5500	CPVC, Schedule 80, threaded joints									
5540	90° elbow, 1/4"	1 Plum	32	.250	Ea.	12.80	9.95		22.75	30.50
5560	1/2"		30.30	.264		7.40	10.50		17.90	25.50
5570	3/4"		26	.308		11.10	12.25		23.35	32
5580	1"		22.70	.352		15.60	14.05		29.65	40
5590	1-1/4"		20.20	.396		30	15.75		45.75	59

22 11 Facility Water Distribution

22 11 13 – Facility Water Distribution Piping

22 11 13.76 Pipe Fittings, Plastic	Crew	Daily Output	Labor-Hours	Unit	Material	2018 Bare Costs Labor	Equipment	Total	Total Incl O&P
5600	1-1/2"	1 Plum	18.20	.440	Ea.	32.50	17.50		50	64.50
5610	2"	Q-1	33.10	.483		43.50	17.30		60.80	76
5730	Coupling, 1/4"	1 Plum	32	.250		16.30	9.95		26.25	34.50
5732	1/2"		30.30	.264		13.40	10.50		23.90	32
5734	3/4"		26	.308		21.50	12.25		33.75	44
5736	1"		22.70	.352		24.50	14.05		38.55	50
5738	1-1/4"		20.20	.396		26	15.75		41.75	54.50
5740	1-1/2"		18.20	.440		28	17.50		45.50	60
5742	2"	Q-1	33.10	.483		33	17.30		50.30	65
5900	CPVC, Schedule 80, socket joints									
5904	90° elbow, 1/4"	1 Plum	32	.250	Ea.	12.15	9.95		22.10	30
5906	1/2"		30.30	.264		4.77	10.50		15.27	22.50
5908	3/4"		26	.308		6.10	12.25		18.35	26.50
5910	1"		22.70	.352		9.65	14.05		23.70	33.50
5912	1-1/4"		20.20	.396		21	15.75		36.75	49
5914	1-1/2"		18.20	.440		23.50	17.50		41	54.50
5916	2"	Q-1	33.10	.483		28	17.30		45.30	59.50
5930	45° elbow, 1/4"	1 Plum	32	.250		18.10	9.95		28.05	36.50
5932	1/2"		30.30	.264		5.85	10.50		16.35	23.50
5934	3/4"		26	.308		8.40	12.25		20.65	29.50
5936	1"		22.70	.352		13.40	14.05		27.45	38
5938	1-1/4"		20.20	.396		26.50	15.75		42.25	55
5940	1-1/2"		18.20	.440		27	17.50		44.50	58.50
5942	2"	Q-1	33.10	.483		30.50	17.30		47.80	62
5990	Coupling, 1/4"	1 Plum	32	.250		12.95	9.95		22.90	30.50
5992	1/2"		30.30	.264		5.05	10.50		15.55	23
5994	3/4"		26	.308		7.05	12.25		19.30	28
5996	1"		22.70	.352		9.50	14.05		23.55	33.50
5998	1-1/4"		20.20	.396		14.20	15.75		29.95	41.50
6000	1-1/2"		18.20	.440		17.85	17.50		35.35	48.50
6002	2"	Q-1	33.10	.483		21	17.30		38.30	51.50

22 11 19 – Domestic Water Piping Specialties

22 11 19.38 Water Supply Meters

		Crew	Daily Output	Labor-Hours	Unit	Material	Labor	Equipment	Total	Total Incl O&P
0010	**WATER SUPPLY METERS**									
2000	Domestic/commercial, bronze									
2020	Threaded									
2060	5/8" diameter, to 20 GPM	1 Plum	16	.500	Ea.	50.50	19.90		70.40	88.50
2080	3/4" diameter, to 30 GPM		14	.571		92	22.50		114.50	139
2100	1" diameter, to 50 GPM		12	.667		140	26.50		166.50	198

22 11 19.42 Backflow Preventers

		Crew	Daily Output	Labor-Hours	Unit	Material	Labor	Equipment	Total	Total Incl O&P
0010	**BACKFLOW PREVENTERS**, Includes valves									
0020	and four test cocks, corrosion resistant, automatic operation									
4000	Reduced pressure principle									
4100	Threaded, bronze, valves are ball									
4120	3/4" pipe size	1 Plum	16	.500	Ea.	455	19.90		474.90	535

22 11 19.50 Vacuum Breakers

		Crew	Daily Output	Labor-Hours	Unit	Material	Labor	Equipment	Total	Total Incl O&P
0010	**VACUUM BREAKERS**									
0013	See also backflow preventers Section 22 11 19.42									
1000	Anti-siphon continuous pressure type									
1010	Max. 150 psi - 210°F									
1020	Bronze body									
1030	1/2" size	1 Stpi	24	.333	Ea.	171	13.65		184.65	211

22 11 Facility Water Distribution

22 11 19 – Domestic Water Piping Specialties

22 11 19.50 Vacuum Breakers

		Crew	Daily Output	Labor-Hours	Unit	Material	2018 Bare Costs Labor	Equipment	Total	Total Incl O&P
1040	3/4" size	1 Stpi	20	.400	Ea.	171	16.40		187.40	215
1050	1" size		19	.421		176	17.25		193.25	223
1060	1-1/4" size		15	.533		345	22		367	415
1070	1-1/2" size		13	.615		425	25		450	510
1080	2" size	▼	11	.727	▼	440	30		470	535
1200	Max. 125 psi with atmospheric vent									
1210	Brass, in-line construction									
1220	1/4" size	1 Stpi	24	.333	Ea.	133	13.65		146.65	169
1230	3/8" size	"	24	.333		133	13.65		146.65	169
1260	For polished chrome finish, add				▼	13%				
2000	Anti-siphon, non-continuous pressure type									
2010	Hot or cold water 125 psi - 210°F									
2020	Bronze body									
2030	1/4" size	1 Stpi	24	.333	Ea.	73	13.65		86.65	103
2040	3/8" size		24	.333		73	13.65		86.65	103
2050	1/2" size		24	.333		82.50	13.65		96.15	113
2060	3/4" size		20	.400		98.50	16.40		114.90	135
2070	1" size		19	.421		152	17.25		169.25	197
2080	1-1/4" size		15	.533		267	22		289	330
2090	1-1/2" size		13	.615		315	25		340	385
2100	2" size		11	.727		485	30		515	585
2110	2-1/2" size		8	1		1,400	41		1,441	1,625
2120	3" size	▼	6	1.333	▼	1,850	54.50		1,904.50	2,150
2150	For polished chrome finish, add					50%				

22 11 19.54 Water Hammer Arresters/Shock Absorbers

		Crew	Daily Output	Labor-Hours	Unit	Material	2018 Bare Costs Labor	Equipment	Total	Total Incl O&P
0010	**WATER HAMMER ARRESTERS/SHOCK ABSORBERS**									
0490	Copper									
0500	3/4" male IPS For 1 to 11 fixtures	1 Plum	12	.667	Ea.	28	26.50		54.50	74.50

22 13 Facility Sanitary Sewerage

22 13 16 – Sanitary Waste and Vent Piping

22 13 16.20 Pipe, Cast Iron

		Crew	Daily Output	Labor-Hours	Unit	Material	2018 Bare Costs Labor	Equipment	Total	Total Incl O&P
0010	**PIPE, CAST IRON**, Soil, on clevis hanger assemblies, 5' OC R221113-50									
0020	Single hub, service wt., lead & oakum joints 10' OC									
2120	2" diameter	Q-1	63	.254	L.F.	16.90	9.10		26	33.50
2140	3" diameter		60	.267		20	9.55		29.55	38
2160	4" diameter	▼	55	.291	▼	35	10.40		45.40	55.50
4000	No hub, couplings 10' OC									
4100	1-1/2" diameter	Q-1	71	.225	L.F.	16	8.05		24.05	31
4120	2" diameter		67	.239		17.10	8.55		25.65	33
4140	3" diameter		64	.250		19.60	8.95		28.55	36.50
4160	4" diameter	▼	58	.276	▼	34.50	9.90		44.40	54

22 13 16.30 Pipe Fittings, Cast Iron

		Crew	Daily Output	Labor-Hours	Unit	Material	2018 Bare Costs Labor	Equipment	Total	Total Incl O&P
0010	**PIPE FITTINGS, CAST IRON**, Soil									
0040	Hub and spigot, service weight, lead & oakum joints									
0080	1/4 bend, 2"	Q-1	16	1	Ea.	20.50	36		56.50	81.50
0120	3"		14	1.143		27.50	41		68.50	97.50
0140	4"		13	1.231		43	44		87	120
0340	1/8 bend, 2"		16	1		14.65	36		50.65	75
0350	3"	▼	14	1.143	▼	23	41		64	92.50

For customer support on your Residential Costs with RSMeans data, call 800.448.8182.

571

22 13 16.30 Pipe Fittings, Cast Iron

	22 13 16.30 Pipe Fittings, Cast Iron	Crew	Daily Output	Labor-Hours	Unit	Material	2018 Bare Costs Labor	Equipment	Total	Total Incl O&P
0360	4"	Q-1	13	1.231	Ea.	33.50	44		77.50	110
0500	Sanitary tee, 2"		10	1.600		28.50	57.50		86	126
0540	3"		9	1.778		46.50	63.50		110	156
0620	4"		8	2		57	71.50		128.50	181
5990	No hub									
6000	Cplg. & labor required at joints not incl. in fitting									
6010	price. Add 1 coupling per joint for installed price									
6020	1/4 bend, 1-1/2"				Ea.	10.55			10.55	11.65
6060	2"					11.55			11.55	12.70
6080	3"					16.10			16.10	17.70
6120	4"					24			24	26
6184	1/4 bend, long sweep, 1-1/2"					27			27	29.50
6186	2"					25			25	27.50
6188	3"					30.50			30.50	33.50
6189	4"					48.50			48.50	53.50
6190	5"					94			94	103
6191	6"					107			107	118
6192	8"					291			291	320
6193	10"					585			585	645
6200	1/8 bend, 1-1/2"					8.90			8.90	9.80
6210	2"					9.95			9.95	10.95
6212	3"					13.30			13.30	14.65
6214	4"					17.45			17.45	19.15
6380	Sanitary tee, tapped, 1-1/2"					21			21	23
6382	2" x 1-1/2"					18.60			18.60	20.50
6384	2"					19.95			19.95	22
6386	3" x 2"					29.50			29.50	32.50
6388	3"					51			51	56.50
6390	4" x 1-1/2"					26.50			26.50	29
6392	4" x 2"					30			30	33
6393	4"					30			30	33
6394	6" x 1-1/2"					68.50			68.50	75.50
6396	6" x 2"					70			70	77
6459	Sanitary tee, 1-1/2"					14.85			14.85	16.30
6460	2"					15.90			15.90	17.50
6470	3"					19.60			19.60	21.50
6472	4"					37			37	41
8000	Coupling, standard (by CISPI Mfrs.)									
8020	1-1/2"	Q-1	48	.333	Ea.	16.10	11.95		28.05	37.50
8040	2"		44	.364		16.10	13.05		29.15	39
8080	3"		38	.421		19.25	15.10		34.35	46
8120	4"		33	.485		22.50	17.35		39.85	53.50

22 13 16.50 Shower Drains

	22 13 16.50 Shower Drains	Crew	Daily Output	Labor-Hours	Unit	Material	2018 Bare Costs Labor	Equipment	Total	Total Incl O&P
0010	**SHOWER DRAINS**									
2780	Shower, with strainer, uniform diam. trap, bronze top									
2800	2" and 3" pipe size	Q-1	8	2	Ea.	350	71.50		421.50	505
2820	4" pipe size	"	7	2.286		395	82		477	570
2840	For galvanized body, add					190			190	208

572

For customer support on your Residential Costs with RSMeans data, call 800.448.8182.

22 13 16.60 Traps

		Crew	Daily Output	Labor-Hours	Unit	Material	2018 Bare Costs Labor	Equipment	Total	Total Incl O&P
0010	**TRAPS**									
0030	Cast iron, service weight									
0050	Running P trap, without vent									
1100	2"	Q-1	16	1	Ea.	148	36		184	222
1150	4"	"	13	1.231		148	44		192	236
1160	6"	Q-2	17	1.412		720	48.50		768.50	870
3000	P trap, B&S, 2" pipe size	Q-1	16	1		38.50	36		74.50	102
3040	3" pipe size	"	14	1.143		57.50	41		98.50	131
4700	Copper, drainage, drum trap									
4840	3" x 6" swivel, 1-1/2" pipe size	1 Plum	16	.500	Ea.	275	19.90		294.90	335
5100	P trap, standard pattern									
5200	1-1/4" pipe size	1 Plum	18	.444	Ea.	88.50	17.70		106.20	126
5240	1-1/2" pipe size		17	.471		98.50	18.75		117.25	139
5260	2" pipe size		15	.533		152	21		173	202
5280	3" pipe size		11	.727		475	29		504	575
6710	ABS DWV P trap, solvent weld joint									
6720	1-1/2" pipe size	1 Plum	18	.444	Ea.	10.90	17.70		28.60	41
6722	2" pipe size		17	.471		14.35	18.75		33.10	47
6724	3" pipe size		15	.533		57	21		78	97.50
6726	4" pipe size		14	.571		114	22.50		136.50	163
6732	PVC DWV P trap, solvent weld joint									
6733	1-1/2" pipe size	1 Plum	18	.444	Ea.	9.30	17.70		27	39.50
6734	2" pipe size		17	.471		11.30	18.75		30.05	43.50
6735	3" pipe size		15	.533		38.50	21		59.50	77.50
6736	4" pipe size		14	.571		87	22.50		109.50	134
6860	PVC DWV hub x hub, basin trap, 1-1/4" pipe size		18	.444		51.50	17.70		69.20	85.50
6870	Sink P trap, 1-1/2" pipe size		18	.444		16.30	17.70		34	47
6880	Tubular S trap, 1-1/2" pipe size		17	.471		26.50	18.75		45.25	60
6890	PVC sch. 40 DWV, drum trap									
6900	1-1/2" pipe size	1 Plum	16	.500	Ea.	35.50	19.90		55.40	72
6910	P trap, 1-1/2" pipe size		18	.444		8.80	17.70		26.50	38.50
6920	2" pipe size		17	.471		11.80	18.75		30.55	44
6930	3" pipe size		15	.533		40	21		61	79
6940	4" pipe size		14	.571		91	22.50		113.50	138
6950	P trap w/clean out, 1-1/2" pipe size		18	.444		14.95	17.70		32.65	45.50
6960	2" pipe size		17	.471		24.50	18.75		43.25	58

22 13 16.80 Vent Flashing and Caps

		Crew	Daily Output	Labor-Hours	Unit	Material	2018 Bare Costs Labor	Equipment	Total	Total Incl O&P
0010	**VENT FLASHING AND CAPS**									
0120	Vent caps									
0140	Cast iron									
0160	1-1/4" to 1-1/2" pipe	1 Plum	23	.348	Ea.	37.50	13.85		51.35	64.50
0170	2" to 2-1/8" pipe		22	.364		43.50	14.45		57.95	72
0180	2-1/2" to 3-5/8" pipe		21	.381		49.50	15.15		64.65	79.50
0190	4" to 4-1/8" pipe		19	.421		69	16.75		85.75	104
0200	5" to 6" pipe		17	.471		104	18.75		122.75	146
0300	PVC									
0320	1-1/4" to 1-1/2" pipe	1 Plum	24	.333	Ea.	12.95	13.25		26.20	36
0330	2" to 2-1/8" pipe	"	23	.348	"	13.95	13.85		27.80	38.50
0900	Vent flashing									
1350	Copper with neoprene ring									
1400	1-1/4" pipe	1 Plum	20	.400	Ea.	70.50	15.90		86.40	104
1430	1-1/2" pipe		20	.400		70.50	15.90		86.40	104

22 13 Facility Sanitary Sewerage

22 13 16 – Sanitary Waste and Vent Piping

22 13 16.80 Vent Flashing and Caps		Crew	Daily Output	Labor-Hours	Unit	Material	2018 Bare Costs Labor	Equipment	Total	Total Incl O&P
1440	2" pipe	1 Plum	18	.444	Ea.	70.50	17.70		88.20	107
1450	3" pipe		17	.471		85.50	18.75		104.25	125
1460	4" pipe		16	.500		85.50	19.90		105.40	127
2980	Neoprene, one piece									
3000	1-1/4" pipe	1 Plum	24	.333	Ea.	2.93	13.25		16.18	25
3030	1-1/2" pipe		24	.333		2.93	13.25		16.18	25
3040	2" pipe		23	.348		5.45	13.85		19.30	29
3050	3" pipe		21	.381		6.45	15.15		21.60	32
3060	4" pipe		20	.400		9.55	15.90		25.45	36.50

22 13 19 – Sanitary Waste Piping Specialties

22 13 19.13 Sanitary Drains

		Crew	Daily Output	Labor-Hours	Unit	Material	Labor	Equipment	Total	Total Incl O&P
0010	**SANITARY DRAINS**									
2000	Floor, medium duty, CI, deep flange, 7" diam. top									
2040	2" and 3" pipe size	Q-1	12	1.333	Ea.	230	48		278	330
2080	For galvanized body, add					111			111	122
2120	With polished bronze top					360			360	395

22 14 Facility Storm Drainage

22 14 26 – Facility Storm Drains

22 14 26.13 Roof Drains

		Crew	Daily Output	Labor-Hours	Unit	Material	Labor	Equipment	Total	Total Incl O&P
0010	**ROOF DRAINS**									
3860	Roof, flat metal deck, CI body, 12" CI dome									
3890	3" pipe size	Q-1	14	1.143	Ea.	330	41		371	430

22 14 29 – Sump Pumps

22 14 29.16 Submersible Sump Pumps

		Crew	Daily Output	Labor-Hours	Unit	Material	Labor	Equipment	Total	Total Incl O&P
0010	**SUBMERSIBLE SUMP PUMPS**									
7000	Sump pump, automatic									
7100	Plastic, 1-1/4" discharge, 1/4 HP	1 Plum	6.40	1.250	Ea.	150	50		200	247
7500	Cast iron, 1-1/4" discharge, 1/4 HP	"	6	1.333	"	211	53		264	320

22 31 Domestic Water Softeners

22 31 13 – Residential Domestic Water Softeners

22 31 13.10 Residential Water Softeners

		Crew	Daily Output	Labor-Hours	Unit	Material	Labor	Equipment	Total	Total Incl O&P
0010	**RESIDENTIAL WATER SOFTENERS**									
7350	Water softener, automatic, to 30 grains per gallon	2 Plum	5	3.200	Ea.	370	127		497	615
7400	To 100 grains per gallon	"	4	4	"	840	159		999	1,175

22 33 Electric Domestic Water Heaters

22 33 30 – Residential, Electric Domestic Water Heaters

22 33 30.13 Residential, Small-Capacity Elec. Water Heaters	Crew	Daily Output	Labor-Hours	Unit	Material	2018 Bare Costs Labor	Equipment	Total	Total Incl O&P
0010 **RESIDENTIAL, SMALL-CAPACITY ELECTRIC DOMESTIC WATER HEATERS**									
1000 Residential, electric, glass lined tank, 5 yr., 10 gal., single element	1 Plum	2.30	3.478	Ea.	430	138		568	700
1060 30 gallon, double element		2.20	3.636		1,000	145		1,145	1,350
1080 40 gallon, double element		2	4		1,075	159		1,234	1,425
1100 52 gallon, double element		2	4		1,200	159		1,359	1,575
1120 66 gallon, double element		1.80	4.444		1,650	177		1,827	2,100
1140 80 gallon, double element	↓	1.60	5	↓	1,850	199		2,049	2,375

22 34 Fuel-Fired Domestic Water Heaters

22 34 13 – Instantaneous, Tankless, Gas Domestic Water Heaters

22 34 13.10 Instantaneous, Tankless, Gas Water Heaters

0010 **INSTANTANEOUS, TANKLESS, GAS WATER HEATERS**									
9410 Natural gas/propane, 3.2 GPM [G]	1 Plum	2	4	Ea.	500	159		659	810
9420 6.4 GPM [G]		1.90	4.211		715	168		883	1,050
9430 8.4 GPM [G]		1.80	4.444		855	177		1,032	1,225
9440 9.5 GPM [G]	↓	1.60	5	↓	1,000	199		1,199	1,425

22 34 30 – Residential Gas Domestic Water Heaters

22 34 30.13 Residential, Atmos, Gas Domestic Wtr Heaters

0010 **RESIDENTIAL, ATMOSPHERIC, GAS DOMESTIC WATER HEATERS**									
2000 Gas fired, foam lined tank, 10 yr., vent not incl.									
2040 30 gallon	1 Plum	2	4	Ea.	1,775	159		1,934	2,200
2100 75 gallon	"	1.50	5.333	"	2,700	212		2,912	3,325
3000 Tank leak safety, water & gas shut off see 22 05 23.20 8800									

22 34 46 – Oil-Fired Domestic Water Heaters

22 34 46.10 Residential Oil-Fired Water Heaters

0010 **RESIDENTIAL OIL-FIRED WATER HEATERS**									
3000 Oil fired, glass lined tank, 5 yr., vent not included, 30 gallon	1 Plum	2	4	Ea.	1,275	159		1,434	1,650
3040 50 gallon	"	1.80	4.444	"	1,575	177		1,752	2,025

22 41 Residential Plumbing Fixtures

22 41 13 – Residential Water Closets, Urinals, and Bidets

22 41 13.13 Water Closets

0010 **WATER CLOSETS**									
0150 Tank type, vitreous china, incl. seat, supply pipe w/stop, 1.6 gpf or noted									
0200 Wall hung									
0400 Two piece, close coupled	Q-1	5.30	3.019	Ea.	400	108		508	620
0960 For rough-in, supply, waste, vent and carrier	"	2.73	5.861	"	1,150	210		1,360	1,625
0999 Floor mounted									
1020 One piece, low profile	Q-1	5.30	3.019	Ea.	730	108		838	985
1100 Two piece, close coupled		5.30	3.019		230	108		338	430
1102 Economy		5.30	3.019		126	108		234	315
1110 Two piece, close coupled, dual flush		5.30	3.019		320	108		428	535
1140 Two piece, close coupled, 1.28 gpf, ADA [G]	↓	5.30	3.019	↓	310	108		418	520
1960 For color, add					30%				
1980 For rough-in, supply, waste and vent	Q-1	3.05	5.246	Ea.	360	188		548	705

For customer support on your Residential Costs with RSMeans data, call 800.448.8182.

575

22 41 16.13 Lavatories

		Crew	Daily Output	Labor-Hours	Unit	Material	2018 Bare Costs Labor	2018 Bare Costs Equipment	Total	Total Incl O&P
0010	**LAVATORIES**, With trim, white unless noted otherwise									
0500	Vanity top, porcelain enamel on cast iron									
0600	20" x 18"	Q-1	6.40	2.500	Ea.	300	89.50		389.50	475
0640	33" x 19" oval		6.40	2.500		520	89.50		609.50	720
0720	19" round		6.40	2.500		390	89.50		479.50	575
0860	For color, add					25%				
1000	Cultured marble, 19" x 17", single bowl	Q-1	6.40	2.500	Ea.	127	89.50		216.50	286
1120	25" x 22", single bowl		6.40	2.500		166	89.50		255.50	330
1160	37" x 22", single bowl		6.40	2.500		197	89.50		286.50	365
1580	For color, same price									
1900	Stainless steel, self-rimming, 25" x 22", single bowl, ledge	Q-1	6.40	2.500	Ea.	320	89.50		409.50	495
1960	17" x 22", single bowl		6.40	2.500		305	89.50		394.50	485
2600	Steel, enameled, 20" x 17", single bowl		5.80	2.759		129	99		228	305
2900	Vitreous china, 20" x 16", single bowl		5.40	2.963		213	106		319	410
3200	22" x 13", single bowl		5.40	2.963		220	106		326	415
3580	Rough-in, supply, waste and vent for all above lavatories		2.30	6.957		238	249		487	670
4000	Wall hung									
4040	Porcelain enamel on cast iron, 16" x 14", single bowl	Q-1	8	2	Ea.	445	71.50		516.50	610
4180	20" x 18", single bowl	"	8	2	"	246	71.50		317.50	390
4580	For color, add					30%				
6000	Vitreous china, 18" x 15", single bowl with backsplash	Q-1	7	2.286	Ea.	161	82		243	310
6060	19" x 17", single bowl		7	2.286		123	82		205	270
6960	Rough-in, supply, waste and vent for above lavatories		1.66	9.639		465	345		810	1,075
7000	Pedestal type									
7600	Vitreous china, 27" x 21", white	Q-1	6.60	2.424	Ea.	670	87		757	885
7610	27" x 21", colored		6.60	2.424		855	87		942	1,075
7620	27" x 21", premium color		6.60	2.424		975	87		1,062	1,225
7660	26" x 20", white		6.60	2.424		660	87		747	870
7670	26" x 20", colored		6.60	2.424		840	87		927	1,075
7680	26" x 20", premium color		6.60	2.424		955	87		1,042	1,200
7700	24" x 20", white		6.60	2.424		470	87		557	665
7710	24" x 20", colored		6.60	2.424		595	87		682	800
7720	24" x 20", premium color		6.60	2.424		675	87		762	890
7760	21" x 18", white		6.60	2.424		258	87		345	425
7770	21" x 18", colored		6.60	2.424		278	87		365	450
7990	Rough-in, supply, waste and vent for pedestal lavatories		1.66	9.639		465	345		810	1,075

22 41 16.16 Sinks

		Crew	Daily Output	Labor-Hours	Unit	Material	2018 Bare Costs Labor	2018 Bare Costs Equipment	Total	Total Incl O&P
0010	**SINKS**, With faucets and drain									
2000	Kitchen, counter top style, PE on CI, 24" x 21" single bowl	Q-1	5.60	2.857	Ea.	305	102		407	505
2100	31" x 22" single bowl		5.60	2.857		660	102		762	895
2200	32" x 21" double bowl		4.80	3.333		375	119		494	610
3000	Stainless steel, self rimming, 19" x 18" single bowl		5.60	2.857		615	102		717	845
3100	25" x 22" single bowl		5.60	2.857		680	102		782	920
3200	33" x 22" double bowl		4.80	3.333		985	119		1,104	1,275
3300	43" x 22" double bowl		4.80	3.333		1,150	119		1,269	1,450
4000	Steel, enameled, with ledge, 24" x 21" single bowl		5.60	2.857		525	102		627	750
4100	32" x 21" double bowl		4.80	3.333		540	119		659	790
4960	For color sinks except stainless steel, add					10%				
4980	For rough-in, supply, waste and vent, counter top sinks	Q-1	2.14	7.477		275	268		543	745
5000	Kitchen, raised deck, PE on CI									
5100	32" x 21", dual level, double bowl	Q-1	2.60	6.154	Ea.	450	220		670	855
5790	For rough-in, supply, waste & vent, sinks	"	1.85	8.649	"	275	310		585	815

22 41 Residential Plumbing Fixtures

22 41 19 – Residential Bathtubs

22 41 19.10 Baths

		Crew	Daily Output	Labor-Hours	Unit	Material	2018 Bare Costs Labor	Equipment	Total	Total Incl O&P
0010	**BATHS**									
0100	Tubs, recessed porcelain enamel on cast iron, with trim									
0180	48" x 42"	Q-1	4	4	Ea.	2,800	143		2,943	3,300
0220	72" x 36"	"	3	5.333	"	2,850	191		3,041	3,475
0300	Mat bottom									
0380	5' long	Q-1	4.40	3.636	Ea.	1,175	130		1,305	1,525
0480	Above floor drain, 5' long		4	4		845	143		988	1,175
0560	Corner 48" x 44"		4.40	3.636		2,800	130		2,930	3,300
2000	Enameled formed steel, 4'-6" long		5.80	2.759		490	99		589	705
4600	Module tub & showerwall surround, molded fiberglass									
4610	5' long x 34" wide x 76" high	Q-1	4	4	Ea.	760	143		903	1,075
9600	Rough-in, supply, waste and vent, for all above tubs, add	"	2.07	7.729	"	430	277		707	925

22 41 23 – Residential Showers

22 41 23.20 Showers

		Crew	Daily Output	Labor-Hours	Unit	Material	Labor	Equipment	Total	Total Incl O&P
0010	**SHOWERS**									
1500	Stall, with drain only. Add for valve and door/curtain									
1520	32" square	Q-1	5	3.200	Ea.	1,175	115		1,290	1,475
1530	36" square		4.80	3.333		2,925	119		3,044	3,400
1540	Terrazzo receptor, 32" square		5	3.200		1,350	115		1,465	1,675
1560	36" square		4.80	3.333		1,475	119		1,594	1,825
1580	36" corner angle		4.80	3.333		1,750	119		1,869	2,125
3000	Fiberglass, one piece, with 3 walls, 32" x 32" square		5.50	2.909		345	104		449	550
3100	36" x 36" square		5.50	2.909		400	104		504	610
4200	Rough-in, supply, waste and vent for above showers		2.05	7.805		370	280		650	870

22 41 36 – Residential Laundry Trays

22 41 36.10 Laundry Sinks

		Crew	Daily Output	Labor-Hours	Unit	Material	Labor	Equipment	Total	Total Incl O&P
0010	**LAUNDRY SINKS**, With trim									
0020	Porcelain enamel on cast iron, black iron frame									
0050	24" x 21", single compartment	Q-1	6	2.667	Ea.	595	95.50		690.50	805
0100	26" x 21", single compartment	"	6	2.667	"	620	95.50		715.50	840
3000	Plastic, on wall hanger or legs									
3020	18" x 23", single compartment	Q-1	6.50	2.462	Ea.	145	88		233	305
3100	20" x 24", single compartment		6.50	2.462		165	88		253	325
3200	36" x 23", double compartment		5.50	2.909		213	104		317	405
3300	40" x 24", double compartment		5.50	2.909		287	104		391	485
5000	Stainless steel, counter top, 22" x 17" single compartment		6	2.667		76.50	95.50		172	241
5200	33" x 22", double compartment		5	3.200		91.50	115		206.50	290
9600	Rough-in, supply, waste and vent, for all laundry sinks		2.14	7.477		275	268		543	745

22 41 39 – Residential Faucets, Supplies and Trim

22 41 39.10 Faucets and Fittings

		Crew	Daily Output	Labor-Hours	Unit	Material	Labor	Equipment	Total	Total Incl O&P
0010	**FAUCETS AND FITTINGS**									
0150	Bath, faucets, diverter spout combination, sweat	1 Plum	8	1	Ea.	86	40		126	160
0200	For integral stops, IPS unions, add					111			111	122
0420	Bath, press-bal mix valve w/diverter, spout, shower head, arm/flange	1 Plum	8	1		175	40		215	258
0500	Drain, central lift, 1-1/2" IPS male		20	.400		49	15.90		64.90	80
0600	Trip lever, 1-1/2" IPS male		20	.400		60	15.90		75.90	92
1000	Kitchen sink faucets, top mount, cast spout		10	.800		83.50	32		115.50	144
1100	For spray, add		24	.333		17.10	13.25		30.35	41
1300	Single control lever handle									
1310	With pull out spray									

For customer support on your Residential Costs with RSMeans data, call 800.448.8182.

577

22 41 39.10 Faucets and Fittings

		Crew	Daily Output	Labor-Hours	Unit	Material	2018 Bare Costs Labor	Equipment	Total	Total Incl O&P
1320	Polished chrome	1 Plum	10	.800	Ea.	196	32		228	269
2000	Laundry faucets, shelf type, IPS or copper unions		12	.667		61.50	26.50		88	111
2100	Lavatory faucet, centerset, without drain		10	.800		67	32		99	127
2120	With pop-up drain		6.66	1.201		58.50	48		106.50	143
2210	Porcelain cross handles and pop-up drain									
2220	Polished chrome	1 Plum	6.66	1.201	Ea.	207	48		255	305
2230	Polished brass	"	6.66	1.201	"	310	48		358	425
2260	Single lever handle and pop-up drain									
2280	Satin nickel	1 Plum	6.66	1.201	Ea.	273	48		321	380
2290	Polished chrome		6.66	1.201		197	48		245	296
2800	Self-closing, center set		10	.800		149	32		181	217
4000	Shower by-pass valve with union		18	.444		57	17.70		74.70	92
4200	Shower thermostatic mixing valve, concealed, with shower head trim kit		8	1		350	40		390	450
4220	Shower pressure balancing mixing valve									
4230	With shower head, arm, flange and diverter tub spout									
4240	Chrome	1 Plum	6.14	1.303	Ea.	410	52		462	540
4250	Satin nickel		6.14	1.303		555	52		607	695
4260	Polished graphite		6.14	1.303		555	52		607	695
5000	Sillcock, compact, brass, IPS or copper to hose		24	.333		10.55	13.25		23.80	33.50

22 41 39.70 Washer/Dryer Accessories

		Crew	Daily Output	Labor-Hours	Unit	Material	2018 Bare Costs Labor	Equipment	Total	Total Incl O&P
0010	**WASHER/DRYER ACCESSORIES**									
1020	Valves ball type single lever									
1030	1/2" diam., IPS	1 Plum	21	.381	Ea.	62.50	15.15		77.65	93.50
1040	1/2" diam., solder	"	21	.381	"	62.50	15.15		77.65	93.50
1050	Recessed box, 16 ga., two hose valves and drain									
1060	1/2" size, 1-1/2" drain	1 Plum	18	.444	Ea.	138	17.70		155.70	181
1070	1/2" size, 2" drain	"	17	.471	"	123	18.75		141.75	167
1080	With grounding electric receptacle									
1090	1/2" size, 1-1/2" drain	1 Plum	18	.444	Ea.	151	17.70		168.70	195
1100	1/2" size, 2" drain	"	17	.471	"	163	18.75		181.75	210
1110	With grounding and dryer receptacle									
1120	1/2" size, 1-1/2" drain	1 Plum	18	.444	Ea.	187	17.70		204.70	234
1130	1/2" size, 2" drain	"	17	.471	"	188	18.75		206.75	238
1140	Recessed box, 16 ga., ball valves with single lever and drain									
1150	1/2" size, 1-1/2" drain	1 Plum	19	.421	Ea.	260	16.75		276.75	315
1160	1/2" size, 2" drain	"	18	.444	"	225	17.70		242.70	277
1170	With grounding electric receptacle									
1180	1/2" size, 1-1/2" drain	1 Plum	19	.421	Ea.	259	16.75		275.75	315
1190	1/2" size, 2" drain	"	18	.444	"	250	17.70		267.70	305
1200	With grounding and dryer receptacles									
1210	1/2" size, 1-1/2" drain	1 Plum	19	.421	Ea.	246	16.75		262.75	299
1220	1/2" size, 2" drain	"	18	.444	"	273	17.70		290.70	330
1300	Recessed box, 20 ga., two hose valves and drain (economy type)									
1310	1/2" size, 1-1/2" drain	1 Plum	19	.421	Ea.	107	16.75		123.75	146
1320	1/2" size, 2" drain		18	.444		100	17.70		117.70	139
1330	Box with drain only		24	.333		62	13.25		75.25	90.50
1340	1/2" size, 1-1/2" ABS/PVC drain		19	.421		124	16.75		140.75	164
1350	1/2" size, 2" ABS/PVC drain		18	.444		127	17.70		144.70	169
1352	Box with drain and 15 A receptacle		24	.333		61	13.25		74.25	89
1360	1/2" size, 2" drain ABS/PVC, 15 A receptacle		24	.333		140	13.25		153.25	176
1400	Wall mounted									
1410	1/2" size, 1-1/2" plastic drain	1 Plum	19	.421	Ea.	30.50	16.75		47.25	61

22 41 Residential Plumbing Fixtures

22 41 39 – Residential Faucets, Supplies and Trim

22 41 39.70 Washer/Dryer Accessories	Crew	Daily Output	Labor-Hours	Unit	Material	2018 Bare Costs Labor	Equipment	Total	Total Incl O&P
1420	1/2" size, 2" plastic drain	1 Plum	18	.444	Ea.	17.30	17.70		35	48
1500	Dryer vent kit									
1510	8' flex duct, clamps and outside hood	1 Plum	20	.400	Ea.	12.95	15.90		28.85	40.50
1980	Rough-in, supply, waste, and vent for washer boxes		3.46	2.310		310	92		402	495
9605	Washing machine valve assembly, hot & cold water supply, recessed		8	1		78.50	40		118.50	152
9610	Washing machine valve assembly, hot & cold water supply, mounted	▼	8	1	▼	62.50	40		102.50	134

22 42 Commercial Plumbing Fixtures

22 42 13 – Commercial Water Closets, Urinals, and Bidets

22 42 13.13 Water Closets

		Crew	Daily Output	Labor-Hours	Unit	Material	Labor	Equipment	Total	Total Incl O&P
0010	**WATER CLOSETS**									
3000	Bowl only, with flush valve, seat, 1.6 gpf unless noted									
3100	Wall hung	Q-1	5.80	2.759	Ea.	970	99		1,069	1,250
3200	For rough-in, supply, waste and vent, single WC		2.56	6.250		1,200	224		1,424	1,675
3300	Floor mounted		5.80	2.759	▼	325	99		424	520
3370	For rough-in, supply, waste and vent, single WC	▼	2.84	5.634	▼	395	202		597	765
3390	Floor mounted children's size, 10-3/4" high									
3392	With automatic flush sensor, 1.6 gpf	Q-1	6.20	2.581	Ea.	620	92.50		712.50	835
3396	With automatic flush sensor, 1.28 gpf		6.20	2.581		620	92.50		712.50	835
3400	For rough-in, supply, waste and vent, single WC	▼	2.84	5.634	▼	395	202		597	765

22 42 16 – Commercial Lavatories and Sinks

22 42 16.13 Lavatories

0010	**LAVATORIES**, With trim, white unless noted otherwise									
0020	Commercial lavatories same as residential. See Section 22 41 16									

22 42 16.40 Service Sinks

		Crew	Daily Output	Labor-Hours	Unit	Material	Labor	Equipment	Total	Total Incl O&P
0010	**SERVICE SINKS**									
6650	Service, floor, corner, PE on CI, 28" x 28"	Q-1	4.40	3.636	Ea.	1,025	130		1,155	1,350
6750	Vinyl coated rim guard, add					61.50			61.50	68
6755	Mop sink, molded stone, 22" x 18"	1 Plum	3.33	2.402		485	95.50		580.50	685
6760	Mop sink, molded stone, 24" x 36"		3.33	2.402		259	95.50		354.50	440
6770	Mop sink, molded stone, 24" x 36", w/rim 3 sides	▼	3.33	2.402		285	95.50		380.50	470
6790	For rough-in, supply, waste & vent, floor service sinks	Q-1	1.64	9.756	▼	960	350		1,310	1,625

22 42 39 – Commercial Faucets, Supplies, and Trim

22 42 39.10 Faucets and Fittings

		Crew	Daily Output	Labor-Hours	Unit	Material	Labor	Equipment	Total	Total Incl O&P
0010	**FAUCETS AND FITTINGS**									
2790	Faucets for lavatories									
2800	Self-closing, center set	1 Plum	10	.800	Ea.	149	32		181	217
2810	Automatic sensor and operator, with faucet head		6.15	1.301		480	52		532	615
3000	Service sink faucet, cast spout, pail hook, hose end	▼	14	.571	▼	76	22.50		98.50	121

22 42 39.30 Carriers and Supports

		Crew	Daily Output	Labor-Hours	Unit	Material	Labor	Equipment	Total	Total Incl O&P
0010	**CARRIERS AND SUPPORTS**, For plumbing fixtures									
0600	Plate type with studs, top back plate	1 Plum	7	1.143	Ea.	57	45.50		102.50	138
3000	Lavatory, concealed arm									
3050	Floor mounted, single									
3100	High back fixture	1 Plum	6	1.333	Ea.	605	53		658	755
3200	Flat slab fixture	"	6	1.333	"	520	53		573	660
8200	Water closet, residential									
8220	Vertical centerline, floor mount									
8240	Single, 3" caulk, 2" or 3" vent	1 Plum	6	1.333	Ea.	695	53		748	855

For customer support on your Residential Costs with RSMeans data, call 800.448.8182.

579

22 42 Commercial Plumbing Fixtures

22 42 39 – Commercial Faucets, Supplies, and Trim

22 42 39.30 Carriers and Supports	Crew	Daily Output	Labor-Hours	Unit	Material	2018 Bare Costs Labor	Equipment	Total	Total Incl O&P
8260	4" caulk, 2" or 4" vent	1 Plum	6	1.333	Ea.	895	53		948	1,075

22 51 Swimming Pool Plumbing Systems

22 51 19 – Swimming Pool Water Treatment Equipment

22 51 19.50 Swimming Pool Filtration Equipment

		Crew	Daily Output	Labor-Hours	Unit	Material	2018 Bare Costs Labor	Equipment	Total	Total Incl O&P
0010	**SWIMMING POOL FILTRATION EQUIPMENT**									
0900	Filter system, sand or diatomite type, incl. pump, 6,000 gal./hr.	2 Plum	1.80	8.889	Total	2,125	355		2,480	2,925
1020	Add for chlorination system, 800 S.F. pool	"	3	5.333	Ea.	177	212		389	545

Estimating Tips

The labor adjustment factors listed in Subdivision 22 01 02.20 also apply to Division 23.

23 10 00 Facility Fuel Systems

- The prices in this subdivision for above- and below-ground storage tanks do not include foundations or hold-down slabs, unless noted. The estimator should refer to Divisions 3 and 31 for foundation system pricing. In addition to the foundations, required tank accessories, such as tank gauges, leak detection devices, and additional manholes and piping, must be added to the tank prices.

23 50 00 Central Heating Equipment

- When estimating the cost of an HVAC system, check to see who is responsible for providing and installing the temperature control system. It is possible to overlook controls, assuming that they would be included in the electrical estimate.

- When looking up a boiler, be careful on specified capacity. Some manufacturers rate their products on output while others use input.

- Include HVAC insulation for pipe, boiler, and duct (wrap and liner).

- Be careful when looking up mechanical items to get the correct pressure rating and connection type (thread, weld, flange).

23 70 00 Central HVAC Equipment

- Combination heating and cooling units are sized by the air conditioning requirements. (See Reference No. R236000-20 for the preliminary sizing guide.)

- A ton of air conditioning is nominally 400 CFM.

- Rectangular duct is taken off by the linear foot for each size, but its cost is usually estimated by the pound. Remember that SMACNA standards now base duct on internal pressure.

- Prefabricated duct is estimated and purchased like pipe: straight sections and fittings.

- Note that cranes or other lifting equipment are not included on any lines in Division 23. For example, if a crane is required to lift a heavy piece of pipe into place high above a gym floor, or to put a rooftop unit on the roof of a four-story building, etc., it must be added. Due to the potential for extreme variation—from nothing additional required to a major crane or helicopter—we feel that including a nominal amount for "lifting contingency" would be useless and detract from the accuracy of the estimate. When using equipment rental cost data from RSMeans, do not forget to include the cost of the operator(s).

Reference Numbers

Reference numbers are shown at the beginning of some major classifications. These numbers refer to related items in the Reference Section. The reference information may be an estimating procedure, an alternate pricing method, or technical information.

Note: Not all subdivisions listed here necessarily appear. ■

Did you know?

RSMeans data is available through our online application with 24/7 access:

- Search for unit prices by keyword
- Leverage the most up-to-date data
- Build and export estimates

Try it free for 30 days!
www.rsmeans.com/2018freetrial

23 05 05 – Selective Demolition for HVAC

23 05 05.10 HVAC Demolition

		Crew	Daily Output	Labor-Hours	Unit	Material	2018 Bare Costs Labor	Equipment	Total	Total Incl O&P
0010	**HVAC DEMOLITION**									
0100	Air conditioner, split unit, 3 ton	Q-5	2	8	Ea.		295		295	485
0150	Package unit, 3 ton	Q-6	3	8	"		284		284	465
0260	Baseboard, hydronic fin tube, 1/2"	Q-5	117	.137	L.F.		5.05		5.05	8.30
0298	Boilers									
0300	Electric, up thru 148 kW	Q-19	2	12	Ea.		455		455	750
0310	150 thru 518 kW	"	1	24			910		910	1,500
0320	550 thru 2,000 kW	Q-21	.40	80			3,100		3,100	5,075
0330	2,070 kW and up	"	.30	107			4,125		4,125	6,775
0340	Gas and/or oil, up thru 150 MBH	Q-7	2.20	14.545			535		535	880
0350	160 thru 2,000 MBH		.80	40			1,475		1,475	2,425
0360	2,100 thru 4,500 MBH		.50	64			2,350		2,350	3,875
0370	4,600 thru 7,000 MBH		.30	107			3,925		3,925	6,475
0390	12,200 thru 25,000 MBH		.12	267			9,825		9,825	16,200
1000	Ductwork, 4" high, 8" wide	1 Clab	200	.040	L.F.		1.07		1.07	1.78
1100	6" high, 8" wide		165	.048			1.29		1.29	2.15
1200	10" high, 12" wide		125	.064			1.71		1.71	2.84
1300	12"-14" high, 16"-18" wide		85	.094			2.51		2.51	4.18
1500	30" high, 36" wide		56	.143			3.81		3.81	6.35
2200	Furnace, electric	Q-20	2	10	Ea.		355		355	590
2300	Gas or oil, under 120 MBH	Q-9	4	4			138		138	229
2340	Over 120 MBH	"	3	5.333			184		184	305
2800	Heat pump, package unit, 3 ton	Q-5	2.40	6.667			246		246	405
2840	Split unit, 3 ton		2	8			295		295	485
2950	Tank, steel, oil, 275 gal., above ground		10	1.600			59		59	97
2960	Remove and reset		3	5.333			197		197	325
5090	Remove refrigerant from system	1 Stpi	40	.200	Lb.		8.20		8.20	13.50

23 07 13 – Duct Insulation

23 07 13.10 Duct Thermal Insulation

		Crew	Daily Output	Labor-Hours	Unit	Material	2018 Bare Costs Labor	Equipment	Total	Total Incl O&P
0010	**DUCT THERMAL INSULATION**									
3000	Ductwork									
3020	Blanket type, fiberglass, flexible									
3030	Fire rated for grease and hazardous exhaust ducts									
3060	1-1/2" thick	Q-14	84	.190	S.F.	4.54	6.20		10.74	15.45
3090	Fire rated for plenums									
3100	1/2" x 24" x 25'	Q-14	1.94	8.247	Roll	167	268		435	640
3110	1/2" x 24" x 25'		98	.163	S.F.	3.35	5.30		8.65	12.65
3120	1/2" x 48" x 25'		1.04	15.385	Roll	335	500		835	1,225
3126	1/2" x 48" x 25'		104	.154	S.F.	3.35	4.99		8.34	12.15
3140	FSK vapor barrier wrap, .75 lb. density									
3160	1" thick [G]	Q-14	350	.046	S.F.	.22	1.48		1.70	2.75
3170	1-1/2" thick [G]	"	320	.050	"	.27	1.62		1.89	3.05
3210	Vinyl jacket, same as FSK									

582

For customer support on your Residential Costs with RSMeans data, call 800.448.8182.

23 09 Instrumentation and Control for HVAC

23 09 53 – Pneumatic and Electric Control System for HVAC

23 09 53.10 Control Components		Crew	Daily Output	Labor-Hours	Unit	Material	2018 Bare Costs Labor	Equipment	Total	Total Incl O&P
0010	**CONTROL COMPONENTS**									
5000	Thermostats									
5030	Manual	1 Stpi	8	1	Ea.	43	41		84	115
5040	1 set back, electric, timed	G	8	1		76.50	41		117.50	152
5050	2 set back, electric, timed	G	8	1		228	41		269	320

23 13 Facility Fuel-Storage Tanks

23 13 13 – Facility Underground Fuel-Oil, Storage Tanks

23 13 13.09 Single-Wall Steel Fuel-Oil Tanks

		Crew	Daily Output	Labor-Hours	Unit	Material	2018 Bare Costs Labor	Equipment	Total	Total Incl O&P
0010	**SINGLE-WALL STEEL FUEL-OIL TANKS**									
5000	Tanks, steel ugnd., sti-p3, not incl. hold-down bars									
5500	Excavation, pad, pumps and piping not included									
5510	Single wall, 500 gallon capacity, 7 ga. shell	Q-5	2.70	5.926	Ea.	1,800	218		2,018	2,350
5520	1,000 gallon capacity, 7 ga. shell	"	2.50	6.400		2,525	236		2,761	3,175
5530	2,000 gallon capacity, 1/4" thick shell	Q-7	4.60	6.957		3,275	256		3,531	4,025
5535	2,500 gallon capacity, 7 ga. shell	Q-5	3	5.333		4,900	197		5,097	5,700
5610	25,000 gallon capacity, 3/8" thick shell	Q-7	1.30	24.615		26,500	905		27,405	30,700
5630	40,000 gallon capacity, 3/8" thick shell		.90	35.556		42,400	1,300		43,700	48,900
5640	50,000 gallon capacity, 3/8" thick shell		.80	40		47,100	1,475		48,575	54,500

23 13 13.23 Glass-Fiber-Reinfcd-Plastic, Fuel-Oil, Storage

		Crew	Daily Output	Labor-Hours	Unit	Material	2018 Bare Costs Labor	Equipment	Total	Total Incl O&P
0010	**GLASS-FIBER-REINFCD-PLASTIC, UNDERGRND. FUEL-OIL, STORAGE**									
0210	Fiberglass, underground, single wall, UL listed, not including									
0220	manway or hold-down strap									
0230	1,000 gallon capacity	Q-5	2.46	6.504	Ea.	4,775	240		5,015	5,650
0240	2,000 gallon capacity	Q-7	4.57	7.002		6,900	258		7,158	8,000
0245	3,000 gallon capacity		3.90	8.205		8,250	300		8,550	9,575
0255	5,000 gallon capacity		3.20	10		10,600	370		10,970	12,300
0500	For manway, fittings and hold-downs, add					20%	15%			
2210	Fiberglass, underground, single wall, UL listed, including									
2220	hold-down straps, no manways									
2230	1,000 gallon capacity	Q-5	1.88	8.511	Ea.	5,250	315		5,565	6,325
2240	2,000 gallon capacity	Q-7	3.55	9.014	"	7,375	330		7,705	8,675

23 13 23 – Facility Aboveground Fuel-Oil, Storage Tanks

23 13 23.16 Steel

		Crew	Daily Output	Labor-Hours	Unit	Material	2018 Bare Costs Labor	Equipment	Total	Total Incl O&P
3001	**STEEL**, storage, above ground, including supports, coating									
3020	fittings, not including foundation, pumps or piping									
3040	Single wall, 275 gallon	Q-5	5	3.200	Ea.	515	118		633	760
3060	550 gallon	"	2.70	5.926		4,325	218		4,543	5,125
3080	1,000 gallon	Q-7	5	6.400		7,125	236		7,361	8,225
3320	Double wall, 500 gallon capacity	Q-5	2.40	6.667		1,750	246		1,996	2,325
3330	2,000 gallon capacity	Q-7	4.15	7.711		5,700	284		5,984	6,750
3340	4,000 gallon capacity		3.60	8.889		12,500	330		12,830	14,300
3350	6,000 gallon capacity		2.40	13.333		14,200	490		14,690	16,400
3360	8,000 gallon capacity		2	16		16,800	590		17,390	19,500
3370	10,000 gallon capacity		1.80	17.778		30,000	655		30,655	34,100
3380	15,000 gallon capacity		1.50	21.333		40,000	785		40,785	45,300
3390	20,000 gallon capacity		1.30	24.615		43,000	905		43,905	48,800
3400	25,000 gallon capacity		1.15	27.826		53,500	1,025		54,525	60,500
3410	30,000 gallon capacity		1	32		64,500	1,175		65,675	73,000

For customer support on your Residential Costs with RSMeans data, call 800.448.8182.

583

23 13 Facility Fuel-Storage Tanks

23 13 23 – Facility Aboveground Fuel-Oil, Storage Tanks

23 13 23.26 Horizontal, Conc., Abvgrd Fuel-Oil, Stor. Tanks	Crew	Daily Output	Labor-Hours	Unit	Material	2018 Bare Costs Labor	Equipment	Total	Total Incl O&P
0010 **HORIZONTAL, CONCRETE, ABOVEGROUND FUEL-OIL, STORAGE TANKS**									
0050 Concrete, storage, aboveground, including pad & pump									
0100 500 gallon	F-3	2	20	Ea.	9,950	640	248	10,838	12,200
0200 1,000 gallon	"	2	20	"	13,900	640	248	14,788	16,600

23 21 Hydronic Piping and Pumps

23 21 20 – Hydronic HVAC Piping Specialties

23 21 20.46 Expansion Tanks

	Crew	Daily Output	Labor-Hours	Unit	Material	2018 Bare Costs Labor	Equipment	Total	Total Incl O&P
0010 **EXPANSION TANKS**									
1507 Underground fuel-oil storage tanks, see Section 23 13 13									
2000 Steel, liquid expansion, ASME, painted, 15 gallon capacity	Q-5	17	.941	Ea.	720	34.50		754.50	845
2040 30 gallon capacity		12	1.333		830	49		879	995
3000 Steel ASME expansion, rubber diaphragm, 19 gal. cap. accept.		12	1.333		2,725	49		2,774	3,075
3020 31 gallon capacity		8	2		3,025	73.50		3,098.50	3,475

23 21 23 – Hydronic Pumps

23 21 23.13 In-Line Centrifugal Hydronic Pumps

	Crew	Daily Output	Labor-Hours	Unit	Material	2018 Bare Costs Labor	Equipment	Total	Total Incl O&P
0010 **IN-LINE CENTRIFUGAL HYDRONIC PUMPS**									
0600 Bronze, sweat connections, 1/40 HP, in line									
0640 3/4" size	Q-1	16	1	Ea.	246	36		282	330
1000 Flange connection, 3/4" to 1-1/2" size									
1040 1/12 HP	Q-1	6	2.667	Ea.	645	95.50		740.50	865
1060 1/8 HP		6	2.667		1,100	95.50		1,195.50	1,375
2101 Pumps, circulating, 3/4" to 1-1/2" size, 1/3 HP		6	2.667		930	95.50		1,025.50	1,175

23 23 Refrigerant Piping

23 23 16 – Refrigerant Piping Specialties

23 23 16.16 Refrigerant Line Sets

	Crew	Daily Output	Labor-Hours	Unit	Material	2018 Bare Costs Labor	Equipment	Total	Total Incl O&P
0010 **REFRIGERANT LINE SETS**, Standard									
0100 Copper tube									
0110 1/2" insulation, both tubes									
0120 Combination 1/4" and 1/2" tubes									
0130 10' set	Q-5	42	.381	Ea.	47.50	14.05		61.55	75
0135 15' set		42	.381		76.50	14.05		90.55	107
0140 20' set		40	.400		74	14.75		88.75	106
0150 30' set		37	.432		105	15.95		120.95	141
0160 40' set		35	.457		130	16.85		146.85	172
0170 50' set		32	.500		162	18.45		180.45	209
0180 100' set		22	.727		350	27		377	430
0300 Combination 1/4" and 3/4" tubes									
0310 10' set	Q-5	40	.400	Ea.	53	14.75		67.75	83
0320 20' set		38	.421		94	15.50		109.50	129
0330 30' set		35	.457		140	16.85		156.85	182
0340 40' set		33	.485		184	17.85		201.85	232
0350 50' set		30	.533		231	19.65		250.65	287
0380 100' set		20	.800		525	29.50		554.50	630
0500 Combination 3/8" & 3/4" tubes									
0510 10' set	Q-5	28	.571	Ea.	61	21		82	102
0520 20' set		36	.444		95	16.40		111.40	131

23 23 Refrigerant Piping

23 23 16 – Refrigerant Piping Specialties

23 23 16.16 Refrigerant Line Sets		Crew	Daily Output	Labor-Hours	Unit	Material	2018 Bare Costs Labor	Equipment	Total	Total Incl O&P
0530	30' set	Q-5	34	.471	Ea.	129	17.35		146.35	171
0540	40' set		31	.516		170	19		189	219
0550	50' set		28	.571		199	21		220	254
0580	100' set		18	.889		590	33		623	700
0700	Combination 3/8" & 1-1/8" tubes									
0710	10' set	Q-5	36	.444	Ea.	107	16.40		123.40	144
0720	20' set		33	.485		178	17.85		195.85	226
0730	30' set		31	.516		240	19		259	295
0740	40' set		28	.571		360	21		381	430
0750	50' set		26	.615		350	22.50		372.50	425
0900	Combination 1/2" & 3/4" tubes									
0910	10' set	Q-5	37	.432	Ea.	64.50	15.95		80.45	96.50
0920	20' set		35	.457		114	16.85		130.85	154
0930	30' set		33	.485		172	17.85		189.85	219
0940	40' set		30	.533		226	19.65		245.65	282
0950	50' set		27	.593		284	22		306	350
0980	100' set		17	.941		640	34.50		674.50	760
2100	Combination 1/2" & 1-1/8" tubes									
2110	10' set	Q-5	35	.457	Ea.	110	16.85		126.85	149
2120	20' set		31	.516		188	19		207	239
2130	30' set		29	.552		283	20.50		303.50	345
2140	40' set		25	.640		380	23.50		403.50	460
2150	50' set		14	1.143		465	42		507	580
2300	For 1" thick insulation add					30%	15%			
3000	Refrigerant line sets, min-split, flared									
3100	Combination 1/4" & 3/8" tubes									
3120	15' set	Q-5	41	.390	Ea.	78	14.40		92.40	109
3140	25' set		38.50	.416		111	15.30		126.30	147
3160	35' set		37	.432		138	15.95		153.95	178
3180	50' set		30	.533		182	19.65		201.65	233
3200	Combination 1/4" & 1/2" tubes									
3220	15' set	Q-5	41	.390	Ea.	85	14.40		99.40	117
3240	25' set		38.50	.416		113	15.30		128.30	149
3260	35' set		37	.432		144	15.95		159.95	184
3280	50' set		30	.533		190	19.65		209.65	242

23 31 HVAC Ducts and Casings

23 31 13 – Metal Ducts

23 31 13.13 Rectangular Metal Ducts

		Crew	Daily Output	Labor-Hours	Unit	Material	Labor	Equipment	Total	Total Incl O&P
0010	**RECTANGULAR METAL DUCTS**									
0020	Fabricated rectangular, includes fittings, joints, supports,									
0021	allowance for flexible connections and field sketches.									
0030	Does not include "as-built dwgs." or insulation.									
0031	NOTE: Fabrication and installation are combined									
0040	as LABOR cost. Approx. 25% fittings assumed.									
0042	Fabrication/Inst. is to commercial quality standards									
0043	(SMACNA or equiv.) for structure, sealing, leak testing, etc.									
0100	Aluminum, alloy 3003-H14, under 100 lb.	Q-10	75	.320	Lb.	3.19	11.40		14.59	22.50
0110	100 to 500 lb.		80	.300		1.88	10.70		12.58	19.90
0120	500 to 1,000 lb.		95	.253		1.81	9		10.81	17
0140	1,000 to 2,000 lb.		120	.200		1.77	7.15		8.92	13.85

23 31 HVAC Ducts and Casings

23 31 13 – Metal Ducts

23 31 13.13 Rectangular Metal Ducts		Crew	Daily Output	Labor-Hours	Unit	Material	2018 Bare Costs Labor	Equipment	Total	Total Incl O&P
0500	Galvanized steel, under 200 lb.	Q-10	235	.102	Lb.	.57	3.65		4.22	6.70
0520	200 to 500 lb.		245	.098		.56	3.50		4.06	6.45
0540	500 to 1,000 lb.	↓	255	.094	↓	.55	3.36		3.91	6.20

23 33 Air Duct Accessories

23 33 13 – Dampers

23 33 13.13 Volume-Control Dampers

		Crew	Daily Output	Labor-Hours	Unit	Material	Labor	Equipment	Total	Total Incl O&P
0010	**VOLUME-CONTROL DAMPERS**									
6000	12" x 12"	1 Shee	21	.381	Ea.	39.50	14.55		54.05	68
8000	Multi-blade dampers, parallel blade									
8100	8" x 8"	1 Shee	24	.333	Ea.	110	12.75		122.75	142

23 33 13.16 Fire Dampers

		Crew	Daily Output	Labor-Hours	Unit	Material	Labor	Equipment	Total	Total Incl O&P
0010	**FIRE DAMPERS**									
3000	Fire damper, curtain type, 1-1/2 hr. rated, vertical, 6" x 6"	1 Shee	24	.333	Ea.	32.50	12.75		45.25	56.50
3020	8" x 6"	"	22	.364	"	32.50	13.90		46.40	58.50

23 33 46 – Flexible Ducts

23 33 46.10 Flexible Air Ducts

			Crew	Daily Output	Labor-Hours	Unit	Material	Labor	Equipment	Total	Total Incl O&P
0010	**FLEXIBLE AIR DUCTS**										
1300	Flexible, coated fiberglass fabric on corr. resist. metal helix										
1400	pressure to 12" (WG) UL-181										
1500	Noninsulated, 3" diameter		Q-9	400	.040	L.F.	1.20	1.38		2.58	3.61
1540	5" diameter			320	.050		1.38	1.72		3.10	4.39
1560	6" diameter			280	.057		1.60	1.97		3.57	5.05
1580	7" diameter			240	.067		1.78	2.30		4.08	5.80
1900	Insulated, 1" thick, PE jacket, 3" diameter	G		380	.042		2.62	1.45		4.07	5.30
1910	4" diameter	G		340	.047		2.80	1.62		4.42	5.80
1920	5" diameter	G		300	.053		2.93	1.84		4.77	6.30
1940	6" diameter	G		260	.062		3.11	2.12		5.23	6.95
1960	7" diameter	G		220	.073		3.60	2.50		6.10	8.15
1980	8" diameter	G		180	.089		3.80	3.06		6.86	9.30
2040	12" diameter	G	↓	100	.160	↓	5.25	5.50		10.75	15

23 33 53 – Duct Liners

23 33 53.10 Duct Liner Board

			Crew	Daily Output	Labor-Hours	Unit	Material	Labor	Equipment	Total	Total Incl O&P
0010	**DUCT LINER BOARD**										
3490	Board type, fiberglass liner, 3 lb. density										
3940	Board type, non-fibrous foam										
3950	Temperature, bacteria and fungi resistant										
3960	1" thick	G	Q-14	150	.107	S.F.	2.57	3.46		6.03	8.70
3970	1-1/2" thick	G		130	.123		4.03	3.99		8.02	11.20
3980	2" thick	G	↓	120	.133	↓	4.02	4.33		8.35	11.70

23 34 HVAC Fans

23 34 23 – HVAC Power Ventilators

23 34 23.10 HVAC Power Circulators and Ventilators	Crew	Daily Output	Labor-Hours	Unit	Material	2018 Bare Costs Labor	Equipment	Total	Total Incl O&P
0010 **HVAC POWER CIRCULATORS AND VENTILATORS**									
6650 Residential, bath exhaust, grille, back draft damper									
6660 50 CFM	Q-20	24	.833	Ea.	52.50	29.50		82	107
6670 110 CFM		22	.909		104	32.50		136.50	168
6900 Kitchen exhaust, grille, complete, 160 CFM		22	.909		105	32.50		137.50	170
6910 180 CFM		20	1		106	35.50		141.50	176
6920 270 CFM		18	1.111		230	39.50		269.50	320
6930 350 CFM	↓	16	1.250	↓	145	44.50		189.50	234
6940 Residential roof jacks and wall caps									
6944 Wall cap with back draft damper									
6946 3" & 4" diam. round duct	1 Shee	11	.727	Ea.	25	28		53	74
6948 6" diam. round duct	"	11	.727	"	72.50	28		100.50	126
6958 Roof jack with bird screen and back draft damper									
6960 3" & 4" diam. round duct	1 Shee	11	.727	Ea.	24.50	28		52.50	73.50
6962 3-1/4" x 10" rectangular duct	"	10	.800	"	44	30.50		74.50	99.50
8020 Attic, roof type									
8030 Aluminum dome, damper & curb									
8080 12" diameter, 1,000 CFM (gravity)	1 Elec	10	.800	Ea.	600	32		632	715
8090 16" diameter, 1,500 CFM (gravity)		9	.889		725	35.50		760.50	860
8100 20" diameter, 2,500 CFM (gravity)	↓	8	1	↓	890	40		930	1,050
8160 Plastic, ABS dome									
8180 1,050 CFM	1 Elec	14	.571	Ea.	177	23		200	233
8200 1,600 CFM	"	12	.667	"	265	27		292	335
8240 Attic, wall type, with shutter, one speed									
8250 12" diameter, 1,000 CFM	1 Elec	14	.571	Ea.	380	23		403	460
8260 14" diameter, 1,500 CFM		12	.667		415	27		442	500
8270 16" diameter, 2,000 CFM	↓	9	.889	↓	465	35.50		500.50	575
8290 Whole house, wall type, with shutter, one speed									
8300 30" diameter, 4,800 CFM	1 Elec	7	1.143	Ea.	1,000	46		1,046	1,175
8310 36" diameter, 7,000 CFM		6	1.333		1,075	53.50		1,128.50	1,300
8320 42" diameter, 10,000 CFM		5	1.600		1,225	64		1,289	1,450
8330 48" diameter, 16,000 CFM	↓	4	2		1,525	80.50		1,605.50	1,800
8340 For two speed, add				↓	91			91	100
8350 Whole house, lay-down type, with shutter, one speed									
8360 30" diameter, 4,500 CFM	1 Elec	8	1	Ea.	1,075	40		1,115	1,250
8370 36" diameter, 6,500 CFM		7	1.143		1,150	46		1,196	1,325
8380 42" diameter, 9,000 CFM		6	1.333		1,250	53.50		1,303.50	1,475
8390 48" diameter, 12,000 CFM	↓	5	1.600		1,425	64		1,489	1,675
8440 For two speed, add					68.50			68.50	75.50
8450 For 12 hour timer switch, add	1 Elec	32	.250	↓	68.50	10.05		78.55	92

For customer support on your Residential Costs with RSMeans data, call 800.448.8182.

587

23 37 13.10 Diffusers

	Crew	Daily Output	Labor-Hours	Unit	Material	2018 Bare Costs Labor	2018 Bare Costs Equipment	Total	Total Incl O&P
0010 **DIFFUSERS**, Aluminum, opposed blade damper unless noted									
0100 Ceiling, linear, also for sidewall									
0120 2" wide	1 Shee	32	.250	L.F.	17.45	9.55		27	35
0160 4" wide		26	.308	"	23.50	11.75		35.25	45
0500 Perforated, 24" x 24" lay-in panel size, 6" x 6"		16	.500	Ea.	158	19.15	.	177.15	206
0520 8" x 8"		15	.533		166	20.50		186.50	217
0530 9" x 9"		14	.571		169	22		191	223
0590 16" x 16"		11	.727		197	28		225	264
1000 Rectangular, 1 to 4 way blow, 6" x 6"		16	.500		42	19.15		61.15	78
1010 8" x 8"		15	.533		58.50	20.50		79	98
1014 9" x 9"		15	.533		52.50	20.50		73	91.50
1016 10" x 10"		15	.533		80.50	20.50		101	123
1020 12" x 6"		15	.533		72	20.50		92.50	114
1040 12" x 9"		14	.571		76.50	22		98.50	121
1060 12" x 12"		12	.667		70.50	25.50		96	120
1070 14" x 6"		13	.615		78.50	23.50		102	125
1074 14" x 14"		12	.667		129	25.50		154.50	185
1150 18" x 18"		9	.889		114	34		148	182
1170 24" x 12"		10	.800		165	30.50		195.50	232
1180 24" x 24"		7	1.143		272	43.50		315.50	370
1500 Round, butterfly damper, steel, diffuser size, 6" diameter		18	.444		11	17		28	40.50
1520 8" diameter		16	.500		11.70	19.15		30.85	45
2000 T-bar mounting, 24" x 24" lay-in frame, 6" x 6"		16	.500		69.50	19.15		88.65	109
2020 8" x 8"		14	.571		69.50	22		91.50	113
2040 12" x 12"		12	.667		85.50	25.50		111	137
2060 16" x 16"		11	.727		106	28		134	163
2080 18" x 18"		10	.800		117	30.50		147.50	180
6000 For steel diffusers instead of aluminum, deduct					10%				

23 37 13.30 Grilles

	Crew	Daily Output	Labor-Hours	Unit	Material	2018 Bare Costs Labor	2018 Bare Costs Equipment	Total	Total Incl O&P
0010 **GRILLES**									
0020 Aluminum, unless noted otherwise									
1000 Air return, steel, 6" x 6"	1 Shee	26	.308	Ea.	19.80	11.75		31.55	41.50
1020 10" x 6"		24	.333		19.80	12.75		32.55	43
1080 16" x 8"		22	.364		28	13.90		41.90	54
1100 12" x 12"		22	.364		28	13.90		41.90	54
1120 24" x 12"		18	.444		37	17		54	69
1180 16" x 16"		22	.364		36	13.90		49.90	62.50

23 37 13.60 Registers

	Crew	Daily Output	Labor-Hours	Unit	Material	2018 Bare Costs Labor	2018 Bare Costs Equipment	Total	Total Incl O&P
0010 **REGISTERS**									
0980 Air supply									
3000 Baseboard, hand adj. damper, enameled steel									
3012 8" x 6"	1 Shee	26	.308	Ea.	6.85	11.75		18.60	27
3020 10" x 6"		24	.333		7.45	12.75		20.20	29
3040 12" x 5"		23	.348		6.05	13.30		19.35	28.50
3060 12" x 6"		23	.348		8.10	13.30		21.40	31
4000 Floor, toe operated damper, enameled steel									
4020 4" x 8"	1 Shee	32	.250	Ea.	12.70	9.55		22.25	30
4040 4" x 12"	"	26	.308	"	14.90	11.75		26.65	36
4300 Spiral pipe supply register									
4310 Steel, with air scoop									
4320 4" x 12", for 8" thru 13" diameter duct	1 Shee	25	.320	Ea.	78	12.25		90.25	107
4330 4" x 18", for 8" thru 13" diameter duct		18	.444		91.50	17		108.50	129

For customer support on your Residential Costs with RSMeans data, call 800.448.8182.

23 37 Air Outlets and Inlets

23 37 13 – Diffusers, Registers, and Grilles

23 37 13.60 Registers		Crew	Daily Output	Labor-Hours	Unit	Material	2018 Bare Costs Labor	2018 Bare Costs Equipment	Total	Total Incl O&P
4340	6" x 12", for 14" thru 21" diameter duct	1 Shee	19	.421	Ea.	83	16.10		99.10	119
4350	6" x 16", for 14" thru 21" diameter duct		18	.444		95	17		112	134
4360	6" x 20", for 14" thru 21" diameter duct		17	.471		104	18		122	144
4370	6" x 24", for 14" thru 21" diameter duct		16	.500		123	19.15		142.15	167
4380	8" x 16", for 22" thru 31" diameter duct		19	.421		99.50	16.10		115.60	137
4390	8" x 18", for 22" thru 31" diameter duct		18	.444		104	17		121	143
4400	8" x 24", for 22" thru 31" diameter duct	↓	15	.533	↓	128	20.50		148.50	175

23 41 Particulate Air Filtration

23 41 13 – Panel Air Filters

23 41 13.10 Panel Type Air Filters

0010	**PANEL TYPE AIR FILTERS**									
2950	Mechanical media filtration units									
3000	High efficiency type, with frame, non-supported	G				MCFM	35		35	38.50
3100	Supported type	G				"	55		55	60.50
5500	Throwaway glass or paper media type, 12" x 36" x 1"					Ea.	2.39		2.39	2.63

23 41 16 – Renewable-Media Air Filters

23 41 16.10 Disposable Media Air Filters

0010	**DISPOSABLE MEDIA AIR FILTERS**									
5000	Renewable disposable roll					C.S.F.	5.25		5.25	5.75

23 41 19 – Washable Air Filters

23 41 19.10 Permanent Air Filters

0010	**PERMANENT AIR FILTERS**									
4500	Permanent washable	G				MCFM	25		25	27.50

23 41 23 – Extended Surface Filters

23 41 23.10 Expanded Surface Filters

0010	**EXPANDED SURFACE FILTERS**									
4000	Medium efficiency, extended surface	G				MCFM	6		6	6.60

23 42 Gas-Phase Air Filtration

23 42 13 – Activated-Carbon Air Filtration

23 42 13.10 Charcoal Type Air Filtration

0010	**CHARCOAL TYPE AIR FILTRATION**									
0050	Activated charcoal type, full flow					MCFM	650		650	715
0060	Full flow, impregnated media 12" deep						225		225	248
0070	HEPA filter & frame for field erection						450		450	495
0080	HEPA filter-diffuser, ceiling install.				↓		350		350	385

For customer support on your Residential Costs with RSMeans data, call 800.448.8182.

589

23 43 Electronic Air Cleaners

23 43 13 – Washable Electronic Air Cleaners

23 43 13.10 Electronic Air Cleaners	Crew	Daily Output	Labor-Hours	Unit	Material	2018 Bare Costs Labor	Equipment	Total	Total Incl O&P
0010 **ELECTRONIC AIR CLEANERS**									
2000 Electronic air cleaner, duct mounted									
2150 1,000 CFM	1 Shee	4	2	Ea.	430	76.50		506.50	600
2200 1,200 CFM		3.80	2.105		505	80.50		585.50	690
2250 1,400 CFM	↓	3.60	2.222	↓	530	85		615	725

23 51 Breechings, Chimneys, and Stacks

23 51 23 – Gas Vents

23 51 23.10 Gas Chimney Vents

	Crew	Daily Output	Labor-Hours	Unit	Material	2018 Bare Costs Labor	Equipment	Total	Total Incl O&P
0010 **GAS CHIMNEY VENTS**, Prefab metal, UL listed									
0020 Gas, double wall, galvanized steel									
0080 3" diameter	Q-9	72	.222	V.L.F.	6.60	7.65		14.25	20
0100 4" diameter	"	68	.235	"	8.15	8.10		16.25	22.50

23 52 Heating Boilers

23 52 13 – Electric Boilers

23 52 13.10 Electric Boilers, ASME

	Crew	Daily Output	Labor-Hours	Unit	Material	2018 Bare Costs Labor	Equipment	Total	Total Incl O&P
0010 **ELECTRIC BOILERS, ASME**, Standard controls and trim									
1000 Steam, 6 KW, 20.5 MBH	Q-19	1.20	20	Ea.	4,250	760		5,010	5,925
1160 60 KW, 205 MBH		1	24		6,925	910		7,835	9,100
2000 Hot water, 7.5 KW, 25.6 MBH		1.30	18.462		5,125	700		5,825	6,800
2040 30 KW, 102 MBH		1.20	20		5,525	760		6,285	7,325
2060 45 KW, 164 MBH	↓	1.20	20	↓	5,625	760		6,385	7,425

23 52 23 – Cast-Iron Boilers

23 52 23.20 Gas-Fired Boilers

	Crew	Daily Output	Labor-Hours	Unit	Material	2018 Bare Costs Labor	Equipment	Total	Total Incl O&P
0010 **GAS-FIRED BOILERS**, Natural or propane, standard controls, packaged									
1000 Cast iron, with insulated jacket									
3000 Hot water, gross output, 80 MBH	Q-7	1.46	21.918	Ea.	1,925	810		2,735	3,450
3020 100 MBH	"	1.35	23.704		2,450	875		3,325	4,150
7000 For tankless water heater, add					10%				
7050 For additional zone valves up to 312 MBH, add				↓	196			196	216

23 52 23.30 Gas/Oil Fired Boilers

	Crew	Daily Output	Labor-Hours	Unit	Material	2018 Bare Costs Labor	Equipment	Total	Total Incl O&P
0010 **GAS/OIL FIRED BOILERS**, Combination with burners and controls, packaged									
1000 Cast iron with insulated jacket									
2000 Steam, gross output, 720 MBH	Q-7	.43	74.074	Ea.	14,200	2,725		16,925	20,100
2900 Hot water, gross output									
2910 200 MBH	Q-6	.62	39.024	Ea.	10,200	1,375		11,575	13,600
2920 300 MBH		.49	49.080		10,200	1,750		11,950	14,200
2930 400 MBH		.41	57.971		12,000	2,050		14,050	16,600
2940 500 MBH	↓	.36	67.039		12,900	2,375		15,275	18,100
3000 584 MBH	Q-7	.44	72.072	↓	14,300	2,650		16,950	20,100

23 52 23.40 Oil-Fired Boilers

	Crew	Daily Output	Labor-Hours	Unit	Material	2018 Bare Costs Labor	Equipment	Total	Total Incl O&P
0010 **OIL-FIRED BOILERS**, Standard controls, flame retention burner, packaged									
1000 Cast iron, with insulated flush jacket									
2000 Steam, gross output, 109 MBH	Q-7	1.20	26.667	Ea.	2,225	985		3,210	4,075
2060 207 MBH	"	.90	35.556	"	3,025	1,300		4,325	5,475
3000 Hot water, same price as steam									

23 52 Heating Boilers

23 52 26 – Steel Boilers

23 52 26.40 Oil-Fired Boilers

		Crew	Daily Output	Labor-Hours	Unit	Material	2018 Bare Costs Labor	Equipment	Total	Total Incl O&P
0010	**OIL-FIRED BOILERS**, Standard controls, flame retention burner									
5000	Steel, with insulated flush jacket									
7000	Hot water, gross output, 103 MBH	Q-6	1.60	15	Ea.	1,875	530		2,405	2,950
7020	122 MBH		1.45	16.506		2,000	585		2,585	3,175
7060	168 MBH		1.30	18.405		2,250	655		2,905	3,550
7080	225 MBH	↓	1.22	19.704	↓	3,250	700		3,950	4,725

23 52 28 – Swimming Pool Boilers

23 52 28.10 Swimming Pool Heaters

		Crew	Daily Output	Labor-Hours	Unit	Material	Labor	Equipment	Total	Total Incl O&P
0010	**SWIMMING POOL HEATERS**, Not including wiring, external									
0020	piping, base or pad									
0160	Gas fired, input, 155 MBH	Q-6	1.50	16	Ea.	1,975	570		2,545	3,100
0200	199 MBH		1	24		2,100	850		2,950	3,700
0280	500 MBH	↓	.40	60		8,725	2,125		10,850	13,100
2000	Electric, 12 KW, 4,800 gallon pool	Q-19	3	8		2,100	305		2,405	2,825
2020	15 KW, 7,200 gallon pool		2.80	8.571		2,150	325		2,475	2,875
2040	24 KW, 9,600 gallon pool		2.40	10		2,475	380		2,855	3,350
2100	57 KW, 24,000 gallon pool	↓	1.20	20	↓	3,650	760		4,410	5,250

23 52 88 – Burners

23 52 88.10 Replacement Type Burners

		Crew	Daily Output	Labor-Hours	Unit	Material	Labor	Equipment	Total	Total Incl O&P
0010	**REPLACEMENT TYPE BURNERS**									
0990	Residential, conversion, gas fired, LP or natural									
1000	Gun type, atmospheric input 50 to 225 MBH	Q-1	2.50	6.400	Ea.	1,000	229		1,229	1,475
1020	100 to 400 MBH	"	2	8		1,675	287		1,962	2,325
1025	Burner, gas, 100 to 400 MBH [G]					1,675			1,675	1,850
1040	300 to 1,000 MBH	Q-1	1.70	9.412	↓	5,100	335		5,435	6,150

23 54 Furnaces

23 54 13 – Electric-Resistance Furnaces

23 54 13.10 Electric Furnaces

		Crew	Daily Output	Labor-Hours	Unit	Material	Labor	Equipment	Total	Total Incl O&P
0010	**ELECTRIC FURNACES**, Hot air, blowers, std. controls									
0011	not including gas, oil or flue piping									
1000	Electric, UL listed									
1070	10.2 MBH	Q-20	4.40	4.545	Ea.	335	162		497	640
1080	17.1 MBH		4.60	4.348		415	155		570	710
1100	34.1 MBH	↓	4.40	4.545	↓	520	162		682	840

23 54 16 – Fuel-Fired Furnaces

23 54 16.13 Gas-Fired Furnaces

		Crew	Daily Output	Labor-Hours	Unit	Material	Labor	Equipment	Total	Total Incl O&P
0010	**GAS-FIRED FURNACES**									
3000	Gas, AGA certified, upflow, direct drive models									
3020	45 MBH input	Q-9	4	4	Ea.	645	138		783	940
3040	60 MBH input		3.80	4.211		640	145		785	945
3060	75 MBH input		3.60	4.444		710	153		863	1,050
3100	100 MBH input		3.20	5		740	172		912	1,100
3120	125 MBH input		3	5.333		770	184		954	1,150
3130	150 MBH input		2.80	5.714		1,550	197		1,747	2,025
3140	200 MBH input		2.60	6.154		3,175	212		3,387	3,850
4000	For starter plenum, add	↓	16	1	↓	94	34.50		128.50	161

For customer support on your Residential Costs with RSMeans data, call 800.448.8182.

591

23 54 Furnaces

23 54 16 – Fuel-Fired Furnaces

23 54 16.16 Oil-Fired Furnaces		Crew	Daily Output	Labor-Hours	Unit	Material	2018 Bare Costs Labor	Equipment	Total	Total Incl O&P
0010	**OIL-FIRED FURNACES**									
6000	Oil, UL listed, atomizing gun type burner									
6020	56 MBH output	Q-9	3.60	4.444	Ea.	2,825	153		2,978	3,350
6030	84 MBH output		3.50	4.571		2,850	157		3,007	3,400
6040	95 MBH output		3.40	4.706		2,875	162		3,037	3,425
6060	134 MBH output		3.20	5		2,825	172		2,997	3,375
6080	151 MBH output		3	5.333		3,100	184		3,284	3,725
6100	200 MBH input	↓	2.60	6.154	↓	3,550	212		3,762	4,250

23 54 16.21 Solid Fuel-Fired Furnaces

			Crew	Daily Output	Labor-Hours	Unit	Material	Labor	Equipment	Total	Total Incl O&P
0010	**SOLID FUEL-FIRED FURNACES**										
6020	Wood fired furnaces										
6030	Includes hot water coil, thermostat, and auto draft control										
6040	24" long firebox	G	Q-9	4	4	Ea.	4,375	138		4,513	5,050
6050	30" long firebox	G		3.60	4.444		5,100	153		5,253	5,875
6060	With fireplace glass doors	G	↓	3.20	5	↓	6,450	172		6,622	7,375
6200	Wood/oil fired furnaces, includes two thermostats										
6210	Includes hot water coil and auto draft control										
6240	24" long firebox	G	Q-9	3.40	4.706	Ea.	5,375	162		5,537	6,175
6250	30" long firebox	G		3	5.333		5,975	184		6,159	6,875
6260	With fireplace glass doors	G	↓	2.80	5.714	↓	7,325	197		7,522	8,375
6400	Wood/gas fired furnaces, includes two thermostats										
6410	Includes hot water coil and auto draft control										
6440	24" long firebox	G	Q-9	2.80	5.714	Ea.	6,000	197		6,197	6,925
6450	30" long firebox	G		2.40	6.667		6,700	230		6,930	7,750
6460	With fireplace glass doors	G	↓	2	8	↓	8,025	275		8,300	9,275
6600	Wood/oil/gas fired furnaces, optional accessories										
6610	Hot air plenum		Q-9	16	1	Ea.	108	34.50		142.50	177
6620	Safety heat dump			24	.667		94	23		117	141
6630	Auto air intake			18	.889		148	30.50		178.50	214
6640	Cold air return package	↓	14	1.143		159	39.50		198.50	241
6650	Wood fork				↓		51			51	56
6700	Wood fired outdoor furnace										
6740	24" long firebox	G	Q-9	3.80	4.211	Ea.	4,000	145		4,145	4,650
6760	Wood fired outdoor furnace, optional accessories										
6770	Chimney section, stainless steel, 6" ID x 3' long	G	Q-9	36	.444	Ea.	176	15.30		191.30	220
6780	Chimney cap, stainless steel	G	"	40	.400	"	104	13.75		117.75	137
6800	Wood fired hot water furnace										
6820	Includes 200 gal. hot water storage, thermostat, and auto draft control										
6840	30" long firebox	G	Q-9	2.10	7.619	Ea.	8,350	262		8,612	9,625
6850	Water to air heat exchanger										
6870	Includes mounting kit and blower relay										
6880	140 MBH, 18.75" W x 18.75" L		Q-9	7.50	2.133	Ea.	375	73.50		448.50	535
6890	200 MBH, 24" W x 24" L		"	7	2.286	"	535	78.50		613.50	715
6900	Water to water heat exchanger										
6940	100 MBH		Q-9	6.50	2.462	Ea.	380	85		465	555
6960	290 MBH		"	6	2.667	"	545	92		637	750
7000	Optional accessories										
7010	Large volume circulation pump (2 included)		Q-9	14	1.143	Ea.	246	39.50		285.50	335
7020	Air bleed fittings (package)			24	.667		51.50	23		74.50	94.50
7030	Domestic water preheater			6	2.667		222	92		314	400
7040	Smoke pipe kit	↓	4	4	↓	92	138		230	330

23 54 Furnaces

23 54 24 – Furnace Components for Cooling

23 54 24.10 Furnace Components and Combinations	Crew	Daily Output	Labor-Hours	Unit	Material	2018 Bare Costs Labor	Equipment	Total	Total Incl O&P
0010 **FURNACE COMPONENTS AND COMBINATIONS**									
0080 Coils, A.C. evaporator, for gas or oil furnaces									
0090 Add-on, with holding charge									
0100 Upflow									
0120 1-1/2 ton cooling	Q-5	4	4	Ea.	209	147		356	475
0130 2 ton cooling		3.70	4.324		263	159		422	550
0140 3 ton cooling		3.30	4.848		400	179		579	735
0150 4 ton cooling		3	5.333		550	197		747	930
0160 5 ton cooling	▼	2.70	5.926	▼	590	218		808	1,000
0300 Downflow									
0330 2-1/2 ton cooling	Q-5	3	5.333	Ea.	315	197		512	670
0340 3-1/2 ton cooling		2.60	6.154		465	227		692	885
0350 5 ton cooling	▼	2.20	7.273	▼	590	268		858	1,100
0600 Horizontal									
0630 2 ton cooling	Q-5	3.90	4.103	Ea.	385	151		536	675
0640 3 ton cooling		3.50	4.571		430	168		598	750
0650 4 ton cooling		3.20	5		480	184		664	830
0660 5 ton cooling	▼	2.90	5.517	▼	480	203		683	860
2000 Cased evaporator coils for air handlers									
2100 1-1/2 ton cooling	Q-5	4.40	3.636	Ea.	310	134		444	560
2110 2 ton cooling		4.10	3.902		355	144		499	625
2120 2-1/2 ton cooling		3.90	4.103		375	151		526	665
2130 3 ton cooling		3.70	4.324		420	159		579	720
2140 3-1/2 ton cooling		3.50	4.571		520	168		688	850
2150 4 ton cooling		3.20	5		610	184		794	975
2160 5 ton cooling	▼	2.90	5.517	▼	600	203		803	995
3010 Air handler, modular									
3100 With cased evaporator cooling coil									
3120 1-1/2 ton cooling	Q-5	3.80	4.211	Ea.	965	155		1,120	1,325
3130 2 ton cooling		3.50	4.571		965	168		1,133	1,350
3140 2-1/2 ton cooling		3.30	4.848		1,025	179		1,204	1,425
3150 3 ton cooling		3.10	5.161		1,175	190		1,365	1,625
3160 3-1/2 ton cooling		2.90	5.517		1,450	203		1,653	1,925
3170 4 ton cooling		2.50	6.400		1,450	236		1,686	2,000
3180 5 ton cooling	▼	2.10	7.619	▼	1,750	281		2,031	2,375
3500 With no cooling coil									
3520 1-1/2 ton coil size	Q-5	12	1.333	Ea.	1,100	49		1,149	1,275
3530 2 ton coil size		10	1.600		1,175	59		1,234	1,400
3540 2-1/2 ton coil size		10	1.600		1,225	59		1,284	1,425
3554 3 ton coil size		9	1.778		1,200	65.50		1,265.50	1,425
3560 3-1/2 ton coil size		9	1.778		1,250	65.50		1,315.50	1,475
3570 4 ton coil size		8.50	1.882		1,325	69.50		1,394.50	1,575
3580 5 ton coil size	▼	8	2	▼	1,425	73.50		1,498.50	1,675
4000 With heater									
4120 5 kW, 17.1 MBH	Q-5	16	1	Ea.	455	37		492	560
4130 7.5 kW, 25.6 MBH		15.60	1.026		1,100	38		1,138	1,250
4140 10 kW, 34.2 MBH		15.20	1.053		1,175	39		1,214	1,375
4150 12.5 kW, 42.7 MBH		14.80	1.081		1,550	40		1,590	1,775
4160 15 kW, 51.2 MBH		14.40	1.111		1,700	41		1,741	1,950
4170 25 kW, 85.4 MBH		14	1.143		2,250	42		2,292	2,550
4180 30 kW, 102 MBH	▼	13	1.231	▼	2,600	45.50		2,645.50	2,950

For customer support on your Residential Costs with RSMeans data, call 800.448.8182.

593

23 62 Packaged Compressor and Condenser Units

23 62 13 – Packaged Air-Cooled Refrigerant Compressor and Condenser Units

23 62 13.10 Packaged Air-Cooled Refrig. Condensing Units	Crew	Daily Output	Labor-Hours	Unit	Material	2018 Bare Costs Labor	Equipment	Total	Total Incl O&P
0010 **PACKAGED AIR-COOLED REFRIGERANT CONDENSING UNITS**									
0020 Condensing unit									
0030 Air cooled, compressor, standard controls									
0050 1.5 ton	Q-5	2.50	6.400	Ea.	1,025	236		1,261	1,525
0100 2 ton		2.10	7.619		1,050	281		1,331	1,600
0200 2.5 ton		1.70	9.412		1,350	345		1,695	2,075
0300 3 ton		1.30	12.308		1,350	455		1,805	2,225
0350 3.5 ton		1.10	14.545		1,575	535		2,110	2,600
0400 4 ton		.90	17.778		1,775	655		2,430	3,025
0500 5 ton		.60	26.667		2,150	985		3,135	4,000

23 74 Packaged Outdoor HVAC Equipment

23 74 33 – Dedicated Outdoor-Air Units

23 74 33.10 Rooftop Air Conditioners

	Crew	Daily Output	Labor-Hours	Unit	Material	2018 Bare Costs Labor	Equipment	Total	Total Incl O&P
0010 **ROOFTOP AIR CONDITIONERS**, Standard controls, curb, economizer									
1000 Single zone, electric cool, gas heat									
1140 5 ton cooling, 112 MBH heating	Q-5	.56	28.521	Ea.	4,375	1,050		5,425	6,525
1150 7.5 ton cooling, 170 MBH heating		.50	32.258		5,825	1,200		7,025	8,350
1156 8.5 ton cooling, 170 MBH heating		.46	34.783		7,075	1,275		8,350	9,875
1160 10 ton cooling, 200 MBH heating	Q-6	.67	35.982		9,150	1,275		10,425	12,200

23 81 Decentralized Unitary HVAC Equipment

23 81 13 – Packaged Terminal Air-Conditioners

23 81 13.10 Packaged Cabinet Type Air-Conditioners

	Crew	Daily Output	Labor-Hours	Unit	Material	2018 Bare Costs Labor	Equipment	Total	Total Incl O&P
0010 **PACKAGED CABINET TYPE AIR-CONDITIONERS**, Cabinet, wall sleeve,									
0100 louver, electric heat, thermostat, manual changeover, 208 V									
0200 6,000 BTUH cooling, 8,800 BTU heat	Q-5	6	2.667	Ea.	720	98.50		818.50	950
0220 9,000 BTUH cooling, 13,900 BTU heat		5	3.200		940	118		1,058	1,225
0240 12,000 BTUH cooling, 13,900 BTU heat		4	4		1,450	147		1,597	1,850
0260 15,000 BTUH cooling, 13,900 BTU heat		3	5.333		1,400	197		1,597	1,875

23 81 19 – Self-Contained Air-Conditioners

23 81 19.10 Window Unit Air Conditioners

	Crew	Daily Output	Labor-Hours	Unit	Material	2018 Bare Costs Labor	Equipment	Total	Total Incl O&P
0010 **WINDOW UNIT AIR CONDITIONERS**									
4000 Portable/window, 15 amp, 125 V grounded receptacle required									
4060 5,000 BTUH	1 Carp	8	1	Ea.	297	35		332	385
4340 6,000 BTUH		8	1		249	35		284	330
4480 8,000 BTUH		6	1.333		400	46.50		446.50	520
4500 10,000 BTUH		6	1.333		680	46.50		726.50	825
4520 12,000 BTUH	L-2	8	2		1,925	61.50		1,986.50	2,200
4600 Window/thru-the-wall, 15 amp, 230 V grounded receptacle required									
4780 18,000 BTUH	L-2	6	2.667	Ea.	745	82		827	955
4940 25,000 BTUH		4	4		895	123		1,018	1,200
4960 29,000 BTUH		4	4		1,075	123		1,198	1,375

23 81 Decentralized Unitary HVAC Equipment

23 81 19 – Self-Contained Air-Conditioners

23 81 19.20 Self-Contained Single Package

		Crew	Daily Output	Labor-Hours	Unit	Material	2018 Bare Costs Labor	Equipment	Total	Total Incl O&P
0010	**SELF-CONTAINED SINGLE PACKAGE**									
0100	Air cooled, for free blow or duct, not incl. remote condenser									
0110	Constant volume									
0200	3 ton cooling	Q-5	1	16	Ea.	3,750	590		4,340	5,100
0210	4 ton cooling	"	.80	20	"	4,100	735		4,835	5,725
1000	Water cooled for free blow or duct, not including tower									
1010	Constant volume									
1100	3 ton cooling	Q-6	1	24	Ea.	3,700	850		4,550	5,475
1300	For hot water or steam heat coils, add					"	12%	10%			

23 81 26 – Split-System Air-Conditioners

23 81 26.10 Split Ductless Systems

		Crew	Daily Output	Labor-Hours	Unit	Material	2018 Bare Costs Labor	Equipment	Total	Total Incl O&P
0010	**SPLIT DUCTLESS SYSTEMS**									
0100	Cooling only, single zone									
0110	Wall mount									
0120	3/4 ton cooling	Q-5	2	8	Ea.	1,100	295		1,395	1,700
0130	1 ton cooling		1.80	8.889		1,225	330		1,555	1,900
0140	1-1/2 ton cooling		1.60	10		1,975	370		2,345	2,775
0150	2 ton cooling		1.40	11.429		2,250	420		2,670	3,175
1000	Ceiling mount									
1020	2 ton cooling	Q-5	1.40	11.429	Ea.	2,000	420		2,420	2,900
1030	3 ton cooling	"	1.20	13.333	"	2,550	490		3,040	3,600
3000	Multizone									
3010	Wall mount									
3020	2 @ 3/4 ton cooling	Q-5	1.80	8.889	Ea.	3,375	330		3,705	4,250
5000	Cooling/Heating									
5010	Wall mount									
5110	1 ton cooling	Q-5	1.70	9.412	Ea.	1,300	345		1,645	2,000
5120	1-1/2 ton cooling	"	1.50	10.667	"	1,975	395		2,370	2,825
7000	Accessories for all split ductless systems									
7010	Add for ambient frost control	Q-5	8	2	Ea.	126	73.50		199.50	260
7020	Add for tube/wiring kit (line sets)									
7030	15' kit	Q-5	32	.500	Ea.	88	18.45		106.45	128
7036	25' kit		28	.571		133	21		154	181
7040	35' kit		24	.667		200	24.50		224.50	261
7050	50' kit		20	.800		202	29.50		231.50	271

23 81 43 – Air-Source Unitary Heat Pumps

23 81 43.10 Air-Source Heat Pumps

		Crew	Daily Output	Labor-Hours	Unit	Material	2018 Bare Costs Labor	Equipment	Total	Total Incl O&P
0010	**AIR-SOURCE HEAT PUMPS**, Not including interconnecting tubing									
1000	Air to air, split system, not including curbs, pads, fan coil and ductwork									
1012	Outside condensing unit only, for fan coil see Section 23 82 19.10									
1020	2 ton cooling, 8.5 MBH heat @ 0°F	Q-5	2	8	Ea.	1,650	295		1,945	2,275
1054	4 ton cooling, 24 MBH heat @ 0°F	"	.80	20	"	2,450	735		3,185	3,925
1500	Single package, not including curbs, pads, or plenums									
1520	2 ton cooling, 6.5 MBH heat @ 0°F	Q-5	1.50	10.667	Ea.	3,200	395		3,595	4,175
1580	4 ton cooling, 13 MBH heat @ 0°F	"	.96	16.667	"	4,300	615		4,915	5,725

For customer support on your Residential Costs with RSMeans data, call 800.448.8182.

595

23 81 Decentralized Unitary HVAC Equipment

23 81 46 – Water-Source Unitary Heat Pumps

23 81 46.10 Water Source Heat Pumps

	Crew	Daily Output	Labor-Hours	Unit	Material	2018 Bare Costs Labor	2018 Bare Costs Equipment	Total	Total Incl O&P
0010 **WATER SOURCE HEAT PUMPS**, Not incl. connecting tubing or water source									
2000 Water source to air, single package									
2100 1 ton cooling, 13 MBH heat @ 75°F	Q-5	2	8	Ea.	1,900	295		2,195	2,550
2200 4 ton cooling, 31 MBH heat @ 75°F	"	1.20	13.333	"	3,000	490		3,490	4,100

23 82 Convection Heating and Cooling Units

23 82 19 – Fan Coil Units

23 82 19.10 Fan Coil Air Conditioning

	Crew	Daily Output	Labor-Hours	Unit	Material	Labor	Equipment	Total	Total Incl O&P
0010 **FAN COIL AIR CONDITIONING**									
0030 Fan coil AC, cabinet mounted, filters and controls									
0100 Chilled water, 1/2 ton cooling	Q-5	8	2	Ea.	555	73.50		628.50	730
0110 3/4 ton cooling		7	2.286		680	84		764	890
0120 1 ton cooling	↓	6	2.667	↓	830	98.50		928.50	1,075

23 82 29 – Radiators

23 82 29.10 Hydronic Heating

	Crew	Daily Output	Labor-Hours	Unit	Material	Labor	Equipment	Total	Total Incl O&P
0010 **HYDRONIC HEATING**, Terminal units, not incl. main supply pipe									
1000 Radiation									
1100 Panel, baseboard, C.I., including supports, no covers	Q-5	46	.348	L.F.	44.50	12.80		57.30	70
3000 Radiators, cast iron									
3100 Free standing or wall hung, 6 tube, 25" high	Q-5	96	.167	Section	62	6.15		68.15	78
3200 4 tube, 19" high	"	96	.167	"	37.50	6.15		43.65	51.50
3250 Adj. brackets, 2 per wall radiator up to 30 sections	1 Stpi	32	.250	Ea.	65	10.25		75.25	88

23 82 36 – Finned-Tube Radiation Heaters

23 82 36.10 Finned Tube Radiation

	Crew	Daily Output	Labor-Hours	Unit	Material	Labor	Equipment	Total	Total Incl O&P
0010 **FINNED TUBE RADIATION**, Terminal units, not incl. main supply pipe									
1310 Baseboard, pkgd, 1/2" copper tube, alum. fin, 7" high	Q-5	60	.267	L.F.	11.60	9.85		21.45	29
1320 3/4" copper tube, alum. fin, 7" high		58	.276		7.85	10.15		18	25.50
1340 1" copper tube, alum. fin, 8-7/8" high		56	.286		21	10.55		31.55	40.50
1360 1-1/4" copper tube, alum. fin, 8-7/8" high	↓	54	.296	↓	31.50	10.90		42.40	52.50
1500 Note: fin tube may also require corners, caps, etc.									

23 83 Radiant Heating Units

23 83 16 – Radiant-Heating Hydronic Piping

23 83 16.10 Radiant Floor Heating

	Crew	Daily Output	Labor-Hours	Unit	Material	Labor	Equipment	Total	Total Incl O&P
0010 **RADIANT FLOOR HEATING**									
0100 Tubing, PEX (cross-linked polyethylene)									
0110 Oxygen barrier type for systems with ferrous materials									
0120 1/2"	Q-5	800	.020	L.F.	1.05	.74		1.79	2.37
0130 3/4"		535	.030		1.48	1.10		2.58	3.44
0140 1"	↓	400	.040	↓	2.31	1.47		3.78	4.97
0200 Non barrier type for ferrous free systems									
0210 1/2"	Q-5	800	.020	L.F.	.50	.74		1.24	1.76
0220 3/4"		535	.030		.92	1.10		2.02	2.82
0230 1"	↓	400	.040	↓	1.58	1.47		3.05	4.17
1000 Manifolds									
1110 Brass									
1120 With supply and return valves, flow meter, thermometer,									
1122 auto air vent and drain/fill valve.									

23 83 16 – Radiant-Heating Hydronic Piping

23 83 16.10 Radiant Floor Heating		Crew	Daily Output	Labor-Hours	Unit	Material	2018 Bare Costs Labor	Equipment	Total	Total Incl O&P
1130	1", 2 circuit	Q-5	14	1.143	Ea.	298	42		340	400
1140	1", 3 circuit		13.50	1.185		340	43.50		383.50	445
1150	1", 4 circuit		13	1.231		370	45.50		415.50	480
1154	1", 5 circuit		12.50	1.280		440	47		487	565
1158	1", 6 circuit		12	1.333		480	49		529	610
1162	1", 7 circuit		11.50	1.391		525	51.50		576.50	665
1166	1", 8 circuit		11	1.455		565	53.50		618.50	715
1172	1", 9 circuit		10.50	1.524		625	56		681	785
1174	1", 10 circuit		10	1.600		670	59		729	835
1178	1", 11 circuit		9.50	1.684		700	62		762	870
1182	1", 12 circuit	▼	9	1.778	▼	775	65.50		840.50	960
1610	Copper manifold header (cut to size)									
1620	1" header, 12 circuit 1/2" sweat outlets	Q-5	3.33	4.805	Ea.	102	177		279	405
1630	1-1/4" header, 12 circuit 1/2" sweat outlets		3.20	5		118	184		302	435
1640	1-1/4" header, 12 circuit 3/4" sweat outlets		3	5.333		128	197		325	465
1650	1-1/2" header, 12 circuit 3/4" sweat outlets		3.10	5.161		153	190		343	485
1660	2" header, 12 circuit 3/4" sweat outlets	▼	2.90	5.517	▼	226	203		429	585
3000	Valves									
3110	Thermostatic zone valve actuator with end switch	Q-5	40	.400	Ea.	44.50	14.75		59.25	73.50
3114	Thermostatic zone valve actuator	"	36	.444	"	91	16.40		107.40	127
3120	Motorized straight zone valve with operator complete									
3130	3/4"	Q-5	35	.457	Ea.	146	16.85		162.85	189
3140	1"		32	.500		158	18.45		176.45	205
3150	1-1/4"	▼	29.60	.541	▼	200	19.90		219.90	253
3500	4 way mixing valve, manual, brass									
3530	1"	Q-5	13.30	1.203	Ea.	203	44.50		247.50	296
3540	1-1/4"		11.40	1.404		220	51.50		271.50	325
3550	1-1/2"		11	1.455		281	53.50		334.50	400
3560	2"		10.60	1.509		395	55.50		450.50	525
3800	Mixing valve motor, 4 way for valves, 1" and 1-1/4"		34	.471		405	17.35		422.35	475
3810	Mixing valve motor, 4 way for valves, 1-1/2" and 2"	▼	30	.533	▼	400	19.65		419.65	480
5000	Radiant floor heating, zone control panel									
5120	4 zone actuator valve control, expandable	Q-5	20	.800	Ea.	159	29.50		188.50	224
5130	6 zone actuator valve control, expandable		18	.889		253	33		286	330
6070	Thermal track, straight panel for long continuous runs, 5.333 S.F.		40	.400		31	14.75		45.75	58.50
6080	Thermal track, utility panel, for direction reverse at run end, 5.333 S.F.		40	.400		31	14.75		45.75	58.50
6090	Combination panel, for direction reverse plus straight run, 5.333 S.F.	▼	40	.400	▼	31	14.75		45.75	58.50
7000	PEX tubing fittings									
7100	Compression type									
7116	Coupling									
7120	1/2" x 1/2"	1 Stpi	27	.296	Ea.	6.75	12.15		18.90	27.50
7124	3/4" x 3/4"	"	23	.348	"	13.05	14.25		27.30	38
7130	Adapter									
7132	1/2" x female sweat 1/2"	1 Stpi	27	.296	Ea.	4.37	12.15		16.52	25
7134	1/2" x female sweat 3/4"		26	.308		4.89	12.60		17.49	26
7136	5/8" x female sweat 3/4"	▼	24	.333	▼	7	13.65		20.65	30
7140	Elbow									
7142	1/2" x female sweat 1/2"	1 Stpi	27	.296	Ea.	6.70	12.15		18.85	27.50
7144	1/2" x female sweat 3/4"		26	.308		7.80	12.60		20.40	29
7146	5/8" x female sweat 3/4"	▼	24	.333	▼	8.75	13.65		22.40	32
7200	Insert type									
7206	PEX x male NPT									
7210	1/2" x 1/2"	1 Stpi	29	.276	Ea.	2.73	11.30		14.03	21.50

For customer support on your Residential Costs with RSMeans data, call 800.448.8182.

597

23 83 16 – Radiant-Heating Hydronic Piping

23 83 16.10 Radiant Floor Heating		Crew	Daily Output	Labor-Hours	Unit	Material	2018 Bare Costs Labor	Equipment	Total	Total Incl O&P
7220	3/4" x 3/4"	1 Stpi	27	.296	Ea.	4.02	12.15		16.17	24.50
7230	1" x 1"	↓	26	.308	↓	6.80	12.60		19.40	28
7300	PEX coupling									
7310	1/2" x 1/2"	1 Stpi	30	.267	Ea.	.54	10.90		11.44	18.55
7320	3/4" x 3/4"		29	.276		.76	11.30		12.06	19.45
7330	1" x 1"	↓	28	.286	↓	1.35	11.70		13.05	20.50
7400	PEX stainless crimp ring									
7410	1/2" x 1/2"	1 Stpi	86	.093	Ea.	.46	3.81		4.27	6.75
7420	3/4" x 3/4"		84	.095		.63	3.90		4.53	7.10
7430	1" x 1"	↓	82	.098	↓	.90	4		4.90	7.60

23 83 33 – Electric Radiant Heaters

23 83 33.10 Electric Heating

		Crew	Daily Output	Labor-Hours	Unit	Material	2018 Bare Costs Labor	Equipment	Total	Total Incl O&P
0010	**ELECTRIC HEATING**, not incl. conduit or feed wiring									
1100	Rule of thumb: Baseboard units, including control	1 Elec	4.40	1.818	kW	109	73		182	238
1300	Baseboard heaters, 2' long, 350 watt		8	1	Ea.	27.50	40		67.50	96
1400	3' long, 750 watt		8	1		31.50	40		71.50	100
1600	4' long, 1,000 watt		6.70	1.194		37	48		85	120
1800	5' long, 935 watt		5.70	1.404		46.50	56.50		103	143
2000	6' long, 1,500 watt		5	1.600		51.50	64		115.50	162
2400	8' long, 2,000 watt		4	2		62.50	80.50		143	200
2800	10' long, 1,875 watt	↓	3.30	2.424	↓	154	97.50		251.50	330
2950	Wall heaters with fan, 120 to 277 volt									
3170	1,000 watt	1 Elec	6	1.333	Ea.	93	53.50		146.50	190
3180	1,250 watt		5	1.600		100	64		164	215
3190	1,500 watt		4	2		100	80.50		180.50	241
3600	Thermostats, integral		16	.500		27.50	20		47.50	63
3800	Line voltage, 1 pole		8	1		15.70	40		55.70	83
5000	Radiant heating ceiling panels, 2' x 4', 500 watt		16	.500		380	20		400	455
5050	750 watt		16	.500		480	20		500	565
5300	Infrared quartz heaters, 120 volts, 1,000 watt		6.70	1.194		335	48		383	450
5350	1,500 watt		5	1.600		335	64		399	475
5400	240 volts, 1,500 watt		5	1.600		335	64		399	475
5450	2,000 watt		4	2		335	80.50		415.50	500
5500	3,000 watt	↓	3	2.667	↓	385	107		492	595

Estimating Tips
26 05 00 Common Work Results for Electrical

- Conduit should be taken off in three main categories—power distribution, branch power, and branch lighting—so the estimator can concentrate on systems and components, therefore making it easier to ensure all items have been accounted for.

- For cost modifications for elevated conduit installation, add the percentages to labor according to the height of installation and only to the quantities exceeding the different height levels, not to the total conduit quantities. Refer to 26 01 02.20 for labor adjustment factors.

- Remember that aluminum wiring of equal ampacity is larger in diameter than copper and may require larger conduit.

- If more than three wires at a time are being pulled, deduct percentages from the labor hours of that grouping of wires.

- When taking off grounding systems, identify separately the type and size of wire, and list each unique type of ground connection.

- The estimator should take the weights of materials into consideration when completing a takeoff. Topics to consider include: How will the materials be supported? What methods of support are available? How high will the support structure have to reach? Will the final support structure be able to withstand the total burden? Is the support material included or separate from the fixture, equipment, and material specified?

- Do not overlook the costs for equipment used in the installation. If scaffolding or highlifts are available in the field, contractors may use them in lieu of the proposed ladders and rolling staging.

26 20 00 Low-Voltage Electrical Transmission

- Supports and concrete pads may be shown on drawings for the larger equipment, or the support system may be only a piece of plywood for the back of a panelboard. In either case, they must be included in the costs.

26 40 00 Electrical and Cathodic Protection

- When taking off cathodic protection systems, identify the type and size of cable, and list each unique type of anode connection.

26 50 00 Lighting

- Fixtures should be taken off room by room using the fixture schedule, specifications, and the ceiling plan. For large concentrations of lighting fixtures in the same area, deduct the percentages from labor hours.

Reference Numbers

Reference numbers are shown at the beginning of some major classifications. These numbers refer to related items in the Reference Section. The reference information may be an estimating procedure, an alternate pricing method, or technical information.

Note: Not all subdivisions listed here necessarily appear. ■

Did you know?

RSMeans data is available through our online application with 24/7 access:

- Search for unit prices by keyword
- Leverage the most up-to-date data
- Build and export estimates

Try it free for 30 days!
www.rsmeans.com/2018freetrial

26 05 05.10 Electrical Demolition	Crew	Daily Output	Labor-Hours	Unit	Material	2018 Bare Costs Labor	Equipment	Total	Total Incl O&P
0010	**ELECTRICAL DEMOLITION**									
0020	Electrical demolition, conduit to 10' high, incl. fittings & hangers									
0100	Rigid galvanized steel, 1/2" to 1" diameter	1 Elec	242	.033	L.F.		1.33		1.33	2.17
0120	1-1/4" to 2"	"	200	.040	"		1.61		1.61	2.63
0270	Armored cable (BX) avg. 50' runs									
0280	#14, 2 wire	1 Elec	690	.012	L.F.		.47		.47	.76
0290	#14, 3 wire		571	.014			.56		.56	.92
0300	#12, 2 wire		605	.013			.53		.53	.87
0310	#12, 3 wire		514	.016			.62		.62	1.02
0320	#10, 2 wire		514	.016			.62		.62	1.02
0330	#10, 3 wire		425	.019			.76		.76	1.24
0340	#8, 3 wire		342	.023			.94		.94	1.54
0350	Non metallic sheathed cable (Romex)									
0360	#14, 2 wire	1 Elec	720	.011	L.F.		.45		.45	.73
0370	#14, 3 wire		657	.012			.49		.49	.80
0380	#12, 2 wire		629	.013			.51		.51	.84
0390	#10, 3 wire		450	.018			.71		.71	1.17
0400	Wiremold raceway, including fittings & hangers									
0420	No. 3000	1 Elec	250	.032	L.F.		1.28		1.28	2.10
0440	No. 4000		217	.037			1.48		1.48	2.42
0460	No. 6000		166	.048			1.93		1.93	3.16
0462	Plugmold with receptacle		114	.070			2.82		2.82	4.61
0465	Telephone/power pole		12	.667	Ea.		27		27	44
0470	Non-metallic, straight section		480	.017	L.F.		.67		.67	1.09
0500	Channels, steel, including fittings & hangers									
0520	3/4" x 1-1/2"	1 Elec	308	.026	L.F.		1.04		1.04	1.70
0540	1-1/2" x 1-1/2"		269	.030			1.19		1.19	1.95
0560	1-1/2" x 1-7/8"		229	.035			1.40		1.40	2.29
1210	Panel boards, incl. removal of all breakers,									
1220	conduit terminations & wire connections									
1230	3 wire, 120/240 V, 100A, to 20 circuits	1 Elec	2.60	3.077	Ea.		124		124	202
1240	200 amps, to 42 circuits	2 Elec	2.60	6.154			247		247	405
1241	225 amps, to 42 circuits		5.60	2.857			115		115	188
1250	400 amps, to 42 circuits		2.20	7.273			292		292	475
1260	4 wire, 120/208 V, 125A, to 20 circuits	1 Elec	2.40	3.333			134		134	219
1270	200 amps, to 42 circuits	2 Elec	2.40	6.667			268		268	440
1720	Junction boxes, 4" sq. & oct.	1 Elec	80	.100			4.02		4.02	6.55
1740	Handy box		107	.075			3		3	4.91
1760	Switch box		107	.075			3		3	4.91
1780	Receptacle & switch plates		257	.031			1.25		1.25	2.04
1800	Wire, THW-THWN-THHN, removed from									
1810	in place conduit, to 10' high									
1830	#14	1 Elec	65	.123	C.L.F.		4.94		4.94	8.10
1840	#12		55	.145			5.85		5.85	9.55
1850	#10		45.50	.176			7.05		7.05	11.55
2000	Interior fluorescent fixtures, incl. supports									
2010	& whips, to 10' high									
2100	Recessed drop-in 2' x 2', 2 lamp	2 Elec	35	.457	Ea.		18.35		18.35	30
2110	2' x 2', 4 lamp		30	.533			21.50		21.50	35
2140	2' x 4', 4 lamp		30	.533			21.50		21.50	35
2180	Surface mount, acrylic lens & hinged frame									
2220	2' x 2', 2 lamp	2 Elec	44	.364	Ea.		14.60		14.60	24
2260	2' x 4', 4 lamp	"	33	.485	"		19.45		19.45	32

26 05 05 – Selective Demolition for Electrical

26 05 05.10 Electrical Demolition	Crew	Daily Output	Labor-Hours	Unit	Material	2018 Bare Costs Labor	Equipment	Total	Total Incl O&P
2300	Strip fixtures, surface mount									
2320	4' long, 1 lamp	2 Elec	53	.302	Ea.		12.10		12.10	19.80
2380	8' long, 2 lamp	"	40	.400	"		16.05		16.05	26.50
2460	Interior incandescent, surface, ceiling									
2470	or wall mount, to 12' high									
2480	Metal cylinder type, 75 Watt	2 Elec	62	.258	Ea.		10.35		10.35	16.95
2600	Exterior fixtures, incandescent, wall mount									
2620	100 Watt	2 Elec	50	.320	Ea.		12.85		12.85	21
3000	Ceiling fan, tear out and remove	1 Elec	24	.333	"		13.40		13.40	22

26 05 19 – Low-Voltage Electrical Power Conductors and Cables

26 05 19.20 Armored Cable

		Crew	Daily Output	Labor-Hours	Unit	Material	Labor	Equipment	Total	Total Incl O&P
0010	**ARMORED CABLE**									
0051	600 volt, copper (BX), #14, 2 conductor, solid	1 Elec	240	.033	L.F.	.41	1.34		1.75	2.64
0101	3 conductor, solid		200	.040		.65	1.61		2.26	3.34
0151	#12, 2 conductor, solid		210	.038		.42	1.53		1.95	2.96
0201	3 conductor, solid		180	.044		.71	1.78		2.49	3.70
0251	#10, 2 conductor, solid		180	.044		.81	1.78		2.59	3.81
0301	3 conductor, solid		150	.053		1.10	2.14		3.24	4.71
0351	3 conductor, stranded		120	.067		2.44	2.68		5.12	7.05

26 05 19.55 Non-Metallic Sheathed Cable

		Crew	Daily Output	Labor-Hours	Unit	Material	Labor	Equipment	Total	Total Incl O&P
0010	**NON-METALLIC SHEATHED CABLE** 600 volt									
0100	Copper with ground wire (Romex)									
0151	#14, 2 wire	1 Elec	250	.032	L.F.	.18	1.28		1.46	2.30
0201	3 wire		230	.035		.26	1.40		1.66	2.56
0251	#12, 2 wire		220	.036		.24	1.46		1.70	2.65
0301	3 wire		200	.040		.40	1.61		2.01	3.07
0351	#10, 2 wire		200	.040		.41	1.61		2.02	3.08
0401	3 wire		140	.057		.63	2.29		2.92	4.44
0451	#8, 3 conductor		130	.062		1.17	2.47		3.64	5.35
0501	#6, 3 wire		120	.067		1.97	2.68		4.65	6.55
0550	SE type SER aluminum cable, 3 RHW and									
0601	1 bare neutral, 3 #8 & 1 #8	1 Elec	150	.053	L.F.	.57	2.14		2.71	4.12
0651	3 #6 & 1 #6	"	130	.062		.67	2.47		3.14	4.78
0701	3 #4 & 1 #6	2 Elec	220	.073		.71	2.92		3.63	5.55
0751	3 #2 & 1 #4		200	.080		1.24	3.21		4.45	6.60
0801	3 #1/0 & 1 #2		180	.089		1.82	3.57		5.39	7.85
0851	3 #2/0 & 1 #1		160	.100		2.12	4.02		6.14	8.90
0901	3 #4/0 & 1 #2/0		140	.114		2.95	4.59		7.54	10.75
2401	SEU service entrance cable, copper 2 conductors, #8 + #8 neutral	1 Elec	150	.053		1.01	2.14		3.15	4.61
2601	#6 + #8 neutral		130	.062		1.27	2.47		3.74	5.45
2801	#6 + #6 neutral		130	.062		1.49	2.47		3.96	5.65
3001	#4 + #6 neutral	2 Elec	220	.073		1.95	2.92		4.87	6.90
3201	#4 + #4 neutral		220	.073		2.13	2.92		5.05	7.10
3401	#3 + #5 neutral		210	.076		3.33	3.06		6.39	8.65
6500	Service entrance cap for copper SEU									
6600	100 amp	1 Elec	12	.667	Ea.	8.10	27		35.10	53
6700	150 amp		10	.800		10.90	32		42.90	64.50
6800	200 amp		8	1		17.80	40		57.80	85

26 05 19.90 Wire

		Crew	Daily Output	Labor-Hours	Unit	Material	Labor	Equipment	Total	Total Incl O&P
0010	**WIRE**, normal installation conditions in wireway, conduit, cable tray									
0021	600 volt, copper type THW, solid, #14	1 Elec	1300	.006	L.F.	.06	.25		.31	.46
0031	#12		1100	.007		.10	.29		.39	.59

For customer support on your Residential Costs with RSMeans data, call 800.448.8182.

601

26 05 19 – Low-Voltage Electrical Power Conductors and Cables

26 05 19.90 Wire

		Crew	Daily Output	Labor-Hours	Unit	Material	2018 Bare Costs Labor	Equipment	Total	Total Incl O&P
0041	#10	1 Elec	1000	.008	L.F.	.16	.32		.48	.70
0050	Stranded, #14		13	.615	C.L.F.	8.60	24.50		33.10	50
0100	#12		11	.727		13.20	29		42.20	62.50
0120	#10		10	.800		21	32		53	75.50
0161	#6		650	.012	L.F.	.43	.49		.92	1.29
0181	#4	2 Elec	1060	.015		.69	.61		1.30	1.75
0201	#3		1000	.016		1.05	.64		1.69	2.20
0221	#2		900	.018		1.02	.71		1.73	2.29
0241	#1		800	.020		1.36	.80		2.16	2.80
0261	1/0		660	.024		2.16	.97		3.13	3.97
0281	2/0		580	.028		2.05	1.11		3.16	4.07
0301	3/0		500	.032		2.35	1.28		3.63	4.68
0351	4/0		440	.036		4.33	1.46		5.79	7.15

26 05 26 – Grounding and Bonding for Electrical Systems

26 05 26.80 Grounding

		Crew	Daily Output	Labor-Hours	Unit	Material	2018 Bare Costs Labor	Equipment	Total	Total Incl O&P
0010	**GROUNDING**									
0030	Rod, copper clad, 8' long, 1/2" diameter	1 Elec	5.50	1.455	Ea.	19.90	58.50		78.40	118
0040	5/8" diameter		5.50	1.455		21.50	58.50		80	119
0050	3/4" diameter		5.30	1.509		35	60.50		95.50	138
0080	10' long, 1/2" diameter		4.80	1.667		22.50	67		89.50	134
0090	5/8" diameter		4.60	1.739		24	70		94	140
0100	3/4" diameter		4.40	1.818		42	73		115	166
0130	15' long, 3/4" diameter		4	2		49	80.50		129.50	185
0261	Wire, ground bare armored, #8-1 conductor		200	.040	L.F.	.70	1.61		2.31	3.40
0271	#6-1 conductor		180	.044	"	.85	1.78		2.63	3.85
0390	Bare copper wire, stranded, #8		11	.727	C.L.F.	43	29		72	95.50
0401	Bare copper, #6 wire		1000	.008	L.F.	.41	.32		.73	.98
0601	#2 stranded	2 Elec	1000	.016	"	.89	.64		1.53	2.03
1800	Water pipe ground clamps, heavy duty									
2000	Bronze, 1/2" to 1" diameter	1 Elec	8	1	Ea.	29.50	40		69.50	98
2100	1-1/4" to 2" diameter		8	1		28.50	40		68.50	97
2200	2-1/2" to 3" diameter		6	1.333		68	53.50		121.50	162

26 05 33 – Raceway and Boxes for Electrical Systems

26 05 33.13 Conduit

		Crew	Daily Output	Labor-Hours	Unit	Material	2018 Bare Costs Labor	Equipment	Total	Total Incl O&P
0010	**CONDUIT** To 10' high, includes 2 terminations, 2 elbows,									
0020	11 beam clamps, and 11 couplings per 100 L.F.									
1750	Rigid galvanized steel, 1/2" diameter	1 Elec	90	.089	L.F.	2.31	3.57		5.88	8.40
1770	3/4" diameter		80	.100		4.57	4.02		8.59	11.60
1800	1" diameter		65	.123		6.75	4.94		11.69	15.55
1830	1-1/4" diameter		60	.133		4.74	5.35		10.09	13.95
1850	1-1/2" diameter		55	.145		7.90	5.85		13.75	18.25
1870	2" diameter		45	.178		9.95	7.15		17.10	22.50
5000	Electric metallic tubing (EMT), 1/2" diameter		170	.047		.80	1.89		2.69	3.96
5020	3/4" diameter		130	.062		1.13	2.47		3.60	5.30
5040	1" diameter		115	.070		1.84	2.79		4.63	6.60
5060	1-1/4" diameter		100	.080		2.99	3.21		6.20	8.55
5080	1-1/2" diameter		90	.089		2.82	3.57		6.39	8.95
9100	PVC, schedule 40, 1/2" diameter		190	.042		.89	1.69		2.58	3.74
9110	3/4" diameter		145	.055		1	2.22		3.22	4.73
9120	1" diameter		125	.064		1.52	2.57		4.09	5.85
9130	1-1/4" diameter		110	.073		1.80	2.92		4.72	6.75
9140	1-1/2" diameter		100	.080		2.09	3.21		5.30	7.55

26 05 33 – Raceway and Boxes for Electrical Systems

26 05 33.13 Conduit

		Crew	Daily Output	Labor-Hours	Unit	Material	2018 Bare Costs Labor	Equipment	Total	Total Incl O&P
9150	2" diameter	1 Elec	90	.089	L.F.	2.72	3.57		6.29	8.85
9995	Do not include labor when adding couplings to a fitting installation									

26 05 33.16 Boxes for Electrical Systems

		Crew	Daily Output	Labor-Hours	Unit	Material	2018 Bare Costs Labor	Equipment	Total	Total Incl O&P
0010	**BOXES FOR ELECTRICAL SYSTEMS**									
0021	Pressed steel, octagon, 4"	1 Elec	18	.444	Ea.	2.86	17.85		20.71	32
0060	Covers, blank		64	.125		.88	5		5.88	9.15
0100	Extension rings		40	.200		3.68	8.05		11.73	17.20
0151	Square 4"		18	.444		5.25	17.85		23.10	35
0200	Extension rings		40	.200		3.79	8.05		11.84	17.30
0250	Covers, blank		64	.125		.67	5		5.67	8.95
0300	Plaster rings		64	.125		1.54	5		6.54	9.90
0651	Switchbox		24	.333		4.61	13.40		18.01	27
1100	Concrete, floor, 1 gang		5.30	1.509		94.50	60.50		155	203

26 05 33.17 Outlet Boxes, Plastic

		Crew	Daily Output	Labor-Hours	Unit	Material	2018 Bare Costs Labor	Equipment	Total	Total Incl O&P
0010	**OUTLET BOXES, PLASTIC**									
0051	4" diameter, round, with 2 mounting nails	1 Elec	23	.348	Ea.	2.73	13.95		16.68	26
0101	Bar hanger mounted		23	.348		5.30	13.95		19.25	29
0201	Square with 2 mounting nails		23	.348		5.30	13.95		19.25	29
0300	Plaster ring		64	.125		2.02	5		7.02	10.40
0401	Switch box with 2 mounting nails, 1 gang		27	.296		3.03	11.90		14.93	23
0501	2 gang		23	.348		3.70	13.95		17.65	27
0601	3 gang		18	.444		4.92	17.85		22.77	34.50

26 05 33.18 Pull Boxes

		Crew	Daily Output	Labor-Hours	Unit	Material	2018 Bare Costs Labor	Equipment	Total	Total Incl O&P
0010	**PULL BOXES**									
0100	Steel, pull box, NEMA 1, type SC, 6" W x 6" H x 4" D	1 Elec	8	1	Ea.	9.75	40		49.75	76
0200	8" W x 8" H x 4" D		8	1		12.55	40		52.55	79.50
0300	10" W x 12" H x 6" D		5.30	1.509		28.50	60.50		89	130

26 05 33.25 Conduit Fittings for Rigid Galvanized Steel

		Crew	Daily Output	Labor-Hours	Unit	Material	2018 Bare Costs Labor	Equipment	Total	Total Incl O&P
0010	**CONDUIT FITTINGS FOR RIGID GALVANIZED STEEL**									
2280	LB, LR or LL fittings & covers, 1/2" diameter	1 Elec	16	.500	Ea.	7.10	20		27.10	41
2290	3/4" diameter		13	.615		8.35	24.50		32.85	49.50
2300	1" diameter		11	.727		13.05	29		42.05	62.50
2330	1-1/4" diameter		8	1		19.45	40		59.45	87
2350	1-1/2" diameter		6	1.333		25	53.50		78.50	115
2370	2" diameter		5	1.600		40.50	64		104.50	150
5280	Service entrance cap, 1/2" diameter		16	.500		4.99	20		24.99	38.50
5300	3/4" diameter		13	.615		5.95	24.50		30.45	47
5320	1" diameter		10	.800		4.88	32		36.88	58
5340	1-1/4" diameter		8	1		4.87	40		44.87	71
5360	1-1/2" diameter		6.50	1.231		9.85	49.50		59.35	92
5380	2" diameter		5.50	1.455		20	58.50		78.50	118

26 05 33.35 Flexible Metallic Conduit

		Crew	Daily Output	Labor-Hours	Unit	Material	2018 Bare Costs Labor	Equipment	Total	Total Incl O&P
0010	**FLEXIBLE METALLIC CONDUIT**									
0050	Steel, 3/8" diameter	1 Elec	200	.040	L.F.	.38	1.61		1.99	3.05
0100	1/2" diameter		200	.040		.42	1.61		2.03	3.09
0200	3/4" diameter		160	.050		.58	2.01		2.59	3.92
0250	1" diameter		100	.080		1	3.21		4.21	6.35
0300	1-1/4" diameter		70	.114		1.39	4.59		5.98	9.05
0350	1-1/2" diameter		50	.160		2.26	6.40		8.66	13
0370	2" diameter		40	.200		2.79	8.05		10.84	16.20

26 05 39 – Underfloor Raceways for Electrical Systems

26 05 39.30 Conduit In Concrete Slab	Crew	Daily Output	Labor-Hours	Unit	Material	2018 Bare Costs Labor	Equipment	Total	Total Incl O&P
0010 **CONDUIT IN CONCRETE SLAB** Including terminations,									
0020 fittings and supports									
3230 PVC, schedule 40, 1/2" diameter	1 Elec	270	.030	L.F.	.49	1.19		1.68	2.48
3250 3/4" diameter		230	.035		.56	1.40		1.96	2.90
3270 1" diameter		200	.040		.77	1.61		2.38	3.47
3300 1-1/4" diameter		170	.047		1.01	1.89		2.90	4.20
3330 1-1/2" diameter		140	.057		1.21	2.29		3.50	5.10
3350 2" diameter		120	.067		1.52	2.68		4.20	6.05
4350 Rigid galvanized steel, 1/2" diameter		200	.040		1.99	1.61		3.60	4.82
4400 3/4" diameter		170	.047		4.27	1.89		6.16	7.80
4450 1" diameter		130	.062		6.60	2.47		9.07	11.30
4500 1-1/4" diameter		110	.073		4.21	2.92		7.13	9.40
4600 1-1/2" diameter		100	.080		7.30	3.21		10.51	13.30
4800 2" diameter		90	.089		9.05	3.57		12.62	15.80

26 05 39.40 Conduit In Trench

	Crew	Daily Output	Labor-Hours	Unit	Material	Labor	Equipment	Total	Total Incl O&P
0010 **CONDUIT IN TRENCH** Includes terminations and fittings									
0020 Does not include excavation or backfill, see Section 31 23 16									
0200 Rigid galvanized steel, 2" diameter	1 Elec	150	.053	L.F.	8.70	2.14		10.84	13.10
0400 2-1/2" diameter	"	100	.080		10.70	3.21		13.91	17
0600 3" diameter	2 Elec	160	.100		12.55	4.02		16.57	20.50
0800 3-1/2" diameter	"	140	.114		16.55	4.59		21.14	25.50

26 05 80 – Wiring Connections

26 05 80.10 Motor Connections

	Crew	Daily Output	Labor-Hours	Unit	Material	Labor	Equipment	Total	Total Incl O&P
0010 **MOTOR CONNECTIONS**									
0020 Flexible conduit and fittings, 115 volt, 1 phase, up to 1 HP motor	1 Elec	8	1	Ea.	4.87	40		44.87	71

26 05 90 – Residential Applications

26 05 90.10 Residential Wiring

	Crew	Daily Output	Labor-Hours	Unit	Material	Labor	Equipment	Total	Total Incl O&P
0010 **RESIDENTIAL WIRING**									
0020 20' avg. runs and #14/2 wiring incl. unless otherwise noted									
1000 Service & panel, includes 24' SE-AL cable, service eye, meter,									
1010 Socket, panel board, main bkr., ground rod, 15 or 20 amp									
1020 1-pole circuit breakers, and misc. hardware									
1100 100 amp, with 10 branch breakers	1 Elec	1.19	6.723	Ea.	325	270		595	800
1110 With PVC conduit and wire		.92	8.696		360	350		710	965
1120 With RGS conduit and wire		.73	10.959		560	440		1,000	1,325
1150 150 amp, with 14 branch breakers		1.03	7.767		850	310		1,160	1,450
1170 With PVC conduit and wire		.82	9.756		920	390		1,310	1,650
1180 With RGS conduit and wire		.67	11.940		1,175	480		1,655	2,075
1200 200 amp, with 18 branch breakers	2 Elec	1.80	8.889		1,000	355		1,355	1,675
1220 With PVC conduit and wire		1.46	10.959		1,075	440		1,515	1,900
1230 With RGS conduit and wire		1.24	12.903		1,400	520		1,920	2,400
1800 Lightning surge suppressor	1 Elec	32	.250		76	10.05		86.05	100
2000 Switch devices									
2100 Single pole, 15 amp, ivory, with a 1-gang box, cover plate,									
2110 Type NM (Romex) cable	1 Elec	17.10	.468	Ea.	15.50	18.80		34.30	47.50
2120 Type MC cable		14.30	.559		25	22.50		47.50	64
2130 EMT & wire		5.71	1.401		36	56.50		92.50	132
2150 3-way, #14/3, type NM cable		14.55	.550		9.80	22		31.80	47
2170 Type MC cable		12.31	.650		22.50	26		48.50	67.50
2180 EMT & wire		5	1.600		30	64		94	138
2200 4-way, #14/3, type NM cable		14.55	.550		18.45	22		40.45	56.50

604

For customer support on your Residential Costs with RSMeans data, call 800.448.8182.

26 05 90.10 Residential Wiring		Crew	Daily Output	Labor-Hours	Unit	Material	2018 Bare Costs Labor	2018 Bare Costs Equipment	Total	Total Incl O&P
2220	Type MC cable	1 Elec	12.31	.650	Ea.	31.50	26		57.50	77
2230	EMT & wire		5	1.600		38.50	64		102.50	148
2250	S.P., 20 amp, #12/2, type NM cable		13.33	.600		11.10	24		35.10	51.50
2270	Type MC cable		11.43	.700		19.65	28		47.65	67.50
2280	EMT & wire		4.85	1.649		33.50	66		99.50	145
2290	S.P. rotary dimmer, 600 W, no wiring		17	.471		30	18.90		48.90	64
2300	S.P. rotary dimmer, 600 W, type NM cable		14.55	.550		33.50	22		55.50	73
2320	Type MC cable		12.31	.650		43.50	26		69.50	90
2330	EMT & wire		5	1.600		55	64		119	166
2350	3-way rotary dimmer, type NM cable		13.33	.600		21.50	24		45.50	63
2370	Type MC cable		11.43	.700		31	28		59	80
2380	EMT & wire	▼	4.85	1.649	▼	43	66		109	155
2400	Interval timer wall switch, 20 amp, 1-30 min., #12/2									
2410	Type NM cable	1 Elec	14.55	.550	Ea.	59.50	22		81.50	102
2420	Type MC cable		12.31	.650		65.50	26		91.50	115
2430	EMT & wire	▼	5	1.600	▼	82	64		146	195
2500	Decorator style									
2510	S.P., 15 amp, type NM cable	1 Elec	17.10	.468	Ea.	19.35	18.80		38.15	52
2520	Type MC cable		14.30	.559		29	22.50		51.50	68.50
2530	EMT & wire		5.71	1.401		39.50	56.50		96	136
2550	3-way, #14/3, type NM cable		14.55	.550		13.70	22		35.70	51
2570	Type MC cable		12.31	.650		26.50	26		52.50	71.50
2580	EMT & wire		5	1.600		33.50	64		97.50	142
2600	4-way, #14/3, type NM cable		14.55	.550		22.50	22		44.50	60.50
2620	Type MC cable		12.31	.650		35	26		61	81
2630	EMT & wire		5	1.600		42.50	64		106.50	152
2650	S.P., 20 amp, #12/2, type NM cable		13.33	.600		14.95	24		38.95	56
2670	Type MC cable		11.43	.700		23.50	28		51.50	72
2680	EMT & wire		4.85	1.649		37.50	66		103.50	150
2700	S.P., slide dimmer, type NM cable		17.10	.468		30.50	18.80		49.30	64
2720	Type MC cable		14.30	.559		40	22.50		62.50	80.50
2730	EMT & wire		5.71	1.401		52	56.50		108.50	149
2750	S.P., touch dimmer, type NM cable		17.10	.468		46.50	18.80		65.30	82
2770	Type MC cable		14.30	.559		56	22.50		78.50	98.50
2780	EMT & wire		5.71	1.401		68	56.50		124.50	167
2800	3-way touch dimmer, type NM cable		13.33	.600		49.50	24		73.50	93.50
2820	Type MC cable		11.43	.700		59	28		87	111
2830	EMT & wire	▼	4.85	1.649	▼	71	66		137	186
3000	Combination devices									
3100	S.P. switch/15 amp recpt., ivory, 1-gang box, plate									
3110	Type NM cable	1 Elec	11.43	.700	Ea.	20.50	28		48.50	68.50
3120	Type MC cable		10	.800		30	32		62	85.50
3130	EMT & wire		4.40	1.818		42	73		115	165
3150	S.P. switch/pilot light, type NM cable		11.43	.700		21	28		49	69
3170	Type MC cable		10	.800		30.50	32		62.50	86
3180	EMT & wire		4.43	1.806		42.50	72.50		115	166
3190	2-S.P. switches, 2-#14/2, no wiring		14	.571		12.85	23		35.85	51.50
3200	2-S.P. switches, 2-#14/2, type NM cables		10	.800		22	32		54	77
3220	Type MC cable		8.89	.900		36.50	36		72.50	99
3230	EMT & wire		4.10	1.951		43.50	78.50		122	176
3250	3-way switch/15 amp recpt., #14/3, type NM cable		10	.800		26.50	32		58.50	82
3270	Type MC cable		8.89	.900		39.50	36		75.50	103
3280	EMT & wire	▼	4.10	1.951	▼	46.50	78.50		125	180

26 05 90.10 Residential Wiring	Crew	Daily Output	Labor-Hours	Unit	Material	2018 Bare Costs Labor	2018 Bare Costs Equipment	Total	Total Incl O&P
3300	2-3 way switches, 2-#14/3, type NM cables	1 Elec	8.89	.900	Ea.	34.50	36		70.50	97
3320	Type MC cable		8	1		55.50	40		95.50	127
3330	EMT & wire		4	2		53	80.50		133.50	190
3350	S.P. switch/20 amp recpt., #12/2, type NM cable		10	.800		28	32		60	83.50
3370	Type MC cable		8.89	.900		34	36		70	96.50
3380	EMT & wire		4.10	1.951		50.50	78.50		129	184
3400	Decorator style									
3410	S.P. switch/15 amp recpt., type NM cable	1 Elec	11.43	.700	Ea.	24	28		52	72.50
3420	Type MC cable		10	.800		34	32		66	89.50
3430	EMT & wire		4.40	1.818		45.50	73		118.50	170
3450	S.P. switch/pilot light, type NM cable		11.43	.700		25	28		53	73.50
3470	Type MC cable		10	.800		34.50	32		66.50	90.50
3480	EMT & wire		4.40	1.818		46.50	73		119.50	170
3500	2-S.P. switches, 2-#14/2, type NM cables		10	.800		26	32		58	81
3520	Type MC cable		8.89	.900		40	36		76	103
3530	EMT & wire		4.10	1.951		47.50	78.50		126	180
3550	3-way/15 amp recpt., #14/3, type NM cable		10	.800		30.50	32		62.50	86
3570	Type MC cable		8.89	.900		43.50	36		79.50	107
3580	EMT & wire		4.10	1.951		50.50	78.50		129	184
3650	2-3 way switches, 2-#14/3, type NM cables		8.89	.900		38.50	36		74.50	102
3670	Type MC cable		8	1		59	40		99	131
3680	EMT & wire		4	2		57	80.50		137.50	194
3700	S.P. switch/20 amp recpt., #12/2, type NM cable		10	.800		32	32		64	87.50
3720	Type MC cable		8.89	.900		38	36		74	101
3730	EMT & wire		4.10	1.951		54.50	78.50		133	188
4000	Receptacle devices									
4010	Duplex outlet, 15 amp recpt., ivory, 1-gang box, plate									
4015	Type NM cable	1 Elec	14.55	.550	Ea.	8.40	22		30.40	45
4020	Type MC cable		12.31	.650		17.95	26		43.95	62.50
4030	EMT & wire		5.33	1.501		28.50	60.50		89	130
4050	With #12/2, type NM cable		12.31	.650		9.50	26		35.50	53
4070	Type MC cable		10.67	.750		18.05	30		48.05	69
4080	EMT & wire		4.71	1.699		32	68		100	148
4100	20 amp recpt., #12/2, type NM cable		12.31	.650		17.25	26		43.25	61.50
4120	Type MC cable		10.67	.750		26	30		56	77.50
4130	EMT & wire		4.71	1.699		40	68		108	156
4140	For GFI see Section 26 05 90.10 line 4300 below									
4150	Decorator style, 15 amp recpt., type NM cable	1 Elec	14.55	.550	Ea.	12.25	22		34.25	49.50
4170	Type MC cable		12.31	.650		22	26		48	66.50
4180	EMT & wire		5.33	1.501		32.50	60.50		93	135
4200	With #12/2, type NM cable		12.31	.650		13.35	26		39.35	57
4220	Type MC cable		10.67	.750		22	30		52	73
4230	EMT & wire		4.71	1.699		36	68		104	152
4250	20 amp recpt., #12/2, type NM cable		12.31	.650		21	26		47	66
4270	Type MC cable		10.67	.750		29.50	30		59.50	81.50
4280	EMT & wire		4.71	1.699		44	68		112	160
4300	GFI, 15 amp recpt., type NM cable		12.31	.650		21	26		47	65.50
4320	Type MC cable		10.67	.750		30.50	30		60.50	82.50
4330	EMT & wire		4.71	1.699		41	68		109	157
4350	GFI with #12/2, type NM cable		10.67	.750		22	30		52	73
4370	Type MC cable		9.20	.870		30.50	35		65.50	90.50
4380	EMT & wire		4.21	1.900		44.50	76.50		121	174
4400	20 amp recpt., #12/2, type NM cable		10.67	.750		48.50	30		78.50	102

26 05 90.10 Residential Wiring	Crew	Daily Output	Labor-Hours	Unit	Material	2018 Bare Costs Labor	Equipment	Total	Total Incl O&P
4420	Type MC cable	1 Elec	9.20	.870	Ea.	57	35		92	120
4430	EMT & wire		4.21	1.900		71	76.50		147.50	203
4500	Weather-proof cover for above receptacles, add	▼	32	.250	▼	2.10	10.05		12.15	18.70
4550	Air conditioner outlet, 20 amp-240 volt recpt.									
4560	30' of #12/2, 2 pole circuit breaker									
4570	Type NM cable	1 Elec	10	.800	Ea.	58	32		90	117
4580	Type MC cable		9	.889		68.50	35.50		104	134
4590	EMT & wire		4	2		81	80.50		161.50	221
4600	Decorator style, type NM cable		10	.800		63	32		95	122
4620	Type MC cable		9	.889		73	35.50		108.50	139
4630	EMT & wire	▼	4	2	▼	86	80.50		166.50	226
4650	Dryer outlet, 30 amp-240 volt recpt., 20' of #10/3									
4660	2 pole circuit breaker									
4670	Type NM cable	1 Elec	6.41	1.248	Ea.	53.50	50		103.50	141
4680	Type MC cable		5.71	1.401		62.50	56.50		119	161
4690	EMT & wire	▼	3.48	2.299	▼	75	92.50		167.50	233
4700	Range outlet, 50 amp-240 volt recpt., 30' of #8/3									
4710	Type NM cable	1 Elec	4.21	1.900	Ea.	82	76.50		158.50	215
4720	Type MC cable		4	2		128	80.50		208.50	272
4730	EMT & wire		2.96	2.703		106	109		215	293
4750	Central vacuum outlet, type NM cable		6.40	1.250		50	50		100	137
4770	Type MC cable		5.71	1.401		64.50	56.50		121	163
4780	EMT & wire	▼	3.48	2.299	▼	83	92.50		175.50	242
4800	30 amp-110 volt locking recpt., #10/2 circ. bkr.									
4810	Type NM cable	1 Elec	6.20	1.290	Ea.	58	52		110	149
4820	Type MC cable		5.40	1.481		76	59.50		135.50	181
4830	EMT & wire	▼	3.20	2.500	▼	93.50	100		193.50	267
4900	Low voltage outlets									
4910	Telephone recpt., 20' of 4/C phone wire	1 Elec	26	.308	Ea.	8	12.35		20.35	29
4920	TV recpt., 20' of RG59U coax wire, F type connector	"	16	.500	"	15.55	20		35.55	50
4950	Door bell chime, transformer, 2 buttons, 60' of bellwire									
4970	Economy model	1 Elec	11.50	.696	Ea.	52	28		80	103
4980	Custom model		11.50	.696		94.50	28		122.50	150
4990	Luxury model, 3 buttons	▼	9.50	.842	▼	185	34		219	260
6000	Lighting outlets									
6050	Wire only (for fixture), type NM cable	1 Elec	32	.250	Ea.	5.75	10.05		15.80	23
6070	Type MC cable		24	.333		11.70	13.40		25.10	35
6080	EMT & wire		10	.800		21.50	32		53.50	76.50
6100	Box (4"), and wire (for fixture), type NM cable		25	.320		14.55	12.85		27.40	37
6120	Type MC cable		20	.400		20.50	16.05		36.55	49
6130	EMT & wire	▼	11	.727	▼	30.50	29		59.50	81.50
6200	Fixtures (use with line 6050 or 6100 above)									
6210	Canopy style, economy grade	1 Elec	40	.200	Ea.	23	8.05		31.05	38.50
6220	Custom grade		40	.200		53.50	8.05		61.55	72
6250	Dining room chandelier, economy grade		19	.421		72	16.90		88.90	107
6260	Custom grade		19	.421		305	16.90		321.90	365
6270	Luxury grade		15	.533		955	21.50		976.50	1,075
6310	Kitchen fixture (fluorescent), economy grade		30	.267		71.50	10.70		82.20	96.50
6320	Custom grade		25	.320		137	12.85		149.85	172
6350	Outdoor, wall mounted, economy grade		30	.267		28.50	10.70		39.20	49
6360	Custom grade		30	.267		121	10.70		131.70	151
6370	Luxury grade		25	.320		248	12.85		260.85	294
6410	Outdoor PAR floodlights, 1 lamp, 150 watt	▼	20	.400	▼	26	16.05		42.05	55

26 05 Common Work Results for Electrical

26 05 90 – Residential Applications

26 05 90.10 Residential Wiring	Crew	Daily Output	Labor-Hours	Unit	Material	2018 Bare Costs Labor	Equipment	Total	Total Incl O&P
6420	2 lamp, 150 watt each	1 Elec	20	.400	Ea.	45.50	16.05		61.55	76.50
6425	Motion sensing, 2 lamp, 150 watt each		20	.400		88.50	16.05		104.55	124
6430	For infrared security sensor, add		32	.250		99.50	10.05		109.55	126
6450	Outdoor, quartz-halogen, 300 watt flood		20	.400		33	16.05		49.05	63
6600	Recessed downlight, round, pre-wired, 50 or 75 watt trim		30	.267		62.50	10.70		73.20	86
6610	With shower light trim		30	.267		86	10.70		96.70	112
6620	With wall washer trim		28	.286		79.50	11.45		90.95	106
6630	With eye-ball trim		28	.286		66.50	11.45		77.95	92
6700	Porcelain lamp holder		40	.200		2.92	8.05		10.97	16.35
6710	With pull switch		40	.200		11.30	8.05		19.35	25.50
6750	Fluorescent strip, 2-20 watt tube, wrap around diffuser, 24"		24	.333		46	13.40		59.40	73
6760	1-34 watt tube, 48"		24	.333		105	13.40		118.40	138
6770	2-34 watt tubes, 48"		20	.400		122	16.05		138.05	161
6800	Bathroom heat lamp, 1-250 watt		28	.286		31	11.45		42.45	53.50
6810	2-250 watt lamps	▼	28	.286	▼	73	11.45		84.45	99.50
6820	For timer switch, see Section 26 05 90.10 line 2400									
6900	Outdoor post lamp, incl. post, fixture, 35' of #14/2									
6910	Type NM cable	1 Elec	3.50	2.286	Ea.	300	92		392	480
6920	Photo-eye, add		27	.296		27.50	11.90		39.40	49.50
6950	Clock dial time switch, 24 hr., w/enclosure, type NM cable		11.43	.700		68.50	28		96.50	121
6970	Type MC cable		11	.727		78	29		107	134
6980	EMT & wire	▼	4.85	1.649	▼	88.50	66		154.50	206
7000	Alarm systems									
7050	Smoke detectors, box, #14/3, type NM cable	1 Elec	14.55	.550	Ea.	33.50	22		55.50	73
7070	Type MC cable		12.31	.650		44	26		70	90.50
7080	EMT & wire	▼	5	1.600		51	64		115	161
7090	For relay output to security system, add				▼	10.75			10.75	11.85
8000	Residential equipment									
8050	Disposal hook-up, incl. switch, outlet box, 3' of flex									
8060	20 amp-1 pole circ. bkr., and 25' of #12/2									
8070	Type NM cable	1 Elec	10	.800	Ea.	28	32		60	83.50
8080	Type MC cable		8	1		37.50	40		77.50	107
8090	EMT & wire	▼	5	1.600	▼	54.50	64		118.50	165
8100	Trash compactor or dishwasher hook-up, incl. outlet box,									
8110	3' of flex, 15 amp-1 pole circ. bkr., and 25' of #14/2									
8130	Type MC cable	1 Elec	8	1	Ea.	26.50	40		66.50	94.50
8140	EMT & wire	"	5	1.600	"	41	64		105	150
8150	Hot water sink dispenser hook-up, use line 8100									
8200	Vent/exhaust fan hook-up, type NM cable	1 Elec	32	.250	Ea.	5.75	10.05		15.80	23
8220	Type MC cable		24	.333		11.70	13.40		25.10	35
8230	EMT & wire	▼	10	.800	▼	21.50	32		53.50	76.50
8250	Bathroom vent fan, 50 CFM (use with above hook-up)									
8260	Economy model	1 Elec	15	.533	Ea.	19.40	21.50		40.90	56.50
8270	Low noise model		15	.533		44	21.50		65.50	83.50
8280	Custom model	▼	12	.667	▼	121	27		148	177
8300	Bathroom or kitchen vent fan, 110 CFM									
8310	Economy model	1 Elec	15	.533	Ea.	68	21.50		89.50	110
8320	Low noise model	"	15	.533	"	92	21.50		113.50	136
8350	Paddle fan, variable speed (w/o lights)									
8360	Economy model (AC motor)	1 Elec	10	.800	Ea.	110	32		142	174
8362	With light kit		10	.800		149	32		181	217
8370	Custom model (AC motor)		10	.800		264	32		296	345
8372	With light kit		10	.800		305	32		337	390

608

26 05 Common Work Results for Electrical

26 05 90 – Residential Applications

26 05 90.10 Residential Wiring

		Crew	Daily Output	Labor-Hours	Unit	Material	2018 Bare Costs Labor	Equipment	Total	Total Incl O&P
8380	Luxury model (DC motor)	1 Elec	8	1	Ea.	315	40		355	415
8382	With light kit		8	1		355	40		395	455
8390	Remote speed switch for above, add	↓	12	.667	↓	33.50	27		60.50	81
8500	Whole house exhaust fan, ceiling mount, 36", variable speed									
8510	Remote switch, incl. shutters, 20 amp-1 pole circ. bkr.									
8520	30' of #12/2, type NM cable	1 Elec	4	2	Ea.	1,225	80.50		1,305.50	1,475
8530	Type MC cable		3.50	2.286		1,250	92		1,342	1,525
8540	EMT & wire	↓	3	2.667	↓	1,250	107		1,357	1,550
8600	Whirlpool tub hook-up, incl. timer switch, outlet box									
8610	3' of flex, 20 amp-1 pole GFI circ. bkr.									
8620	30' of #12/2, type NM cable	1 Elec	5	1.600	Ea.	125	64		189	243
8630	Type MC cable		4.20	1.905		131	76.50		207.50	269
8640	EMT & wire	↓	3.40	2.353	↓	145	94.50		239.50	315
8650	Hot water heater hook-up, incl. 1-2 pole circ. bkr., box;									
8660	3' of flex, 20' of #10/2, type NM cable	1 Elec	5	1.600	Ea.	28.50	64		92.50	137
8670	Type MC cable		4.20	1.905		41	76.50		117.50	170
8680	EMT & wire	↓	3.40	2.353	↓	47.50	94.50		142	207
9000	Heating/air conditioning									
9050	Furnace/boiler hook-up, incl. firestat, local on-off switch									
9060	Emergency switch, and 40' of type NM cable	1 Elec	4	2	Ea.	52	80.50		132.50	188
9070	Type MC cable		3.50	2.286		66	92		158	223
9080	EMT & wire	↓	1.50	5.333	↓	89	214		303	450
9100	Air conditioner hook-up, incl. local 60 amp disc. switch									
9110	3' sealtite, 40 amp, 2 pole circuit breaker									
9130	40' of #8/2, type NM cable	1 Elec	3.50	2.286	Ea.	134	92		226	297
9140	Type MC cable		3	2.667		201	107		308	395
9150	EMT & wire	↓	1.30	6.154	↓	182	247		429	605
9200	Heat pump hook-up, 1-40 & 1-100 amp 2 pole circ. bkr.									
9210	Local disconnect switch, 3' sealtite									
9220	40' of #8/2 & 30' of #3/2									
9230	Type NM cable	1 Elec	1.30	6.154	Ea.	475	247		722	925
9240	Type MC cable		1.08	7.407		525	297		822	1,050
9250	EMT & wire	↓	.94	8.511	↓	490	340		830	1,100
9500	Thermostat hook-up, using low voltage wire									
9520	Heating only, 25' of #18-3	1 Elec	24	.333	Ea.	5.70	13.40		19.10	28.50
9530	Heating/cooling, 25' of #18-4	"	20	.400	"	7.15	16.05		23.20	34.50

26 24 Switchboards and Panelboards

26 24 16 – Panelboards

26 24 16.10 Load Centers

		Crew	Daily Output	Labor-Hours	Unit	Material	2018 Bare Costs Labor	Equipment	Total	Total Incl O&P
0010	**LOAD CENTERS** (residential type)									
0100	3 wire, 120/240 V, 1 phase, including 1 pole plug-in breakers									
0200	100 amp main lugs, indoor, 8 circuits	1 Elec	1.40	5.714	Ea.	85.50	229		314.50	470
0300	12 circuits		1.20	6.667		145	268		413	600
0400	Rainproof, 8 circuits		1.40	5.714		108	229		337	495
0500	12 circuits	↓	1.20	6.667		141	268		409	595
0600	200 amp main lugs, indoor, 16 circuits	R-1A	1.80	8.889		205	320		525	750
0700	20 circuits		1.50	10.667		189	385		574	840
0800	24 circuits		1.30	12.308		255	445		700	1,000
1200	Rainproof, 16 circuits		1.80	8.889		207	320		527	750
1300	20 circuits		1.50	10.667		288	385		673	945

For customer support on your Residential Costs with RSMeans data, call 800.448.8182.

609

26 24 Switchboards and Panelboards

26 24 16 – Panelboards

26 24 16.10 Load Centers	Crew	Daily Output	Labor-Hours	Unit	Material	2018 Bare Costs Labor	Equipment	Total	Total Incl O&P
1400	24 circuits	R-1A	1.30	12.308	Ea.	282	445		727	1,025

26 24 16.20 Panelboard and Load Center Circuit Breakers

		Crew	Daily Output	Labor-Hours	Unit	Material	Labor	Equipment	Total	Total Incl O&P
0010	**PANELBOARD AND LOAD CENTER CIRCUIT BREAKERS**									
2000	Plug-in panel or load center, 120/240 volt, to 60 amp, 1 pole	1 Elec	12	.667	Ea.	6.15	27		33.15	51
2004	Circuit breaker, 120/240 volt, 20 A, 1 pole with NM cable		6.50	1.231		10.90	49.50		60.40	93
2006	30 A, 1 pole with NM cable		6.50	1.231		10.90	49.50		60.40	93
2010	2 pole		9	.889		26	35.50		61.50	87
2014	50 A, 2 pole with NM cable		5.50	1.455		30.50	58.50		89	129
2020	3 pole		7.50	1.067		85.50	43		128.50	164
2030	100 amp, 2 pole		6	1.333		104	53.50		157.50	203
2040	3 pole		4.50	1.778		117	71.50		188.50	246
2050	150-200 amp, 2 pole		3	2.667		218	107		325	415
2060	Plug-in tandem, 120/240 V, 2-15 A, 1 pole		11	.727		21	29		50	71
2070	1-15 A & 1-20 A		11	.727		17.60	29		46.60	67.50
2080	2-20 A		11	.727		21	29		50	71
2300	Ground fault, 240 volt, 30 amp, 1 pole		7	1.143		87.50	46		133.50	172
2310	2 pole		6	1.333		156	53.50		209.50	260

26 27 Low-Voltage Distribution Equipment

26 27 13 – Electricity Metering

26 27 13.10 Meter Centers and Sockets

		Crew	Daily Output	Labor-Hours	Unit	Material	Labor	Equipment	Total	Total Incl O&P
0010	**METER CENTERS AND SOCKETS**									
0100	Sockets, single position, 4 terminal, 100 amp	1 Elec	3.20	2.500	Ea.	45	100		145	214
0200	150 amp		2.30	3.478		62	140		202	296
0300	200 amp		1.90	4.211		96	169		265	380
0500	Double position, 4 terminal, 100 amp		2.80	2.857		212	115		327	420
0600	150 amp		2.10	3.810		249	153		402	525
0700	200 amp		1.70	4.706		515	189		704	875
1100	Meter centers and sockets, three phase, single pos, 7 terminal, 100 amp		2.80	2.857		105	115		220	305
1200	200 amp		2.10	3.810		214	153		367	485
1400	400 amp		1.70	4.706		810	189		999	1,200
2590	Basic meter device									
2600	1P 3W 120/240 V 4 jaw 125A sockets, 3 meter	2 Elec	1	16	Ea.	360	640		1,000	1,450
2610	4 meter		.90	17.778		450	715		1,165	1,675
2620	5 meter		.80	20		515	805		1,320	1,900
2630	6 meter		.60	26.667		560	1,075		1,635	2,375
2640	7 meter		.56	28.571		1,400	1,150		2,550	3,425
2660	10 meter		.48	33.333		1,925	1,350		3,275	4,325
2680	Rainproof 1P 3W 120/240 V 4 jaw 125A sockets									
2690	3 meter	2 Elec	1	16	Ea.	640	640		1,280	1,750
2710	6 meter		.60	26.667		1,100	1,075		2,175	2,975
2730	8 meter		.52	30.769		1,525	1,225		2,750	3,725
2750	1P 3W 120/240 V 4 jaw sockets									
2760	with 125A circuit breaker, 3 meter	2 Elec	1	16	Ea.	1,200	640		1,840	2,375
2780	5 meter		.80	20		1,900	805		2,705	3,400
2800	7 meter		.56	28.571		2,725	1,150		3,875	4,850
2820	10 meter		.48	33.333		3,800	1,350		5,150	6,375
2830	Rainproof 1P 3W 120/240 V 4 jaw sockets									
2840	with 125A circuit breaker, 3 meter	2 Elec	1	16	Ea.	1,200	640		1,840	2,375
2870	6 meter		.60	26.667		2,225	1,075		3,300	4,200
2890	8 meter		.52	30.769		3,025	1,225		4,250	5,350

26 27 Low-Voltage Distribution Equipment

26 27 13 – Electricity Metering

26 27 13.10 Meter Centers and Sockets	Crew	Daily Output	Labor-Hours	Unit	Material	2018 Bare Costs Labor	Equipment	Total	Total Incl O&P
3250	1P 3W 120/240 V 4 jaw sockets									
3260	with 200A circuit breaker, 3 meter	2 Elec	1	16	Ea.	1,800	640		2,440	3,025
3290	6 meter		.60	26.667		3,600	1,075		4,675	5,700
3310	8 meter		.56	28.571		4,875	1,150		6,025	7,225
3330	Rainproof 1P 3W 120/240 V 4 jaw sockets									
3350	with 200A circuit breaker, 3 meter	2 Elec	1	16	Ea.	1,800	640		2,440	3,025
3380	6 meter		.60	26.667		3,600	1,075		4,675	5,700
3400	8 meter		.52	30.769		4,875	1,225		6,100	7,375

26 27 23 – Indoor Service Poles

26 27 23.40 Surface Raceway

		Crew	Daily Output	Labor-Hours	Unit	Material	Labor	Equipment	Total	Total Incl O&P
0010	**SURFACE RACEWAY**									
0090	Metal, straight section									
0100	No. 500	1 Elec	100	.080	L.F.	1.07	3.21		4.28	6.45
0110	No. 700		100	.080		1.19	3.21		4.40	6.55
0200	No. 1000		90	.089		1.45	3.57		5.02	7.45
0400	No. 1500, small pancake		90	.089		2.20	3.57		5.77	8.25
0600	No. 2000, base & cover, blank		90	.089		2.24	3.57		5.81	8.30
0800	No. 3000, base & cover, blank		75	.107		4.28	4.28		8.56	11.70
2400	Fittings, elbows, No. 500		40	.200	Ea.	1.94	8.05		9.99	15.30
2800	Elbow cover, No. 2000		40	.200		3.64	8.05		11.69	17.15
2880	Tee, No. 500		42	.190		3.74	7.65		11.39	16.60
2900	No. 2000		27	.296		12.10	11.90		24	33
3000	Switch box, No. 500		16	.500		12.50	20		32.50	47
3400	Telephone outlet, No. 1500		16	.500		14.50	20		34.50	49
3600	Junction box, No. 1500		16	.500		10.30	20		30.30	44.50
3800	Plugmold wired sections, No. 2000									
4000	1 circuit, 6 outlets, 3' long	1 Elec	8	1	Ea.	38	40		78	108
4100	2 circuits, 8 outlets, 6' long	"	5.30	1.509	"	62.50	60.50		123	168

26 27 26 – Wiring Devices

26 27 26.10 Low Voltage Switching

		Crew	Daily Output	Labor-Hours	Unit	Material	Labor	Equipment	Total	Total Incl O&P
0010	**LOW VOLTAGE SWITCHING**									
3600	Relays, 120 V or 277 V standard	1 Elec	12	.667	Ea.	41	27		68	89
3800	Flush switch, standard		40	.200		11.50	8.05		19.55	26
4000	Interchangeable		40	.200		15.05	8.05		23.10	29.50
4100	Surface switch, standard		40	.200		8.20	8.05		16.25	22
4200	Transformer 115 V to 25 V		12	.667		128	27		155	185
4400	Master control, 12 circuit, manual		4	2		128	80.50		208.50	272
4500	25 circuit, motorized		4	2		141	80.50		221.50	286
4600	Rectifier, silicon		12	.667		46	27		73	94.50
4800	Switchplates, 1 gang, 1, 2 or 3 switch, plastic		80	.100		5	4.02		9.02	12.05
5000	Stainless steel		80	.100		11.35	4.02		15.37	19.05
5400	2 gang, 3 switch, stainless steel		53	.151		23	6.05		29.05	35
5500	4 switch, plastic		53	.151		10.45	6.05		16.50	21.50
5600	2 gang, 4 switch, stainless steel		53	.151		22	6.05		28.05	34
5700	6 switch, stainless steel		53	.151		43.50	6.05		49.55	58
5800	3 gang, 9 switch, stainless steel		32	.250		65.50	10.05		75.55	88.50

26 27 26.20 Wiring Devices Elements

		Crew	Daily Output	Labor-Hours	Unit	Material	Labor	Equipment	Total	Total Incl O&P
0010	**WIRING DEVICES ELEMENTS**									
0200	Toggle switch, quiet type, single pole, 15 amp	1 Elec	40	.200	Ea.	.49	8.05		8.54	13.70
0600	3 way, 15 amp		23	.348		1.52	13.95		15.47	24.50
0900	4 way, 15 amp		15	.533		10.70	21.50		32.20	47

For customer support on your Residential Costs with RSMeans data, call 800.448.8182.

611

26 27 Low-Voltage Distribution Equipment

26 27 26 – Wiring Devices

26 27 26.20 Wiring Devices Elements

		Crew	Daily Output	Labor-Hours	Unit	Material	2018 Bare Costs Labor	2018 Bare Costs Equipment	Total	Total Incl O&P
1650	Dimmer switch, 120 volt, incandescent, 600 watt, 1 pole **G**	1 Elec	16	.500	Ea.	22.50	20		42.50	58
2460	Receptacle, duplex, 120 volt, grounded, 15 amp		40	.200		1.30	8.05		9.35	14.60
2470	20 amp		27	.296		9.05	11.90		20.95	29.50
2490	Dryer, 30 amp		15	.533		4.44	21.50		25.94	40
2500	Range, 50 amp		11	.727		10.60	29		39.60	59.50
2600	Wall plates, stainless steel, 1 gang		80	.100		2.56	4.02		6.58	9.35
2800	2 gang		53	.151		4.33	6.05		10.38	14.65
3200	Lampholder, keyless		26	.308		16.85	12.35		29.20	38.50
3400	Pullchain with receptacle	↓	22	.364	↓	20.50	14.60		35.10	46.50

26 27 73 – Door Chimes

26 27 73.10 Doorbell System

		Crew	Daily Output	Labor-Hours	Unit	Material	2018 Bare Costs Labor	2018 Bare Costs Equipment	Total	Total Incl O&P
0010	**DOORBELL SYSTEM**, incl. transformer, button & signal									
0100	6" bell	1 Elec	4	2	Ea.	139	80.50		219.50	284
0200	Buzzer		4	2		109	80.50		189.50	251
1000	Door chimes, 2 notes		16	.500		30	20		50	66
1020	with ambient light		12	.667		120	27		147	176
1100	Tube type, 3 tube system		12	.667		226	27		253	292
1180	4 tube system		10	.800		480	32		512	585
1900	For transformer & button, add		5	1.600		14.20	64		78.20	121
3000	For push button only	↓	24	.333		.94	13.40		14.34	23

26 28 Low-Voltage Circuit Protective Devices

26 28 16 – Enclosed Switches and Circuit Breakers

26 28 16.10 Circuit Breakers

		Crew	Daily Output	Labor-Hours	Unit	Material	2018 Bare Costs Labor	2018 Bare Costs Equipment	Total	Total Incl O&P
0010	**CIRCUIT BREAKERS** (in enclosure)									
0100	Enclosed (NEMA 1), 600 volt, 3 pole, 30 amp	1 Elec	3.20	2.500	Ea.	520	100		620	735
0200	60 amp		2.80	2.857		635	115		750	890
0400	100 amp	↓	2.30	3.478	↓	730	140		870	1,025

26 28 16.20 Safety Switches

		Crew	Daily Output	Labor-Hours	Unit	Material	2018 Bare Costs Labor	2018 Bare Costs Equipment	Total	Total Incl O&P
0010	**SAFETY SWITCHES**									
0100	General duty 240 volt, 3 pole NEMA 1, fusible, 30 amp	1 Elec	3.20	2.500	Ea.	55.50	100		155.50	225
0200	60 amp		2.30	3.478		94	140		234	330
0300	100 amp		1.90	4.211		159	169		328	450
0400	200 amp	↓	1.30	6.154		345	247		592	785
0500	400 amp	2 Elec	1.80	8.889		905	355		1,260	1,575
9010	Disc. switch, 600 volt 3 pole fusible, 30 amp, to 10 HP motor	1 Elec	3.20	2.500		355	100		455	555
9050	60 amp, to 30 HP motor		2.30	3.478		815	140		955	1,125
9070	100 amp, to 60 HP motor	↓	1.90	4.211	↓	815	169		984	1,175

26 28 16.40 Time Switches

		Crew	Daily Output	Labor-Hours	Unit	Material	2018 Bare Costs Labor	2018 Bare Costs Equipment	Total	Total Incl O&P
0010	**TIME SWITCHES**									
0100	Single pole, single throw, 24 hour dial	1 Elec	4	2	Ea.	125	80.50		205.50	268
0200	24 hour dial with reserve power		3.60	2.222		565	89		654	765
0300	Astronomic dial		3.60	2.222		240	89		329	410
0400	Astronomic dial with reserve power		3.30	2.424		665	97.50		762.50	890
0500	7 day calendar dial		3.30	2.424		140	97.50		237.50	315
0600	7 day calendar dial with reserve power		3.20	2.500		253	100		353	440
0700	Photo cell 2,000 watt	↓	8	1	↓	27	40		67	95.50

26 32 Packaged Generator Assemblies

26 32 13 – Engine Generators

26 32 13.16 Gas-Engine-Driven Generator Sets

		Crew	Daily Output	Labor-Hours	Unit	Material	2018 Bare Costs Labor	Equipment	Total	Total Incl O&P
0010	**GAS-ENGINE-DRIVEN GENERATOR SETS**									
0020	Gas or gasoline operated, includes battery,									
0050	charger, & muffler									
0200	7.5 kW	R-3	.83	24.096	Ea.	8,550	955	156	9,661	11,200
0300	11.5 kW		.71	28.169		12,100	1,125	183	13,408	15,300
0400	20 kW		.63	31.746		14,300	1,250	206	15,756	18,000
0500	35 kW		.55	36.364		17,000	1,450	236	18,686	21,300

26 33 Battery Equipment

26 33 43 – Battery Chargers

26 33 43.55 Electric Vehicle Charging

			Crew	Daily Output	Labor-Hours	Unit	Material	2018 Bare Costs Labor	Equipment	Total	Total Incl O&P
0010	**ELECTRIC VEHICLE CHARGING**										
0020	Level 2, wall mounted										
2100	Light duty, hard wired	G	1 Elec	20.48	.391	Ea.	570	15.70		585.70	650
2110	plug in	G	"	30.72	.260	"	570	10.45		580.45	640

26 36 Transfer Switches

26 36 23 – Automatic Transfer Switches

26 36 23.10 Automatic Transfer Switch Devices

		Crew	Daily Output	Labor-Hours	Unit	Material	2018 Bare Costs Labor	Equipment	Total	Total Incl O&P
0010	**AUTOMATIC TRANSFER SWITCH DEVICES**									
0100	Switches, enclosed 480 volt, 3 pole, 30 amp	1 Elec	2.30	3.478	Ea.	2,950	140		3,090	3,475
0200	60 amp	"	1.90	4.211	"	2,950	169		3,119	3,525

26 41 Facility Lightning Protection

26 41 13 – Lightning Protection for Structures

26 41 13.13 Lightning Protection for Buildings

		Crew	Daily Output	Labor-Hours	Unit	Material	2018 Bare Costs Labor	Equipment	Total	Total Incl O&P
0010	**LIGHTNING PROTECTION FOR BUILDINGS**									
0200	Air terminals & base, copper									
0400	3/8" diameter x 10" (to 75' high)	1 Elec	8	1	Ea.	22.50	40		62.50	90
1000	Aluminum, 1/2" diameter x 12" (to 75' high)		8	1		17.85	40		57.85	85
1020	1/2" diameter x 24"		7.30	1.096		19.50	44		63.50	93.50
1040	1/2" diameter x 60"		6.70	1.194		26.50	48		74.50	108
2000	Cable, copper, 220 lb. per thousand ft. (to 75' high)		320	.025	L.F.	3.03	1		4.03	4.97
2500	Aluminum, 101 lb. per thousand ft. (to 75' high)		280	.029	"	.95	1.15		2.10	2.93
3000	Arrester, 175 volt AC to ground		8	1	Ea.	149	40		189	230

For customer support on your Residential Costs with RSMeans data, call 800.448.8182.

613

26 51 13 – Interior Lighting Fixtures, Lamps, and Ballasts

26 51 13.50 Interior Lighting Fixtures		Crew	Daily Output	Labor-Hours	Unit	Material	2018 Bare Costs Labor	Equipment	Total	Total Incl O&P
0010	**INTERIOR LIGHTING FIXTURES** Including lamps, mounting									
0030	hardware and connections									
0100	Fluorescent, C.W. lamps, troffer, recess mounted in grid, RS									
0130	Grid ceiling mount									
0200	Acrylic lens, 1' W x 4' L, two 40 watt	1 Elec	5.70	1.404	Ea.	51.50	56.50		108	149
0300	2' W x 2' L, two U40 watt		5.70	1.404		55.50	56.50		112	153
0600	2' W x 4' L, four 40 watt	↓	4.70	1.702	↓	62	68.50		130.50	181
1000	Surface mounted, RS									
1030	Acrylic lens with hinged & latched door frame									
1100	1' W x 4' L, two 40 watt	1 Elec	7	1.143	Ea.	65.50	46		111.50	148
1200	2' W x 2' L, two U40 watt		7	1.143		70.50	46		116.50	153
1500	2' W x 4' L, four 40 watt		5.30	1.509		85	60.50		145.50	193
1501	2' W x 4' L, six 40 watt T8	↓	5.20	1.538	↓	85	62		147	195
2100	Strip fixture									
2200	4' long, one 40 watt, RS	1 Elec	8.50	.941	Ea.	31	38		69	96
2300	4' long, two 40 watt, RS	"	8	1		43	40		83	113
2600	8' long, one 75 watt, SL	2 Elec	13.40	1.194		55	48		103	139
2700	8' long, two 75 watt, SL	"	12.40	1.290	↓	66.50	52		118.50	158
4450	Incandescent, high hat can, round alzak reflector, prewired									
4470	100 watt	1 Elec	8	1	Ea.	64.50	40		104.50	137
4480	150 watt		8	1		105	40		145	181
4500	300 watt	↓	6.70	1.194	↓	242	48		290	345
5200	Ceiling, surface mounted, opal glass drum									
5300	8", one 60 watt lamp	1 Elec	10	.800	Ea.	65.50	32		97.50	125
5400	10", two 60 watt lamps		8	1		72.50	40		112.50	146
5500	12", four 60 watt lamps		6.70	1.194		103	48		151	192
6900	Mirror light, fluorescent, RS, acrylic enclosure, two 40 watt		8	1		116	40		156	194
6910	One 40 watt		8	1		96	40		136	172
6920	One 20 watt	↓	12	.667	↓	82.50	27		109.50	135

26 51 13.55 Interior LED Fixtures

26 51 13.55 Interior LED Fixtures			Crew	Daily Output	Labor-Hours	Unit	Material	2018 Bare Costs Labor	Equipment	Total	Total Incl O&P
0010	**INTERIOR LED FIXTURES** Incl. lamps and mounting hardware										
0100	Downlight, recess mounted, 7.5" diameter, 25 watt	G	1 Elec	8	1	Ea.	335	40		375	430
0120	10" diameter, 36 watt	G		8	1		360	40		400	460
0160	cylinder, 10 watts	G		8	1		103	40		143	179
0180	20 watts	G		8	1		129	40		169	208
1000	Troffer, recess mounted, 2' x 4', 3,200 lumens	G		5.30	1.509		138	60.50		198.50	250
1010	4,800 lumens	G		5	1.600		179	64		243	300
1020	6,400 lumens	G		4.70	1.702		198	68.50		266.50	330
1100	Troffer retrofit lamp, 38 watt	G		21	.381		69.50	15.30		84.80	102
1110	60 watt	G		20	.400		141	16.05		157.05	183
1120	100 watt	G		18	.444		206	17.85		223.85	255
1200	Troffer, volumetric recess mounted, 2' x 2'	G		5.70	1.404		305	56.50		361.50	425
2000	Strip, surface mounted, one light bar 4' long, 3,500 K	G		8.50	.941		260	38		298	350
2010	5,000 K	G		8	1		260	40		300	350
2020	Two light bar 4' long, 5,000 K	G		7	1.143		410	46		456	525
3000	Linear, suspended mounted, one light bar 4' long, 37 watt	G		6.70	1.194		153	48		201	247
3010	One light bar 8' long, 74 watt	G	2 Elec	12.20	1.311		283	52.50		335.50	395
3020	Two light bar 4' long, 74 watt	G	1 Elec	5.70	1.404		305	56.50		361.50	425
3030	Two light bar 8' long, 148 watt	G	2 Elec	8.80	1.818		350	73		423	505
4000	High bay, surface mounted, round, 150 watts	G		5.41	2.959		425	119		544	665
4010	2 bars, 164 watts	G		5.41	2.959		385	119		504	620
4020	3 bars, 246 watts	G	↓	5.01	3.197	↓	515	128		643	775

26 51 Interior Lighting

26 51 13 – Interior Lighting Fixtures, Lamps, and Ballasts

26 51 13.55 Interior LED Fixtures	Crew	Daily Output	Labor-Hours	Unit	Material	2018 Bare Costs Labor	Equipment	Total	Total Incl O&P
4030	4 bars, 328 watts	G	2 Elec	4.60	3.478	Ea.	715	140		855	1,025
4040	5 bars, 410 watts	G	3 Elec	4.20	5.716		810	229		1,039	1,275
4050	6 bars, 492 watts	G		3.80	6.324		910	254		1,164	1,425
4060	7 bars, 574 watts	G		3.39	7.075		990	284		1,274	1,575
4070	8 bars, 656 watts	G		2.99	8.029		1,050	320		1,370	1,675
5000	Track, lighthead, 6 watt	G	1 Elec	32	.250		54.50	10.05		64.55	76.50
5010	9 watt	G	"	32	.250		61.50	10.05		71.55	84
6000	Garage, surface mounted, 103 watts	G	2 Elec	6.50	2.462		970	99		1,069	1,225
6100	pendent mounted, 80 watts	G		6.50	2.462		660	99		759	885
6200	95 watts	G		6.50	2.462		780	99		879	1,025
6300	125 watts	G		6.50	2.462		820	99		919	1,075

26 51 13.70 Residential Fixtures

		Crew	Daily Output	Labor-Hours	Unit	Material	Labor	Equipment	Total	Total Incl O&P
0010	**RESIDENTIAL FIXTURES**									
0400	Fluorescent, interior, surface, circline, 32 watt & 40 watt	1 Elec	20	.400	Ea.	155	16.05		171.05	198
0500	2' x 2', two U-tube 32 watt T8		8	1		129	40		169	208
0700	Shallow under cabinet, two 20 watt		16	.500		71.50	20		91.50	112
0900	Wall mounted, 4' L, two 32 watt T8, with baffle		10	.800		133	32		165	199
2000	Incandescent, exterior lantern, wall mounted, 60 watt		16	.500		60.50	20		80.50	99.50
2100	Post light, 150 W, with 7' post		4	2		270	80.50		350.50	430
2500	Lamp holder, weatherproof with 150 W PAR		16	.500		33.50	20		53.50	69.50
2550	With reflector and guard		12	.667		59.50	27		86.50	110
2600	Interior pendent, globe with shade, 150 W		20	.400		201	16.05		217.05	248

26 55 Special Purpose Lighting

26 55 59 – Display Lighting

26 55 59.10 Track Lighting	Crew	Daily Output	Labor-Hours	Unit	Material	Labor	Equipment	Total	Total Incl O&P
0010	**TRACK LIGHTING**									
0100	8' section	2 Elec	10.60	1.509	Ea.	70.50	60.50		131	177
0300	3 circuits, 4' section	1 Elec	6.70	1.194		94	48		142	183
0400	8' section	2 Elec	10.60	1.509		133	60.50		193.50	245
0500	12' section	"	8.80	1.818		159	73		232	294
1000	Feed kit, surface mounting	1 Elec	16	.500		14.80	20		34.80	49.50
1100	End cover		24	.333		5.55	13.40		18.95	28
1200	Feed kit, stem mounting, 1 circuit		16	.500		51.50	20		71.50	89.50
1300	3 circuit		16	.500		51.50	20		71.50	89.50
2000	Electrical joiner, for continuous runs, 1 circuit		32	.250		33.50	10.05		43.55	53
2100	3 circuit		32	.250		61	10.05		71.05	83.50
2200	Fixtures, spotlight, 75 W PAR halogen		16	.500		51	20		71	89
2210	50 W MR16 halogen		16	.500		171	20		191	221
3000	Wall washer, 250 W tungsten halogen		16	.500		132	20		152	178
3100	Low voltage, 25/50 W, 1 circuit		16	.500		133	20		153	179
3120	3 circuit		16	.500		198	20		218	251

For customer support on your Residential Costs with RSMeans data, call 800.448.8182.

615

26 56 Exterior Lighting

26 56 13 – Lighting Poles and Standards

26 56 13.10 Lighting Poles

		Crew	Daily Output	Labor-Hours	Unit	Material	2018 Bare Costs Labor	Equipment	Total	Total Incl O&P
0010	**LIGHTING POLES**									
6420	Wood pole, 4-1/2" x 5-1/8", 8' high	1 Elec	6	1.333	Ea.	375	53.50		428.50	505
6440	12' high		5.70	1.404		540	56.50		596.50	685
6460	20' high		4	2		765	80.50		845.50	975

26 56 23 – Area Lighting

26 56 23.10 Exterior Fixtures

		Crew	Daily Output	Labor-Hours	Unit	Material	2018 Bare Costs Labor	Equipment	Total	Total Incl O&P
0010	**EXTERIOR FIXTURES** With lamps									
0400	Quartz, 500 watt	1 Elec	5.30	1.509	Ea.	61.50	60.50		122	167
1100	Wall pack, low pressure sodium, 35 watt		4	2		191	80.50		271.50	340
1150	55 watt		4	2		227	80.50		307.50	380

26 56 26 – Landscape Lighting

26 56 26.20 Landscape Fixtures

		Crew	Daily Output	Labor-Hours	Unit	Material	2018 Bare Costs Labor	Equipment	Total	Total Incl O&P
0010	**LANDSCAPE FIXTURES**									
7380	Landscape recessed uplight, incl. housing, ballast, transformer									
7390	& reflector, not incl. conduit, wire, trench									
7420	Incandescent, 250 watt	1 Elec	5	1.600	Ea.	655	64		719	825
7440	Quartz, 250 watt	"	5	1.600	"	620	64		684	785

26 56 26.50 Landscape LED Fixtures

		Crew	Daily Output	Labor-Hours	Unit	Material	2018 Bare Costs Labor	Equipment	Total	Total Incl O&P
0010	**LANDSCAPE LED FIXTURES**									
0100	12 volt alum bullet hooded-BLK	1 Elec	5	1.600	Ea.	96	64		160	211
0200	12 volt alum bullet hooded-BRZ		5	1.600		96	64		160	211
0300	12 volt alum bullet hooded-GRN		5	1.600		96	64		160	211
1000	12 volt alum large bullet hooded-BLK		5	1.600		71	64		135	183
1100	12 volt alum large bullet hooded-BRZ		5	1.600		71	64		135	183
1200	12 volt alum large bullet hooded-GRN		5	1.600		71	64		135	183
2000	12 volt large bullet landscape light fixture		5	1.600		71	64		135	183
2100	12 volt alum light large bullet		5	1.600		71	64		135	183
2200	12 volt alum bullet light		5	1.600		71	64		135	183

26 56 33 – Walkway Lighting

26 56 33.10 Walkway Luminaire

		Crew	Daily Output	Labor-Hours	Unit	Material	2018 Bare Costs Labor	Equipment	Total	Total Incl O&P
0010	**WALKWAY LUMINAIRE**									
6500	Bollard light, lamp & ballast, 42" high with polycarbonate lens									
7200	Incandescent, 150 watt	1 Elec	3	2.667	Ea.	670	107		777	910

26 61 Lighting Systems and Accessories

26 61 23 – Lamps Applications

26 61 23.10 Lamps

		Crew	Daily Output	Labor-Hours	Unit	Material	2018 Bare Costs Labor	Equipment	Total	Total Incl O&P
0010	**LAMPS**									
0081	Fluorescent, rapid start, cool white, 2' long, 20 watt	1 Elec	100	.080	Ea.	3.05	3.21		6.26	8.60
0101	4' long, 40 watt		90	.089		2.56	3.57		6.13	8.65
1351	High pressure sodium, 70 watt		30	.267		16.40	10.70		27.10	35.50
1371	150 watt		30	.267		16.75	10.70		27.45	36

Estimating Tips
27 20 00 Data Communications
27 30 00 Voice Communications
27 40 00 Audio-Video Communications

- When estimating material costs for special systems, it is always prudent to obtain manufacturers' quotations for equipment prices and special installation requirements that may affect the total cost.

- For cost modifications for elevated tray installation, add the percentages to labor according to the height of the installation and only to the quantities exceeding the different height levels, not to the total tray quantities. Refer to 26 01 02.20 for labor adjustment factors.

- Do not overlook the costs for equipment used in the installation. If scissor lifts and boom lifts are available in the field, contractors may use them in lieu of the proposed ladders and rolling staging.

Reference Numbers
Reference numbers are shown at the beginning of some major classifications. These numbers refer to related items in the Reference Section. The reference information may be an estimating procedure, an alternate pricing method, or technical information.

Note: Not all subdivisions listed here necessarily appear. ■

Did you know?

RSMeans data is available through our online application with 24/7 access:

- Search for unit prices by keyword
- Leverage the most up-to-date data
- Build and export estimates

Try it free for 30 days!
www.rsmeans.com/2018freetrial

27 13 Communications Backbone Cabling

27 13 23 – Communications Optical Fiber Backbone Cabling

27 13 23.13 Communications Optical Fiber

		Crew	Daily Output	Labor-Hours	Unit	Material	2018 Bare Costs Labor	Equipment	Total	Total Incl O&P
0010	**COMMUNICATIONS OPTICAL FIBER**									
0040	Specialized tools & techniques cause installation costs to vary.									
0070	Fiber optic, cable, bulk simplex, single mode	1 Elec	8	1	C.L.F.	22.50	40		62.50	90.50
0080	Multi mode		8	1		29.50	40		69.50	98
0090	4 strand, single mode		7.34	1.090		38	44		82	113
0095	Multi mode		7.34	1.090		50.50	44		94.50	127
0100	12 strand, single mode		6.67	1.199		72.50	48		120.50	158
0105	Multi mode		6.67	1.199		96.50	48		144.50	185
0150	Jumper				Ea.	33			33	36.50
0200	Pigtail					29			29	32
0300	Connector	1 Elec	24	.333		26	13.40		39.40	50.50
0350	Finger splice		32	.250		38	10.05		48.05	58.50
0400	Transceiver (low cost bi-directional)		8	1		430	40		470	540
0450	Rack housing, 4 rack spaces, 12 panels (144 fibers)		2	4		560	161		721	880
1000	Cable, 62.5 microns, direct burial, 4 fiber	R-15	1200	.040	L.F.	.91	1.58	.24	2.73	3.84
1020	Indoor, 2 fiber	R-19	1000	.020		.42	.81		1.23	1.78
1040	Outdoor, aerial/duct	"	1670	.012		.65	.48		1.13	1.51
1060	50 microns, direct burial, 8 fiber	R-22	4000	.009		1.27	.34		1.61	1.96
1080	12 fiber		4000	.009		2.08	.34		2.42	2.85
1100	Indoor, 12 fiber		759	.049		1.98	1.81		3.79	5.15
1120	Connectors, 62.5 micron cable, transmission	R-19	40	.500	Ea.	14.35	20		34.35	49
1140	Cable splice		40	.500		17.50	20		37.50	52.50
1160	125 micron cable, transmission		16	1.250		15.10	50.50		65.60	99
1180	Receiver, 1.2 mile range		20	1		240	40.50		280.50	330
1200	1.9 mile range		20	1		226	40.50		266.50	315
1220	6.2 mile range		5	4		282	161		443	575
1240	Transmitter, 1.2 mile range		20	1		264	40.50		304.50	355
1260	1.9 mile range		20	1		299	40.50		339.50	395
1280	6.2 mile range		5	4		380	161		541	685
1300	Modem, 1.2 mile range		5	4		173	161		334	455
1320	6.2 mile range		5	4		315	161		476	615
1340	1.9 mile range, 12 channel		5	4		2,000	161		2,161	2,475
1360	Repeater, 1.2 mile range		10	2		370	80.50		450.50	540
1380	1.9 mile range		10	2		475	80.50		555.50	655
1400	6.2 mile range		5	4		920	161		1,081	1,275
1420	1.2 mile range, digital		5	4		440	161		601	745

27 41 Audio-Video Systems

27 41 33 – Master Antenna Television Systems

27 41 33.10 TV Systems

		Crew	Daily Output	Labor-Hours	Unit	Material	2018 Bare Costs Labor	Equipment	Total	Total Incl O&P
0010	**TV SYSTEMS,** not including rough-in wires, cables & conduits									
0100	Master TV antenna system									
0200	VHF reception & distribution, 12 outlets	1 Elec	6	1.333	Outlet	114	53.50		167.50	214
0800	VHF & UHF reception & distribution, 12 outlets		6	1.333	"	249	53.50		302.50	360
5000	Antenna, small		6	1.333	Ea.	55	53.50		108.50	148
5100	Large		4	2		231	80.50		311.50	385
5110	Rotor unit		8	1		63	40		103	135
5120	Single booster		8	1		31.50	40		71.50	100
5130	Antenna pole, 10'		3.20	2.500		20	100		120	186
6100	Satellite TV system	2 Elec	1	16		2,150	640		2,790	3,425
6110	Dish, mesh, 10' diam.	"	2.40	6.667		1,675	268		1,943	2,300

For customer support on your Residential Costs with RSMeans data, call 800.448.8182.

27 41 Audio-Video Systems

27 41 33 – Master Antenna Television Systems

27 41 33.10 TV Systems	Crew	Daily Output	Labor-Hours	Unit	Material	2018 Bare Costs Labor	Equipment	Total	Total Incl O&P
6111	Two way RF/IF tapeoff	1 Elec	36	.222	Ea.	3.59	8.90		12.49	18.55
6112	Two way RF/IF splitter		24	.333		13.70	13.40		27.10	37
6113	Line amplifier		24	.333		13.10	13.40		26.50	36.50
6114	Line splitters		36	.222		4.95	8.90		13.85	20
6115	Line multi switches		8	1		655	40		695	785
6120	Motor unit		2.40	3.333		259	134		393	505
7000	Home theater, widescreen, 42", high definition, TV		10.24	.781		400	31.50		431.50	490
7050	Flat wall mount bracket		10.24	.781		76	31.50		107.50	135
7100	7 channel home theater receiver		10.24	.781		385	31.50		416.50	475
7200	Home theater speakers		10.24	.781	Set	216	31.50		247.50	289
7300	Home theater programmable remote		10.24	.781	Ea.	249	31.50		280.50	325
8000	Main video splitter		4	2		3,225	80.50		3,305.50	3,675
8010	Video distribution units		4	2		140	80.50		220.50	285

Division Notes

	CREW	DAILY OUTPUT	LABOR-HOURS	UNIT	BARE COSTS				TOTAL INCL O&P
					MAT.	LABOR	EQUIP.	TOTAL	

Estimating Tips

- When estimating material costs for electronic safety and security systems, it is always prudent to obtain manufacturers' quotations for equipment prices and special installation requirements that may affect the total cost.

- Fire alarm systems consist of control panels, annunciator panels, batteries with rack, charger, and fire alarm actuating and indicating devices. Some fire alarm systems include speakers, telephone lines, door closer controls, and other components. Be careful not to overlook the costs related to installation for these items. Also be aware of costs for integrated automation instrumentation and terminal devices, control equipment, control wiring, and programming. Insurance underwriters may have specific requirements for the type of materials to be installed or design requirements based on the hazard to be protected. Local jurisdictions may have requirements not covered by code. It is advisable to be aware of any special conditions.

- Security equipment includes items such as CCTV, access control, and other detection and identification systems to perform alert and alarm functions. Be sure to consider the costs related to installation for this security equipment, such as for integrated automation instrumentation and terminal devices, control equipment, control wiring, and programming.

Reference Numbers

Reference numbers are shown at the beginning of some major classifications. These numbers refer to related items in the Reference Section. The reference information may be an estimating procedure, an alternate pricing method, or technical information.

Note: Not all subdivisions listed here necessarily appear. ■

Did you know?

RSMeans data is available through our online application with 24/7 access:

- Search for unit prices by keyword
- Leverage the most up-to-date data
- Build and export estimates

Try it free for 30 days!
www.rsmeans.com/2018freetrial

28 31 Intrusion Detection

28 31 16 – Intrusion Detection Systems Infrastructure

28 31 16.50 Intrusion Detection	Crew	Daily Output	Labor-Hours	Unit	Material	2018 Bare Costs Labor	Equipment	Total	Total Incl O&P
0010 **INTRUSION DETECTION**, not including wires & conduits									
0100 Burglar alarm, battery operated, mechanical trigger	1 Elec	4	2	Ea.	280	80.50		360.50	440
0200 Electrical trigger		4	2		335	80.50		415.50	500
0400 For outside key control, add		8	1		87	40		127	161
0600 For remote signaling circuitry, add		8	1		138	40		178	218
0800 Card reader, flush type, standard		2.70	2.963		810	119		929	1,075
1000 Multi-code		2.70	2.963		1,200	119		1,319	1,525
1200 Door switches, hinge switch		5.30	1.509		62.50	60.50		123	168
1400 Magnetic switch		5.30	1.509		100	60.50		160.50	209
2800 Ultrasonic motion detector, 12 V		2.30	3.478		194	140		334	440
3000 Infrared photoelectric detector		4	2		146	80.50		226.50	292
3200 Passive infrared detector		4	2		236	80.50		316.50	390
3420 Switchmats, 30" x 5'		5.30	1.509		105	60.50		165.50	215
3440 30" x 25'		4	2		205	80.50		285.50	355
3460 Police connect panel		4	2		272	80.50		352.50	430
3480 Telephone dialer		5.30	1.509		390	60.50		450.50	525
3500 Alarm bell		4	2		105	80.50		185.50	247
3520 Siren		4	2		147	80.50		227.50	293

28 46 Fire Detection and Alarm

28 46 11 – Fire Sensors and Detectors

28 46 11.21 Carbon-Monoxide Detection Sensors

	Crew	Daily Output	Labor-Hours	Unit	Material	Labor	Equipment	Total	Total Incl O&P
0010 **CARBON-MONOXIDE DETECTION SENSORS**									
8400 Smoke and carbon monoxide alarm battery operated photoelectric low profile	1 Elec	24	.333	Ea.	54.50	13.40		67.90	81.50
8410 low profile photoelectric battery powered		24	.333		28.50	13.40		41.90	53.50
8420 photoelectric low profile sealed lithium		24	.333		52	13.40		65.40	79
8430 Photoelectric low profile sealed lithium smoke and CO with voice combo		24	.333		38	13.40		51.40	64
8500 Carbon monoxide sensor, wall mount 1Mod 1 relay output smoke & heat		24	.333		490	13.40		503.40	560
8700 Carbon monoxide detector, battery operated, wall mounted		16	.500		52	20		72	90
8710 Hardwired, wall and ceiling mounted		8	1		99.50	40		139.50	175
8720 Duct mounted		8	1		345	40		385	440

28 46 11.27 Other Sensors

	Crew	Daily Output	Labor-Hours	Unit	Material	Labor	Equipment	Total	Total Incl O&P
0010 **OTHER SENSORS**									
5200 Smoke detector, ceiling type	1 Elec	6.20	1.290	Ea.	120	52		172	217
5240 Smoke detector addressable type		6	1.333		224	53.50		277.50	335
5420 Duct addressable type		3.20	2.500		515	100		615	730
8300 Smoke alarm with integrated strobe light 120 V 16DB 60 fpm flash rate		16	.500		103	20		123	146
8310 Photoelectric smoke detector with strobe 120 V 90 DB ceiling mount		12	.667		158	27		185	218
8320 120 V, 90 DB wall mount		12	.667		166	27		193	226

28 46 21 – Fire Alarm

28 46 21.50 Alarm Panels and Devices

	Crew	Daily Output	Labor-Hours	Unit	Material	Labor	Equipment	Total	Total Incl O&P
0010 **ALARM PANELS AND DEVICES**, not including wires & conduits									
5600 Strobe and horn	1 Elec	5.30	1.509	Ea.	154	60.50		214.50	268
5610 Strobe and horn (ADA type)		5.30	1.509		154	60.50		214.50	268
5620 Visual alarm (ADA type)		6.70	1.194		104	48		152	193
5800 Fire alarm horn		6.70	1.194		61.50	48		109.50	146
6600 Drill switch		8	1		375	40		415	475
6800 Master box		2.70	2.963		6,475	119		6,594	7,325
7000 Break glass station		8	1		57.50	40		97.50	129
7800 Remote annunciator, 8 zone lamp		1.80	4.444		213	178		391	525

28 46 Fire Detection and Alarm

28 46 21 – Fire Alarm

28 46 21.50 Alarm Panels and Devices	Crew	Daily Output	Labor-Hours	Unit	Material	2018 Bare Costs Labor	Equipment	Total	Total Incl O&P
8000	12 zone lamp	2 Elec	2.60	6.154	Ea.	415	247		662	865
8200	16 zone lamp	"	2.20	7.273	↓	375	292		667	885

For customer support on your Residential Costs with RSMeans data, call 800.448.8182.

623

Division Notes

	CREW	DAILY OUTPUT	LABOR-HOURS	UNIT	BARE COSTS				TOTAL INCL O&P
					MAT.	LABOR	EQUIP.	TOTAL	

Estimating Tips
31 05 00 Common Work Results for Earthwork

- Estimating the actual cost of performing earthwork requires careful consideration of the variables involved. This includes items such as type of soil, whether water will be encountered, dewatering, whether banks need bracing, disposal of excavated earth, and length of haul to fill or spoil sites, etc. If the project has large quantities of cut or fill, consider raising or lowering the site to reduce costs, while paying close attention to the effect on site drainage and utilities.

- If the project has large quantities of fill, creating a borrow pit on the site can significantly lower the costs.

- It is very important to consider what time of year the project is scheduled for completion. Bad weather can create large cost overruns from dewatering, site repair, and lost productivity from cold weather.

Reference Numbers

Reference numbers are shown at the beginning of some major classifications. These numbers refer to related items in the Reference Section. The reference information may be an estimating procedure, an alternate pricing method, or technical information.

Note: Not all subdivisions listed here necessarily appear. ■

Did you know?

RSMeans data is available through our online application with 24/7 access:

- Search for unit prices by keyword
- Leverage the most up-to-date data
- Build and export estimates

Try it free for 30 days!
www.rsmeans.com/2018freetrial

31 05 Common Work Results for Earthwork

31 05 13 – Soils for Earthwork

31 05 13.10 Borrow	Crew	Daily Output	Labor-Hours	Unit	Material	2018 Bare Costs Labor	Equipment	Total	Total Incl O&P
0010 **BORROW**									
0020 Spread, 200 HP dozer, no compaction, 2 mile RT haul									
0200 Common borrow	B-15	600	.047	C.Y.	12.75	1.52	3.93	18.20	21

31 05 16 – Aggregates for Earthwork

31 05 16.10 Borrow	Crew	Daily Output	Labor-Hours	Unit	Material	2018 Bare Costs Labor	Equipment	Total	Total Incl O&P
0010 **BORROW**									
0020 Spread, with 200 HP dozer, no compaction, 2 mile RT haul									
0100 Bank run gravel	B-15	600	.047	L.C.Y.	18.50	1.52	3.93	23.95	27.50
0300 Crushed stone (1.40 tons per C.Y.), 1-1/2"		600	.047		28	1.52	3.93	33.45	37.50
0320 3/4"		600	.047		28	1.52	3.93	33.45	37.50
0340 1/2"		600	.047		29.50	1.52	3.93	34.95	39.50
0360 3/8"		600	.047		32.50	1.52	3.93	37.95	43
0400 Sand, washed, concrete		600	.047		36	1.52	3.93	41.45	46.50
0500 Dead or bank sand		600	.047		18.15	1.52	3.93	23.60	27

31 11 Clearing and Grubbing

31 11 10 – Clearing and Grubbing Land

31 11 10.10 Clear and Grub Site	Crew	Daily Output	Labor-Hours	Unit	Material	2018 Bare Costs Labor	Equipment	Total	Total Incl O&P
0010 **CLEAR AND GRUB SITE**									
0020 Cut & chip light trees to 6" diam.	B-7	1	48	Acre		1,375	1,700	3,075	4,125
0150 Grub stumps and remove	B-30	2	12			405	990	1,395	1,775
0200 Cut & chip medium trees to 12" diam.	B-7	.70	68.571			1,975	2,425	4,400	5,925
0250 Grub stumps and remove	B-30	1	24			810	1,975	2,785	3,500
0300 Cut & chip heavy trees to 24" diam.	B-7	.30	160			4,600	5,625	10,225	13,800
0350 Grub stumps and remove	B-30	.50	48			1,625	3,950	5,575	7,000
0400 If burning is allowed, deduct cut & chip								40%	40%

31 13 Selective Tree and Shrub Removal and Trimming

31 13 13 – Selective Tree and Shrub Removal

31 13 13.10 Selective Clearing	Crew	Daily Output	Labor-Hours	Unit	Material	2018 Bare Costs Labor	Equipment	Total	Total Incl O&P
0010 **SELECTIVE CLEARING**									
0020 Clearing brush with brush saw	A-1C	.25	32	Acre		855	111	966	1,550
0100 By hand	1 Clab	.12	66.667			1,775		1,775	2,950
0300 With dozer, ball and chain, light clearing	B-11A	2	8			253	635	888	1,125
0400 Medium clearing	"	1.50	10.667			335	850	1,185	1,500

31 14 Earth Stripping and Stockpiling

31 14 13 – Soil Stripping and Stockpiling

31 14 13.23 Topsoil Stripping and Stockpiling	Crew	Daily Output	Labor-Hours	Unit	Material	2018 Bare Costs Labor	Equipment	Total	Total Incl O&P
0010 **TOPSOIL STRIPPING AND STOCKPILING**									
1400 Loam or topsoil, remove and stockpile on site									
1420 6" deep, 200' haul	B-10B	865	.009	C.Y.		.34	1.47	1.81	2.18
1430 300' haul		520	.015			.56	2.45	3.01	3.62
1440 500' haul		225	.036			1.30	5.65	6.95	8.35
1450 Alternate method: 6" deep, 200' haul		5090	.002	S.Y.		.06	.25	.31	.36
1460 500' haul		1325	.006	"		.22	.96	1.18	1.42
1500 Loam or topsoil, remove/stockpile on site									

31 14 Earth Stripping and Stockpiling

31 14 13 – Soil Stripping and Stockpiling

31 14 13.23 Topsoil Stripping and Stockpiling	Crew	Daily Output	Labor-Hours	Unit	Material	2018 Bare Costs Labor	Equipment	Total	Total Incl O&P
1510	By hand, 6" deep, 50' haul, less than 100 S.Y.	B-1	100	.240	S.Y.		6.55		6.55	10.90
1520	By skid steer, 6" deep, 100' haul, 101-500 S.Y.	B-62	500	.048			1.42	.35	1.77	2.73
1530	100' haul, 501-900 S.Y.	"	900	.027			.79	.19	.98	1.52
1540	200' haul, 901-1,100 S.Y.	B-63	1000	.040			1.07	.17	1.24	1.97
1550	By dozer, 200' haul, 1,101-4,000 S.Y.	B-10B	4000	.002	▼		.07	.32	.39	.47

31 22 Grading

31 22 16 – Fine Grading

31 22 16.10 Finish Grading

		Crew	Daily Output	Labor-Hours	Unit	Material	2018 Bare Costs Labor	Equipment	Total	Total Incl O&P
0010	**FINISH GRADING**									
0012	Finish grading area to be paved with grader, small area	B-11L	400	.040	S.Y.		1.27	1.61	2.88	3.86
0100	Large area		2000	.008			.25	.32	.57	.77
0200	Grade subgrade for base course, roadways	▼	3500	.005			.14	.18	.32	.44
1020	For large parking lots	B-32C	5000	.010			.31	.43	.74	.98
1050	For small irregular areas	"	2000	.024			.77	1.07	1.84	2.45
1100	Fine grade for slab on grade, machine	B-11L	1040	.015			.49	.62	1.11	1.48
1150	Hand grading	B-18	700	.034			.94	.06	1	1.62
1200	Fine grade granular base for sidewalks and bikeways	B-62	1200	.020	▼		.59	.15	.74	1.14
2550	Hand grade select gravel	2 Clab	60	.267	C.S.F.		7.10		7.10	11.85
3000	Hand grade select gravel, including compaction, 4" deep	B-18	555	.043	S.Y.		1.18	.07	1.25	2.05
3100	6" deep		400	.060			1.64	.10	1.74	2.84
3120	8" deep	▼	300	.080			2.19	.14	2.33	3.79
3300	Finishing grading slopes, gentle	B-11L	8900	.002			.06	.07	.13	.17
3310	Steep slopes	"	7100	.002	▼		.07	.09	.16	.22

31 23 Excavation and Fill

31 23 16 – Excavation

31 23 16.13 Excavating, Trench

		Crew	Daily Output	Labor-Hours	Unit	Material	2018 Bare Costs Labor	Equipment	Total	Total Incl O&P
0010	**EXCAVATING, TRENCH**									
0011	Or continuous footing									
0050	1' to 4' deep, 3/8 C.Y. excavator	B-11C	150	.107	B.C.Y.		3.37	2.08	5.45	7.90
0060	1/2 C.Y. excavator	B-11M	200	.080			2.53	1.92	4.45	6.30
0090	4' to 6' deep, 1/2 C.Y. excavator	"	200	.080			2.53	1.92	4.45	6.30
0100	5/8 C.Y. excavator	B-12Q	250	.064			2.04	2.31	4.35	5.90
0300	1/2 C.Y. excavator, truck mounted	B-12J	200	.080			2.55	4.35	6.90	9
1352	4' to 6' deep, 1/2 C.Y. excavator w/trench box	B-13H	188	.085			2.71	5.05	7.76	10.05
1354	5/8 C.Y. excavator	"	235	.068			2.17	4.05	6.22	8.05
1400	By hand with pick and shovel 2' to 6' deep, light soil	1 Clab	8	1			26.50		26.50	44.50
1500	Heavy soil	"	4	2	▼		53.50		53.50	89
5020	Loam & sandy clay with no sheeting or dewatering included									
5050	1' to 4' deep, 3/8 C.Y. tractor loader/backhoe	B-11C	162	.099	B.C.Y.		3.12	1.93	5.05	7.25
5060	1/2 C.Y. excavator	B-11M	216	.074			2.34	1.78	4.12	5.85
5080	4' to 6' deep, 1/2 C.Y. excavator	"	216	.074			2.34	1.78	4.12	5.85
5090	5/8 C.Y. excavator	B-12Q	276	.058			1.85	2.09	3.94	5.35
5130	1/2 C.Y. excavator, truck mounted	B-12J	216	.074			2.36	4.03	6.39	8.35
5352	4' to 6' deep, 1/2 C.Y. excavator w/trench box	B-13H	205	.078			2.49	4.64	7.13	9.20
5354	5/8 C.Y. excavator	"	257	.062	▼		1.98	3.70	5.68	7.35
6020	Sand & gravel with no sheeting or dewatering included									
6050	1' to 4' deep, 3/8 C.Y. excavator	B-11C	165	.097	B.C.Y.		3.07	1.89	4.96	7.15

For customer support on your Residential Costs with RSMeans data, call 800.448.8182.

627

31 23 Excavation and Fill

31 23 16 – Excavation

31 23 16.13 Excavating, Trench

		Crew	Daily Output	Labor-Hours	Unit	Material	2018 Bare Costs Labor	Equipment	Total	Total Incl O&P
6060	1/2 C.Y. excavator	B-11M	220	.073	B.C.Y.		2.30	1.75	4.05	5.75
6080	4' to 6' deep, 1/2 C.Y. excavator	"	220	.073			2.30	1.75	4.05	5.75
6090	5/8 C.Y. excavator	B-12Q	275	.058			1.85	2.10	3.95	5.40
6130	1/2 C.Y. excavator, truck mounted	B-12J	220	.073			2.32	3.95	6.27	8.20
6352	4' to 6' deep, 1/2 C.Y. excavator w/trench box	B-13H	209	.077			2.44	4.55	6.99	9.05
6354	5/8 C.Y. excavator	"	261	.061	▼		1.95	3.64	5.59	7.25
7020	Dense hard clay with no sheeting or dewatering included									
7050	1' to 4' deep, 3/8 C.Y. excavator	B-11C	132	.121	B.C.Y.		3.83	2.37	6.20	8.95
7060	1/2 C.Y. excavator	B-11M	176	.091			2.88	2.19	5.07	7.15
7080	4' to 6' deep, 1/2 C.Y. excavator	"	176	.091			2.88	2.19	5.07	7.15
7090	5/8 C.Y. excavator	B-12Q	220	.073			2.32	2.62	4.94	6.70
7130	1/2 C.Y. excavator, truck mounted	B-12J	176	.091	▼		2.90	4.94	7.84	10.25

31 23 16.14 Excavating, Utility Trench

		Crew	Daily Output	Labor-Hours	Unit	Material	2018 Bare Costs Labor	Equipment	Total	Total Incl O&P
0010	**EXCAVATING, UTILITY TRENCH**									
0011	Common earth									
0050	Trenching with chain trencher, 12 HP, operator walking									
0100	4" wide trench, 12" deep	B-53	800	.010	L.F.		.27	.08	.35	.53
1000	Backfill by hand including compaction, add									
1050	4" wide trench, 12" deep	A-1G	800	.010	L.F.		.27	.07	.34	.51

31 23 16.16 Structural Excavation for Minor Structures

		Crew	Daily Output	Labor-Hours	Unit	Material	2018 Bare Costs Labor	Equipment	Total	Total Incl O&P
0010	**STRUCTURAL EXCAVATION FOR MINOR STRUCTURES**									
0015	Hand, pits to 6' deep, sandy soil	1 Clab	8	1	B.C.Y.		26.50		26.50	44.50
0100	Heavy soil or clay		4	2			53.50		53.50	89
1100	Hand loading trucks from stock pile, sandy soil		12	.667			17.80		17.80	29.50
1300	Heavy soil or clay	▼	8	1	▼		26.50		26.50	44.50
1500	For wet or muck hand excavation, add to above								50%	50%

31 23 16.42 Excavating, Bulk Bank Measure

		Crew	Daily Output	Labor-Hours	Unit	Material	2018 Bare Costs Labor	Equipment	Total	Total Incl O&P
0010	**EXCAVATING, BULK BANK MEASURE**									
0011	Common earth piled									
0020	For loading onto trucks, add								15%	15%
0200	Excavator, hydraulic, crawler mtd., 1 C.Y. cap. = 100 C.Y./hr.	B-12A	800	.020	B.C.Y.		.64	.93	1.57	2.07
0310	Wheel mounted, 1/2 C.Y. cap. = 40 C.Y./hr.	B-12E	320	.050			1.59	1.36	2.95	4.14
1200	Front end loader, track mtd., 1-1/2 C.Y. cap. = 70 C.Y./hr.	B-10N	560	.014			.52	1.04	1.56	2
1500	Wheel mounted, 3/4 C.Y. cap. = 45 C.Y./hr.	B-10R	360	.022	▼		.81	.81	1.62	2.23
5000	Excavating, bulk bank measure, sandy clay & loam piled									
5020	For loading onto trucks, add								15%	15%
5100	Excavator, hydraulic, crawler mtd., 1 C.Y. cap. = 120 C.Y./hr.	B-12A	960	.017	B.C.Y.		.53	.77	1.30	1.73
5610	Wheel mounted, 1/2 C.Y. cap. = 44 C.Y./hr.	B-12E	352	.045	"		1.45	1.24	2.69	3.76
8000	For hauling excavated material, see Section 31 23 23.20									

31 23 23 – Fill

31 23 23.13 Backfill

		Crew	Daily Output	Labor-Hours	Unit	Material	2018 Bare Costs Labor	Equipment	Total	Total Incl O&P
0010	**BACKFILL**									
0015	By hand, no compaction, light soil	1 Clab	14	.571	L.C.Y.		15.25		15.25	25.50
0100	Heavy soil		11	.727	"		19.40		19.40	32.50
0300	Compaction in 6" layers, hand tamp, add to above	▼	20.60	.388	E.C.Y.		10.35		10.35	17.25
0400	Roller compaction operator walking, add	B-10A	100	.080			2.92	1.80	4.72	6.80
0500	Air tamp, add	B-9D	190	.211			5.70	1.42	7.12	11.05
0600	Vibrating plate, add	A-1D	60	.133			3.56	.53	4.09	6.50
0800	Compaction in 12" layers, hand tamp, add to above	1 Clab	34	.235	▼		6.30		6.30	10.45
1300	Dozer backfilling, bulk, up to 300' haul, no compaction	B-10B	1200	.007	L.C.Y.		.24	1.06	1.30	1.57
1400	Air tamped, add	B-11B	80	.200	E.C.Y.		6.20	3.56	9.76	14.20

628

For customer support on your Residential Costs with RSMeans data, call 800.448.8182.

31 23 Excavation and Fill

31 23 23 – Fill

31 23 23.16 Fill By Borrow and Utility Bedding

	31 23 23.16 Fill By Borrow and Utility Bedding	Crew	Daily Output	Labor-Hours	Unit	Material	2018 Bare Costs Labor	Equipment	Total	Total Incl O&P
0010	**FILL BY BORROW AND UTILITY BEDDING**									
0049	Utility bedding, for pipe & conduit, not incl. compaction									
0050	Crushed or screened bank run gravel	B-6	150	.160	L.C.Y.	21	4.74	2.08	27.82	33
0100	Crushed stone 3/4" to 1/2"		150	.160		28	4.74	2.08	34.82	40.50
0200	Sand, dead or bank	↓	150	.160	↓	18.15	4.74	2.08	24.97	30
0500	Compacting bedding in trench	A-1D	90	.089	E.C.Y.		2.37	.35	2.72	4.34
0600	If material source exceeds 2 miles, add for extra mileage.									
0610	See Section 31 23 23.20 for hauling mileage add.									

31 23 23.17 General Fill

		Crew	Daily Output	Labor-Hours	Unit	Material	2018 Bare Costs Labor	Equipment	Total	Total Incl O&P
0010	**GENERAL FILL**									
0011	Spread dumped material, no compaction									
0020	By dozer	B-10B	1000	.008	L.C.Y.		.29	1.27	1.56	1.88
0100	By hand	1 Clab	12	.667	"		17.80		17.80	29.50
0500	Gravel fill, compacted, under floor slabs, 4" deep	B-37	10000	.005	S.F.	.40	.14	.02	.56	.69
0600	6" deep		8600	.006		.60	.16	.02	.78	.94
0700	9" deep		7200	.007		1	.19	.02	1.21	1.44
0800	12" deep		6000	.008	↓	1.40	.23	.03	1.66	1.95
1000	Alternate pricing method, 4" deep		120	.400	E.C.Y.	30	11.40	1.26	42.66	53.50
1100	6" deep		160	.300		30	8.55	.95	39.50	48
1200	9" deep		200	.240		30	6.85	.76	37.61	45
1300	12" deep	↓	220	.218	↓	30	6.20	.69	36.89	44

31 23 23.20 Hauling

		Crew	Daily Output	Labor-Hours	Unit	Material	2018 Bare Costs Labor	Equipment	Total	Total Incl O&P
0010	**HAULING**									
0011	Excavated or borrow, loose cubic yards									
0012	no loading equipment, including hauling, waiting, loading/dumping									
0013	time per cycle (wait, load, travel, unload or dump & return)									
0014	8 C.Y. truck, 15 MPH avg., cycle 0.5 miles, 10 min. wait/ld./uld.	B-34A	320	.025	L.C.Y.		.81	1.06	1.87	2.49
0016	cycle 1 mile		272	.029			.95	1.24	2.19	2.93
0018	cycle 2 miles		208	.038			1.24	1.63	2.87	3.83
0020	cycle 4 miles		144	.056			1.79	2.35	4.14	5.55
0022	cycle 6 miles		112	.071			2.30	3.02	5.32	7.10
0024	cycle 8 miles		88	.091			2.93	3.85	6.78	9.05
0026	20 MPH avg., cycle 0.5 mile		336	.024			.77	1.01	1.78	2.37
0028	cycle 1 mile		296	.027			.87	1.14	2.01	2.69
0030	cycle 2 miles		240	.033			1.07	1.41	2.48	3.32
0032	cycle 4 miles		176	.045			1.46	1.92	3.38	4.53
0034	cycle 6 miles		136	.059			1.89	2.49	4.38	5.85
0036	cycle 8 miles		112	.071			2.30	3.02	5.32	7.10
0044	25 MPH avg., cycle 4 miles		192	.042			1.34	1.76	3.10	4.15
0046	cycle 6 miles		160	.050			1.61	2.12	3.73	4.98
0048	cycle 8 miles		128	.063			2.01	2.65	4.66	6.20
0050	30 MPH avg., cycle 4 miles		216	.037			1.19	1.57	2.76	3.68
0052	cycle 6 miles		176	.045			1.46	1.92	3.38	4.53
0054	cycle 8 miles		144	.056			1.79	2.35	4.14	5.55
0114	15 MPH avg., cycle 0.5 mile, 15 min. wait/ld./uld.		224	.036			1.15	1.51	2.66	3.55
0116	cycle 1 mile		200	.040			1.29	1.69	2.98	3.98
0118	cycle 2 miles		168	.048			1.53	2.02	3.55	4.74
0120	cycle 4 miles		120	.067			2.15	2.82	4.97	6.65
0122	cycle 6 miles		96	.083			2.68	3.53	6.21	8.30
0124	cycle 8 miles		80	.100			3.22	4.23	7.45	9.95
0126	20 MPH avg., cycle 0.5 mile		232	.034			1.11	1.46	2.57	3.44
0128	cycle 1 mile	↓	208	.038	↓		1.24	1.63	2.87	3.83

31 23 23.20 Hauling

		Crew	Daily Output	Labor-Hours	Unit	Material	2018 Bare Costs Labor	2018 Bare Costs Equipment	Total	Total Incl O&P
0130	cycle 2 miles	B-34A	184	.043	L.C.Y.		1.40	1.84	3.24	4.32
0132	cycle 4 miles		144	.056			1.79	2.35	4.14	5.55
0134	cycle 6 miles		112	.071			2.30	3.02	5.32	7.10
0136	cycle 8 miles		96	.083			2.68	3.53	6.21	8.30
0144	25 MPH avg., cycle 4 miles		152	.053			1.69	2.23	3.92	5.25
0146	cycle 6 miles		128	.063			2.01	2.65	4.66	6.20
0148	cycle 8 miles		112	.071			2.30	3.02	5.32	7.10
0150	30 MPH avg., cycle 4 miles		168	.048			1.53	2.02	3.55	4.74
0152	cycle 6 miles		144	.056			1.79	2.35	4.14	5.55
0154	cycle 8 miles		120	.067			2.15	2.82	4.97	6.65
0214	15 MPH avg., cycle 0.5 mile, 20 min. wait/ld./uld.		176	.045			1.46	1.92	3.38	4.53
0216	cycle 1 mile		160	.050			1.61	2.12	3.73	4.98
0218	cycle 2 miles		136	.059			1.89	2.49	4.38	5.85
0220	cycle 4 miles		104	.077			2.48	3.26	5.74	7.65
0222	cycle 6 miles		88	.091			2.93	3.85	6.78	9.05
0224	cycle 8 miles		72	.111			3.58	4.70	8.28	11.05
0226	20 MPH avg., cycle 0.5 mile		176	.045			1.46	1.92	3.38	4.53
0228	cycle 1 mile		168	.048			1.53	2.02	3.55	4.74
0230	cycle 2 miles		144	.056			1.79	2.35	4.14	5.55
0232	cycle 4 miles		120	.067			2.15	2.82	4.97	6.65
0234	cycle 6 miles		96	.083			2.68	3.53	6.21	8.30
0236	cycle 8 miles		88	.091			2.93	3.85	6.78	9.05
0244	25 MPH avg., cycle 4 miles		128	.063			2.01	2.65	4.66	6.20
0246	cycle 6 miles		112	.071			2.30	3.02	5.32	7.10
0248	cycle 8 miles		96	.083			2.68	3.53	6.21	8.30
0250	30 MPH avg., cycle 4 miles		136	.059			1.89	2.49	4.38	5.85
0252	cycle 6 miles		120	.067			2.15	2.82	4.97	6.65
0254	cycle 8 miles		104	.077			2.48	3.26	5.74	7.65
0314	15 MPH avg., cycle 0.5 mile, 25 min. wait/ld./uld.		144	.056			1.79	2.35	4.14	5.55
0316	cycle 1 mile		128	.063			2.01	2.65	4.66	6.20
0318	cycle 2 miles		112	.071			2.30	3.02	5.32	7.10
0320	cycle 4 miles		96	.083			2.68	3.53	6.21	8.30
0322	cycle 6 miles		80	.100			3.22	4.23	7.45	9.95
0324	cycle 8 miles		64	.125			4.03	5.30	9.33	12.45
0326	20 MPH avg., cycle 0.5 mile		144	.056			1.79	2.35	4.14	5.55
0328	cycle 1 mile		136	.059			1.89	2.49	4.38	5.85
0330	cycle 2 miles		120	.067			2.15	2.82	4.97	6.65
0332	cycle 4 miles		104	.077			2.48	3.26	5.74	7.65
0334	cycle 6 miles		88	.091			2.93	3.85	6.78	9.05
0336	cycle 8 miles		80	.100			3.22	4.23	7.45	9.95
0344	25 MPH avg., cycle 4 miles		112	.071			2.30	3.02	5.32	7.10
0346	cycle 6 miles		96	.083			2.68	3.53	6.21	8.30
0348	cycle 8 miles		88	.091			2.93	3.85	6.78	9.05
0350	30 MPH avg., cycle 4 miles		112	.071			2.30	3.02	5.32	7.10
0352	cycle 6 miles		104	.077			2.48	3.26	5.74	7.65
0354	cycle 8 miles		96	.083			2.68	3.53	6.21	8.30
0414	15 MPH avg., cycle 0.5 mile, 30 min. wait/ld./uld.		120	.067			2.15	2.82	4.97	6.65
0416	cycle 1 mile		112	.071			2.30	3.02	5.32	7.10
0418	cycle 2 miles		96	.083			2.68	3.53	6.21	8.30
0420	cycle 4 miles		80	.100			3.22	4.23	7.45	9.95
0422	cycle 6 miles		72	.111			3.58	4.70	8.28	11.05
0424	cycle 8 miles		64	.125			4.03	5.30	9.33	12.45
0426	20 MPH avg., cycle 0.5 mile		120	.067			2.15	2.82	4.97	6.65

31 23 23.20 Hauling	Crew	Daily Output	Labor-Hours	Unit	Material	2018 Bare Costs Labor	2018 Bare Costs Equipment	Total	Total Incl O&P
0428	cycle 1 mile	B-34A	112	.071	L.C.Y.		2.30	3.02	5.32	7.10
0430	cycle 2 miles		104	.077			2.48	3.26	5.74	7.65
0432	cycle 4 miles		88	.091			2.93	3.85	6.78	9.05
0434	cycle 6 miles		80	.100			3.22	4.23	7.45	9.95
0436	cycle 8 miles		72	.111			3.58	4.70	8.28	11.05
0444	25 MPH avg., cycle 4 miles		96	.083			2.68	3.53	6.21	8.30
0446	cycle 6 miles		88	.091			2.93	3.85	6.78	9.05
0448	cycle 8 miles		80	.100			3.22	4.23	7.45	9.95
0450	30 MPH avg., cycle 4 miles		96	.083			2.68	3.53	6.21	8.30
0452	cycle 6 miles		88	.091			2.93	3.85	6.78	9.05
0454	cycle 8 miles		80	.100			3.22	4.23	7.45	9.95
0514	15 MPH avg., cycle 0.5 mile, 35 min. wait/ld./uld.		104	.077			2.48	3.26	5.74	7.65
0516	cycle 1 mile		96	.083			2.68	3.53	6.21	8.30
0518	cycle 2 miles		88	.091			2.93	3.85	6.78	9.05
0520	cycle 4 miles		72	.111			3.58	4.70	8.28	11.05
0522	cycle 6 miles		64	.125			4.03	5.30	9.33	12.45
0524	cycle 8 miles		56	.143			4.60	6.05	10.65	14.20
0526	20 MPH avg., cycle 0.5 mile		104	.077			2.48	3.26	5.74	7.65
0528	cycle 1 mile		96	.083			2.68	3.53	6.21	8.30
0530	cycle 2 miles		96	.083			2.68	3.53	6.21	8.30
0532	cycle 4 miles		80	.100			3.22	4.23	7.45	9.95
0534	cycle 6 miles		72	.111			3.58	4.70	8.28	11.05
0536	cycle 8 miles		64	.125			4.03	5.30	9.33	12.45
0544	25 MPH avg., cycle 4 miles		88	.091			2.93	3.85	6.78	9.05
0546	cycle 6 miles		80	.100			3.22	4.23	7.45	9.95
0548	cycle 8 miles		72	.111			3.58	4.70	8.28	11.05
0550	30 MPH avg., cycle 4 miles		88	.091			2.93	3.85	6.78	9.05
0552	cycle 6 miles		80	.100			3.22	4.23	7.45	9.95
0554	cycle 8 miles		72	.111			3.58	4.70	8.28	11.05
1014	12 C.Y. truck, cycle 0.5 mile, 15 MPH avg., 15 min. wait/ld./uld.	B-34B	336	.024			.77	1.62	2.39	3.04
1016	cycle 1 mile		300	.027			.86	1.81	2.67	3.40
1018	cycle 2 miles		252	.032			1.02	2.15	3.17	4.05
1020	cycle 4 miles		180	.044			1.43	3.02	4.45	5.70
1022	cycle 6 miles		144	.056			1.79	3.77	5.56	7.10
1024	cycle 8 miles		120	.067			2.15	4.52	6.67	8.50
1025	cycle 10 miles		96	.083			2.68	5.65	8.33	10.60
1026	20 MPH avg., cycle 0.5 mile		348	.023			.74	1.56	2.30	2.94
1028	cycle 1 mile		312	.026			.83	1.74	2.57	3.27
1030	cycle 2 miles		276	.029			.93	1.97	2.90	3.70
1032	cycle 4 miles		216	.037			1.19	2.51	3.70	4.72
1034	cycle 6 miles		168	.048			1.53	3.23	4.76	6.05
1036	cycle 8 miles		144	.056			1.79	3.77	5.56	7.10
1038	cycle 10 miles		120	.067			2.15	4.52	6.67	8.50
1040	25 MPH avg., cycle 4 miles		228	.035			1.13	2.38	3.51	4.48
1042	cycle 6 miles		192	.042			1.34	2.83	4.17	5.30
1044	cycle 8 miles		168	.048			1.53	3.23	4.76	6.05
1046	cycle 10 miles		144	.056			1.79	3.77	5.56	7.10
1050	30 MPH avg., cycle 4 miles		252	.032			1.02	2.15	3.17	4.05
1052	cycle 6 miles		216	.037			1.19	2.51	3.70	4.72
1054	cycle 8 miles		180	.044			1.43	3.02	4.45	5.70
1056	cycle 10 miles		156	.051			1.65	3.48	5.13	6.55
1060	35 MPH avg., cycle 4 miles		264	.030			.98	2.06	3.04	3.87
1062	cycle 6 miles		228	.035			1.13	2.38	3.51	4.48

31 23 23.20 Hauling		Crew	Daily Output	Labor-Hours	Unit	Material	2018 Bare Costs Labor	Equipment	Total	Total Incl O&P
1064	cycle 8 miles	B-34B	204	.039	L.C.Y.		1.26	2.66	3.92	5
1066	cycle 10 miles		180	.044			1.43	3.02	4.45	5.70
1068	cycle 20 miles		120	.067			2.15	4.52	6.67	8.50
1069	cycle 30 miles		84	.095			3.07	6.45	9.52	12.15
1070	cycle 40 miles		72	.111			3.58	7.55	11.13	14.20
1072	40 MPH avg., cycle 6 miles		240	.033			1.07	2.26	3.33	4.26
1074	cycle 8 miles		216	.037			1.19	2.51	3.70	4.72
1076	cycle 10 miles		192	.042			1.34	2.83	4.17	5.30
1078	cycle 20 miles		120	.067			2.15	4.52	6.67	8.50
1080	cycle 30 miles		96	.083			2.68	5.65	8.33	10.60
1082	cycle 40 miles		72	.111			3.58	7.55	11.13	14.20
1084	cycle 50 miles		60	.133			4.29	9.05	13.34	17
1094	45 MPH avg., cycle 8 miles		216	.037			1.19	2.51	3.70	4.72
1096	cycle 10 miles		204	.039			1.26	2.66	3.92	5
1098	cycle 20 miles		132	.061			1.95	4.11	6.06	7.75
1100	cycle 30 miles		108	.074			2.39	5.05	7.44	9.50
1102	cycle 40 miles		84	.095			3.07	6.45	9.52	12.15
1104	cycle 50 miles		72	.111			3.58	7.55	11.13	14.20
1106	50 MPH avg., cycle 10 miles		216	.037			1.19	2.51	3.70	4.72
1108	cycle 20 miles		144	.056			1.79	3.77	5.56	7.10
1110	cycle 30 miles		108	.074			2.39	5.05	7.44	9.50
1112	cycle 40 miles		84	.095			3.07	6.45	9.52	12.15
1114	cycle 50 miles		72	.111			3.58	7.55	11.13	14.20
1214	15 MPH avg., cycle 0.5 mile, 20 min. wait/ld./uld.		264	.030			.98	2.06	3.04	3.87
1216	cycle 1 mile		240	.033			1.07	2.26	3.33	4.26
1218	cycle 2 miles		204	.039			1.26	2.66	3.92	5
1220	cycle 4 miles		156	.051			1.65	3.48	5.13	6.55
1222	cycle 6 miles		132	.061			1.95	4.11	6.06	7.75
1224	cycle 8 miles		108	.074			2.39	5.05	7.44	9.50
1225	cycle 10 miles		96	.083			2.68	5.65	8.33	10.60
1226	20 MPH avg., cycle 0.5 mile		264	.030			.98	2.06	3.04	3.87
1228	cycle 1 mile		252	.032			1.02	2.15	3.17	4.05
1230	cycle 2 miles		216	.037			1.19	2.51	3.70	4.72
1232	cycle 4 miles		180	.044			1.43	3.02	4.45	5.70
1234	cycle 6 miles		144	.056			1.79	3.77	5.56	7.10
1236	cycle 8 miles		132	.061			1.95	4.11	6.06	7.75
1238	cycle 10 miles		108	.074			2.39	5.05	7.44	9.50
1240	25 MPH avg., cycle 4 miles		192	.042			1.34	2.83	4.17	5.30
1242	cycle 6 miles		168	.048			1.53	3.23	4.76	6.05
1244	cycle 8 miles		144	.056			1.79	3.77	5.56	7.10
1246	cycle 10 miles		132	.061			1.95	4.11	6.06	7.75
1250	30 MPH avg., cycle 4 miles		204	.039			1.26	2.66	3.92	5
1252	cycle 6 miles		180	.044			1.43	3.02	4.45	5.70
1254	cycle 8 miles		156	.051			1.65	3.48	5.13	6.55
1256	cycle 10 miles		144	.056			1.79	3.77	5.56	7.10
1260	35 MPH avg., cycle 4 miles		216	.037			1.19	2.51	3.70	4.72
1262	cycle 6 miles		192	.042			1.34	2.83	4.17	5.30
1264	cycle 8 miles		168	.048			1.53	3.23	4.76	6.05
1266	cycle 10 miles		156	.051			1.65	3.48	5.13	6.55
1268	cycle 20 miles		108	.074			2.39	5.05	7.44	9.50
1269	cycle 30 miles		72	.111			3.58	7.55	11.13	14.20
1270	cycle 40 miles		60	.133			4.29	9.05	13.34	17
1272	40 MPH avg., cycle 6 miles		192	.042			1.34	2.83	4.17	5.30

31 23 23 – Fill

31 23 23.20 Hauling	Crew	Daily Output	Labor-Hours	Unit	Material	2018 Bare Costs Labor	2018 Bare Costs Equipment	Total	Total Incl O&P
1274	cycle 8 miles	B-34B	180	.044	L.C.Y.		1.43	3.02	4.45	5.70
1276	cycle 10 miles		156	.051			1.65	3.48	5.13	6.55
1278	cycle 20 miles		108	.074			2.39	5.05	7.44	9.50
1280	cycle 30 miles		84	.095			3.07	6.45	9.52	12.15
1282	cycle 40 miles		72	.111			3.58	7.55	11.13	14.20
1284	cycle 50 miles		60	.133			4.29	9.05	13.34	17
1294	45 MPH avg., cycle 8 miles		180	.044			1.43	3.02	4.45	5.70
1296	cycle 10 miles		168	.048			1.53	3.23	4.76	6.05
1298	cycle 20 miles		120	.067			2.15	4.52	6.67	8.50
1300	cycle 30 miles		96	.083			2.68	5.65	8.33	10.60
1302	cycle 40 miles		72	.111			3.58	7.55	11.13	14.20
1304	cycle 50 miles		60	.133			4.29	9.05	13.34	17
1306	50 MPH avg., cycle 10 miles		180	.044			1.43	3.02	4.45	5.70
1308	cycle 20 miles		132	.061			1.95	4.11	6.06	7.75
1310	cycle 30 miles		96	.083			2.68	5.65	8.33	10.60
1312	cycle 40 miles		84	.095			3.07	6.45	9.52	12.15
1314	cycle 50 miles		72	.111			3.58	7.55	11.13	14.20
1414	15 MPH avg., cycle 0.5 mile, 25 min. wait/ld./uld.		204	.039			1.26	2.66	3.92	5
1416	cycle 1 mile		192	.042			1.34	2.83	4.17	5.30
1418	cycle 2 miles		168	.048			1.53	3.23	4.76	6.05
1420	cycle 4 miles		132	.061			1.95	4.11	6.06	7.75
1422	cycle 6 miles		120	.067			2.15	4.52	6.67	8.50
1424	cycle 8 miles		96	.083			2.68	5.65	8.33	10.60
1425	cycle 10 miles		84	.095			3.07	6.45	9.52	12.15
1426	20 MPH avg., cycle 0.5 mile		216	.037			1.19	2.51	3.70	4.72
1428	cycle 1 mile		204	.039			1.26	2.66	3.92	5
1430	cycle 2 miles		180	.044			1.43	3.02	4.45	5.70
1432	cycle 4 miles		156	.051			1.65	3.48	5.13	6.55
1434	cycle 6 miles		132	.061			1.95	4.11	6.06	7.75
1436	cycle 8 miles		120	.067			2.15	4.52	6.67	8.50
1438	cycle 10 miles		96	.083			2.68	5.65	8.33	10.60
1440	25 MPH avg., cycle 4 miles		168	.048			1.53	3.23	4.76	6.05
1442	cycle 6 miles		144	.056			1.79	3.77	5.56	7.10
1444	cycle 8 miles		132	.061			1.95	4.11	6.06	7.75
1446	cycle 10 miles		108	.074			2.39	5.05	7.44	9.50
1450	30 MPH avg., cycle 4 miles		168	.048			1.53	3.23	4.76	6.05
1452	cycle 6 miles		156	.051			1.65	3.48	5.13	6.55
1454	cycle 8 miles		132	.061			1.95	4.11	6.06	7.75
1456	cycle 10 miles		120	.067			2.15	4.52	6.67	8.50
1460	35 MPH avg., cycle 4 miles		180	.044			1.43	3.02	4.45	5.70
1462	cycle 6 miles		156	.051			1.65	3.48	5.13	6.55
1464	cycle 8 miles		144	.056			1.79	3.77	5.56	7.10
1466	cycle 10 miles		132	.061			1.95	4.11	6.06	7.75
1468	cycle 20 miles		96	.083			2.68	5.65	8.33	10.60
1469	cycle 30 miles		72	.111			3.58	7.55	11.13	14.20
1470	cycle 40 miles		60	.133			4.29	9.05	13.34	17
1472	40 MPH avg., cycle 6 miles		168	.048			1.53	3.23	4.76	6.05
1474	cycle 8 miles		156	.051			1.65	3.48	5.13	6.55
1476	cycle 10 miles		144	.056			1.79	3.77	5.56	7.10
1478	cycle 20 miles		96	.083			2.68	5.65	8.33	10.60
1480	cycle 30 miles		84	.095			3.07	6.45	9.52	12.15
1482	cycle 40 miles		60	.133			4.29	9.05	13.34	17
1484	cycle 50 miles		60	.133			4.29	9.05	13.34	17

633

For customer support on your Residential Costs with RSMeans data, call 800.448.8182.

31 23 23.20 Hauling	Crew	Daily Output	Labor-Hours	Unit	Material	2018 Bare Costs Labor	2018 Bare Costs Equipment	Total	Total Incl O&P
1494	45 MPH avg., cycle 8 miles	B-34B	156	.051	L.C.Y.		1.65	3.48	5.13	6.55
1496	cycle 10 miles		144	.056			1.79	3.77	5.56	7.10
1498	cycle 20 miles		108	.074			2.39	5.05	7.44	9.50
1500	cycle 30 miles		84	.095			3.07	6.45	9.52	12.15
1502	cycle 40 miles		72	.111			3.58	7.55	11.13	14.20
1504	cycle 50 miles		60	.133			4.29	9.05	13.34	17
1506	50 MPH avg., cycle 10 miles		156	.051			1.65	3.48	5.13	6.55
1508	cycle 20 miles		120	.067			2.15	4.52	6.67	8.50
1510	cycle 30 miles		96	.083			2.68	5.65	8.33	10.60
1512	cycle 40 miles		72	.111			3.58	7.55	11.13	14.20
1514	cycle 50 miles		60	.133			4.29	9.05	13.34	17
1614	15 MPH avg., cycle 0.5 mile, 30 min. wait/ld./uld.		180	.044			1.43	3.02	4.45	5.70
1616	cycle 1 mile		168	.048			1.53	3.23	4.76	6.05
1618	cycle 2 miles		144	.056			1.79	3.77	5.56	7.10
1620	cycle 4 miles		120	.067			2.15	4.52	6.67	8.50
1622	cycle 6 miles		108	.074			2.39	5.05	7.44	9.50
1624	cycle 8 miles		84	.095			3.07	6.45	9.52	12.15
1625	cycle 10 miles		84	.095			3.07	6.45	9.52	12.15
1626	20 MPH avg., cycle 0.5 mile		180	.044			1.43	3.02	4.45	5.70
1628	cycle 1 mile		168	.048			1.53	3.23	4.76	6.05
1630	cycle 2 miles		156	.051			1.65	3.48	5.13	6.55
1632	cycle 4 miles		132	.061			1.95	4.11	6.06	7.75
1634	cycle 6 miles		120	.067			2.15	4.52	6.67	8.50
1636	cycle 8 miles		108	.074			2.39	5.05	7.44	9.50
1638	cycle 10 miles		96	.083			2.68	5.65	8.33	10.60
1640	25 MPH avg., cycle 4 miles		144	.056			1.79	3.77	5.56	7.10
1642	cycle 6 miles		132	.061			1.95	4.11	6.06	7.75
1644	cycle 8 miles		108	.074			2.39	5.05	7.44	9.50
1646	cycle 10 miles		108	.074			2.39	5.05	7.44	9.50
1650	30 MPH avg., cycle 4 miles		144	.056			1.79	3.77	5.56	7.10
1652	cycle 6 miles		132	.061			1.95	4.11	6.06	7.75
1654	cycle 8 miles		120	.067			2.15	4.52	6.67	8.50
1656	cycle 10 miles		108	.074			2.39	5.05	7.44	9.50
1660	35 MPH avg., cycle 4 miles		156	.051			1.65	3.48	5.13	6.55
1662	cycle 6 miles		144	.056			1.79	3.77	5.56	7.10
1664	cycle 8 miles		132	.061			1.95	4.11	6.06	7.75
1666	cycle 10 miles		120	.067			2.15	4.52	6.67	8.50
1668	cycle 20 miles		84	.095			3.07	6.45	9.52	12.15
1669	cycle 30 miles		72	.111			3.58	7.55	11.13	14.20
1670	cycle 40 miles		60	.133			4.29	9.05	13.34	17
1672	40 MPH avg., cycle 6 miles		144	.056			1.79	3.77	5.56	7.10
1674	cycle 8 miles		132	.061			1.95	4.11	6.06	7.75
1676	cycle 10 miles		120	.067			2.15	4.52	6.67	8.50
1678	cycle 20 miles		96	.083			2.68	5.65	8.33	10.60
1680	cycle 30 miles		72	.111			3.58	7.55	11.13	14.20
1682	cycle 40 miles		60	.133			4.29	9.05	13.34	17
1684	cycle 50 miles		48	.167			5.35	11.30	16.65	21.50
1694	45 MPH avg., cycle 8 miles		144	.056			1.79	3.77	5.56	7.10
1696	cycle 10 miles		132	.061			1.95	4.11	6.06	7.75
1698	cycle 20 miles		96	.083			2.68	5.65	8.33	10.60
1700	cycle 30 miles		84	.095			3.07	6.45	9.52	12.15
1702	cycle 40 miles		60	.133			4.29	9.05	13.34	17
1704	cycle 50 miles		60	.133			4.29	9.05	13.34	17

31 23 23.20 Hauling	Crew	Daily Output	Labor-Hours	Unit	Material	2018 Bare Costs Labor	2018 Bare Costs Equipment	Total	Total Incl O&P
1706 50 MPH avg., cycle 10 miles	B-34B	132	.061	L.C.Y.		1.95	4.11	6.06	7.75
1708 cycle 20 miles		108	.074			2.39	5.05	7.44	9.50
1710 cycle 30 miles		84	.095			3.07	6.45	9.52	12.15
1712 cycle 40 miles		72	.111			3.58	7.55	11.13	14.20
1714 cycle 50 miles		60	.133			4.29	9.05	13.34	17
2000 Hauling, 8 C.Y. truck, small project cost per hour	B-34A	8	1	Hr.		32	42.50	74.50	99.50
2100 12 C.Y. truck	B-34B	8	1			32	68	100	128
2150 16.5 C.Y. truck	B-34C	8	1			32	76	108	137
2175 18 C.Y. 8 wheel truck	B-34I	8	1			32	88.50	120.50	150
2200 20 C.Y. truck	B-34D	8	1			32	78	110	139
9014 18 C.Y. truck, 8 wheels,15 min. wait/ld./uld.,15 MPH, cycle 0.5 mi.	B-34I	504	.016	L.C.Y.		.51	1.40	1.91	2.38
9016 cycle 1 mile		450	.018			.57	1.57	2.14	2.67
9018 cycle 2 miles		378	.021			.68	1.87	2.55	3.17
9020 cycle 4 miles		270	.030			.95	2.61	3.56	4.45
9022 cycle 6 miles		216	.037			1.19	3.27	4.46	5.55
9024 cycle 8 miles		180	.044			1.43	3.92	5.35	6.65
9025 cycle 10 miles		144	.056			1.79	4.90	6.69	8.35
9026 20 MPH avg., cycle 0.5 mile		522	.015			.49	1.35	1.84	2.30
9028 cycle 1 mile		468	.017			.55	1.51	2.06	2.57
9030 cycle 2 miles		414	.019			.62	1.71	2.33	2.90
9032 cycle 4 miles		324	.025			.80	2.18	2.98	3.71
9034 cycle 6 miles		252	.032			1.02	2.80	3.82	4.76
9036 cycle 8 miles		216	.037			1.19	3.27	4.46	5.55
9038 cycle 10 miles		180	.044			1.43	3.92	5.35	6.65
9040 25 MPH avg., cycle 4 miles		342	.023			.75	2.06	2.81	3.51
9042 cycle 6 miles		288	.028			.89	2.45	3.34	4.17
9044 cycle 8 miles		252	.032			1.02	2.80	3.82	4.76
9046 cycle 10 miles		216	.037			1.19	3.27	4.46	5.55
9050 30 MPH avg., cycle 4 miles		378	.021			.68	1.87	2.55	3.17
9052 cycle 6 miles		324	.025			.80	2.18	2.98	3.71
9054 cycle 8 miles		270	.030			.95	2.61	3.56	4.45
9056 cycle 10 miles		234	.034			1.10	3.02	4.12	5.15
9060 35 MPH avg., cycle 4 miles		396	.020			.65	1.78	2.43	3.03
9062 cycle 6 miles		342	.023			.75	2.06	2.81	3.51
9064 cycle 8 miles		288	.028			.89	2.45	3.34	4.17
9066 cycle 10 miles		270	.030			.95	2.61	3.56	4.45
9068 cycle 20 miles		162	.049			1.59	4.36	5.95	7.40
9070 cycle 30 miles		126	.063			2.04	5.60	7.64	9.50
9072 cycle 40 miles		90	.089			2.86	7.85	10.71	13.35
9074 40 MPH avg., cycle 6 miles		360	.022			.72	1.96	2.68	3.34
9076 cycle 8 miles		324	.025			.80	2.18	2.98	3.71
9078 cycle 10 miles		288	.028			.89	2.45	3.34	4.17
9080 cycle 20 miles		180	.044			1.43	3.92	5.35	6.65
9082 cycle 30 miles		144	.056			1.79	4.90	6.69	8.35
9084 cycle 40 miles		108	.074			2.39	6.55	8.94	11.15
9086 cycle 50 miles		90	.089			2.86	7.85	10.71	13.35
9094 45 MPH avg., cycle 8 miles		324	.025			.80	2.18	2.98	3.71
9096 cycle 10 miles		306	.026			.84	2.31	3.15	3.93
9098 cycle 20 miles		198	.040			1.30	3.57	4.87	6.05
9100 cycle 30 miles		144	.056			1.79	4.90	6.69	8.35
9102 cycle 40 miles		126	.063			2.04	5.60	7.64	9.50
9104 cycle 50 miles		108	.074			2.39	6.55	8.94	11.15
9106 50 MPH avg., cycle 10 miles		324	.025			.80	2.18	2.98	3.71

31 23 23.20 Hauling		Crew	Daily Output	Labor-Hours	Unit	Material	2018 Bare Costs Labor	2018 Bare Costs Equipment	Total	Total Incl O&P
9108	cycle 20 miles	B-34I	216	.037	L.C.Y.		1.19	3.27	4.46	5.55
9110	cycle 30 miles		162	.049			1.59	4.36	5.95	7.40
9112	cycle 40 miles		126	.063			2.04	5.60	7.64	9.50
9114	cycle 50 miles		108	.074			2.39	6.55	8.94	11.15
9214	20 min. wait/ld./uld.,15 MPH, cycle 0.5 mi.		396	.020			.65	1.78	2.43	3.03
9216	cycle 1 mile		360	.022			.72	1.96	2.68	3.34
9218	cycle 2 miles		306	.026			.84	2.31	3.15	3.93
9220	cycle 4 miles		234	.034			1.10	3.02	4.12	5.15
9222	cycle 6 miles		198	.040			1.30	3.57	4.87	6.05
9224	cycle 8 miles		162	.049			1.59	4.36	5.95	7.40
9225	cycle 10 miles		144	.056			1.79	4.90	6.69	8.35
9226	20 MPH avg., cycle 0.5 mile		396	.020			.65	1.78	2.43	3.03
9228	cycle 1 mile		378	.021			.68	1.87	2.55	3.17
9230	cycle 2 miles		324	.025			.80	2.18	2.98	3.71
9232	cycle 4 miles		270	.030			.95	2.61	3.56	4.45
9234	cycle 6 miles		216	.037			1.19	3.27	4.46	5.55
9236	cycle 8 miles		198	.040			1.30	3.57	4.87	6.05
9238	cycle 10 miles		162	.049			1.59	4.36	5.95	7.40
9240	25 MPH avg., cycle 4 miles		288	.028			.89	2.45	3.34	4.17
9242	cycle 6 miles		252	.032			1.02	2.80	3.82	4.76
9244	cycle 8 miles		216	.037			1.19	3.27	4.46	5.55
9246	cycle 10 miles		198	.040			1.30	3.57	4.87	6.05
9250	30 MPH avg., cycle 4 miles		306	.026			.84	2.31	3.15	3.93
9252	cycle 6 miles		270	.030			.95	2.61	3.56	4.45
9254	cycle 8 miles		234	.034			1.10	3.02	4.12	5.15
9256	cycle 10 miles		216	.037			1.19	3.27	4.46	5.55
9260	35 MPH avg., cycle 4 miles		324	.025			.80	2.18	2.98	3.71
9262	cycle 6 miles		288	.028			.89	2.45	3.34	4.17
9264	cycle 8 miles		252	.032			1.02	2.80	3.82	4.76
9266	cycle 10 miles		234	.034			1.10	3.02	4.12	5.15
9268	cycle 20 miles		162	.049			1.59	4.36	5.95	7.40
9270	cycle 30 miles		108	.074			2.39	6.55	8.94	11.15
9272	cycle 40 miles		90	.089			2.86	7.85	10.71	13.35
9274	40 MPH avg., cycle 6 miles		288	.028			.89	2.45	3.34	4.17
9276	cycle 8 miles		270	.030			.95	2.61	3.56	4.45
9278	cycle 10 miles		234	.034			1.10	3.02	4.12	5.15
9280	cycle 20 miles		162	.049			1.59	4.36	5.95	7.40
9282	cycle 30 miles		126	.063			2.04	5.60	7.64	9.50
9284	cycle 40 miles		108	.074			2.39	6.55	8.94	11.15
9286	cycle 50 miles		90	.089			2.86	7.85	10.71	13.35
9294	45 MPH avg., cycle 8 miles		270	.030			.95	2.61	3.56	4.45
9296	cycle 10 miles		252	.032			1.02	2.80	3.82	4.76
9298	cycle 20 miles		180	.044			1.43	3.92	5.35	6.65
9300	cycle 30 miles		144	.056			1.79	4.90	6.69	8.35
9302	cycle 40 miles		108	.074			2.39	6.55	8.94	11.15
9304	cycle 50 miles		90	.089			2.86	7.85	10.71	13.35
9306	50 MPH avg., cycle 10 miles		270	.030			.95	2.61	3.56	4.45
9308	cycle 20 miles		198	.040			1.30	3.57	4.87	6.05
9310	cycle 30 miles		144	.056			1.79	4.90	6.69	8.35
9312	cycle 40 miles		126	.063			2.04	5.60	7.64	9.50
9314	cycle 50 miles		108	.074			2.39	6.55	8.94	11.15
9414	25 min. wait/ld./uld.,15 MPH, cycle 0.5 mi.		306	.026			.84	2.31	3.15	3.93
9416	cycle 1 mile		288	.028			.89	2.45	3.34	4.17

For customer support on your Residential Costs with RSMeans data, call 800.448.8182.

31 23 23 – Fill

31 23 23.20 Hauling		Crew	Daily Output	Labor-Hours	Unit	Material	2018 Bare Costs Labor	2018 Bare Costs Equipment	Total	Total Incl O&P
9418	cycle 2 miles	B-34I	252	.032	L.C.Y.		1.02	2.80	3.82	4.76
9420	cycle 4 miles		198	.040			1.30	3.57	4.87	6.05
9422	cycle 6 miles		180	.044			1.43	3.92	5.35	6.65
9424	cycle 8 miles		144	.056			1.79	4.90	6.69	8.35
9425	cycle 10 miles		126	.063			2.04	5.60	7.64	9.50
9426	20 MPH avg., cycle 0.5 mile		324	.025			.80	2.18	2.98	3.71
9428	cycle 1 mile		306	.026			.84	2.31	3.15	3.93
9430	cycle 2 miles		270	.030			.95	2.61	3.56	4.45
9432	cycle 4 miles		234	.034			1.10	3.02	4.12	5.15
9434	cycle 6 miles		198	.040			1.30	3.57	4.87	6.05
9436	cycle 8 miles		180	.044			1.43	3.92	5.35	6.65
9438	cycle 10 miles		144	.056			1.79	4.90	6.69	8.35
9440	25 MPH avg., cycle 4 miles		252	.032			1.02	2.80	3.82	4.76
9442	cycle 6 miles		216	.037			1.19	3.27	4.46	5.55
9444	cycle 8 miles		198	.040			1.30	3.57	4.87	6.05
9446	cycle 10 miles		180	.044			1.43	3.92	5.35	6.65
9450	30 MPH avg., cycle 4 miles		252	.032			1.02	2.80	3.82	4.76
9452	cycle 6 miles		234	.034			1.10	3.02	4.12	5.15
9454	cycle 8 miles		198	.040			1.30	3.57	4.87	6.05
9456	cycle 10 miles		180	.044			1.43	3.92	5.35	6.65
9460	35 MPH avg., cycle 4 miles		270	.030			.95	2.61	3.56	4.45
9462	cycle 6 miles		234	.034			1.10	3.02	4.12	5.15
9464	cycle 8 miles		216	.037			1.19	3.27	4.46	5.55
9466	cycle 10 miles		198	.040			1.30	3.57	4.87	6.05
9468	cycle 20 miles		144	.056			1.79	4.90	6.69	8.35
9470	cycle 30 miles		108	.074			2.39	6.55	8.94	11.15
9472	cycle 40 miles		90	.089			2.86	7.85	10.71	13.35
9474	40 MPH avg., cycle 6 miles		252	.032			1.02	2.80	3.82	4.76
9476	cycle 8 miles		234	.034			1.10	3.02	4.12	5.15
9478	cycle 10 miles		216	.037			1.19	3.27	4.46	5.55
9480	cycle 20 miles		144	.056			1.79	4.90	6.69	8.35
9482	cycle 30 miles		126	.063			2.04	5.60	7.64	9.50
9484	cycle 40 miles		90	.089			2.86	7.85	10.71	13.35
9486	cycle 50 miles		90	.089			2.86	7.85	10.71	13.35
9494	45 MPH avg., cycle 8 miles		234	.034			1.10	3.02	4.12	5.15
9496	cycle 10 miles		216	.037			1.19	3.27	4.46	5.55
9498	cycle 20 miles		162	.049			1.59	4.36	5.95	7.40
9500	cycle 30 miles		126	.063			2.04	5.60	7.64	9.50
9502	cycle 40 miles		108	.074			2.39	6.55	8.94	11.15
9504	cycle 50 miles		90	.089			2.86	7.85	10.71	13.35
9506	50 MPH avg., cycle 10 miles		234	.034			1.10	3.02	4.12	5.15
9508	cycle 20 miles		180	.044			1.43	3.92	5.35	6.65
9510	cycle 30 miles		144	.056			1.79	4.90	6.69	8.35
9512	cycle 40 miles		108	.074			2.39	6.55	8.94	11.15
9514	cycle 50 miles		90	.089			2.86	7.85	10.71	13.35
9614	30 min. wait/ld./uld.,15 MPH, cycle 0.5 mi.		270	.030			.95	2.61	3.56	4.45
9616	cycle 1 mile		252	.032			1.02	2.80	3.82	4.76
9618	cycle 2 miles		216	.037			1.19	3.27	4.46	5.55
9620	cycle 4 miles		180	.044			1.43	3.92	5.35	6.65
9622	cycle 6 miles		162	.049			1.59	4.36	5.95	7.40
9624	cycle 8 miles		126	.063			2.04	5.60	7.64	9.50
9625	cycle 10 miles		126	.063			2.04	5.60	7.64	9.50
9626	20 MPH avg., cycle 0.5 mile		270	.030			.95	2.61	3.56	4.45

31 23 23.20 Hauling

		Crew	Daily Output	Labor-Hours	Unit	Material	Labor	Equipment	Total	Total Incl O&P
							2018 Bare Costs			
9628	cycle 1 mile	B-34I	252	.032	L.C.Y.		1.02	2.80	3.82	4.76
9630	cycle 2 miles		234	.034			1.10	3.02	4.12	5.15
9632	cycle 4 miles		198	.040			1.30	3.57	4.87	6.05
9634	cycle 6 miles		180	.044			1.43	3.92	5.35	6.65
9636	cycle 8 miles		162	.049			1.59	4.36	5.95	7.40
9638	cycle 10 miles		144	.056			1.79	4.90	6.69	8.35
9640	25 MPH avg., cycle 4 miles		216	.037			1.19	3.27	4.46	5.55
9642	cycle 6 miles		198	.040			1.30	3.57	4.87	6.05
9644	cycle 8 miles		180	.044			1.43	3.92	5.35	6.65
9646	cycle 10 miles		162	.049			1.59	4.36	5.95	7.40
9650	30 MPH avg., cycle 4 miles		216	.037			1.19	3.27	4.46	5.55
9652	cycle 6 miles		198	.040			1.30	3.57	4.87	6.05
9654	cycle 8 miles		180	.044			1.43	3.92	5.35	6.65
9656	cycle 10 miles		162	.049			1.59	4.36	5.95	7.40
9660	35 MPH avg., cycle 4 miles		234	.034			1.10	3.02	4.12	5.15
9662	cycle 6 miles		216	.037			1.19	3.27	4.46	5.55
9664	cycle 8 miles		198	.040			1.30	3.57	4.87	6.05
9666	cycle 10 miles		180	.044			1.43	3.92	5.35	6.65
9668	cycle 20 miles		126	.063			2.04	5.60	7.64	9.50
9670	cycle 30 miles		108	.074			2.39	6.55	8.94	11.15
9672	cycle 40 miles		90	.089			2.86	7.85	10.71	13.35
9674	40 MPH avg., cycle 6 miles		216	.037	L.C.Y.		1.19	3.27	4.46	5.55
9676	cycle 8 miles		198	.040			1.30	3.57	4.87	6.05
9678	cycle 10 miles		180	.044			1.43	3.92	5.35	6.65
9680	cycle 20 miles		144	.056			1.79	4.90	6.69	8.35
9682	cycle 30 miles		108	.074			2.39	6.55	8.94	11.15
9684	cycle 40 miles		90	.089			2.86	7.85	10.71	13.35
9686	cycle 50 miles		72	.111			3.58	9.80	13.38	16.70
9694	45 MPH avg., cycle 8 miles		216	.037			1.19	3.27	4.46	5.55
9696	cycle 10 miles		198	.040			1.30	3.57	4.87	6.05
9698	cycle 20 miles		144	.056			1.79	4.90	6.69	8.35
9700	cycle 30 miles		126	.063			2.04	5.60	7.64	9.50
9702	cycle 40 miles		108	.074			2.39	6.55	8.94	11.15
9704	cycle 50 miles		90	.089			2.86	7.85	10.71	13.35
9706	50 MPH avg., cycle 10 miles		198	.040			1.30	3.57	4.87	6.05
9708	cycle 20 miles		162	.049			1.59	4.36	5.95	7.40
9710	cycle 30 miles		126	.063			2.04	5.60	7.64	9.50
9712	cycle 40 miles		108	.074			2.39	6.55	8.94	11.15
9714	cycle 50 miles		90	.089			2.86	7.85	10.71	13.35

31 23 23.24 Compaction, Structural

		Crew	Daily Output	Labor-Hours	Unit	Material	Labor	Equipment	Total	Total Incl O&P
0010	**COMPACTION, STRUCTURAL**									
0020	Steel wheel tandem roller, 5 tons	B-10E	8	1	Hr.		36.50	18.90	55.40	81.50
0050	Air tamp, 6" to 8" lifts, common fill	B-9	250	.160	E.C.Y.		4.34	.94	5.28	8.25
0060	Select fill	"	300	.133			3.61	.78	4.39	6.85
0600	Vibratory plate, 8" lifts, common fill	A-1D	200	.040			1.07	.16	1.23	1.95
0700	Select fill	"	216	.037			.99	.15	1.14	1.80

31 25 Erosion and Sedimentation Controls

31 25 14 – Stabilization Measures for Erosion and Sedimentation Control

31 25 14.16 Rolled Erosion Control Mats and Blankets

	31 25 14.16 Rolled Erosion Control Mats and Blankets		Crew	Daily Output	Labor-Hours	Unit	Material	2018 Bare Costs Labor	Equipment	Total	Total Incl O&P
0010	**ROLLED EROSION CONTROL MATS AND BLANKETS**										
0020	Jute mesh, 100 S.Y. per roll, 4' wide, stapled	G	B-80A	2400	.010	S.Y.	.91	.27	.10	1.28	1.55
0100	Plastic netting, stapled, 2" x 1" mesh, 20 mil	G	B-1	2500	.010		.29	.26		.55	.76
0120	Revegetation mat, webbed	G	2 Clab	1000	.016		2.91	.43		3.34	3.91
0200	Polypropylene mesh, stapled, 6.5 oz./S.Y.	G	B-1	2500	.010		1.63	.26		1.89	2.23
0300	Tobacco netting, or jute mesh #2, stapled	G	"	2500	.010		.25	.26		.51	.72
1000	Silt fence, install and maintain, remove	G	B-62	1300	.018	L.F.	.46	.55	.13	1.14	1.57
1100	Allow 10% per month for maintenance; 6-month max life										

31 31 Soil Treatment

31 31 16 – Termite Control

31 31 16.13 Chemical Termite Control

	31 31 16.13 Chemical Termite Control		Crew	Daily Output	Labor-Hours	Unit	Material	2018 Bare Costs Labor	Equipment	Total	Total Incl O&P
0010	**CHEMICAL TERMITE CONTROL**										
0020	Slab and walls, residential		1 Skwk	1200	.007	SF Flr.	.32	.24		.56	.75
0030	SS mesh, no chemicals, avg 1,400 S.F. home, min	G		1000	.008	"	.32	.29		.61	.83
0400	Insecticides for termite control, minimum			14.20	.563	Gal.	69	20		89	109
0500	Maximum			11	.727	"	118	26		144	174

For customer support on your Residential Costs with RSMeans data, call 800.448.8182.

639

Division Notes

		CREW	DAILY OUTPUT	LABOR-HOURS	UNIT	BARE COSTS				TOTAL INCL O&P
						MAT.	LABOR	EQUIP.	TOTAL	

Estimating Tips

32 01 00 Operations and Maintenance of Exterior Improvements

- Recycling of asphalt pavement is becoming very popular and is an alternative to removal and replacement. It can be a good value engineering proposal if removed pavement can be recycled, either at the project site or at another site that is reasonably close to the project site. Sections on repair of flexible and rigid pavement are included.

32 10 00 Bases, Ballasts, and Paving

- When estimating paving, keep in mind the project schedule. Also note that prices for asphalt and concrete are generally higher in the cold seasons. Lines for pavement markings, including tactile warning systems and fence lines, are included.

32 90 00 Planting

- The timing of planting and guarantee specifications often dictate the costs for establishing tree and shrub growth and a stand of grass or ground cover. Establish the work performance schedule to coincide with the local planting season. Maintenance and growth guarantees can add from 20%–100% to the total landscaping cost and can be contractually cumbersome. The cost to replace trees and shrubs can be as high as 5% of the total cost, depending on the planting zone, soil conditions, and time of year.

Reference Numbers

Reference numbers are shown at the beginning of some major classifications. These numbers refer to related items in the Reference Section. The reference information may be an estimating procedure, an alternate pricing method, or technical information.

Note: Not all subdivisions listed here necessarily appear. ■

32 01 Operation and Maintenance of Exterior Improvements

32 01 13 – Flexible Paving Surface Treatment

32 01 13.66 Fog Seal

		Crew	Daily Output	Labor-Hours	Unit	Material	2018 Bare Costs Labor	Equipment	Total	Total Incl O&P
0010	**FOG SEAL**									
0012	Sealcoating, 2 coat coal tar pitch emulsion over 10,000 S.Y.	B-45	5000	.003	S.Y.	.85	.09	.16	1.10	1.28
0030	1,000 to 10,000 S.Y.	"	3000	.005		.85	.16	.27	1.28	1.49
0100	Under 1,000 S.Y.	B-1	1050	.023		.85	.63		1.48	1.98
0300	Petroleum resistant, over 10,000 S.Y.	B-45	5000	.003		1.32	.09	.16	1.57	1.79
0320	1,000 to 10,000 S.Y.	"	3000	.005		1.32	.16	.27	1.75	2
0400	Under 1,000 S.Y.	B-1	1050	.023	↓	1.32	.63		1.95	2.49

32 06 Schedules for Exterior Improvements

32 06 10 – Schedules for Bases, Ballasts, and Paving

32 06 10.10 Sidewalks, Driveways and Patios

		Crew	Daily Output	Labor-Hours	Unit	Material	2018 Bare Costs Labor	Equipment	Total	Total Incl O&P
0010	**SIDEWALKS, DRIVEWAYS AND PATIOS** No base									
0021	Asphaltic concrete, 2" thick	B-37	6480	.007	S.F.	.76	.21	.02	.99	1.22
0101	2-1/2" thick	"	5950	.008	"	.97	.23	.03	1.23	1.48
0300	Concrete, 3,000 psi, CIP, 6 x 6 - W1.4 x W1.4 mesh,									
0310	broomed finish, no base, 4" thick	B-24	600	.040	S.F.	2.01	1.28		3.29	4.33
0350	5" thick		545	.044		2.68	1.41		4.09	5.30
0400	6" thick	↓	510	.047		3.13	1.51		4.64	5.95
0450	For bank run gravel base, 4" thick, add	B-18	2500	.010		.43	.26	.02	.71	.93
0520	8" thick, add	"	1600	.015		.87	.41	.03	1.31	1.67
1000	Crushed stone, 1" thick, white marble	2 Clab	1700	.009		.48	.25		.73	.95
1050	Bluestone	"	1700	.009		.20	.25		.45	.64
1700	Redwood, prefabricated, 4' x 4' sections	2 Carp	316	.051		4.77	1.77		6.54	8.20
1750	Redwood planks, 1" thick, on sleepers	"	240	.067	↓	4.77	2.33		7.10	9.10
2250	Stone dust, 4" thick	B-62	900	.027	S.Y.	3.97	.79	.19	4.95	5.90

32 06 10.20 Steps

		Crew	Daily Output	Labor-Hours	Unit	Material	2018 Bare Costs Labor	Equipment	Total	Total Incl O&P
0010	**STEPS**									
0011	Incl. excav., borrow & concrete base as required									
0100	Brick steps	B-24	35	.686	LF Riser	17.25	22		39.25	55.50
0200	Railroad ties	2 Clab	25	.640		3.55	17.10		20.65	32.50
0300	Bluestone treads, 12" x 2" or 12" x 1-1/2"	B-24	30	.800	↓	41.50	25.50		67	88
0600	Precast concrete, see Section 03 41 23.50									
4025	Steel edge strips, incl. stakes, 1/4" x 5"	B-1	390	.062	L.F.	4.54	1.68		6.22	7.80
4050	Edging, landscape timber or railroad ties, 6" x 8"	2 Carp	170	.094	"	2.36	3.29		5.65	8.05

32 11 Base Courses

32 11 23 – Aggregate Base Courses

32 11 23.23 Base Course Drainage Layers

		Crew	Daily Output	Labor-Hours	Unit	Material	2018 Bare Costs Labor	Equipment	Total	Total Incl O&P
0010	**BASE COURSE DRAINAGE LAYERS**									
0011	For roadways and large areas									
0051	3/4" stone compacted to 3" deep	B-36	36000	.001	S.F.	.30	.03	.04	.37	.44
0101	6" deep		35100	.001		.60	.04	.04	.68	.77
0201	9" deep		25875	.002		.88	.05	.06	.99	1.12
0305	12" deep		21150	.002		1.51	.06	.07	1.64	1.84
0306	Crushed 1-1/2" stone base, compacted to 4" deep		47000	.001		.06	.03	.03	.12	.14
0307	6" deep		35100	.001		.75	.04	.04	.83	.94
0308	8" deep		27000	.001		1	.05	.06	1.11	1.24
0309	12" deep	↓	16200	.002	↓	1.51	.08	.10	1.69	1.89
0350	Bank run gravel, spread and compacted									

32 11 Base Courses

32 11 23 – Aggregate Base Courses

32 11 23.23 Base Course Drainage Layers

		Crew	Daily Output	Labor-Hours	Unit	Material	2018 Bare Costs Labor	Equipment	Total	Total Incl O&P
0371	6" deep	B-32	54000	.001	S.F.	.40	.02	.04	.46	.51
0391	9" deep		39600	.001		.58	.03	.05	.66	.75
0401	12" deep	↓	32400	.001	↓	.80	.03	.07	.90	1.01
6900	For small and irregular areas, add						50%	50%		

32 11 26 – Asphaltic Base Courses

32 11 26.19 Bituminous-Stabilized Base Courses

		Crew	Daily Output	Labor-Hours	Unit	Material	2018 Bare Costs Labor	Equipment	Total	Total Incl O&P
0010	**BITUMINOUS-STABILIZED BASE COURSES**									
0020	And large paved areas									
0700	Liquid application to gravel base, asphalt emulsion	B-45	6000	.003	Gal.	4.50	.08	.13	4.71	5.25
0800	Prime and seal, cut back asphalt		6000	.003	"	5.30	.08	.13	5.51	6.15
1000	Macadam penetration crushed stone, 2 gal./S.Y., 4" thick		6000	.003	S.Y.	9	.08	.13	9.21	10.20
1100	6" thick, 3 gal./S.Y.		4000	.004		13.50	.12	.20	13.82	15.25
1200	8" thick, 4 gal./S.Y.	↓	3000	.005	↓	18	.16	.27	18.43	20.50
8900	For small and irregular areas, add						50%	50%		

32 12 Flexible Paving

32 12 16 – Asphalt Paving

32 12 16.14 Paving Asphaltic Concrete

		Crew	Daily Output	Labor-Hours	Unit	Material	2018 Bare Costs Labor	Equipment	Total	Total Incl O&P
0010	**PAVING ASPHALTIC CONCRETE**									
0020	6" stone base, 2" binder course, 1" topping	B-25C	9000	.005	S.F.	1.82	.16	.26	2.24	2.56
0025	2" binder course, 2" topping		9000	.005		2.25	.16	.26	2.67	3.04
0030	3" binder course, 2" topping		9000	.005		2.65	.16	.26	3.07	3.47
0035	4" binder course, 2" topping		9000	.005		3.04	.16	.26	3.46	3.90
0040	1-1/2" binder course, 1" topping		9000	.005		1.63	.16	.26	2.05	2.35
0042	3" binder course, 1" topping		9000	.005		2.22	.16	.26	2.64	3
0045	3" binder course, 3" topping		9000	.005		3.08	.16	.26	3.50	3.95
0050	4" binder course, 3" topping		9000	.005		3.47	.16	.26	3.89	4.38
0055	4" binder course, 4" topping		9000	.005		3.90	.16	.26	4.32	4.84
0300	Binder course, 1-1/2" thick		35000	.001		.59	.04	.07	.70	.79
0400	2" thick		25000	.002		.76	.06	.09	.91	1.04
0500	3" thick		15000	.003		1.18	.10	.16	1.44	1.63
0600	4" thick		10800	.004		1.54	.13	.22	1.89	2.16
0800	Sand finish course, 3/4" thick		41000	.001		.31	.04	.06	.41	.46
0900	1" thick	↓	34000	.001		.38	.04	.07	.49	.57
1000	Fill pot holes, hot mix, 2" thick	B-16	4200	.008		.81	.22	.13	1.16	1.39
1100	4" thick		3500	.009		1.18	.26	.16	1.60	1.90
1120	6" thick	↓	3100	.010		1.59	.29	.18	2.06	2.42
1140	Cold patch, 2" thick	B-51	3000	.016		.92	.45	.06	1.43	1.82
1160	4" thick		2700	.018		1.75	.50	.07	2.32	2.82
1180	6" thick	↓	1900	.025	↓	2.72	.70	.10	3.52	4.27

For customer support on your Residential Costs with RSMeans data, call 800.448.8182.

643

32 13 Rigid Paving

32 13 13 – Concrete Paving

32 13 13.25 Concrete Pavement, Highways

		Crew	Daily Output	Labor-Hours	Unit	Material	2018 Bare Costs Labor	2018 Bare Costs Equipment	Total	Total Incl O&P
0010	**CONCRETE PAVEMENT, HIGHWAYS**									
0015	Including joints, finishing and curing									
0021	Fixed form, 12' pass, unreinforced, 6" thick	B-26	18000	.005	S.F.	3.78	.15	.17	4.10	4.59
0101	8" thick	"	13500	.007		5.10	.20	.22	5.52	6.15
0701	Finishing, broom finish small areas	2 Cefi	1215	.013	↓		.46		.46	.74

32 14 Unit Paving

32 14 13 – Precast Concrete Unit Paving

32 14 13.18 Precast Concrete Plantable Pavers

		Crew	Daily Output	Labor-Hours	Unit	Material	2018 Bare Costs Labor	2018 Bare Costs Equipment	Total	Total Incl O&P
0010	**PRECAST CONCRETE PLANTABLE PAVERS** (50% grass)									
0300	3/4" crushed stone base for plantable pavers, 6" depth	B-62	1000	.024	S.Y.	4.22	.71	.17	5.10	6
0400	8" depth		900	.027		5.60	.79	.19	6.58	7.70
0500	10" depth		800	.030		7.05	.89	.22	8.16	9.45
0600	12" depth	↓	700	.034	↓	8.40	1.02	.25	9.67	11.20
0700	Hydro seeding plantable pavers	B-81A	20	.800	M.S.F.	11.40	23.50	17.95	52.85	71
0800	Apply fertilizer and seed to plantable pavers	1 Clab	8	1	"	41.50	26.50		68	90

32 14 16 – Brick Unit Paving

32 14 16.10 Brick Paving

		Crew	Daily Output	Labor-Hours	Unit	Material	2018 Bare Costs Labor	2018 Bare Costs Equipment	Total	Total Incl O&P
0010	**BRICK PAVING**									
0012	4" x 8" x 1-1/2", without joints (4.5 bricks/S.F.)	D-1	110	.145	S.F.	2.57	4.61		7.18	10.50
0100	Grouted, 3/8" joint (3.9 bricks/S.F.)		90	.178		2.08	5.65		7.73	11.70
0200	4" x 8" x 2-1/4", without joints (4.5 bricks/S.F.)		110	.145		2.40	4.61		7.01	10.35
0300	Grouted, 3/8" joint (3.9 bricks/S.F.)		90	.178		2.08	5.65		7.73	11.70
0455	Pervious brick paving, 4" x 8" x 3-1/4", without joints (4.5 bricks/S.F.)	↓	110	.145		3.60	4.61		8.21	11.65
0500	Bedding, asphalt, 3/4" thick	B-25	5130	.017		.65	.51	.52	1.68	2.13
0540	Course washed sand bed, 1" thick	B-18	5000	.005		.35	.13	.01	.49	.62
0580	Mortar, 1" thick	D-1	300	.053		.65	1.69		2.34	3.53
0620	2" thick		200	.080		1.29	2.53		3.82	5.65
1500	Brick on 1" thick sand bed laid flat, 4.5/S.F.		100	.160		2.85	5.05		7.90	11.60
2000	Brick pavers, laid on edge, 7.2/S.F.	↓	70	.229	↓	4.46	7.25		11.71	17

32 14 23 – Asphalt Unit Paving

32 14 23.10 Asphalt Blocks

		Crew	Daily Output	Labor-Hours	Unit	Material	2018 Bare Costs Labor	2018 Bare Costs Equipment	Total	Total Incl O&P
0010	**ASPHALT BLOCKS**									
0020	Rectangular, 6" x 12" x 1-1/4", w/bed & neopr. adhesive	D-1	135	.119	S.F.	8.55	3.75		12.30	15.65
0100	3" thick		130	.123		12	3.90		15.90	19.70
0300	Hexagonal tile, 8" wide, 1-1/4" thick		135	.119		8.55	3.75		12.30	15.65
0400	2" thick		130	.123		12	3.90		15.90	19.70
0500	Square, 8" x 8", 1-1/4" thick		135	.119		8.55	3.75		12.30	15.65
0600	2" thick	↓	130	.123	↓	12	3.90		15.90	19.70

32 14 40 – Stone Paving

32 14 40.10 Stone Pavers

		Crew	Daily Output	Labor-Hours	Unit	Material	2018 Bare Costs Labor	2018 Bare Costs Equipment	Total	Total Incl O&P
0010	**STONE PAVERS**									
1100	Flagging, bluestone, irregular, 1" thick,	D-1	81	.198	S.F.	10.45	6.25		16.70	22
1150	Snapped random rectangular, 1" thick		92	.174		15.85	5.50		21.35	26.50
1200	1-1/2" thick		85	.188		19.05	5.95		25	31
1250	2" thick		83	.193		22	6.10		28.10	34.50
1300	Slate, natural cleft, irregular, 3/4" thick		92	.174		9.45	5.50		14.95	19.60
1310	1" thick		85	.188		11.05	5.95		17	22
1351	Random rectangular, gauged, 1/2" thick	↓	105	.152	↓	20.50	4.83		25.33	30.50

32 14 Unit Paving

32 14 40 – Stone Paving

32 14 40.10 Stone Pavers

32 14 40.10 Stone Pavers	Crew	Daily Output	Labor-Hours	Unit	Material	2018 Bare Costs Labor	Equipment	Total	Total Incl O&P
1400	Random rectangular, butt joint, gauged, 1/4" thick	D-1	150	.107	S.F.	22	3.38		25.38	30
1450	For sand rubbed finish, add				▼	9.45			9.45	10.40
1500	For interior setting, add								25%	25%
1550	Granite blocks, 3-1/2" x 3-1/2" x 3-1/2"	D-1	92	.174	S.F.	19.30	5.50		24.80	30

32 16 Curbs, Gutters, Sidewalks, and Driveways

32 16 13 – Curbs and Gutters

32 16 13.13 Cast-in-Place Concrete Curbs and Gutters

32 16 13.13 Cast-in-Place Concrete Curbs and Gutters	Crew	Daily Output	Labor-Hours	Unit	Material	2018 Bare Costs Labor	Equipment	Total	Total Incl O&P
0010	**CAST-IN-PLACE CONCRETE CURBS AND GUTTERS**									
0290	Forms only, no concrete									
0300	Concrete, wood forms, 6" x 18", straight	C-2	500	.096	L.F.	2.93	2.99		5.92	8.20
0400	6" x 18", radius	"	200	.240		3.05	7.45		10.50	15.80
0404	Concrete, wood forms, 6" x 18", straight & concrete	C-2A	500	.096		6.30	3.25		9.55	12.35
0406	6" x 18", radius	"	200	.240	▼	6.45	8.15		14.60	20.50

32 16 13.23 Precast Concrete Curbs and Gutters

32 16 13.23 Precast Concrete Curbs and Gutters	Crew	Daily Output	Labor-Hours	Unit	Material	2018 Bare Costs Labor	Equipment	Total	Total Incl O&P
0010	**PRECAST CONCRETE CURBS AND GUTTERS**									
0550	Precast, 6" x 18", straight	B-29	700	.069	L.F.	9.35	1.97	1.24	12.56	14.90
0600	6" x 18", radius	"	325	.148	"	10.55	4.25	2.68	17.48	21.50

32 16 13.33 Asphalt Curbs

32 16 13.33 Asphalt Curbs	Crew	Daily Output	Labor-Hours	Unit	Material	2018 Bare Costs Labor	Equipment	Total	Total Incl O&P
0010	**ASPHALT CURBS**									
0012	Curbs, asphaltic, machine formed, 8" wide, 6" high, 40 L.F./ton	B-27	1000	.032	L.F.	1.63	.87	.32	2.82	3.60
0100	8" wide, 8" high, 30 L.F./ton		900	.036		2.18	.97	.35	3.50	4.40
0150	Asphaltic berm, 12" W, 3" to 6" H, 35 L.F./ton, before pavement	▼	700	.046		.04	1.24	.45	1.73	2.62
0200	12" W, 1-1/2" to 4" H, 60 L.F./ton, laid with pavement	B-2	1050	.038	▼	.02	1.03		1.05	1.75

32 16 13.43 Stone Curbs

32 16 13.43 Stone Curbs	Crew	Daily Output	Labor-Hours	Unit	Material	2018 Bare Costs Labor	Equipment	Total	Total Incl O&P
0010	**STONE CURBS**									
1000	Granite, split face, straight, 5" x 16"	D-13	275	.175	L.F.	15.20	5.85	1.32	22.37	28
1100	6" x 18"	"	250	.192		19.95	6.40	1.46	27.81	34.50
1300	Radius curbing, 6" x 18", over 10' radius	B-29	260	.185	▼	24.50	5.30	3.35	33.15	39.50
1400	Corners, 2' radius	"	80	.600	Ea.	82	17.25	10.85	110.10	131
1600	Edging, 4-1/2" x 12", straight	D-13	300	.160	L.F.	7.60	5.35	1.21	14.16	18.60
1800	Curb inlets (guttermouth) straight	B-29	41	1.171	Ea.	182	33.50	21	236.50	281
2000	Indian granite (Belgian block)									
2100	Jumbo, 10-1/2" x 7-1/2" x 4", grey	D-1	150	.107	L.F.	8.35	3.38		11.73	14.85
2150	Pink		150	.107		8.40	3.38		11.78	14.90
2200	Regular, 9" x 4-1/2" x 4-1/2", grey		160	.100		4.45	3.17		7.62	10.20
2250	Pink		160	.100		5.85	3.17		9.02	11.75
2300	Cubes, 4" x 4" x 4", grey		175	.091		3.62	2.90		6.52	8.85
2350	Pink		175	.091		3.72	2.90		6.62	8.95
2400	6" x 6" x 6", pink	▼	155	.103	▼	12.70	3.27		15.97	19.40
2500	Alternate pricing method for Indian granite									
2550	Jumbo, 10-1/2" x 7-1/2" x 4" (30 lb.), grey				Ton	475			475	525
2600	Pink					490			490	540
2650	Regular, 9" x 4-1/2" x 4-1/2" (20 lb.), grey					315			315	345
2700	Pink					410			410	450
2750	Cubes, 4" x 4" x 4" (5 lb.), grey					440			440	485
2800	Pink					480			480	525
2850	6" x 6" x 6" (25 lb.), pink					495			495	545
2900	For pallets, add				▼	22			22	24

For customer support on your Residential Costs with RSMeans data, call 800.448.8182.

645

32 31 Fences and Gates

32 31 13 – Chain Link Fences and Gates

32 31 13.15 Chain Link Fence

32 31 13.15 Chain Link Fence	Crew	Daily Output	Labor-Hours	Unit	Material	2018 Bare Costs Labor	Equipment	Total	Total Incl O&P
0010	**CHAIN LINK FENCE**									
0020	1-5/8" post 10' OC, 1-3/8" top rail, 2" corner post galv. stl., 3' high	B-1	185	.130	L.F.	10.75	3.55		14.30	17.75
0050	4' high		170	.141		10.25	3.86		14.11	17.75
0100	6' high		115	.209	↓	12.25	5.70		17.95	23
0150	Add for gate 3' wide, 1-3/8" frame 3' high		12	2	Ea.	98.50	54.50		153	199
0170	4' high		10	2.400		117	65.50		182.50	238
0190	6' high		10	2.400		141	65.50		206.50	264
0200	Add for gate 4' wide, 1-3/8" frame 3' high		9	2.667		112	73		185	244
0220	4' high		9	2.667		122	73		195	255
0240	6' high		8	3	↓	145	82		227	297
0350	Aluminized steel, 9 ga. wire, 3' high		185	.130	L.F.	8.75	3.55		12.30	15.55
0380	4' high		170	.141		9.25	3.86		13.11	16.65
0400	6' high		115	.209	↓	12.25	5.70		17.95	23
0450	Add for gate 3' wide, 1-3/8" frame 3' high		12	2	Ea.	150	54.50		204.50	256
0470	4' high		10	2.400		160	65.50		225.50	285
0490	6' high		10	2.400		200	65.50		265.50	330
0500	Add for gate 4' wide, 1-3/8" frame 3' high		10	2.400		155	65.50		220.50	279
0520	4' high		9	2.667		143	73		216	278
0540	6' high		8	3	↓	193	82		275	350
0620	Vinyl covered 9 ga. wire, 3' high		185	.130	L.F.	8.75	3.55		12.30	15.55
0640	4' high		170	.141		7.25	3.86		11.11	14.45
0660	6' high		115	.209	↓	9.75	5.70		15.45	20.50
0720	Add for gate 3' wide, 1-3/8" frame 3' high		12	2	Ea.	107	54.50		161.50	209
0740	4' high		10	2.400		125	65.50		190.50	247
0760	6' high		10	2.400		155	65.50		220.50	280
0780	Add for gate 4' wide, 1-3/8" frame 3' high		10	2.400		120	65.50		185.50	241
0800	4' high		9	2.667		130	73		203	264
0820	6' high	↓	8	3	↓	163	82		245	315
0860	Tennis courts, 11 ga. wire, 2-1/2" post 10' OC, 1-5/8" top rail									
0900	2-1/2" corner post, 10' high	B-1	95	.253	L.F.	9	6.90		15.90	21.50
0920	12' high		80	.300	"	9.25	8.20		17.45	24
1000	Add for gate 3' wide, 1-5/8" frame 10' high		10	2.400	Ea.	232	65.50		297.50	365
1040	Aluminized, 11 ga. wire 10' high		95	.253	L.F.	11	6.90		17.90	23.50
1100	12' high		80	.300	"	12	8.20		20.20	27
1140	Add for gate 3' wide, 1-5/8" frame, 10' high		10	2.400	Ea.	165	65.50		230.50	291
1250	Vinyl covered 11 ga. wire, 10' high		95	.253	L.F.	9.50	6.90		16.40	22
1300	12' high		80	.300	"	11.35	8.20		19.55	26
1400	Add for gate 3' wide, 1-3/8" frame, 10' high	↓	10	2.400	Ea.	325	65.50		390.50	470

32 31 13.80 Residential Chain Link Gate

32 31 13.80 Residential Chain Link Gate	Crew	Daily Output	Labor-Hours	Unit	Material	Labor	Equipment	Total	Total Incl O&P
0010	**RESIDENTIAL CHAIN LINK GATE**									
0110	Residential 4' gate, single incl. hardware and concrete	B-80C	10	2.400	Ea.	123	68	19.60	210.60	270
0120	5'		10	2.400		133	68	19.60	220.60	281
0130	6'		10	2.400		143	68	19.60	230.60	293
0510	Residential 4' gate, double incl. hardware and concrete		10	2.400		219	68	19.60	306.60	375
0520	5'		10	2.400		234	68	19.60	321.60	395
0530	6'	↓	10	2.400	↓	273	68	19.60	360.60	435

32 31 13.82 Internal Chain Link Gate

32 31 13.82 Internal Chain Link Gate	Crew	Daily Output	Labor-Hours	Unit	Material	Labor	Equipment	Total	Total Incl O&P
0010	**INTERNAL CHAIN LINK GATE**									
0110	Internal 6' gate, single incl. post flange, hardware and concrete	B-80C	10	2.400	Ea.	269	68	19.60	356.60	430
0120	8'		10	2.400		305	68	19.60	392.60	470
0130	10'		10	2.400		420	68	19.60	507.60	600
0510	Internal 6' gate, double incl. post flange, hardware and concrete	↓	10	2.400	↓	490	68	19.60	577.60	675

32 31 Fences and Gates

32 31 13 – Chain Link Fences and Gates

32 31 13.82 Internal Chain Link Gate

		Crew	Daily Output	Labor-Hours	Unit	Material	2018 Bare Costs Labor	Equipment	Total	Total Incl O&P
0520	8'	B-80C	10	2.400	Ea.	565	68	19.60	652.60	755
0530	10'	↓	10	2.400	↓	710	68	19.60	797.60	920

32 31 13.84 Industrial Chain Link Gate

		Crew	Daily Output	Labor-Hours	Unit	Material	Labor	Equipment	Total	Total Incl O&P
0010	**INDUSTRIAL CHAIN LINK GATE**									
0110	Industrial 8' gate, single incl. hardware and concrete	B-80C	10	2.400	Ea.	460	68	19.60	547.60	640
0120	10'		10	2.400		520	68	19.60	607.60	710
0510	Industrial 8' gate, double incl. hardware and concrete		10	2.400		715	68	19.60	802.60	920
0520	10'	↓	10	2.400	↓	815	68	19.60	902.60	1,025

32 31 13.88 Chain Link Transom

		Crew	Daily Output	Labor-Hours	Unit	Material	Labor	Equipment	Total	Total Incl O&P
0010	**CHAIN LINK TRANSOM**									
0110	Add for, single transom, 3' wide, incl. components & hardware	B-80C	10	2.400	Ea.	114	68	19.60	201.60	260
0120	Add for, double transom, 6' wide, incl. components & hardware	"	10	2.400	"	122	68	19.60	209.60	269

32 31 23 – Plastic Fences and Gates

32 31 23.10 Fence, Vinyl

		Crew	Daily Output	Labor-Hours	Unit	Material	Labor	Equipment	Total	Total Incl O&P
0010	**FENCE, VINYL**									
0011	White, steel reinforced, stainless steel fasteners									
0020	Picket, 4" x 4" posts @ 6'-0" OC, 3' high	B-1	140	.171	L.F.	24.50	4.69		29.19	34.50
0030	4' high		130	.185		26.50	5.05		31.55	37.50
0040	5' high		120	.200		29	5.45		34.45	41
0100	Board (semi-privacy), 5" x 5" posts @ 7'-6" OC, 5' high		130	.185		27	5.05		32.05	38
0120	6' high		125	.192		29	5.25		34.25	41
0200	Basket weave, 5" x 5" posts @ 7'-6" OC, 5' high		160	.150		26	4.11		30.11	35.50
0220	6' high		150	.160		29	4.38		33.38	39
0300	Privacy, 5" x 5" posts @ 7'-6" OC, 5' high		130	.185		25	5.05		30.05	36
0320	6' high		150	.160	↓	29	4.38		33.38	39
0350	Gate, 5' high		9	2.667	Ea.	315	73		388	470
0360	6' high		9	2.667		360	73		433	520
0400	For posts set in concrete, add		25	.960	↓	10	26.50		36.50	54.50
0500	Post and rail fence, 2 rail		150	.160	L.F.	6	4.38		10.38	13.90
0510	3 rail		150	.160		7.75	4.38		12.13	15.85
0515	4 rail	↓	150	.160	↓	10.05	4.38		14.43	18.35

32 31 26 – Wire Fences and Gates

32 31 26.10 Fences, Misc. Metal

		Crew	Daily Output	Labor-Hours	Unit	Material	Labor	Equipment	Total	Total Incl O&P
0010	**FENCES, MISC. METAL**									
0012	Chicken wire, posts @ 4', 1" mesh, 4' high	B-80C	410	.059	L.F.	3.50	1.66	.48	5.64	7.15
0100	2" mesh, 6' high		350	.069		3.93	1.94	.56	6.43	8.15
0200	Galv. steel, 12 ga., 2" x 4" mesh, posts 5' OC, 3' high		300	.080		2.70	2.27	.65	5.62	7.45
0300	5' high		300	.080		3.28	2.27	.65	6.20	8.10
0400	14 ga., 1" x 2" mesh, 3' high		300	.080		3.32	2.27	.65	6.24	8.10
0500	5' high	↓	300	.080	↓	4.44	2.27	.65	7.36	9.35
1000	Kennel fencing, 1-1/2" mesh, 6' long, 3'-6" wide, 6'-2" high	2 Clab	4	4	Ea.	495	107		602	725
1050	12' long		4	4		695	107		802	945
1200	Top covers, 1-1/2" mesh, 6' long		15	1.067		132	28.50		160.50	193
1250	12' long	↓	12	1.333	↓	185	35.50		220.50	263

For customer support on your Residential Costs with RSMeans data, call 800.448.8182.

647

32 31 29.10 Fence, Wood	Crew	Daily Output	Labor-Hours	Unit	Material	2018 Bare Costs Labor	Equipment	Total	Total Incl O&P
0010 **FENCE, WOOD**									
0011 Basket weave, 3/8" x 4" boards, 2" x 4"									
0020 stringers on spreaders, 4" x 4" posts									
0050 No. 1 cedar, 6' high	B-80C	160	.150	L.F.	26	4.25	1.22	31.47	37
0070 Treated pine, 6' high		150	.160		37	4.53	1.31	42.84	49.50
0090 Vertical weave, 6' high		145	.166		22.50	4.69	1.35	28.54	34
0200 Board fence, 1" x 4" boards, 2" x 4" rails, 4" x 4" post									
0220 Preservative treated, 2 rail, 3' high	B-80C	145	.166	L.F.	10.30	4.69	1.35	16.34	20.50
0240 4' high		135	.178		11.85	5.05	1.45	18.35	23
0260 3 rail, 5' high		130	.185		12.40	5.25	1.51	19.16	24
0300 6' high		125	.192		15.30	5.45	1.57	22.32	27.50
0320 No. 2 grade western cedar, 2 rail, 3' high		145	.166		12.35	4.69	1.35	18.39	23
0340 4' high		135	.178		11.75	5.05	1.45	18.25	23
0360 3 rail, 5' high		130	.185		13.85	5.25	1.51	20.61	25.50
0400 6' high		125	.192		14.75	5.45	1.57	21.77	27
0420 No. 1 grade cedar, 2 rail, 3' high		145	.166		13.30	4.69	1.35	19.34	24
0440 4' high		135	.178		14.70	5.05	1.45	21.20	26
0460 3 rail, 5' high		130	.185		17.40	5.25	1.51	24.16	29.50
0500 6' high		125	.192		21.50	5.45	1.57	28.52	34
0540 Shadow box, 1" x 6" board, 2" x 4" rail, 4" x 4" post									
0560 Pine, pressure treated, 3 rail, 6' high	B-80C	150	.160	L.F.	24	4.53	1.31	29.84	35.50
0600 Gate, 3'-6" wide		8	3	Ea.	135	85	24.50	244.50	315
0620 No. 1 cedar, 3 rail, 4' high		130	.185	L.F.	18.85	5.25	1.51	25.61	31.50
0640 6' high		125	.192		26	5.45	1.57	33.02	39
0860 Open rail fence, split rails, 2 rail, 3' high, no. 1 cedar		160	.150		9.75	4.25	1.22	15.22	19.15
0870 No. 2 cedar		160	.150		8.15	4.25	1.22	13.62	17.35
0880 3 rail, 4' high, no. 1 cedar		150	.160		12.15	4.53	1.31	17.99	22.50
0890 No. 2 cedar		150	.160		8	4.53	1.31	13.84	17.70
0920 Rustic rails, 2 rail, 3' high, no. 1 cedar		160	.150		12.30	4.25	1.22	17.77	22
0930 No. 2 cedar		160	.150		11.15	4.25	1.22	16.62	20.50
0940 3 rail, 4' high		150	.160		11.80	4.53	1.31	17.64	22
0950 No. 2 cedar		150	.160		8.05	4.53	1.31	13.89	17.80
0960 Picket fence, gothic, pressure treated pine									
1000 2 rail, 3' high	B-80C	140	.171	L.F.	7.80	4.86	1.40	14.06	18.20
1020 3 rail, 4' high		130	.185	"	8.90	5.25	1.51	15.66	20
1040 Gate, 3'-6" wide		9	2.667	Ea.	74	75.50	22	171.50	231
1060 No. 2 cedar, 2 rail, 3' high		140	.171	L.F.	9	4.86	1.40	15.26	19.50
1100 3 rail, 4' high		130	.185	"	9.10	5.25	1.51	15.86	20.50
1120 Gate, 3'-6" wide		9	2.667	Ea.	79	75.50	22	176.50	236
1140 No. 1 cedar, 2 rail, 3' high		140	.171	L.F.	13.70	4.86	1.40	19.96	24.50
1160 3 rail, 4' high		130	.185		17.85	5.25	1.51	24.61	30
1200 Rustic picket, molded pine, 2 rail, 3' high		140	.171		8.50	4.86	1.40	14.76	18.95
1220 No. 1 cedar, 2 rail, 3' high		140	.171		10.40	4.86	1.40	16.66	21
1240 Stockade fence, no. 1 cedar, 3-1/4" rails, 6' high		160	.150		13.15	4.25	1.22	18.62	23
1260 8' high		155	.155		18.10	4.39	1.26	23.75	28.50
1300 No. 2 cedar, treated wood rails, 6' high		160	.150		13.45	4.25	1.22	18.92	23
1320 Gate, 3'-6" wide		8	3	Ea.	89.50	85	24.50	199	267
1360 Treated pine, treated rails, 6' high		160	.150	L.F.	13.95	4.25	1.22	19.42	24
1400 8' high		150	.160	"	19.80	4.53	1.31	25.64	31

32 31 Fences and Gates

32 31 29 – Wood Fences and Gates

32 31 29.20 Fence, Wood Rail	Crew	Daily Output	Labor-Hours	Unit	Material	2018 Bare Costs Labor	Equipment	Total	Total Incl O&P
0010 **FENCE, WOOD RAIL**									
0012 Picket, No. 2 cedar, Gothic, 2 rail, 3' high	B-1	160	.150	L.F.	8.10	4.11		12.21	15.75
0050 Gate, 3'-6" wide	B-80C	9	2.667	Ea.	78	75.50	22	175.50	235
0400 3 rail, 4' high		150	.160	L.F.	9.10	4.53	1.31	14.94	18.95
0500 Gate, 3'-6" wide		9	2.667	Ea.	95	75.50	22	192.50	253
1200 Stockade, No. 2 cedar, treated wood rails, 6' high		160	.150	L.F.	9.10	4.25	1.22	14.57	18.40
1250 Gate, 3' wide		9	2.667	Ea.	97	75.50	22	194.50	256
1300 No. 1 cedar, 3-1/4" cedar rails, 6' high		160	.150	L.F.	19.25	4.25	1.22	24.72	29.50
1500 Gate, 3' wide		9	2.667	Ea.	227	75.50	22	324.50	400
2700 Prefabricated redwood or cedar, 4' high		160	.150	L.F.	15.90	4.25	1.22	21.37	26
2800 6' high		150	.160		23.50	4.53	1.31	29.34	35
3300 Board, shadow box, 1" x 6", treated pine, 6' high		160	.150		12.65	4.25	1.22	18.12	22.50
3400 No. 1 cedar, 6' high		150	.160		25	4.53	1.31	30.84	36.50
3900 Basket weave, No. 1 cedar, 6' high	↓	160	.150	↓	34	4.25	1.22	39.47	46
4200 Gate, 3'-6" wide	B-1	9	2.667	Ea.	178	73		251	315
5000 Fence rail, redwood, 2" x 4", merch. grade, 8'	"	2400	.010	L.F.	2.53	.27		2.80	3.24

32 32 Retaining Walls

32 32 13 – Cast-in-Place Concrete Retaining Walls

32 32 13.10 Retaining Walls, Cast Concrete

	Crew	Daily Output	Labor-Hours	Unit	Material	Labor	Equipment	Total	Total Incl O&P
0010 **RETAINING WALLS, CAST CONCRETE**									
1800 Concrete gravity wall with vertical face including excavation & backfill									
1850 No reinforcing									
1900 6' high, level embankment	C-17C	36	2.306	L.F.	89	83.50	15.80	188.30	254
2000 33° slope embankment	"	32	2.594	"	103	94	17.80	214.80	290
2800 Reinforced concrete cantilever, incl. excavation, backfill & reinf.									
2900 6' high, 33° slope embankment	C-17C	35	2.371	L.F.	80.50	86	16.25	182.75	249

32 32 23 – Segmental Retaining Walls

32 32 23.13 Segmental Conc. Unit Masonry Retaining Walls

	Crew	Daily Output	Labor-Hours	Unit	Material	Labor	Equipment	Total	Total Incl O&P
0010 **SEGMENTAL CONC. UNIT MASONRY RETAINING WALLS**									
7100 Segmental retaining wall system, incl. pins and void fill									
7120 base and backfill not included									
7140 Large unit, 8" high x 18" wide x 20" deep, 3 plane split	B-62	300	.080	S.F.	12.75	2.37	.58	15.70	18.60
7150 Straight split		300	.080		12.85	2.37	.58	15.80	18.65
7160 Medium, lt. wt., 8" high x 18" wide x 12" deep, 3 plane split		400	.060		6.30	1.78	.44	8.52	10.30
7170 Straight split		400	.060		9.65	1.78	.44	11.87	14.05
7180 Small unit, 4" x 18" x 10" deep, 3 plane split		400	.060		14.60	1.78	.44	16.82	19.45
7190 Straight split		400	.060		10.65	1.78	.44	12.87	15.10
7200 Cap unit, 3 plane split		300	.080		13.40	2.37	.58	16.35	19.25
7210 Cap unit, straight split		300	.080		13.40	2.37	.58	16.35	19.25
7250 Geo-grid soil reinforcement 4' x 50'	2 Clab	22500	.001		.68	.02		.70	.78
7255 Geo-grid soil reinforcement 6' x 150'	"	22500	.001	↓	.54	.02		.56	.62

32 32 26 – Metal Crib Retaining Walls

32 32 26.10 Metal Bin Retaining Walls

	Crew	Daily Output	Labor-Hours	Unit	Material	Labor	Equipment	Total	Total Incl O&P
0010 **METAL BIN RETAINING WALLS**									
0011 Aluminized steel bin, excavation									
0020 and backfill not included, 10' wide									
0100 4' high, 5.5' deep	B-13	650	.074	S.F.	28	2.12	.91	31.03	35.50
0200 8' high, 5.5' deep	↓	615	.078	↓	32	2.24	.96	35.20	40.50

For customer support on your Residential Costs with RSMeans data, call 800.448.8182.

649

32 32 Retaining Walls

32 32 26 – Metal Crib Retaining Walls

32 32 26.10 Metal Bin Retaining Walls	Crew	Daily Output	Labor-Hours	Unit	Material	2018 Bare Costs Labor	Equipment	Total	Total Incl O&P
0300	10' high, 7.7' deep	B-13	580	.083	S.F.	35.50	2.38	1.02	38.90	44.50
0400	12' high, 7.7' deep		530	.091		38.50	2.60	1.11	42.21	48
0500	16' high, 7.7' deep	↓	515	.093	↓	40.50	2.68	1.15	44.33	50.50

32 32 29 – Timber Retaining Walls

32 32 29.10 Landscape Timber Retaining Walls

		Crew	Daily Output	Labor-Hours	Unit	Material	Labor	Equipment	Total	Total Incl O&P
0010	**LANDSCAPE TIMBER RETAINING WALLS**									
0100	Treated timbers, 6" x 6"	1 Clab	265	.030	L.F.	2.47	.81		3.28	4.06
0110	6" x 8'	"	200	.040	"	5.40	1.07		6.47	7.70
0120	Drilling holes in timbers for fastening, 1/2"	1 Carp	450	.018	Inch		.62		.62	1.03
0130	5/8"	"	450	.018	"		.62		.62	1.03
0140	Reinforcing rods for fastening, 1/2"	1 Clab	312	.026	L.F.	.35	.68		1.03	1.53
0150	5/8"	"	312	.026	"	.55	.68		1.23	1.74
0160	Reinforcing fabric	2 Clab	2500	.006	S.Y.	2.13	.17		2.30	2.62
0170	Gravel backfill		28	.571	C.Y.	16.80	15.25		32.05	44
0180	Perforated pipe, 4" diameter with silt sock	↓	1200	.013	L.F.	1	.36		1.36	1.69
0190	Galvanized 60d common nails	1 Clab	625	.013	Ea.	.18	.34		.52	.77
0200	20d common nails	"	3800	.002	"	.04	.06		.10	.13

32 32 53 – Stone Retaining Walls

32 32 53.10 Retaining Walls, Stone

		Crew	Daily Output	Labor-Hours	Unit	Material	Labor	Equipment	Total	Total Incl O&P
0010	**RETAINING WALLS, STONE**									
0015	Including excavation, concrete footing and									
0020	stone 3' below grade. Price is exposed face area.									
0200	Decorative random stone, to 6' high, 1'-6" thick, dry set	D-1	35	.457	S.F.	69	14.50		83.50	99.50
0300	Mortar set		40	.400		71	12.65		83.65	99
0500	Cut stone, to 6' high, 1'-6" thick, dry set		35	.457		71.50	14.50		86	103
0600	Mortar set		40	.400		72	12.65		84.65	100
0800	Random stone, 6' to 10' high, 2' thick, dry set		45	.356		77.50	11.25		88.75	104
0900	Mortar set		50	.320		81	10.15		91.15	106
1100	Cut stone, 6' to 10' high, 2' thick, dry set		45	.356		78.50	11.25		89.75	105
1200	Mortar set	↓	50	.320	↓	81.50	10.15		91.65	106

32 33 Site Furnishings

32 33 33 – Site Manufactured Planters

32 33 33.10 Planters

		Crew	Daily Output	Labor-Hours	Unit	Material	Labor	Equipment	Total	Total Incl O&P
0010	**PLANTERS**									
0012	Concrete, sandblasted, precast, 48" diameter, 24" high	2 Clab	15	1.067	Ea.	660	28.50		688.50	775
0300	Fiberglass, circular, 36" diameter, 24" high		15	1.067		690	28.50		718.50	810
1200	Wood, square, 48" side, 24" high		15	1.067		1,625	28.50		1,653.50	1,850
1300	Circular, 48" diameter, 30" high		10	1.600		1,100	42.50		1,142.50	1,300
1600	Planter/bench, 72"	↓	5	3.200	↓	3,675	85.50		3,760.50	4,200

32 33 43 – Site Seating and Tables

32 33 43.13 Site Seating

		Crew	Daily Output	Labor-Hours	Unit	Material	Labor	Equipment	Total	Total Incl O&P
0010	**SITE SEATING**									
0012	Seating, benches, park, precast conc., w/backs, wood rails, 4' long	2 Clab	5	3.200	Ea.	635	85.50		720.50	840
0100	8' long		4	4		1,075	107		1,182	1,350
0500	Steel barstock pedestals w/backs, 2" x 3" wood rails, 4' long		10	1.600		1,350	42.50		1,392.50	1,575
0510	8' long		7	2.286		1,725	61		1,786	2,000
0800	Cast iron pedestals, back & arms, wood slats, 4' long		8	2		435	53.50		488.50	570
0820	8' long	↓	5	3.200	↓	1,100	85.50		1,185.50	1,350

32 33 Site Furnishings

32 33 43 – Site Seating and Tables

32 33 43.13 Site Seating	Crew	Daily Output	Labor-Hours	Unit	Material	2018 Bare Costs Labor	Equipment	Total	Total Incl O&P
1700	Steel frame, fir seat, 10' long	2 Clab	10	1.600	Ea.	395	42.50		437.50	505

32 84 Planting Irrigation

32 84 23 – Underground Sprinklers

32 84 23.10 Sprinkler Irrigation System

		Crew	Daily Output	Labor-Hours	Unit	Material	2018 Bare Costs Labor	Equipment	Total	Total Incl O&P
0010	**SPRINKLER IRRIGATION SYSTEM**									
0011	For lawns									
0800	Residential system, custom, 1" supply	B-20	2000	.012	S.F.	.25	.33		.58	.83
0900	1-1/2" supply	"	1800	.013	"	.48	.36		.84	1.14

32 91 Planting Preparation

32 91 13 – Soil Preparation

32 91 13.16 Mulching

		Crew	Daily Output	Labor-Hours	Unit	Material	2018 Bare Costs Labor	Equipment	Total	Total Incl O&P
0010	**MULCHING**									
0100	Aged barks, 3" deep, hand spread	1 Clab	100	.080	S.Y.	3.80	2.14		5.94	7.75
0150	Skid steer loader	B-63	13.50	2.963	M.S.F.	420	79	12.90	511.90	610
0200	Hay, 1" deep, hand spread	1 Clab	475	.017	S.Y.	.45	.45		.90	1.25
0250	Power mulcher, small	B-64	180	.089	M.S.F.	50	2.59	1.82	54.41	61.50
0350	Large	B-65	530	.030	"	50	.88	.86	51.74	57.50
0400	Humus peat, 1" deep, hand spread	1 Clab	700	.011	S.Y.	2.96	.31		3.27	3.77
0450	Push spreader	"	2500	.003	"	2.96	.09		3.05	3.40
0550	Tractor spreader	B-66	700	.011	M.S.F.	330	.40	.35	330.75	360
0600	Oat straw, 1" deep, hand spread	1 Clab	475	.017	S.Y.	.56	.45		1.01	1.37
0650	Power mulcher, small	B-64	180	.089	M.S.F.	62	2.59	1.82	66.41	75
0700	Large	B-65	530	.030	"	62	.88	.86	63.74	71
0750	Add for asphaltic emulsion	B-45	1770	.009	Gal.	5.80	.27	.45	6.52	7.35
0800	Peat moss, 1" deep, hand spread	1 Clab	900	.009	S.Y.	4.90	.24		5.14	5.80
0850	Push spreader	"	2500	.003	"	4.90	.09		4.99	5.55
0950	Tractor spreader	B-66	700	.011	M.S.F.	545	.40	.35	545.75	600
1000	Polyethylene film, 6 mil	2 Clab	2000	.008	S.Y.	.53	.21		.74	.94
1100	Redwood nuggets, 3" deep, hand spread	1 Clab	150	.053	"	3.12	1.42		4.54	5.80
1150	Skid steer loader	B-63	13.50	2.963	M.S.F.	345	79	12.90	436.90	525
1200	Stone mulch, hand spread, ceramic chips, economy	1 Clab	125	.064	S.Y.	7	1.71		8.71	10.55
1250	Deluxe	"	95	.084	"	10.60	2.25		12.85	15.40
1300	Granite chips	B-1	10	2.400	C.Y.	72.50	65.50		138	189
1400	Marble chips		10	2.400		215	65.50		280.50	345
1600	Pea gravel		28	.857		111	23.50		134.50	161
1700	Quartz		10	2.400		190	65.50		255.50	320
1800	Tar paper, 15 lb. felt	1 Clab	800	.010	S.Y.	.48	.27		.75	.97
1900	Wood chips, 2" deep, hand spread	"	220	.036	"	1.58	.97		2.55	3.35
1950	Skid steer loader	B-63	20.30	1.970	M.S.F.	176	52.50	8.55	237.05	290

32 91 13.26 Planting Beds

		Crew	Daily Output	Labor-Hours	Unit	Material	2018 Bare Costs Labor	Equipment	Total	Total Incl O&P
0010	**PLANTING BEDS**									
0100	Backfill planting pit, by hand, on site topsoil	2 Clab	18	.889	C.Y.		23.50		23.50	39.50
0200	Prepared planting mix, by hand	"	24	.667			17.80		17.80	29.50
0300	Skid steer loader, on site topsoil	B-62	340	.071			2.09	.51	2.60	4.02
0400	Prepared planting mix	"	410	.059			1.73	.42	2.15	3.34
1000	Excavate planting pit, by hand, sandy soil	2 Clab	16	1			26.50		26.50	44.50
1100	Heavy soil or clay	"	8	2			53.50		53.50	89

For customer support on your Residential Costs with RSMeans data, call 800.448.8182.

651

32 91 Planting Preparation

32 91 13 – Soil Preparation

32 91 13.26 Planting Beds

		Crew	Daily Output	Labor-Hours	Unit	Material	2018 Bare Costs Labor	Equipment	Total	Total Incl O&P
1200	1/2 C.Y. backhoe, sandy soil	B-11C	150	.107	C.Y.		3.37	2.08	5.45	7.90
1300	Heavy soil or clay	"	115	.139			4.40	2.72	7.12	10.30
2000	Mix planting soil, incl. loam, manure, peat, by hand	2 Clab	60	.267		44	7.10		51.10	60.50
2100	Skid steer loader	B-62	150	.160		44	4.74	1.16	49.90	57.50
3000	Pile sod, skid steer loader	"	2800	.009	S.Y.		.25	.06	.31	.49
3100	By hand	2 Clab	400	.040			1.07		1.07	1.78
4000	Remove sod, F.E. loader	B-10S	2000	.004			.15	.17	.32	.43
4100	Sod cutter	B-12K	3200	.005			.16	.39	.55	.69
4200	By hand	2 Clab	240	.067			1.78		1.78	2.96

32 91 19 – Landscape Grading

32 91 19.13 Topsoil Placement and Grading

		Crew	Daily Output	Labor-Hours	Unit	Material	2018 Bare Costs Labor	Equipment	Total	Total Incl O&P
0010	**TOPSOIL PLACEMENT AND GRADING**									
0300	Fine grade, base course for paving, see Section 32 11 23.23									
0701	Furnish and place, truck dumped, unscreened, 4" deep	B-10S	12000	.001	S.F.	.42	.02	.03	.47	.53
0801	6" deep	"	7400	.001	"	.49	.04	.05	.58	.66
0900	Fine grading and seeding, incl. lime, fertilizer & seed,									
1001	With equipment	B-14	9000	.005	S.F.	.07	.15	.03	.25	.37

32 92 Turf and Grasses

32 92 19 – Seeding

32 92 19.13 Mechanical Seeding

		Crew	Daily Output	Labor-Hours	Unit	Material	2018 Bare Costs Labor	Equipment	Total	Total Incl O&P
0010	**MECHANICAL SEEDING**									
0020	Mechanical seeding, 215 lb./acre	B-66	1.50	5.333	Acre	560	189	165	914	1,100
0100	44 lb./M.S.Y.	"	2500	.003	S.Y.	.20	.11	.10	.41	.52
0101	44 lb./M.S.Y.	1 Clab	13950	.001	S.F.	.02	.02		.04	.05
0300	Fine grading and seeding incl. lime, fertilizer & seed,									
0310	with equipment	B-14	1000	.048	S.Y.	.45	1.37	.31	2.13	3.11
0400	Fertilizer hand push spreader, 35 lb./M.S.F.	1 Clab	200	.040	M.S.F.	9.85	1.07		10.92	12.65
0600	Limestone hand push spreader, 50 lb./M.S.F.		180	.044		4.65	1.19		5.84	7.05
0800	Grass seed hand push spreader, 4.5 lb./M.S.F.		180	.044		22	1.19		23.19	26.50
1000	Hydro or air seeding for large areas, incl. seed and fertilizer	B-81	8900	.002	S.Y.	.50	.05	.06	.61	.71
1100	With wood fiber mulch added	"	8900	.002	"	2	.05	.06	2.11	2.36
1300	Seed only, over 100 lb., field seed, minimum				Lb.	1.85			1.85	2.04
1400	Maximum					1.75			1.75	1.93
1500	Lawn seed, minimum					1.45			1.45	1.60
1600	Maximum					2.50			2.50	2.75
1800	Aerial operations, seeding only, field seed	B-58	50	.480	Acre	600	14.20	63.50	677.70	755
1900	Lawn seed		50	.480		470	14.20	63.50	547.70	615
2100	Seed and liquid fertilizer, field seed		50	.480		675	14.20	63.50	752.70	840
2200	Lawn seed		50	.480		545	14.20	63.50	622.70	695

32 92 23 – Sodding

32 92 23.10 Sodding Systems

		Crew	Daily Output	Labor-Hours	Unit	Material	2018 Bare Costs Labor	Equipment	Total	Total Incl O&P
0010	**SODDING SYSTEMS**									
0020	Sodding, 1" deep, bluegrass sod, on level ground, over 8 M.S.F.	B-63	22	1.818	M.S.F.	246	48.50	7.90	302.40	360
0200	4 M.S.F.		17	2.353		240	63	10.25	313.25	380
0300	1,000 S.F.		13.50	2.963		290	79	12.90	381.90	465
0500	Sloped ground, over 8 M.S.F.		6	6.667		246	178	29	453	600
0600	4 M.S.F.		5	8		240	214	35	489	660
0700	1,000 S.F.		4	10		290	267	43.50	600.50	815
1000	Bent grass sod, on level ground, over 6 M.S.F.		20	2		257	53.50	8.70	319.20	380

32 92 Turf and Grasses

32 92 23 – Sodding

32 92 23.10 Sodding Systems		Crew	Daily Output	Labor-Hours	Unit	Material	2018 Bare Costs Labor	Equipment	Total	Total Incl O&P
1100	3 M.S.F.	B-63	18	2.222	M.S.F.	271	59.50	9.65	340.15	405
1200	Sodding 1,000 S.F. or less		14	2.857		297	76.50	12.45	385.95	465
1500	Sloped ground, over 6 M.S.F.		15	2.667		257	71	11.60	339.60	415
1600	3 M.S.F.		13.50	2.963		271	79	12.90	362.90	445
1700	1,000 S.F.	↓	12	3.333	↓	297	89	14.50	400.50	490

32 93 Plants

32 93 13 – Ground Covers

32 93 13.10 Ground Cover Plants

		Crew	Daily Output	Labor-Hours	Unit	Material	2018 Bare Costs Labor	Equipment	Total	Total Incl O&P
0010	**GROUND COVER PLANTS**									
0012	Plants, pachysandra, in prepared beds	B-1	15	1.600	C	79.50	44		123.50	161
0200	Vinca minor, 1 yr., bare root, in prepared beds		12	2	"	70	54.50		124.50	168
0600	Stone chips, in 50 lb. bags, Georgia marble		520	.046	Bag	4.80	1.26		6.06	7.40
0700	Onyx gemstone		260	.092		16	2.53		18.53	22
0800	Quartz		260	.092	↓	16	2.53		18.53	22
0900	Pea gravel, truckload lots	↓	28	.857	Ton	30	23.50		53.50	72.50

32 93 33 – Shrubs

32 93 33.10 Shrubs and Trees

		Crew	Daily Output	Labor-Hours	Unit	Material	2018 Bare Costs Labor	Equipment	Total	Total Incl O&P
0010	**SHRUBS AND TREES**									
0011	Evergreen, in prepared beds, B&B									
0100	Arborvitae pyramidal, 4'-5'	B-17	30	1.067	Ea.	103	32.50	21.50	157	191
0150	Globe, 12"-15"	B-1	96	.250		26	6.85		32.85	40
0300	Cedar, blue, 8'-10'	B-17	18	1.778		239	54	36	329	390
0500	Hemlock, Canadian, 2-1/2'-3'	B-1	36	.667		32.50	18.25		50.75	66
0550	Holly, Savannah, 8'-10' H		9.68	2.479		300	68		368	445
0600	Juniper, andorra, 18"-24"		80	.300		50	8.20		58.20	68.50
0620	Wiltoni, 15"-18"	↓	80	.300		26.50	8.20		34.70	43
0640	Skyrocket, 4-1/2'-5'	B-17	55	.582		110	17.60	11.85	139.45	163
0660	Blue pfitzer, 2'-2-1/2'	B-1	44	.545		39.50	14.95		54.45	68.50
0680	Ketleerie, 2-1/2'-3'		50	.480		56	13.15		69.15	83.50
0700	Pine, black, 2-1/2'-3'		50	.480		62	13.15		75.15	90.50
0720	Mugo, 18"-24"	↓	60	.400		56	10.95		66.95	80
0740	White, 4'-5'	B-17	75	.427		53	12.90	8.70	74.60	89.50
0800	Spruce, blue, 18"-24"	B-1	60	.400		69	10.95		79.95	94
0840	Norway, 4'-5'	B-17	75	.427		86	12.90	8.70	107.60	126
0900	Yew, denisforma, 12"-15"	B-1	60	.400		36.50	10.95		47.45	58.50
1000	Capitata, 18"-24"		30	.800		34	22		56	74
1100	Hicksi, 2'-2-1/2'	↓	30	.800	↓	103	22		125	151

32 93 33.20 Shrubs

		Crew	Daily Output	Labor-Hours	Unit	Material	2018 Bare Costs Labor	Equipment	Total	Total Incl O&P
0010	**SHRUBS**									
0011	Broadleaf Evergreen, planted in prepared beds									
0100	Andromeda, 15"-18", cont	B-1	96	.250	Ea.	33.50	6.85		40.35	48.50
0200	Azalea, 15"-18", cont		96	.250		30.50	6.85		37.35	45.50
0300	Barberry, 9"-12", cont		130	.185		18.70	5.05		23.75	29
0400	Boxwood, 15"-18", B&B		96	.250		44.50	6.85		51.35	60.50
0500	Euonymus, emerald gaiety, 12"-15", cont		115	.209		24.50	5.70		30.20	36.50
0600	Holly, 15"-18", B&B		96	.250		40	6.85		46.85	55.50
0900	Mount laurel, 18"-24", B&B		80	.300		72.50	8.20		80.70	93.50
1000	Paxistema, 9"-12" H		130	.185		22	5.05		27.05	32.50
1100	Rhododendron, 18"-24", cont	↓	48	.500	↓	38.50	13.70		52.20	65.50

For customer support on your Residential Costs with RSMeans data, call 800.448.8182.

653

32 93 Plants

32 93 33 – Shrubs

32 93 33.20 Shrubs

		Crew	Daily Output	Labor-Hours	Unit	Material	2018 Bare Costs Labor	Equipment	Total	Total Incl O&P
1200	Rosemary, 1 gal. cont	B-1	600	.040	Ea.	18.55	1.09		19.64	22.50
2000	Deciduous, planted in prepared beds, amelanchier, 2'-3', B&B		57	.421		123	11.50		134.50	154
2100	Azalea, 15"-18", B&B		96	.250		30.50	6.85		37.35	45
2300	Bayberry, 2'-3', B&B		57	.421		30	11.50		41.50	52
2600	Cotoneaster, 15"-18", B&B		80	.300		27	8.20		35.20	43.50
2800	Dogwood, 3'-4', B&B	B-17	40	.800		33	24	16.25	73.25	94.50
2900	Euonymus, alatus compacta, 15"-18", cont	B-1	80	.300		27	8.20		35.20	43
3200	Forsythia, 2'-3', cont	"	60	.400		17.80	10.95		28.75	38
3300	Hibiscus, 3'-4', B&B	B-17	75	.427		42	12.90	8.70	63.60	77
3400	Honeysuckle, 3'-4', B&B	B-1	60	.400		27.50	10.95		38.45	48.50
3500	Hydrangea, 2'-3', B&B	"	57	.421		31	11.50		42.50	53
3600	Lilac, 3'-4', B&B	B-17	40	.800		27.50	24	16.25	67.75	88
3900	Privet, bare root, 18"-24"	B-1	80	.300		15.15	8.20		23.35	30.50
4100	Quince, 2'-3', B&B	"	57	.421		29	11.50		40.50	51
4200	Russian olive, 3'-4', B&B	B-17	75	.427		25	12.90	8.70	46.60	58.50
4400	Spirea, 3'-4', B&B	B-1	70	.343		20.50	9.40		29.90	38
4500	Viburnum, 3'-4', B&B	B-17	40	.800		26	24	16.25	66.25	86.50

32 93 43 – Trees

32 93 43.20 Trees

			Crew	Daily Output	Labor-Hours	Unit	Material	2018 Bare Costs Labor	Equipment	Total	Total Incl O&P
0010	**TREES**										
0011	Deciduous, in prep. beds, balled & burlapped (B&B)										
0100	Ash, 2" caliper	G	B-17	8	4	Ea.	200	121	81.50	402.50	510
0200	Beech, 5'-6'	G		50	.640		210	19.35	13	242.35	277
0300	Birch, 6'-8', 3 stems	G		20	1.600		167	48.50	32.50	248	300
0500	Crabapple, 6'-8'	G		20	1.600		139	48.50	32.50	220	269
0600	Dogwood, 4'-5'	G		40	.800		140	24	16.25	180.25	212
0700	Eastern redbud, 4'-5'	G		40	.800		148	24	16.25	188.25	221
0800	Elm, 8'-10'	G		20	1.600		325	48.50	32.50	406	475
0900	Ginkgo, 6'-7'	G		24	1.333		148	40.50	27	215.50	260
1000	Hawthorn, 8'-10', 1" caliper	G		20	1.600		161	48.50	32.50	242	293
1100	Honeylocust, 10'-12', 1-1/2" caliper	G		10	3.200		206	97	65	368	460
1300	Larch, 8'	G		32	1		128	30.50	20.50	179	213
1400	Linden, 8'-10', 1" caliper	G		20	1.600		144	48.50	32.50	225	274
1500	Magnolia, 4'-5'	G		20	1.600		115	48.50	32.50	196	243
1600	Maple, red, 8'-10', 1-1/2" caliper	G		10	3.200		199	97	65	361	450
1700	Mountain ash, 8'-10', 1" caliper	G		16	2		177	60.50	40.50	278	340
1800	Oak, 2-1/2"-3" caliper	G		6	5.333		320	161	108	589	740
2100	Planetree, 9'-11', 1-1/4" caliper	G		10	3.200		266	97	65	428	525
2200	Plum, 6'-8', 1" caliper	G		20	1.600		79	48.50	32.50	160	203
2300	Poplar, 9'-11', 1-1/4" caliper	G		10	3.200		100	97	65	262	340
2500	Sumac, 2'-3'	G		75	.427		44.50	12.90	8.70	66.10	80
2700	Tulip, 5'-6'	G		40	.800		46	24	16.25	86.25	108
2800	Willow, 6'-8', 1" caliper	G		20	1.600		97	48.50	32.50	178	223

32 94 Planting Accessories

32 94 13 – Landscape Edging

32 94 13.20 Edging		Crew	Daily Output	Labor-Hours	Unit	Material	2018 Bare Costs Labor	Equipment	Total	Total Incl O&P
0010	**EDGING**									
0050	Aluminum alloy, including stakes, 1/8" x 4", mill finish	B-1	390	.062	L.F.	2.21	1.68		3.89	5.25
0051	Black paint		390	.062		2.56	1.68		4.24	5.60
0052	Black anodized		390	.062		2.96	1.68		4.64	6.05
0100	Brick, set horizontally, 1-1/2 bricks per L.F.	D-1	370	.043		1.39	1.37		2.76	3.82
0150	Set vertically, 3 bricks per L.F.	"	135	.119		3.54	3.75		7.29	10.15
0200	Corrugated aluminum, roll, 4" wide	1 Carp	650	.012		2.19	.43		2.62	3.13
0250	6" wide	"	550	.015		2.74	.51		3.25	3.86
0600	Railroad ties, 6" x 8"	2 Carp	170	.094		2.36	3.29		5.65	8.05
0650	7" x 9"		136	.118		2.62	4.11		6.73	9.75
0750	Redwood 2" x 4"		330	.048		2.26	1.69		3.95	5.30
0800	Steel edge strips, incl. stakes, 1/4" x 5"	B-1	390	.062		4.54	1.68		6.22	7.80
0850	3/16" x 4"	"	390	.062		3.59	1.68		5.27	6.75

32 94 50 – Tree Guying

32 94 50.10 Tree Guying Systems

		Crew	Daily Output	Labor-Hours	Unit	Material	Labor	Equipment	Total	Total Incl O&P
0010	**TREE GUYING SYSTEMS**									
0015	Tree guying including stakes, guy wire and wrap									
0100	Less than 3" caliper, 2 stakes	2 Clab	35	.457	Ea.	13.10	12.20		25.30	35
0200	3" to 4" caliper, 3 stakes	"	21	.762	"	19.85	20.50		40.35	56
1000	Including arrowhead anchor, cable, turnbuckles and wrap									
1100	Less than 3" caliper, 3" anchors	2 Clab	20	.800	Ea.	21	21.50		42.50	59
1200	3" to 6" caliper, 4" anchors		15	1.067		31.50	28.50		60	82.50
1300	6" caliper, 6" anchors		12	1.333		23.50	35.50		59	85
1400	8" caliper, 8" anchors		9	1.778		116	47.50		163.50	207

32 96 Transplanting

32 96 23 – Plant and Bulb Transplanting

32 96 23.23 Planting

		Crew	Daily Output	Labor-Hours	Unit	Material	Labor	Equipment	Total	Total Incl O&P
0010	**PLANTING**									
0012	Moving shrubs on site, 12" ball	B-62	28	.857	Ea.		25.50	6.20	31.70	49
0100	24" ball	"	22	1.091	"		32.50	7.90	40.40	62

32 96 23.43 Moving Trees

		Crew	Daily Output	Labor-Hours	Unit	Material	Labor	Equipment	Total	Total Incl O&P
0010	**MOVING TREES**, On site									
0300	Moving trees on site, 36" ball	B-6	3.75	6.400	Ea.		189	83.50	272.50	405
0400	60" ball	"	1	24	"		710	310	1,020	1,525

For customer support on your Residential Costs with RSMeans data, call 800.448.8182.

655

Division Notes

		CREW	DAILY OUTPUT	LABOR-HOURS	UNIT	BARE COSTS				TOTAL INCL O&P
						MAT.	LABOR	EQUIP.	TOTAL	

Estimating Tips
33 10 00 Water Utilities
33 30 00 Sanitary Sewerage Utilities
33 40 00 Storm Drainage Utilities

- Never assume that the water, sewer, and drainage lines will go in at the early stages of the project. Consider the site access needs before dividing the site in half with open trenches, loose pipe, and machinery obstructions. Always inspect the site to establish that the site drawings are complete. Check off all existing utilities on your drawings as you locate them. Be especially careful with underground utilities because appurtenances are sometimes buried during regrading or repaving operations. If you find any discrepancies, mark up the site plan for further research. Differing site conditions can be very costly if discovered later in the project.

- See also Section 33 01 00 for restoration of pipe where removal/replacement may be undesirable. Use of new types of piping materials can reduce the overall project cost. Owners/design engineers should consider the installing contractor as a valuable source of current information on utility products and local conditions that could lead to significant cost savings.

Reference Numbers

Reference numbers are shown at the beginning of some major classifications. These numbers refer to related items in the Reference Section. The reference information may be an estimating procedure, an alternate pricing method, or technical information.

Note: Not all subdivisions listed here necessarily appear. ■

---

33 05 Common Work Results for Utilities

33 05 07 – Trenchless Installation of Utility Piping

33 05 07.36 Microtunneling

33 05 07.36 Microtunneling	Crew	Daily Output	Labor-Hours	Unit	Material	2018 Bare Costs Labor	Equipment	Total	Total Incl O&P
0010 **MICROTUNNELING**									
0011 Not including excavation, backfill, shoring,									
0020 or dewatering, average 50'/day, slurry method									
0100 24" to 48" outside diameter, minimum				L.F.				965	965
0110 Adverse conditions, add				"				500	500
1000 Rent microtunneling machine, average monthly lease				Month				97,500	107,000
1010 Operating technician				Day				630	690
1100 Mobilization and demobilization, minimum				Job				41,200	45,900
1110 Maximum				"				445,500	490,500

33 05 61 – Concrete Manholes

33 05 61.10 Storm Drainage Manholes, Frames and Covers

	Crew	Daily Output	Labor-Hours	Unit	Material	2018 Bare Costs Labor	Equipment	Total	Total Incl O&P
0010 **STORM DRAINAGE MANHOLES, FRAMES & COVERS**									
0020 Excludes footing, excavation, backfill (See line items for frame & cover)									
0050 Brick, 4' inside diameter, 4' deep	D-1	1	16	Ea.	590	505		1,095	1,500
1110 Precast, 4' ID, 4' deep	B-22	4.10	7.317	"	845	214	47.50	1,106.50	1,325

33 05 63 – Concrete Vaults and Chambers

33 05 63.13 Precast Concrete Utility Structures

	Crew	Daily Output	Labor-Hours	Unit	Material	2018 Bare Costs Labor	Equipment	Total	Total Incl O&P
0010 **PRECAST CONCRETE UTILITY STRUCTURES**, 6" thick									
0050 5' x 10' x 6' high, ID	B-13	2	24	Ea.	1,800	690	295	2,785	3,450
0350 Hand hole, precast concrete, 1-1/2" thick									
0400 1'-0" x 2'-0" x 1'-9", ID, light duty	B-1	4	6	Ea.	430	164		594	745
0450 4'-6" x 3'-2" x 2'-0", OD, heavy duty	B-6	3	8	"	1,525	237	104	1,866	2,175

33 11 Groundwater Sources

33 11 13 – Potable Water Supply Wells

33 11 13.10 Wells and Accessories

	Crew	Daily Output	Labor-Hours	Unit	Material	2018 Bare Costs Labor	Equipment	Total	Total Incl O&P
0010 **WELLS & ACCESSORIES**									
0011 Domestic									
0100 Drilled, 4" to 6" diameter	B-23	120	.333	L.F.		9.05	22.50	31.55	39.50
1500 Pumps, installed in wells to 100' deep, 4" submersible									
1520 3/4 HP	Q-1	2.66	6.015	Ea.	845	216		1,061	1,275
1600 1 HP	"	2.29	6.987	"	910	250		1,160	1,400

33 14 Water Utility Transmission and Distribution

33 14 13 – Public Water Utility Distribution Piping

33 14 13.15 Water Supply, Ductile Iron Pipe

	Crew	Daily Output	Labor-Hours	Unit	Material	2018 Bare Costs Labor	Equipment	Total	Total Incl O&P
0010 **WATER SUPPLY, DUCTILE IRON PIPE**									
0020 Not including excavation or backfill									
2000 Pipe, class 50 water piping, 18' lengths									
2020 Mechanical joint, 4" diameter	B-21A	200	.200	L.F.	30.50	6.55	1.82	38.87	46.50
2040 6" diameter		160	.250		37.50	8.20	2.27	47.97	57
3000 Push-on joint, 4" diameter		400	.100		21	3.28	.91	25.19	29.50
3020 6" diameter		333.33	.120		21.50	3.94	1.09	26.53	31.50
8000 Piping, fittings, mechanical joint, AWWA C110									
8006 90° bend, 4" diameter	B-20A	16	2	Ea.	170	63.50		233.50	292
8020 6" diameter		12.80	2.500		255	79.50		334.50	410
8200 Wye or tee, 4" diameter		10.67	2.999		380	95.50		475.50	575
8220 6" diameter		8.53	3.751		570	119		689	820

33 14 Water Utility Transmission and Distribution

33 14 13 – Public Water Utility Distribution Piping

33 14 13.15 Water Supply, Ductile Iron Pipe

		Crew	Daily Output	Labor-Hours	Unit	Material	2018 Bare Costs Labor	Equipment	Total	Total Incl O&P
8398	45° bend, 4" diameter	B-20A	16	2	Ea.	211	63.50		274.50	335
8400	6" diameter		12.80	2.500		253	79.50		332.50	410
8405	8" diameter		10.67	2.999		365	95.50		460.50	560
8450	Decreaser, 6" x 4" diameter		14.22	2.250		232	71.50		303.50	375
8460	8" x 6" diameter		11.64	2.749		350	87.50		437.50	530
8550	Piping, butterfly valves, cast iron									
8560	4" diameter	B-20	6	4	Ea.	405	109		514	625
8700	Joint restraint, ductile iron mechanical joints									
8710	4" diameter	B-20A	32	1	Ea.	31	32		63	86.50
8720	6" diameter		25.60	1.250		37.50	39.50		77	107
8730	8" diameter		21.33	1.500		56	47.50		103.50	141
8740	10" diameter		18.28	1.751		87	55.50		142.50	188
8750	12" diameter		16.84	1.900		114	60.50		174.50	226
8760	14" diameter		16	2		129	63.50		192.50	247
8770	16" diameter		11.64	2.749		155	87.50		242.50	315
8780	18" diameter		11.03	2.901		217	92		309	390
8785	20" diameter		9.14	3.501		267	111		378	480
8790	24" diameter		7.53	4.250		370	135		505	630
9600	Steel sleeve with tap, 4" diameter	B-20	3	8		440	219		659	850
9620	6" diameter	"	2	12		510	330		840	1,100

33 14 13.25 Water Supply, Polyvinyl Chloride Pipe

		Crew	Daily Output	Labor-Hours	Unit	Material	2018 Bare Costs Labor	Equipment	Total	Total Incl O&P
0010	**WATER SUPPLY, POLYVINYL CHLORIDE PIPE**									
2100	PVC pipe, Class 150, 1-1/2" diameter	Q-1A	750	.013	L.F.	.52	.54		1.06	1.45
2120	2" diameter		686	.015		.91	.59		1.50	1.96
2140	2-1/2" diameter		500	.020		1.18	.80		1.98	2.62
2160	3" diameter	B-20	430	.056		1.59	1.53		3.12	4.29
8700	PVC pipe, joint restraint									
8710	4" diameter	B-20A	32	1	Ea.	44.50	32		76.50	102
8720	6" diameter		25.60	1.250		55	39.50		94.50	127
8730	8" diameter		21.33	1.500		81.50	47.50		129	169
8740	10" diameter		18.28	1.751		147	55.50		202.50	253
8750	12" diameter		16.84	1.900		154	60.50		214.50	269
8760	14" diameter		16	2		218	63.50		281.50	345
8770	16" diameter		11.64	2.749		293	87.50		380.50	470
8780	18" diameter		11.03	2.901		365	92		457	550
8785	20" diameter		9.14	3.501		465	111		576	700
8790	24" diameter		7.53	4.250		540	135		675	820

33 14 17 – Site Water Utility Service Laterals

33 14 17.15 Tapping, Crosses and Sleeves

		Crew	Daily Output	Labor-Hours	Unit	Material	2018 Bare Costs Labor	Equipment	Total	Total Incl O&P
0010	**TAPPING, CROSSES AND SLEEVES**									
4000	Drill and tap pressurized main (labor only)									
4100	6" main, 1" to 2" service	Q-1	3	5.333	Ea.		191		191	315
4150	8" main, 1" to 2" service	"	2.75	5.818	"		208		208	345
4500	Tap and insert gate valve									
4600	8" main, 4" branch	B-21	3.20	8.750	Ea.		252	40.50	292.50	465
4650	6" branch		2.70	10.370			298	48	346	550
4700	10" main, 4" branch		2.70	10.370			298	48	346	550
4750	6" branch		2.35	11.915			345	55	400	630
4800	12" main, 6" branch		2.35	11.915			345	55	400	630

For customer support on your Residential Costs with RSMeans data, call 800.448.8182.

659

33 31 Sanitary Sewerage Piping

33 31 11 – Public Sanitary Sewerage Gravity Piping

33 31 11.15 Sewage Collection, Concrete Pipe

33 31 11.15 Sewage Collection, Concrete Pipe	Crew	Daily Output	Labor-Hours	Unit	Material	2018 Bare Costs Labor	Equipment	Total	Total Incl O&P
0010 **SEWAGE COLLECTION, CONCRETE PIPE**									
0020 See Section 33 41 13.60 for sewage/drainage collection, concrete pipe									

33 31 11.25 Sewage Collection, Polyvinyl Chloride Pipe

33 31 11.25	Crew	Daily Output	Labor-Hours	Unit	Material	2018 Bare Costs Labor	Equipment	Total	Total Incl O&P
0010 **SEWAGE COLLECTION, POLYVINYL CHLORIDE PIPE**									
0020 Not including excavation or backfill									
2000 20' lengths, SDR 35, B&S, 4" diameter	B-20	375	.064	L.F.	1.64	1.75		3.39	4.71
2040 6" diameter		350	.069		3.49	1.88		5.37	6.95
2080 13' lengths, SDR 35, B&S, 8" diameter	↓	335	.072		6.50	1.96		8.46	10.40
2120 10" diameter	B-21	330	.085		11.35	2.44	.39	14.18	16.95
4000 Piping, DWV PVC, no exc./bkfill., 10' L, Sch 40, 4" diameter	B-20	375	.064		3.74	1.75		5.49	7
4010 6" diameter		350	.069		8.10	1.88		9.98	12
4020 8" diameter		335	.072	↓	12.75	1.96		14.71	17.25
4030 Fittings, 1/4 bend DWV PVC, 4" diameter		19	1.263	Ea.	23	34.50		57.50	82.50
4040 6" diameter		12	2		105	54.50		159.50	206
4050 8" diameter		11	2.182		105	59.50		164.50	215
4060 1/8 bend DWV PVC, 4" diameter		19	1.263		19.05	34.50		53.55	78.50
4070 6" diameter		12	2		71	54.50		125.50	169
4080 8" diameter		11	2.182		116	59.50		175.50	228
4090 Tee DWV PVC, 4" diameter		12	2		31	54.50		85.50	126
4100 6" diameter		10	2.400		125	65.50		190.50	247
4110 8" diameter	↓	9	2.667	↓	277	73		350	425

33 34 Onsite Wastewater Disposal

33 34 13 – Septic Tanks

33 34 13.13 Concrete Septic Tanks

33 34 13.13 Concrete Septic Tanks	Crew	Daily Output	Labor-Hours	Unit	Material	2018 Bare Costs Labor	Equipment	Total	Total Incl O&P
0010 **CONCRETE SEPTIC TANKS**									
0011 Not including excavation or piping									
0015 Septic tanks, precast, 1,000 gallon	B-21	8	3.500	Ea.	1,025	101	16.20	1,142.20	1,325
0060 1,500 gallon		7	4		1,600	115	18.50	1,733.50	1,975
0100 2,000 gallon	↓	5	5.600		2,275	161	26	2,462	2,800
0900 Concrete riser 24" x 12" with standard lid	1 Clab	6	1.333		87	35.50		122.50	155
0905 24" x 12" with heavy duty lid		6	1.333		110	35.50		145.50	180
0910 24" x 8" with standard lid		6	1.333		81	35.50		116.50	148
0915 24" x 8" with heavy duty lid		6	1.333		104	35.50		139.50	174
0917 24" x 12" extension		12	.667		58.50	17.80		76.30	94
0920 24" x 8" extension		12	.667		53	17.80		70.80	87.50
0950 HDPE riser 20" x 12" with standard lid		8	1		135	26.50		161.50	193
0951 HDPE 20" x 12" with heavy duty lid		8	1		150	26.50		176.50	209
0955 HDPE 24" x 12" with standard lid		8	1		119	26.50		145.50	176
0956 HDPE 24" x 12" with heavy duty lid		8	1		145	26.50		171.50	204
0960 HDPE 20" x 6" extension		48	.167		30	4.45		34.45	40.50
0962 HDPE 20" x 12" extension		48	.167		55	4.45		59.45	68
0965 HDPE 24" x 6" extension		48	.167		29.50	4.45		33.95	40
0967 HDPE 24" x 12" extension	↓	48	.167		47.50	4.45		51.95	59.50
1150 Leaching field chambers, 13' x 3'-7" x 1'-4", standard	B-13	16	3		500	86.50	37	623.50	735
1420 Leaching pit, precast concrete, 6' diameter, 3' deep	B-21	4.70	5.957	↓	815	171	27.50	1,013.50	1,200

33 34 13.33 Polyethylene Septic Tanks

33 34 13.33 Polyethylene Septic Tanks	Crew	Daily Output	Labor-Hours	Unit	Material	2018 Bare Costs Labor	Equipment	Total	Total Incl O&P
0010 **POLYETHYLENE SEPTIC TANKS**									
0015 High density polyethylene, 1,000 gallon	B-21	8	3.500	Ea.	1,325	101	16.20	1,442.20	1,625
0020 1,250 gallon	↓	8	3.500		1,200	101	16.20	1,317.20	1,475

For customer support on your Residential Costs with RSMeans data, call 800.448.8182.

33 34 13 – Septic Tanks

33 34 13.33 Polyethylene Septic Tanks	Crew	Daily Output	Labor-Hours	Unit	Material	2018 Bare Costs Labor	Equipment	Total	Total Incl O&P
0025 1,500 gallon	B-21	7	4	Ea.	1,350	115	18.50	1,483.50	1,675

33 34 16 – Septic Tank Effluent Filters

33 34 16.13 Septic Tank Gravity Effluent Filters

	Crew	Daily Output	Labor-Hours	Unit	Material	Labor	Equipment	Total	Total Incl O&P
0010 **SEPTIC TANK GRAVITY EFFLUENT FILTERS**									
3000 Effluent filter, 4" diameter	1 Skwk	8	1	Ea.	39.50	36		75.50	104
3020 6" diameter	"	7	1.143	"	55	41		96	129

33 34 51 – Drainage Field Systems

33 34 51.10 Drainage Field Excavation and Fill

	Crew	Daily Output	Labor-Hours	Unit	Material	Labor	Equipment	Total	Total Incl O&P
0010 **DRAINAGE FIELD EXCAVATION AND FILL**									
2200 Septic tank & drainage field excavation with 3/4 C.Y. backhoe	B-12F	145	.110	C.Y.		3.52	4.62	8.14	10.90
2400 4' trench for disposal field, 3/4 C.Y. backhoe	"	335	.048	L.F.		1.52	2	3.52	4.72
2600 Gravel fill, run of bank	B-6	150	.160	C.Y.	16.80	4.74	2.08	23.62	28.50
2800 Crushed stone, 3/4"	"	150	.160	"	36	4.74	2.08	42.82	49.50

33 34 51.13 Utility Septic Tank Tile Drainage Field

	Crew	Daily Output	Labor-Hours	Unit	Material	Labor	Equipment	Total	Total Incl O&P
0010 **UTILITY SEPTIC TANK TILE DRAINAGE FIELD**									
0015 Distribution box, concrete, 5 outlets	2 Clab	20	.800	Ea.	92.50	21.50		114	138
0020 7 outlets		16	1		92.50	26.50		119	147
0025 9 outlets		8	2		550	53.50		603.50	695
0115 Distribution boxes, HDPE, 5 outlets		20	.800		75	21.50		96.50	118
0117 6 outlets		15	1.067		75	28.50		103.50	130
0118 7 outlets		15	1.067		75	28.50		103.50	130
0120 8 outlets		10	1.600		79	42.50		121.50	158
0240 Distribution boxes, outlet flow leveler	1 Clab	50	.160		2.22	4.27		6.49	9.55
0300 Precast concrete, galley, 4' x 4' x 4'	B-21	16	1.750		236	50.50	8.10	294.60	350
0350 HDPE infiltration chamber 12" H x 15" W	2 Clab	300	.053	L.F.	6.95	1.42		8.37	10
0351 12" H x 15" W end cap	1 Clab	32	.250	Ea.	18.65	6.70		25.35	31.50
0355 chamber 12" H x 22" W	2 Clab	300	.053	L.F.	6.45	1.42		7.87	9.45
0356 12" H x 22" W end cap	1 Clab	32	.250	Ea.	16.80	6.70		23.50	29.50
0360 chamber 13" H x 34" W	2 Clab	300	.053	L.F.	14.15	1.42		15.57	17.95
0361 13" H x 34" W end cap	1 Clab	32	.250	Ea.	50.50	6.70		57.20	66.50
0365 chamber 16" H x 34" W	2 Clab	300	.053	L.F.	19.05	1.42		20.47	23.50
0366 16" H x 34" W end cap	1 Clab	32	.250	Ea.	15.70	6.70		22.40	28.50
0370 chamber 8" H x 16" W	2 Clab	300	.053	L.F.	9.85	1.42		11.27	13.15
0371 8" H x 16" W end cap	1 Clab	32	.250	Ea.	11	6.70		17.70	23

33 41 Subdrainage

33 41 16 – Subdrainage Piping

33 41 16.25 Piping, Subdrainage, Corrugated Metal

	Crew	Daily Output	Labor-Hours	Unit	Material	Labor	Equipment	Total	Total Incl O&P
0010 **PIPING, SUBDRAINAGE, CORRUGATED METAL**									
0021 Not including excavation and backfill									
2010 Aluminum, perforated									
2020 6" diameter, 18 ga.	B-20	380	.063	L.F.	6.65	1.73		8.38	10.20
2200 8" diameter, 16 ga.	"	370	.065		8.25	1.78		10.03	12.05
2220 10" diameter, 16 ga.	B-21	360	.078		10.30	2.24	.36	12.90	15.45
3000 Uncoated galvanized, perforated									
3020 6" diameter, 18 ga.	B-20	380	.063	L.F.	6.45	1.73		8.18	10
3200 8" diameter, 16 ga.	"	370	.065		7.85	1.78		9.63	11.60
3220 10" diameter, 16 ga.	B-21	360	.078		8.30	2.24	.36	10.90	13.25
3240 12" diameter, 16 ga.	"	285	.098		9.25	2.82	.45	12.52	15.40
4000 Steel, perforated, asphalt coated									

For customer support on your Residential Costs with RSMeans data, call 800.448.8182.

661

33 41 Subdrainage

33 41 16 – Subdrainage Piping

33 41 16.25 Piping, Subdrainage, Corrugated Metal	Crew	Daily Output	Labor-Hours	Unit	Material	2018 Bare Costs Labor	Equipment	Total	Total Incl O&P
4020	6" diameter, 18 ga.	B-20	380	.063	L.F.	6.65	1.73		8.38	10.20
4030	8" diameter, 18 ga.	"	370	.065		8.25	1.78		10.03	12.05
4040	10" diameter, 16 ga.	B-21	360	.078		10.40	2.24	.36	13	15.55
4050	12" diameter, 16 ga.		285	.098		11.65	2.82	.45	14.92	18.05
4060	18" diameter, 16 ga.	▼	205	.137	▼	18.90	3.93	.63	23.46	28

33 42 Stormwater Conveyance

33 42 11 – Stormwater Gravity Piping

33 42 11.60 Sewage/Drainage Collection, Concrete Pipe

		Crew	Daily Output	Labor-Hours	Unit	Material	2018 Bare Costs Labor	Equipment	Total	Total Incl O&P
0010	**SEWAGE/DRAINAGE COLLECTION, CONCRETE PIPE**									
0020	Not including excavation or backfill									
1020	8" diameter	B-14	224	.214	L.F.	8.30	6.10	1.40	15.80	21
1030	10" diameter	"	216	.222	"	9.20	6.35	1.45	17	22
3780	Concrete slotted pipe, class 4 mortar joint									
3800	12" diameter	B-21	168	.167	L.F.	31	4.79	.77	36.56	43
3840	18" diameter	"	152	.184	"	35	5.30	.85	41.15	48
3900	Concrete slotted pipe, Class 4 O-ring joint									
3940	12" diameter	B-21	168	.167	L.F.	28	4.79	.77	33.56	39.50
3960	18" diameter	"	152	.184	"	32	5.30	.85	38.15	44.50

33 42 33 – Stormwater Curbside Drains and Inlets

33 42 33.13 Catch Basins

		Crew	Daily Output	Labor-Hours	Unit	Material	2018 Bare Costs Labor	Equipment	Total	Total Incl O&P
0010	**CATCH BASINS**									
0011	Not including footing & excavation									
1600	Frames & grates, C.I., 24" square, 500 lb.	B-6	7.80	3.077	Ea.	355	91	40	486	585

33 52 Hydrocarbon Transmission and Distribution

33 52 16 – Gas Hydrocarbon Piping

33 52 16.13 Steel Natural Gas Piping

		Crew	Daily Output	Labor-Hours	Unit	Material	2018 Bare Costs Labor	Equipment	Total	Total Incl O&P
0010	**STEEL NATURAL GAS PIPING**									
0020	Not including excavation or backfill, tar coated and wrapped									
4000	Pipe schedule 40, plain end									
4040	1" diameter	Q-4	300	.107	L.F.	5.20	4.03	.20	9.43	12.55
4080	2" diameter	"	280	.114	"	8.15	4.32	.21	12.68	16.30

33 52 16.20 Piping, Gas Service and Distribution, P.E.

		Crew	Daily Output	Labor-Hours	Unit	Material	2018 Bare Costs Labor	Equipment	Total	Total Incl O&P
0010	**PIPING, GAS SERVICE AND DISTRIBUTION, POLYETHYLENE**									
0020	Not including excavation or backfill									
1000	60 psi coils, compression coupling @ 100', 1/2" diameter, SDR 11	B-20A	608	.053	L.F.	.50	1.67		2.17	3.31
1010	1" diameter, SDR 11		544	.059		1.16	1.87		3.03	4.36
1040	1-1/4" diameter, SDR 11		544	.059		1.64	1.87		3.51	4.89
1100	2" diameter, SDR 11		488	.066		2.55	2.08		4.63	6.25
1160	3" diameter, SDR 11	▼	408	.078		5.35	2.49		7.84	10
1500	60 psi 40' joints with coupling, 3" diameter, SDR 11	B-21A	408	.098		7.30	3.22	.89	11.41	14.30
1540	4" diameter, SDR 11		352	.114		13	3.73	1.03	17.76	21.50
1600	6" diameter, SDR 11		328	.122		31	4	1.11	36.11	42
1640	8" diameter, SDR 11	▼	272	.147	▼	44.50	4.83	1.34	50.67	58.50

Estimating Tips

- When estimating costs for the installation of electrical power generation equipment, factors to review include access to the job site, access and setting up at the installation site, required connections, uncrating pads, anchors, leveling, final assembly of the components, and temporary protection from physical damage, such as environmental exposure.

- Be aware of the costs of equipment supports, concrete pads, and vibration isolators. Cross-reference them against other trades' specifications. Also, review site and structural drawings for items that must be included in the estimates.

- It is important to include items that are not documented in the plans and specifications but must be priced. These items include, but are not limited to, testing, dust protection, roof penetration, core drilling concrete floors and walls, patching, cleanup, and final adjustments. Add a contingency or allowance for utility company fees for power hookups, if needed.

- The project size and scope of electrical power generation equipment will have a significant impact on cost. The intent of RSMeans cost data is to provide a benchmark cost so that owners, engineers, and electrical contractors will have a comfortable number with which to start a project. Additionally, there are many websites available to use for research and to obtain a vendor's quote to finalize costs.

Reference Numbers

Reference numbers are shown at the beginning of some major classifications. These numbers refer to related items in the Reference Section. The reference information may be an estimating procedure, an alternate pricing method, or technical information.

Note: Not all subdivisions listed here necessarily appear. ■

48 15 13.50 Wind Turbines and Components	Crew	Daily Output	Labor-Hours	Unit	Material	2018 Bare Costs Labor	Equipment	Total	Total Incl O&P
0010 **WIND TURBINES & COMPONENTS**									
0500 Complete system, grid connected									
1000 20 kW, 31' diam., incl. labor & material	G			System				49,900	49,900
1010 Enhanced	G							55,000	55,000
1500 10 kW, 23' diam., incl. labor & material	G							74,000	74,000
2000 2.4 kW, 12' diam., incl. labor & material	G							18,000	18,000
2900 Component system									
3200 1,000 W, 9' diam.	G	1 Elec 2.05	3.902	Ea.	1,500	157		1,657	1,900
3400 Mounting hardware									
3500 30' guyed tower kit	G	2 Clab 5.12	3.125	Ea.	395	83.50		478.50	570
3505 3' galvanized helical earth screw	G	1 Clab 8	1		48	26.50		74.50	97
3510 Attic mount kit	G	1 Rofc 2.56	3.125		202	93.50		295.50	390
3520 Roof mount kit	G	1 Clab 3.41	2.346		292	62.50		354.50	425
8900 Equipment									
9100 DC to AC inverter for, 48 V, 4,000 W	G	1 Elec 2	4	Ea.	2,275	161		2,436	2,775

Reference Section

All the reference information is in one section, making it easy to find what you need to know and easy to use the data set on a daily basis. This section is visually identified by a vertical black bar on the page edges.

In this Reference Section, we've included Equipment Rental Costs, a listing of rental and operating costs; Crew Listings, a full listing of all crews, equipment, and their costs; Location Factors for adjusting costs to the region you are in; Reference Tables, where you will find explanations, estimating information and procedures, and technical data; an explanation of all the Abbreviations in the data set; and sample Estimating Forms.

Table of Contents

Estimating Tips

- This section contains the average costs to rent and operate hundreds of pieces of construction equipment. This is useful information when estimating the time and material requirements of any particular operation in order to establish a unit or total cost. Bare equipment costs shown on a unit cost line include not only rental, but also operating costs for equipment under normal use.

Rental Costs

- Equipment rental rates are obtained from the following industry sources throughout North America: contractors, suppliers, dealers, manufacturers, and distributors.

- Rental rates vary throughout the country, with larger cities generally having lower rates. Lease plans for new equipment are available for periods in excess of six months, with a percentage of payments applying toward purchase.

- Monthly rental rates vary from 2% to 5% of the purchase price of the equipment depending on the anticipated life of the equipment and its wearing parts.

- Weekly rental rates are about 1/3 of the monthly rates, and daily rental rates are about 1/3 of the weekly rate.

- Rental rates can also be treated as reimbursement costs for contractor-owned equipment. Owned equipment costs include depreciation, loan payments, interest, taxes, insurance, storage, and major repairs.

Operating Costs

- The operating costs include parts and labor for routine servicing, such as the repair and replacement of pumps, filters, and worn lines. Normal operating expendables, such as fuel, lubricants, tires, and electricity (where applicable), are also included.

- Extraordinary operating expendables with highly variable wear patterns, such as diamond bits and blades, are excluded. These costs can be found as material costs in the Unit Price section.

- The hourly operating costs listed do not include the operator's wages.

Equipment Cost/Day

- Any power equipment required by a crew is shown in the Crew Listings with a daily cost.

- This daily cost of equipment needed by a crew includes both the rental cost and the operating cost and is based on dividing the weekly rental rate by 5 (number of working days in the week), and then adding the hourly operating cost times 8 (the number of hours in a day). This "Equipment Cost/Day" is shown in the far right column of the Equipment Rental section.

- If equipment is needed for only one or two days, it is best to develop your own cost by including components for daily rent and hourly operating costs. This is important when the listed Crew for a task does not contain the equipment needed, such as a crane for lifting mechanical heating/cooling equipment up onto a roof.

- If the quantity of work is less than the crew's Daily Output shown for a Unit Price line item that includes a bare unit equipment cost, the recommendation is to estimate one day's rental cost and operating cost for equipment shown in the Crew Listing for that line item.

Mobilization, Demobilization Costs

- The cost to move construction equipment from an equipment yard or rental company to the job site and back again is not included in equipment rental costs listed in the Reference Section, nor in the bare equipment cost of any unit price line item, nor in any equipment costs shown in the Crew Listings.

- Mobilization (to the site) and demobilization (from the site) costs can be found in the Unit Price section.

- If a piece of equipment is already at the job site, it is not appropriate to utilize mobilization, demobilization costs again in an estimate. ∎

01 54 33 | Equipment Rental

		UNIT	HOURLY OPER. COST	RENT PER DAY	RENT PER WEEK	RENT PER MONTH	EQUIPMENT COST/DAY	
10	**0010**	**CONCRETE EQUIPMENT RENTAL** without operators `R015433 -10`						**10**
	0200	Bucket, concrete lightweight, 1/2 C.Y.	Ea.	.85	24.50	73	219	21.40	
	0300	1 C.Y.		.95	28.50	85	255	24.60	
	0400	1-1/2 C.Y.		1.20	38.50	115	345	32.60	
	0500	2 C.Y.		1.30	46.50	140	420	38.40	
	0580	8 C.Y.		6.35	263	790	2,375	208.80	
	0600	Cart, concrete, self-propelled, operator walking, 10 C.F.		2.85	58.50	175	525	57.80	
	0700	Operator riding, 18 C.F.		4.80	98.50	295	885	97.40	
	0800	Conveyer for concrete, portable, gas, 16" wide, 26' long		10.60	130	390	1,175	162.80	
	0900	46' long		11.00	155	465	1,400	181	
	1000	56' long		11.15	163	490	1,475	187.20	
	1100	Core drill, electric, 2-1/2 H.P., 1" to 8" bit diameter		1.56	58.50	175	525	47.50	
	1150	11 H.P., 8" to 18" cores		5.40	115	345	1,025	112.20	
	1200	Finisher, concrete floor, gas, riding trowel, 96" wide		9.65	148	445	1,325	166.20	
	1300	Gas, walk-behind, 3 blade, 36" trowel		2.15	23	69	207	31	
	1400	4 blade, 48" trowel		3.05	28	84	252	41.20	
	1500	Float, hand-operated (Bull float), 48" wide		.08	13.65	41	123	8.85	
	1570	Curb builder, 14 H.P., gas, single screw		13.90	275	825	2,475	276.20	
	1590	Double screw		14.80	330	995	2,975	317.40	
	1600	Floor grinder, concrete and terrazzo, electric, 22" path		2.95	183	550	1,650	133.60	
	1700	Edger, concrete, electric, 7" path		1.12	56.50	170	510	42.95	
	1750	Vacuum pick-up system for floor grinders, wet/dry		1.62	90	270	810	66.95	
	1800	Mixer, powered, mortar and concrete, gas, 6 C.F., 18 H.P.		7.40	123	370	1,100	133.20	
	1900	10 C.F., 25 H.P.		9.00	148	445	1,325	161	
	2000	16 C.F.		9.35	172	515	1,550	177.80	
	2100	Concrete, stationary, tilt drum, 2 C.Y.		7.40	242	725	2,175	204.20	
	2120	Pump, concrete, truck mounted, 4" line, 80' boom		29.20	1,075	3,240	9,725	881.60	
	2140	5" line, 110' boom		36.60	1,375	4,115	12,300	1,116	
	2160	Mud jack, 50 C.F. per hr.		6.45	128	385	1,150	128.60	
	2180	225 C.F. per hr.		8.80	147	440	1,325	158.40	
	2190	Shotcrete pump rig, 12 C.Y./hr.		13.95	225	675	2,025	246.60	
	2200	35 C.Y./hr.		15.65	242	725	2,175	270.20	
	2600	Saw, concrete, manual, gas, 18 H.P.		5.40	46.50	140	420	71.20	
	2650	Self-propelled, gas, 30 H.P.		7.55	70	210	630	102.40	
	2675	V-groove crack chaser, manual, gas, 6 H.P.		1.80	18.35	55	165	25.40	
	2700	Vibrators, concrete, electric, 60 cycle, 2 H.P.		.47	9	27	81	9.15	
	2800	3 H.P.		.60	11.65	35	105	11.80	
	2900	Gas engine, 5 H.P.		1.50	16.35	49	147	21.80	
	3000	8 H.P.		2.00	16	48	144	25.60	
	3050	Vibrating screed, gas engine, 8 H.P.		2.82	88	264	790	75.35	
	3120	Concrete transit mixer, 6 x 4, 250 H.P., 8 C.Y., rear discharge		48.75	600	1,805	5,425	751	
	3200	Front discharge		56.60	735	2,205	6,625	893.80	
	3300	6 x 6, 285 H.P., 12 C.Y., rear discharge		55.65	700	2,095	6,275	864.20	
	3400	Front discharge		57.95	735	2,210	6,625	905.60	
20	**0010**	**EARTHWORK EQUIPMENT RENTAL** without operators `R015433 -10`						**20**
	0040	Aggregate spreader, push type, 8' to 12' wide	Ea.	2.60	26.50	80	240	36.80	
	0045	Tailgate type, 8' wide		2.55	33.50	100	300	40.40	
	0055	Earth auger, truck mounted, for fence & sign posts, utility poles		12.60	455	1,365	4,100	373.80	
	0060	For borings and monitoring wells		42.60	695	2,090	6,275	758.80	
	0070	Portable, trailer mounted		2.30	33	99	297	38.20	
	0075	Truck mounted, for caissons, water wells		85.75	2,925	8,790	26,400	2,444	
	0080	Horizontal boring machine, 12" to 36" diameter, 45 H.P.		22.70	197	590	1,775	299.60	
	0090	12" to 48" diameter, 65 H.P.		30.60	340	1,020	3,050	448.80	
	0095	Auger, for fence posts, gas engine, hand held		.45	6.35	19	57	7.40	
	0100	Excavator, diesel hydraulic, crawler mounted, 1/2 C.Y. cap.		21.30	440	1,325	3,975	435.40	
	0120	5/8 C.Y. capacity		28.60	580	1,740	5,225	576.80	
	0140	3/4 C.Y. capacity		31.95	690	2,070	6,200	669.60	
	0150	1 C.Y. capacity		39.90	705	2,115	6,350	742.20	

For customer support on your Residential Costs with RSMeans data, call 800.448.8182.

01 54 33 | Equipment Rental

		UNIT	HOURLY OPER. COST	RENT PER DAY	RENT PER WEEK	RENT PER MONTH	EQUIPMENT COST/DAY
0200	1-1/2 C.Y. capacity	Ea.	47.45	855	2,570	7,700	893.60
0300	2 C.Y. capacity		54.15	1,025	3,095	9,275	1,052
0320	2-1/2 C.Y. capacity		78.90	1,300	3,900	11,700	1,411
0325	3-1/2 C.Y. capacity		113.65	2,150	6,430	19,300	2,195
0330	4-1/2 C.Y. capacity		143.70	2,650	7,925	23,800	2,735
0335	6 C.Y. capacity		181.40	3,400	10,180	30,500	3,487
0340	7 C.Y. capacity		175.20	3,125	9,365	28,100	3,275
0342	Excavator attachments, bucket thumbs		3.35	252	755	2,275	177.80
0345	Grapples		3.05	220	660	1,975	156.40
0346	Hydraulic hammer for boom mounting, 4000 ft lb.		13.20	365	1,095	3,275	324.60
0347	5000 ft lb.		15.50	450	1,350	4,050	394
0348	8000 ft lb.		22.85	655	1,970	5,900	576.80
0349	12,000 ft lb.		24.90	785	2,350	7,050	669.20
0350	Gradall type, truck mounted, 3 ton @ 15' radius, 5/8 C.Y.		44.95	850	2,550	7,650	869.60
0370	1 C.Y. capacity		60.20	1,275	3,805	11,400	1,243
0400	Backhoe-loader, 40 to 45 H.P., 5/8 C.Y. capacity		12.25	248	745	2,225	247
0450	45 H.P. to 60 H.P., 3/4 C.Y. capacity		17.80	283	850	2,550	312.40
0460	80 H.P., 1-1/4 C.Y. capacity		19.50	380	1,145	3,425	385
0470	112 H.P., 1-1/2 C.Y. capacity		31.90	605	1,820	5,450	619.20
0482	Backhoe-loader attachment, compactor, 20,000 lb.		6.25	148	445	1,325	139
0485	Hydraulic hammer, 750 ft lb.		3.55	102	305	915	89.40
0486	Hydraulic hammer, 1200 ft lb.		6.55	208	625	1,875	177.40
0500	Brush chipper, gas engine, 6" cutter head, 35 H.P.		9.15	110	330	990	139.20
0550	Diesel engine, 12" cutter head, 130 H.P.		24.00	335	1,005	3,025	393
0600	15" cutter head, 165 H.P.		26.05	400	1,200	3,600	448.40
0750	Bucket, clamshell, general purpose, 3/8 C.Y.		1.40	40	120	360	35.20
0800	1/2 C.Y.		1.50	48.50	145	435	41
0850	3/4 C.Y.		1.65	56.50	170	510	47.20
0900	1 C.Y.		1.70	61.50	185	555	50.60
0950	1-1/2 C.Y.		2.75	85	255	765	73
1000	2 C.Y.		2.90	93.50	280	840	79.20
1010	Bucket, dragline, medium duty, 1/2 C.Y.		.80	24.50	73	219	21
1020	3/4 C.Y.		.80	25.50	77	231	21.80
1030	1 C.Y.		.85	27	81	243	23
1040	1-1/2 C.Y.		1.30	41.50	125	375	35.40
1050	2 C.Y.		1.35	45	135	405	37.80
1070	3 C.Y.		2.10	65	195	585	55.80
1200	Compactor, manually guided 2-drum vibratory smooth roller, 7.5 H.P.		7.20	203	610	1,825	179.60
1250	Rammer/tamper, gas, 8"		2.25	46.50	140	420	46
1260	15"		2.50	53.50	160	480	52
1300	Vibratory plate, gas, 18" plate, 3000 lb. blow		2.15	24.50	73	219	31.80
1350	21" plate, 5000 lb. blow		2.60	33	99	297	40.60
1370	Curb builder/extruder, 14 H.P., gas, single screw		13.90	275	825	2,475	276.20
1390	Double screw		14.80	330	995	2,975	317.40
1500	Disc harrow attachment, for tractor		.47	78.50	235	705	50.75
1810	Feller buncher, shearing & accumulating trees, 100 H.P.		40.55	810	2,430	7,300	810.40
1860	Grader, self-propelled, 25,000 lb.		32.75	745	2,235	6,700	709
1910	30,000 lb.		32.35	640	1,925	5,775	643.80
1920	40,000 lb.		54.75	1,250	3,765	11,300	1,191
1930	55,000 lb.		67.80	1,650	4,975	14,900	1,537
1950	Hammer, pavement breaker, self-propelled, diesel, 1000 to 1250 lb.		28.40	460	1,380	4,150	503.20
2000	1300 to 1500 lb.		42.60	920	2,760	8,275	892.80
2050	Pile driving hammer, steam or air, 4150 ft lb. @ 225 bpm		11.80	565	1,695	5,075	433.40
2100	8750 ft lb. @ 145 bpm		14.30	805	2,410	7,225	596.40
2150	15,000 ft lb. @ 60 bpm		14.65	845	2,530	7,600	623.20
2200	24,450 ft lb. @ 111 bpm		15.65	935	2,800	8,400	685.20
2250	Leads, 60' high for pile driving hammers up to 20,000 ft lb.		3.60	84.50	253	760	79.40
2300	90' high for hammers over 20,000 ft lb.		5.35	148	444	1,325	131.60

01 54 33 | Equipment Rental

		UNIT	HOURLY OPER. COST	RENT PER DAY	RENT PER WEEK	RENT PER MONTH	EQUIPMENT COST/DAY
2350	Diesel type hammer, 22,400 ft lb.	Ea.	18.25	470	1,405	4,225	427
2400	41,300 ft lb.		26.50	600	1,800	5,400	572
2450	141,000 ft lb.		41.80	945	2,840	8,525	902.40
2500	Vib. elec. hammer/extractor, 200 kW diesel generator, 34 H.P.		40.35	695	2,080	6,250	738.80
2550	80 H.P.		70.10	1,000	3,000	9,000	1,161
2600	150 H.P.		129.40	1,925	5,780	17,300	2,191
2800	Log chipper, up to 22" diameter, 600 H.P.		46.10	660	1,980	5,950	764.80
2850	Logger, for skidding & stacking logs, 150 H.P.		44.05	830	2,485	7,450	849.40
2860	Mulcher, diesel powered, trailer mounted		17.10	220	660	1,975	268.80
2900	Rake, spring tooth, with tractor		14.54	360	1,075	3,225	331.30
3000	Roller, vibratory, tandem, smooth drum, 20 H.P.		7.65	150	450	1,350	151.20
3050	35 H.P.		10.05	252	755	2,275	231.40
3100	Towed type vibratory compactor, smooth drum, 50 H.P.		25.25	370	1,105	3,325	423
3150	Sheepsfoot, 50 H.P.		25.35	370	1,115	3,350	425.80
3170	Landfill compactor, 220 H.P.		71.35	1,575	4,710	14,100	1,513
3200	Pneumatic tire roller, 80 H.P.		12.90	390	1,175	3,525	338.20
3250	120 H.P.		19.45	645	1,930	5,800	541.60
3300	Sheepsfoot vibratory roller, 240 H.P.		62.05	1,375	4,100	12,300	1,316
3320	340 H.P.		83.10	2,025	6,075	18,200	1,880
3350	Smooth drum vibratory roller, 75 H.P.		23.30	650	1,950	5,850	576.40
3400	125 H.P.		27.60	730	2,195	6,575	659.80
3410	Rotary mower, brush, 60", with tractor		19.00	340	1,025	3,075	357
3420	Rototiller, walk-behind, gas, 5 H.P.		2.21	76.50	229	685	63.50
3422	8 H.P.		2.80	87	261	785	74.60
3440	Scrapers, towed type, 7 C.Y. capacity		6.30	122	365	1,100	123.40
3450	10 C.Y. capacity		7.10	165	495	1,475	155.80
3500	15 C.Y. capacity		7.60	192	575	1,725	175.80
3525	Self-propelled, single engine, 14 C.Y. capacity		131.25	2,400	7,200	21,600	2,490
3550	Dual engine, 21 C.Y. capacity		137.25	2,275	6,850	20,600	2,468
3600	31 C.Y. capacity		184.55	3,500	10,500	31,500	3,576
3640	44 C.Y. capacity		225.40	4,425	13,270	39,800	4,457
3650	Elevating type, single engine, 11 C.Y. capacity		61.05	1,100	3,335	10,000	1,155
3700	22 C.Y. capacity		114.65	2,275	6,835	20,500	2,284
3710	Screening plant, 110 H.P. w/5' x 10' screen		20.55	385	1,160	3,475	396.40
3720	5' x 16' screen		25.96	495	1,480	4,450	503.70
3850	Shovel, crawler-mounted, front-loading, 7 C.Y. capacity		204.00	3,750	11,240	33,700	3,880
3855	12 C.Y. capacity		332.00	5,200	15,567	46,700	5,769
3860	Shovel/backhoe bucket, 1/2 C.Y.		2.70	71.50	215	645	64.60
3870	3/4 C.Y.		2.75	78.50	235	705	69
3880	1 C.Y.		2.85	88.50	265	795	75.80
3890	1-1/2 C.Y.		3.00	102	305	915	85
3910	3 C.Y.		3.35	137	410	1,225	108.80
3950	Stump chipper, 18" deep, 30 H.P.		6.93	214	643	1,925	184.05
4110	Dozer, crawler, torque converter, diesel 80 H.P.		25.05	440	1,320	3,950	464.40
4150	105 H.P.		34.25	555	1,670	5,000	608
4200	140 H.P.		42.20	845	2,540	7,625	845.60
4260	200 H.P.		61.65	1,300	3,900	11,700	1,273
4310	300 H.P.		82.45	1,950	5,845	17,500	1,829
4360	410 H.P.		109.35	2,350	7,080	21,200	2,291
4370	500 H.P.		137.25	2,875	8,600	25,800	2,818
4380	700 H.P.		229.75	5,275	15,805	47,400	4,999
4400	Loader, crawler, torque conv., diesel, 1-1/2 C.Y., 80 H.P.		29.60	570	1,715	5,150	579.80
4450	1-1/2 to 1-3/4 C.Y., 95 H.P.		30.55	670	2,005	6,025	645.40
4510	1-3/4 to 2-1/4 C.Y., 130 H.P.		47.50	1,000	2,995	8,975	979
4530	2-1/2 to 3-1/4 C.Y., 190 H.P.		57.55	1,250	3,720	11,200	1,204
4560	3-1/2 to 5 C.Y., 275 H.P.		71.55	1,475	4,435	13,300	1,459
4610	Front end loader, 4WD, articulated frame, diesel, 1 to 1-1/4 C.Y., 70 H.P.		16.10	270	810	2,425	290.80
4620	1-1/2 to 1-3/4 C.Y., 95 H.P.		19.35	315	940	2,825	342.80

01 54 33 | Equipment Rental

		UNIT	HOURLY OPER. COST	RENT PER DAY	RENT PER WEEK	RENT PER MONTH	EQUIPMENT COST/DAY	
20	4650	1-3/4 to 2 C.Y., 130 H.P.	Ea.	20.50	380	1,140	3,425	392	**20**
	4710	2-1/2 to 3-1/2 C.Y., 145 H.P.		29.15	485	1,450	4,350	523.20	
	4730	3 to 4-1/2 C.Y., 185 H.P.		32.20	530	1,585	4,750	574.60	
	4760	5-1/4 to 5-3/4 C.Y., 270 H.P.		53.15	930	2,785	8,350	982.20	
	4810	7 to 9 C.Y., 475 H.P.		90.60	1,750	5,275	15,800	1,780	
	4870	9 to 11 C.Y., 620 H.P.		131.70	2,625	7,880	23,600	2,630	
	4880	Skid-steer loader, wheeled, 10 C.F., 30 H.P. gas		9.50	163	490	1,475	174	
	4890	1 C.Y., 78 H.P., diesel		18.35	365	1,090	3,275	364.80	
	4892	Skid-steer attachment, auger		.82	136	408	1,225	88.15	
	4893	Backhoe		.74	123	369	1,100	79.70	
	4894	Broom		.71	118	355	1,075	76.70	
	4895	Forks		.16	27	81	243	17.50	
	4896	Grapple		.69	115	346	1,050	74.70	
	4897	Concrete hammer		1.05	175	526	1,575	113.60	
	4898	Tree spade		.60	100	300	900	64.80	
	4899	Trencher		.69	115	344	1,025	74.30	
	4900	Trencher, chain, boom type, gas, operator walking, 12 H.P.		4.10	48.50	145	435	61.80	
	4910	Operator riding, 40 H.P.		16.30	345	1,030	3,100	336.40	
	5000	Wheel type, diesel, 4' deep, 12" wide		69.35	900	2,700	8,100	1,095	
	5100	6' deep, 20" wide		87.50	2,100	6,285	18,900	1,957	
	5150	Chain type, diesel, 5' deep, 8" wide		16.30	345	1,030	3,100	336.40	
	5200	Diesel, 8' deep, 16" wide		95.25	2,150	6,430	19,300	2,048	
	5202	Rock trencher, wheel type, 6" wide x 18" deep		43.65	925	2,775	8,325	904.20	
	5206	Chain type, 18" wide x 7' deep		106.10	3,050	9,160	27,500	2,681	
	5210	Tree spade, self-propelled		14.24	390	1,167	3,500	347.30	
	5250	Truck, dump, 2-axle, 12 ton, 8 C.Y. payload, 220 H.P.		23.95	245	735	2,200	338.60	
	5300	Three axle dump, 16 ton, 12 C.Y. payload, 400 H.P.		41.60	350	1,050	3,150	542.80	
	5310	Four axle dump, 25 ton, 18 C.Y. payload, 450 H.P.		50.00	510	1,530	4,600	706	
	5350	Dump trailer only, rear dump, 16-1/2 C.Y.		5.75	145	435	1,300	133	
	5400	20 C.Y.		6.20	163	490	1,475	147.60	
	5450	Flatbed, single axle, 1-1/2 ton rating		18.30	70	210	630	188.40	
	5500	3 ton rating		22.25	100	300	900	238	
	5550	Off highway rear dump, 25 ton capacity		61.80	1,425	4,245	12,700	1,343	
	5600	35 ton capacity		66.10	1,525	4,595	13,800	1,448	
	5610	50 ton capacity		81.20	1,675	5,060	15,200	1,662	
	5620	65 ton capacity		84.95	1,950	5,820	17,500	1,844	
	5630	100 ton capacity		119.60	2,850	8,560	25,700	2,669	
	6000	Vibratory plow, 25 H.P., walking		6.80	61.50	185	555	91.40	
40	0010	**GENERAL EQUIPMENT RENTAL** without operators [R015433-10]							**40**
	0020	Aerial lift, scissor type, to 20' high, 1200 lb. capacity, electric	Ea.	3.40	51.50	155	465	58.20	
	0030	To 30' high, 1200 lb. capacity		3.85	68.50	205	615	71.80	
	0040	Over 30' high, 1500 lb. capacity		5.15	122	365	1,100	114.20	
	0070	Articulating boom, to 45' high, 500 lb. capacity, diesel [R015433-15]		9.70	273	820	2,450	241.60	
	0075	To 60' high, 500 lb. capacity		13.70	470	1,410	4,225	391.60	
	0080	To 80' high, 500 lb. capacity		16.10	560	1,685	5,050	465.80	
	0085	To 125' high, 500 lb. capacity		18.40	780	2,335	7,000	614.20	
	0100	Telescoping boom to 40' high, 500 lb. capacity, diesel		11.60	315	950	2,850	282.80	
	0105	To 45' high, 500 lb. capacity		12.40	320	958	2,875	290.80	
	0110	To 60' high, 500 lb. capacity		16.20	540	1,625	4,875	454.60	
	0115	To 80' high, 500 lb. capacity		21.70	625	1,875	5,625	548.60	
	0120	To 100' high, 500 lb. capacity		28.80	805	2,420	7,250	714.40	
	0125	To 120' high, 500 lb. capacity		29.25	845	2,530	7,600	740	
	0195	Air compressor, portable, 6.5 CFM, electric		.91	13	39	117	15.10	
	0196	Gasoline		.66	19.65	59	177	17.10	
	0200	Towed type, gas engine, 60 CFM		9.30	51.50	155	465	105.40	
	0300	160 CFM		10.65	53.50	160	480	117.20	
	0400	Diesel engine, rotary screw, 250 CFM		12.05	118	355	1,075	167.40	
	0500	365 CFM		15.80	142	425	1,275	211.40	

For customer support on your Residential Costs with RSMeans data, call 800.448.8182.

671

		UNIT	HOURLY OPER. COST	RENT PER DAY	RENT PER WEEK	RENT PER MONTH	EQUIPMENT COST/DAY
0550	450 CFM	Ea.	19.70	177	530	1,600	263.60
0600	600 CFM		33.90	243	730	2,200	417.20
0700	750 CFM		34.10	252	755	2,275	423.80
0930	Air tools, breaker, pavement, 60 lb.		.55	10.35	31	93	10.60
0940	80 lb.		.55	10.65	32	96	10.80
0950	Drills, hand (jackhammer), 65 lb.		.65	17.65	53	159	15.80
0960	Track or wagon, swing boom, 4" drifter		54.80	925	2,775	8,325	993.40
0970	5" drifter		63.45	1,100	3,325	9,975	1,173
0975	Track mounted quarry drill, 6" diameter drill		104.15	1,650	4,945	14,800	1,822
0980	Dust control per drill		1.04	24.50	74	222	23.10
0990	Hammer, chipping, 12 lb.		.60	27	81	243	21
1000	Hose, air with couplings, 50' long, 3/4" diameter		.07	11.35	34	102	7.35
1100	1" diameter		.08	13	39	117	8.45
1200	1-1/2" diameter		.21	35	105	315	22.70
1300	2" diameter		.24	40	120	360	25.90
1400	2-1/2" diameter		.35	58.50	175	525	37.80
1410	3" diameter		.40	66.50	200	600	43.20
1450	Drill, steel, 7/8" x 2'		.08	13.65	41	123	8.85
1460	7/8" x 6'		.11	17.65	53	159	11.50
1520	Moil points		.03	4.67	14	42	3.05
1525	Pneumatic nailer w/accessories		.48	32	96	288	23.05
1530	Sheeting driver for 60 lb. breaker		.04	7.35	22	66	4.70
1540	For 90 lb. breaker		.15	9.65	29	87	7
1550	Spade, 25 lb.		.50	7.35	22	66	8.40
1560	Tamper, single, 35 lb.		.59	39.50	118	355	28.30
1570	Triple, 140 lb.		.89	59	177	530	42.50
1580	Wrenches, impact, air powered, up to 3/4" bolt		.45	13	39	117	11.40
1590	Up to 1-1/4" bolt		.55	23.50	71	213	18.60
1600	Barricades, barrels, reflectorized, 1 to 99 barrels		.03	5.35	16	48	3.45
1610	100 to 200 barrels		.02	4.13	12.40	37	2.65
1620	Barrels with flashers, 1 to 99 barrels		.04	6	18	54	3.90
1630	100 to 200 barrels		.03	4.80	14.40	43	3.10
1640	Barrels with steady burn type C lights		.05	8	24	72	5.20
1650	Illuminated board, trailer mounted, with generator		3.30	133	400	1,200	106.40
1670	Portable barricade, stock, with flashers, 1 to 6 units		.04	6	18	54	3.90
1680	25 to 50 units		.03	5.60	16.80	50.50	3.60
1685	Butt fusion machine, wheeled, 1.5 H.P. electric, 2" - 8" diameter pipe		2.63	167	500	1,500	121.05
1690	Tracked, 20 H.P. diesel, 4" - 12" diameter pipe		11.21	560	1,680	5,050	425.70
1695	83 H.P. diesel, 8" - 24" diameter pipe		49.46	2,525	7,560	22,700	1,908
1700	Carts, brick, gas engine, 1000 lb. capacity		2.95	61.50	185	555	60.60
1800	1500 lb., 7-1/2' lift		3.00	65	195	585	63
1822	Dehumidifier, medium, 6 lb./hr., 150 CFM		1.16	72.50	218	655	52.90
1824	Large, 18 lb./hr., 600 CFM		2.20	138	413	1,250	100.20
1830	Distributor, asphalt, trailer mounted, 2000 gal., 38 H.P. diesel		10.75	350	1,050	3,150	296
1840	3000 gal., 38 H.P. diesel		12.35	380	1,140	3,425	326.80
1850	Drill, rotary hammer, electric		1.12	27	81	243	25.15
1860	Carbide bit, 1-1/2" diameter, add to electric rotary hammer		.03	5	15	45	3.25
1865	Rotary, crawler, 250 H.P.		136.15	2,225	6,690	20,100	2,427
1870	Emulsion sprayer, 65 gal., 5 H.P. gas engine		2.77	103	309	925	83.95
1880	200 gal., 5 H.P. engine		7.30	172	515	1,550	161.40
1900	Floor auto-scrubbing machine, walk-behind, 28" path		5.41	350	1,055	3,175	254.30
1930	Floodlight, mercury vapor, or quartz, on tripod, 1000 watt		.46	22	66	198	16.90
1940	2000 watt		.63	27.50	82	246	21.45
1950	Floodlights, trailer mounted with generator, 1 - 300 watt light		3.60	76.50	230	690	74.80
1960	2 - 1000 watt lights		4.50	102	305	915	97
2000	4 - 300 watt lights		4.25	96.50	290	870	92
2005	Foam spray rig, incl. box trailer, compressor, generator, proportioner		23.78	515	1,545	4,625	499.25
2015	Forklift, pneumatic tire, rough terr, straight mast, 5000 lb, 12' lift, gas		19.00	212	635	1,900	279

		UNIT	HOURLY OPER. COST	RENT PER DAY	RENT PER WEEK	RENT PER MONTH	EQUIPMENT COST/DAY	
40	2025	8000 lb., 12' lift	Ea.	22.75	283	850	2,550	352	40
	2030	5000 lb., 12' lift, diesel		15.70	237	710	2,125	267.60	
	2035	8000 lb., 12' lift, diesel		16.75	268	805	2,425	295	
	2045	All terrain, telescoping boom, diesel, 5000 lb., 10' reach, 19' lift		17.25	325	980	2,950	334	
	2055	6600 lb., 29' reach, 42' lift		21.10	380	1,140	3,425	396.80	
	2065	10,000 lb., 31' reach, 45' lift		23.65	490	1,475	4,425	484.20	
	2070	Cushion tire, smooth floor, gas, 5000 lb. capacity		8.25	76.50	230	690	112	
	2075	8000 lb. capacity		11.40	95	285	855	148.20	
	2085	Diesel, 5000 lb. capacity		7.75	83.50	250	750	112	
	2090	12,000 lb. capacity		12.05	130	390	1,175	174.40	
	2095	20,000 lb. capacity		17.00	165	495	1,475	235	
	2100	Generator, electric, gas engine, 1.5 kW to 3 kW		2.70	11.35	34	102	28.40	
	2200	5 kW		3.35	14.35	43	129	35.40	
	2300	10 kW		6.25	35	105	315	71	
	2400	25 kW		7.60	86.50	260	780	112.80	
	2500	Diesel engine, 20 kW		9.20	76.50	230	690	119.60	
	2600	50 kW		15.85	100	300	900	186.80	
	2700	100 kW		28.20	137	410	1,225	307.60	
	2800	250 kW		56.05	260	780	2,350	604.40	
	2850	Hammer, hydraulic, for mounting on boom, to 500 ft lb.		2.85	86.50	260	780	74.80	
	2860	1000 ft lb.		4.70	133	400	1,200	117.60	
	2900	Heaters, space, oil or electric, 50 MBH		1.47	8	24	72	16.55	
	3000	100 MBH		2.73	11.35	34	102	28.65	
	3100	300 MBH		7.84	40	120	360	86.70	
	3150	500 MBH		12.75	45	135	405	129	
	3200	Hose, water, suction with coupling, 20' long, 2" diameter		.02	3	9	27	1.95	
	3210	3" diameter		.03	4.33	13	39	2.85	
	3220	4" diameter		.03	5	15	45	3.25	
	3230	6" diameter		.11	17.65	53	159	11.50	
	3240	8" diameter		.28	46.50	140	420	30.25	
	3250	Discharge hose with coupling, 50' long, 2" diameter		.01	1.33	4	12	.90	
	3260	3" diameter		.01	2.33	7	21	1.50	
	3270	4" diameter		.02	3.67	11	33	2.35	
	3280	6" diameter		.06	9.35	28	84	6.10	
	3290	8" diameter		.24	40	120	360	25.90	
	3295	Insulation blower		.83	6	18	54	10.25	
	3300	Ladders, extension type, 16' to 36' long		.18	30	90	270	19.45	
	3400	40' to 60' long		.67	112	335	1,000	72.35	
	3405	Lance for cutting concrete		2.23	58.50	176	530	53.05	
	3407	Lawn mower, rotary, 22", 5 H.P.		1.15	25	75	225	24.20	
	3408	48" self-propelled		2.86	90	270	810	76.90	
	3410	Level, electronic, automatic, with tripod and leveling rod		1.05	70	210	630	50.40	
	3430	Laser type, for pipe and sewer line and grade		2.13	142	425	1,275	102.05	
	3440	Rotating beam for interior control		.90	60	180	540	43.20	
	3460	Builder's optical transit, with tripod and rod		.10	16.35	49	147	10.60	
	3500	Light towers, towable, with diesel generator, 2000 watt		4.25	96.50	290	870	92	
	3600	4000 watt		4.50	102	305	915	97	
	3700	Mixer, powered, plaster and mortar, 6 C.F., 7 H.P.		2.05	20.50	62	186	28.80	
	3800	10 C.F., 9 H.P.		2.20	33.50	100	300	37.60	
	3850	Nailer, pneumatic		.48	32	96	288	23.05	
	3900	Paint sprayers complete, 8 CFM		.94	62.50	188	565	45.10	
	4000	17 CFM		1.69	112	337	1,000	80.90	
	4020	Pavers, bituminous, rubber tires, 8' wide, 50 H.P., diesel		31.55	550	1,645	4,925	581.40	
	4030	10' wide, 150 H.P.		96.50	1,875	5,655	17,000	1,903	
	4050	Crawler, 8' wide, 100 H.P., diesel		87.60	2,025	6,105	18,300	1,922	
	4060	10' wide, 150 H.P.		104.50	2,350	7,015	21,000	2,239	
	4070	Concrete paver, 12' to 24' wide, 250 H.P.		87.45	1,625	4,875	14,600	1,675	
	4080	Placer-spreader-trimmer, 24' wide, 300 H.P.		117.20	2,375	7,115	21,300	2,361	

01 54 33 | Equipment Rental

		UNIT	HOURLY OPER. COST	RENT PER DAY	RENT PER WEEK	RENT PER MONTH	EQUIPMENT COST/DAY	
4100	Pump, centrifugal gas pump, 1-1/2" diameter, 65 GPM	Ea.	3.90	53.50	160	480	63.20	40
4200	2" diameter, 130 GPM		5.00	63.50	190	570	78	
4300	3" diameter, 250 GPM		5.15	63.50	190	570	79.20	
4400	6" diameter, 1500 GPM		22.30	197	590	1,775	296.40	
4500	Submersible electric pump, 1-1/4" diameter, 55 GPM		.41	17.65	53	159	13.90	
4600	1-1/2" diameter, 83 GPM		.45	20.50	61	183	15.80	
4700	2" diameter, 120 GPM		1.65	25.50	76	228	28.40	
4800	3" diameter, 300 GPM		2.94	45	135	405	50.50	
4900	4" diameter, 560 GPM		14.70	167	500	1,500	217.60	
5000	6" diameter, 1590 GPM		21.94	218	655	1,975	306.50	
5100	Diaphragm pump, gas, single, 1-1/2" diameter		1.12	54.50	164	490	41.75	
5200	2" diameter		4.00	68.50	205	615	73	
5300	3" diameter		4.05	68.50	205	615	73.40	
5400	Double, 4" diameter		5.85	113	340	1,025	114.80	
5450	Pressure washer 5 GPM, 3000 psi		3.95	53.50	160	480	63.60	
5460	7 GPM, 3000 psi		4.90	63.50	190	570	77.20	
5500	Trash pump, self-priming, gas, 2" diameter		3.80	23.50	70	210	44.40	
5600	Diesel, 4" diameter		6.95	95	285	855	112.60	
5650	Diesel, 6" diameter		16.90	167	500	1,500	235.20	
5655	Grout pump		19.50	275	825	2,475	321	
5700	Salamanders, L.P. gas fired, 100,000 BTU		2.93	14	42	126	31.85	
5705	50,000 BTU		1.67	11.35	34	102	20.15	
5720	Sandblaster, portable, open top, 3 C.F. capacity		.60	27	81	243	21	
5730	6 C.F. capacity		1.00	40	120	360	32	
5740	Accessories for above		.14	22.50	68	204	14.70	
5750	Sander, floor		.77	17.65	53	159	16.75	
5760	Edger		.52	15	45	135	13.15	
5800	Saw, chain, gas engine, 18" long		1.80	22.50	67	201	27.80	
5900	Hydraulic powered, 36" long		.80	66.50	200	600	46.40	
5950	60" long		.80	68.50	205	615	47.40	
6000	Masonry, table mounted, 14" diameter, 5 H.P.		1.32	56.50	170	510	44.55	
6050	Portable cut-off, 8 H.P.		1.85	33.50	100	300	34.80	
6100	Circular, hand held, electric, 7-1/4" diameter		.23	5	15	45	4.85	
6200	12" diameter		.23	8	24	72	6.65	
6250	Wall saw, w/hydraulic power, 10 H.P.		3.30	33.50	100	300	46.40	
6275	Shot blaster, walk-behind, 20" wide		4.85	293	880	2,650	214.80	
6280	Sidewalk broom, walk-behind		2.39	85	255	765	70.10	
6300	Steam cleaner, 100 gallons per hour		3.35	80	240	720	74.80	
6310	200 gallons per hour		4.40	96.50	290	870	93.20	
6340	Tar kettle/pot, 400 gallons		15.15	76.50	230	690	167.20	
6350	Torch, cutting, acetylene-oxygen, 150' hose, excludes gases		.45	15	45	135	12.60	
6360	Hourly operating cost includes tips and gas		21.00				168	
6410	Toilet, portable chemical		.13	22	66	198	14.25	
6420	Recycle flush type		.16	27	81	243	17.50	
6430	Toilet, fresh water flush, garden hose,		.19	32.50	97	291	20.90	
6440	Hoisted, non-flush, for high rise		.16	26.50	79	237	17.10	
6465	Tractor, farm with attachment		17.80	340	1,025	3,075	347.40	
6480	Trailers, platform, flush deck, 2 axle, 3 ton capacity		1.60	21	63	189	25.40	
6500	25 ton capacity		6.25	138	415	1,250	133	
6600	40 ton capacity		8.00	193	580	1,750	180	
6700	3 axle, 50 ton capacity		8.65	215	645	1,925	198.20	
6800	75 ton capacity		10.90	285	855	2,575	258.20	
6810	Trailer mounted cable reel for high voltage line work		5.79	276	827	2,475	211.70	
6820	Trailer mounted cable tensioning rig		11.48	545	1,640	4,925	419.85	
6830	Cable pulling rig		72.98	3,075	9,210	27,600	2,426	
6850	Portable cable/wire puller, 8000 lb. max pulling capacity		3.72	167	502	1,500	130.15	
6900	Water tank trailer, engine driven discharge, 5000 gallons		7.20	150	450	1,350	147.60	
6925	10,000 gallons		9.70	207	620	1,850	201.60	

			UNIT	HOURLY OPER. COST	RENT PER DAY	RENT PER WEEK	RENT PER MONTH	EQUIPMENT COST/DAY	
40	6950	Water truck, off highway, 6000 gallons	Ea.	70.16	805	2,420	7,250	1,045	**40**
	7010	Tram car for high voltage line work, powered, 2 conductor		6.85	150	449	1,350	144.60	
	7020	Transit (builder's level) with tripod		.10	16.35	49	147	10.60	
	7030	Trench box, 3000 lb., 6' x 8'		.56	93.50	280	840	60.50	
	7040	7200 lb., 6' x 20'		.75	125	375	1,125	81	
	7050	8000 lb., 8' x 16'		1.08	180	540	1,625	116.65	
	7060	9500 lb., 8' x 20'		1.21	201	603	1,800	130.30	
	7065	11,000 lb., 8' x 24'		1.27	211	633	1,900	136.75	
	7070	12,000 lb., 10' x 20'		1.50	251	752	2,250	162.40	
	7100	Truck, pickup, 3/4 ton, 2 wheel drive		9.90	60	180	540	115.20	
	7200	4 wheel drive		10.20	75	225	675	126.60	
	7250	Crew carrier, 9 passenger		14.00	90	270	810	166	
	7290	Flat bed truck, 20,000 lb. GVW		14.90	130	390	1,175	197.20	
	7300	Tractor, 4 x 2, 220 H.P.		21.00	203	610	1,825	290	
	7410	330 H.P.		30.80	280	840	2,525	414.40	
	7500	6 x 4, 380 H.P.		35.15	325	975	2,925	476.20	
	7600	450 H.P.		43.30	395	1,185	3,550	583.40	
	7610	Tractor, with A frame, boom and winch, 225 H.P.		24.10	282	845	2,525	361.80	
	7620	Vacuum truck, hazardous material, 2500 gallons		12.85	305	910	2,725	284.80	
	7625	5000 gallons		13.11	425	1,270	3,800	358.90	
	7650	Vacuum, HEPA, 16 gallon, wet/dry		.90	18	54	162	18	
	7655	55 gallon, wet/dry		.81	27	81	243	22.70	
	7660	Water tank, portable		.74	123	370	1,100	79.90	
	7690	Sewer/catch basin vacuum, 14 C.Y., 1500 gallons		17.59	635	1,910	5,725	522.70	
	7700	Welder, electric, 200 amp		3.99	16.35	49	147	41.70	
	7800	300 amp		5.90	20	60	180	59.20	
	7900	Gas engine, 200 amp		9.10	24.50	74	222	87.60	
	8000	300 amp		10.35	26	78	234	98.40	
	8100	Wheelbarrow, any size		.06	10.65	32	96	6.90	
	8200	Wrecking ball, 4000 lb.		2.45	71.50	215	645	62.60	
50	0010	**HIGHWAY EQUIPMENT RENTAL** without operators							**50**
	0050	Asphalt batch plant, portable drum mixer, 100 ton/hr. [R015433-10]	Ea.	85.49	1,500	4,505	13,500	1,585	
	0060	200 ton/hr.		97.81	1,600	4,800	14,400	1,742	
	0070	300 ton/hr.		116.21	1,875	5,625	16,900	2,055	
	0100	Backhoe attachment, long stick, up to 185 H.P., 10.5' long		.37	24.50	73	219	17.55	
	0140	Up to 250 H.P., 12' long		.41	27	81	243	19.50	
	0180	Over 250 H.P., 15' long		.56	37	111	335	26.70	
	0200	Special dipper arm, up to 100 H.P., 32' long		1.14	75.50	227	680	54.50	
	0240	Over 100 H.P., 33' long		1.42	94.50	284	850	68.15	
	0280	Catch basin/sewer cleaning truck, 3 ton, 9 C.Y., 1000 gal.		35.10	405	1,210	3,625	522.80	
	0300	Concrete batch plant, portable, electric, 200 C.Y./hr.		24.34	545	1,630	4,900	520.70	
	0520	Grader/dozer attachment, ripper/scarifier, rear mounted, up to 135 H.P.		3.15	61.50	185	555	62.20	
	0540	Up to 180 H.P.		4.10	91.50	275	825	87.80	
	0580	Up to 250 H.P.		5.70	145	435	1,300	132.60	
	0700	Pvmt. removal bucket, for hyd. excavator, up to 90 H.P.		2.10	56.50	170	510	50.80	
	0740	Up to 200 H.P.		2.25	71.50	215	645	61	
	0780	Over 200 H.P.		2.45	88.50	265	795	72.60	
	0900	Aggregate spreader, self-propelled, 187 H.P.		50.00	730	2,185	6,550	837	
	1000	Chemical spreader, 3 C.Y.		3.15	45	135	405	52.20	
	1900	Hammermill, traveling, 250 H.P.		68.23	2,200	6,620	19,900	1,870	
	2000	Horizontal borer, 3" diameter, 13 H.P. gas driven		5.50	56.50	170	510	78	
	2150	Horizontal directional drill, 20,000 lb. thrust, 78 H.P. diesel		27.50	680	2,045	6,125	629	
	2160	30,000 lb. thrust, 115 H.P.		33.65	1,050	3,135	9,400	896.20	
	2170	50,000 lb. thrust, 170 H.P.		48.35	1,325	4,005	12,000	1,188	
	2190	Mud trailer for HDD, 1500 gallons, 175 H.P., gas		24.10	158	475	1,425	287.80	
	2200	Hydromulcher, diesel, 3000 gallon, for truck mounting		16.35	253	760	2,275	282.80	
	2300	Gas, 600 gallon		7.40	103	310	930	121.20	
	2400	Joint & crack cleaner, walk behind, 25 H.P.		3.10	51.50	155	465	55.80	

01 54 33 | Equipment Rental

		UNIT	HOURLY OPER. COST	RENT PER DAY	RENT PER WEEK	RENT PER MONTH	EQUIPMENT COST/DAY	
50	2500	Filler, trailer mounted, 400 gallons, 20 H.P.	Ea.	8.40	218	655	1,975	198.20	**50**
	3000	Paint striper, self-propelled, 40 gallon, 22 H.P.		6.75	162	485	1,450	151	
	3100	120 gallon, 120 H.P.		18.90	405	1,220	3,650	395.20	
	3200	Post drivers, 6" I-Beam frame, for truck mounting		12.45	390	1,175	3,525	334.60	
	3400	Road sweeper, self-propelled, 8' wide, 90 H.P.		35.95	670	2,005	6,025	688.60	
	3450	Road sweeper, vacuum assisted, 4 C.Y., 220 gallons		56.05	655	1,960	5,875	840.40	
	4000	Road mixer, self-propelled, 130 H.P.		45.95	800	2,405	7,225	848.60	
	4100	310 H.P.		75.55	2,150	6,425	19,300	1,889	
	4220	Cold mix paver, incl. pug mill and bitumen tank, 165 H.P.		94.60	2,300	6,915	20,700	2,140	
	4240	Pavement brush, towed		3.40	96.50	290	870	85.20	
	4250	Paver, asphalt, wheel or crawler, 130 H.P., diesel		94.25	2,275	6,845	20,500	2,123	
	4300	Paver, road widener, gas, 1' to 6', 67 H.P.		46.65	940	2,825	8,475	938.20	
	4400	Diesel, 2' to 14', 88 H.P.		56.75	1,125	3,355	10,100	1,125	
	4600	Slipform pavers, curb and gutter, 2 track, 75 H.P.		56.30	1,200	3,615	10,800	1,173	
	4700	4 track, 165 H.P.		36.95	825	2,470	7,400	789.60	
	4800	Median barrier, 215 H.P.		57.45	1,275	3,805	11,400	1,221	
	4901	Trailer, low bed, 75 ton capacity		11.05	268	805	2,425	249.40	
	5000	Road planer, walk behind, 10" cutting width, 10 H.P.		2.50	33.50	100	300	40	
	5100	Self-propelled, 12" cutting width, 64 H.P.		8.00	115	345	1,025	133	
	5120	Traffic line remover, metal ball blaster, truck mounted, 115 H.P.		46.70	800	2,395	7,175	852.60	
	5140	Grinder, truck mounted, 115 H.P.		51.05	850	2,555	7,675	919.40	
	5160	Walk-behind, 11 H.P.		3.55	55	165	495	61.40	
	5200	Pavement profiler, 4' to 6' wide, 450 H.P.		218.90	3,450	10,350	31,100	3,821	
	5300	8' to 10' wide, 750 H.P.		336.50	4,550	13,635	40,900	5,419	
	5400	Roadway plate, steel, 1" x 8' x 20'		.09	14.35	43	129	9.30	
	5600	Stabilizer, self-propelled, 150 H.P.		41.50	680	2,045	6,125	741	
	5700	310 H.P.		77.65	1,825	5,485	16,500	1,718	
	5800	Striper, truck mounted, 120 gallon paint, 460 H.P.		48.50	505	1,510	4,525	690	
	5900	Thermal paint heating kettle, 115 gallons		7.73	26.50	80	240	77.85	
	6000	Tar kettle, 330 gallon, trailer mounted		11.65	60	180	540	129.20	
	7000	Tunnel locomotive, diesel, 8 to 12 ton		29.90	600	1,800	5,400	599.20	
	7005	Electric, 10 ton		28.40	685	2,060	6,175	639.20	
	7010	Muck cars, 1/2 C.Y. capacity		2.25	26	78	234	33.60	
	7020	1 C.Y. capacity		2.45	33.50	100	300	39.60	
	7030	2 C.Y. capacity		2.60	38.50	115	345	43.80	
	7040	Side dump, 2 C.Y. capacity		2.80	46.50	140	420	50.40	
	7050	3 C.Y. capacity		3.80	51.50	155	465	61.40	
	7060	5 C.Y. capacity		5.50	66.50	200	600	84	
	7100	Ventilating blower for tunnel, 7-1/2 H.P.		2.16	51.50	155	465	48.30	
	7110	10 H.P.		2.39	53.50	160	480	51.10	
	7120	20 H.P.		3.56	69.50	208	625	70.10	
	7140	40 H.P.		6.02	80	240	720	96.15	
	7160	60 H.P.		8.75	98.50	295	885	129	
	7175	75 H.P.		10.31	153	460	1,375	174.50	
	7180	200 H.P.		20.73	305	920	2,750	349.85	
	7800	Windrow loader, elevating		54.10	1,350	4,045	12,100	1,242	
60	0010	**LIFTING AND HOISTING EQUIPMENT RENTAL** without operators							**60**
	0150	Crane, flatbed mounted, 3 ton capacity	Ea.	14.10	205	615	1,850	235.80	
	0200	Crane, climbing, 106' jib, 6000 lb. capacity, 410 fpm		41.13	1,750	5,260	15,800	1,381	
	0300	101' jib, 10,250 lb. capacity, 270 fpm		48.18	2,225	6,670	20,000	1,719	
	0500	Tower, static, 130' high, 106' jib, 6200 lb. capacity at 400 fpm		45.23	2,025	6,080	18,200	1,578	
	0520	Mini crawler spider crane, up to 24" wide, 1990 lb. lifting capacity		12.54	755	2,265	6,800	553.30	
	0525	Up to 30" wide, 6450 lb. lifting capacity		14.57	840	2,520	7,550	620.55	
	0530	Up to 52" wide, 6680 lb. lifting capacity		23.17	1,325	3,960	11,900	977.35	
	0535	Up to 55" wide, 8920 lb. lifting capacity		25.87	1,500	4,500	13,500	1,107	
	0540	Up to 66" wide, 13,350 lb. lifting capacity		35.03	2,050	6,120	18,400	1,504	
	0600	Crawler mounted, lattice boom, 1/2 C.Y., 15 tons at 12' radius		37.10	885	2,660	7,975	828.80	
	0700	3/4 C.Y., 20 tons at 12' radius		49.46	1,100	3,320	9,950	1,060	

Reference boxes: R015433 -10 (at row 0010); R312316 -45 (at row 0200)

01 54 33 | Equipment Rental

		UNIT	HOURLY OPER. COST	RENT PER DAY	RENT PER WEEK	RENT PER MONTH	EQUIPMENT COST/DAY
60	0800	1 C.Y., 25 tons at 12' radius	Ea.	65.95	1,375	4,100	12,300	1,348	60
	0900	1-1/2 C.Y., 40 tons at 12' radius		66.45	1,400	4,190	12,600	1,370
	1000	2 C.Y., 50 tons at 12' radius		88.90	2,050	6,145	18,400	1,940
	1100	3 C.Y., 75 tons at 12' radius		76.00	1,825	5,500	16,500	1,708
	1200	100 ton capacity, 60' boom		86.05	1,975	5,920	17,800	1,872
	1300	165 ton capacity, 60' boom		105.00	2,325	6,970	20,900	2,234
	1400	200 ton capacity, 70' boom		140.85	3,125	9,385	28,200	3,004
	1500	350 ton capacity, 80' boom		182.50	4,125	12,375	37,100	3,935
	1600	Truck mounted, lattice boom, 6 x 4, 20 tons at 10' radius		37.84	1,300	3,900	11,700	1,083
	1700	25 tons at 10' radius		40.92	1,425	4,240	12,700	1,175
	1800	8 x 4, 30 tons at 10' radius		44.29	1,500	4,520	13,600	1,258
	1900	40 tons at 12' radius		47.09	1,575	4,720	14,200	1,321
	2000	60 tons at 15' radius		52.56	1,675	5,000	15,000	1,420
	2050	82 tons at 15' radius		58.46	1,775	5,340	16,000	1,536
	2100	90 tons at 15' radius		65.33	1,950	5,820	17,500	1,687
	2200	115 tons at 15' radius		73.60	2,175	6,500	19,500	1,889
	2300	150 tons at 18' radius		80.95	2,275	6,845	20,500	2,017
	2350	165 tons at 18' radius		85.95	2,425	7,260	21,800	2,140
	2400	Truck mounted, hydraulic, 12 ton capacity		31.00	415	1,240	3,725	496
	2500	25 ton capacity		37.45	485	1,455	4,375	590.60
	2550	33 ton capacity		50.70	890	2,675	8,025	940.60
	2560	40 ton capacity		51.00	905	2,710	8,125	950
	2600	55 ton capacity		56.60	900	2,705	8,125	993.80
	2700	80 ton capacity		77.85	1,500	4,475	13,400	1,518
	2720	100 ton capacity		76.80	1,525	4,595	13,800	1,533
	2740	120 ton capacity		101.90	1,825	5,460	16,400	1,907
	2760	150 ton capacity		107.85	1,975	5,960	17,900	2,055
	2800	Self-propelled, 4 x 4, with telescoping boom, 5 ton		15.00	232	695	2,075	259
	2900	12-1/2 ton capacity		20.45	335	1,000	3,000	363.60
	3000	15 ton capacity		33.95	520	1,560	4,675	583.60
	3050	20 ton capacity		25.40	615	1,840	5,525	571.20
	3100	25 ton capacity		37.30	615	1,850	5,550	668.40
	3150	40 ton capacity		44.10	635	1,910	5,725	734.80
	3200	Derricks, guy, 20 ton capacity, 60' boom, 75' mast		23.07	430	1,288	3,875	442.15
	3300	100' boom, 115' mast		36.55	735	2,210	6,625	734.40
	3400	Stiffleg, 20 ton capacity, 70' boom, 37' mast		25.74	555	1,670	5,000	539.90
	3500	100' boom, 47' mast		39.84	895	2,680	8,050	854.70
	3550	Helicopter, small, lift to 1250 lb. maximum, w/pilot		97.10	3,475	10,400	31,200	2,857
	3600	Hoists, chain type, overhead, manual, 3/4 ton		.15	.33	1	3	1.40
	3900	10 ton		.80	6	18	54	10
	4000	Hoist and tower, 5000 lb. cap., portable electric, 40' high		5.19	247	742	2,225	189.90
	4100	For each added 10' section, add		.12	19.35	58	174	12.55
	4200	Hoist and single tubular tower, 5000 lb. electric, 100' high		7.03	345	1,036	3,100	263.45
	4300	For each added 6'-6" section, add		.20	33.50	101	305	21.80
	4400	Hoist and double tubular tower, 5000 lb., 100' high		7.56	380	1,141	3,425	288.70
	4500	For each added 6'-6" section, add		.22	37	111	335	23.95
	4550	Hoist and tower, mast type, 6000 lb., 100' high		8.14	395	1,183	3,550	301.70
	4570	For each added 10' section, add		.14	22.50	68	204	14.70
	4600	Hoist and tower, personnel, electric, 2000 lb., 100' @ 125 fpm		17.23	1,050	3,150	9,450	767.85
	4700	3000 lb., 100' @ 200 fpm		19.70	1,200	3,570	10,700	871.60
	4800	3000 lb., 150' @ 300 fpm		21.85	1,325	4,000	12,000	974.80
	4900	4000 lb., 100' @ 300 fpm		22.62	1,350	4,080	12,200	996.95
	5000	6000 lb., 100' @ 275 fpm	▼	24.32	1,425	4,270	12,800	1,049
	5100	For added heights up to 500', add	L.F.	.01	1.67	5	15	1.10
	5200	Jacks, hydraulic, 20 ton	Ea.	.05	2	6	18	1.60
	5500	100 ton		.40	12	36	108	10.40
	6100	Jacks, hydraulic, climbing w/50' jackrods, control console, 30 ton cap.		2.13	142	426	1,275	102.25
	6150	For each added 10' jackrod section, add	▼	.05	3.33	10	30	2.40

01 54 33 | Equipment Rental

		UNIT	HOURLY OPER. COST	RENT PER DAY	RENT PER WEEK	RENT PER MONTH	EQUIPMENT COST/DAY	
60	6300	50 ton capacity	Ea.	3.43	228	685	2,050	164.45	**60**
	6350	For each added 10' jackrod section, add		.06	4	12	36	2.90	
	6500	125 ton capacity		8.95	595	1,790	5,375	429.60	
	6550	For each added 10' jackrod section, add		.61	40.50	121	365	29.10	
	6600	Cable jack, 10 ton capacity with 200' cable		1.79	119	357	1,075	85.70	
	6650	For each added 50' of cable, add		.22	14.35	43	129	10.35	
70	0010	**WELLPOINT EQUIPMENT RENTAL** without operators	R015433 -10						**70**
	0020	Based on 2 months rental							
	0100	Combination jetting & wellpoint pump, 60 H.P. diesel	Ea.	15.83	350	1,057	3,175	338.05	
	0200	High pressure gas jet pump, 200 H.P., 300 psi	"	34.42	300	903	2,700	455.95	
	0300	Discharge pipe, 8" diameter	L.F.	.01	.57	1.71	5.15	.40	
	0350	12" diameter		.01	.84	2.53	7.60	.60	
	0400	Header pipe, flows up to 150 GPM, 4" diameter		.01	.52	1.56	4.68	.40	
	0500	400 GPM, 6" diameter		.01	.61	1.83	5.50	.45	
	0600	800 GPM, 8" diameter		.01	.84	2.53	7.60	.60	
	0700	1500 GPM, 10" diameter		.01	.89	2.66	8	.60	
	0800	2500 GPM, 12" diameter		.03	1.68	5.03	15.10	1.25	
	0900	4500 GPM, 16" diameter		.03	2.15	6.44	19.30	1.55	
	0950	For quick coupling aluminum and plastic pipe, add		.03	2.22	6.67	20	1.55	
	1100	Wellpoint, 25' long, with fittings & riser pipe, 1-1/2" or 2" diameter	Ea.	.07	4.44	13.31	40	3.20	
	1200	Wellpoint pump, diesel powered, 4" suction, 20 H.P.		7.07	203	609	1,825	178.35	
	1300	6" suction, 30 H.P.		9.51	252	756	2,275	227.30	
	1400	8" suction, 40 H.P.		12.87	345	1,036	3,100	310.15	
	1500	10" suction, 75 H.P.		19.01	405	1,211	3,625	394.30	
	1600	12" suction, 100 H.P.		27.56	645	1,930	5,800	606.50	
	1700	12" suction, 175 H.P.		39.50	710	2,130	6,400	742	
80	0010	**MARINE EQUIPMENT RENTAL** without operators	R015433 -10						**80**
	0200	Barge, 400 ton, 30' wide x 90' long	Ea.	18.05	1,150	3,455	10,400	835.40	
	0240	800 ton, 45' wide x 90' long		21.95	1,425	4,240	12,700	1,024	
	2000	Tugboat, diesel, 100 H.P.		28.80	228	685	2,050	367.40	
	2040	250 H.P.		54.40	410	1,225	3,675	680.20	
	2080	380 H.P.		122.45	1,225	3,685	11,100	1,717	
	3000	Small work boat, gas, 16-foot, 50 H.P.		12.50	63.50	190	570	138	
	4000	Large, diesel, 48-foot, 200 H.P.		74.90	1,300	3,930	11,800	1,385	

Crew No.	Bare Costs Hr.	Bare Costs Daily	Incl. Subs O&P Hr.	Incl. Subs O&P Daily	Cost Per Labor-Hour Bare Costs	Cost Per Labor-Hour Incl. O&P
Crew A-1						
1 Building Laborer	$26.70	$213.60	$44.40	$355.20	$26.70	$44.40
1 Concrete Saw, Gas Manual		71.20		78.32	8.90	9.79
8 L.H., Daily Totals		$284.80		$433.52	$35.60	$54.19
Crew A-1A						
1 Skilled Worker	$35.80	$286.40	$59.80	$478.40	$35.80	$59.80
1 Shot Blaster, 20"		214.80		236.28	26.85	29.54
8 L.H., Daily Totals		$501.20		$714.68	$62.65	$89.33
Crew A-1B						
1 Building Laborer	$26.70	$213.60	$44.40	$355.20	$26.70	$44.40
1 Concrete Saw		102.40		112.64	12.80	14.08
8 L.H., Daily Totals		$316.00		$467.84	$39.50	$58.48
Crew A-1C						
1 Building Laborer	$26.70	$213.60	$44.40	$355.20	$26.70	$44.40
1 Chain Saw, Gas, 18"		27.80		30.58	3.48	3.82
8 L.H., Daily Totals		$241.40		$385.78	$30.18	$48.22
Crew A-1D						
1 Building Laborer	$26.70	$213.60	$44.40	$355.20	$26.70	$44.40
1 Vibrating Plate, Gas, 18"		31.80		34.98	3.98	4.37
8 L.H., Daily Totals		$245.40		$390.18	$30.68	$48.77
Crew A-1E						
1 Building Laborer	$26.70	$213.60	$44.40	$355.20	$26.70	$44.40
1 Vibrating Plate, Gas, 21"		40.60		44.66	5.08	5.58
8 L.H., Daily Totals		$254.20		$399.86	$31.77	$49.98
Crew A-1F						
1 Building Laborer	$26.70	$213.60	$44.40	$355.20	$26.70	$44.40
1 Rammer/Tamper, Gas, 8"		46.00		50.60	5.75	6.33
8 L.H., Daily Totals		$259.60		$405.80	$32.45	$50.73
Crew A-1G						
1 Building Laborer	$26.70	$213.60	$44.40	$355.20	$26.70	$44.40
1 Rammer/Tamper, Gas, 15"		52.00		57.20	6.50	7.15
8 L.H., Daily Totals		$265.60		$412.40	$33.20	$51.55
Crew A-1H						
1 Building Laborer	$26.70	$213.60	$44.40	$355.20	$26.70	$44.40
1 Exterior Steam Cleaner		74.80		82.28	9.35	10.29
8 L.H., Daily Totals		$288.40		$437.48	$36.05	$54.69
Crew A-1J						
1 Building Laborer	$26.70	$213.60	$44.40	$355.20	$26.70	$44.40
1 Cultivator, Walk-Behind, 5 H.P.		63.50		69.85	7.94	8.73
8 L.H., Daily Totals		$277.10		$425.05	$34.64	$53.13
Crew A-1K						
1 Building Laborer	$26.70	$213.60	$44.40	$355.20	$26.70	$44.40
1 Cultivator, Walk-Behind, 8 H.P.		74.60		82.06	9.32	10.26
8 L.H., Daily Totals		$288.20		$437.26	$36.02	$54.66
Crew A-1M						
1 Building Laborer	$26.70	$213.60	$44.40	$355.20	$26.70	$44.40
1 Snow Blower, Walk-Behind		70.10		77.11	8.76	9.64
8 L.H., Daily Totals		$283.70		$432.31	$35.46	$54.04

Crew No.	Bare Costs Hr.	Bare Costs Daily	Incl. Subs O&P Hr.	Incl. Subs O&P Daily	Cost Per Labor-Hour Bare Costs	Cost Per Labor-Hour Incl. O&P
Crew A-2						
2 Laborers	$26.70	$427.20	$44.40	$710.40	$28.33	$46.93
1 Truck Driver (light)	31.60	252.80	52.00	416.00		
1 Flatbed Truck, Gas, 1.5 Ton		188.40		207.24	7.85	8.63
24 L.H., Daily Totals		$868.40		$1333.64	$36.18	$55.57
Crew A-2A						
2 Laborers	$26.70	$427.20	$44.40	$710.40	$28.33	$46.93
1 Truck Driver (light)	31.60	252.80	52.00	416.00		
1 Flatbed Truck, Gas, 1.5 Ton		188.40		207.24		
1 Concrete Saw		102.40		112.64	12.12	13.33
24 L.H., Daily Totals		$970.80		$1446.28	$40.45	$60.26
Crew A-2B						
1 Truck Driver (light)	$31.60	$252.80	$52.00	$416.00	$31.60	$52.00
1 Flatbed Truck, Gas, 1.5 Ton		188.40		207.24	23.55	25.91
8 L.H., Daily Totals		$441.20		$623.24	$55.15	$77.91
Crew A-3A						
1 Equip. Oper. (light)	$35.40	$283.20	$58.35	$466.80	$35.40	$58.35
1 Pickup Truck, 4x4, 3/4 Ton		126.60		139.26	15.82	17.41
8 L.H., Daily Totals		$409.80		$606.06	$51.23	$75.76
Crew A-3B						
1 Equip. Oper. (medium)	$36.55	$292.40	$60.25	$482.00	$34.38	$56.63
1 Truck Driver (heavy)	32.20	257.60	53.00	424.00		
1 Dump Truck, 12 C.Y., 400 H.P.		542.80		597.08		
1 F.E. Loader, W.M., 2.5 C.Y.		523.20		575.52	66.63	73.29
16 L.H., Daily Totals		$1616.00		$2078.60	$101.00	$129.91
Crew A-3C						
1 Equip. Oper. (light)	$35.40	$283.20	$58.35	$466.80	$35.40	$58.35
1 Loader, Skid Steer, 78 H.P.		364.80		401.28	45.60	50.16
8 L.H., Daily Totals		$648.00		$868.08	$81.00	$108.51
Crew A-3D						
1 Truck Driver (light)	$31.60	$252.80	$52.00	$416.00	$31.60	$52.00
1 Pickup Truck, 4x4, 3/4 Ton		126.60		139.26		
1 Flatbed Trailer, 25 Ton		133.00		146.30	32.45	35.70
8 L.H., Daily Totals		$512.40		$701.56	$64.05	$87.69
Crew A-3E						
1 Equip. Oper. (crane)	$37.05	$296.40	$61.05	$488.40	$34.63	$57.02
1 Truck Driver (heavy)	32.20	257.60	53.00	424.00		
1 Pickup Truck, 4x4, 3/4 Ton		126.60		139.26	7.91	8.70
16 L.H., Daily Totals		$680.60		$1051.66	$42.54	$65.73
Crew A-3F						
1 Equip. Oper. (crane)	$37.05	$296.40	$61.05	$488.40	$34.63	$57.02
1 Truck Driver (heavy)	32.20	257.60	53.00	424.00		
1 Pickup Truck, 4x4, 3/4 Ton		126.60		139.26		
1 Truck Tractor, 6x4, 380 H.P.		476.20		523.82		
1 Lowbed Trailer, 75 Ton		249.40		274.34	53.26	58.59
16 L.H., Daily Totals		$1406.20		$1849.82	$87.89	$115.61

Crew A-3G

Crew No.	Bare Costs Hr.	Daily	Incl. Subs O&P Hr.	Daily	Cost Per Labor-Hour Bare Costs	Incl. O&P
1 Equip. Oper. (crane)	$37.05	$296.40	$61.05	$488.40	$34.63	$57.02
1 Truck Driver (heavy)	32.20	257.60	53.00	424.00		
1 Pickup Truck, 4x4, 3/4 Ton		126.60		139.26		
1 Truck Tractor, 6x4, 450 H.P.		583.40		641.74		
1 Lowbed Trailer, 75 Ton		249.40		274.34	59.96	65.96
16 L.H., Daily Totals		$1513.40		$1967.74	$94.59	$122.98

Crew A-3H

Crew No.	Bare Costs Hr.	Daily	Incl. Subs O&P Hr.	Daily	Cost Per Labor-Hour Bare Costs	Incl. O&P
1 Equip. Oper. (crane)	$37.05	$296.40	$61.05	$488.40	$37.05	$61.05
1 Hyd. Crane, 12 Ton (Daily)		628.60		691.46	78.58	86.43
8 L.H., Daily Totals		$925.00		$1179.86	$115.63	$147.48

Crew A-3I

Crew No.	Bare Costs Hr.	Daily	Incl. Subs O&P Hr.	Daily	Cost Per Labor-Hour Bare Costs	Incl. O&P
1 Equip. Oper. (crane)	$37.05	$296.40	$61.05	$488.40	$37.05	$61.05
1 Hyd. Crane, 25 Ton (Daily)		759.40		835.34	94.92	104.42
8 L.H., Daily Totals		$1055.80		$1323.74	$131.97	$165.47

Crew A-3J

Crew No.	Bare Costs Hr.	Daily	Incl. Subs O&P Hr.	Daily	Cost Per Labor-Hour Bare Costs	Incl. O&P
1 Equip. Oper. (crane)	$37.05	$296.40	$61.05	$488.40	$37.05	$61.05
1 Hyd. Crane, 40 Ton (Daily)		1313.00		1444.30	164.13	180.54
8 L.H., Daily Totals		$1609.40		$1932.70	$201.18	$241.59

Crew A-3K

Crew No.	Bare Costs Hr.	Daily	Incl. Subs O&P Hr.	Daily	Cost Per Labor-Hour Bare Costs	Incl. O&P
1 Equip. Oper. (crane)	$37.05	$296.40	$61.05	$488.40	$34.80	$57.35
1 Equip. Oper. (oiler)	32.55	260.40	53.65	429.20		
1 Hyd. Crane, 55 Ton (Daily)		1353.00		1488.30		
1 P/U Truck, 3/4 Ton (Daily)		139.20		153.12	93.26	102.59
16 L.H., Daily Totals		$2049.00		$2559.02	$128.06	$159.94

Crew A-3L

Crew No.	Bare Costs Hr.	Daily	Incl. Subs O&P Hr.	Daily	Cost Per Labor-Hour Bare Costs	Incl. O&P
1 Equip. Oper. (crane)	$37.05	$296.40	$61.05	$488.40	$34.80	$57.35
1 Equip. Oper. (oiler)	32.55	260.40	53.65	429.20		
1 Hyd. Crane, 80 Ton (Daily)		2113.00		2324.30		
1 P/U Truck, 3/4 Ton (Daily)		139.20		153.12	140.76	154.84
16 L.H., Daily Totals		$2809.00		$3395.02	$175.56	$212.19

Crew A-3M

Crew No.	Bare Costs Hr.	Daily	Incl. Subs O&P Hr.	Daily	Cost Per Labor-Hour Bare Costs	Incl. O&P
1 Equip. Oper. (crane)	$37.05	$296.40	$61.05	$488.40	$34.80	$57.35
1 Equip. Oper. (oiler)	32.55	260.40	53.65	429.20		
1 Hyd. Crane, 100 Ton (Daily)		2144.00		2358.40		
1 P/U Truck, 3/4 Ton (Daily)		139.20		153.12	142.70	156.97
16 L.H., Daily Totals		$2840.00		$3429.12	$177.50	$214.32

Crew A-3N

Crew No.	Bare Costs Hr.	Daily	Incl. Subs O&P Hr.	Daily	Cost Per Labor-Hour Bare Costs	Incl. O&P
1 Equip. Oper. (crane)	$37.05	$296.40	$61.05	$488.40	$37.05	$61.05
1 Tower Crane (monthly)		1180.00		1298.00	147.50	162.25
8 L.H., Daily Totals		$1476.40		$1786.40	$184.55	$223.30

Crew A-3P

Crew No.	Bare Costs Hr.	Daily	Incl. Subs O&P Hr.	Daily	Cost Per Labor-Hour Bare Costs	Incl. O&P
1 Equip. Oper. (light)	$35.40	$283.20	$58.35	$466.80	$35.40	$58.35
1 A.T. Forklift, 31' reach, 45' lift		484.20		532.62	60.52	66.58
8 L.H., Daily Totals		$767.40		$999.42	$95.92	$124.93

Crew A-3Q

Crew No.	Bare Costs Hr.	Daily	Incl. Subs O&P Hr.	Daily	Cost Per Labor-Hour Bare Costs	Incl. O&P
1 Equip. Oper. (light)	$35.40	$283.20	$58.35	$466.80	$35.40	$58.35
1 Pickup Truck, 4x4, 3/4 Ton		126.60		139.26		
1 Flatbed Trailer, 3 Ton		25.40		27.94	19.00	20.90
8 L.H., Daily Totals		$435.20		$634.00	$54.40	$79.25

Crew A-3R

Crew No.	Bare Costs Hr.	Daily	Incl. Subs O&P Hr.	Daily	Cost Per Labor-Hour Bare Costs	Incl. O&P
1 Equip. Oper. (light)	$35.40	$283.20	$58.35	$466.80	$35.40	$58.35
1 Forklift, Smooth Floor, 8,000 Lb.		148.20		163.02	18.52	20.38
8 L.H., Daily Totals		$431.40		$629.82	$53.92	$78.73

Crew A-4

Crew No.	Bare Costs Hr.	Daily	Incl. Subs O&P Hr.	Daily	Cost Per Labor-Hour Bare Costs	Incl. O&P
2 Carpenters	$34.95	$559.20	$58.10	$929.60	$33.08	$54.83
1 Painter, Ordinary	29.35	234.80	48.30	386.40		
24 L.H., Daily Totals		$794.00		$1316.00	$33.08	$54.83

Crew A-5

Crew No.	Bare Costs Hr.	Daily	Incl. Subs O&P Hr.	Daily	Cost Per Labor-Hour Bare Costs	Incl. O&P
2 Laborers	$26.70	$427.20	$44.40	$710.40	$27.24	$45.24
.25 Truck Driver (light)	31.60	63.20	52.00	104.00		
.25 Flatbed Truck, Gas, 1.5 Ton		47.10		51.81	2.62	2.88
18 L.H., Daily Totals		$537.50		$866.21	$29.86	$48.12

Crew A-6

Crew No.	Bare Costs Hr.	Daily	Incl. Subs O&P Hr.	Daily	Cost Per Labor-Hour Bare Costs	Incl. O&P
1 Instrument Man	$35.80	$286.40	$59.80	$478.40	$34.88	$57.98
1 Rodman/Chainman	33.95	271.60	56.15	449.20		
1 Level, Electronic		50.40		55.44	3.15	3.46
16 L.H., Daily Totals		$608.40		$983.04	$38.02	$61.44

Crew A-7

Crew No.	Bare Costs Hr.	Daily	Incl. Subs O&P Hr.	Daily	Cost Per Labor-Hour Bare Costs	Incl. O&P
1 Chief of Party	$41.40	$331.20	$68.50	$548.00	$37.05	$61.48
1 Instrument Man	35.80	286.40	59.80	478.40		
1 Rodman/Chainman	33.95	271.60	56.15	449.20		
1 Level, Electronic		50.40		55.44	2.10	2.31
24 L.H., Daily Totals		$939.60		$1531.04	$39.15	$63.79

Crew A-8

Crew No.	Bare Costs Hr.	Daily	Incl. Subs O&P Hr.	Daily	Cost Per Labor-Hour Bare Costs	Incl. O&P
1 Chief of Party	$41.40	$331.20	$68.50	$548.00	$36.27	$60.15
1 Instrument Man	35.80	286.40	59.80	478.40		
2 Rodmen/Chainmen	33.95	543.20	56.15	898.40		
1 Level, Electronic		50.40		55.44	1.58	1.73
32 L.H., Daily Totals		$1211.20		$1980.24	$37.85	$61.88

Crew A-9

Crew No.	Bare Costs Hr.	Daily	Incl. Subs O&P Hr.	Daily	Cost Per Labor-Hour Bare Costs	Incl. O&P
1 Asbestos Foreman	$36.55	$292.40	$61.85	$494.80	$36.11	$61.11
7 Asbestos Workers	36.05	2018.80	61.00	3416.00		
64 L.H., Daily Totals		$2311.20		$3910.80	$36.11	$61.11

Crew A-10A

Crew No.	Bare Costs Hr.	Daily	Incl. Subs O&P Hr.	Daily	Cost Per Labor-Hour Bare Costs	Incl. O&P
1 Asbestos Foreman	$36.55	$292.40	$61.85	$494.80	$36.22	$61.28
2 Asbestos Workers	36.05	576.80	61.00	976.00		
24 L.H., Daily Totals		$869.20		$1470.80	$36.22	$61.28

Crew A-10B

Crew No.	Bare Costs Hr.	Daily	Incl. Subs O&P Hr.	Daily	Cost Per Labor-Hour Bare Costs	Incl. O&P
1 Asbestos Foreman	$36.55	$292.40	$61.85	$494.80	$36.17	$61.21
3 Asbestos Workers	36.05	865.20	61.00	1464.00		
32 L.H., Daily Totals		$1157.60		$1958.80	$36.17	$61.21

Crew A-10C

Crew No.	Bare Costs Hr.	Daily	Incl. Subs O&P Hr.	Daily	Cost Per Labor-Hour Bare Costs	Incl. O&P
3 Asbestos Workers	$36.05	$865.20	$61.00	$1464.00	$36.05	$61.00
1 Flatbed Truck, Gas, 1.5 Ton		188.40		207.24	7.85	8.63
24 L.H., Daily Totals		$1053.60		$1671.24	$43.90	$69.64

Crew A-10D

Crew No.	Bare Costs Hr.	Daily	Incl. Subs O&P Hr.	Daily	Cost Per Labor-Hour Bare Costs	Incl. O&P
2 Asbestos Workers	$36.05	$576.80	$61.00	$976.00	$35.42	$59.17
1 Equip. Oper. (crane)	37.05	296.40	61.05	488.40		
1 Equip. Oper. (oiler)	32.55	260.40	53.65	429.20		
1 Hydraulic Crane, 33 Ton		940.60		1034.66	29.39	32.33
32 L.H., Daily Totals		$2074.20		$2928.26	$64.82	$91.51

Crew A-11

Crew No.	Bare Costs Hr.	Daily	Incl. Subs O&P Hr.	Daily	Cost Per Labor-Hour Bare Costs	Incl. O&P
1 Asbestos Foreman	$36.55	$292.40	$61.85	$494.80	$36.11	$61.11
7 Asbestos Workers	36.05	2018.80	61.00	3416.00		
2 Chip. Hammers, 12 Lb., Elec.		42.00		46.20	0.66	0.72
64 L.H., Daily Totals		$2353.20		$3957.00	$36.77	$61.83

Crew A-12

Crew No.	Bare Costs Hr.	Daily	Incl. Subs O&P Hr.	Daily	Cost Per Labor-Hour Bare Costs	Incl. O&P
1 Asbestos Foreman	$36.55	$292.40	$61.85	$494.80	$36.11	$61.11
7 Asbestos Workers	36.05	2018.80	61.00	3416.00		
1 Trk-Mtd Vac, 14 CY, 1500 Gal.		522.70		574.97		
1 Flatbed Truck, 20,000 GVW		197.20		216.92	11.25	12.37
64 L.H., Daily Totals		$3031.10		$4702.69	$47.36	$73.48

Crew A-13

Crew No.	Bare Costs Hr.	Daily	Incl. Subs O&P Hr.	Daily	Cost Per Labor-Hour Bare Costs	Incl. O&P
1 Equip. Oper. (light)	$35.40	$283.20	$58.35	$466.80	$35.40	$58.35
1 Trk-Mtd Vac, 14 CY, 1500 Gal.		522.70		574.97		
1 Flatbed Truck, 20,000 GVW		197.20		216.92	89.99	98.99
8 L.H., Daily Totals		$1003.10		$1258.69	$125.39	$157.34

Crew B-1

Crew No.	Bare Costs Hr.	Daily	Incl. Subs O&P Hr.	Daily	Cost Per Labor-Hour Bare Costs	Incl. O&P
1 Labor Foreman (outside)	$28.70	$229.60	$47.75	$382.00	$27.37	$45.52
2 Laborers	26.70	427.20	44.40	710.40		
24 L.H., Daily Totals		$656.80		$1092.40	$27.37	$45.52

Crew B-1A

Crew No.	Bare Costs Hr.	Daily	Incl. Subs O&P Hr.	Daily	Cost Per Labor-Hour Bare Costs	Incl. O&P
1 Labor Foreman (outside)	$28.70	$229.60	$47.75	$382.00	$27.37	$45.52
2 Laborers	26.70	427.20	44.40	710.40		
2 Cutting Torches		25.20		27.72		
2 Sets of Gases		336.00		369.60	15.05	16.56
24 L.H., Daily Totals		$1018.00		$1489.72	$42.42	$62.07

Crew B-1B

Crew No.	Bare Costs Hr.	Daily	Incl. Subs O&P Hr.	Daily	Cost Per Labor-Hour Bare Costs	Incl. O&P
1 Labor Foreman (outside)	$28.70	$229.60	$47.75	$382.00	$29.79	$49.40
2 Laborers	26.70	427.20	44.40	710.40		
1 Equip. Oper. (crane)	37.05	296.40	61.05	488.40		
2 Cutting Torches		25.20		27.72		
2 Sets of Gases		336.00		369.60		
1 Hyd. Crane, 12 Ton		496.00		545.60	26.79	29.47
32 L.H., Daily Totals		$1810.40		$2523.72	$56.58	$78.87

Crew B-1C

Crew No.	Bare Costs Hr.	Daily	Incl. Subs O&P Hr.	Daily	Cost Per Labor-Hour Bare Costs	Incl. O&P
1 Labor Foreman (outside)	$28.70	$229.60	$47.75	$382.00	$27.37	$45.52
2 Laborers	26.70	427.20	44.40	710.40		
1 Telescoping Boom Lift, to 60'		454.60		500.06	18.94	20.84
24 L.H., Daily Totals		$1111.40		$1592.46	$46.31	$66.35

Crew B-1D

Crew No.	Bare Costs Hr.	Daily	Incl. Subs O&P Hr.	Daily	Cost Per Labor-Hour Bare Costs	Incl. O&P
2 Laborers	$26.70	$427.20	$44.40	$710.40	$26.70	$44.40
1 Small Work Boat, Gas, 50 H.P.		138.00		151.80		
1 Pressure Washer, 7 GPM		77.20		84.92	13.45	14.80
16 L.H., Daily Totals		$642.40		$947.12	$40.15	$59.20

Crew B-1E

Crew No.	Bare Costs Hr.	Daily	Incl. Subs O&P Hr.	Daily	Cost Per Labor-Hour Bare Costs	Incl. O&P
1 Labor Foreman (outside)	$28.70	$229.60	$47.75	$382.00	$27.20	$45.24
3 Laborers	26.70	640.80	44.40	1065.60		
1 Work Boat, Diesel, 200 H.P.		1385.00		1523.50		
2 Pressure Washers, 7 GPM		154.40		169.84	48.11	52.92
32 L.H., Daily Totals		$2409.80		$3140.94	$75.31	$98.15

Crew B-1F

Crew No.	Bare Costs Hr.	Daily	Incl. Subs O&P Hr.	Daily	Cost Per Labor-Hour Bare Costs	Incl. O&P
2 Skilled Workers	$35.80	$572.80	$59.80	$956.80	$32.77	$54.67
1 Laborer	26.70	213.60	44.40	355.20		
1 Small Work Boat, Gas, 50 H.P.		138.00		151.80		
1 Pressure Washer, 7 GPM		77.20		84.92	8.97	9.86
24 L.H., Daily Totals		$1001.60		$1548.72	$41.73	$64.53

Crew B-1G

Crew No.	Bare Costs Hr.	Daily	Incl. Subs O&P Hr.	Daily	Cost Per Labor-Hour Bare Costs	Incl. O&P
2 Laborers	$26.70	$427.20	$44.40	$710.40	$26.70	$44.40
1 Small Work Boat, Gas, 50 H.P.		138.00		151.80	8.63	9.49
16 L.H., Daily Totals		$565.20		$862.20	$35.33	$53.89

Crew B-1H

Crew No.	Bare Costs Hr.	Daily	Incl. Subs O&P Hr.	Daily	Cost Per Labor-Hour Bare Costs	Incl. O&P
2 Skilled Workers	$35.80	$572.80	$59.80	$956.80	$32.77	$54.67
1 Laborer	26.70	213.60	44.40	355.20		
1 Small Work Boat, Gas, 50 H.P.		138.00		151.80	5.75	6.33
24 L.H., Daily Totals		$924.40		$1463.80	$38.52	$60.99

Crew B-1J

Crew No.	Bare Costs Hr.	Daily	Incl. Subs O&P Hr.	Daily	Cost Per Labor-Hour Bare Costs	Incl. O&P
1 Labor Foreman (inside)	$27.20	$217.60	$45.25	$362.00	$26.95	$44.83
1 Laborer	26.70	213.60	44.40	355.20		
16 L.H., Daily Totals		$431.20		$717.20	$26.95	$44.83

Crew B-1K

Crew No.	Bare Costs Hr.	Daily	Incl. Subs O&P Hr.	Daily	Cost Per Labor-Hour Bare Costs	Incl. O&P
1 Carpenter Foreman (inside)	$35.45	$283.60	$58.95	$471.60	$35.20	$58.52
1 Carpenter	34.95	279.60	58.10	464.80		
16 L.H., Daily Totals		$563.20		$936.40	$35.20	$58.52

Crew B-2

Crew No.	Bare Costs Hr.	Daily	Incl. Subs O&P Hr.	Daily	Cost Per Labor-Hour Bare Costs	Incl. O&P
1 Labor Foreman (outside)	$28.70	$229.60	$47.75	$382.00	$27.10	$45.07
4 Laborers	26.70	854.40	44.40	1420.80		
40 L.H., Daily Totals		$1084.00		$1802.80	$27.10	$45.07

Crew B-2A

Crew No.	Bare Costs Hr.	Daily	Incl. Subs O&P Hr.	Daily	Cost Per Labor-Hour Bare Costs	Incl. O&P
1 Labor Foreman (outside)	$28.70	$229.60	$47.75	$382.00	$27.37	$45.52
2 Laborers	26.70	427.20	44.40	710.40		
1 Telescoping Boom Lift, to 60'		454.60		500.06	18.94	20.84
24 L.H., Daily Totals		$1111.40		$1592.46	$46.31	$66.35

Crew B-3

Crew No.	Bare Costs Hr.	Daily	Incl. Subs O&P Hr.	Daily	Cost Per Labor-Hour Bare Costs	Incl. O&P
1 Labor Foreman (outside)	$28.70	$229.60	$47.75	$382.00	$30.51	$50.47
2 Laborers	26.70	427.20	44.40	710.40		
1 Equip. Oper. (medium)	36.55	292.40	60.25	482.00		
2 Truck Drivers (heavy)	32.20	515.20	53.00	848.00		
1 Crawler Loader, 3 C.Y.		1204.00		1324.40		
2 Dump Trucks, 12 C.Y., 400 H.P.		1085.60		1194.16	47.70	52.47
48 L.H., Daily Totals		$3754.00		$4940.96	$78.21	$102.94

Crew B-3A

Crew No.	Bare Costs Hr.	Daily	Incl. Subs O&P Hr.	Daily	Cost Per Labor-Hour Bare Costs	Incl. O&P
4 Laborers	$26.70	$854.40	$44.40	$1420.80	$28.67	$47.57
1 Equip. Oper. (medium)	36.55	292.40	60.25	482.00		
1 Hyd. Excavator, 1.5 C.Y.		893.60		982.96	22.34	24.57
40 L.H., Daily Totals		$2040.40		$2885.76	$51.01	$72.14

Crew B-3B

Crew No.	Hr.	Daily	Hr.	Daily	Bare Costs	Incl. O&P
2 Laborers	$26.70	$427.20	$44.40	$710.40	$30.54	$50.51
1 Equip. Oper. (medium)	36.55	292.40	60.25	482.00		
1 Truck Driver (heavy)	32.20	257.60	53.00	424.00		
1 Backhoe Loader, 80 H.P.		385.00		423.50		
1 Dump Truck, 12 C.Y., 400 H.P.		542.80		597.08	28.99	31.89
32 L.H., Daily Totals		$1905.00		$2636.98	$59.53	$82.41

Crew B-3C

Crew No.	Hr.	Daily	Hr.	Daily	Bare Costs	Incl. O&P
3 Laborers	$26.70	$640.80	$44.40	$1065.60	$29.16	$48.36
1 Equip. Oper. (medium)	36.55	292.40	60.25	482.00		
1 Crawler Loader, 4 C.Y.		1459.00		1604.90	45.59	50.15
32 L.H., Daily Totals		$2392.20		$3152.50	$74.76	$98.52

Crew B-4

Crew No.	Hr.	Daily	Hr.	Daily	Bare Costs	Incl. O&P
1 Labor Foreman (outside)	$28.70	$229.60	$47.75	$382.00	$27.95	$46.39
4 Laborers	26.70	854.40	44.40	1420.80		
1 Truck Driver (heavy)	32.20	257.60	53.00	424.00		
1 Truck Tractor, 220 H.P.		290.00		319.00		
1 Flatbed Trailer, 40 Ton		180.00		198.00	9.79	10.77
48 L.H., Daily Totals		$1811.60		$2743.80	$37.74	$57.16

Crew B-5

Crew No.	Hr.	Daily	Hr.	Daily	Bare Costs	Incl. O&P
1 Labor Foreman (outside)	$28.70	$229.60	$47.75	$382.00	$29.07	$48.24
3 Laborers	26.70	640.80	44.40	1065.60		
1 Equip. Oper. (medium)	36.55	292.40	60.25	482.00		
1 Air Compressor, 250 cfm		167.40		184.14		
2 Breakers, Pavement, 60 lb.		21.20		23.32		
2 -50' Air Hoses, 1.5"		45.40		49.94		
1 Crawler Loader, 3 C.Y.		1204.00		1324.40	35.95	39.55
40 L.H., Daily Totals		$2600.80		$3511.40	$65.02	$87.78

Crew B-5A

Crew No.	Hr.	Daily	Hr.	Daily	Bare Costs	Incl. O&P
1 Labor Foreman (outside)	$28.70	$229.60	$47.75	$382.00	$30.15	$49.92
6 Laborers	26.70	1281.60	44.40	2131.20		
2 Equip. Oper. (medium)	36.55	584.80	60.25	964.00		
1 Equip. Oper. (light)	35.40	283.20	58.35	466.80		
2 Truck Drivers (heavy)	32.20	515.20	53.00	848.00		
1 Air Compressor, 365 cfm		211.40		232.54		
2 Breakers, Pavement, 60 lb.		21.20		23.32		
8 -50' Air Hoses, 1"		67.60		74.36		
2 Dump Trucks, 8 C.Y., 220 H.P.		677.20		744.92	10.18	11.20
96 L.H., Daily Totals		$3871.80		$5867.14	$40.33	$61.12

Crew B-5B

Crew No.	Hr.	Daily	Hr.	Daily	Bare Costs	Incl. O&P
1 Powderman	$35.80	$286.40	$59.80	$478.40	$34.25	$56.55
2 Equip. Oper. (medium)	36.55	584.80	60.25	964.00		
3 Truck Drivers (heavy)	32.20	772.80	53.00	1272.00		
1 F.E. Loader, W.M., 2.5 C.Y.		523.20		575.52		
3 Dump Trucks, 12 C.Y., 400 H.P.		1628.40		1791.24		
1 Air Compressor, 365 cfm		211.40		232.54	49.23	54.15
48 L.H., Daily Totals		$4007.00		$5313.70	$83.48	$110.70

Crew B-5C

Crew No.	Hr.	Daily	Hr.	Daily	Bare Costs	Incl. O&P
3 Laborers	$26.70	$640.80	$44.40	$1065.60	$31.33	$51.77
1 Equip. Oper. (medium)	36.55	292.40	60.25	482.00		
2 Truck Drivers (heavy)	32.20	515.20	53.00	848.00		
1 Equip. Oper. (crane)	37.05	296.40	61.05	488.40		
1 Equip. Oper. (oiler)	32.55	260.40	53.65	429.20		
2 Dump Trucks, 12 C.Y., 400 H.P.		1085.60		1194.16		
1 Crawler Loader, 4 C.Y.		1459.00		1604.90		
1 S.P. Crane, 4x4, 25 Ton		668.40		735.24	50.20	55.22
64 L.H., Daily Totals		$5218.20		$6847.50	$81.53	$106.99

Crew B-5D

Crew No.	Hr.	Daily	Hr.	Daily	Bare Costs	Incl. O&P
1 Labor Foreman (outside)	$28.70	$229.60	$47.75	$382.00	$29.59	$49.03
3 Laborers	26.70	640.80	44.40	1065.60		
1 Equip. Oper. (medium)	36.55	292.40	60.25	482.00		
1 Truck Driver (heavy)	32.20	257.60	53.00	424.00		
1 Air Compressor, 250 cfm		167.40		184.14		
2 Breakers, Pavement, 60 lb.		21.20		23.32		
2 -50' Air Hoses, 1.5"		45.40		49.94		
1 Crawler Loader, 3 C.Y.		1204.00		1324.40		
1 Dump Truck, 12 C.Y., 400 H.P.		542.80		597.08	41.27	45.39
48 L.H., Daily Totals		$3401.20		$4532.48	$70.86	$94.43

Crew B-6

Crew No.	Hr.	Daily	Hr.	Daily	Bare Costs	Incl. O&P
2 Laborers	$26.70	$427.20	$44.40	$710.40	$29.60	$49.05
1 Equip. Oper. (light)	35.40	283.20	58.35	466.80		
1 Backhoe Loader, 48 H.P.		312.40		343.64	13.02	14.32
24 L.H., Daily Totals		$1022.80		$1520.84	$42.62	$63.37

Crew B-6B

Crew No.	Hr.	Daily	Hr.	Daily	Bare Costs	Incl. O&P
2 Labor Foremen (outside)	$28.70	$459.20	$47.75	$764.00	$27.37	$45.52
4 Laborers	26.70	854.40	44.40	1420.80		
1 S.P. Crane, 4x4, 5 Ton		259.00		284.90		
1 Flatbed Truck, Gas, 1.5 Ton		188.40		207.24		
1 Butt Fusion Mach., 4"-12" diam.		425.70		468.27	18.19	20.01
48 L.H., Daily Totals		$2186.70		$3145.21	$45.56	$65.53

Crew B-6C

Crew No.	Hr.	Daily	Hr.	Daily	Bare Costs	Incl. O&P
2 Labor Foremen (outside)	$28.70	$459.20	$47.75	$764.00	$27.37	$45.52
4 Laborers	26.70	854.40	44.40	1420.80		
1 S.P. Crane, 4x4, 12 Ton		363.60		399.96		
1 Flatbed Truck, Gas, 3 Ton		238.00		261.80		
1 Butt Fusion Mach., 8"-24" diam.		1908.00		2098.80	52.28	57.51
48 L.H., Daily Totals		$3823.20		$4945.36	$79.65	$103.03

Crew B-7

Crew No.	Hr.	Daily	Hr.	Daily	Bare Costs	Incl. O&P
1 Labor Foreman (outside)	$28.70	$229.60	$47.75	$382.00	$28.68	$47.60
4 Laborers	26.70	854.40	44.40	1420.80		
1 Equip. Oper. (medium)	36.55	292.40	60.25	482.00		
1 Brush Chipper, 12", 130 H.P.		393.00		432.30		
1 Crawler Loader, 3 C.Y.		1204.00		1324.40		
2 Chain Saws, Gas, 36" Long		92.80		102.08	35.20	38.72
48 L.H., Daily Totals		$3066.20		$4143.58	$63.88	$86.32

Crew B-7A

Crew No.	Hr.	Daily	Hr.	Daily	Bare Costs	Incl. O&P
2 Laborers	$26.70	$427.20	$44.40	$710.40	$29.60	$49.05
1 Equip. Oper. (light)	35.40	283.20	58.35	466.80		
1 Rake w/Tractor		331.30		364.43		
2 Chain Saws, Gas, 18"		55.60		61.16	16.12	17.73
24 L.H., Daily Totals		$1097.30		$1602.79	$45.72	$66.78

For customer support on your Residential Costs with RSMeans data, call 800.448.8182.

Crew No.	Bare Costs		Incl. Subs O&P		Cost Per Labor-Hour	
Crew B-7B	Hr.	Daily	Hr.	Daily	Bare Costs	Incl. O&P
1 Labor Foreman (outside)	$28.70	$229.60	$47.75	$382.00	$29.18	$48.37
4 Laborers	26.70	854.40	44.40	1420.80		
1 Equip. Oper. (medium)	36.55	292.40	60.25	482.00		
1 Truck Driver (heavy)	32.20	257.60	53.00	424.00		
1 Brush Chipper, 12", 130 H.P.		393.00		432.30		
1 Crawler Loader, 3 C.Y.		1204.00		1324.40		
2 Chain Saws, Gas, 36" Long		92.80		102.08		
1 Dump Truck, 8 C.Y., 220 H.P.		338.60		372.46	36.22	39.84
56 L.H., Daily Totals		$3662.40		$4940.04	$65.40	$88.22
Crew B-7C	Hr.	Daily	Hr.	Daily	Bare Costs	Incl. O&P
1 Labor Foreman (outside)	$28.70	$229.60	$47.75	$382.00	$29.18	$48.37
4 Laborers	26.70	854.40	44.40	1420.80		
1 Equip. Oper. (medium)	36.55	292.40	60.25	482.00		
1 Truck Driver (heavy)	32.20	257.60	53.00	424.00		
1 Brush Chipper, 12", 130 H.P.		393.00		432.30		
1 Crawler Loader, 3 C.Y.		1204.00		1324.40		
2 Chain Saws, Gas, 36" Long		92.80		102.08		
1 Dump Truck, 12 C.Y., 400 H.P.		542.80		597.08	39.87	43.85
56 L.H., Daily Totals		$3866.60		$5164.66	$69.05	$92.23
Crew B-8	Hr.	Daily	Hr.	Daily	Bare Costs	Incl. O&P
1 Labor Foreman (outside)	$28.70	$229.60	$47.75	$382.00	$31.37	$51.86
2 Laborers	26.70	427.20	44.40	710.40		
2 Equip. Oper. (medium)	36.55	584.80	60.25	964.00		
2 Truck Drivers (heavy)	32.20	515.20	53.00	848.00		
1 Hyd. Crane, 25 Ton		590.60		649.66		
1 Crawler Loader, 3 C.Y.		1204.00		1324.40		
2 Dump Trucks, 12 C.Y., 400 H.P.		1085.60		1194.16	51.43	56.58
56 L.H., Daily Totals		$4637.00		$6072.62	$82.80	$108.44
Crew B-9	Hr.	Daily	Hr.	Daily	Bare Costs	Incl. O&P
1 Labor Foreman (outside)	$28.70	$229.60	$47.75	$382.00	$27.10	$45.07
4 Laborers	26.70	854.40	44.40	1420.80		
1 Air Compressor, 250 cfm		167.40		184.14		
2 Breakers, Pavement, 60 lb.		21.20		23.32		
2 -50' Air Hoses, 1.5"		45.40		49.94	5.85	6.43
40 L.H., Daily Totals		$1318.00		$2060.20	$32.95	$51.51
Crew B-9A	Hr.	Daily	Hr.	Daily	Bare Costs	Incl. O&P
2 Laborers	$26.70	$427.20	$44.40	$710.40	$28.53	$47.27
1 Truck Driver (heavy)	32.20	257.60	53.00	424.00		
1 Water Tank Trailer, 5000 Gal.		147.60		162.36		
1 Truck Tractor, 220 H.P.		290.00		319.00		
2 -50' Discharge Hoses, 3"		3.00		3.30	18.36	20.19
24 L.H., Daily Totals		$1125.40		$1619.06	$46.89	$67.46
Crew B-9B	Hr.	Daily	Hr.	Daily	Bare Costs	Incl. O&P
2 Laborers	$26.70	$427.20	$44.40	$710.40	$28.53	$47.27
1 Truck Driver (heavy)	32.20	257.60	53.00	424.00		
2 -50' Discharge Hoses, 3"		3.00		3.30		
1 Water Tank Trailer, 5000 Gal.		147.60		162.36		
1 Truck Tractor, 220 H.P.		290.00		319.00		
1 Pressure Washer		63.60		69.96	21.01	23.11
24 L.H., Daily Totals		$1189.00		$1689.02	$49.54	$70.38

Crew No.	Bare Costs		Incl. Subs O&P		Cost Per Labor-Hour	
Crew B-9D	Hr.	Daily	Hr.	Daily	Bare Costs	Incl. O&P
1 Labor Foreman (outside)	$28.70	$229.60	$47.75	$382.00	$27.10	$45.07
4 Common Laborers	26.70	854.40	44.40	1420.80		
1 Air Compressor, 250 cfm		167.40		184.14		
2 -50' Air Hoses, 1.5"		45.40		49.94		
2 Air Powered Tampers		56.60		62.26	6.74	7.41
40 L.H., Daily Totals		$1353.40		$2099.14	$33.84	$52.48
Crew B-10	Hr.	Daily	Hr.	Daily	Bare Costs	Incl. O&P
1 Equip. Oper. (medium)	36.55	292.40	60.25	482.00	36.55	60.25
8 L.H., Daily Totals		$292.40		$482.00	$36.55	$60.25
Crew B-10A	Hr.	Daily	Hr.	Daily	Bare Costs	Incl. O&P
1 Equip. Oper. (medium)	$36.55	$292.40	$60.25	$482.00	$36.55	$60.25
1 Roller, 2-Drum, W.B., 7.5 H.P.		179.60		197.56	22.45	24.70
8 L.H., Daily Totals		$472.00		$679.56	$59.00	$84.94
Crew B-10B	Hr.	Daily	Hr.	Daily	Bare Costs	Incl. O&P
1 Equip. Oper. (medium)	$36.55	$292.40	$60.25	$482.00	$36.55	$60.25
1 Dozer, 200 H.P.		1273.00		1400.30	159.13	175.04
8 L.H., Daily Totals		$1565.40		$1882.30	$195.68	$235.29
Crew B-10C	Hr.	Daily	Hr.	Daily	Bare Costs	Incl. O&P
1 Equip. Oper. (medium)	$36.55	$292.40	$60.25	$482.00	$36.55	$60.25
1 Dozer, 200 H.P.		1273.00		1400.30		
1 Vibratory Roller, Towed, 23 Ton		423.00		465.30	212.00	233.20
8 L.H., Daily Totals		$1988.40		$2347.60	$248.55	$293.45
Crew B-10D	Hr.	Daily	Hr.	Daily	Bare Costs	Incl. O&P
1 Equip. Oper. (medium)	$36.55	$292.40	$60.25	$482.00	$36.55	$60.25
1 Dozer, 200 H.P.		1273.00		1400.30		
1 Sheepsft. Roller, Towed		425.80		468.38	212.35	233.59
8 L.H., Daily Totals		$1991.20		$2350.68	$248.90	$293.83
Crew B-10E	Hr.	Daily	Hr.	Daily	Bare Costs	Incl. O&P
1 Equip. Oper. (medium)	$36.55	$292.40	$60.25	$482.00	$36.55	$60.25
1 Tandem Roller, 5 Ton		151.20		166.32	18.90	20.79
8 L.H., Daily Totals		$443.60		$648.32	$55.45	$81.04
Crew B-10F	Hr.	Daily	Hr.	Daily	Bare Costs	Incl. O&P
1 Equip. Oper. (medium)	$36.55	$292.40	$60.25	$482.00	$36.55	$60.25
1 Tandem Roller, 10 Ton		231.40		254.54	28.93	31.82
8 L.H., Daily Totals		$523.80		$736.54	$65.47	$92.07
Crew B-10G	Hr.	Daily	Hr.	Daily	Bare Costs	Incl. O&P
1 Equip. Oper. (medium)	$36.55	$292.40	$60.25	$482.00	$36.55	$60.25
1 Sheepsfoot Roller, 240 H.P.		1316.00		1447.60	164.50	180.95
8 L.H., Daily Totals		$1608.40		$1929.60	$201.05	$241.20
Crew B-10H	Hr.	Daily	Hr.	Daily	Bare Costs	Incl. O&P
1 Equip. Oper. (medium)	$36.55	$292.40	$60.25	$482.00	$36.55	$60.25
1 Diaphragm Water Pump, 2"		73.00		80.30		
1 -20' Suction Hose, 2"		1.95		2.15		
2 -50' Discharge Hoses, 2"		1.80		1.98	9.59	10.55
8 L.H., Daily Totals		$369.15		$566.42	$46.14	$70.80

Crews - Residential

Crew No.	Bare Costs		Incl. Subs O&P		Cost Per Labor-Hour	

Crew B-10I

Crew B-10I	Hr.	Daily	Hr.	Daily	Bare Costs	Incl. O&P
1 Equip. Oper. (medium)	$36.55	$292.40	$60.25	$482.00	$36.55	$60.25
1 Diaphragm Water Pump, 4"		114.80		126.28		
1 -20' Suction Hose, 4"		3.25		3.58		
2 -50' Discharge Hoses, 4"		4.70		5.17	15.34	16.88
8 L.H., Daily Totals		$415.15		$617.02	$51.89	$77.13

Crew B-10J	Hr.	Daily	Hr.	Daily	Bare Costs	Incl. O&P
1 Equip. Oper. (medium)	$36.55	$292.40	$60.25	$482.00	$36.55	$60.25
1 Centrifugal Water Pump, 3"		79.20		87.12		
1 -20' Suction Hose, 3"		2.85		3.13		
2 -50' Discharge Hoses, 3"		3.00		3.30	10.63	11.69
8 L.H., Daily Totals		$377.45		$575.55	$47.18	$71.94

Crew B-10K	Hr.	Daily	Hr.	Daily	Bare Costs	Incl. O&P
1 Equip. Oper. (medium)	$36.55	$292.40	$60.25	$482.00	$36.55	$60.25
1 Centr. Water Pump, 6"		296.40		326.04		
1 -20' Suction Hose, 6"		11.50		12.65		
2 -50' Discharge Hoses, 6"		12.20		13.42	40.01	44.01
8 L.H., Daily Totals		$612.50		$834.11	$76.56	$104.26

Crew B-10L	Hr.	Daily	Hr.	Daily	Bare Costs	Incl. O&P
1 Equip. Oper. (medium)	$36.55	$292.40	$60.25	$482.00	$36.55	$60.25
1 Dozer, 80 H.P.		464.40		510.84	58.05	63.85
8 L.H., Daily Totals		$756.80		$992.84	$94.60	$124.11

Crew B-10M	Hr.	Daily	Hr.	Daily	Bare Costs	Incl. O&P
1 Equip. Oper. (medium)	$36.55	$292.40	$60.25	$482.00	$36.55	$60.25
1 Dozer, 300 H.P.		1829.00		2011.90	228.63	251.49
8 L.H., Daily Totals		$2121.40		$2493.90	$265.18	$311.74

Crew B-10N	Hr.	Daily	Hr.	Daily	Bare Costs	Incl. O&P
1 Equip. Oper. (medium)	$36.55	$292.40	$60.25	$482.00	$36.55	$60.25
1 F.E. Loader, T.M., 1.5 C.Y.		579.80		637.78	72.47	79.72
8 L.H., Daily Totals		$872.20		$1119.78	$109.03	$139.97

Crew B-10O	Hr.	Daily	Hr.	Daily	Bare Costs	Incl. O&P
1 Equip. Oper. (medium)	$36.55	$292.40	$60.25	$482.00	$36.55	$60.25
1 F.E. Loader, T.M., 2.25 C.Y.		979.00		1076.90	122.38	134.61
8 L.H., Daily Totals		$1271.40		$1558.90	$158.93	$194.86

Crew B-10P	Hr.	Daily	Hr.	Daily	Bare Costs	Incl. O&P
1 Equip. Oper. (medium)	$36.55	$292.40	$60.25	$482.00	$36.55	$60.25
1 Crawler Loader, 3 C.Y.		1204.00		1324.40	150.50	165.55
8 L.H., Daily Totals		$1496.40		$1806.40	$187.05	$225.80

Crew B-10Q	Hr.	Daily	Hr.	Daily	Bare Costs	Incl. O&P
1 Equip. Oper. (medium)	$36.55	$292.40	$60.25	$482.00	$36.55	$60.25
1 Crawler Loader, 4 C.Y.		1459.00		1604.90	182.38	200.61
8 L.H., Daily Totals		$1751.40		$2086.90	$218.93	$260.86

Crew B-10R	Hr.	Daily	Hr.	Daily	Bare Costs	Incl. O&P
1 Equip. Oper. (medium)	$36.55	$292.40	$60.25	$482.00	$36.55	$60.25
1 F.E. Loader, W.M., 1 C.Y.		290.80		319.88	36.35	39.98
8 L.H., Daily Totals		$583.20		$801.88	$72.90	$100.24

Crew B-10S	Hr.	Daily	Hr.	Daily	Bare Costs	Incl. O&P
1 Equip. Oper. (medium)	$36.55	$292.40	$60.25	$482.00	$36.55	$60.25
1 F.E. Loader, W.M., 1.5 C.Y.		342.80		377.08	42.85	47.13
8 L.H., Daily Totals		$635.20		$859.08	$79.40	$107.39

Crew B-10T	Hr.	Daily	Hr.	Daily	Bare Costs	Incl. O&P
1 Equip. Oper. (medium)	$36.55	$292.40	$60.25	$482.00	$36.55	$60.25
1 F.E. Loader, W.M., 2.5 C.Y.		523.20		575.52	65.40	71.94
8 L.H., Daily Totals		$815.60		$1057.52	$101.95	$132.19

Crew B-10U	Hr.	Daily	Hr.	Daily	Bare Costs	Incl. O&P
1 Equip. Oper. (medium)	$36.55	$292.40	$60.25	$482.00	$36.55	$60.25
1 F.E. Loader, W.M., 5.5 C.Y.		982.20		1080.42	122.78	135.05
8 L.H., Daily Totals		$1274.60		$1562.42	$159.32	$195.30

Crew B-10V	Hr.	Daily	Hr.	Daily	Bare Costs	Incl. O&P
1 Equip. Oper. (medium)	$36.55	$292.40	$60.25	$482.00	$36.55	$60.25
1 Dozer, 700 H.P.		4999.00		5498.90	624.88	687.36
8 L.H., Daily Totals		$5291.40		$5980.90	$661.42	$747.61

Crew B-10W	Hr.	Daily	Hr.	Daily	Bare Costs	Incl. O&P
1 Equip. Oper. (medium)	$36.55	$292.40	$60.25	$482.00	$36.55	$60.25
1 Dozer, 105 H.P.		608.00		668.80	76.00	83.60
8 L.H., Daily Totals		$900.40		$1150.80	$112.55	$143.85

Crew B-10X	Hr.	Daily	Hr.	Daily	Bare Costs	Incl. O&P
1 Equip. Oper. (medium)	$36.55	$292.40	$60.25	$482.00	$36.55	$60.25
1 Dozer, 410 H.P.		2291.00		2520.10	286.38	315.01
8 L.H., Daily Totals		$2583.40		$3002.10	$322.93	$375.26

Crew B-10Y	Hr.	Daily	Hr.	Daily	Bare Costs	Incl. O&P
1 Equip. Oper. (medium)	$36.55	$292.40	$60.25	$482.00	$36.55	$60.25
1 Vibr. Roller, Towed, 12 Ton		576.40		634.04	72.05	79.25
8 L.H., Daily Totals		$868.80		$1116.04	$108.60	$139.51

Crew B-11A	Hr.	Daily	Hr.	Daily	Bare Costs	Incl. O&P
1 Equipment Oper. (med.)	$36.55	$292.40	$60.25	$482.00	$31.63	$52.33
1 Laborer	26.70	213.60	44.40	355.20		
1 Dozer, 200 H.P.		1273.00		1400.30	79.56	87.52
16 L.H., Daily Totals		$1779.00		$2237.50	$111.19	$139.84

Crew B-11B	Hr.	Daily	Hr.	Daily	Bare Costs	Incl. O&P
1 Equipment Oper. (light)	$35.40	$283.20	$58.35	$466.80	$31.05	$51.38
1 Laborer	26.70	213.60	44.40	355.20		
1 Air Powered Tamper		28.30		31.13		
1 Air Compressor, 365 cfm		211.40		232.54		
2 -50' Air Hoses, 1.5"		45.40		49.94	17.82	19.60
16 L.H., Daily Totals		$781.90		$1135.61	$48.87	$70.98

Crew B-11C	Hr.	Daily	Hr.	Daily	Bare Costs	Incl. O&P
1 Equipment Oper. (med.)	$36.55	$292.40	$60.25	$482.00	$31.63	$52.33
1 Laborer	26.70	213.60	44.40	355.20		
1 Backhoe Loader, 48 H.P.		312.40		343.64	19.52	21.48
16 L.H., Daily Totals		$818.40		$1180.84	$51.15	$73.80

Crew B-11K	Hr.	Daily	Hr.	Daily	Bare Costs	Incl. O&P
1 Equipment Oper. (med.)	$36.55	$292.40	$60.25	$482.00	$31.63	$52.33
1 Laborer	26.70	213.60	44.40	355.20		
1 Trencher, Chain Type, 8' D		2048.00		2252.80	128.00	140.80
16 L.H., Daily Totals		$2554.00		$3090.00	$159.63	$193.13

Crew No.	Bare Costs		Incl. Subs O&P		Cost Per Labor-Hour	
Crew B-11L	Hr.	Daily	Hr.	Daily	Bare Costs	Incl. O&P
1 Equipment Oper. (med.)	$36.55	$292.40	$60.25	$482.00	$31.63	$52.33
1 Laborer	26.70	213.60	44.40	355.20		
1 Grader, 30,000 Lbs.		643.80		708.18	40.24	44.26
16 L.H., Daily Totals		$1149.80		$1545.38	$71.86	$96.59

Crew No.	Bare Costs		Incl. Subs O&P		Cost Per Labor-Hour	
Crew B-11M	Hr.	Daily	Hr.	Daily	Bare Costs	Incl. O&P
1 Equipment Oper. (med.)	$36.55	$292.40	$60.25	$482.00	$31.63	$52.33
1 Laborer	26.70	213.60	44.40	355.20		
1 Backhoe Loader, 80 H.P.		385.00		423.50	24.06	26.47
16 L.H., Daily Totals		$891.00		$1260.70	$55.69	$78.79

Crew No.	Bare Costs		Incl. Subs O&P		Cost Per Labor-Hour	
Crew B-11W	Hr.	Daily	Hr.	Daily	Bare Costs	Incl. O&P
1 Equipment Operator (med.)	$36.55	$292.40	$60.25	$482.00	$32.10	$52.89
1 Common Laborer	26.70	213.60	44.40	355.20		
10 Truck Drivers (heavy)	32.20	2576.00	53.00	4240.00		
1 Dozer, 200 H.P.		1273.00		1400.30		
1 Vibratory Roller, Towed, 23 Ton		423.00		465.30		
10 Dump Trucks, 8 C.Y., 220 H.P.		3386.00		3724.60	52.94	58.23
96 L.H., Daily Totals		$8164.00		$10667.40	$85.04	$111.12

Crew No.	Bare Costs		Incl. Subs O&P		Cost Per Labor-Hour	
Crew B-11Y	Hr.	Daily	Hr.	Daily	Bare Costs	Incl. O&P
1 Labor Foreman (outside)	$28.70	$229.60	$47.75	$382.00	$30.21	$50.06
5 Common Laborers	26.70	1068.00	44.40	1776.00		
3 Equipment Operators (med.)	36.55	877.20	60.25	1446.00		
1 Dozer, 80 H.P.		464.40		510.84		
2 Rollers, 2-Drums, W.B., 7.5 H.P.		359.20		395.12		
4 Vibrating Plates, Gas, 21"		162.40		178.64	13.69	15.06
72 L.H., Daily Totals		$3160.80		$4688.60	$43.90	$65.12

Crew No.	Bare Costs		Incl. Subs O&P		Cost Per Labor-Hour	
Crew B-12A	Hr.	Daily	Hr.	Daily	Bare Costs	Incl. O&P
1 Equip. Oper. (crane)	$37.05	$296.40	$61.05	$488.40	$31.88	$52.73
1 Laborer	26.70	213.60	44.40	355.20		
1 Hyd. Excavator, 1 C.Y.		742.20		816.42	46.39	51.03
16 L.H., Daily Totals		$1252.20		$1660.02	$78.26	$103.75

Crew No.	Bare Costs		Incl. Subs O&P		Cost Per Labor-Hour	
Crew B-12B	Hr.	Daily	Hr.	Daily	Bare Costs	Incl. O&P
1 Equip. Oper. (crane)	$37.05	$296.40	$61.05	$488.40	$31.88	$52.73
1 Laborer	26.70	213.60	44.40	355.20		
1 Hyd. Excavator, 1.5 C.Y.		893.60		982.96	55.85	61.44
16 L.H., Daily Totals		$1403.60		$1826.56	$87.72	$114.16

Crew No.	Bare Costs		Incl. Subs O&P		Cost Per Labor-Hour	
Crew B-12C	Hr.	Daily	Hr.	Daily	Bare Costs	Incl. O&P
1 Equip. Oper. (crane)	$37.05	$296.40	$61.05	$488.40	$31.88	$52.73
1 Laborer	26.70	213.60	44.40	355.20		
1 Hyd. Excavator, 2 C.Y.		1052.00		1157.20	65.75	72.33
16 L.H., Daily Totals		$1562.00		$2000.80	$97.63	$125.05

Crew No.	Bare Costs		Incl. Subs O&P		Cost Per Labor-Hour	
Crew B-12D	Hr.	Daily	Hr.	Daily	Bare Costs	Incl. O&P
1 Equip. Oper. (crane)	$37.05	$296.40	$61.05	$488.40	$31.88	$52.73
1 Laborer	26.70	213.60	44.40	355.20		
1 Hyd. Excavator, 3.5 C.Y.		2195.00		2414.50	137.19	150.91
16 L.H., Daily Totals		$2705.00		$3258.10	$169.06	$203.63

Crew No.	Bare Costs		Incl. Subs O&P		Cost Per Labor-Hour	
Crew B-12E	Hr.	Daily	Hr.	Daily	Bare Costs	Incl. O&P
1 Equip. Oper. (crane)	$37.05	$296.40	$61.05	$488.40	$31.88	$52.73
1 Laborer	26.70	213.60	44.40	355.20		
1 Hyd. Excavator, .5 C.Y.		435.40		478.94	27.21	29.93
16 L.H., Daily Totals		$945.40		$1322.54	$59.09	$82.66

Crew No.	Bare Costs		Incl. Subs O&P		Cost Per Labor-Hour	
Crew B-12F	Hr.	Daily	Hr.	Daily	Bare Costs	Incl. O&P
1 Equip. Oper. (crane)	$37.05	$296.40	$61.05	$488.40	$31.88	$52.73
1 Laborer	26.70	213.60	44.40	355.20		
1 Hyd. Excavator, .75 C.Y.		669.60		736.56	41.85	46.03
16 L.H., Daily Totals		$1179.60		$1580.16	$73.72	$98.76

Crew No.	Bare Costs		Incl. Subs O&P		Cost Per Labor-Hour	
Crew B-12G	Hr.	Daily	Hr.	Daily	Bare Costs	Incl. O&P
1 Equip. Oper. (crane)	$37.05	$296.40	$61.05	$488.40	$31.88	$52.73
1 Laborer	26.70	213.60	44.40	355.20		
1 Crawler Crane, 15 Ton		828.80		911.68		
1 Clamshell Bucket, .5 C.Y.		41.00		45.10	54.36	59.80
16 L.H., Daily Totals		$1379.80		$1800.38	$86.24	$112.52

Crew No.	Bare Costs		Incl. Subs O&P		Cost Per Labor-Hour	
Crew B-12H	Hr.	Daily	Hr.	Daily	Bare Costs	Incl. O&P
1 Equip. Oper. (crane)	$37.05	$296.40	$61.05	$488.40	$31.88	$52.73
1 Laborer	26.70	213.60	44.40	355.20		
1 Crawler Crane, 25 Ton		1348.00		1482.80		
1 Clamshell Bucket, 1 C.Y.		50.60		55.66	87.41	96.15
16 L.H., Daily Totals		$1908.60		$2382.06	$119.29	$148.88

Crew No.	Bare Costs		Incl. Subs O&P		Cost Per Labor-Hour	
Crew B-12I	Hr.	Daily	Hr.	Daily	Bare Costs	Incl. O&P
1 Equip. Oper. (crane)	$37.05	$296.40	$61.05	$488.40	$31.88	$52.73
1 Laborer	26.70	213.60	44.40	355.20		
1 Crawler Crane, 20 Ton		1060.00		1166.00		
1 Dragline Bucket, .75 C.Y.		21.80		23.98	67.61	74.37
16 L.H., Daily Totals		$1591.80		$2033.58	$99.49	$127.10

Crew No.	Bare Costs		Incl. Subs O&P		Cost Per Labor-Hour	
Crew B-12J	Hr.	Daily	Hr.	Daily	Bare Costs	Incl. O&P
1 Equip. Oper. (crane)	$37.05	$296.40	$61.05	$488.40	$31.88	$52.73
1 Laborer	26.70	213.60	44.40	355.20		
1 Gradall, 5/8 C.Y.		869.60		956.56	54.35	59.78
16 L.H., Daily Totals		$1379.60		$1800.16	$86.22	$112.51

Crew No.	Bare Costs		Incl. Subs O&P		Cost Per Labor-Hour	
Crew B-12K	Hr.	Daily	Hr.	Daily	Bare Costs	Incl. O&P
1 Equip. Oper. (crane)	$37.05	$296.40	$61.05	$488.40	$31.88	$52.73
1 Laborer	26.70	213.60	44.40	355.20		
1 Gradall, 3 Ton, 1 C.Y.		1243.00		1367.30	77.69	85.46
16 L.H., Daily Totals		$1753.00		$2210.90	$109.56	$138.18

Crew No.	Bare Costs		Incl. Subs O&P		Cost Per Labor-Hour	
Crew B-12L	Hr.	Daily	Hr.	Daily	Bare Costs	Incl. O&P
1 Equip. Oper. (crane)	$37.05	$296.40	$61.05	$488.40	$31.88	$52.73
1 Laborer	26.70	213.60	44.40	355.20		
1 Crawler Crane, 15 Ton		828.80		911.68		
1 F.E. Attachment, .5 C.Y.		64.60		71.06	55.84	61.42
16 L.H., Daily Totals		$1403.40		$1826.34	$87.71	$114.15

Crew No.	Bare Costs		Incl. Subs O&P		Cost Per Labor-Hour	
Crew B-12M	Hr.	Daily	Hr.	Daily	Bare Costs	Incl. O&P
1 Equip. Oper. (crane)	$37.05	$296.40	$61.05	$488.40	$31.88	$52.73
1 Laborer	26.70	213.60	44.40	355.20		
1 Crawler Crane, 20 Ton		1060.00		1166.00		
1 F.E. Attachment, .75 C.Y.		69.00		75.90	70.56	77.62
16 L.H., Daily Totals		$1639.00		$2085.50	$102.44	$130.34

Crew No.	Bare Costs		Incl. Subs O&P		Cost Per Labor-Hour	
Crew B-12N	Hr.	Daily	Hr.	Daily	Bare Costs	Incl. O&P
1 Equip. Oper. (crane)	$37.05	$296.40	$61.05	$488.40	$31.88	$52.73
1 Laborer	26.70	213.60	44.40	355.20		
1 Crawler Crane, 25 Ton		1348.00		1482.80		
1 F.E. Attachment, 1 C.Y.		75.80		83.38	88.99	97.89
16 L.H., Daily Totals		$1933.80		$2409.78	$120.86	$150.61

Crew B-120

Crew No.	Bare Costs Hr.	Daily	Incl. Subs O&P Hr.	Daily	Cost Per Labor-Hour Bare Costs	Incl. O&P
1 Equip. Oper. (crane)	$37.05	$296.40	$61.05	$488.40	$31.88	$52.73
1 Laborer	26.70	213.60	44.40	355.20		
1 Crawler Crane, 40 Ton		1370.00		1507.00		
1 F.E. Attachment, 1.5 C.Y.		85.00		93.50	90.94	100.03
16 L.H., Daily Totals		$1965.00		$2444.10	$122.81	$152.76

Crew B-12P

	Hr.	Daily	Hr.	Daily	Bare Costs	Incl. O&P
1 Equip. Oper. (crane)	$37.05	$296.40	$61.05	$488.40	$31.88	$52.73
1 Laborer	26.70	213.60	44.40	355.20		
1 Crawler Crane, 40 Ton		1370.00		1507.00		
1 Dragline Bucket, 1.5 C.Y.		35.40		38.94	87.84	96.62
16 L.H., Daily Totals		$1915.40		$2389.54	$119.71	$149.35

Crew B-12Q

	Hr.	Daily	Hr.	Daily	Bare Costs	Incl. O&P
1 Equip. Oper. (crane)	$37.05	$296.40	$61.05	$488.40	$31.88	$52.73
1 Laborer	26.70	213.60	44.40	355.20		
1 Hyd. Excavator, 5/8 C.Y.		576.80		634.48	36.05	39.66
16 L.H., Daily Totals		$1086.80		$1478.08	$67.92	$92.38

Crew B-12S

	Hr.	Daily	Hr.	Daily	Bare Costs	Incl. O&P
1 Equip. Oper. (crane)	$37.05	$296.40	$61.05	$488.40	$31.88	$52.73
1 Laborer	26.70	213.60	44.40	355.20		
1 Hyd. Excavator, 2.5 C.Y.		1411.00		1552.10	88.19	97.01
16 L.H., Daily Totals		$1921.00		$2395.70	$120.06	$149.73

Crew B-12T

	Hr.	Daily	Hr.	Daily	Bare Costs	Incl. O&P
1 Equip. Oper. (crane)	$37.05	$296.40	$61.05	$488.40	$31.88	$52.73
1 Laborer	26.70	213.60	44.40	355.20		
1 Crawler Crane, 75 Ton		1708.00		1878.80		
1 F.E. Attachment, 3 C.Y.		108.80		119.68	113.55	124.91
16 L.H., Daily Totals		$2326.80		$2842.08	$145.43	$177.63

Crew B-12V

	Hr.	Daily	Hr.	Daily	Bare Costs	Incl. O&P
1 Equip. Oper. (crane)	$37.05	$296.40	$61.05	$488.40	$31.88	$52.73
1 Laborer	26.70	213.60	44.40	355.20		
1 Crawler Crane, 75 Ton		1708.00		1878.80		
1 Dragline Bucket, 3 C.Y.		55.80		61.38	110.24	121.26
16 L.H., Daily Totals		$2273.80		$2783.78	$142.11	$173.99

Crew B-12Y

	Hr.	Daily	Hr.	Daily	Bare Costs	Incl. O&P
1 Equip. Oper. (crane)	$37.05	$296.40	$61.05	$488.40	$30.15	$49.95
2 Laborers	26.70	427.20	44.40	710.40		
1 Hyd. Excavator, 3.5 C.Y.		2195.00		2414.50	91.46	100.60
24 L.H., Daily Totals		$2918.60		$3613.30	$121.61	$150.55

Crew B-12Z

	Hr.	Daily	Hr.	Daily	Bare Costs	Incl. O&P
1 Equip. Oper. (crane)	$37.05	$296.40	$61.05	$488.40	$30.15	$49.95
2 Laborers	26.70	427.20	44.40	710.40		
1 Hyd. Excavator, 2.5 C.Y.		1411.00		1552.10	58.79	64.67
24 L.H., Daily Totals		$2134.60		$2750.90	$88.94	$114.62

Crew B-13

	Hr.	Daily	Hr.	Daily	Bare Costs	Incl. O&P
1 Labor Foreman (outside)	$28.70	$229.60	$47.75	$382.00	$28.76	$47.73
4 Laborers	26.70	854.40	44.40	1420.80		
1 Equip. Oper. (crane)	37.05	296.40	61.05	488.40		
1 Hyd. Crane, 25 Ton		590.60		649.66	12.30	13.53
48 L.H., Daily Totals		$1971.00		$2940.86	$41.06	$61.27

Crew B-13A

Crew No.	Bare Costs Hr.	Daily	Incl. Subs O&P Hr.	Daily	Cost Per Labor-Hour Bare Costs	Incl. O&P
1 Labor Foreman (outside)	$28.70	$229.60	$47.75	$382.00	$31.37	$51.86
2 Laborers	26.70	427.20	44.40	710.40		
2 Equipment Operators (med.)	36.55	584.80	60.25	964.00		
2 Truck Drivers (heavy)	32.20	515.20	53.00	848.00		
1 Crawler Crane, 75 Ton		1708.00		1878.80		
1 Crawler Loader, 4 C.Y.		1459.00		1604.90		
2 Dump Trucks, 8 C.Y., 220 H.P.		677.20		744.92	68.65	75.51
56 L.H., Daily Totals		$5601.00		$7133.02	$100.02	$127.38

Crew B-13B

	Hr.	Daily	Hr.	Daily	Bare Costs	Incl. O&P
1 Labor Foreman (outside)	$28.70	$229.60	$47.75	$382.00	$29.30	$48.58
4 Laborers	26.70	854.40	44.40	1420.80		
1 Equip. Oper. (crane)	37.05	296.40	61.05	488.40		
1 Equip. Oper. (oiler)	32.55	260.40	53.65	429.20		
1 Hyd. Crane, 55 Ton		993.80		1093.18	17.75	19.52
56 L.H., Daily Totals		$2634.60		$3813.58	$47.05	$68.10

Crew B-13C

	Hr.	Daily	Hr.	Daily	Bare Costs	Incl. O&P
1 Labor Foreman (outside)	$28.70	$229.60	$47.75	$382.00	$29.30	$48.58
4 Laborers	26.70	854.40	44.40	1420.80		
1 Equip. Oper. (crane)	37.05	296.40	61.05	488.40		
1 Equip. Oper. (oiler)	32.55	260.40	53.65	429.20		
1 Crawler Crane, 100 Ton		1872.00		2059.20	33.43	36.77
56 L.H., Daily Totals		$3512.80		$4779.60	$62.73	$85.35

Crew B-13D

	Hr.	Daily	Hr.	Daily	Bare Costs	Incl. O&P
1 Laborer	$26.70	$213.60	$44.40	$355.20	$31.88	$52.73
1 Equip. Oper. (crane)	37.05	296.40	61.05	488.40		
1 Hyd. Excavator, 1 C.Y.		742.20		816.42		
1 Trench Box		81.00		89.10	51.45	56.59
16 L.H., Daily Totals		$1333.20		$1749.12	$83.33	$109.32

Crew B-13E

	Hr.	Daily	Hr.	Daily	Bare Costs	Incl. O&P
1 Laborer	$26.70	$213.60	$44.40	$355.20	$31.88	$52.73
1 Equip. Oper. (crane)	37.05	296.40	61.05	488.40		
1 Hyd. Excavator, 1.5 C.Y.		893.60		982.96		
1 Trench Box		81.00		89.10	60.91	67.00
16 L.H., Daily Totals		$1484.60		$1915.66	$92.79	$119.73

Crew B-13F

	Hr.	Daily	Hr.	Daily	Bare Costs	Incl. O&P
1 Laborer	$26.70	$213.60	$44.40	$355.20	$31.88	$52.73
1 Equip. Oper. (crane)	37.05	296.40	61.05	488.40		
1 Hyd. Excavator, 3.5 C.Y.		2195.00		2414.50		
1 Trench Box		81.00		89.10	142.25	156.47
16 L.H., Daily Totals		$2786.00		$3347.20	$174.13	$209.20

Crew B-13G

	Hr.	Daily	Hr.	Daily	Bare Costs	Incl. O&P
1 Laborer	$26.70	$213.60	$44.40	$355.20	$31.88	$52.73
1 Equip. Oper. (crane)	37.05	296.40	61.05	488.40		
1 Hyd. Excavator, .75 C.Y.		669.60		736.56		
1 Trench Box		81.00		89.10	46.91	51.60
16 L.H., Daily Totals		$1260.60		$1669.26	$78.79	$104.33

Crew B-13H

	Hr.	Daily	Hr.	Daily	Bare Costs	Incl. O&P
1 Laborer	$26.70	$213.60	$44.40	$355.20	$31.88	$52.73
1 Equip. Oper. (crane)	37.05	296.40	61.05	488.40		
1 Gradall, 5/8 C.Y.		869.60		956.56		
1 Trench Box		81.00		89.10	59.41	65.35
16 L.H., Daily Totals		$1460.60		$1889.26	$91.29	$118.08

Crew No.	Bare Costs		Incl. Subs O&P		Cost Per Labor-Hour	

Crew B-13I

	Hr.	Daily	Hr.	Daily	Bare Costs	Incl. O&P
1 Laborer	$26.70	$213.60	$44.40	$355.20	$31.88	$52.73
1 Equip. Oper. (crane)	37.05	296.40	61.05	488.40		
1 Gradall, 3 Ton, 1 C.Y.		1243.00		1367.30		
1 Trench Box		81.00		89.10	82.75	91.03
16 L.H., Daily Totals		$1834.00		$2300.00	$114.63	$143.75

Crew B-13J

	Hr.	Daily	Hr.	Daily	Bare Costs	Incl. O&P
1 Laborer	$26.70	$213.60	$44.40	$355.20	$31.88	$52.73
1 Equip. Oper. (crane)	37.05	296.40	61.05	488.40		
1 Hyd. Excavator, 2.5 C.Y.		1411.00		1552.10		
1 Trench Box		81.00		89.10	93.25	102.58
16 L.H., Daily Totals		$2002.00		$2484.80	$125.13	$155.30

Crew B-13K

	Hr.	Daily	Hr.	Daily	Bare Costs	Incl. O&P
2 Equip. Opers. (crane)	$37.05	$592.80	$61.05	$976.80	$37.05	$61.05
1 Hyd. Excavator, .75 C.Y.		669.60		736.56		
1 Hyd. Hammer, 4000 ft-lb		324.60		357.06		
1 Hyd. Excavator, .75 C.Y.		669.60		736.56	103.99	114.39
16 L.H., Daily Totals		$2256.60		$2806.98	$141.04	$175.44

Crew B-13L

	Hr.	Daily	Hr.	Daily	Bare Costs	Incl. O&P
2 Equip. Opers. (crane)	$37.05	$592.80	$61.05	$976.80	$37.05	$61.05
1 Hyd. Excavator, 1.5 C.Y.		893.60		982.96		
1 Hyd. Hammer, 5000 ft-lb		394.00		433.40		
1 Hyd. Excavator, .75 C.Y.		669.60		736.56	122.33	134.56
16 L.H., Daily Totals		$2550.00		$3129.72	$159.38	$195.61

Crew B-13M

	Hr.	Daily	Hr.	Daily	Bare Costs	Incl. O&P
2 Equip. Opers. (crane)	$37.05	$592.80	$61.05	$976.80	$37.05	$61.05
1 Hyd. Excavator, 2.5 C.Y.		1411.00		1552.10		
1 Hyd. Hammer, 8000 ft-lb		576.80		634.48		
1 Hyd. Excavator, 1.5 C.Y.		893.60		982.96	180.09	198.10
16 L.H., Daily Totals		$3474.20		$4146.34	$217.14	$259.15

Crew B-13N

	Hr.	Daily	Hr.	Daily	Bare Costs	Incl. O&P
2 Equip. Opers. (crane)	$37.05	$592.80	$61.05	$976.80	$37.05	$61.05
1 Hyd. Excavator, 3.5 C.Y.		2195.00		2414.50		
1 Hyd. Hammer, 12,000 ft-lb		669.20		736.12		
1 Hyd. Excavator, 1.5 C.Y.		893.60		982.96	234.86	258.35
16 L.H., Daily Totals		$4350.60		$5110.38	$271.91	$319.40

Crew B-14

	Hr.	Daily	Hr.	Daily	Bare Costs	Incl. O&P
1 Labor Foreman (outside)	$28.70	$229.60	$47.75	$382.00	$28.48	$47.28
4 Laborers	26.70	854.40	44.40	1420.80		
1 Equip. Oper. (light)	35.40	283.20	58.35	466.80		
1 Backhoe Loader, 48 H.P.		312.40		343.64	6.51	7.16
48 L.H., Daily Totals		$1679.60		$2613.24	$34.99	$54.44

Crew B-14A

	Hr.	Daily	Hr.	Daily	Bare Costs	Incl. O&P
1 Equip. Oper. (crane)	$37.05	$296.40	$61.05	$488.40	$33.60	$55.50
.5 Laborer	26.70	106.80	44.40	177.60		
1 Hyd. Excavator, 4.5 C.Y.		2735.00		3008.50	227.92	250.71
12 L.H., Daily Totals		$3138.20		$3674.50	$261.52	$306.21

Crew B-14B

	Hr.	Daily	Hr.	Daily	Bare Costs	Incl. O&P
1 Equip. Oper. (crane)	$37.05	$296.40	$61.05	$488.40	$33.60	$55.50
.5 Laborer	26.70	106.80	44.40	177.60		
1 Hyd. Excavator, 6 C.Y.		3487.00		3835.70	290.58	319.64
12 L.H., Daily Totals		$3890.20		$4501.70	$324.18	$375.14

Crew B-14C

	Hr.	Daily	Hr.	Daily	Bare Costs	Incl. O&P
1 Equip. Oper. (crane)	$37.05	$296.40	$61.05	$488.40	$33.60	$55.50
.5 Laborer	26.70	106.80	44.40	177.60		
1 Hyd. Excavator, 7 C.Y.		3275.00		3602.50	272.92	300.21
12 L.H., Daily Totals		$3678.20		$4268.50	$306.52	$355.71

Crew B-14F

	Hr.	Daily	Hr.	Daily	Bare Costs	Incl. O&P
1 Equip. Oper. (crane)	$37.05	$296.40	$61.05	$488.40	$33.60	$55.50
.5 Laborer	26.70	106.80	44.40	177.60		
1 Hyd. Shovel, 7 C.Y.		3880.00		4268.00	323.33	355.67
12 L.H., Daily Totals		$4283.20		$4934.00	$356.93	$411.17

Crew B-14G

	Hr.	Daily	Hr.	Daily	Bare Costs	Incl. O&P
1 Equip. Oper. (crane)	$37.05	$296.40	$61.05	$488.40	$33.60	$55.50
.5 Laborer	26.70	106.80	44.40	177.60		
1 Hyd. Shovel, 12 C.Y.		5769.00		6345.90	480.75	528.83
12 L.H., Daily Totals		$6172.20		$7011.90	$514.35	$584.33

Crew B-14J

	Hr.	Daily	Hr.	Daily	Bare Costs	Incl. O&P
1 Equip. Oper. (medium)	$36.55	$292.40	$60.25	$482.00	$33.27	$54.97
.5 Laborer	26.70	106.80	44.40	177.60		
1 F.E. Loader, 8 C.Y.		1780.00		1958.00	148.33	163.17
12 L.H., Daily Totals		$2179.20		$2617.60	$181.60	$218.13

Crew B-14K

	Hr.	Daily	Hr.	Daily	Bare Costs	Incl. O&P
1 Equip. Oper. (medium)	$36.55	$292.40	$60.25	$482.00	$33.27	$54.97
.5 Laborer	26.70	106.80	44.40	177.60		
1 F.E. Loader, 10 C.Y.		2630.00		2893.00	219.17	241.08
12 L.H., Daily Totals		$3029.20		$3552.60	$252.43	$296.05

Crew B-15

	Hr.	Daily	Hr.	Daily	Bare Costs	Incl. O&P
1 Equipment Oper. (med.)	$36.55	$292.40	$60.25	$482.00	$32.66	$53.84
.5 Laborer	26.70	106.80	44.40	177.60		
2 Truck Drivers (heavy)	32.20	515.20	53.00	848.00		
2 Dump Trucks, 12 C.Y., 400 H.P.		1085.60		1194.16		
1 Dozer, 200 H.P.		1273.00		1400.30	84.24	92.66
28 L.H., Daily Totals		$3273.00		$4102.06	$116.89	$146.50

Crew B-16

	Hr.	Daily	Hr.	Daily	Bare Costs	Incl. O&P
1 Labor Foreman (outside)	$28.70	$229.60	$47.75	$382.00	$28.57	$47.39
2 Laborers	26.70	427.20	44.40	710.40		
1 Truck Driver (heavy)	32.20	257.60	53.00	424.00		
1 Dump Truck, 12 C.Y., 400 H.P.		542.80		597.08	16.96	18.66
32 L.H., Daily Totals		$1457.20		$2113.48	$45.54	$66.05

Crew B-17

	Hr.	Daily	Hr.	Daily	Bare Costs	Incl. O&P
2 Laborers	$26.70	$427.20	$44.40	$710.40	$30.25	$50.04
1 Equip. Oper. (light)	35.40	283.20	58.35	466.80		
1 Truck Driver (heavy)	32.20	257.60	53.00	424.00		
1 Backhoe Loader, 48 H.P.		312.40		343.64		
1 Dump Truck, 8 C.Y., 220 H.P.		338.60		372.46	20.34	22.38
32 L.H., Daily Totals		$1619.00		$2317.30	$50.59	$72.42

Crew B-17A

	Hr.	Daily	Hr.	Daily	Bare Costs	Incl. O&P
2 Labor Foremen (outside)	$28.70	$459.20	$47.75	$764.00	$29.12	$48.48
6 Laborers	26.70	1281.60	44.40	2131.20		
1 Skilled Worker Foreman (out)	37.80	302.40	63.15	505.20		
1 Skilled Worker	35.80	286.40	59.80	478.40		
80 L.H., Daily Totals		$2329.60		$3878.80	$29.12	$48.48

Crew B-17B	Hr.	Daily	Hr.	Daily	Bare Costs	Incl. O&P
2 Laborers	$26.70	$427.20	$44.40	$710.40	$30.25	$50.04
1 Equip. Oper. (light)	35.40	283.20	58.35	466.80		
1 Truck Driver (heavy)	32.20	257.60	53.00	424.00		
1 Backhoe Loader, 48 H.P.		312.40		343.64		
1 Dump Truck, 12 C.Y., 400 H.P.		542.80		597.08	26.73	29.40
32 L.H., Daily Totals		$1823.20		$2541.92	$56.98	$79.44

Crew B-18	Hr.	Daily	Hr.	Daily	Bare Costs	Incl. O&P
1 Labor Foreman (outside)	$28.70	$229.60	$47.75	$382.00	$27.37	$45.52
2 Laborers	26.70	427.20	44.40	710.40		
1 Vibrating Plate, Gas, 21"		40.60		44.66	1.69	1.86
24 L.H., Daily Totals		$697.40		$1137.06	$29.06	$47.38

Crew B-19	Hr.	Daily	Hr.	Daily	Bare Costs	Incl. O&P
1 Pile Driver Foreman (outside)	$37.40	$299.20	$64.20	$513.60	$34.68	$58.98
4 Pile Drivers	35.40	1132.80	60.80	1945.60		
1 Equip. Oper. (crane)	37.05	296.40	61.05	488.40		
1 Building Laborer	26.70	213.60	44.40	355.20		
1 Crawler Crane, 40 Ton		1370.00		1507.00		
1 Lead, 90' High		131.60		144.76		
1 Hammer, Diesel, 22k ft-lb		427.00		469.70	34.44	37.88
56 L.H., Daily Totals		$3870.60		$5424.26	$69.12	$96.86

Crew B-19A	Hr.	Daily	Hr.	Daily	Bare Costs	Incl. O&P
1 Pile Driver Foreman (outside)	$37.40	$299.20	$64.20	$513.60	$34.68	$58.98
4 Pile Drivers	35.40	1132.80	60.80	1945.60		
1 Equip. Oper. (crane)	37.05	296.40	61.05	488.40		
1 Common Laborer	26.70	213.60	44.40	355.20		
1 Crawler Crane, 75 Ton		1708.00		1878.80		
1 Lead, 90' High		131.60		144.76		
1 Hammer, Diesel, 41k ft-lb		572.00		629.20	43.06	47.37
56 L.H., Daily Totals		$4353.60		$5955.56	$77.74	$106.35

Crew B-19B	Hr.	Daily	Hr.	Daily	Bare Costs	Incl. O&P
1 Pile Driver Foreman (outside)	$37.40	$299.20	$64.20	$513.60	$34.68	$58.98
4 Pile Drivers	35.40	1132.80	60.80	1945.60		
1 Equip. Oper. (crane)	37.05	296.40	61.05	488.40		
1 Common Laborer	26.70	213.60	44.40	355.20		
1 Crawler Crane, 40 Ton		1370.00		1507.00		
1 Lead, 90' High		131.60		144.76		
1 Hammer, Diesel, 22k ft-lb		427.00		469.70		
1 Barge, 400 Ton		835.40		918.94	49.36	54.29
56 L.H., Daily Totals		$4706.00		$6343.20	$84.04	$113.27

Crew B-19C	Hr.	Daily	Hr.	Daily	Bare Costs	Incl. O&P
1 Pile Driver Foreman (outside)	$37.40	$299.20	$64.20	$513.60	$34.68	$58.98
4 Pile Drivers	35.40	1132.80	60.80	1945.60		
1 Equip. Oper. (crane)	37.05	296.40	61.05	488.40		
1 Common Laborer	26.70	213.60	44.40	355.20		
1 Crawler Crane, 75 Ton		1708.00		1878.80		
1 Lead, 90' High		131.60		144.76		
1 Hammer, Diesel, 41k ft-lb		572.00		629.20		
1 Barge, 400 Ton		835.40		918.94	57.98	63.78
56 L.H., Daily Totals		$5189.00		$6874.50	$92.66	$122.76

Crew B-20	Hr.	Daily	Hr.	Daily	Bare Costs	Incl. O&P
1 Labor Foreman (outside)	$28.70	$229.60	$47.75	$382.00	$27.37	$45.52
2 Laborers	26.70	427.20	44.40	710.40		
24 L.H., Daily Totals		$656.80		$1092.40	$27.37	$45.52

Crew B-20A	Hr.	Daily	Hr.	Daily	Bare Costs	Incl. O&P
1 Labor Foreman (outside)	$28.70	$229.60	$47.75	$382.00	$31.76	$52.52
1 Laborer	26.70	213.60	44.40	355.20		
1 Plumber	39.80	318.40	65.50	524.00		
1 Plumber Apprentice	31.85	254.80	52.45	419.60		
32 L.H., Daily Totals		$1016.40		$1680.80	$31.76	$52.52

Crew B-21	Hr.	Daily	Hr.	Daily	Bare Costs	Incl. O&P
1 Labor Foreman (outside)	$28.70	$229.60	$47.75	$382.00	$28.75	$47.74
2 Laborers	26.70	427.20	44.40	710.40		
.5 Equip. Oper. (crane)	37.05	148.20	61.05	244.20		
.5 S.P. Crane, 4x4, 5 Ton		129.50		142.45	4.63	5.09
28 L.H., Daily Totals		$934.50		$1479.05	$33.38	$52.82

Crew B-21A	Hr.	Daily	Hr.	Daily	Bare Costs	Incl. O&P
1 Labor Foreman (outside)	$28.70	$229.60	$47.75	$382.00	$32.82	$54.23
1 Laborer	26.70	213.60	44.40	355.20		
1 Plumber	39.80	318.40	65.50	524.00		
1 Plumber Apprentice	31.85	254.80	52.45	419.60		
1 Equip. Oper. (crane)	37.05	296.40	61.05	488.40		
1 S.P. Crane, 4x4, 12 Ton		363.60		399.96	9.09	10.00
40 L.H., Daily Totals		$1676.40		$2569.16	$41.91	$64.23

Crew B-21B	Hr.	Daily	Hr.	Daily	Bare Costs	Incl. O&P
1 Labor Foreman (outside)	$28.70	$229.60	$47.75	$382.00	$29.17	$48.40
3 Laborers	26.70	640.80	44.40	1065.60		
1 Equip. Oper. (crane)	37.05	296.40	61.05	488.40		
1 Hyd. Crane, 12 Ton		496.00		545.60	12.40	13.64
40 L.H., Daily Totals		$1662.80		$2481.60	$41.57	$62.04

Crew B-21C	Hr.	Daily	Hr.	Daily	Bare Costs	Incl. O&P
1 Labor Foreman (outside)	$28.70	$229.60	$47.75	$382.00	$29.30	$48.58
4 Laborers	26.70	854.40	44.40	1420.80		
1 Equip. Oper. (crane)	37.05	296.40	61.05	488.40		
1 Equip. Oper. (oiler)	32.55	260.40	53.65	429.20		
2 Cutting Torches		25.20		27.72		
2 Sets of Gases		336.00		369.60		
1 Lattice Boom Crane, 90 Ton		1687.00		1855.70	36.58	40.23
56 L.H., Daily Totals		$3689.00		$4973.42	$65.88	$88.81

Crew B-22	Hr.	Daily	Hr.	Daily	Bare Costs	Incl. O&P
1 Labor Foreman (outside)	$28.70	$229.60	$47.75	$382.00	$29.30	$48.62
2 Laborers	26.70	427.20	44.40	710.40		
.75 Equip. Oper. (crane)	37.05	222.30	61.05	366.30		
.75 S.P. Crane, 4x4, 5 Ton		194.25		213.68	6.47	7.12
30 L.H., Daily Totals		$1073.35		$1672.38	$35.78	$55.75

Crew B-22A	Hr.	Daily	Hr.	Daily	Bare Costs	Incl. O&P
1 Labor Foreman (outside)	$28.70	$229.60	$47.75	$382.00	$30.99	$51.48
1 Skilled Worker	35.80	286.40	59.80	478.40		
2 Laborers	26.70	427.20	44.40	710.40		
1 Equipment Operator, Crane	37.05	296.40	61.05	488.40		
1 S.P. Crane, 4x4, 5 Ton		259.00		284.90		
1 Butt Fusion Mach., 4"-12" diam.		425.70		468.27	17.12	18.83
40 L.H., Daily Totals		$1924.30		$2812.37	$48.11	$70.31

Crew No.	Bare Costs		Incl. Subs O&P		Cost Per Labor-Hour	

Crew B-22B

	Hr.	Daily	Hr.	Daily	Bare Costs	Incl. O&P
1 Labor Foreman (outside)	$28.70	$229.60	$47.75	$382.00	$30.99	$51.48
1 Skilled Worker	35.80	286.40	59.80	478.40		
2 Laborers	26.70	427.20	44.40	710.40		
1 Equip. Oper. (crane)	37.05	296.40	61.05	488.40		
1 S.P. Crane, 4x4, 5 Ton		259.00		284.90		
1 Butt Fusion Mach., 8"-24" diam.		1908.00		2098.80	54.17	59.59
40 L.H., Daily Totals		$3406.60		$4442.90	$85.17	$111.07

Crew B-22C

	Hr.	Daily	Hr.	Daily	Bare Costs	Incl. O&P
1 Skilled Worker	$35.80	$286.40	$59.80	$478.40	$31.25	$52.10
1 Laborer	26.70	213.60	44.40	355.20		
1 Butt Fusion Mach., 2"-8" diam.		121.05		133.16	7.57	8.32
16 L.H., Daily Totals		$621.05		$966.76	$38.82	$60.42

Crew B-23

	Hr.	Daily	Hr.	Daily	Bare Costs	Incl. O&P
1 Labor Foreman (outside)	$28.70	$229.60	$47.75	$382.00	$27.10	$45.07
4 Laborers	26.70	854.40	44.40	1420.80		
1 Drill Rig, Truck-Mounted		2444.00		2688.40		
1 Flatbed Truck, Gas, 3 Ton		238.00		261.80	67.05	73.75
40 L.H., Daily Totals		$3766.00		$4753.00	$94.15	$118.83

Crew B-23A

	Hr.	Daily	Hr.	Daily	Bare Costs	Incl. O&P
1 Labor Foreman (outside)	$28.70	$229.60	$47.75	$382.00	$30.65	$50.80
1 Laborer	26.70	213.60	44.40	355.20		
1 Equip. Oper. (medium)	36.55	292.40	60.25	482.00		
1 Drill Rig, Truck-Mounted		2444.00		2688.40		
1 Pickup Truck, 3/4 Ton		115.20		126.72	106.63	117.30
24 L.H., Daily Totals		$3294.80		$4034.32	$137.28	$168.10

Crew B-23B

	Hr.	Daily	Hr.	Daily	Bare Costs	Incl. O&P
1 Labor Foreman (outside)	$28.70	$229.60	$47.75	$382.00	$30.65	$50.80
1 Laborer	26.70	213.60	44.40	355.20		
1 Equip. Oper. (medium)	36.55	292.40	60.25	482.00		
1 Drill Rig, Truck-Mounted		2444.00		2688.40		
1 Pickup Truck, 3/4 Ton		115.20		126.72		
1 Centr. Water Pump, 6"		296.40		326.04	118.98	130.88
24 L.H., Daily Totals		$3591.20		$4360.36	$149.63	$181.68

Crew B-24

	Hr.	Daily	Hr.	Daily	Bare Costs	Incl. O&P
1 Cement Finisher	$34.70	$277.60	$56.30	$450.40	$32.12	$52.93
1 Laborer	26.70	213.60	44.40	355.20		
1 Carpenter	34.95	279.60	58.10	464.80		
24 L.H., Daily Totals		$770.80		$1270.40	$32.12	$52.93

Crew B-25

	Hr.	Daily	Hr.	Daily	Bare Costs	Incl. O&P
1 Labor Foreman (outside)	$28.70	$229.60	$47.75	$382.00	$29.57	$49.03
7 Laborers	26.70	1495.20	44.40	2486.40		
3 Equip. Oper. (medium)	36.55	877.20	60.25	1446.00		
1 Asphalt Paver, 130 H.P.		2123.00		2335.30		
1 Tandem Roller, 10 Ton		231.40		254.54		
1 Roller, Pneum. Whl., 12 Ton		338.20		372.02	30.60	33.66
88 L.H., Daily Totals		$5294.60		$7276.26	$60.17	$82.68

Crew B-25B

	Hr.	Daily	Hr.	Daily	Bare Costs	Incl. O&P
1 Labor Foreman (outside)	$28.70	$229.60	$47.75	$382.00	$30.15	$49.96
7 Laborers	26.70	1495.20	44.40	2486.40		
4 Equip. Oper. (medium)	36.55	1169.60	60.25	1928.00		
1 Asphalt Paver, 130 H.P.		2123.00		2335.30		
2 Tandem Rollers, 10 Ton		462.80		509.08		
1 Roller, Pneum. Whl., 12 Ton		338.20		372.02	30.46	33.50
96 L.H., Daily Totals		$5818.40		$8012.80	$60.61	$83.47

Crew B-25C

	Hr.	Daily	Hr.	Daily	Bare Costs	Incl. O&P
1 Labor Foreman (outside)	$28.70	$229.60	$47.75	$382.00	$30.32	$50.24
3 Laborers	26.70	640.80	44.40	1065.60		
2 Equip. Oper. (medium)	36.55	584.80	60.25	964.00		
1 Asphalt Paver, 130 H.P.		2123.00		2335.30		
1 Tandem Roller, 10 Ton		231.40		254.54	49.05	53.95
48 L.H., Daily Totals		$3809.60		$5001.44	$79.37	$104.20

Crew B-25D

	Hr.	Daily	Hr.	Daily	Bare Costs	Incl. O&P
1 Labor Foreman (outside)	$28.70	$229.60	$47.75	$382.00	$30.48	$50.50
3 Laborers	26.70	640.80	44.40	1065.60		
2.125 Equip. Oper. (medium)	36.55	621.35	60.25	1024.25		
.125 Truck Driver (heavy)	32.20	32.20	53.00	53.00		
.125 Truck Tractor, 6x4, 380 H.P.		59.52		65.48		
.125 Dist. Tanker, 3000 Gallon		40.85		44.94		
1 Asphalt Paver, 130 H.P.		2123.00		2335.30		
1 Tandem Roller, 10 Ton		231.40		254.54	49.10	54.01
50 L.H., Daily Totals		$3978.72		$5225.10	$79.57	$104.50

Crew B-25E

	Hr.	Daily	Hr.	Daily	Bare Costs	Incl. O&P
1 Labor Foreman (outside)	$28.70	$229.60	$47.75	$382.00	$30.63	$50.73
3 Laborers	26.70	640.80	44.40	1065.60		
2.250 Equip. Oper. (medium)	36.55	657.90	60.25	1084.50		
.25 Truck Driver (heavy)	32.20	64.40	53.00	106.00		
.25 Truck Tractor, 6x4, 380 H.P.		119.05		130.96		
.25 Dist. Tanker, 3000 Gallon		81.70		89.87		
1 Asphalt Paver, 130 H.P.		2123.00		2335.30		
1 Tandem Roller, 10 Ton		231.40		254.54	49.14	54.05
52 L.H., Daily Totals		$4147.85		$5448.77	$79.77	$104.78

Crew B-26

	Hr.	Daily	Hr.	Daily	Bare Costs	Incl. O&P
1 Labor Foreman (outside)	$28.70	$229.60	$47.75	$382.00	$30.50	$50.51
6 Laborers	26.70	1281.60	44.40	2131.20		
2 Equip. Oper. (medium)	36.55	584.80	60.25	964.00		
1 Rodman (reinf.)	38.80	310.40	64.70	517.60		
1 Cement Finisher	34.70	277.60	56.30	450.40		
1 Grader, 30,000 Lbs.		643.80		708.18		
1 Paving Mach. & Equip.		2361.00		2597.10	34.15	37.56
88 L.H., Daily Totals		$5688.80		$7750.48	$64.65	$88.07

Crew B-26A

	Hr.	Daily	Hr.	Daily	Bare Costs	Incl. O&P
1 Labor Foreman (outside)	$28.70	$229.60	$47.75	$382.00	$30.50	$50.51
6 Laborers	26.70	1281.60	44.40	2131.20		
2 Equip. Oper. (medium)	36.55	584.80	60.25	964.00		
1 Rodman (reinf.)	38.80	310.40	64.70	517.60		
1 Cement Finisher	34.70	277.60	56.30	450.40		
1 Grader, 30,000 Lbs.		643.80		708.18		
1 Paving Mach. & Equip.		2361.00		2597.10		
1 Concrete Saw		102.40		112.64	35.31	38.84
88 L.H., Daily Totals		$5791.20		$7863.12	$65.81	$89.35

Crew B-26B

	Hr.	Daily	Hr.	Daily	Bare Costs	Incl. O&P
1 Labor Foreman (outside)	$28.70	$229.60	$47.75	$382.00	$31.00	$51.33
6 Laborers	26.70	1281.60	44.40	2131.20		
3 Equip. Oper. (medium)	36.55	877.20	60.25	1446.00		
1 Rodman (reinf.)	38.80	310.40	64.70	517.60		
1 Cement Finisher	34.70	277.60	56.30	450.40		
1 Grader, 30,000 Lbs.		643.80		708.18		
1 Paving Mach. & Equip.		2361.00		2597.10		
1 Concrete Pump, 110' Boom		1116.00		1227.60	42.92	47.22
96 L.H., Daily Totals		$7097.20		$9460.08	$73.93	$98.54

Crew No.	Bare Costs		Incl. Subs O&P		Cost Per Labor-Hour	
Crew B-26C	Hr.	Daily	Hr.	Daily	Bare Costs	Incl. O&P
1 Labor Foreman (outside)	$28.70	$229.60	$47.75	$382.00	$29.90	$49.54
6 Laborers	26.70	1281.60	44.40	2131.20		
1 Equip. Oper. (medium)	36.55	292.40	60.25	482.00		
1 Rodman (reinf.)	38.80	310.40	64.70	517.60		
1 Cement Finisher	34.70	277.60	56.30	450.40		
1 Paving Mach. & Equip.		2361.00		2597.10		
1 Concrete Saw		102.40		112.64	30.79	33.87
80 L.H., Daily Totals		$4855.00		$6672.94	$60.69	$83.41
Crew B-27	Hr.	Daily	Hr.	Daily	Bare Costs	Incl. O&P
1 Labor Foreman (outside)	$28.70	$229.60	$47.75	$382.00	$27.20	$45.24
3 Laborers	26.70	640.80	44.40	1065.60		
1 Berm Machine		317.40		349.14	9.92	10.91
32 L.H., Daily Totals		$1187.80		$1796.74	$37.12	$56.15
Crew B-28	Hr.	Daily	Hr.	Daily	Bare Costs	Incl. O&P
2 Carpenters	$34.95	$559.20	$58.10	$929.60	$32.20	$53.53
1 Laborer	26.70	213.60	44.40	355.20		
24 L.H., Daily Totals		$772.80		$1284.80	$32.20	$53.53
Crew B-29	Hr.	Daily	Hr.	Daily	Bare Costs	Incl. O&P
1 Labor Foreman (outside)	$28.70	$229.60	$47.75	$382.00	$28.76	$47.73
4 Laborers	26.70	854.40	44.40	1420.80		
1 Equip. Oper. (crane)	37.05	296.40	61.05	488.40		
1 Gradall, 5/8 C.Y.		869.60		956.56	18.12	19.93
48 L.H., Daily Totals		$2250.00		$3247.76	$46.88	$67.66
Crew B-30	Hr.	Daily	Hr.	Daily	Bare Costs	Incl. O&P
1 Equip. Oper. (medium)	$36.55	$292.40	$60.25	$482.00	$33.65	$55.42
2 Truck Drivers (heavy)	32.20	515.20	53.00	848.00		
1 Hyd. Excavator, 1.5 C.Y.		893.60		982.96		
2 Dump Trucks, 12 C.Y., 400 H.P.		1085.60		1194.16	82.47	90.71
24 L.H., Daily Totals		$2786.80		$3507.12	$116.12	$146.13
Crew B-31	Hr.	Daily	Hr.	Daily	Bare Costs	Incl. O&P
1 Labor Foreman (outside)	$28.70	$229.60	$47.75	$382.00	$27.10	$45.07
4 Laborers	26.70	854.40	44.40	1420.80		
1 Air Compressor, 250 cfm		167.40		184.14		
1 Sheeting Driver		7.00		7.70		
2 -50' Air Hoses, 1.5"		45.40		49.94	5.50	6.04
40 L.H., Daily Totals		$1303.80		$2044.58	$32.59	$51.11
Crew B-32	Hr.	Daily	Hr.	Daily	Bare Costs	Incl. O&P
1 Laborer	$26.70	$213.60	$44.40	$355.20	$34.09	$56.29
3 Equip. Oper. (medium)	36.55	877.20	60.25	1446.00		
1 Grader, 30,000 Lbs.		643.80		708.18		
1 Tandem Roller, 10 Ton		231.40		254.54		
1 Dozer, 200 H.P.		1273.00		1400.30	67.13	73.84
32 L.H., Daily Totals		$3239.00		$4164.22	$101.22	$130.13
Crew B-32A	Hr.	Daily	Hr.	Daily	Bare Costs	Incl. O&P
1 Laborer	$26.70	$213.60	$44.40	$355.20	$33.27	$54.97
2 Equip. Oper. (medium)	36.55	584.80	60.25	964.00		
1 Grader, 30,000 Lbs.		643.80		708.18		
1 Roller, Vibratory, 25 Ton		659.80		725.78	54.32	59.75
24 L.H., Daily Totals		$2102.00		$2753.16	$87.58	$114.72

Crew No.	Bare Costs		Incl. Subs O&P		Cost Per Labor-Hour	
Crew B-32B	Hr.	Daily	Hr.	Daily	Bare Costs	Incl. O&P
1 Laborer	$26.70	$213.60	$44.40	$355.20	$33.27	$54.97
2 Equip. Oper. (medium)	36.55	584.80	60.25	964.00		
1 Dozer, 200 H.P.		1273.00		1400.30		
1 Roller, Vibratory, 25 Ton		659.80		725.78	80.53	88.59
24 L.H., Daily Totals		$2731.20		$3445.28	$113.80	$143.55
Crew B-32C	Hr.	Daily	Hr.	Daily	Bare Costs	Incl. O&P
1 Labor Foreman (outside)	$28.70	$229.60	$47.75	$382.00	$31.96	$52.88
2 Laborers	26.70	427.20	44.40	710.40		
3 Equip. Oper. (medium)	36.55	877.20	60.25	1446.00		
1 Grader, 30,000 Lbs.		643.80		708.18		
1 Tandem Roller, 10 Ton		231.40		254.54		
1 Dozer, 200 H.P.		1273.00		1400.30	44.75	49.23
48 L.H., Daily Totals		$3682.20		$4901.42	$76.71	$102.11
Crew B-33A	Hr.	Daily	Hr.	Daily	Bare Costs	Incl. O&P
1 Equip. Oper. (medium)	$36.55	$292.40	$60.25	$482.00	$36.55	$60.25
.25 Equip. Oper. (medium)	36.55	73.10	60.25	120.50		
1 Scraper, Towed, 7 C.Y.		123.40		135.74		
1.250 Dozers, 300 H.P.		2286.25		2514.88	240.97	265.06
10 L.H., Daily Totals		$2775.15		$3253.11	$277.51	$325.31
Crew B-33B	Hr.	Daily	Hr.	Daily	Bare Costs	Incl. O&P
1 Equip. Oper. (medium)	$36.55	$292.40	$60.25	$482.00	$36.55	$60.25
.25 Equip. Oper. (medium)	36.55	73.10	60.25	120.50		
1 Scraper, Towed, 10 C.Y.		155.80		171.38		
1.250 Dozers, 300 H.P.		2286.25		2514.88	244.21	268.63
10 L.H., Daily Totals		$2807.55		$3288.76	$280.76	$328.88
Crew B-33C	Hr.	Daily	Hr.	Daily	Bare Costs	Incl. O&P
1 Equip. Oper. (medium)	$36.55	$292.40	$60.25	$482.00	$36.55	$60.25
.25 Equip. Oper. (medium)	36.55	73.10	60.25	120.50		
1 Scraper, Towed, 15 C.Y.		175.80		193.38		
1.250 Dozers, 300 H.P.		2286.25		2514.88	246.21	270.83
10 L.H., Daily Totals		$2827.55		$3310.76	$282.76	$331.08
Crew B-33D	Hr.	Daily	Hr.	Daily	Bare Costs	Incl. O&P
1 Equip. Oper. (medium)	$36.55	$292.40	$60.25	$482.00	$36.55	$60.25
.25 Equip. Oper. (medium)	36.55	73.10	60.25	120.50		
1 S.P. Scraper, 14 C.Y.		2490.00		2739.00		
.25 Dozer, 300 H.P.		457.25		502.98	294.73	324.20
10 L.H., Daily Totals		$3312.75		$3844.47	$331.27	$384.45
Crew B-33E	Hr.	Daily	Hr.	Daily	Bare Costs	Incl. O&P
1 Equip. Oper. (medium)	$36.55	$292.40	$60.25	$482.00	$36.55	$60.25
.25 Equip. Oper. (medium)	36.55	73.10	60.25	120.50		
1 S.P. Scraper, 21 C.Y.		2468.00		2714.80		
.25 Dozer, 300 H.P.		457.25		502.98	292.52	321.78
10 L.H., Daily Totals		$3290.75		$3820.28	$329.07	$382.03
Crew B-33F	Hr.	Daily	Hr.	Daily	Bare Costs	Incl. O&P
1 Equip. Oper. (medium)	$36.55	$292.40	$60.25	$482.00	$36.55	$60.25
.25 Equip. Oper. (medium)	36.55	73.10	60.25	120.50		
1 Elev. Scraper, 11 C.Y.		1155.00		1270.50		
.25 Dozer, 300 H.P.		457.25		502.98	161.22	177.35
10 L.H., Daily Totals		$1977.75		$2375.97	$197.78	$237.60

Crew No.	Bare Costs		Incl. Subs O&P		Cost Per Labor-Hour	
Crew B-33G	Hr.	Daily	Hr.	Daily	Bare Costs	Incl. O&P
1 Equip. Oper. (medium)	$36.55	$292.40	$60.25	$482.00	$36.55	$60.25
.25 Equip. Oper. (medium)	36.55	73.10	60.25	120.50		
1 Elev. Scraper, 22 C.Y.		2284.00		2512.40		
.25 Dozer, 300 H.P.		457.25		502.98	274.13	301.54
10 L.H., Daily Totals		$3106.75		$3617.88	$310.68	$361.79

Crew No.	Bare Costs		Incl. Subs O&P		Cost Per Labor-Hour	
Crew B-33K	Hr.	Daily	Hr.	Daily	Bare Costs	Incl. O&P
1 Equipment Operator (med.)	$36.55	$292.40	$60.25	$482.00	$33.74	$55.72
.25 Equipment Operator (med.)	36.55	73.10	60.25	120.50		
.5 Laborer	26.70	106.80	44.40	177.60		
1 S.P. Scraper, 31 C.Y.		3576.00		3933.60		
.25 Dozer, 410 H.P.		572.75		630.02	296.34	325.97
14 L.H., Daily Totals		$4621.05		$5343.73	$330.07	$381.69

Crew No.	Bare Costs		Incl. Subs O&P		Cost Per Labor-Hour	
Crew B-34A	Hr.	Daily	Hr.	Daily	Bare Costs	Incl. O&P
1 Truck Driver (heavy)	$32.20	$257.60	$53.00	$424.00	$32.20	$53.00
1 Dump Truck, 8 C.Y., 220 H.P.		338.60		372.46	42.33	46.56
8 L.H., Daily Totals		$596.20		$796.46	$74.53	$99.56

Crew No.	Bare Costs		Incl. Subs O&P		Cost Per Labor-Hour	
Crew B-34B	Hr.	Daily	Hr.	Daily	Bare Costs	Incl. O&P
1 Truck Driver (heavy)	$32.20	$257.60	$53.00	$424.00	$32.20	$53.00
1 Dump Truck, 12 C.Y., 400 H.P.		542.80		597.08	67.85	74.64
8 L.H., Daily Totals		$800.40		$1021.08	$100.05	$127.64

Crew No.	Bare Costs		Incl. Subs O&P		Cost Per Labor-Hour	
Crew B-34C	Hr.	Daily	Hr.	Daily	Bare Costs	Incl. O&P
1 Truck Driver (heavy)	$32.20	$257.60	$53.00	$424.00	$32.20	$53.00
1 Truck Tractor, 6x4, 380 H.P.		476.20		523.82		
1 Dump Trailer, 16.5 C.Y.		133.00		146.30	76.15	83.77
8 L.H., Daily Totals		$866.80		$1094.12	$108.35	$136.76

Crew No.	Bare Costs		Incl. Subs O&P		Cost Per Labor-Hour	
Crew B-34D	Hr.	Daily	Hr.	Daily	Bare Costs	Incl. O&P
1 Truck Driver (heavy)	$32.20	$257.60	$53.00	$424.00	$32.20	$53.00
1 Truck Tractor, 6x4, 380 H.P.		476.20		523.82		
1 Dump Trailer, 20 C.Y.		147.60		162.36	77.97	85.77
8 L.H., Daily Totals		$881.40		$1110.18	$110.18	$138.77

Crew No.	Bare Costs		Incl. Subs O&P		Cost Per Labor-Hour	
Crew B-34E	Hr.	Daily	Hr.	Daily	Bare Costs	Incl. O&P
1 Truck Driver (heavy)	$32.20	$257.60	$53.00	$424.00	$32.20	$53.00
1 Dump Truck, Off Hwy., 25 Ton		1343.00		1477.30	167.88	184.66
8 L.H., Daily Totals		$1600.60		$1901.30	$200.07	$237.66

Crew No.	Bare Costs		Incl. Subs O&P		Cost Per Labor-Hour	
Crew B-34F	Hr.	Daily	Hr.	Daily	Bare Costs	Incl. O&P
1 Truck Driver (heavy)	$32.20	$257.60	$53.00	$424.00	$32.20	$53.00
1 Dump Truck, Off Hwy., 35 Ton		1448.00		1592.80	181.00	199.10
8 L.H., Daily Totals		$1705.60		$2016.80	$213.20	$252.10

Crew No.	Bare Costs		Incl. Subs O&P		Cost Per Labor-Hour	
Crew B-34G	Hr.	Daily	Hr.	Daily	Bare Costs	Incl. O&P
1 Truck Driver (heavy)	$32.20	$257.60	$53.00	$424.00	$32.20	$53.00
1 Dump Truck, Off Hwy., 50 Ton		1662.00		1828.20	207.75	228.53
8 L.H., Daily Totals		$1919.60		$2252.20	$239.95	$281.52

Crew No.	Bare Costs		Incl. Subs O&P		Cost Per Labor-Hour	
Crew B-34H	Hr.	Daily	Hr.	Daily	Bare Costs	Incl. O&P
1 Truck Driver (heavy)	$32.20	$257.60	$53.00	$424.00	$32.20	$53.00
1 Dump Truck, Off Hwy., 65 Ton		1844.00		2028.40	230.50	253.55
8 L.H., Daily Totals		$2101.60		$2452.40	$262.70	$306.55

Crew No.	Bare Costs		Incl. Subs O&P		Cost Per Labor-Hour	
Crew B-34I	Hr.	Daily	Hr.	Daily	Bare Costs	Incl. O&P
1 Truck Driver (heavy)	$32.20	$257.60	$53.00	$424.00	$32.20	$53.00
1 Dump Truck, 18 C.Y., 450 H.P.		706.00		776.60	88.25	97.08
8 L.H., Daily Totals		$963.60		$1200.60	$120.45	$150.07

Crew No.	Bare Costs		Incl. Subs O&P		Cost Per Labor-Hour	
Crew B-34J	Hr.	Daily	Hr.	Daily	Bare Costs	Incl. O&P
1 Truck Driver (heavy)	$32.20	$257.60	$53.00	$424.00	$32.20	$53.00
1 Dump Truck, Off Hwy., 100 Ton		2669.00		2935.90	333.63	366.99
8 L.H., Daily Totals		$2926.60		$3359.90	$365.82	$419.99

Crew No.	Bare Costs		Incl. Subs O&P		Cost Per Labor-Hour	
Crew B-34K	Hr.	Daily	Hr.	Daily	Bare Costs	Incl. O&P
1 Truck Driver (heavy)	$32.20	$257.60	$53.00	$424.00	$32.20	$53.00
1 Truck Tractor, 6x4, 450 H.P.		583.40		641.74		
1 Lowbed Trailer, 75 Ton		249.40		274.34	104.10	114.51
8 L.H., Daily Totals		$1090.40		$1340.08	$136.30	$167.51

Crew No.	Bare Costs		Incl. Subs O&P		Cost Per Labor-Hour	
Crew B-34L	Hr.	Daily	Hr.	Daily	Bare Costs	Incl. O&P
1 Equip. Oper. (light)	$35.40	$283.20	$58.35	$466.80	$35.40	$58.35
1 Flatbed Truck, Gas, 1.5 Ton		188.40		207.24	23.55	25.91
8 L.H., Daily Totals		$471.60		$674.04	$58.95	$84.25

Crew No.	Bare Costs		Incl. Subs O&P		Cost Per Labor-Hour	
Crew B-34M	Hr.	Daily	Hr.	Daily	Bare Costs	Incl. O&P
1 Equip. Oper. (light)	$35.40	$283.20	$58.35	$466.80	$35.40	$58.35
1 Flatbed Truck, Gas, 3 Ton		238.00		261.80	29.75	32.73
8 L.H., Daily Totals		$521.20		$728.60	$65.15	$91.08

Crew No.	Bare Costs		Incl. Subs O&P		Cost Per Labor-Hour	
Crew B-34N	Hr.	Daily	Hr.	Daily	Bare Costs	Incl. O&P
1 Truck Driver (heavy)	$32.20	$257.60	$53.00	$424.00	$34.38	$56.63
1 Equip. Oper. (medium)	36.55	292.40	60.25	482.00		
1 Truck Tractor, 6x4, 380 H.P.		476.20		523.82		
1 Flatbed Trailer, 40 Ton		180.00		198.00	41.01	45.11
16 L.H., Daily Totals		$1206.20		$1627.82	$75.39	$101.74

Crew No.	Bare Costs		Incl. Subs O&P		Cost Per Labor-Hour	
Crew B-34P	Hr.	Daily	Hr.	Daily	Bare Costs	Incl. O&P
1 Pipe Fitter	$40.95	$327.60	$67.40	$539.20	$36.37	$59.88
1 Truck Driver (light)	31.60	252.80	52.00	416.00		
1 Equip. Oper. (medium)	36.55	292.40	60.25	482.00		
1 Flatbed Truck, Gas, 3 Ton		238.00		261.80		
1 Backhoe Loader, 48 H.P.		312.40		343.64	22.93	25.23
24 L.H., Daily Totals		$1423.20		$2042.64	$59.30	$85.11

Crew No.	Bare Costs		Incl. Subs O&P		Cost Per Labor-Hour	
Crew B-34Q	Hr.	Daily	Hr.	Daily	Bare Costs	Incl. O&P
1 Pipe Fitter	$40.95	$327.60	$67.40	$539.20	$36.53	$60.15
1 Truck Driver (light)	31.60	252.80	52.00	416.00		
1 Equip. Oper. (crane)	37.05	296.40	61.05	488.40		
1 Flatbed Trailer, 25 Ton		133.00		146.30		
1 Dump Truck, 8 C.Y., 220 H.P.		338.60		372.46		
1 Hyd. Crane, 25 Ton		590.60		649.66	44.26	48.68
24 L.H., Daily Totals		$1939.00		$2612.02	$80.79	$108.83

Crew No.	Bare Costs		Incl. Subs O&P		Cost Per Labor-Hour	
Crew B-34R	Hr.	Daily	Hr.	Daily	Bare Costs	Incl. O&P
1 Pipe Fitter	$40.95	$327.60	$67.40	$539.20	$36.53	$60.15
1 Truck Driver (light)	31.60	252.80	52.00	416.00		
1 Equip. Oper. (crane)	37.05	296.40	61.05	488.40		
1 Flatbed Trailer, 25 Ton		133.00		146.30		
1 Dump Truck, 8 C.Y., 220 H.P.		338.60		372.46		
1 Hyd. Crane, 25 Ton		590.60		649.66		
1 Hyd. Excavator, 1 C.Y.		742.20		816.42	75.18	82.70
24 L.H., Daily Totals		$2681.20		$3428.44	$111.72	$142.85

Crew No.	Bare Costs		Incl. Subs O&P		Cost Per Labor-Hour	
Crew B-34S	Hr.	Daily	Hr.	Daily	Bare Costs	Incl. O&P
2 Pipe Fitters	$40.95	$655.20	$67.40	$1078.40	$37.79	$62.21
1 Truck Driver (heavy)	32.20	257.60	53.00	424.00		
1 Equip. Oper. (crane)	37.05	296.40	61.05	488.40		
1 Flatbed Trailer, 40 Ton		180.00		198.00		
1 Truck Tractor, 6x4, 380 H.P.		476.20		523.82		
1 Hyd. Crane, 80 Ton		1518.00		1669.80		
1 Hyd. Excavator, 2 C.Y.		1052.00		1157.20	100.82	110.90
32 L.H., Daily Totals		$4435.40		$5539.62	$138.61	$173.11
Crew B-34T	Hr.	Daily	Hr.	Daily	Bare Costs	Incl. O&P
2 Pipe Fitters	$40.95	$655.20	$67.40	$1078.40	$37.79	$62.21
1 Truck Driver (heavy)	32.20	257.60	53.00	424.00		
1 Equip. Oper. (crane)	37.05	296.40	61.05	488.40		
1 Flatbed Trailer, 40 Ton		180.00		198.00		
1 Truck Tractor, 6x4, 380 H.P.		476.20		523.82		
1 Hyd. Crane, 80 Ton		1518.00		1669.80	67.94	74.74
32 L.H., Daily Totals		$3383.40		$4382.42	$105.73	$136.95
Crew B-34U	Hr.	Daily	Hr.	Daily	Bare Costs	Incl. O&P
1 Truck Driver (heavy)	$32.20	$257.60	$53.00	$424.00	$33.80	$55.67
1 Equip. Oper. (light)	35.40	283.20	58.35	466.80		
1 Truck Tractor, 220 H.P.		290.00		319.00		
1 Flatbed Trailer, 25 Ton		133.00		146.30	26.44	29.08
16 L.H., Daily Totals		$963.80		$1356.10	$60.24	$84.76
Crew B-34V	Hr.	Daily	Hr.	Daily	Bare Costs	Incl. O&P
1 Truck Driver (heavy)	$32.20	$257.60	$53.00	$424.00	$34.88	$57.47
1 Equip. Oper. (crane)	37.05	296.40	61.05	488.40		
1 Equip. Oper. (light)	35.40	283.20	58.35	466.80		
1 Truck Tractor, 6x4, 450 H.P.		583.40		641.74		
1 Equipment Trailer, 50 Ton		198.20		218.02		
1 Pickup Truck, 4x4, 3/4 Ton		126.60		139.26	37.84	41.63
24 L.H., Daily Totals		$1745.40		$2378.22	$72.72	$99.09
Crew B-34W	Hr.	Daily	Hr.	Daily	Bare Costs	Incl. O&P
5 Truck Drivers (heavy)	$32.20	$1288.00	$53.00	$2120.00	$33.22	$54.74
2 Equip. Opers. (crane)	37.05	592.80	61.05	976.80		
1 Equip. Oper. (mechanic)	37.15	297.20	61.20	489.60		
1 Laborer	26.70	213.60	44.40	355.20		
4 Truck Tractors, 6x4, 380 H.P.		1904.80		2095.28		
2 Equipment Trailers, 50 Ton		396.40		436.04		
2 Flatbed Trailers, 40 Ton		360.00		396.00		
1 Pickup Truck, 4x4, 3/4 Ton		126.60		139.26		
1 S.P. Crane, 4x4, 20 Ton		571.20		628.32	46.65	51.32
72 L.H., Daily Totals		$5750.60		$7636.50	$79.87	$106.06
Crew B-35	Hr.	Daily	Hr.	Daily	Bare Costs	Incl. O&P
1 Labor Foreman (outside)	$28.70	$229.60	$47.75	$382.00	$33.61	$55.70
1 Skilled Worker	35.80	286.40	59.80	478.40		
1 Welder (plumber)	39.80	318.40	65.50	524.00		
1 Laborer	26.70	213.60	44.40	355.20		
1 Equip. Oper. (crane)	37.05	296.40	61.05	488.40		
1 Welder, Electric, 300 amp		59.20		65.12		
1 Hyd. Excavator, .75 C.Y.		669.60		736.56	18.22	20.04
40 L.H., Daily Totals		$2073.20		$3029.68	$51.83	$75.74

Crew No.	Bare Costs		Incl. Subs O&P		Cost Per Labor-Hour	
Crew B-35A	Hr.	Daily	Hr.	Daily	Bare Costs	Incl. O&P
1 Labor Foreman (outside)	$28.70	$229.60	$47.75	$382.00	$32.47	$53.79
2 Laborers	26.70	427.20	44.40	710.40		
1 Skilled Worker	35.80	286.40	59.80	478.40		
1 Welder (plumber)	39.80	318.40	65.50	524.00		
1 Equip. Oper. (crane)	37.05	296.40	61.05	488.40		
1 Equip. Oper. (oiler)	32.55	260.40	53.65	429.20		
1 Welder, Gas Engine, 300 amp		98.40		108.24		
1 Crawler Crane, 75 Ton		1708.00		1878.80	32.26	35.48
56 L.H., Daily Totals		$3624.80		$4999.44	$64.73	$89.28
Crew B-36	Hr.	Daily	Hr.	Daily	Bare Costs	Incl. O&P
1 Labor Foreman (outside)	$28.70	$229.60	$47.75	$382.00	$31.04	$51.41
2 Laborers	26.70	427.20	44.40	710.40		
2 Equip. Oper. (medium)	36.55	584.80	60.25	964.00		
1 Dozer, 200 H.P.		1273.00		1400.30		
1 Aggregate Spreader		36.80		40.48		
1 Tandem Roller, 10 Ton		231.40		254.54	38.53	42.38
40 L.H., Daily Totals		$2782.80		$3751.72	$69.57	$93.79
Crew B-36A	Hr.	Daily	Hr.	Daily	Bare Costs	Incl. O&P
1 Labor Foreman (outside)	$28.70	$229.60	$47.75	$382.00	$32.61	$53.94
2 Laborers	26.70	427.20	44.40	710.40		
4 Equip. Oper. (medium)	36.55	1169.60	60.25	1928.00		
1 Dozer, 200 H.P.		1273.00		1400.30		
1 Aggregate Spreader		36.80		40.48		
1 Tandem Roller, 10 Ton		231.40		254.54		
1 Roller, Pneum. Whl., 12 Ton		338.20		372.02	33.56	36.92
56 L.H., Daily Totals		$3705.80		$5087.74	$66.17	$90.85
Crew B-36B	Hr.	Daily	Hr.	Daily	Bare Costs	Incl. O&P
1 Labor Foreman (outside)	$28.70	$229.60	$47.75	$382.00	$32.56	$53.82
2 Laborers	26.70	427.20	44.40	710.40		
4 Equip. Oper. (medium)	36.55	1169.60	60.25	1928.00		
1 Truck Driver (heavy)	32.20	257.60	53.00	424.00		
1 Grader, 30,000 Lbs.		643.80		708.18		
1 F.E. Loader, Crl, 1.5 C.Y.		645.40		709.94		
1 Dozer, 300 H.P.		1829.00		2011.90		
1 Roller, Vibratory, 25 Ton		659.80		725.78		
1 Truck Tractor, 6x4, 450 H.P.		583.40		641.74		
1 Water Tank Trailer, 5000 Gal.		147.60		162.36	70.45	77.50
64 L.H., Daily Totals		$6593.00		$8404.30	$103.02	$131.32
Crew B-36C	Hr.	Daily	Hr.	Daily	Bare Costs	Incl. O&P
1 Labor Foreman (outside)	$28.70	$229.60	$47.75	$382.00	$34.11	$56.30
3 Equip. Oper. (medium)	36.55	877.20	60.25	1446.00		
1 Truck Driver (heavy)	32.20	257.60	53.00	424.00		
1 Grader, 30,000 Lbs.		643.80		708.18		
1 Dozer, 300 H.P.		1829.00		2011.90		
1 Roller, Vibratory, 25 Ton		659.80		725.78		
1 Truck Tractor, 6x4, 450 H.P.		583.40		641.74		
1 Water Tank Trailer, 5000 Gal.		147.60		162.36	96.59	106.25
40 L.H., Daily Totals		$5228.00		$6501.96	$130.70	$162.55

Crew No.	Bare Costs		Incl. Subs O&P		Cost Per Labor-Hour	

Crew B-36E

	Hr.	Daily	Hr.	Daily	Bare Costs	Incl. O&P
1 Labor Foreman (outside)	$28.70	$229.60	$47.75	$382.00	$34.52	$56.96
4 Equip. Oper. (medium)	36.55	1169.60	60.25	1928.00		
1 Truck Driver (heavy)	32.20	257.60	53.00	424.00		
1 Grader, 30,000 Lbs.		643.80		708.18		
1 Dozer, 300 H.P.		1829.00		2011.90		
1 Roller, Vibratory, 25 Ton		659.80		725.78		
1 Truck Tractor, 6x4, 380 H.P.		476.20		523.82		
1 Dist. Tanker, 3000 Gallon		326.80		359.48	81.99	90.19
48 L.H., Daily Totals		$5592.40		$7063.16	$116.51	$147.15

Crew B-37

	Hr.	Daily	Hr.	Daily	Bare Costs	Incl. O&P
1 Labor Foreman (outside)	$28.70	$229.60	$47.75	$382.00	$28.48	$47.28
4 Laborers	26.70	854.40	44.40	1420.80		
1 Equip. Oper. (light)	35.40	283.20	58.35	466.80		
1 Tandem Roller, 5 Ton		151.20		166.32	3.15	3.46
48 L.H., Daily Totals		$1518.40		$2435.92	$31.63	$50.75

Crew B-37A

	Hr.	Daily	Hr.	Daily	Bare Costs	Incl. O&P
2 Laborers	$26.70	$427.20	$44.40	$710.40	$28.33	$46.93
1 Truck Driver (light)	31.60	252.80	52.00	416.00		
1 Flatbed Truck, Gas, 1.5 Ton		188.40		207.24		
1 Tar Kettle, T.M.		129.20		142.12	13.23	14.56
24 L.H., Daily Totals		$997.60		$1475.76	$41.57	$61.49

Crew B-37B

	Hr.	Daily	Hr.	Daily	Bare Costs	Incl. O&P
3 Laborers	$26.70	$640.80	$44.40	$1065.60	$27.93	$46.30
1 Truck Driver (light)	31.60	252.80	52.00	416.00		
1 Flatbed Truck, Gas, 1.5 Ton		188.40		207.24		
1 Tar Kettle, T.M.		129.20		142.12	9.93	10.92
32 L.H., Daily Totals		$1211.20		$1830.96	$37.85	$57.22

Crew B-37C

	Hr.	Daily	Hr.	Daily	Bare Costs	Incl. O&P
2 Laborers	$26.70	$427.20	$44.40	$710.40	$29.15	$48.20
2 Truck Drivers (light)	31.60	505.60	52.00	832.00		
2 Flatbed Trucks, Gas, 1.5 Ton		376.80		414.48		
1 Tar Kettle, T.M.		129.20		142.12	15.81	17.39
32 L.H., Daily Totals		$1438.80		$2099.00	$44.96	$65.59

Crew B-37D

	Hr.	Daily	Hr.	Daily	Bare Costs	Incl. O&P
1 Laborer	$26.70	$213.60	$44.40	$355.20	$29.15	$48.20
1 Truck Driver (light)	31.60	252.80	52.00	416.00		
1 Pickup Truck, 3/4 Ton		115.20		126.72	7.20	7.92
16 L.H., Daily Totals		$581.60		$897.92	$36.35	$56.12

Crew B-37E

	Hr.	Daily	Hr.	Daily	Bare Costs	Incl. O&P
3 Laborers	$26.70	$640.80	$44.40	$1065.60	$30.75	$50.83
1 Equip. Oper. (light)	35.40	283.20	58.35	466.80		
1 Equip. Oper. (medium)	36.55	292.40	60.25	482.00		
2 Truck Drivers (light)	31.60	505.60	52.00	832.00		
4 Barrels w/Flasher		15.60		17.16		
1 Concrete Saw		102.40		112.64		
1 Rotary Hammer Drill		25.15		27.66		
1 Hammer Drill Bit		3.25		3.58		
1 Loader, Skid Steer, 30 H.P.		174.00		191.40		
1 Conc. Hammer Attach.		113.60		124.96		
1 Vibrating Plate, Gas, 18"		31.80		34.98		
2 Flatbed Trucks, Gas, 1.5 Ton		376.80		414.48	15.05	16.55
56 L.H., Daily Totals		$2564.60		$3773.26	$45.80	$67.38

Crew B-37F

	Hr.	Daily	Hr.	Daily	Bare Costs	Incl. O&P
3 Laborers	$26.70	$640.80	$44.40	$1065.60	$27.93	$46.30
1 Truck Driver (light)	31.60	252.80	52.00	416.00		
4 Barrels w/Flasher		15.60		17.16		
1 Concrete Mixer, 10 C.F.		161.00		177.10		
1 Air Compressor, 60 cfm		105.40		115.94		
1 -50' Air Hose, 3/4"		7.35		8.09		
1 Spade (Chipper)		8.40		9.24		
1 Flatbed Truck, Gas, 1.5 Ton		188.40		207.24	15.19	16.71
32 L.H., Daily Totals		$1379.75		$2016.37	$43.12	$63.01

Crew B-37G

	Hr.	Daily	Hr.	Daily	Bare Costs	Incl. O&P
1 Labor Foreman (outside)	$28.70	$229.60	$47.75	$382.00	$28.48	$47.28
4 Laborers	26.70	854.40	44.40	1420.80		
1 Equip. Oper. (light)	35.40	283.20	58.35	466.80		
1 Berm Machine		317.40		349.14		
1 Tandem Roller, 5 Ton		151.20		166.32	9.76	10.74
48 L.H., Daily Totals		$1835.80		$2785.06	$38.25	$58.02

Crew B-37H

	Hr.	Daily	Hr.	Daily	Bare Costs	Incl. O&P
1 Labor Foreman (outside)	$28.70	$229.60	$47.75	$382.00	$28.48	$47.28
4 Laborers	26.70	854.40	44.40	1420.80		
1 Equip. Oper. (light)	35.40	283.20	58.35	466.80		
1 Tandem Roller, 5 Ton		151.20		166.32		
1 Flatbed Truck, Gas, 1.5 Ton		188.40		207.24		
1 Tar Kettle, T.M.		129.20		142.12	9.77	10.74
48 L.H., Daily Totals		$1836.00		$2785.28	$38.25	$58.03

Crew B-37I

	Hr.	Daily	Hr.	Daily	Bare Costs	Incl. O&P
3 Laborers	$26.70	$640.80	$44.40	$1065.60	$30.75	$50.83
1 Equip. Oper. (light)	35.40	283.20	58.35	466.80		
1 Equip. Oper. (medium)	36.55	292.40	60.25	482.00		
2 Truck Drivers (light)	31.60	505.60	52.00	832.00		
4 Barrels w/Flasher		15.60		17.16		
1 Concrete Saw		102.40		112.64		
1 Rotary Hammer Drill		25.15		27.66		
1 Hammer Drill Bit		3.25		3.58		
1 Air Compressor, 60 cfm		105.40		115.94		
1 -50' Air Hose, 3/4"		7.35		8.09		
1 Spade (Chipper)		8.40		9.24		
1 Loader, Skid Steer, 30 H.P.		174.00		191.40		
1 Conc. Hammer Attach.		113.60		124.96		
1 Concrete Mixer, 10 C.F.		161.00		177.10		
1 Vibrating Plate, Gas, 18"		31.80		34.98		
2 Flatbed Trucks, Gas, 1.5 Ton		376.80		414.48	20.08	22.09
56 L.H., Daily Totals		$2846.75		$4083.63	$50.83	$72.92

Crew B-37J

	Hr.	Daily	Hr.	Daily	Bare Costs	Incl. O&P
1 Labor Foreman (outside)	$28.70	$229.60	$47.75	$382.00	$28.48	$47.28
4 Laborers	26.70	854.40	44.40	1420.80		
1 Equip. Oper. (light)	35.40	283.20	58.35	466.80		
1 Air Compressor, 60 cfm		105.40		115.94		
1 -50' Air Hose, 3/4"		7.35		8.09		
2 Concrete Mixers, 10 C.F.		322.00		354.20		
2 Flatbed Trucks, Gas, 1.5 Ton		376.80		414.48		
1 Shot Blaster, 20"		214.80		236.28	21.38	23.52
48 L.H., Daily Totals		$2393.55		$3398.59	$49.87	$70.80

Crew No.	Bare Costs		Incl. Subs O&P		Cost Per Labor-Hour	
Crew B-37K	**Hr.**	**Daily**	**Hr.**	**Daily**	**Bare Costs**	**Incl. O&P**
1 Labor Foreman (outside)	$28.70	$229.60	$47.75	$382.00	$28.48	$47.28
4 Laborers	26.70	854.40	44.40	1420.80		
1 Equip. Oper. (light)	35.40	283.20	58.35	466.80		
1 Air Compressor, 60 cfm		105.40		115.94		
1 -50' Air Hose, 3/4"		7.35		8.09		
2 Flatbed Trucks, Gas, 1.5 Ton		376.80		414.48		
1 Shot Blaster, 20"		214.80		236.28	14.67	16.14
48 L.H., Daily Totals		$2071.55		$3044.39	$43.16	$63.42
Crew B-38	**Hr.**	**Daily**	**Hr.**	**Daily**	**Bare Costs**	**Incl. O&P**
2 Laborers	$26.70	$427.20	$44.40	$710.40	$29.60	$49.05
1 Equip. Oper. (light)	35.40	283.20	58.35	466.80		
1 Backhoe Loader, 48 H.P.		312.40		343.64		
1 Hyd. Hammer (1200 lb.)		177.40		195.14		
1 F.E. Loader, W.M., 4 C.Y.		574.60		632.06		
1 Pvmt. Rem. Bucket		61.00		67.10	46.89	51.58
24 L.H., Daily Totals		$1835.80		$2415.14	$76.49	$100.63
Crew B-39	**Hr.**	**Daily**	**Hr.**	**Daily**	**Bare Costs**	**Incl. O&P**
1 Labor Foreman (outside)	$28.70	$229.60	$47.75	$382.00	$27.03	$44.96
5 Laborers	26.70	1068.00	44.40	1776.00		
1 Air Compressor, 250 cfm		167.40		184.14		
2 Breakers, Pavement, 60 lb.		21.20		23.32		
2 -50' Air Hoses, 1.5"		45.40		49.94	4.88	5.36
48 L.H., Daily Totals		$1531.60		$2415.40	$31.91	$50.32
Crew B-40	**Hr.**	**Daily**	**Hr.**	**Daily**	**Bare Costs**	**Incl. O&P**
1 Pile Driver Foreman (outside)	$37.40	$299.20	$64.20	$513.60	$34.68	$58.98
4 Pile Drivers	35.40	1132.80	60.80	1945.60		
1 Building Laborer	26.70	213.60	44.40	355.20		
1 Equip. Oper. (crane)	37.05	296.40	61.05	488.40		
1 Crawler Crane, 40 Ton		1370.00		1507.00		
1 Vibratory Hammer & Gen.		2191.00		2410.10	63.59	69.95
56 L.H., Daily Totals		$5503.00		$7219.90	$98.27	$128.93
Crew B-40B	**Hr.**	**Daily**	**Hr.**	**Daily**	**Bare Costs**	**Incl. O&P**
1 Labor Foreman (outside)	$28.70	$229.60	$47.75	$382.00	$29.73	$49.27
3 Laborers	26.70	640.80	44.40	1065.60		
1 Equip. Oper. (crane)	37.05	296.40	61.05	488.40		
1 Equip. Oper. (oiler)	32.55	260.40	53.65	429.20		
1 Lattice Boom Crane, 40 Ton		1321.00		1453.10	27.52	30.27
48 L.H., Daily Totals		$2748.20		$3818.30	$57.25	$79.55
Crew B-41	**Hr.**	**Daily**	**Hr.**	**Daily**	**Bare Costs**	**Incl. O&P**
1 Labor Foreman (outside)	$28.70	$229.60	$47.75	$382.00	$27.80	$46.19
4 Laborers	26.70	854.40	44.40	1420.80		
.25 Equip. Oper. (crane)	37.05	74.10	61.05	122.10		
.25 Equip. Oper. (oiler)	32.55	65.10	53.65	107.30		
.25 Crawler Crane, 40 Ton		342.50		376.75	7.78	8.56
44 L.H., Daily Totals		$1565.70		$2408.95	$35.58	$54.75
Crew B-42	**Hr.**	**Daily**	**Hr.**	**Daily**	**Bare Costs**	**Incl. O&P**
1 Labor Foreman (outside)	$28.70	$229.60	$47.75	$382.00	$30.34	$50.27
4 Laborers	26.70	854.40	44.40	1420.80		
1 Equip. Oper. (crane)	37.05	296.40	61.05	488.40		
1 Welder	39.80	318.40	65.50	524.00		
1 Hyd. Crane, 25 Ton		590.60		649.66		
1 Welder, Gas Engine, 300 amp		98.40		108.24		
1 Horz. Boring Csg. Mch.		448.80		493.68	20.32	22.35
56 L.H., Daily Totals		$2836.60		$4066.78	$50.65	$72.62

Crew No.	Bare Costs		Incl. Subs O&P		Cost Per Labor-Hour	
Crew B-43	**Hr.**	**Daily**	**Hr.**	**Daily**	**Bare Costs**	**Incl. O&P**
1 Labor Foreman (outside)	$28.70	$229.60	$47.75	$382.00	$27.10	$45.07
4 Laborers	26.70	854.40	44.40	1420.80		
1 Drill Rig, Truck-Mounted		2444.00		2688.40	61.10	67.21
40 L.H., Daily Totals		$3528.00		$4491.20	$88.20	$112.28
Crew B-44	**Hr.**	**Daily**	**Hr.**	**Daily**	**Bare Costs**	**Incl. O&P**
1 Pile Driver Foreman (outside)	$37.40	$299.20	$64.20	$513.60	$33.68	$57.16
4 Pile Drivers	35.40	1132.80	60.80	1945.60		
1 Equip. Oper. (crane)	37.05	296.40	61.05	488.40		
2 Laborers	26.70	427.20	44.40	710.40		
1 Crawler Crane, 40 Ton		1370.00		1507.00		
1 Lead, 60' High		79.40		87.34		
1 Hammer, Diesel, 15K ft.-lbs.		623.20		685.52	32.38	35.62
64 L.H., Daily Totals		$4228.20		$5937.86	$66.07	$92.78
Crew B-45	**Hr.**	**Daily**	**Hr.**	**Daily**	**Bare Costs**	**Incl. O&P**
1 Building Laborer	$26.70	$213.60	$44.40	$355.20	$29.45	$48.70
1 Truck Driver (heavy)	32.20	257.60	53.00	424.00		
1 Dist. Tanker, 3000 Gallon		326.80		359.48		
1 Truck Tractor, 6x4, 380 H.P.		476.20		523.82	50.19	55.21
16 L.H., Daily Totals		$1274.20		$1662.50	$79.64	$103.91
Crew B-46	**Hr.**	**Daily**	**Hr.**	**Daily**	**Bare Costs**	**Incl. O&P**
1 Pile Driver Foreman (outside)	$37.40	$299.20	$64.20	$513.60	$31.38	$53.17
2 Pile Drivers	35.40	566.40	60.80	972.80		
3 Laborers	26.70	640.80	44.40	1065.60		
1 Chain Saw, Gas, 36" Long		46.40		51.04	0.97	1.06
48 L.H., Daily Totals		$1552.80		$2603.04	$32.35	$54.23
Crew B-47	**Hr.**	**Daily**	**Hr.**	**Daily**	**Bare Costs**	**Incl. O&P**
1 Blast Foreman (outside)	$28.70	$229.60	$47.75	$382.00	$27.70	$46.08
1 Driller	26.70	213.60	44.40	355.20		
1 Air Track Drill, 4"		993.40		1092.74		
1 Air Compressor, 600 cfm		417.20		458.92		
2 -50' Air Hoses, 3"		86.40		95.04	93.56	102.92
16 L.H., Daily Totals		$1940.20		$2383.90	$121.26	$148.99
Crew B-47A	**Hr.**	**Daily**	**Hr.**	**Daily**	**Bare Costs**	**Incl. O&P**
1 Drilling Foreman (outside)	$28.70	$229.60	$47.75	$382.00	$32.77	$54.15
1 Equip. Oper. (heavy)	37.05	296.40	61.05	488.40		
1 Equip. Oper. (oiler)	32.55	260.40	53.65	429.20		
1 Air Track Drill, 5"		1173.00		1290.30	48.88	53.76
24 L.H., Daily Totals		$1959.40		$2589.90	$81.64	$107.91
Crew B-47C	**Hr.**	**Daily**	**Hr.**	**Daily**	**Bare Costs**	**Incl. O&P**
1 Laborer	$26.70	$213.60	$44.40	$355.20	$31.05	$51.38
1 Equip. Oper. (light)	35.40	283.20	58.35	466.80		
1 Air Compressor, 750 cfm		423.80		466.18		
2 -50' Air Hoses, 3"		86.40		95.04		
1 Air Track Drill, 4"		993.40		1092.74	93.97	103.37
16 L.H., Daily Totals		$2000.40		$2475.96	$125.03	$154.75
Crew B-47E	**Hr.**	**Daily**	**Hr.**	**Daily**	**Bare Costs**	**Incl. O&P**
1 Labor Foreman (outside)	$28.70	$229.60	$47.75	$382.00	$27.20	$45.24
3 Laborers	26.70	640.80	44.40	1065.60		
1 Flatbed Truck, Gas, 3 Ton		238.00		261.80	7.44	8.18
32 L.H., Daily Totals		$1108.40		$1709.40	$34.64	$53.42

For customer support on your Residential Costs with RSMeans data, call 800.448.8182.

Crew No.	Bare Costs		Incl. Subs O&P		Cost Per Labor-Hour	

Crew B-47G	Hr.	Daily	Hr.	Daily	Bare Costs	Incl. O&P
1 Labor Foreman (outside)	$28.70	$229.60	$47.75	$382.00	$27.37	$45.52
2 Laborers	26.70	427.20	44.40	710.40		
1 Air Track Drill, 4"		993.40		1092.74		
1 Air Compressor, 600 cfm		417.20		458.92		
2 -50' Air Hoses, 3"		86.40		95.04		
1 Gunite Pump Rig		321.00		353.10	75.75	83.33
24 L.H., Daily Totals		$2474.80		$3092.20	$103.12	$128.84

Crew B-47H	Hr.	Daily	Hr.	Daily	Bare Costs	Incl. O&P
1 Skilled Worker Foreman (out)	$37.80	$302.40	$63.15	$505.20	$36.30	$60.64
3 Skilled Workers	35.80	859.20	59.80	1435.20		
1 Flatbed Truck, Gas, 3 Ton		238.00		261.80	7.44	8.18
32 L.H., Daily Totals		$1399.60		$2202.20	$43.74	$68.82

Crew B-48	Hr.	Daily	Hr.	Daily	Bare Costs	Incl. O&P
1 Labor Foreman (outside)	$28.70	$229.60	$47.75	$382.00	$28.76	$47.73
4 Laborers	26.70	854.40	44.40	1420.80		
1 Equip. Oper. (crane)	37.05	296.40	61.05	488.40		
1 Centr. Water Pump, 6"		296.40		326.04		
1 -20' Suction Hose, 6"		11.50		12.65		
1 -50' Discharge Hose, 6"		6.10		6.71		
1 Drill Rig, Truck-Mounted		2444.00		2688.40	57.46	63.20
48 L.H., Daily Totals		$4138.40		$5325.00	$86.22	$110.94

Crew B-49	Hr.	Daily	Hr.	Daily	Bare Costs	Incl. O&P
1 Labor Foreman (outside)	$28.70	$229.60	$47.75	$382.00	$30.01	$50.27
5 Laborers	26.70	1068.00	44.40	1776.00		
1 Equip. Oper. (crane)	37.05	296.40	61.05	488.40		
2 Pile Drivers	35.40	566.40	60.80	972.80		
1 Hyd. Crane, 25 Ton		590.60		649.66		
1 Centr. Water Pump, 6"		296.40		326.04		
1 -20' Suction Hose, 6"		11.50		12.65		
1 -50' Discharge Hose, 6"		6.10		6.71		
1 Drill Rig, Truck-Mounted		2444.00		2688.40	46.51	51.16
72 L.H., Daily Totals		$5509.00		$7302.66	$76.51	$101.43

Crew B-50	Hr.	Daily	Hr.	Daily	Bare Costs	Incl. O&P
1 Pile Driver Foreman (outside)	$37.40	$299.20	$64.20	$513.60	$32.33	$54.77
6 Pile Drivers	35.40	1699.20	60.80	2918.40		
1 Equip. Oper. (crane)	37.05	296.40	61.05	488.40		
5 Laborers	26.70	1068.00	44.40	1776.00		
1 Crawler Crane, 40 Ton		1370.00		1507.00		
1 Lead, 60' High		79.40		87.34		
1 Hammer, Diesel, 15K ft.-lbs.		623.20		685.52		
1 Air Compressor, 600 cfm		417.20		458.92		
2 -50' Air Hoses, 3"		86.40		95.04		
1 Chain Saw, Gas, 36" Long		46.40		51.04	25.22	27.74
104 L.H., Daily Totals		$5985.40		$8581.26	$57.55	$82.51

Crew B-51	Hr.	Daily	Hr.	Daily	Bare Costs	Incl. O&P
1 Labor Foreman (outside)	$28.70	$229.60	$47.75	$382.00	$27.85	$46.23
4 Laborers	26.70	854.40	44.40	1420.80		
1 Truck Driver (light)	31.60	252.80	52.00	416.00		
1 Flatbed Truck, Gas, 1.5 Ton		188.40		207.24	3.92	4.32
48 L.H., Daily Totals		$1525.20		$2426.04	$31.77	$50.54

Crew B-52	Hr.	Daily	Hr.	Daily	Bare Costs	Incl. O&P
1 Labor Foreman (outside)	$28.70	$229.60	$47.75	$382.00	$29.73	$49.42
1 Carpenter	34.95	279.60	58.10	464.80		
4 Laborers	26.70	854.40	44.40	1420.80		
.5 Rodman (reinf.)	38.80	155.20	64.70	258.80		
.5 Equip. Oper. (medium)	36.55	146.20	60.25	241.00		
.5 Crawler Loader, 3 C.Y.		602.00		662.20	10.75	11.82
56 L.H., Daily Totals		$2267.00		$3429.60	$40.48	$61.24

Crew B-53	Hr.	Daily	Hr.	Daily	Bare Costs	Incl. O&P
1 Building Laborer	$26.70	$213.60	$44.40	$355.20	$26.70	$44.40
1 Trencher, Chain, 12 H.P.		61.80		67.98	7.72	8.50
8 L.H., Daily Totals		$275.40		$423.18	$34.42	$52.90

Crew B-54	Hr.	Daily	Hr.	Daily	Bare Costs	Incl. O&P
1 Equip. Oper. (light)	$35.40	$283.20	$58.35	$466.80	$35.40	$58.35
1 Trencher, Chain, 40 H.P.		336.40		370.04	42.05	46.26
8 L.H., Daily Totals		$619.60		$836.84	$77.45	$104.61

Crew B-54A	Hr.	Daily	Hr.	Daily	Bare Costs	Incl. O&P
.17 Labor Foreman (outside)	$28.70	$39.03	$47.75	$64.94	$35.41	$58.43
1 Equipment Operator (med.)	36.55	292.40	60.25	482.00		
1 Wheel Trencher, 67 H.P.		1095.00		1204.50	116.99	128.69
9.36 L.H., Daily Totals		$1426.43		$1751.44	$152.40	$187.12

Crew B-54B	Hr.	Daily	Hr.	Daily	Bare Costs	Incl. O&P
.25 Labor Foreman (outside)	$28.70	$57.40	$47.75	$95.50	$34.98	$57.75
1 Equipment Operator (med.)	36.55	292.40	60.25	482.00		
1 Wheel Trencher, 150 H.P.		1957.00		2152.70	195.70	215.27
10 L.H., Daily Totals		$2306.80		$2730.20	$230.68	$273.02

Crew B-54D	Hr.	Daily	Hr.	Daily	Bare Costs	Incl. O&P
1 Laborer	$26.70	$213.60	$44.40	$355.20	$31.63	$52.33
1 Equipment Operator (med.)	36.55	292.40	60.25	482.00		
1 Rock Trencher, 6" Width		904.20		994.62	56.51	62.16
16 L.H., Daily Totals		$1410.20		$1831.82	$88.14	$114.49

Crew B-54E	Hr.	Daily	Hr.	Daily	Bare Costs	Incl. O&P
1 Laborer	$26.70	$213.60	$44.40	$355.20	$31.63	$52.33
1 Equipment Operator (med.)	36.55	292.40	60.25	482.00		
1 Rock Trencher, 18" Width		2681.00		2949.10	167.56	184.32
16 L.H., Daily Totals		$3187.00		$3786.30	$199.19	$236.64

Crew B-55	Hr.	Daily	Hr.	Daily	Bare Costs	Incl. O&P
1 Laborer	$26.70	$213.60	$44.40	$355.20	$29.15	$48.20
1 Truck Driver (light)	31.60	252.80	52.00	416.00		
1 Truck-Mounted Earth Auger		758.80		834.68		
1 Flatbed Truck, Gas, 3 Ton		238.00		261.80	62.30	68.53
16 L.H., Daily Totals		$1463.20		$1867.68	$91.45	$116.73

Crew B-56	Hr.	Daily	Hr.	Daily	Bare Costs	Incl. O&P
2 Laborers	$26.70	$427.20	$44.40	$710.40	$26.70	$44.40
1 Air Track Drill, 4"		993.40		1092.74		
1 Air Compressor, 600 cfm		417.20		458.92		
1 -50' Air Hose, 3"		43.20		47.52	90.86	99.95
16 L.H., Daily Totals		$1881.00		$2309.58	$117.56	$144.35

For customer support on your Residential Costs with RSMeans data, call 800.448.8182.

695

Crew B-57

Crew No.	Bare Costs Hr.	Daily	Incl. Subs O&P Hr.	Daily	Cost Per Labor-Hour Bare Costs	Incl. O&P
1 Labor Foreman (outside)	$28.70	$229.60	$47.75	$382.00	$29.17	$48.40
3 Laborers	26.70	640.80	44.40	1065.60		
1 Equip. Oper. (crane)	37.05	296.40	61.05	488.40		
1 Crawler Crane, 25 Ton		1348.00		1482.80		
1 Clamshell Bucket, 1 C.Y.		50.60		55.66		
1 Centr. Water Pump, 6"		296.40		326.04		
1 -20' Suction Hose, 6"		11.50		12.65		
20 -50' Discharge Hoses, 6"		122.00		134.20	45.71	50.28
40 L.H., Daily Totals		$2995.30		$3947.35	$74.88	$98.68

Crew B-58

Crew No.	Bare Costs Hr.	Daily	Incl. Subs O&P Hr.	Daily	Bare Costs	Incl. O&P
2 Laborers	$26.70	$427.20	$44.40	$710.40	$29.60	$49.05
1 Equip. Oper. (light)	35.40	283.20	58.35	466.80		
1 Backhoe Loader, 48 H.P.		312.40		343.64		
1 Small Helicopter, w/ Pilot		2857.00		3142.70	132.06	145.26
24 L.H., Daily Totals		$3879.80		$4663.54	$161.66	$194.31

Crew B-59

Crew No.	Bare Costs Hr.	Daily	Incl. Subs O&P Hr.	Daily	Bare Costs	Incl. O&P
1 Truck Driver (heavy)	$32.20	$257.60	$53.00	$424.00	$32.20	$53.00
1 Truck Tractor, 220 H.P.		290.00		319.00		
1 Water Tank Trailer, 5000 Gal.		147.60		162.36	54.70	60.17
8 L.H., Daily Totals		$695.20		$905.36	$86.90	$113.17

Crew B-60

Crew No.	Bare Costs Hr.	Daily	Incl. Subs O&P Hr.	Daily	Bare Costs	Incl. O&P
1 Labor Foreman (outside)	$28.70	$229.60	$47.75	$382.00	$30.21	$50.06
3 Laborers	26.70	640.80	44.40	1065.60		
1 Equip. Oper. (crane)	37.05	296.40	61.05	488.40		
1 Equip. Oper. (light)	35.40	283.20	58.35	466.80		
1 Crawler Crane, 40 Ton		1370.00		1507.00		
1 Lead, 60' High		79.40		87.34		
1 Hammer, Diesel, 15K ft.-lbs.		623.20		685.52		
1 Backhoe Loader, 48 H.P.		312.40		343.64	49.69	54.66
48 L.H., Daily Totals		$3835.00		$5026.30	$79.90	$104.71

Crew B-61

Crew No.	Bare Costs Hr.	Daily	Incl. Subs O&P Hr.	Daily	Bare Costs	Incl. O&P
1 Labor Foreman (outside)	$28.70	$229.60	$47.75	$382.00	$27.10	$45.07
4 Laborers	26.70	854.40	44.40	1420.80		
1 Cement Mixer, 2 C.Y.		204.20		224.62		
1 Air Compressor, 160 cfm		117.20		128.92	8.04	8.84
40 L.H., Daily Totals		$1405.40		$2156.34	$35.13	$53.91

Crew B-62

Crew No.	Bare Costs Hr.	Daily	Incl. Subs O&P Hr.	Daily	Bare Costs	Incl. O&P
2 Laborers	$26.70	$427.20	$44.40	$710.40	$29.60	$49.05
1 Equip. Oper. (light)	35.40	283.20	58.35	466.80		
1 Loader, Skid Steer, 30 H.P.		174.00		191.40	7.25	7.97
24 L.H., Daily Totals		$884.40		$1368.60	$36.85	$57.02

Crew B-62A

Crew No.	Bare Costs Hr.	Daily	Incl. Subs O&P Hr.	Daily	Bare Costs	Incl. O&P
2 Laborers	$26.70	$427.20	$44.40	$710.40	$29.60	$49.05
1 Equip. Oper. (light)	35.40	283.20	58.35	466.80		
1 Loader, Skid Steer, 30 H.P.		174.00		191.40		
1 Trencher Attachment		74.30		81.73	10.35	11.38
24 L.H., Daily Totals		$958.70		$1450.33	$39.95	$60.43

Crew B-63

Crew No.	Bare Costs Hr.	Daily	Incl. Subs O&P Hr.	Daily	Bare Costs	Incl. O&P
5 Laborers	$26.70	$1068.00	$44.40	$1776.00	$26.70	$44.40
1 Loader, Skid Steer, 30 H.P.		174.00		191.40	4.35	4.79
40 L.H., Daily Totals		$1242.00		$1967.40	$31.05	$49.19

Crew B-63B

Crew No.	Bare Costs Hr.	Daily	Incl. Subs O&P Hr.	Daily	Bare Costs	Incl. O&P
1 Labor Foreman (inside)	$27.20	$217.60	$45.25	$362.00	$29.00	$48.10
2 Laborers	26.70	427.20	44.40	710.40		
1 Equip. Oper. (light)	35.40	283.20	58.35	466.80		
1 Loader, Skid Steer, 78 H.P.		364.80		401.28	11.40	12.54
32 L.H., Daily Totals		$1292.80		$1940.48	$40.40	$60.64

Crew B-64

Crew No.	Bare Costs Hr.	Daily	Incl. Subs O&P Hr.	Daily	Bare Costs	Incl. O&P
1 Laborer	$26.70	$213.60	$44.40	$355.20	$29.15	$48.20
1 Truck Driver (light)	31.60	252.80	52.00	416.00		
1 Power Mulcher (small)		139.20		153.12		
1 Flatbed Truck, Gas, 1.5 Ton		188.40		207.24	20.48	22.52
16 L.H., Daily Totals		$794.00		$1131.56	$49.63	$70.72

Crew B-65

Crew No.	Bare Costs Hr.	Daily	Incl. Subs O&P Hr.	Daily	Bare Costs	Incl. O&P
1 Laborer	$26.70	$213.60	$44.40	$355.20	$29.15	$48.20
1 Truck Driver (light)	31.60	252.80	52.00	416.00		
1 Power Mulcher (Large)		268.80		295.68		
1 Flatbed Truck, Gas, 1.5 Ton		188.40		207.24	28.57	31.43
16 L.H., Daily Totals		$923.60		$1274.12	$57.73	$79.63

Crew B-66

Crew No.	Bare Costs Hr.	Daily	Incl. Subs O&P Hr.	Daily	Bare Costs	Incl. O&P
1 Equip. Oper. (light)	$35.40	$283.20	$58.35	$466.80	$35.40	$58.35
1 Loader-Backhoe, 40 H.P.		247.00		271.70	30.88	33.96
8 L.H., Daily Totals		$530.20		$738.50	$66.28	$92.31

Crew B-67

Crew No.	Bare Costs Hr.	Daily	Incl. Subs O&P Hr.	Daily	Bare Costs	Incl. O&P
1 Millwright	$36.95	$295.60	$59.10	$472.80	$36.17	$58.73
1 Equip. Oper. (light)	35.40	283.20	58.35	466.80		
1 R.T. Forklift, 5,000 Lb., diesel		267.60		294.36	16.73	18.40
16 L.H., Daily Totals		$846.40		$1233.96	$52.90	$77.12

Crew B-67B

Crew No.	Bare Costs Hr.	Daily	Incl. Subs O&P Hr.	Daily	Bare Costs	Incl. O&P
1 Millwright Foreman (inside)	$37.45	$299.60	$59.90	$479.20	$37.20	$59.50
1 Millwright	36.95	295.60	59.10	472.80		
16 L.H., Daily Totals		$595.20		$952.00	$37.20	$59.50

Crew B-68

Crew No.	Bare Costs Hr.	Daily	Incl. Subs O&P Hr.	Daily	Bare Costs	Incl. O&P
2 Millwrights	$36.95	$591.20	$59.10	$945.60	$36.43	$58.85
1 Equip. Oper. (light)	35.40	283.20	58.35	466.80		
1 R.T. Forklift, 5,000 Lb., diesel		267.60		294.36	11.15	12.27
24 L.H., Daily Totals		$1142.00		$1706.76	$47.58	$71.11

Crew B-68A

Crew No.	Bare Costs Hr.	Daily	Incl. Subs O&P Hr.	Daily	Bare Costs	Incl. O&P
1 Millwright Foreman (inside)	$37.45	$299.60	$59.90	$479.20	$37.12	$59.37
2 Millwrights	36.95	591.20	59.10	945.60		
1 Forklift, Smooth Floor, 8,000 Lb.		148.20		163.02	6.17	6.79
24 L.H., Daily Totals		$1039.00		$1587.82	$43.29	$66.16

Crew B-68B

Crew No.	Bare Costs Hr.	Daily	Incl. Subs O&P Hr.	Daily	Bare Costs	Incl. O&P
1 Millwright Foreman (inside)	$37.45	$299.60	$59.90	$479.20	$38.75	$62.91
2 Millwrights	36.95	591.20	59.10	945.60		
2 Electricians	40.15	642.40	65.65	1050.40		
2 Plumbers	39.80	636.80	65.50	1048.00		
1 R.T. Forklift, 5,000 Lb., gas		279.00		306.90	4.98	5.48
56 L.H., Daily Totals		$2449.00		$3830.10	$43.73	$68.39

For customer support on your Residential Costs with RSMeans data, call 800.448.8182.

Crew No.	Bare Costs		Incl. Subs O&P		Cost Per Labor-Hour	

Crew B-68C	Hr.	Daily	Hr.	Daily	Bare Costs	Incl. O&P
1 Millwright Foreman (inside)	$37.45	$299.60	$59.90	$479.20	$38.59	$62.54
1 Millwright	36.95	295.60	59.10	472.80		
1 Electrician	40.15	321.20	65.65	525.20		
1 Plumber	39.80	318.40	65.50	524.00		
1 R.T. Forklift, 5,000 Lb., gas		279.00		306.90	8.72	9.59
32 L.H., Daily Totals		$1513.80		$2308.10	$47.31	$72.13

Crew B-68D	Hr.	Daily	Hr.	Daily	Bare Costs	Incl. O&P
1 Labor Foreman (inside)	$27.20	$217.60	$45.25	$362.00	$29.77	$49.33
1 Laborer	26.70	213.60	44.40	355.20		
1 Equip. Oper. (light)	35.40	283.20	58.35	466.80		
1 R.T. Forklift, 5,000 Lb., gas		279.00		306.90	11.63	12.79
24 L.H., Daily Totals		$993.40		$1490.90	$41.39	$62.12

Crew B-68E	Hr.	Daily	Hr.	Daily	Bare Costs	Incl. O&P
1 Struc. Steel Foreman (inside)	$39.30	$314.40	$69.75	$558.00	$38.90	$69.03
3 Struc. Steel Workers	38.80	931.20	68.85	1652.40		
1 Welder	38.80	310.40	68.85	550.80		
1 Forklift, Smooth Floor, 8,000 Lb.		148.20		163.02	3.71	4.08
40 L.H., Daily Totals		$1704.20		$2924.22	$42.60	$73.11

Crew B-68F	Hr.	Daily	Hr.	Daily	Bare Costs	Incl. O&P
1 Skilled Worker Foreman (out)	$37.80	$302.40	$63.15	$505.20	$36.47	$60.92
2 Skilled Workers	35.80	572.80	59.80	956.80		
1 R.T. Forklift, 5,000 Lb., gas		279.00		306.90	11.63	12.79
24 L.H., Daily Totals		$1154.20		$1768.90	$48.09	$73.70

Crew B-68G	Hr.	Daily	Hr.	Daily	Bare Costs	Incl. O&P
2 Structural Steel Workers	$38.80	$620.80	$68.85	$1101.60	$38.80	$68.85
1 R.T. Forklift, 5,000 Lb., gas		279.00		306.90	17.44	19.18
16 L.H., Daily Totals		$899.80		$1408.50	$56.24	$88.03

Crew B-69	Hr.	Daily	Hr.	Daily	Bare Costs	Incl. O&P
1 Labor Foreman (outside)	$28.70	$229.60	$47.75	$382.00	$29.73	$49.27
3 Laborers	26.70	640.80	44.40	1065.60		
1 Equip. Oper. (crane)	37.05	296.40	61.05	488.40		
1 Equip. Oper. (oiler)	32.55	260.40	53.65	429.20		
1 Hyd. Crane, 80 Ton		1518.00		1669.80	31.63	34.79
48 L.H., Daily Totals		$2945.20		$4035.00	$61.36	$84.06

Crew B-69A	Hr.	Daily	Hr.	Daily	Bare Costs	Incl. O&P
1 Labor Foreman (outside)	$28.70	$229.60	$47.75	$382.00	$30.01	$49.58
3 Laborers	26.70	640.80	44.40	1065.60		
1 Equip. Oper. (medium)	36.55	292.40	60.25	482.00		
1 Concrete Finisher	34.70	277.60	56.30	450.40		
1 Curb/Gutter Paver, 2-Track		1173.00		1290.30	24.44	26.88
48 L.H., Daily Totals		$2613.40		$3670.30	$54.45	$76.46

Crew B-69B	Hr.	Daily	Hr.	Daily	Bare Costs	Incl. O&P
1 Labor Foreman (outside)	$28.70	$229.60	$47.75	$382.00	$30.01	$49.58
3 Laborers	26.70	640.80	44.40	1065.60		
1 Equip. Oper. (medium)	36.55	292.40	60.25	482.00		
1 Cement Finisher	34.70	277.60	56.30	450.40		
1 Curb/Gutter Paver, 4-Track		789.60		868.56	16.45	18.09
48 L.H., Daily Totals		$2230.00		$3248.56	$46.46	$67.68

Crew B-70	Hr.	Daily	Hr.	Daily	Bare Costs	Incl. O&P
1 Labor Foreman (outside)	$28.70	$229.60	$47.75	$382.00	$31.21	$51.67
3 Laborers	26.70	640.80	44.40	1065.60		
3 Equip. Oper. (medium)	36.55	877.20	60.25	1446.00		
1 Grader, 30,000 Lbs.		643.80		708.18		
1 Ripper, Beam & 1 Shank		87.80		96.58		
1 Road Sweeper, S.P., 8' wide		688.60		757.46		
1 F.E. Loader, W.M., 1.5 C.Y.		342.80		377.08	31.48	34.63
56 L.H., Daily Totals		$3510.60		$4832.90	$62.69	$86.30

Crew B-71	Hr.	Daily	Hr.	Daily	Bare Costs	Incl. O&P
1 Labor Foreman (outside)	$28.70	$229.60	$47.75	$382.00	$31.21	$51.67
3 Laborers	26.70	640.80	44.40	1065.60		
3 Equip. Oper. (medium)	36.55	877.20	60.25	1446.00		
1 Pvmt. Profiler, 750 H.P.		5419.00		5960.90		
1 Road Sweeper, S.P., 8' wide		688.60		757.46		
1 F.E. Loader, W.M., 1.5 C.Y.		342.80		377.08	115.19	126.70
56 L.H., Daily Totals		$8198.00		$9989.04	$146.39	$178.38

Crew B-72	Hr.	Daily	Hr.	Daily	Bare Costs	Incl. O&P
1 Labor Foreman (outside)	$28.70	$229.60	$47.75	$382.00	$31.88	$52.74
3 Laborers	26.70	640.80	44.40	1065.60		
4 Equip. Oper. (medium)	36.55	1169.60	60.25	1928.00		
1 Pvmt. Profiler, 750 H.P.		5419.00		5960.90		
1 Hammermill, 250 H.P.		1870.00		2057.00		
1 Windrow Loader		1242.00		1366.20		
1 Mix Paver, 165 H.P.		2140.00		2354.00		
1 Roller, Pneum. Whl., 12 Ton		338.20		372.02	172.02	189.22
64 L.H., Daily Totals		$13049.20		$15485.72	$203.89	$241.96

Crew B-73	Hr.	Daily	Hr.	Daily	Bare Costs	Incl. O&P
1 Labor Foreman (outside)	$28.70	$229.60	$47.75	$382.00	$33.11	$54.73
2 Laborers	26.70	427.20	44.40	710.40		
5 Equip. Oper. (medium)	36.55	1462.00	60.25	2410.00		
1 Road Mixer, 310 H.P.		1889.00		2077.90		
1 Tandem Roller, 10 Ton		231.40		254.54		
1 Hammermill, 250 H.P.		1870.00		2057.00		
1 Grader, 30,000 Lbs.		643.80		708.18		
.5 F.E. Loader, W.M., 1.5 C.Y.		171.40		188.54		
.5 Truck Tractor, 220 H.P.		145.00		159.50		
.5 Water Tank Trailer, 5000 Gal.		73.80		81.18	78.51	86.36
64 L.H., Daily Totals		$7143.20		$9029.24	$111.61	$141.08

Crew B-74	Hr.	Daily	Hr.	Daily	Bare Costs	Incl. O&P
1 Labor Foreman (outside)	$28.70	$229.60	$47.75	$382.00	$33.25	$54.89
1 Laborer	26.70	213.60	44.40	355.20		
4 Equip. Oper. (medium)	36.55	1169.60	60.25	1928.00		
2 Truck Drivers (heavy)	32.20	515.20	53.00	848.00		
1 Grader, 30,000 Lbs.		643.80		708.18		
1 Ripper, Beam & 1 Shank		87.80		96.58		
2 Stabilizers, 310 H.P.		3436.00		3779.60		
1 Flatbed Truck, Gas, 3 Ton		238.00		261.80		
1 Chem. Spreader, Towed		52.20		57.42		
1 Roller, Vibratory, 25 Ton		659.80		725.78		
1 Water Tank Trailer, 5000 Gal.		147.60		162.36		
1 Truck Tractor, 220 H.P.		290.00		319.00	86.80	95.48
64 L.H., Daily Totals		$7683.20		$9623.92	$120.05	$150.37

Crew No.	Bare Costs		Incl. Subs O&P		Cost Per Labor-Hour	
Crew B-75	Hr.	Daily	Hr.	Daily	Bare Costs	Incl. O&P
1 Labor Foreman (outside)	$28.70	$229.60	$47.75	$382.00	$33.40	$55.16
1 Laborer	26.70	213.60	44.40	355.20		
4 Equip. Oper. (medium)	36.55	1169.60	60.25	1928.00		
1 Truck Driver (heavy)	32.20	257.60	53.00	424.00		
1 Grader, 30,000 Lbs.		643.80		708.18		
1 Ripper, Beam & 1 Shank		87.80		96.58		
2 Stabilizers, 310 H.P.		3436.00		3779.60		
1 Dist. Tanker, 3000 Gallon		326.80		359.48		
1 Truck Tractor, 6x4, 380 H.P.		476.20		523.82		
1 Roller, Vibratory, 25 Ton		659.80		725.78	100.54	110.60
56 L.H., Daily Totals		$7500.80		$9282.64	$133.94	$165.76

Crew No.	Bare Costs		Incl. Subs O&P		Cost Per Labor-Hour	
Crew B-76	Hr.	Daily	Hr.	Daily	Bare Costs	Incl. O&P
1 Dock Builder Foreman (outside)	$37.40	$299.20	$64.20	$513.60	$35.67	$60.44
5 Dock Builders	35.40	1416.00	60.80	2432.00		
2 Equip. Oper. (crane)	37.05	592.80	61.05	976.80		
1 Equip. Oper. (oiler)	32.55	260.40	53.65	429.20		
1 Crawler Crane, 50 Ton		1940.00		2134.00		
1 Barge, 400 Ton		835.40		918.94		
1 Hammer, Diesel, 15K ft.-lbs.		623.20		685.52		
1 Lead, 60' High		79.40		87.34		
1 Air Compressor, 600 cfm		417.20		458.92		
2 -50' Air Hoses, 3"		86.40		95.04	55.30	60.83
72 L.H., Daily Totals		$6550.00		$8731.36	$90.97	$121.27

Crew No.	Bare Costs		Incl. Subs O&P		Cost Per Labor-Hour	
Crew B-76A	Hr.	Daily	Hr.	Daily	Bare Costs	Incl. O&P
1 Labor Foreman (outside)	$28.70	$229.60	$47.75	$382.00	$28.98	$48.06
5 Laborers	26.70	1068.00	44.40	1776.00		
1 Equip. Oper. (crane)	37.05	296.40	61.05	488.40		
1 Equip. Oper. (oiler)	32.55	260.40	53.65	429.20		
1 Crawler Crane, 50 Ton		1940.00		2134.00		
1 Barge, 400 Ton		835.40		918.94	43.37	47.70
64 L.H., Daily Totals		$4629.80		$6128.54	$72.34	$95.76

Crew No.	Bare Costs		Incl. Subs O&P		Cost Per Labor-Hour	
Crew B-77	Hr.	Daily	Hr.	Daily	Bare Costs	Incl. O&P
1 Labor Foreman (outside)	$28.70	$229.60	$47.75	$382.00	$28.08	$46.59
3 Laborers	26.70	640.80	44.40	1065.60		
1 Truck Driver (light)	31.60	252.80	52.00	416.00		
1 Crack Cleaner, 25 H.P.		55.80		61.38		
1 Crack Filler, Trailer Mtd.		198.20		218.02		
1 Flatbed Truck, Gas, 3 Ton		238.00		261.80	12.30	13.53
40 L.H., Daily Totals		$1615.20		$2404.80	$40.38	$60.12

Crew No.	Bare Costs		Incl. Subs O&P		Cost Per Labor-Hour	
Crew B-78	Hr.	Daily	Hr.	Daily	Bare Costs	Incl. O&P
1 Labor Foreman (outside)	$28.70	$229.60	$47.75	$382.00	$27.10	$45.07
4 Laborers	26.70	854.40	44.40	1420.80		
1 Paint Striper, S.P., 40 Gallon		151.00		166.10		
1 Flatbed Truck, Gas, 3 Ton		238.00		261.80		
1 Pickup Truck, 3/4 Ton		115.20		126.72	12.61	13.87
40 L.H., Daily Totals		$1588.20		$2357.42	$39.70	$58.94

Crew No.	Bare Costs		Incl. Subs O&P		Cost Per Labor-Hour	
Crew B-78B	Hr.	Daily	Hr.	Daily	Bare Costs	Incl. O&P
2 Laborers	$26.70	$427.20	$44.40	$710.40	$27.67	$45.95
.25 Equip. Oper. (light)	35.40	70.80	58.35	116.70		
1 Pickup Truck, 3/4 Ton		115.20		126.72		
1 Line Rem.,11 H.P.,Walk Behind		61.40		67.54		
.25 Road Sweeper, S.P., 8' wide		172.15		189.37	19.38	21.31
18 L.H., Daily Totals		$846.75		$1210.72	$47.04	$67.26

Crew No.	Bare Costs		Incl. Subs O&P		Cost Per Labor-Hour	
Crew B-78C	Hr.	Daily	Hr.	Daily	Bare Costs	Incl. O&P
1 Labor Foreman (outside)	$28.70	$229.60	$47.75	$382.00	$27.85	$46.23
4 Laborers	26.70	854.40	44.40	1420.80		
1 Truck Driver (light)	31.60	252.80	52.00	416.00		
1 Paint Striper, T.M., 120 Gal.		690.00		759.00		
1 Flatbed Truck, Gas, 3 Ton		238.00		261.80		
1 Pickup Truck, 3/4 Ton		115.20		126.72	21.73	23.91
48 L.H., Daily Totals		$2380.00		$3366.32	$49.58	$70.13

Crew No.	Bare Costs		Incl. Subs O&P		Cost Per Labor-Hour	
Crew B-78D	Hr.	Daily	Hr.	Daily	Bare Costs	Incl. O&P
2 Labor Foremen (outside)	$28.70	$459.20	$47.75	$764.00	$27.59	$45.83
7 Laborers	26.70	1495.20	44.40	2486.40		
1 Truck Driver (light)	31.60	252.80	52.00	416.00		
1 Paint Striper, T.M., 120 Gal.		690.00		759.00		
1 Flatbed Truck, Gas, 3 Ton		238.00		261.80		
3 Pickup Trucks, 3/4 Ton		345.60		380.16		
1 Air Compressor, 60 cfm		105.40		115.94		
1 -50' Air Hose, 3/4"		7.35		8.09		
1 Breaker, Pavement, 60 lb.		10.60		11.66	17.46	19.21
80 L.H., Daily Totals		$3604.15		$5203.05	$45.05	$65.04

Crew No.	Bare Costs		Incl. Subs O&P		Cost Per Labor-Hour	
Crew B-78E	Hr.	Daily	Hr.	Daily	Bare Costs	Incl. O&P
2 Labor Foremen (outside)	$28.70	$459.20	$47.75	$764.00	$27.44	$45.59
9 Laborers	26.70	1922.40	44.40	3196.80		
1 Truck Driver (light)	31.60	252.80	52.00	416.00		
1 Paint Striper, T.M., 120 Gal.		690.00		759.00		
1 Flatbed Truck, Gas, 3 Ton		238.00		261.80		
4 Pickup Trucks, 3/4 Ton		460.80		506.88		
2 Air Compressors, 60 cfm		210.80		231.88		
2 -50' Air Hoses, 3/4"		14.70		16.17		
2 Breakers, Pavement, 60 lb.		21.20		23.32	17.04	18.74
96 L.H., Daily Totals		$4269.90		$6175.85	$44.48	$64.33

Crew No.	Bare Costs		Incl. Subs O&P		Cost Per Labor-Hour	
Crew B-78F	Hr.	Daily	Hr.	Daily	Bare Costs	Incl. O&P
2 Labor Foremen (outside)	$28.70	$459.20	$47.75	$764.00	$27.34	$45.42
11 Laborers	26.70	2349.60	44.40	3907.20		
1 Truck Driver (light)	31.60	252.80	52.00	416.00		
1 Paint Striper, T.M., 120 Gal.		690.00		759.00		
1 Flatbed Truck, Gas, 3 Ton		238.00		261.80		
7 Pickup Trucks, 3/4 Ton		806.40		887.04		
3 Air Compressors, 60 cfm		316.20		347.82		
3 -50' Air Hoses, 3/4"		22.05		24.25		
3 Breakers, Pavement, 60 lb.		31.80		34.98	18.79	20.67
112 L.H., Daily Totals		$5166.05		$7402.10	$46.13	$66.09

Crew No.	Bare Costs		Incl. Subs O&P		Cost Per Labor-Hour	
Crew B-79	Hr.	Daily	Hr.	Daily	Bare Costs	Incl. O&P
1 Labor Foreman (outside)	$28.70	$229.60	$47.75	$382.00	$28.08	$46.59
3 Laborers	26.70	640.80	44.40	1065.60		
1 Truck Driver (light)	31.60	252.80	52.00	416.00		
1 Paint Striper, T.M., 120 Gal.		690.00		759.00		
1 Heating Kettle, 115 Gallon		77.85		85.64		
1 Flatbed Truck, Gas, 3 Ton		238.00		261.80		
2 Pickup Trucks, 3/4 Ton		230.40		253.44	30.91	34.00
40 L.H., Daily Totals		$2359.45		$3223.47	$58.99	$80.59

Crew No.	Bare Costs		Incl. Subs O&P		Cost Per Labor-Hour	
Crew B-79B	Hr.	Daily	Hr.	Daily	Bare Costs	Incl. O&P
1 Laborer	$26.70	$213.60	$44.40	$355.20	$26.70	$44.40
1 Set of Gases		168.00		184.80	21.00	23.10
8 L.H., Daily Totals		$381.60		$540.00	$47.70	$67.50

For customer support on your Residential Costs with RSMeans data, call 800.448.8182.

Crew No.	Bare Costs		Incl. Subs O&P		Cost Per Labor-Hour	
Crew B-79C	Hr.	Daily	Hr.	Daily	Bare Costs	Incl. O&P
1 Labor Foreman (outside)	$28.70	$229.60	$47.75	$382.00	$27.69	$45.96
5 Laborers	26.70	1068.00	44.40	1776.00		
1 Truck Driver (light)	31.60	252.80	52.00	416.00		
1 Paint Striper, T.M., 120 Gal.		690.00		759.00		
1 Heating Kettle, 115 Gallon		77.85		85.64		
1 Flatbed Truck, Gas, 3 Ton		238.00		261.80		
3 Pickup Trucks, 3/4 Ton		345.60		380.16		
1 Air Compressor, 60 cfm		105.40		115.94		
1 -50' Air Hose, 3/4"		7.35		8.09		
1 Breaker, Pavement, 60 lb.		10.60		11.66	26.34	28.97
56 L.H., Daily Totals		$3025.20		$4196.28	$54.02	$74.93

Crew No.	Bare Costs		Incl. Subs O&P		Cost Per Labor-Hour	
Crew B-79D	Hr.	Daily	Hr.	Daily	Bare Costs	Incl. O&P
2 Labor Foremen (outside)	$28.70	$459.20	$47.75	$764.00	$27.81	$46.19
5 Laborers	26.70	1068.00	44.40	1776.00		
1 Truck Driver (light)	31.60	252.80	52.00	416.00		
1 Paint Striper, T.M., 120 Gal.		690.00		759.00		
1 Heating Kettle, 115 Gallon		77.85		85.64		
1 Flatbed Truck, Gas, 3 Ton		238.00		261.80		
4 Pickup Trucks, 3/4 Ton		460.80		506.88		
1 Air Compressor, 60 cfm		105.40		115.94		
1 -50' Air Hose, 3/4"		7.35		8.09		
1 Breaker, Pavement, 60 lb.		10.60		11.66	24.84	27.33
64 L.H., Daily Totals		$3370.00		$4705.00	$52.66	$73.52

Crew No.	Bare Costs		Incl. Subs O&P		Cost Per Labor-Hour	
Crew B-79E	Hr.	Daily	Hr.	Daily	Bare Costs	Incl. O&P
2 Labor Foremen (outside)	$28.70	$459.20	$47.75	$764.00	$27.59	$45.83
7 Laborers	26.70	1495.20	44.40	2486.40		
1 Truck Driver (light)	31.60	252.80	52.00	416.00		
1 Paint Striper, T.M., 120 Gal.		690.00		759.00		
1 Heating Kettle, 115 Gallon		77.85		85.64		
1 Flatbed Truck, Gas, 3 Ton		238.00		261.80		
5 Pickup Trucks, 3/4 Ton		576.00		633.60		
2 Air Compressors, 60 cfm		210.80		231.88		
2 -50' Air Hoses, 3/4"		14.70		16.17		
2 Breakers, Pavement, 60 lb.		21.20		23.32	22.86	25.14
80 L.H., Daily Totals		$4035.75		$5677.81	$50.45	$70.97

Crew No.	Bare Costs		Incl. Subs O&P		Cost Per Labor-Hour	
Crew B-80	Hr.	Daily	Hr.	Daily	Bare Costs	Incl. O&P
1 Labor Foreman (outside)	$28.70	$229.60	$47.75	$382.00	$27.37	$45.52
2 Laborers	26.70	427.20	44.40	710.40		
1 Flatbed Truck, Gas, 3 Ton		238.00		261.80		
1 Earth Auger, Truck-Mtd.		373.80		411.18	25.49	28.04
24 L.H., Daily Totals		$1268.60		$1765.38	$52.86	$73.56

Crew No.	Bare Costs		Incl. Subs O&P		Cost Per Labor-Hour	
Crew B-80A	Hr.	Daily	Hr.	Daily	Bare Costs	Incl. O&P
3 Laborers	$26.70	$640.80	$44.40	$1065.60	$26.70	$44.40
1 Flatbed Truck, Gas, 3 Ton		238.00		261.80	9.92	10.91
24 L.H., Daily Totals		$878.80		$1327.40	$36.62	$55.31

Crew No.	Bare Costs		Incl. Subs O&P		Cost Per Labor-Hour	
Crew B-80B	Hr.	Daily	Hr.	Daily	Bare Costs	Incl. O&P
3 Laborers	$26.70	$640.80	$44.40	$1065.60	$28.88	$47.89
1 Equip. Oper. (light)	35.40	283.20	58.35	466.80		
1 Crane, Flatbed Mounted, 3 Ton		235.80		259.38	7.37	8.11
32 L.H., Daily Totals		$1159.80		$1791.78	$36.24	$55.99

Crew No.	Bare Costs		Incl. Subs O&P		Cost Per Labor-Hour	
Crew B-80C	Hr.	Daily	Hr.	Daily	Bare Costs	Incl. O&P
2 Laborers	$26.70	$427.20	$44.40	$710.40	$28.33	$46.93
1 Truck Driver (light)	31.60	252.80	52.00	416.00		
1 Flatbed Truck, Gas, 1.5 Ton		188.40		207.24		
1 Manual Fence Post Auger, Gas		7.40		8.14	8.16	8.97
24 L.H., Daily Totals		$875.80		$1341.78	$36.49	$55.91

Crew No.	Bare Costs		Incl. Subs O&P		Cost Per Labor-Hour	
Crew B-81	Hr.	Daily	Hr.	Daily	Bare Costs	Incl. O&P
1 Laborer	$26.70	$213.60	$44.40	$355.20	$29.45	$48.70
1 Truck Driver (heavy)	32.20	257.60	53.00	424.00		
1 Hydromulcher, T.M., 3000 Gal.		282.80		311.08		
1 Truck Tractor, 220 H.P.		290.00		319.00	35.80	39.38
16 L.H., Daily Totals		$1044.00		$1409.28	$65.25	$88.08

Crew No.	Bare Costs		Incl. Subs O&P		Cost Per Labor-Hour	
Crew B-81A	Hr.	Daily	Hr.	Daily	Bare Costs	Incl. O&P
1 Laborer	$26.70	$213.60	$44.40	$355.20	$29.15	$48.20
1 Truck Driver (light)	31.60	252.80	52.00	416.00		
1 Hydromulcher, T.M., 600 Gal.		121.20		133.32		
1 Flatbed Truck, Gas, 3 Ton		238.00		261.80	22.45	24.70
16 L.H., Daily Totals		$825.60		$1166.32	$51.60	$72.89

Crew No.	Bare Costs		Incl. Subs O&P		Cost Per Labor-Hour	
Crew B-82	Hr.	Daily	Hr.	Daily	Bare Costs	Incl. O&P
1 Laborer	$26.70	$213.60	$44.40	$355.20	$31.05	$51.38
1 Equip. Oper. (light)	35.40	283.20	58.35	466.80		
1 Horiz. Borer, 6 H.P.		78.00		85.80	4.88	5.36
16 L.H., Daily Totals		$574.80		$907.80	$35.92	$56.74

Crew No.	Bare Costs		Incl. Subs O&P		Cost Per Labor-Hour	
Crew B-82A	Hr.	Daily	Hr.	Daily	Bare Costs	Incl. O&P
2 Laborers	$26.70	$427.20	$44.40	$710.40	$31.05	$51.38
2 Equip. Opers. (light)	35.40	566.40	58.35	933.60		
2 Dump Trucks, 8 C.Y., 220 H.P.		677.20		744.92		
1 Flatbed Trailer, 25 Ton		133.00		146.30		
1 Horiz. Dir. Drill, 20k lb. Thrust		629.00		691.90		
1 Mud Trailer for HDD, 1500 Gal.		287.80		316.58		
1 Pickup Truck, 4x4, 3/4 Ton		126.60		139.26		
1 Flatbed Trailer, 3 Ton		25.40		27.94		
1 Loader, Skid Steer, 78 H.P.		364.80		401.28	70.12	77.13
32 L.H., Daily Totals		$3237.40		$4112.18	$101.17	$128.51

Crew No.	Bare Costs		Incl. Subs O&P		Cost Per Labor-Hour	
Crew B-82B	Hr.	Daily	Hr.	Daily	Bare Costs	Incl. O&P
2 Laborers	$26.70	$427.20	$44.40	$710.40	$31.05	$51.38
2 Equip. Opers. (light)	35.40	566.40	58.35	933.60		
2 Dump Trucks, 8 C.Y., 220 H.P.		677.20		744.92		
1 Flatbed Trailer, 25 Ton		133.00		146.30		
1 Horiz. Dir. Drill, 30k lb. Thrust		896.20		985.82		
1 Mud Trailer for HDD, 1500 Gal.		287.80		316.58		
1 Pickup Truck, 4x4, 3/4 Ton		126.60		139.26		
1 Flatbed Trailer, 3 Ton		25.40		27.94		
1 Loader, Skid Steer, 78 H.P.		364.80		401.28	78.47	86.32
32 L.H., Daily Totals		$3504.60		$4406.10	$109.52	$137.69

Crew No.	Bare Costs		Incl. Subs O&P		Cost Per Labor-Hour	
Crew B-82C	Hr.	Daily	Hr.	Daily	Bare Costs	Incl. O&P
2 Laborers	$26.70	$427.20	$44.40	$710.40	$31.05	$51.38
2 Equip. Opers. (light)	35.40	566.40	58.35	933.60		
2 Dump Trucks, 8 C.Y., 220 H.P.		677.20		744.92		
1 Flatbed Trailer, 25 Ton		133.00		146.30		
1 Horiz. Dir. Drill, 50k lb. Thrust		1188.00		1306.80		
1 Mud Trailer for HDD, 1500 Gal.		287.80		316.58		
1 Pickup Truck, 4x4, 3/4 Ton		126.60		139.26		
1 Flatbed Trailer, 3 Ton		25.40		27.94		
1 Loader, Skid Steer, 78 H.P.		364.80		401.28	87.59	96.35
32 L.H., Daily Totals		$3796.40		$4727.08	$118.64	$147.72

Crew No.	Bare Costs Hr.	Daily	Incl. Subs O&P Hr.	Daily	Cost Per Labor-Hour Bare Costs	Incl. O&P
Crew B-82D	Hr.	Daily	Hr.	Daily	Bare Costs	Incl. O&P
1 Equip. Oper. (light)	$35.40	$283.20	$58.35	$466.80	$35.40	$58.35
1 Mud Trailer for HDD, 1500 Gal.		287.80		316.58	35.98	39.57
8 L.H., Daily Totals		$571.00		$783.38	$71.38	$97.92
Crew B-83	Hr.	Daily	Hr.	Daily	Bare Costs	Incl. O&P
1 Tugboat Captain	$36.55	$292.40	$60.25	$482.00	$31.63	$52.33
1 Tugboat Hand	26.70	213.60	44.40	355.20		
1 Tugboat, 250 H.P.		680.20		748.22	42.51	46.76
16 L.H., Daily Totals		$1186.20		$1585.42	$74.14	$99.09
Crew B-84	Hr.	Daily	Hr.	Daily	Bare Costs	Incl. O&P
1 Equip. Oper. (medium)	$36.55	$292.40	$60.25	$482.00	$36.55	$60.25
1 Rotary Mower/Tractor		357.00		392.70	44.63	49.09
8 L.H., Daily Totals		$649.40		$874.70	$81.17	$109.34
Crew B-85	Hr.	Daily	Hr.	Daily	Bare Costs	Incl. O&P
3 Laborers	$26.70	$640.80	$44.40	$1065.60	$29.77	$49.29
1 Equip. Oper. (medium)	36.55	292.40	60.25	482.00		
1 Truck Driver (heavy)	32.20	257.60	53.00	424.00		
1 Telescoping Boom Lift, to 80'		548.60		603.46		
1 Brush Chipper, 12", 130 H.P.		393.00		432.30		
1 Pruning Saw, Rotary		6.65		7.32	23.71	26.08
40 L.H., Daily Totals		$2139.05		$3014.68	$53.48	$75.37
Crew B-86	Hr.	Daily	Hr.	Daily	Bare Costs	Incl. O&P
1 Equip. Oper. (medium)	$36.55	$292.40	$60.25	$482.00	$36.55	$60.25
1 Stump Chipper, S.P.		184.05		202.46	23.01	25.31
8 L.H., Daily Totals		$476.45		$684.46	$59.56	$85.56
Crew B-86A	Hr.	Daily	Hr.	Daily	Bare Costs	Incl. O&P
1 Equip. Oper. (medium)	$36.55	$292.40	$60.25	$482.00	$36.55	$60.25
1 Grader, 30,000 Lbs.		643.80		708.18	80.47	88.52
8 L.H., Daily Totals		$936.20		$1190.18	$117.03	$148.77
Crew B-86B	Hr.	Daily	Hr.	Daily	Bare Costs	Incl. O&P
1 Equip. Oper. (medium)	$36.55	$292.40	$60.25	$482.00	$36.55	$60.25
1 Dozer, 200 H.P.		1273.00		1400.30	159.13	175.04
8 L.H., Daily Totals		$1565.40		$1882.30	$195.68	$235.29
Crew B-87	Hr.	Daily	Hr.	Daily	Bare Costs	Incl. O&P
1 Laborer	$26.70	$213.60	$44.40	$355.20	$34.58	$57.08
4 Equip. Oper. (medium)	36.55	1169.60	60.25	1928.00		
2 Feller Bunchers, 100 H.P.		1620.80		1782.88		
1 Log Chipper, 22" Tree		764.80		841.28		
1 Dozer, 105 H.P.		608.00		668.80		
1 Chain Saw, Gas, 36" Long		46.40		51.04	76.00	83.60
40 L.H., Daily Totals		$4423.20		$5627.20	$110.58	$140.68
Crew B-88	Hr.	Daily	Hr.	Daily	Bare Costs	Incl. O&P
1 Laborer	$26.70	$213.60	$44.40	$355.20	$35.14	$57.99
6 Equip. Oper. (medium)	36.55	1754.40	60.25	2892.00		
2 Feller Bunchers, 100 H.P.		1620.80		1782.88		
1 Log Chipper, 22" Tree		764.80		841.28		
2 Log Skidders, 50 H.P.		1698.80		1868.68		
1 Dozer, 105 H.P.		608.00		668.80		
1 Chain Saw, Gas, 36" Long		46.40		51.04	84.62	93.08
56 L.H., Daily Totals		$6706.80		$8459.88	$119.76	$151.07

Crew No.	Bare Costs Hr.	Daily	Incl. Subs O&P Hr.	Daily	Cost Per Labor-Hour Bare Costs	Incl. O&P
Crew B-89	Hr.	Daily	Hr.	Daily	Bare Costs	Incl. O&P
1 Skilled Worker	$35.80	$286.40	$59.80	$478.40	$31.25	$52.10
1 Building Laborer	26.70	213.60	44.40	355.20		
1 Flatbed Truck, Gas, 3 Ton		238.00		261.80		
1 Concrete Saw		102.40		112.64		
1 Water Tank, 65 Gal.		79.90		87.89	26.27	28.90
16 L.H., Daily Totals		$920.30		$1295.93	$57.52	$81.00
Crew B-89A	Hr.	Daily	Hr.	Daily	Bare Costs	Incl. O&P
1 Skilled Worker	$35.80	$286.40	$59.80	$478.40	$31.25	$52.10
1 Laborer	26.70	213.60	44.40	355.20		
1 Core Drill (Large)		112.20		123.42	7.01	7.71
16 L.H., Daily Totals		$612.20		$957.02	$38.26	$59.81
Crew B-89B	Hr.	Daily	Hr.	Daily	Bare Costs	Incl. O&P
1 Equip. Oper. (light)	$35.40	$283.20	$58.35	$466.80	$33.50	$55.17
1 Truck Driver (light)	31.60	252.80	52.00	416.00		
1 Wall Saw, Hydraulic, 10 H.P.		46.40		51.04		
1 Generator, Diesel, 100 kW		307.60		338.36		
1 Water Tank, 65 Gal.		79.90		87.89		
1 Flatbed Truck, Gas, 3 Ton		238.00		261.80	41.99	46.19
16 L.H., Daily Totals		$1207.90		$1621.89	$75.49	$101.37
Crew B-90	Hr.	Daily	Hr.	Daily	Bare Costs	Incl. O&P
1 Labor Foreman (outside)	$28.70	$229.60	$47.75	$382.00	$30.50	$50.46
3 Laborers	26.70	640.80	44.40	1065.60		
2 Equip. Oper. (light)	35.40	566.40	58.35	933.60		
2 Truck Drivers (heavy)	32.20	515.20	53.00	848.00		
1 Road Mixer, 310 H.P.		1889.00		2077.90		
1 Dist. Truck, 2000 Gal.		296.00		325.60	34.14	37.55
64 L.H., Daily Totals		$4137.00		$5632.70	$64.64	$88.01
Crew B-90A	Hr.	Daily	Hr.	Daily	Bare Costs	Incl. O&P
1 Labor Foreman (outside)	$28.70	$229.60	$47.75	$382.00	$32.61	$53.94
2 Laborers	26.70	427.20	44.40	710.40		
4 Equip. Oper. (medium)	36.55	1169.60	60.25	1928.00		
2 Graders, 30,000 Lbs.		1287.60		1416.36		
1 Tandem Roller, 10 Ton		231.40		254.54		
1 Roller, Pneum. Whl., 12 Ton		338.20		372.02	33.16	36.48
56 L.H., Daily Totals		$3683.60		$5063.32	$65.78	$90.42
Crew B-90B	Hr.	Daily	Hr.	Daily	Bare Costs	Incl. O&P
1 Labor Foreman (outside)	$28.70	$229.60	$47.75	$382.00	$31.96	$52.88
2 Laborers	26.70	427.20	44.40	710.40		
3 Equip. Oper. (medium)	36.55	877.20	60.25	1446.00		
1 Roller, Pneum. Whl., 12 Ton		338.20		372.02		
1 Road Mixer, 310 H.P.		1889.00		2077.90	46.40	51.04
48 L.H., Daily Totals		$3761.20		$4988.32	$78.36	$103.92
Crew B-90C	Hr.	Daily	Hr.	Daily	Bare Costs	Incl. O&P
1 Labor Foreman (outside)	$28.70	$229.60	$47.75	$382.00	$31.07	$51.37
4 Laborers	26.70	854.40	44.40	1420.80		
3 Equip. Oper. (medium)	36.55	877.20	60.25	1446.00		
3 Truck Drivers (heavy)	32.20	772.80	53.00	1272.00		
3 Road Mixers, 310 H.P.		5667.00		6233.70	64.40	70.84
88 L.H., Daily Totals		$8401.00		$10754.50	$95.47	$122.21

Crew No.	Bare Costs		Incl. Subs O&P		Cost Per Labor-Hour	
Crew B-90D	Hr.	Daily	Hr.	Daily	Bare Costs	Incl. O&P
1 Labor Foreman (outside)	$28.70	$229.60	$47.75	$382.00	$30.40	$50.30
6 Laborers	26.70	1281.60	44.40	2131.20		
3 Equip. Oper. (medium)	36.55	877.20	60.25	1446.00		
3 Truck Drivers (heavy)	32.20	772.80	53.00	1272.00		
3 Road Mixers, 310 H.P.		5667.00		6233.70	54.49	59.94
104 L.H., Daily Totals		$8828.20		$11464.90	$84.89	$110.24
Crew B-90E	Hr.	Daily	Hr.	Daily	Bare Costs	Incl. O&P
1 Labor Foreman (outside)	$28.70	$229.60	$47.75	$382.00	$30.82	$51.01
4 Laborers	26.70	854.40	44.40	1420.80		
3 Equip. Oper. (medium)	36.55	877.20	60.25	1446.00		
1 Truck Driver (heavy)	32.20	257.60	53.00	424.00		
1 Road Mixer, 310 H.P.		1889.00		2077.90	26.24	28.86
72 L.H., Daily Totals		$4107.80		$5750.70	$57.05	$79.87
Crew B-91	Hr.	Daily	Hr.	Daily	Bare Costs	Incl. O&P
1 Labor Foreman (outside)	$28.70	$229.60	$47.75	$382.00	$32.56	$53.82
2 Laborers	26.70	427.20	44.40	710.40		
4 Equip. Oper. (medium)	36.55	1169.60	60.25	1928.00		
1 Truck Driver (heavy)	32.20	257.60	53.00	424.00		
1 Dist. Tanker, 3000 Gallon		326.80		359.48		
1 Truck Tractor, 6x4, 380 H.P.		476.20		523.82		
1 Aggreg. Spreader, S.P.		837.00		920.70		
1 Roller, Pneum. Whl., 12 Ton		338.20		372.02		
1 Tandem Roller, 10 Ton		231.40		254.54	34.52	37.98
64 L.H., Daily Totals		$4293.60		$5874.96	$67.09	$91.80
Crew B-91B	Hr.	Daily	Hr.	Daily	Bare Costs	Incl. O&P
1 Laborer	$26.70	$213.60	$44.40	$355.20	$31.63	$52.33
1 Equipment Oper. (med.)	36.55	292.40	60.25	482.00		
1 Road Sweeper, Vac. Assist.		840.40		924.44	52.52	57.78
16 L.H., Daily Totals		$1346.40		$1761.64	$84.15	$110.10
Crew B-91C	Hr.	Daily	Hr.	Daily	Bare Costs	Incl. O&P
1 Laborer	$26.70	$213.60	$44.40	$355.20	$29.15	$48.20
1 Truck Driver (light)	31.60	252.80	52.00	416.00		
1 Catch Basin Cleaning Truck		522.80		575.08	32.67	35.94
16 L.H., Daily Totals		$989.20		$1346.28	$61.83	$84.14
Crew B-91D	Hr.	Daily	Hr.	Daily	Bare Costs	Incl. O&P
1 Labor Foreman (outside)	$28.70	$229.60	$47.75	$382.00	$31.49	$52.08
5 Laborers	26.70	1068.00	44.40	1776.00		
5 Equip. Oper. (medium)	36.55	1462.00	60.25	2410.00		
2 Truck Drivers (heavy)	32.20	515.20	53.00	848.00		
1 Aggreg. Spreader, S.P.		837.00		920.70		
2 Truck Tractors, 6x4, 380 H.P.		952.40		1047.64		
2 Dist. Tankers, 3000 Gallon		653.60		718.96		
2 Pavement Brushes, Towed		170.40		187.44		
2 Rollers Pneum. Whl., 12 Ton		676.40		744.04	31.63	34.80
104 L.H., Daily Totals		$6564.60		$9034.78	$63.12	$86.87
Crew B-92	Hr.	Daily	Hr.	Daily	Bare Costs	Incl. O&P
1 Labor Foreman (outside)	$28.70	$229.60	$47.75	$382.00	$27.20	$45.24
3 Laborers	26.70	640.80	44.40	1065.60		
1 Crack Cleaner, 25 H.P.		55.80		61.38		
1 Air Compressor, 60 cfm		105.40		115.94		
1 Tar Kettle, T.M.		129.20		142.12		
1 Flatbed Truck, Gas, 3 Ton		238.00		261.80	16.51	18.16
32 L.H., Daily Totals		$1398.80		$2028.84	$43.71	$63.40

Crew No.	Bare Costs		Incl. Subs O&P		Cost Per Labor-Hour	
Crew B-93	Hr.	Daily	Hr.	Daily	Bare Costs	Incl. O&P
1 Equip. Oper. (medium)	$36.55	$292.40	$60.25	$482.00	$36.55	$60.25
1 Feller Buncher, 100 H.P.		810.40		891.44	101.30	111.43
8 L.H., Daily Totals		$1102.80		$1373.44	$137.85	$171.68
Crew B-94A	Hr.	Daily	Hr.	Daily	Bare Costs	Incl. O&P
1 Laborer	$26.70	$213.60	$44.40	$355.20	$26.70	$44.40
1 Diaphragm Water Pump, 2"		73.00		80.30		
1 -20' Suction Hose, 2"		1.95		2.15		
2 -50' Discharge Hoses, 2"		1.80		1.98	9.59	10.55
8 L.H., Daily Totals		$290.35		$439.63	$36.29	$54.95
Crew B-94B	Hr.	Daily	Hr.	Daily	Bare Costs	Incl. O&P
1 Laborer	$26.70	$213.60	$44.40	$355.20	$26.70	$44.40
1 Diaphragm Water Pump, 4"		114.80		126.28		
1 -20' Suction Hose, 4"		3.25		3.58		
2 -50' Discharge Hoses, 4"		4.70		5.17	15.34	16.88
8 L.H., Daily Totals		$336.35		$490.23	$42.04	$61.28
Crew B-94C	Hr.	Daily	Hr.	Daily	Bare Costs	Incl. O&P
1 Laborer	$26.70	$213.60	$44.40	$355.20	$26.70	$44.40
1 Centrifugal Water Pump, 3"		79.20		87.12		
1 -20' Suction Hose, 3"		2.85		3.13		
2 -50' Discharge Hoses, 3"		3.00		3.30	10.63	11.69
8 L.H., Daily Totals		$298.65		$448.76	$37.33	$56.09
Crew B-94D	Hr.	Daily	Hr.	Daily	Bare Costs	Incl. O&P
1 Laborer	$26.70	$213.60	$44.40	$355.20	$26.70	$44.40
1 Centr. Water Pump, 6"		296.40		326.04		
1 -20' Suction Hose, 6"		11.50		12.65		
2 -50' Discharge Hoses, 6"		12.20		13.42	40.01	44.01
8 L.H., Daily Totals		$533.70		$707.31	$66.71	$88.41
Crew C-1	Hr.	Daily	Hr.	Daily	Bare Costs	Incl. O&P
2 Carpenters	$34.95	$559.20	$58.10	$929.60	$30.79	$51.33
1 Carpenter Helper	26.55	212.40	44.70	357.60		
1 Laborer	26.70	213.60	44.40	355.20		
32 L.H., Daily Totals		$985.20		$1642.40	$30.79	$51.33
Crew C-2	Hr.	Daily	Hr.	Daily	Bare Costs	Incl. O&P
1 Carpenter Foreman (outside)	$36.95	$295.60	$61.45	$491.60	$31.11	$51.91
2 Carpenters	34.95	559.20	58.10	929.60		
2 Carpenter Helpers	26.55	424.80	44.70	715.20		
1 Laborer	26.70	213.60	44.40	355.20		
48 L.H., Daily Totals		$1493.20		$2491.60	$31.11	$51.91
Crew C-2A	Hr.	Daily	Hr.	Daily	Bare Costs	Incl. O&P
1 Carpenter Foreman (outside)	$36.95	$295.60	$61.45	$491.60	$33.87	$56.08
3 Carpenters	34.95	838.80	58.10	1394.40		
1 Cement Finisher	34.70	277.60	56.30	450.40		
1 Laborer	26.70	213.60	44.40	355.20		
48 L.H., Daily Totals		$1625.60		$2691.60	$33.87	$56.08
Crew C-3	Hr.	Daily	Hr.	Daily	Bare Costs	Incl. O&P
1 Rodman Foreman (outside)	$40.80	$326.40	$68.00	$544.00	$34.09	$56.71
3 Rodmen (reinf.)	38.80	931.20	64.70	1552.80		
1 Equip. Oper. (light)	35.40	283.20	58.35	466.80		
3 Laborers	26.70	640.80	44.40	1065.60		
3 Stressing Equipment		31.20		34.32		
.5 Grouting Equipment		79.20		87.12	1.73	1.90
64 L.H., Daily Totals		$2292.00		$3750.64	$35.81	$58.60

For customer support on your Residential Costs with RSMeans data, call 800.448.8182.

Crew No.	Bare Costs		Incl. Subs O&P		Cost Per Labor-Hour	
Crew C-4	Hr.	Daily	Hr.	Daily	Bare Costs	Incl. O&P
1 Rodman Foreman (outside)	$40.80	$326.40	$68.00	$544.00	$36.27	$60.45
2 Rodmen (reinf.)	38.80	620.80	64.70	1035.20		
1 Building Laborer	26.70	213.60	44.40	355.20		
3 Stressing Equipment		31.20		34.32	0.97	1.07
32 L.H., Daily Totals		$1192.00		$1968.72	$37.25	$61.52
Crew C-4A	Hr.	Daily	Hr.	Daily	Bare Costs	Incl. O&P
2 Rodmen (reinf.)	$38.80	$620.80	$64.70	$1035.20	$38.80	$64.70
4 Stressing Equipment		41.60		45.76	2.60	2.86
16 L.H., Daily Totals		$662.40		$1080.96	$41.40	$67.56
Crew C-5	Hr.	Daily	Hr.	Daily	Bare Costs	Incl. O&P
1 Rodman Foreman (outside)	$40.80	$326.40	$68.00	$544.00	$34.81	$57.88
2 Rodmen (reinf.)	38.80	620.80	64.70	1035.20		
1 Equip. Oper. (crane)	37.05	296.40	61.05	488.40		
2 Building Laborers	26.70	427.20	44.40	710.40		
1 Hyd. Crane, 25 Ton		590.60		649.66	12.30	13.53
48 L.H., Daily Totals		$2261.40		$3427.66	$47.11	$71.41
Crew C-6	Hr.	Daily	Hr.	Daily	Bare Costs	Incl. O&P
1 Labor Foreman (outside)	$28.70	$229.60	$47.75	$382.00	$28.37	$46.94
4 Laborers	26.70	854.40	44.40	1420.80		
1 Cement Finisher	34.70	277.60	56.30	450.40		
2 Gas Engine Vibrators		51.20		56.32	1.07	1.17
48 L.H., Daily Totals		$1412.80		$2309.52	$29.43	$48.12
Crew C-7	Hr.	Daily	Hr.	Daily	Bare Costs	Incl. O&P
1 Labor Foreman (outside)	$28.70	$229.60	$47.75	$382.00	$29.56	$48.88
5 Laborers	26.70	1068.00	44.40	1776.00		
1 Cement Finisher	34.70	277.60	56.30	450.40		
1 Equip. Oper. (medium)	36.55	292.40	60.25	482.00		
1 Equip. Oper. (oiler)	32.55	260.40	53.65	429.20		
2 Gas Engine Vibrators		51.20		56.32		
1 Concrete Bucket, 1 C.Y.		24.60		27.06		
1 Hyd. Crane, 55 Ton		993.80		1093.18	14.86	16.34
72 L.H., Daily Totals		$3197.60		$4696.16	$44.41	$65.22
Crew C-8	Hr.	Daily	Hr.	Daily	Bare Costs	Incl. O&P
1 Labor Foreman (outside)	$28.70	$229.60	$47.75	$382.00	$30.68	$50.54
3 Laborers	26.70	640.80	44.40	1065.60		
2 Cement Finishers	34.70	555.20	56.30	900.80		
1 Equip. Oper. (medium)	36.55	292.40	60.25	482.00		
1 Concrete Pump (Small)		881.60		969.76	15.74	17.32
56 L.H., Daily Totals		$2599.60		$3800.16	$46.42	$67.86
Crew C-8A	Hr.	Daily	Hr.	Daily	Bare Costs	Incl. O&P
1 Labor Foreman (outside)	$28.70	$229.60	$47.75	$382.00	$29.70	$48.92
3 Laborers	26.70	640.80	44.40	1065.60		
2 Cement Finishers	34.70	555.20	56.30	900.80		
48 L.H., Daily Totals		$1425.60		$2348.40	$29.70	$48.92
Crew C-8B	Hr.	Daily	Hr.	Daily	Bare Costs	Incl. O&P
1 Labor Foreman (outside)	$28.70	$229.60	$47.75	$382.00	$29.07	$48.24
3 Laborers	26.70	640.80	44.40	1065.60		
1 Equip. Oper. (medium)	36.55	292.40	60.25	482.00		
1 Vibrating Power Screed		75.35		82.89		
1 Roller, Vibratory, 25 Ton		659.80		725.78		
1 Dozer, 200 H.P.		1273.00		1400.30	50.20	55.22
40 L.H., Daily Totals		$3170.95		$4138.56	$79.27	$103.46

Crew No.	Bare Costs		Incl. Subs O&P		Cost Per Labor-Hour	
Crew C-8C	Hr.	Daily	Hr.	Daily	Bare Costs	Incl. O&P
1 Labor Foreman (outside)	$28.70	$229.60	$47.75	$382.00	$30.01	$49.58
3 Laborers	26.70	640.80	44.40	1065.60		
1 Cement Finisher	34.70	277.60	56.30	450.40		
1 Equip. Oper. (medium)	36.55	292.40	60.25	482.00		
1 Shotcrete Rig, 12 C.Y./hr		246.60		271.26		
1 Air Compressor, 160 cfm		117.20		128.92		
4 -50' Air Hoses, 1"		33.80		37.18		
4 -50' Air Hoses, 2"		103.60		113.96	10.44	11.49
48 L.H., Daily Totals		$1941.60		$2931.32	$40.45	$61.07
Crew C-8D	Hr.	Daily	Hr.	Daily	Bare Costs	Incl. O&P
1 Labor Foreman (outside)	$28.70	$229.60	$47.75	$382.00	$31.38	$51.70
1 Laborer	26.70	213.60	44.40	355.20		
1 Cement Finisher	34.70	277.60	56.30	450.40		
1 Equipment Oper. (light)	35.40	283.20	58.35	466.80		
1 Air Compressor, 250 cfm		167.40		184.14		
2 -50' Air Hoses, 1"		16.90		18.59	5.76	6.34
32 L.H., Daily Totals		$1188.30		$1857.13	$37.13	$58.04
Crew C-8E	Hr.	Daily	Hr.	Daily	Bare Costs	Incl. O&P
1 Labor Foreman (outside)	$28.70	$229.60	$47.75	$382.00	$29.82	$49.27
3 Laborers	26.70	640.80	44.40	1065.60		
1 Cement Finisher	34.70	277.60	56.30	450.40		
1 Equipment Oper. (light)	35.40	283.20	58.35	466.80		
1 Shotcrete Rig, 35 C.Y./hr		270.20		297.22		
1 Air Compressor, 250 cfm		167.40		184.14		
4 -50' Air Hoses, 1"		33.80		37.18		
4 -50' Air Hoses, 2"		103.60		113.96	11.98	13.18
48 L.H., Daily Totals		$2006.20		$2997.30	$41.80	$62.44
Crew C-10	Hr.	Daily	Hr.	Daily	Bare Costs	Incl. O&P
1 Laborer	$26.70	$213.60	$44.40	$355.20	$32.03	$52.33
2 Cement Finishers	34.70	555.20	56.30	900.80		
24 L.H., Daily Totals		$768.80		$1256.00	$32.03	$52.33
Crew C-10B	Hr.	Daily	Hr.	Daily	Bare Costs	Incl. O&P
3 Laborers	$26.70	$640.80	$44.40	$1065.60	$29.90	$49.16
2 Cement Finishers	34.70	555.20	56.30	900.80		
1 Concrete Mixer, 10 C.F.		161.00		177.10		
2 Trowels, 48" Walk-Behind		82.40		90.64	6.09	6.69
40 L.H., Daily Totals		$1439.40		$2234.14	$35.98	$55.85
Crew C-10C	Hr.	Daily	Hr.	Daily	Bare Costs	Incl. O&P
1 Laborer	$26.70	$213.60	$44.40	$355.20	$32.03	$52.33
2 Cement Finishers	34.70	555.20	56.30	900.80		
1 Trowel, 48" Walk-Behind		41.20		45.32	1.72	1.89
24 L.H., Daily Totals		$810.00		$1301.32	$33.75	$54.22
Crew C-10D	Hr.	Daily	Hr.	Daily	Bare Costs	Incl. O&P
1 Laborer	$26.70	$213.60	$44.40	$355.20	$32.03	$52.33
2 Cement Finishers	34.70	555.20	56.30	900.80		
1 Vibrating Power Screed		75.35		82.89		
1 Trowel, 48" Walk-Behind		41.20		45.32	4.86	5.34
24 L.H., Daily Totals		$885.35		$1384.20	$36.89	$57.68
Crew C-10E	Hr.	Daily	Hr.	Daily	Bare Costs	Incl. O&P
1 Laborer	$26.70	$213.60	$44.40	$355.20	$32.03	$52.33
2 Cement Finishers	34.70	555.20	56.30	900.80		
1 Vibrating Power Screed		75.35		82.89		
1 Cement Trowel, 96" Ride-On		166.20		182.82	10.06	11.07
24 L.H., Daily Totals		$1010.35		$1521.70	$42.10	$63.40

For customer support on your Residential Costs with RSMeans data, call 800.448.8182.

Crew No.	Bare Costs		Incl. Subs O&P		Cost Per Labor-Hour	

Crew C-10F

Crew No.	Hr.	Daily	Hr.	Daily	Bare Costs	Incl. O&P
1 Laborer	$26.70	$213.60	$44.40	$355.20	$32.03	$52.33
2 Cement Finishers	34.70	555.20	56.30	900.80		
1 Telescoping Boom Lift, to 60'		454.60		500.06	18.94	20.84
24 L.H., Daily Totals		$1223.40		$1756.06	$50.98	$73.17

Crew C-11

Crew No.	Hr.	Daily	Hr.	Daily	Bare Costs	Incl. O&P
1 Skilled Worker Foreman	$37.80	$302.40	$63.15	$505.20	$36.26	$60.46
5 Skilled Workers	35.80	1432.00	59.80	2392.00		
1 Equip. Oper. (crane)	37.05	296.40	61.05	488.40		
1 Lattice Boom Crane, 150 Ton		2017.00		2218.70	36.02	39.62
56 L.H., Daily Totals		$4047.80		$5604.30	$72.28	$100.08

Crew C-12

Crew No.	Hr.	Daily	Hr.	Daily	Bare Costs	Incl. O&P
1 Carpenter Foreman (outside)	$36.95	$295.60	$61.45	$491.60	$34.26	$56.87
3 Carpenters	34.95	838.80	58.10	1394.40		
1 Laborer	26.70	213.60	44.40	355.20		
1 Equip. Oper. (crane)	37.05	296.40	61.05	488.40		
1 Hyd. Crane, 12 Ton		496.00		545.60	10.33	11.37
48 L.H., Daily Totals		$2140.40		$3275.20	$44.59	$68.23

Crew C-13

Crew No.	Hr.	Daily	Hr.	Daily	Bare Costs	Incl. O&P
2 Struc. Steel Workers	$38.80	$620.80	$68.85	$1101.60	$37.52	$65.27
1 Carpenter	34.95	279.60	58.10	464.80		
1 Welder, Gas Engine, 300 amp		98.40		108.24	4.10	4.51
24 L.H., Daily Totals		$998.80		$1674.64	$41.62	$69.78

Crew C-14

Crew No.	Hr.	Daily	Hr.	Daily	Bare Costs	Incl. O&P
1 Carpenter Foreman (outside)	$36.95	$295.60	$61.45	$491.60	$31.70	$52.66
3 Carpenters	34.95	838.80	58.10	1394.40		
2 Carpenter Helpers	26.55	424.80	44.70	715.20		
4 Laborers	26.70	854.40	44.40	1420.80		
2 Rodmen (reinf.)	38.80	620.80	64.70	1035.20		
2 Rodman Helpers	26.55	424.80	44.70	715.20		
2 Cement Finishers	34.70	555.20	56.30	900.80		
1 Equip. Oper. (crane)	37.05	296.40	61.05	488.40		
1 Hyd. Crane, 80 Ton		1518.00		1669.80	11.16	12.28
136 L.H., Daily Totals		$5828.80		$8831.40	$42.86	$64.94

Crew C-14A

Crew No.	Hr.	Daily	Hr.	Daily	Bare Costs	Incl. O&P
1 Carpenter Foreman (outside)	$36.95	$295.60	$61.45	$491.60	$35.04	$58.21
16 Carpenters	34.95	4473.60	58.10	7436.80		
4 Rodmen (reinf.)	38.80	1241.60	64.70	2070.40		
2 Laborers	26.70	427.20	44.40	710.40		
1 Cement Finisher	34.70	277.60	56.30	450.40		
1 Equip. Oper. (medium)	36.55	292.40	60.25	482.00		
1 Gas Engine Vibrator		25.60		28.16		
1 Concrete Pump (Small)		881.60		969.76	4.54	4.99
200 L.H., Daily Totals		$7915.20		$12639.52	$39.58	$63.20

Crew C-14B

Crew No.	Hr.	Daily	Hr.	Daily	Bare Costs	Incl. O&P
1 Carpenter Foreman (outside)	$36.95	$295.60	$61.45	$491.60	$35.03	$58.13
16 Carpenters	34.95	4473.60	58.10	7436.80		
4 Rodmen (reinf.)	38.80	1241.60	64.70	2070.40		
2 Laborers	26.70	427.20	44.40	710.40		
2 Cement Finishers	34.70	555.20	56.30	900.80		
1 Equip. Oper. (medium)	36.55	292.40	60.25	482.00		
1 Gas Engine Vibrator		25.60		28.16		
1 Concrete Pump (Small)		881.60		969.76	4.36	4.80
208 L.H., Daily Totals		$8192.80		$13089.92	$39.39	$62.93

Crew C-14C

Crew No.	Hr.	Daily	Hr.	Daily	Bare Costs	Incl. O&P
1 Carpenter Foreman (outside)	$36.95	$295.60	$61.45	$491.60	$33.27	$55.24
6 Carpenters	34.95	1677.60	58.10	2788.80		
2 Rodmen (reinf.)	38.80	620.80	64.70	1035.20		
4 Laborers	26.70	854.40	44.40	1420.80		
1 Cement Finisher	34.70	277.60	56.30	450.40		
1 Gas Engine Vibrator		25.60		28.16	0.23	0.25
112 L.H., Daily Totals		$3751.60		$6214.96	$33.50	$55.49

Crew C-14D

Crew No.	Hr.	Daily	Hr.	Daily	Bare Costs	Incl. O&P
1 Carpenter Foreman (outside)	$36.95	$295.60	$61.45	$491.60	$34.73	$57.68
18 Carpenters	34.95	5032.80	58.10	8366.40		
2 Rodmen (reinf.)	38.80	620.80	64.70	1035.20		
2 Laborers	26.70	427.20	44.40	710.40		
1 Cement Finisher	34.70	277.60	56.30	450.40		
1 Equip. Oper. (medium)	36.55	292.40	60.25	482.00		
1 Gas Engine Vibrator		25.60		28.16		
1 Concrete Pump (Small)		881.60		969.76	4.54	4.99
200 L.H., Daily Totals		$7853.60		$12533.92	$39.27	$62.67

Crew C-14E

Crew No.	Hr.	Daily	Hr.	Daily	Bare Costs	Incl. O&P
1 Carpenter Foreman (outside)	$36.95	$295.60	$61.45	$491.60	$34.26	$56.90
2 Carpenters	34.95	559.20	58.10	929.60		
4 Rodmen (reinf.)	38.80	1241.60	64.70	2070.40		
3 Laborers	26.70	640.80	44.40	1065.60		
1 Cement Finisher	34.70	277.60	56.30	450.40		
1 Gas Engine Vibrator		25.60		28.16	0.29	0.32
88 L.H., Daily Totals		$3040.40		$5035.76	$34.55	$57.22

Crew C-14F

Crew No.	Hr.	Daily	Hr.	Daily	Bare Costs	Incl. O&P
1 Labor Foreman (outside)	$28.70	$229.60	$47.75	$382.00	$32.26	$52.71
2 Laborers	26.70	427.20	44.40	710.40		
6 Cement Finishers	34.70	1665.60	56.30	2702.40		
1 Gas Engine Vibrator		25.60		28.16	0.36	0.39
72 L.H., Daily Totals		$2348.00		$3822.96	$32.61	$53.10

Crew C-14G

Crew No.	Hr.	Daily	Hr.	Daily	Bare Costs	Incl. O&P
1 Labor Foreman (outside)	$28.70	$229.60	$47.75	$382.00	$31.56	$51.68
2 Laborers	26.70	427.20	44.40	710.40		
4 Cement Finishers	34.70	1110.40	56.30	1801.60		
1 Gas Engine Vibrator		25.60		28.16	0.46	0.50
56 L.H., Daily Totals		$1792.80		$2922.16	$32.01	$52.18

Crew C-14H

Crew No.	Hr.	Daily	Hr.	Daily	Bare Costs	Incl. O&P
1 Carpenter Foreman (outside)	$36.95	$295.60	$61.45	$491.60	$34.51	$57.17
2 Carpenters	34.95	559.20	58.10	929.60		
1 Rodman (reinf.)	38.80	310.40	64.70	517.60		
1 Laborer	26.70	213.60	44.40	355.20		
1 Cement Finisher	34.70	277.60	56.30	450.40		
1 Gas Engine Vibrator		25.60		28.16	0.53	0.59
48 L.H., Daily Totals		$1682.00		$2772.56	$35.04	$57.76

Crew C-14L

Crew No.	Hr.	Daily	Hr.	Daily	Bare Costs	Incl. O&P
1 Carpenter Foreman (outside)	$36.95	$295.60	$61.45	$491.60	$32.35	$53.66
6 Carpenters	34.95	1677.60	58.10	2788.80		
4 Laborers	26.70	854.40	44.40	1420.80		
1 Cement Finisher	34.70	277.60	56.30	450.40		
1 Gas Engine Vibrator		25.60		28.16	0.27	0.29
96 L.H., Daily Totals		$3130.80		$5179.76	$32.61	$53.96

703

For customer support on your Residential Costs with RSMeans data, call 800.448.8182.

Crew C-14M

	Bare Costs Hr.	Daily	Incl. Subs O&P Hr.	Daily	Cost Per Labor-Hour Bare Costs	Incl. O&P
1 Carpenter Foreman (outside)	$36.95	$295.60	$61.45	$491.60	$33.79	$55.96
2 Carpenters	34.95	559.20	58.10	929.60		
1 Rodman (reinf.)	38.80	310.40	64.70	517.60		
2 Laborers	26.70	427.20	44.40	710.40		
1 Cement Finisher	34.70	277.60	56.30	450.40		
1 Equip. Oper. (medium)	36.55	292.40	60.25	482.00		
1 Gas Engine Vibrator		25.60		28.16		
1 Concrete Pump (Small)		881.60		969.76	14.18	15.59
64 L.H., Daily Totals		$3069.60		$4579.52	$47.96	$71.56

Crew C-15

	Bare Costs Hr.	Daily	Incl. Subs O&P Hr.	Daily	Cost Per Labor-Hour Bare Costs	Incl. O&P
1 Carpenter Foreman (outside)	$36.95	$295.60	$61.45	$491.60	$32.79	$54.24
2 Carpenters	34.95	559.20	58.10	929.60		
3 Laborers	26.70	640.80	44.40	1065.60		
2 Cement Finishers	34.70	555.20	56.30	900.80		
1 Rodman (reinf.)	38.80	310.40	64.70	517.60		
72 L.H., Daily Totals		$2361.20		$3905.20	$32.79	$54.24

Crew C-16

	Bare Costs Hr.	Daily	Incl. Subs O&P Hr.	Daily	Cost Per Labor-Hour Bare Costs	Incl. O&P
1 Labor Foreman (outside)	$28.70	$229.60	$47.75	$382.00	$30.68	$50.54
3 Laborers	26.70	640.80	44.40	1065.60		
2 Cement Finishers	34.70	555.20	56.30	900.80		
1 Equip. Oper. (medium)	36.55	292.40	60.25	482.00		
1 Gunite Pump Rig		321.00		353.10		
2 -50' Air Hoses, 3/4"		14.70		16.17		
2 -50' Air Hoses, 2"		51.80		56.98	6.92	7.61
56 L.H., Daily Totals		$2105.50		$3256.65	$37.60	$58.15

Crew C-16A

	Bare Costs Hr.	Daily	Incl. Subs O&P Hr.	Daily	Cost Per Labor-Hour Bare Costs	Incl. O&P
1 Laborer	$26.70	$213.60	$44.40	$355.20	$33.16	$54.31
2 Cement Finishers	34.70	555.20	56.30	900.80		
1 Equip. Oper. (medium)	36.55	292.40	60.25	482.00		
1 Gunite Pump Rig		321.00		353.10		
2 -50' Air Hoses, 3/4"		14.70		16.17		
2 -50' Air Hoses, 2"		51.80		56.98		
1 Telescoping Boom Lift, to 60'		454.60		500.06	26.32	28.95
32 L.H., Daily Totals		$1903.30		$2664.31	$59.48	$83.26

Crew C-17

	Bare Costs Hr.	Daily	Incl. Subs O&P Hr.	Daily	Cost Per Labor-Hour Bare Costs	Incl. O&P
2 Skilled Worker Foremen (out)	$37.80	$604.80	$63.15	$1010.40	$36.20	$60.47
8 Skilled Workers	35.80	2291.20	59.80	3827.20		
80 L.H., Daily Totals		$2896.00		$4837.60	$36.20	$60.47

Crew C-17A

	Bare Costs Hr.	Daily	Incl. Subs O&P Hr.	Daily	Cost Per Labor-Hour Bare Costs	Incl. O&P
2 Skilled Worker Foremen (out)	$37.80	$604.80	$63.15	$1010.40	$36.21	$60.48
8 Skilled Workers	35.80	2291.20	59.80	3827.20		
.125 Equip. Oper. (crane)	37.05	37.05	61.05	61.05		
.125 Hyd. Crane, 80 Ton		189.75		208.72	2.34	2.58
81 L.H., Daily Totals		$3122.80		$5107.38	$38.55	$63.05

Crew C-17B

	Bare Costs Hr.	Daily	Incl. Subs O&P Hr.	Daily	Cost Per Labor-Hour Bare Costs	Incl. O&P
2 Skilled Worker Foremen (out)	$37.80	$604.80	$63.15	$1010.40	$36.22	$60.48
8 Skilled Workers	35.80	2291.20	59.80	3827.20		
.25 Equip. Oper. (crane)	37.05	74.10	61.05	122.10		
.25 Hyd. Crane, 80 Ton		379.50		417.45		
.25 Trowel, 48" Walk-Behind		10.30		11.33	4.75	5.23
82 L.H., Daily Totals		$3359.90		$5388.48	$40.97	$65.71

Crew C-17C

	Bare Costs Hr.	Daily	Incl. Subs O&P Hr.	Daily	Cost Per Labor-Hour Bare Costs	Incl. O&P
2 Skilled Worker Foremen (out)	$37.80	$604.80	$63.15	$1010.40	$36.23	$60.49
8 Skilled Workers	35.80	2291.20	59.80	3827.20		
.375 Equip. Oper. (crane)	37.05	111.15	61.05	183.15		
.375 Hyd. Crane, 80 Ton		569.25		626.17	6.86	7.54
83 L.H., Daily Totals		$3576.40		$5646.93	$43.09	$68.04

Crew C-17D

	Bare Costs Hr.	Daily	Incl. Subs O&P Hr.	Daily	Cost Per Labor-Hour Bare Costs	Incl. O&P
2 Skilled Worker Foremen (out)	$37.80	$604.80	$63.15	$1010.40	$36.24	$60.50
8 Skilled Workers	35.80	2291.20	59.80	3827.20		
.5 Equip. Oper. (crane)	37.05	148.20	61.05	244.20		
.5 Hyd. Crane, 80 Ton		759.00		834.90	9.04	9.94
84 L.H., Daily Totals		$3803.20		$5916.70	$45.28	$70.44

Crew C-17E

	Bare Costs Hr.	Daily	Incl. Subs O&P Hr.	Daily	Cost Per Labor-Hour Bare Costs	Incl. O&P
2 Skilled Worker Foremen (out)	$37.80	$604.80	$63.15	$1010.40	$36.20	$60.47
8 Skilled Workers	35.80	2291.20	59.80	3827.20		
1 Hyd. Jack with Rods		102.25		112.47	1.28	1.41
80 L.H., Daily Totals		$2998.25		$4950.07	$37.48	$61.88

Crew C-18

	Bare Costs Hr.	Daily	Incl. Subs O&P Hr.	Daily	Cost Per Labor-Hour Bare Costs	Incl. O&P
.125 Labor Foreman (outside)	$28.70	$28.70	$47.75	$47.75	$26.92	$44.77
1 Laborer	26.70	213.60	44.40	355.20		
1 Concrete Cart, 10 C.F.		57.80		63.58	6.42	7.06
9 L.H., Daily Totals		$300.10		$466.53	$33.34	$51.84

Crew C-19

	Bare Costs Hr.	Daily	Incl. Subs O&P Hr.	Daily	Cost Per Labor-Hour Bare Costs	Incl. O&P
.125 Labor Foreman (outside)	$28.70	$28.70	$47.75	$47.75	$26.92	$44.77
1 Laborer	26.70	213.60	44.40	355.20		
1 Concrete Cart, 18 C.F.		97.40		107.14	10.82	11.90
9 L.H., Daily Totals		$339.70		$510.09	$37.74	$56.68

Crew C-20

	Bare Costs Hr.	Daily	Incl. Subs O&P Hr.	Daily	Cost Per Labor-Hour Bare Costs	Incl. O&P
1 Labor Foreman (outside)	$28.70	$229.60	$47.75	$382.00	$29.18	$48.29
5 Laborers	26.70	1068.00	44.40	1776.00		
1 Cement Finisher	34.70	277.60	56.30	450.40		
1 Equip. Oper. (medium)	36.55	292.40	60.25	482.00		
2 Gas Engine Vibrators		51.20		56.32		
1 Concrete Pump (Small)		881.60		969.76	14.57	16.03
64 L.H., Daily Totals		$2800.40		$4116.48	$43.76	$64.32

Crew C-21

	Bare Costs Hr.	Daily	Incl. Subs O&P Hr.	Daily	Cost Per Labor-Hour Bare Costs	Incl. O&P
1 Labor Foreman (outside)	$28.70	$229.60	$47.75	$382.00	$29.18	$48.29
5 Laborers	26.70	1068.00	44.40	1776.00		
1 Cement Finisher	34.70	277.60	56.30	450.40		
1 Equip. Oper. (medium)	36.55	292.40	60.25	482.00		
2 Gas Engine Vibrators		51.20		56.32		
1 Concrete Conveyer		187.20		205.92	3.73	4.10
64 L.H., Daily Totals		$2106.00		$3352.64	$32.91	$52.38

Crew C-22

	Bare Costs Hr.	Daily	Incl. Subs O&P Hr.	Daily	Cost Per Labor-Hour Bare Costs	Incl. O&P
1 Rodman Foreman (outside)	$40.80	$326.40	$68.00	$544.00	$38.99	$64.98
4 Rodmen (reinf.)	38.80	1241.60	64.70	2070.40		
.125 Equip. Oper. (crane)	37.05	37.05	61.05	61.05		
.125 Equip. Oper. (oiler)	32.55	32.55	53.65	53.65		
.125 Hyd. Crane, 25 Ton		73.83		81.21	1.76	1.93
42 L.H., Daily Totals		$1711.43		$2810.31	$40.75	$66.91

Crews - Residential

Crew No.	Bare Costs		Incl. Subs O&P		Cost Per Labor-Hour	
Crew C-23	Hr.	Daily	Hr.	Daily	Bare Costs	Incl. O&P
2 Skilled Worker Foremen (out)	$37.80	$604.80	$63.15	$1010.40	$36.00	$59.98
6 Skilled Workers	35.80	1718.40	59.80	2870.40		
1 Equip. Oper. (crane)	37.05	296.40	61.05	488.40		
1 Equip. Oper. (oiler)	32.55	260.40	53.65	429.20		
1 Lattice Boom Crane, 90 Ton		1687.00		1855.70	21.09	23.20
80 L.H., Daily Totals		$4567.00		$6654.10	$57.09	$83.18
Crew C-24	Hr.	Daily	Hr.	Daily	Bare Costs	Incl. O&P
2 Skilled Worker Foremen (out)	$37.80	$604.80	$63.15	$1010.40	$36.00	$59.98
6 Skilled Workers	35.80	1718.40	59.80	2870.40		
1 Equip. Oper. (crane)	37.05	296.40	61.05	488.40		
1 Equip. Oper. (oiler)	32.55	260.40	53.65	429.20		
1 Lattice Boom Crane, 150 Ton		2017.00		2218.70	25.21	27.73
80 L.H., Daily Totals		$4897.00		$7017.10	$61.21	$87.71
Crew C-25	Hr.	Daily	Hr.	Daily	Bare Costs	Incl. O&P
2 Rodmen (reinf.)	$38.80	$620.80	$64.70	$1035.20	$30.57	$52.73
2 Rodmen Helpers	22.35	357.60	40.75	652.00		
32 L.H., Daily Totals		$978.40		$1687.20	$30.57	$52.73
Crew C-27	Hr.	Daily	Hr.	Daily	Bare Costs	Incl. O&P
2 Cement Finishers	$34.70	$555.20	$56.30	$900.80	$34.70	$56.30
1 Concrete Saw		102.40		112.64	6.40	7.04
16 L.H., Daily Totals		$657.60		$1013.44	$41.10	$63.34
Crew C-28	Hr.	Daily	Hr.	Daily	Bare Costs	Incl. O&P
1 Cement Finisher	$34.70	$277.60	$56.30	$450.40	$34.70	$56.30
1 Portable Air Compressor, Gas		17.10		18.81	2.14	2.35
8 L.H., Daily Totals		$294.70		$469.21	$36.84	$58.65
Crew C-29	Hr.	Daily	Hr.	Daily	Bare Costs	Incl. O&P
1 Laborer	$26.70	$213.60	$44.40	$355.20	$26.70	$44.40
1 Pressure Washer		63.60		69.96	7.95	8.74
8 L.H., Daily Totals		$277.20		$425.16	$34.65	$53.15
Crew C-30	Hr.	Daily	Hr.	Daily	Bare Costs	Incl. O&P
1 Laborer	$26.70	$213.60	$44.40	$355.20	$26.70	$44.40
1 Concrete Mixer, 10 C.F.		161.00		177.10	20.13	22.14
8 L.H., Daily Totals		$374.60		$532.30	$46.83	$66.54
Crew C-31	Hr.	Daily	Hr.	Daily	Bare Costs	Incl. O&P
1 Cement Finisher	$34.70	$277.60	$56.30	$450.40	$34.70	$56.30
1 Grout Pump		321.00		353.10	40.13	44.14
8 L.H., Daily Totals		$598.60		$803.50	$74.83	$100.44
Crew C-32	Hr.	Daily	Hr.	Daily	Bare Costs	Incl. O&P
1 Cement Finisher	$34.70	$277.60	$56.30	$450.40	$30.70	$50.35
1 Laborer	26.70	213.60	44.40	355.20		
1 Crack Chaser Saw, Gas, 6 H.P.		25.40		27.94		
1 Vacuum Pick-Up System		66.95		73.64	5.77	6.35
16 L.H., Daily Totals		$583.55		$907.18	$36.47	$56.70
Crew D-1	Hr.	Daily	Hr.	Daily	Bare Costs	Incl. O&P
1 Bricklayer	$34.70	$277.60	$58.00	$464.00	$31.68	$52.92
1 Bricklayer Helper	28.65	229.20	47.85	382.80		
16 L.H., Daily Totals		$506.80		$846.80	$31.68	$52.92

Crew No.	Bare Costs		Incl. Subs O&P		Cost Per Labor-Hour	
Crew D-2	Hr.	Daily	Hr.	Daily	Bare Costs	Incl. O&P
3 Bricklayers	$34.70	$832.80	$58.00	$1392.00	$32.28	$53.94
2 Bricklayer Helpers	28.65	458.40	47.85	765.60		
40 L.H., Daily Totals		$1291.20		$2157.60	$32.28	$53.94
Crew D-3	Hr.	Daily	Hr.	Daily	Bare Costs	Incl. O&P
3 Bricklayers	$34.70	$832.80	$58.00	$1392.00	$32.41	$54.14
2 Bricklayer Helpers	28.65	458.40	47.85	765.60		
.25 Carpenter	34.95	69.90	58.10	116.20		
42 L.H., Daily Totals		$1361.10		$2273.80	$32.41	$54.14
Crew D-4	Hr.	Daily	Hr.	Daily	Bare Costs	Incl. O&P
1 Bricklayer	$34.70	$277.60	$58.00	$464.00	$29.47	$49.19
3 Bricklayer Helpers	28.65	687.60	47.85	1148.40		
1 Building Laborer	26.70	213.60	44.40	355.20		
1 Grout Pump, 50 C.F./hr.		128.60		141.46	3.21	3.54
40 L.H., Daily Totals		$1307.40		$2109.06	$32.69	$52.73
Crew D-5	Hr.	Daily	Hr.	Daily	Bare Costs	Incl. O&P
1 Block Mason Helper	28.65	229.20	47.85	382.80	28.65	47.85
8 L.H., Daily Totals		$229.20		$382.80	$28.65	$47.85
Crew D-6	Hr.	Daily	Hr.	Daily	Bare Costs	Incl. O&P
3 Bricklayers	$34.70	$832.80	$58.00	$1392.00	$31.68	$52.92
3 Bricklayer Helpers	28.65	687.60	47.85	1148.40		
48 L.H., Daily Totals		$1520.40		$2540.40	$31.68	$52.92
Crew D-7	Hr.	Daily	Hr.	Daily	Bare Costs	Incl. O&P
1 Tile Layer	$33.75	$270.00	$54.70	$437.60	$29.90	$48.45
1 Tile Layer Helper	26.05	208.40	42.20	337.60		
16 L.H., Daily Totals		$478.40		$775.20	$29.90	$48.45
Crew D-8	Hr.	Daily	Hr.	Daily	Bare Costs	Incl. O&P
3 Bricklayers	$34.70	$832.80	$58.00	$1392.00	$32.28	$53.94
2 Bricklayer Helpers	28.65	458.40	47.85	765.60		
40 L.H., Daily Totals		$1291.20		$2157.60	$32.28	$53.94
Crew D-9	Hr.	Daily	Hr.	Daily	Bare Costs	Incl. O&P
3 Bricklayers	$34.70	$832.80	$58.00	$1392.00	$31.68	$52.92
3 Bricklayer Helpers	28.65	687.60	47.85	1148.40		
48 L.H., Daily Totals		$1520.40		$2540.40	$31.68	$52.92
Crew D-10	Hr.	Daily	Hr.	Daily	Bare Costs	Incl. O&P
1 Bricklayer Foreman (outside)	$36.70	$293.60	$61.35	$490.80	$34.27	$57.06
1 Bricklayer	34.70	277.60	58.00	464.00		
1 Bricklayer Helper	28.65	229.20	47.85	382.80		
1 Equip. Oper. (crane)	37.05	296.40	61.05	488.40		
1 S.P. Crane, 4x4, 12 Ton		363.60		399.96	11.36	12.50
32 L.H., Daily Totals		$1460.40		$2225.96	$45.64	$69.56
Crew D-11	Hr.	Daily	Hr.	Daily	Bare Costs	Incl. O&P
2 Bricklayers	$34.70	$555.20	$58.00	$928.00	$32.68	$54.62
1 Bricklayer Helper	28.65	229.20	47.85	382.80		
24 L.H., Daily Totals		$784.40		$1310.80	$32.68	$54.62
Crew D-12	Hr.	Daily	Hr.	Daily	Bare Costs	Incl. O&P
2 Bricklayers	$34.70	$555.20	$58.00	$928.00	$31.68	$52.92
2 Bricklayer Helpers	28.65	458.40	47.85	765.60		
32 L.H., Daily Totals		$1013.60		$1693.60	$31.68	$52.92

For customer support on your Residential Costs with RSMeans data, call 800.448.8182.

Crew No.	Bare Costs		Incl. Subs O&P		Cost Per Labor-Hour	
	Hr.	Daily	Hr.	Daily	Bare Costs	Incl. O&P
Crew D-13					Bare Costs	Incl. O&P
1 Bricklayer Foreman (outside)	$36.70	$293.60	$61.35	$490.80	$33.41	$55.68
2 Bricklayers	34.70	555.20	58.00	928.00		
2 Bricklayer Helpers	28.65	458.40	47.85	765.60		
1 Equip. Oper. (crane)	37.05	296.40	61.05	488.40		
1 S.P. Crane, 4x4, 12 Ton		363.60		399.96	7.58	8.33
48 L.H., Daily Totals		$1967.20		$3072.76	$40.98	$64.02
Crew D-14					Bare Costs	Incl. O&P
3 Bricklayers	$34.70	$832.80	$58.00	$1392.00	$33.19	$55.46
1 Bricklayer Helper	28.65	229.20	47.85	382.80		
32 L.H., Daily Totals		$1062.00		$1774.80	$33.19	$55.46
Crew E-1					Bare Costs	Incl. O&P
2 Struc. Steel Workers	$38.80	$620.80	$68.85	$1101.60	$38.80	$68.85
1 Welder, Gas Engine, 300 amp		98.40		108.24	6.15	6.76
16 L.H., Daily Totals		$719.20		$1209.84	$44.95	$75.61
Crew E-2					Bare Costs	Incl. O&P
1 Struc. Steel Foreman (outside)	$40.80	$326.40	$72.40	$579.20	$38.84	$68.14
4 Struc. Steel Workers	38.80	1241.60	68.85	2203.20		
1 Equip. Oper. (crane)	37.05	296.40	61.05	488.40		
1 Lattice Boom Crane, 90 Ton		1687.00		1855.70	35.15	38.66
48 L.H., Daily Totals		$3551.40		$5126.50	$73.99	$106.80
Crew E-3					Bare Costs	Incl. O&P
1 Struc. Steel Foreman (outside)	$40.80	$326.40	$72.40	$579.20	$39.47	$70.03
2 Struc. Steel Workers	38.80	620.80	68.85	1101.60		
1 Welder, Gas Engine, 300 amp		98.40		108.24	4.10	4.51
24 L.H., Daily Totals		$1045.60		$1789.04	$43.57	$74.54
Crew E-3A					Bare Costs	Incl. O&P
1 Struc. Steel Foreman (outside)	$40.80	$326.40	$72.40	$579.20	$39.47	$70.03
2 Struc. Steel Workers	38.80	620.80	68.85	1101.60		
1 Welder, Gas Engine, 300 amp		98.40		108.24		
1 Telescoping Boom Lift, to 40'		282.80		311.08	15.88	17.47
24 L.H., Daily Totals		$1328.40		$2100.12	$55.35	$87.50
Crew E-4					Bare Costs	Incl. O&P
1 Struc. Steel Foreman (outside)	$40.80	$326.40	$72.40	$579.20	$39.30	$69.74
3 Struc. Steel Workers	38.80	931.20	68.85	1652.40		
1 Welder, Gas Engine, 300 amp		98.40		108.24	3.08	3.38
32 L.H., Daily Totals		$1356.00		$2339.84	$42.38	$73.12
Crew E-5					Bare Costs	Incl. O&P
1 Struc. Steel Foreman (outside)	$40.80	$326.40	$72.40	$579.20	$38.83	$68.38
7 Struc. Steel Workers	38.80	2172.80	68.85	3855.60		
1 Equip. Oper. (crane)	37.05	296.40	61.05	488.40		
1 Lattice Boom Crane, 90 Ton		1687.00		1855.70		
1 Welder, Gas Engine, 300 amp		98.40		108.24	24.80	27.28
72 L.H., Daily Totals		$4581.00		$6887.14	$63.63	$95.65

Crew No.	Bare Costs		Incl. Subs O&P		Cost Per Labor-Hour	
	Hr.	Daily	Hr.	Daily	Bare Costs	Incl. O&P
Crew E-6					Bare Costs	Incl. O&P
1 Struc. Steel Foreman (outside)	$40.80	$326.40	$72.40	$579.20	$38.59	$67.87
12 Struc. Steel Workers	38.80	3724.80	68.85	6609.60		
1 Equip. Oper. (crane)	37.05	296.40	61.05	488.40		
1 Equip. Oper. (light)	35.40	283.20	58.35	466.80		
1 Lattice Boom Crane, 90 Ton		1687.00		1855.70		
1 Welder, Gas Engine, 300 amp		98.40		108.24		
1 Air Compressor, 160 cfm		117.20		128.92		
2 Impact Wrenches		37.20		40.92	16.16	17.78
120 L.H., Daily Totals		$6570.60		$10277.78	$54.76	$85.65
Crew E-7					Bare Costs	Incl. O&P
1 Struc. Steel Foreman (outside)	$40.80	$326.40	$72.40	$579.20	$38.83	$68.38
7 Struc. Steel Workers	38.80	2172.80	68.85	3855.60		
1 Equip. Oper. (crane)	37.05	296.40	61.05	488.40		
1 Lattice Boom Crane, 90 Ton		1687.00		1855.70		
2 Welders, Gas Engine, 300 amp		196.80		216.48	26.16	28.78
72 L.H., Daily Totals		$4679.40		$6995.38	$64.99	$97.16
Crew E-8					Bare Costs	Incl. O&P
1 Struc. Steel Foreman (outside)	$40.80	$326.40	$72.40	$579.20	$38.82	$68.46
9 Struc. Steel Workers	38.80	2793.60	68.85	4957.20		
1 Equip. Oper. (crane)	37.05	296.40	61.05	488.40		
1 Lattice Boom Crane, 90 Ton		1687.00		1855.70		
4 Welders, Gas Engine, 300 amp		393.60		432.96	23.64	26.01
88 L.H., Daily Totals		$5497.00		$8313.46	$62.47	$94.47
Crew E-9					Bare Costs	Incl. O&P
2 Struc. Steel Foremen (outside)	$40.80	$652.80	$72.40	$1158.40	$38.46	$67.42
5 Struc. Steel Workers	38.80	1552.00	68.85	2754.00		
1 Welder Foreman (outside)	40.80	326.40	72.40	579.20		
5 Welders	38.80	1552.00	68.85	2754.00		
1 Equip. Oper. (crane)	37.05	296.40	61.05	488.40		
1 Equip. Oper. (oiler)	32.55	260.40	53.65	429.20		
1 Equip. Oper. (light)	35.40	283.20	58.35	466.80		
1 Lattice Boom Crane, 90 Ton		1687.00		1855.70		
5 Welders, Gas Engine, 300 amp		492.00		541.20	17.02	18.73
128 L.H., Daily Totals		$7102.20		$11026.90	$55.49	$86.15
Crew E-10					Bare Costs	Incl. O&P
1 Struc. Steel Foreman (outside)	$40.80	$326.40	$72.40	$579.20	$39.47	$70.03
2 Struc. Steel Workers	38.80	620.80	68.85	1101.60		
1 Welder, Gas Engine, 300 amp		98.40		108.24		
1 Flatbed Truck, Gas, 3 Ton		238.00		261.80	14.02	15.42
24 L.H., Daily Totals		$1283.60		$2050.84	$53.48	$85.45
Crew E-11					Bare Costs	Incl. O&P
2 Painters, Struc. Steel	$29.60	$473.60	$53.45	$855.20	$30.32	$52.41
1 Building Laborer	26.70	213.60	44.40	355.20		
1 Equip. Oper. (light)	35.40	283.20	58.35	466.80		
1 Air Compressor, 250 cfm		167.40		184.14		
1 Sandblaster, Portable, 3 C.F.		21.00		23.10		
1 Set Sand Blasting Accessories		14.70		16.17	6.35	6.98
32 L.H., Daily Totals		$1173.50		$1900.61	$36.67	$59.39

Crew No.	Bare Costs		Incl. Subs O&P		Cost Per Labor-Hour	
Crew E-11A	**Hr.**	**Daily**	**Hr.**	**Daily**	**Bare Costs**	**Incl. O&P**
2 Painters, Struc. Steel	$29.60	$473.60	$53.45	$855.20	$30.32	$52.41
1 Building Laborer	26.70	213.60	44.40	355.20		
1 Equip. Oper. (light)	35.40	283.20	58.35	466.80		
1 Air Compressor, 250 cfm		167.40		184.14		
1 Sandblaster, Portable, 3 C.F.		21.00		23.10		
1 Set Sand Blasting Accessories		14.70		16.17		
1 Telescoping Boom Lift, to 60'		454.60		500.06	20.55	22.61
32 L.H., Daily Totals		$1628.10		$2400.67	$50.88	$75.02
Crew E-11B	**Hr.**	**Daily**	**Hr.**	**Daily**	**Bare Costs**	**Incl. O&P**
2 Painters, Struc. Steel	$29.60	$473.60	$53.45	$855.20	$28.63	$50.43
1 Building Laborer	26.70	213.60	44.40	355.20		
2 Paint Sprayers, 8 C.F.M.		90.20		99.22		
1 Telescoping Boom Lift, to 60'		454.60		500.06	22.70	24.97
24 L.H., Daily Totals		$1232.00		$1809.68	$51.33	$75.40
Crew E-12	**Hr.**	**Daily**	**Hr.**	**Daily**	**Bare Costs**	**Incl. O&P**
1 Welder Foreman (outside)	$40.80	$326.40	$72.40	$579.20	$38.10	$65.38
1 Equip. Oper. (light)	35.40	283.20	58.35	466.80		
1 Welder, Gas Engine, 300 amp		98.40		108.24	6.15	6.76
16 L.H., Daily Totals		$708.00		$1154.24	$44.25	$72.14
Crew E-13	**Hr.**	**Daily**	**Hr.**	**Daily**	**Bare Costs**	**Incl. O&P**
1 Welder Foreman (outside)	$40.80	$326.40	$72.40	$579.20	$39.00	$67.72
.5 Equip. Oper. (light)	35.40	141.60	58.35	233.40		
1 Welder, Gas Engine, 300 amp		98.40		108.24	8.20	9.02
12 L.H., Daily Totals		$566.40		$920.84	$47.20	$76.74
Crew E-14	**Hr.**	**Daily**	**Hr.**	**Daily**	**Bare Costs**	**Incl. O&P**
1 Struc. Steel Worker	$38.80	$310.40	$68.85	$550.80	$38.80	$68.85
1 Welder, Gas Engine, 300 amp		98.40		108.24	12.30	13.53
8 L.H., Daily Totals		$408.80		$659.04	$51.10	$82.38
Crew E-16	**Hr.**	**Daily**	**Hr.**	**Daily**	**Bare Costs**	**Incl. O&P**
1 Welder Foreman (outside)	$40.80	$326.40	$72.40	$579.20	$39.80	$70.63
1 Welder	38.80	310.40	68.85	550.80		
1 Welder, Gas Engine, 300 amp		98.40		108.24	6.15	6.76
16 L.H., Daily Totals		$735.20		$1238.24	$45.95	$77.39
Crew E-17	**Hr.**	**Daily**	**Hr.**	**Daily**	**Bare Costs**	**Incl. O&P**
1 Struc. Steel Foreman (outside)	$40.80	$326.40	$72.40	$579.20	$39.80	$70.63
1 Structural Steel Worker	38.80	310.40	68.85	550.80		
16 L.H., Daily Totals		$636.80		$1130.00	$39.80	$70.63
Crew E-18	**Hr.**	**Daily**	**Hr.**	**Daily**	**Bare Costs**	**Incl. O&P**
1 Struc. Steel Foreman (outside)	$40.80	$326.40	$72.40	$579.20	$38.75	$67.84
3 Structural Steel Workers	38.80	931.20	68.85	1652.40		
1 Equipment Operator (med.)	36.55	292.40	60.25	482.00		
1 Lattice Boom Crane, 20 Ton		1083.00		1191.30	27.07	29.78
40 L.H., Daily Totals		$2633.00		$3904.90	$65.83	$97.62
Crew E-19	**Hr.**	**Daily**	**Hr.**	**Daily**	**Bare Costs**	**Incl. O&P**
1 Struc. Steel Foreman (outside)	$40.80	$326.40	$72.40	$579.20	$38.33	$66.53
1 Structural Steel Worker	38.80	310.40	68.85	550.80		
1 Equip. Oper. (light)	35.40	283.20	58.35	466.80		
1 Lattice Boom Crane, 20 Ton		1083.00		1191.30	45.13	49.64
24 L.H., Daily Totals		$2003.00		$2788.10	$83.46	$116.17

Crew No.	Bare Costs		Incl. Subs O&P		Cost Per Labor-Hour	
Crew E-20	**Hr.**	**Daily**	**Hr.**	**Daily**	**Bare Costs**	**Incl. O&P**
1 Struc. Steel Foreman (outside)	$40.80	$326.40	$72.40	$579.20	$38.05	$66.42
5 Structural Steel Workers	38.80	1552.00	68.85	2754.00		
1 Equip. Oper. (crane)	37.05	296.40	61.05	488.40		
1 Equip. Oper. (oiler)	32.55	260.40	53.65	429.20		
1 Lattice Boom Crane, 40 Ton		1321.00		1453.10	20.64	22.70
64 L.H., Daily Totals		$3756.20		$5703.90	$58.69	$89.12
Crew E-22	**Hr.**	**Daily**	**Hr.**	**Daily**	**Bare Costs**	**Incl. O&P**
1 Skilled Worker Foreman (out)	$37.80	$302.40	$63.15	$505.20	$36.47	$60.92
2 Skilled Workers	35.80	572.80	59.80	956.80		
24 L.H., Daily Totals		$875.20		$1462.00	$36.47	$60.92
Crew E-24	**Hr.**	**Daily**	**Hr.**	**Daily**	**Bare Costs**	**Incl. O&P**
3 Structural Steel Workers	$38.80	$931.20	$68.85	$1652.40	$38.24	$66.70
1 Equipment Operator (med.)	36.55	292.40	60.25	482.00		
1 Hyd. Crane, 25 Ton		590.60		649.66	18.46	20.30
32 L.H., Daily Totals		$1814.20		$2784.06	$56.69	$87.00
Crew E-25	**Hr.**	**Daily**	**Hr.**	**Daily**	**Bare Costs**	**Incl. O&P**
1 Welder	$38.80	$310.40	$68.85	$550.80	$38.80	$68.85
1 Cutting Torch		12.60		13.86	1.58	1.73
8 L.H., Daily Totals		$323.00		$564.66	$40.38	$70.58
Crew E-26	**Hr.**	**Daily**	**Hr.**	**Daily**	**Bare Costs**	**Incl. O&P**
1 Struc. Steel Foreman (outside)	$40.80	$326.40	$72.40	$579.20	$39.54	$69.40
1 Struc. Steel Worker	38.80	310.40	68.85	550.80		
1 Welder	38.80	310.40	68.85	550.80		
.25 Electrician	40.15	80.30	65.65	131.30		
.25 Plumber	39.80	79.60	65.50	131.00		
1 Welder, Gas Engine, 300 amp		98.40		108.24	3.51	3.87
28 L.H., Daily Totals		$1205.50		$2051.34	$43.05	$73.26
Crew E-27	**Hr.**	**Daily**	**Hr.**	**Daily**	**Bare Costs**	**Incl. O&P**
1 Struc. Steel Foreman (outside)	$40.80	$326.40	$72.40	$579.20	$38.83	$68.32
6 Struc. Steel Workers	38.80	1862.40	68.85	3304.80		
1 Equip. Oper. (crane)	37.05	296.40	61.05	488.40		
1 Hyd. Crane, 12 Ton		496.00		545.60		
1 Hyd. Crane, 80 Ton		1518.00		1669.80	31.47	34.62
64 L.H., Daily Totals		$4499.20		$6587.80	$70.30	$102.93
Crew F-3	**Hr.**	**Daily**	**Hr.**	**Daily**	**Bare Costs**	**Incl. O&P**
2 Carpenters	$34.95	$559.20	$58.10	$929.60	$32.01	$53.33
2 Carpenter Helpers	26.55	424.80	44.70	715.20		
1 Equip. Oper. (crane)	37.05	296.40	61.05	488.40		
1 Hyd. Crane, 12 Ton		496.00		545.60	12.40	13.64
40 L.H., Daily Totals		$1776.40		$2678.80	$44.41	$66.97
Crew F-4	**Hr.**	**Daily**	**Hr.**	**Daily**	**Bare Costs**	**Incl. O&P**
2 Carpenters	$34.95	$559.20	$58.10	$929.60	$32.01	$53.33
2 Carpenter Helpers	26.55	424.80	44.70	715.20		
1 Equip. Oper. (crane)	37.05	296.40	61.05	488.40		
1 Hyd. Crane, 55 Ton		993.80		1093.18	24.84	27.33
40 L.H., Daily Totals		$2274.20		$3226.38	$56.85	$80.66
Crew F-5	**Hr.**	**Daily**	**Hr.**	**Daily**	**Bare Costs**	**Incl. O&P**
2 Carpenters	$34.95	$559.20	$58.10	$929.60	$30.75	$51.40
2 Carpenter Helpers	26.55	424.80	44.70	715.20		
32 L.H., Daily Totals		$984.00		$1644.80	$30.75	$51.40

For customer support on your Residential Costs with RSMeans data, call 800.448.8182.

Crew No.	Bare Costs		Incl. Subs O&P		Cost Per Labor-Hour	
Crew F-6	Hr.	Daily	Hr.	Daily	Bare Costs	Incl. O&P
2 Carpenters	$34.95	$559.20	$58.10	$929.60	$32.07	$53.21
2 Building Laborers	26.70	427.20	44.40	710.40		
1 Equip. Oper. (crane)	37.05	296.40	61.05	488.40		
1 Hyd. Crane, 12 Ton		496.00		545.60	12.40	13.64
40 L.H., Daily Totals		$1778.80		$2674.00	$44.47	$66.85
Crew F-7	Hr.	Daily	Hr.	Daily	Bare Costs	Incl. O&P
2 Carpenters	$34.95	$559.20	$58.10	$929.60	$30.82	$51.25
2 Building Laborers	26.70	427.20	44.40	710.40		
32 L.H., Daily Totals		$986.40		$1640.00	$30.82	$51.25
Crew G-1	Hr.	Daily	Hr.	Daily	Bare Costs	Incl. O&P
1 Roofer Foreman (outside)	$31.90	$255.20	$58.15	$465.20	$28.03	$51.09
4 Roofers Composition	29.90	956.80	54.50	1744.00		
2 Roofer Helpers	22.35	357.60	40.75	652.00		
1 Application Equipment		181.00		199.10		
1 Tar Kettle/Pot		167.20		183.92		
1 Crew Truck		166.00		182.60	9.18	10.10
56 L.H., Daily Totals		$2083.80		$3426.82	$37.21	$61.19
Crew G-2	Hr.	Daily	Hr.	Daily	Bare Costs	Incl. O&P
1 Plasterer	$33.90	$271.20	$55.60	$444.80	$29.87	$49.18
1 Plasterer Helper	29.00	232.00	47.55	380.40		
1 Building Laborer	26.70	213.60	44.40	355.20		
1 Grout Pump, 50 C.F./hr.		128.60		141.46	5.36	5.89
24 L.H., Daily Totals		$845.40		$1321.86	$35.23	$55.08
Crew G-2A	Hr.	Daily	Hr.	Daily	Bare Costs	Incl. O&P
1 Roofer Composition	$29.90	$239.20	$54.50	$436.00	$26.32	$46.55
1 Roofer Helper	22.35	178.80	40.75	326.00		
1 Building Laborer	26.70	213.60	44.40	355.20		
1 Foam Spray Rig, Trailer-Mtd.		499.25		549.17		
1 Pickup Truck, 3/4 Ton		115.20		126.72	25.60	28.16
24 L.H., Daily Totals		$1246.05		$1793.10	$51.92	$74.71
Crew G-3	Hr.	Daily	Hr.	Daily	Bare Costs	Incl. O&P
2 Sheet Metal Workers	$38.25	$612.00	$63.70	$1019.20	$32.48	$54.05
2 Building Laborers	26.70	427.20	44.40	710.40		
32 L.H., Daily Totals		$1039.20		$1729.60	$32.48	$54.05
Crew G-4	Hr.	Daily	Hr.	Daily	Bare Costs	Incl. O&P
1 Labor Foreman (outside)	$28.70	$229.60	$47.75	$382.00	$27.37	$45.52
2 Building Laborers	26.70	427.20	44.40	710.40		
1 Flatbed Truck, Gas, 1.5 Ton		188.40		207.24		
1 Air Compressor, 160 cfm		117.20		128.92	12.73	14.01
24 L.H., Daily Totals		$962.40		$1428.56	$40.10	$59.52
Crew G-5	Hr.	Daily	Hr.	Daily	Bare Costs	Incl. O&P
1 Roofer Foreman (outside)	$31.90	$255.20	$58.15	$465.20	$27.28	$49.73
2 Roofers Composition	29.90	478.40	54.50	872.00		
2 Roofer Helpers	22.35	357.60	40.75	652.00		
1 Application Equipment		181.00		199.10	4.53	4.98
40 L.H., Daily Totals		$1272.20		$2188.30	$31.81	$54.71
Crew G-6A	Hr.	Daily	Hr.	Daily	Bare Costs	Incl. O&P
2 Roofers Composition	$29.90	$478.40	$54.50	$872.00	$29.90	$54.50
1 Small Compressor, Electric		15.10		16.61		
2 Pneumatic Nailers		46.10		50.71	3.83	4.21
16 L.H., Daily Totals		$539.60		$939.32	$33.73	$58.71

Crew No.	Bare Costs		Incl. Subs O&P		Cost Per Labor-Hour	
Crew G-7	Hr.	Daily	Hr.	Daily	Bare Costs	Incl. O&P
1 Carpenter	$34.95	$279.60	$58.10	$464.80	$34.95	$58.10
1 Small Compressor, Electric		15.10		16.61		
1 Pneumatic Nailer		23.05		25.36	4.77	5.25
8 L.H., Daily Totals		$317.75		$506.76	$39.72	$63.35
Crew H-1	Hr.	Daily	Hr.	Daily	Bare Costs	Incl. O&P
2 Glaziers	$33.95	$543.20	$56.15	$898.40	$36.38	$62.50
2 Struc. Steel Workers	38.80	620.80	68.85	1101.60		
32 L.H., Daily Totals		$1164.00		$2000.00	$36.38	$62.50
Crew H-2	Hr.	Daily	Hr.	Daily	Bare Costs	Incl. O&P
2 Glaziers	$33.95	$543.20	$56.15	$898.40	$31.53	$52.23
1 Building Laborer	26.70	213.60	44.40	355.20		
24 L.H., Daily Totals		$756.80		$1253.60	$31.53	$52.23
Crew H-3	Hr.	Daily	Hr.	Daily	Bare Costs	Incl. O&P
1 Glazier	$33.95	$271.60	$56.15	$449.20	$30.25	$50.42
1 Helper	26.55	212.40	44.70	357.60		
16 L.H., Daily Totals		$484.00		$806.80	$30.25	$50.42
Crew H-4	Hr.	Daily	Hr.	Daily	Bare Costs	Incl. O&P
1 Carpenter	$34.95	$279.60	$58.10	$464.80	$32.63	$54.25
1 Carpenter Helper	26.55	212.40	44.70	357.60		
.5 Electrician	40.15	160.60	65.65	262.60		
20 L.H., Daily Totals		$652.60		$1085.00	$32.63	$54.25
Crew J-1	Hr.	Daily	Hr.	Daily	Bare Costs	Incl. O&P
3 Plasterers	$33.90	$813.60	$55.60	$1334.40	$31.94	$52.38
2 Plasterer Helpers	29.00	464.00	47.55	760.80		
1 Mixing Machine, 6 C.F.		133.20		146.52	3.33	3.66
40 L.H., Daily Totals		$1410.80		$2241.72	$35.27	$56.04
Crew J-2	Hr.	Daily	Hr.	Daily	Bare Costs	Incl. O&P
3 Plasterers	$33.90	$813.60	$55.60	$1334.40	$32.35	$52.92
2 Plasterer Helpers	29.00	464.00	47.55	760.80		
1 Lather	34.40	275.20	55.65	445.20		
1 Mixing Machine, 6 C.F.		133.20		146.52	2.77	3.05
48 L.H., Daily Totals		$1686.00		$2686.92	$35.13	$55.98
Crew J-3	Hr.	Daily	Hr.	Daily	Bare Costs	Incl. O&P
1 Terrazzo Worker	$33.40	$267.20	$54.10	$432.80	$30.98	$50.17
1 Terrazzo Helper	28.55	228.40	46.25	370.00		
1 Floor Grinder, 22" Path		133.60		146.96		
1 Terrazzo Mixer		177.80		195.58	19.46	21.41
16 L.H., Daily Totals		$807.00		$1145.34	$50.44	$71.58
Crew J-4	Hr.	Daily	Hr.	Daily	Bare Costs	Incl. O&P
2 Cement Finishers	$34.70	$555.20	$56.30	$900.80	$32.03	$52.33
1 Laborer	26.70	213.60	44.40	355.20		
1 Floor Grinder, 22" Path		133.60		146.96		
1 Floor Edger, 7" Path		42.95		47.24		
1 Vacuum Pick-Up System		66.95		73.64	10.15	11.16
24 L.H., Daily Totals		$1012.30		$1523.85	$42.18	$63.49

For customer support on your Residential Costs with RSMeans data, call 800.448.8182.

Crew J-4A

Crew No.	Bare Costs Hr.	Daily	Incl. Subs O&P Hr.	Daily	Cost Per Labor-Hour Bare Costs	Incl. O&P
2 Cement Finishers	$34.70	$555.20	$56.30	$900.80	$30.70	$50.35
2 Laborers	26.70	427.20	44.40	710.40		
1 Floor Grinder, 22" Path		133.60		146.96		
1 Floor Edger, 7" Path		42.95		47.24		
1 Vacuum Pick-Up System		66.95		73.64		
1 Floor Auto Scrubber		254.30		279.73	15.56	17.11
32 L.H., Daily Totals		$1480.20		$2158.78	$46.26	$67.46

Crew J-4B

Crew No.	Bare Costs Hr.	Daily	Incl. Subs O&P Hr.	Daily	Cost Per Labor-Hour Bare Costs	Incl. O&P
1 Laborer	$26.70	$213.60	$44.40	$355.20	$26.70	$44.40
1 Floor Auto Scrubber		254.30		279.73	31.79	34.97
8 L.H., Daily Totals		$467.90		$634.93	$58.49	$79.37

Crew J-6

Crew No.	Bare Costs Hr.	Daily	Incl. Subs O&P Hr.	Daily	Cost Per Labor-Hour Bare Costs	Incl. O&P
2 Painters	$29.35	$469.60	$48.30	$772.80	$30.20	$49.84
1 Building Laborer	26.70	213.60	44.40	355.20		
1 Equip. Oper. (light)	35.40	283.20	58.35	466.80		
1 Air Compressor, 250 cfm		167.40		184.14		
1 Sandblaster, Portable, 3 C.F.		21.00		23.10		
1 Set Sand Blasting Accessories		14.70		16.17	6.35	6.98
32 L.H., Daily Totals		$1169.50		$1818.21	$36.55	$56.82

Crew J-7

Crew No.	Bare Costs Hr.	Daily	Incl. Subs O&P Hr.	Daily	Cost Per Labor-Hour Bare Costs	Incl. O&P
2 Painters	$29.35	$469.60	$48.30	$772.80	$29.35	$48.30
1 Floor Belt Sander		16.75		18.43		
1 Floor Sanding Edger		13.15		14.47	1.87	2.06
16 L.H., Daily Totals		$499.50		$805.69	$31.22	$50.36

Crew K-1

Crew No.	Bare Costs Hr.	Daily	Incl. Subs O&P Hr.	Daily	Cost Per Labor-Hour Bare Costs	Incl. O&P
1 Carpenter	$34.95	$279.60	$58.10	$464.80	$33.27	$55.05
1 Truck Driver (light)	31.60	252.80	52.00	416.00		
1 Flatbed Truck, Gas, 3 Ton		238.00		261.80	14.88	16.36
16 L.H., Daily Totals		$770.40		$1142.60	$48.15	$71.41

Crew K-2

Crew No.	Bare Costs Hr.	Daily	Incl. Subs O&P Hr.	Daily	Cost Per Labor-Hour Bare Costs	Incl. O&P
1 Struc. Steel Foreman (outside)	$40.80	$326.40	$72.40	$579.20	$37.07	$64.42
1 Struc. Steel Worker	38.80	310.40	68.85	550.80		
1 Truck Driver (light)	31.60	252.80	52.00	416.00		
1 Flatbed Truck, Gas, 3 Ton		238.00		261.80	9.92	10.91
24 L.H., Daily Totals		$1127.60		$1807.80	$46.98	$75.33

Crew L-1

Crew No.	Bare Costs Hr.	Daily	Incl. Subs O&P Hr.	Daily	Cost Per Labor-Hour Bare Costs	Incl. O&P
.25 Electrician	$40.15	$80.30	$65.65	$131.30	$39.87	$65.53
1 Plumber	39.80	318.40	65.50	524.00		
10 L.H., Daily Totals		$398.70		$655.30	$39.87	$65.53

Crew L-2

Crew No.	Bare Costs Hr.	Daily	Incl. Subs O&P Hr.	Daily	Cost Per Labor-Hour Bare Costs	Incl. O&P
1 Carpenter	$34.95	$279.60	$58.10	$464.80	$30.75	$51.40
1 Carpenter Helper	26.55	212.40	44.70	357.60		
16 L.H., Daily Totals		$492.00		$822.40	$30.75	$51.40

Crew L-3

Crew No.	Bare Costs Hr.	Daily	Incl. Subs O&P Hr.	Daily	Cost Per Labor-Hour Bare Costs	Incl. O&P
1 Carpenter	$34.95	$279.60	$58.10	$464.80	$35.99	$59.61
.25 Electrician	40.15	80.30	65.65	131.30		
10 L.H., Daily Totals		$359.90		$596.10	$35.99	$59.61

Crew L-3A

Crew No.	Bare Costs Hr.	Daily	Incl. Subs O&P Hr.	Daily	Cost Per Labor-Hour Bare Costs	Incl. O&P
1 Carpenter Foreman (outside)	$36.95	$295.60	$61.45	$491.60	$37.38	$62.20
.5 Sheet Metal Worker	38.25	153.00	63.70	254.80		
12 L.H., Daily Totals		$448.60		$746.40	$37.38	$62.20

Crew L-4

Crew No.	Bare Costs Hr.	Daily	Incl. Subs O&P Hr.	Daily	Cost Per Labor-Hour Bare Costs	Incl. O&P
1 Skilled Worker	$35.80	$286.40	$59.80	$478.40	$31.18	$52.25
1 Helper	26.55	212.40	44.70	357.60		
16 L.H., Daily Totals		$498.80		$836.00	$31.18	$52.25

Crew L-5

Crew No.	Bare Costs Hr.	Daily	Incl. Subs O&P Hr.	Daily	Cost Per Labor-Hour Bare Costs	Incl. O&P
1 Struc. Steel Foreman (outside)	$40.80	$326.40	$72.40	$579.20	$38.84	$68.24
5 Struc. Steel Workers	38.80	1552.00	68.85	2754.00		
1 Equip. Oper. (crane)	37.05	296.40	61.05	488.40		
1 Hyd. Crane, 25 Ton		590.60		649.66	10.55	11.60
56 L.H., Daily Totals		$2765.40		$4471.26	$49.38	$79.84

Crew L-5A

Crew No.	Bare Costs Hr.	Daily	Incl. Subs O&P Hr.	Daily	Cost Per Labor-Hour Bare Costs	Incl. O&P
1 Struc. Steel Foreman (outside)	$40.80	$326.40	$72.40	$579.20	$38.86	$67.79
2 Structural Steel Workers	38.80	620.80	68.85	1101.60		
1 Equip. Oper. (crane)	37.05	296.40	61.05	488.40		
1 S.P. Crane, 4x4, 25 Ton		668.40		735.24	20.89	22.98
32 L.H., Daily Totals		$1912.00		$2904.44	$59.75	$90.76

Crew L-5B

Crew No.	Bare Costs Hr.	Daily	Incl. Subs O&P Hr.	Daily	Cost Per Labor-Hour Bare Costs	Incl. O&P
1 Struc. Steel Foreman (outside)	$40.80	$326.40	$72.40	$579.20	$38.26	$64.63
2 Structural Steel Workers	38.80	620.80	68.85	1101.60		
2 Electricians	40.15	642.40	65.65	1050.40		
2 Steamfitters/Pipefitters	40.95	655.20	67.40	1078.40		
1 Equip. Oper. (crane)	37.05	296.40	61.05	488.40		
1 Common Laborer	26.70	213.60	44.40	355.20		
1 Hyd. Crane, 80 Ton		1518.00		1669.80	21.08	23.19
72 L.H., Daily Totals		$4272.80		$6323.00	$59.34	$87.82

Crew L-6

Crew No.	Bare Costs Hr.	Daily	Incl. Subs O&P Hr.	Daily	Cost Per Labor-Hour Bare Costs	Incl. O&P
1 Plumber	$39.80	$318.40	$65.50	$524.00	$39.92	$65.55
.5 Electrician	40.15	160.60	65.65	262.60		
12 L.H., Daily Totals		$479.00		$786.60	$39.92	$65.55

Crew L-7

Crew No.	Bare Costs Hr.	Daily	Incl. Subs O&P Hr.	Daily	Cost Per Labor-Hour Bare Costs	Incl. O&P
1 Carpenter	$34.95	$279.60	$58.10	$464.80	$30.18	$50.43
2 Carpenter Helpers	26.55	424.80	44.70	715.20		
.25 Electrician	40.15	80.30	65.65	131.30		
26 L.H., Daily Totals		$784.70		$1311.30	$30.18	$50.43

Crew L-8

Crew No.	Bare Costs Hr.	Daily	Incl. Subs O&P Hr.	Daily	Cost Per Labor-Hour Bare Costs	Incl. O&P
1 Carpenter	$34.95	$279.60	$58.10	$464.80	$32.56	$54.22
1 Carpenter Helper	26.55	212.40	44.70	357.60		
.5 Plumber	39.80	159.20	65.50	262.00		
20 L.H., Daily Totals		$651.20		$1084.40	$32.56	$54.22

Crew L-9

Crew No.	Bare Costs Hr.	Daily	Incl. Subs O&P Hr.	Daily	Cost Per Labor-Hour Bare Costs	Incl. O&P
1 Skilled Worker Foreman	$37.80	$302.40	$63.15	$505.20	$32.62	$54.48
1 Skilled Worker	35.80	286.40	59.80	478.40		
2 Helpers	26.55	424.80	44.70	715.20		
.5 Electrician	40.15	160.60	65.65	262.60		
36 L.H., Daily Totals		$1174.20		$1961.40	$32.62	$54.48

Crew L-10

Crew No.	Bare Costs Hr.	Daily	Incl. Subs O&P Hr.	Daily	Cost Per Labor-Hour Bare Costs	Incl. O&P
1 Struc. Steel Foreman (outside)	$40.80	$326.40	$72.40	$579.20	$38.88	$67.43
1 Structural Steel Worker	38.80	310.40	68.85	550.80		
1 Equip. Oper. (crane)	37.05	296.40	61.05	488.40		
1 Hyd. Crane, 12 Ton		496.00		545.60	20.67	22.73
24 L.H., Daily Totals		$1429.20		$2164.00	$59.55	$90.17

Crew L-11

Crew No.	Bare Costs Hr.	Daily	Incl. Subs O&P Hr.	Daily	Cost Per Labor-Hour Bare Costs	Incl. O&P
2 Wreckers	$27.10	$433.60	$46.40	$742.40	$31.66	$53.05
1 Equip. Oper. (crane)	37.05	296.40	61.05	488.40		
1 Equip. Oper. (light)	35.40	283.20	58.35	466.80		
1 Hyd. Excavator, 2.5 C.Y.		1411.00		1552.10		
1 Loader, Skid Steer, 78 H.P.		364.80		401.28	55.49	61.04
32 L.H., Daily Totals		$2789.00		$3650.98	$87.16	$114.09

Crew M-1

Crew No.	Hr.	Daily	Hr.	Daily	Bare Costs	Incl. O&P
3 Elevator Constructors	$54.25	$1302.00	$88.45	$2122.80	$51.54	$84.03
1 Elevator Apprentice	43.40	347.20	70.75	566.00		
5 Hand Tools		50.00		55.00	1.56	1.72
32 L.H., Daily Totals		$1699.20		$2743.80	$53.10	$85.74

Crew M-3

Crew No.	Hr.	Daily	Hr.	Daily	Bare Costs	Incl. O&P
1 Electrician Foreman (outside)	$42.15	$337.20	$68.90	$551.20	$41.33	$67.66
1 Common Laborer	26.70	213.60	44.40	355.20		
.25 Equipment Operator (med.)	36.55	73.10	60.25	120.50		
1 Elevator Constructor	54.25	434.00	88.45	707.60		
1 Elevator Apprentice	43.40	347.20	70.75	566.00		
.25 S.P. Crane, 4x4, 20 Ton		142.80		157.08	4.20	4.62
34 L.H., Daily Totals		$1547.90		$2457.58	$45.53	$72.28

Crew M-4

Crew No.	Hr.	Daily	Hr.	Daily	Bare Costs	Incl. O&P
1 Electrician Foreman (outside)	$42.15	$337.20	$68.90	$551.20	$40.87	$66.93
1 Common Laborer	26.70	213.60	44.40	355.20		
.25 Equipment Operator, Crane	37.05	74.10	61.05	122.10		
.25 Equip. Oper. (oiler)	32.55	65.10	53.65	107.30		
1 Elevator Constructor	54.25	434.00	88.45	707.60		
1 Elevator Apprentice	43.40	347.20	70.75	566.00		
.25 S.P. Crane, 4x4, 40 Ton		183.70		202.07	5.10	5.61
36 L.H., Daily Totals		$1654.90		$2611.47	$45.97	$72.54

Crew Q-1

Crew No.	Hr.	Daily	Hr.	Daily	Bare Costs	Incl. O&P
1 Plumber	$39.80	$318.40	$65.50	$524.00	$35.83	$58.98
1 Plumber Apprentice	31.85	254.80	52.45	419.60		
16 L.H., Daily Totals		$573.20		$943.60	$35.83	$58.98

Crew Q-1A

Crew No.	Hr.	Daily	Hr.	Daily	Bare Costs	Incl. O&P
.25 Plumber Foreman (outside)	$41.80	$83.60	$68.80	$137.60	$40.20	$66.16
1 Plumber	39.80	318.40	65.50	524.00		
10 L.H., Daily Totals		$402.00		$661.60	$40.20	$66.16

Crew Q-1C

Crew No.	Hr.	Daily	Hr.	Daily	Bare Costs	Incl. O&P
1 Plumber	$39.80	$318.40	$65.50	$524.00	$36.07	$59.40
1 Plumber Apprentice	31.85	254.80	52.45	419.60		
1 Equip. Oper. (medium)	36.55	292.40	60.25	482.00		
1 Trencher, Chain Type, 8' D		2048.00		2252.80	85.33	93.87
24 L.H., Daily Totals		$2913.60		$3678.40	$121.40	$153.27

Crew Q-2

Crew No.	Hr.	Daily	Hr.	Daily	Bare Costs	Incl. O&P
1 Plumber	$39.80	$318.40	$65.50	$524.00	$34.50	$56.80
2 Plumber Apprentices	31.85	509.60	52.45	839.20		
24 L.H., Daily Totals		$828.00		$1363.20	$34.50	$56.80

Crew Q-3

Crew No.	Hr.	Daily	Hr.	Daily	Bare Costs	Incl. O&P
2 Plumbers	$39.80	$636.80	$65.50	$1048.00	$35.83	$58.98
2 Plumber Apprentices	31.85	509.60	52.45	839.20		
32 L.H., Daily Totals		$1146.40		$1887.20	$35.83	$58.98

Crew Q-4

Crew No.	Hr.	Daily	Hr.	Daily	Bare Costs	Incl. O&P
2 Plumbers	$39.80	$636.80	$65.50	$1048.00	$37.81	$62.24
1 Welder (plumber)	39.80	318.40	65.50	524.00		
1 Plumber Apprentice	31.85	254.80	52.45	419.60		
1 Welder, Electric, 300 amp		59.20		65.12	1.85	2.04
32 L.H., Daily Totals		$1269.20		$2056.72	$39.66	$64.27

Crew Q-5

Crew No.	Hr.	Daily	Hr.	Daily	Bare Costs	Incl. O&P
1 Steamfitter	$40.95	$327.60	$67.40	$539.20	$36.85	$60.65
1 Steamfitter Apprentice	32.75	262.00	53.90	431.20		
16 L.H., Daily Totals		$589.60		$970.40	$36.85	$60.65

Crew Q-6

Crew No.	Hr.	Daily	Hr.	Daily	Bare Costs	Incl. O&P
1 Steamfitter	$40.95	$327.60	$67.40	$539.20	$35.48	$58.40
2 Steamfitter Apprentices	32.75	524.00	53.90	862.40		
24 L.H., Daily Totals		$851.60		$1401.60	$35.48	$58.40

Crew Q-7

Crew No.	Hr.	Daily	Hr.	Daily	Bare Costs	Incl. O&P
2 Steamfitters	$40.95	$655.20	$67.40	$1078.40	$36.85	$60.65
2 Steamfitter Apprentices	32.75	524.00	53.90	862.40		
32 L.H., Daily Totals		$1179.20		$1940.80	$36.85	$60.65

Crew Q-8

Crew No.	Hr.	Daily	Hr.	Daily	Bare Costs	Incl. O&P
2 Steamfitters	$40.95	$655.20	$67.40	$1078.40	$38.90	$64.03
1 Welder (steamfitter)	40.95	327.60	67.40	539.20		
1 Steamfitter Apprentice	32.75	262.00	53.90	431.20		
1 Welder, Electric, 300 amp		59.20		65.12	1.85	2.04
32 L.H., Daily Totals		$1304.00		$2113.92	$40.75	$66.06

Crew Q-9

Crew No.	Hr.	Daily	Hr.	Daily	Bare Costs	Incl. O&P
1 Sheet Metal Worker	$38.25	$306.00	$63.70	$509.60	$34.42	$57.35
1 Sheet Metal Apprentice	30.60	244.80	51.00	408.00		
16 L.H., Daily Totals		$550.80		$917.60	$34.42	$57.35

Crew Q-10

Crew No.	Hr.	Daily	Hr.	Daily	Bare Costs	Incl. O&P
2 Sheet Metal Workers	$38.25	$612.00	$63.70	$1019.20	$35.70	$59.47
1 Sheet Metal Apprentice	30.60	244.80	51.00	408.00		
24 L.H., Daily Totals		$856.80		$1427.20	$35.70	$59.47

Crew Q-11

Crew No.	Hr.	Daily	Hr.	Daily	Bare Costs	Incl. O&P
2 Sheet Metal Workers	$38.25	$612.00	$63.70	$1019.20	$34.42	$57.35
2 Sheet Metal Apprentices	30.60	489.60	51.00	816.00		
32 L.H., Daily Totals		$1101.60		$1835.20	$34.42	$57.35

Crew Q-12

Crew No.	Hr.	Daily	Hr.	Daily	Bare Costs	Incl. O&P
1 Sprinkler Installer	$38.90	$311.20	$64.20	$513.60	$35.00	$57.75
1 Sprinkler Apprentice	31.10	248.80	51.30	410.40		
16 L.H., Daily Totals		$560.00		$924.00	$35.00	$57.75

Crew Q-13

Crew No.	Hr.	Daily	Hr.	Daily	Bare Costs	Incl. O&P
2 Sprinkler Installers	$38.90	$622.40	$64.20	$1027.20	$35.00	$57.75
2 Sprinkler Apprentices	31.10	497.60	51.30	820.80		
32 L.H., Daily Totals		$1120.00		$1848.00	$35.00	$57.75

Crew Q-14

Crew No.	Hr.	Daily	Hr.	Daily	Bare Costs	Incl. O&P
1 Asbestos Worker	$36.05	$288.40	$61.00	$488.00	$32.45	$54.90
1 Asbestos Apprentice	28.85	230.80	48.80	390.40		
16 L.H., Daily Totals		$519.20		$878.40	$32.45	$54.90

Crews - Residential

Crew Q-15

Crew No.	Bare Costs Hr.	Daily	Incl. Subs O&P Hr.	Daily	Cost Per Labor-Hour Bare Costs	Incl. O&P
1 Plumber	$39.80	$318.40	$65.50	$524.00	$35.83	$58.98
1 Plumber Apprentice	31.85	254.80	52.45	419.60		
1 Welder, Electric, 300 amp		59.20		65.12	3.70	4.07
16 L.H., Daily Totals		$632.40		$1008.72	$39.52	$63.05

Crew Q-16

Crew No.	Bare Costs Hr.	Daily	Incl. Subs O&P Hr.	Daily	Cost Per Labor-Hour Bare Costs	Incl. O&P
2 Plumbers	$39.80	$636.80	$65.50	$1048.00	$37.15	$61.15
1 Plumber Apprentice	31.85	254.80	52.45	419.60		
1 Welder, Electric, 300 amp		59.20		65.12	2.47	2.71
24 L.H., Daily Totals		$950.80		$1532.72	$39.62	$63.86

Crew Q-17

Crew No.	Bare Costs Hr.	Daily	Incl. Subs O&P Hr.	Daily	Cost Per Labor-Hour Bare Costs	Incl. O&P
1 Steamfitter	$40.95	$327.60	$67.40	$539.20	$36.85	$60.65
1 Steamfitter Apprentice	32.75	262.00	53.90	431.20		
1 Welder, Electric, 300 amp		59.20		65.12	3.70	4.07
16 L.H., Daily Totals		$648.80		$1035.52	$40.55	$64.72

Crew Q-17A

Crew No.	Bare Costs Hr.	Daily	Incl. Subs O&P Hr.	Daily	Cost Per Labor-Hour Bare Costs	Incl. O&P
1 Steamfitter	$40.95	$327.60	$67.40	$539.20	$36.92	$60.78
1 Steamfitter Apprentice	32.75	262.00	53.90	431.20		
1 Equip. Oper. (crane)	37.05	296.40	61.05	488.40		
1 Hyd. Crane, 12 Ton		496.00		545.60		
1 Welder, Electric, 300 amp		59.20		65.12	23.13	25.45
24 L.H., Daily Totals		$1441.20		$2069.52	$60.05	$86.23

Crew Q-18

Crew No.	Bare Costs Hr.	Daily	Incl. Subs O&P Hr.	Daily	Cost Per Labor-Hour Bare Costs	Incl. O&P
2 Steamfitters	$40.95	$655.20	$67.40	$1078.40	$38.22	$62.90
1 Steamfitter Apprentice	32.75	262.00	53.90	431.20		
1 Welder, Electric, 300 amp		59.20		65.12	2.47	2.71
24 L.H., Daily Totals		$976.40		$1574.72	$40.68	$65.61

Crew Q-19

Crew No.	Bare Costs Hr.	Daily	Incl. Subs O&P Hr.	Daily	Cost Per Labor-Hour Bare Costs	Incl. O&P
1 Steamfitter	$40.95	$327.60	$67.40	$539.20	$37.95	$62.32
1 Steamfitter Apprentice	32.75	262.00	53.90	431.20		
1 Electrician	40.15	321.20	65.65	525.20		
24 L.H., Daily Totals		$910.80		$1495.60	$37.95	$62.32

Crew Q-20

Crew No.	Bare Costs Hr.	Daily	Incl. Subs O&P Hr.	Daily	Cost Per Labor-Hour Bare Costs	Incl. O&P
1 Sheet Metal Worker	$38.25	$306.00	$63.70	$509.60	$35.57	$59.01
1 Sheet Metal Apprentice	30.60	244.80	51.00	408.00		
.5 Electrician	40.15	160.60	65.65	262.60		
20 L.H., Daily Totals		$711.40		$1180.20	$35.57	$59.01

Crew Q-21

Crew No.	Bare Costs Hr.	Daily	Incl. Subs O&P Hr.	Daily	Cost Per Labor-Hour Bare Costs	Incl. O&P
2 Steamfitters	$40.95	$655.20	$67.40	$1078.40	$38.70	$63.59
1 Steamfitter Apprentice	32.75	262.00	53.90	431.20		
1 Electrician	40.15	321.20	65.65	525.20		
32 L.H., Daily Totals		$1238.40		$2034.80	$38.70	$63.59

Crew Q-22

Crew No.	Bare Costs Hr.	Daily	Incl. Subs O&P Hr.	Daily	Cost Per Labor-Hour Bare Costs	Incl. O&P
1 Plumber	$39.80	$318.40	$65.50	$524.00	$35.83	$58.98
1 Plumber Apprentice	31.85	254.80	52.45	419.60		
1 Hyd. Crane, 12 Ton		496.00		545.60	31.00	34.10
16 L.H., Daily Totals		$1069.20		$1489.20	$66.83	$93.08

Crew Q-22A

Crew No.	Bare Costs Hr.	Daily	Incl. Subs O&P Hr.	Daily	Cost Per Labor-Hour Bare Costs	Incl. O&P
1 Plumber	$39.80	$318.40	$65.50	$524.00	$33.85	$55.85
1 Plumber Apprentice	31.85	254.80	52.45	419.60		
1 Laborer	26.70	213.60	44.40	355.20		
1 Equip. Oper. (crane)	37.05	296.40	61.05	488.40		
1 Hyd. Crane, 12 Ton		496.00		545.60	15.50	17.05
32 L.H., Daily Totals		$1579.20		$2332.80	$49.35	$72.90

Crew Q-23

Crew No.	Bare Costs Hr.	Daily	Incl. Subs O&P Hr.	Daily	Cost Per Labor-Hour Bare Costs	Incl. O&P
1 Plumber Foreman (outside)	$41.80	$334.40	$68.80	$550.40	$39.38	$64.85
1 Plumber	39.80	318.40	65.50	524.00		
1 Equip. Oper. (medium)	36.55	292.40	60.25	482.00		
1 Lattice Boom Crane, 20 Ton		1083.00		1191.30	45.13	49.64
24 L.H., Daily Totals		$2028.20		$2747.70	$84.51	$114.49

Crew R-1

Crew No.	Bare Costs Hr.	Daily	Incl. Subs O&P Hr.	Daily	Cost Per Labor-Hour Bare Costs	Incl. O&P
1 Electrician Foreman	$40.65	$325.20	$66.45	$531.60	$37.55	$61.40
3 Electricians	40.15	963.60	65.65	1575.60		
2 Electrician Apprentices	32.10	513.60	52.50	840.00		
48 L.H., Daily Totals		$1802.40		$2947.20	$37.55	$61.40

Crew R-1A

Crew No.	Bare Costs Hr.	Daily	Incl. Subs O&P Hr.	Daily	Cost Per Labor-Hour Bare Costs	Incl. O&P
1 Electrician	$40.15	$321.20	$65.65	$525.20	$36.13	$59.08
1 Electrician Apprentice	32.10	256.80	52.50	420.00		
16 L.H., Daily Totals		$578.00		$945.20	$36.13	$59.08

Crew R-1B

Crew No.	Bare Costs Hr.	Daily	Incl. Subs O&P Hr.	Daily	Cost Per Labor-Hour Bare Costs	Incl. O&P
1 Electrician	$40.15	$321.20	$65.65	$525.20	$34.78	$56.88
2 Electrician Apprentices	32.10	513.60	52.50	840.00		
24 L.H., Daily Totals		$834.80		$1365.20	$34.78	$56.88

Crew R-1C

Crew No.	Bare Costs Hr.	Daily	Incl. Subs O&P Hr.	Daily	Cost Per Labor-Hour Bare Costs	Incl. O&P
2 Electricians	$40.15	$642.40	$65.65	$1050.40	$36.13	$59.08
2 Electrician Apprentices	32.10	513.60	52.50	840.00		
1 Portable Cable Puller, 8000 lb.		130.15		143.16	4.07	4.47
32 L.H., Daily Totals		$1286.15		$2033.57	$40.19	$63.55

Crew R-2

Crew No.	Bare Costs Hr.	Daily	Incl. Subs O&P Hr.	Daily	Cost Per Labor-Hour Bare Costs	Incl. O&P
1 Electrician Foreman	$40.65	$325.20	$66.45	$531.60	$37.48	$61.35
3 Electricians	40.15	963.60	65.65	1575.60		
2 Electrician Apprentices	32.10	513.60	52.50	840.00		
1 Equip. Oper. (crane)	37.05	296.40	61.05	488.40		
1 S.P. Crane, 4x4, 5 Ton		259.00		284.90	4.63	5.09
56 L.H., Daily Totals		$2357.80		$3720.50	$42.10	$66.44

Crew R-3

Crew No.	Bare Costs Hr.	Daily	Incl. Subs O&P Hr.	Daily	Cost Per Labor-Hour Bare Costs	Incl. O&P
1 Electrician Foreman	$40.65	$325.20	$66.45	$531.60	$39.73	$65.05
1 Electrician	40.15	321.20	65.65	525.20		
.5 Equip. Oper. (crane)	37.05	148.20	61.05	244.20		
.5 S.P. Crane, 4x4, 5 Ton		129.50		142.45	6.47	7.12
20 L.H., Daily Totals		$924.10		$1443.45	$46.20	$72.17

Crew R-4

Crew No.	Bare Costs Hr.	Daily	Incl. Subs O&P Hr.	Daily	Cost Per Labor-Hour Bare Costs	Incl. O&P
1 Struc. Steel Foreman (outside)	$40.80	$326.40	$72.40	$579.20	$39.47	$68.92
3 Struc. Steel Workers	38.80	931.20	68.85	1652.40		
1 Electrician	40.15	321.20	65.65	525.20		
1 Welder, Gas Engine, 300 amp		98.40		108.24	2.46	2.71
40 L.H., Daily Totals		$1677.20		$2865.04	$41.93	$71.63

For customer support on your Residential Costs with RSMeans data, call 800.448.8182.

Crew No.	Bare Costs		Incl. Subs O&P		Cost Per Labor-Hour	

Crew R-5

Crew R-5	Hr.	Daily	Hr.	Daily	Bare Costs	Incl. O&P
1 Electrician Foreman	$40.65	$325.20	$66.45	$531.60	$35.25	$58.10
4 Electrician Linemen	40.15	1284.80	65.65	2100.80		
2 Electrician Operators	40.15	642.40	65.65	1050.40		
4 Electrician Groundmen	26.55	849.60	44.70	1430.40		
1 Crew Truck		166.00		182.60		
1 Flatbed Truck, 20,000 GVW		197.20		216.92		
1 Pickup Truck, 3/4 Ton		115.20		126.72		
.2 Hyd. Crane, 55 Ton		198.76		218.64		
.2 Hyd. Crane, 12 Ton		99.20		109.12		
.2 Earth Auger, Truck-Mtd.		74.76		82.24		
1 Tractor w/Winch		361.80		397.98	13.78	15.16
88 L.H., Daily Totals		$4314.92		$6447.41	$49.03	$73.27

Crew R-6

Crew R-6	Hr.	Daily	Hr.	Daily	Bare Costs	Incl. O&P
1 Electrician Foreman	$40.65	$325.20	$66.45	$531.60	$35.25	$58.10
4 Electrician Linemen	40.15	1284.80	65.65	2100.80		
2 Electrician Operators	40.15	642.40	65.65	1050.40		
4 Electrician Groundmen	26.55	849.60	44.70	1430.40		
1 Crew Truck		166.00		182.60		
1 Flatbed Truck, 20,000 GVW		197.20		216.92		
1 Pickup Truck, 3/4 Ton		115.20		126.72		
.2 Hyd. Crane, 55 Ton		198.76		218.64		
.2 Hyd. Crane, 12 Ton		99.20		109.12		
.2 Earth Auger, Truck-Mtd.		74.76		82.24		
1 Tractor w/Winch		361.80		397.98		
3 Cable Trailers		635.10		698.61		
.5 Tensioning Rig		209.93		230.92		
.5 Cable Pulling Rig		1213.00		1334.30	37.17	40.89
88 L.H., Daily Totals		$6372.94		$8711.24	$72.42	$98.99

Crew R-7

Crew R-7	Hr.	Daily	Hr.	Daily	Bare Costs	Incl. O&P
1 Electrician Foreman	$40.65	$325.20	$66.45	$531.60	$28.90	$48.33
5 Electrician Groundmen	26.55	1062.00	44.70	1788.00		
1 Crew Truck		166.00		182.60	3.46	3.80
48 L.H., Daily Totals		$1553.20		$2502.20	$32.36	$52.13

Crew R-8

Crew R-8	Hr.	Daily	Hr.	Daily	Bare Costs	Incl. O&P
1 Electrician Foreman	$40.65	$325.20	$66.45	$531.60	$35.70	$58.80
3 Electrician Linemen	40.15	963.60	65.65	1575.60		
2 Electrician Groundmen	26.55	424.80	44.70	715.20		
1 Pickup Truck, 3/4 Ton		115.20		126.72		
1 Crew Truck		166.00		182.60	5.86	6.44
48 L.H., Daily Totals		$1994.80		$3131.72	$41.56	$65.24

Crew R-9

Crew R-9	Hr.	Daily	Hr.	Daily	Bare Costs	Incl. O&P
1 Electrician Foreman	$40.65	$325.20	$66.45	$531.60	$33.41	$55.27
1 Electrician Lineman	40.15	321.20	65.65	525.20		
2 Electrician Operators	40.15	642.40	65.65	1050.40		
4 Electrician Groundmen	26.55	849.60	44.70	1430.40		
1 Pickup Truck, 3/4 Ton		115.20		126.72		
1 Crew Truck		166.00		182.60	4.39	4.83
64 L.H., Daily Totals		$2419.60		$3846.92	$37.81	$60.11

Crew R-10

Crew R-10	Hr.	Daily	Hr.	Daily	Bare Costs	Incl. O&P
1 Electrician Foreman	$40.65	$325.20	$66.45	$531.60	$37.97	$62.29
4 Electrician Linemen	40.15	1284.80	65.65	2100.80		
1 Electrician Groundman	26.55	212.40	44.70	357.60		
1 Crew Truck		166.00		182.60		
3 Tram Cars		433.80		477.18	12.50	13.75
48 L.H., Daily Totals		$2422.20		$3649.78	$50.46	$76.04

Crew R-11

Crew R-11	Hr.	Daily	Hr.	Daily	Bare Costs	Incl. O&P
1 Electrician Foreman	$40.65	$325.20	$66.45	$531.60	$37.86	$62.07
4 Electricians	40.15	1284.80	65.65	2100.80		
1 Equip. Oper. (crane)	37.05	296.40	61.05	488.40		
1 Common Laborer	26.70	213.60	44.40	355.20		
1 Crew Truck		166.00		182.60		
1 Hyd. Crane, 12 Ton		496.00		545.60	11.82	13.00
56 L.H., Daily Totals		$2782.00		$4204.20	$49.68	$75.08

Crew R-12

Crew R-12	Hr.	Daily	Hr.	Daily	Bare Costs	Incl. O&P
1 Carpenter Foreman (inside)	$35.45	$283.60	$58.95	$471.60	$32.49	$54.37
4 Carpenters	34.95	1118.40	58.10	1859.20		
4 Common Laborers	26.70	854.40	44.40	1420.80		
1 Equip. Oper. (medium)	36.55	292.40	60.25	482.00		
1 Steel Worker	38.80	310.40	68.85	550.80		
1 Dozer, 200 H.P.		1273.00		1400.30		
1 Pickup Truck, 3/4 Ton		115.20		126.72	15.78	17.35
88 L.H., Daily Totals		$4247.40		$6311.42	$48.27	$71.72

Crew R-13

Crew R-13	Hr.	Daily	Hr.	Daily	Bare Costs	Incl. O&P
1 Electrician Foreman	$40.65	$325.20	$66.45	$531.60	$38.65	$63.30
3 Electricians	40.15	963.60	65.65	1575.60		
.25 Equip. Oper. (crane)	37.05	74.10	61.05	122.10		
1 Equipment Oiler	32.55	260.40	53.65	429.20		
.25 Hydraulic Crane, 33 Ton		235.15		258.67	5.60	6.16
42 L.H., Daily Totals		$1858.45		$2917.17	$44.25	$69.46

Crew R-15

Crew R-15	Hr.	Daily	Hr.	Daily	Bare Costs	Incl. O&P
1 Electrician Foreman	$40.65	$325.20	$66.45	$531.60	$39.44	$64.57
4 Electricians	40.15	1284.80	65.65	2100.80		
1 Equipment Oper. (light)	35.40	283.20	58.35	466.80		
1 Telescoping Boom Lift, to 40'		282.80		311.08	5.89	6.48
48 L.H., Daily Totals		$2176.00		$3410.28	$45.33	$71.05

Crew R-15A

Crew R-15A	Hr.	Daily	Hr.	Daily	Bare Costs	Incl. O&P
1 Electrician Foreman	$40.65	$325.20	$66.45	$531.60	$34.96	$57.48
2 Electricians	40.15	642.40	65.65	1050.40		
2 Common Laborers	26.70	427.20	44.40	710.40		
1 Equip. Oper. (light)	35.40	283.20	58.35	466.80		
1 Telescoping Boom Lift, to 40'		282.80		311.08	5.89	6.48
48 L.H., Daily Totals		$1960.80		$3070.28	$40.85	$63.96

Crew R-18

Crew R-18	Hr.	Daily	Hr.	Daily	Bare Costs	Incl. O&P
.25 Electrician Foreman	$40.65	$81.30	$66.45	$132.90	$35.23	$57.62
1 Electrician	40.15	321.20	65.65	525.20		
2 Electrician Apprentices	32.10	513.60	52.50	840.00		
26 L.H., Daily Totals		$916.10		$1498.10	$35.23	$57.62

Crew R-19

Crew R-19	Hr.	Daily	Hr.	Daily	Bare Costs	Incl. O&P
.5 Electrician Foreman	$40.65	$162.60	$66.45	$265.80	$40.25	$65.81
2 Electricians	40.15	642.40	65.65	1050.40		
20 L.H., Daily Totals		$805.00		$1316.20	$40.25	$65.81

Crew R-21

Crew R-21	Hr.	Daily	Hr.	Daily	Bare Costs	Incl. O&P
1 Electrician Foreman	$40.65	$325.20	$66.45	$531.60	$40.18	$65.71
3 Electricians	40.15	963.60	65.65	1575.60		
.1 Equip. Oper. (medium)	36.55	29.24	60.25	48.20		
.1 S.P. Crane, 4x4, 25 Ton		66.84		73.52	2.04	2.24
32.8 L.H., Daily Totals		$1384.88		$2228.92	$42.22	$67.95

Crews - Residential

Crew No.	Bare Costs		Incl. Subs O & P		Cost Per Labor-Hour	

Crew R-22	Hr.	Daily	Hr.	Daily	Bare Costs	Incl. O&P
.66 Electrician Foreman	$40.65	$214.63	$66.45	$350.86	$36.77	$60.12
2 Electricians	40.15	642.40	65.65	1050.40		
2 Electrician Apprentices	32.10	513.60	52.50	840.00		
37.28 L.H., Daily Totals		$1370.63		$2241.26	$36.77	$60.12

Crew R-30	Hr.	Daily	Hr.	Daily	Bare Costs	Incl. O&P
.25 Electrician Foreman (outside)	$42.15	$84.30	$68.90	$137.80	$32.03	$52.82
1 Electrician	40.15	321.20	65.65	525.20		
2 Laborers (Semi-Skilled)	26.70	427.20	44.40	710.40		
26 L.H., Daily Totals		$832.70		$1373.40	$32.03	$52.82

For customer support on your Residential Costs with RSMeans data, call 800.448.8182.

Costs shown in *Residential Costs with RSMeans data* are based on national averages for materials and installation. To adjust these costs to a specific location, simply multiply the base cost by the factor for that city. The data is arranged alphabetically by state and postal zip code numbers. For a city not listed, use the factor for a nearby city with similar economic characteristics.

STATE	CITY	Residential
ALABAMA		
350-352	Birmingham	.83
354	Tuscaloosa	.84
355	Jasper	.83
356	Decatur	.83
357-358	Huntsville	.83
359	Gadsden	.82
360-361	Montgomery	.82
362	Anniston	.81
363	Dothan	.84
364	Evergreen	.81
365-366	Mobile	.83
367	Selma	.82
368	Phenix City	.83
369	Butler	.81
ALASKA		
995-996	Anchorage	1.22
997	Fairbanks	1.25
998	Juneau	1.23
999	Ketchikan	1.26
ARIZONA		
850,853	Phoenix	.87
851,852	Mesa/Tempe	.86
855	Globe	.85
856-857	Tucson	.85
859	Show Low	.86
860	Flagstaff	.87
863	Prescott	.86
864	Kingman	.85
865	Chambers	.87
ARKANSAS		
716	Pine Bluff	.78
717	Camden	.76
718	Texarkana	.78
719	Hot Springs	.75
720-722	Little Rock	.80
723	West Memphis	.78
724	Jonesboro	.77
725	Batesville	.75
726	Harrison	.76
727	Fayetteville	.74
728	Russellville	.75
729	Fort Smith	.81
CALIFORNIA		
900-902	Los Angeles	1.15
903-905	Inglewood	1.12
906-908	Long Beach	1.12
910-912	Pasadena	1.11
913-916	Van Nuys	1.14
917-918	Alhambra	1.15
919-921	San Diego	1.10
922	Palm Springs	1.11
923-924	San Bernardino	1.14
925	Riverside	1.14
926-927	Santa Ana	1.13
928	Anaheim	1.14
930	Oxnard	1.14
931	Santa Barbara	1.14
932-933	Bakersfield	1.11
934	San Luis Obispo	1.15
935	Mojave	1.12
936-938	Fresno	1.18
939	Salinas	1.21
940-941	San Francisco	1.33
942,956-958	Sacramento	1.19
943	Palo Alto	1.27
944	San Mateo	1.30
945	Vallejo	1.23
946	Oakland	1.29
947	Berkeley	1.32
948	Richmond	1.31
949	San Rafael	1.30
950	Santa Cruz	1.25
951	San Jose	1.30
952	Stockton	1.19
953	Modesto	1.18

STATE	CITY	Residential
CALIFORNIA (CONT'D)		
954	Santa Rosa	1.27
955	Eureka	1.23
959	Marysville	1.20
960	Redding	1.24
961	Susanville	1.21
COLORADO		
800-802	Denver	.88
803	Boulder	.90
804	Golden	.85
805	Fort Collins	.88
806	Greeley	.88
807	Fort Morgan	.85
808-809	Colorado Springs	.84
810	Pueblo	.84
811	Alamosa	.84
812	Salida	.82
813	Durango	.86
814	Montrose	.81
815	Grand Junction	.92
816	Glenwood Springs	.81
CONNECTICUT		
060	New Britain	1.11
061	Hartford	1.09
062	Willimantic	1.11
063	New London	1.10
064	Meriden	1.10
065	New Haven	1.11
066	Bridgeport	1.12
067	Waterbury	1.11
068	Norwalk	1.11
069	Stamford	1.12
D.C.		
200-205	Washington	.92
DELAWARE		
197	Newark	1.02
198	Wilmington	1.01
199	Dover	1.02
FLORIDA		
320,322	Jacksonville	.80
321	Daytona Beach	.83
323	Tallahassee	.80
324	Panama City	.81
325	Pensacola	.84
326,344	Gainesville	.80
327-328,347	Orlando	.81
329	Melbourne	.83
330-332,340	Miami	.79
333	Fort Lauderdale	.80
334,349	West Palm Beach	.80
335-336,346	Tampa	.82
337	St. Petersburg	.81
338	Lakeland	.79
339,341	Fort Myers	.79
342	Sarasota	.83
GEORGIA		
300-303,399	Atlanta	.88
304	Statesboro	.79
305	Gainesville	.82
306	Athens	.81
307	Dalton	.84
308-309	Augusta	.87
310-312	Macon	.84
313-314	Savannah	.85
315	Waycross	.81
316	Valdosta	.75
317,398	Albany	.84
318-319	Columbus	.84
HAWAII		
967	Hilo	1.19
968	Honolulu	1.21

Location Factors - Residential

STATE	CITY	Residential		STATE	CITY	Residential
STATES & POSS.				**KENTUCKY (CONT'D)**		
969	Guam	.95		406	Frankfort	.86
				407-409	Corbin	.82
IDAHO				410	Covington	.85
832	Pocatello	.89		411-412	Ashland	.90
833	Twin Falls	.89		413-414	Campton	.87
834	Idaho Falls	.87		415-416	Pikeville	.86
835	Lewiston	.98		417-418	Hazard	.86
836-837	Boise	.90		420	Paducah	.88
838	Coeur d'Alene	.98		421-422	Bowling Green	.87
				423	Owensboro	.88
ILLINOIS				424	Henderson	.88
600-603	North Suburban	1.22		425-426	Somerset	.84
604	Joliet	1.24		427	Elizabethtown	.84
605	South Suburban	1.22				
606-608	Chicago	1.26		**LOUISIANA**		
609	Kankakee	1.16		700-701	New Orleans	.86
610-611	Rockford	1.13		703	Thibodaux	.82
612	Rock Island	.98		704	Hammond	.78
613	La Salle	1.12		705	Lafayette	.86
614	Galesburg	1.04		706	Lake Charles	.87
615-616	Peoria	1.07		707-708	Baton Rouge	.86
617	Bloomington	1.04		710-711	Shreveport	.83
618-619	Champaign	1.05		712	Monroe	.81
620-622	East St. Louis	1.01		713-714	Alexandria	.82
623	Quincy	1.03				
624	Effingham	1.04		**MAINE**		
625	Decatur	1.03		039	Kittery	.88
626-627	Springfield	1.04		040-041	Portland	.92
628	Centralia	1.01		042	Lewiston	.92
629	Carbondale	1.00		043	Augusta	.88
				044	Bangor	.91
INDIANA				045	Bath	.88
460	Anderson	.90		046	Machias	.87
461-462	Indianapolis	.92		047	Houlton	.88
463-464	Gary	1.04		048	Rockland	.87
465-466	South Bend	.90		049	Waterville	.87
467-468	Fort Wayne	.88				
469	Kokomo	.90		**MARYLAND**		
470	Lawrenceburg	.87		206	Waldorf	.90
471	New Albany	.87		207-208	College Park	.89
472	Columbus	.91		209	Silver Spring	.90
473	Muncie	.91		210-212	Baltimore	.92
474	Bloomington	.93		214	Annapolis	.90
475	Washington	.91		215	Cumberland	.91
476-477	Evansville	.90		216	Easton	.87
478	Terre Haute	.90		217	Hagerstown	.91
479	Lafayette	.92		218	Salisbury	.85
				219	Elkton	.93
IOWA						
500-503,509	Des Moines	.92		**MASSACHUSETTS**		
504	Mason City	.85		010-011	Springfield	1.05
505	Fort Dodge	.84		012	Pittsfield	1.04
506-507	Waterloo	.88		013	Greenfield	1.03
508	Creston	.88		014	Fitchburg	1.11
510-511	Sioux City	.89		015-016	Worcester	1.12
512	Sibley	.75		017	Framingham	1.13
513	Spencer	.76		018	Lowell	1.14
514	Carroll	.87		019	Lawrence	1.14
515	Council Bluffs	.87		020-022, 024	Boston	1.19
516	Shenandoah	.81		023	Brockton	1.12
520	Dubuque	.89		025	Buzzards Bay	1.10
521	Decorah	.87		026	Hyannis	1.09
522-524	Cedar Rapids	.94		027	New Bedford	1.11
525	Ottumwa	.87				
526	Burlington	.90		**MICHIGAN**		
527-528	Davenport	.99		480,483	Royal Oak	.99
				481	Ann Arbor	1.00
KANSAS				482	Detroit	1.02
660-662	Kansas City	1.00		484-485	Flint	.93
664-666	Topeka	.87		486	Saginaw	.90
667	Fort Scott	.91		487	Bay City	.90
668	Emporia	.84		488-489	Lansing	.90
669	Belleville	.82		490	Battle Creek	.88
670-672	Wichita	.82		491	Kalamazoo	.88
673	Independence	.90		492	Jackson	.90
674	Salina	.81		493,495	Grand Rapids	.88
675	Hutchinson	.80		494	Muskegon	.86
676	Hays	.83		496	Traverse City	.82
677	Colby	.86		497	Gaylord	.86
678	Dodge City	.83		498-499	Iron Mountain	.86
679	Liberal	.82				
				MINNESOTA		
KENTUCKY				550-551	Saint Paul	1.06
400-402	Louisville	.87		553-555	Minneapolis	1.07
403-405	Lexington	.87		556-558	Duluth	.99

For customer support on your Residential Costs with RSMeans data, call 800.448.8182.

Location Factors - Residential

STATE	CITY	Residential
MINNESOTA (CONT'D)		
559	Rochester	.99
560	Mankato	.96
561	Windom	.91
562	Willmar	.93
563	St. Cloud	1.01
564	Brainerd	.92
565	Detroit Lakes	.90
566	Bemidji	.91
567	Thief River Falls	.90
MISSISSIPPI		
386	Clarksdale	.73
387	Greenville	.81
388	Tupelo	.74
389	Greenwood	.77
390-392	Jackson	.82
393	Meridian	.80
394	Laurel	.78
395	Biloxi	.84
396	McComb	.74
397	Columbus	.75
MISSOURI		
630-631	St. Louis	1.03
633	Bowling Green	.97
634	Hannibal	.95
635	Kirksville	.91
636	Flat River	.97
637	Cape Girardeau	.91
638	Sikeston	.89
639	Poplar Bluff	.89
640-641	Kansas City	1.02
644-645	St. Joseph	.97
646	Chillicothe	.96
647	Harrisonville	.99
648	Joplin	.90
650-651	Jefferson City	.93
652	Columbia	.91
653	Sedalia	.90
654-655	Rolla	.96
656-658	Springfield	.88
MONTANA		
590-591	Billings	.87
592	Wolf Point	.86
593	Miles City	.87
594	Great Falls	.87
595	Havre	.83
596	Helena	.87
597	Butte	.84
598	Missoula	.84
599	Kalispell	.84
NEBRASKA		
680-681	Omaha	.90
683-685	Lincoln	.90
686	Columbus	.89
687	Norfolk	.89
688	Grand Island	.87
689	Hastings	.89
690	McCook	.83
691	North Platte	.86
692	Valentine	.83
693	Alliance	.82
NEVADA		
889-891	Las Vegas	1.02
893	Ely	1.00
894-895	Reno	.90
897	Carson City	.91
898	Elko	.93
NEW HAMPSHIRE		
030	Nashua	.98
031	Manchester	.97
032-033	Concord	.97
034	Keene	.91
035	Littleton	.94
036	Charleston	.89
037	Claremont	.89
038	Portsmouth	.95

STATE	CITY	Residential
NEW JERSEY		
070-071	Newark	1.20
072	Elizabeth	1.22
073	Jersey City	1.18
074-075	Paterson	1.20
076	Hackensack	1.19
077	Long Branch	1.19
078	Dover	1.19
079	Summit	1.20
080,083	Vineland	1.18
081	Camden	1.19
082,084	Atlantic City	1.22
085-086	Trenton	1.19
087	Point Pleasant	1.19
088-089	New Brunswick	1.22
NEW MEXICO		
870-872	Albuquerque	.83
873	Gallup	.82
874	Farmington	.83
875	Santa Fe	.84
877	Las Vegas	.82
878	Socorro	.82
879	Truth/Consequences	.81
880	Las Cruces	.82
881	Clovis	.83
882	Roswell	.83
883	Carrizozo	.83
884	Tucumcari	.84
NEW YORK		
100-102	New York	1.40
103	Staten Island	1.32
104	Bronx	1.36
105	Mount Vernon	1.17
106	White Plains	1.20
107	Yonkers	1.24
108	New Rochelle	1.21
109	Suffern	1.19
110	Queens	1.37
111	Long Island City	1.40
112	Brooklyn	1.38
113	Flushing	1.39
114	Jamaica	1.38
115,117,118	Hicksville	1.26
116	Far Rockaway	1.38
119	Riverhead	1.28
120-122	Albany	1.02
123	Schenectady	1.04
124	Kingston	1.13
125-126	Poughkeepsie	1.25
127	Monticello	1.14
128	Glens Falls	.96
129	Plattsburgh	1.01
130-132	Syracuse	.99
133-135	Utica	.97
136	Watertown	.95
137-139	Binghamton	.99
140-142	Buffalo	1.09
143	Niagara Falls	1.05
144-146	Rochester	1.01
147	Jamestown	.93
148-149	Elmira	.96
NORTH CAROLINA		
270,272-274	Greensboro	.94
271	Winston-Salem	.94
275-276	Raleigh	.93
277	Durham	.94
278	Rocky Mount	.92
279	Elizabeth City	.91
280	Gastonia	.96
281-282	Charlotte	.94
283	Fayetteville	.94
284	Wilmington	.93
285	Kinston	.92
286	Hickory	.93
287-288	Asheville	.94
289	Murphy	.95
NORTH DAKOTA		
580-581	Fargo	.90
582	Grand Forks	.88
583	Devils Lake	.90
584	Jamestown	.88
585	Bismarck	.89

For customer support on your Residential Costs with RSMeans data, call 800.448.8182.

Location Factors - Residential

STATE	CITY	Residential
NORTH DAKOTA (CONT'D)		
586	Dickinson	.88
587	Minot	.87
588	Williston	.88
OHIO		
430-432	Columbus	.91
433	Marion	.90
434-436	Toledo	.96
437-438	Zanesville	.88
439	Steubenville	.93
440	Lorain	.94
441	Cleveland	.97
442-443	Akron	.96
444-445	Youngstown	.93
446-447	Canton	.92
448-449	Mansfield	.90
450	Hamilton	.89
451-452	Cincinnati	.91
453-454	Dayton	.89
455	Springfield	.90
456	Chillicothe	.93
457	Athens	.91
458	Lima	.89
OKLAHOMA		
730-731	Oklahoma City	.81
734	Ardmore	.82
735	Lawton	.83
736	Clinton	.82
737	Enid	.83
738	Woodward	.78
739	Guymon	.82
740-741	Tulsa	.82
743	Miami	.80
744	Muskogee	.80
745	McAlester	.74
746	Ponca City	.80
747	Durant	.79
748	Shawnee	.79
749	Poteau	.79
OREGON		
970-972	Portland	1.00
973	Salem	.99
974	Eugene	1.00
975	Medford	.98
976	Klamath Falls	.99
977	Bend	1.01
978	Pendleton	.99
979	Vale	.97
PENNSYLVANIA		
150-152	Pittsburgh	1.01
153	Washington	.96
154	Uniontown	.93
155	Bedford	.90
156	Greensburg	.96
157	Indiana	.94
158	Dubois	.91
159	Johnstown	.91
160	Butler	.92
161	New Castle	.93
162	Kittanning	.93
163	Oil City	.93
164-165	Erie	.94
166	Altoona	.90
167	Bradford	.96
168	State College	.92
169	Wellsboro	.92
170-171	Harrisburg	.95
172	Chambersburg	.89
173-174	York	.91
175-176	Lancaster	.94
177	Williamsport	.90
178	Sunbury	.92
179	Pottsville	.92
180	Lehigh Valley	1.06
181	Allentown	1.03
182	Hazleton	.91
183	Stroudsburg	.96
184-185	Scranton	.97
186-187	Wilkes-Barre	.94
188	Montrose	.93
189	Doylestown	1.07

STATE	CITY	Residential
PENNSYLVANIA (CONT'D)		
190-191	Philadelphia	1.17
193	Westchester	1.09
194	Norristown	1.09
195-196	Reading	.98
PUERTO RICO		
009	San Juan	.77
RHODE ISLAND		
028	Newport	1.07
029	Providence	1.08
SOUTH CAROLINA		
290-292	Columbia	.95
293	Spartanburg	.95
294	Charleston	.96
295	Florence	.92
296	Greenville	.94
297	Rock Hill	.94
298	Aiken	.95
299	Beaufort	.85
SOUTH DAKOTA		
570-571	Sioux Falls	.87
572	Watertown	.76
573	Mitchell	.78
574	Aberdeen	.79
575	Pierre	.83
576	Mobridge	.78
577	Rapid City	.86
TENNESSEE		
370-372	Nashville	.83
373-374	Chattanooga	.83
375,380-381	Memphis	.84
376	Johnson City	.76
377-379	Knoxville	.80
382	McKenzie	.72
383	Jackson	.75
384	Columbia	.79
385	Cookeville	.70
TEXAS		
750	McKinney	.80
751	Waxahackie	.80
752-753	Dallas	.83
754	Greenville	.80
755	Texarkana	.82
756	Longview	.80
757	Tyler	.82
758	Palestine	.78
759	Lufkin	.79
760-761	Fort Worth	.82
762	Denton	.84
763	Wichita Falls	.84
764	Eastland	.82
765	Temple	.80
766-767	Waco	.84
768	Brownwood	.80
769	San Angelo	.78
770-772	Houston	.83
773	Huntsville	.80
774	Wharton	.80
775	Galveston	.82
776-777	Beaumont	.85
778	Bryan	.78
779	Victoria	.81
780	Laredo	.80
781-782	San Antonio	.80
783-784	Corpus Christi	.83
785	McAllen	.84
786-787	Austin	.79
788	Del Rio	.81
789	Giddings	.80
790-791	Amarillo	.79
792	Childress	.81
793-794	Lubbock	.80
795-796	Abilene	.82
797	Midland	.84
798-799,885	El Paso	.79
UTAH		
840-841	Salt Lake City	.84
842,844	Ogden	.82
843	Logan	.83

For customer support on your Residential Costs with RSMeans data, call 800.448.8182.

STATE	CITY	Residential
UTAH (CONT'D)		
845	Price	.84
846-847	Provo	.83
VERMONT		
050	White River Jct.	.96
051	Bellows Falls	1.01
052	Bennington	1.00
053	Brattleboro	1.01
054	Burlington	.93
056	Montpelier	.99
057	Rutland	.92
058	St. Johnsbury	.97
059	Guildhall	.97
VIRGINIA		
220-221	Fairfax	1.01
222	Arlington	1.03
223	Alexandria	1.04
224-225	Fredericksburg	1.00
226	Winchester	1.01
227	Culpeper	1.00
228	Harrisonburg	.86
229	Charlottesville	.88
230-232	Richmond	1.03
233-235	Norfolk	.95
236	Newport News	.95
237	Portsmouth	.87
238	Petersburg	.95
239	Farmville	.85
240-241	Roanoke	1.00
242	Bristol	.91
243	Pulaski	.83
244	Staunton	.86
245	Lynchburg	.98
246	Grundy	.81
WASHINGTON		
980-981,987	Seattle	1.05
982	Everett	1.06
983-984	Tacoma	1.03
985	Olympia	1.01
986	Vancouver	.98
988	Wenatchee	.95
989	Yakima	1.00
990-992	Spokane	.97
993	Richland	.97
994	Clarkston	.94
WEST VIRGINIA		
247-248	Bluefield	.93
249	Lewisburg	.92
250-253	Charleston	.95
254	Martinsburg	.89
255-257	Huntington	.97
258-259	Beckley	.93
260	Wheeling	.93
261	Parkersburg	.91
262	Buckhannon	.92
263-264	Clarksburg	.92
265	Morgantown	.92
266	Gassaway	.92
267	Romney	.89
268	Petersburg	.89
WISCONSIN		
530,532	Milwaukee	1.05
531	Kenosha	1.05
534	Racine	1.04
535	Beloit	.98
537	Madison	.98
538	Lancaster	.97
539	Portage	.94
540	New Richmond	.95
541-543	Green Bay	1.03
544	Wausau	.96
545	Rhinelander	.93
546	La Crosse	.97
547	Eau Claire	.99
548	Superior	.95
549	Oshkosh	.93
WYOMING		
820	Cheyenne	.86
821	Yellowstone Nat. Pk.	.83
822	Wheatland	.79

STATE	CITY	Residential
WYOMING (CONT'D)		
823	Rawlins	.86
824	Worland	.83
825	Riverton	.82
826	Casper	.83
827	Newcastle	.85
828	Sheridan	.86
829-831	Rock Springs	.88
CANADIAN FACTORS (reflect Canadian currency)		
ALBERTA		
	Calgary	1.07
	Edmonton	1.07
	Fort McMurray	1.10
	Lethbridge	1.07
	Lloydminster	1.02
	Medicine Hat	1.03
	Red Deer	1.03
BRITISH COLUMBIA		
	Kamloops	1.00
	Prince George	.99
	Vancouver	1.01
	Victoria	1.01
MANITOBA		
	Brandon	1.07
	Portage la Prairie	.97
	Winnipeg	.95
NEW BRUNSWICK		
	Bathurst	.89
	Dalhousie	.90
	Fredericton	.94
	Moncton	.91
	Newcastle	.90
	St. John	.99
NEWFOUNDLAND		
	Corner Brook	1.06
	St. Johns	1.04
NORTHWEST TERRITORIES		
	Yellowknife	1.12
NOVA SCOTIA		
	Bridgewater	.93
	Dartmouth	1.02
	Halifax	1.01
	New Glasgow	1.01
	Sydney	1.00
	Truro	.93
	Yarmouth	1.01
ONTARIO		
	Barrie	1.10
	Brantford	1.09
	Cornwall	1.08
	Hamilton	1.06
	Kingston	1.08
	Kitchener	1.03
	London	1.05
	North Bay	1.16
	Oshawa	1.06
	Ottawa	1.06
	Owen Sound	1.09
	Peterborough	1.07
	Sarnia	1.10
	Sault Ste. Marie	1.04
	St. Catharines	1.04
	Sudbury	1.02
	Thunder Bay	1.07
	Timmins	1.06
	Toronto	1.08
	Windsor	1.06
PRINCE EDWARD ISLAND		
	Charlottetown	.89
	Summerside	.95
QUEBEC		
	Cap-de-la-Madeleine	1.07
	Charlesbourg	1.07
	Chicoutimi	1.10
	Gatineau	1.06
	Granby	1.06

718

Location Factors - Residential

STATE	CITY	Residential
QUEBEC (CONT'D)		
	Hull	1.06
	Joliette	1.07
	Laval	1.06
	Montreal	1.05
	Quebec City	1.06
	Rimouski	1.10
	Rouyn-Noranda	1.06
	Saint-Hyacinthe	1.06
	Sherbrooke	1.06
	Sorel	1.07
	Saint-Jerome	1.06
	Trois-Rivieres	1.17
SASKATCHEWAN		
	Moose Jaw	.89
	Prince Albert	.88
	Regina	1.08
	Saskatoon	1.04
YUKON		
	Whitehorse	1.03

For customer support on your Residential Costs with RSMeans data, call 800.448.8182.

R011105-05   Tips for Accurate Estimating

1. Use pre-printed or columnar forms for orderly sequence of dimensions and locations and for recording telephone quotations.

2. Use only the front side of each paper or form except for certain pre-printed summary forms.

3. Be consistent in listing dimensions: For example, length x width x height. This helps in rechecking to ensure that, the total length of partitions is appropriate for the building area.

4. Use printed (rather than measured) dimensions where given.

5. Add up multiple printed dimensions for a single entry where possible.

6. Measure all other dimensions carefully.

7. Use each set of dimensions to calculate multiple related quantities.

8. Convert foot and inch measurements to decimal feet when listing. Memorize decimal equivalents to .01 parts of a foot (1/8″ equals approximately .01′).

9. Do not "round off" quantities until the final summary.

10. Mark drawings with different colors as items are taken off.

11. Keep similar items together, different items separate.

12. Identify location and drawing numbers to aid in future checking for completeness.

13. Measure or list everything on the drawings or mentioned in the specifications.

14. It may be necessary to list items not called for to make the job complete.

15. Be alert for: Notes on plans such as N.T.S. (not to scale); changes in scale throughout the drawings; reduced size drawings; discrepancies between the specifications and the drawings.

16. Develop a consistent pattern of performing an estimate. For example:
    a. Start the quantity takeoff at the lower floor and move to the next higher floor.
    b. Proceed from the main section of the building to the wings.
    c. Proceed from south to north or vice versa, clockwise or counterclockwise.
    d. Take off floor plan quantities first, elevations next, then detail drawings.

17. List all gross dimensions that can be either used again for different quantities, or used as a rough check of other quantities for verification (exterior perimeter, gross floor area, individual floor areas, etc.).

18. Utilize design symmetry or repetition (repetitive floors, repetitive wings, symmetrical design around a center line, similar room layouts, etc.). Note: Extreme caution is needed here so as not to omit or duplicate an area.

19. Do not convert units until the final total is obtained. For instance, when estimating concrete work, keep all units to the nearest cubic foot, then summarize and convert to cubic yards.

20. When figuring alternatives, it is best to total all items involved in the basic system, then total all items involved in the alternates. Therefore you work with positive numbers in all cases. When adds and deducts are used, it is often confusing whether to add or subtract a portion of an item; especially on a complicated or involved alternate.

R011105-50   Metric Conversion Factors

Description: This table is primarily for converting customary U.S. units in the left hand column to SI metric units in the right hand column. In addition, conversion factors for some commonly encountered Canadian and non-SI metric units are included.

	If You Know		Multiply By		To Find
Length	Inches	x	25.4[a]	=	Millimeters
	Feet	x	0.3048[a]	=	Meters
	Yards	x	0.9144[a]	=	Meters
	Miles (statute)	x	1.609	=	Kilometers
Area	Square inches	x	645.2	=	Square millimeters
	Square feet	x	0.0929	=	Square meters
	Square yards	x	0.8361	=	Square meters
Volume (Capacity)	Cubic inches	x	16,387	=	Cubic millimeters
	Cubic feet	x	0.02832	=	Cubic meters
	Cubic yards	x	0.7646	=	Cubic meters
	Gallons (U.S. liquids)[b]	x	0.003785	=	Cubic meters[c]
	Gallons (Canadian liquid)[b]	x	0.004546	=	Cubic meters[c]
	Ounces (U.S. liquid)[b]	x	29.57	=	Milliliters[c, d]
	Quarts (U.S. liquid)[b]	x	0.9464	=	Liters[c, d]
	Gallons (U.S. liquid)[b]	x	3.785	=	Liters[c, d]
Force	Kilograms force[d]	x	9.807	=	Newtons
	Pounds force	x	4.448	=	Newtons
	Pounds force	x	0.4536	=	Kilograms force[d]
	Kips	x	4448	=	Newtons
	Kips	x	453.6	=	Kilograms force[d]
Pressure, Stress, Strength (Force per unit area)	Kilograms force per square centimeter[d]	x	0.09807	=	Megapascals
	Pounds force per square inch (psi)	x	0.006895	=	Megapascals
	Kips per square inch	x	6.895	=	Megapascals
	Pounds force per square inch (psi)	x	0.07031	=	Kilograms force per square centimeter[d]
	Pounds force per square foot	x	47.88	=	Pascals
	Pounds force per square foot	x	4.882	=	Kilograms force per square meter[d]
Flow	Cubic feet per minute	x	0.4719	=	Liters per second
	Gallons per minute	x	0.0631	=	Liters per second
	Gallons per hour	x	1.05	=	Milliliters per second
Bending Moment Or Torque	Inch-pounds force	x	0.01152	=	Meter-kilograms force[d]
	Inch-pounds force	x	0.1130	=	Newton-meters
	Foot-pounds force	x	0.1383	=	Meter-kilograms force[d]
	Foot-pounds force	x	1.356	=	Newton-meters
	Meter-kilograms force[d]	x	9.807	=	Newton-meters
Mass	Ounces (avoirdupois)	x	28.35	=	Grams
	Pounds (avoirdupois)	x	0.4536	=	Kilograms
	Tons (metric)	x	1000	=	Kilograms
	Tons, short (2000 pounds)	x	907.2	=	Kilograms
	Tons, short (2000 pounds)	x	0.9072	=	Megagrams[e]
Mass per Unit Volume	Pounds mass per cubic foot	x	16.02	=	Kilograms per cubic meter
	Pounds mass per cubic yard	x	0.5933	=	Kilograms per cubic meter
	Pounds mass per gallon (U.S. liquid)[b]	x	119.8	=	Kilograms per cubic meter
	Pounds mass per gallon (Canadian liquid)[b]	x	99.78	=	Kilograms per cubic meter
Temperature	Degrees Fahrenheit	(F-32)/1.8		=	Degrees Celsius
	Degrees Fahrenheit	(F+459.67)/1.8		=	Degrees Kelvin
	Degrees Celsius	C+273.15		=	Degrees Kelvin

[a]The factor given is exact
[b]One U.S. gallon = 0.8327 Canadian gallon
[c]1 liter = 1000 milliliters = 1000 cubic centimeters
 1 cubic decimeter = 0.001 cubic meter

[d]Metric but not SI unit
[e]Called "tonne" in England and
 "metric ton" in other metric countries

721

R011105-60   Weights and Measures

Measures of Length
1 Mile = 1760 Yards = 5280 Feet
1 Yard = 3 Feet = 36 inches
1 Foot = 12 Inches
1 Mil = 0.001 Inch
1 Fathom = 2 Yards = 6 Feet
1 Rod = 5.5 Yards = 16.5 Feet
1 Hand = 4 Inches
1 Span = 9 Inches
1 Micro-inch = One Millionth Inch or 0.000001 Inch
1 Micron = One Millionth Meter + 0.00003937 Inch

Surveyor's Measure
1 Mile = 8 Furlongs = 80 Chains
1 Furlong = 10 Chains = 220 Yards
1 Chain = 4 Rods = 22 Yards = 66 Feet = 100 Links
1 Link = 7.92 Inches

Square Measure
1 Square Mile = 640 Acres = 6400 Square Chains
1 Acre = 10 Square Chains = 4840 Square Yards =
  43,560 Sq. Ft.
1 Square Chain = 16 Square Rods = 484 Square Yards =
  4356 Sq. Ft.
1 Square Rod = 30.25 Square Yards = 272.25 Square Feet = 625 Square
  Lines
1 Square Yard = 9 Square Feet
1 Square Foot = 144 Square Inches
An Acre equals a Square 208.7 Feet per Side

Cubic Measure
1 Cubic Yard = 27 Cubic Feet
1 Cubic Foot = 1728 Cubic Inches
1 Cord of Wood = 4 x 4 x 8 Feet = 128 Cubic Feet
1 Perch of Masonry = 16½ x 1½ x 1 Foot = 24.75 Cubic Feet

Avoirdupois or Commercial Weight
1 Gross or Long Ton = 2240 Pounds
1 Net or Short Ton = 2000 Pounds
1 Pound = 16 Ounces = 7000 Grains
1 Ounce = 16 Drachms = 437.5 Grains
1 Stone = 14 Pounds

Power
1 British Thermal Unit per Hour = 0.2931 Watts
1 Ton (Refrigeration) = 3.517 Kilowatts
1 Horsepower (Boiler) = 9.81 Kilowatts
1 Horsepower (550 ft-lb/s) = 0.746 Kilowatts

Shipping Measure
For Measuring Internal Capacity of a Vessel:
  1 Register Ton = 100 Cubic Feet

For Measurement of Cargo:
  Approximately 40 Cubic Feet of Merchandise is considered a Shipping
  Ton, unless that bulk would weigh more than 2000 Pounds, in which case
  Freight Charge may be based upon weight.

40 Cubic Feet = 32.143 U.S. Bushels = 31.16 Imp. Bushels

Liquid Measure
1 Imperial Gallon = 1.2009 U.S. Gallon = 277.42 Cu. In.
1 Cubic Foot = 7.48 U.S. Gallons

R011110-10   Architectural Fees

Tabulated below are typical percentage fees by project size, for good professional architectural service. Fees may vary from those listed depending upon degree of design difficulty and economic conditions in any particular area.

Rates can be interpolated horizontally and vertically. Various portions of the same project requiring different rates should be adjusted proportionately. For alterations, add 50% to the fee for the first $500,000 of project cost and add 25% to the fee for project cost over $500,000.

Architectural fees tabulated below include Structural, Mechanical and Electrical Engineering Fees. They do not include the fees for special consultants such as kitchen planning, security, acoustical, interior design, etc.

Civil Engineering fees are included in the Architectural fee for project sites requiring minimal design such as city sites. However, separate Civil Engineering fees must be added when utility connections require design, drainage calculations are needed, stepped foundations are required, or provisions are required to protect adjacent wetlands.

Building Types	Total Project Size in Thousands of Dollars						
	100	250	500	1,000	5,000	10,000	50,000
Factories, garages, warehouses, repetitive housing	9.0%	8.0%	7.0%	6.2%	5.3%	4.9%	4.5%
Apartments, banks, schools, libraries, offices, municipal buildings	12.2	12.3	9.2	8.0	7.0	6.6	6.2
Churches, hospitals, homes, laboratories, museums, research	15.0	13.6	12.7	11.9	9.5	8.8	8.0
Memorials, monumental work, decorative furnishings	—	16.0	14.5	13.1	10.0	9.0	8.3

R012909-80 Sales Tax by State

State sales tax on materials is tabulated below (5 states have no sales tax). Many states allow local jurisdictions, such as a county or city, to levy additional sales tax.

Some projects may be sales tax exempt, particularly those constructed with public funds.

State	Tax (%)	State	Tax (%)	State	Tax (%)	State	Tax (%)
Alabama	4	Illinois	6.25	Montana	0	Rhode Island	7
Alaska	0	Indiana	7	Nebraska	5.5	South Carolina	6
Arizona	5.6	Iowa	6	Nevada	6.85	South Dakota	5
Arkansas	6.5	Kansas	6.5	New Hampshire	0	Tennessee	7
California	7.25	Kentucky	6	New Jersey	7	Texas	6.25
Colorado	2.9	Louisiana	4	New Mexico	5.125	Utah	5.95
Connecticut	6.35	Maine	5.5	New York	4	Vermont	6
Delaware	0	Maryland	6	North Carolina	4.75	Virginia	5.3
District of Columbia	5.75	Massachusetts	6.25	North Dakota	5	Washington	6.5
Florida	6	Michigan	6	Ohio	5.75	West Virginia	6
Georgia	4	Minnesota	6.875	Oklahoma	4.5	Wisconsin	5
Hawaii	4	Mississippi	7	Oregon	0	Wyoming	4
Idaho	6	Missouri	4.225	Pennsylvania	6	Average	5.11 %

Sales Tax by Province (Canada)

GST - a value-added tax, which the government imposes on most goods and services provided in or imported into Canada. PST - a retail sales tax, which five of the provinces impose on the prices of most goods and some services. QST - a value-added tax, similar to the federal GST, which Quebec imposes. HST - Three provinces have combined their retail sales taxes with the federal GST into one harmonized tax.

Province	PST (%)	QST (%)	GST(%)	HST(%)
Alberta	0	0	5	0
British Columbia	7	0	5	0
Manitoba	8	0	5	0
New Brunswick	0	0	0	15
Newfoundland	0	0	0	15
Northwest Territories	0	0	5	0
Nova Scotia	0	0	0	15
Ontario	0	0	0	13
Prince Edward Island	0	0	0	15
Quebec	0	9.975	5	0
Saskatchewan	6	0	5	0
Yukon	0	0	5	0

R012909-85 Unemployment Taxes and Social Security Taxes

State unemployment tax rates vary not only from state to state, but also with the experience rating of the contractor. The federal unemployment tax rate is 6.0% of the first $7,000 of wages. This is reduced by a credit of up to 5.4% for timely payment to the state. The minimum federal unemployment tax is 0.6% after all credits.

Social security (FICA) for 2018 is estimated at time of publication to be 7.65% of wages up to $127,200.

General Requirements — R0129 Payment Procedures

R012909-86   Unemployment Tax by State

Information is from the U.S. Department of Labor, state unemployment tax rates.

State	Tax (%)	State	Tax (%)	State	Tax (%)	State	Tax (%)
Alabama	6.74	Illinois	7.75	Montana	6.12	Rhode Island	9.79
Alaska	5.4	Indiana	7.474	Nebraska	5.4	South Carolina	5.46
Arizona	8.91	Iowa	8	Nevada	5.4	South Dakota	9.5
Arkansas	6.0	Kansas	7.6	New Hampshire	7.5	Tennessee	10.0
California	6.2	Kentucky	10.0	New Jersey	5.8	Texas	7.5
Colorado	8.9	Louisiana	6.2	New Mexico	5.4	Utah	7.2
Connecticut	6.8	Maine	5.4	New York	8.5	Vermont	8.4
Delaware	8.0	Maryland	7.50	North Carolina	5.76	Virginia	6.27
District of Columbia	7	Massachusetts	11.13	North Dakota	10.72	Washington	5.7
Florida	5.4	Michigan	10.3	Ohio	8.7	West Virginia	7.5
Georgia	5.4	Minnesota	9.0	Oklahoma	5.5	Wisconsin	12.0
Hawaii	5.6	Mississippi	5.4	Oregon	5.4	Wyoming	8.8
Idaho	5.4	Missouri	9.75	Pennsylvania	10.89	Median	7.47 %

General Requirements — R0131 Project Management & Coordination

R013113-40   Builder's Risk Insurance

Builder's risk insurance is insurance on a building during construction. Premiums are paid by the owner or the contractor. Blasting, collapse and underground insurance would raise total insurance costs.

R013113-50   General Contractor's Overhead

There are two distinct types of overhead on a construction project: Project overhead and main office overhead. Project overhead includes those costs at a construction site not directly associated with the installation of construction materials. Examples of project overhead costs include the following:

1. Superintendent
2. Construction office and storage trailers
3. Temporary sanitary facilities
4. Temporary utilities
5. Security fencing
6. Photographs
7. Cleanup
8. Performance and payment bonds

The above project overhead items are also referred to as general requirements and therefore are estimated in Division 1. Division 1 is the first division listed in the CSI MasterFormat but it is usually the last division estimated. The sum of the costs in Divisions 1 through 49 is referred to as the sum of the direct costs.

All construction projects also include indirect costs. The primary components of indirect costs are the contractor's main office overhead and profit. The amount of the main office overhead expense varies depending on the following:

1. Owner's compensation
2. Project managers' and estimators' wages
3. Clerical support wages
4. Office rent and utilities
5. Corporate legal and accounting costs
6. Advertising
7. Automobile expenses
8. Association dues
9. Travel and entertainment expenses

These costs are usually calculated as a percentage of annual sales volume. This percentage can range from 35% for a small contractor doing less than $500,000 to 5% for a large contractor with sales in excess of $100 million.

R013113-55   Installing Contractor's Overhead

Installing contractors (subcontractors) also incur costs for general requirements and main office overhead.

Included within the total incl. overhead and profit costs is a percent mark-up for overhead that includes:
1. Compensation and benefits for office staff and project managers
2. Office rent, utilities, business equipment, and maintenance
3. Corporate legal and accounting costs

4. Advertising
5. Vehicle expenses (for office staff and project managers)
6. Association dues
7. Travel, entertainment
8. Insurance
9. Small tools and equipment

R013113-60   Workers' Compensation Insurance Rates by Trade

The table below tabulates the national averages for workers' compensation insurance rates by trade and type of building. The average "Insurance Rate" is multiplied by the "% of Building Cost" for each trade. This produces the "Workers' Compensation" cost by % of total labor cost, to be added for each trade by building type to determine the weighted average workers' compensation rate for the building types analyzed.

Trade	Insurance Rate (% Labor Cost) Range	Average	% of Building Cost Office Bldgs.	Schools & Apts.	Mfg.	Workers' Compensation Office Bldgs.	Schools & Apts.	Mfg.
Excavation, Grading, etc.	2.7 % to 20.1%	8.5%	4.8%	4.9%	4.5%	0.41%	0.42%	0.38%
Piles & Foundations	5.3 to 29.8	13.4	7.1	5.2	8.7	0.95	0.70	1.17
Concrete	4.1 to 28.0	11.8	5.0	14.8	3.7	0.59	1.75	0.44
Masonry	3.9 to 49.3	13.8	6.9	7.5	1.9	0.95	1.04	0.26
Structural Steel	5.3 to 59.1	21.2	10.7	3.9	17.6	2.27	0.83	3.73
Miscellaneous & Ornamental Metals	3.3 to 24.4	10.6	2.8	4.0	3.6	0.30	0.42	0.38
Carpentry & Millwork	4.4 to 32.4	13.0	3.7	4.0	0.5	0.48	0.52	0.07
Metal or Composition Siding	5.5 to 107.2	19.0	2.3	0.3	4.3	0.44	0.06	0.82
Roofing	5.5 to 120.3	29.0	2.3	2.6	3.1	0.67	0.75	0.90
Doors & Hardware	3.2 to 32.4	11.0	0.9	1.4	0.4	0.10	0.15	0.04
Sash & Glazing	4.7 to 25.5	12.1	3.5	4.0	1.0	0.42	0.48	0.12
Lath & Plaster	3.0 to 31.6	10.7	3.3	6.9	0.8	0.35	0.74	0.09
Tile, Marble & Floors	2.7 to 18.3	8.7	2.6	3.0	0.5	0.23	0.26	0.04
Acoustical Ceilings	2.4 to 46.3	8.5	2.4	0.2	0.3	0.20	0.02	0.03
Painting	3.3 to 38.8	11.2	1.5	1.6	1.6	0.17	0.18	0.18
Interior Partitions	4.4 to 32.4	13.0	3.9	4.3	4.4	0.51	0.56	0.57
Miscellaneous Items	2.3 to 97.7	11.2	5.2	3.7	9.7	0.58	0.42	1.09
Elevators	1.3 to 13.7	4.7	2.1	1.1	2.2	0.10	0.05	0.10
Sprinklers	2.0 to 15.5	6.7	0.5	—	2.0	0.03	—	0.13
Plumbing	1.7 to 14.0	6.3	4.9	7.2	5.2	0.31	0.45	0.33
Heat., Vent., Air Conditioning	3.3 to 17.8	8.3	13.5	11.0	12.9	1.12	0.91	1.07
Electrical	1.9 to 11.6	5.2	10.1	8.4	11.1	0.53	0.44	0.58
Total	1.3 % to 120.3%	—	100.0%	100.0%	100.0%	11.71%	11.15%	12.52%

Overall Weighted Average   11.79%

Workers' Compensation Insurance Rates by States

The table below lists the weighted average Workers' Compensation base rate for each state with a factor comparing this with the national average of 11.8%.

State	Weighted Average	Factor	State	Weighted Average	Factor	State	Weighted Average	Factor
Alabama	15.0%	127	Kentucky	10.4%	88	North Dakota	6.2%	53
Alaska	10.4	88	Louisiana	18.7	158	Ohio	7.2	61
Arizona	9.6	81	Maine	10.4	88	Oklahoma	8.9	75
Arkansas	7.0	59	Maryland	11.3	96	Oregon	9.3	79
California	22.2	188	Massachusetts	11.2	95	Pennsylvania	21.1	179
Colorado	7.5	64	Michigan	8.2	69	Rhode Island	13.7	116
Connecticut	17.5	148	Minnesota	16.9	143	South Carolina	16.5	140
Delaware	13.9	118	Mississippi	11.8	100	South Dakota	11.8	100
District of Columbia	9.1	77	Missouri	12.4	105	Tennessee	8.6	73
Florida	11.1	94	Montana	8.8	75	Texas	6.6	56
Georgia	31.9	270	Nebraska	13.5	114	Utah	7.4	63
Hawaii	8.5	72	Nevada	7.5	64	Vermont	10.9	92
Idaho	9.4	80	New Hampshire	12.0	102	Virginia	6.9	58
Illinois	21.1	179	New Jersey	14.8	125	Washington	9.1	77
Indiana	4.1	35	New Mexico	13.3	113	West Virginia	4.5	38
Iowa	13.7	116	New York	19.2	163	Wisconsin	12.2	103
Kansas	6.5	55	North Carolina	15.8	134	Wyoming	5.6	47

Weighted Average for U.S. is   11.8% of payroll = 100%

The weighted average skilled worker rate for 35 trades is 11.8%. For bidding purposes, apply the full value of Workers' Compensation directly to total labor costs, or if labor is 38%, materials 42% and overhead and profit 20% of total cost, carry 38/80 x 11.8% = 6.0% of cost (before overhead and profit)

into overhead. Rates vary not only from state to state but also with the experience rating of the contractor.

Rates are the most current available at the time of publication.

R015423-10   Steel Tubular Scaffolding

On new construction, tubular scaffolding is efficient up to 60' high or five stories. Above this it is usually better to use a hung scaffolding if construction permits. Swing scaffolding operations may interfere with tenants. In this case, the tubular is more practical at all heights.

In repairing or cleaning the front of an existing building the cost of tubular scaffolding per S.F. of building front increases as the height increases above the first tier. The first tier cost is relatively high due to leveling and alignment.

The minimum efficient crew for erecting and dismantling is three workers. They can set up and remove 18 frame sections per day up to 5 stories high. For 6 to 12 stories high, a crew of four is most efficient. Use two or more on top and two on the bottom for handing up or hoisting. They can

also set up and remove 18 frame sections per day. At 7' horizontal spacing, this will run about 800 S.F. per day of erecting and dismantling. Time for placing and removing planks must be added to the above. A crew of three can place and remove 72 planks per day up to 5 stories. For over 5 stories, a crew of four can place and remove 80 planks per day.

The table below shows the number of pieces required to erect tubular steel scaffolding for 1000 S.F. of building frontage. This area is made up of a scaffolding system that is 12 frames (11 bays) long by 2 frames high.

For jobs under twenty-five frames, add 50% to rental cost. Rental rates will be lower for jobs over three months duration. Large quantities for long periods can reduce rental rates by 20%.

Description of Component	Number of Pieces for 1000 S.F. of Building Front	Unit
5' Wide Standard Frame, 6'-4" High	24	Ea.
Leveling Jack & Plate	24	
Cross Brace	44	
Side Arm Bracket, 21"	12	
Guardrail Post	12	
Guardrail, 7' section	22	
Stairway Section	2	
Stairway Starter Bar	1	
Stairway Inside Handrail	2	
Stairway Outside Handrail	2	
Walk-Thru Frame Guardrail	2	

Scaffolding is often used as falsework over 15' high during construction of cast-in-place concrete beams and slabs. Two foot wide scaffolding is generally used for heavy beam construction. The span between frames depends upon the load to be carried with a maximum span of 5'.

Heavy duty shoring frames with a capacity of 10,000#/leg can be spaced up to 10' O.C. depending upon form support design and loading.

Scaffolding used as horizontal shoring requires less than half the material required with conventional shoring.

On new construction, erection is done by carpenters.

Rolling towers supporting horizontal shores can reduce labor and speed the job. For maintenance work, catwalks with spans up to 70' can be supported by the rolling towers.

R015423-20   Pump Staging

Pump staging is generally not available for rent. The table below shows the number of pieces required to erect pump staging for 2400 S.F. of building

frontage. This area is made up of a pump jack system that is 3 poles (2 bays) wide by 2 poles high.

Item	Number of Pieces for 2400 S.F. of Building Front	Unit
Aluminum pole section, 24' long	6	Ea.
Aluminum splice joint, 6' long	3	
Aluminum foldable brace	3	
Aluminum pump jack	3	
Aluminum support for workbench/back safety rail	3	
Aluminum scaffold plank/workbench, 14" wide x 24' long	4	
Safety net, 22' long	2	
Aluminum plank end safety rail	2	

The cost in place for this 2400 S.F. will depend on how many uses are realized during the life of the equipment.

R015436-50   Mobilization

Costs to move rented construction equipment to a job site from an equipment dealer's or contractor's yard (mobilization) or off the job site (demobilization) are not included in the rental or operating rates, nor in the equipment cost on a unit price line or in a crew listing. These costs can be found consolidated in the Mobilization section of the data and elsewhere in particular site work sections. If a piece of

equipment is already on the job site, it is not appropriate to include mob/demob costs in a new estimate that requires use of that equipment. The following table identifies approximate sizes of rented construction equipment that would be hauled on a towed trailer. Because this listing is not all-encompassing, the user can infer as to what size trailer might be required for a piece of equipment not listed.

3-ton Trailer	20-ton Trailer	40-ton Trailer	50-ton Trailer
20 H.P. Excavator	110 H.P. Excavator	200 H.P. Excavator	270 H.P. Excavator
50 H.P. Skid Steer	165 H.P. Dozer	300 H.P. Dozer	Small Crawler Crane
35 H.P. Roller	150 H.P. Roller	400 H.P. Scraper	500 H.P. Scraper
40 H.P. Trencher	Backhoe	450 H.P. Art. Dump Truck	500 H.P. Art. Dump Truck

727

R024119-10   Demolition Defined

Whole Building Demolition - Demolition of the whole building with no concern for any particular building element, component, or material type being demolished. This type of demolition is accomplished with large pieces of construction equipment that break up the structure, load it into trucks and haul it to a disposal site, but disposal or dump fees are not included. Demolition of below-grade foundation elements, such as footings, foundation walls, grade beams, slabs on grade, etc., is not included. Certain mechanical equipment containing flammable liquids or ozone-depleting refrigerants, electric lighting elements, communication equipment components, and other building elements may contain hazardous waste, and must be removed, either selectively or carefully, as hazardous waste before the building can be demolished.

Foundation Demolition - Demolition of below-grade foundation footings, foundation walls, grade beams, and slabs on grade. This type of demolition is accomplished by hand or pneumatic hand tools, and does not include saw cutting, or handling, loading, hauling, or disposal of the debris.

Gutting - Removal of building interior finishes and electrical/mechanical systems down to the load-bearing and sub-floor elements of the rough building frame, with no concern for any particular building element, component, or material type being demolished. This type of demolition is accomplished by hand or pneumatic hand tools, and includes loading into trucks, but not hauling, disposal or dump fees, scaffolding, or shoring. Certain mechanical equipment containing flammable liquids or ozone-depleting refrigerants, electric lighting elements, communication equipment components, and other building elements may contain hazardous waste, and must be removed, either selectively or carefully, as hazardous waste, before the building is gutted.

Selective Demolition - Demolition of a selected building element, component, or finish, with some concern for surrounding or adjacent elements, components, or finishes (see the first Subdivision (s) at the beginning of appropriate Divisions). This type of demolition is accomplished by hand or pneumatic hand tools, and does not include handling, loading, storing, hauling, or disposal of the debris, scaffolding, or shoring. "Gutting" methods may be used in order to save time, but damage that is caused to surrounding or adjacent elements, components, or finishes may have to be repaired at a later time.

Careful Removal - Removal of a piece of service equipment, building element or component, or material type, with great concern for both the removed item and surrounding or adjacent elements, components or finishes. The purpose of careful removal may be to protect the removed item for later re-use, preserve a higher salvage value of the removed item, or replace an item while taking care to protect surrounding or adjacent elements, components, connections, or finishes from cosmetic and/or structural damage. An approximation of the time required to perform this type of removal is 1/3 to 1/2 the time it would take to install a new item of like kind. This type of removal is accomplished by hand or pneumatic hand tools, and does not include loading, hauling, or storing the removed item, scaffolding, shoring, or lifting equipment.

Cutout Demolition - Demolition of a small quantity of floor, wall, roof, or other assembly, with concern for the appearance and structural integrity of the surrounding materials. This type of demolition is accomplished by hand or pneumatic hand tools, and does not include saw cutting, handling, loading, hauling, or disposal of debris, scaffolding, or shoring.

Rubbish Handling - Work activities that involve handling, loading or hauling of debris. Generally, the cost of rubbish handling must be added to the cost of all types of demolition, with the exception of whole building demolition.

Minor Site Demolition - Demolition of site elements outside the footprint of a building. This type of demolition is accomplished by hand or pneumatic hand tools, or with larger pieces of construction equipment, and may include loading a removed item onto a truck (check the Crew for equipment used). It does not include saw cutting, hauling or disposal of debris, and, sometimes, handling or loading.

R024119-20   Dumpsters

Dumpster rental costs on construction sites are presented in two ways.

The cost per week rental includes the delivery of the dumpster; its pulling or emptying once per week, and its final removal. The assumption is made that the dumpster contractor could choose to empty a dumpster by simply bringing in an empty unit and removing the full one. These costs also include the disposal of the materials in the dumpster.

The Alternate Pricing can be used when actual planned conditions are not approximated by the weekly numbers. For example, these lines can be used when a dumpster is needed for 4 weeks and will need to be emptied 2 or 3 times per week. Conversely the Alternate Pricing lines can be used when a dumpster will be rented for several weeks or months but needs to be emptied only a few times over this period.

R040130-10 Cleaning Face Brick

On smooth brick a person can clean 70 S.F. an hour; on rough brick 50 S.F. per hour. Use one gallon muriatic acid to 20 gallons of water for 1000

S.F. Do not use acid solution until wall is at least seven days old, but a mild soap solution may be used after two days.

Time has been allowed for clean-up in brick prices.

R040513-10 Cement Mortar (material only)

Type N - 1:1:6 mix by volume. Use everywhere above grade except as noted below. - 1:3 mix using conventional masonry cement which saves handling two separate bagged materials.

Type M - 1:1/4:3 mix by volume, or 1 part cement, 1/4 (10% by wt.) lime, 3 parts sand. Use for heavy loads and where earthquakes or hurricanes may occur. Also for reinforced brick, sewers, manholes and everywhere below grade.

Mix Proportions by Volume and Compressive Strength of Mortar

Where Used	Mortar Type	Portland Cement	Masonry Cement	Hydrated Lime	Masonry Sand	Compressive Strength @ 28 days
Plain Masonry	M	1	1	—	6	
		1	—	1/4	3	2500 psi
	S	1/2	1	—	4	
		1	—	1/4 to 1/2	4	1800 psi
	N	—	1	—	3	
		1	—	1/2 to 1-1/4	6	750 psi
	O	—	1	—	3	
		1	—	1-1/4 to 2-1/2	9	350 psi
	K	1	—	2-1/2 to 4	12	75 psi
Reinforced Masonry	PM	1	1	—	6	2500 psi
	PL	1	—	1/4 to 1/2	4	2500 psi

Note: The total aggregate should be between 2.25 to 3 times the sum of the cement and lime used.

The labor cost to mix the mortar is included in the productivity and labor cost of unit price lines in unit cost sections for brickwork, blockwork and stonework.

The material cost of mixed mortar is included in the material cost of those same unit price lines and includes the cost of renting and operating a 10 C.F. mixer at the rate of 200 C.F. per day.

There are two types of mortar color used. One type is the inert additive type with about 100 lbs. per M brick as the typical quantity required. These colors are also available in smaller-batch-sized bags (1 lb. to 15 lb.) which can be placed directly into the mixer without measuring. The other type is premixed and replaces the masonry cement. Dark green color has the highest cost.

R040519-50 Masonry Reinforcing

Horizontal joint reinforcing helps prevent wall cracks where wall movement may occur and in many locations is required by code. Horizontal joint reinforcing is generally not considered to be structural reinforcing and an unreinforced wall may still contain joint reinforcing.

Reinforcing strips come in 10′ and 12′ lengths and in truss and ladder shapes, with and without drips. Field labor runs between 2.7 to 5.3 hours per 1000 L.F. for wall thicknesses up to 12″.

The wire meets ASTM A82 for cold drawn steel wire and the typical size is 9 ga. sides and ties with 3/16″ diameter also available. Typical finish is mill galvanized with zinc coating at .10 oz. per S.F. Class I (.40 oz. per S.F.) and Class III (.80 oz. per S.F.) are also available, as is hot dipped galvanizing at 1.50 oz. per S.F.

R042110-10 Economy in Bricklaying

Have adequate supervision. Be sure bricklayers are always supplied with materials so there is no waiting. Place experienced bricklayers at corners and openings.

Use only screened sand for mortar. Otherwise, labor time will be wasted picking out pebbles. Use seamless metal tubs for mortar as they do not leak or catch the trowel. Locate stack and mortar for easy wheeling.

Have brick delivered for stacking. This makes for faster handling, reduces chipping and breakage, and requires less storage space. Many dealers will deliver select common in 2' x 3' x 4' pallets or face brick packaged. This affords quick handling with a crane or forklift and easy tonging in units of ten, which reduces waste.

Use wider bricks for one wythe wall construction. Keep scaffolding away from the wall to allow mortar to fall clear and not stain the wall.

On large jobs develop specialized crews for each type of masonry unit.

Consider designing for prefabricated panel construction on high rise projects.

Avoid excessive corners or openings. Each opening adds about 50% to the labor cost for area of opening.

Bolting stone panels and using window frames as stops reduce labor costs and speed up erection.

R042110-20 Common and Face Brick

Common building brick manufactured according to ASTM C62 and facing brick manufactured according to ASTM C216 are the two standard bricks available for general building use.

Building brick is made in three grades: SW, where high resistance to damage caused by cyclic freezing is required; MW, where moderate resistance to cyclic freezing is needed; and NW, where little resistance to cyclic freezing is needed. Facing brick is made in only the two grades SW and MW. Additionally, facing brick is available in three types: FBS, for general use; FBX, for general use where a higher degree of precision and lower permissible variation in size than FBS are needed; and FBA, for general use to produce characteristic architectural effects resulting from non-uniformity in size and texture of the units.

In figuring the material cost of brickwork, an allowance of 25% mortar waste and 3% brick breakage was included. If bricks are delivered palletized

with 280 to 300 per pallet, or packaged, allow only 1-1/2% for breakage. Packaged or palletized delivery is practical when a job is big enough to have a crane or other equipment available to handle a package of brick. This is so on all industrial work but not always true on small commercial buildings.

The use of buff and gray face is increasing, and there is a continuing trend to the Norman, Roman, Jumbo and SCR brick.

Common red clay brick for backup is not used that often. Concrete block is the most usual backup material with occasional use of sand lime or cement brick. Building brick is commonly used in solid walls for strength and as a fire stop.

Brick panels built on the ground and then crane erected to the upper floors have proven to be economical. This allows the work to be done under cover and without scaffolding.

R042110-50 Brick, Block & Mortar Quantities

Running Bond							For Other Bonds Standard Size Add to S.F. Quantities in Table to Left		
Number of Brick per S.F. of Wall - Single Wythe with 3/8" Joints					C.F. of Mortar per M Bricks, Waste Included				
Type Brick	Nominal Size (incl. mortar) L H W		Modular Coursing	Number of Brick per S.F.	3/8" Joint	1/2" Joint	Bond Type	Description	Factor
Standard	8 x 2-2/3 x 4		3C=8"	6.75	10.3	12.9	Common	full header every fifth course	+20%
Economy	8 x 4 x 4		1C=4"	4.50	11.4	14.6		full header every sixth course	+16.7%
Engineer	8 x 3-1/5 x 4		5C=16"	5.63	10.6	13.6	English	full header every second course	+50%
Fire	9 x 2-1/2 x 4-1/2		2C=5"	6.40	550 # Fireclay	—	Flemish	alternate headers every course	+33.3%
Jumbo	12 x 4 x 6 or 8		1C=4"	3.00	23.8	30.8		every sixth course	+5.6%
Norman	12 x 2-2/3 x 4		3C=8"	4.50	14.0	17.9	Header = W x H exposed		+100%
Norwegian	12 x 3-1/5 x 4		5C=16"	3.75	14.6	18.6	Rowlock = H x W exposed		+100%
Roman	12 x 2 x 4		2C=4"	6.00	13.4	17.0	Rowlock stretcher = L x W exposed		+33.3%
SCR	12 x 2-2/3 x 6		3C=8"	4.50	21.8	28.0	Soldier = H x L exposed		—
Utility	12 x 4 x 4		1C=4"	3.00	15.4	19.6	Sailor = W x L exposed		-33.3%

Concrete Blocks Nominal Size		Approximate Weight per S.F.		Blocks per 100 S.F.	Mortar per M block, waste included	
		Standard	Lightweight		Partitions	Back up
2"	x 8" x 16"	20 PSF	15 PSF	113	27 C.F.	36 C.F.
4"		30	20		41	51
6"		42	30		56	66
8"		55	38		72	82
10"		70	47		87	97
12"		85	55		102	112

Brick & Mortar Quantities
©Brick Industry Association. 2009 Feb. Technical Notes on
Brick Construction 10:
   Dimensioning and Estimating Brick Masonry. Reston (VA): BIA. Table 1
   Modular Brick Sizes and Table 4 Quantity Estimates for Brick Masonry.

Masonry — R0422 Concrete Unit Masonry

R042210-20 Concrete Block

The material cost of special block such as corner, jamb and head block can be figured at the same price as ordinary block of equal size. Labor on specials is about the same as equal-sized regular block.

Bond beams and 16" high lintel blocks are more expensive than regular units of equal size. Lintel blocks are 8" long and either 8" or 16" high.

Use of a motorized mortar spreader box will speed construction of continuous walls.

Hollow non-load-bearing units are made according to ASTM C129 and hollow load-bearing units according to ASTM C90.

Metals — R0531 Steel Decking

R053100-10 Decking Descriptions

General - All Deck Products

A steel deck is made by cold forming structural grade sheet steel into a repeating pattern of parallel ribs. The strength and stiffness of the panels are the result of the ribs and the material properties of the steel. Deck lengths can be varied to suit job conditions, but because of shipping considerations, are usually less than 40 feet. Standard deck width varies with the product used but full sheets are usually 12", 18", 24", 30", or 36". The deck is typically furnished in a standard width with the ends cut square. Any cutting for width, such as at openings or for angular fit, is done at the job site.

The deck is typically attached to the building frame with arc puddle welds, self-drilling screws, or powder or pneumatically driven pins. Sheet to sheet fastening is done with screws, button punching (crimping), or welds.

Composite Floor Deck

After installation and adequate fastening, a floor deck serves several purposes. It (a) acts as a working platform, (b) stabilizes the frame, (c) serves as a concrete form for the slab, and (d) reinforces the slab to carry the design loads applied during the life of the building. Composite decks are distinguished by the presence of shear connector devices as part of the deck. These devices are designed to mechanically lock the concrete and deck together so that the concrete and the deck work together to carry subsequent floor loads. These shear connector devices can be rolled-in embossments, lugs, holes, or wires welded to the panels. The deck profile can also be used to interlock concrete and steel.

Composite deck finishes are either galvanized (zinc coated) or phosphatized/painted. Galvanized deck has a zinc coating on both the top and bottom surfaces. The phosphatized/painted deck has a bare (phosphatized) top surface that will come into contact with the concrete. This bare top surface can be expected to develop rust before the concrete is placed. The bottom side of the deck has a primer coat of paint.

A composite floor deck is normally installed so the panel ends do not overlap on the supporting beams. Shear lugs or panel profile shapes often prevent a tight metal to metal fit if the panel ends overlap; the air gap caused by overlapping will prevent proper fusion with the structural steel supports when the panel end laps are shear stud welded.

Adequate end bearing of the deck must be obtained as shown on the drawings. If bearing is actually less in the field than shown on the drawings, further investigation is required.

Roof Deck

A roof deck is not designed to act compositely with other materials. A roof deck acts alone in transferring horizontal and vertical loads into the building frame. Roof deck rib openings are usually narrower than floor deck rib openings. This provides adequate support of the rigid thermal insulation board.

A roof deck is typically installed to endlap approximately 2" over supports. However, it can be butted (or lapped more than 2") to solve field fit problems. Since designers frequently use the installed deck system as part of the horizontal bracing system (the deck as a diaphragm), any fastening substitution or change should be approved by the designer. Continuous perimeter support of the deck is necessary to limit edge deflection in the finished roof and may be required for diaphragm shear transfer.

Standard roof deck finishes are galvanized or primer painted. The standard factory applied paint for roof decks is a primer paint and is not intended to weather for extended periods of time. Field painting or touching up of abrasions and deterioration of the primer coat or other protective finishes is the responsibility of the contractor.

Cellular Deck

A cellular deck is made by attaching a bottom steel sheet to a roof deck or composite floor deck panel. A cellular deck can be used in the same manner as a floor deck. Electrical, telephone, and data wires are easily run through the chase created between the deck panel and the bottom sheet.

When used as part of the electrical distribution system, the cellular deck must be installed so that the ribs line up and create a smooth cell transition at abutting ends. The joint that occurs at butting cell ends must be taped or otherwise sealed to prevent wet concrete from seeping into the cell. Cell interiors must be free of welding burrs, or other sharp intrusions, to prevent damage to wires.

When used as a roof deck, the bottom flat plate is usually left exposed to view. Care must be maintained during erection to keep good alignment and prevent damage.

A cellular deck is sometimes used with the flat plate on the top side to provide a flat working surface. Installation of the deck for this purpose requires special methods for attachment to the frame because the flat plate, now on the top, can prevent direct access to the deck material that is bearing on the structural steel. It may be advisable to treat the flat top surface to prevent slipping.

A cellular deck is always furnished galvanized or painted over galvanized.

Form Deck

A form deck can be any floor or roof deck product used as a concrete form. Connections to the frame are by the same methods used to anchor floor and roof decks. Welding washers are recommended when welding a deck that is less than 20 gauge thickness.

A form deck is furnished galvanized, prime painted, or uncoated. A galvanized deck must be used for those roof deck systems where a form deck is used to carry a lightweight insulating concrete fill.

R061110-30 Lumber Product Material Prices

The price of forest products fluctuates widely from location to location and from season to season depending upon economic conditions. The bare material prices in the unit cost sections of the data set show the National Average material prices in effect Jan. 1 of this data year. It must be noted that lumber prices in general may change significantly during the year.

Availability of certain items depends upon geographic location and must be checked prior to firm-price bidding.

R061636-20 Plywood

There are two types of plywood used in construction: interior, which is moisture-resistant but not waterproofed, and exterior, which is waterproofed.

The grade of the exterior surface of the plywood sheets is designated by the first letter: A, for smooth surface with patches allowed; B, for solid surface with patches and plugs allowed; C, which may be surface plugged or may have knot holes up to 1″ wide; and D, which is used only for interior type plywood and may have knot holes up to 2-1/2″ wide. "Structural Grade" is specifically designed for engineered applications such as box beams. All CC & DD grades have roof and floor spans marked on them.

Underlayment-grade plywood runs from 1/4″ to 1-1/4″ thick. Thicknesses 5/8″ and over have optional tongue and groove joints which eliminate the need for blocking the edges. Underlayment 19/32″ and over may be referred to as Sturd-i-Floor.

The price of plywood can fluctuate widely due to geographic and economic conditions.

Typical uses for various plywood grades are as follows:

AA-AD Interior — cupboards, shelving, paneling, furniture

BB Plyform — concrete form plywood

CDX — wall and roof sheathing

Structural — box beams, girders, stressed skin panels

AA-AC Exterior — fences, signs, siding, soffits, etc.

Underlayment — base for resilient floor coverings

Overlaid HDO — high density for concrete forms & highway signs

Overlaid MDO — medium density for painting, siding, soffits & signs

303 Siding — exterior siding, textured, striated, embossed, etc.

R073126-20 Roof Slate

16″, 18″ and 20″ are standard lengths, and slate usually comes in random widths. For standard 3/16″ thickness use 1-1/2″ copper nails. Allow for 3% breakage.

R075213-30 Modified Bitumen Roofing

The cost of modified bitumen roofing is highly dependent on the type of installation that is planned. Installation is based on the type of modifier used in the bitumen. The two most popular modifiers are atactic polypropylene (APP) and styrene butadiene styrene (SBS). The modifiers are added to heated bitumen during the manufacturing process to change its characteristics. A polyethylene, polyester or fiberglass reinforcing sheet is then sandwiched between layers of this bitumen. When completed, the result is a pre-assembled, built-up roof that has increased elasticity and weatherability. Some manufacturers include a surfacing material such as ceramic or mineral granules, metal particles or sand.

The preferred method of adhering SBS-modified bitumen roofing to the substrate is with hot-mopped asphalt (much the same as built-up roofing). This installation method requires a tar kettle/pot to heat the asphalt, as well as the labor, tools and equipment necessary to distribute and spread the hot asphalt.

The alternative method for applying APP and SBS modified bitumen is as follows. A skilled installer uses a torch to melt a small pool of bitumen off the membrane. This pool must form across the entire roll for proper adhesion. The installer must unroll the roofing at a pace slow enough to melt the bitumen, but fast enough to prevent damage to the rest of the membrane.

Modified bitumen roofing provides the advantages of both built-up and single-ply roofing. Labor costs are reduced over those of built-up roofing because only a single ply is necessary. The elasticity of single-ply roofing is attained with the reinforcing sheet and polymer modifiers. Modifieds have some self-healing characteristics and because of their multi-layer construction, they offer the reliability and safety of built-up roofing.

R081313-20 Steel Door Selection Guide

Standard steel doors are classified into four levels, as recommended by the Steel Door Institute in the chart below. Each of the four levels offers a range of construction models and designs to meet architectural requirements for preference and appearance, including full flush, seamless, and stile & rail. Recommended minimum gauge requirements are also included.

For complete standard steel door construction specifications and available sizes, refer to the Steel Door Institute Technical Data Series, ANSI A250.8-98 (SDI-100), and ANSI A250.4-94 Test Procedure and Acceptance Criteria for Physical Endurance of Steel Door and Hardware Reinforcements.

Level		Model	Construction	For Full Flush or Seamless		
				Min. Gauge	Thickness (in)	Thickness (mm)
I	Standard Duty	1	Full Flush	20	0.032	0.8
		2	Seamless			
II	Heavy Duty	1	Full Flush	18	0.042	1.0
		2	Seamless			
III	Extra Heavy Duty	1	Full Flush	16	0.053	1.3
		2	Seamless			
		3	*Stile & Rail			
IV	Maximum Duty	1	Full Flush	14	0.067	1.6
		2	Seamless			

*Stiles & rails are 16 gauge; flush panels, when specified, are 18 gauge.

R085216-10 Window Estimates

To ensure a complete window estimate, be sure to include the material and labor costs for each window, as well as the material and labor costs for an interior wood trim set.

R085313-20 Replacement Windows

Replacement windows are typically measured per United Inch.

United Inches are calculated by rounding the width and height of the window opening up to the nearest inch, then adding the two figures.

The labor cost for replacement windows includes removal of sash, existing sash balance or weights, parting bead where necessary and installation of new window.

Debris hauling and dump fees are not included.

Reference Tables

R087110-10   Hardware Finishes

This table describes hardware finishes used throughout the industry. It also shows the base metal and the respective symbols in the three predominate systems of identification. Many of these are used in pricing descriptions in Division Eight.

US″	BMHA*	CDN^	Base	Description
US P	600	CP	Steel	Primed for Painting
US 1B	601	C1B	Steel	Bright Black Japanned
US 2C	602	C2C	Steel	Zinc Plated
US 2G	603	C2G	Steel	Zinc Plated
US 3	605	C3	Brass	Bright Brass, Clear Coated
US 4	606	C4	Brass	Satin Brass, Clear Coated
US 5	609	C5	Brass	Satin Brass, Blackened, Satin Relieved, Clear Coated
US 7	610	C7	Brass	Satin Brass, Blackened, Bright Relieved, Clear Coated
US 9	611	C9	Bronze	Bright Bronze, Clear Coated
US 10	612	C10	Bronze	Satin Bronze, Clear Coated
US 10A	641	C10A	Steel	Antiqued Bronze, Oiled and Lacquered
US 10B	613	C10B	Bronze	Antiqued Bronze, Oiled
US 11	616	C11	Bronze	Satin Bronze, Blackened, Satin Relieved, Clear Coated
US 14	618	C14	Brass/Bronze	Bright Nickel Plated, Clear Coated
US 15	619	C15	Brass/Bronze	Satin Nickel, Clear Coated
US 15A	620	C15A	Brass/Bronze	Satin Nickel Plated, Blackened, Satin Relieved, Clear Coated
US 17A	621	C17A	Brass/Bronze	Nickel Plated, Blackened, Relieved, Clear Coated
US 19	622	C19	Brass/Bronze	Flat Black Coated
US 20	623	C20	Brass/Bronze	Statuary Bronze, Light
US 20A	624	C20A	Brass/Bronze	Statuary Bronze, Dark
US 26	625	C26	Brass/Bronze	Bright Chromium
US 26D	626	C26D	Brass/Bronze	Satin Chromium
US 20	627	C27	Aluminum	Satin Aluminum Clear
US 28	628	C28	Aluminum	Anodized Dull Aluminum
US 32	629	C32	Stainless Steel	Bright Stainless Steel
US 32D	630	C32D	Stainless Steel	Stainless Steel
US 3	632	C3	Steel	Bright Brass Plated, Clear Coated
US 4	633	C4	Steel	Satin Brass, Clear Coated
US 7	636	C7	Steel	Satin Brass Plated, Blackened, Bright Relieved, Clear Coated
US 9	637	C9	Steel	Bright Bronze Plated, Clear Coated
US 5	638	C5	Steel	Satin Brass Plated, Blackened, Bright Relieved, Clear Coated
US 10	639	C10	Steel	Satin Bronze Plated, Clear Coated
US 10B	640	C10B	Steel	Antique Bronze, Oiled
US 10A	641	C10A	Steel	Antiqued Bronze, Oiled and Lacquered
US 11	643	C11	Steel	Satin Bronze Plated, Blackened, Bright Relieved, Clear Coated
US 14	645	C14	Steel	Bright Nickel Plated, Clear Coated
US 15	646	C15	Steel	Satin Nickel Plated, Clear Coated
US 15A	647	C15A	Steel	Nickel Plated, Blackened, Bright Relieved, Clear Coated
US 17A	648	C17A	Steel	Nickel Plated, Blackened, Relieved, Clear Coated
US 20	649	C20	Steel	Statuary Bronze, Light
US 20A	650	C20A	Steel	Statuary Bronze, Dark
US 26	651	C26	Steel	Bright Chromium Plated
US 26D	652	C26D	Steel	Satin Chromium Plated

* - BMHA Builders Hardware Manufacturing Association
″ - US Equivalent
^ - Canadian Equivalent
Japanning is imitating Asian lacquer work

R092000-50 Lath, Plaster and Gypsum Board

Gypsum board lath is available in 3/8″ thick x 16″ wide x 4′ long sheets as a base material for multi-layer plaster applications. It is also available as a base for either multi-layer or veneer plaster applications in 1/2″ and 5/8″ thick—4′ wide x 8′, 10′ or 12′ long sheets. Fasteners are screws or blued ring shank nails for wood framing and screws for metal framing.

Metal lath is available in diamond mesh patterns with flat or self-furring profiles. Paper backing is available for applications where excessive plaster waste needs to be avoided. A slotted mesh ribbed lath should be used in areas where the span between structural supports is greater than normal. Most metal lath comes in 27″ x 96″ sheets. Diamond mesh weighs 1.75, 2.5 or 3.4 pounds per square yard, slotted mesh lath weighs 2.75 or 3.4 pounds per square yard. Metal lath can be nailed, screwed or tied in place.

Many **accessories** are available. Corner beads, flat reinforcing strips, casing beads, control and expansion joints, furring brackets and channels are some examples. Note that accessories are not included in plaster or stucco line items.

Plaster is defined as a material or combination of materials that when mixed with a suitable amount of water, forms a plastic mass or paste. When applied to a surface, the paste adheres to it and subsequently hardens, preserving in a rigid state the form or texture imposed during the period of elasticity.

Gypsum plaster is made from ground calcined gypsum. It is mixed with aggregates and water for use as a base coat plaster.

Vermiculite plaster is a fire-retardant plaster covering used on steel beams, concrete slabs and other heavy construction materials. Vermiculite is a group name for certain clay minerals, hydrous silicates or aluminum, magnesium and iron that have been expanded by heat.

Perlite plaster is a plaster using perlite as an aggregate instead of sand. Perlite is a volcanic glass that has been expanded by heat.

Gauging plaster is a mix of gypsum plaster and lime putty that when applied produces a quick drying finish coat.

Veneer plaster is a one or two component gypsum plaster used as a thin finish coat over special gypsum board.

Keenes cement is a white cementitious material manufactured from gypsum that has been burned at a high temperature and ground to a fine powder. Alum is added to accelerate the set. The resulting plaster is hard and strong and accepts and maintains a high polish, hence it is used as a finishing plaster.

Stucco is a Portland cement based plaster used primarily as an exterior finish.

Plaster is used on both interior and exterior surfaces. Generally it is applied in multiple-coat systems. A three-coat system uses the terms scratch, brown and finish to identify each coat. A two-coat system uses base and finish to describe each coat. Each type of plaster and application system has attributes that are chosen by the designer to best fit the intended use.

Gypsum Plaster Quantities for 100 S.Y.	2 Coat, 5/8″ Thick		3 Coat, 3/4″ Thick		
	Base	Finish	Scratch	Brown	Finish
	1:3 Mix	2:1 Mix	1:2 Mix	1:3 Mix	2:1 Mix
Gypsum plaster	1,300 lb.		1,350 lb.	650 lb.	
Sand	1.75 C.Y.		1.85 C.Y.	1.35 C.Y.	
Finish hydrated lime		340 lb.			340 lb.
Gauging plaster		170 lb.			170 lb.

Vermiculite or Perlite Plaster Quantities for 100 S.Y.	2 Coat, 5/8″ Thick		3 Coat, 3/4″ Thick		
	Base	Finish	Scratch	Brown	Finish
Gypsum plaster	1,250 lb.		1,450 lb.	800 lb.	
Vermiculite or perlite	7.8 bags		8.0 bags	3.3 bags	
Finish hydrated lime		340 lb.			340 lb.
Gauging plaster		170 lb.			170 lb.

Stucco–Three-Coat System Quantities for 100 S.Y.	On Wood Frame	On Masonry
Portland cement	29 bags	21 bags
Sand	2.6 C.Y.	2.0 C.Y.
Hydrated lime	180 lb.	120 lb.

R092910-10   Levels of Gypsum Board Finish

In the past, contract documents often used phrases such as "industry standard" and "workmanlike finish" to specify the expected quality of gypsum board wall and ceiling installations. The vagueness of these descriptions led to unacceptable work and disputes.

In order to resolve this problem, four major trade associations concerned with the manufacture, erection, finish and decoration of gypsum board wall and ceiling systems developed an industry-wide *Recommended Levels of Gypsum Board Finish*.

The finish of gypsum board walls and ceilings for specific final decoration is dependent on a number of factors. A primary consideration is the location of the surface and the degree of decorative treatment desired. Painted and unpainted surfaces in warehouses and other areas where appearance is normally not critical may simply require the taping of wallboard joints and 'spotting' of fastener heads. Blemish-free, smooth, monolithic surfaces often intended for painted and decorated walls and ceilings in habitated structures, ranging from single-family dwellings through monumental buildings, require additional finishing prior to the application of the final decoration.

Other factors to be considered in determining the level of finish of the gypsum board surface are (1) the type of angle of surface illumination (both natural and artificial lighting), and (2) the paint and method of application or the type and finish of wallcovering specified as the final decoration. Critical lighting conditions, gloss paints, and thin wall coverings require a higher level of gypsum board finish than heavily textured surfaces which are subsequently painted or surfaces which are to be decorated with heavy grade wall coverings.

The following descriptions were developed by the Association of the Wall and Ceiling Industries-International (AWCI), Ceiling & Interior Systems Construction Association (CISCA), Gypsum Association (GA), and Painting and Decorating Contractors of America (PDCA) as a guide.

Level 0: Used in temporary construction or wherever the final decoration has not been determined. Unfinished. No taping, finishing or corner beads are required. Also could be used where non-predecorated panels will be used in demountable-type partitions that are to be painted as a final finish. .

Level 1: Frequently used in plenum areas above ceilings, in attics, in areas where the assembly would generally be concealed, or in building service corridors and other areas not normally open to public view. Some degree of sound and smoke control is provided; in some geographic areas, this level is referred to as "fire-taping," although this level of finish does not typically meet fire-resistant assembly requirements. Where a fire resistance rating is required for the gypsum board assembly, details of construction should be in accordance with reports of fire tests of assemblies that have met the requirements of the fire rating acceptable.

All joints and interior angles shall have tape embedded in joint compound. Accessories are optional at specifier discretion in corridors and other areas with pedestrian traffic. Tape and fastener heads need not be covered with joint compound. Surface shall be free of excess joint compound. Tool marks and ridges are acceptable.

Level 2: It may be specified for standard gypsum board surfaces in garages, warehouse storage, or other similar areas where surface appearance is not of primary importance.

All joints and interior angles shall have tape embedded in joint compound and shall be immediately wiped with a joint knife or trowel, leaving a thin coating of joint compound over all joints and interior angles. Fastener heads and accessories shall be covered with a coat of joint compound. Surface shall be free of excess joint compound. Tool marks and ridges are acceptable.

Level 3: Typically used in areas receiving heavy texture (spray or hand applied) finishes before final painting, or where commercial-grade (heavy duty) wall coverings are to be applied as the final decoration. This level of finish should not be used where smooth painted surfaces or where lighter weight wall coverings are specified. The prepared surface shall be coated with a drywall primer prior to the application of final finishes.

All joints and interior angles shall have tape embedded in joint compound and shall be immediately wiped with a joint knife or trowel, leaving a thin coating of joint compound over all joints and interior angles. One additional coat of joint compound shall be applied over all joints and interior angles. Fastener heads and accessories shall be covered with two separate coats of joint compound. All joint compounds shall be smooth and free of tool marks and ridges. The prepared surface shall be covered with a drywall primer prior to the application of the final decoration.

Level 4: This level should be used where residential grade (light duty) wall coverings, flat paints, or light textures are to be applied. The prepared surface shall be coated with a drywall primer prior to the application of final finishes. Release agents for wall coverings are specifically formulated to minimize damage if coverings are subsequently removed.

The weight, texture, and sheen level of the wall covering material selected should be taken into consideration when specifying wall coverings over this level of drywall treatment. Joints and fasteners must be sufficiently concealed if the wall covering material is lightweight, contains limited pattern, has a glossy finish, or has any combination of these features. In critical lighting areas, flat paints applied over light textures tend to reduce joint photographing. Gloss, semi-gloss, and enamel paints are not recommended over this level of finish.

All joints and interior angles shall have tape embedded in joint compound and shall be immediately wiped with a joint knife or trowel, leaving a thin coating of joint compound over all joints and interior angles. In addition, two separate coats of joint compound shall be applied over all flat joints and one separate coat of joint compound applied over interior angles. Fastener heads and accessories shall be covered with three separate coats of joint compound. All joint compounds shall be smooth and free of tool marks and ridges. The prepared surface shall be covered with a drywall primer like Sheetrock first coat prior to the application of the final decoration.

Level 5: The highest quality finish is the most effective method to provide a uniform surface and minimize the possibility of joint photographing and of fasteners showing through the final decoration. This level of finish is required where gloss, semi-gloss, or enamel is specified; when flat joints are specified over an untextured surface; or where critical lighting conditions occur. The prepared surface shall be coated with a drywall primer prior to the application of the final decoration.

All joints and interior angles shall have tape embedded in joint compound and be immediately wiped with a joint knife or trowel, leaving a thin coating of joint compound over all joints and interior angles. Two separate coats of joint compound shall be applied over all flat joints and one separate coat of joint compound applied over interior angles. Fastener heads and accessories shall be covered with three separate coats of joint compound.

A thin skim coat of joint compound shall be trowel applied to the entire surface. Excess compound is immediately troweled off, leaving a film or skim coating of compound completely covering the paper. As an alternative to a skim coat, a material manufactured especially for this purpose may be applied such as Sheetrock Tuff-Hide primer surfacer. The surface must be smooth and free of tool marks and ridges. The prepared surface shall be covered with a drywall primer prior to the application of the final decoration.

Finishes — R0972 Wall Coverings

R097223-10   Wall Covering

The table below lists the quantities required for 100 S.F. of wall covering.

Description	Medium-Priced Paper	Expensive Paper
Paper	1.6 dbl. rolls	1.6 dbl. rolls
Wall sizing	0.25 gallon	0.25 gallon
Vinyl wall paste	0.6 gallon	0.6 gallon
Apply sizing	0.3 hour	0.3 hour
Apply paper	1.2 hours	1.5 hours

Most wallpapers now come in double rolls only.
To remove old paper, allow 1.3 hours per 100 S.F.

Finishes — R0991 Painting

R099100-10   Painting Estimating Techniques

Proper estimating methodology is needed to obtain an accurate painting estimate. There is no known reliable shortcut or square foot method. The following steps should be followed:

- List all surfaces to be painted, with an accurate quantity (area) of each. Items having similar surface condition, finish, application method and accessibility may be grouped together.

- List all the tasks required for each surface to be painted, including surface preparation, masking, and protection of adjacent surfaces. Surface preparation may include minor repairs, washing, sanding and puttying.

- Select the proper Means line for each task. Review and consider all adjustments to labor and materials for type of paint and location of work. Apply the height adjustment carefully. For instance, when applying the adjustment for work over 8' high to a wall that is 12' high, apply the adjustment only to the area between 8' and 12' high, and not to the entire wall.

When applying more than one percent (%) adjustment, apply each to the base cost of the data, rather than applying one percentage adjustment on top of the other.

When estimating the cost of painting walls and ceilings remember to add the brushwork for all cut-ins at inside corners and around windows and doors as a LF measure. One linear foot of cut-in with a brush equals one square foot of painting.

All items for spray painting include the labor for roll-back.

Deduct for openings greater than 100 SF or openings that extend from floor to ceiling and are greater than 5' wide. Do not deduct small openings.

The cost of brushes, rollers, ladders and spray equipment are considered part of a painting contractor's overhead, and should not be added to the estimate. The cost of rented equipment such as scaffolding and swing staging should be added to the estimate.

R099100-20   Painting

Item	Coat	One Gallon Covers			In 8 Hours a Laborer Covers			Labor-Hours per 100 S.F.		
		Brush	Roller	Spray	Brush	Roller	Spray	Brush	Roller	Spray
Paint wood siding	prime	250 S.F.	225 S.F.	290 S.F.	1150 S.F.	1300 S.F.	2275 S.F.	.695	.615	.351
	others	270	250	290	1300	1625	2600	.615	.492	.307
Paint exterior trim	prime	400	—	—	650	—	—	1.230	—	—
	1st	475	—	—	800	—	—	1.000	—	—
	2nd	520	—	—	975	—	—	.820	—	—
Paint shingle siding	prime	270	255	300	650	975	1950	1.230	.820	.410
	others	360	340	380	800	1150	2275	1.000	.695	.351
Stain shingle siding	1st	180	170	200	750	1125	2250	1.068	.711	.355
	2nd	270	250	290	900	1325	2600	.888	.603	.307
Paint brick masonry	prime	180	135	160	750	800	1800	1.066	1.000	.444
	1st	270	225	290	815	975	2275	.981	.820	.351
	2nd	340	305	360	815	1150	2925	.981	.695	.273
Paint interior plaster or drywall	prime	400	380	495	1150	2000	3250	.695	.400	.246
	others	450	425	495	1300	2300	4000	.615	.347	.200
Paint interior doors and windows	prime	400	—	—	650	—	—	1.230	—	—
	1st	425	—	—	800	—	—	1.000	—	—
	2nd	450	—	—	975	—	—	.820	—	—

R131113-20   Swimming Pools

Pool prices given per square foot of surface area include pool structure, filter and chlorination equipment, pumps, related piping, ladders/steps, maintenance kit, skimmer and vacuum system. Decks and electrical service to equipment are not included.

Residential in-ground pool construction can be divided into two categories: vinyl lined and gunite. Vinyl lined pool walls are constructed of different materials including wood, concrete, plastic or metal. The bottom is often graded with sand over which the vinyl liner is installed. Vermiculite or soil cement bottoms may be substituted for an added cost.

Gunite pool construction is used both in residential and municipal installations. These structures are steel reinforced for strength and finished with a white cement limestone plaster.

Municipal pools will have a higher cost because plumbing codes require more expensive materials, chlorination equipment and higher filtration rates.

Municipal pools greater than 1,800 S.F. require gutter systems to control waves. This gutter may be formed into the concrete wall. Often a vinyl/stainless steel gutter or gutter/wall system is specified, which will raise the pool cost.

Competition pools usually require tile bottoms and sides with contrasting lane striping, which will also raise the pool cost.

Plumbing | R2211 Facility Water Distribution

R221113-50 Pipe Material Considerations

1. Malleable fittings should be used for gas service.
2. Malleable fittings are used where there are stresses/strains due to expansion and vibration.
3. Cast fittings may be broken as an aid to disassembling heating lines frozen by long use, temperature and minerals.
4. A cast iron pipe is extensively used for underground and submerged service.
5. Type M (light wall) copper tubing is available in hard temper only and is used for nonpressure and less severe applications than K and L.

6. Type L (medium wall) copper tubing, available hard or soft for interior service.
7. Type K (heavy wall) copper tubing, available in hard or soft temper for use where conditions are severe. For underground and interior service.
8. Hard drawn tubing requires fewer hangers or supports but should not be bent. Silver brazed fittings are recommended, but soft solder is normally used.
9. Type DMV (very light wall) copper tubing designed for drainage, waste and vent plus other non-critical pressure services.

Domestic/Imported Pipe and Fittings Costs

The prices shown in this publication for steel/cast iron pipe and steel, cast iron, and malleable iron fittings are based on domestic production sold at the normal trade discounts. The above listed items of foreign manufacture may be available at prices 1/3 to 1/2 of those shown. Some imported items after minor machining or finishing operations are being sold as domestic to further complicate the system.

Caution: Most pipe prices in this data set also include a coupling and pipe hangers which for the larger sizes can add significantly to the per foot cost and should be taken into account when comparing "book cost" with the quoted supplier's cost.

Exterior Improvements | R3292 Turf & Grasses

R329219-50 Seeding

The type of grass is determined by light, shade and moisture content of soil plus intended use. Fertilizer should be disked 4" before seeding. For steep slopes disk five tons of mulch and lay two tons of hay or straw on surface per acre after seeding. Surface mulch can be staked, lightly disked or tar emulsion sprayed. Material for mulch can be wood chips, peat moss, partially rotted hay or straw, wood fibers and sprayed emulsions. Hemp seed blankets with fertilizer are also available. For spring seeding, watering is necessary. Late fall seeding may have to be reseeded in the spring. Hydraulic seeding, power mulching, and aerial seeding can be used on large areas.

R331113-80   Piping Designations

There are several systems currently in use to describe pipe and fittings. The following paragraphs will help to identify and clarify classifications of piping systems used for water distribution.

Piping may be classified by schedule. Piping schedules include 5S, 10S, 10, 20, 30, Standard, 40, 60, Extra Strong, 80, 100, 120, 140, 160 and Double Extra Strong. These schedules are dependent upon the pipe wall thickness. The wall thickness of a particular schedule may vary with pipe size.

Ductile iron pipe for water distribution is classified by Pressure Classes such as Class 150, 200, 250, 300 and 350. These classes are actually the rated water working pressure of the pipe in pounds per square inch (psi). The pipe in these pressure classes is designed to withstand the rated water working pressure plus a surge allowance of 100 psi.

The American Water Works Association (AWWA) provides standards for various types of **plastic pipe.** C-900 is the specification for polyvinyl chloride (PVC) piping used for water distribution in sizes ranging from 4″ through 12″. C-901 is the specification for polyethylene (PE) pressure pipe, tubing and fittings used for water distribution in sizes ranging from 1/2″ through 3″. C-905 is the specification for PVC piping sizes 14″ and greater.

PVC pressure-rated pipe is identified using the standard dimensional ratio (SDR) method. This method is defined by the American Society for Testing and Materials (ASTM) Standard D 2241. This pipe is available in SDR numbers 64, 41, 32.5, 26, 21, 17, and 13.5. A pipe with an SDR of 64 will have the thinnest wall while a pipe with an SDR of 13.5 will have the thickest wall. When the pressure rating (PR) of a pipe is given in psi, it is based on a line supplying water at 73 degrees F.

The National Sanitation Foundation (NSF) seal of approval is applied to products that can be used with potable water. These products have been tested to ANSI/NSF Standard 14.

Valves and strainers are classified by American National Standards Institute (ANSI) Classes. These Classes are 125, 150, 200, 250, 300, 400, 600, 900, 1500 and 2500. Within each class there is an operating pressure range dependent upon temperature. Design parameters should be compared to the appropriate material dependent, pressure-temperature rating chart for accurate valve selection.

Abbreviations

A	Area Square Feet; Ampere	Brk., brk	Brick	Csc	Cosecant
AAFES	Army and Air Force Exchange Service	brkt	Bracket	C.S.F.	Hundred Square Feet
		Brs.	Brass	CSI	Construction Specifications Institute
ABS	Acrylonitrile Butadiene Stryrene; Asbestos Bonded Steel	Brz.	Bronze		
		Bsn.	Basin	CT	Current Transformer
A.C., AC	Alternating Current; Air-Conditioning; Asbestos Cement; Plywood Grade A & C	Btr.	Better	CTS	Copper Tube Size
		BTU	British Thermal Unit	Cu	Copper, Cubic
		BTUH	BTU per Hour	Cu. Ft.	Cubic Foot
		Bu.	Bushels	cw	Continuous Wave
ACI	American Concrete Institute	BUR	Built-up Roofing	C.W.	Cool White; Cold Water
ACR	Air Conditioning Refrigeration	BX	Interlocked Armored Cable	Cwt.	100 Pounds
ADA	Americans with Disabilities Act	°C	Degree Centigrade	C.W.X.	Cool White Deluxe
AD	Plywood, Grade A & D	c	Conductivity, Copper Sweat	C.Y.	Cubic Yard (27 cubic feet)
Addit.	Additional	C	Hundred; Centigrade	C.Y./Hr.	Cubic Yard per Hour
Adh.	Adhesive	C/C	Center to Center, Cedar on Cedar	Cyl.	Cylinder
Adj.	Adjustable	C-C	Center to Center	d	Penny (nail size)
af	Audio-frequency	Cab	Cabinet	D	Deep; Depth; Discharge
AFFF	Aqueous Film Forming Foam	Cair.	Air Tool Laborer	Dis., Disch.	Discharge
AFUE	Annual Fuel Utilization Efficiency	Cal.	Caliper	Db	Decibel
AGA	American Gas Association	Calc	Calculated	Dbl.	Double
Agg.	Aggregate	Cap.	Capacity	DC	Direct Current
A.H., Ah	Ampere Hours	Carp.	Carpenter	DDC	Direct Digital Control
A hr.	Ampere-hour	C.B.	Circuit Breaker	Demob.	Demobilization
A.H.U., AHU	Air Handling Unit	C.C.A.	Chromate Copper Arsenate	d.f.t.	Dry Film Thickness
A.I.A.	American Institute of Architects	C.C.F.	Hundred Cubic Feet	d.f.u.	Drainage Fixture Units
AIC	Ampere Interrupting Capacity	cd	Candela	D.H.	Double Hung
Allow.	Allowance	cd/sf	Candela per Square Foot	DHW	Domestic Hot Water
alt., alt	Alternate	CD	Grade of Plywood Face & Back	DI	Ductile Iron
Alum.	Aluminum	CDX	Plywood, Grade C & D, exterior glue	Diag.	Diagonal
a.m.	Ante Meridiem			Diam., Dia	Diameter
Amp.	Ampere	Cefi.	Cement Finisher	Distrib.	Distribution
Anod.	Anodized	Cem.	Cement	Div.	Division
ANSI	American National Standards Institute	CF	Hundred Feet	Dk.	Deck
		C.F.	Cubic Feet	D.L.	Dead Load; Diesel
APA	American Plywood Association	CFM	Cubic Feet per Minute	DLH	Deep Long Span Bar Joist
Approx.	Approximate	CFRP	Carbon Fiber Reinforced Plastic	dlx	Deluxe
Apt.	Apartment	c.g.	Center of Gravity	Do.	Ditto
Asb.	Asbestos	CHW	Chilled Water; Commercial Hot Water	DOP	Dioctyl Phthalate Penetration Test (Air Filters)
A.S.B.C.	American Standard Building Code				
Asbe.	Asbestos Worker	C.I., CI	Cast Iron	Dp., dp	Depth
ASCE	American Society of Civil Engineers	C.I.P., CIP	Cast in Place	D.P.S.T.	Double Pole, Single Throw
A.S.H.R.A.E.	American Society of Heating, Refrig. & AC Engineers	Circ.	Circuit	Dr.	Drive
		C.L.	Carload Lot	DR	Dimension Ratio
ASME	American Society of Mechanical Engineers	CL	Chain Link	Drink.	Drinking
		Clab.	Common Laborer	D.S.	Double Strength
ASTM	American Society for Testing and Materials	Clam	Common Maintenance Laborer	D.S.A.	Double Strength A Grade
		C.L.F.	Hundred Linear Feet	D.S.B.	Double Strength B Grade
Attchmt.	Attachment	CLF	Current Limiting Fuse	Dty.	Duty
Avg., Ave.	Average	CLP	Cross Linked Polyethylene	DWV	Drain Waste Vent
AWG	American Wire Gauge	cm	Centimeter	DX	Deluxe White, Direct Expansion
AWWA	American Water Works Assoc.	CMP	Corr. Metal Pipe	dyn	Dyne
Bbl.	Barrel	CMU	Concrete Masonry Unit	e	Eccentricity
B&B, BB	Grade B and Better; Balled & Burlapped	CN	Change Notice	E	Equipment Only; East; Emissivity
		Col.	Column	Ea.	Each
B&S	Bell and Spigot	CO₂	Carbon Dioxide	EB	Encased Burial
B.&W.	Black and White	Comb.	Combination	Econ.	Economy
b.c.c.	Body-centered Cubic	comm.	Commercial, Communication	E.C.Y	Embankment Cubic Yards
B.C.Y.	Bank Cubic Yards	Compr.	Compressor	EDP	Electronic Data Processing
BE	Bevel End	Conc.	Concrete	EIFS	Exterior Insulation Finish System
B.F.	Board Feet	Cont., cont	Continuous; Continued, Container	E.D.R.	Equiv. Direct Radiation
Bg. cem.	Bag of Cement	Corkbd.	Cork Board	Eq.	Equation
BHP	Boiler Horsepower; Brake Horsepower	Corr.	Corrugated	EL	Elevation
		Cos	Cosine	Elec.	Electrician; Electrical
B.I.	Black Iron	Cot	Cotangent	Elev.	Elevator; Elevating
bidir.	bidirectional	Cov.	Cover	EMT	Electrical Metallic Conduit; Thin Wall Conduit
Bit., Bitum.	Bituminous	C/P	Cedar on Paneling		
Bit., Conc.	Bituminous Concrete	CPA	Control Point Adjustment	Eng.	Engine, Engineered
Bk.	Backed	Cplg.	Coupling	EPDM	Ethylene Propylene Diene Monomer
Bkrs.	Breakers	CPM	Critical Path Method		
Bldg., bldg	Building	CPVC	Chlorinated Polyvinyl Chloride	EPS	Expanded Polystyrene
Blk.	Block	C.Pr.	Hundred Pair	Eqhv.	Equip. Oper., Heavy
Bm.	Beam	CRC	Cold Rolled Channel	Eqlt.	Equip. Oper., Light
Boil.	Boilermaker	Creos.	Creosote	Eqmd.	Equip. Oper., Medium
bpm	Blows per Minute	Crpt.	Carpet & Linoleum Layer	Eqmm.	Equip. Oper., Master Mechanic
BR	Bedroom	CRT	Cathode-ray Tube	Eqol.	Equip. Oper., Oilers
Brg., brng.	Bearing	CS	Carbon Steel, Constant Shear Bar Joist	Equip.	Equipment
Brhe.	Bricklayer Helper			ERW	Electric Resistance Welded
Bric.	Bricklayer				

741

Abbreviation	Meaning
E.S.	Energy Saver
Est.	Estimated
esu	Electrostatic Units
E.W.	Each Way
EWT	Entering Water Temperature
Excav.	Excavation
excl	Excluding
Exp., exp	Expansion, Exposure
Ext., ext	Exterior; Extension
Extru.	Extrusion
f.	Fiber Stress
F	Fahrenheit; Female; Fill
Fab., fab	Fabricated; Fabric
FBGS	Fiberglass
F.C.	Footcandles
f.c.c.	Face-centered Cubic
f'c.	Compressive Stress in Concrete; Extreme Compressive Stress
F.E.	Front End
FEP	Fluorinated Ethylene Propylene (Teflon)
F.G.	Flat Grain
F.H.A.	Federal Housing Administration
Fig.	Figure
Fin.	Finished
FIPS	Female Iron Pipe Size
Fixt.	Fixture
FJP	Finger jointed and primed
Fl. Oz.	Fluid Ounces
Flr.	Floor
Flrs.	Floors
FM	Frequency Modulation; Factory Mutual
Fmg.	Framing
FM/UL	Factory Mutual/Underwriters Labs
Fdn.	Foundation
FNPT	Female National Pipe Thread
Fori.	Foreman, Inside
Foro.	Foreman, Outside
Fount.	Fountain
fpm	Feet per Minute
FPT	Female Pipe Thread
Fr	Frame
F.R.	Fire Rating
FRK	Foil Reinforced Kraft
FSK	Foil/Scrim/Kraft
FRP	Fiberglass Reinforced Plastic
FS	Forged Steel
FSC	Cast Body; Cast Switch Box
Ft., ft	Foot; Feet
Ftng.	Fitting
Ftg.	Footing
Ft lb	Foot Pound
Furn.	Furniture
FVNR	Full Voltage Non-Reversing
FVR	Full Voltage Reversing
FXM	Female by Male
Fy.	Minimum Yield Stress of Steel
g	Gram
G	Gauss
Ga.	Gauge
Gal., gal.	Gallon
Galv., galv	Galvanized
GC/MS	Gas Chromatograph/Mass Spectrometer
Gen.	General
GFI	Ground Fault Interrupter
GFRC	Glass Fiber Reinforced Concrete
Glaz.	Glazier
GPD	Gallons per Day
gpf	Gallon per Flush
GPH	Gallons per Hour
gpm, GPM	Gallons per Minute
GR	Grade
Gran.	Granular
Grnd.	Ground
GVW	Gross Vehicle Weight
GWB	Gypsum Wall Board
H	High Henry
HC	High Capacity
H.D., HD	Heavy Duty; High Density
H.D.O.	High Density Overlaid
HDPE	High Density Polyethylene Plastic
Hdr.	Header
Hdwe.	Hardware
H.I.D., HID	High Intensity Discharge
Help.	Helper Average
HEPA	High Efficiency Particulate Air Filter
Hg	Mercury
HIC	High Interrupting Capacity
HM	Hollow Metal
HMWPE	High Molecular Weight Polyethylene
HO	High Output
Horiz.	Horizontal
H.P., HP	Horsepower; High Pressure
H.P.F.	High Power Factor
Hr.	Hour
Hrs./Day	Hours per Day
HSC	High Short Circuit
Ht.	Height
Htg.	Heating
Htrs.	Heaters
HVAC	Heating, Ventilation & Air-Conditioning
Hvy.	Heavy
HW	Hot Water
Hyd.; Hydr.	Hydraulic
Hz	Hertz (cycles)
I.	Moment of Inertia
IBC	International Building Code
I.C.	Interrupting Capacity
ID	Inside Diameter
I.D.	Inside Dimension; Identification
I.F.	Inside Frosted
I.M.C.	Intermediate Metal Conduit
In.	Inch
Incan.	Incandescent
Incl.	Included; Including
Int.	Interior
Inst.	Installation
Insul., insul	Insulation/Insulated
I.P.	Iron Pipe
I.P.S., IPS	Iron Pipe Size
IPT	Iron Pipe Threaded
I.W.	Indirect Waste
J	Joule
J.I.C.	Joint Industrial Council
K	Thousand; Thousand Pounds; Heavy Wall Copper Tubing, Kelvin
K.A.H.	Thousand Amp. Hours
kcmil	Thousand Circular Mils
KD	Knock Down
K.D.A.T.	Kiln Dried After Treatment
kg	Kilogram
kG	Kilogauss
kgf	Kilogram Force
kHz	Kilohertz
Kip	1000 Pounds
KJ	Kilojoule
K.L.	Effective Length Factor
K.L.F.	Kips per Linear Foot
Km	Kilometer
KO	Knock Out
K.S.F.	Kips per Square Foot
K.S.I.	Kips per Square Inch
kV	Kilovolt
kVA	Kilovolt Ampere
kVAR	Kilovar (Reactance)
KW	Kilowatt
KWh	Kilowatt-hour
L	Labor Only; Length; Long; Medium Wall Copper Tubing
Lab.	Labor
lat	Latitude
Lath.	Lather
Lav.	Lavatory
lb.; #	Pound
L.B., LB	Load Bearing; L Conduit Body
L. & E.	Labor & Equipment
lb./hr.	Pounds per Hour
lb./L.F.	Pounds per Linear Foot
lbf/sq.in.	Pound-force per Square Inch
L.C.L.	Less than Carload Lot
L.C.Y.	Loose Cubic Yard
Ld.	Load
LE	Lead Equivalent
LED	Light Emitting Diode
L.F.	Linear Foot
L.F. Hdr	Linear Feet of Header
L.F. Nose	Linear Foot of Stair Nosing
L.F. Rsr	Linear Foot of Stair Riser
Lg.	Long; Length; Large
L & H	Light and Heat
LH	Long Span Bar Joist
L.H.	Labor Hours
L.L., LL	Live Load
L.L.D.	Lamp Lumen Depreciation
lm	Lumen
lm/sf	Lumen per Square Foot
lm/W	Lumen per Watt
LOA	Length Over All
log	Logarithm
L-O-L	Lateralolet
long.	Longitude
L.P., LP	Liquefied Petroleum; Low Pressure
L.P.F.	Low Power Factor
LR	Long Radius
L.S.	Lump Sum
Lt.	Light
Lt. Ga.	Light Gauge
L.T.L.	Less than Truckload Lot
Lt. Wt.	Lightweight
L.V.	Low Voltage
M	Thousand; Material; Male; Light Wall Copper Tubing
M²CA	Meters Squared Contact Area
m/hr.; M.H.	Man-hour
mA	Milliampere
Mach.	Machine
Mag. Str.	Magnetic Starter
Maint.	Maintenance
Marb.	Marble Setter
Mat; Mat'l.	Material
Max.	Maximum
MBF	Thousand Board Feet
MBH	Thousand BTU's per hr.
MC	Metal Clad Cable
MCC	Motor Control Center
M.C.F.	Thousand Cubic Feet
MCFM	Thousand Cubic Feet per Minute
M.C.M.	Thousand Circular Mils
MCP	Motor Circuit Protector
MD	Medium Duty
MDF	Medium-density fibreboard
M.D.O.	Medium Density Overlaid
Med.	Medium
MF	Thousand Feet
M.F.B.M.	Thousand Feet Board Measure
Mfg.	Manufacturing
Mfrs.	Manufacturers
mg	Milligram
MGD	Million Gallons per Day
MGPH	Million Gallons per Hour
MH, M.H.	Manhole; Metal Halide; Man-Hour
MHz	Megahertz
Mi.	Mile
MI	Malleable Iron; Mineral Insulated
MIPS	Male Iron Pipe Size
mj	Mechanical Joint
m	Meter
mm	Millimeter
Mill.	Millwright
Min., min.	Minimum, Minute

742

Misc.	Miscellaneous	PCM	Phase Contrast Microscopy	SBS	Styrene Butadiere Styrene
ml	Milliliter, Mainline	PDCA	Painting and Decorating	SC	Screw Cover
M.L.F.	Thousand Linear Feet		Contractors of America	SCFM	Standard Cubic Feet per Minute
Mo.	Month	P.E., PE	Professional Engineer;	Scaf.	Scaffold
Mobil.	Mobilization		Porcelain Enamel;	Sch., Sched.	Schedule
Mog.	Mogul Base		Polyethylene; Plain End	S.C.R.	Modular Brick
MPH	Miles per Hour	P.E.C.I.	Porcelain Enamel on Cast Iron	S.D.	Sound Deadening
MPT	Male Pipe Thread	Perf.	Perforated	SDR	Standard Dimension Ratio
MRGWB	Moisture Resistant Gypsum	PEX	Cross Linked Polyethylene	S.E.	Surfaced Edge
	Wallboard	Ph.	Phase	Sel.	Select
MRT	Mile Round Trip	P.I.	Pressure Injected	SER, SEU	Service Entrance Cable
ms	Millisecond	Pile.	Pile Driver	S.F.	Square Foot
M.S.F.	Thousand Square Feet	Pkg.	Package	S.F.C.A.	Square Foot Contact Area
Mstz.	Mosaic & Terrazzo Worker	Pl.	Plate	S.F. Flr.	Square Foot of Floor
M.S.Y.	Thousand Square Yards	Plah.	Plasterer Helper	S.F.G.	Square Foot of Ground
Mtd., mtd., mtd	Mounted	Plas.	Plasterer	S.F. Hor.	Square Foot Horizontal
Mthe.	Mosaic & Terrazzo Helper	plf	Pounds Per Linear Foot	SFR	Square Feet of Radiation
Mtng.	Mounting	Pluh.	Plumber Helper	S.F. Shlf.	Square Foot of Shelf
Mult.	Multi; Multiply	Plum.	Plumber	S4S	Surface 4 Sides
MUTCD	Manual on Uniform Traffic Control	Ply.	Plywood	Shee.	Sheet Metal Worker
	Devices	p.m.	Post Meridiem	Sin.	Sine
M.V.A.	Million Volt Amperes	Pntd.	Painted	Skwk.	Skilled Worker
M.V.A.R.	Million Volt Amperes Reactance	Pord.	Painter, Ordinary	SL	Saran Lined
MV	Megavolt	pp	Pages	S.L.	Slimline
MW	Megawatt	PP, PPL	Polypropylene	Sldr.	Solder
MXM	Male by Male	P.P.M.	Parts per Million	SLH	Super Long Span Bar Joist
MYD	Thousand Yards	Pr.	Pair	S.N.	Solid Neutral
N	Natural; North	P.E.S.B.	Pre-engineered Steel Building	SO	Stranded with oil resistant inside
nA	Nanoampere	Prefab.	Prefabricated		insulation
NA	Not Available; Not Applicable	Prefin.	Prefinished	S-O-L	Socketolet
N.B.C.	National Building Code	Prop.	Propelled	sp	Standpipe
NC	Normally Closed	PSF, psf	Pounds per Square Foot	S.P.	Static Pressure; Single Pole; Self-
NEMA	National Electrical Manufacturers	PSI, psi	Pounds per Square Inch		Propelled
	Assoc.	PSIG	Pounds per Square Inch Gauge	Spri.	Sprinkler Installer
NEHB	Bolted Circuit Breaker to 600V.	PSP	Plastic Sewer Pipe	spwg	Static Pressure Water Gauge
NFPA	National Fire Protection Association	Pspr.	Painter, Spray	S.P.D.T.	Single Pole, Double Throw
NLB	Non-Load-Bearing	Psst.	Painter, Structural Steel	SPF	Spruce Pine Fir; Sprayed
NM	Non-Metallic Cable	P.T.	Potential Transformer		Polyurethane Foam
nm	Nanometer	P. & T.	Pressure & Temperature	S.P.S.T.	Single Pole, Single Throw
No.	Number	Ptd.	Painted	SPT	Standard Pipe Thread
NO	Normally Open	Ptns.	Partitions	Sq.	Square; 100 Square Feet
N.O.C.	Not Otherwise Classified	Pu	Ultimate Load	Sq. Hd.	Square Head
Nose.	Nosing	PVC	Polyvinyl Chloride	Sq. In.	Square Inch
NPT	National Pipe Thread	Pvmt.	Pavement	S.S.	Single Strength; Stainless Steel
NQOD	Combination Plug-on/Bolt on	PRV	Pressure Relief Valve	S.S.B.	Single Strength B Grade
	Circuit Breaker to 240V.	Pwr.	Power	sst, ss	Stainless Steel
N.R.C., NRC	Noise Reduction Coefficient/	Q	Quantity Heat Flow	Sswk.	Structural Steel Worker
	Nuclear Regulator Commission	Qt.	Quart	Sswl.	Structural Steel Welder
N.R.S.	Non Rising Stem	Quan., Qty.	Quantity	St.; Stl.	Steel
ns	Nanosecond	Q.C.	Quick Coupling	STC	Sound Transmission Coefficient
NTP	Notice to Proceed	r	Radius of Gyration	Std.	Standard
nW	Nanowatt	R	Resistance	Stg.	Staging
OB	Opposing Blade	R.C.P.	Reinforced Concrete Pipe	STK	Select Tight Knot
OC	On Center	Rect.	Rectangle	STP	Standard Temperature & Pressure
OD	Outside Diameter	recpt.	Receptacle	Stpi.	Steamfitter, Pipefitter
O.D.	Outside Dimension	Reg.	Regular	Str.	Strength; Starter; Straight
ODS	Overhead Distribution System	Reinf.	Reinforced	Strd.	Stranded
O.G.	Ogee	Req'd.	Required	Struct.	Structural
O.H.	Overhead	Res.	Resistant	Sty.	Story
O&P	Overhead and Profit	Resi.	Residential	Subj.	Subject
Oper.	Operator	RF	Radio Frequency	Subs.	Subcontractors
Opng.	Opening	RFID	Radio-frequency Identification	Surf.	Surface
Orna.	Ornamental	Rgh.	Rough	Sw.	Switch
OSB	Oriented Strand Board	RGS	Rigid Galvanized Steel	Swbd.	Switchboard
OS&Y	Outside Screw and Yoke	RHW	Rubber, Heat & Water Resistant;	S.Y.	Square Yard
OSHA	Occupational Safety and Health		Residential Hot Water	Syn.	Synthetic
	Act	rms	Root Mean Square	S.Y.P.	Southern Yellow Pine
Ovhd.	Overhead	Rnd.	Round	Sys.	System
OWG	Oil, Water or Gas	Rodm.	Rodman	t.	Thickness
Oz.	Ounce	Rofc.	Roofer, Composition	T	Temperature; Ton
P.	Pole; Applied Load; Projection	Rofp.	Roofer, Precast	Tan	Tangent
p.	Page	Rohe.	Roofer Helpers (Composition)	T.C.	Terra Cotta
Pape.	Paperhanger	Rots.	Roofer, Tile & Slate	T & C	Threaded and Coupled
P.A.P.R.	Powered Air Purifying Respirator	R.O.W.	Right of Way	T.D.	Temperature Difference
PAR	Parabolic Reflector	RPM	Revolutions per Minute	TDD	Telecommunications Device for
P.B., PB	Push Button	R.S.	Rapid Start		the Deaf
Pc., Pcs.	Piece, Pieces	Rsr	Riser	T.E.M.	Transmission Electron Microscopy
P.C.	Portland Cement; Power Connector	RT	Round Trip	temp	Temperature, Tempered, Temporary
P.C.F.	Pounds per Cubic Foot	S.	Suction; Single Entrance; South	TFFN	Nylon Jacketed Wire

743

TFE	Tetrafluoroethylene (Teflon)	U.L., UL	Underwriters Laboratory	w/	With
T. & G.	Tongue & Groove;	Uld.	Unloading	W.C., WC	Water Column; Water Closet
	Tar & Gravel	Unfin.	Unfinished	W.F.	Wide Flange
Th., Thk.	Thick	UPS	Uninterruptible Power Supply	W.G.	Water Gauge
Thn.	Thin	URD	Underground Residential	Wldg.	Welding
Thrded	Threaded		Distribution	W. Mile	Wire Mile
Tilf.	Tile Layer, Floor	US	United States	W-O-L	Weldolet
Tilh.	Tile Layer, Helper	USGBC	U.S. Green Building Council	W.R.	Water Resistant
THHN	Nylon Jacketed Wire	USP	United States Primed	Wrck.	Wrecker
THW.	Insulated Strand Wire	UTMCD	Uniform Traffic Manual For Control	WSFU	Water Supply Fixture Unit
THWN	Nylon Jacketed Wire		Devices	W.S.P.	Water, Steam, Petroleum
T.L., TL	Truckload	UTP	Unshielded Twisted Pair	WT., Wt.	Weight
T.M.	Track Mounted	V	Volt	WWF	Welded Wire Fabric
Tot.	Total	VA	Volt Amperes	XFER	Transfer
T-O-L	Threadolet	VAT	Vinyl Asbestos Tile	XFMR	Transformer
tmpd	Tempered	V.C.T.	Vinyl Composition Tile	XHD	Extra Heavy Duty
TPO	Thermoplastic Polyolefin	VAV	Variable Air Volume	XHHW	Cross-Linked Polyethylene Wire
T.S.	Trigger Start	VC	Veneer Core	XLPE	Insulation
Tr.	Trade	VDC	Volts Direct Current	XLP	Cross-linked Polyethylene
Transf.	Transformer	Vent.	Ventilation	Xport	Transport
Trhv.	Truck Driver, Heavy	Vert.	Vertical	Y	Wye
Trlr	Trailer	V.F.	Vinyl Faced	yd	Yard
Trlt.	Truck Driver, Light	V.G.	Vertical Grain	yr	Year
TTY	Teletypewriter	VHF	Very High Frequency	Δ	Delta
TV	Television	VHO	Very High Output	%	Percent
T.W.	Thermoplastic Water Resistant	Vib.	Vibrating	~	Approximately
	Wire	VLF	Vertical Linear Foot	Ø	Phase; diameter
UCI	Uniform Construction Index	VOC	Volatile Organic Compound	@	At
UF	Underground Feeder	Vol.	Volume	#	Pound; Number
UGND	Underground Feeder	VRP	Vinyl Reinforced Polyester	<	Less Than
UHF	Ultra High Frequency	W	Wire; Watt; Wide; West	>	Greater Than
U.I.	United Inch			Z	Zone

RESIDENTIAL
COST ESTIMATE

OWNER'S NAME: _____ APPRAISER: _____

RESIDENCE ADDRESS: _____ PROJECT: _____

CITY, STATE, ZIP CODE: _____ DATE: _____

CLASS OF CONSTRUCTION	RESIDENCE TYPE	CONFIGURATION	EXTERIOR WALL SYSTEM
☐ ECONOMY	☐ 1 STORY	☐ DETACHED	☐ WOOD SIDING—WOOD FRAME
☐ AVERAGE	☐ 1-1/2 STORY	☐ TOWN/ROW HOUSE	☐ BRICK VENEER—WOOD FRAME
☐ CUSTOM	☐ 2 STORY	☐ SEMI-DETACHED	☐ STUCCO ON WOOD FRAME
☐ LUXURY	☐ 2-1/2 STORY		☐ PAINTED CONCRETE BLOCK
	☐ 3 STORY	OCCUPANCY	☐ SOLID MASONRY (AVERAGE & CUSTOM)
	☐ BI-LEVEL	☐ ONE STORY	☐ STONE VENEER—WOOD FRAME
	☐ TRI-LEVEL	☐ TWO FAMILY	☐ SOLID BRICK (LUXURY)
		☐ THREE FAMILY	☐ SOLID STONE (LUXURY)
		☐ OTHER _____	

*LIVING AREA (Main Building)

First Level _____ S.F.
Second Level _____ S.F.
Third Level _____ S.F.

Total _____ S.F.

*LIVING AREA (Wing or Ell) (   )

First Level _____ S.F.
Second Level _____ S.F.
Third Level _____ S.F.

Total _____ S.F.

*LIVING AREA (Wing or Ell) (   )

First Level _____ S.F.
Second Level _____ S.F.
Third Level _____ S.F.

Total _____ S.F.

*Basement Area is not part of living area.

MAIN BUILDING	COSTS PER S.F. LIVING AREA
Cost per Square Foot of Living Area, from Page _____	$
Basement Addition: _____ % Finished, _____ % Unfinished	+
Roof Cover Adjustment: _____ Type, Page _____ (Add or Deduct)	( )
Central Air Conditioning: ☐ Separate Ducts ☐ Heating Ducts, Page _____	+
Heating System Adjustment: _____ Type, Page _____ (Add or Deduct)	( )
Main Building: Adjusted Cost per S.F. of Living Area	$

MAIN BUILDING TOTAL COST $ _____ /S.F. x _____ S.F. x _____ = $ _____
Cost per S.F. Living Area   Living Area   Town/Row House Multiplier (Use 1 for Detached)   TOTAL COST

WING OR ELL ( ) _____ STORY	COSTS PER S.F. LIVING AREA
Cost per Square Foot of Living Area, from Page _____	$
Basement Addition: _____ % Finished, _____ % Unfinished	+
Roof Cover Adjustment: _____ Type, Page _____ (Add or Deduct)	( )
Central Air Conditioning: ☐ Separate Ducts ☐ Heating Ducts, Page _____	+
Heating System Adjustment: _____ Type, Page _____ (Add or Deduct)	( )
Wing or Ell ( ): Adjusted Cost per S.F. of Living Area	$

WING OR ELL (   ) TOTAL COST $ _____ /S.F. x _____ S.F. = $ _____
Cost per S.F. Living Area   Living Area   TOTAL COST

WING OR ELL ( ) _____ STORY	COSTS PER S.F. LIVING AREA
Cost per Square Foot of Living Area, from Page _____	$
Basement Addition: _____ % Finished, _____ % Unfinished	+
Roof Cover Adjustment: _____ Type, Page _____ (Add or Deduct)	( )
Central Air Conditioning: ☐ Separate Ducts ☐ Heating Ducts, Page _____	+
Heating System Adjustment: _____ Type, Page _____ (Add or Deduct)	( )
Wing or Ell ( ): Adjusted Cost per S.F. of Living Area	$

WING OR ELL (   ) TOTAL COST $ _____ /S.F. x _____ S.F. = $ _____
Cost per S.F. Living Area   Living Area   TOTAL COST

TOTAL THIS PAGE   [   ]

Page 1 of 2

RESIDENTIAL COST ESTIMATE

		QUANTITY	UNIT COST	
Total Page 1				$
Additional Bathrooms: _____ Full, _____ Half				
Finished Attic: _____ Ft. x _____ Ft.		S.F.		+
Breezeway: ☐ Open ☐ Enclosed _____ Ft. x _____ Ft.		S.F.		+
Covered Porch: ☐ Open ☐ Enclosed _____ Ft. x _____ Ft.		S.F.		+
Fireplace: ☐ Interior Chimney ☐ Exterior Chimney ☐ No. of Flues ☐ Additional Fireplaces				+
Appliances:				+
Kitchen Cabinets Adjustment: (+/−)				
☐ Garage ☐ Carport: _____ Car(s) Description _____ (+/−)				
Miscellaneous:				+

ADJUSTED TOTAL BUILDING COST $ _____

REPLACEMENT COST	
ADJUSTED TOTAL BUILDING COST	$ _____
Site Improvements	
(A) Paving & Sidewalks	$ _____
(B) Landscaping	$ _____
(C) Fences	$ _____
(D) Swimming Pool	$ _____
(E) Miscellaneous	$ _____
TOTAL	$ _____
Location Factor	x _____
Location Replacement Cost	$ _____
Depreciation	−$ _____
LOCAL DEPRECIATED COST	$ _____

INSURANCE COST	
ADJUSTED TOTAL BUILDING COST	$ _____
Insurance Exclusions	
(A) Footings, Site Work, Underground Piping	−$ _____
(B) Architects' Fees	−$ _____
Total Building Cost Less Exclusion	$ _____
Location Factor	x _____
LOCAL INSURABLE REPLACEMENT COST	$ _____

SKETCH AND ADDITIONAL CALCULATIONS

For customer support on your Residential Costs with RSMeans data, call 800.448.8182.

Index

Index

For customer support on your Residential Costs with RSMeans data, call 800.448.8182.

753

Index

Index

For customer support on your Residential Costs with RSMeans data, call 800.448.8182.

For customer support on your Residential Costs with RSMeans data, call 800.448.8182.

Index

For customer support on your Residential Costs with RSMeans data, call 800.448.8182.

Other Data & Services
A tradition of excellence in construction cost information
and services since 1942

Table of Contents
Annual Cost Guides
Online Estimating Solution
Seminars and Professional Development

For more information visit our website at www.RSMeans.com

Unit prices according to the latest MasterFormat®

Cost Data Selection Guide

The following table provides definitive information on the content of each cost data publication. The number of lines of data provided in each unit price or assemblies division, as well as the number of crews, is listed for each data set. The presence of other elements such as reference tables, square foot models, equipment rental costs, historical cost indexes, and city cost indexes, is also indicated. You can use the table to help select the RSMeans data set that has the quantity and type of information you most need in your work.

Unit Cost Divisions	Building Construction	Mechanical	Electrical	Commercial Renovation	Square Foot	Site Work Landsc.	Green Building	Interior	Concrete Masonry	Open Shop	Heavy Construction	Light Commercial	Facilities Construction	Plumbing	Residential
1	590	411	428	530	0	524	200	331	473	589	527	273	1065	421	178
2	777	280	86	733	0	991	207	399	214	776	733	481	1220	287	274
3	1688	340	230	1081	0	1480	986	354	2034	1688	1690	482	1788	316	389
4	960	21	0	920	0	725	180	615	1158	928	615	533	1175	0	447
5	1901	158	155	1093	0	852	1799	1106	729	1901	1037	979	1918	204	746
6	2453	18	18	2111	0	110	589	1528	281	2449	123	2141	2125	22	2661
7	1596	215	128	1634	0	580	763	532	523	1593	26	1329	1697	227	1049
8	2140	80	3	2733	0	255	1140	1813	105	2142	0	2328	2966	0	1552
9	2107	86	45	1931	0	309	455	2193	412	2050	15	1756	2356	54	1521
10	1090	17	10	685	0	234	32	899	136	1090	34	589	1181	237	224
11	1097	201	166	541	0	135	56	925	29	1064	0	231	1117	164	110
12	548	0	2	299	0	219	147	1551	14	515	0	273	1574	23	217
13	744	149	158	253	0	366	125	254	78	720	267	109	760	115	104
14	273	36	0	223	0	0	0	257	0	273	0	12	293	16	6
21	130	0	41	37	0	0	0	296	0	130	0	121	668	688	259
22	1165	7557	160	1226	0	1572	1063	849	20	1154	1681	875	7505	9414	719
23	1194	6995	581	938	0	157	898	787	38	1177	110	886	5235	1917	480
25	0	0	14	14	0	0	0	0	0	0	0	0	0	0	0
26	1512	491	10456	1293	0	811	644	1159	55	1438	600	1360	10237	399	636
27	94	0	447	101	0	0	0	71	0	94	39	67	388	0	56
28	143	79	223	124	0	0	28	97	0	127	0	70	209	57	41
31	1510	733	610	806	0	3265	288	7	1217	1455	3276	604	1569	660	613
32	836	49	8	905	0	4472	353	405	314	808	1889	440	1751	142	487
33	1246	1076	534	252	0	3021	38	0	237	523	3058	128	1698	2085	154
34	107	0	47	4	0	190	0	0	31	62	212	0	136	0	0
35	18	0	0	0	0	327	0	0	0	18	442	0	84	0	0
41	62	0	0	33	0	8	0	22	0	61	31	0	68	14	0
44	75	79	0	0	0	0	0	0	0	0	0	0	75	75	0
46	23	16	0	0	0	274	261	0	0	23	264	0	33	33	0
48	10	0	38	2	0	0	23	0	0	10	17	10	23	0	10
Totals	26089	19087	14588	20502	0	20877	10275	16450	8098	24858	16686	16077	50914	17570	12933

Assem Div	Building Construction	Mechanical	Electrical	Commercial Renovation	Square Foot	Site Work Landscape	Assemblies	Green Building	Interior	Concrete Masonry	Heavy Construction	Light Commercial	Facilities Construction	Plumbing	Asm Div	Residential
A		15	0	188	164	577	598	0	0	536	571	154	24	0	1	378
B		0	0	848	2554	0	5661	56	329	1976	368	2094	174	0	2	211
C		0	0	647	954	0	1334	0	1642	146	0	844	255	0	3	588
D		1057	941	712	1859	72	2538	330	825	0	0	1345	1105	1088	4	851
E		0	0	86	261	0	301	0	5	0	0	258	5	0	5	391
F		0	0	0	114	0	143	0	0	0	0	114	0	0	6	357
G		527	447	318	312	3377	792	0	0	534	1349	205	293	677	7	307
															8	760
															9	80
															10	0
															11	0
															12	0
Totals		1599	1388	2799	6218	4026	11367	386	2801	3192	2288	5014	1856	1765		3923

Reference Section	Building Construction Costs	Mechanical	Electrical	Commercial Renovation	Square Foot	Site Work Landscape	Assem.	Green Building	Interior	Concrete Masonry	Open Shop	Heavy Construction	Light Commercial	Facilities Construction	Plumbing	Resi.
Reference Tables	yes	yes	yes	yes	no	yes	yes	yes	yes	yes	yes	yes	yes	yes	yes	yes
Models					111		25						50			28
Crews	578	578	578	556		578		578	578	578	555	578	555	556	578	555
Equipment Rental Costs	yes	yes	yes	yes		yes		yes	yes	yes	yes	yes	yes	yes	yes	yes
Historical Cost Indexes	yes	yes	yes	yes	yes	yes	yes	yes	yes	yes	yes	yes	yes	yes	yes	no
City Cost Indexes	yes	yes	yes	yes	yes	yes	yes	yes	yes	yes	yes	yes	yes	yes	yes	yes

Our Online Estimating Solution

Competitive Cost Estimates Made Easy

Our online estimating solution is a web-based service that provides accurate and up-to-date cost information to help you build competitive estimates or budgets in less time.

Quick, intuitive, easy to use and automatically updated, you'll gain instant access to hundreds of thousands of material, labor, and equipment costs from RSMeans' comprehensive database, delivering the information you need to build budgets and competitive estimates every time.

With our online estimating solutions, you can perform quick searches to locate specific costs and adjust costs to reflect prices in your geographic area. Tag and store your favorites for fast access to your frequently used line items and assemblies and clone estimates to save time. System notifications will alert you as updated data becomes available. This data is automatically updated throughout the year.

Our visual, interactive estimating features help you create, manage, save and share estimates with ease! You'll enjoy increased flexibility with customizable advanced reports. Easily edit custom report templates and import your company logo onto your estimates.

	Core	Advanced	Complete
Unit Prices	✓	✓	✓
Assemblies	⊘	✓	✓
Sq Foot Models	⊘	⊘	✓
Editable Sq Foot Models	⊘	⊘	✓
Editable Assembly Components	⊘	✓	✓
Custom Cost Data	⊘	✓	✓
User Defined Components	⊘	✓	✓
Advanced Reporting & Customization	⊘	✓	✓
Union Labor Type	✓	✓	✓

Continue to check our website at www.RSMeans.com for more product offerings.

Estimate with Precision

Find everything you need to develop complete, accurate estimates.

- Verified costs for construction materials
- Equipment rental costs
- Crew sizing, labor hours and labor rates
- Localized costs for U.S. and Canada

Save Time & Increase Efficiency

Make cost estimating and calculating faster and easier than ever with secure, online estimating tools.

- Quickly locate costs in the searchable database
- Create estimates in minutes with RSMeans cost lines
- Tag and store favorites for fast access to frequently used items

Improve Planning & Decision-Making

Back your estimates with complete, accurate and up-to-date cost data for informed business decisions.

- Verify construction costs from third parties
- Check validity of subcontractor proposals
- Evaluate material and assembly alternatives

Increase Profits

Use our online estimating solution to estimate projects quickly and accurately, so you can gain an edge over your competition.

- Create accurate and competitive bids
- Minimize the risk of cost overruns
- Reduce variability

Access the data online

Search for unit prices by keyword　　Leverage the most up-to-date data　　Build and export estimates

Try it free for 30 days　　www.rsmeans.com/2018freetrial

2018 Seminar Schedule　　📞 877-620-6245

Note: call for exact dates, locations, and details as some cities are subject to change.

Location	Dates	Location	Dates
Seattle, WA	January and August	San Francisco, CA	June
Dallas/Ft. Worth, TX	January	Bethesda, MD	June
Austin, TX	February	El Segundo, CA	August
Anchorage, AK	March and September	Dallas, TX	September
Las Vegas, NV	March	Raleigh, NC	October
New Orleans, LA	March	Salt Lake City, UT	October
Washington, DC	April and September	Baltimore, MD	November
Phoenix, AZ	April	Orlando, FL	November
Toronto	May	San Diego, CA	December
Denver, CO	May	San Antonio, TX	December

Gordian also offers a suite of online RSMeans data self-paced offerings. Check our website www.RSMeans.com for more information.

Self-Paced Professional Development Courses

Training on how to use RSMeans data and estimating tools, as well as Professional Development courses on industry topics are now offered in a convenient self-paced format. These courses are on-demand and allow more flexibility to learn around your busy schedule, while saving the cost of travel and time.

Current course offerings include:

Facilities Construction Estimating—our best-selling live class now available as an on-demand training course! Let the subject matter experts of construction estimating—the RSMeans Engineering Staff—walk you through the basics and much more of estimating for renovation and facilities construction.

RSMeansOnline.com Training—learn the ins and outs of the flagship delivery method of RSMeans data!

The Construction Process—how much do you and your team really know about the ins and outs of the "contract-side" of a construction project? This self-paced course will clarify best practices for items such as schedules, change orders, and project closeout.

These self-paced training courses can be completed over the course of 45 days and are comprised of multiple lessons with documentation, video presentation, software simulation, assessment quizzes and certificate of completion.

Site Work Estimating with RSMeans data

This new one-day program focuses directly on site work costs, a unique portion of most construction projects that often is the wild card in determining whether you have developed a good estimate or not. Accurately scoping, quantifying, and pricing site preparation, underground utility work, and improvements to exterior site elements are often the most difficult estimating tasks on any project. The program takes the participant from preparing a never-developed site through underground utility installation, pad preparation, paving and sidewalks, and landscaping. Attendees will use the full array of site work cost data through the RSMeans online program and participate in exercises to strengthen their estimating skills.

Some of what you'll learn:
  Evaluation of site work and understanding site scope of work.
  Site work estimating topics including: site clearing, grading, excavation, disposal and trucking of materials, erosion control devices, backfill and compaction, underground utilities, paving, sidewalks, fences & gates, and seeding & planting.
  Unit price site work estimates—Correct use of RSMeans site work cost data to develop a cost estimate.
  Using and modifying assemblies—Save valuable time when estimating site work activities using custom assemblies.

Who should attend: Engineers, contractors, estimators, project managers, owner's representatives, and others who are concerned with the proper preparation and/or evaluation of site work estimates.

Please bring a laptop with ability to access the internet.

Training for our Online Estimating Solution

Construction estimating is vital to the decision-making process at each state of every project. Our online solution works the way you do. It's systematic, flexible and intuitive. In this one-day class you will see how you can estimate any phase of any project faster and better.

Some of what you'll learn:
- Customizing our online estimating solution
- Making the most of RSMeans "Circle Reference" numbers
- How to integrate your cost data
- Generating reports, exporting estimates to MS Excel, sharing, collaborating and more

Also offered as a self-paced or on-site training program!

Facilities Construction Estimating

In this two-day course, professionals working in facilities management can get help with their daily challenges to establish budgets for all phases of a project.

Some of what you'll learn:
- Determining the full scope of a project
- Identifying the scope of risks and opportunities
- Creative solutions to estimating issues
- Organizing estimates for presentation and discussion
- Special techniques for repair/remodel and maintenance projects
- Negotiating project change orders

Who should attend: facility managers, engineers, contractors, facility tradespeople, planners, and project managers.

Construction Cost Estimating: Concepts and Practice

This one-day introductory course to improve estimating skills and effectiveness starts with the details of interpreting bid documents and ends with the summary of the estimate and bid submission.

Some of what you'll learn:
- Using the plans and specifications to create estimates
- The takeoff process—deriving all tasks with correct quantities
- Developing pricing using various sources; how subcontractor pricing fits in
- Summarizing the estimate to arrive at the final number
- Formulas for area and cubic measure, adding waste and adjusting productivity to specific projects
- Evaluating subcontractors' proposals and prices
- Adding insurance and bonds
- Understanding how labor costs are calculated
- Submitting bids and proposals

Who should attend: project managers, architects, engineers, owners' representatives, contractors, and anyone who's responsible for budgeting or estimating construction projects.

Maintenance & Repair Estimating for Facilities

This two-day course teaches attendees how to plan, budget, and estimate the cost of ongoing and preventive maintenance and repair for existing buildings and grounds.

Some of what you'll learn:
- The most financially favorable maintenance, repair, and replacement scheduling and estimating
- Auditing and value engineering facilities
- Preventive planning and facilities upgrading
- Determining both in-house and contract-out service costs
- Annual, asset-protecting M&R plan

Who should attend: facility managers, maintenance supervisors, buildings and grounds superintendents, plant managers, planners, estimators, and others involved in facilities planning and budgeting.

Practical Project Management for Construction Professionals

In this two-day course, acquire the essential knowledge and develop the skills to effectively and efficiently execute the day-to-day responsibilities of the construction project manager.

Some of what you'll learn:
- General conditions of the construction contract
- Contract modifications: change orders and construction change directives
- Negotiations with subcontractors and vendors
- Effective writing: notification and communications
- Dispute resolution: claims and liens

Who should attend: architects, engineers, owners' representatives, and project managers.

Mechanical & Electrical Estimating

This two-day course teaches attendees how to prepare more accurate and complete mechanical/electrical estimates, avoid the pitfalls of omission and double-counting, and understand the composition and rationale within the RSMeans mechanical/electrical database.

Some of what you'll learn:
- The unique way mechanical and electrical systems are interrelated
- M&E estimates—conceptual, planning, budgeting, and bidding stages
- Order of magnitude, square foot, assemblies, and unit price estimating
- Comparative cost analysis of equipment and design alternatives

Who should attend: architects, engineers, facilities managers, mechanical and electrical contractors, and others who need a highly reliable method for developing, understanding, and evaluating mechanical and electrical contracts.

Unit Price Estimating

This interactive two-day seminar teaches attendees how to interpret project information and process it into final, detailed estimates with the greatest accuracy level.

The most important credential an estimator can take to the job is the ability to visualize construction and estimate accurately.

Some of what you'll learn:
- Interpreting the design in terms of cost
- The most detailed, time-tested methodology for accurate pricing
- Key cost drivers—material, labor, equipment, staging, and subcontracts
- Understanding direct and indirect costs for accurate job cost accounting and change order management

Who should attend: corporate and government estimators and purchasers, architects, engineers, and others who need to produce accurate project estimates.

Life Cycle Cost Estimating for Facility Asset Managers

Life Cycle Cost Estimating will take the attendee through choosing the correct RSMeans database to use and then correctly applying RSMeans data to their specific life cycle application. Conceptual estimating through RSMeans new building models, conceptual estimating of major existing building projects through RSMeans renovation models, pricing specific renovation elements, estimating repair, replacement and preventive maintenance costs today and forward up to 30 years will be covered.

Some of what you'll learn:
- Cost implications of managing assets
- Planning projects and initial & life cycle costs
- How to use RSMeans data online

Who should attend: facilities owners and managers and anyone involved in the financial side of the decision making process in the planning, design, procurement, and operation of facility real assets.

Please bring a laptop with ability to access the internet.

Training for our CD Estimating Solution

This one-day course helps users become more familiar with the functionality of the CD. Each menu, icon, screen, and function found in the program is explained in depth. Time is devoted to hands-on estimating exercises.

Some of what you'll learn:
- Searching the database using all navigation methods
- Exporting RSMeans data to your preferred spreadsheet format
- Viewing crews, assembly components, and much more
- Automatically regionalizing the database

This training session requires you to bring a laptop computer to class.

When you register for this course you will receive an outline for your laptop requirements.

Also offered as a self-paced or on-site training program!

Assessing Scope of Work for Facilities Construction Estimating

This two-day practical training program addresses the vital importance of understanding the scope of projects in order to produce accurate cost estimates for facility repair and remodeling.

Some of what you'll learn:
- Discussions of site visits, plans/specs, record drawings of facilities, and site-specific lists
- Review of CSI divisions, including means, methods, materials, and the challenges of scoping each topic
- Exercises in scope identification and scope writing for accurate estimating of projects
- Hands-on exercises that require scope, take-off, and pricing

Who should attend: corporate and government estimators, planners, facility managers, and others who need to produce accurate project estimates.

Facilities Estimating Using the CD

This two-day class combines hands-on skill-building with best estimating practices and real-life problems. You will learn key concepts, tips, pointers, and guidelines to save time and avoid cost oversights and errors.

Some of what you'll learn:
- Estimating process concepts
- Customizing and adapting RSMeans cost data
- Establishing scope of work to account for all known variables
- Budget estimating: when, why, and how
- Site visits: what to look for and what you can't afford to overlook
- How to estimate repair and remodeling variables

This training session requires you to bring a laptop computer to class.

Who should attend: facility managers, architects, engineers, contractors, facility tradespeople, planners, project managers, and anyone involved with JOC, SABRE, or IDIQ.

Building Systems and the Construction Process

This one-day course was written to assist novices and those outside the industry in obtaining a solid understanding of the construction process - from both a building systems and construction administration approach.

Some of what you'll learn:
- Various systems used and how components come together to create a building
- Start with foundation and end with the physical systems of the structure such as HVAC and Electrical
- Focus on the process from start of design through project closeout

This training session requires you to bring a laptop computer to class.

Who should attend: building professionals or novices to help make the crossover to the construction industry; suited for anyone responsible for providing high level oversight on construction projects.

Professional Development

For more information visit our website at www.RSMeans.com

Registration Information

Register early and save up to $100!
Register 30 days before the start date of a seminar and save $100 off your total fee. Note: This discount can be applied only once per order. It cannot be applied to team discount registrations or any other special offer.

How to register
By Phone
Register by phone at 877-620-6245

Online
Register online at
www.RSMeans.com/products/seminars.aspx

Note: Purchase Orders or Credits Cards are required to register.

Two-day seminar registration fee - $1,045.

One-Day Construction Cost Estimating or Building Systems and the Construction Process - $630.

Government pricing
All federal government employees save off the regular seminar price. Other promotional discounts cannot be combined with the government discount.

Team discount program
For over five attendee registrations. Call for pricing: 781-422-5115

Refund policy
Cancellations will be accepted up to ten business days prior to the seminar start. There are no refunds for cancellations received later than ten working days prior to the first day of the seminar. A $150 processing fee will be applied for all cancellations. Written notice of the cancellation is required. Substitutions can be made at any time before the session starts. No-shows are subject to the full seminar fee.

Note: Pricing subject to change.

AACE approved courses
Many seminars described and offered here have been approved for 14 hours (1.4 recertification credits) of credit by the AACE International Certification Board toward meeting the continuing education requirements for recertification as a Certified Cost Engineer/Certified Cost Consultant.

AIA Continuing Education
We are registered with the AIA Continuing Education System (AIA/CES) and are committed to developing quality learning activities in accordance with the CES criteria. Many seminars meet the AIA/CES criteria for Quality Level 2. AIA members may receive 14 learning units (LUs) for each two-day RSMeans course.

Daily course schedule
The first day of each seminar session begins at 8:30 a.m. and ends at 4:30 p.m. The second day begins at 8:00 a.m. and ends at 4:00 p.m. Participants are urged to bring a hand-held calculator since many actual problems will be worked out in each session.

Continental breakfast
Your registration includes the cost of a continental breakfast and a morning and afternoon refreshment break. These informal segments allow you to discuss topics of mutual interest with other seminar attendees. (You are free to make your own lunch and dinner arrangements.)

Hotel/transportation arrangements
We arrange to hold a block of rooms at most host hotels. To take advantage of special group rates when making your reservation, be sure to mention that you are attending the RSMeans Institute data seminar. You are, of course, free to stay at the lodging place of your choice. (Hotel reservations and transportation arrangements should be made directly by seminar attendees.)

Important
Class sizes are limited, so please register as soon as possible.

Professional Development